An Introduction to Linear Algebra for Science and Engineering

Daniel Norman **Dan Wolczuk**
University of Waterloo

Third Edition

Pearson Canada Inc., 26 Prince Andrew Place, North York, Ontario M3C 2H4.

9780134682631

1 20

Library and Archives Canada Cataloguing in Publication

Norman, Daniel, 1938-, author

Introduction to linear algebra for science and engineering / Daniel Norman, Dan Wolczuk, University of Waterloo. – Third edition.

ISBN 978-0-13-468263-1 (softcover)

1. Algebras, Linear–Textbooks. 2. Textbooks. I. Wolczuk, Dan, 1972-, author II. Title.

QA184.2.N67 2018 512'.5 C2018-906600-8

Table of Contents

A Note to Students

Linear Algebra – What Is It?

Welcome to the third edition of *An Introduction to Linear Algebra for Science and Engineering*! Linear algebra is essentially the study of vectors, matrices, and linear mappings, and is now an extremely important topic in mathematics. Its application and usefulness in a variety of different areas is undeniable. It encompasses technological innovation, economic decision making, industry development, and scientific research. We are literally surrounded by applications of linear algebra.

Most people who have learned linear algebra and calculus believe that the ideas of elementary calculus (such as limits and integrals) are more difficult than those of introductory linear algebra, and that most problems encountered in calculus courses are harder than those found in linear algebra courses. So, at least by this comparison, linear algebra is not hard. Still, some students find learning linear algebra challenging. We think two factors contribute to the difficulty some students have.

First, students do not always see what linear algebra is good for. This is why it is important to read the applications in the text–even if you do not understand them completely. They will give you some sense of where linear algebra fits into the broader picture.

Second, mathematics is often mistakenly seen as a collection of recipes for solving standard problems. Students are often uncomfortable with the fact that linear algebra is "abstract" and includes a lot of "theory." However, students need to realize that there will be no long-term payoff in simply memorizing the recipes–computers carry them out far faster and more accurately than any human. That being said, practicing the procedures on specific examples is often an important step towards a much more important goal: understanding the *concepts* used in linear algebra to formulate and solve problems, and learning to interpret the results of calculations. Such understanding requires us to come to terms with some theory. In this text, when working through the examples and exercises – which are often small – keep in mind that when you do apply these ideas later, you may very well have a million variables and a million equations, but the theory and methods remain constant. For example, Google's PageRank system uses a matrix that has thirty billion columns and thirty billion rows – you do not want to do that by hand! **When you are solving computational problems, always try to observe how your work relates to the theory you have learned.**

Mathematics is useful in so many areas because it is *abstract*: the same good idea can unlock the problems of control engineers, civil engineers, physicists, social scientists, and mathematicians because the idea has been abstracted from a particular setting. One technique solves many problems because someone has established a *theory* of how to deal with these kinds of problems. *Definitions* are the way we try to capture important ideas, and *theorems* are how we summarize useful general facts about the kind of problems we are studying. *Proofs* not only show us that a statement is true; they can help us understand the statement, give us practice using important ideas, and make it easier to learn a given subject. In particular, proofs show us how ideas are tied together, so we do not have to memorize too many disconnected facts.

Many of the concepts introduced in linear algebra are natural and easy, but some may seem unnatural and "technical" to beginners. Do not avoid these seemingly more difficult ideas; use examples and theorems to see how these ideas are an essential part of the story of linear algebra. By learning the "vocabulary" and "grammar" of linear algebra, you will be equipping yourself with concepts and techniques that mathematicians, engineers, and scientists find invaluable for tackling an extraordinarily rich variety of problems.

Linear Algebra – Who Needs It?

Mathematicians

Linear algebra and its applications are a subject of continuing research. Linear algebra is vital to mathematics because it provides essential ideas and tools in areas as diverse as abstract algebra, differential equations, calculus of functions of several variables, differential geometry, functional analysis, and numerical analysis.

Engineers

Suppose you become a control engineer and have to design or upgrade an automatic control system. The system may be controlling a manufacturing process, or perhaps an airplane landing system. You will probably start with a linear model of the system, requiring linear algebra for its solution. To include feedback control, your system must take account of many measurements (for the example of the airplane, position, velocity, pitch, etc.), and it will have to assess this information very rapidly in order to determine the correct control responses. A standard part of such a control system is a Kalman-Bucy filter, which is not so much a piece of hardware as a piece of mathematical machinery for doing the required calculations. Linear algebra is an essential part of the Kalman-Bucy filter.

If you become a structural engineer or a mechanical engineer, you may be concerned with the problem of vibrations in structures or machinery. To understand the problem, you will have to know about eigenvalues and eigenvectors and how they determine the normal modes of oscillation. Eigenvalues and eigenvectors are some of the central topics in linear algebra.

An electrical engineer will need linear algebra to analyze circuits and systems; a civil engineer will need linear algebra to determine internal forces in static structures and to understand principal axes of strain.

In addition to these fairly specific uses, engineers will also find that they need to know linear algebra to understand systems of differential equations and some aspects of the calculus of functions of two or more variables. Moreover, the ideas and techniques of linear algebra are central to numerical techniques for solving problems of heat and fluid flow, which are major concerns in mechanical engineering. Also, the ideas of linear algebra underlie advanced techniques such as Laplace transforms and Fourier analysis.

Physicists

Linear algebra is important in physics, partly for the reasons described above. In addition, it is vital in applications such as the inertia tensor in general rotating motion. Linear algebra is an absolutely essential tool in quantum physics (where, for example, energy levels may be determined as eigenvalues of linear operators) and relativity (where understanding change of coordinates is one of the central issues).

Life and Social Scientists

Input-output models, described by matrices, are often used in economics and other social sciences. Similar ideas can be used in modeling populations where one needs to keep track of sub-populations (generations, for example, or genotypes). In all sciences, statistical analysis of data is of a great importance, and much of this analysis uses linear algebra. For example, the method of least squares (for regression) can be understood in terms of projections in linear algebra.

Managers and Other Professionals

All managers need to make decisions about the best allocation of resources. Enormous amounts of computer time around the world are devoted to linear programming algorithms that solve such allocation problems. In industry, the same sorts of techniques are used in production, networking, and many other areas.

Who needs linear algebra? Almost every mathematician, engineer, scientist, economist, manager, or professional will find linear algebra an important and useful. So, who needs linear algebra? You do!

Will these applications be explained in this book?

Unfortunately, most of these applications require too much specialized background to be included in a first-year linear algebra book. To give you an idea of how some of these concepts are applied, a wide variety of applications are mentioned throughout the text. You will get to see many more applications of linear algebra in your future courses.

How To Make the Most of This Book: SQ3R

The SQ3R reading technique was developed by Francis Robinson to help students read textbooks more effectively. Here is a brief summary of this powerful method for learning. It is easy to learn more about this and other similar strategies online.

Survey: Quickly skim over the section. Make note of any heading or boldface words. Read over the definitions, the statement of theorems, and the statement of examples or exercises (do not read proofs or solutions at this time). Also, briefly examine the figures.

Question: Make a purpose for your reading by writing down general questions about the headings, boldface words, definitions, or theorems that you surveyed. For example, a couple of questions for Section 1.1 could be:
How do we use vectors in $\mathbb{R}^2$ *and* $\mathbb{R}^3$*?*
How does this material relate to what I have previously learned?
What is the relationship between vectors in $\mathbb{R}^2$ *and directed line segments?*
What are the similarities and differences between vectors and lines in $\mathbb{R}^2$ *and in* $\mathbb{R}^3$*?*

Read: Read the material in chunks of about one to two pages. Read carefully and look for the answers to your questions as well as key concepts and supporting details. *Take the time to solve the mid-section exercises before reading past them. Also, try to solve examples before reading the solutions, and try to figure out the proofs before you read them.* If you are not able to solve them, look carefully through the provided solution to figure out the step where you got stuck.

Recall: As you finish each chunk, put the book aside and summarize the important details of what you have just read. Write down the answers to any questions that you made and write down any further questions that you have. Think critically about how well you have understood the concepts, and if necessary, go back and reread a part or do some relevant end of section problems.

Review: This is an ongoing process. Once you complete an entire section, go back and review your notes and questions from the entire section. Test your understanding by trying to solve the end-of-section problems without referring to the book or your notes. Repeat this again when you finish an entire chapter and then again in the future as necessary.

Yes, you are going to find that this makes the reading go much slower for the first couple of chapters. However, students who use this technique consistently report that they feel that they end up spending a lot less time studying for the course as they learn the material so much better at the beginning, which makes future concepts much easier to learn.

A Note to Instructors

Welcome to the third edition of *An Introduction to Linear Algebra for Science and Engineering*! Thanks to the feedback I have received from students and instructors as well as my own research into the science of teaching and learning, I am very excited to present to you this new and improved version of the text. Overall, I believe the modifications I have made complement my overall approach to teaching. I believe in introducing the students slowly to difficult concepts and helping students learn these concepts more deeply by exposing them to the same concepts multiple times over spaced intervals.

One aspect of teaching linear algebra that I find fascinating is that so many different approaches can be used effectively. Typically, the biggest difference between most calculus textbooks is whether they have early or late transcendentals. However, linear algebra textbooks and courses can be done in a wide variety of orders. For example, in China it is not uncommon to begin an introductory linear algebra course with determinants and not cover solving systems of linear equations until after matrices and general vector spaces. Examination of the advantages and disadvantages of a variety of these methods has led me to my current approach.

It is well known that students of linear algebra typically find the computational problems easy but have great difficulty in understanding or applying the abstract concepts and the theory. However, with my approach, I find not only that very few students have trouble with concepts like general vector spaces but that they also retain their mastery of the linear algebra content in their upper year courses.

Although I have found my approach to be very successful with my students, I see the value in a multitude of other ways of organizing an introductory linear algebra course. Therefore, I have tried to write this book to accommodate a variety of orders. See Using This Text To Teach Linear Algebra below.

Changes to the Third Edition

- Some of the content has been reordered to make even better use of the spacing effect. The spacing effect is a well known and extensively studied effect from psychology, which states that students learn concepts better if they are exposed to the same concept multiple times over spaced intervals as opposed to learning it all at once. See:

  Dempster, F.N. (1988). *The spacing effect: A case study in the failure to apply the results of psychological research.* American Psychologist, 43(8), 627–634.

  Fain, R. J., Hieb, J. L., Ralston, P. A., Lyle, K. B. (2015, June), *Can the Spacing Effect Improve the Effectiveness of a Math Intervention Course for Engineering Students?* Paper presented at 2015 ASEE Annual Conference & Exposition, Seattle, Washington.

- The number and type of applications has been greatly increased and are used either to motivate the need for certain concepts or definitions in linear algebra, or to demonstrate how some linear algebra concepts are used in applications.

- A greater emphasis has been placed on the geometry of many concepts. In particular, Chapter 1 has been reorganized to focus on the geometry of linear algebra in $\mathbb{R}^2$ and $\mathbb{R}^3$ before exploring $\mathbb{R}^n$.

- Numerous small changes have been made to improve student comprehension.

Approach and Organization

Students of linear algebra typically have little trouble with computational questions, but they often struggle with abstract concepts and proofs. This is problematic because computers perform the computations in the vast majority of real world applications of linear algebra. Human users, meanwhile, must apply the theory to transform a given problem into a linear algebra context, input the data properly, and interpret the result correctly.

The approach of this book is both to use the spacing effect and to mix theory and computations throughout the course. Additionally, it uses real world applications to both motivate and explain the usefulness of some of the seemingly abstract concepts, and it uses the geometry of linear algebra in $\mathbb{R}^2$ and $\mathbb{R}^3$ to help students visualize many of the concepts. The benefits of this approach are as follows:

- It prevents students from mistaking linear algebra as very easy and very computational early in the course, and then getting overwhelmed by abstract concepts and theories later.

- It allows important linear algebra concepts to be developed and extended more slowly.

- It encourages students to use computational problems to help them understand the theory of linear algebra rather than blindly memorize algorithms.

- It helps students understand the concepts and why they are useful.

One example of this approach is our treatment of the concepts of spanning and linear independence. They are both introduced in Section 1.2 in $\mathbb{R}^2$ and $\mathbb{R}^3$, where they are motivated in a geometrical context. They are expanded to vectors in $\mathbb{R}^n$ in Section 1.4, and used again for matrices in Section 3.1 and polynomials in Section 4.1, before they are finally extended to general vector spaces in Section 4.2.

Other features of the text's organization include

- The idea of linear mappings is introduced early in a geometrical context, and is used to explain aspects of matrix multiplication, matrix inversion, features of systems of linear equations, and the geometry of eigenvalues and eigenvectors. Geometrical transformations provide intuitively satisfying illustrations of important concepts.

- Topics are ordered to give students a chance to work with concepts in a simpler setting before using them in a much more involved or abstract setting. For example, before reaching the definition of a vector space in Section 4.2, students will have seen the ten vector space axioms and the concepts of linear independence and spanning for three different vectors spaces, and will have had some experience in working with bases and dimensions. Thus, instead of being bombarded with new concepts at the introduction of general vector spaces, students will just be generalizing concepts with which they are already familiar.

Pedagogical Features

Since mathematics is best learned by doing, the following pedagogical elements are included in the text:

- A selection of routine mid-section exercises are provided, with answers included in the back of the book. These allow students to use and test their understanding of one concept before moving onto other concepts in the section.

- Practice problems are provided for students at the end of each section. See "A Note on the Exercises and Problems" below.

Applications

Often the applications of linear algebra are not as transparent, concise, or approachable as those of elementary calculus. Most convincing applications of linear algebra require a fairly lengthy buildup of background, which would be inappropriate in a linear algebra text. However, without some of these applications, many students would find it difficult to remain motivated to learn linear algebra. An additional difficultly is that the applications of linear algebra are so varied that there is very little agreement on which applications should be covered.

In this text we briefly discuss a few applications to give students some exposure to how linear algebra is applied.

List of Applications

- Force vectors in physics (Sections 1.1, 1.3)
- Bravais lattice (Section 1.2)
- Graphing quadratic forms (Sections 1.2, 6.2, 8.3)
- Acceleration due to forces (Section 1.3)
- Area and volume (Sections 1.3, 1.5, 5.4)
- Minimum distance from a point to a plane (Section 1.5)
- Best approximation (Section 1.5)
- Forces and moments (Section 2.1)
- Flow through a network (Sections 2.1, 2.4, 3.1)
- Spring-mass systems (Sections 2.4, 3.1, 3.5, 6.1)
- Electrical circuits (Sections 2.4, 9.2)
- Partial fraction decompositions (Section 2.4)
- Balancing chemical equations (Section 2.4)
- Planar trusses (Section 2.4)
- Linear programming (Section 2.4)
- Magic squares (Chapter 4 Review)
- Systems of Linear Difference Equations (Section 6.2)
- Markov processes (Section 6.3)
- Differential equations (Section 6.3)
- Curve of best fit (Section 7.3)

- Overdetermined systems (Section 7.3)
- Fourier series (Section 7.5)
- Small deformations (Sections 6.2, 8.4)
- Inertia tensor (Section 8.4)
- Effective rank (Section 8.5)
- Image compression (Section 8.5)

A wide variety of additional applications are mentioned throughout the text.

A Note on the Exercises and Problems

Most sections contain mid-section exercises. The purpose of these exercises is to give students a way of checking their understanding of some concepts before proceeding to further concepts in the section. Thus, when reading through a chapter, a student should always complete each exercise before continuing to read the rest of the chapter.

At the end of each section, problems are divided into A, B, and C Problems.

The A Problems are practice problems and are intended to provide a sufficient variety and number of standard computational problems and the odd theoretical problem for students to master the techniques of the course; answers are provided at the back of the text. Full solutions are available in the Student Solutions Manual.

The B Problems are homework problems. They are generally identical to the A Problems, with no answers provided, and can be used by by instructors for homework. In a few cases, the B Problems are not exactly parallel to the A Problems.

The C Problems usually require students to work with general cases, to write simple arguments, or to invent examples. These are important aspects of mastering mathematical ideas, and all students should attempt at least some of these–and not get discouraged if they make slow progress. With effort most students will be able to solve many of these problems and will benefit greatly in the understanding of the concepts and connections in doing so.

In addition to the mid-section exercises and end-of-section problems, there is a sample Chapter Quiz in the Chapter Review at the end of each chapter. Students should be aware that their instructors may have a different idea of what constitutes an appropriate test on this material.

At the end of each chapter, there are some Further Problems; these are similar to the C Problems and provide an extended investigation of certain ideas or applications of linear algebra. Further Problems are intended for advanced students who wish to challenge themselves and explore additional concepts.

Using This Text To Teach Linear Algebra

There are many different approaches to teaching linear algebra. Although we suggest covering the chapters in order, the text has been written to try to accommodate a variety of approaches.

Early Vector Spaces We believe that it is very beneficial to introduce general vector spaces immediately after students have gained some experience in working with a few specific examples of vector spaces. Students find it easier to generalize the concepts of spanning, linear independence, bases, dimension, and linear mappings while the earlier specific cases are still fresh in their minds. Additionally, we feel that it can be unhelpful to students to have determinants available too soon. Some students are far too eager to latch onto mindless algorithms involving determinants (for example, to check linear independence of three vectors in three-dimensional space), rather than actually come to terms with the defining ideas. Lastly, this allows eigenvalues, eigenvectors, and diagonalization to be focused on later in the course. I personally find that if diagonalization is taught too soon, students will focus mainly on being able to diagonalize small matrices by hand, which causes the importance of diagonalization to be lost.

Early Systems of Linear Equations For courses that begin with solving systems of linear questions, the first two sections of Chapter 2 may be covered prior to covering Chapter 1 content.

Early Determinants and Diagonalization Some reviewers have commented that they want to be able to cover determinants and diagonalization before abstract vectors spaces and that in some introductory courses abstract vector spaces may be omitted entirely. Thus, this text has been written so that Chapter 5, Chapter 6, most of Chapter 7, and Chapter 8 may be taught prior to Chapter 4 (note that all required information about subspaces, bases, and dimension for diagonalization of matrices over $\mathbb{R}$ is covered in Chapters 1, 2, and 3). Moreover, we have made sure that there is a very natural flow from matrix inverses and elementary matrices at the end of Chapter 3 to determinants in Chapter 5.

Early Complex Numbers Some introductory linear algebra courses include the use of complex numbers from the beginning. We have written Chapter 9 so that the sections of Chapter 9 may be covered immediately after covering the relevant material over $\mathbb{R}$.

A Matrix-Oriented Course For both options above, the text is organized so that sections or subsections involving linear mappings may be omitted without loss of continuity.

MyLab Math

MyLab Math and MathXL are online learning resources available to instructors and students using *An Introduction to Linear Algebra for Science and Engineering.*

MyLab Math provides engaging experiences that personalize, stimulate, and measure learning for each student. MyLab's comprehensive **online gradebook** automatically tracks your students' results on tests, quizzes, homework, and in the study plan. The homework and practice exercises in MyLab Math are correlated to the exercises in the textbook, and MyLab provides **immediate, helpful feedback** when students enter incorrect answers. The **study plan** can be assigned or used for individual practice and is personalized to each student, tracking areas for improvement as students navigate problems. With over 100 questions (all algorithmic) added to the third edition, MyLab Math for *An Introduction to Linear Algebra for Science and Engineering* is a well-equipped resource that can help improve individual students' performance.

To learn more about how MyLab combines proven learning applications with powerful assessment, visit www.pearson.com/mylab or contact your Pearson representative.

A Personal Note

The third edition of *An Introduction to Linear Algebra for Science and Engineering* is meant to engage students and pique their curiosity, as well as provide a template for instructors. I am constantly fascinated by the countless potential applications of linear algebra in everyday life, and I intend for this textbook to be approachable to all. I will not pretend that mathematical prerequisites and previous knowledge are not required. However, the approach taken in this textbook encourages the reader to explore a variety of concepts and provides exposure to an extensive amount of mathematical knowledge. Linear algebra is an exciting discipline. My hope is that those reading this book will share in my enthusiasm.

Acknowledgments

Thanks are expressed to:

Agnieszka Wolczuk for her support and encouragement.

Mike La Croix for all of the amazing figures in the text, and for his assistance in editing, formatting, and LaTeX'ing.

Peiyao Zeng, Daniel Yu, Adam Radek Martinez, Bruno Verdugo Paredes, and Alex Liao for proof-reading and their many valuable comments and suggestions.

Stephen New, Paul McGrath, Ken McCay, Paul Kates, and many other of my colleagues who have helped me become a better instructor.

To all of the reviewers whose comments, corrections, and recommendations have resulted in many positive improvements.

Charlotte Morrison-Reed for all of her hard work in making the third edition of this text possible and for her suggestions and editing.

A very special thank you to Daniel Norman and all those who contributed to the first and second editions.

Dan Wolczuk
University of Waterloo

CHAPTER 1

Euclidean Vector Spaces

CHAPTER OUTLINE

Some of the material in this chapter will be familiar to many students, but some ideas that are introduced here will be new to most. In this chapter we will look at operations on and important concepts related to vectors. We will also look at some applications of vectors in the familiar setting of Euclidean space. Most of these concepts will later be extended to more general settings. A firm understanding of the material from this chapter will help greatly in understanding the topics in the rest of this book.

1.1 Vectors in $\mathbb{R}^2$ and $\mathbb{R}^3$

We begin by considering the two-dimensional plane in Cartesian coordinates. Choose an origin O and two mutually perpendicular axes, called the x_1-axis and the x_2-axis, as shown in Figure 1.1.1. Any point P in the plane can be uniquely identified by the 2-tuple (p_1, p_2), called the **coordinates** of P. In particular, p_1 is the distance from P to the x_2-axis, with p_1 positive if P is to the right of this axis and negative if P is to the left, and p_2 is the distance from P to the x_1-axis, with p_2 positive if P is above this axis and negative if P is below. You have already learned how to plot graphs of equations in this plane.

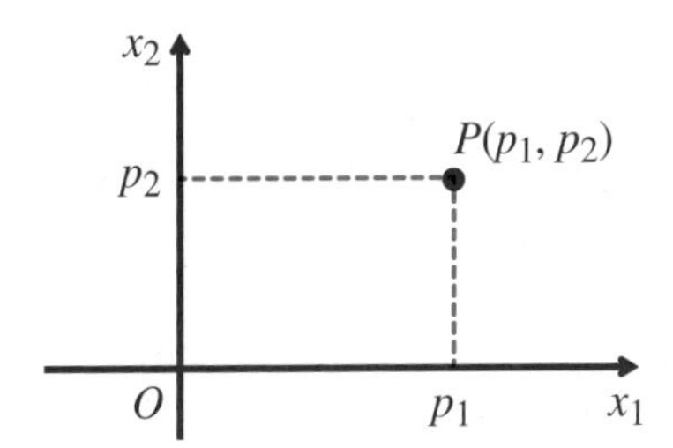

Figure 1.1.1 Coordinates in the plane.

For applications in many areas of mathematics, and in many subjects such as physics, chemistry, economics, and engineering, it is useful to view points more abstractly. In particular, we will view them as **vectors** and provide rules for adding them and multiplying them by constants.

Definition
$\mathbb{R}^2$

We let $\mathbb{R}^2$ denote the set of all vectors of the form $\begin{bmatrix} x_1 \\ x_2 \end{bmatrix}$, where x_1 and x_2 are real numbers called the **components** of the vector. Mathematically, we write

$$\mathbb{R}^2 = \left\{ \begin{bmatrix} x_1 \\ x_2 \end{bmatrix} \mid x_1, x_2 \in \mathbb{R} \right\}$$

We say two vectors $\begin{bmatrix} x_1 \\ x_2 \end{bmatrix}, \begin{bmatrix} y_1 \\ y_2 \end{bmatrix} \in \mathbb{R}^2$ are **equal** if $x_1 = y_1$ and $x_2 = y_2$. We write

$$\begin{bmatrix} x_1 \\ x_2 \end{bmatrix} = \begin{bmatrix} y_1 \\ y_2 \end{bmatrix}$$

Although we are viewing the elements of $\mathbb{R}^2$ as vectors, we can still interpret these geometrically as points. That is, the vector $\vec{p} = \begin{bmatrix} p_1 \\ p_2 \end{bmatrix}$ can be interpreted as the point $P(p_1, p_2)$. Graphically, this is often represented by drawing an arrow from $(0, 0)$ to (p_1, p_2), as shown in Figure 1.1.2. Note, that the point $(0, 0)$ and the points between $(0, 0)$ and (p_1, p_2) should not be thought of as points "on the vector." The representation of a vector as an arrow is particularly common in physics; force and acceleration are vector quantities that can conveniently be represented by an arrow of suitable magnitude and direction.

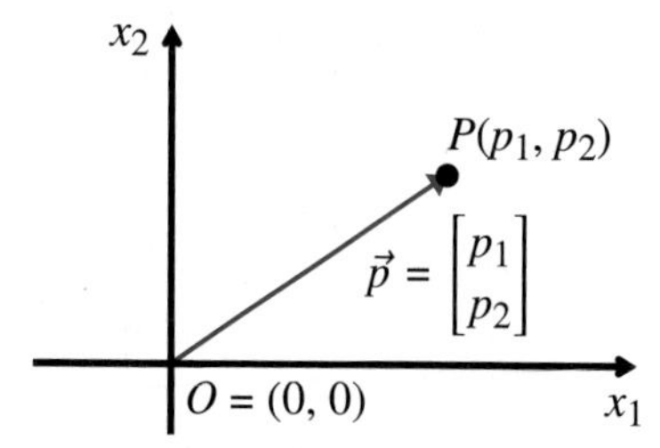

Figure 1.1.2 Graphical representation of a vector.

EXAMPLE 1.1.1

An object on a frictionless surface is being pulled by two strings with force and direction as given in the diagram.

(a) Represent each force as a vector in $\mathbb{R}^2$.
(b) Represent the net force being applied to the object as a vector in $\mathbb{R}^2$.

Solution: (a) The force F_1 has $150N$ of horizontal force and $0N$ of vertical force. Thus, we can represent this with the vector

$$\vec{F}_1 = \begin{bmatrix} 150 \\ 0 \end{bmatrix}$$

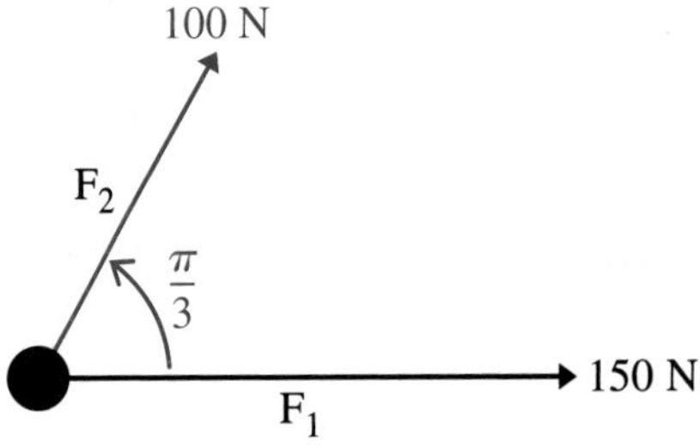

The force F_2 has horizontal component $100\cos\frac{\pi}{3} = 50$ N and vertical component $100\sin\frac{\pi}{3} = 50\sqrt{3}$ N. Therefore, we can represent this with the vector

$$\vec{F}_2 = \begin{bmatrix} 50 \\ 50\sqrt{3} \end{bmatrix}$$

(b) We know from physics that to get the net force we add the horizontal components of the forces together and we add the vertical components of the forces together. Thus, the net horizontal component is $150N + 50N = 200N$. The net vertical force is $0N + 50\sqrt{3}N = 50\sqrt{3}N$. We can represent this as the vector

$$\vec{F} = \begin{bmatrix} 200 \\ 50\sqrt{3} \end{bmatrix}$$

The example shows that in physics we add vectors by adding their corresponding components. Similarly, we find that in physics we multiply a vector by a scalar by multiplying each component of the vector by the scalar.

Since we want our generalized concept of vectors to be able to help us solve physical problems like these and more, we define addition and scalar multiplication of vectors in $\mathbb{R}^2$ to match.

Definition
Addition and Scalar Multiplication in $\mathbb{R}^2$

Let $\vec{x} = \begin{bmatrix} x_1 \\ x_2 \end{bmatrix}, \vec{y} = \begin{bmatrix} y_1 \\ y_2 \end{bmatrix} \in \mathbb{R}^2$. We define **addition** of vectors by

$$\vec{x} + \vec{y} = \begin{bmatrix} x_1 \\ x_2 \end{bmatrix} + \begin{bmatrix} y_1 \\ y_2 \end{bmatrix} = \begin{bmatrix} x_1 + y_1 \\ x_2 + y_2 \end{bmatrix}$$

We define **scalar multiplication** of $\vec{x}$ by a factor of $t \in \mathbb{R}$, called a **scalar**, by

$$t\vec{x} = t\begin{bmatrix} x_1 \\ x_2 \end{bmatrix} = \begin{bmatrix} tx_1 \\ tx_2 \end{bmatrix}$$

Remark

It is important to note that $\vec{x} - \vec{y}$ is to be interpreted as $\vec{x} + (-1)\vec{y}$.

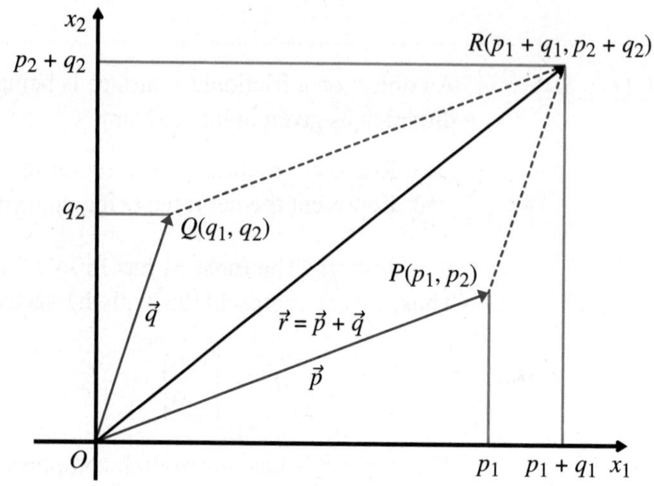

Figure 1.1.3 Addition of vectors $\vec{p}$ and $\vec{q}$.

The addition of two vectors is illustrated in Figure 1.1.3: construct a parallelogram with vectors $\vec{p}$ and $\vec{q}$ as adjacent sides; then $\vec{p} + \vec{q}$ is the vector corresponding to the vertex of the parallelogram opposite to the origin. Observe that the components really are added according to the definition. This is often called the **parallelogram rule for addition**.

EXAMPLE 1.1.2

Let $\vec{x} = \begin{bmatrix} -2 \\ 3 \end{bmatrix}, \vec{y} = \begin{bmatrix} 5 \\ 1 \end{bmatrix} \in \mathbb{R}^2$. Calculate $\vec{x} + \vec{y}$.

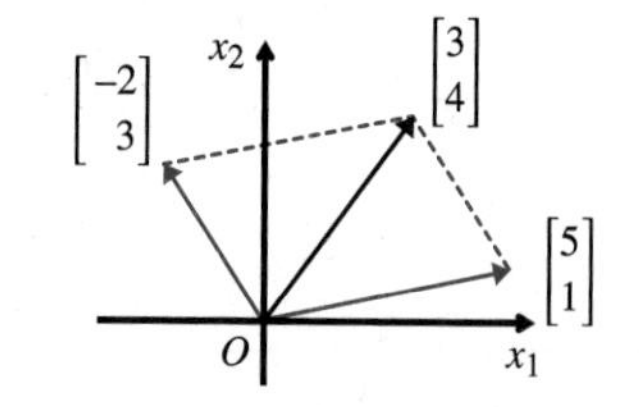

Solution: We have $\vec{x} + \vec{y} = \begin{bmatrix} -2 + 5 \\ 3 + 1 \end{bmatrix} = \begin{bmatrix} 3 \\ 4 \end{bmatrix}$.

Scalar multiplication is illustrated in Figure 1.1.4. Observe that multiplication by a negative scalar reverses the direction of the vector.

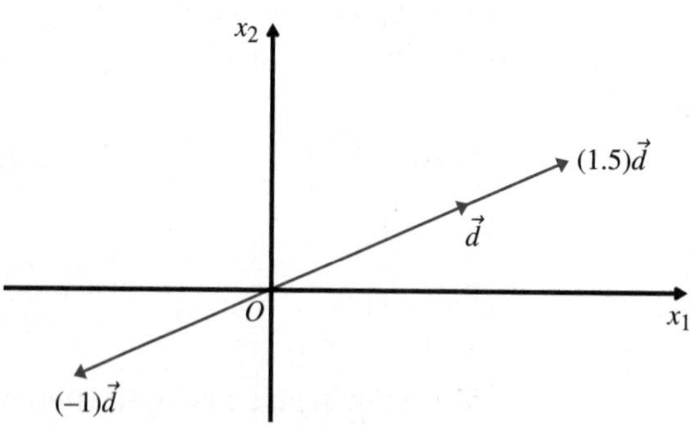

Figure 1.1.4 Scalar multiplication of the vector $\vec{d}$.

EXAMPLE 1.1.3

Let $\vec{u} = \begin{bmatrix} 3 \\ 1 \end{bmatrix}, \vec{v} = \begin{bmatrix} -2 \\ 3 \end{bmatrix}, \vec{w} = \begin{bmatrix} 0 \\ -1 \end{bmatrix} \in \mathbb{R}^2$. Calculate $\vec{u} + \vec{v}$, $3\vec{w}$, and $2\vec{v} - \vec{w}$.

Solution: We get

$$\vec{u} + \vec{v} = \begin{bmatrix} 3 \\ 1 \end{bmatrix} + \begin{bmatrix} -2 \\ 3 \end{bmatrix} = \begin{bmatrix} 1 \\ 4 \end{bmatrix}$$

$$3\vec{w} = 3\begin{bmatrix} 0 \\ -1 \end{bmatrix} = \begin{bmatrix} 0 \\ -3 \end{bmatrix}$$

$$2\vec{v} - \vec{w} = 2\begin{bmatrix} -2 \\ 3 \end{bmatrix} + (-1)\begin{bmatrix} 0 \\ -1 \end{bmatrix} = \begin{bmatrix} -4 \\ 6 \end{bmatrix} + \begin{bmatrix} 0 \\ 1 \end{bmatrix} = \begin{bmatrix} -4 \\ 7 \end{bmatrix}$$

EXERCISE 1.1.1

Let $\vec{u} = \begin{bmatrix} 1 \\ -1 \end{bmatrix}, \vec{v} = \begin{bmatrix} 2 \\ 1 \end{bmatrix}, \vec{w} = \begin{bmatrix} 0 \\ 1 \end{bmatrix} \in \mathbb{R}^2$. Calculate each of the following and illustrate with a sketch.

(a) $\vec{u} + \vec{w}$ (b) $-\vec{v}$ (c) $(\vec{u} + \vec{w}) - \vec{v}$

We will frequently look at sums of scalar multiples of vectors. So, we make the following definition.

Definition
Linear Combination

Let $\vec{v}_1, \dots, \vec{v}_k \in \mathbb{R}^2$ and $c_1, \dots, c_k \in \mathbb{R}$. We call the sum $c_1\vec{v}_1 + \cdots + c_k\vec{v}_k$ a **linear combination** of the vectors $\vec{v}_1, \dots, \vec{v}_k$.

It is important to observe that $\mathbb{R}^2$ has the property that any linear combination of vectors in $\mathbb{R}^2$ is a vector in $\mathbb{R}^2$ (combining properties V1, V6 in Theorem 1.1.1 below). Although this property is clear for $\mathbb{R}^2$, it does not hold for most subsets of $\mathbb{R}^2$. As we will see in Section 1.4, in linear algebra, we are mostly interested in sets that have this property.

Theorem 1.1.1

For all $\vec{w}, \vec{x}, \vec{y} \in \mathbb{R}^2$ and $s, t \in \mathbb{R}$ we have

V1 $\vec{x} + \vec{y} \in \mathbb{R}^2$ (closed under addition)
V2 $\vec{x} + \vec{y} = \vec{y} + \vec{x}$ (addition is commutative)
V3 $(\vec{x} + \vec{y}) + \vec{w} = \vec{x} + (\vec{y} + \vec{w})$ (addition is associative)
V4 There exists a vector $\vec{0} \in \mathbb{R}^2$ such that $\vec{z} + \vec{0} = \vec{z}$ for all $\vec{z} \in \mathbb{R}^2$ (zero vector)
V5 For each $\vec{x} \in \mathbb{R}^2$ there exists a vector $-\vec{x} \in \mathbb{R}^2$ such that $\vec{x} + (-\vec{x}) = \vec{0}$ (additive inverses)
V6 $s\vec{x} \in \mathbb{R}^2$ (closed under scalar multiplication)
V7 $s(t\vec{x}) = (st)\vec{x}$ (scalar multiplication is associative)
V8 $(s + t)\vec{x} = s\vec{x} + t\vec{x}$ (a distributive law)
V9 $s(\vec{x} + \vec{y}) = s\vec{x} + s\vec{y}$ (another distributive law)
V10 $1\vec{x} = \vec{x}$ (scalar multiplicative identity)

Observe that the zero vector from property V4 is the vector $\vec{0} = \begin{bmatrix} 0 \\ 0 \end{bmatrix}$, and the additive inverse of $\vec{x}$ from V5 is $-\vec{x} = (-1)\vec{x}$.

The Vector Equation of a Line in $\mathbb{R}^2$

In Figure 1.1.4, it is apparent that the set of all multiples of a non-zero vector $\vec{d}$ creates a line through the origin. We make this our definition of a line in $\mathbb{R}^2$: a **line through the origin in** $\mathbb{R}^2$ is a set of the form

$$\{t\vec{d} \mid t \in \mathbb{R}\}$$

Often we do not use formal set notation but simply write a **vector equation** of the line:

$$\vec{x} = t\vec{d}, \quad t \in \mathbb{R}$$

The non-zero vector $\vec{d}$ is called a **direction vector** of the line.

Similarly, we define a **line through** $\vec{p}$ **with direction vector** $\vec{d} \neq \vec{0}$ to be the set

$$\{\vec{p} + t\vec{d} \mid t \in \mathbb{R}\}$$

which has vector equation

$$\vec{x} = \vec{p} + t\vec{d}, \quad t \in \mathbb{R}$$

This line is parallel to the line with equation $\vec{x} = t\vec{d}, t \in \mathbb{R}$ because of the parallelogram rule for addition. As shown in Figure 1.1.5, each point on the line through $\vec{p}$ can be obtained from a corresponding point on the line $\vec{x} = t\vec{d}, t \in \mathbb{R}$ by adding the vector $\vec{p}$. We say that the line has been **translated** by $\vec{p}$. More generally, two lines are parallel if the direction vector of one line is a non-zero scalar multiple of the direction vector of the other line.

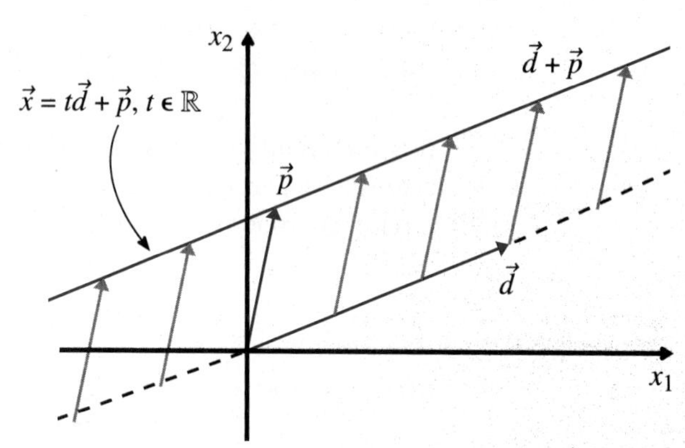

Figure 1.1.5 The line with vector equation $\vec{x} = t\vec{d} + \vec{p}, t \in \mathbb{R}$.

EXAMPLE 1.1.4

A vector equation of the line through the point $P(2, -3)$ with direction vector $\begin{bmatrix} -4 \\ 5 \end{bmatrix}$ is

$$\vec{x} = \begin{bmatrix} 2 \\ -3 \end{bmatrix} + t\begin{bmatrix} -4 \\ 5 \end{bmatrix}, \quad t \in \mathbb{R}$$

EXAMPLE 1.1.5

Write a vector equation of the line through $P(1, 2)$ parallel to the line with vector equation

$$\vec{x} = t\begin{bmatrix}2\\3\end{bmatrix}, \quad t \in \mathbb{R}$$

Solution: Since they are parallel, we can choose the same direction vector. Hence, a vector equation of the line is

$$\vec{x} = \begin{bmatrix}1\\2\end{bmatrix} + t\begin{bmatrix}2\\3\end{bmatrix}, \quad t \in \mathbb{R}$$

EXERCISE 1.1.2

Write a vector equation of a line through $P(0, 0)$ parallel to the line

$$\vec{x} = \begin{bmatrix}4\\-3\end{bmatrix} + t\begin{bmatrix}3\\-1\end{bmatrix}, \quad t \in \mathbb{R}$$

Sometimes the components of a vector equation are written separately. In particular, expanding a vector equation $\vec{x} = \vec{p} + t\vec{d}, t \in \mathbb{R}$ we get

$$\begin{bmatrix}x_1\\x_2\end{bmatrix} = \begin{bmatrix}p_1\\p_2\end{bmatrix} + t\begin{bmatrix}d_1\\d_2\end{bmatrix} = \begin{bmatrix}p_1 + td_1\\p_2 + td_2\end{bmatrix}$$

Comparing entries, we get **parametric equations** of the line:

$$\begin{cases} x_1 = p_1 + td_1 \\ x_2 = p_2 + td_2, \end{cases} \quad t \in \mathbb{R}$$

The familiar **scalar equation** of the line is obtained by eliminating the parameter t. Provided that $d_1 \neq 0$ we solve the first equation for t to get

$$\frac{x_1 - p_1}{d_1} = t$$

Substituting this into the second equation gives the scalar equation

$$x_2 = p_2 + \frac{d_2}{d_1}(x_1 - p_1) \tag{1.1}$$

What can you say about the line if $d_1 = 0$?

EXAMPLE 1.1.6

Write a vector equation, a scalar equation, and parametric equations of the line passing through the point $P(3, 4)$ with direction vector $\begin{bmatrix}-5\\1\end{bmatrix}$.

Solution: A vector equation is $\begin{bmatrix}x_1\\x_2\end{bmatrix} = \begin{bmatrix}3\\4\end{bmatrix} + t\begin{bmatrix}-5\\1\end{bmatrix}, t \in \mathbb{R}$.

So, parametric equations are $\begin{cases} x_1 = 3 - 5t \\ x_2 = 4 + t, \end{cases} \quad t \in \mathbb{R}.$

Hence, a scalar equation is $x_2 = 4 - \frac{1}{5}(x_1 - 3)$.

Directed Line Segments

For dealing with certain geometrical problems, it is useful to introduce **directed line segments**. We denote the directed line segment from point P to point Q by $\vec{PQ}$ as in Figure 1.1.6. We think of it as an "arrow" starting at P and pointing towards Q. We shall identify directed line segments from the origin O with the corresponding vectors; we write $\vec{OP} = \vec{p}$, $\vec{OQ} = \vec{q}$, and so on. A directed line segment that starts at the origin is called the **position vector** of the point.

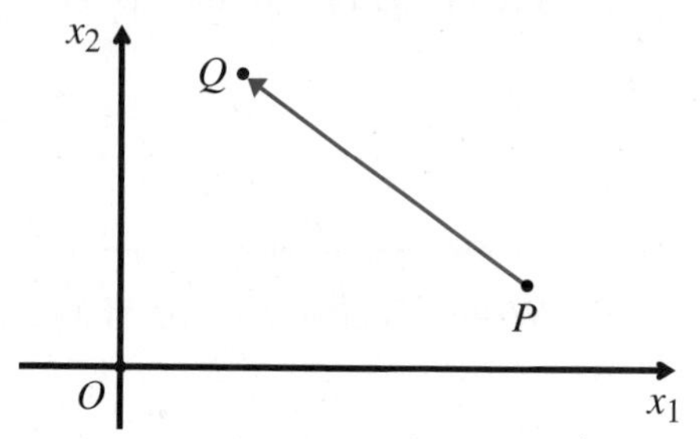

Figure 1.1.6 The directed line segment $\vec{PQ}$ from P to Q.

For many problems, we are interested only in the direction and length of the directed line segment; we are not interested in the point where it is located. For example, in Figure 1.1.3 on page 4, we may wish to treat the line segment $\vec{QR}$ as if it were the same as $\vec{OP}$. Taking our cue from this example, for arbitrary points P, Q, R in $\mathbb{R}^2$, we define $\vec{QR}$ to be **equivalent** to $\vec{OP}$ if $\vec{r} - \vec{q} = \vec{p}$. In this case, we have used one directed line segment $\vec{OP}$ starting from the origin in our definition.

More generally, for arbitrary points Q, R, S, and T in $\mathbb{R}^2$, we define $\vec{QR}$ to be equivalent to $\vec{ST}$ if they are both equivalent to the same $\vec{OP}$ for some P. That is, if

$$\vec{r} - \vec{q} = \vec{p} \text{ and } \vec{t} - \vec{s} = \vec{p} \text{ for the same } \vec{p}$$

We can abbreviate this by simply requiring that

$$\vec{r} - \vec{q} = \vec{t} - \vec{s}$$

EXAMPLE 1.1.7

For points $Q(1,3)$, $R(6,-1)$, $S(-2,4)$, and $T(3,0)$, we have that $\vec{QR}$ is equivalent to $\vec{ST}$ because

$$\vec{r} - \vec{q} = \begin{bmatrix} 6 \\ -1 \end{bmatrix} - \begin{bmatrix} 1 \\ 3 \end{bmatrix} = \begin{bmatrix} 5 \\ -4 \end{bmatrix} = \begin{bmatrix} 3 \\ 0 \end{bmatrix} - \begin{bmatrix} -2 \\ 4 \end{bmatrix} = \vec{t} - \vec{s}$$

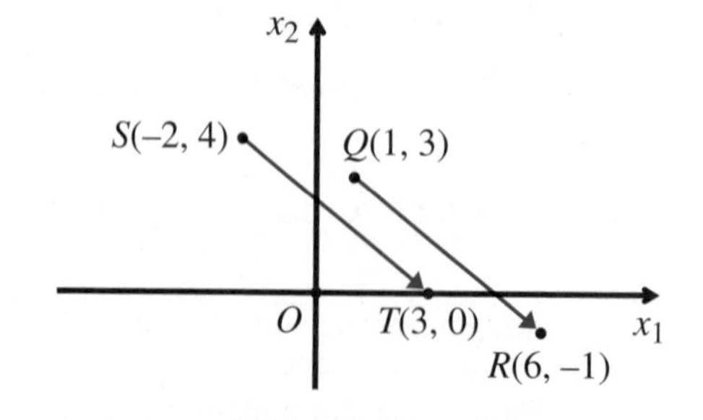

In some problems, where it is not necessary to distinguish between equivalent directed line segments, we "identify" them (that is, we treat them as the same object) and write $\vec{PQ} = \vec{RS}$. Indeed, we identify them with the corresponding line segment starting at the origin, so in Example 1.1.7 we write $\vec{QR} = \vec{ST} = \begin{bmatrix} 5 \\ -4 \end{bmatrix}$.

Remark

Writing $\vec{QR} = \vec{ST}$ is a bit sloppy—an abuse of notation—because $\vec{QR}$ is not really the same object as $\vec{ST}$. However, introducing the precise language of "equivalence classes" and more careful notation with directed line segments is not helpful at this stage. By introducing directed line segments, we are encouraged to think about vectors that are located at arbitrary points in space. This is helpful in solving some geometrical problems, as we shall see below.

EXAMPLE 1.1.8

Find a vector equation of the line through $P(1,2)$ and $Q(3,-1)$.

Solution: A direction vector of the line is

$$\vec{PQ} = \vec{q} - \vec{p} = \begin{bmatrix} 3 \\ -1 \end{bmatrix} - \begin{bmatrix} 1 \\ 2 \end{bmatrix} = \begin{bmatrix} 2 \\ -3 \end{bmatrix}$$

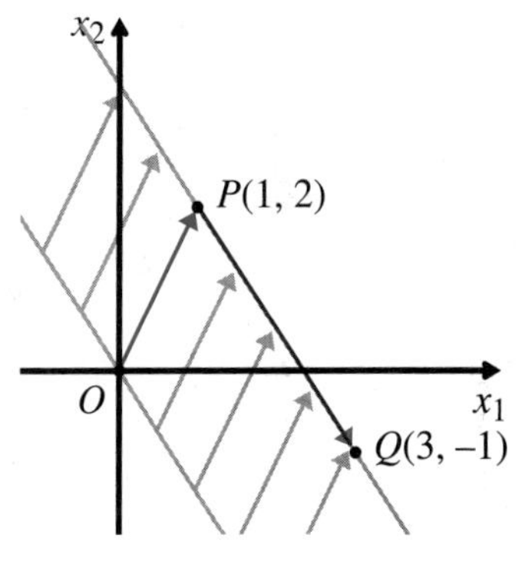

Hence, a vector equation of the line with direction $\vec{PQ}$ that passes through $P(1,2)$ is

$$\vec{x} = \vec{p} + t\vec{PQ} = \begin{bmatrix} 1 \\ 2 \end{bmatrix} + t\begin{bmatrix} 2 \\ -3 \end{bmatrix}, \quad t \in \mathbb{R}$$

Observe in the example above that we would have the same line if we started at the second point and "moved" toward the first point—or even if we took a direction vector in the opposite direction. Thus, the same line is described by the vector equations

$$\vec{x} = \begin{bmatrix} 3 \\ -1 \end{bmatrix} + r\begin{bmatrix} -2 \\ 3 \end{bmatrix}, \quad r \in \mathbb{R}$$

$$\vec{x} = \begin{bmatrix} 3 \\ -1 \end{bmatrix} + s\begin{bmatrix} 2 \\ -3 \end{bmatrix}, \quad s \in \mathbb{R}$$

$$\vec{x} = \begin{bmatrix} 1 \\ 2 \end{bmatrix} + t\begin{bmatrix} -2 \\ 3 \end{bmatrix}, \quad t \in \mathbb{R}$$

In fact, there are infinitely many descriptions of a line: we may choose any point on the line, and we may use any non-zero scalar multiple of the direction vector.

EXERCISE 1.1.3

Find a vector equation of the line through $P(1,1)$ and $Q(-2,2)$.

Vectors, Lines, and Planes in $\mathbb{R}^3$

Everything we have done so far works perfectly well in three dimensions. We choose an origin O and three mutually perpendicular axes, as shown in Figure 1.1.7. The x_1-axis is usually pictured coming out of the page (or screen), the x_2-axis to the right, and the x_3-axis towards the top of the picture.

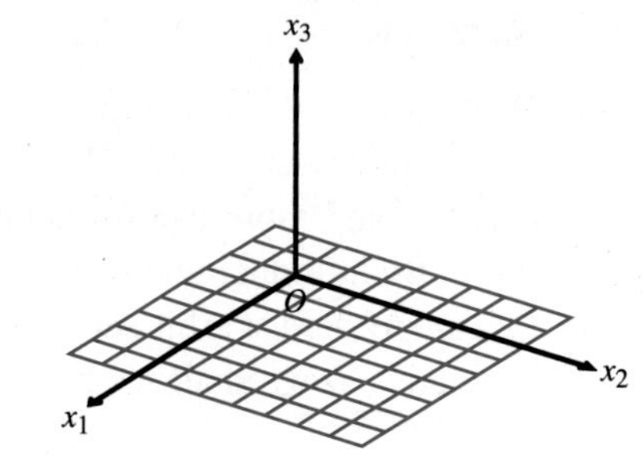

Figure 1.1.7 The positive coordinate axes in $\mathbb{R}^3$.

It should be noted that we are adopting the convention that the coordinate axes form a **right-handed system**. One way to visualize a right-handed system is to spread out the thumb, index finger, and middle finger of your right hand. The thumb is the x_1-axis; the index finger is the x_2-axis; and the middle finger is the x_3-axis. See Figure 1.1.8.

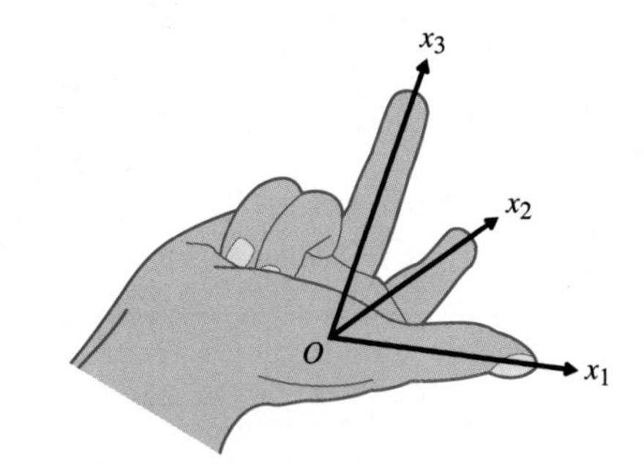

Figure 1.1.8 Identifying a right-handed system.

We now define $\mathbb{R}^3$ to be the three-dimensional analog of $\mathbb{R}^2$.

Definition
$\mathbb{R}^3$

We define $\mathbb{R}^3$ to be the set of all vectors of the form $\begin{bmatrix} x_1 \\ x_2 \\ x_3 \end{bmatrix}$, with $x_1, x_2, x_3 \in \mathbb{R}$.

Mathematically,

$$\mathbb{R}^3 = \left\{ \begin{bmatrix} x_1 \\ x_2 \\ x_3 \end{bmatrix} \mid x_1, x_2, x_3 \in \mathbb{R} \right\}$$

We say two vectors $\vec{x} = \begin{bmatrix} x_1 \\ x_2 \\ x_3 \end{bmatrix}, \vec{y} = \begin{bmatrix} y_1 \\ y_2 \\ y_3 \end{bmatrix}$ are **equal** and write $\vec{x} = \vec{y}$ if $x_i = y_i$, for $i = 1, 2, 3$.

Definition
Addition and Scalar Multiplication in $\mathbb{R}^3$

Let $\vec{x} = \begin{bmatrix} x_1 \\ x_2 \\ x_3 \end{bmatrix}, \vec{y} = \begin{bmatrix} y_1 \\ y_2 \\ y_3 \end{bmatrix} \in \mathbb{R}^3$. We define **addition** of vectors by

$$\vec{x} + \vec{y} = \begin{bmatrix} x_1 \\ x_2 \\ x_3 \end{bmatrix} + \begin{bmatrix} y_1 \\ y_2 \\ y_3 \end{bmatrix} = \begin{bmatrix} x_1 + y_1 \\ x_2 + y_2 \\ x_3 + y_3 \end{bmatrix}$$

We define the **scalar multiplication** of $\vec{x}$ by a **scalar** $t \in \mathbb{R}$ by

$$t\vec{x} = t\begin{bmatrix} x_1 \\ x_2 \\ x_3 \end{bmatrix} = \begin{bmatrix} tx_1 \\ tx_2 \\ tx_3 \end{bmatrix}$$

Addition still follows the parallelogram rule. It may help you to visualize this if you realize that two vectors in $\mathbb{R}^3$ must lie within a plane in $\mathbb{R}^3$ so that the two-dimensional picture is still valid. See Figure 1.1.9.

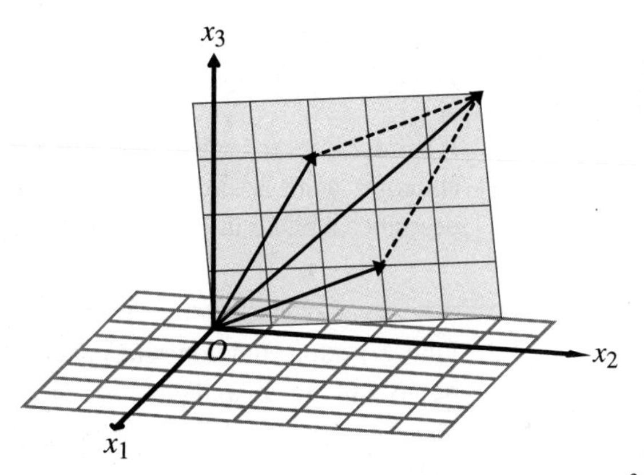

Figure 1.1.9 Two-dimensional parallelogram rule in $\mathbb{R}^3$.

EXAMPLE 1.1.9

Let $\vec{u} = \begin{bmatrix} 1 \\ 1 \\ -1 \end{bmatrix}, \vec{v} = \begin{bmatrix} -2 \\ 1 \\ 2 \end{bmatrix}, \vec{w} = \begin{bmatrix} 1 \\ 0 \\ 1 \end{bmatrix} \in \mathbb{R}^3$. Calculate $\vec{v} + \vec{u}$, $-\vec{w}$, and $-\vec{v} + 2\vec{w} - \vec{u}$.

Solution: We have

$$\vec{v} + \vec{u} = \begin{bmatrix} -2 \\ 1 \\ 2 \end{bmatrix} + \begin{bmatrix} 1 \\ 1 \\ -1 \end{bmatrix} = \begin{bmatrix} -1 \\ 2 \\ 1 \end{bmatrix}$$

$$-\vec{w} = -\begin{bmatrix} 1 \\ 0 \\ 1 \end{bmatrix} = \begin{bmatrix} -1 \\ 0 \\ -1 \end{bmatrix}$$

$$-\vec{v} + 2\vec{w} - \vec{u} = -\begin{bmatrix} -2 \\ 1 \\ 2 \end{bmatrix} + 2\begin{bmatrix} 1 \\ 0 \\ 1 \end{bmatrix} - \begin{bmatrix} 1 \\ 1 \\ -1 \end{bmatrix} = \begin{bmatrix} 2 \\ -1 \\ -2 \end{bmatrix} + \begin{bmatrix} 2 \\ 0 \\ 2 \end{bmatrix} + \begin{bmatrix} -1 \\ -1 \\ 1 \end{bmatrix} = \begin{bmatrix} 3 \\ -2 \\ 1 \end{bmatrix}$$

EXERCISE 1.1.4

Let $\vec{x} = \begin{bmatrix} 2 \\ 1 \\ -5 \end{bmatrix}, \vec{y} = \begin{bmatrix} 1 \\ 2 \\ 1 \end{bmatrix}, \vec{z} = \begin{bmatrix} 3 \\ 0 \\ -1 \end{bmatrix} \in \mathbb{R}^3$. Calculate $2\vec{x} - \vec{y} + 3\vec{z}$.

As before, we call a sum of scalar multiples of vectors in $\mathbb{R}^3$ a linear combination. Moreover, of course, vectors in $\mathbb{R}^3$ satisfy all the same properties in Theorem 1.1.1 replacing $\mathbb{R}^2$ by $\mathbb{R}^3$ in properties V1, V4, V5, and V6.

The zero vector in $\mathbb{R}^3$ is $\vec{0} = \begin{bmatrix} 0 \\ 0 \\ 0 \end{bmatrix}$ and the additive inverse of $\vec{x} \in \mathbb{R}^3$ is $-\vec{x} = (-1)\vec{x}$.

Directed line segments are the same in three-dimensional space as in the two-dimensional case.

The line through the point P in $\mathbb{R}^3$ (corresponding to a vector $\vec{p}$) with direction vector $\vec{d} \neq \vec{0}$ can be described by a vector equation:

$$\vec{x} = \vec{p} + t\vec{d}, \quad t \in \mathbb{R}$$

CONNECTION

It is important to realize that a line in $\mathbb{R}^3$ cannot be described by a single scalar equation, as in $\mathbb{R}^2$. We shall see in Section 1.3 that a single scalar equation in $\mathbb{R}^3$ describes a plane in $\mathbb{R}^3$.

EXAMPLE 1.1.10

Find a vector equation and parametric equations of the line that passes through the points $P(1, 5, -2)$ and $Q(4, -1, 3)$.

Solution: A direction vector is $\vec{PQ} = \begin{bmatrix} 4-1 \\ -1-5 \\ 3-(-2) \end{bmatrix} = \begin{bmatrix} 3 \\ -6 \\ 5 \end{bmatrix}$. Hence, a vector equation of the line is

$$\vec{x} = \begin{bmatrix} 1 \\ 5 \\ -2 \end{bmatrix} + t\begin{bmatrix} 3 \\ -6 \\ 5 \end{bmatrix}, \quad t \in \mathbb{R}$$

Hence, we have

$$\begin{bmatrix} x_1 \\ x_2 \\ x_3 \end{bmatrix} = \begin{bmatrix} 1+3t \\ 5-6t \\ -2+5t \end{bmatrix}, \quad t \in \mathbb{R}$$

Consequently, corresponding parametric equations are

$$\begin{cases} x_1 = 1 + 3t \\ x_2 = 5 - 6t \\ x_3 = -2 + 5t, \end{cases} \quad t \in \mathbb{R}$$

EXERCISE 1.1.5

Find a vector equation and parametric equations of the line that passes through the points $P(1,2,2)$ and $Q(1,-2,3)$.

Let $\vec{u}$ and $\vec{v}$ be vectors in $\mathbb{R}^3$ that are not scalar multiples of each other. This implies that the sets $\{t\vec{u} \mid t \in \mathbb{R}\}$ and $\{s\vec{v} \mid s \in \mathbb{R}\}$ are both lines in $\mathbb{R}^3$ through the origin in different directions. Thus, the set of all possible linear combinations of $\vec{u}$ and $\vec{v}$ forms a two-dimensional plane. That is, the set

$$\{t\vec{u} + s\vec{v} \mid s,t \in \mathbb{R}\}$$

is a **plane through the origin in** $\mathbb{R}^3$. As we did with lines, we could say that

$$\{\vec{p} + t\vec{u} + s\vec{v} \mid s,t \in \mathbb{R}\}$$

is a **plane through** $\vec{p}$ **in** $\mathbb{R}^3$ and that

$$\vec{x} = \vec{p} + t\vec{u} + s\vec{v}, \quad s,t \in \mathbb{R}$$

is a **vector equation** for the plane. It is very important to note that if either $\vec{u}$ or $\vec{v}$ is a scalar multiple of the other, then the set $\{t\vec{u} + s\vec{v} \mid s,t \in \mathbb{R}\}$ would *not* be a plane.

EXAMPLE 1.1.11

Determine which of the following vectors are in the plane with vector equation

$$\vec{x} = s\begin{bmatrix}1\\0\\1\end{bmatrix} + t\begin{bmatrix}1\\2\\2\end{bmatrix}, \quad s,t \in \mathbb{R}$$

(a) $\vec{p} = \begin{bmatrix}5/2\\1\\3\end{bmatrix}$ (b) $\vec{q} = \begin{bmatrix}1\\1\\1\end{bmatrix}$

Solution: (a) The vector $\vec{p}$ is in the plane if and only if there are scalars $s,t \in \mathbb{R}$ such that

$$\begin{bmatrix}5/2\\1\\3\end{bmatrix} = s\begin{bmatrix}1\\0\\1\end{bmatrix} + t\begin{bmatrix}1\\2\\2\end{bmatrix}$$

Performing the linear combination on the right-hand side gives

$$\begin{bmatrix}5/2\\1\\3\end{bmatrix} = \begin{bmatrix}s+t\\2t\\s+2t\end{bmatrix}$$

For these vectors to be equal we must have

$$s + t = 5/2, \qquad 2t = 1, \qquad s + 2t = 3$$

EXAMPLE 1.1.11 (continued)

We find that the values $t = 1/2$ and $s = 2$ satisfies all three equations. Hence, we have found that $\vec{p}$ is in the plane. In particular,

$$\begin{bmatrix} 5/2 \\ 1 \\ 3 \end{bmatrix} = 2\begin{bmatrix} 1 \\ 0 \\ 1 \end{bmatrix} + \frac{1}{2}\begin{bmatrix} 1 \\ 2 \\ 2 \end{bmatrix}$$

(b) We need to determine whether there exists $s, t \in \mathbb{R}$ such that

$$\begin{bmatrix} 1 \\ 1 \\ 1 \end{bmatrix} = s\begin{bmatrix} 1 \\ 0 \\ 1 \end{bmatrix} + t\begin{bmatrix} 1 \\ 2 \\ 2 \end{bmatrix}$$

Performing the linear combination on the right-hand side gives

$$\begin{bmatrix} 1 \\ 1 \\ 1 \end{bmatrix} = \begin{bmatrix} s+t \\ 2t \\ s+2t \end{bmatrix}$$

Since vectors are equal only if they have equal components, for these vectors to be equal we must have

$$s + t = 1, \qquad 2t = 1, \qquad s + 2t = 1$$

The middle equation shows us that we must have $t = 1/2$. However, then we would need $s = 1/2$ for the first equation and $s = 0$ for the third equation. Therefore, there is no value of s and t that satisfies all three equations. Thus, $\vec{q}$ is not a linear combination of $\begin{bmatrix} 1 \\ 0 \\ 1 \end{bmatrix}$ and $\begin{bmatrix} 1 \\ 2 \\ 2 \end{bmatrix}$, so $\vec{q}$ is not in the plane.

EXERCISE 1.1.6

Consider the plane in $\mathbb{R}^3$ with vector equation

$$\vec{x} = \begin{bmatrix} 1 \\ 1 \\ 1 \end{bmatrix} + s\begin{bmatrix} 1 \\ 1 \\ 0 \end{bmatrix} + t\begin{bmatrix} 1 \\ 0 \\ 1 \end{bmatrix}, \quad s, t \in \mathbb{R}$$

Find two vectors $\vec{u}, \vec{v} \in \mathbb{R}^3$ that are in the plane, and find a vector $\vec{w} \in \mathbb{R}^3$ that is not in the plane.

EXERCISE 1.1.7

Find two non-zero vectors $\vec{u}$ and $\vec{v}$ in $\mathbb{R}^3$ such that $\{s\vec{u} + t\vec{v} \mid s, t \in \mathbb{R}\}$ is a line in $\mathbb{R}^3$.

PROBLEMS 1.1

Practice Problems

In Problems A1–A4, compute the given linear combination in $\mathbb{R}^2$ and illustrate with a sketch.

A1 $\begin{bmatrix}1\\4\end{bmatrix}+\begin{bmatrix}2\\3\end{bmatrix}$

A2 $\begin{bmatrix}3\\2\end{bmatrix}-\begin{bmatrix}4\\1\end{bmatrix}$

A3 $3\begin{bmatrix}-1\\4\end{bmatrix}$

A4 $2\begin{bmatrix}2\\1\end{bmatrix}-2\begin{bmatrix}3\\-1\end{bmatrix}$

In Problems A5–A10, compute the given linear combination in $\mathbb{R}^2$.

A5 $\begin{bmatrix}4\\-2\end{bmatrix}+\begin{bmatrix}-1\\3\end{bmatrix}$

A6 $\begin{bmatrix}-3\\-4\end{bmatrix}-\begin{bmatrix}-2\\5\end{bmatrix}$

A7 $-2\begin{bmatrix}3\\-2\end{bmatrix}$

A8 $\frac{1}{2}\begin{bmatrix}2\\6\end{bmatrix}+\frac{1}{3}\begin{bmatrix}4\\3\end{bmatrix}$

A9 $\frac{2}{3}\begin{bmatrix}3\\1\end{bmatrix}-2\begin{bmatrix}1/4\\1/3\end{bmatrix}$

A10 $\sqrt{2}\begin{bmatrix}\sqrt{2}\\\sqrt{3}\end{bmatrix}+3\begin{bmatrix}1\\\sqrt{6}\end{bmatrix}$

In Problems A11–A16, compute the given linear combination in $\mathbb{R}^3$.

A11 $\begin{bmatrix}2\\3\\4\end{bmatrix}-\begin{bmatrix}5\\1\\-2\end{bmatrix}$

A12 $\begin{bmatrix}2\\1\\-6\end{bmatrix}+\begin{bmatrix}-3\\1\\-4\end{bmatrix}$

A13 $-6\begin{bmatrix}4\\-5\\-6\end{bmatrix}$

A14 $-2\begin{bmatrix}-5\\1\\1\end{bmatrix}+3\begin{bmatrix}-1\\0\\-1\end{bmatrix}$

A15 $2\begin{bmatrix}2/3\\-1/3\\2\end{bmatrix}+\frac{1}{3}\begin{bmatrix}3\\-2\\1\end{bmatrix}$

A16 $\sqrt{2}\begin{bmatrix}1\\1\\1\end{bmatrix}+\pi\begin{bmatrix}-1\\0\\1\end{bmatrix}$

A17 Let $\vec{v}=\begin{bmatrix}1\\2\\-2\end{bmatrix}$ and $\vec{w}=\begin{bmatrix}2\\-1\\3\end{bmatrix}$. Determine

(a) $2\vec{v}-3\vec{w}$
(b) $-3(\vec{v}+2\vec{w})+5\vec{v}$
(c) $\vec{u}$ such that $\vec{w}-2\vec{u}=3\vec{v}$
(d) $\vec{u}$ such that $\vec{u}-3\vec{v}=2\vec{u}$

A18 Let $\vec{v}=\begin{bmatrix}3\\1\\1\end{bmatrix}$ and $\vec{w}=\begin{bmatrix}5\\-1\\-2\end{bmatrix}$. Determine

(a) $\frac{1}{2}\vec{v}+\frac{1}{2}\vec{w}$
(b) $2(\vec{v}+\vec{w})-(2\vec{v}-3\vec{w})$
(c) $\vec{u}$ such that $\vec{w}-\vec{u}=2\vec{v}$
(d) $\vec{u}$ such that $\frac{1}{2}\vec{u}+\frac{1}{3}\vec{v}=\vec{w}$

A19 Consider the points $P(2,3,1)$, $Q(3,1,-2)$, $R(1,4,0)$, and $S(-5,1,5)$. Determine $\vec{PQ}$, $\vec{PR}$, $\vec{PS}$, $\vec{QR}$, and $\vec{SR}$. Verify that $\vec{PQ}+\vec{QR}=\vec{PR}=\vec{PS}+\vec{SR}$.

For Problems A20–A23, write a vector equation for the line passing through the given point with the given direction vector.

A20 $P(3,4)$, $\vec{d}=\begin{bmatrix}-5\\1\end{bmatrix}$

A21 $P(2,3)$, $\vec{d}=\begin{bmatrix}-4\\-6\end{bmatrix}$

A22 $P(2,0,5)$, $\vec{d}=\begin{bmatrix}4\\-2\\-11\end{bmatrix}$

A23 $P(4,1,5)$, $\vec{d}=\begin{bmatrix}-2\\1\\2\end{bmatrix}$

For Problems A24–A28, write a vector equation for the line that passes through the given points.

A24 $P(-1,2)$, $Q(2,-3)$

A25 $P(4,1)$, $Q(-2,-1)$

A26 $P(1,3,-5)$, $Q(-2,1,0)$

A27 $P(-2,1,1)$, $Q(4,2,2)$

A28 $P\left(\frac{1}{2},\frac{1}{4},1\right)$, $Q\left(-1,1,\frac{1}{3}\right)$

For Problems A29–A32, determine parametric equations and a scalar equation for the line that passes through the given points.

A29 $P(-1,2)$, $Q(2,-3)$

A30 $P(1,1)$, $Q(2,2)$

A31 $P(1,0)$, $Q(3,0)$

A32 $P(1,3)$, $Q(-1,5)$

A33 (a) A set of points is **collinear** if all the points lie on the same line. By considering directed line segments, give a general method for determining whether a given set of three points is collinear.
(b) Determine whether the points $P(1,2)$, $Q(4,1)$, and $R(-5,4)$ are collinear. Show how you decide.
(c) Determine whether the points $S(1,0,1)$, $T(3,-2,3)$, and $U(-3,4,-1)$ are collinear. Show how you decide.

A34 Prove properties V2 and V8 of Theorem 1.1.1.

A35 Consider the object from Example 1.1.1. If the force F_1 is tripled to $450N$ and the force F_2 is halved to $50N$, then what is the vector representing the net force being applied to the object?

Homework Problems

In Problems B1–B4, compute the given linear combination and illustrate with a sketch.

B1 $\begin{bmatrix}1\\2\end{bmatrix}+\begin{bmatrix}-1\\2\end{bmatrix}$

B2 $\begin{bmatrix}-3\\4\end{bmatrix}-\begin{bmatrix}2\\1\end{bmatrix}$

B3 $2\begin{bmatrix}1\\3\end{bmatrix}$

B4 $3\begin{bmatrix}2\\5\end{bmatrix}-\begin{bmatrix}1\\-1\end{bmatrix}$

In Problems B5–B9, compute the given linear combination in $\mathbb{R}^2$.

B5 $\begin{bmatrix}7\\11\end{bmatrix}+\begin{bmatrix}-2\\4\end{bmatrix}$

B6 $\begin{bmatrix}2\\-5\end{bmatrix}-\begin{bmatrix}3\\-3\end{bmatrix}$

B7 $5\begin{bmatrix}3\\-2\end{bmatrix}$

B8 $\frac{3}{4}\begin{bmatrix}2\\6\end{bmatrix}-\frac{1}{4}\begin{bmatrix}3\\-1\end{bmatrix}$

B9 $\sqrt{2}\begin{bmatrix}\sqrt{2}\\2\end{bmatrix}+\sqrt{6}\begin{bmatrix}1\\-\sqrt{3}\end{bmatrix}-\begin{bmatrix}\sqrt{6}\\\sqrt{2}\end{bmatrix}$

In Problems B10–B16, compute the given linear combination in $\mathbb{R}^3$.

B10 $\begin{bmatrix}3\\2\\-1\end{bmatrix}-\begin{bmatrix}3\\5\\8\end{bmatrix}$

B11 $\begin{bmatrix}4\\2\\2\end{bmatrix}+\begin{bmatrix}-3\\2\\-3\end{bmatrix}$

B12 $\begin{bmatrix}1\\2\\1\end{bmatrix}+\begin{bmatrix}3\\1\\0\end{bmatrix}+\begin{bmatrix}0\\1\\1\end{bmatrix}$

B13 $3\begin{bmatrix}1\\2\\5\end{bmatrix}$

B14 $0\begin{bmatrix}2\\-5\\-1\end{bmatrix}$

B15 $\frac{2}{3}\begin{bmatrix}3\\5\\-1\end{bmatrix}+\frac{1}{3}\begin{bmatrix}-6\\-10\\2\end{bmatrix}$

B16 $(1+\sqrt{2})\begin{bmatrix}1+\sqrt{2}\\0\\\sqrt{2}-1\end{bmatrix}-\begin{bmatrix}\sqrt{2}\\0\\2\end{bmatrix}$

B17 Let $\vec{v}=\begin{bmatrix}2\\4\\3\end{bmatrix}$ and $\vec{w}=\begin{bmatrix}2\\-2\\-1\end{bmatrix}$. Determine

(a) $3\vec{v}-2\vec{w}$
(b) $-2(\vec{v}-2\vec{w})+3\vec{w}$
(c) $\vec{u}$ such that $\vec{w}+\vec{u}=2\vec{v}$
(d) $\vec{u}$ such that $2\vec{u}+3\vec{w}=-\vec{v}$

B18 Let $\vec{v}=\begin{bmatrix}1\\3\\-1\end{bmatrix}$ and $\vec{w}=\begin{bmatrix}2\\4\\3\end{bmatrix}$. Determine

(a) $3\vec{v}+4\vec{w}$
(b) $-\frac{1}{2}\vec{v}+\frac{3}{4}\vec{w}$
(c) $\vec{u}$ such that $2\vec{v}+\vec{u}=\vec{v}$
(d) $\vec{u}$ such that $3\vec{u}-2\vec{w}=\vec{v}$

For Problems B19 and B20, determine $\vec{PQ}$, $\vec{PR}$, $\vec{PS}$, $\vec{QR}$, and $\vec{SR}$, and verify that $\vec{PQ}+\vec{QR}=\vec{PR}=\vec{PS}+\vec{SR}$.

B19 $P(2,3,2)$, $Q(5,4,1)$, $R(-2,3,-1)$, $S(7,-3,4)$

B20 $P(-2,3,-1)$, $Q(4,5,1)$, $R(-2,-1,0)$, and $S(3,1,-1)$

For Problems B21–B26, write a vector equation for the line passing through the given point with the given direction vector.

B21 $P(2,-1)$, $\vec{d}=\begin{bmatrix}-3\\2\end{bmatrix}$

B22 $P(0,0,0)$, $\vec{d}=\begin{bmatrix}2\\2\\1\end{bmatrix}$

B23 $P(3,1)$, $\vec{d}=\begin{bmatrix}1\\2\end{bmatrix}$

B24 $P(1,-1,2)$, $\vec{d}=\begin{bmatrix}1\\1\\0\end{bmatrix}$

B25 $P(1,1,1)$, $\vec{d}=\begin{bmatrix}1\\0\\1\end{bmatrix}$

B26 $P(-2,3,1)$, $\vec{d}=\begin{bmatrix}2\\3\\1\end{bmatrix}$

For Problems B27–B32, write a vector equation for the line that passes through the given points.

B27 $P(2,4)$, $Q(1,2)$

B28 $P(-2,5)$, $Q(-1,-1)$

B29 $P(1,3,2)$, $Q(0,0,0)$

B30 $P(0,1,4)$, $Q(-1,2,2)$

B31 $P(-2,6,1)$, $Q(-2,5,1)$

B32 $P\left(1,2,\frac{1}{2}\right)$, $Q\left(\frac{1}{2},\frac{1}{3},0\right)$

For Problems B33–B38, determine parametric equations and a scalar equation for the line that passes through the given points.

B33 $P(2,5)$, $Q(3,3)$

B34 $P(3,-1)$, $Q(6,1)$

B35 $P(0,3)$, $Q(1,-5)$

B36 $P(-3,1)$, $Q(4,1)$

B37 $P(2,0)$, $Q(0,-3)$

B38 $P(5,-2)$, $Q(6,3)$

For Problems B39–B41, use the solution from Problem A33 (a) to determine whether the given points are collinear. Show how you decide.

B39 $P(1,1)$, $Q(4,3)$, $R(-5,-3)$

B40 $P(2,-1,2)$, $Q(3,2,3)$, $R(1,-4,0)$

B41 $S(0,4,4)$, $T(-1,5,6)$, $U(4,0,-4)$

Conceptual Problems

C1 Let $\vec{v} = \begin{bmatrix} 1 \\ 1 \end{bmatrix}$ and $\vec{w} = \begin{bmatrix} 1 \\ -1 \end{bmatrix}$.

(a) Find real numbers t_1 and t_2 such that $t_1\vec{v} + t_2\vec{w} = \begin{bmatrix} 3 \\ -2 \end{bmatrix}$. Illustrate with a sketch.

(b) Find real numbers t_1 and t_2 such that $t_1\vec{v} + t_2\vec{w} = \begin{bmatrix} x_1 \\ x_2 \end{bmatrix}$ for any $x_1, x_2 \in \mathbb{R}$.

(c) Use your result in part (b) to find real numbers t_1 and t_2 such that $t_1\vec{v}_1 + t_2\vec{v}_2 = \begin{bmatrix} \sqrt{2} \\ \pi \end{bmatrix}$.

C2 Let P, Q, and R be points in $\mathbb{R}^2$ corresponding to vectors $\vec{p}$, $\vec{q}$, and $\vec{r}$ respectively.

(a) Explain in terms of directed line segments why

$$\vec{PQ} + \vec{QR} + \vec{RP} = \vec{0}$$

(b) Verify the equation of part (a) by expressing $\vec{PQ}$, $\vec{QR}$, and $\vec{RP}$ in terms of $\vec{p}$, $\vec{q}$, and $\vec{r}$.

For Problems C3 and C4, let $\vec{x}$ and $\vec{y}$ be vectors in $\mathbb{R}^3$ and $s, t \in \mathbb{R}$.

C3 Prove that $s(t\vec{x}) = (st)\vec{x}$

C4 Prove that $s(\vec{x} + \vec{y}) = s\vec{x} + s\vec{y}$

C5 Let $\vec{p}$ and $\vec{d} \neq \vec{0}$ be vectors in $\mathbb{R}^2$. Prove that $\vec{x} = \vec{p} + t\vec{d}$, $t \in \mathbb{R}$, is a line in $\mathbb{R}^2$ passing through the origin if and only if $\vec{p}$ is a scalar multiple of $\vec{d}$.

C6 Let $\vec{p}, \vec{u}, \vec{v} \in \mathbb{R}^3$ such that $\vec{u}$ and $\vec{v}$ are not scalar multiples of each other. Prove that $\vec{x} = \vec{p} + s\vec{u} + t\vec{v}$, $s, t \in \mathbb{R}$ is a plane in $\mathbb{R}^3$ passing through the origin if and only if $\vec{p}$ is a linear combination of $\vec{u}$ and $\vec{v}$.

C7 Let O, Q, P, and R be the corner points of a parallelogram (see Figure 1.1.3). Prove that the two diagonals of the parallelogram $\vec{OR}$ and $\vec{PQ}$ bisect each other.

C8 Let $A(a_1, a_2)$ and $B(b_1, b_2)$ be points in $\mathbb{R}^2$. Find the coordinates of the point 1/3 of the way from the point A to the point B.

C9 Consider the plane in $\mathbb{R}^3$ with vector equation

$$\vec{x} = \begin{bmatrix} 2 \\ 1 \\ 0 \end{bmatrix} + s\begin{bmatrix} 1 \\ 2 \\ 3 \end{bmatrix} + t\begin{bmatrix} 1 \\ 1 \\ 2 \end{bmatrix}, \quad s, t \in \mathbb{R}$$

(a) Find parametric equations for the plane.

(b) Use the parametric equations you found in (a) to find a scalar equation for the plane.

C10 We have seen how to use a vector equation of a line to find parametric equations, and how to use parametric equations to find a scalar equation of the line. In this exercise, we will perform these steps in reverse. Let $ax_1 + bx_2 = c$ be the scalar equation of a line in $\mathbb{R}^2$ with a and b both non-zero.

(a) Find parametric equations for the line by setting $x_2 = t$ and solving for x_1.

(b) Substitute the parametric equations into the vector $\vec{x} = \begin{bmatrix} x_1 \\ x_2 \end{bmatrix}$ and use operations on vectors to write $\vec{x}$ in the form $\vec{x} = \vec{p} + t\vec{d}$, $t \in \mathbb{R}$.

(c) Find a vector equation of the line $2x_1 + 3x_2 = 5$.

(d) Find a vector equation of the line $x_1 = 3$.

C11 Let L be a line in $\mathbb{R}^2$ with vector equation $\vec{x} = t\begin{bmatrix} d_1 \\ d_2 \end{bmatrix}$, $t \in \mathbb{R}$. Prove that a point $P(p_1, p_2)$ is on the line L if and only if $p_1 d_2 = p_2 d_1$.

C12 Show that if two lines in $\mathbb{R}^2$ are not parallel to each other, then they must have a point of intersection. (Hint: Use the result of Problem C11.)

1.2 Spanning and Linear Independence in $\mathbb{R}^2$ and $\mathbb{R}^3$

In this section we will give a preview of some important concepts in linear algebra. We will use the geometry of $\mathbb{R}^2$ and $\mathbb{R}^3$ to help you visualize and understand these concepts.

Spanning in $\mathbb{R}^2$ and $\mathbb{R}^3$

We saw in the previous section that lines in $\mathbb{R}^2$ and $\mathbb{R}^3$ that pass through the origin have the form

$$\vec{x} = t\vec{d}, \quad t \in \mathbb{R}$$

for some non-zero direction vector $\vec{d}$. That is, such a line is the set of all possible scalar multiples of $\vec{d}$.

Similarly, we saw that planes in $\mathbb{R}^3$ that pass through the origin have the form

$$\vec{x} = s\vec{u} + t\vec{v}, \quad s, t \in \mathbb{R}$$

where neither $\vec{u}$ nor $\vec{v}$ is a scalar multiple of the other. Hence, such a plane is the set of all possible linear combinations of $\vec{u}$ and $\vec{v}$.

Sets of all possible linear combinations (or scalar multiples in the case of a single vector) are extremely important in linear algebra. We make the following definition.

Definition
Span in $\mathbb{R}^2$

Let $B = \{\vec{v}_1, \dots, \vec{v}_k\}$ be a set of vectors in $\mathbb{R}^2$ or a set of vectors in $\mathbb{R}^3$. We define the **span** of B, denoted Span B, to be the set of all possible linear combinations of the vectors in B. Mathematically,

$$\operatorname{Span} B = \{c_1\vec{v}_1 + \cdots + c_k\vec{v}_k \mid c_1, \dots, c_k \in \mathbb{R}\}$$

A **vector equation** for Span B is

$$\vec{x} = c_1\vec{v}_1 + \cdots + c_k\vec{v}_k, \quad c_1, \dots, c_k \in \mathbb{R}$$

If $S = \operatorname{Span} B$, then we say that B **spans** S, that B is a **spanning set** for S, and that S is **spanned** by B.

EXAMPLE 1.2.1

Describe the set spanned by $\left\{\begin{bmatrix}1\\2\end{bmatrix}\right\}$ geometrically.

Solution: A vector equation for the spanned set is

$$\vec{x} = s\begin{bmatrix}1\\2\end{bmatrix}, \quad s \in \mathbb{R}$$

Thus, the spanned set is a line in $\mathbb{R}^2$ through the origin with direction vector $\begin{bmatrix}1\\2\end{bmatrix}$.

EXAMPLE 1.2.2

Is the vector $\begin{bmatrix} 3 \\ 1 \end{bmatrix}$ in Span $\left\{ \begin{bmatrix} 1 \\ 2 \end{bmatrix}, \begin{bmatrix} -1 \\ 1 \end{bmatrix} \right\}$?

Solution: Using the definition of span, the vector $\begin{bmatrix} 3 \\ 1 \end{bmatrix}$ is in the spanned set if it can be written as a linear combination of the vectors in the spanning set. That is, we need to determine whether there exists $c_1, c_2 \in \mathbb{R}$ such that

$$\begin{bmatrix} 3 \\ 1 \end{bmatrix} = c_1 \begin{bmatrix} 1 \\ 2 \end{bmatrix} + c_2 \begin{bmatrix} -1 \\ 1 \end{bmatrix}$$

Performing operations on vectors on the right-hand side gives

$$\begin{bmatrix} 3 \\ 1 \end{bmatrix} = \begin{bmatrix} c_1 - c_2 \\ 2c_1 + c_2 \end{bmatrix}$$

Since vectors are equal if and only if their corresponding entries are equal, we get that this vector equation implies

$$\begin{aligned} 3 &= c_1 - c_2 \\ 1 &= 2c_1 + c_2 \end{aligned}$$

Adding the equations gives $4 = 3c_1$ and so $c_1 = \frac{4}{3}$.
Substituting this into either equation gives $c_2 = -\frac{5}{3}$. Hence, we have that

$$\begin{bmatrix} 3 \\ 1 \end{bmatrix} = \frac{4}{3} \begin{bmatrix} 1 \\ 2 \end{bmatrix} - \frac{5}{3} \begin{bmatrix} -1 \\ 1 \end{bmatrix}$$

Thus, by definition, $\begin{bmatrix} 3 \\ 1 \end{bmatrix} \in \text{Span} \left\{ \begin{bmatrix} 1 \\ 2 \end{bmatrix}, \begin{bmatrix} -1 \\ 1 \end{bmatrix} \right\}$.

EXAMPLE 1.2.3

Let $\vec{e}_1 = \begin{bmatrix} 1 \\ 0 \end{bmatrix}$ and $\vec{e}_2 = \begin{bmatrix} 0 \\ 1 \end{bmatrix}$. Show that $\text{Span}\{\vec{e}_1, \vec{e}_2\} = \mathbb{R}^2$.

Solution: We need to show that every vector in $\mathbb{R}^2$ can be written as a linear combination of the vectors $\vec{e}_1$ and $\vec{e}_2$. We pick a general vector $\vec{x} = \begin{bmatrix} x_1 \\ x_2 \end{bmatrix}$ in $\mathbb{R}^2$. We need to determine whether there exists $c_1, c_2 \in \mathbb{R}$ such that

$$\begin{bmatrix} x_1 \\ x_2 \end{bmatrix} = c_1 \begin{bmatrix} 1 \\ 0 \end{bmatrix} + c_2 \begin{bmatrix} 0 \\ 1 \end{bmatrix}$$

We observe that we can take $c_1 = x_1$ and $c_2 = x_2$. That is, we have

$$\begin{bmatrix} x_1 \\ x_2 \end{bmatrix} = x_1 \begin{bmatrix} 1 \\ 0 \end{bmatrix} + x_2 \begin{bmatrix} 0 \\ 1 \end{bmatrix}$$

So, $\text{Span}\{\vec{e}_1, \vec{e}_2\} = \mathbb{R}^2$.

We have just shown that every vector in $\mathbb{R}^2$ can be written as a unique linear combination of the vectors $\vec{e}_1$ and $\vec{e}_2$. This is not surprising since Span$\{\vec{e}_1\}$ is the x_1-axis and Span$\{\vec{e}_2\}$ is the x_2-axis. In particular, when we write $\begin{bmatrix} x_1 \\ x_2 \end{bmatrix} = x_1\vec{e}_1 + x_2\vec{e}_2$ we are really just writing the vector $\vec{x}$ in terms of its standard coordinates. We call the set $\{\vec{e}_1, \vec{e}_2\}$ the **standard basis** for $\mathbb{R}^2$.

Remark

In physics and engineering, it is common to use the notation $\mathbf{i} = \begin{bmatrix} 1 \\ 0 \end{bmatrix}$ and $\mathbf{j} = \begin{bmatrix} 0 \\ 1 \end{bmatrix}$ instead of $\vec{e}_1$ and $\vec{e}_2$.

Using the definition of span, we have that if $\vec{d} \in \mathbb{R}^2$ with $\vec{d} \neq \vec{0}$, then geometrically Span$\{\vec{d}\}$ is a line through the origin in $\mathbb{R}^2$. If $\vec{u}, \vec{v} \in \mathbb{R}^2$ with $\vec{u} \neq \vec{0}$ and $\vec{v} \neq \vec{0}$, then what is Span$\{\vec{u}, \vec{v}\}$ geometrically? It is tempting to say that the set $\{\vec{u}, \vec{v}\}$ would span $\mathbb{R}^2$. However, as demonstrated in the next example, this does not have to be true.

EXAMPLE 1.2.4

Describe Span $\left\{ \begin{bmatrix} 3 \\ 2 \end{bmatrix}, \begin{bmatrix} 6 \\ 4 \end{bmatrix} \right\}$ geometrically.

Solution: Using the definition of span, a vector equation of the spanned set is

$$\vec{x} = s\begin{bmatrix} 3 \\ 2 \end{bmatrix} + t\begin{bmatrix} 6 \\ 4 \end{bmatrix}, \quad s, t \in \mathbb{R}$$

Observe that we can rewrite this as

$$\begin{aligned} \vec{x} &= s\begin{bmatrix} 3 \\ 2 \end{bmatrix} + (2t)\begin{bmatrix} 3 \\ 2 \end{bmatrix}, \quad s, t \in \mathbb{R} \\ &= (s + 2t)\begin{bmatrix} 3 \\ 2 \end{bmatrix}, \quad s, t \in \mathbb{R} \end{aligned}$$

Since $c = s + 2t$ can take any real value, the spanned set is a line through the origin with direction vector $\begin{bmatrix} 3 \\ 2 \end{bmatrix}$.

Hence, before we can describe a spanned set geometrically, we must first see if we can simplify the spanning set.

EXAMPLE 1.2.5

Describe Span $\left\{ \begin{bmatrix} 3 \\ 1 \\ -3 \end{bmatrix}, \begin{bmatrix} 0 \\ -1 \\ 1 \end{bmatrix}, \begin{bmatrix} 3 \\ 0 \\ -2 \end{bmatrix} \right\}$ geometrically.

Solution: By definition, a vector equation for the spanned set is

$$\vec{x} = c_1 \begin{bmatrix} 3 \\ 1 \\ -3 \end{bmatrix} + c_2 \begin{bmatrix} 0 \\ -1 \\ 1 \end{bmatrix} + c_3 \begin{bmatrix} 3 \\ 0 \\ -2 \end{bmatrix}, \quad c_1, c_2, c_3 \in \mathbb{R}$$

We observe that $\begin{bmatrix} 3 \\ 1 \\ -3 \end{bmatrix} + \begin{bmatrix} 0 \\ -1 \\ 1 \end{bmatrix} = \begin{bmatrix} 3 \\ 0 \\ -2 \end{bmatrix}$. Hence, we can rewrite the vector equation as

$$\begin{aligned} \vec{x} &= c_1 \begin{bmatrix} 3 \\ 1 \\ -3 \end{bmatrix} + c_2 \begin{bmatrix} 0 \\ -1 \\ 1 \end{bmatrix} + c_3 \left(\begin{bmatrix} 3 \\ 1 \\ -3 \end{bmatrix} + \begin{bmatrix} 0 \\ -1 \\ 1 \end{bmatrix} \right), \quad c_1, c_2, c_3 \in \mathbb{R} \\ &= (c_1 + c_3) \begin{bmatrix} 3 \\ 1 \\ -3 \end{bmatrix} + (c_2 + c_3) \begin{bmatrix} 0 \\ -1 \\ 1 \end{bmatrix}, \quad c_1, c_2, c_3 \in \mathbb{R} \end{aligned}$$

Since $\begin{bmatrix} 3 \\ 1 \\ -3 \end{bmatrix}$ and $\begin{bmatrix} 0 \\ -1 \\ 1 \end{bmatrix}$ are not scalar multiples of each other, we cannot simplify the vector equation any more. Thus, the set is a plane with vector equation

$$\vec{x} = s \begin{bmatrix} 3 \\ 1 \\ -3 \end{bmatrix} + t \begin{bmatrix} 0 \\ -1 \\ 1 \end{bmatrix}, \quad s, t \in \mathbb{R}$$

In Example 1.2.4, we used the fact that the second vector was a scalar multiple of the first to simplify the vector equation. In Example 1.2.5, we used the fact that the third vector could be written as a linear combination of the first two vectors to simplify the spanning set. Rather than having to perform these steps each time, we create a theorem to help us.

Theorem 1.2.1

Let $\mathcal{B} = \{\vec{v}_1, \dots, \vec{v}_k\}$ be a set of vectors in $\mathbb{R}^2$ or a set of vectors in $\mathbb{R}^3$. Some vector $\vec{v}_i$, $1 \leq i \leq k$, can be written as a linear combination of $\vec{v}_1, \dots, \vec{v}_{i-1}, \vec{v}_{i+1}, \dots, \vec{v}_k$ if and only if

$$\text{Span}\{\vec{v}_1, \dots, \vec{v}_k\} = \text{Span}\{\vec{v}_1, \dots, \vec{v}_{i-1}, \vec{v}_{i+1}, \dots, \vec{v}_k\}$$

This theorem shows that if one vector $\vec{v}_i$ in the spanning set can be written as a linear combination of the other vectors, then $\vec{v}_i$ can be removed from the spanning set without changing the set that is being spanned.

EXERCISE 1.2.1

Use Theorem 1.2.1 to find a simplified spanning set for each of the following sets.

(a) Span $\left\{\begin{bmatrix}1\\1\end{bmatrix}, \begin{bmatrix}-1\\1\end{bmatrix}\right\}$ (b) Span $\left\{\begin{bmatrix}-1\\2\\1\end{bmatrix}, \begin{bmatrix}-2\\4\\2\end{bmatrix}, \begin{bmatrix}1\\-2\\0\end{bmatrix}\right\}$

Linear Independence and Bases in $\mathbb{R}^2$ and $\mathbb{R}^3$

Examples 1.2.4 and 1.2.5 show that it is important to identify if a spanning set is as simple as possible. For example, if $\vec{v}_1, \vec{v}_2, \vec{v}_3 \in \mathbb{R}^3$, then it is impossible to determine the geometric interpretation of Span$\{\vec{v}_1, \vec{v}_2, \vec{v}_3\}$ without knowing if one of the vectors in the spanning set can be removed using Theorem 1.2.1. We now look at a mathematical way of determining if one vector in a set can be written as a linear combination of the others.

Definition
Linearly Dependent
Linearly Independent

Let $\mathcal{B} = \{\vec{v}_1, \dots, \vec{v}_k\}$ be a set of vectors in $\mathbb{R}^2$ or a set of vectors in $\mathbb{R}^3$. The set $\mathcal{B}$ is said to be **linearly dependent** if there exist real coefficients $c_1, \dots, c_k$ not all zero such that

$$c_1\vec{v}_1 + \cdots + c_k\vec{v}_k = \vec{0}$$

The set $\mathcal{B}$ is said to be **linearly independent** if the only solution to

$$c_1\vec{v}_1 + \cdots + c_k\vec{v}_k = \vec{0}$$

is $c_1 = c_2 = \cdots = c_k = 0$ (called the **trivial solution**).

EXAMPLE 1.2.6

Determine whether the set $\mathcal{B} = \left\{\begin{bmatrix}1\\0\\1\end{bmatrix}, \begin{bmatrix}0\\1\\1\end{bmatrix}, \begin{bmatrix}1\\1\\0\end{bmatrix}\right\}$ in $\mathbb{R}^3$ is linearly independent.

Solution: By definition, we need to find all solutions of the equation

$$\begin{bmatrix}0\\0\\0\end{bmatrix} = c_1\begin{bmatrix}1\\0\\1\end{bmatrix} + c_2\begin{bmatrix}0\\1\\1\end{bmatrix} + c_3\begin{bmatrix}1\\1\\0\end{bmatrix}$$

Performing the linear combination on the right-hand side gives

$$\begin{bmatrix}0\\0\\0\end{bmatrix} = \begin{bmatrix}c_1 + c_3\\c_2 + c_3\\c_1 + c_2\end{bmatrix}$$

Comparing entries gives the system of equations

$$c_1 + c_3 = 0, \qquad c_2 + c_3 = 0, \qquad c_1 + c_2 = 0$$

Adding the first to the second and then subtracting the third gives $2c_3 = 0$. Hence, $c_3 = 0$ which then implies $c_1 = c_2 = 0$ from the first and second equations.

Since $c_1 = c_2 = c_3 = 0$ is the only solution, the set is linearly independent.

EXAMPLE 1.2.7

Determine whether the set $C = \left\{ \begin{bmatrix} 1 \\ 1 \end{bmatrix}, \begin{bmatrix} 1 \\ 0 \end{bmatrix}, \begin{bmatrix} 2 \\ 2 \end{bmatrix} \right\}$ is linearly independent or linearly dependent.

Solution: We consider the equation

$$\begin{bmatrix} 0 \\ 0 \end{bmatrix} = c_1 \begin{bmatrix} 1 \\ 1 \end{bmatrix} + c_2 \begin{bmatrix} 1 \\ 0 \end{bmatrix} + c_3 \begin{bmatrix} 2 \\ 2 \end{bmatrix}$$

We observe that taking $c_1 = -2$, $c_2 = 0$, and $c_3 = 1$ satisfies the equation. Hence, by definition, the set C is linearly dependent.

As desired, the definition of linear independence/linear dependence gives us the following theorem.

Theorem 1.2.2

Let $\mathcal{B} = \{\vec{v}_1, \dots, \vec{v}_k\}$ be a set of vectors in $\mathbb{R}^2$ or a set of vectors in $\mathbb{R}^3$. The set $\mathcal{B}$ is linearly dependent if and only if $\vec{v}_i \in \text{Span}\{\vec{v}_1, \dots, \vec{v}_{i-1}, \vec{v}_{i+1}, \dots, \vec{v}_k\}$ for some i, $1 \leq i \leq k$.

Theorem 1.2.2 tells us that a set $\mathcal{B}$ is linearly independent if and only if none of the vectors in $\mathcal{B}$ can be written as a linear combination of the others. That is, the simplest spanning set $\mathcal{B}$ for a given set S is one that is linearly independent. Hence, we make the following definition.

Definition
Basis of $\mathbb{R}^2$
Basis of $\mathbb{R}^3$

Let $\mathcal{B} = \{\vec{v}_1, \vec{v}_2\}$ be a set in $\mathbb{R}^2$. If $\mathcal{B}$ is linearly independent and Span $\mathcal{B} = \mathbb{R}^2$, then the set $\mathcal{B}$ is called a **basis** of $\mathbb{R}^2$.

Let $\mathcal{B} = \{\vec{v}_1, \vec{v}_2, \vec{v}_3\}$ be a set in $\mathbb{R}^3$. If $\mathcal{B}$ is linearly independent and Span $\mathcal{B} = \mathbb{R}^3$, then the set $\mathcal{B}$ is called a **basis** of $\mathbb{R}^3$.

Remarks

1. The plural of basis is **bases**. As we will see, both $\mathbb{R}^2$ and $\mathbb{R}^3$ have infinitely many bases.

2. Here we are relying on our geometric intuition to say that all bases of $\mathbb{R}^2$ have exactly two vectors and all bases of $\mathbb{R}^3$ have exactly three vectors. In Chapter 2, we will mathematically prove this assertion.

We saw in Example 1.2.3 that the set $\{\vec{e}_1, \vec{e}_2\} = \left\{ \begin{bmatrix} 1 \\ 0 \end{bmatrix}, \begin{bmatrix} 0 \\ 1 \end{bmatrix} \right\}$ is the standard basis for $\mathbb{R}^2$. We now look at the standard basis $\{\vec{e}_1, \vec{e}_2, \vec{e}_3\} = \left\{ \begin{bmatrix} 1 \\ 0 \\ 0 \end{bmatrix}, \begin{bmatrix} 0 \\ 1 \\ 0 \end{bmatrix}, \begin{bmatrix} 0 \\ 0 \\ 1 \end{bmatrix} \right\}$ for $\mathbb{R}^3$.

EXAMPLE 1.2.8

Prove that the set $\mathcal{B} = \{\vec{e}_1, \vec{e}_2, \vec{e}_3\}$ is a basis for $\mathbb{R}^3$.

Solution: To show that $\mathcal{B}$ is a basis, we need to prove that it is linearly independent and spans $\mathbb{R}^3$.

Linear Independence: Consider

$$\begin{bmatrix}0\\0\\0\end{bmatrix} = c_1\begin{bmatrix}1\\0\\0\end{bmatrix} + c_2\begin{bmatrix}0\\1\\0\end{bmatrix} + c_3\begin{bmatrix}0\\0\\1\end{bmatrix} = \begin{bmatrix}c_1\\c_2\\c_3\end{bmatrix}$$

Comparing entries, we get that $c_1 = c_2 = c_3 = 0$. Therefore, $\mathcal{B}$ is linearly independent.

Spanning: Let $\begin{bmatrix}x_1\\x_2\\x_3\end{bmatrix}$ be any vector in $\mathbb{R}^3$. Observe that we have

$$\begin{bmatrix}x_1\\x_2\\x_3\end{bmatrix} = x_1\begin{bmatrix}1\\0\\0\end{bmatrix} + x_2\begin{bmatrix}0\\1\\0\end{bmatrix} + x_3\begin{bmatrix}0\\0\\1\end{bmatrix}$$

Hence, $\operatorname{Span}\mathcal{B} = \mathbb{R}^3$.

Since $\mathcal{B}$ is a linearly independent spanning set for $\mathbb{R}^3$, it is a basis for $\mathbb{R}^3$.

EXAMPLE 1.2.9

Prove that the set $C = \left\{\begin{bmatrix}-1\\2\end{bmatrix}, \begin{bmatrix}1\\1\end{bmatrix}\right\}$ is a basis for $\mathbb{R}^2$.

Solution: We need to show that $\operatorname{Span} C = \mathbb{R}^2$ and that C is linearly independent.

Spanning: Let $\begin{bmatrix}x_1\\x_2\end{bmatrix} \in \mathbb{R}^2$ and consider

$$\begin{bmatrix}x_1\\x_2\end{bmatrix} = c_1\begin{bmatrix}-1\\2\end{bmatrix} + c_2\begin{bmatrix}1\\1\end{bmatrix} = \begin{bmatrix}-c_1 + c_2\\2c_1 + c_2\end{bmatrix} \tag{1.2}$$

Comparing entries, we get two equations in two unknowns

$$\begin{aligned}-c_1 + c_2 &= x_1\\ 2c_1 + c_2 &= x_2\end{aligned}$$

Solving gives $c_1 = \frac{1}{3}(-x_1 + x_2)$ and $c_2 = \frac{1}{3}(2x_1 + x_2)$. Hence, we have that

$$\frac{1}{3}(-x_1 + x_2)\begin{bmatrix}-1\\2\end{bmatrix} + \frac{1}{3}(2x_1 + x_2)\begin{bmatrix}1\\1\end{bmatrix} = \begin{bmatrix}x_1\\x_2\end{bmatrix}$$

Thus, $\operatorname{Span} C = \mathbb{R}^2$.

Linear Independence: Take $x_1 = x_2 = 0$ in equation (1.2). Our general solution to that equation says that the only solution is $c_1 = \frac{1}{3}(-0 + 0) = 0$ and $c_2 = \frac{1}{3}(2(0) + 0) = 0$. So, C is also linearly independent. Therefore, C is a basis for $\mathbb{R}^2$.

Graphically, the basis in Example 1.2.9 represents a different set of coordinate axes. Figure 1.2.1 shows how these vectors still form a grid covering all points in $\mathbb{R}^2$.

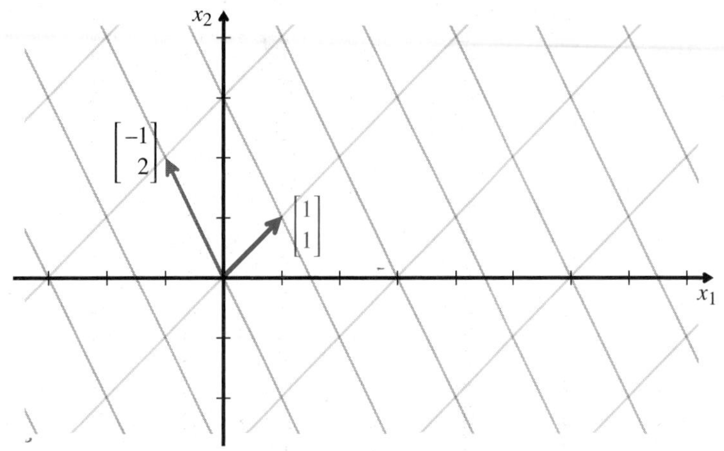

Figure 1.2.1 Geometric representation of the basis $\mathcal{B}$ in Example 1.2.9.

Observe that the coefficients $c_1 = \frac{1}{3}(-x_1 + x_2)$ and $c_2 = \frac{1}{3}(2x_1 + x_2)$ represent the location of the vector $\vec{x}$ on this grid system in the same way that the values x_1 and x_2 represent the location of the vector $\vec{x}$ on the standard coordinate axes.

Definition
Coordinates in $\mathbb{R}^2$

Let $\mathcal{B} = \{\vec{v}_1, \vec{v}_2\}$ be a basis for $\mathbb{R}^2$ and let $\vec{x} \in \mathbb{R}^2$. If $\vec{x} = c_1\vec{v}_1 + c_2\vec{v}_2$, then the scalars c_1 and c_2 are called the **coordinates of $\vec{x}$ with respect to the basis $\mathcal{B}$.**

EXAMPLE 1.2.10

Find the coordinates of $\vec{x} = \begin{bmatrix} 4 \\ 1 \end{bmatrix}$ with respect to the basis $C = \left\{ \begin{bmatrix} -1 \\ 2 \end{bmatrix}, \begin{bmatrix} 1 \\ 1 \end{bmatrix} \right\}$.

Solution: From our work in Example 1.2.9 we have that the coordinates are

$$c_1 = \frac{1}{3}(-4 + 1) = -1, \qquad c_2 = \frac{1}{3}(2(4) + 1) = 3$$

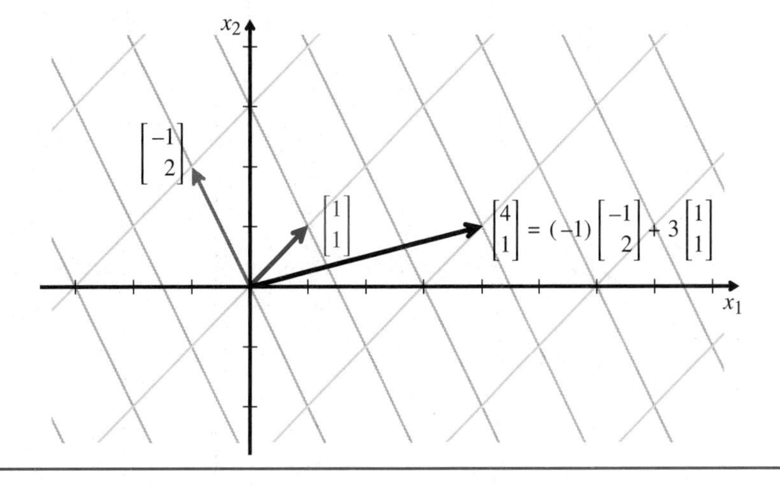

EXAMPLE 1.2.11

Given that $\mathcal{B} = \left\{\begin{bmatrix}1\\2\end{bmatrix}, \begin{bmatrix}2\\1\end{bmatrix}\right\}$ is a basis for $\mathbb{R}^2$, find the coordinates of $\vec{x} = \begin{bmatrix}4\\1\end{bmatrix}$ with respect to $\mathcal{B}$.

Solution: We need to find $c_1, c_2 \in \mathbb{R}$ such that

$$\begin{bmatrix}4\\1\end{bmatrix} = c_1\begin{bmatrix}1\\2\end{bmatrix} + c_2\begin{bmatrix}2\\1\end{bmatrix} = \begin{bmatrix}c_1 + 2c_2\\2c_1 + c_2\end{bmatrix}$$

Comparing entries gives the system of two equations in two unknowns

$$\begin{aligned}4 &= c_1 + 2c_2\\1 &= 2c_1 + c_2\end{aligned}$$

Solving, we find that the coordinates of $\vec{x}$ with respect to the basis $\mathcal{B}$ are $c_1 = -2/3$ and $c_2 = 7/3$.

Applications of Non-Standard Bases

Although we generally use the standard basis in $\mathbb{R}^2$ and $\mathbb{R}^3$, there are applications in which the naturally occurring grid system does not line up with the standard coordinate axes. For example, in crystallography, atoms in a monoclinic crystal are depicted in a Bravais lattice such that the angle between two of the axes is not 90°. Figure 1.2.2 shows a two-dimensional Bravais lattice depicting a monoclinic crystal.

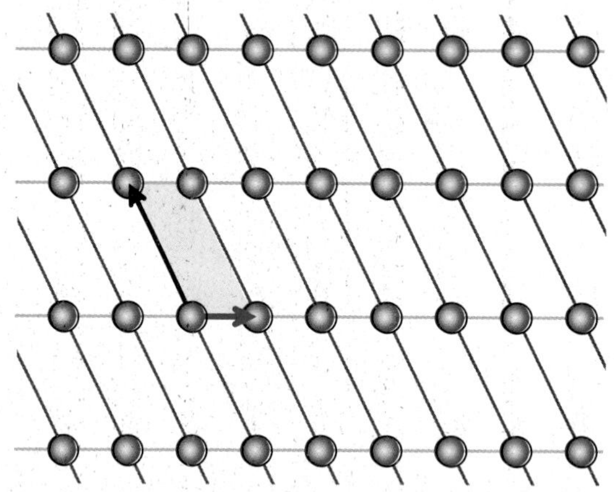

Figure 1.2.2 Two dimensional Bravais lattice of a monoclinic crystal.

Here is another example demonstrating when it can be advantageous to use an alternate coordinate system.

EXAMPLE 1.2.12

Consider the ellipse

$$17x_1^2 + 12x_1x_2 + 8x_2^2 = 2500$$

This equation, which is written in terms of standard coordinates, does not look easy to use.

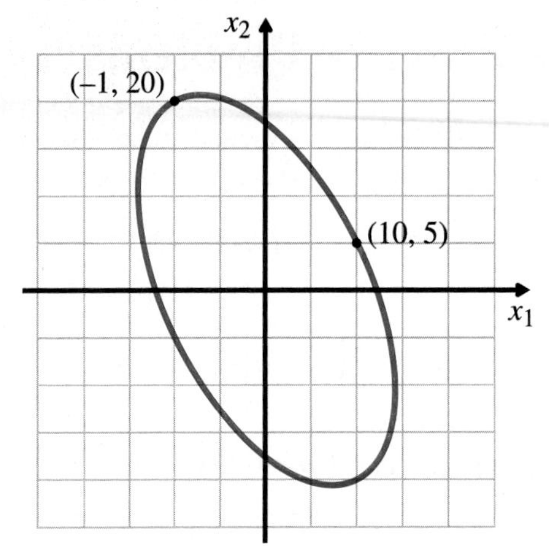

However, if we sketch the ellipse on a new grid system based on the basis

$$\mathcal{B} = \left\{ \begin{bmatrix} 10 \\ 5 \end{bmatrix}, \begin{bmatrix} -5 \\ 10 \end{bmatrix} \right\}$$

then we see that in this grid system the equation of the ellipse is

$$\frac{y_1^2}{1^2} + \frac{y_2^2}{2^2} = 1 \Rightarrow 4y_1^2 + y_2^2 = 4$$

which is much easier to work with.

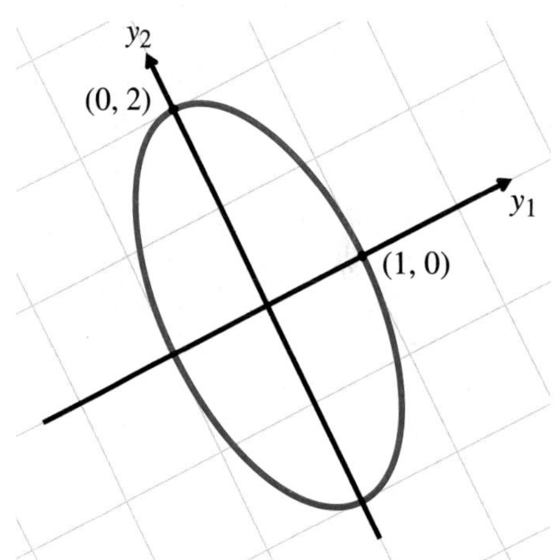

CONNECTION

A natural question to ask is how we found the basis $\mathcal{B}$ in Example 1.2.12 that corresponded to the ellipse. This procedure, covered in Section 8.3, requires many of the concepts from Chapters 2, 3, 5, 6, and 8 to understand.

EXERCISE 1.2.2

(a) If $\vec{u}$ is a non-zero vector in $\mathbb{R}^3$, then what is the geometric interpretation of $S_1 = \text{Span}\{\vec{u}\}$?
(b) If $\vec{u}, \vec{v}$ are non-zero vectors in $\mathbb{R}^3$, then what is the geometric interpretation of $\text{Span}\{\vec{u}, \vec{v}\}$?
(c) If $\vec{u}, \vec{v}$ are non-zero vectors in $\mathbb{R}^3$ such that $\{\vec{u}, \vec{v}\}$ is linearly independent, then what is the geometric interpretation of $\text{Span}\{\vec{u}, \vec{v}\}$?
(d) If $\vec{u}, \vec{v}, \vec{w}$ are non-zero vectors in $\mathbb{R}^3$, then what is the geometric interpretation of $\text{Span}\{\vec{u}, \vec{v}, \vec{w}\}$?

PROBLEMS 1.2

Practice Problems

For Problems A1–A6, determine whether the vector $\vec{x}$ is in Span $\mathcal{B}$. If so, write $\vec{x}$ as a linear combination of the vectors in $\mathcal{B}$.

A1 $\vec{x} = \begin{bmatrix} 3 \\ 1 \end{bmatrix}, \mathcal{B} = \left\{ \begin{bmatrix} 1 \\ 3 \end{bmatrix}, \begin{bmatrix} 1 \\ -1 \end{bmatrix} \right\}$

A2 $\vec{x} = \begin{bmatrix} 8 \\ -4 \end{bmatrix}, \mathcal{B} = \left\{ \begin{bmatrix} -2 \\ 1 \end{bmatrix} \right\}$

A3 $\vec{x} = \begin{bmatrix} 6 \\ 3 \end{bmatrix}, \mathcal{B} = \left\{ \begin{bmatrix} -2 \\ 1 \end{bmatrix} \right\}$

A4 $\vec{x} = \begin{bmatrix} 2 \\ 5 \end{bmatrix}, \mathcal{B} = \left\{ \begin{bmatrix} 2 \\ -1 \end{bmatrix}, \begin{bmatrix} 1 \\ 2 \end{bmatrix} \right\}$

A5 $\vec{x} = \begin{bmatrix} 1 \\ 2 \\ -1 \end{bmatrix}, \mathcal{B} = \left\{ \begin{bmatrix} 1 \\ 1 \\ 0 \end{bmatrix}, \begin{bmatrix} 0 \\ 1 \\ 1 \end{bmatrix}, \begin{bmatrix} 1 \\ 0 \\ 1 \end{bmatrix} \right\}$

A6 $\vec{x} = \begin{bmatrix} 0 \\ 1 \\ 3 \end{bmatrix}, \mathcal{B} = \left\{ \begin{bmatrix} 1 \\ 2 \\ 2 \end{bmatrix}, \begin{bmatrix} 0 \\ -1 \\ 1 \end{bmatrix}, \begin{bmatrix} 1 \\ 0 \\ 4 \end{bmatrix} \right\}$

For Problems A7–A14, determine whether the set is linearly independent. If it is linearly dependent, write one of the vectors in the set as a linear combination of the others.

A7 $\left\{ \begin{bmatrix} 1 \\ 2 \end{bmatrix}, \begin{bmatrix} 1 \\ 3 \end{bmatrix}, \begin{bmatrix} 1 \\ 4 \end{bmatrix} \right\}$

A8 $\left\{ \begin{bmatrix} 3 \\ 1 \end{bmatrix}, \begin{bmatrix} -1 \\ 3 \end{bmatrix} \right\}$

A9 $\left\{ \begin{bmatrix} 1 \\ 1 \end{bmatrix}, \begin{bmatrix} 1 \\ 0 \end{bmatrix} \right\}$

A10 $\left\{ \begin{bmatrix} 2 \\ 3 \end{bmatrix}, \begin{bmatrix} -4 \\ -6 \end{bmatrix} \right\}$

A11 $\left\{ \begin{bmatrix} 1 \\ 2 \\ 1 \end{bmatrix} \right\}$

A12 $\left\{ \begin{bmatrix} 1 \\ -3 \\ -2 \end{bmatrix}, \begin{bmatrix} 4 \\ 6 \\ 1 \end{bmatrix}, \begin{bmatrix} 0 \\ 0 \\ 0 \end{bmatrix} \right\}$

A13 $\left\{ \begin{bmatrix} 1 \\ 1 \\ 0 \end{bmatrix}, \begin{bmatrix} 1 \\ 2 \\ -1 \end{bmatrix}, \begin{bmatrix} -2 \\ -4 \\ 2 \end{bmatrix} \right\}$

A14 $\left\{ \begin{bmatrix} 1 \\ -2 \\ 1 \end{bmatrix}, \begin{bmatrix} 2 \\ 3 \\ 4 \end{bmatrix}, \begin{bmatrix} 0 \\ -1 \\ -2 \end{bmatrix} \right\}$

For Problems A15–A20, describe the set geometrically and write a simplified vector equation for the set.

A15 Span $\left\{ \begin{bmatrix} 1 \\ 0 \end{bmatrix} \right\}$

A16 Span $\left\{ \begin{bmatrix} -1 \\ 1 \end{bmatrix}, \begin{bmatrix} 2 \\ -2 \end{bmatrix} \right\}$

A17 Span $\left\{ \begin{bmatrix} 1 \\ -3 \\ 1 \end{bmatrix}, \begin{bmatrix} -2 \\ 6 \\ -2 \end{bmatrix} \right\}$

A18 $\left\{ \begin{bmatrix} 1 \\ -3 \\ 1 \end{bmatrix}, \begin{bmatrix} -2 \\ 6 \\ -2 \end{bmatrix} \right\}$

A19 Span $\left\{ \begin{bmatrix} 1 \\ 0 \\ -2 \end{bmatrix}, \begin{bmatrix} 2 \\ 1 \\ -1 \end{bmatrix} \right\}$

A20 Span $\left\{ \begin{bmatrix} 0 \\ 0 \\ 0 \end{bmatrix} \right\}$

For Problems A21–A26, determine whether the set forms a basis for $\mathbb{R}^2$.

A21 $\mathcal{B} = \left\{ \begin{bmatrix} 3 \\ 2 \end{bmatrix} \right\}$

A22 $\mathcal{B} = \left\{ \begin{bmatrix} 2 \\ 3 \end{bmatrix}, \begin{bmatrix} 1 \\ 0 \end{bmatrix} \right\}$

A23 $\mathcal{B} = \left\{ \begin{bmatrix} 2 \\ 1 \end{bmatrix}, \begin{bmatrix} 0 \\ 0 \end{bmatrix} \right\}$

A24 $\mathcal{B} = \left\{ \begin{bmatrix} 2 \\ 2 \end{bmatrix}, \begin{bmatrix} -3 \\ -3 \end{bmatrix} \right\}$

A25 $\mathcal{B} = \left\{ \begin{bmatrix} -1 \\ 1 \end{bmatrix}, \begin{bmatrix} 1 \\ 3 \end{bmatrix} \right\}$

A26 $\mathcal{B} = \left\{ \begin{bmatrix} -1 \\ 1 \end{bmatrix}, \begin{bmatrix} 1 \\ 3 \end{bmatrix}, \begin{bmatrix} 0 \\ 2 \end{bmatrix} \right\}$

For Problems A27–A30, determine whether the set forms a basis for $\mathbb{R}^3$.

A27 $\mathcal{B} = \left\{ \begin{bmatrix} 1 \\ 2 \\ 1 \end{bmatrix}, \begin{bmatrix} 0 \\ 0 \\ 0 \end{bmatrix}, \begin{bmatrix} 1 \\ 4 \\ 3 \end{bmatrix} \right\}$

A28 $\mathcal{B} = \left\{ \begin{bmatrix} -1 \\ 2 \\ -1 \end{bmatrix}, \begin{bmatrix} 1 \\ 1 \\ 2 \end{bmatrix} \right\}$

A29 $\mathcal{B} = \left\{ \begin{bmatrix} 1 \\ 0 \\ 1 \end{bmatrix}, \begin{bmatrix} 1 \\ 0 \\ 0 \end{bmatrix}, \begin{bmatrix} 0 \\ 1 \\ 2 \end{bmatrix} \right\}$

A30 $\mathcal{B} = \left\{ \begin{bmatrix} 1 \\ 0 \\ 1 \end{bmatrix}, \begin{bmatrix} 1 \\ 1 \\ 1 \end{bmatrix}, \begin{bmatrix} 1 \\ 1 \\ 0 \end{bmatrix} \right\}$

For Problems A31–A33:

(a) Prove that $\mathcal{B}$ is a basis for $\mathbb{R}^2$.

(b) Find the coordinates of $\vec{e}_1 = \begin{bmatrix} 1 \\ 0 \end{bmatrix}$, $\vec{e}_2 = \begin{bmatrix} 0 \\ 1 \end{bmatrix}$, and $\vec{x} = \begin{bmatrix} 1 \\ 3 \end{bmatrix}$ with respect to $\mathcal{B}$.

A31 $\mathcal{B} = \left\{ \begin{bmatrix} 1 \\ 0 \end{bmatrix}, \begin{bmatrix} 1 \\ 1 \end{bmatrix} \right\}$

A32 $\mathcal{B} = \left\{ \begin{bmatrix} 1 \\ 1 \end{bmatrix}, \begin{bmatrix} 1 \\ -1 \end{bmatrix} \right\}$

A33 $\mathcal{B} = \left\{ \begin{bmatrix} 1 \\ 2 \end{bmatrix}, \begin{bmatrix} -1 \\ -1 \end{bmatrix} \right\}$

A34 Let $\{\vec{v}_1, \vec{v}_2\}$ be a set of vectors in $\mathbb{R}^3$. Prove that $\{\vec{v}_1, \vec{v}_2\}$ is linearly independent if and only if neither $\vec{v}_1$ nor $\vec{v}_2$ is a scalar multiple of the other.

A35 Let $\{\vec{v}_1, \vec{v}_2\}$ be a set of vectors in $\mathbb{R}^3$. Prove that $\text{Span}\{\vec{v}_1, \vec{v}_2\} = \text{Span}\{\vec{v}_1, t\vec{v}_2\}$ for any $t \in \mathbb{R}, t \neq 0$.

Homework Problems

For Problems B1–B6, determine whether the vector $\vec{x}$ is in Span $\mathcal{B}$. If so, write $\vec{x}$ as a linear combination of the vectors in $\mathcal{B}$.

B1 $\vec{x} = \begin{bmatrix}3\\2\end{bmatrix}, \mathcal{B} = \left\{\begin{bmatrix}1\\1\end{bmatrix}, \begin{bmatrix}1\\-1\end{bmatrix}\right\}$

B2 $\vec{x} = \begin{bmatrix}7\\3\end{bmatrix}, \mathcal{B} = \left\{\begin{bmatrix}3\\5\end{bmatrix}\right\}$

B3 $\vec{x} = \begin{bmatrix}2\\-2\end{bmatrix}, \mathcal{B} = \left\{\begin{bmatrix}-3\\3\end{bmatrix}\right\}$

B4 $\vec{x} = \begin{bmatrix}1\\0\end{bmatrix}, \mathcal{B} = \left\{\begin{bmatrix}2\\-1\end{bmatrix}, \begin{bmatrix}1\\2\end{bmatrix}\right\}$

B5 $\vec{x} = \begin{bmatrix}3\\1\\4\end{bmatrix}, \mathcal{B} = \left\{\begin{bmatrix}1\\1\\0\end{bmatrix}, \begin{bmatrix}0\\1\\1\end{bmatrix}, \begin{bmatrix}1\\0\\1\end{bmatrix}\right\}$

B6 $\vec{x} = \begin{bmatrix}-3\\5\\7\end{bmatrix}, \mathcal{B} = \left\{\begin{bmatrix}1\\2\\2\end{bmatrix}, \begin{bmatrix}0\\-1\\1\end{bmatrix}, \begin{bmatrix}1\\0\\4\end{bmatrix}\right\}$

For Problems B7–B14, determine whether the set is linearly independent. If it is linearly dependent, write one of the vectors in the set as a linear combination of the others.

B7 $\left\{\begin{bmatrix}1\\1\end{bmatrix}, \begin{bmatrix}3\\5\end{bmatrix}, \begin{bmatrix}2\\1\end{bmatrix}\right\}$

B8 $\left\{\begin{bmatrix}0\\0\end{bmatrix}, \begin{bmatrix}8\\3\end{bmatrix}\right\}$

B9 $\left\{\begin{bmatrix}-3\\6\end{bmatrix}, \begin{bmatrix}1\\3\end{bmatrix}\right\}$

B10 $\left\{\begin{bmatrix}-2\\5\end{bmatrix}, \begin{bmatrix}4\\-10\end{bmatrix}\right\}$

B11 $\left\{\begin{bmatrix}5\\3\end{bmatrix}\right\}$

B12 $\left\{\begin{bmatrix}0\\0\\0\end{bmatrix}, \begin{bmatrix}1\\-3\\2\end{bmatrix}\right\}$

B13 $\left\{\begin{bmatrix}2\\-1\\1\end{bmatrix}, \begin{bmatrix}4\\-2\\2\end{bmatrix}, \begin{bmatrix}1\\5\\3\end{bmatrix}\right\}$

B14 $\left\{\begin{bmatrix}4\\2\\1\end{bmatrix}, \begin{bmatrix}2\\6\\3\end{bmatrix}, \begin{bmatrix}3\\4\\2\end{bmatrix}\right\}$

For Problems B15–B20, describe the set geometrically and write a simplified vector equation for the set.

B15 $\operatorname{Span}\left\{\begin{bmatrix}1\\0\end{bmatrix}\right\}$

B16 $\operatorname{Span}\left\{\begin{bmatrix}1\\1\end{bmatrix}, \begin{bmatrix}2\\3\end{bmatrix}\right\}$

B17 $\operatorname{Span}\left\{\begin{bmatrix}2\\4\end{bmatrix}, \begin{bmatrix}1\\2\end{bmatrix}, \begin{bmatrix}-3\\-6\end{bmatrix}\right\}$

B18 $\operatorname{Span}\left\{\begin{bmatrix}0\\0\\0\end{bmatrix}\right\}$

B19 $\operatorname{Span}\left\{\begin{bmatrix}3\\1\\1\end{bmatrix}, \begin{bmatrix}-6\\-2\\-2\end{bmatrix}\right\}$

B20 $\operatorname{Span}\left\{\begin{bmatrix}2\\1\\3\end{bmatrix}\right\}$

For Problems B21–B24, determine whether the set forms a basis for $\mathbb{R}^2$.

B21 $\left\{\begin{bmatrix}1\\1\end{bmatrix}, \begin{bmatrix}3\\2\end{bmatrix}\right\}$

B22 $\left\{\begin{bmatrix}2\\4\end{bmatrix}, \begin{bmatrix}-1\\-2\end{bmatrix}\right\}$

B23 $\left\{\begin{bmatrix}5\\6\end{bmatrix}, \begin{bmatrix}4\\3\end{bmatrix}\right\}$

B24 $\left\{\begin{bmatrix}1\\-2\end{bmatrix}, \begin{bmatrix}4\\2\end{bmatrix}\right\}$

For Problems B25–B28, determine whether the set forms a basis for $\mathbb{R}^3$.

B25 $\left\{\begin{bmatrix}-1\\2\\1\end{bmatrix}, \begin{bmatrix}5\\-2\\1\end{bmatrix}\right\}$

B26 $\left\{\begin{bmatrix}1\\1\\3\end{bmatrix}, \begin{bmatrix}1\\-1\\0\end{bmatrix}, \begin{bmatrix}1\\1\\0\end{bmatrix}\right\}$

B27 $\left\{\begin{bmatrix}2\\1\\2\end{bmatrix}, \begin{bmatrix}1\\0\\1\end{bmatrix}, \begin{bmatrix}0\\1\\0\end{bmatrix}\right\}$

B28 $\left\{\begin{bmatrix}-1\\1\\1\end{bmatrix}, \begin{bmatrix}2\\1\\1\end{bmatrix}, \begin{bmatrix}1\\1\\-1\end{bmatrix}\right\}$

B29 Let $\mathcal{B} = \left\{\begin{bmatrix}1\\-1\end{bmatrix}, \begin{bmatrix}0\\1\end{bmatrix}\right\}$.

(a) Prove that $\mathcal{B}$ is a basis for $\mathbb{R}^2$.

(b) Find the coordinates of $\vec{e}_1 = \begin{bmatrix}1\\0\end{bmatrix}$, $\vec{e}_2 = \begin{bmatrix}0\\1\end{bmatrix}$, and $\vec{x} = \begin{bmatrix}2\\1\end{bmatrix}$ with respect to $\mathcal{B}$.

For Problems B30–B33:

(a) Prove that $\mathcal{B}$ is a basis for $\mathbb{R}^2$.

(b) Find the coordinates of $\vec{e}_1 = \begin{bmatrix}1\\0\end{bmatrix}$, $\vec{e}_2 = \begin{bmatrix}0\\1\end{bmatrix}$, and $\vec{x} = \begin{bmatrix}1\\3\end{bmatrix}$ with respect to $\mathcal{B}$.

B30 $\mathcal{B} = \left\{\begin{bmatrix}2\\1\end{bmatrix}, \begin{bmatrix}1\\3\end{bmatrix}\right\}$

B31 $\mathcal{B} = \left\{\begin{bmatrix}2\\0\end{bmatrix}, \begin{bmatrix}1\\3\end{bmatrix}\right\}$

B32 $\mathcal{B} = \left\{\begin{bmatrix}1\\-2\end{bmatrix}, \begin{bmatrix}2\\1\end{bmatrix}\right\}$

B33 $\mathcal{B} = \left\{\begin{bmatrix}-2\\-1\end{bmatrix}, \begin{bmatrix}-3\\5\end{bmatrix}\right\}$

1.3 Length and Angles in $\mathbb{R}^2$ and $\mathbb{R}^3$

In many physical applications, we are given measurements in terms of angles and magnitudes. We must convert this data into vectors so that we can apply the tools of linear algebra to solve problems. For example, we may need to find a vector representing the path (and speed) of a plane flying northwest at 1300 km/h. To do this, we need to identify the length of a vector and the angle between two vectors. In this section, we see how we can calculate both of these quantities with the dot product operator.

Length, Angles, and Dot Products in $\mathbb{R}^2$

The length of a vector in $\mathbb{R}^2$ is defined by the usual distance formula (that is, Pythagoras' Theorem), as in Figure 1.3.1.

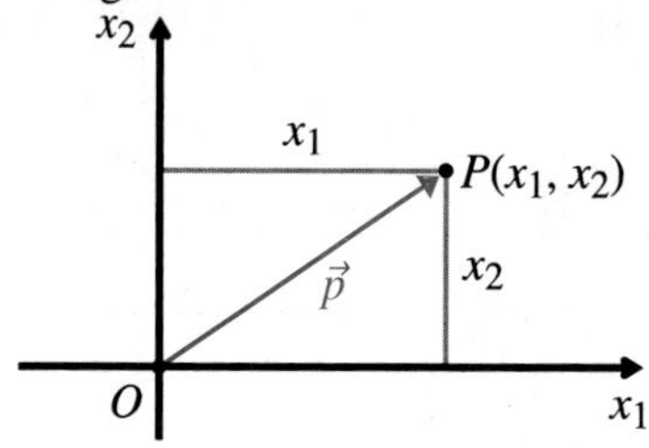

Figure 1.3.1 Length in $\mathbb{R}^2$.

Definition
Length in $\mathbb{R}^2$

If $\vec{x} = \begin{bmatrix} x_1 \\ x_2 \end{bmatrix} \in \mathbb{R}^2$, its **length** is defined to be $\|\vec{x}\| = \sqrt{x_1^2 + x_2^2}$.

EXAMPLE 1.3.1

An object that weighs 10kg is being pulled by two strings with force and direction as given in the diagram. Newton's second law of motion says that the force F acting on the object is equal to its mass m times its acceleration a. What is the resulting acceleration of the object?

Solution: In Example 1.1.1 we found that the net force being applied to the object is $\vec{F} = \begin{bmatrix} 200 \\ 50\sqrt{3} \end{bmatrix}$. The total amount of force being applied will be the length of this vector:

$$\begin{aligned} \|\vec{F}\| &= \sqrt{(200)^2 + (50\sqrt{3})^2} \\ &= \sqrt{47500} \\ &\approx 218 \end{aligned}$$

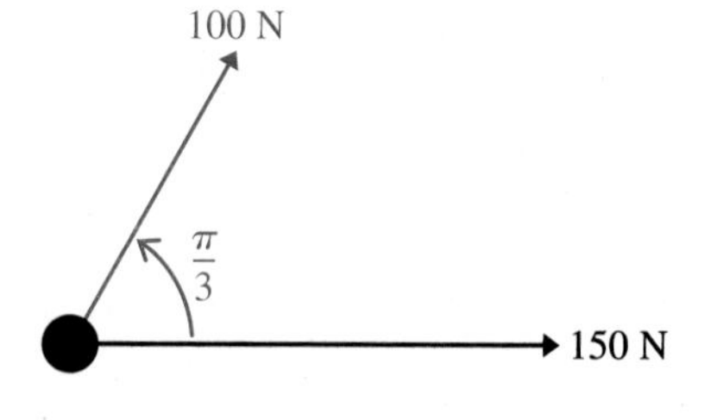

Hence, the total amount of force is about $F = 218$ N. Thus, we find that the acceleration is approximately

$$a = \frac{F}{m} = \frac{218}{10} = 21.8 m/s^2$$

In the example above, we may also be interested in finding the angle from the horizontal at which the object is moving.

Theorem 1.3.1 If $\vec{p}, \vec{q} \in \mathbb{R}^2$ and θ is an angle between $\vec{p}$ and $\vec{q}$, then

$$p_1q_1 + p_2q_2 = \|\vec{p}\|\,\|\vec{q}\|\cos\theta$$

Proof: Consider the figure below.

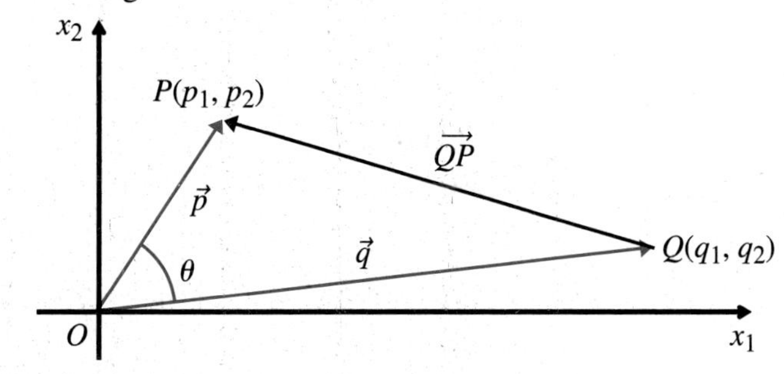

The Law of Cosines gives

$$\|\overrightarrow{QP}\|^2 = \|\vec{p}\|^2 + \|\vec{q}\|^2 - 2\|\vec{p}\|\,\|\vec{q}\|\cos\theta \tag{1.3}$$

Substituting $\vec{p} = \begin{bmatrix} p_1 \\ p_2 \end{bmatrix}$, $\vec{q} = \begin{bmatrix} q_1 \\ q_2 \end{bmatrix}$, $\overrightarrow{QP} = \begin{bmatrix} p_1 - q_1 \\ p_2 - q_2 \end{bmatrix}$ into (1.3) and simplifying gives

$$p_1q_1 + p_2q_2 = \|\vec{p}\|\,\|\vec{q}\|\cos\theta$$

■

Remark

When solving for θ we usually choose θ such that $0 \leq \theta \leq \pi$.

EXAMPLE 1.3.2 Find the angle θ at which the object in Example 1.3.1 is moving.

Solution: We need to calculate the angle between the net force vector and either of the initial force vectors. We have $\vec{F}_1 = \begin{bmatrix} 150 \\ 0 \end{bmatrix}$ and $\vec{F} = \begin{bmatrix} 200 \\ 50\sqrt{3} \end{bmatrix}$. Thus, an angle θ between them satisfies

$$\begin{aligned} 150(200) + 0(50\sqrt{3}) &= \|\vec{F}_1\|\,\|\vec{F}\|\cos\theta \\ 150(200) &\approx 150(218)\cos\theta \\ \frac{100}{109} &\approx \cos\theta \end{aligned}$$

We get that $\theta \approx 0.41$ radians.

EXAMPLE 1.3.3

Find the angle in $\mathbb{R}^2$ between $\vec{v} = \begin{bmatrix} 1 \\ -2 \end{bmatrix}$ and $\vec{w} = \begin{bmatrix} 2 \\ 1 \end{bmatrix}$.

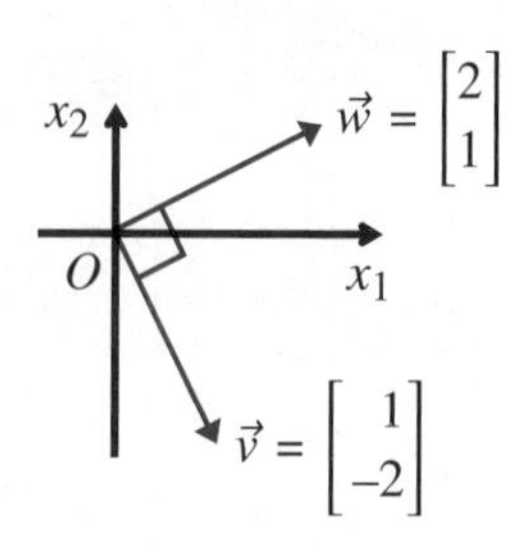

Solution: We have $v_1w_1 + v_2w_2 = 1(2) + (-2)(1) = 0$. Hence, $\cos\theta = \dfrac{0}{\|\vec{v}\|\,\|\vec{w}\|} = 0$. Thus, $\theta = \dfrac{\pi}{2}$ radians. That is, $\vec{v}$ and $\vec{w}$ are perpendicular to each other.

The formula $p_1q_1 + p_2q_2$ on the left-hand side of the equation for the angle between vectors in Theorem 1.3.1 matches with what we use in the equation for the length of a vector. So, we make the following definition.

Definition
Dot Product in $\mathbb{R}^2$

Let $\vec{x} = \begin{bmatrix} x_1 \\ x_2 \end{bmatrix}$ and $\vec{y} = \begin{bmatrix} y_1 \\ y_2 \end{bmatrix}$ be vectors in $\mathbb{R}^2$. The **dot product** of $\vec{x}$ and $\vec{y}$, denoted $\vec{x} \cdot \vec{y}$, is defined by

$$\vec{x} \cdot \vec{y} = x_1y_1 + x_2y_2$$

Using this definition, we can rewrite our formulas for angles and length as

$$\|\vec{x}\| = \sqrt{\vec{x} \cdot \vec{x}}$$
$$\vec{x} \cdot \vec{y} = \|\vec{x}\|\,\|\vec{y}\| \cos\theta$$

Using the dot product and what we saw in Example 1.3.3, we make the following definition.

Definition
Orthogonal Vectors in $\mathbb{R}^2$

Two vectors $\vec{x}$ and $\vec{y}$ in $\mathbb{R}^2$ are **orthogonal** to each other if and only if $\vec{x} \cdot \vec{y} = 0$.

EXAMPLE 1.3.4

Show that $\begin{bmatrix} 1 \\ 2 \end{bmatrix}$ and $\begin{bmatrix} -2 \\ 1 \end{bmatrix}$ are orthogonal.

Solution: We have $\begin{bmatrix} 1 \\ 2 \end{bmatrix} \cdot \begin{bmatrix} -2 \\ 1 \end{bmatrix} = 1(-2) + 2(1) = 0$. Hence, by definition, they are orthogonal.

Remark

Notice that the definition of orthogonality implies that the zero vector $\vec{0}$ is orthogonal to every vector in $\mathbb{R}^2$.

Length, Angles, and Dot Products in $\mathbb{R}^3$

To define length and angles in $\mathbb{R}^3$, we repeat what we did in $\mathbb{R}^2$. This is easiest if we begin by defining the dot product in $\mathbb{R}^3$.

Definition
Dot Product in $\mathbb{R}^3$

Let $\vec{x} = \begin{bmatrix} x_1 \\ x_2 \\ x_3 \end{bmatrix}$ and $\vec{y} = \begin{bmatrix} y_1 \\ y_2 \\ y_3 \end{bmatrix}$ be vectors in $\mathbb{R}^3$. The **dot product** of $\vec{x}$ and $\vec{y}$, denoted $\vec{x} \cdot \vec{y}$, is defined by

$$\vec{x} \cdot \vec{y} = x_1y_1 + x_2y_2 + x_3y_3$$

EXAMPLE 1.3.5

Calculate the dot product of the following pairs of vectors.

(a) $\vec{x} = \begin{bmatrix} 1 \\ -2 \\ 1 \end{bmatrix}, \vec{y} = \begin{bmatrix} 2 \\ 1 \\ 3 \end{bmatrix}$ (b) $\vec{u} = \begin{bmatrix} 1/2 \\ 0 \\ 1/2 \end{bmatrix}, \vec{v} = \begin{bmatrix} 1 \\ 5 \\ -1 \end{bmatrix}$

Solution: (a) We have $\vec{x} \cdot \vec{y} = 1(2) + (-2)(1) + 1(3) = 3$.

(b) We have $\vec{u} \cdot \vec{v} = \frac{1}{2}(1) + 0(5) + \frac{1}{2}(-1) = 0$.

The dot product has the following important properties.

Theorem 1.3.2

If $\vec{x}, \vec{y}, \vec{z}$ are vectors in $\mathbb{R}^2$ or $\mathbb{R}^3$, and $s, t \in \mathbb{R}$, then

(1) $\vec{x} \cdot \vec{x} \geq 0$
(2) $\vec{x} \cdot \vec{x} = 0$ if and only if $\vec{x} = \vec{0}$
(3) $\vec{x} \cdot \vec{y} = \vec{y} \cdot \vec{x}$
(4) $\vec{x} \cdot (s\vec{y} + t\vec{z}) = s(\vec{x} \cdot \vec{y}) + t(\vec{x} \cdot \vec{z})$

For vectors in $\mathbb{R}^3$, the formula for the length can be obtained from a two-step calculation using the formula in $\mathbb{R}^2$. Consider the points $X(x_1, x_2, x_3)$ and $P(x_1, x_2, 0)$. Observe that OPX is a right triangle, so that

$$\|\vec{x}\|^2 = \|\vec{OP}\|^2 + \|\vec{PX}\|^2 = (x_1^2 + x_2^2) + x_3^2$$

Definition
Length in $\mathbb{R}^3$

If $\vec{x} = \begin{bmatrix} x_1 \\ x_2 \\ x_3 \end{bmatrix} \in \mathbb{R}^3$, its **length** is defined to be

$$\|\vec{x}\| = \sqrt{\vec{x} \cdot \vec{x}}$$

One immediate application of this formula is to calculate the distance between two points. In particular, if we have points P and Q, then the distance between them is the length of the directed line segment $\vec{PQ}$.

EXAMPLE 1.3.6

Find the distance between the points $P(-1,3,4)$ and $Q(2,-5,1)$ in $\mathbb{R}^3$.

Solution: We have $\vec{PQ} = \begin{bmatrix} 2-(-1) \\ -5-3 \\ 1-4 \end{bmatrix} = \begin{bmatrix} 3 \\ -8 \\ -3 \end{bmatrix}$. Hence, the distance between the two points is

$$\|\vec{PQ}\| = \sqrt{\vec{PQ}\cdot\vec{PQ}} = \sqrt{3^2+(-8)^2+(-3)^2} = \sqrt{82}$$

We also find that if $\vec{p}$ and $\vec{q}$ are vectors in $\mathbb{R}^3$ and θ is an angle between them, then

$$\vec{p}\cdot\vec{q} = \|\vec{p}\|\,\|\vec{q}\|\cos\theta$$

EXAMPLE 1.3.7

Find the angle in $\mathbb{R}^3$ between $\vec{v} = \begin{bmatrix} 1 \\ 4 \\ -2 \end{bmatrix}$ and $\vec{w} = \begin{bmatrix} 3 \\ -1 \\ 4 \end{bmatrix}$.

Solution: We have

$$\begin{aligned} \vec{v}\cdot\vec{w} &= 1(3)+4(-1)+(-2)(4) = -9 \\ \|\vec{v}\| &= \sqrt{1^2+4^2+(-2)^2} = \sqrt{21} \\ \|\vec{w}\| &= \sqrt{3^2+(-1)^2+4^2} = \sqrt{26} \end{aligned}$$

Hence,

$$\cos\theta = \frac{-9}{\sqrt{21}\sqrt{26}} \approx -0.38516$$

So, $\theta \approx 1.966$ radians.

EXERCISE 1.3.1

Find the angle in $\mathbb{R}^3$ between $\vec{v} = \begin{bmatrix} 1 \\ 2 \\ -1 \end{bmatrix}$ and $\vec{w} = \begin{bmatrix} 1 \\ -1 \\ -1 \end{bmatrix}$.

If the angle θ between $\vec{x}$ and $\vec{y}$ in $\mathbb{R}^3$ is $\frac{\pi}{2}$ radians, then, as in $\mathbb{R}^2$, we get

$$\vec{x}\cdot\vec{y} = \|\vec{x}\|\,\|\vec{y}\|\cos\frac{\pi}{2} = 0$$

Definition
Orthogonal Vectors in $\mathbb{R}^3$

Two vectors $\vec{x}$ and $\vec{y}$ in $\mathbb{R}^3$ are **orthogonal** to each other if and only if $\vec{x}\cdot\vec{y} = 0$.

EXAMPLE 1.3.8

Show that $\vec{x} = \begin{bmatrix} 3 \\ -6 \\ 5 \end{bmatrix}$ and $\vec{y} = \begin{bmatrix} 9 \\ 2 \\ -3 \end{bmatrix}$ are orthogonal.

Solution: We have $\vec{x} \cdot \vec{y} = 3(9) + (-6)(2) + 5(-3) = 0$ as required.

Scalar Equations of Planes

We saw in Section 1.1 that a plane can be described by the vector equation $\vec{x} = \vec{p} + t_1\vec{v}_1 + t_2\vec{v}_2, t_1, t_2 \in \mathbb{R}$, where $\{\vec{v}_1, \vec{v}_2\}$ is linearly independent. In many problems, it is more useful to have a **scalar equation** that represents the plane. We now look at how to use the dot product and orthogonality to find such an equation.

Suppose that we want to find an equation of the plane that passes through the point $P(p_1, p_2, p_3)$. Suppose that we can find a non-zero vector $\vec{n} = \begin{bmatrix} n_1 \\ n_2 \\ n_3 \end{bmatrix}$, called a **normal vector** of the plane, that is orthogonal to any directed line segment $\vec{PQ}$ lying in the plane. (That is, $\vec{n}$ is orthogonal to $\vec{PQ}$ for any point Q in the plane; see Figure 1.3.2.) To find the equation of this plane, let $X(x_1, x_2, x_3)$ be any point on the plane. Then $\vec{n}$ is orthogonal to $\vec{PX}$, so

$$0 = \vec{n} \cdot \vec{PX} = \vec{n} \cdot (\vec{x} - \vec{p}) = n_1(x_1 - p_1) + n_2(x_2 - p_2) + n_3(x_3 - p_3)$$

This equation, which must be satisfied by the coordinates of a point X in the plane, can be written in the form

$$n_1x_1 + n_2x_2 + n_3x_3 = d, \quad \text{where} \quad d = n_1p_1 + n_2p_2 + n_3p_3 = \vec{n} \cdot \vec{p}$$

This is a scalar equation of this plane. For computational purposes, the form $\vec{n} \cdot (\vec{x} - \vec{p}) = 0$ is often easiest to use.

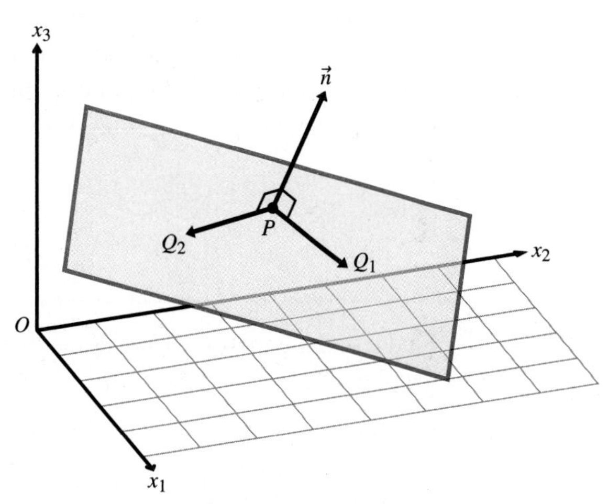

Figure 1.3.2 The normal $\vec{n}$ is orthogonal to every directed line segment lying in the plane.

EXAMPLE 1.3.9

Find a scalar equation of the plane that passes through the point $P(2, 3, -1)$ and has normal vector $\vec{n} = \begin{bmatrix} 1 \\ -4 \\ 1 \end{bmatrix}$.

Solution: The equation is

$$\vec{n} \cdot (\vec{x} - \vec{p}) = \begin{bmatrix} 1 \\ -4 \\ 1 \end{bmatrix} \cdot \begin{bmatrix} x_1 - 2 \\ x_2 - 3 \\ x_3 + 1 \end{bmatrix} = 0$$

or

$$x_1 - 4x_2 + x_3 = 1(2) + (-4)(3) + 1(-1) = -11$$

Note that any non-zero scalar multiple of $\vec{n}$ will also be a normal vector for the plane. Moreover, our argument above works in reverse. That is, if $n_1x_1 + n_2x_2 + n_3x_3 = d$ is a scalar equation of the plane, then $\vec{n} = \begin{bmatrix} n_1 \\ n_2 \\ n_3 \end{bmatrix}$ is a normal vector for the plane.

EXAMPLE 1.3.10

Find a normal vector of the plane with scalar equation $x_1 - 2x_3 = 5$.

Solution: We should think of this equation as $x_1 + 0x_2 - 2x_3 = 5$. Hence, a normal vector for the plane is $\vec{n} = \begin{bmatrix} 1 \\ 0 \\ -2 \end{bmatrix}$.

Two planes are defined to be **parallel** if the normal vector to one plane is a non-zero scalar multiple of the normal vector of the other plane. Thus, for example, the plane $x_1 + 2x_2 - x_3 = 1$ is parallel to the plane $2x_1 + 4x_2 - 2x_3 = 7$.

Two planes are **orthogonal** to each other if their normal vectors are orthogonal. For example, the plane $x_1 + x_2 + x_3 = 0$ is orthogonal to the plane $x_1 + x_2 - 2x_3 = 0$ since

$$\begin{bmatrix} 1 \\ 1 \\ 1 \end{bmatrix} \cdot \begin{bmatrix} 1 \\ 1 \\ -2 \end{bmatrix} = 0$$

EXAMPLE 1.3.11

Find a scalar equation of the plane that contains the point $P(2, 4, -1)$ and is parallel to the plane $2x_1 + 3x_2 - 5x_3 = 6$.

Solution: A scalar equation of the desired plane can have the form $2x_1 + 3x_2 - 5x_3 = d$ since the planes are parallel. The plane must pass through P, so we find that a scalar equation of the plane is

$$2x_1 + 3x_2 - 5x_3 = 2(2) + 3(4) - 5(-1) = 21$$

EXAMPLE 1.3.12

Find a scalar equation of a plane that contains the point $P(3,-1,3)$ and is orthogonal to the plane $x_1 - 2x_2 + 4x_3 = 2$.

Solution: Any normal vector of the desired plane must be orthogonal to $\begin{bmatrix} 1 \\ -2 \\ 4 \end{bmatrix}$. We pick $\begin{bmatrix} 0 \\ 2 \\ 1 \end{bmatrix}$. Thus, a scalar equation of this plane can have the form $2x_2 + x_3 = d$. Since the plane passes through P, we find that a scalar equation of this plane is

$$2x_2 + x_3 = 0(3) + 2(-1) + 1(3) = 1$$

EXERCISE 1.3.2

Find a scalar equation of the plane that contains the point $P(1,2,3)$ and is parallel to the plane $x_1 - 3x_2 - 2x_3 = -5$.

It is important to note that we can reverse the reasoning that leads to the scalar equation of the plane in order to identify the set of points that satisfies the equation

$$n_1x_1 + n_2x_2 + n_3x_3 = d$$

If $n_1 \neq 0$, we can solve this equation for x_1 to get

$$x_1 = \frac{d}{n_1} - \frac{n_2}{n_1}x_2 - \frac{n_3}{n_1}x_3$$

Hence, every vector $\vec{x}$ in the plane satisfies the vector equation

$$\vec{x} = \begin{bmatrix} x_1 \\ x_2 \\ x_3 \end{bmatrix} = \begin{bmatrix} \frac{d}{n_1} - \frac{n_2}{n_1}x_2 - \frac{n_3}{n_1}x_3 \\ x_2 \\ x_3 \end{bmatrix} = \begin{bmatrix} d/n_1 \\ 0 \\ 0 \end{bmatrix} + x_2 \begin{bmatrix} -n_2/n_1 \\ 1 \\ 0 \end{bmatrix} + x_3 \begin{bmatrix} -n_3/n_1 \\ 0 \\ 1 \end{bmatrix}, \quad x_2, x_3 \in \mathbb{R}$$

Observe that this is a vector equation of a plane through the point $P(d/n_1, 0, 0)$. If $n_2 \neq 0$ or $n_3 \neq 0$, we could instead solve the scalar equation of the plane for x_2 or x_3 respectively to get alternate vector equations for the plane.

EXAMPLE 1.3.13

Find a vector equation of the plane in $\mathbb{R}^3$ that satisfies $5x_1 - 6x_2 + 7x_3 = 11$.

Solution: Solving the scalar equation of the plane for x_1 gives

$$x_1 = \frac{11}{5} + \frac{6}{5}x_2 - \frac{7}{5}x_3$$

Hence, a vector equation is

$$\vec{x} = \begin{bmatrix} \frac{11}{5} + \frac{6}{5}x_2 - \frac{7}{5}x_3 \\ x_2 \\ x_3 \end{bmatrix} = \begin{bmatrix} 11/5 \\ 0 \\ 0 \end{bmatrix} + x_2 \begin{bmatrix} 6/5 \\ 1 \\ 0 \end{bmatrix} + x_3 \begin{bmatrix} -7/5 \\ 0 \\ 1 \end{bmatrix}, \quad x_2, x_3 \in \mathbb{R}$$

EXERCISE 1.3.3 Find a vector equation of the plane in $\mathbb{R}^3$ that satisfies $2x_1 + 3x_2 - x_3 = 0$.

At first, one may wonder why we would want both a scalar equation and a vector equation of a plane. Depending on what we need, both representations have their uses. For example, the scalar equation makes it very easy to check if a given point $Q(q_1, q_2, q_3)$ is in the plane. We just have to verify whether or not

$$n_1q_1 + n_2q_2 + n_3q_3 = d$$

On the other hand, if we want to generate 100 points in the plane, the vector equation

$$\vec{x} = \vec{p} + s\vec{u} + t\vec{v}, \quad s, t \in \mathbb{R}$$

is more suitable. We just need to pick 100 different pairs of s and t.

Cross Products

Given a pair of vectors $\vec{u}$ and $\vec{v}$ in $\mathbb{R}^3$, how can we find a third vector $\vec{w}$ that is orthogonal to both $\vec{u}$ and $\vec{v}$? This problem arises naturally in many ways. For example, in physics, it is observed that the force on an electrically charged particle moving in a magnetic field is in the direction orthogonal to the velocity of the particle and to the vector describing the magnetic field.

Let $\vec{u}, \vec{v} \in \mathbb{R}^3$. If $\vec{w}$ is orthogonal to both $\vec{u}$ and $\vec{v}$, it must satisfy the equations

$$\begin{aligned} \vec{u} \cdot \vec{w} &= u_1w_1 + u_2w_2 + u_3w_3 = 0 \\ \vec{v} \cdot \vec{w} &= v_1w_1 + v_2w_2 + v_3w_3 = 0 \end{aligned}$$

In Chapter 2, we shall develop systematic methods for solving such equations for w_1, w_2, w_3. For the present, we simply give a solution:

$$\vec{w} = \begin{bmatrix} u_2v_3 - u_3v_2 \\ u_3v_1 - u_1v_3 \\ u_1v_2 - u_2v_1 \end{bmatrix}$$

Definition
Cross Product

The **cross product** of vectors $\vec{u} = \begin{bmatrix} u_1 \\ u_2 \\ u_3 \end{bmatrix}$ and $\vec{v} = \begin{bmatrix} v_1 \\ v_2 \\ v_3 \end{bmatrix}$ in $\mathbb{R}^3$ is defined by

$$\vec{u} \times \vec{v} = \begin{bmatrix} u_2v_3 - u_3v_2 \\ u_3v_1 - u_1v_3 \\ u_1v_2 - u_2v_1 \end{bmatrix}$$

EXAMPLE 1.3.14

Calculate the cross product of $\begin{bmatrix} 2 \\ 3 \\ 5 \end{bmatrix}$ and $\begin{bmatrix} -1 \\ 1 \\ 2 \end{bmatrix}$ in $\mathbb{R}^3$.

Solution: $\begin{bmatrix} 2 \\ 3 \\ 5 \end{bmatrix} \times \begin{bmatrix} -1 \\ 1 \\ 2 \end{bmatrix} = \begin{bmatrix} 6-5 \\ -5-4 \\ 2-(-3) \end{bmatrix} = \begin{bmatrix} 1 \\ -9 \\ 5 \end{bmatrix}$.

Remarks

1. Unlike the dot product of two vectors, which is a scalar, the cross product of two vectors in $\mathbb{R}^3$ is itself a new vector.
2. The cross product is a construction that is defined only in $\mathbb{R}^3$. (There is a generalization to higher dimensions, but it is considerably more complicated, and it will not be considered in this book.)

The formula for the cross product is a little awkward to remember, but there are many tricks for remembering it. One way is to write the components of $\vec{u}$ in a row above the components of $\vec{v}$:

$$\begin{matrix} u_1 & u_2 & u_3 \\ v_1 & v_2 & v_3 \end{matrix}$$

Then, for the first entry in $\vec{u} \times \vec{v}$, we cover the first column and calculate the difference of the products of the cross-terms:

$$\begin{vmatrix} u_2 & u_3 \\ v_2 & v_3 \end{vmatrix} \Rightarrow u_2v_3 - u_3v_2$$

For the second entry in $\vec{u} \times \vec{v}$, we cover the second column and take the negative of the difference of the products of the cross-terms:

$$-\begin{vmatrix} u_1 & u_3 \\ v_1 & v_3 \end{vmatrix} \Rightarrow -(u_1v_3 - u_3v_1)$$

Similarly, for the third entry, we cover the third column and calculate the difference of the products of the cross-terms:

$$\begin{vmatrix} u_1 & u_2 \\ v_1 & v_2 \end{vmatrix} \Rightarrow u_1v_2 - u_2v_1$$

Note carefully that the second term must be given a minus sign in order for this procedure to provide the correct answer. Since the formula can be difficult to remember, we recommend checking the answer by verifying that it is orthogonal to both $\vec{u}$ and $\vec{v}$.

EXERCISE 1.3.4

Calculate the cross product of $\begin{bmatrix} 3 \\ -2 \\ 1 \end{bmatrix}$ and $\begin{bmatrix} 2 \\ 3 \\ 7 \end{bmatrix}$.

By construction, $\vec{u} \times \vec{v}$ is orthogonal to $\vec{u}$ and $\vec{v}$, so the direction of $\vec{u} \times \vec{v}$ is known except for the sign: does it point "up" or "down"? The general rule is as follows: the three vectors $\vec{u}$, $\vec{v}$, and $\vec{u} \times \vec{v}$, taken in this order, form a right-handed system (see page 10). Let us see how this works for simple cases.

EXERCISE 1.3.5

Let $\vec{e}_1$, $\vec{e}_2$, and $\vec{e}_3$ be the standard basis vectors in $\mathbb{R}^3$. Verify that

$$\vec{e}_1 \times \vec{e}_2 = \vec{e}_3, \quad \vec{e}_2 \times \vec{e}_3 = \vec{e}_1, \quad \vec{e}_3 \times \vec{e}_1 = \vec{e}_2$$

but

$$\vec{e}_2 \times \vec{e}_1 = -\vec{e}_3, \quad \vec{e}_3 \times \vec{e}_2 = -\vec{e}_1, \quad \vec{e}_1 \times \vec{e}_3 = -\vec{e}_2$$

Check that in every case, the three vectors taken in order form a right-handed system.

These simple examples also suggest some of the general properties of the cross product.

Theorem 1.3.3

For $\vec{x}, \vec{y}, \vec{z} \in \mathbb{R}^3$ and $t \in \mathbb{R}$, we have

(1) $\vec{x} \times \vec{y} = -\vec{y} \times \vec{x}$
(2) $\vec{x} \times \vec{x} = \vec{0}$
(3) $\vec{x} \times (\vec{y} + \vec{z}) = \vec{x} \times \vec{y} + \vec{x} \times \vec{z}$
(4) $(t\vec{x}) \times \vec{y} = t(\vec{x} \times \vec{y}) = \vec{x} \times (t\vec{y})$
(5) $\vec{x} \times \vec{y} = \vec{0}$ if and only if either $\vec{x} = \vec{0}$ or $\vec{y}$ is a scalar multiple of $\vec{x}$
(6) If $\vec{n} = \vec{x} \times \vec{y}$, then for any $\vec{w} \in \operatorname{Span}\{\vec{x}, \vec{y}\}$ we have $\vec{w} \cdot \vec{n} = 0$

Proof: These properties follow easily from the definition of the cross product and are left to the reader.

One rule we might expect does not in fact hold. In general,

$$\vec{x} \times (\vec{y} \times \vec{z}) \neq (\vec{x} \times \vec{y}) \times \vec{z}$$

This means that the parentheses cannot be omitted in a cross product. (There are formulas available for these triple-vector products, but we shall not need them. See Problem F3 in Further Problems at the end of this chapter.)

Applications of the Cross Product

Finding the Normal to a Plane

In Section 1.1, the vector equation of a plane was given in the form $\vec{x} = \vec{p} + s\vec{u} + t\vec{v}$, where $\{\vec{u}, \vec{v}\}$ is linearly independent. By definition, a normal vector $\vec{n}$ must be a non-zero vector orthogonal to both $\vec{u}$ and $\vec{v}$. Therefore, we can take $\vec{n} = \vec{u} \times \vec{v}$.

EXAMPLE 1.3.15

The lines $\vec{x} = \begin{bmatrix} 1 \\ 3 \\ 2 \end{bmatrix} + s \begin{bmatrix} 1 \\ 0 \\ 2 \end{bmatrix}$, $s \in \mathbb{R}$ and $\vec{x} = \begin{bmatrix} 1 \\ 3 \\ 2 \end{bmatrix} + t \begin{bmatrix} -1 \\ 2 \\ 1 \end{bmatrix}$, $t \in \mathbb{R}$ must lie in a common plane since they have the point $(1, 3, 2)$ in common. Find a scalar equation of the plane that contains these lines.

Solution: To find a normal vector for the plane we take the cross product of the direction vectors for the lines. We get

$$\vec{n} = \begin{bmatrix} 1 \\ 0 \\ 2 \end{bmatrix} \times \begin{bmatrix} -1 \\ 2 \\ 1 \end{bmatrix} = \begin{bmatrix} -4 \\ -3 \\ 2 \end{bmatrix}$$

Therefore, since the plane passes through $P(1, 3, 2)$, we find that a scalar equation of the plane is

$$-4x_1 - 3x_2 + 2x_3 = (-4)(1) + (-3)(3) + 2(2) = -9$$

EXAMPLE 1.3.16

Find a scalar equation of the plane that contains the three points $P(1, -2, 1)$, $Q(2, -2, -1)$, and $R(4, 1, 1)$.

Solution: Since P, Q, and R lie in the plane, then so do the directed line segments $\vec{PQ}$ and $\vec{PR}$. Hence, the normal to the plane is given by

$$\vec{n} = \vec{PQ} \times \vec{PR} = \begin{bmatrix} 1 \\ 0 \\ -2 \end{bmatrix} \times \begin{bmatrix} 3 \\ 3 \\ 0 \end{bmatrix} = \begin{bmatrix} 6 \\ -6 \\ 3 \end{bmatrix}$$

Since the plane passes through P, we find that a scalar equation of the plane is

$$6x_1 - 6x_2 + 3x_3 = (6)(1) + (-6)(-2) + 3(1) = 21, \quad \text{or} \quad 2x_1 - 2x_2 + x_3 = 7$$

EXERCISE 1.3.6

Find a scalar equation of the plane with vector equation

$$\vec{x} = \begin{bmatrix} 1 \\ 0 \\ -3 \end{bmatrix} + s \begin{bmatrix} 2 \\ 1 \\ -1 \end{bmatrix} + t \begin{bmatrix} 1 \\ 2 \\ 2 \end{bmatrix}, \quad s, t \in \mathbb{R}$$

The Length of the Cross Product

Given $\vec{u}$ and $\vec{v}$, the direction of their cross product is known. What is the length of the cross product of $\vec{u}$ and $\vec{v}$? We give the answer in the following theorem.

Theorem 1.3.4 Let $\vec{u}, \vec{v} \in \mathbb{R}^3$ and θ be the angle between $\vec{u}$ and $\vec{v}$, then $\|\vec{u} \times \vec{v}\| = \|\vec{u}\| \, \|\vec{v}\| \sin\theta$.

Proof: We give an outline of the proof. We have

$$\|\vec{u} \times \vec{v}\|^2 = (u_2v_3 - u_3v_2)^2 + (u_3v_1 - u_1v_3)^2 + (u_1v_2 - u_2v_1)^2$$

Expand by the binomial theorem and then add and subtract the term $(u_1^2v_1^2 + u_2^2v_2^2 + u_3^2v_3^2)$. The resulting terms can be arranged so as to be seen to be equal to

$$(u_1^2 + u_2^2 + u_3^2)(v_1^2 + v_2^2 + v_3^2) - (u_1v_1 + u_2v_2 + u_3v_3)^2$$

Thus,

$$\begin{aligned}\|\vec{u} \times \vec{v}\|^2 &= \|\vec{u}\|^2\|\vec{v}\|^2 - (\vec{u} \cdot \vec{v})^2 = \|\vec{u}\|^2\|\vec{v}\|^2 - \|\vec{u}\|^2\|\vec{v}\|^2\cos^2\theta \\ &= \|\vec{u}\|^2\|\vec{v}\|^2(1 - \cos^2\theta) = \|\vec{u}\|^2\|\vec{v}\|^2\sin^2\theta\end{aligned}$$

and the result follows. ∎

To interpret this formula, consider Figure 1.3.3. Assuming $\{\vec{u}, \vec{v}\}$ is linearly independent, then the vectors $\vec{u}$ and $\vec{v}$ determine a parallelogram. Take the length of $\vec{u}$ to be the base of the parallelogram. From trigonometry, we know that the length of the altitude is $\|\vec{v}\| \sin\theta$. So, the area of the parallelogram is

$$(\text{base}) \times (\text{altitude}) = \|\vec{u}\| \, \|\vec{v}\| \sin\theta = \|\vec{u} \times \vec{v}\|$$

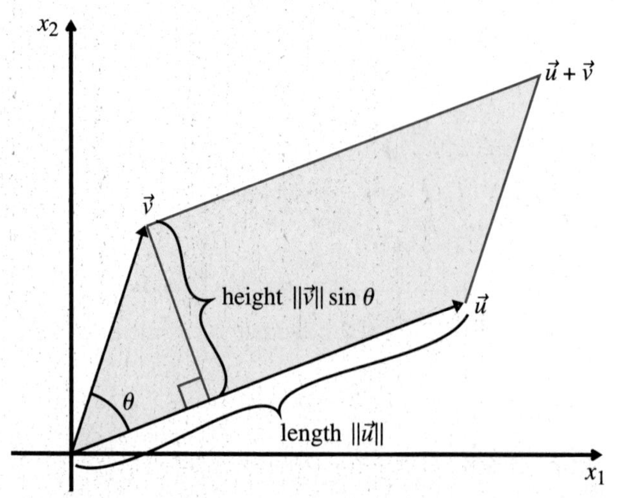

Figure 1.3.3 The area of the parallelogram is $\|\vec{u}\| \, \|\vec{v}\| \sin\theta$.

EXAMPLE 1.3.17

Find the area of the parallelogram determined by $\vec{u} = \begin{bmatrix} 1 \\ -2 \\ 1 \end{bmatrix}$ and $\vec{v} = \begin{bmatrix} -1 \\ 1 \\ 1 \end{bmatrix}$.

Solution: By Theorem 1.3.4 and our work above, we find that the area is

$$\|\vec{u} \times \vec{v}\| = \left\| \begin{bmatrix} -3 \\ -2 \\ -1 \end{bmatrix} \right\| = \sqrt{(-3)^2 + (-2)^2 + (-1)^2} = \sqrt{14}$$

EXERCISE 1.3.7

Find the area of the parallelogram determined by $\vec{u} = \begin{bmatrix} 0 \\ 2 \\ -1 \end{bmatrix}$ and $\vec{v} = \begin{bmatrix} 1 \\ 2 \\ -2 \end{bmatrix}$.

Finding the Line of Intersection of Two Planes

Unless two planes in $\mathbb{R}^3$ are parallel, their intersection will be a line. The direction vector of this line lies in both planes, so it is perpendicular to both of the normals. It can therefore be obtained as the cross product of the two normals. Once we find a point that lies on this line, we can write the vector equation of the line.

EXAMPLE 1.3.18

Find a vector equation of the line of intersection of the two planes $x_1 + x_2 - 2x_3 = 3$ and $2x_1 - x_2 + 3x_3 = 6$.

Solution: The normal vectors of the planes are $\begin{bmatrix} 1 \\ 1 \\ -2 \end{bmatrix}$ and $\begin{bmatrix} 2 \\ -1 \\ 3 \end{bmatrix}$. Hence, a direction vector of the line of intersection is

$$\vec{d} = \begin{bmatrix} 1 \\ 1 \\ -2 \end{bmatrix} \times \begin{bmatrix} 2 \\ -1 \\ 3 \end{bmatrix} = \begin{bmatrix} 1 \\ -7 \\ -3 \end{bmatrix}$$

One easy way to find a point on the line is to let $x_3 = 0$ and then solve the remaining equations $x_1 + x_2 = 3$ and $2x_1 - x_2 = 6$. The solution is $x_1 = 3$ and $x_2 = 0$. Hence, a vector equation of the line of intersection is

$$\vec{x} = \begin{bmatrix} 3 \\ 0 \\ 0 \end{bmatrix} + t \begin{bmatrix} 1 \\ -7 \\ -3 \end{bmatrix}, \quad t \in \mathbb{R}$$

EXERCISE 1.3.8

Find a vector equation of the line of intersection of the two planes $-x_1 - 2x_2 + x_3 = -2$ and $2x_1 + x_2 - 2x_3 = 1$.

PROBLEMS 1.3

Practice Problems

For Problems A1–A6, calculate the length of the vector.

A1 $\begin{bmatrix} 2 \\ -5 \end{bmatrix}$ **A2** $\begin{bmatrix} 2/\sqrt{29} \\ -5/\sqrt{29} \end{bmatrix}$ **A3** $\begin{bmatrix} 1 \\ 0 \\ -1 \end{bmatrix}$

A4 $\begin{bmatrix} 2 \\ 3 \\ -2 \end{bmatrix}$ **A5** $\begin{bmatrix} 1 \\ 1/5 \\ -3 \end{bmatrix}$ **A6** $\begin{bmatrix} 1/\sqrt{3} \\ 1/\sqrt{3} \\ -1/\sqrt{3} \end{bmatrix}$

For Problems A7–A10, calculate the distance from P to Q.

A7 $P(2,3), Q(-4,1)$ **A8** $P(1,1,-2), Q(-3,1,1)$

A9 $P(4,-6,1), Q(-3,5,1)$ **A10** $P(2,1,1), Q(4,6,-2)$

For Problems A11–A16, determine whether the pair of vectors is orthogonal.

A11 $\begin{bmatrix} 1 \\ 3 \\ 2 \end{bmatrix}, \begin{bmatrix} 2 \\ -2 \\ 2 \end{bmatrix}$ **A12** $\begin{bmatrix} -3 \\ 1 \\ 7 \end{bmatrix}, \begin{bmatrix} 2 \\ -1 \\ 1 \end{bmatrix}$

A13 $\begin{bmatrix} 2 \\ 1 \\ 1 \end{bmatrix}, \begin{bmatrix} -1 \\ 4 \\ 2 \end{bmatrix}$ **A14** $\begin{bmatrix} 4 \\ 1 \\ 0 \end{bmatrix}, \begin{bmatrix} -1 \\ 4 \\ 3 \end{bmatrix}$

A15 $\begin{bmatrix} 0 \\ 0 \\ 0 \end{bmatrix}, \begin{bmatrix} x_1 \\ x_2 \\ x_3 \end{bmatrix}$ **A16** $\begin{bmatrix} 1/3 \\ 2/3 \\ -1/3 \end{bmatrix}, \begin{bmatrix} 3/2 \\ 0 \\ -3/2 \end{bmatrix}$

For Problems A17–A20, determine all values of k for which the pair of vectors is orthogonal.

A17 $\begin{bmatrix} 3 \\ -1 \end{bmatrix}, \begin{bmatrix} 2 \\ k \end{bmatrix}$ **A18** $\begin{bmatrix} 3 \\ -1 \end{bmatrix}, \begin{bmatrix} k \\ k^2 \end{bmatrix}$

A19 $\begin{bmatrix} 1 \\ 2 \\ 3 \end{bmatrix}, \begin{bmatrix} 3 \\ -k \\ k \end{bmatrix}$ **A20** $\begin{bmatrix} 1 \\ 2 \\ 3 \end{bmatrix}, \begin{bmatrix} k \\ k \\ -k \end{bmatrix}$

For Problems A21–A23, find a scalar equation of the plane that contains the given point with the given normal vector.

A21 $P(-1,2,-3), \vec{n} = \begin{bmatrix} 2 \\ 4 \\ -1 \end{bmatrix}$ **A22** $P(2,5,4), \vec{n} = \begin{bmatrix} 3 \\ 0 \\ 5 \end{bmatrix}$

A23 $P(1,-1,1), \vec{n} = \begin{bmatrix} 3 \\ -4 \\ 1 \end{bmatrix}$

For Problems A24–A29, calculate the cross product.

A24 $\begin{bmatrix} 1 \\ -5 \\ 2 \end{bmatrix} \times \begin{bmatrix} -2 \\ 1 \\ 5 \end{bmatrix}$ **A25** $\begin{bmatrix} 2 \\ -3 \\ -5 \end{bmatrix} \times \begin{bmatrix} 4 \\ -2 \\ 7 \end{bmatrix}$

A26 $\begin{bmatrix} -1 \\ 0 \\ -1 \end{bmatrix} \times \begin{bmatrix} 0 \\ 4 \\ 5 \end{bmatrix}$ **A27** $\begin{bmatrix} 1 \\ 2 \\ 0 \end{bmatrix} \times \begin{bmatrix} -1 \\ -3 \\ 0 \end{bmatrix}$

A28 $\begin{bmatrix} 4 \\ -2 \\ 6 \end{bmatrix} \times \begin{bmatrix} -2 \\ 1 \\ -3 \end{bmatrix}$ **A29** $\begin{bmatrix} 3 \\ 1 \\ 3 \end{bmatrix} \times \begin{bmatrix} 3 \\ 1 \\ 3 \end{bmatrix}$

A30 Let $\vec{u} = \begin{bmatrix} -1 \\ 4 \\ 2 \end{bmatrix}$, $\vec{v} = \begin{bmatrix} 3 \\ 1 \\ -1 \end{bmatrix}$, and $\vec{w} = \begin{bmatrix} 2 \\ -3 \\ -1 \end{bmatrix}$. Check by calculation that the following general properties hold.

(a) $\vec{u} \times \vec{u} = \vec{0}$

(b) $\vec{u} \times \vec{v} = -\vec{v} \times \vec{u}$

(c) $\vec{u} \times 3\vec{w} = 3(\vec{u} \times \vec{w})$

(d) $\vec{u} \times (\vec{v} + \vec{w}) = \vec{u} \times \vec{v} + \vec{u} \times \vec{w}$

(e) $\vec{u} \cdot (\vec{v} \times \vec{w}) = \vec{w} \cdot (\vec{u} \times \vec{v})$

(f) $\vec{u} \cdot (\vec{v} \times \vec{w}) = -\vec{v} \cdot (\vec{u} \times \vec{w})$

For Problems A31–A34, determine a scalar equation of the plane with the given vector equation.

A31 $\vec{x} = \begin{bmatrix} 1 \\ 4 \\ 7 \end{bmatrix} + s\begin{bmatrix} 2 \\ 3 \\ -1 \end{bmatrix} + t\begin{bmatrix} 4 \\ 1 \\ 0 \end{bmatrix}, \quad s,t \in \mathbb{R}$

A32 $\vec{x} = \begin{bmatrix} 2 \\ 3 \\ -1 \end{bmatrix} + s\begin{bmatrix} 1 \\ 1 \\ 0 \end{bmatrix} + t\begin{bmatrix} -2 \\ 1 \\ 2 \end{bmatrix}, \quad s,t \in \mathbb{R}$

A33 $\vec{x} = \begin{bmatrix} 1 \\ -1 \\ 3 \end{bmatrix} + s\begin{bmatrix} 2 \\ -2 \\ 1 \end{bmatrix} + t\begin{bmatrix} 0 \\ 3 \\ 1 \end{bmatrix}, \quad s,t \in \mathbb{R}$

A34 $\vec{x} = s\begin{bmatrix} 1 \\ 3 \\ 2 \end{bmatrix} + t\begin{bmatrix} -2 \\ 4 \\ -3 \end{bmatrix}, \quad s,t \in \mathbb{R}$

For Problems A35–A40, determine a vector equation of the plane with the given scalar equation.

A35 $2x_1 - 3x_2 + x_3 = 0$ **A36** $4x_1 + x_2 - 2x_3 = 5$

A37 $x_1 + 2x_2 + 2x_3 = 1$ **A38** $3x_1 + 5x_2 - 4x_3 = 7$

A39 $6x_1 - 3x_2 + 9x_3 = 0$ **A40** $2x_1 + x_2 + 3x_3 = 3$

For Problems A41–A46, determine a scalar equation of the plane that contains the set of points.

A41 $P(2,1,5), Q(4,-3,2), R(2,6,-1)$

A42 $P(3,1,4), Q(-2,0,2), R(1,4,-1)$

A43 $P(-1,4,2), Q(3,1,-1), R(2,-3,-1)$

A44 $P(1,0,1), Q(-1,0,1), R(0,0,0)$

A45 $P(0,2,1), Q(3,-1,1), R(1,3,0)$

A46 $P(1,5,-3), Q(2,6,-1), R(1,0,1)$

For Problems A47–A49, find a scalar equation of the plane through the given point and parallel to the given plane.

A47 $P(1,-3,-1)$, $2x_1 - 3x_2 + 5x_3 = 17$

A48 $P(0,-2,4)$, $x_2 = 0$

A49 $P(1,2,1)$, $x_1 - x_2 + 3x_3 = 5$

For Problems A50–A53, determine a vector equation of the line of intersection of the given planes.

A50 $x_1 + 3x_2 - x_3 = 5$ and $2x_1 - 5x_2 + x_3 = 7$

A51 $2x_1 - 3x_3 = 7$ and $x_2 + 2x_3 = 4$

A52 $x_1 - 2x_2 + x_3 = 1$ and $3x_1 + 4x_2 - x_3 = 5$

A53 $x_1 - 2x_2 + x_3 = 0$ and $3x_1 + 4x_2 - x_3 = 0$

For Problems A54–A56, calculate the area of the parallelogram determined by each pair of vectors.

A54 $\begin{bmatrix}1\\2\\1\end{bmatrix}, \begin{bmatrix}2\\3\\-1\end{bmatrix}$ **A55** $\begin{bmatrix}1\\0\\1\end{bmatrix}, \begin{bmatrix}1\\1\\4\end{bmatrix}$ **A56** $\begin{bmatrix}-3\\1\end{bmatrix}, \begin{bmatrix}4\\3\end{bmatrix}$

(Hint: For A56, think of the vectors as $\begin{bmatrix}-3\\1\\0\end{bmatrix}, \begin{bmatrix}4\\3\\0\end{bmatrix} \in \mathbb{R}^3$.)

A57 What does it mean, geometrically, if $\vec{u} \cdot (\vec{v} \times \vec{w}) = 0$?

A58 Show that $(\vec{u} - \vec{v}) \times (\vec{u} + \vec{v}) = 2(\vec{u} \times \vec{v})$.

Homework Problems

For Problems B1–B6, calculate the length of the vector.

B1 $\begin{bmatrix}1\\-4\end{bmatrix}$ **B2** $\begin{bmatrix}2\\3\end{bmatrix}$ **B3** $\begin{bmatrix}0\\0\\0\end{bmatrix}$

B4 $\begin{bmatrix}2/3\\1/3\\2/3\end{bmatrix}$ **B5** $\begin{bmatrix}-1/\sqrt{6}\\-2/\sqrt{6}\\1/\sqrt{6}\end{bmatrix}$ **B6** $\begin{bmatrix}2/\sqrt{6}\\2/\sqrt{6}\\1/\sqrt{6}\end{bmatrix}$

For Problems B7–B13, calculate the distance from P to Q.

B7 $P(3,1)$, $Q(-2,2)$ **B8** $P(2,5)$, $Q(3,9)$

B9 $P(1,0)$, $Q(-3,5)$ **B10** $P(3,-1,3)$, $Q(6,-2,4)$

B11 $P(7,3,-5)$, $Q(9,-1,-3)$

B12 $P(4,0,2)$, $Q(5,-2,-1)$

B13 $P(1,-2,1)$, $Q(3,5,-1)$

For Problems B14–B19, determine whether the pair of vectors is orthogonal.

B14 $\begin{bmatrix}9\\-3\end{bmatrix}, \begin{bmatrix}2\\-6\end{bmatrix}$ **B15** $\begin{bmatrix}4\\6\end{bmatrix}, \begin{bmatrix}3\\2\end{bmatrix}$

B16 $\begin{bmatrix}4\\3\\-5\end{bmatrix}, \begin{bmatrix}-4\\-3\\-5\end{bmatrix}$ **B17** $\begin{bmatrix}2\\1\\-1\end{bmatrix}, \begin{bmatrix}-2\\0\\2\end{bmatrix}$

B18 $\begin{bmatrix}1\\2\\1\end{bmatrix}, \begin{bmatrix}-3\\5\\1\end{bmatrix}$ **B19** $\begin{bmatrix}1\\3\\8\end{bmatrix}, \begin{bmatrix}-2\\-2\\1\end{bmatrix}$

For Problems B20–B23, determine all values of k for which the pair of vectors is orthogonal.

B20 $\begin{bmatrix}1\\2\end{bmatrix}, \begin{bmatrix}k\\2k\end{bmatrix}$ **B21** $\begin{bmatrix}3\\2\end{bmatrix}, \begin{bmatrix}2k\\k^2\end{bmatrix}$

B22 $\begin{bmatrix}1\\3\\-1\end{bmatrix}, \begin{bmatrix}k\\2k\\2\end{bmatrix}$ **B23** $\begin{bmatrix}2\\1\\1\end{bmatrix}, \begin{bmatrix}k\\3k\\-k^2\end{bmatrix}$

For Problems B24–B28, find a scalar equation of the plane that contains the given point with the given normal vector.

B24 $P(3,9,2)$, $\vec{n} = \begin{bmatrix}1\\-1\\5\end{bmatrix}$ **B25** $P(4,3,1)$, $\vec{n} = \begin{bmatrix}3\\3\\-4\end{bmatrix}$

B26 $P(0,2,1)$, $\vec{n} = \begin{bmatrix}0\\-4\\-2\end{bmatrix}$ **B27** $P(1,3,1)$, $\vec{n} = \begin{bmatrix}1\\3\\1\end{bmatrix}$

B28 $P(0,0,0)$, $\vec{n} = \begin{bmatrix}5\\-6\\3\end{bmatrix}$

For Problems B29–B34, calculate the cross product.

B29 $\begin{bmatrix}2\\-1\\1\end{bmatrix} \times \begin{bmatrix}-6\\3\\-6\end{bmatrix}$ **B30** $\begin{bmatrix}3\\1\\-2\end{bmatrix} \times \begin{bmatrix}-1\\3\\-1\end{bmatrix}$

B31 $\begin{bmatrix}1\\2\\1\end{bmatrix} \times \begin{bmatrix}2\\4\\2\end{bmatrix}$ **B32** $\begin{bmatrix}1\\3\\-2\end{bmatrix} \times \begin{bmatrix}5\\-1\\1\end{bmatrix}$

B33 $\begin{bmatrix}5\\-1\\1\end{bmatrix} \times \begin{bmatrix}1\\3\\-2\end{bmatrix}$ **B34** $\begin{bmatrix}2\\4\\6\end{bmatrix} \times \begin{bmatrix}4\\2\\2\end{bmatrix}$

B35 Let $\vec{u} = \begin{bmatrix} 2 \\ -1 \\ 2 \end{bmatrix}$, $\vec{v} = \begin{bmatrix} 1 \\ -1 \\ 2 \end{bmatrix}$, and $\vec{w} = \begin{bmatrix} 1 \\ 0 \\ 3 \end{bmatrix}$. Check by calculation that the following general properties hold.

(a) $\vec{u} \times \vec{u} = \vec{0}$

(b) $\vec{u} \times \vec{v} = -\vec{v} \times \vec{u}$

(c) $\vec{u} \times 2\vec{w} = 2(\vec{u} \times \vec{w})$

(d) $\vec{u} \times (\vec{v} + \vec{w}) = \vec{u} \times \vec{v} + \vec{u} \times \vec{w}$

(e) $\vec{u} \cdot (\vec{v} \times \vec{w}) = \vec{w} \cdot (\vec{u} \times \vec{v})$

(f) $\vec{u} \cdot (\vec{v} \times \vec{w}) = -\vec{v} \cdot (\vec{u} \times \vec{w})$

For Problems B36–B41, determine a scalar equation of the plane with the given vector equation.

B36 $\vec{x} = \begin{bmatrix} 1 \\ 7 \\ 3 \end{bmatrix} + s\begin{bmatrix} 1 \\ -3 \\ -2 \end{bmatrix} + t\begin{bmatrix} 1 \\ 2 \\ 2 \end{bmatrix}, \quad s, t \in \mathbb{R}$

B37 $\vec{x} = \begin{bmatrix} 1 \\ 3 \\ 3 \end{bmatrix} + s\begin{bmatrix} 2 \\ 1 \\ -1 \end{bmatrix} + t\begin{bmatrix} 1 \\ -2 \\ 3 \end{bmatrix}, \quad s, t \in \mathbb{R}$

B38 $\vec{x} = \begin{bmatrix} 1 \\ 1 \\ -1 \end{bmatrix} + s\begin{bmatrix} 3 \\ 2 \\ 1 \end{bmatrix} + t\begin{bmatrix} 1 \\ 0 \\ 1 \end{bmatrix}, \quad s, t \in \mathbb{R}$

B39 $\vec{x} = s\begin{bmatrix} 1 \\ 5 \\ -4 \end{bmatrix} + t\begin{bmatrix} 3 \\ 1 \\ -1 \end{bmatrix}, \quad s, t \in \mathbb{R}$

B40 $\vec{x} = s\begin{bmatrix} 1 \\ 0 \\ -1 \end{bmatrix} + t\begin{bmatrix} 2 \\ 1 \\ -2 \end{bmatrix}, \quad s, t \in \mathbb{R}$

B41 $\vec{x} = s\begin{bmatrix} -2 \\ -1 \\ -4 \end{bmatrix} + t\begin{bmatrix} -1 \\ 1 \\ -1 \end{bmatrix}, \quad s, t \in \mathbb{R}$

For Problems B42–B47, determine a vector equation of the plane with the given scalar equation.

B42 $-x_1 + 5x_2 + 2x_3 = 2$

B43 $x_2 + x_3 = 1$

B44 $2x_1 + 2x_2 + 4x_3 = 0$

B45 $x_1 + x_2 - x_3 = 6$

B46 $3x_1 + 5x_2 - x_3 = 0$

B47 $-2x_1 - 2x_2 + 3x_3 = 4$

For Problems B48–B53, determine a scalar equation of the plane that contains the given points.

B48 $P(6, 3, 2), Q(4, 3, 3), R(9, 2, 6)$

B49 $P(4, 3, -2), Q(2, 1, 5), R(3, 1, 1)$

B50 $P(0, 2, 1), Q(4, 1, 0), R(2, 2, 0)$

B51 $P(3, 1, 2), Q(-2, -6, -1), R(1, 1, 1)$

B52 $P(9, 2, -1), Q(8, 3, 3), R(7, 3, 1)$

B53 $P(-3, 0, 3), Q(5, 5, 1), R(-2, -2, 0)$

For Problems B54–B59, find a scalar equation of the plane through the given point and parallel to the given plane.

B54 $P(2, 4, -3), 4x_1 + x_2 + 2x_3 = 2$

B55 $P(1, -2, 6), -x_1 + 2x_2 - 3x_3 = 3$

B56 $P(3, 1, 2), 2x_1 + 3x_3 = 1$

B57 $P(1, 1, 0), -x_1 - 5x_2 + 3x_3 = 5$

B58 $P(0, 0, 0), 2x_1 + 3x_2 - 4x_3 = 1$

B59 $P(-5, 2, 8), 4x_1 + 2x_2 + 2x_3 = 5$

For Problems B60–B64, determine a vector equation of the line of intersection of the given planes.

B60 $x_1 + 2x_2 + x_3 = 1$ and $2x_1 - 3x_2 + x_3 = 4$

B61 $-x_1 + 2x_2 - 2x_3 = 1$ and $x_1 + 2x_2 - x_3 = 2$

B62 $x_1 - 2x_2 + x_3 = 1$ and $3x_1 + x_2 - x_3 = 4$

B63 $2x_1 + 3x_2 + x_3 = 5$ and $4x_1 - 2x_2 + x_3 = 6$

B64 $5x_1 + 2x_2 - x_3 = 0$ and $4x_1 + x_2 - 3x_3 = 2$

For Problems B65–B70, calculate the area of the parallelogram determined by the given vectors.

B65 $\begin{bmatrix} 1 \\ 1 \\ 2 \end{bmatrix}, \begin{bmatrix} 3 \\ 2 \\ -1 \end{bmatrix}$

B66 $\begin{bmatrix} 4 \\ 2 \\ 3 \end{bmatrix}, \begin{bmatrix} 1 \\ 2 \\ 2 \end{bmatrix}$

B67 $\begin{bmatrix} 1 \\ 1 \\ -3 \end{bmatrix}, \begin{bmatrix} 3 \\ 1 \\ 1 \end{bmatrix}$

B68 $\begin{bmatrix} 2 \\ 3 \end{bmatrix}, \begin{bmatrix} -5 \\ 2 \end{bmatrix}$

B69 $\begin{bmatrix} 3 \\ 1 \end{bmatrix}, \begin{bmatrix} 4 \\ 2 \end{bmatrix}$

B70 $\begin{bmatrix} 1 \\ 1 \end{bmatrix}, \begin{bmatrix} -1 \\ 5 \end{bmatrix}$

(Hint: For Problems B54–B59 use the same method as for Problem A56.)

Conceptual Problems

C1 (a) Using geometrical arguments in $\mathbb{R}^3$, what can you say about the vectors $\vec{p}$, $\vec{n}$, and $\vec{d}$ if the line with vector equation $\vec{x} = \vec{p} + t\vec{d}$, $t \in \mathbb{R}$ and the plane with scalar equation $\vec{n} \cdot \vec{x} = k$ have no point of intersection?

(b) Confirm your answer in part (a) by determining when it is possible to find a value of the parameter t that gives a point of intersection.

C2 Let $\vec{x}, \vec{y}, \vec{z} \in \mathbb{R}^3$ and $s, t \in \mathbb{R}$. Prove the following properties of the dot product.

(a) $\vec{x} \cdot \vec{x} \geq 0$

(b) $\vec{x} \cdot \vec{x} = 0$ if and only if $\vec{x} = \vec{0}$

(c) $\vec{x} \cdot \vec{y} = \vec{y} \cdot \vec{x}$

(d) $\vec{x} \cdot (s\vec{y} + t\vec{z}) = s(\vec{x} \cdot \vec{y}) + t(\vec{x} \cdot \vec{z})$

C3 Prove that the zero vector $\vec{0}$ in $\mathbb{R}^2$ is orthogonal to every vector $\vec{x} \in \mathbb{R}^2$ in two ways.

(a) By directly calculating $\vec{x} \cdot \vec{0}$.

(b) By using property (4) in Theorem 1.3.2.

C4 Determine an equation of the set of points in $\mathbb{R}^3$ that are equidistant from points P and Q. Explain why the set is a plane, and determine its normal vector.

C5 Find a scalar equation of the plane such that each point of the plane is equidistant from the points $P(2, 2, 5)$ and $Q(-3, 4, 1)$ in two ways.

(a) Write and simplify the equation $\|\vec{PX}\| = \|\vec{QX}\|$.

(b) Determine a point on the plane and the normal vector by geometrical arguments.

C6 Let $\vec{x}, \vec{y}, \vec{z} \in \mathbb{R}^3$. Consider the following statement: "If $\vec{x} \cdot \vec{y} = \vec{x} \cdot \vec{z}$, then $\vec{y} = \vec{z}$."

(a) If the statement is true, prove it. If the statement is false, provide a counterexample.

(b) If we specify $\vec{x} \neq \vec{0}$, does that change the result?

C7 Show that if X is a point in $\mathbb{R}^3$ on the line through P and Q, then $\vec{x} \times (\vec{q} - \vec{p}) = \vec{p} \times \vec{q}$, where $\vec{x} = \overrightarrow{OX}$, $\vec{p} = \overrightarrow{OP}$, and $\vec{q} = \overrightarrow{OQ}$.

C8 (a) Let $\vec{n}$ be a vector in $\mathbb{R}^3$ such that $\|\vec{n}\| = 1$ (called a **unit vector**). Let α be the angle between $\vec{n}$ and the x_1-axis; let β be the angle between $\vec{n}$ and the x_2-axis; and let γ be the angle between $\vec{n}$ and the x_3-axis. Explain why

$$\vec{n} = \begin{bmatrix} \cos\alpha \\ \cos\beta \\ \cos\gamma \end{bmatrix}$$

(Hint: Take the dot product of $\vec{n}$ with the standard basis vectors.)

Because of this equation, the components n_1, n_2, n_3 are sometimes called the **direction cosines**.

(b) Explain why $\cos^2\alpha + \cos^2\beta + \cos^2\gamma = 1$.

(c) Give a two-dimensional version of the direction cosines, and explain the connection to the identity $\cos^2\theta + \sin^2\theta = 1$.

C9 Let $\vec{u}, \vec{v}, \vec{w} \in \mathbb{R}^3$. Consider the following statement: "If $\vec{u} \neq \vec{0}$, and $\vec{u} \times \vec{v} = \vec{u} \times \vec{w}$, then $\vec{v} = \vec{w}$."
If the statement is true, prove it. If it is false, give a counterexample.

C10 Let $\vec{u}, \vec{v}, \vec{w} \in \mathbb{R}^3$. Explain why $\vec{u} \times (\vec{v} \times \vec{w})$ must be a vector that satisfies the vector equation $\vec{x} = s\vec{v} + t\vec{w}$.

C11 Give an example of distinct vectors $\vec{u}$, $\vec{v}$, and $\vec{w}$ in $\mathbb{R}^3$ such that

(a) $\vec{u} \times (\vec{v} \times \vec{w}) = (\vec{u} \times \vec{v}) \times \vec{w}$

(b) $\vec{u} \times (\vec{v} \times \vec{w}) \neq (\vec{u} \times \vec{v}) \times \vec{w}$

C12 Prove that if $\vec{x}$ and $\vec{y}$ are non-zero orthogonal vectors in $\mathbb{R}^2$, then $\{\vec{x}, \vec{y}\}$ is linearly independent. (Hint: Use the definition of linear independence, and take the dot product of both sides with respect to $\vec{x}$.)

C13 A set $\mathcal{B} = \{\vec{v}_1, \vec{v}_2\}$ is said to be an **orthogonal basis** for $\mathbb{R}^2$ if $\mathcal{B}$ is a basis for $\mathbb{R}^2$, and $\vec{v}_1$ and $\vec{v}_2$ are orthogonal. Show that the coordinates of any vector $\vec{x} \in \mathbb{R}^2$ with respect to $\mathcal{B}$ are $c_1 = \dfrac{\vec{x} \cdot \vec{v}_1}{\|\vec{v}_1\|^2}$ and $c_2 = \dfrac{\vec{x} \cdot \vec{v}_2}{\|\vec{v}_2\|^2}$.

1.4 Vectors in $\mathbb{R}^n$

We now extend the ideas from Sections 1.1 and 1.2 to n-dimensional Euclidean space.

Applications of Vectors in $\mathbb{R}^n$

Students sometimes do not see the point in discussing n-dimensional space because it does not seem to correspond to any physical realistic geometry. It is important to realize that a vector in $\mathbb{R}^n$ does not have to refer to an object in n-dimensional space. Here are just a few uses of vectors in $\mathbb{R}^n$.

String Theory Some scientists working in string theory work with vectors in six-, ten-, or eleven-dimensional space.

Position of a Rigid Object To discuss the position of a particle, an engineer needs to specify its position (3 variables) and the direction it is pointing (3 more variables); the engineer therefore uses a vector with 6 components.

Economic Models An economist seeking to model the Canadian economy uses many variables. One standard model has more than 1500 variables.

Population Age Distribution A biologist may use a vector with n components where the i-th component of the vector is the number of individuals in a given age class.

Analog Signal Sampling To convert an analog signal into digital form, the average intensity level of the signal over n equally spaced time intervals is recorded into a vector with n components.

Addition and Scalar Multiplication of Vectors in $\mathbb{R}^n$

Definition
$\mathbb{R}^n$

$\mathbb{R}^n$ is the set of all vectors of the form $\begin{bmatrix} x_1 \\ \vdots \\ x_n \end{bmatrix}$, where $x_i \in \mathbb{R}$. Mathematically,

$$\mathbb{R}^n = \left\{ \begin{bmatrix} x_1 \\ \vdots \\ x_n \end{bmatrix} \mid x_1, \dots, x_n \in \mathbb{R} \right\}$$

If $\vec{x} = \begin{bmatrix} x_1 \\ \vdots \\ x_n \end{bmatrix}$ and $\vec{y} = \begin{bmatrix} y_1 \\ \vdots \\ y_n \end{bmatrix}$ are two vectors in $\mathbb{R}^n$ such that $x_i = y_i$ for $1 \leq i \leq n$, then we say that $\vec{x}$ and $\vec{y}$ are **equal** and write $\vec{x} = \vec{y}$.

As we have seen already in $\mathbb{R}^2$ and $\mathbb{R}^3$, we write vectors in $\mathbb{R}^n$ as columns. These are sometimes called **column vectors.**

Definition
Addition and Scalar Multiplication in $\mathbb{R}^n$

If $\vec{x} = \begin{bmatrix} x_1 \\ \vdots \\ x_n \end{bmatrix}, \vec{y} = \begin{bmatrix} y_1 \\ \vdots \\ y_n \end{bmatrix} \in \mathbb{R}^n$, then we define **addition** of vectors by

$$\vec{x} + \vec{y} = \begin{bmatrix} x_1 \\ \vdots \\ x_n \end{bmatrix} + \begin{bmatrix} y_1 \\ \vdots \\ y_n \end{bmatrix} = \begin{bmatrix} x_1 + y_1 \\ \vdots \\ x_n + y_n \end{bmatrix}$$

We define **scalar multiplication** of a vector $\vec{x}$ by a scalar $t \in \mathbb{R}$ by

$$t\vec{x} = t\begin{bmatrix} x_1 \\ \vdots \\ x_n \end{bmatrix} = \begin{bmatrix} tx_1 \\ \vdots \\ tx_n \end{bmatrix}$$

As in $\mathbb{R}^2$ and $\mathbb{R}^3$, we call a sum of scalar multiples of vectors in $\mathbb{R}^n$ a **linear combination**, and by $\vec{x} - \vec{y}$ we mean $\vec{x} + (-1)\vec{y}$.

EXAMPLE 1.4.1

Let $\vec{u} = \begin{bmatrix} 1 \\ 2 \\ -5 \\ 1 \end{bmatrix}, \vec{v} = \begin{bmatrix} 2 \\ 1 \\ -5 \\ 4 \end{bmatrix} \in \mathbb{R}^4$. Calculate the linear combination $2\vec{u} - 3\vec{v}$.

Solution: We get

$$2\vec{u} - 3\vec{v} = 2\begin{bmatrix} 1 \\ 2 \\ -5 \\ 1 \end{bmatrix} + (-3)\begin{bmatrix} 2 \\ 1 \\ -5 \\ 4 \end{bmatrix} = \begin{bmatrix} 2 \\ 4 \\ -10 \\ 2 \end{bmatrix} + \begin{bmatrix} -6 \\ -3 \\ 15 \\ -12 \end{bmatrix} = \begin{bmatrix} -4 \\ 1 \\ 5 \\ -10 \end{bmatrix}$$

Of course, we get exactly the same properties that we saw in Theorem 1.1.1.

Theorem 1.4.1

For all $\vec{w}, \vec{x}, \vec{y} \in \mathbb{R}^n$ and $s, t \in \mathbb{R}$ we have

V1 $\vec{x} + \vec{y} \in \mathbb{R}^n$ (closed under addition)
V2 $\vec{x} + \vec{y} = \vec{y} + \vec{x}$ (addition is commutative)
V3 $(\vec{x} + \vec{y}) + \vec{w} = \vec{x} + (\vec{y} + \vec{w})$ (addition is associative)
V4 There exists a vector $\vec{0} \in \mathbb{R}^n$ such that $\vec{z} + \vec{0} = \vec{z}$ for all $\vec{z} \in \mathbb{R}^n$ (zero vector)
V5 For each $\vec{x} \in \mathbb{R}^n$ there exists a vector $-\vec{x} \in \mathbb{R}^n$ such that $\vec{x} + (-\vec{x}) = \vec{0}$ (additive inverses)
V6 $t\vec{x} \in \mathbb{R}^n$ (closed under scalar multiplication)
V7 $s(t\vec{x}) = (st)\vec{x}$ (scalar multiplication is associative)
V8 $(s + t)\vec{x} = s\vec{x} + t\vec{x}$ (a distributive law)
V9 $t(\vec{x} + \vec{y}) = t\vec{x} + t\vec{y}$ (another distributive law)
V10 $1\vec{x} = \vec{x}$ (scalar multiplicative identity)

Proof: We will prove property V2 and leave the other proofs to the reader.

For V2,

$$\vec{x} + \vec{y} = \begin{bmatrix} x_1 + y_1 \\ \vdots \\ x_n + y_n \end{bmatrix} = \begin{bmatrix} y_1 + x_1 \\ \vdots \\ y_n + x_n \end{bmatrix} = \vec{y} + \vec{x}$$

■

EXERCISE 1.4.1 Prove properties V5 and V7 from Theorem 1.4.1.

Observe that properties V2, V3, V7, V8, V9, and V10 from Theorem 1.4.1 refer only to the operations of addition and scalar multiplication, while the other properties, V1, V4, V5, and V6, are about the relationship between the operations and the set $\mathbb{R}^n$. These facts should be clear in the proof of the theorem. Moreover, we see that the zero vector of $\mathbb{R}^n$ is the vector $\vec{0} = \begin{bmatrix} 0 \\ \vdots \\ 0 \end{bmatrix}$, and the additive inverse of $\vec{x}$ is $-\vec{x} = (-1)\vec{x}$.

Subspaces

Properties V1 and V6 from Theorem 1.4.1 show that $\mathbb{R}^n$ is **closed under linear combinations**. That is, if $\vec{v}_1, \ldots, \vec{v}_k \in \mathbb{R}^n$, then $c_1\vec{v}_1 + \cdots + c_k\vec{v}_k$ is also a vector in $\mathbb{R}^n$ for any $c_1, \ldots, c_k \in \mathbb{R}$. As previously mentioned, not all subsets of $\mathbb{R}^n$ are closed under linear combinations.

EXAMPLE 1.4.2 Consider the line L in $\mathbb{R}^2$ defined by

$$L = \left\{ \begin{bmatrix} 1 \\ 2 \end{bmatrix} + t \begin{bmatrix} 1 \\ 0 \end{bmatrix} \mid t \in \mathbb{R} \right\}$$

Show that L is not closed under linear combinations.

Solution: If we take two vectors in the line, say

$$\vec{x} = \begin{bmatrix} 1 \\ 2 \end{bmatrix} + 0 \begin{bmatrix} 1 \\ 0 \end{bmatrix} = \begin{bmatrix} 1 \\ 2 \end{bmatrix}, \quad \text{and} \quad \vec{y} = \begin{bmatrix} 1 \\ 2 \end{bmatrix} + 1 \begin{bmatrix} 1 \\ 0 \end{bmatrix} = \begin{bmatrix} 2 \\ 2 \end{bmatrix}$$

We get $\vec{x} + \vec{y} = \begin{bmatrix} 3 \\ 4 \end{bmatrix}$. However, we see that there is no value of $t \in \mathbb{R}$ such that

$$\begin{bmatrix} 3 \\ 4 \end{bmatrix} = \begin{bmatrix} 1 \\ 2 \end{bmatrix} + t \begin{bmatrix} 1 \\ 0 \end{bmatrix}$$

Consequently, $\begin{bmatrix} 3 \\ 4 \end{bmatrix}$ is not a vector on the line. So, this set is not closed under linear combinations.

We are most interested in non-empty subsets of $\mathbb{R}^n$ that are closed under linear combinations.

Definition
Subspace

A non-empty subset S of $\mathbb{R}^n$ is called a **subspace** of $\mathbb{R}^n$ if for all vectors $\vec{x}, \vec{y} \in S$ and $s, t \in \mathbb{R}$ we have

$$s\vec{x} + t\vec{y} \in S$$

The definition requires that a subspace S be non-empty. In particular, it follows from the definition that if we pick any two vectors $\vec{x}, \vec{y} \in S$ and take $s = t = 0$, then $\vec{0} = 0\vec{x} + 0\vec{y} \in S$. Hence, every subspace S of $\mathbb{R}^n$ contains the zero vector. This fact provides an easy method for disqualifying any subsets that do not contain the zero vector as subspaces. For instance, as we saw in Example 1.4.2, a line in $\mathbb{R}^2$ cannot be a subspace if it does not pass through the origin. Thus, when checking to determine if a set S is non-empty, it makes sense to first check if $\vec{0} \in S$.

It is easy to see that the set $\{\vec{0}\}$ consisting of only the zero vector in $\mathbb{R}^n$ is a subspace of $\mathbb{R}^n$; this is called the **trivial subspace**. Additionally, $\mathbb{R}^n$ is a subspace of itself. We will see throughout the text that other subspaces arise naturally in linear algebra.

EXAMPLE 1.4.3

Show that $T = \left\{ \begin{bmatrix} x_1 \\ x_2 \\ x_3 \end{bmatrix} \mid x_1 - x_2 + x_3 = 0 \right\}$ is a subspace of $\mathbb{R}^3$.

Solution: By definition, T is a subset of $\mathbb{R}^3$ and we have that $\vec{0} = \begin{bmatrix} 0 \\ 0 \\ 0 \end{bmatrix} \in T$ since taking $x_1 = 0, x_2 = 0$, and $x_3 = 0$ satisfies $x_1 - x_2 + x_3 = 0$.

Let $\vec{x} = \begin{bmatrix} x_1 \\ x_2 \\ x_3 \end{bmatrix}, \vec{y} = \begin{bmatrix} y_1 \\ y_2 \\ y_3 \end{bmatrix} \in T$. Then they must satisfy the condition of the set, so $x_1 - x_2 + x_3 = 0$ and $y_1 - y_2 + y_3 = 0$.

We must show that $s\vec{x} + t\vec{y}$ satisfies the condition on T. We have

$$s\vec{x} + t\vec{y} = \begin{bmatrix} sx_1 + ty_1 \\ sx_2 + ty_2 \\ sx_3 + ty_3 \end{bmatrix}$$

and

$$(sx_1 + ty_1) - (sx_2 + ty_2) + (sx_3 + ty_3) = s(x_1 - x_2 + x_3) + t(y_1 - y_2 + y_3) = s(0) + t(0) = 0$$

Hence, $s\vec{x} + t\vec{y} \in T$.

Therefore, by definition, T is a subspace of $\mathbb{R}^3$.

EXAMPLE 1.4.4

Show that $U = \left\{ \begin{bmatrix} x_1 \\ x_2 \end{bmatrix} \mid x_1x_2 = 0 \right\}$ is not a subspace of $\mathbb{R}^2$.

Solution: To show that U is not a subspace, we just need to give one example showing that U does not satisfy the definition of a subspace.

Observe that $\vec{x} = \begin{bmatrix} 1 \\ 0 \end{bmatrix}$ and $\vec{y} = \begin{bmatrix} 0 \\ 1 \end{bmatrix}$ are both in U, but $\vec{x} + \vec{y} = \begin{bmatrix} 1 \\ 1 \end{bmatrix} \notin U$, since $1(1) \neq 0$.

Thus, U is not a subspace of $\mathbb{R}^2$.

EXERCISE 1.4.2

Show that $S = \left\{ \begin{bmatrix} x_1 \\ x_2 \end{bmatrix} \mid 2x_1 = x_2 \right\}$ is a subspace of $\mathbb{R}^2$ and $T = \left\{ \begin{bmatrix} x_1 \\ x_2 \end{bmatrix} \mid x_1 + x_2 = 2 \right\}$ is not a subspace of $\mathbb{R}^2$.

EXERCISE 1.4.3

Prove that if P is a plane in $\mathbb{R}^3$ with vector equation

$$\vec{x} = a\vec{v}_1 + b\vec{v}_2, \quad a, b \in \mathbb{R}$$

then P is a subspace of $\mathbb{R}^3$.

Spanning Sets and Linear Independence

It can be shown that the only subspaces of $\mathbb{R}^2$ are $\{\vec{0}\}$, lines through the origin, and $\mathbb{R}^2$ itself. Similarly, the only subspaces of $\mathbb{R}^3$ are $\{\vec{0}\}$, lines through the origin, planes through the origin, and $\mathbb{R}^3$ itself. We see that all of these sets can be described as the span of a set of vectors. Thus, we also extend our definition of spanning to $\mathbb{R}^n$.

Definition
Span in $\mathbb{R}^n$

Let $B = \{\vec{v}_1, \dots, \vec{v}_k\}$ be a set of vectors in $\mathbb{R}^n$. We define the **span** of B, denoted Span B, to be the set of all possible linear combinations of the vectors in B. Mathematically,

$$\operatorname{Span} B = \{c_1\vec{v}_1 + \cdots + c_k\vec{v}_k \mid c_1, \dots, c_k \in \mathbb{R}\}$$

A **vector equation** for Span B is

$$\vec{x} = c_1\vec{v}_1 + \cdots + c_k\vec{v}_k, \quad c_1, \dots, c_k \in \mathbb{R}$$

If $S = \operatorname{Span} B$, then we say that B **spans** S, that B is a **spanning set** for S, and that S is **spanned** by B.

The following theorem says that, like in $\mathbb{R}^2$ and $\mathbb{R}^3$, all spanned sets are subspaces.

Theorem 1.4.2 If $\{\vec{v}_1, \dots, \vec{v}_k\}$ is a set of vectors in $\mathbb{R}^n$ and $S = \text{Span}\{\vec{v}_1, \dots, \vec{v}_k\}$, then S is a subspace of $\mathbb{R}^n$.

Proof: By properties V1 and V6 of Theorem 1.4.1, $t_1\vec{v}_1 + \cdots + t_k\vec{v}_k \in \mathbb{R}^n$, so S is a subset of $\mathbb{R}^n$.

Taking $t_i = 0$ for $1 \leq i \leq k$, we get $\vec{0} = 0\vec{v}_1 + \cdots + 0\vec{v}_k \in S$, so S is non-empty.

Let $\vec{x}, \vec{y} \in S$. Then, for some real numbers c_i and d_i, $1 \leq i \leq k$, $\vec{x} = c_1\vec{v}_1 + \cdots + c_k\vec{v}_k$ and $\vec{y} = d_1\vec{v}_1 + \cdots + d_k\vec{v}_k$. It follows that

$$\begin{aligned} s\vec{x} + t\vec{y} &= s(c_1\vec{v}_1 + \cdots + c_k\vec{v}_k) + t(d_1\vec{v}_1 + \cdots + d_k\vec{v}_k) \\ &= (sc_1 + td_1)\vec{v}_1 + \cdots + (sc_k + td_k)\vec{v}_k \end{aligned}$$

so, $s\vec{x} + t\vec{y} \in S$ since $(sc_i + td_i) \in \mathbb{R}$.

Therefore, S is a subspace of $\mathbb{R}^n$. ■

To simplify spanning sets in $\mathbb{R}^n$, we use the following theorem, which corresponds to what we saw in Section 1.2.

Theorem 1.4.3 Let $\mathcal{B} = \{\vec{v}_1, \dots, \vec{v}_k\}$ be a set of vectors in $\mathbb{R}^n$. Some vector $\vec{v}_i$, $1 \leq i \leq k$, can be written as a linear combination of $\vec{v}_1, \dots, \vec{v}_{i-1}, \vec{v}_{i+1}, \dots, \vec{v}_k$ if and only if

$$\text{Span}\{\vec{v}_1, \dots, \vec{v}_k\} = \text{Span}\{\vec{v}_1, \dots, \vec{v}_{i-1}, \vec{v}_{i+1}, \dots, \vec{v}_k\}$$

As before, we define linear independence so that a spanning set is as simple as possible if and only if it is linearly independent.

Definition
Linearly Dependent
Linearly Independent

Let $\mathcal{B} = \{\vec{v}_1, \dots, \vec{v}_k\}$ be a set of vectors in $\mathbb{R}^n$. The set $\mathcal{B}$ is said to be **linearly dependent** if there exist real coefficients $t_1, \dots, t_k$ not all zero such that

$$t_1\vec{v}_1 + \cdots + t_k\vec{v}_k = \vec{0}$$

The set $\mathcal{B}$ is said to be **linearly independent** if the only solution to

$$t_1\vec{v}_1 + \cdots + t_k\vec{v}_k = \vec{0}$$

is $t_1 = t_2 = \cdots = t_k = 0$.

EXAMPLE 1.4.5

Prove that the set $\mathcal{B} = \left\{ \begin{bmatrix} 1 \\ 0 \\ 0 \\ 1 \end{bmatrix}, \begin{bmatrix} 0 \\ 1 \\ 0 \\ -2 \end{bmatrix}, \begin{bmatrix} 0 \\ 1 \\ 1 \\ -1 \end{bmatrix} \right\}$ is linearly independent.

Solution: Consider the vector equation

$$\begin{bmatrix} 0 \\ 0 \\ 0 \\ 0 \end{bmatrix} = t_1 \begin{bmatrix} 1 \\ 0 \\ 0 \\ 1 \end{bmatrix} + t_2 \begin{bmatrix} 0 \\ 1 \\ 0 \\ -2 \end{bmatrix} + t_3 \begin{bmatrix} 0 \\ 1 \\ 1 \\ -1 \end{bmatrix} = \begin{bmatrix} t_1 \\ t_2 + t_3 \\ t_3 \\ t_1 - 2t_2 - t_3 \end{bmatrix}$$

Comparing entries gives the system of equations

$$t_1 = 0, \quad t_2 + t_3 = 0, \quad t_3 = 0, \quad t_1 - 2t_2 - t_3 = 0$$

Solving the system gives $t_1 = t_2 = t_3 = 0$. Hence, $\mathcal{B}$ is linearly independent.

CONNECTION

Observe that determining whether a set $\{\vec{v}_1, \dots, \vec{v}_k\}$ in $\mathbb{R}^n$ is linearly dependent or linearly independent requires determining solutions of the vector equation

$$t_1\vec{v}_1 + \cdots + t_k\vec{v}_k = \vec{0}$$

Similarly, determining whether a vector $\vec{b} \in \mathbb{R}^n$ is in a spanned set Span$\{\vec{w}_1, \dots, \vec{w}_\ell\}$ in $\mathbb{R}^n$ requires determining if the vector equation

$$s_1\vec{w}_1 + \cdots + s_k\vec{w}_k = \vec{b}$$

has a solution. Both of these vector equations actually represents n equations (one for each entry of the vectors) in the k unknown scalars. In the next chapter, we will look at how to efficiently solve such systems of equations.

We can now extend our geometrical concepts of lines and planes to $\mathbb{R}^n$ for $n > 3$. To match what we did in $\mathbb{R}^2$ and $\mathbb{R}^3$, we make the following definitions.

Definition
Line in $\mathbb{R}^n$

Let $\vec{p}, \vec{v} \in \mathbb{R}^n$ with $\vec{v} \neq \vec{0}$. We call the set with vector equation

$$\vec{x} = \vec{p} + t_1\vec{v}, \quad t_1 \in \mathbb{R}$$

a **line** in $\mathbb{R}^n$ that passes through $\vec{p}$.

Definition
Plane in $\mathbb{R}^n$

Let $\vec{v}_1, \vec{v}_2, \vec{p} \in \mathbb{R}^n$, with $\{\vec{v}_1, \vec{v}_2\}$ being a linearly independent set. The set with vector equation

$$\vec{x} = \vec{p} + t_1\vec{v}_1 + t_2\vec{v}_2, \quad t_1, t_2 \in \mathbb{R}$$

is called a **plane** in $\mathbb{R}^n$ that passes through $\vec{p}$.

Definition
Hyperplane in $\mathbb{R}^n$

Let $\vec{v}_1, \dots, \vec{v}_{n-1}, \vec{p} \in \mathbb{R}^n$, with $\{\vec{v}_1, \dots, \vec{v}_{n-1}\}$ being linearly independent. The set with vector equation

$$\vec{x} = \vec{p} + t_1\vec{v}_1 + \cdots + t_{n-1}\vec{v}_{n-1}, \quad t_1, \dots, t_{n-1} \in \mathbb{R}$$

is called a **hyperplane** in $\mathbb{R}^n$ that passes through $\vec{p}$.

EXAMPLE 1.4.6

Let $\mathcal{B} = \left\{\begin{bmatrix}1\\0\\0\\1\end{bmatrix}, \begin{bmatrix}0\\1\\0\\-2\end{bmatrix}, \begin{bmatrix}0\\1\\1\\-1\end{bmatrix}\right\}$. Since $\mathcal{B}$ is linearly independent, the set $\operatorname{Span}\mathcal{B}$ is a hyperplane in $\mathbb{R}^4$.

Bases of Subspaces of $\mathbb{R}^n$

In Section 1.2, we defined a basis for $\mathbb{R}^2$ to be a set $\mathcal{B}$ such that $\operatorname{Span}\mathcal{B} = \mathbb{R}^2$ and $\mathcal{B}$ is linearly independent. We now generalize the concept of a basis to any subspace of $\mathbb{R}^n$.

Definition
Basis of a Subspace

Let $\mathcal{B} = \{\vec{v}_1, \dots, \vec{v}_k\}$ be a set in $\mathbb{R}^n$, and let S be a non-trivial subspace of $\mathbb{R}^n$. If $\mathcal{B}$ is linearly independent and $\operatorname{Span}\mathcal{B} = S$, then the set $\mathcal{B}$ is called a **basis** for S.
A basis for the trivial subspace $\{\vec{0}\}$ is defined to be the empty set.

EXAMPLE 1.4.7

Prove that $\mathcal{B} = \left\{\begin{bmatrix}1\\0\\2\end{bmatrix}, \begin{bmatrix}0\\1\\3\end{bmatrix}\right\}$ is a basis for the plane P in $\mathbb{R}^3$ with scalar equation $2x_1 + 3x_2 - x_3 = 0$.

Solution: By definition of P, every $\vec{x} = \begin{bmatrix}x_1\\x_2\\x_3\end{bmatrix} \in P$ satisfies $2x_1 + 3x_2 - x_3 = 0$. Solving this for x_3 gives $x_3 = 2x_1 + 3x_2$. Consider

$$\begin{bmatrix}x_1\\x_2\\2x_1 + 3x_2\end{bmatrix} = c_1\begin{bmatrix}1\\0\\2\end{bmatrix} + c_2\begin{bmatrix}0\\1\\3\end{bmatrix} = \begin{bmatrix}c_1\\c_2\\2c_1 + 3c_2\end{bmatrix}$$

Solving gives $c_1 = x_1, c_2 = x_2$. Thus, $\mathcal{B}$ spans P. Now consider

$$\begin{bmatrix}0\\0\\0\end{bmatrix} = c_1\begin{bmatrix}1\\0\\2\end{bmatrix} + c_2\begin{bmatrix}0\\1\\3\end{bmatrix} = \begin{bmatrix}c_1\\c_2\\2c_1 + 3c_2\end{bmatrix}$$

Comparing entries, we get that $c_1 = c_2 = 0$. Hence, $\mathcal{B}$ is also linearly independent.

Since $\mathcal{B}$ is linearly independent and spans P, it is a basis for P.

We can think of the basis $\mathcal{B} = \left\{ \begin{bmatrix} 1 \\ 0 \\ 2 \end{bmatrix}, \begin{bmatrix} 0 \\ 1 \\ 3 \end{bmatrix} \right\}$ for P in Example 1.4.7 in exactly the same way as for any basis of $\mathbb{R}^2$. That is, the lines Span $\left\{ \begin{bmatrix} 1 \\ 0 \\ 2 \end{bmatrix} \right\}$ and Span $\left\{ \begin{bmatrix} 0 \\ 1 \\ 3 \end{bmatrix} \right\}$ form coordinate axes for the plane as in Figure 1.4.1.

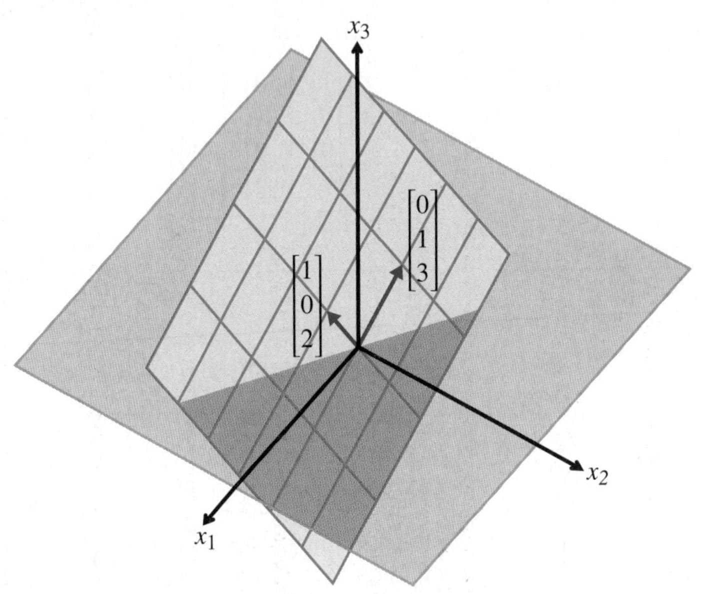

Figure 1.4.1 A basis for P.

The standard basis for $\mathbb{R}^n$ matches what we saw for $\mathbb{R}^2$ and $\mathbb{R}^3$.

Definition
Standard Basis for $\mathbb{R}^n$

In $\mathbb{R}^n$, let $\vec{e}_i$ represent the vector whose i-th component is 1 and all other components are 0. The set $\{\vec{e}_1, \dots, \vec{e}_n\}$ is called the **standard basis for** $\mathbb{R}^n$.

EXERCISE 1.4.4

State the standard basis for $\mathbb{R}^4$. Prove that it is linearly independent, and show that it is a spanning set for $\mathbb{R}^4$.

As we saw in Section 1.2, a basis $\mathcal{B}$ defines a coordinate system for the subspace S. In particular, because $\mathcal{B}$ spans S, it means that every vector $\vec{x} \in S$ can be written as a linear combination of the vectors in $\mathcal{B}$. Moreover, since $\mathcal{B}$ is linearly independent, it means that there is only one such linear combination for each $\vec{x} \in S$. We state this as a theorem.

Theorem 1.4.4

If $\mathcal{B} = \{\vec{v}_1, \dots, \vec{v}_k\}$ is a basis for a subspace S of $\mathbb{R}^n$, then for each $\vec{x} \in S$ there exists unique scalars $c_1, \dots, c_k \in \mathbb{R}$ such that

$$\vec{x} = c_1\vec{v}_1 + \cdots + c_k\vec{v}_k$$

PROBLEMS 1.4

Practice Problems

For Problems A1–A4, compute the linear combination.

A1 $\begin{bmatrix}1\\3\\2\\-1\end{bmatrix}+2\begin{bmatrix}2\\3\\-1\\1\end{bmatrix}$

A2 $\begin{bmatrix}1\\-2\\5\\1\end{bmatrix}-3\begin{bmatrix}-1\\1\\1\\2\end{bmatrix}+2\begin{bmatrix}3\\-1\\4\\0\end{bmatrix}$

A3 $\begin{bmatrix}3\\-4\\-1\\2\\1\end{bmatrix}+\begin{bmatrix}5\\2\\2\\4\\3\end{bmatrix}-\begin{bmatrix}2\\-2\\-3\\1\\1\end{bmatrix}$

A4 $2\begin{bmatrix}1\\2\\1\\0\\-1\end{bmatrix}+2\begin{bmatrix}2\\-2\\1\\2\\1\end{bmatrix}-3\begin{bmatrix}2\\0\\1\\1\\1\end{bmatrix}$

For Problems A5–A10, determine whether the set is a subspace of the appropriate $\mathbb{R}^n$.

A5 $\text{Span}\left\{\begin{bmatrix}1\\2\end{bmatrix}\right\}$

A6 $\left\{\vec{x}\in\mathbb{R}^3 \mid x_1^2-x_2^2=x_3\right\}$

A7 $\left\{\vec{x}\in\mathbb{R}^3 \mid x_1=x_3\right\}$

A8 $\left\{\vec{x}\in\mathbb{R}^2 \mid x_1+x_2=0\right\}$

A9 $\left\{\vec{x}\in\mathbb{R}^3 \mid x_1x_2=x_3\right\}$

A10 $\vec{x}=\begin{bmatrix}2\\2\\2\end{bmatrix}+t_1\begin{bmatrix}1\\1\\1\end{bmatrix}, t\in\mathbb{R}$

For Problems A11–A16, determine whether the set is a subspace of $\mathbb{R}^4$.

A11 $\left\{\vec{x}\in\mathbb{R}^4 \mid x_1+x_2+x_3+x_4=0\right\}$

A12 $\{\vec{v}_1\in\mathbb{R}^4\}$, where $\vec{v}_1\neq\vec{0}$

A13 $\left\{\vec{x}\in\mathbb{R}^4 \mid x_1+2x_3=5, x_1-3x_4=0\right\}$

A14 $\left\{\vec{x}\in\mathbb{R}^4 \mid x_1=x_3x_4, x_2-x_4=0\right\}$

A15 $\left\{\vec{x}\in\mathbb{R}^4 \mid 2x_1=3x_4, x_2-5x_3=0\right\}$

A16 $\left\{\vec{x}\in\mathbb{R}^4 \mid x_1+x_2=-x_4, x_3=2\right\}$

For Problems A17–A20, show that the set is linearly dependent by writing a non-trivial linear combination of the vectors that equals the zero vector.

A17 $\left\{\begin{bmatrix}0\\0\\0\\0\end{bmatrix},\begin{bmatrix}1\\0\\1\\2\end{bmatrix},\begin{bmatrix}2\\1\\-1\\-3\end{bmatrix}\right\}$

A18 $\left\{\begin{bmatrix}2\\-1\\3\\2\end{bmatrix},\begin{bmatrix}0\\2\\1\\-1\end{bmatrix},\begin{bmatrix}0\\4\\2\\-2\end{bmatrix}\right\}$

A19 $\left\{\begin{bmatrix}1\\1\\0\\2\end{bmatrix},\begin{bmatrix}1\\1\\1\\1\end{bmatrix},\begin{bmatrix}2\\2\\1\\3\end{bmatrix}\right\}$

A20 $\left\{\begin{bmatrix}1\\1\\-2\\3\end{bmatrix},\begin{bmatrix}1\\2\\1\\3\end{bmatrix},\begin{bmatrix}1\\-1\\-8\\3\end{bmatrix}\right\}$

For Problems A21–A24, determine whether the set is linearly independent. If it is linearly dependent, write one of the vectors in the set as a linear combination of the others.

A21 $\left\{\begin{bmatrix}1\\1\\0\\3\end{bmatrix},\begin{bmatrix}1\\2\\-1\\1\end{bmatrix},\begin{bmatrix}-2\\-4\\2\\-2\end{bmatrix}\right\}$

A22 $\left\{\begin{bmatrix}1\\1\\2\\1\end{bmatrix},\begin{bmatrix}2\\2\\4\\2\end{bmatrix},\begin{bmatrix}1\\0\\1\\0\end{bmatrix}\right\}$

A23 $\left\{\begin{bmatrix}1\\1\\0\\1\end{bmatrix},\begin{bmatrix}0\\1\\1\\1\end{bmatrix}\right\}$

A24 $\left\{\begin{bmatrix}3\\2\\1\\2\end{bmatrix},\begin{bmatrix}4\\4\\-5\\0\end{bmatrix},\begin{bmatrix}3\\3\\-2\\1\end{bmatrix}\right\}$

For Problems A25–A27, prove that B is a basis for the plane in $\mathbb{R}^3$ with the given scalar equation.

A25 $B=\left\{\begin{bmatrix}1\\0\\-2\end{bmatrix},\begin{bmatrix}0\\1\\-1\end{bmatrix}\right\}$, $2x_1+x_2+x_3=0$.

A26 $B=\left\{\begin{bmatrix}1\\-3\\0\end{bmatrix},\begin{bmatrix}0\\2\\1\end{bmatrix}\right\}$, $3x_1+x_2-2x_3=0$.

A27 $B=\left\{\begin{bmatrix}1\\0\\3/2\end{bmatrix},\begin{bmatrix}0\\1\\1/2\end{bmatrix}\right\}$, $3x_1+x_2-2x_3=0$.

A28 Prove that $B=\left\{\begin{bmatrix}1\\0\\0\\1\end{bmatrix},\begin{bmatrix}0\\1\\0\\1\end{bmatrix},\begin{bmatrix}0\\0\\1\\1\end{bmatrix}\right\}$ is a basis for the hyperplane P in $\mathbb{R}^4$ with scalar equation $x_1+x_2+x_3-x_4=0$.

For Problems A29–A32, determine whether the set represents a line, a plane, or a hyperplane in $\mathbb{R}^4$. Give a basis for the subspace.

A29 $\text{Span}\left\{\begin{bmatrix}1\\0\\1\\1\end{bmatrix},\begin{bmatrix}1\\2\\1\\3\end{bmatrix}\right\}$

A30 $\text{Span}\left\{\begin{bmatrix}1\\0\\0\\0\end{bmatrix},\begin{bmatrix}0\\1\\0\\0\end{bmatrix},\begin{bmatrix}0\\0\\0\\1\end{bmatrix}\right\}$

A31 $\text{Span}\left\{\begin{bmatrix}3\\1\\-1\\0\end{bmatrix},\begin{bmatrix}0\\0\\0\\0\end{bmatrix},\begin{bmatrix}6\\2\\-2\\0\end{bmatrix}\right\}$

A32 $\text{Span}\left\{\begin{bmatrix}1\\1\\0\\2\end{bmatrix},\begin{bmatrix}1\\0\\0\\-1\end{bmatrix},\begin{bmatrix}2\\1\\0\\1\end{bmatrix}\right\}$

A33 Let $\vec{p}, \vec{d} \in \mathbb{R}^n$. Prove that $\vec{x} = \vec{p} + t\vec{d}, t \in \mathbb{R}$ is a subspace of $\mathbb{R}^n$ if and only if $\vec{p}$ is a scalar multiple of $\vec{d}$.

A34 Suppose that $\mathcal{B} = \{\vec{v}_1, \dots, \vec{v}_k\}$ is a linearly independent set in $\mathbb{R}^n$. Prove that any non-empty subset of $\mathcal{B}$ is linearly independent.

A35 Let $\vec{v}_1, \dots, \vec{v}_k$ be vectors in $\mathbb{R}^n$.
(a) Prove if $\text{Span}\{\vec{v}_1, \dots, \vec{v}_k\} = \text{Span}\{\vec{v}_1, \dots, \vec{v}_{k-1}\}$, then $\vec{v}_k$ can be written as a linear combination of $\vec{v}_1, \dots, \vec{v}_{k-1}$.
(b) Prove if $\vec{v}_k$ can be written as a linear combination of $\vec{v}_1, \dots, \vec{v}_{k-1}$, then $\text{Span}\{\vec{v}_1, \dots, \vec{v}_k\} = \text{Span}\{\vec{v}_1, \dots, \vec{v}_{k-1}\}$.

A36 A factory produces thingamajiggers and whatchamacallits. It takes 3kg of steel, 2L of whipped cream, 4 nails, and 3 calculators to create a thingamajigger, while it takes 1kg of steel, 10L of whipped cream, 2 nails, and 5 calculators to create a whatchamacallit. We can represent the amount of each building material for thingamajiggers and whatchamacallits as the vectors $\vec{t} = \begin{bmatrix} 3 \\ 2 \\ 4 \\ 3 \end{bmatrix}$ and $\vec{w} = \begin{bmatrix} 1 \\ 10 \\ 2 \\ 5 \end{bmatrix}$. What does the linear combination $100\vec{t} + 250\vec{w}$ represent in this situation?

Homework Problems

For Problems B1–B4, compute the linear combination.

B1 $\begin{bmatrix} 3 \\ 2 \\ 4 \\ 1 \end{bmatrix} + 2\begin{bmatrix} 3 \\ 1 \\ 6 \\ 11 \end{bmatrix}$

B2 $\begin{bmatrix} 5 \\ 1 \\ 7 \\ 3 \\ 2 \end{bmatrix} - 3\begin{bmatrix} 1 \\ -1 \\ 2 \\ 4 \\ 1 \end{bmatrix} + \begin{bmatrix} 5 \\ 2 \\ 4 \\ 6 \\ 0 \end{bmatrix}$

B3 $\begin{bmatrix} 3 \\ -3 \\ 1 \\ 4 \\ -2 \end{bmatrix} - \begin{bmatrix} -1 \\ 2 \\ 6 \\ 1 \\ -2 \end{bmatrix}$

B4 $5\begin{bmatrix} -2 \\ 1 \\ 1 \\ 2 \\ 3 \end{bmatrix} + \begin{bmatrix} 6 \\ 4 \\ 1 \\ 2 \\ 5 \end{bmatrix} + \begin{bmatrix} -1 \\ -7 \\ 6 \\ -7 \\ 0 \end{bmatrix}$

For Problems B5–B14, determine whether the set is a subspace of the appropriate $\mathbb{R}^n$.

B5 $\{\vec{x} \in \mathbb{R}^2 \mid x_1 + 3x_2 = 0\}$ **B6** $\{\vec{x} \in \mathbb{R}^2 \mid x_1 + x_2 \leq 0\}$

B7 $\{\vec{x} \in \mathbb{R}^2 \mid x_1^2 = x_2^3\}$ **B8** $\text{Span}\left\{\begin{bmatrix} 1 \\ 2 \\ 3 \end{bmatrix}, \begin{bmatrix} 2 \\ 4 \\ 6 \end{bmatrix}\right\}$

B9 $\{\vec{x} \in \mathbb{R}^3 \mid x_2 = 1\}$

B10 $\{\vec{x} \in \mathbb{R}^3 \mid 3x_1 + 3x_2 - 2x_3 = 0\}$

B11 $\{\vec{x} \in \mathbb{R}^3 \mid x_1 - 2x_2 = 3, x_1 + x_2 = 0\}$

B12 $\{\vec{x} \in \mathbb{R}^3 \mid x_1 - 2x_2 = 0, x_1 + x_2 = 0\}$

B13 $\vec{x} = \begin{bmatrix} 1 \\ 0 \\ 0 \end{bmatrix} + t_1\begin{bmatrix} 2 \\ 1 \\ 2 \end{bmatrix} + t_2\begin{bmatrix} 1 \\ 1 \\ 1 \end{bmatrix}, t_1, t_2 \in \mathbb{R}$

B14 $\vec{x} = \begin{bmatrix} 2 \\ 1 \\ 5 \end{bmatrix} + t_1\begin{bmatrix} -3 \\ 1 \\ 3 \end{bmatrix} + t_2\begin{bmatrix} 1 \\ -2 \\ -8 \end{bmatrix}, t_1, t_2 \in \mathbb{R}$

For Problems B15–B18, determine whether the set is a subspace of $\mathbb{R}^4$.

B15 $\{\vec{x} \in \mathbb{R}^4 \mid x_1 + x_2 + 3x_4 = 0, 3x_2 = 2x_4\}$

B16 $\{\vec{x} \in \mathbb{R}^4 \mid x_1 + 2x_2 - x_3 = x_4\}$

B17 $\{\vec{x} \in \mathbb{R}^4 \mid x_1 + x_2 - 3x_3 = 1, x_1 = x_4\}$

B18 $\{\vec{x} \in \mathbb{R}^4 \mid x_1 = 2x_3 - x_4, x_1 - 3x_4 = 0\}$

For Problems B19–B22, show that the set is linearly dependent by writing a non-trivial linear combination of the vectors that equals the zero vector.

B19 $\left\{\begin{bmatrix} 1 \\ 2 \\ 5 \\ 1 \end{bmatrix}, \begin{bmatrix} 2 \\ -1 \\ 1 \\ 1 \end{bmatrix}, \begin{bmatrix} 3 \\ 1 \\ 6 \\ 2 \end{bmatrix}, \begin{bmatrix} 1 \\ 1 \\ 1 \\ 1 \end{bmatrix}\right\}$

B20 $\left\{\begin{bmatrix} 4 \\ 1 \\ -2 \\ 1 \end{bmatrix}, \begin{bmatrix} 3 \\ 1 \\ 1 \\ 2 \end{bmatrix}, \begin{bmatrix} 3 \\ 0 \\ -9 \\ -3 \end{bmatrix}\right\}$

B21 $\left\{\begin{bmatrix} 1 \\ -1 \\ 1 \\ -1 \end{bmatrix}, \begin{bmatrix} 2 \\ 2 \\ 1 \\ 1 \end{bmatrix}, \begin{bmatrix} 1 \\ 1 \\ 1 \\ 1 \end{bmatrix}, \begin{bmatrix} -2 \\ -2 \\ -1 \\ -1 \end{bmatrix}\right\}$

B22 $\left\{\begin{bmatrix} 1 \\ 1 \\ 2 \\ 5 \end{bmatrix}, \begin{bmatrix} 3 \\ 1 \\ 5 \\ 2 \end{bmatrix}, \begin{bmatrix} 0 \\ 0 \\ 0 \\ 0 \end{bmatrix}\right\}$

For Problems B23–B26, determine whether the set is linearly independent. If it is linearly dependent, write one of the vectors in the set as a linear combination of the others.

B23 $\left\{\begin{bmatrix} 1 \\ 2 \\ -1 \\ 0 \end{bmatrix}, \begin{bmatrix} 2 \\ 3 \\ 4 \\ 0 \end{bmatrix}, \begin{bmatrix} 0 \\ -1 \\ -2 \\ 0 \end{bmatrix}\right\}$

B24 $\left\{\begin{bmatrix} 1 \\ 1 \\ 0 \\ 0 \end{bmatrix}, \begin{bmatrix} 0 \\ 0 \\ 1 \\ 1 \end{bmatrix}, \begin{bmatrix} 0 \\ 1 \\ 1 \\ 0 \end{bmatrix}\right\}$

B25 $\left\{\begin{bmatrix} 2 \\ 1 \\ 1 \\ 3 \end{bmatrix}, \begin{bmatrix} 1 \\ -1 \\ 0 \\ 4 \end{bmatrix}, \begin{bmatrix} 3 \\ 0 \\ 1 \\ 7 \end{bmatrix}\right\}$

B26 $\left\{\begin{bmatrix} -2 \\ 5 \\ 1 \\ 4 \end{bmatrix}, \begin{bmatrix} 2 \\ 3 \\ 1 \\ 2 \end{bmatrix}, \begin{bmatrix} 0 \\ 4 \\ 1 \\ 3 \end{bmatrix}\right\}$

B27 Prove that $B = \left\{ \begin{bmatrix} 1 \\ 0 \\ -1 \end{bmatrix}, \begin{bmatrix} 0 \\ 1 \\ 3 \end{bmatrix} \right\}$ is a basis for the plane P in $\mathbb{R}^3$ with scalar equation $x_1 - 3x_2 + x_3 = 0$.

B28 Prove that $B = \left\{ \begin{bmatrix} 3/2 \\ 1 \\ 0 \end{bmatrix}, \begin{bmatrix} 5/2 \\ 0 \\ 1 \end{bmatrix} \right\}$ is a basis for the plane P in $\mathbb{R}^3$ with scalar equation $-2x_1 + 3x_2 + 5x_3 = 0$.

B29 Prove that $B = \left\{ \begin{bmatrix} 1 \\ 0 \\ 0 \\ 2 \end{bmatrix}, \begin{bmatrix} 0 \\ 1 \\ 0 \\ 0 \end{bmatrix}, \begin{bmatrix} 0 \\ 0 \\ 1 \\ -3 \end{bmatrix} \right\}$ is a basis for the hyperplane P in $\mathbb{R}^4$ with scalar equation $2x_1 - 3x_3 - x_4 = 0$.

For Problems B30–B35, determine whether the set represents a line, a plane, or a hyperplane in $\mathbb{R}^4$. Give a basis for the subspace.

B30 $\text{Span}\left\{ \begin{bmatrix} 1 \\ 3 \\ 1 \\ 2 \end{bmatrix}, \begin{bmatrix} 4 \\ 1 \\ 3 \\ -2 \end{bmatrix} \right\}$

B31 $\text{Span}\left\{ \begin{bmatrix} 1 \\ 2 \\ -2 \\ 1 \end{bmatrix}, \begin{bmatrix} 3 \\ 2 \\ 4 \\ -1 \end{bmatrix}, \begin{bmatrix} 2 \\ 2 \\ 1 \\ 0 \end{bmatrix} \right\}$

B32 $\text{Span}\left\{ \begin{bmatrix} 1 \\ 2 \\ 2 \\ 0 \end{bmatrix}, \begin{bmatrix} -1 \\ -2 \\ 2 \\ 0 \end{bmatrix}, \begin{bmatrix} 3 \\ 1 \\ 0 \\ 1 \end{bmatrix} \right\}$

B33 $\text{Span}\left\{ \begin{bmatrix} 2 \\ 4 \\ -2 \\ 6 \end{bmatrix}, \begin{bmatrix} -1 \\ -2 \\ 1 \\ -3 \end{bmatrix}, \begin{bmatrix} 0 \\ 0 \\ 0 \\ 0 \end{bmatrix} \right\}$

B34 $\text{Span}\left\{ \begin{bmatrix} 1 \\ 0 \\ 1 \\ 1 \end{bmatrix}, \begin{bmatrix} 1 \\ 0 \\ 2 \\ 1 \end{bmatrix}, \begin{bmatrix} 3 \\ 1 \\ 0 \\ 0 \end{bmatrix} \right\}$

B35 $\text{Span}\left\{ \begin{bmatrix} 1 \\ 2 \\ 1 \\ 1 \end{bmatrix}, \begin{bmatrix} 2 \\ 4 \\ 2 \\ 2 \end{bmatrix}, \begin{bmatrix} 3 \\ 6 \\ 3 \\ 3 \end{bmatrix} \right\}$

Conceptual Problems

C1 Prove property V8 from Theorem 1.4.1.

C2 Prove property V9 from Theorem 1.4.1.

C3 Prove if $\mathcal{B} = \{\vec{v}_1, \dots, \vec{v}_k\}$ is a basis for a subspace S of $\mathbb{R}^n$, then for each $\vec{x} \in S$, there exists unique scalars $c_1, \dots, c_k \in \mathbb{R}$ such that $x = c_1\vec{v}_1 + \cdots + c_k\vec{v}_k$.

C4 Let $\mathcal{B} = \{\vec{v}_1, \dots, \vec{v}_k\}$ be a set of vectors in $\mathbb{R}^n$. Prove that if $\vec{v}_i = \vec{0}$ for some i, then $\mathcal{B}$ is linearly dependent.

C5 Let U and V be subspaces of $\mathbb{R}^n$.

(a) Prove that the intersection of U and V is a subspace of $\mathbb{R}^n$.

(b) Give an example to show that the union of two subspaces of $\mathbb{R}^n$ does not have to be a subspace of $\mathbb{R}^n$.

(c) Define $U + V = \{\vec{u} + \vec{v} \mid \vec{u} \in U, \vec{v} \in V\}$. Prove that $U + V$ is a subspace of $\mathbb{R}^n$.

C6 Pick vectors $\vec{p}, \vec{v}_1, \vec{v}_2$, and $\vec{v}_3$ in $\mathbb{R}^4$ such that the vector equation $\vec{x} = \vec{p} + t_1\vec{v}_1 + t_2\vec{v}_2 + t_3\vec{v}_3$

(a) is a hyperplane not passing through the origin.

(b) is a plane passing through the origin.

(c) is the point $(1, 3, 1, 1)$.

(d) is a line passing through the origin.

C7 Let $\vec{v}_1, \vec{v}_2 \in \mathbb{R}^n$, and let s and t be fixed real numbers with $t \neq 0$. Prove that

$$\text{Span}\{\vec{v}_1, \vec{v}_2\} = \text{Span}\{\vec{v}_1, s\vec{v}_1 + t\vec{v}_2\}$$

C8 Explain the difference between a subset of $\mathbb{R}^n$ and a subspace of $\mathbb{R}^n$.

For Problems C9–C14, given that $\vec{v}_1, \vec{v}_2, \vec{v}_3 \in \mathbb{R}^n$, state whether each of the following statements is true or false. If the statement is true, explain briefly. If the statement is false, give a counterexample.

C9 If $\vec{v}_2 = t\vec{v}_1$ for some real number t, then $\{\vec{v}_1, \vec{v}_2\}$ is linearly dependent.

C10 If $\vec{v}_1$ is not a scalar multiple of $\vec{v}_2$, then $\{\vec{v}_1, \vec{v}_2\}$ is linearly independent.

C11 If $\{\vec{v}_1, \vec{v}_2, \vec{v}_3\}$ is linearly dependent, then $\vec{v}_1$ can be written as a linear combination of $\vec{v}_2$ and $\vec{v}_3$.

C12 If $\vec{v}_1$ can be written as a linear combination of $\vec{v}_2$ and $\vec{v}_3$, then $\{\vec{v}_1, \vec{v}_2, \vec{v}_3\}$ is linearly dependent.

C13 $\{\vec{v}_1\}$ is not a subspace of $\mathbb{R}^n$.

C14 $\text{Span}\{\vec{v}_1\}$ is a subspace of $\mathbb{R}^n$.

1.5 Dot Products and Projections in $\mathbb{R}^n$

We now extend everything we did with dot products, lengths, and orthogonality in $\mathbb{R}^2$ and $\mathbb{R}^3$ to $\mathbb{R}^n$. We will use these concepts to define projections in $\mathbb{R}^n$.

Definition
Dot Product

Let $\vec{x} = \begin{bmatrix} x_1 \\ \vdots \\ x_n \end{bmatrix}, \vec{y} = \begin{bmatrix} y_1 \\ \vdots \\ y_n \end{bmatrix} \in \mathbb{R}^n$. We define the **dot product** of $\vec{x}$ and $\vec{y}$ by

$$\vec{x} \cdot \vec{y} = x_1y_1 + x_2y_2 + \cdots + x_ny_n$$

CONNECTION

The dot product is also sometimes called the **scalar product** or the **standard inner product**. We will look at other inner products in Section 7.4.

EXAMPLE 1.5.1

Let $\vec{x} = \begin{bmatrix} 1 \\ 2 \\ -1 \\ 3 \end{bmatrix}$ and $\vec{y} = \begin{bmatrix} -2 \\ 5 \\ 0 \\ -4 \end{bmatrix}$. Calculate $\vec{x} \cdot \vec{y}$.

Solution: We have $\vec{x} \cdot \vec{y} = 1(-2) + 2(5) + (-1)(0) + 3(-4) = -4$.

From this definition, some important properties follow.

Theorem 1.5.1

Let $\vec{x}, \vec{y}, \vec{z} \in \mathbb{R}^n$ and $s, t \in \mathbb{R}$. Then,

(1) $\vec{x} \cdot \vec{x} \geq 0$
(2) $\vec{x} \cdot \vec{x} = 0$ if and only if $\vec{x} = \vec{0}$
(3) $\vec{x} \cdot \vec{y} = \vec{y} \cdot \vec{x}$
(4) $\vec{x} \cdot (s\vec{y} + t\vec{z}) = s(\vec{x} \cdot \vec{y}) + t(\vec{x} \cdot \vec{z})$

Proof: We leave the proof of these properties to the reader.

Because of property (1), we can now define the length of a vector in $\mathbb{R}^n$. The word **norm** is often used as a synonym for **length** when we are speaking of vectors.

Definition
Norm

Let $\vec{x} = \begin{bmatrix} x_1 \\ \vdots \\ x_n \end{bmatrix}$. We define the **norm**, or **length**, of $\vec{x}$ by

$$\|\vec{x}\| = \sqrt{\vec{x} \cdot \vec{x}} = \sqrt{x_1^2 + \cdots + x_n^2}$$

EXAMPLE 1.5.2

Let $\vec{x} = \begin{bmatrix} 2 \\ 1 \\ 3 \\ -1 \end{bmatrix}$ and $\vec{y} = \begin{bmatrix} 1/3 \\ -2/3 \\ 0 \\ -2/3 \end{bmatrix}$. Find $\|\vec{x}\|$ and $\|\vec{y}\|$.

Solution: We have

$$\|\vec{x}\| = \sqrt{2^2 + 1^2 + 3^2 + (-1)^2} = \sqrt{15}$$
$$\|\vec{y}\| = \sqrt{(1/3)^2 + (-2/3)^2 + 0^2 + (-2/3)^2} = \sqrt{1/9 + 4/9 + 0 + 4/9} = 1$$

EXERCISE 1.5.1

Let $\vec{x} = \begin{bmatrix} 1 \\ 2 \\ 1 \end{bmatrix}$ and let $\vec{y} = \frac{1}{\|\vec{x}\|}\vec{x}$. Determine $\|\vec{x}\|$ and $\|\vec{y}\|$.

We now give some important properties of the norm in $\mathbb{R}^n$.

Theorem 1.5.2

Let $\vec{x}, \vec{y} \in \mathbb{R}^n$ and $t \in \mathbb{R}$. Then

(1) $\|\vec{x}\| \geq 0$, and $\|\vec{x}\| = 0$ if and only if $\vec{x} = \vec{0}$
(2) $\|t\vec{x}\| = |t|\,\|\vec{x}\|$
(3) $|\vec{x} \cdot \vec{y}| \leq \|\vec{x}\|\,\|\vec{y}\|$, with equality if and only if $\{\vec{x}, \vec{y}\}$ is linearly dependent
(4) $\|\vec{x} + \vec{y}\| \leq \|\vec{x}\| + \|\vec{y}\|$

Remark

Property (3) is called the **Cauchy-Schwarz Inequality**.
Property (4) is called the **Triangle Inequality**.

EXERCISE 1.5.2

Prove that the vector $\hat{x} = \frac{1}{\|\vec{x}\|}\vec{x}$ is parallel to $\vec{x}$ and satisfies $\|\hat{x}\| = 1$.

Definition
Unit Vector

A vector $\vec{x} \in \mathbb{R}^n$ such that $\|\vec{x}\| = 1$ is called a **unit vector**.

We will see that unit vectors can be very useful. We often want to find a unit vector that has the same direction as a given vector $\vec{x}$. Using the result in Exercise 1.5.2, we see that we can use the vector

$$\hat{x} = \frac{1}{\|\vec{x}\|}\vec{x}$$

We could now define angles between vectors $\vec{x}$ and $\vec{y}$ in $\mathbb{R}^n$ by matching what we did in $\mathbb{R}^2$ and $\mathbb{R}^3$. However, in linear algebra we are generally interested only in whether two vectors are orthogonal.

Definition
Orthogonal

Two vectors $\vec{x}$ and $\vec{y}$ in $\mathbb{R}^n$ are **orthogonal** to each other if and only if $\vec{x} \cdot \vec{y} = 0$.

Notice that this definition implies that $\vec{0}$ is orthogonal to every vector in $\mathbb{R}^n$.

EXAMPLE 1.5.3

Let $\vec{v} = \begin{bmatrix} 1 \\ 0 \\ 3 \\ -2 \end{bmatrix}$, $\vec{w} = \begin{bmatrix} 2 \\ 3 \\ 0 \\ 1 \end{bmatrix}$, and $\vec{z} = \begin{bmatrix} -1 \\ -1 \\ 1 \\ 2 \end{bmatrix}$. Show that $\vec{v}$ is orthogonal to $\vec{w}$ but $\vec{v}$ is not orthogonal to $\vec{z}$.

Solution: We have $\vec{v} \cdot \vec{w} = 1(2) + 0(3) + 3(0) + (-2)(1) = 0$, so they are orthogonal. $\vec{v} \cdot \vec{z} = 1(-1) + 0(-1) + 3(1) + (-2)(2) = -2$, so they are not orthogonal.

The Scalar Equation of a Hyperplane

We can repeat the argument we used to find a scalar equation of a plane in Section 1.3 to find a scalar equation of a hyperplane in $\mathbb{R}^n$. In particular, if we have a vector $\vec{m}$ that is orthogonal to any directed line segment $\vec{PQ}$ lying in the hyperplane, then for any point $X(x_1, \dots, x_n)$ in the hyperplane, we have

$$0 = \vec{m} \cdot \vec{PX}$$

As before, we can rearrange this as

$$\begin{aligned} 0 &= \vec{m} \cdot (\vec{x} - \vec{p}) \\ 0 &= \vec{m} \cdot \vec{x} - \vec{m} \cdot \vec{p} \\ \vec{m} \cdot \vec{x} &= \vec{m} \cdot \vec{p} \\ m_1x_1 + \cdots + m_nx_n &= \vec{m} \cdot \vec{p} \end{aligned}$$

Thus, we see that a single scalar equation in $\mathbb{R}^n$ represents a hyperplane in $\mathbb{R}^n$.

EXAMPLE 1.5.4

Find a scalar equation of the hyperplane in $\mathbb{R}^4$ that has normal vector $\vec{m} = \begin{bmatrix} 2 \\ 3 \\ -2 \\ 1 \end{bmatrix}$ and passes through the point $P(1, 0, 2, -1)$.

Solution: The equation is

$$2x_1 + 3x_2 - 2x_3 + x_4 = 2(1) + 3(0) + (-2)(2) + 1(-1) = -3$$

Projections in $\mathbb{R}^n$

The idea of a projection is one of the most important applications of the dot product. Suppose that we want to know how much of a given vector $\vec{y}$ is in the direction of some other given vector $\vec{x}$ (see Figure 1.5.1). In elementary physics, this is exactly what is required when a force is "resolved" into its components along certain directions (for example, into its vertical and horizontal components). When we define projections, it is helpful to think of examples in two or three dimensions, but the ideas do not really depend on whether the vectors are in $\mathbb{R}^2$, $\mathbb{R}^3$, or $\mathbb{R}^n$.

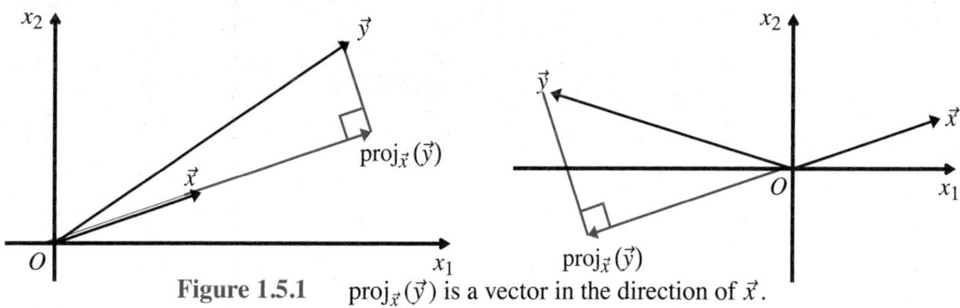

Figure 1.5.1 $\mathrm{proj}_{\vec{x}}(\vec{y})$ is a vector in the direction of $\vec{x}$.

First, let us consider the case where $\vec{x} = \vec{e}_1$ in $\mathbb{R}^2$. How much of an arbitrary vector $\vec{y} = \begin{bmatrix} y_1 \\ y_2 \end{bmatrix}$ points along $\vec{x}$? Clearly, the part of $\vec{y}$ that is in the direction of $\vec{x}$ is $\begin{bmatrix} y_1 \\ 0 \end{bmatrix} = y_1\vec{e}_1 = (\vec{x} \cdot \vec{y})\vec{x}$. This will be called the **projection of $\vec{y}$ onto $\vec{x}$** and is denoted $\mathrm{proj}_{\vec{x}}(\vec{y})$.

Next, consider the case where $\vec{x} \in \mathbb{R}^2$ has arbitrary direction and is a unit vector. First, draw the line through the origin with direction vector $\vec{x}$. Now, draw the line perpendicular to this line that passes through the point (y_1, y_2). This forms a right triangle, as in Figure 1.5.1. The projection of $\vec{y}$ onto $\vec{x}$ is the portion of the triangle that lies on the line with direction $\vec{x}$. Thus, the resulting projection is a scalar multiple of $\vec{x}$, that is $\mathrm{proj}_{\vec{x}}(\vec{y}) = k\vec{x}$. We need to determine the value of k. To do this, let $\vec{z}$ denote the vector from $\mathrm{proj}_{\vec{x}}(\vec{y})$ to $\vec{y}$. Then, by definition, $\vec{z}$ is orthogonal to $\vec{x}$ and we can write

$$\vec{y} = \vec{z} + \mathrm{proj}_{\vec{x}}(\vec{y}) = \vec{z} + k\vec{x}$$

We now employ a very useful and common trick which is to take the dot product of $\vec{y}$ with $\vec{x}$:

$$\vec{x} \cdot \vec{y} = \vec{x} \cdot (\vec{z} + k\vec{x}) = \vec{x} \cdot \vec{z} + \vec{x} \cdot (k\vec{x}) = 0 + k(\vec{x} \cdot \vec{x}) = k\|\vec{x}\|^2 = k$$

since $\vec{x}$ is orthogonal to $\vec{z}$ and is a unit vector. Thus,

$$\mathrm{proj}_{\vec{x}}(\vec{y}) = (\vec{x} \cdot \vec{y})\vec{x}$$

EXAMPLE 1.5.5

Find the projection of $\vec{u} = \begin{bmatrix} -3 \\ 1 \end{bmatrix}$ onto the unit vector $\vec{v} = \begin{bmatrix} 1/\sqrt{2} \\ 1/\sqrt{2} \end{bmatrix}$.

Solution: We have

$$\begin{aligned} \text{proj}_{\vec{v}}(\vec{u}) &= (\vec{v} \cdot \vec{u})\vec{v} \\ &= \left(\frac{-3}{\sqrt{2}} + \frac{1}{\sqrt{2}} \right) \begin{bmatrix} 1/\sqrt{2} \\ 1/\sqrt{2} \end{bmatrix} \\ &= \frac{-2}{\sqrt{2}} \begin{bmatrix} 1/\sqrt{2} \\ 1/\sqrt{2} \end{bmatrix} \\ &= \begin{bmatrix} -1 \\ -1 \end{bmatrix} \end{aligned}$$

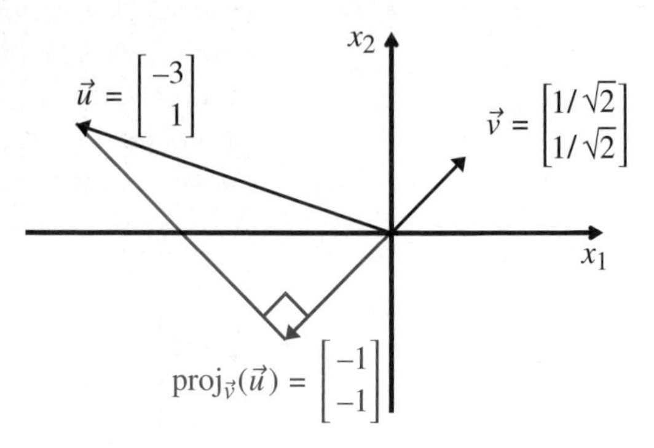

If $\vec{x} \in \mathbb{R}^2$ is an arbitrary non-zero vector, then a unit vector in the direction of $\vec{x}$ is $\hat{x} = \frac{\vec{x}}{\|\vec{x}\|}$. Hence, we find that the projection of $\vec{y}$ onto $\vec{x}$ is

$$\text{proj}_{\vec{x}}(\vec{y}) = \text{proj}_{\hat{x}}(\vec{y}) = (\hat{x} \cdot \vec{y})\hat{x} = \left(\frac{\vec{x}}{\|\vec{x}\|} \cdot \vec{y} \right) \frac{\vec{x}}{\|\vec{x}\|} = \frac{\vec{x} \cdot \vec{y}}{\|\vec{x}\|^2}\vec{x}$$

To match this result, we make the following definition for vectors in $\mathbb{R}^n$.

Definition
Projection

For any vectors $\vec{y}, \vec{x}$ in $\mathbb{R}^n$, with $\vec{x} \neq \vec{0}$, we define the **projection** of $\vec{y}$ onto $\vec{x}$ by

$$\text{proj}_{\vec{x}}(\vec{y}) = \frac{\vec{x} \cdot \vec{y}}{\|\vec{x}\|^2}\vec{x}$$

EXAMPLE 1.5.6

Let $\vec{v} = \begin{bmatrix} 4 \\ 3 \\ -1 \end{bmatrix}$ and $\vec{u} = \begin{bmatrix} -2 \\ 5 \\ 3 \end{bmatrix}$. Determine $\text{proj}_{\vec{v}}(\vec{u})$ and $\text{proj}_{\vec{u}}(\vec{v})$.

Solution: We have

$$\text{proj}_{\vec{v}}(\vec{u}) = \frac{\vec{v} \cdot \vec{u}}{\|\vec{v}\|^2}\vec{v} = \frac{(4)(-2) + 3(5) + (-1)(3)}{4^2 + 3^2 + (-1)^2}\vec{v} = \frac{4}{26}\begin{bmatrix} 4 \\ 3 \\ -1 \end{bmatrix} = \begin{bmatrix} 8/13 \\ 6/13 \\ -2/13 \end{bmatrix}$$

$$\text{proj}_{\vec{u}}(\vec{v}) = \frac{\vec{u} \cdot \vec{v}}{\|\vec{u}\|^2}\vec{u} = \frac{(-2)(4) + 5(3) + 3(-1)}{(-2)^2 + 5^2 + 3^2}\vec{u} = \frac{4}{38}\begin{bmatrix} -2 \\ 5 \\ 3 \end{bmatrix} = \begin{bmatrix} -4/19 \\ 10/19 \\ 6/19 \end{bmatrix}$$

Remarks

1. This example illustrates that, in general, $\text{proj}_{\vec{x}}(\vec{y}) \neq \text{proj}_{\vec{y}}(\vec{x})$. Of course, we should not expect equality, because $\text{proj}_{\vec{x}}(\vec{y})$ is in the direction of $\vec{x}$, whereas $\text{proj}_{\vec{y}}(\vec{x})$ is in the direction of $\vec{y}$.

2. Observe that for any $\vec{x} \in \mathbb{R}^n$, we can consider $\text{proj}_{\vec{x}}$ a function whose domain and codomain are $\mathbb{R}^n$. To indicate this, we can write $\text{proj}_{\vec{x}} : \mathbb{R}^n \to \mathbb{R}^n$. Since the output of this function is a vector, we call it a **vector-valued function**.

EXAMPLE 1.5.7

A 30 kg rail cart with a wind sail is on a frictionless track that runs northeast and southwest. A wind applies a force vector of $\vec{F} = \begin{bmatrix} 100 \\ 50 \end{bmatrix}$ to the cart. Calculate the acceleration of the cart.

Solution: Since the rail cart is on a track, only the amount of force in the direction of the track will create acceleration in the direction of the track. The track has direction vector $\vec{d} = \begin{bmatrix} 1 \\ 1 \end{bmatrix}$. Thus, the force vector in that direction is

$$\text{proj}_{\vec{d}}(\vec{F}) = \frac{\vec{d} \cdot \vec{F}}{\|\vec{d}\|^2}\vec{d} = \frac{150}{2}\begin{bmatrix} 1 \\ 1 \end{bmatrix} = \begin{bmatrix} 75 \\ 75 \end{bmatrix}$$

Thus, the amount of force in the direction of the track is

$$\left\|\text{proj}_{\vec{d}}(\vec{F})\right\| = \sqrt{(75)^2 + (75)^2} \approx 106N$$

Consequently, the acceleration of the cart along the track will be $a = \dfrac{F}{m} \approx 3.53m/s^2$.

The Perpendicular Part

When you resolve a force in physics, you often not only want the component of the force in the direction of a given vector $\vec{x}$, but also the component of the force perpendicular to $\vec{x}$.

We begin by restating the problem. In $\mathbb{R}^n$, given a non-zero vector $\vec{x}$, express any $\vec{y} \in \mathbb{R}^n$ as the sum of a vector parallel to $\vec{x}$ and a vector orthogonal to $\vec{x}$. That is, write $\vec{y} = \vec{w} + \vec{z}$, where $\vec{w} = c\vec{x}$ for some $c \in \mathbb{R}$ and $\vec{z} \cdot \vec{x} = 0$.

We use the same trick we did $\mathbb{R}^2$. Taking the dot product of $\vec{x}$ and $\vec{y}$ gives

$$\vec{x} \cdot \vec{y} = \vec{x} \cdot (\vec{z} + \vec{w}) = \vec{x} \cdot \vec{z} + \vec{x} \cdot (c\vec{x}) = 0 + c(\vec{x} \cdot \vec{x}) = c\|\vec{x}\|^2$$

Therefore, $c = \dfrac{\vec{x} \cdot \vec{y}}{\|\vec{x}\|^2}$, so in fact, $\vec{w} = c\vec{x} = \text{proj}_{\vec{x}}(\vec{y})$, as we might have expected. One bonus of approaching the problem this way is that it is now clear that this is the only way to choose $\vec{w}$ to satisfy the problem.

Next, since $\vec{y} = \text{proj}_{\vec{x}}(\vec{y}) + \vec{z}$, it follows that $\vec{z} = \vec{y} - \text{proj}_{\vec{x}}(\vec{y})$. Is this $\vec{z}$ really orthogonal to $\vec{x}$? To check, calculate

$$\begin{aligned} \vec{x} \cdot \vec{z} &= \vec{x} \cdot (\vec{y} - \text{proj}_{\vec{x}}(\vec{y})) \\ &= \vec{x} \cdot \vec{y} - \left(\frac{\vec{x} \cdot \vec{y}}{\|\vec{x}\|^2}\vec{x}\right) \cdot \vec{x} \\ &= \vec{x} \cdot \vec{y} - \left(\frac{\vec{x} \cdot \vec{y}}{\|\vec{x}\|^2}\right)(\vec{x} \cdot \vec{x}) \\ &= \vec{x} \cdot \vec{y} - \left(\frac{\vec{x} \cdot \vec{y}}{\|\vec{x}\|^2}\right)\|\vec{x}\|^2 \\ &= \vec{x} \cdot \vec{y} - \vec{x} \cdot \vec{y} = 0 \end{aligned}$$

So, $\vec{z}$ is orthogonal to $\vec{x}$, as required. Since it is often useful to construct a vector $\vec{z}$ in this way, we introduce a name for it.

Definition
Perpendicular of a Projection

For any vectors $\vec{x}, \vec{y} \in \mathbb{R}^n$, with $\vec{x} \neq \vec{0}$, define the **projection of $\vec{y}$ perpendicular to $\vec{x}$** to be

$$\text{perp}_{\vec{x}}(\vec{y}) = \vec{y} - \text{proj}_{\vec{x}}(\vec{y})$$

Notice that $\text{perp}_{\vec{x}}(\vec{y})$ is again a vector-valued function on $\mathbb{R}^n$. Also observe that $\vec{y} = \text{proj}_{\vec{x}}(\vec{y}) + \text{perp}_{\vec{x}}(\vec{y})$. See Figure 1.5.2.

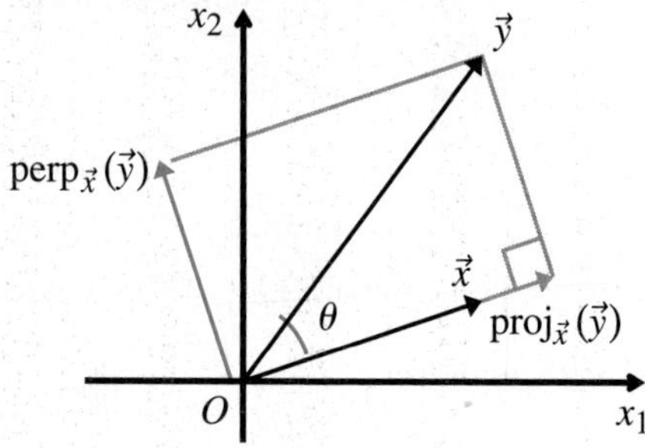

Figure 1.5.2 $\text{perp}_{\vec{x}}(\vec{y})$ is perpendicular to $\vec{x}$, and $\text{proj}_{\vec{x}}(\vec{y}) + \text{perp}_{\vec{x}}(\vec{y}) = \vec{y}$.

EXAMPLE 1.5.8

Let $\vec{v} = \begin{bmatrix} 2 \\ 1 \\ 1 \end{bmatrix}$ and $\vec{u} = \begin{bmatrix} 1 \\ 5 \\ 1 \end{bmatrix}$. Determine $\text{proj}_{\vec{v}}(\vec{u})$ and $\text{perp}_{\vec{v}}(\vec{u})$.

Solution:

$$\text{proj}_{\vec{v}}(\vec{u}) = \frac{\vec{v} \cdot \vec{u}}{\|\vec{v}\|^2}\vec{v} = \frac{8}{6}\begin{bmatrix} 2 \\ 1 \\ 1 \end{bmatrix} = \begin{bmatrix} 8/3 \\ 4/3 \\ 4/3 \end{bmatrix}$$

$$\text{perp}_{\vec{v}}(\vec{u}) = \vec{u} - \text{proj}_{\vec{v}}(\vec{u}) = \begin{bmatrix} 1 \\ 5 \\ 1 \end{bmatrix} - \begin{bmatrix} 8/3 \\ 4/3 \\ 4/3 \end{bmatrix} = \begin{bmatrix} -5/3 \\ 11/3 \\ -1/3 \end{bmatrix}$$

EXERCISE 1.5.3

Let $\vec{v} = \begin{bmatrix} 3 \\ 1 \\ 2 \end{bmatrix}$ and $\vec{u} = \begin{bmatrix} 1 \\ -2 \\ 0 \end{bmatrix}$. Determine $\text{proj}_{\vec{v}}(\vec{u})$ and $\text{perp}_{\vec{v}}(\vec{u})$.

Two Properties of Projections

Projections will appear several times in this book, and some of their special properties are important. The first is called the **linearity property**, and the second is called the **projection property**. Let $\vec{x} \in \mathbb{R}^n$ with $\vec{x} \neq \vec{0}$, then

(L1) $\text{proj}_{\vec{x}}(s\vec{y} + t\vec{z}) = s\,\text{proj}_{\vec{x}}(\vec{y}) + t\,\text{proj}_{\vec{x}}(\vec{z})$ for all $\vec{y}, \vec{z} \in \mathbb{R}^n$, $s, t \in \mathbb{R}$
(L2) $\text{proj}_{\vec{x}}(\text{proj}_{\vec{x}}(\vec{y})) = \text{proj}_{\vec{x}}(\vec{y})$, for all $\vec{y}$ in $\mathbb{R}^n$

EXERCISE 1.5.4 Verify that properties (L1) and (L2) are true.

It follows that $\text{perp}_{\vec{x}}$ also satisfies the corresponding equations. We shall see that $\text{proj}_{\vec{x}}$ and $\text{perp}_{\vec{x}}$ are just two cases amongst the many functions that satisfy the linearity property.

Applications of Projections

Minimum Distance What is the distance from a point $Q(q_1, q_2)$ to the line with vector equation $\vec{x} = \vec{p} + t\vec{d}$, $t \in \mathbb{R}$? In this and similar problems, *distance* always means the minimum distance. Geometrically, we see that the minimum distance is found along a line segment perpendicular to the given line through a point P on the line. A formal proof that minimum distance requires perpendicularity can be given by using the Pythagorean Theorem. (See Problem C9.)

To answer the question, take *any* point on the line $\vec{x} = \vec{p} + t\vec{d}$, $t \in \mathbb{R}$. The obvious choice is $P(p_1, p_2)$ corresponding to $\vec{p}$. From Figure 1.5.3, we see that the required distance is the length $\text{perp}_{\vec{d}}\left(\vec{PQ}\right)$.

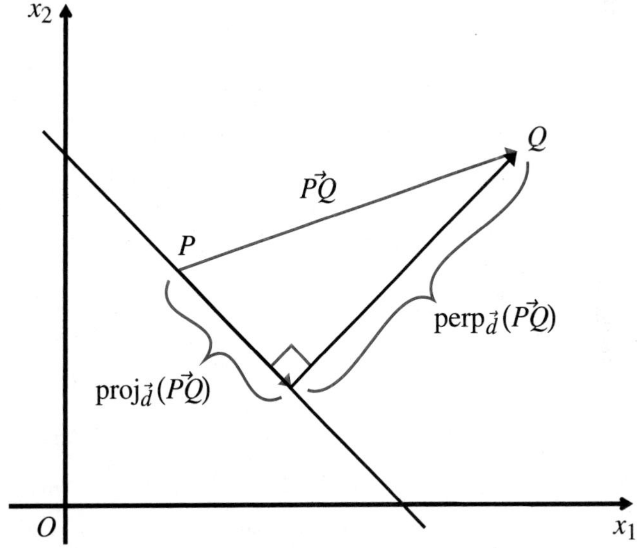

Figure 1.5.3 The distance from Q to the line $\vec{x} = \vec{p} + t\vec{d}$, $t \in \mathbb{R}$ is $\|\text{perp}_{\vec{d}}\left(\vec{PQ}\right)\|$.

EXAMPLE 1.5.9

Find the distance from $Q(4,3)$ to the line $\vec{x} = \begin{bmatrix}1\\2\end{bmatrix} + t\begin{bmatrix}-1\\1\end{bmatrix}, t \in \mathbb{R}$.

Solution: We pick the point $P(1,2)$ on the line. Then, $\vec{PQ} = \begin{bmatrix}4-1\\3-2\end{bmatrix} = \begin{bmatrix}3\\1\end{bmatrix}$. So, the distance is

$$\begin{aligned}\|\operatorname{perp}_{\vec{d}}(\vec{PQ})\| &= \|\vec{PQ} - \operatorname{proj}_{\vec{d}}(\vec{PQ})\| \\ &= \left\|\begin{bmatrix}3\\1\end{bmatrix} - \left(\frac{-3+1}{1+1}\right)\begin{bmatrix}-1\\1\end{bmatrix}\right\| \\ &= \left\|\begin{bmatrix}3\\1\end{bmatrix} + \begin{bmatrix}-1\\1\end{bmatrix}\right\| = \left\|\begin{bmatrix}2\\2\end{bmatrix}\right\| = 2\sqrt{2}\end{aligned}$$

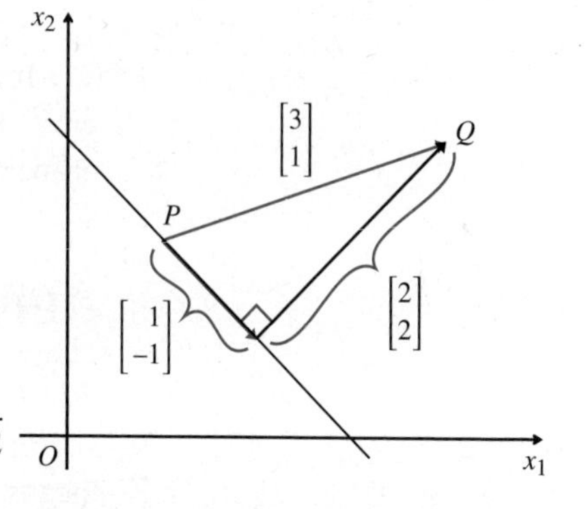

Notice that in this problem and similar problems, we take advantage of the fact that the direction vector $\vec{d}$ can be thought of as "starting" at any point. When $\operatorname{perp}_{\vec{d}}(\vec{AB})$ is calculated, both vectors are "located" at point P. When projections were originally defined, it was implicitly assumed that all vectors were located at the origin. Now, it is apparent that the definitions make sense as long as all vectors in the calculation are located at the same point.

We now look at the similar problem of finding the distance from a point $Q(q_1, q_2, q_3)$ to a plane in $\mathbb{R}^3$ with normal vector $\vec{n}$. If P is any point in the plane, then $\operatorname{proj}_{\vec{n}}(\vec{PQ})$ is the directed line segment from the plane to the point Q that is perpendicular to the plane. Hence, $\|\operatorname{proj}_{\vec{n}}(\vec{PQ})\|$ is the distance from Q to the plane. Moreover, $\operatorname{perp}_{\vec{n}}(\vec{PQ})$ is a directed line segment lying in the plane. In particular, it is the projection of $\vec{PQ}$ onto the plane. See Figure 1.5.4.

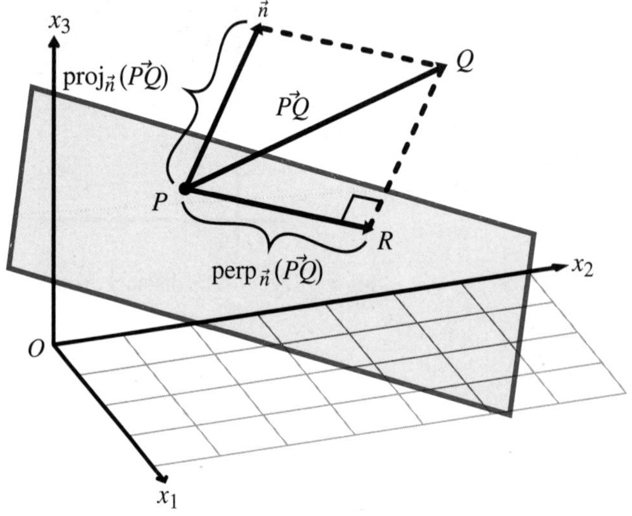

Figure 1.5.4 $\operatorname{proj}_{\vec{n}}(\vec{PQ})$ and $\operatorname{perp}_{\vec{n}}(\vec{PQ})$, where $\vec{n}$ is normal to the plane.

EXAMPLE 1.5.10

What is the distance from $Q(q_1, q_2, q_3)$ to a plane in $\mathbb{R}^3$ with equation $n_1x_1 + n_2x_2 + n_3x_3 = d$?

Solution: Assuming that $n_1 \neq 0$, we pick $P(d/n_1, 0, 0)$ to be our point in the plane. Thus, the distance is

$$\begin{aligned}
\| \text{proj}_{\vec{n}} \left(\vec{PQ}\right) \| &= \left\| \frac{(\vec{q} - \vec{p}) \cdot \vec{n}}{\|\vec{n}\|^2} \vec{n} \right\| \\
&= \left| \frac{(\vec{q} - \vec{p}) \cdot \vec{n}}{\|\vec{n}\|^2} \right| \|\vec{n}\| \\
&= \left| \frac{(\vec{q} - \vec{p}) \cdot \vec{n}}{\|\vec{n}\|} \right| \\
&= \left| \frac{(q_1 - d/n_1)n_1 + q_2n_2 + q_3n_3}{\sqrt{n_1^2 + n_2^2 + n_3^2}} \right| \\
&= \left| \frac{q_1n_1 + q_2n_2 + q_3n_3 - d}{\sqrt{n_1^2 + n_2^2 + n_3^2}} \right|
\end{aligned}$$

This is a standard formula for this distance problem. However, the lengths of projections along or perpendicular to a suitable vector can be used for all of these problems. It is better to learn to use this powerful and versatile idea, as illustrated in the problems above, than to memorize complicated formulas.

Finding the Nearest Point In some applications, we need to determine the point in the plane that is nearest to the point Q. Let us call this point R, as in Figure 1.5.4. Then we can determine R by observing that

$$\vec{OR} = \vec{OP} + \vec{PR} = \vec{OP} + \text{perp}_{\vec{n}} \left(\vec{PQ}\right)$$

However, we get an easier calculation if we observe from the figure that

$$\vec{OR} = \vec{OQ} + \vec{QR} = \vec{OQ} + \text{proj}_{\vec{n}} \left(\vec{QP}\right)$$

Notice that we need $\vec{QP}$ here instead of $\vec{PQ}$. Problem C10 asks you to check that these two calculations of $\vec{OR}$ are consistent.

If the plane in this problem passes through the origin, then we may take $P = O$, and the point in the plane that is closest to Q is given by $\text{perp}_{\vec{n}}(\vec{q})$.

EXAMPLE 1.5.11

Find the point on the plane $x_1 - 2x_2 + 2x_3 = 5$ that is closest to the point $Q(2, 1, 1)$.

Solution: We pick $P(1, -1, 1)$ to be the point on the plane. Then $\vec{QP} = \begin{bmatrix} -1 \\ -2 \\ 0 \end{bmatrix}$, and we find that the point on the plane closest to Q is

$$\vec{OR} = \vec{q} + \text{proj}_{\vec{n}} \left(\vec{QP}\right) = \vec{q} + \frac{\vec{n} \cdot \vec{QP}}{\|\vec{n}\|^2} \vec{n} = \begin{bmatrix} 2 \\ 1 \\ 1 \end{bmatrix} + \frac{3}{9} \begin{bmatrix} 1 \\ -2 \\ 2 \end{bmatrix} = \begin{bmatrix} 7/3 \\ 1/3 \\ 5/3 \end{bmatrix}$$

The Scalar Triple Product and Volumes in $\mathbb{R}^3$ The three vectors $\vec{u}$, $\vec{v}$, and $\vec{w}$ in $\mathbb{R}^3$ may be taken to be the three adjacent edges of a parallelepiped (see Figure 1.5.5). Is there an expression for the volume of the parallelepiped in terms of the three vectors? To obtain such a formula, observe that the parallelogram determined by $\vec{u}$ and $\vec{v}$ can be regarded as the base of the solid of the parallelepiped. This base has area $\|\vec{u} \times \vec{v}\|$. With respect to this base, the altitude of the solid is the length of the amount of $\vec{w}$ in the direction of the normal vector $\vec{n} = \vec{u} \times \vec{v}$ to the base.

$$\text{altitude} = \|\operatorname{proj}_{\vec{n}}(\vec{w})\| = \frac{|\vec{n} \cdot \vec{w}|}{\|\vec{n}\|} = \frac{|(\vec{u} \times \vec{v}) \cdot \vec{w}|}{\|\vec{u} \times \vec{v}\|}$$

To get the volume, multiply this altitude by the area of the base to get

$$\text{volume of the parallelepiped} = \frac{|(\vec{u} \times \vec{v}) \cdot \vec{w}|}{\|\vec{u} \times \vec{v}\|} \times \|\vec{u} \times \vec{v}\| = |(\vec{u} \times \vec{v}) \cdot \vec{w}|$$

The product $(\vec{u} \times \vec{v}) \cdot \vec{w}$ is called the **scalar triple product** of $\vec{w}$, $\vec{u}$, and $\vec{v}$. Notice that the result is a real number (a scalar).

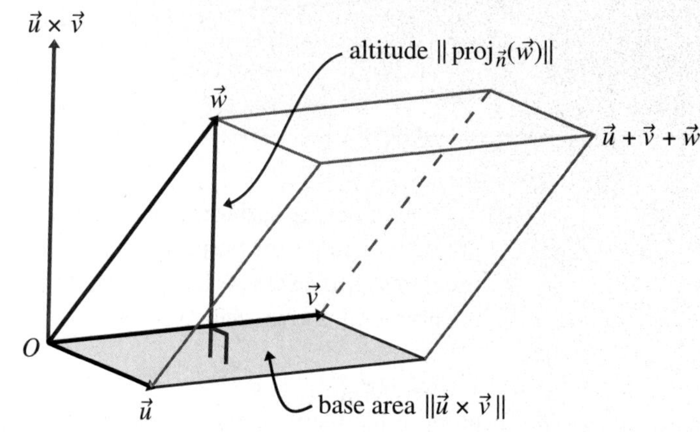

Figure 1.5.5 The parallelepiped with edges $\vec{u}$, $\vec{v}$, $\vec{w}$ has volume $|(\vec{u} \times \vec{v}) \cdot \vec{w}|$.

The sign of the scalar triple product also has an interpretation. Recall that the ordered triple of vectors $\{\vec{u}, \vec{v}, \vec{u} \times \vec{v}\}$ is right-handed; we can think of $\vec{u} \times \vec{v}$ as the "upwards" normal vector to the plane with vector equation $\vec{x} = s\vec{u} + t\vec{v}$, $s, t \in \mathbb{R}$. Some other vector $\vec{w}$ is then "upwards," and $\{\vec{u}, \vec{v}, \vec{w}\}$ (in that order) is right-handed, if and only if the scalar triple product is positive. If the triple scalar product is negative, then $\{\vec{u}, \vec{v}, \vec{w}\}$ is a left-handed system.

It is often useful to note that $(\vec{u} \times \vec{v}) \cdot \vec{w} = (\vec{v} \times \vec{w}) \cdot \vec{u} = (\vec{w} \times \vec{u}) \cdot \vec{v}$. This is straightforward but tedious to verify.

EXAMPLE 1.5.12

Find the volume of the parallelepiped determined by the vectors $\begin{bmatrix} 2 \\ 1 \\ 1 \end{bmatrix}, \begin{bmatrix} -1 \\ -1 \\ 3 \end{bmatrix}$, and $\begin{bmatrix} 1 \\ 0 \\ 2 \end{bmatrix}$.

Solution: The volume V is

$$V = \left| \left(\begin{bmatrix} 2 \\ 1 \\ 1 \end{bmatrix} \times \begin{bmatrix} -1 \\ -1 \\ 3 \end{bmatrix} \right) \cdot \begin{bmatrix} 1 \\ 0 \\ 2 \end{bmatrix} \right| = \left| \begin{bmatrix} 4 \\ -7 \\ -1 \end{bmatrix} \cdot \begin{bmatrix} 1 \\ 0 \\ 2 \end{bmatrix} \right| = 2$$

Best Approximation Finding the point Q on a line L that is closest to a given point P does not have to be interpreted geometrically. For example, it can be used to find the best approximation of a given set of data. For now we look at just one example.

EXAMPLE 1.5.13

Hooke's Law states that the force F applied to a spring is proportional to the distance x that the spring is stretched. That is, $F = kx$ where k is a constant called the *spring constant*. To determine the spring constant for a given spring, some physics students experiment with different forces on the spring. They find that forces of 2 N, 3 N, 5 N, and 6 N stretch the spring by 0.25 m, 0.40 m, 0.55 m, and 0.75 m respectively. Thus, by Hooke's Law, they find that

$$\begin{aligned} 2 &= 0.25k \\ 3 &= 0.40k \\ 5 &= 0.55k \\ 6 &= 0.75k \end{aligned}$$

Not surprisingly, due to measurement error, each equation gives a different value of k. To solve this problem, we interpret it in terms of vectors. Let $\vec{p} = \begin{bmatrix} 2 \\ 3 \\ 5 \\ 6 \end{bmatrix}$ and $\vec{d} = \begin{bmatrix} 0.25 \\ 0.40 \\ 0.55 \\ 0.75 \end{bmatrix}$.

We want to find the value of k that makes the vector $k\vec{d}$ closest to the point $P(2, 3, 5, 6)$. Observe that we can interpret $k\vec{d}$ as the line L with vector equation

$$\vec{x} = k \begin{bmatrix} 0.25 \\ 0.40 \\ 0.55 \\ 0.75 \end{bmatrix}, \quad k \in \mathbb{R}$$

The vector on L that is closest to P is the projection of P onto L. Moreover, we know that the coefficient k of the projection is

$$k = \frac{\vec{d} \cdot \vec{p}}{\|\vec{d}\|^2} = \frac{8.95}{1.0875} \approx 8.23$$

Thus, based on the data, the best approximation of k would be $k \approx 8.23$.

CONNECTION

In this section, we have only looked at how to do projections onto lines in $\mathbb{R}^n$ and onto planes in $\mathbb{R}^3$. In Chapter 7 we will extend this to finding projections onto subspaces of $\mathbb{R}^n$. This will also allow us to find more general equations of best fit via the method of least squares.

PROBLEMS 1.5

Practice Problems

For Problems A1–A3, calculate the dot product.

A1 $\begin{bmatrix} 5 \\ 3 \\ -6 \\ 1 \end{bmatrix} \cdot \begin{bmatrix} 3 \\ 2 \\ 4 \\ 0 \end{bmatrix}$ **A2** $\begin{bmatrix} 1 \\ -2 \\ -2 \\ 4 \end{bmatrix} \cdot \begin{bmatrix} 2 \\ 1/2 \\ 1/2 \\ -1 \end{bmatrix}$ **A3** $\begin{bmatrix} 1 \\ 4 \\ -1 \\ 1 \end{bmatrix} \cdot \begin{bmatrix} 2 \\ -1 \\ -1 \\ 1 \end{bmatrix}$

For Problems A4–A6, find the length.

A4 $\left\| \begin{bmatrix} \sqrt{2} \\ 1 \\ -\sqrt{2} \\ -1 \end{bmatrix} \right\|$ **A5** $\left\| \begin{bmatrix} 1/2 \\ 1/2 \\ 1/2 \\ 1/2 \end{bmatrix} \right\|$ **A6** $\left\| \begin{bmatrix} 1 \\ 2 \\ -1 \\ 3 \end{bmatrix} \right\|$

For Problems A7–A12, find a unit vector in the direction of $\vec{x}$.

A7 $\vec{x} = \begin{bmatrix} 1 \\ 2 \\ 5 \end{bmatrix}$ **A8** $\vec{x} = \begin{bmatrix} 3 \\ -2 \\ -1 \\ 1 \end{bmatrix}$ **A9** $\vec{x} = \begin{bmatrix} -2 \\ 1 \\ 0 \\ 1 \end{bmatrix}$

A10 $\vec{x} = \begin{bmatrix} 1 \\ 2 \\ 5 \\ -3 \end{bmatrix}$ **A11** $\vec{x} = \begin{bmatrix} 1/2 \\ 1/2 \\ 1/2 \\ 1/2 \end{bmatrix}$ **A12** $\vec{x} = \begin{bmatrix} 1 \\ 0 \\ 1 \\ 0 \\ 1 \end{bmatrix}$

For Problems A13 and A14, verify the triangle inequality and the Cauchy-Schwarz inequality for the given vectors.

A13 $\vec{x} = \begin{bmatrix} 4 \\ 3 \\ 1 \end{bmatrix}, \vec{y} = \begin{bmatrix} 2 \\ 1 \\ 5 \end{bmatrix}$ **A14** $\vec{x} = \begin{bmatrix} 1 \\ -1 \\ 2 \end{bmatrix}, \vec{y} = \begin{bmatrix} -3 \\ 2 \\ 4 \end{bmatrix}$

For Problems A15–A20, determine a scalar equation of the hyperplane that passes through the given point with the given normal vector.

A15 $P(1,1,-1), \vec{n} = \begin{bmatrix} 3 \\ 1 \\ 4 \end{bmatrix}$ **A16** $P(2,-2,0,1), \vec{n} = \begin{bmatrix} 0 \\ 1 \\ 3 \\ 3 \end{bmatrix}$

A17 $P(2,1,1,5), \vec{n} = \begin{bmatrix} 3 \\ -2 \\ -5 \\ 1 \end{bmatrix}$ **A18** $P(3,1,0,7), \vec{n} = \begin{bmatrix} 2 \\ -4 \\ 1 \\ -3 \end{bmatrix}$

A19 $P(0,0,0,0), \vec{n} = \begin{bmatrix} 1 \\ -4 \\ 5 \\ -2 \end{bmatrix}$ **A20** $P(1,0,1,2,1), \vec{n} = \begin{bmatrix} 0 \\ 1 \\ 2 \\ 1 \\ 1 \end{bmatrix}$

For Problems A21–A25, determine a normal vector of the hyperplane.

A21 $2x_1 + x_2 = 3$ in $\mathbb{R}^2$

A22 $3x_1 - 2x_2 + 3x_3 = 7$ in $\mathbb{R}^3$

A23 $-4x_1 + 3x_2 - 5x_3 - 6 = 0$ in $\mathbb{R}^3$

A24 $x_1 - x_2 + 2x_3 - 3x_4 = 5$ in $\mathbb{R}^4$

A25 $x_1 + x_2 - x_3 + 2x_4 - x_5 = 0$ in $\mathbb{R}^5$

For Problems A26–A31, determine $\text{proj}_{\vec{v}}(\vec{u})$ and $\text{perp}_{\vec{v}}(\vec{u})$. Check your results by verifying that $\vec{v} \cdot \text{perp}_{\vec{v}}(\vec{u}) = 0$ and $\text{proj}_{\vec{v}}(\vec{u}) + \text{perp}_{\vec{v}}(\vec{u}) = \vec{u}$.

A26 $\vec{v} = \begin{bmatrix} 0 \\ 1 \end{bmatrix}, \vec{u} = \begin{bmatrix} 3 \\ -5 \end{bmatrix}$ **A27** $\vec{v} = \begin{bmatrix} 3/5 \\ 4/5 \end{bmatrix}, \vec{u} = \begin{bmatrix} -4 \\ 6 \end{bmatrix}$

A28 $\vec{v} = \begin{bmatrix} 0 \\ 1 \\ 0 \end{bmatrix}, \vec{u} = \begin{bmatrix} -3 \\ 5 \\ 2 \end{bmatrix}$ **A29** $\vec{v} = \begin{bmatrix} 1/3 \\ -2/3 \\ 2/3 \end{bmatrix}, \vec{u} = \begin{bmatrix} 4 \\ 1 \\ -3 \end{bmatrix}$

A30 $\vec{v} = \begin{bmatrix} 1 \\ 1 \\ 0 \\ -2 \end{bmatrix}, \vec{u} = \begin{bmatrix} -1 \\ -1 \\ 2 \\ -1 \end{bmatrix}$ **A31** $\vec{v} = \begin{bmatrix} 1 \\ 0 \\ 0 \\ 1 \end{bmatrix}, \vec{u} = \begin{bmatrix} 2 \\ 3 \\ 2 \\ -3 \end{bmatrix}$

For Problems A32–A37, determine $\text{proj}_{\vec{v}}(\vec{u})$ and $\text{perp}_{\vec{v}}(\vec{u})$.

A32 $\vec{v} = \begin{bmatrix} 1 \\ 1 \end{bmatrix}, \vec{u} = \begin{bmatrix} 3 \\ -3 \end{bmatrix}$ **A33** $\vec{v} = \begin{bmatrix} 2 \\ 3 \\ -2 \end{bmatrix}, \vec{u} = \begin{bmatrix} 4 \\ -1 \\ 3 \end{bmatrix}$

A34 $\vec{v} = \begin{bmatrix} -2 \\ 1 \\ -1 \end{bmatrix}, \vec{u} = \begin{bmatrix} 5 \\ -1 \\ 3 \end{bmatrix}$ **A35** $\vec{v} = \begin{bmatrix} 1 \\ 1 \\ -2 \end{bmatrix}, \vec{u} = \begin{bmatrix} 4 \\ 1 \\ -2 \end{bmatrix}$

A36 $\vec{v} = \begin{bmatrix} -1 \\ 2 \\ 1 \\ -3 \end{bmatrix}, \vec{u} = \begin{bmatrix} 2 \\ -1 \\ 2 \\ 1 \end{bmatrix}$ **A37** $\vec{v} = \begin{bmatrix} 2 \\ 0 \\ 1 \\ 1 \end{bmatrix}, \vec{u} = \begin{bmatrix} -1 \\ 2 \\ -1 \\ 2 \end{bmatrix}$

For Problems A38 and A39:

(a) Determine a unit vector in the direction of $\vec{u}$.
(b) Calculate $\text{proj}_{\vec{u}}(\vec{F})$.
(c) Calculate $\text{perp}_{\vec{u}}(\vec{F})$.

A38 $\vec{u} = \begin{bmatrix} 2 \\ 6 \\ 3 \end{bmatrix}, \vec{F} = \begin{bmatrix} 10 \\ 18 \\ -6 \end{bmatrix}$ **A39** $\vec{u} = \begin{bmatrix} 3 \\ 1 \\ -2 \end{bmatrix}, \vec{F} = \begin{bmatrix} 3 \\ 11 \\ 2 \end{bmatrix}$

For Problems A40–A43, use a projection to find the point on the line that is closest to the given point, and find the distance from the point to the line.

A40 $Q(0,0)$, line $\vec{x} = \begin{bmatrix} 1 \\ 4 \end{bmatrix} + t\begin{bmatrix} -2 \\ 2 \end{bmatrix}, \quad t \in \mathbb{R}$

A41 $Q(2,5)$, line $\vec{x} = \begin{bmatrix} 3 \\ 7 \end{bmatrix} + t\begin{bmatrix} 1 \\ -4 \end{bmatrix}, \quad t \in \mathbb{R}$

A42 $Q(1,0,1)$, line $\vec{x} = \begin{bmatrix} 2 \\ 2 \\ -1 \end{bmatrix} + t\begin{bmatrix} 1 \\ -2 \\ 1 \end{bmatrix}, \quad t \in \mathbb{R}$

A43 $Q(2,3,2)$, line $\vec{x} = \begin{bmatrix} 1 \\ 1 \\ -1 \end{bmatrix} + t\begin{bmatrix} 1 \\ 4 \\ 1 \end{bmatrix}, \quad t \in \mathbb{R}$

For Problems A44–A50, use a projection to find the distance from the point to the plane in $\mathbb{R}^3$.

A44 $Q(2,3,1)$, plane $3x_1 - x_2 + 4x_3 = 5$

A45 $Q(-2,3,-1)$, plane $2x_1 - 3x_2 - 5x_3 = 5$

A46 $Q(0,2,-1)$, plane $2x_1 - x_3 = 5$

A47 $Q(-1,-1,1)$, plane $2x_1 - x_2 - x_3 = 4$

A48 $Q(1,0,1)$, plane $x_1 + x_2 + 3x_3 = 7$

A49 $Q(0,0,2)$, plane $2x_1 + x_2 - 4x_3 = 5$

A50 $Q(2,-1,2)$, plane $x_1 - x_2 - x_3 = 6$

For Problems A51–A54, use a projection to determine the point in the hyperplane that is closest to the given point.

A51 $Q(1,0,0,1)$, hyperplane $2x_1 - x_2 + x_3 + x_4 = 0$

A52 $Q(1,2,1,3)$, hyperplane $x_1 - 2x_2 + 3x_3 = 1$

A53 $Q(2,4,3,4)$, hyperplane $3x_1 - x_2 + 4x_3 + x_4 = 0$

A54 $Q(-1,3,2,-2)$, hyperplane $x_1 + 2x_2 + x_3 - x_4 = 4$

For Problems A55–A58, find the volume of the parallelepiped determined by the given vectors.

A55 $\begin{bmatrix} 1 \\ 0 \\ 1 \end{bmatrix}, \begin{bmatrix} 0 \\ 1 \\ 1 \end{bmatrix}, \begin{bmatrix} 0 \\ 0 \\ 1 \end{bmatrix}$

A56 $\begin{bmatrix} 4 \\ 1 \\ -1 \end{bmatrix}, \begin{bmatrix} -1 \\ 5 \\ 2 \end{bmatrix}, \begin{bmatrix} 1 \\ 1 \\ 6 \end{bmatrix}$

A57 $\begin{bmatrix} -2 \\ 1 \\ 2 \end{bmatrix}, \begin{bmatrix} 3 \\ 1 \\ 2 \end{bmatrix}, \begin{bmatrix} 0 \\ 2 \\ 5 \end{bmatrix}$

A58 $\begin{bmatrix} 1 \\ 5 \\ -3 \end{bmatrix}, \begin{bmatrix} 1 \\ 0 \\ -1 \end{bmatrix}, \begin{bmatrix} 3 \\ 0 \\ 4 \end{bmatrix}$

A59 To determine the spring constant for a given spring, some physics students apply different forces on the spring. They find that forces of 3.0 N, 6.5 N, and 9.0 N stretch the spring by 1.0 cm, 2.0 cm, and 3.0 cm respectively. Approximate the spring constant k using the method outlined in Example 1.5.13.

Homework Problems

For Problems B1–B3, calculate the dot product.

B1 $\begin{bmatrix} 1 \\ 2 \\ -1 \\ 3 \end{bmatrix} \cdot \begin{bmatrix} 3 \\ -2 \\ 1 \\ 0 \end{bmatrix}$

B2 $\begin{bmatrix} 4 \\ 2 \\ -5 \\ 1 \end{bmatrix} \cdot \begin{bmatrix} 3 \\ 0 \\ -2 \\ 1 \end{bmatrix}$

B3 $\begin{bmatrix} 2 \\ 2 \\ 5 \\ 1 \end{bmatrix} \cdot \begin{bmatrix} -2 \\ 1 \\ -1 \\ 4 \end{bmatrix}$

For Problems B4–B6, find the length.

B4 $\left\| \begin{bmatrix} 1 \\ 2 \\ 2 \\ 3 \end{bmatrix} \right\|$

B5 $\left\| \begin{bmatrix} 3 \\ 1 \\ 1 \\ 4 \end{bmatrix} \right\|$

B6 $\left\| \begin{bmatrix} 2/3 \\ 1/3 \\ 1/3 \\ 2/3 \end{bmatrix} \right\|$

For Problems B7 and B8, evaluate the expression.

B7 $\left(2\begin{bmatrix} 1 \\ 2 \\ -1 \\ 3 \end{bmatrix}\right) \cdot \left(3\begin{bmatrix} 3 \\ -2 \\ 1 \\ 0 \end{bmatrix}\right)$

B8 $\left(\begin{bmatrix} 3 \\ 5 \\ 11 \\ 3 \end{bmatrix} \cdot \begin{bmatrix} -2 \\ 1 \\ -2 \\ 1 \end{bmatrix}\right)\begin{bmatrix} 1 \\ 2 \\ 0 \\ -1 \end{bmatrix}$

For Problems B9–B14, find a unit vector in the direction of $\vec{x}$.

B9 $\vec{x} = \begin{bmatrix} 1 \\ 1 \\ -2 \end{bmatrix}$

B10 $\vec{x} = \begin{bmatrix} 5 \\ 0 \\ 0 \\ 1 \end{bmatrix}$

B11 $\vec{x} = \begin{bmatrix} 3 \\ 2 \\ -1 \\ 0 \end{bmatrix}$

B12 $\vec{x} = \begin{bmatrix} -2 \\ 1 \\ -2 \\ 3 \end{bmatrix}$

B13 $\vec{x} = \begin{bmatrix} 1/3 \\ 1/2 \\ 1/6 \\ 0 \end{bmatrix}$

B14 $\vec{x} = \begin{bmatrix} 1 \\ 1 \\ 1 \\ 1 \\ 1 \end{bmatrix}$

For Problems B15 and B16, verify the triangle inequality and the Cauchy-Schwarz inequality for the given vectors.

B15 $\vec{x} = \begin{bmatrix} -1 \\ 2 \\ 4 \end{bmatrix}, \vec{y} = \begin{bmatrix} -3 \\ 1 \\ 5 \end{bmatrix}$

B16 $\vec{x} = \begin{bmatrix} 3 \\ 1 \\ -2 \end{bmatrix}, \vec{y} = \begin{bmatrix} 2 \\ 2 \\ 2 \end{bmatrix}$

For Problems B17–B20 determine a scalar equation of the hyperplane that passes through the given point with the given normal vector.

B17 $P(1,2,3,5), \vec{n} = \begin{bmatrix} 2 \\ 2 \\ 6 \\ -1 \end{bmatrix}$ **B18** $P(3,1,4,1), \vec{n} = \begin{bmatrix} 1 \\ 5 \\ 9 \\ 2 \end{bmatrix}$

B19 $P(2,1,2,1), \vec{n} = \begin{bmatrix} 2 \\ 1 \\ 2 \\ 1 \end{bmatrix}$ **B20** $P(1,2,0,1), \vec{n} = \begin{bmatrix} 0 \\ 1 \\ 2 \\ 1 \end{bmatrix}$

For Problems B21–B26 determine a normal vector of the hyperplane.

B21 $3x_1 + x_2 = 0$ in $\mathbb{R}^2$

B22 $x_1 + 2x_2 + 7x_3 = 1$ in $\mathbb{R}^3$

B23 $3x_1 - 5x_2 + x_3 - x_4 = 4$ in $\mathbb{R}^4$

B24 $x_1 - 3x_3 + 9x_4 = 15$ in $\mathbb{R}^4$

B25 $2x_1 + x_3 + 3x_5 = 2$ in $\mathbb{R}^5$

B26 $-2x_1 - x_2 - 2x_3 + 2x_4 - 2x_5 = 0$ in $\mathbb{R}^5$

For Problems B27–B34 determine $\text{proj}_{\vec{v}}(\vec{u})$ and $\text{perp}_{\vec{v}}(\vec{u})$. Check your results by verifying that $\vec{v} \cdot \text{perp}_{\vec{v}}(\vec{u}) = 0$ and $\text{proj}_{\vec{v}}(\vec{u}) + \text{perp}_{\vec{v}}(\vec{u}) = \vec{u}$.

B27 $\vec{v} = \begin{bmatrix} 1 \\ 1 \end{bmatrix}, \vec{u} = \begin{bmatrix} 1 \\ 2 \end{bmatrix}$ **B28** $\vec{v} = \begin{bmatrix} -2 \\ 3 \end{bmatrix}, \vec{u} = \begin{bmatrix} 4 \\ -6 \end{bmatrix}$

B29 $\vec{v} = \begin{bmatrix} 4 \\ 3 \end{bmatrix}, \vec{u} = \begin{bmatrix} 1 \\ 0 \end{bmatrix}$ **B30** $\vec{v} = \begin{bmatrix} -4/5 \\ 3/5 \end{bmatrix}, \vec{u} = \begin{bmatrix} -2 \\ 5 \end{bmatrix}$

B31 $\vec{v} = \begin{bmatrix} 1 \\ 0 \\ 1 \end{bmatrix}, \vec{u} = \begin{bmatrix} 2 \\ -4 \\ 7 \end{bmatrix}$ **B32** $\vec{v} = \begin{bmatrix} 2 \\ -1 \\ 2 \end{bmatrix}, \vec{u} = \begin{bmatrix} -2 \\ 3 \\ 2 \end{bmatrix}$

B33 $\vec{v} = \begin{bmatrix} 1 \\ 0 \\ 1 \\ 1 \end{bmatrix}, \vec{u} = \begin{bmatrix} 3 \\ 3 \\ -4 \\ 2 \end{bmatrix}$ **B34** $\vec{v} = \begin{bmatrix} 7 \\ 2 \\ -1 \\ 3 \end{bmatrix}, \vec{u} = \begin{bmatrix} -1 \\ 3 \\ 2 \\ 1 \end{bmatrix}$

For Problems B35–B42 determine $\text{proj}_{\vec{v}}(\vec{u})$ and $\text{perp}_{\vec{v}}(\vec{u})$.

B35 $\vec{v} = \begin{bmatrix} 1 \\ -1 \end{bmatrix} \vec{u} = \begin{bmatrix} 3 \\ 5 \end{bmatrix}$ **B36** $\vec{v} = \begin{bmatrix} 1 \\ 0 \end{bmatrix}, \vec{u} = \begin{bmatrix} 5 \\ 3 \end{bmatrix}$

B37 $\vec{v} = \begin{bmatrix} 2 \\ 2 \end{bmatrix} \vec{u} = \begin{bmatrix} 3 \\ -1 \end{bmatrix}$ **B38** $\vec{v} = \begin{bmatrix} 1 \\ 3 \end{bmatrix}, \vec{u} = \begin{bmatrix} 5 \\ 3 \end{bmatrix}$

B39 $\vec{v} = \begin{bmatrix} 2 \\ -2 \\ 3 \end{bmatrix}, \vec{u} = \begin{bmatrix} 4 \\ 1 \\ -2 \end{bmatrix}$ **B40** $\vec{v} = \begin{bmatrix} 1 \\ 1 \\ -1 \end{bmatrix}, \vec{u} = \begin{bmatrix} 4 \\ 1 \\ 2 \end{bmatrix}$

B41 $\vec{v} = \begin{bmatrix} 1 \\ 2 \\ -1 \end{bmatrix}, \vec{u} = \begin{bmatrix} 3 \\ 3 \\ -3 \end{bmatrix}$ **B42** $\vec{v} = \begin{bmatrix} 2 \\ 0 \\ 1 \\ 2 \end{bmatrix}, \vec{u} = \begin{bmatrix} -1 \\ 2 \\ -1 \\ 2 \end{bmatrix}$

For Problems B43 and B44

(a) Determine a unit vector in the direction of $\vec{u}$.

(b) Calculate $\text{proj}_{\vec{u}}(\vec{F})$.

(c) Calculate $\text{perp}_{\vec{u}}(\vec{F})$.

B43 $\vec{u} = \begin{bmatrix} 2 \\ 1 \\ 2 \end{bmatrix}, \vec{F} = \begin{bmatrix} -3 \\ 5 \\ 2 \end{bmatrix}$ **B44** $\vec{u} = \begin{bmatrix} 1 \\ 3 \\ -3 \end{bmatrix}, \vec{F} = \begin{bmatrix} 4 \\ 3 \\ 5 \end{bmatrix}$

For Problems B45–B48 use a projection to find the point on the line that is closest to the given point, and find the distance from the point to the line.

B45 $Q(4,-5)$, line $\vec{x} = \begin{bmatrix} 2 \\ -4 \end{bmatrix} + t\begin{bmatrix} 3 \\ -4 \end{bmatrix}, \quad t \in \mathbb{R}$

B46 $Q(0,0,1)$, line $\vec{x} = \begin{bmatrix} 2 \\ 1 \\ 0 \end{bmatrix} + t\begin{bmatrix} -1 \\ 2 \\ 2 \end{bmatrix}, \quad t \in \mathbb{R}$

B47 $Q(2,-2,1)$, line $\vec{x} = \begin{bmatrix} 1 \\ -1 \\ -1 \end{bmatrix} + t\begin{bmatrix} 1 \\ 1 \\ 1 \end{bmatrix}, \quad t \in \mathbb{R}$

B48 $Q(3,2,0)$, line $\vec{x} = \begin{bmatrix} 4 \\ 1 \\ -2 \end{bmatrix} + t\begin{bmatrix} 2 \\ 1 \\ 4 \end{bmatrix}, \quad t \in \mathbb{R}$

For Problems B49–B52 use a projection to find the distance from the point to the plane in $\mathbb{R}^3$.

B49 $Q(3,5,2)$, plane $2x_1 - 3x_2 - 5x_3 = 7$

B50 $Q(1,0,1)$, plane $2x_1 + x_2 - 4x_3 = 5$

B51 $Q(2,6,2)$, plane $x_1 - x_2 - x_3 = 6$

B52 $Q(0,0,0)$, plane $x_1 + 2x_2 - x_3 = 4$

For Problems B53–B56 use a projection to determine the point in the hyperplane that is closest to the given point.

B53 $Q(2,1,0,-1)$, hyperplane $2x_1 + 2x_3 + 3x_4 = 0$

B54 $Q(1,3,0,1)$, hyperplane $2x_1 - 2x_2 + x_3 + 3x_4 = 0$

B55 $Q(3,1,2,6)$, hyperplane $3x_1 - x_2 - x_3 + x_4 = 3$

B56 $Q(3,1,3,0)$, hyperplane $2x_1 + x_2 + 4x_3 + 3x_4 = 4$

For Problems B57–B60 find the volume of the parallelepiped determined by the given vectors.

B57 $\begin{bmatrix} 5 \\ 1 \\ 1 \end{bmatrix}, \begin{bmatrix} 2 \\ 0 \\ 2 \end{bmatrix}, \begin{bmatrix} 1 \\ 2 \\ -6 \end{bmatrix}$ **B58** $\begin{bmatrix} 3 \\ 1 \\ 1 \end{bmatrix}, \begin{bmatrix} 2 \\ 1 \\ 4 \end{bmatrix}, \begin{bmatrix} -2 \\ 1 \\ -5 \end{bmatrix}$

B59 $\begin{bmatrix} 3 \\ 1 \\ 0 \end{bmatrix}, \begin{bmatrix} 2 \\ 3 \\ 3 \end{bmatrix}, \begin{bmatrix} 1 \\ 4 \\ -1 \end{bmatrix}$ **B60** $\begin{bmatrix} 6 \\ 1 \\ 5 \end{bmatrix}, \begin{bmatrix} 7 \\ 0 \\ 5 \end{bmatrix}, \begin{bmatrix} 2 \\ 2 \\ -4 \end{bmatrix}$

Conceptual Problems

C1 Consider the statement "If $\vec{u}, \vec{v}, \vec{w} \in \mathbb{R}^n$ such that $\vec{u} \cdot \vec{v} = \vec{u} \cdot \vec{w}$, then $\vec{v} = \vec{w}$."
(a) If the statement is true, prove it. If it is false, provide a counterexample.
(b) If we specify $\vec{u} \neq \vec{0}$, does this change the result?

C2 Prove that, as a consequence of the triangle inequality, $\big| \|\vec{x}\| - \|\vec{y}\| \big| \leq \|\vec{x} - \vec{y}\|$. (Hint: $\|\vec{x}\| = \|\vec{x} - \vec{y} + \vec{y}\|$.)

C3 Let $\vec{v}_1$ and $\vec{v}_2$ be orthogonal vectors in $\mathbb{R}^n$. Prove that $\|\vec{v}_1 + \vec{v}_2\|^2 = \|\vec{v}_1\|^2 + \|\vec{v}_2\|^2$.

C4 Prove that if $\vec{x} \in \mathbb{R}^n$ with $\vec{x} \neq \vec{0}$, then $\dfrac{1}{\|\vec{x}\|}\vec{x}$ is a unit vector.

C5 Let $\{\vec{v}_1, \dots, \vec{v}_k\}$ be a set of non-zero vectors in $\mathbb{R}^n$ such that all of the vectors are mutually orthogonal. That is, $\vec{v}_i \cdot \vec{v}_j = 0$ for all $i \neq j$. Prove that $\{\vec{v}_1, \dots, \vec{v}_k\}$ is linearly independent.

C6 Let S be any set of vectors in $\mathbb{R}^n$. Let $S^\perp$ be the set of all vectors that are orthogonal to every vector in S. That is,

$$S^\perp = \{\vec{w} \in \mathbb{R}^n \mid \vec{v} \cdot \vec{w} = 0 \text{ for all } \vec{v} \in S\}$$

Show that $S^\perp$ is a subspace of $\mathbb{R}^n$.

C7 (a) Given $\vec{u}$ and $\vec{v}$ in $\mathbb{R}^3$ with $\vec{u} \neq \vec{0}$ and $\vec{v} \neq \vec{0}$, verify that the composite map $C : \mathbb{R}^3 \to \mathbb{R}^3$ defined by $C(\vec{x}) = \text{proj}_{\vec{u}}(\text{proj}_{\vec{v}}\,\vec{x})$ also has the linearity property (L1).
(b) Suppose that $C(\vec{x}) = \vec{0}$ for all $\vec{x} \in \mathbb{R}^3$, where C is defined as in part (a). What can you say about $\vec{u}$ and $\vec{v}$? Explain.

C8 By the linearity property (L1), we know that $\text{proj}_{\vec{u}}(-\vec{x}) = -\text{proj}_{\vec{u}}\,\vec{x}$. Check, and explain geometrically, that $\text{proj}_{-\vec{u}}(\vec{x}) = \text{proj}_{\vec{u}}(\vec{x})$.

C9 (a) (Pythagorean Theorem) Use the fact that $\|\vec{x}\|^2 = \vec{x} \cdot \vec{x}$ to prove that $\|\vec{x} + \vec{y}\|^2 = \|\vec{x}\|^2 + \|\vec{y}\|^2$ if and only if $\vec{x} \cdot \vec{y} = 0$.
(b) Let ℓ be the line in $\mathbb{R}^n$ with vector equation $\vec{x} = t\vec{d}$ and let P be any point that is not on ℓ. Prove that for any point Q on the line, the smallest value of $\|\vec{p} - \vec{q}\|^2$ is obtained when $\vec{q} = \text{proj}_{\vec{d}}(\vec{p})$ (that is, when $\vec{p} - \vec{q}$ is perpendicular to $\vec{d}$). (Hint: Consider $\|\vec{p} - \vec{q}\| = \|\vec{p} - \text{proj}_{\vec{d}}(\vec{p}) + \text{proj}_{\vec{d}}(\vec{p}) - \vec{q}\|$.)

C10 By using the definition of $\text{perp}_{\vec{n}}$ and the fact that $\vec{PQ} = -\vec{QP}$, show that

$$\vec{OP} + \text{perp}_{\vec{n}}\left(\vec{PQ}\right) = \vec{OQ} + \text{proj}_{\vec{n}}\left(\vec{QP}\right)$$

C11 (a) Let $\vec{u} = \begin{bmatrix} 1 \\ 1 \\ -1 \end{bmatrix}$ and $\vec{x} = \begin{bmatrix} 2 \\ 5 \\ 3 \end{bmatrix}$ be vectors in $\mathbb{R}^3$. Show that $\text{proj}_{\vec{u}}(\text{perp}_{\vec{u}}(\vec{x})) = \vec{0}$.
(b) For any $\vec{u} \in \mathbb{R}^3$, prove algebraically that for any $\vec{x} \in \mathbb{R}^3$, $\text{proj}_{\vec{u}}(\text{perp}_{\vec{u}}(\vec{x})) = \vec{0}$.
(c) Explain geometrically why $\text{proj}_{\vec{u}}(\text{perp}_{\vec{u}}(\vec{x})) = \vec{0}$ for every $\vec{x} \in \mathbb{R}^3$.

C12 A set $\{\vec{v}_1, \dots, \vec{v}_k\}$ is said to be an **orthonormal set** if each vector in the set is orthogonal to every other vector in the set, and each vector is a unit vector.
(a) Prove that the standard basis for $\mathbb{R}^3$ is an orthonormal set.
(b) Prove that any orthonormal set is a basis for the set it spans.

CHAPTER REVIEW
Suggestions for Student Review

Organizing your own review is an important step towards mastering new material. It is much more valuable than memorizing someone else's list of key ideas. To retain new concepts as useful tools, you must be able to state definitions and make connections between various ideas and techniques. You should also be able to give (or, even better, create) instructive examples. The suggestions below are not intended to be an exhaustive checklist; instead, they suggest the kinds of activities and questioning that will help you gain a confident grasp of the material.

1 Find some person or persons to talk with about mathematics. There is lots of evidence that this is the best way to learn. Be sure you do your share of asking and answering. Note that a little bit of embarrassment is a small price for learning. Also, be sure to get lots of practice in writing answers independently. Looking for how you will need to apply linear algebra in your chosen field can be extremely helpful and motivating.

2 Draw pictures to illustrate addition, subtraction, and scalar multiplication of vectors. (Section 1.1)

3 Explain how you find a vector equation for a line, and make up examples to show why the vector equation of a line is not unique. (Section 1.1)

4 How do you determine if a vector is on a given line or plane? If you were asked to find 10 points on a given plane, how would you do it? (Sections 1.1, 1.3)

5 What are the differences and similarities between vectors in $\mathbb{R}^2$ or $\mathbb{R}^3$ and directed line segments? Make sure to discuss the different applications. (Section 1.1)

6 State the formal definition of spanning. Can a line in $\mathbb{R}^3$ that does not pass through the origin have a spanning set? What are the theorems related to spanning, and how do we use them? (Sections 1.2, 1.4)

7 State the formal definition of linear independence. Explain the connection between the formal definition of linear dependence and an intuitive geometric understanding of linear dependence. Why is linear independence important when looking at spanning sets? (Sections 1.2, 1.4)

8 State the formal definition of a basis. What does a basis represent geometrically? What is the difference between a basis and a spanning set? What is the importance of that difference? (Sections 1.2, 1.4)

9 State the relation (or relations) between the length and angles in $\mathbb{R}^3$ and the dot product in $\mathbb{R}^3$. Use examples to illustrate. (Sections 1.3, 1.5)

10 What are the properties of dot products and lengths? What are some applications of dot products? (Sections 1.3, 1.5)

11 State the important algebraic and geometric properties of the cross product. What are some applications of the cross product? (Section 1.3)

12 State the definition of a subspace of $\mathbb{R}^n$. Give examples of subspaces in $\mathbb{R}^3$ that are lines, planes, and all of $\mathbb{R}^3$. Show that there is only one subspace in $\mathbb{R}^3$ that does not have infinitely many vectors in it. (Section 1.4)

13 Show that the subspace spanned by three vectors in $\mathbb{R}^3$ can either be a point, a line, a plane, or all of $\mathbb{R}^3$, by giving examples. Explain how this relates to the concept of linear independence. (Section 1.4)

14 Let $\{\vec{v}_1, \vec{v}_2\}$ be a linearly independent spanning set for a subspace S of $\mathbb{R}^3$. Explain how you could construct other spanning sets and other linearly independent spanning sets for S. (Section 1.4)

15 Explain how the projection onto a vector $\vec{v}$ is defined in terms of the dot product. Illustrate with a picture. Define the part of a vector $\vec{x}$ perpendicular to $\vec{v}$ and verify (in the general case) that it is perpendicular to $\vec{v}$. (Section 1.5)

16 Explain with a picture how projections help us to solve the minimum distance problem. (Section 1.5)

17 Discuss the role of a normal vector to a plane or hyperplane in determining a scalar equation of the plane or hyperplane. Explain how you can get from a scalar equation of a plane to a vector equation for the plane and from a vector equation of the plane to the scalar equation. (Sections 1.3, 1.5)

Chapter Quiz

Note: Your instructor may have different ideas of an appropriate level of difficulty for a test on this material.

E1 Calculate $\begin{bmatrix} 3 \\ -2 \end{bmatrix} + 2\begin{bmatrix} -1 \\ 2 \end{bmatrix}$ and illustrate with a sketch.

E2 Let $\vec{x} = \begin{bmatrix} 1 \\ 2 \\ 0 \end{bmatrix}$ and $\vec{y} = \begin{bmatrix} -3 \\ 1 \\ 1 \end{bmatrix}$. Determine a unit vector that is orthogonal to both $\vec{x}$ and $\vec{y}$.

E3 Let $\vec{u} = \begin{bmatrix} 3 \\ 1 \\ 2 \\ -1 \end{bmatrix}$ and $\vec{v} = \begin{bmatrix} 1 \\ 0 \\ 2 \\ 3 \end{bmatrix}$ be vectors in $\mathbb{R}^4$. Calculate $\text{proj}_{\vec{u}}(\vec{v})$ and $\text{perp}_{\vec{u}}(\vec{v})$.

E4 Determine a vector equation of the line passing through points $P(-2, 1, -4)$ and $Q(5, -2, 1)$.

E5 Determine a vector equation of the plane in $\mathbb{R}^3$ that satisfies $x_1 - 2x_3 = 3$.

E6 Determine a scalar equation of the plane that contains the points $P(1, -1, 0)$, $Q(3, 1, -2)$, and $R(-4, 1, 6)$.

E7 Describe Span $\left\{ \begin{bmatrix} 2 \\ 6 \\ 4 \end{bmatrix}, \begin{bmatrix} 1 \\ 3 \\ 3 \end{bmatrix}, \begin{bmatrix} 3 \\ 9 \\ 7 \end{bmatrix} \right\}$ geometrically and give a basis for the spanned set.

E8 Determine whether the set $\left\{ \begin{bmatrix} 1 \\ 2 \\ 1 \end{bmatrix}, \begin{bmatrix} 1 \\ -1 \\ 3 \end{bmatrix}, \begin{bmatrix} 2 \\ 0 \\ 1 \end{bmatrix} \right\}$ is linearly independent or linearly dependent.

E9 Let $B = \left\{ \begin{bmatrix} 1 \\ 2 \end{bmatrix}, \begin{bmatrix} -1 \\ 2 \end{bmatrix} \right\}$.

(a) Prove that B is a basis for $\mathbb{R}^2$.

(b) Find the coordinates of $\vec{x} = \begin{bmatrix} 3 \\ 5 \end{bmatrix}$ with respect to the basis B.

(c) Find the coordinates of $\vec{y} = \begin{bmatrix} 6 \\ 10 \end{bmatrix}$ with respect to the basis B.

E10 Let $S = \left\{ \begin{bmatrix} x_1 \\ x_2 \end{bmatrix} \in \mathbb{R}^2 \mid x_2 = 3 - 5x_1 \right\}$. Determine whether S is a subspace of $\mathbb{R}^2$.

E11 Prove that $S = \left\{ \begin{bmatrix} x_1 \\ x_2 \\ x_3 \end{bmatrix} \in \mathbb{R}^3 \mid a_1x_2 + a_2x_2 + a_3x_3 = d \right\}$ is a subspace of $\mathbb{R}^3$ for any real numbers a_1, a_2, a_3 if and only if $d = 0$.

E12 Prove that $B = \left\{ \begin{bmatrix} 1 \\ 0 \\ -1 \end{bmatrix}, \begin{bmatrix} 0 \\ 1 \\ 3 \end{bmatrix} \right\}$ is a basis of the plane P in $\mathbb{R}^3$ with scalar equation $x_1 - 3x_2 + x_3 = 0$.

E13 Find the point on the line $\vec{x} = t\begin{bmatrix} 3 \\ -2 \\ 3 \end{bmatrix}, t \in \mathbb{R}$ that is closest to the point $P(2, 3, 4)$. Illustrate your method of calculation with a sketch.

E14 Find the point on the hyperplane $x_1 + x_2 + x_3 + x_4 = 1$ that is closest to the point $P(3, -2, 0, 2)$, and determine the distance from the point to the plane.

E15 Prove that the volume of the parallelepiped determined by $\vec{u}$, $\vec{v}$, and $\vec{w}$ has the same volume as the parallelepiped determined by $(\vec{u} + k\vec{v})$, $\vec{v}$, and $\vec{w}$.

For Problems E16–E22, determine whether the statement is true, and if so, explain briefly. If false, give a counterexample. Each statement is to be interpreted in $\mathbb{R}^3$.

E16 Any three distinct points lie in exactly one plane.

E17 The subspace spanned by a single non-zero vector is a line passing through the origin.

E18 If $\vec{x} = s\vec{v}_1 + t\vec{v}_2, s, t \in \mathbb{R}$ is a plane, then $\{\vec{v}_1, \vec{v}_2\}$ is a basis for the plane.

E19 The dot product of a vector with itself cannot be zero.

E20 For any vectors $\vec{x}$ and $\vec{y}$, $\text{proj}_{\vec{x}}(\vec{y}) = \text{proj}_{\vec{y}}(\vec{x})$.

E21 For any vectors $\vec{x}$ and $\vec{y}$, the set $\{\text{proj}_{\vec{x}}(\vec{y}), \text{perp}_{\vec{x}}(\vec{y})\}$ is linearly independent.

E22 The area of the parallelogram determined by $\vec{u}$ and $\vec{v}$ is the same as the area of the parallelogram determined by $\vec{u}$ and $(\vec{v} + 3\vec{u})$.

Further Problems

These problems are intended to be a little more challenging than the problems at the end of each section. Some explore topics beyond the material discussed in the text, and some preview topics that will appear later in the text.

F1 Consider the statement "If $\vec{u} \neq \vec{0}$, and both $\vec{u} \cdot \vec{v} = \vec{u} \cdot \vec{w}$ and $\vec{u} \times \vec{v} = \vec{u} \times \vec{w}$, then $\vec{v} = \vec{w}$." Either prove the statement or give a counterexample.

F2 Suppose that $\vec{u}$ and $\vec{v}$ are orthogonal unit vectors in $\mathbb{R}^3$. Prove that for every $\vec{x} \in \mathbb{R}^3$,

$$\text{perp}_{\vec{u}\times\vec{v}}(\vec{x}) = \text{proj}_{\vec{u}}(\vec{x}) + \text{proj}_{\vec{v}}(\vec{x})$$

F3 In Section 1.3 Problem C10, you were asked to show that $\vec{u} \times (\vec{v} \times \vec{w}) = s\vec{v} + t\vec{w}$ for some $s, t \in \mathbb{R}$.

(a) By direct calculation, prove that

$$\vec{u} \times (\vec{v} \times \vec{w}) = (\vec{u} \cdot \vec{w})\vec{v} - (\vec{u} \cdot \vec{v})\vec{w}$$

(b) Prove that

$$\vec{u} \times (\vec{v} \times \vec{w}) + \vec{v} \times (\vec{w} \times \vec{u}) + \vec{w} \times (\vec{u} \times \vec{v}) = \vec{0}$$

F4 What condition is required on the values of $a, b, c, d \in \mathbb{R}$ so that $\mathcal{B} = \left\{ \begin{bmatrix} a \\ b \end{bmatrix}, \begin{bmatrix} c \\ d \end{bmatrix} \right\}$ is a basis for $\mathbb{R}^2$.

F5 Let $\{\vec{v}_1, \vec{v}_2\}$ be a basis for a plane P in $\mathbb{R}^3$.

(a) Find a vector $\vec{w}$ such that $\vec{w}$ is orthogonal to $\vec{v}_1$ and $\{\vec{v}_1, \vec{w}\}$ is also a basis for P. (The set $\{\vec{v}_1, \vec{w}\}$ is called an **orthogonal basis** for P.)

(b) Find an orthogonal basis for $\mathbb{R}^3$ that includes the vectors $\vec{v}_1$ and $\vec{w}$. That is, find another vector $\vec{y} \in \mathbb{R}^3$ such that $B = \{\vec{v}_1, \vec{w}, \vec{y}\}$ is a basis for $\mathbb{R}^3$.

(c) Find a formula for the coordinates of any vector $\vec{x} \in \mathbb{R}^3$ with respect to the basis B. These are called the coordinates with respect to an orthogonal basis.

F6 Let $\mathbb{U}, \mathbb{W}$ be subspaces of $\mathbb{R}^n$ such that $\mathbb{U} \cap \mathbb{W} = \{\vec{0}\}$. Define a subset of $\mathbb{R}^n$ by

$$\mathbb{U} \oplus \mathbb{W} = \{\vec{u} + \vec{w} \mid \vec{u} \in \mathbb{U} \text{ and } \vec{w} \in \mathbb{W}\}$$

(a) Prove that $\mathbb{U} \oplus \mathbb{W}$ is a subspace of $\mathbb{R}^n$.

(b) Prove that if $\{\vec{u}_1, \ldots, \vec{u}_k\}$ is a basis for $\mathbb{U}$ and $\{\vec{w}_1, \ldots, \vec{w}_\ell\}$ is a basis for $\mathbb{W}$, then $\{\vec{u}_1, \ldots, \vec{u}_k, \vec{w}_1, \ldots, \vec{w}_\ell\}$ is a basis for $\mathbb{U} \oplus \mathbb{W}$.

F7 Prove that

(a) $\vec{u} \cdot \vec{v} = \frac{1}{4}\|\vec{u} + \vec{v}\|^2 - \frac{1}{4}\|\vec{u} - \vec{v}\|^2$

(b) $\|\vec{u} + \vec{v}\|^2 + \|\vec{u} - \vec{v}\|^2 = 2\|\vec{u}\|^2 + 2\|\vec{v}\|^2$

(c) Interpret (a) and (b) in terms of a parallelogram determined by vectors $\vec{u}$ and $\vec{v}$.

F8 Show that if P, Q, and R are collinear points and $\vec{OP} = \vec{p}$, $\vec{OQ} = \vec{q}$, and $\vec{OR} = \vec{r}$, then

$$(\vec{p} \times \vec{q}) + (\vec{q} \times \vec{r}) + (\vec{r} \times \vec{p}) = \vec{0}$$

F9 In $\mathbb{R}^2$, two lines fail to have a point of intersection only if they are parallel. However, in $\mathbb{R}^3$, a pair of lines can fail to have a point of intersection even if they are not parallel. Two such lines in $\mathbb{R}^3$ are called **skew**.

(a) Observe that if two lines are skew, then they do not lie in a common plane. Show that two skew lines do lie in parallel planes.

(b) Find the distance between the skew lines

$$\vec{x} = \begin{bmatrix} 1 \\ 4 \\ 2 \end{bmatrix} + s\begin{bmatrix} 2 \\ 0 \\ 1 \end{bmatrix}, s \in \mathbb{R} \text{ and } \vec{x} = \begin{bmatrix} 2 \\ -3 \\ 1 \end{bmatrix} + t\begin{bmatrix} 1 \\ 1 \\ 3 \end{bmatrix}, \quad t \in \mathbb{R}$$

CHAPTER 2

Systems of Linear Equations

CHAPTER OUTLINE

In Chapter 1 there were several times that we needed to find a vector $\vec{x}$ in $\mathbb{R}^n$ that simultaneously satisfied several linear equations. For example, when determining whether a set was linearly independent, and when deriving the formula for the cross product. In such cases, we used a **system of linear equations**. *Such systems arise frequently in almost every conceivable area where mathematics is applied: in analyzing stresses in complicated structures; in allocating resources or managing inventory; in determining appropriate controls to guide aircraft or robots; and as a fundamental tool in the numerical analysis of the flow of fluids or heat.*

You have previously learned how to solve small systems of linear equations with substitution and elimination. However, in real-life problems, it is possible for a system of linear equations to have thousands of equations and thousands of variables. Hence, we want to develop some theory that will allow us to solve such problems quickly. In this chapter, we will see that substitution and elimination can be represented by row reduction of a matrix to its reduced row echelon form. This is a fundamental procedure in linear algebra. Obtaining and interpreting the reduced row echelon form of a matrix will play an important role in almost everything we do in the rest of this book.

2.1 Systems of Linear Equations and Elimination

Definition
Linear Equation

A **linear equation** in n variables $x_1, \dots, x_n$ is an equation that can be written in the form

$$a_1x_1 + a_2x_2 + a_3x_3 + \cdots + a_nx_n = b \tag{2.1}$$

The numbers $a_1, \dots, a_n$ are called the **coefficients** of the equation, and b is usually referred to as "the right-hand side" or "the constant term." The x_i are the unknowns or variables to be solved for.

CONNECTION

1. In Chapters 1 – 8 the coefficients and constant term will be real numbers. In Chapter 9, we will look at linear equations where the coefficients and constant term are a little more 'complex'.

2. From our work in Section 1.5, we know that a linear equation in n variables with real coefficients and constant term geometrically represents a hyperplane in $\mathbb{R}^n$.

EXAMPLE 2.1.1

The equation $x_1 + 2x_2 = 4$ is linear.

The equation $x_1 + \sqrt{3}x_2 - 1 = \pi x_3$ is also linear as it can be written in the form $x_1 + \sqrt{3}x_2 - \pi x_3 = 1$.

The equations $x_1^2 - x_2 = 1$ and $x_1 x_2 = 0$ are both not linear.

Definition
System of Linear Equations

A set of m linear equations in the same variables $x_1, \ldots, x_n$ is called a **system of m linear equations in n variables**.

A general **system of m linear equations in n variables** is written in the form

$$\begin{aligned} a_{11}x_1 + a_{12}x_2 + \cdots + a_{1n}x_n &= b_1 \\ a_{21}x_1 + a_{22}x_2 + \cdots + a_{2n}x_n &= b_2 \\ &\vdots \\ a_{m1}x_1 + a_{m2}x_2 + \cdots + a_{mn}x_n &= b_m \end{aligned}$$

Note that for each coefficient a_{ij}, the first index i indicates in which equation the coefficient appears. The second index j indicates which variable the coefficient multiplies. That is, a_{ij} is the coefficient of x_j in the i-th equation. The index i on the constant term b_i indicates which equation the constant appears in.

Definition
Solution of a System
Solution Set

A vector $\vec{s} = \begin{bmatrix} s_1 \\ \vdots \\ s_n \end{bmatrix} \in \mathbb{R}^n$ is called a **solution** of a system of m linear equations in n variables if all m equations are satisfied when we set $x_i = s_i$ for $1 \leq i \leq n$. The set of all solutions of a system of linear equations is called the **solution set** of the system.

Definition
Consistent
Inconsistent

If a system of linear equations has at least one solution, then it is said to be **consistent**. Otherwise, it is said to be **inconsistent**.

Observe that geometrically a system of m linear equations in n variables represents m hyperplanes in $\mathbb{R}^n$. A solution of the system is a vector in $\mathbb{R}^n$ which lies on all m hyperplanes. The system is inconsistent if all m hyperplanes do not share a point of intersection.

EXAMPLE 2.1.2 The system of 2 equations in 3 variables

$$\begin{aligned} x_1 + 2x_2 - x_3 &= 1 \\ 3x_1 + 6x_2 - 3x_3 &= 2 \end{aligned}$$

does not have any solutions since the two corresponding planes are parallel. Hence, the system is inconsistent.

EXAMPLE 2.1.3 The system of 3 linear equations in 2 variables

$$\begin{aligned} x_1 - x_2 &= -5 \\ 2x_1 + 3x_2 &= 0 \\ -x_1 + 4x_2 &= 11 \end{aligned}$$

is consistent since if we take $x_1 = -3$ and $x_2 = 2$, then we get

$$\begin{aligned} -3 - 2 &= -5 \\ 2(-3) + 3(2) &= 0 \\ -(-3) + 4(2) &= 11 \end{aligned}$$

It can be shown this is the only solution, so the solution set is $\left\{ \begin{bmatrix} -3 \\ 2 \end{bmatrix} \right\}$.

EXERCISE 2.1.1 Sketch the lines $x_1 - x_2 = -5$, $2x_1 + 3x_2 = 0$, and $-x_1 + 4x_2 = 11$ on a single graph to verify the result of Example 2.1.3 geometrically.

EXAMPLE 2.1.4 It can be shown that the solution set of the system of 2 equations in 3 variables

$$\begin{aligned} x_1 - 2x_2 &= 3 \\ x_1 + x_2 + 3x_3 &= 9 \end{aligned}$$

has vector equation

$$\vec{x} = \begin{bmatrix} 7 \\ 2 \\ 0 \end{bmatrix} + t \begin{bmatrix} 2 \\ 1 \\ -1 \end{bmatrix}, \quad t \in \mathbb{R}$$

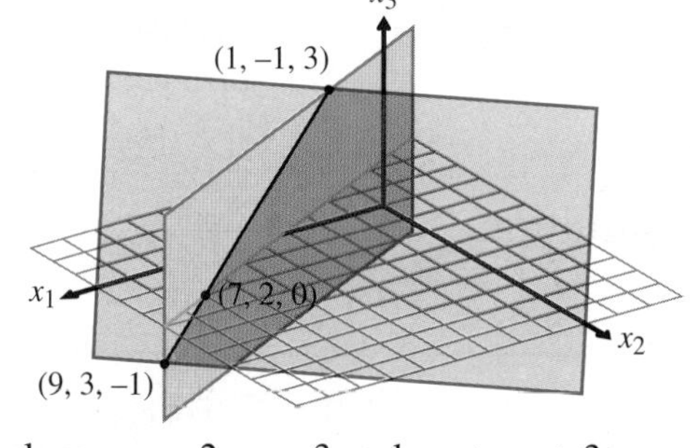

Geometrically, this means that the planes $x_1 - 2x_2 = 3$ and $x_1 + x_2 + 3x_3 = 9$ intersect in the line with the given vector equation.

Remark

It may look like the first equation in Example 2.1.4 has only 2 variables. However, we always assume that all the equations have the same variables. Thus, we interpret the first equation as

$$x_1 - 2x_2 + 0x_3 = 0$$

EXERCISE 2.1.2

Verify that every vector on the line with vector equation $\vec{x} = \begin{bmatrix} 7 \\ 2 \\ 0 \end{bmatrix} + t\begin{bmatrix} 2 \\ 1 \\ -1 \end{bmatrix}$, $t \in \mathbb{R}$ is indeed a solution of the system of linear equations in Example 2.1.4.

To illustrate the possibilities, consider a system of three linear equations in three unknowns. Each equation represents a plane in $\mathbb{R}^3$ which we will label as P_1, P_2, P_3. A solution of the system determines a point of intersection of the three planes. Figure 2.1.1 illustrates an inconsistent system: there is no point common to all three planes. Figure 2.1.2 illustrates a unique solution: all three planes intersect in exactly one point. Figure 2.1.3 demonstrates a case where there are infinitely many solutions.

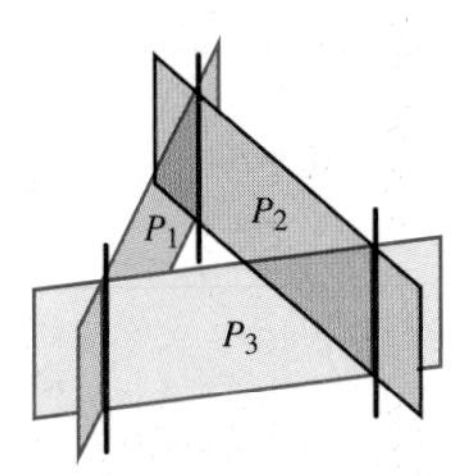

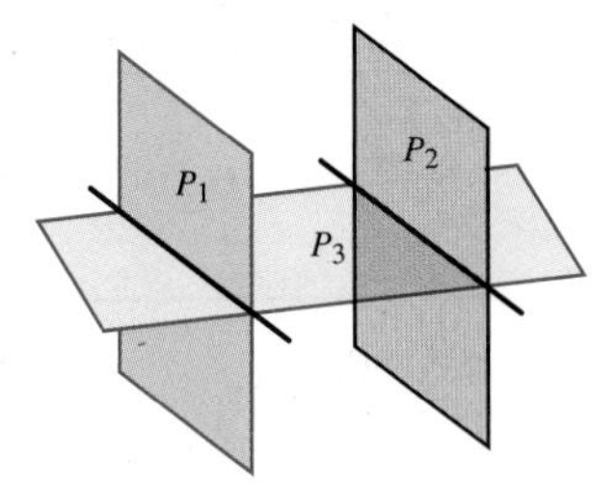

Figure 2.1.1 Two cases where three planes have no common point of intersection: the corresponding system is inconsistent.

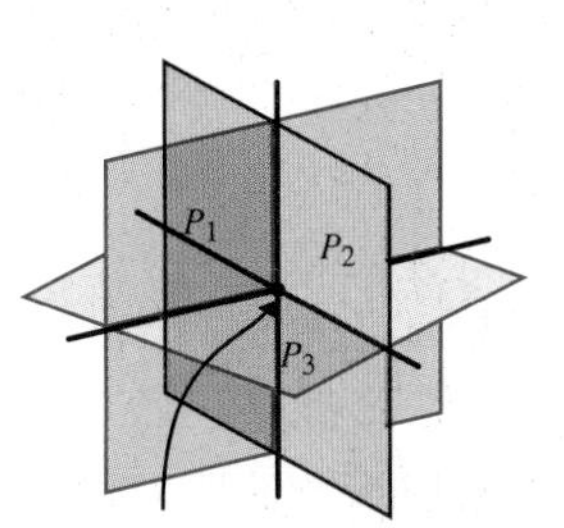

Figure 2.1.2 Three planes with one intersection point: the corresponding system of equations has a unique solution.

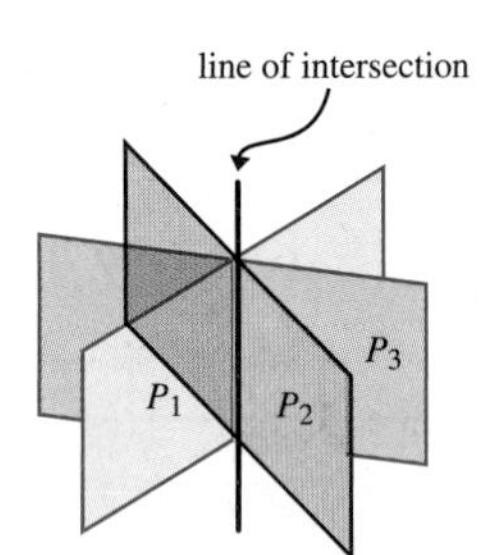

Figure 2.1.3 Three planes that meet in a common line: the corresponding system has infinitely many solutions.

EXERCISE 2.1.3

For each of the four pictures above, create a system of 3 linear equations in 3 unknowns which will be graphically represented by the picture.

These are, in fact, the only three possibilities for any system of linear equations. That is, every system of linear equations is either inconsistent, consistent with a unique solution, or consistent with infinitely many solutions. The following theorem shows that if a system of linear equations has two solutions, then it must have infinitely many solutions. In particular, every vector lying on the line that passes through the two vectors is also a solution of the system.

Theorem 2.1.1 If a system of linear equations has two distinct solutions $\vec{s}$ and $\vec{t}$, then $\vec{x} = \vec{s} + c(\vec{s} - \vec{t})$ is a distinct solution for each $c \in \mathbb{R}$.

You are asked to prove this theorem in Problem C4.

EXAMPLE 2.1.5 We have now seen that the system of linear equations

$$\begin{aligned} x_1 - x_2 &= 3 \\ 2x_1 + x_2 &= 1 \end{aligned}$$

can be interpreted geometrically as a pair of lines in $\mathbb{R}^2$ and its solution $(4/3, -5/3)$ viewed as the point of intersection.

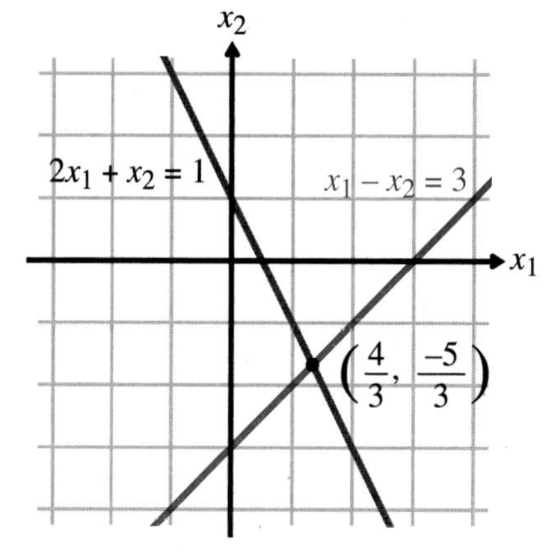

However, we do have an alternate view. From Example 1.2.2 on page 19, we see that this system also represents trying to write the vector $\vec{b} = \begin{bmatrix} 3 \\ 1 \end{bmatrix}$ and a linear combination of $\vec{u} = \begin{bmatrix} 1 \\ 2 \end{bmatrix}$ and $\vec{v} = \begin{bmatrix} -1 \\ 1 \end{bmatrix}$. That is, the solution $(4/3, -5/3)$ tells us that

$$\frac{4}{3}\begin{bmatrix} 1 \\ 2 \end{bmatrix} - \frac{5}{3}\begin{bmatrix} -1 \\ 1 \end{bmatrix} = \begin{bmatrix} 3 \\ 1 \end{bmatrix}$$

This is represented in the diagram.

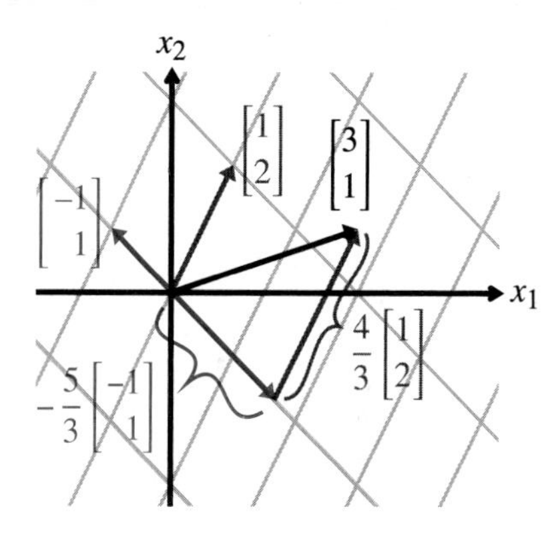

Solving Systems of Linear Equations

We now want to establish a standard procedure for determining the solution set of any system of linear equations. To do this, we begin by carefully analyzing how we solve a small system of equations using substitution and elimination.

EXAMPLE 2.1.6 Find all solutions of the system of linear equations

$$\begin{aligned} x_1 + x_2 - 2x_3 &= 4 \\ x_1 + 3x_2 - x_3 &= 7 \\ 2x_1 + x_2 - 5x_3 &= 7 \end{aligned}$$

Solution: To solve this system by elimination, we begin by eliminating x_1 from all equations except the first one.

EXAMPLE 2.1.6
(continued)

Add (−1) times the first equation to the second equation. The first and third equations are unchanged, so the system is now

$$\begin{aligned} x_1 + x_2 - 2x_3 &= 4 \\ 2x_2 + x_3 &= 3 \\ 2x_1 + x_2 - 5x_3 &= 7 \end{aligned}$$

Note two important things about this step. First, if x_1, x_2, x_3 satisfy the original system, then they certainly satisfy the revised system after the step. This follows from the rule of arithmetic that if $P = Q$ and $R = S$, then $P+R = Q+S$. So, when we add a multiple of one equation to another equation and both are satisfied, the resulting equation is satisfied. Thus, the revised system has the same solution set as the original system.

Add (−2) times the first equation to the third equation.

$$\begin{aligned} x_1 + x_2 - 2x_3 &= 4 \\ 2x_2 + x_3 &= 3 \\ -x_2 - x_3 &= -1 \end{aligned}$$

Again, this step does not change the solution set. Now that x_1 has been eliminated from all equations except the first one, we leave the first equation alone and turn our attention to x_2.

Although we will not modify or use the first equation in the next several steps, we keep writing the entire system after each step. This is important because it leads to a good general procedure for dealing with large systems.

It is convenient, but not necessary, to work with an equation in which x_2 has the coefficient 1. We could multiply the second equation by $1/2$. However, to avoid fractions, we instead just swap the order of the equations.

Interchange the second and third equations. This step definitely does not change the solution set.

$$\begin{aligned} x_1 + x_2 - 2x_3 &= 4 \\ -x_2 - x_3 &= -1 \\ 2x_2 + x_3 &= 3 \end{aligned}$$

Multiply the second equation by (-1). This does not change the solution set.

$$\begin{aligned} x_1 + x_2 - 2x_3 &= 4 \\ x_2 + x_3 &= 1 \\ 2x_2 + x_3 &= 3 \end{aligned}$$

Add (−2) times the second equation to the third equation. This does not change the solution set.

$$\begin{aligned} x_1 + x_2 - 2x_3 &= 4 \\ x_2 + x_3 &= 1 \\ -x_3 &= 1 \end{aligned}$$

In the third equation, all variables except x_3 have been eliminated; by elimination, we have solved for x_3. Using similar steps, we could continue and eliminate x_3 from the second and first equations and x_2 from the first equation. However, it is often a much simpler task to complete the solution process by **back-substitution**.

EXAMPLE 2.1.6
(continued)

We have $x_3 = -1$. Substitute this value into the second equation to find that

$$x_2 = 1 - x_3 = 1 - (-1) = 2$$

Next, substitute both these values into the first equation to obtain

$$x_1 = 4 - x_2 + 2x_3 = 4 - 2 + 2(-1) = 0$$

Thus, the only solution of this system is

$$\begin{bmatrix} x_1 \\ x_2 \\ x_3 \end{bmatrix} = \begin{bmatrix} 0 \\ 2 \\ -1 \end{bmatrix}$$

But, all of operations did not change the solution set, so this solution is also the unique solution of the original system.

Observe that we can easily check that $\begin{bmatrix} 0 \\ 2 \\ -1 \end{bmatrix}$ satisfies the original system of equations:

$$\begin{aligned} 0 + 2 - 2(-1) &= 4 \\ 0 + 3(2) - (-1) &= 7 \\ 2(0) + 2 - 5(-1) &= 7 \end{aligned}$$

It is important to observe the form of the equations in our final system. The first variable with a non-zero coefficient in each equation, called a **leading variable**, does not appear in any equation below it. Also, the leading variable in the second equation is to the right of the leading variable in the first, and the leading variable in the third is to the right of the leading variable in the second.

The system solved in Example 2.1.6 is a simple one. However, the solution procedure introduces all the steps that are needed in the process of elimination.

Types of Steps in Elimination

(1) Multiply one equation by a non-zero constant.

(2) Interchange two equations.

(3) Add a multiple of one equation to another equation.

Warning! Do *not* combine steps of type (1) and type (3) into one step of the form "Add a multiple of one equation to a multiple of another equation." Although such a combination would not lead to errors in this chapter, it can lead to errors when we apply these ideas in Chapter 5.

Observe that after each step we actually have a new system of linear equations to solve. The idea is that each new system has the same solution set as the original and is easier to solve. Moreover, observe that each elimination step is reversible, so that we can always return to the original system from the new system. Hence, we make the following definition.

Definition
Equivalent

Two systems of linear equations that have the same solution set are said to be **equivalent**.

EXERCISE 2.1.4

Solve the system below by using substitution and elimination. Clearly show/explain all of your steps used in solving the system, and describe your general procedure.

$$\begin{aligned} 2x_1 + 3x_2 &= 11 \\ 3x_1 + 6x_2 &= 7 \end{aligned}$$

EXAMPLE 2.1.7

Determine the solution set of the system of 2 linear equations in 4 variables

$$\begin{aligned} x_1 + 2x_3 + x_4 &= 14 \\ x_1 + 3x_3 + 3x_4 &= 19 \end{aligned}$$

Solution: First, notice that neither equation contains x_2. This may seem peculiar, but it happens in some applications that one of the variables of interest does not appear in any of the linear equations. If it truly is one of the variables of the problem, ignoring it is incorrect. We rewrite the equations to make it explicit:

$$\begin{aligned} x_1 + 0x_2 + 2x_3 + x_4 &= 14 \\ x_1 + 0x_2 + 3x_3 + 3x_4 &= 19 \end{aligned}$$

As in Example 2.1.6, we want our leading variable in the first equation to be to the left of the leading variable in the second equation, and we want the leading variable to be eliminated from the second equation. Thus, we use a type (3) step to eliminate x_1 from the second equation.
Add (-1) times the first equation to the second equation:

$$\begin{aligned} x_1 + 0x_2 + 2x_3 + x_4 &= 14 \\ x_3 + 2x_4 &= 5 \end{aligned}$$

Observe that x_2 is not shown in the second equation because the leading variable must have a non-zero coefficient. Moreover, we have already finished our elimination procedure as we have our desired form. The solution can now be completed by back-substitution.

Note that the equations do not completely determine both x_3 and x_4: one of them can be chosen arbitrarily, and the equations can still be satisfied. For consistency, we always choose the variables that do not appear as a leading variable in any equation to be the ones that will be chosen arbitrarily. We will call these **free variables**.

Thus, in the revised system, we see that neither x_2 nor x_4 appears as a leading variable in any equation. Therefore, x_2 and x_4 are the free variables and may be chosen arbitrarily (for example, $x_4 = t \in \mathbb{R}$ and $x_2 = s \in \mathbb{R}$). Then the second equation can be solved for the leading variable x_3:

$$x_3 = 5 - 2x_4 = 5 - 2t$$

Now, solve the first equation for its leading variable x_1:

$$x_1 = 14 - 2x_3 - x_4 = 14 - 2(5 - 2t) - t = 4 + 3t$$

EXAMPLE 2.1.7
(continued)

Thus, the solution set of the system is

$$\begin{bmatrix} x_1 \\ x_2 \\ x_3 \\ x_4 \end{bmatrix} = \begin{bmatrix} 4+3t \\ s \\ 5-2t \\ t \end{bmatrix}, \quad s,t \in \mathbb{R}$$

In this case, there are infinitely many solutions because for each value of s and for each value of t that we choose, we get a different solution. We say that this equation is the **general solution** of the system, and we call s and t the **parameters** of the general solution. For many purposes, it is useful to recognize that this solution can be split into a constant part, a part in t, and a part in s:

$$\begin{bmatrix} x_1 \\ x_2 \\ x_3 \\ x_4 \end{bmatrix} = \begin{bmatrix} 4 \\ 0 \\ 5 \\ 0 \end{bmatrix} + s\begin{bmatrix} 0 \\ 1 \\ 0 \\ 0 \end{bmatrix} + t\begin{bmatrix} 3 \\ 0 \\ -2 \\ 1 \end{bmatrix}, \quad s,t \in \mathbb{R}$$

This will be the standard format for displaying general solutions. It is acceptable to leave x_2 in the place of s and x_4 in the place of t, but then you *must* say $x_2, x_4 \in \mathbb{R}$. Observe that one immediate advantage of this form is that we can instantly see the geometric interpretation of the solution. The intersection of the two hyperplanes $x_1 + 2x_3 + x_4 = 14$ and $x_1 + 3x_3 + 3x_4 = 19$ in $\mathbb{R}^4$ is the plane in $\mathbb{R}^4$ that passes through $P(4, 0, 5, 0)$ with vector equation

$$\vec{x} = \begin{bmatrix} 4 \\ 0 \\ 5 \\ 0 \end{bmatrix} + s\begin{bmatrix} 0 \\ 1 \\ 0 \\ 0 \end{bmatrix} + t\begin{bmatrix} 3 \\ 0 \\ -2 \\ 1 \end{bmatrix}, \quad s,t \in \mathbb{R}$$

The solution procedure we have introduced is known as **Gaussian elimination with back-substitution**. A slight variation of this procedure is introduced in the next section.

EXERCISE 2.1.5

Find the general solution to the system of linear equations

$$\begin{aligned} 2x_1 + 4x_2 + 0x_3 &= 12 \\ x_1 + 2x_2 - x_3 &= 4 \end{aligned}$$

Use the general solution to find three different solutions of the system.

CONNECTION

In Chapter 1 (see page 37) when we were using a scalar equation of a plane to find a vector equation of the plane, we were really just solving a system of 1 linear equation in 3 variables. Observe that the procedure we used there is exactly the same as what we are now doing.

The Matrix Representation of a System of Linear Equations

After you have solved a few systems of equations using elimination, you may realize that you could write the solution faster if you could omit the letters x_1, x_2, and so on—as long as you could keep the coefficients lined up properly. To do this, we write out the coefficients in a rectangular array called a **matrix**.

Definition
Augmented Matrix
Coefficient Matrix

A general linear system of m equations in n unknowns can be represented by the matrix

$$\left[\begin{array}{cccccc|c} a_{11} & a_{12} & \cdots & a_{1j} & \cdots & a_{1n} & b_1 \\ a_{21} & a_{22} & \cdots & a_{2j} & \cdots & a_{2n} & b_2 \\ \vdots & \vdots & & \vdots & & \vdots & \vdots \\ a_{i1} & a_{i2} & \cdots & a_{ij} & \cdots & a_{in} & b_i \\ \vdots & \vdots & & \vdots & & \vdots & \vdots \\ a_{m1} & a_{m2} & \cdots & a_{mj} & \cdots & a_{mn} & b_m \end{array}\right]$$

where the coefficient a_{ij} appears in the i-th row and j-th column of the coefficient matrix. This is called the **augmented matrix** of the system. It is augmented because it includes as its last column the right-hand side of the equations. The matrix without this last column is called the **coefficient matrix** of the system:

$$\begin{bmatrix} a_{11} & a_{12} & \cdots & a_{1j} & \cdots & a_{1n} \\ a_{21} & a_{22} & \cdots & a_{2j} & \cdots & a_{2n} \\ \vdots & \vdots & & \vdots & & \vdots \\ a_{i1} & a_{i2} & \cdots & a_{ij} & \cdots & a_{in} \\ \vdots & \vdots & & \vdots & & \vdots \\ a_{m1} & a_{m2} & \cdots & a_{mj} & \cdots & a_{mn} \end{bmatrix}$$

For convenience, we sometimes denote the augmented matrix of a system with coefficient matrix A and right-hand side $\vec{b} = \begin{bmatrix} b_1 \\ \vdots \\ b_m \end{bmatrix}$ by $\left[A \mid \vec{b}\right]$. In Chapter 3, we will develop an even better way of representing a system of linear equations.

EXAMPLE 2.1.8

Write the coefficient matrix and augmented matrix for the following system:

$$\begin{aligned} 3x_1 + 8x_2 - 18x_3 + x_4 &= 35 \\ x_1 + 2x_2 - 4x_3 \qquad &= 11 \\ x_1 + 3x_2 - 7x_3 + x_4 &= 10 \end{aligned}$$

Solution: The coefficient matrix is formed by writing the coefficients of each equation as the rows of the matrix. Thus, we get the matrix

$$A = \begin{bmatrix} 3 & 8 & -18 & 1 \\ 1 & 2 & -4 & 0 \\ 1 & 3 & -7 & 1 \end{bmatrix}$$

EXAMPLE 2.1.8
(continued)

For the augmented matrix, we just add the right-hand side as the last column. We get

$$\left[\begin{array}{cccc|c} 3 & 8 & -18 & 1 & 35 \\ 1 & 2 & -4 & 0 & 11 \\ 1 & 3 & -7 & 1 & 10 \end{array}\right]$$

EXAMPLE 2.1.9

Write the system of linear equations that has the augmented matrix

$$\left[\begin{array}{ccc|c} 1 & 0 & 2 & 3 \\ 0 & -1 & 1 & 1 \\ 0 & 0 & 1 & -2 \end{array}\right]$$

Solution: The rows of the matrix tell us the coefficients and constant terms of each equation. We get the system

$$\begin{aligned} x_1 \qquad + 2x_3 &= 3 \\ -x_2 + x_3 &= 1 \\ x_3 &= -2 \end{aligned}$$

Remark

Another way to view the coefficient matrix is to see that the j-th column of the coefficient matrix is the vector containing all the coefficients of x_j. We will use this interpretation throughout the text.

Since each row in the augmented matrix corresponds to an equation in the system of linear equations, performing operations on the equations of the system corresponds to performing the same operations on the rows of the matrix. Thus, the steps in elimination correspond to the following elementary row operations.

Definition
Elementary Row Operations

The three **elementary row operations (EROs)** are:
(1) Multiply one row by a non-zero constant.
(2) Interchange two rows.
(3) Add a multiple of one row to another row.

As with the steps in elimination, we do not combine operations of type (1) and type (3) into one operation.

The process of performing elementary row operations on a matrix to bring it into some simpler form is called **row reduction**.

Recall that if a system of equations is obtained from another system by one or more of the elimination steps, the systems are said to be equivalent. For matrices, if the matrix M is row reduced into a matrix N by a sequence of elementary row operations, then we say that M is **row equivalent** to N and we write $M \sim N$. Note that it would be incorrect to use $=$ or $\Rightarrow$ instead of $\sim$.

Just as elimination steps are reversible, so are elementary row operations. It follows that if M is row equivalent to N, then N is row equivalent to M, so we may say that M and N are row equivalent. It also follows that if A is row equivalent to B and B is row equivalent to C, then A is row equivalent to C.

We often use short hand notation to indicate which operations we are using:

cR_i indicates multiplying the i-th row by $c \neq 0$.

$R_i \updownarrow R_j$ indicates swapping the i-th row and the j-th row.

$R_i + cR_j$ indicates adding c times the j-th row to the i-th row.

As one becomes confident with row reducing, one may omit these indicators. However, including them can make checking the steps easier, is required for one concept in Chapter 5, and instructors may require them in work submitted for grading.

Theorem 2.1.2 If the augmented matrices $\left[A_1 \mid \vec{b}_1\right]$ and $\left[A \mid \vec{b}\right]$ are row equivalent, then the systems of linear equations associated with each augmented matrix are equivalent.

EXAMPLE 2.1.10 Rewrite the elimination steps for the system in Example 2.1.6 in matrix form.

Solution: The augmented matrix for the system is

$$\left[\begin{array}{ccc|c} 1 & 1 & -2 & 4 \\ 1 & 3 & -1 & 7 \\ 2 & 1 & -5 & 7 \end{array}\right]$$

The first step in the elimination was to add (-1) times the first equation to the second. Here we add (-1) multiplied by the first row to the second. We write

$$\left[\begin{array}{ccc|c} 1 & 1 & -2 & 4 \\ 1 & 3 & -1 & 7 \\ 2 & 1 & -5 & 7 \end{array}\right] R_2 - 1R_1 \sim \left[\begin{array}{ccc|c} 1 & 1 & -2 & 4 \\ 0 & 2 & 1 & 3 \\ 2 & 1 & -5 & 7 \end{array}\right]$$

The remaining steps are

$$\left[\begin{array}{ccc|c} 1 & 1 & -2 & 4 \\ 0 & 2 & 1 & 3 \\ 2 & 1 & -5 & 7 \end{array}\right] R_3 - 2R_1 \sim \left[\begin{array}{ccc|c} 1 & 1 & -2 & 4 \\ 0 & 2 & 1 & 3 \\ 0 & -1 & -1 & -1 \end{array}\right] R_2 \updownarrow R_3$$

$$\sim \left[\begin{array}{ccc|c} 1 & 1 & -2 & 4 \\ 0 & -1 & -1 & -1 \\ 0 & 2 & 1 & 3 \end{array}\right] (-1)R_2 \sim \left[\begin{array}{ccc|c} 1 & 1 & -2 & 4 \\ 0 & 1 & 1 & 1 \\ 0 & 2 & 1 & 3 \end{array}\right] R_3 - 2R_2 \sim \left[\begin{array}{ccc|c} 1 & 1 & -2 & 4 \\ 0 & 1 & 1 & 1 \\ 0 & 0 & -1 & 1 \end{array}\right]$$

All the elementary row operations corresponding to the elimination in Example 2.1.6 have been performed. Observe that the last matrix is the augmented matrix for the final system of linear equations that we obtained in Example 2.1.6.

EXAMPLE 2.1.11 Write the matrix representation of the elimination in Example 2.1.7.

Solution: We have

$$\left[\begin{array}{cccc|c} 1 & 0 & 2 & 1 & 14 \\ 1 & 0 & 3 & 3 & 19 \end{array}\right] R_2 + (-1)R_1 \sim \left[\begin{array}{cccc|c} 1 & 0 & 2 & 1 & 14 \\ 0 & 0 & 1 & 2 & 5 \end{array}\right]$$

EXERCISE 2.1.6 Write out the matrix representation of the elimination used in Exercise 2.1.5.

In the next example, we will solve a system of linear equations using Gaussian elimination with back-substitution entirely in matrix form.

EXAMPLE 2.1.12 Find the general solution of the system

$$\begin{aligned} 3x_1 + 8x_2 - 18x_3 + x_4 &= 35 \\ x_1 + 2x_2 - 4x_3 \quad\quad &= 11 \\ x_1 + 3x_2 - 7x_3 + x_4 &= 10 \end{aligned}$$

Solution: Write the augmented matrix of the system and row reduce:

$$\left[\begin{array}{cccc|c} 3 & 8 & -18 & 1 & 35 \\ 1 & 2 & -4 & 0 & 11 \\ 1 & 3 & -7 & 1 & 10 \end{array}\right] \begin{array}{l} R_1 \updownarrow R_2 \\ \\ \\ \end{array} \sim \left[\begin{array}{cccc|c} 1 & 2 & -4 & 0 & 11 \\ 3 & 8 & -18 & 1 & 35 \\ 1 & 3 & -7 & 1 & 10 \end{array}\right] \begin{array}{l} \\ R_2 - 3R_1 \\ \\ \end{array} \sim$$

$$\left[\begin{array}{cccc|c} 1 & 2 & -4 & 0 & 11 \\ 0 & 2 & -6 & 1 & 2 \\ 1 & 3 & -7 & 1 & 10 \end{array}\right] \begin{array}{l} \\ \\ R_3 - 1R_1 \end{array} \sim \left[\begin{array}{cccc|c} 1 & 2 & -4 & 0 & 11 \\ 0 & 2 & -6 & 1 & 2 \\ 0 & 1 & -3 & 1 & -1 \end{array}\right] \begin{array}{l} \\ R_2 \updownarrow R_3 \\ \\ \end{array} \sim$$

$$\left[\begin{array}{cccc|c} 1 & 2 & -4 & 0 & 11 \\ 0 & 1 & -3 & 1 & -1 \\ 0 & 2 & -6 & 1 & 2 \end{array}\right] \begin{array}{l} \\ \\ R_3 - 2R_2 \end{array} \sim \left[\begin{array}{cccc|c} 1 & 2 & -4 & 0 & 11 \\ 0 & 1 & -3 & 1 & -1 \\ 0 & 0 & 0 & -1 & 4 \end{array}\right]$$

To find the general solution, we now interpret the final matrix as the augmented matrix of the equivalent system. We get the system

$$\begin{aligned} x_1 + 2x_2 - 4x_3 \quad\quad &= 11 \\ x_2 - 3x_3 + x_4 &= -1 \\ -x_4 &= 4 \end{aligned}$$

Since x_3 is a free variable, we let $x_3 = t \in \mathbb{R}$. Then we use back-substitution to get

$$\begin{aligned} x_4 &= -4 \\ x_2 &= -1 + 3x_3 - x_4 = 3 + 3t \\ x_1 &= 11 - 2x_2 + 4x_3 = 5 - 2t \end{aligned}$$

Thus, the general solution is

$$\begin{bmatrix} x_1 \\ x_2 \\ x_3 \\ x_4 \end{bmatrix} = \begin{bmatrix} 5-2t \\ 3+3t \\ t \\ -4 \end{bmatrix} = \begin{bmatrix} 5 \\ 3 \\ 0 \\ -4 \end{bmatrix} + t \begin{bmatrix} -2 \\ 3 \\ 1 \\ 0 \end{bmatrix}, \quad t \in \mathbb{R}$$

Check this solution by substituting these values for x_1, x_2, x_3, x_4 into the original equations.

Observe that there are many different ways that we could choose to row reduce the augmented matrix in any of these examples. For instance, in Example 2.1.12 we could interchange row 1 and row 3 instead of interchanging row 1 and row 2. Alternatively, we could use the elementary row operations $R_2 - \frac{1}{3}R_1$ and $R_3 - \frac{1}{3}R_1$ to eliminate the non-zero entries beneath the first leading variable. It is natural to ask if there is a way of determining which elementary row operations will work the best. Unfortunately, there is no such algorithm for doing these by hand. However, we will give a basic algorithm for row reducing a matrix into the "proper" form. We start by defining this form.

Row Echelon Form

Based on how we used elimination to solve the system of equations, we define the following form of a matrix.

Definition
Row Echelon Form (REF)

A matrix is in **row echelon form (REF)** if

(1) A zero row (all entries in the row are zero) must appear below all rows that contain a non-zero entry.

(2) When two non-zero rows are compared, the first non-zero entry, called the leading entry, in the upper row is to the left of the leading entry in the lower row.

If a matrix A is row equivalent to a matrix R in row echelon form, then we say that R is a row echelon form of A.

Remark

It follows from these properties that all entries in a column beneath a leading entry must be 0. For otherwise, (1) or (2) would be violated.

EXAMPLE 2.1.13 Determine which of the following matrices are in row echelon form. For each matrix that is not in row echelon form, explain why it is not in row echelon form.

$$\text{(a)} \begin{bmatrix} 1 & 1 & -2 & 4 \\ 0 & 1 & 1 & 1 \\ 0 & 0 & 1 & -1 \end{bmatrix} \qquad \text{(b)} \begin{bmatrix} 0 & 1 & 0 & 1 & 2 \\ 0 & 0 & 0 & -3 & 1 \\ 0 & 0 & 0 & 0 & 0 \end{bmatrix}$$

$$\text{(c)} \begin{bmatrix} 1 & 0 & 0 & 0 \\ 0 & 0 & 2 & -3 \\ 0 & 2 & 3 & 3 \end{bmatrix} \qquad \text{(d)} \begin{bmatrix} 1 & 1 & 2 & -1 & 1 \\ 1 & 3 & 1 & 4 & -2 \end{bmatrix}$$

Solution: The matrices in (a) and (b) are both in REF. The matrix in (c) is not in REF since the leading entry in the second row is to the right of the leading entry in the third row. The matrix in (d) is not in REF since there is a non-zero entry beneath the leading entry in the first row.

Algorithm 2.1.1 Gaussian Elimination

Any matrix can be row reduced to row echelon form by using the following steps:

1. Working from the left, find the first column of the matrix that contains some non-zero entry. Interchange rows (if necessary) so that the top entry in the column is non-zero. Of course, the column may contain multiple non-zero entries. You can use any of these non-zero entries, but some choices will make your calculations considerably easier than others; see the Remarks on page 95. We will call this entry a **pivot**.
2. Use elementary row operations of type (3) to make all entries beneath the pivot into zeros.
3. Repeat Steps 1 and 2 on the submatrix consisting of all rows below the row with the most recently obtained pivot until you reach the bottom row or there are no remaining non-zero rows.

EXAMPLE 2.1.14

Use Gaussian Elimination to bring the matrix $A = \begin{bmatrix} 2 & 0 & 4 & 4 \\ -1 & 0 & -1 & 0 \\ -3 & 0 & 0 & 6 \end{bmatrix}$ into REF.

Solution: We begin by considering the first column. The top entry of the first column is non-zero, so this becomes our pivot.

$$\begin{bmatrix} 2 & 0 & 4 & 4 \\ -1 & 0 & -1 & 0 \\ -3 & 0 & 0 & 6 \end{bmatrix}$$

We now put zeros beneath the pivot using $R_2 + \frac{1}{2}R_1$ and $R_3 + \frac{3}{2}R_1$.

$$\begin{bmatrix} 2 & 0 & 4 & 4 \\ -1 & 0 & -1 & 0 \\ -3 & 0 & 0 & 6 \end{bmatrix} \begin{matrix} \\ R_2 + \frac{1}{2}R_1 \\ \\ \end{matrix} \sim \begin{bmatrix} 2 & 0 & 4 & 4 \\ 0 & 0 & 1 & 2 \\ -3 & 0 & 0 & 6 \end{bmatrix} \begin{matrix} \\ \\ R_3 + \frac{3}{2}R_1 \end{matrix} \sim \begin{bmatrix} 2 & 0 & 4 & 4 \\ 0 & 0 & 1 & 2 \\ 0 & 0 & 6 & 12 \end{bmatrix}$$

We now temporarily ignore the first row and repeat the procedure.

$$\begin{bmatrix} \not2 & \not0 & \not4 & \not4 \\ 0 & 0 & 1 & 2 \\ 0 & 0 & 6 & 12 \end{bmatrix}$$

The first two columns of the submatrix contain only zeros, so we move to the third column. Its top entry is non-zero, so that becomes our pivot. We use $R_3 - 6R_2$ to put a zero beneath it.

$$\begin{bmatrix} 2 & 0 & 4 & 4 \\ 0 & 0 & 1 & 2 \\ 0 & 0 & 6 & 12 \end{bmatrix} \begin{matrix} \\ \\ R_3 - 6R_2 \end{matrix} \sim \begin{bmatrix} 2 & 0 & 4 & 4 \\ 0 & 0 & 1 & 2 \\ 0 & 0 & 0 & 0 \end{bmatrix}$$

Since we only have zero rows beneath the most recent pivot, the matrix is now in row echelon form.

EXAMPLE 2.1.15 Use Gaussian Elimination to row reduce the augmented matrix of the following system to row echelon form. Determine all solutions of the system.

$$\begin{aligned} x_2 + x_3 &= 2 \\ x_1 + x_2 \quad &= 1 \\ x_1 + 2x_2 + x_3 &= -2 \end{aligned}$$

Solution: We first write the augmented matrix of the system.

$$\left[\begin{array}{ccc|c} 0 & 1 & 1 & 2 \\ 1 & 1 & 0 & 1 \\ 1 & 2 & 1 & -2 \end{array}\right]$$

We begin by considering the first column. To make the top entry in the first column non-zero, we exchange row 1 and row 2

$$\left[\begin{array}{ccc|c} 0 & 1 & 1 & 2 \\ 1 & 1 & 0 & 1 \\ 1 & 2 & 1 & -2 \end{array}\right] R_1 \updownarrow R_2 \sim \left[\begin{array}{ccc|c} 1 & 1 & 0 & 1 \\ 0 & 1 & 1 & 2 \\ 1 & 2 & 1 & -2 \end{array}\right]$$

We now have our first pivot. The second row already has a zero beneath this pivot, so we just use $R_3 - R_1$ so that all entries beneath the pivot are 0.

$$\left[\begin{array}{ccc|c} 1 & 1 & 0 & 1 \\ 0 & 1 & 1 & 2 \\ 1 & 2 & 1 & -2 \end{array}\right] \begin{array}{c} \\ \\ R_3 - R_1 \end{array} \sim \left[\begin{array}{ccc|c} 1 & 1 & 0 & 1 \\ 0 & 1 & 1 & 2 \\ 0 & 1 & 1 & -3 \end{array}\right]$$

We now temporarily ignore the first row and repeat the procedure.

$$\left[\begin{array}{ccc|c} \not{1} & \not{1} & \not{0} & \not{1} \\ 0 & 1 & 1 & 2 \\ 0 & 1 & 1 & -3 \end{array}\right]$$

We see that the first column of the submatrix contains all zeros, so we move to the next column. The entry in the top of the next column is non-zero, so this becomes our next pivot. We use $R_3 - R_2$ to place a zero beneath it.

$$\left[\begin{array}{ccc|c} 1 & 1 & 0 & 1 \\ 0 & 1 & 1 & 2 \\ 0 & 1 & 1 & -3 \end{array}\right] \begin{array}{c} \\ \\ R_3 - R_2 \end{array} \sim \left[\begin{array}{ccc|c} 1 & 1 & 0 & 1 \\ 0 & 1 & 1 & 2 \\ 0 & 0 & 0 & -5 \end{array}\right]$$

Since we have arrived at the bottom row, the matrix is in row echelon form.

Observe that when we write the system of linear equations represented by this augmented matrix, we get

$$\begin{aligned} x_1 + x_2 \quad &= 1 \\ x_2 + x_3 &= 2 \\ 0 &= -5 \end{aligned}$$

Clearly, the last equation is impossible. This means we cannot find values of x_1, x_2, x_3 that satisfy all three equations. Consequently this system, and hence the original system, has no solution.

Remarks

1. Although the previous algorithm will always work, it is not necessarily the fastest or easiest method to use for any particular matrix. In principle, it does not matter which non-zero entry is chosen as the pivot in the procedure just described. In practice, it can have considerable impact on the amount of work required and on the accuracy of the result. The ability to row reduce a general matrix to REF by hand quickly and efficiently comes only with a considerable amount of practice. Note that for hand calculations on simple integer examples, it is sensible to go to some trouble to postpone fractions because avoiding fractions may reduce both the effort required and the chance of making errors.

2. Every matrix, except the matrix containing all zeros, has infinitely many row echelon forms that are all row equivalent. However, it can be shown that any two row echelon forms for the same matrix A must agree on the position of the leading entries. (This fact may seem obvious, but it is not easy to prove. It follows from Problem F2 in the Chapter 4 Further Problems.)

Row echelon form allows us to answer questions about consistency and uniqueness. In particular, we have the following theorem.

Theorem 2.1.3 Suppose that the augmented matrix $\left[A \mid \vec{b}\right]$ of a system of linear equations is row equivalent to $\left[R \mid \vec{c}\right]$, which is in row echelon form.

(1) The given system is inconsistent if and only if some row of $\left[R \mid \vec{c}\right]$ is of the form $\left[\begin{array}{cccc|c} 0 & 0 & \cdots & 0 & c \end{array}\right]$, with $c \neq 0$.

(2) If the system is consistent, there are two possibilities. Either the number of pivots in R is equal to the number of variables in the system and the system has a unique solution, or the number of pivots is less than the number of variables and the system has infinitely many solutions.

Proof: (1) If $\left[R \mid \vec{c}\right]$ contains a row of the form $\left[\begin{array}{cccc|c} 0 & 0 & \cdots & 0 & c \end{array}\right]$, where $c \neq 0$, then this corresponds to the equation $0 = c$, which clearly has no solution. Hence, the system is inconsistent. On the other hand, if it contains no such row, then each row must either be of the form $\left[\begin{array}{cccc|c} 0 & 0 & \cdots & 0 & 0 \end{array}\right]$, which corresponds to an equation satisfied by any values of $x_1, \dots, x_n$, or else contains a pivot. We may ignore the rows that consist entirely of zeros, leaving only rows with pivots. In the latter case, the corresponding system can be solved by assigning arbitrary values to the free variables and then determining the remaining variables by back-substitution. Thus, if there is no row of the form $\left[\begin{array}{cccc|c} 0 & 0 & \cdots & 0 & c \end{array}\right]$, the system is consistent.

(2) Now consider the case of a consistent system. The number of leading variables cannot be greater than the number of columns in the coefficient matrix; if it is equal, then each variable is a leading variable and thus is determined uniquely by the system corresponding to $\left[R \mid \vec{c}\right]$. If some variables are not leading variables, then they are free variables, and they may be chosen arbitrarily. Hence, there are infinitely many solutions. ■

Remark

As we will see later in the text, sometimes we are only interested in whether a system is consistent or inconsistent or in how many solutions a system has. We may not necessarily be interested in finding a particular solution. In these cases, Theorem 2.1.3, or the System-Rank Theorem in Section 2.2, can be very useful.

Some Shortcuts and Some Bad Moves

When carrying out elementary row operations, you may get weary of rewriting the matrix every time. Fortunately, we can combine some elementary row operations in one rewriting. For example,

$$\left[\begin{array}{ccc|c} 1 & 1 & -2 & 4 \\ 1 & 3 & -1 & 7 \\ 2 & 1 & -5 & 7 \end{array}\right] \begin{array}{c} \\ R_2 - R_1 \\ R_3 - 2R_1 \end{array} \sim \left[\begin{array}{ccc|c} 1 & 1 & -2 & 4 \\ 0 & 2 & 1 & 3 \\ 0 & -1 & -1 & -1 \end{array}\right]$$

Choosing one particular row (in this case, the first row) and adding multiples of it to several other rows is perfectly acceptable. There are other elementary row operations that can be combined, but these should not be used until one is extremely comfortable with row reducing. This is because some combinations of steps do cause errors. For example,

$$\begin{bmatrix} 1 & 1 & 3 \\ 1 & 2 & 4 \end{bmatrix} \begin{array}{c} R_1 - R_2 \\ R_2 - R_1 \end{array} \sim \begin{bmatrix} 0 & -1 & -1 \\ 0 & 1 & 1 \end{bmatrix} \qquad (\textit{WRONG!})$$

This is nonsense because the final matrix should have a leading 1 in the first column. By performing one elementary row operation, we change one row; thereafter, we must use that row in its new changed form. Thus, when performing multiple elementary row operations in one step, make sure that you are not modifying a row that you are using in another elementary row operation.

EXERCISE 2.1.7

Use Gaussian Elimination to bring the matrix $A = \begin{bmatrix} 0 & 2 & 1 & -1 \\ 1 & 2 & -1 & 1 \\ -1 & 4 & 4 & 1 \end{bmatrix}$ into a row echelon form.

Applications

To illustrate the application of systems of equations we give a couple simple examples. In Section 2.4 we discuss more applications from physics/engineering.

EXAMPLE 2.1.16

A boy has a jar full of coins. Altogether there are 180 nickels, dimes, and quarters. The number of dimes is one-half of the total number of nickels and quarters. The value of the coins is \$16.00. How many of each kind of coin does he have?

Solution: Let n be the number of nickels, d the number of dimes, and q the number of quarters. Then

$$n + d + q = 180$$

The second piece of information we are given is that

$$d = \frac{1}{2}(n + q)$$

We rewrite this into standard form for a linear equation:

$$n - 2d + q = 0$$

Finally, we have the value of the coins, in cents:

$$5n + 10d + 25q = 1600$$

Thus, n, d, and q satisfy the system of linear equations:

$$\begin{aligned} n \quad + d \quad + q &= 180 \\ n \quad - 2d \quad + q &= 0 \\ 5n + 10d + 25q &= 1600 \end{aligned}$$

Write the augmented matrix and row reduce:

$$\left[\begin{array}{ccc|c} 1 & 1 & 1 & 180 \\ 1 & -2 & 1 & 0 \\ 5 & 10 & 25 & 1600 \end{array}\right] \begin{array}{c} \\ R_2 - R_1 \\ R_3 - 5R_1 \end{array} \sim \left[\begin{array}{ccc|c} 1 & 1 & 1 & 180 \\ 0 & -3 & 0 & -180 \\ 0 & 5 & 20 & 700 \end{array}\right] \begin{array}{c} \\ (-1/3)R_2 \\ (1/5)R_3 \end{array} \sim$$

$$\left[\begin{array}{ccc|c} 1 & 1 & 1 & 180 \\ 0 & 1 & 0 & 60 \\ 0 & 1 & 4 & 140 \end{array}\right] \begin{array}{c} \\ \\ R_3 - R_2 \end{array} \sim \left[\begin{array}{ccc|c} 1 & 1 & 1 & 180 \\ 0 & 1 & 0 & 60 \\ 0 & 0 & 4 & 80 \end{array}\right]$$

According to Theorem 2.1.3, the system is consistent with a unique solution. In particular, writing the final augmented matrix as a system of equations, we get

$$\begin{aligned} n + d + q &= 180 \\ d &= 60 \\ 4q &= 80 \end{aligned}$$

So, by back-substitution, we get $q = 20$, $d = 60$, $n = 180 - d - q = 100$. Hence, the boy has 100 nickels, 60 dimes, and 20 quarters.

EXAMPLE 2.1.17 Determine the parabola $y = a + bx + cx^2$ that passes through the points $(1, 1)$, $(-1, 7)$, and $(2, 4)$.

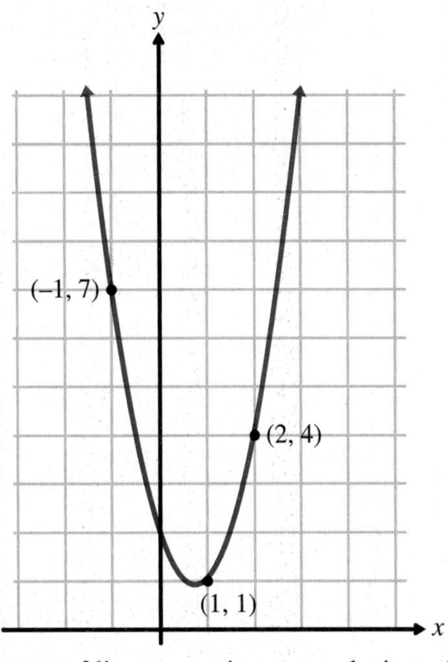

Solution: To create a system of linear equations, we substitute the x and y values of each point into the general equation of the parabola $y = a + bx + cx^2$. We get

$$\begin{aligned} 1 &= a + b(1) + c(1)^2 = a + b + c \\ 7 &= a + b(-1) + c(-1)^2 = a - b + c \\ 4 &= a + b(2) + c(2)^2 = a + 2b + 4c \end{aligned}$$

As usual, we write the corresponding augmented matrix and row reduce.

$$\left[\begin{array}{ccc|c} 1 & 1 & 1 & 1 \\ 1 & -1 & 1 & 7 \\ 1 & 2 & 4 & 4 \end{array}\right] \begin{array}{l} \\ R_2 - R_1 \\ R_3 - R_1 \end{array} \sim \left[\begin{array}{ccc|c} 1 & 1 & 1 & 1 \\ 0 & -2 & 0 & 6 \\ 0 & 1 & 3 & 3 \end{array}\right] \begin{array}{l} \\ -(1/2)R_2 \\ \\ \end{array} \sim$$

$$\left[\begin{array}{ccc|c} 1 & 1 & 1 & 1 \\ 0 & 1 & 0 & -3 \\ 0 & 1 & 3 & 3 \end{array}\right] \begin{array}{l} \\ \\ R_3 - R_2 \end{array} \sim \left[\begin{array}{ccc|c} 1 & 1 & 1 & 1 \\ 0 & 1 & 0 & -3 \\ 0 & 0 & 3 & 6 \end{array}\right]$$

By back-substitution we get $c = 2$, $b = -3$, and $a = 1 - b - c = 1 - (-3) - 2 = 2$. Hence, the parabola is

$$y = 2 - 3x + 2x^2$$

It is easy to verify the answer by checking to ensure that all three points are on this parabola.

A Remark on Computer Calculations

In computer calculations, the choice of the pivot may affect accuracy. The problem is that real numbers are represented to only a finite number of digits in a computer, so inevitably some round-off or truncation errors can occur. When you are doing a large number of arithmetic operations, these errors can accumulate, and they can be particularly serious if at some stage you subtract two nearly equal numbers. The following example gives some idea of the difficulties that might be encountered.

The system

$$\begin{aligned} 0.1000x_1 + 0.9990x_2 &= 1.000 \\ 0.1000x_1 + 1.000x_2 &= 1.006 \end{aligned}$$

is easily found to have solution $x_2 = 6.000$, $x_1 = -49.94$. Notice that the coefficients were given to four digits. Suppose all entries are rounded to three digits. The system becomes

$$\begin{aligned} 0.100x_1 + 0.999x_2 &= 1.00 \\ 0.100x_1 + 1.00x_2 &= 1.01 \end{aligned}$$

The solution is now $x_2 = 10$, $x_1 = -89.9$. Notice that, despite the fact that there was only a small change in one term on the right-hand side, the resulting solution is not close to the solution of the original problem. Geometrically, this can be understood by observing that the solution is the intersection point of two nearly parallel lines; therefore, a small displacement of one line causes a major shift of the intersection point. Difficulties of this kind may arise in higher-dimensional systems of equations in real applications.

Carefully choosing pivots in computer programs can reduce the error caused by these sorts of problems. However, some matrices are **ill conditioned**; even with high-precision calculations, the solutions produced by computers with such matrices may be unreliable. In applications, the entries in the matrices may be experimentally determined, and small errors in the entries may result in large errors in calculated solutions, no matter how much precision is used in computation. To understand this problem better, you need to know something about sources of error in numerical computation—and more linear algebra. We shall not discuss it further in this book, but you should be aware of the difficulty if you use computers to solve systems of linear equations.

PROBLEMS 2.1

Practice Problems

For Problems A1–A4, solve the system using back-substitution. Write the general solution in standard form.

A1
$$\begin{aligned} x_1 - 3x_2 &= 5 \\ x_2 &= 4 \end{aligned}$$

A2
$$\begin{aligned} x_1 + 2x_2 - x_3 &= 7 \\ x_3 &= 6 \end{aligned}$$

A3
$$\begin{aligned} x_1 + 3x_2 - 2x_3 &= 4 \\ x_2 + 5x_3 &= 2 \\ x_3 &= 2 \end{aligned}$$

A4
$$\begin{aligned} x_1 - 2x_2 + x_3 + 4x_4 &= 7 \\ x_2 \qquad - x_4 &= -3 \\ x_3 + x_4 &= 2 \end{aligned}$$

For Problems A5–A8, determine whether the matrix is in REF. If not, explain why it is not.

A5 $A = \begin{bmatrix} 1 & 2 & 3 & 4 \\ 0 & 1 & -2 & -3 \\ 0 & 0 & 0 & 3 \end{bmatrix}$ **A6** $B = \begin{bmatrix} 0 & 1 & 2 & 3 \\ 0 & 0 & 1 & 1 \\ 0 & 0 & 0 & 0 \end{bmatrix}$

A7 $C = \begin{bmatrix} 1 & -1 & -2 & -3 \\ 0 & 1 & 2 & 0 \\ 0 & 1 & 0 & 3 \end{bmatrix}$ **A8** $D = \begin{bmatrix} 1 & 0 & 2 & 1 \\ 0 & 0 & 0 & 1 \\ 0 & 0 & 1 & 1 \end{bmatrix}$

For Problems A9–A14, row reduce the matrix to REF. Show your steps.

A9 $\begin{bmatrix} 4 & 1 & 1 \\ 1 & -3 & 2 \end{bmatrix}$ **A10** $\begin{bmatrix} 2 & -2 & 5 & 8 \\ 1 & -1 & 2 & 3 \\ -1 & 1 & 0 & 2 \end{bmatrix}$

A11 $\begin{bmatrix} 1 & -1 & -1 \\ 2 & -1 & -2 \\ 5 & 0 & 0 \\ 3 & 4 & 5 \end{bmatrix}$ **A12** $\begin{bmatrix} 2 & 0 & 2 & 0 \\ 1 & 2 & 3 & 4 \\ 1 & 4 & 9 & 16 \\ 3 & 6 & 13 & 20 \end{bmatrix}$

A13 $\begin{bmatrix} 0 & 1 & 2 & 1 \\ 1 & 2 & 1 & 1 \\ 3 & -1 & -4 & 1 \\ 2 & 1 & 3 & 6 \end{bmatrix}$ **A14** $\begin{bmatrix} 3 & 1 & 8 & 2 & 4 \\ 1 & 0 & 3 & 0 & 1 \\ 0 & 2 & -2 & 4 & 3 \\ -4 & 1 & 11 & 3 & 8 \end{bmatrix}$

For Problems A15–A20, the given matrix is an augmented matrix of a system of linear equations. Either show the system is inconsistent or write a vector equation for the solution set.

A15 $\left[\begin{array}{ccc|c} 1 & 2 & -1 & 2 \\ 0 & 1 & 3 & 4 \\ 0 & 0 & 0 & -5 \end{array}\right]$ **A16** $\left[\begin{array}{ccc|c} 1 & 0 & 0 & 2 \\ 0 & 0 & 1 & 3 \\ 0 & 0 & 0 & 0 \end{array}\right]$

A17 $\left[\begin{array}{cccc|c} 1 & 0 & 1 & 0 & 1 \\ 0 & 1 & 1 & 1 & 2 \\ 0 & 0 & 0 & 1 & 3 \end{array}\right]$ **A18** $\left[\begin{array}{cccc|c} 1 & 1 & -1 & 3 & 1 \\ 0 & 0 & 2 & 1 & 3 \\ 0 & 0 & 0 & 1 & -2 \end{array}\right]$

A19 $\left[\begin{array}{cccc|c} 1 & 0 & 1 & -1 & 0 \\ 0 & 1 & 0 & 0 & 0 \\ 0 & 0 & 0 & 0 & 0 \\ 0 & 0 & 0 & 0 & 0 \end{array}\right]$ **A20** $\left[\begin{array}{ccc|c} 1 & -1 & 0 & 2 \\ 0 & 0 & 1 & 2 \\ 0 & 0 & 0 & 1 \end{array}\right]$

For Problems A21–A27:

(a) Write the augmented matrix.
(b) Row reduce the augmented matrix to REF.
(c) Determine whether the system is consistent or inconsistent. If it is consistent, determine the number of parameters in the general solution.
(d) If the system is consistent, write its general solution in standard form.

A21
$$\begin{aligned} 3x_1 - 5x_2 &= 2 \\ x_1 + 2x_2 &= 4 \end{aligned}$$

A22
$$\begin{aligned} x_1 + 2x_2 + x_3 &= 5 \\ 2x_1 - 3x_2 + 2x_3 &= 6 \end{aligned}$$

A23
$$\begin{aligned} x_1 + 2x_2 - 3x_3 &= 8 \\ x_1 + 3x_2 - 5x_3 &= 11 \\ 2x_1 + 5x_2 - 8x_3 &= 19 \end{aligned}$$

A24
$$\begin{aligned} -3x_1 + 6x_2 + 16x_3 &= 36 \\ x_1 - 2x_2 - 5x_3 &= -11 \\ 2x_1 - 3x_2 - 8x_3 &= -17 \end{aligned}$$

A25
$$\begin{aligned} x_1 + 2x_2 - x_3 &= 4 \\ 2x_1 + 5x_2 + x_3 &= 10 \\ 4x_1 + 9x_2 - x_3 &= 19 \end{aligned}$$

A26
$$\begin{aligned} x_1 + 2x_2 - 3x_3 \qquad &= -5 \\ 2x_1 + 4x_2 - 6x_3 + x_4 &= -8 \\ 6x_1 + 13x_2 - 17x_3 + 4x_4 &= -21 \end{aligned}$$

A27
$$\begin{aligned} 2x_2 - 2x_3 \qquad + x_5 &= 2 \\ x_1 + 2x_2 - 3x_3 + x_4 + 4x_5 &= 1 \\ 2x_1 + 4x_2 - 5x_3 + 3x_4 + 8x_5 &= 3 \\ 2x_1 + 5x_2 - 7x_3 + 3x_4 + 10x_5 &= 5 \end{aligned}$$

For Problems A28–A31, find the parabola $y = a + bx + cx^2$ that passes through the given three points.

A28 $(1,3), (2,5), (4,15)$ **A29** $(0,2), (1,-1), (2,-10)$

A30 $(-2,9), (-1,2), (2,17)$ **A31** $(2,7), (-1,1), (0,-3)$

For Problems A32–A33, given that the matrix is an augmented matrix of a system of linear equations, determine the values of a, b, c, d for which the system is consistent. If it is consistent, determine whether it has a unique solution.

A32 $\left[\begin{array}{ccc|c} 2 & 4 & -3 & 6 \\ 0 & b & 7 & 2 \\ 0 & 0 & a & a \end{array}\right]$ **A33** $\left[\begin{array}{cccc|c} 1 & -1 & 4 & -2 & 5 \\ 0 & 1 & 2 & 3 & 4 \\ 0 & 0 & d & 5 & 7 \\ 0 & 0 & 0 & cd & c \end{array}\right]$

A34 A fruit seller has apples, bananas, and oranges. Altogether he has 1500 pieces of fruit. On average, each apple weighs 120 grams, each banana weighs 140 grams, and each orange weighs 160 grams. He can sell apples for 25 cents each, bananas for 20 cents each, and oranges for 30 cents each. If the fruit weighs 208 kilograms, and the total selling price is $380, how many of each kind of fruit does the fruit seller have?

A35 A student is taking courses in algebra, calculus, and physics at a university where grades are given in percentages. To determine her standing for a physics prize, a weighted average is calculated based on 50% of the student's physics grades, 30% of her calculus grade, and 20% of her algebra grade; the weighted average is 84. For an applied mathematics prize, a weighted average based on one-third of each of the three grades is calculated to be 83. For a pure mathematics prize, her average based on 50% of her calculus grade and 50% of her algebra grade is 82.5. What are her grades in the individual courses?

Homework Problems

For Problems B1–B8, solve the system using back-substitution. Write the general solution in standard form.

B1 $\begin{aligned} 2x_1 + 5x_2 &= 6 \\ x_2 &= 2 \end{aligned}$

B2 $\begin{aligned} x_1 - 2x_2 &= -6 \\ x_2 &= 7 \end{aligned}$

B3 $\begin{aligned} x_1 - 4x_2 + x_3 &= 1 \\ x_3 &= -2 \end{aligned}$

B4 $\begin{aligned} 3x_1 - 3x_2 + 2x_3 &= -6 \\ x_2 &= 1 \end{aligned}$

B5 $\begin{aligned} x_1 - 3x_2 + 3x_3 &= 1 \\ 2x_2 + x_3 &= 0 \\ x_3 &= 8 \end{aligned}$

B6 $\begin{aligned} 6x_1 + x_2 - x_3 &= 12 \\ 3x_2 - 6x_3 &= 9 \\ x_3 &= 2 \end{aligned}$

B7 $\begin{aligned} x_1 + \quad +7x_3 &= 4 \\ 4x_2 - 2x_3 &= 2 \\ x_3 &= -3 \end{aligned}$

B8 $\begin{aligned} x_1 + 5x_2 - 2x_3 + x_4 &= 3 \\ x_3 - 2x_4 &= 2 \end{aligned}$

For Problems B9–B12, determine whether the matrix is in REF. If not, explain why it is not.

B9 $A = \begin{bmatrix} 2 & 0 & -1 & 0 \\ 0 & 0 & 2 & 3 \\ 0 & 0 & 0 & 5 \end{bmatrix}$ **B10** $B = \begin{bmatrix} 1 & -4 & 12 & 3 \\ 0 & 0 & 0 & 0 \\ 0 & 0 & 2 & 1 \end{bmatrix}$

B11 $C = \begin{bmatrix} 3 & 3 & 1/4 & 0 \\ 0 & 0 & 4 & 2 \\ 5 & -2 & 3 & 4 \end{bmatrix}$ **B12** $D = \begin{bmatrix} 0 & 8 & 1 & -3 \\ 0 & 0 & 0 & 5 \\ 0 & 0 & 0 & 0 \end{bmatrix}$

For Problems B13–B24, row reduce the matrix to REF. Show your steps.

B13 $\begin{bmatrix} 3 & 6 \\ -2 & 1 \end{bmatrix}$ **B14** $\begin{bmatrix} 8 & -6 \\ -12 & 9 \end{bmatrix}$

B15 $\begin{bmatrix} 1 & 3 & 5 \\ 2 & 4 & 4 \\ -1 & 0 & 4 \end{bmatrix}$ **B16** $\begin{bmatrix} 0 & 2 & 3 \\ 5 & 6 & 9 \\ 1 & 1 & 2 \end{bmatrix}$

B17 $\begin{bmatrix} 1 & 2 & 4 \\ -2 & 0 & -3 \\ 5 & 6 & 7 \end{bmatrix}$ **B18** $\begin{bmatrix} 2 & 0 & 4 \\ 2 & 8 & -2 \\ -6 & -9 & -6 \end{bmatrix}$

B19 $\begin{bmatrix} 0 & -1 & 2 & 1 \\ 1 & 3 & -2 & 5 \\ 2 & 4 & 8 & 5 \end{bmatrix}$ **B20** $\begin{bmatrix} 1 & 3 & -2 & 1 \\ -3 & 1 & 3 & 1 \\ -3 & 11 & 0 & 5 \end{bmatrix}$

B21 $\begin{bmatrix} 3 & 1 & 1 \\ 2 & 2 & -6 \\ 1 & 2 & -8 \\ 5 & 2 & 0 \end{bmatrix}$ **B22** $\begin{bmatrix} 1 & 3 & 4 \\ 2 & 1 & 1 \\ -3 & 1 & 2 \\ 2 & 0 & 1 \end{bmatrix}$

B23 $\begin{bmatrix} 3 & 2 & 1 & 3 \\ -9 & -4 & -1 & -9 \\ 12 & 0 & 1 & 7 \\ 6 & 2 & 5 & 3 \end{bmatrix}$ **B24** $\begin{bmatrix} 0 & 1 & 3 & -5 \\ 3 & 2 & 7 & -6 \\ 1 & 1 & -2 & 7 \\ 0 & -5 & -5 & 5 \end{bmatrix}$

For Problems B25–B30, the given matrix is an augmented matrix of a system of linear equations. Either show the system is inconsistent or write a vector equation for the solution set.

B25 $\left[\begin{array}{ccc|c} 1 & 0 & 3 & 2 \\ 0 & 1 & -2 & -1 \\ 0 & 0 & 0 & 0 \end{array}\right]$ **B26** $\left[\begin{array}{ccc|c} 2 & -1 & 3 & 2 \\ 0 & 2 & 3 & 1 \\ 0 & 0 & 3 & 6 \end{array}\right]$

B27 $\left[\begin{array}{cccc|c} 1 & 2 & -1 & 3 & 0 \\ 0 & 1 & 2 & 4 & 0 \\ 0 & 0 & 0 & 1 & 0 \end{array}\right]$ **B28** $\left[\begin{array}{ccc|c} 4 & -1 & 3 & 5 \\ 0 & 0 & 2 & 7 \\ 0 & 0 & 0 & -3 \end{array}\right]$

B29 $\left[\begin{array}{cccc|c} 2 & 1 & 0 & -1 & 1 \\ 0 & 0 & 2 & -1 & 1 \\ 0 & 0 & 0 & 0 & 0 \end{array}\right]$ **B30** $\left[\begin{array}{cccc|c} 1 & 2 & 3 & 1 & 1 \\ 0 & 0 & 0 & 5 & 1 \\ 0 & 0 & 0 & 4 & 2 \end{array}\right]$

For Problems B31–B37:

(a) Write the augmented matrix.
(b) Row reduce the augmented matrix to REF.
(c) Determine whether the system is consistent or inconsistent. If it is consistent, determine the number of parameters in the general solution.
(d) If the system is consistent, write its general solution in standard form.

B31
$$\begin{aligned} 6x_1 + 3x_2 &= 9 \\ 4x_1 + 2x_2 &= 6 \end{aligned}$$

B32
$$\begin{aligned} 2x_2 + 4x_3 &= 4 \\ x_1 + 5x_2 + 4x_3 &= 8 \end{aligned}$$

B33
$$\begin{aligned} 5x_1 - 2x_2 - x_3 &= 0 \\ -4x_1 + x_2 - x_3 &= 7 \\ x_1 + x_2 + 4x_3 &= 9 \end{aligned}$$

B34
$$\begin{aligned} x_1 + 3x_2 + 3x_3 &= 2 \\ 4x_1 + 5x_2 + 12x_3 &= 1 \\ -2x_1 + 7x_2 + 7x_3 &= -4 \end{aligned}$$

B35
$$\begin{aligned} -x_1 - 2x_2 + x_3 &= 17 \\ x_1 + 2x_2 + 5x_3 &= 1 \\ x_1 + 2x_2 + 9x_3 &= 13 \end{aligned}$$

B36
$$\begin{aligned} x_1 + 4x_2 + 6x_3 + 9x_4 &= 1 \\ 2x_1 + 3x_2 + 7x_3 + 3x_4 &= 2 \\ -2x_1 + x_2 - 3x_3 + 9x_4 &= 1 \end{aligned}$$

B37
$$\begin{aligned} x_1 + 4x_2 + 6x_3 + 9x_4 &= 0 \\ 2x_1 + 3x_2 + 7x_3 + 3x_4 &= 5 \\ -2x_1 + x_2 - 3x_3 + 9x_4 &= -9 \end{aligned}$$

For Problems B38–B41, find the parabola $y = a + bx + cx^2$ that passes through the given three points.

B38 $(-1,8),(1,-2),(2,14)$ **B39** $(-1,9),(0,1),(1,-3)$

B40 $(-2,-5),(1,10),(2,17)$ **B41** $(1,-4),(2,3),(3,16)$

For Problems B42 and B43, given that the matrix is an augmented matrix of a system of linear equations, determine the values of a, b, c, d for which the system is consistent. If it is consistent, determine whether it has a unique solution.

B42 $\left[\begin{array}{ccc|c} 1 & 2 & 1 & 0 \\ 0 & a & 1 & b \\ 0 & 0 & b & 1 \end{array}\right]$ **B43** $\left[\begin{array}{ccc|c} 2 & 3 & 1 & 0 \\ 0 & c & 0 & d \\ 0 & d & 1 & c \end{array}\right]$

B44 A bookkeeper is trying to determine the prices that a manufacturer was charging. He examines old sales slips that show the number of various items shipped and the total price. He finds that 20 armchairs, 10 sofa beds, and 8 double beds cost \$15200; 15 armchairs, 12 sofa beds, and 10 double beds cost \$15700; and 12 armchairs, 20 sofa beds, and 10 double beds cost \$19600. Determine the cost for each item or explain why the sales slips must be in error.

B45 Steady flow through a network can be described by a system of linear equations. Such networks are used to model, for example, traffic along roads, water through pipes, electricity through a circuit, blood through arteries, or fluxes in a metabolic network. We assume the network in the diagram is in equilibrium; the flow into each node equals the flow out. For instance, at the top node, we get

$$\begin{aligned} \text{flow in} &= \text{flow out} \\ f_1 + f_2 &= 40 \end{aligned}$$

(a) Finish creating the corresponding system of linear equations by writing the equation for the other three nodes.
(b) Solve the system to determine how the flows are f_1, f_2, f_3, f_4 when in equilibrium.
(c) What is the physical interpretation of the negative flow?

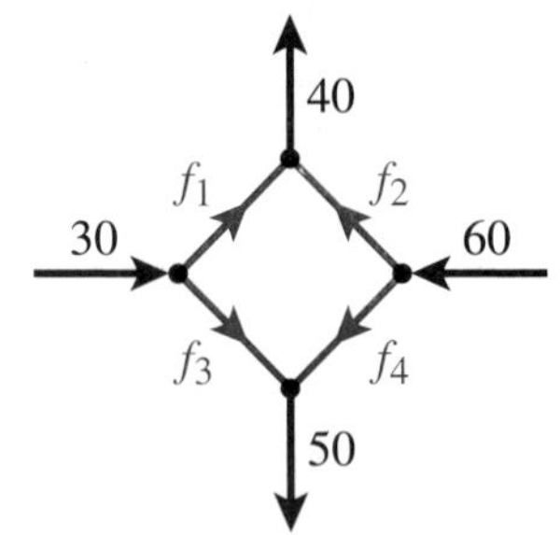

B46 Students at Linear University write a linear algebra examination. An average mark is computed for 100 students in business, an average is computed for 300 students in liberal arts, and an average is computed for 200 students in science. The average of these three averages is 85%. However, the overall average for the 600 students is 86%. Also, the average for the 300 students in business and science is 4 marks higher than the average for the students in liberal arts. Determine the average for each group of students by solving a system of linear equations.

B47 (Requires knowledge of forces and moments.)
A rod 10 m long is pivoted at its centre; it swings in the horizontal plane. Forces of magnitude F_1, F_2, F_3 are applied perpendicular to the rod in the directions indicated by the arrows in the diagram below; F_1 is applied to the left end of the rod, F_2 is applied at a point 2 m to the right of centre, and F_3 at a point 4 m to the right of centre. The total force on the pivot is zero, the moment about the centre is zero, and the sum of the magnitudes of forces is 80 newtons. Write a system of three equations for F_1, F_2, and F_3; write the corresponding augmented matrix; and use the standard procedure to find F_1, F_2, and F_3.

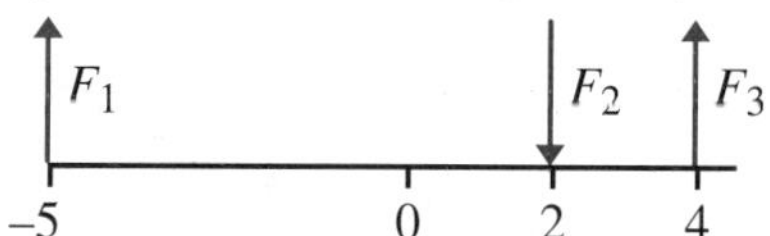

Conceptual Problems

C1 Consider the linear system in x, y, z, and w:

$$\begin{aligned} x + y \quad\quad + w &= b \\ 2x + 3y + z + 5w &= 6 \\ z + w &= 4 \\ 2y + 2z + aw &= 1 \end{aligned}$$

For what values of a and b is the system
(a) Inconsistent?
(b) Consistent with a unique solution?
(c) Consistent with infinitely many solutions?

C2 Recall that two planes $\vec{n}\cdot\vec{x} = c$ and $\vec{m}\cdot\vec{x} = d$ in $\mathbb{R}^3$ are parallel if and only if the normal vector $\vec{m}$ is a non-zero multiple of the normal vector $\vec{n}$. Row reduce a suitable augmented matrix to explain why two parallel planes must either coincide or else have no points in common.

C3 Consider the system

$$\begin{aligned} ax_1 + bx_2 &= 0 \\ cx_1 + dx_2 &= 0 \end{aligned}$$

(a) Explain why the system is consistent.
(b) Prove that if $ad - bc \neq 0$, then the system has a unique solution.
(c) Prove that if the system has a unique solution, then $ad - bc \neq 0$.

C4 Prove Theorem 2.1.1 by
(a) first proving that $\vec{x} = \vec{s} + c(\vec{s} - \vec{t})$ is a solution for each $c \in \mathbb{R}$,
(b) and then proving that if $c_1 \neq c_2$, then

$$\vec{s} + c_1(\vec{s} - \vec{t}) \neq \vec{s} + c_2(\vec{s} - \vec{t})$$

2.2 Reduced Row Echelon Form, Rank, and Homogeneous Systems

The standard basic procedure for determining the solution of a system of linear equations is elimination with back-substitution, as described in Section 2.1. In some situations and applications, however, it is advantageous to carry the elimination steps (elementary row operations) as far as possible to avoid the need for back-substitution.

To see what further elementary row operations might be worthwhile, recall that the Gaussian elimination procedure proceeds by selecting a pivot and using elementary row operations to create zeros beneath the pivot. The only further elimination steps that simplify the system are steps that create zeros above the pivot.

EXAMPLE 2.2.1

In Example 2.1.10 on page 90 we row reduced the augmented matrix for the original system to a row equivalent matrix in row echelon form. That is, we found that

$$\left[\begin{array}{ccc|c} 1 & 1 & -2 & 4 \\ 1 & 3 & -1 & 7 \\ 2 & 1 & -5 & 7 \end{array}\right] \sim \left[\begin{array}{ccc|c} 1 & 1 & -2 & 4 \\ 0 & 1 & 1 & 1 \\ 0 & 0 & -1 & 1 \end{array}\right]$$

Instead of using back-substitution to solve the system as we did in Example 2.1.10, we instead perform the following elementary row operations:

$$\left[\begin{array}{ccc|c} 1 & 1 & -2 & 4 \\ 0 & 1 & 1 & 1 \\ 0 & 0 & -1 & 1 \end{array}\right] \begin{array}{c} \\ \\ (-1)R_3 \end{array} \sim \left[\begin{array}{ccc|c} 1 & 1 & -2 & 4 \\ 0 & 1 & 1 & 1 \\ 0 & 0 & 1 & -1 \end{array}\right] \begin{array}{c} \\ R_2 - R_3 \\ \\ \end{array} \sim$$

$$\left[\begin{array}{ccc|c} 1 & 1 & -2 & 4 \\ 0 & 1 & 0 & 2 \\ 0 & 0 & 1 & -1 \end{array}\right] \begin{array}{c} R_1 + 2R_3 \\ \\ \\ \end{array} \sim \left[\begin{array}{ccc|c} 1 & 1 & 0 & 2 \\ 0 & 1 & 0 & 2 \\ 0 & 0 & 1 & -1 \end{array}\right] \begin{array}{c} R_1 - R_2 \\ \\ \\ \end{array} \sim \left[\begin{array}{ccc|c} 1 & 0 & 0 & 0 \\ 0 & 1 & 0 & 2 \\ 0 & 0 & 1 & -1 \end{array}\right]$$

This is the augmented matrix for the system $x_1 = 0$, $x_2 = 2$, and $x_3 = -1$, which gives us the solution we found in Example 2.1.10.

This system has been solved by **complete elimination**. The leading variable in the j-th equation has been eliminated from every other equation. This procedure is often called **Gauss-Jordan elimination** to distinguish it from Gaussian elimination with back-substitution. Observe that the elementary row operations used in Example 2.2.1 are exactly the same as the operations performed in the back-substitution in Example 2.1.10.

A matrix corresponding to a system on which Gauss-Jordan elimination has been carried out is in a special kind of row echelon form.

Definition
Reduced Row Echelon Form (RREF)

A matrix R is said to be in **reduced row echelon form (RREF)** if
(1) It is in row echelon form.
(2) All leading entries are 1, called a **leading one**.
(3) In a column with a leading one, all the other entries are zeros.
If A is row equivalent to a matrix R in RREF, then we say that R is the **reduced row echelon form** of A.

EXAMPLE 2.2.2

Determine which of the following matrices are in RREF.

(a) $\begin{bmatrix} 1 & 0 & 2 & 0 \\ 0 & 1 & 1 & 0 \\ 0 & 0 & 0 & 1 \end{bmatrix}$ (b) $\begin{bmatrix} 0 & 1 & 0 & 0 \\ 0 & 0 & 0 & 1 \\ 0 & 0 & 0 & 0 \end{bmatrix}$ (c) $\begin{bmatrix} 1 & 0 & 0 & 2 \\ 0 & 1 & 1 & 1 \\ 0 & 0 & 1 & 1 \end{bmatrix}$ (d) $\begin{bmatrix} 0 & 1 & 0 \\ 1 & 0 & 0 \end{bmatrix}$

Solution: The matrices in (a) and (b) are in RREF. The matrix in (c) is not in RREF as there is a non-zero entry above the leading one in the third column. The matrix (d) is not in RREF since the leading one in the second row is to the left of the leading one in the row above it.

EXERCISE 2.2.1

Determine which of the following matrices are in RREF.

(a) $\begin{bmatrix} 1 & 0 & -1 & 1 \\ 0 & 1 & -1 & -1 \\ 0 & 0 & 0 & 1 \end{bmatrix}$ (b) $\begin{bmatrix} 1 & 0 & -1 & 0 \\ 0 & 0 & 0 & 1 \\ 0 & 0 & 0 & 0 \end{bmatrix}$ (c) $\begin{bmatrix} 0 & 2 & 0 & 3 \\ 0 & 0 & 2 & 0 \\ 0 & 0 & 0 & 0 \end{bmatrix}$ (d) $\begin{bmatrix} 1 & 0 & 7 & 0 \\ 0 & 0 & 0 & 0 \\ 0 & 1 & 0 & 1 \end{bmatrix}$

As in the case of row echelon form, it is easy to see that every matrix is row equivalent to a matrix in reduced row echelon form via Gauss-Jordan elimination. However, in this case we get a stronger result.

Theorem 2.2.1 For any given matrix A there is a unique matrix in reduced row echelon form that is row equivalent to A.

You are asked to prove that there is only one matrix in reduced row echelon form that is row equivalent to A in Problem F2 in the Chapter 4 Further Problems.

EXAMPLE 2.2.3

Obtain the matrix in reduced row echelon form that is row equivalent to the matrix

$$A = \begin{bmatrix} 1 & 1 & 2 & -2 & 2 \\ 3 & 3 & 5 & 0 & 2 \end{bmatrix}$$

Solution: We begin by using Gaussian elimination to put the matrix into REF

$$\begin{bmatrix} 1 & 1 & 2 & -2 & 2 \\ 3 & 3 & 5 & 0 & 2 \end{bmatrix} \begin{matrix} \\ R_2 - 3R_1 \end{matrix} \sim \begin{bmatrix} 1 & 1 & 2 & -2 & 2 \\ 0 & 0 & -1 & 6 & -4 \end{bmatrix}$$

We also need to put zeros above pivots and to ensure that each pivot is a 1.

$$\begin{bmatrix} 1 & 1 & 2 & -2 & 2 \\ 0 & 0 & -1 & 6 & -4 \end{bmatrix} \begin{matrix} R_1 + 2R_2 \\ \\ \end{matrix} \sim \begin{bmatrix} 1 & 1 & 0 & 10 & -6 \\ 0 & 0 & -1 & 6 & -4 \end{bmatrix} \begin{matrix} \\ (-1)R_2 \end{matrix} \sim$$
$$\begin{bmatrix} 1 & 1 & 0 & 10 & -6 \\ 0 & 0 & 1 & -6 & 4 \end{bmatrix}$$

This final matrix is in reduced row echelon form.

When row reducing to reduced row echelon form by hand, it seems more natural not to obtain a row echelon form first. Instead, you might first turn any pivot into a leading one and then obtain zeros below and above it moving on to the next leading one. However, for programming a computer to row reduce a matrix, this is a poor strategy because it requires more multiplications and additions than the previous strategy. See Problem F2 at the end of the chapter.

EXAMPLE 2.2.4

Solve the following system of equations by row reducing the augmented matrix to RREF.

$$\begin{aligned} x_1 + x_2 &= -7 \\ 2x_1 + 4x_2 + x_3 &= -16 \\ x_1 + 2x_2 + x_3 &= 9 \end{aligned}$$

Solution: Our first pivot is already a leading one, so we place zeros beneath it.

$$\left[\begin{array}{ccc|c} 1 & 1 & 0 & -7 \\ 2 & 4 & 1 & -16 \\ 1 & 2 & 1 & 9 \end{array}\right] \begin{array}{l} \\ R_2 - 2R_1 \\ R_3 - R_1 \end{array} \sim \left[\begin{array}{ccc|c} 1 & 1 & 0 & -7 \\ 0 & 2 & 1 & -2 \\ 0 & 1 & 1 & 16 \end{array}\right]$$

To make our next pivot a leading one, rather than introducing fractions, we use $R_2 - R_3$.

$$\left[\begin{array}{ccc|c} 1 & 1 & 0 & -7 \\ 0 & 2 & 1 & -2 \\ 0 & 1 & 1 & 16 \end{array}\right] R_2 - R_3 \sim \left[\begin{array}{ccc|c} 1 & 1 & 0 & -7 \\ 0 & 1 & 0 & -18 \\ 0 & 1 & 1 & 16 \end{array}\right]$$

We now need to get zeros above and below this new leading one.

$$\left[\begin{array}{ccc|c} 1 & 1 & 0 & -7 \\ 0 & 1 & 0 & -18 \\ 0 & 1 & 1 & 16 \end{array}\right] \begin{array}{l} R_1 - R_2 \\ \\ R_3 - R_2 \end{array} \sim \left[\begin{array}{ccc|c} 1 & 0 & 0 & 11 \\ 0 & 1 & 0 & -18 \\ 0 & 0 & 1 & 34 \end{array}\right]$$

The matrix is now in reduced row echelon form. The reduced row echelon form corresponds to the system $x_1 = 11, x_2 = -18, x_3 = 34$. Hence, the solution is $\vec{x} = \begin{bmatrix} 11 \\ -18 \\ 34 \end{bmatrix}$.

EXERCISE 2.2.2

Row reduce $A = \begin{bmatrix} 0 & -2 & 2 & 2 \\ 2 & 1 & 1 & 1 \\ 1 & 0 & 1 & 1 \end{bmatrix}$ into reduced row echelon form.

Remark

In general, reducing an augmented matrix to reduced row echelon form to solve a system is not more efficient than the method used in Section 2.1. As previously mentioned, both methods are essentially equivalent for solving small systems by hand.

Rank of a Matrix

Theorem 2.1.3 shows that the number of pivots in a row echelon form of the coefficient matrix of a system of linear equations determines whether the system is consistent or inconsistent. It also determines how many solutions (one or infinitely many) the system has if it is consistent. Thus, we make the following definition.

Definition
Rank

The **rank** of a matrix A is the number of leading ones in its reduced row echelon form and is denoted by $\text{rank}(A)$.

The rank of A is also equal to the number of leading entries in any row echelon form of A. However, since the row echelon form is not unique, it is more tiresome to give clear arguments in terms of row echelon form. In Section 3.4 we shall see a more conceptual way of describing rank.

EXAMPLE 2.2.5

In Example 2.2.3 we saw that the RREF of $A = \begin{bmatrix} 1 & 1 & 2 & -2 & 2 \\ 3 & 3 & 5 & 0 & 2 \end{bmatrix}$ is $\begin{bmatrix} 1 & 1 & 0 & 10 & -6 \\ 0 & 0 & 1 & -6 & 4 \end{bmatrix}$. Thus, $\text{rank}(A) = 2$.

EXAMPLE 2.2.6

In Example 2.2.4 we saw that the RREF of $B = \begin{bmatrix} 1 & 1 & 0 & -7 \\ 2 & 4 & 1 & -16 \\ 1 & 2 & 1 & 9 \end{bmatrix}$ is $\begin{bmatrix} 1 & 0 & 0 & 11 \\ 0 & 1 & 0 & -18 \\ 0 & 0 & 1 & 34 \end{bmatrix}$. Thus, $\text{rank}(B) = 3$.

EXAMPLE 2.2.7

The RREF of $C = \begin{bmatrix} 1 & 1 & 1 \\ 1 & 1 & 1 \end{bmatrix}$ is $\begin{bmatrix} 1 & 1 & 1 \\ 0 & 0 & 0 \end{bmatrix}$. Hence, $\text{rank}(C) = 1$.

EXERCISE 2.2.3

Determine the rank of each of the following matrices:

(a) $A = \begin{bmatrix} 1 & 0 & 1 & 0 \\ 0 & 0 & 1 & 1 \\ 0 & 0 & 0 & 0 \end{bmatrix}$

(b) $B = \begin{bmatrix} 1 & 1 & 0 & 1 \\ 0 & 0 & 0 & 0 \\ 0 & 0 & 0 & 3 \\ 0 & 0 & 0 & 2 \end{bmatrix}$

Theorem 2.2.2 System-Rank Theorem

Let $\left[A \mid \vec{b}\right]$ be a system of m linear equations in n variables.

(1) The system is consistent if and only if the rank of the coefficient matrix A is equal to the rank of the augmented matrix $\left[A \mid \vec{b}\right]$.

(2) If the system $\left[A \mid \vec{b}\right]$ is consistent, then the number of parameters in the general solution is the number of variables minus the rank of the matrix A:

$$\#\text{ of parameters} = n - \text{rank}(A)$$

(3) $\text{rank}(A) = m$ if and only if the system $\left[A \mid \vec{b}\right]$ is consistent for every $\vec{b} \in \mathbb{R}^m$.

Proof: (1): The rank of A is less than the rank of the augmented matrix if and only if there is a row in the RREF of the augmented matrix of the form $\left[\begin{array}{ccc|c} 0 & \cdots & 0 & 1 \end{array}\right]$. This is true if and only if the system is inconsistent.

(2): If the system is consistent, then the free variables are the variables that are not leading variables of any equation in a row echelon form of the matrix. Thus, by definition, there are $n - \text{rank}(A)$ free variables and hence $n - \text{rank}(A)$ parameters in the general solution.

The proof of (3) is left as Problem A40. ■

For property (1) of the System-Rank Theorem, it is important to realize that we consider the rank of the entire matrix, even the augmented part.

EXAMPLE 2.2.8 Consider the system of linear equations

$$\begin{aligned} x_2 + x_3 &= 2 \\ x_1 + x_2 \quad\;\; &= 1 \\ x_1 + 2x_2 + x_3 &= -2 \end{aligned}$$

from Example 2.1.15. If we row reduce the augmented matrix to RREF, we get

$$\left[\begin{array}{ccc|c} 1 & 0 & -1 & 0 \\ 0 & 1 & 1 & 0 \\ 0 & 0 & 0 & 1 \end{array}\right]$$

Hence, the rank of the augmented matrix is 3. If we just row reduce the coefficient matrix to RREF, then we get

$$\begin{bmatrix} 1 & 0 & -1 \\ 0 & 1 & 1 \\ 0 & 0 & 0 \end{bmatrix}$$

Thus, the rank of the coefficient matrix is 2. So, by the System-Rank Theorem (1), the system in inconsistent.

Homogeneous Linear Equations

Frequently, systems of linear equations appear where all of the terms on the right-hand side are zero.

Definition
Homogeneous

A linear equation is **homogeneous** if the right-hand side is zero. A system of linear equations is **homogeneous** if all of the equations of the system are homogeneous.

Since a homogeneous system is a special case of the systems already discussed, no new tools or techniques are needed to solve them. However, we normally work only with the coefficient matrix of a homogeneous system since the last column of the augmented matrix consists entirely of zeros.

EXAMPLE 2.2.9

Find the general solution of the homogeneous system

$$\begin{aligned} 2x_1 + x_2 \qquad &= 0 \\ x_1 + x_2 - x_3 &= 0 \\ -x_2 + 2x_3 &= 0 \end{aligned}$$

Solution: We row reduce the coefficient matrix of the system to RREF:

$$\begin{bmatrix} 2 & 1 & 0 \\ 1 & 1 & -1 \\ 0 & -1 & 2 \end{bmatrix} \begin{matrix} R_1 - R_2 \\ \\ \\ \end{matrix} \sim \begin{bmatrix} 1 & 0 & 1 \\ 1 & 1 & -1 \\ 0 & -1 & 2 \end{bmatrix} \begin{matrix} \\ R_2 - R_1 \\ \\ \end{matrix} \sim$$

$$\begin{bmatrix} 1 & 0 & 1 \\ 0 & 1 & -2 \\ 0 & -1 & 2 \end{bmatrix} \begin{matrix} \\ \\ R_3 + R_2 \end{matrix} \sim \begin{bmatrix} 1 & 0 & 1 \\ 0 & 1 & -2 \\ 0 & 0 & 0 \end{bmatrix}$$

This corresponds to the homogeneous system

$$\begin{aligned} x_1 \qquad + x_3 &= 0 \\ x_2 - 2x_3 &= 0 \end{aligned}$$

Hence, x_3 is a free variable, so we let $x_3 = t \in \mathbb{R}$. Then $x_1 = -x_3 = -t$, $x_2 = 2x_3 = 2t$, and the general solution is

$$\begin{bmatrix} x_1 \\ x_2 \\ x_3 \end{bmatrix} = \begin{bmatrix} -t \\ 2t \\ t \end{bmatrix} = t \begin{bmatrix} -1 \\ 2 \\ 1 \end{bmatrix}, \quad t \in \mathbb{R}$$

Observe that every homogeneous system is consistent as the zero vector $\vec{0}$ will certainly be a solution. We call $\vec{0}$ the **trivial solution**. Thus, as we will see frequently throughout the text, when dealing with homogeneous systems, we are often mostly interested in how many parameters are in the general solution. Of course, for this we can apply the System-Rank Theorem.

EXAMPLE 2.2.10

Determine the number of parameters in the general solution of the homogeneous system

$$\begin{aligned} x_1 + 2x_2 + 2x_3 + x_4 &= 0 \\ 3x_1 + 7x_2 + 7x_3 + 3x_4 &= 0 \\ 2x_1 + 5x_2 + 5x_3 + 2x_4 &= 0 \end{aligned}$$

Solution: We row reduce the coefficient matrix:

$$\begin{bmatrix} 1 & 2 & 2 & 1 \\ 3 & 7 & 7 & 3 \\ 2 & 5 & 5 & 2 \end{bmatrix} \begin{matrix} \\ R_2 - 3R_1 \\ R_3 - 2R_1 \end{matrix} \sim \begin{bmatrix} 1 & 2 & 2 & 1 \\ 0 & 1 & 1 & 0 \\ 0 & 1 & 1 & 0 \end{bmatrix} \begin{matrix} R_1 - 2R_2 \\ \\ R_3 - R_2 \end{matrix} \sim \begin{bmatrix} 1 & 0 & 0 & 1 \\ 0 & 1 & 1 & 0 \\ 0 & 0 & 0 & 0 \end{bmatrix}$$

The rank of the coefficient matrix is 2 and the number of variables is 4. Thus, by the System-Rank Theorem (2), there are $4 - 2 = 2$ parameters in the general solution.

EXERCISE 2.2.4

Write the general solution of the system in Example 2.2.10.

Example 2.2.9 can be interpreted geometrically. It shows that the intersection of the three planes is the line that passes through the origin with vector equation

$$\vec{x} = t\begin{bmatrix} -1 \\ 2 \\ 1 \end{bmatrix}, \quad t \in \mathbb{R}$$

Thus, from our work in Chapter 1, we know that the solution set is a subspace of $\mathbb{R}^3$. Similarly, Example 2.2.10 shows that the three hyperplanes in $\mathbb{R}^5$ intersect in a plane that passes through the origin in $\mathbb{R}^5$. Hence, the solution set of that system is a subspace of $\mathbb{R}^5$. Of course, we can prove this idea in general.

Theorem 2.2.3

The solution set of a homogeneous system of m linear equations in n variables is a subspace of $\mathbb{R}^n$.

You are asked to prove Theorem 2.2.3 in Problem C2.

This allows us to make the following definition.

Definition
Solution Space

The solution set of a homogeneous system is called the **solution space** of the system.

PROBLEMS 2.2

Practice Problems

For Problems A1–A7, determine whether the matrix is in RREF.

A1 $\begin{bmatrix} 2 & 0 & -3 \\ 0 & 2 & 1 \end{bmatrix}$ **A2** $\begin{bmatrix} 1 & 2 & 1 \\ 0 & 0 & 3 \\ 0 & 1 & 5 \end{bmatrix}$ **A3** $\begin{bmatrix} 0 & 1 & 0 \\ 0 & 0 & 1 \\ 0 & 0 & 0 \end{bmatrix}$

A4 $\begin{bmatrix} 1 & -2 & 0 & 0 \\ 0 & 0 & 1 & 0 \\ 0 & 0 & 0 & 1 \end{bmatrix}$ **A5** $\begin{bmatrix} 1 & 0 & -3 & 4 \\ 0 & 1 & 1 & 2 \\ 0 & 0 & 0 & 0 \end{bmatrix}$

A6 $\begin{bmatrix} 0 & 0 & 1 & 2 \\ 1 & -2 & 1 & 3 \\ 0 & 1 & 0 & 0 \end{bmatrix}$ **A7** $\begin{bmatrix} 1 & 0 & 1 & 0 \\ 0 & 1 & 3 & 3 \\ 0 & 0 & 1 & 2 \end{bmatrix}$

For Problems A8–A16, determine the RREF and the rank of the matrix.

A8 $\begin{bmatrix} 2 & 1 \\ 1 & -1 \\ 3 & 2 \end{bmatrix}$ **A9** $\begin{bmatrix} 2 & 0 & 1 \\ 0 & 1 & 2 \\ 1 & 1 & 1 \end{bmatrix}$ **A10** $\begin{bmatrix} 1 & 2 & 3 \\ 2 & 1 & 2 \\ 2 & 3 & 4 \end{bmatrix}$

A11 $\begin{bmatrix} 1 & 0 & -2 \\ 2 & 1 & 2 \\ 2 & 3 & 4 \end{bmatrix}$ **A12** $\begin{bmatrix} 1 & 2 & 1 \\ 1 & 2 & 3 \\ -1 & -2 & 3 \\ 2 & 4 & 3 \end{bmatrix}$

A13 $\begin{bmatrix} 1 & 1 & 1 & 1 \\ 1 & 1 & 1 & 0 \\ 1 & 1 & 0 & 0 \end{bmatrix}$ **A14** $\begin{bmatrix} 2 & -1 & 2 & 8 \\ 1 & -1 & 0 & 2 \\ 3 & -2 & 3 & 13 \end{bmatrix}$

A15 $\begin{bmatrix} 1 & 1 & 0 & 1 \\ 0 & 1 & 1 & 2 \\ 2 & 3 & 1 & 4 \\ 1 & 2 & 3 & 4 \end{bmatrix}$ **A16** $\begin{bmatrix} 0 & 1 & 0 & 2 & 5 \\ 3 & 1 & 8 & 5 & 3 \\ 1 & 0 & 3 & 2 & 1 \\ 2 & 1 & 6 & 7 & 1 \end{bmatrix}$

For Problems A17–A22, the given matrix is the coefficient matrix of a homogeneous system already in RREF. Determine the number of parameters in the general solution and write out the general solution in standard form.

A17 $\begin{bmatrix} 1 & 0 & 2 & 0 \\ 0 & 1 & -1 & 0 \\ 0 & 0 & 0 & 1 \end{bmatrix}$ **A18** $\begin{bmatrix} 0 & 1 & 2 & 0 \\ 0 & 0 & 0 & 1 \\ 0 & 0 & 0 & 0 \end{bmatrix}$

A19 $\begin{bmatrix} 1 & -3 & 2 & 0 \\ 0 & 0 & 0 & 1 \\ 0 & 0 & 0 & 0 \\ 0 & 0 & 0 & 0 \end{bmatrix}$ **A20** $\begin{bmatrix} 1 & 0 & 2 & 0 & 0 \\ 0 & 1 & -1 & 0 & -2 \\ 0 & 0 & 0 & 1 & 1 \end{bmatrix}$

A21 $\begin{bmatrix} 1 & 0 & 0 & 4 & 0 \\ 0 & 0 & 1 & -5 & 0 \\ 0 & 0 & 0 & 0 & 1 \end{bmatrix}$ **A22** $\begin{bmatrix} 1 & 0 & 0 & 0 & 0 \\ 0 & 1 & 1 & 0 & 0 \\ 0 & 0 & 0 & 1 & 0 \\ 0 & 0 & 0 & 0 & 1 \end{bmatrix}$

For Problems A23–A29, solve the system of linear equations by row reducing the corresponding augmented matrix to RREF. Compare your steps with your solutions from Section 2.1 Problems A21–A27.

A23 $\begin{aligned} 3x_1 - 5x_2 &= 2 \\ x_1 + 2x_2 &= 4 \end{aligned}$ **A24** $\begin{aligned} x_1 + 2x_2 + x_3 &= 5 \\ 2x_1 - 3x_2 + 2x_3 &= 6 \end{aligned}$

A25 $\begin{aligned} x_1 + 2x_2 - 3x_3 &= 8 \\ x_1 + 3x_2 - 5x_3 &= 11 \\ 2x_1 + 5x_2 - 8x_3 &= 19 \end{aligned}$ **A26** $\begin{aligned} -3x_1 + 6x_2 + 16x_3 &= 36 \\ x_1 - 2x_2 - 5x_3 &= -11 \\ 2x_1 - 3x_2 - 8x_3 &= -17 \end{aligned}$

A27 $\begin{aligned} x_1 + 2x_2 - x_3 &= 4 \\ 2x_1 + 5x_2 + x_3 &= 10 \\ 4x_1 + 9x_2 - x_3 &= 19 \end{aligned}$

A28 $\begin{aligned} x_1 + 2x_2 - 3x_3 &= -5 \\ 2x_1 + 4x_2 - 6x_3 + x_4 &= -8 \\ 6x_1 + 13x_2 - 17x_3 + 4x_4 &= -21 \end{aligned}$

A29 $\begin{aligned} 2x_2 - 2x_3 + x_5 &= 2 \\ x_1 + 2x_2 - 3x_3 + x_4 + 4x_5 &= 1 \\ 2x_1 + 4x_2 - 5x_3 + 3x_4 + 8x_5 &= 3 \\ 2x_1 + 5x_2 - 7x_3 + 3x_4 + 10x_5 &= 5 \end{aligned}$

For Problems A30–A33, write the coefficient matrix of the system of linear equations. Determine the rank of the coefficient matrix and write out the general solution in standard form.

A30 $\begin{aligned} 2x_2 - 5x_3 &= 0 \\ x_1 + 2x_2 + 3x_3 &= 0 \\ x_1 + 4x_2 - 3x_3 &= 0 \end{aligned}$ **A31** $\begin{aligned} 3x_1 + x_2 - 9x_3 &= 0 \\ x_1 + x_2 - 5x_3 &= 0 \\ 2x_1 + x_2 - 7x_3 &= 0 \end{aligned}$

A32 $\begin{aligned} x_1 - x_2 + 2x_3 - 3x_4 &= 0 \\ 3x_1 - 3x_2 + 8x_3 - 5x_4 &= 0 \\ 2x_1 - 2x_2 + 5x_3 - 4x_4 &= 0 \\ 3x_1 - 3x_2 + 7x_3 - 7x_4 &= 0 \end{aligned}$

A33 $\begin{aligned} x_2 + 2x_3 + 2x_4 &= 0 \\ x_1 + 2x_2 + 5x_3 + 3x_4 - x_5 &= 0 \\ 2x_1 + x_2 + 5x_3 + x_4 - 3x_5 &= 0 \\ x_1 + x_2 + 4x_3 + 2x_4 - 2x_5 &= 0 \end{aligned}$

For Problems A34–A39, solve the system $\left[A \mid \vec{b}\right]$ by row reducing the augmented matrix to RREF. Then, without any further operations, find the general solution to the homogeneous $\left[A \mid \vec{0}\right]$.

A34 $A = \begin{bmatrix} 2 & -1 & 4 \\ 1 & 3 & 0 \\ 1 & 1 & 2 \end{bmatrix}, \vec{b} = \begin{bmatrix} 1 \\ 0 \\ 2 \end{bmatrix}$

A35 $A = \begin{bmatrix} 1 & 7 & 5 \\ 1 & 0 & 5 \\ -1 & 2 & -5 \end{bmatrix}, \vec{b} = \begin{bmatrix} 5 \\ -2 \\ 4 \end{bmatrix}$

A36 $A = \begin{bmatrix} 0 & -1 & 5 & -2 \\ -1 & -1 & -4 & -1 \end{bmatrix}, \vec{b} = \begin{bmatrix} -1 \\ 4 \end{bmatrix}$

A37 $A = \begin{bmatrix} 1 & 0 & -1 & -1 \\ 4 & 3 & 2 & -4 \\ -1 & -4 & -3 & 5 \end{bmatrix}, \vec{b} = \begin{bmatrix} 3 \\ 3 \\ 5 \end{bmatrix}$

A38 $A = \begin{bmatrix} 1 & -1 & 4 & -1 \\ -1 & -2 & 5 & -2 \\ -4 & -1 & 2 & 2 \\ 5 & 4 & 1 & 8 \end{bmatrix}, \vec{b} = \begin{bmatrix} 4 \\ 5 \\ -4 \\ 5 \end{bmatrix}$

A39 $A = \begin{bmatrix} 1 & 1 & 3 & 1 & 4 \\ 4 & 4 & 6 & -8 & 4 \\ 1 & 1 & 4 & -2 & 1 \\ 3 & 3 & 2 & -4 & 5 \end{bmatrix}, \vec{b} = \begin{bmatrix} 2 \\ -4 \\ -6 \\ 6 \end{bmatrix}$

A40 In this problem, we will look at how to prove System-Rank Theorem (3). For each direction of the 'if and only if', we first look at an example to help us figure out how to do the general proof.

(a) Row reduce the coefficient matrix of the system

$$\begin{aligned} x_1 + x_2 + x_3 \quad\quad &= b_1 \\ x_1 + 2x_2 + x_3 - 2x_4 &= b_2 \\ x_1 + 4x_2 + 2x_3 - 7x_4 &= b_3 \end{aligned}$$

and explain how this proves that the system is consistent for all $\vec{b} \in \mathbb{R}^3$.

(b) Prove if rank$(A) = m$, then $\left[A \mid \vec{b}\right]$ is consistent for every $\vec{b} \in \mathbb{R}^m$.

(c) Find a vector $\vec{b} \in \mathbb{R}^3$ such that the following system is inconsistent. To find such a vector $\vec{b}$, think about how to work backwards from the RREF of the coefficient matrix A.

$$\begin{aligned} x_1 + x_2 &= b_1 \\ 2x_1 + 2x_2 &= b_2 \\ 2x_1 + 3x_2 &= b_3 \end{aligned}$$

(d) Prove if rank$(A) < m$, then there exists $\vec{b} \in \mathbb{R}^m$ such that $\left[A \mid \vec{b}\right]$ is inconsistent.

Homework Problems

For Problems B1–B9, determine whether the matrix is in RREF.

B1 $\begin{bmatrix} 1 & -1 & 2 \\ 0 & 0 & 1 \\ 0 & 0 & 0 \end{bmatrix}$ **B2** $\begin{bmatrix} 1 & 0 & -1 \\ 0 & 1 & 0 \\ 0 & 0 & 0 \end{bmatrix}$ **B3** $\begin{bmatrix} 0 & 0 & 1 \\ 0 & 1 & 0 \\ 1 & 0 & 0 \end{bmatrix}$

B4 $\begin{bmatrix} 1 & 0 & -2 & 1 \\ 0 & 0 & 0 & 0 \\ 0 & 1 & 0 & 0 \end{bmatrix}$ **B5** $\begin{bmatrix} 1 & 0 & -3 & 0 \\ 0 & 2 & 0 & 0 \\ 0 & 0 & 0 & 3 \end{bmatrix}$

B6 $\begin{bmatrix} 1 & 0 & 1 & 0 \\ 0 & 1 & 0 & 0 \\ 0 & 0 & 0 & 1 \end{bmatrix}$ **B7** $\begin{bmatrix} 1 & 1 & 0 & 0 \\ 0 & 1 & 1 & 0 \\ 0 & 0 & 1 & 1 \end{bmatrix}$

B8 $\begin{bmatrix} 1 & 0 & 1 & 2 \\ 0 & 1 & 0 & -3 \\ 0 & 0 & 1 & -1 \\ 0 & 0 & 0 & 0 \end{bmatrix}$ **B9** $\begin{bmatrix} 1 & 0 & -3 & 0 \\ 0 & 1 & 6 & 0 \\ 0 & 0 & 0 & 1 \\ 0 & 0 & 0 & 0 \end{bmatrix}$

For Problems B10–B18, determine the RREF and the rank of matrix.

B10 $\begin{bmatrix} 1 & -2 \\ -3 & 6 \\ 4 & -8 \end{bmatrix}$ **B11** $\begin{bmatrix} 1 & 1 & 1 \\ -1 & -2 & 3 \\ 0 & -2 & 8 \end{bmatrix}$ **B12** $\begin{bmatrix} -1 & 1 & 2 \\ 4 & -1 & -3 \\ -3 & -3 & 1 \end{bmatrix}$

B13 $\begin{bmatrix} 2 & 1 & 3 & -1 \\ -4 & 3 & -11 & 3 \\ 6 & 8 & 4 & -3 \end{bmatrix}$ **B14** $\begin{bmatrix} 6 & 9 & 1 & -1 \\ 2 & 3 & -1 & 1 \\ 4 & 6 & 2 & -2 \end{bmatrix}$

B15 $\begin{bmatrix} 2 & 1 & 1 \\ 1 & 2 & 1 \\ 1 & 1 & 2 \end{bmatrix}$ **B16** $\begin{bmatrix} 0 & 1 & -2 & 2 \\ 1 & 2 & 1 & 3 \\ 3 & 1 & 6 & 6 \end{bmatrix}$

B17 $\begin{bmatrix} 1 & -2 & 1 & 1 \\ -2 & 5 & 0 & -5 \\ 6 & -7 & 6 & -9 \\ 3 & -4 & 2 & 2 \end{bmatrix}$ **B18** $\begin{bmatrix} 1 & -2 & 1 & 5 & 2 \\ 0 & 0 & 3 & 6 & 1 \\ -2 & 4 & 5 & 4 & 5 \\ -2 & 4 & 7 & 8 & 1 \end{bmatrix}$

For Problems B19–B24, the given matrix is the coefficient matrix of a homogeneous system already in RREF. Determine the number of parameters in the general solution and write out the general solution in standard form.

B19 $\begin{bmatrix} 1 & 0 & 0 & 5 \\ 0 & 1 & 0 & 3 \\ 0 & 0 & 1 & 0 \end{bmatrix}$ **B20** $\begin{bmatrix} 1 & 0 & 0 & 2 \\ 0 & 0 & 1 & 3 \\ 0 & 0 & 0 & 0 \end{bmatrix}$

B21 $\begin{bmatrix} 1 & 0 & -1 & 4 \\ 0 & 1 & -3 & 5 \\ 0 & 0 & 0 & 0 \end{bmatrix}$ **B22** $\begin{bmatrix} 1 & 0 & -5 & 0 \\ 0 & 1 & -1 & 0 \\ 0 & 0 & 0 & 1 \end{bmatrix}$

B23 $\begin{bmatrix} 0 & 0 & 1 & 0 & 0 \\ 0 & 0 & 0 & 1 & 0 \\ 0 & 0 & 0 & 0 & 1 \end{bmatrix}$ **B24** $\begin{bmatrix} 1 & 0 & 3 & 0 & 8 \\ 0 & 1 & 3 & 0 & -2 \\ 0 & 0 & 0 & 1 & 0 \\ 0 & 0 & 0 & 0 & 0 \end{bmatrix}$

For Problems B25–B31, solve the system of linear equations by row reducing the corresponding augmented matrix to RREF.

B25 $\begin{aligned} 3x_1 + 5x_2 &= 4 \\ 2x_1 + 5x_2 &= -4 \end{aligned}$ **B26** $\begin{aligned} 2x_1 + 3x_2 + 2x_3 &= 1 \\ x_1 + 2x_2 + 6x_3 &= 2 \end{aligned}$

B27 $\begin{aligned} x_1 + 2x_2 + 4x_3 &= 2 \\ x_1 + 3x_2 + 5x_3 &= 1 \\ x_1 + x_2 + 4x_3 &= 6 \end{aligned}$ **B28** $\begin{aligned} 2x_1 + 5x_2 + 5x_3 &= 0 \\ 4x_1 + 7x_2 + x_3 &= 3 \\ -4x_1 - 6x_2 + 2x_3 &= 5 \end{aligned}$

B29 $\begin{aligned} x_2 + 5x_3 - 4x_4 &= -2 \\ x_1 + 2x_2 + 7x_3 - 3x_4 &= -1 \\ 5x_1 + 4x_2 + 5x_3 + 9x_4 &= 7 \end{aligned}$

B30 $\begin{aligned} 3x_1 + 4x_2 + 2x_3 + 4x_4 &= 1 \\ 6x_1 + 8x_2 + 3x_3 + 3x_4 &= 1 \\ -6x_1 - 8x_2 - x_3 + 7x_4 &= 1 \end{aligned}$

B31 $\begin{aligned} 3x_1 + 4x_2 - 5x_3 + 8x_4 + x_5 &= 7 \\ 5x_1 - 2x_2 + 9x_3 + 6x_4 + x_5 &= 5 \\ 2x_1 + 4x_2 - 6x_3 + 7x_4 + x_5 &= 6 \\ 5x_1 + x_2 + 3x_3 + 5x_4 - x_5 &= 1 \end{aligned}$

For Problems B32–B35, write the coefficient matrix of the system of linear equations. Determine the rank of the coefficient matrix and write out the general solution in standard form.

B32 $\begin{aligned} 2x_2 + x_3 &= 0 \\ 5x_1 + 6x_2 + x_3 &= 0 \\ x_1 + 3x_2 + x_3 &= 0 \end{aligned}$ **B33** $\begin{aligned} x_1 + 2x_2 + 4x_3 &= 0 \\ x_1 + 4x_2 + 2x_3 &= 0 \\ x_1 + 3x_2 + 3x_3 &= 0 \end{aligned}$

B34 $\begin{aligned} 3x_1 + 4x_2 - 2x_3 + 7x_4 &= 0 \\ 9x_1 + 5x_2 + x_3 \quad &= 0 \\ -3x_1 + x_2 - 3x_3 + 8x_4 &= 0 \\ 3x_1 + x_2 + x_3 - 2x_4 &= 0 \end{aligned}$

B35 $\begin{aligned} x_1 - 3x_2 + x_3 - 3x_4 - x_5 &= 0 \\ -x_1 + 4x_2 \quad + x_4 + x_5 &= 0 \\ 2x_1 - 13x_2 - 5x_3 + 8x_4 - x_5 &= 0 \\ 3x_1 - 12x_2 + x_3 - 7x_4 - 3x_5 &= 0 \end{aligned}$

For Problems B36–B41, solve the system $\left[A \mid \vec{b}\right]$ by row reducing the augmented matrix to RREF. Then, without any further operations, find the general solution to the homogeneous $\left[A \mid \vec{0}\right]$.

B36 $A = \begin{bmatrix} 2 & 4 & 0 \\ 1 & 4 & 2 \\ 2 & 5 & 3 \end{bmatrix}, \vec{b} = \begin{bmatrix} -8 \\ 6 \\ 11 \end{bmatrix}$

B37 $A = \begin{bmatrix} 4 & 6 & 3 \\ 6 & 9 & 5 \\ -4 & -6 & 2 \end{bmatrix}, \vec{b} = \begin{bmatrix} 7 \\ 3 \\ 3 \end{bmatrix}$

B38 $A = \begin{bmatrix} 5 & 3 & 12 & 13 \\ 2 & 1 & 5 & 6 \end{bmatrix}, \vec{b} = \begin{bmatrix} 14 \\ 5 \end{bmatrix}$

B39 $A = \begin{bmatrix} 1 & 0 & 3 & -1 \\ 4 & 3 & 6 & -4 \\ -1 & -4 & 5 & 5 \end{bmatrix}, \vec{b} = \begin{bmatrix} 1 \\ 4 \\ 7 \end{bmatrix}$

B40 $A = \begin{bmatrix} 1 & 3 & 2 & 1 \\ 2 & 10 & 6 & -4 \\ 1 & 6 & 4 & -3 \\ 1 & 8 & 3 & -8 \end{bmatrix}, \vec{b} = \begin{bmatrix} 3 \\ 4 \\ 4 \\ -7 \end{bmatrix}$

B41 $A = \begin{bmatrix} 1 & 2 & 5 & 1 & -3 \\ 0 & 2 & 2 & 2 & -4 \\ -1 & -4 & -7 & -1 & 9 \\ 0 & 2 & 2 & 3 & -3 \end{bmatrix}, \vec{b} = \begin{bmatrix} 7 \\ 4 \\ -9 \\ 5 \end{bmatrix}$

Conceptual Problems

C1 We want to find a vector $\vec{x} \neq \vec{0}$ in $\mathbb{R}^3$ that is simultaneously orthogonal to given vectors $\vec{a}, \vec{b}, \vec{c} \in \mathbb{R}^3$.
(a) Write equations that must be satisfied by $\vec{x}$.
(b) What condition must be satisfied by the rank of the matrix $A = \begin{bmatrix} a_1 & a_2 & a_3 \\ b_1 & b_2 & b_3 \\ c_1 & c_2 & c_3 \end{bmatrix}$ if there are to be non-trivial solutions? Explain.

C2 Prove the solution set of a homogeneous system of m linear equations in n variables is a subspace of $\mathbb{R}^n$.

C3 (a) Suppose that $A = \begin{bmatrix} 1 & 0 & 2 \\ 0 & 1 & -1 \end{bmatrix}$ is the coefficient matrix of a homogeneous system $\left[A \mid \vec{0} \right]$. Find the general solution of the system and indicate why it describes a line through the origin.
(b) Suppose that a matrix A with two rows and three columns is the coefficient matrix of a homogeneous system. If A has rank 2, then explain why the solution set of the homogeneous system is a line through the origin. What could you say if rank$(A) = 1$?
(c) Let $\vec{u}$, $\vec{v}$, and $\vec{w}$ be three vectors in $\mathbb{R}^4$. Write conditions on a vector $\vec{x} \in \mathbb{R}^4$ such that $\vec{x}$ is orthogonal to $\vec{u}$, $\vec{v}$, and $\vec{w}$. (This should lead to a homogeneous system with coefficient matrix C, whose rows are $\vec{u}$, $\vec{v}$, and $\vec{w}$.) What does the rank of C tell us about the set of vectors $\vec{x}$ that are orthogonal to $\vec{u}$, $\vec{v}$, and $\vec{w}$?

For Problems C4–C6, what can you say about the consistency of the system of m linear equations in n variables and the number of parameters in the general solution?

C4 $m = 5, n = 7$, the rank of the coefficient matrix is 4.

C5 $m = 3, n = 6$, the rank of the coefficient matrix is 3.

C6 $m = 5, n = 4$, the rank of the augmented matrix is 4.

C7 A system of linear equations has augmented matrix $\left[\begin{array}{ccc|c} 1 & a & b & 1 \\ 1 & 1 & 0 & a \\ 1 & 0 & 1 & b \end{array}\right]$. For which values of a and b is the system consistent? Are there values for which there is a unique solution? Determine the general solution.

C8 Consider three planes P_1, P_2, and P_3, with respective equations
$a_{11}x_1 + a_{12}x_2 + a_{13}x_3 = b_1$,
$a_{21}x_1 + a_{22}x_2 + a_{23}x_3 = b_2$, and
$a_{31}x_1 + a_{32}x_2 + a_{33}x_3 = b_3$
The intersections of these planes is illustrated below. Assume that P_1 and P_2 are parallel.

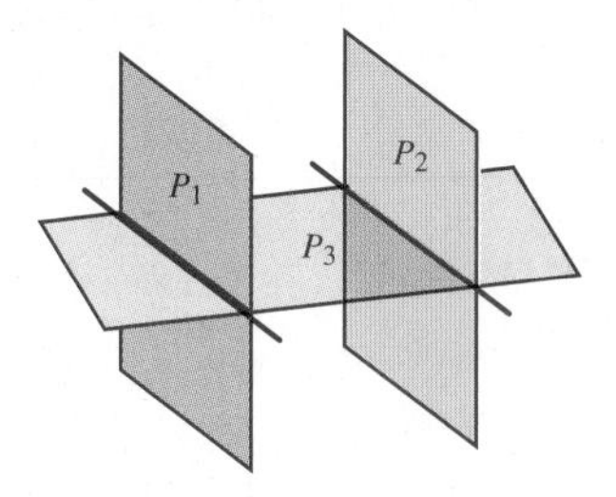

What is the rank of $A = \begin{bmatrix} a_{11} & a_{12} & a_{13} \\ a_{21} & a_{22} & a_{23} \\ a_{31} & a_{32} & a_{33} \end{bmatrix}$?

C9 Prove that if two planes in $\mathbb{R}^3$ intersect, then they intersect either in a line or a plane.

For Problems C10–C12, determine whether the statement is true or false. Justify your answer.

C10 The solution set of a system of linear equations is a subspace of $\mathbb{R}^n$.

C11 If A is the coefficient matrix of a system of m linear equations in n variables where $m < n$, then rank $A = m$.

C12 If a system of m linear equations in n variables $\left[A \mid \vec{b} \right]$ is consistent for every $\vec{b} \in \mathbb{R}^m$ with a unique solution, then $m = n$.

2.3 Application to Spanning and Linear Independence

As discussed at the beginning of this chapter, solving systems of linear equations will play an important role in much of what we do in the rest of the text. Here we will show how to use the methods described in this chapter to solve some of the problems we encountered in Chapter 1.

Spanning Problems

Recall that a vector $\vec{v} \in \mathbb{R}^n$ is in the span of a set $\{\vec{v}_1, \dots, \vec{v}_k\}$ of vectors in $\mathbb{R}^n$ if and only if there exists scalars $t_1, \dots, t_k \in \mathbb{R}$ such that

$$t_1\vec{v}_1 + \cdots + t_k\vec{v}_k = \vec{v}$$

This vector equation actually represents n equations (one for each component of the vectors) in the k unknowns $t_1, \dots, t_k$. Thus, it is easy to establish whether a vector is in the span of a set; we just need to determine whether the corresponding system of linear equations is consistent or not.

EXAMPLE 2.3.1

Determine whether the vector $\vec{v} = \begin{bmatrix} -2 \\ -3 \\ 1 \end{bmatrix}$ is in Span $\left\{ \begin{bmatrix} 1 \\ 1 \\ 1 \end{bmatrix}, \begin{bmatrix} 1 \\ -1 \\ 5 \end{bmatrix}, \begin{bmatrix} 2 \\ 1 \\ 4 \end{bmatrix} \right\}$.

Solution: Consider the vector equation

$$t_1 \begin{bmatrix} 1 \\ 1 \\ 1 \end{bmatrix} + t_2 \begin{bmatrix} 1 \\ -1 \\ 5 \end{bmatrix} + t_3 \begin{bmatrix} 2 \\ 1 \\ 4 \end{bmatrix} = \begin{bmatrix} -2 \\ -3 \\ 1 \end{bmatrix}$$

Simplifying the left-hand side using vector operations, we get

$$\begin{bmatrix} t_1 + t_2 + 2t_3 \\ t_1 - t_2 + t_3 \\ t_1 + 5t_2 + 4t_3 \end{bmatrix} = \begin{bmatrix} -2 \\ -3 \\ 1 \end{bmatrix}$$

Comparing corresponding entries gives the system of linear equations

$$\begin{aligned} t_1 + t_2 + 2t_3 &= -2 \\ t_1 - t_2 + t_3 &= -3 \\ t_1 + 5t_2 + 4t_3 &= 1 \end{aligned}$$

We row reduce the augmented matrix:

$$\left[\begin{array}{ccc|c} 1 & 1 & 2 & -2 \\ 1 & -1 & 1 & -3 \\ 1 & 5 & 4 & 1 \end{array}\right] \begin{array}{c} \\ R_2 - R_1 \\ R_3 - R_1 \end{array} \sim \left[\begin{array}{ccc|c} 1 & 1 & 2 & -2 \\ 0 & -2 & -1 & -1 \\ 0 & 4 & 2 & 3 \end{array}\right] \begin{array}{c} \\ \\ R_3 + 2R_2 \end{array} \sim \left[\begin{array}{ccc|c} 1 & 1 & 2 & -2 \\ 0 & -2 & -1 & -1 \\ 0 & 0 & 0 & 1 \end{array}\right]$$

By Theorem 2.1.3, the system is inconsistent. Hence, $\vec{v}$ is not in the spanned set.

EXAMPLE 2.3.2

Write $\vec{v} = \begin{bmatrix} -1 \\ 1 \\ -1 \end{bmatrix}$ as a linear combination of $\begin{bmatrix} 1 \\ 2 \\ 1 \end{bmatrix}, \begin{bmatrix} -2 \\ 1 \\ 0 \end{bmatrix}$, and $\begin{bmatrix} 1 \\ 1 \\ 1 \end{bmatrix}$.

Solution: We need to find scalars $t_1, t_2, t_3 \in \mathbb{R}$ such that

$$t_1 \begin{bmatrix} 1 \\ 2 \\ 1 \end{bmatrix} + t_2 \begin{bmatrix} -2 \\ 1 \\ 0 \end{bmatrix} + t_3 \begin{bmatrix} 1 \\ 1 \\ 1 \end{bmatrix} = \begin{bmatrix} -1 \\ 1 \\ -1 \end{bmatrix}$$

Simplifying the left-hand side using vector operations, we get

$$\begin{bmatrix} t_1 - 2t_2 + t_3 \\ 2t_1 + t_2 + t_3 \\ t_1 \qquad + t_3 \end{bmatrix} = \begin{bmatrix} -1 \\ 1 \\ -1 \end{bmatrix}$$

This gives the system of linear equations:

$$\begin{aligned} t_1 - 2t_2 + t_3 &= -1 \\ 2t_1 + t_2 + t_3 &= 1 \\ t_1 \qquad + t_3 &= -1 \end{aligned}$$

Row reducing the augmented matrix to RREF gives

$$\left[\begin{array}{ccc|c} 1 & -2 & 1 & -1 \\ 2 & 1 & 1 & 1 \\ 1 & 0 & 1 & -1 \end{array}\right] \sim \left[\begin{array}{ccc|c} 1 & 0 & 0 & 2 \\ 0 & 1 & 0 & 0 \\ 0 & 0 & 1 & -3 \end{array}\right]$$

The solution is $t_1 = 2, t_2 = 0$, and $t_3 = -3$. This tells us that

$$2 \begin{bmatrix} 1 \\ 2 \\ 1 \end{bmatrix} + 0 \begin{bmatrix} -2 \\ 1 \\ 0 \end{bmatrix} - 3 \begin{bmatrix} 1 \\ 1 \\ 1 \end{bmatrix} = \begin{bmatrix} -1 \\ 1 \\ -1 \end{bmatrix}$$

EXERCISE 2.3.1

Determine whether $\vec{v} = \begin{bmatrix} 1 \\ 3 \\ 1 \end{bmatrix}$ is in Span $\left\{ \begin{bmatrix} 1 \\ -3 \\ -3 \end{bmatrix}, \begin{bmatrix} 2 \\ -2 \\ 1 \end{bmatrix}, \begin{bmatrix} -2 \\ 2 \\ -3 \end{bmatrix} \right\}$.

EXAMPLE 2.3.3

Consider the subspace of $\mathbb{R}^4$ defined by $S = \text{Span}\left\{\begin{bmatrix}1\\2\\1\\1\end{bmatrix}, \begin{bmatrix}1\\1\\3\\1\end{bmatrix}, \begin{bmatrix}3\\5\\5\\3\end{bmatrix}\right\}$. Find a homogeneous system of linear equations that defines S.

Solution: A vector $\vec{x} \in \mathbb{R}^4$ is in this set if and only if for some $t_1, t_2, t_3 \in \mathbb{R}$,

$$t_1\begin{bmatrix}1\\2\\1\\1\end{bmatrix} + t_2\begin{bmatrix}1\\1\\3\\1\end{bmatrix} + t_3\begin{bmatrix}3\\5\\5\\3\end{bmatrix} = \begin{bmatrix}x_1\\x_2\\x_3\\x_4\end{bmatrix}$$

Simplifying the left-hand side gives us a system of equations with augmented matrix

$$\left[\begin{array}{ccc|c} 1 & 1 & 3 & x_1 \\ 2 & 1 & 5 & x_2 \\ 1 & 3 & 5 & x_3 \\ 1 & 1 & 3 & x_4 \end{array}\right]$$

Row reducing this matrix to row echelon form gives

$$\left[\begin{array}{ccc|c} 1 & 1 & 3 & x_1 \\ 2 & 1 & 5 & x_2 \\ 1 & 3 & 5 & x_3 \\ 1 & 1 & 3 & x_4 \end{array}\right] \sim \left[\begin{array}{ccc|c} 1 & 1 & 3 & x_1 \\ 0 & 1 & 1 & 2x_1 - x_2 \\ 0 & 0 & 0 & -5x_1 + 2x_2 + x_3 \\ 0 & 0 & 0 & -x_1 + x_4 \end{array}\right]$$

The system is consistent if and only if $-5x_1 + 2x_2 + x_3 = 0$ and $-x_1 + x_4 = 0$. Thus, this homogeneous system of linear equations defines S.

EXAMPLE 2.3.4

Show that $\text{Span}\left\{\begin{bmatrix}-3\\1\\-2\end{bmatrix}, \begin{bmatrix}1\\3\\-3\end{bmatrix}, \begin{bmatrix}2\\-1\\1\end{bmatrix}\right\} = \mathbb{R}^3$.

Solution: Denote the vectors by $\vec{v}_1, \vec{v}_2, \vec{v}_3$. To show that every vector in $\mathbb{R}^3$ can be written as a linear combination of the vectors $\vec{v}_1$, $\vec{v}_2$, and $\vec{v}_3$, we only need to show that the system

$$t_1\vec{v}_1 + t_2\vec{v}_2 + t_3\vec{v}_3 = \vec{b}$$

is consistent for all $\vec{b} \in \mathbb{R}^3$. By the System-Rank Theorem (3), we only need to show that the rank of the coefficient matrix equals the number of rows.

Row reducing the coefficient matrix to RREF gives

$$\begin{bmatrix} -3 & 1 & 2 \\ 1 & 3 & -1 \\ -2 & -3 & 1 \end{bmatrix} \sim \begin{bmatrix} 1 & 0 & 0 \\ 0 & 1 & 0 \\ 0 & 0 & 1 \end{bmatrix}$$

Hence, the rank of the matrix is 3, which equals the number of rows, as required.

We generalize the method used in Example 2.3.4 to get the following important results.

Theorem 2.3.1 A set of k vectors $\{\vec{v}_1, \dots, \vec{v}_k\}$ in $\mathbb{R}^n$ spans $\mathbb{R}^n$ if and only if the rank of the coefficient matrix of the system $t_1\vec{v}_1 + \cdots + t_k\vec{v}_k = \vec{b}$ is n.

Proof: If $\text{Span}\{\vec{v}_1, \dots, \vec{v}_k\} = \mathbb{R}^n$, then every $\vec{b} \in \mathbb{R}^n$ can be written as a linear combination of the vectors $\{\vec{v}_1, \dots, \vec{v}_k\}$. That is, the system of linear equations

$$t_1\vec{v}_1 + \cdots + t_k\vec{v}_k = \vec{b}$$

has a solution for every $\vec{b} \in \mathbb{R}^n$. By the System-Rank Theorem (3), this means that the rank of the coefficient matrix of the system equals n (the number of equations).
On the other hand, if the rank of the coefficient matrix of the system is n, then by the System-Rank Theorem (3) the system is consistent for all $\vec{b} \in \mathbb{R}^n$. Therefore, every $\vec{b} \in \mathbb{R}^n$ can be written as a linear combination of the vectors $\{\vec{v}_1, \dots, \vec{v}_k\}$. Consequently, $\text{Span}\{\vec{v}_1, \dots, \vec{v}_k\} = \mathbb{R}^n$. ■

Theorem 2.3.2 Let $\{\vec{v}_1, \dots, \vec{v}_k\}$ be a set of k vectors in $\mathbb{R}^n$. If $\text{Span}\{\vec{v}_1, \dots, \vec{v}_k\} = \mathbb{R}^n$, then $k \geq n$.

Proof: By Theorem 2.3.1, if $\text{Span}\{\vec{v}_1, \dots, \vec{v}_k\} = \mathbb{R}^n$, then the rank of the coefficient matrix is n. If the matrix has n leading ones, then it must have least n columns to contain the leading ones. Hence, the number of columns, k, must be greater than or equal to n. ■

Linear Independence Problems

Recall that a set of vectors $\{\vec{v}_1, \dots, \vec{v}_k\}$ in $\mathbb{R}^n$ is said to be linearly independent if and only if the only solution to the vector equation

$$t_1\vec{v}_1 + \cdots + t_k\vec{v}_k = \vec{0}$$

is the solution $t_i = 0$ for $1 \leq i \leq k$. By the System-Rank Theorem (2), this is true if and only if the rank of the coefficient matrix of the corresponding homogeneous system is equal to the number of variables k. In particular, if the corresponding homogeneous system has no parameters, then the trivial solution is the only solution.

EXAMPLE 2.3.5

Determine whether the set $\left\{ \begin{bmatrix} 1 \\ 1 \\ 1 \end{bmatrix}, \begin{bmatrix} 1 \\ -1 \\ 5 \end{bmatrix}, \begin{bmatrix} 2 \\ 1 \\ 4 \end{bmatrix}, \begin{bmatrix} -1 \\ -3 \\ 3 \end{bmatrix} \right\}$ is linearly independent in $\mathbb{R}^3$.

Solution: Consider

$$t_1 \begin{bmatrix} 1 \\ 1 \\ 1 \end{bmatrix} + t_2 \begin{bmatrix} 1 \\ -1 \\ 5 \end{bmatrix} + t_3 \begin{bmatrix} 2 \\ 1 \\ 4 \end{bmatrix} + t_4 \begin{bmatrix} -1 \\ -3 \\ 3 \end{bmatrix} = \begin{bmatrix} 0 \\ 0 \\ 0 \end{bmatrix}$$

EXAMPLE 2.3.5
(continued)

Simplifying as above, this gives the homogeneous system with coefficient matrix

$$\begin{bmatrix} 1 & 1 & 2 & -1 \\ 1 & -1 & 1 & -3 \\ 1 & 5 & 4 & 3 \end{bmatrix}$$

Notice that we do not need to row reduce this matrix. By the System-Rank Theorem (2), the number of parameters in the general solution equals the number of variables minus the rank of the matrix. There are 4 variables, but the maximum the rank can be is 3 since there are only 3 rows. Hence, the number of parameters is at least 1, so the system has infinitely many solutions. Therefore, the set is linearly dependent.

EXAMPLE 2.3.6

Let $\vec{v}_1 = \begin{bmatrix} 1 \\ 2 \\ 1 \end{bmatrix}, \vec{v}_2 = \begin{bmatrix} -2 \\ 1 \\ 0 \end{bmatrix}, \vec{v}_3 = \begin{bmatrix} 1 \\ 1 \\ 1 \end{bmatrix} \in \mathbb{R}^3$. Determine whether the set $\{\vec{v}_1, \vec{v}_2, \vec{v}_3\}$ is linearly independent in $\mathbb{R}^3$.

Solution: Consider $t_1\vec{v}_1 + t_2\vec{v}_2 + t_3\vec{v}_3 = \vec{0}$. As above, we find that the coefficient matrix of the corresponding system is $\begin{bmatrix} 1 & -2 & 1 \\ 2 & 1 & 1 \\ 1 & 0 & 1 \end{bmatrix}$. Using the same elementary row operations as in Example 2.3.2, we get

$$\begin{bmatrix} 1 & -2 & 1 \\ 2 & 1 & 1 \\ 1 & 0 & 1 \end{bmatrix} \sim \begin{bmatrix} 1 & 0 & 0 \\ 0 & 1 & 0 \\ 0 & 0 & 1 \end{bmatrix}$$

Therefore, the set is linearly independent since the system has a unique solution.

EXERCISE 2.3.2

Determine whether the set $\left\{ \begin{bmatrix} -1 \\ 1 \\ -3 \end{bmatrix}, \begin{bmatrix} -2 \\ -3 \\ -3 \end{bmatrix}, \begin{bmatrix} 1 \\ 1 \\ 3 \end{bmatrix} \right\}$ is linearly independent or dependent.

We generalize the method used in Examples 2.3.5 and 2.3.6 to prove some important results.

Theorem 2.3.3

A set of vectors $\{\vec{v}_1, \ldots, \vec{v}_k\}$ in $\mathbb{R}^n$ is linearly independent if and only if the rank of the coefficient matrix of the homogeneous system $t_1\vec{v}_1 + \cdots + t_k\vec{v}_k = \vec{0}$ is k.

Proof: If $\{\vec{v}_1, \ldots, \vec{v}_k\}$ is linearly independent, then the system of linear equations

$$c_1\vec{v}_1 + \cdots + c_k\vec{v}_k = \vec{0}$$

has a unique solution. Thus, the rank of the coefficient matrix equals the number of unknowns k by the System-Rank Theorem (2).

On the other hand, if the rank of the coefficient matrix equals k, then the homogeneous system has $k - k = 0$ parameters. Therefore, it has the unique solution $t_1 = \cdots = t_k = 0$, and so the set is linearly independent. ∎

Theorem 2.3.4 If $\{\vec{v}_1, \dots, \vec{v}_k\}$ is a linearly independent set of vectors in $\mathbb{R}^n$, then $k \leq n$.

Proof: By Theorem 2.3.3, if $\{\vec{v}_1, \dots, \vec{v}_k\}$ is linearly independent, then the rank of the coefficient matrix is k. Hence, there must be at least k rows in the matrix to contain the leading ones. Therefore, the number of rows n must be greater than or equal to k. ■

Bases and Dimension of Subspaces

Recall from Section 1.4 that we defined a basis $\mathcal{B}$ of a subspace S of $\mathbb{R}^n$ to be a linearly independent set that spans S. Thus, with our previous tools, we can now easily identify a basis for a subspace. In particular, to show that a set $\mathcal{B}$ of vectors in $\mathbb{R}^n$ is a basis for a subspace S, we just need to show that Span $\mathcal{B} = S$ and $\mathcal{B}$ is linearly independent. We demonstrate this with a couple of examples.

EXAMPLE 2.3.7

Prove that $\mathcal{B} = \left\{ \begin{bmatrix} 1 \\ 1 \\ 2 \end{bmatrix}, \begin{bmatrix} 5 \\ -2 \\ 2 \end{bmatrix}, \begin{bmatrix} -2 \\ 3 \\ 1 \end{bmatrix} \right\}$ is a basis for $\mathbb{R}^3$.

Solution: Consider

$$t_1 \begin{bmatrix} 1 \\ 1 \\ 2 \end{bmatrix} + t_2 \begin{bmatrix} 5 \\ -2 \\ 2 \end{bmatrix} + t_3 \begin{bmatrix} -2 \\ 3 \\ 1 \end{bmatrix} = \vec{v}$$

We find that the coefficient matrix of the corresponding system is

$$\begin{bmatrix} 1 & 5 & -2 \\ 1 & -2 & 3 \\ 2 & 2 & 1 \end{bmatrix}$$

By Theorem 2.3.1, $\mathcal{B}$ spans $\mathbb{R}^3$ if and only if the rank of this matrix equals the number of rows. Moreover, by Theorem 2.3.3, $\mathcal{B}$ is linearly independent if and only if the rank of this matrix equals the number of columns. Hence, we just need to show that the rank of this is matrix is 3. Row reducing the matrix to RREF we get

$$\begin{bmatrix} 1 & 5 & -2 \\ 1 & -2 & 3 \\ 2 & 2 & 1 \end{bmatrix} \sim \begin{bmatrix} 1 & 0 & 0 \\ 0 & 1 & 0 \\ 0 & 0 & 1 \end{bmatrix}$$

Thus, the rank is 3. So, $\mathcal{B}$ is a basis for $\mathbb{R}^3$.

Remark

In Example 2.3.7 we benefited from Theorem 2.3.1 and Theorem 2.3.3 since we were finding a basis for all of $\mathbb{R}^3$. However, this is not always going to be the case. It is important that you do not always just memorize short cuts. It is always necessary to ensure that you have understood the complete concept so that you can solve a variety of problems.

EXAMPLE 2.3.8

Show that $\mathcal{B} = \left\{ \begin{bmatrix} 1 \\ 2 \\ -1 \end{bmatrix}, \begin{bmatrix} 1 \\ 1 \\ 1 \end{bmatrix} \right\}$ is a basis for the plane $-3x_1 + 2x_2 + x_3 = 0$.

Solution: To prove that $\mathcal{B}$ is a basis for the plane, we must show that $\mathcal{B}$ is linearly independent and spans the plane.

We first observe that $\mathcal{B}$ is clearly linearly independent since neither vector is a scalar multiple of the other.

For spanning, first observe that any vector $\vec{x}$ in the plane must satisfy the condition of the plane. Hence, every vector in the plane has the form

$$\vec{x} = \begin{bmatrix} x_1 \\ x_2 \\ 3x_1 - 2x_2 \end{bmatrix}$$

since $x_3 = 3x_1 - 2x_2$. Therefore, we now just need to show that the equation

$$t_1 \begin{bmatrix} 1 \\ 2 \\ -1 \end{bmatrix} + t_2 \begin{bmatrix} 1 \\ 1 \\ 1 \end{bmatrix} = \begin{bmatrix} x_1 \\ x_2 \\ 3x_1 - 2x_2 \end{bmatrix}$$

is always consistent. Row reducing the corresponding augmented matrix gives

$$\left[\begin{array}{cc|c} 1 & 1 & x_1 \\ 2 & 1 & x_2 \\ -1 & 1 & 3x_1 - 2x_2 \end{array}\right] \sim \left[\begin{array}{cc|c} 1 & 1 & x_1 \\ 0 & 1 & 2x_1 - x_2 \\ 0 & 0 & 0 \end{array}\right]$$

The system is consistent for all $x_1, x_2 \in \mathbb{R}$ and hence $\mathcal{B}$ also spans the plane.

Using the method in Example 2.3.7 we get the following useful theorem.

Theorem 2.3.5 A set of vectors $\{\vec{v}_1, \dots, \vec{v}_n\}$ is a basis for $\mathbb{R}^n$ if and only if the rank of the coefficient matrix of $t_1\vec{v}_1 + \cdots + t_n\vec{v}_n = \vec{0}$ is n.

Theorem 2.3.5 gives us a condition to test whether a set of n vectors in $\mathbb{R}^n$ is a basis for $\mathbb{R}^n$. Moreover, Theorem 2.3.1 and Theorem 2.3.3 give us the following theorem.

Theorem 2.3.6 A set of vectors $\{\vec{v}_1, \dots, \vec{v}_n\}$ in $\mathbb{R}^n$ is linearly independent if and only if it spans $\mathbb{R}^n$.

We now want to prove that every basis of a subspace S of $\mathbb{R}^n$ must contain the same number of vectors.

Theorem 2.3.7 Suppose that $\mathcal{B} = \{\vec{v}_1, \dots, \vec{v}_\ell\}$ is a basis for a non-trivial subspace S of $\mathbb{R}^n$ and that $\{\vec{u}_1, \dots, \vec{u}_k\}$ is a set in S. If $k > \ell$, then $\{\vec{u}_1, \dots, \vec{u}_k\}$ is linearly dependent.

Proof: Since each $\vec{u}_i$, $1 \leq i \leq k$, is a vector in S and $\mathcal{B}$ is a basis for S, by Theorem 1.4.4 each $\vec{u}_i$ can be written as a unique linear combination of the vectors in $\mathcal{B}$. We get

$$\begin{aligned}
\vec{u}_1 &= a_{11}\vec{v}_1 + a_{21}\vec{v}_2 + \cdots + a_{\ell 1}\vec{v}_\ell \\
\vec{u}_2 &= a_{12}\vec{v}_1 + a_{22}\vec{v}_2 + \cdots + a_{\ell 2}\vec{v}_\ell \\
&\vdots \\
\vec{u}_k &= a_{1k}\vec{v}_1 + a_{2k}\vec{v}_2 + \cdots + a_{\ell k}\vec{v}_\ell
\end{aligned}$$

Consider the equation

$$\begin{aligned}
\vec{0} &= t_1\vec{u}_1 + \cdots + t_k\vec{u}_k \qquad (2.2) \\
&= t_1(a_{11}\vec{v}_1 + a_{21}\vec{v}_2 + \cdots + a_{\ell 1}\vec{v}_\ell) + \cdots + t_k(a_{1k}\vec{v}_1 + a_{2k}\vec{v}_2 + \cdots + a_{\ell k}\vec{v}_\ell) \\
&= (a_{11}t_1 + \cdots + a_{1k}t_k)\vec{v}_1 + \cdots + (a_{\ell 1}t_1 + \cdots + a_{\ell k}t_k)\vec{v}_\ell
\end{aligned}$$

But, $\{\vec{v}_1, \dots, \vec{v}_\ell\}$ is linearly independent, so the only solution to this equation is

$$\begin{aligned}
a_{11}t_1 + \cdots + a_{1k}t_k &= 0 \\
\vdots \qquad\qquad & \vdots \\
a_{\ell 1}t_1 + \cdots + a_{\ell k}t_k &= 0
\end{aligned}$$

The rank of the coefficient matrix of this homogeneous system is at most ℓ because $\ell < k$. By the System-Rank Theorem (2), the solution space has at least $k - \ell > 0$ parameters. Therefore, there are infinitely many possible $t_1, \dots, t_k$ and so $\{\vec{u}_1, \dots, \vec{u}_k\}$ is linearly dependent since equation (2.2) has infinitely many solutions. ■

Theorem 2.3.8 If $\{\vec{v}_1, \dots, \vec{v}_\ell\}$ and $\{\vec{u}_1, \dots, \vec{u}_k\}$ are both bases of a subspace S of $\mathbb{R}^n$, then $k = \ell$.

Proof: Since $\{\vec{v}_1, \dots, \vec{v}_\ell\}$ is a basis for S and $\{\vec{u}_1, \dots, \vec{u}_k\}$ is linearly independent, by Theorem 2.3.7, we must have $k \leq \ell$. Similarly, since $\{\vec{u}_1, \dots, \vec{u}_k\}$ is a basis and $\{\vec{v}_1, \dots, \vec{v}_\ell\}$ is linearly independent, we must have $\ell \leq k$. Therefore, $\ell = k$, as required. ■

This theorem justifies the following definition.

Definition
Dimension

If S is a non-trivial subspace of $\mathbb{R}^n$ with a basis containing k vectors, then we say that the **dimension** of S is k and write

$$\dim S = k$$

Since a basis for the trivial subspace $\{\vec{0}\}$ of $\mathbb{R}^n$ is the empty set, the dimension of the trivial subspace is 0.

EXAMPLE 2.3.9

Find a basis and the dimension of the solution space of the homogeneous system

$$\begin{aligned} 2x_1 + x_2 \quad &= 0 \\ x_1 + x_2 - x_3 &= 0 \\ -x_2 + 2x_3 &= 0 \end{aligned}$$

Solution: We found in Example 2.2.9 that the general solution of this system is

$$\vec{x} = t\begin{bmatrix} -1 \\ 2 \\ 1 \end{bmatrix}, \quad t \in \mathbb{R}$$

This shows that a spanning set for the solution space is $\mathcal{B} = \left\{\begin{bmatrix} -1 \\ 2 \\ 1 \end{bmatrix}\right\}$. Since $\mathcal{B}$ contains one non-zero vector, it is also linearly independent and hence a basis for the solution space. Since the basis contains 1 vector, we have that the dimension of the solution space is 1.

EXAMPLE 2.3.10

Find a basis and the dimension of the solution space of the homogeneous system

$$\begin{aligned} x_1 + 2x_2 + 2x_3 + x_4 + 4x_5 &= 0 \\ 3x_1 + 7x_2 + 7x_3 + 3x_4 + 13x_5 &= 0 \\ 2x_1 + 5x_2 + 5x_3 + 2x_4 + 9x_5 &= 0 \end{aligned}$$

Solution: We found in Exercise 2.2.4 that the general solution of this system is

$$\vec{x} = t_1\begin{bmatrix} 0 \\ -1 \\ 1 \\ 0 \\ 0 \end{bmatrix} + t_2\begin{bmatrix} -1 \\ 0 \\ 0 \\ 1 \\ 0 \end{bmatrix} + t_3\begin{bmatrix} -2 \\ -1 \\ 0 \\ 0 \\ 1 \end{bmatrix}, \quad t_1, t_2, t_3 \in \mathbb{R}$$

This shows that a spanning set for the solution space is

$$\mathcal{B} = \left\{\begin{bmatrix} 0 \\ -1 \\ 1 \\ 0 \\ 0 \end{bmatrix}, \begin{bmatrix} -1 \\ 0 \\ 0 \\ 1 \\ 0 \end{bmatrix}, \begin{bmatrix} -2 \\ -1 \\ 0 \\ 0 \\ 1 \end{bmatrix}\right\}$$

It is not difficult to verify that $\mathcal{B}$ is also linearly independent. Consequently, $\mathcal{B}$ is a basis for the solution space, and hence the dimension of the solution space is 3.

PROBLEMS 2.3

Practice Problems

A1 Let $B = \left\{ \begin{bmatrix} 1 \\ 0 \\ 1 \\ 1 \end{bmatrix}, \begin{bmatrix} 2 \\ 1 \\ 0 \\ 1 \end{bmatrix}, \begin{bmatrix} -1 \\ 1 \\ 2 \\ 1 \end{bmatrix} \right\}$. For each of the following vectors, either express it as a linear combination of the vectors in B or show that it is not in Span B.

(a) $\begin{bmatrix} -3 \\ 2 \\ 8 \\ 4 \end{bmatrix}$ (b) $\begin{bmatrix} 5 \\ 4 \\ 6 \\ 7 \end{bmatrix}$ (c) $\begin{bmatrix} 2 \\ -2 \\ 1 \\ 1 \end{bmatrix}$

A2 Let $B = \left\{ \begin{bmatrix} 1 \\ -1 \\ 1 \\ 0 \end{bmatrix}, \begin{bmatrix} -1 \\ 1 \\ 0 \\ 2 \end{bmatrix}, \begin{bmatrix} 1 \\ 1 \\ -1 \\ -1 \end{bmatrix} \right\}$. For each of the following vectors, either express it as a linear combination of the vectors in B or show that it is not in Span B.

(a) $\begin{bmatrix} 3 \\ 2 \\ -1 \\ -1 \end{bmatrix}$ (b) $\begin{bmatrix} -7 \\ 3 \\ 0 \\ 8 \end{bmatrix}$ (c) $\begin{bmatrix} 1 \\ 1 \\ 1 \\ 1 \end{bmatrix}$

For Problems A3–A8, find a homogeneous system that defines the given subspace.

A3 Span $\left\{ \begin{bmatrix} 1 \\ 0 \\ 0 \end{bmatrix}, \begin{bmatrix} 0 \\ 1 \\ 0 \end{bmatrix} \right\}$

A4 Span $\left\{ \begin{bmatrix} 2 \\ 1 \\ 0 \end{bmatrix} \right\}$

A5 Span $\left\{ \begin{bmatrix} 1 \\ 1 \\ 2 \end{bmatrix}, \begin{bmatrix} 3 \\ -1 \\ 0 \end{bmatrix} \right\}$

A6 Span $\left\{ \begin{bmatrix} 2 \\ 1 \\ -1 \end{bmatrix}, \begin{bmatrix} 1 \\ -2 \\ 1 \end{bmatrix} \right\}$

A7 Span $\left\{ \begin{bmatrix} 1 \\ 0 \\ 1 \\ 0 \end{bmatrix}, \begin{bmatrix} 2 \\ -1 \\ 1 \\ 1 \end{bmatrix} \right\}$

A8 Span $\left\{ \begin{bmatrix} 1 \\ -1 \\ 1 \\ 2 \end{bmatrix}, \begin{bmatrix} 0 \\ 1 \\ 3 \\ -2 \end{bmatrix}, \begin{bmatrix} -2 \\ 0 \\ 4 \\ -3 \end{bmatrix} \right\}$

For Problems A9 and A10, find a basis and the dimension of the solution space of the given homogeneous system.

A9

$$\begin{aligned} x_1 + 3x_2 + 2x_3 &= 0 \\ 2x_1 + 8x_2 + 6x_3 &= 0 \\ -4x_1 + x_2 + 3x_3 &= 0 \end{aligned}$$

A10

$$\begin{aligned} x_2 + 3x_3 + x_4 &= 0 \\ x_1 + x_2 + 5x_3 \quad &= 0 \\ x_1 + 3x_2 + 11x_3 + 2x_4 &= 0 \end{aligned}$$

For Problems A11–A13, determine whether the given set is a basis for the given plane or hyperplane.

A11 $\left\{ \begin{bmatrix} 1 \\ 1 \\ 2 \end{bmatrix}, \begin{bmatrix} 1 \\ 0 \\ 1 \end{bmatrix} \right\}$ for $x_1 + x_2 - x_3 = 0$

A12 $\left\{ \begin{bmatrix} 1 \\ 1 \\ 1 \end{bmatrix}, \begin{bmatrix} 1 \\ 2 \\ -3 \end{bmatrix} \right\}$ for $2x_1 - 3x_2 + x_3 = 0$

A13 $\left\{ \begin{bmatrix} 1 \\ 1 \\ 0 \\ 0 \end{bmatrix}, \begin{bmatrix} 0 \\ 0 \\ 1 \\ 1 \end{bmatrix}, \begin{bmatrix} 3 \\ 1 \\ 0 \\ 1 \end{bmatrix} \right\}$ for $x_1 - x_2 + 2x_3 - 2x_4 = 0$

For Problems A14–A17, determine whether the set is linearly independent. If the set is linearly dependent, find all linear combinations of the vectors that equal the zero vector.

A14 $\left\{ \begin{bmatrix} 1 \\ 2 \\ 1 \\ -1 \end{bmatrix}, \begin{bmatrix} 1 \\ 2 \\ 3 \\ 1 \end{bmatrix}, \begin{bmatrix} 1 \\ -3 \\ 2 \\ 1 \end{bmatrix} \right\}$

A15 $\left\{ \begin{bmatrix} 1 \\ 0 \\ 1 \\ 0 \end{bmatrix}, \begin{bmatrix} 0 \\ 1 \\ 1 \\ 1 \end{bmatrix}, \begin{bmatrix} 0 \\ 0 \\ 1 \\ 1 \end{bmatrix}, \begin{bmatrix} 3 \\ 2 \\ 6 \\ 3 \end{bmatrix} \right\}$

A16 $\left\{ \begin{bmatrix} 1 \\ 1 \\ 0 \\ 1 \\ 1 \end{bmatrix}, \begin{bmatrix} 2 \\ 3 \\ 1 \\ 3 \\ 3 \end{bmatrix}, \begin{bmatrix} 0 \\ 1 \\ 1 \\ 1 \\ 1 \end{bmatrix} \right\}$

A17 $\left\{ \begin{bmatrix} 0 \\ 1 \\ 1 \\ 0 \end{bmatrix}, \begin{bmatrix} 1 \\ 2 \\ 3 \\ 1 \end{bmatrix}, \begin{bmatrix} -4 \\ 1 \\ 2 \\ 1 \end{bmatrix}, \begin{bmatrix} 3 \\ 1 \\ 2 \\ 0 \end{bmatrix} \right\}$

For Problems A18–A19, determine all values of k such that the given set is linearly independent.

A18 $\left\{ \begin{bmatrix} 1 \\ 0 \\ 1 \\ 0 \end{bmatrix}, \begin{bmatrix} 0 \\ 1 \\ 1 \\ 1 \end{bmatrix}, \begin{bmatrix} 2 \\ -3 \\ -1 \\ k \end{bmatrix} \right\}$

A19 $\left\{ \begin{bmatrix} 1 \\ 1 \\ 1 \\ 2 \end{bmatrix}, \begin{bmatrix} 1 \\ -1 \\ 2 \\ 0 \end{bmatrix}, \begin{bmatrix} -1 \\ 2 \\ k \\ 1 \end{bmatrix} \right\}$

For Problems A20–A23, determine whether the given set is a basis for $\mathbb{R}^3$.

A20 $\left\{ \begin{bmatrix} 1 \\ 1 \\ 2 \end{bmatrix}, \begin{bmatrix} 1 \\ -1 \\ -1 \end{bmatrix}, \begin{bmatrix} 2 \\ 1 \\ 1 \end{bmatrix} \right\}$

A21 $\left\{ \begin{bmatrix} -2 \\ 2 \\ 1 \end{bmatrix}, \begin{bmatrix} 3 \\ -1 \\ 2 \end{bmatrix} \right\}$

A22 $\left\{ \begin{bmatrix} 1 \\ 0 \\ 1 \end{bmatrix}, \begin{bmatrix} -1 \\ 2 \\ 1 \end{bmatrix}, \begin{bmatrix} 1 \\ 3 \\ 5 \end{bmatrix}, \begin{bmatrix} 2 \\ -1 \\ -4 \end{bmatrix} \right\}$

A23 $\left\{ \begin{bmatrix} 1 \\ -1 \\ 1 \end{bmatrix}, \begin{bmatrix} 1 \\ 2 \\ -1 \end{bmatrix}, \begin{bmatrix} 3 \\ 0 \\ 1 \end{bmatrix} \right\}$

Homework Problems

B1 Let $B = \left\{ \begin{bmatrix} 1 \\ -2 \\ 2 \end{bmatrix}, \begin{bmatrix} 2 \\ -3 \\ 1 \end{bmatrix}, \begin{bmatrix} -3 \\ 4 \\ 0 \end{bmatrix} \right\}$. For each of the following vectors, either express it as a linear combination of the vectors in B or show that it is not in Span B.

(a) $\begin{bmatrix} 1 \\ 1 \\ 1 \end{bmatrix}$ (b) $\begin{bmatrix} 7 \\ -9 \\ -1 \end{bmatrix}$ (c) $\begin{bmatrix} 4 \\ -5 \\ -1 \end{bmatrix}$

B2 Let $B = \left\{ \begin{bmatrix} 1 \\ 1 \\ 0 \\ 1 \end{bmatrix}, \begin{bmatrix} 1 \\ 2 \\ 1 \\ 1 \end{bmatrix}, \begin{bmatrix} 2 \\ 2 \\ 1 \\ 3 \end{bmatrix} \right\}$. For each of the following vectors, either express it as a linear combination of the vectors in B or show that it is not in Span B.

(a) $\begin{bmatrix} 1 \\ 4 \\ 2 \\ 1 \end{bmatrix}$ (b) $\begin{bmatrix} 0 \\ -5 \\ -4 \\ 1 \end{bmatrix}$ (c) $\begin{bmatrix} 5 \\ 4 \\ 3 \\ 9 \end{bmatrix}$

B3 Let $B = \left\{ \begin{bmatrix} 0 \\ 2 \\ 0 \\ 2 \end{bmatrix}, \begin{bmatrix} 3 \\ 3 \\ 6 \\ -3 \end{bmatrix}, \begin{bmatrix} -1 \\ -1 \\ -3 \\ 2 \end{bmatrix} \right\}$. For each of the following vectors, either express it as a linear combination of the vectors in B or show that it is not in Span B.

(a) $\begin{bmatrix} 1 \\ 7 \\ 0 \\ 7 \end{bmatrix}$ (b) $\begin{bmatrix} 1 \\ 1 \\ 2 \\ 0 \end{bmatrix}$ (c) $\begin{bmatrix} 3 \\ -5 \\ 3 \\ -8 \end{bmatrix}$

For Problems B4–B11, find a homogeneous system that defines the given subspace.

B4 Span $\left\{ \begin{bmatrix} 1 \\ 0 \\ 1 \end{bmatrix}, \begin{bmatrix} 0 \\ 1 \\ 0 \end{bmatrix} \right\}$

B5 Span $\left\{ \begin{bmatrix} 1 \\ -3 \\ 2 \end{bmatrix} \right\}$

B6 Span $\left\{ \begin{bmatrix} 0 \\ 1 \\ 1 \end{bmatrix}, \begin{bmatrix} 2 \\ -3 \\ 5 \end{bmatrix} \right\}$

B7 Span $\left\{ \begin{bmatrix} 1 \\ 1 \\ 0 \end{bmatrix}, \begin{bmatrix} -1 \\ 1 \\ 1 \end{bmatrix} \right\}$

B8 Span $\left\{ \begin{bmatrix} 1 \\ 1 \\ 1 \end{bmatrix} \right\}$

B9 Span $\left\{ \begin{bmatrix} 1 \\ -3 \\ 2 \end{bmatrix}, \begin{bmatrix} 4 \\ -1 \\ 1 \end{bmatrix} \right\}$

B10 Span $\left\{ \begin{bmatrix} 1 \\ 0 \\ -5 \\ 7 \end{bmatrix}, \begin{bmatrix} 3 \\ 1 \\ -8 \\ 9 \end{bmatrix} \right\}$

B11 Span $\left\{ \begin{bmatrix} 1 \\ 1 \\ 2 \\ 6 \end{bmatrix}, \begin{bmatrix} 2 \\ 3 \\ 4 \\ 9 \end{bmatrix}, \begin{bmatrix} 2 \\ 6 \\ 4 \\ 5 \end{bmatrix} \right\}$

For Problems B12–B15, find a basis and the dimension of the solution space of the given homogeneous system.

B12
$$\begin{aligned} 4x_1 \qquad - 4x_3 &= 0 \\ x_1 + x_2 - x_3 &= 0 \\ 6x_1 + 2x_2 - 5x_3 &= 0 \end{aligned}$$

B13
$$\begin{aligned} 3x_2 + 3x_3 - 2x_4 &= 0 \\ x_1 + 5x_2 + x_3 + 3x_4 &= 0 \\ -x_1 + 2x_2 + 6x_3 + x_4 &= 0 \end{aligned}$$

B14
$$\begin{aligned} x_1 + 2x_2 + 3x_3 - x_4 &= 0 \\ 2x_1 + 4x_2 + 6x_3 + x_4 &= 0 \\ 2x_1 + 4x_2 + 7x_3 + 3x_4 &= 0 \end{aligned}$$

B15
$$\begin{aligned} 3x_1 + 5x_2 + 3x_3 - x_4 &= 0 \\ 5x_1 + 9x_2 + 7x_3 + x_4 &= 0 \\ x_1 + 2x_2 + 2x_3 + x_4 &= 0 \\ 2x_1 + x_2 - 5x_3 - 10x_4 &= 0 \end{aligned}$$

For Problems B16–B21, determine whether the given set is a basis for the given plane or hyperplane.

B16 $\left\{ \begin{bmatrix} 3 \\ 5 \\ 8 \end{bmatrix}, \begin{bmatrix} 2 \\ -5 \\ -3 \end{bmatrix} \right\}$ for $x_1 + x_2 - x_3 = 0$

B17 $\left\{ \begin{bmatrix} 1 \\ -2 \\ 0 \end{bmatrix}, \begin{bmatrix} 1 \\ 1 \\ -6 \end{bmatrix} \right\}$ for $4x_1 + 2x_2 + x_3 = 0$

B18 $\left\{ \begin{bmatrix} 2 \\ 1 \\ 1 \end{bmatrix}, \begin{bmatrix} 6 \\ 3 \\ 3 \end{bmatrix} \right\}$ for $x_1 - 5x_2 + 3x_3 = 0$

B19 $\left\{ \begin{bmatrix} 1 \\ 1 \\ 1 \end{bmatrix}, \begin{bmatrix} 0 \\ 1 \\ 2 \end{bmatrix} \right\}$ for $x_1 + 5x_2 + 2x_3 = 0$

B20 $\left\{ \begin{bmatrix} 1 \\ 1 \\ 0 \\ 2 \end{bmatrix}, \begin{bmatrix} 0 \\ 6 \\ -1 \\ 2 \end{bmatrix} \right\}$ for $3x_1 + x_2 + 2x_3 - 2x_4 = 0$

B21 $\left\{ \begin{bmatrix} 0 \\ 3 \\ 0 \\ 1 \end{bmatrix}, \begin{bmatrix} 5 \\ 0 \\ -1 \\ 0 \end{bmatrix}, \begin{bmatrix} 2 \\ 1 \\ 0 \\ 0 \end{bmatrix} \right\}$ for $x_1 - 2x_2 + 5x_3 + 6x_4 = 0$

For Problems B22–B27, determine whether the set is linearly independent. If the set is linearly dependent, find all linear combinations of the vectors that equal the zero vector.

B22 $\left\{\begin{bmatrix}1\\0\\2\end{bmatrix}, \begin{bmatrix}1\\1\\3\end{bmatrix}, \begin{bmatrix}-1\\4\\2\end{bmatrix}\right\}$

B23 $\left\{\begin{bmatrix}1\\1\\4\end{bmatrix}, \begin{bmatrix}0\\1\\1\end{bmatrix}, \begin{bmatrix}2\\3\\7\end{bmatrix}\right\}$

B24 $\left\{\begin{bmatrix}5\\3\\1\end{bmatrix}, \begin{bmatrix}2\\4\\1\end{bmatrix}, \begin{bmatrix}-7\\7\\1\end{bmatrix}, \begin{bmatrix}1\\9\\2\end{bmatrix}\right\}$

B25 $\left\{\begin{bmatrix}1\\-1\\3\\1\end{bmatrix}, \begin{bmatrix}-1\\1\\2\\1\end{bmatrix}, \begin{bmatrix}-1\\1\\12\\5\end{bmatrix}\right\}$

B26 $\left\{\begin{bmatrix}1\\-1\\0\\-1\end{bmatrix}, \begin{bmatrix}2\\1\\1\\2\end{bmatrix}, \begin{bmatrix}2\\3\\2\\2\end{bmatrix}\right\}$

B27 $\left\{\begin{bmatrix}0\\3\\5\\2\end{bmatrix}, \begin{bmatrix}2\\5\\3\\3\end{bmatrix}, \begin{bmatrix}4\\5\\2\\3\end{bmatrix}, \begin{bmatrix}-1\\3\\8\\2\end{bmatrix}\right\}$

For Problems B28 and B29, determine all values of k such that the given set is linearly independent.

B28 $\left\{\begin{bmatrix}-1\\2\\3\end{bmatrix}, \begin{bmatrix}4\\-7\\-5\end{bmatrix}, \begin{bmatrix}k\\-1\\4\end{bmatrix}\right\}$

B29 $\left\{\begin{bmatrix}-3\\3\\2\\1\end{bmatrix}, \begin{bmatrix}k\\1\\1\\0\end{bmatrix}, \begin{bmatrix}0\\9\\7\\k\end{bmatrix}\right\}$

For Problems B30–B33, determine whether the given set is a basis for $\mathbb{R}^3$.

B30 $\left\{\begin{bmatrix}1\\3\\-4\end{bmatrix}, \begin{bmatrix}2\\-1\\5\end{bmatrix}\right\}$

B31 $\left\{\begin{bmatrix}3\\5\\3\end{bmatrix}, \begin{bmatrix}2\\5\\-2\end{bmatrix}, \begin{bmatrix}4\\-2\\-1\end{bmatrix}\right\}$

B32 $\left\{\begin{bmatrix}1\\7\\3\end{bmatrix}, \begin{bmatrix}0\\1\\-2\end{bmatrix}, \begin{bmatrix}2\\4\\-1\end{bmatrix}, \begin{bmatrix}1\\1\\-5\end{bmatrix}\right\}$

B33 $\left\{\begin{bmatrix}1\\1\\1\end{bmatrix}, \begin{bmatrix}3\\5\\1\end{bmatrix}, \begin{bmatrix}6\\7\\-4\end{bmatrix}\right\}$

Conceptual Problems

C1 Let $B = \{\vec{e}_1, \dots, \vec{e}_n\}$ be the standard basis for $\mathbb{R}^n$. Prove that Span $B = \mathbb{R}^n$ and that B is linearly independent.

C2 Find a basis for the hyperplane in $\mathbb{R}^4$ defined by

$$x_1 + ax_2 + bx_3 + cx_4 = 0$$

C3 Let $B = \{\vec{v}_1, \vec{v}_2\}$ be a basis for a subspace S in $\mathbb{R}^4$ and let $\vec{w}_1, \vec{w}_2, \vec{w}_3 \in S$.

(a) Prove that $\{\vec{w}_1, \vec{w}_2, \vec{w}_3\}$ is linearly dependent.

(b) Find vectors $\vec{x}_1, \vec{x}_2, \vec{x}_3 \in S$ such that $\text{Span}\{\vec{x}_1, \vec{x}_2, \vec{x}_3\} = S$.

(c) Find vectors $\vec{y}_1, \vec{y}_2, \vec{y}_3 \in S$ such that $\text{Span}\{\vec{y}_1, \vec{y}_2, \vec{y}_3\} \neq S$.

C4 Prove if S is a subspace of $\mathbb{R}^n$ and $\dim S = n$, then $S = \mathbb{R}^n$.

C5 Suppose that S is a subspace of $\mathbb{R}^n$ such that

$$\dim S = k$$

(a) Suppose that $B = \{\vec{w}_1, \dots, \vec{w}_\ell\}$ is a set in S. Prove if $\ell < k$, then B does not span S.

(b) Suppose that $C = \{\vec{u}_1, \dots, \vec{u}_k\}$ spans S. Prove that C is linearly independent.

C6 State, with justification, the dimension of the following subspaces of $\mathbb{R}^5$.

(a) A line through the origin.

(b) A plane through the origin.

(c) A hyperplane through the origin.

2.4 Applications of Systems of Linear Equations

We now look at a few applications of systems of linear equations. The first two of these, Spring-Mass Systems and Electric Circuits, will be referred to again in later chapters to demonstrate how our continued development of linear algebra furthers our ability to work with such applications.

Spring-Mass Systems

Consider the spring-mass system depicted in Figure 2.4.1. We want to determine the equilibrium displacements x_1, x_2 (with positive displacement being to the right) of the masses if constant forces f_1, f_2 (with a positive force being to the right) act on each of the masses m_1, m_2 respectively. As indicated, we are assuming that the springs have the corresponding spring constants k_1, k_2, k_3.

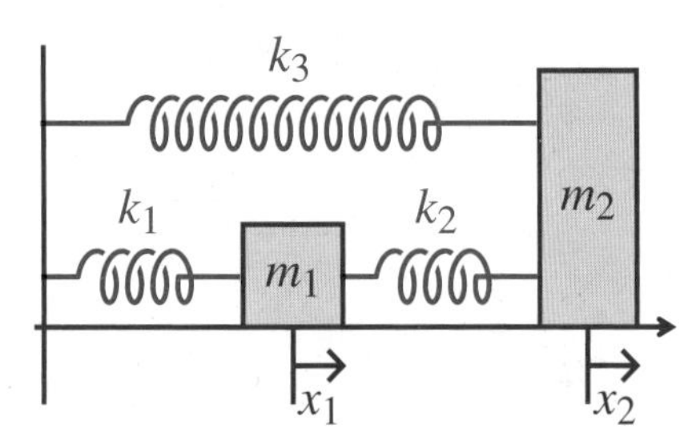

Figure 2.4.1 A spring-mass system.

According to Hooke's Law, we know that the force F required to stretch or compress a spring with spring constant k by a distance x is given by

$$F = kx$$

Using this, and taking into account the direction of force applied to each mass by the spring, we can write an equation for the force acting on each spring.

The mass m_1 is being acted upon by two springs; k_1 and k_2. A displacement of x_1 by mass m_1 will cause the spring k_1 to stretch or compress depending on if x_1 is positive or negative. In either case, by Hooke's Law, the force acting upon mass m_1 by spring k_1 is

$$-k_1 x_1$$

We needed to include the negative sign as if x_1 were positive, then the force will need to be negative as the spring will be trying to pull the mass back.

How much spring k_2 will be stretched or compressed depends on displacements x_1 and x_2. In particular, it will be stretched by $x_2 - x_1$. Hence, the force acting on mass m_1 by spring k_2 is

$$k_2(x_2 - x_1)$$

For the system to be in equilibrium, the sum of the forces acting on mass m_1 must equal 0. Hence, we have

$$f_1 - k_1 x_1 + k_2(x_2 - x_1) = 0$$

Mass m_2 is being acted upon by springs k_2 and k_3. Using similar analysis, we find that we also have

$$f_2 - k_3x_2 - k_2(x_2 - x_1) = 0$$

Rearranging this gives

$$\begin{aligned}(k_1 + k_2)x_1 - k_2x_2 &= f_1\\ -k_2x_1 + (k_2 + k_3)x_2 &= f_2\end{aligned}$$

Therefore, to determine the equilibrium displacements x_1, x_2 we just need to solve this system of linear equations.

EXAMPLE 2.4.1

Consider the spring-mass system in Figure 2.4.1. Assume that spring k_1 has spring constant $k_1 = 2$ N/m, spring k_2 has spring constant $k_2 = 4$ N/m, and spring k_3 has spring constant $k_3 = 3$ N/m. Find the equilibrium displacements x_1, x_2 if forces $f_1 = 10$ N and $f_2 = 5$ N are applied to masses m_1 and m_2 respectively.

Solution: Our work above shows us that we just need to solve the system of linear equations

$$\begin{aligned}(2 + 4)x_1 - 4x_2 &= 10\\ -4x_1 + (4 + 3)x_2 &= 5\end{aligned}$$

Row reducing the corresponding augmented matrix gives

$$\left[\begin{array}{cc|c} 6 & -4 & 10\\ -4 & 7 & 5\end{array}\right] \sim \left[\begin{array}{cc|c} 1 & 0 & 45/13\\ 0 & 1 & 35/13\end{array}\right]$$

Hence, the equilibrium displacements are $x_1 = 45/13$ m and $x_2 = 35/13$ m.

Resistor Circuits in Electricity

The flow of electrical current in simple electrical circuits is described by simple linear laws. In an electrical circuit, the **current** has a direction and therefore has a sign attached to it; **voltage** is also a signed quantity; **resistance** is a positive scalar. The laws for electrical circuits are discussed next.

Ohm's Law

If an electrical current of magnitude I is flowing through a resistor with resistance R, then the drop in the voltage across the resistor is $V = IR$. The filament in a light bulb and the heating element of an electrical heater are familiar examples of electrical resistors. (See Figure 2.4.2.)

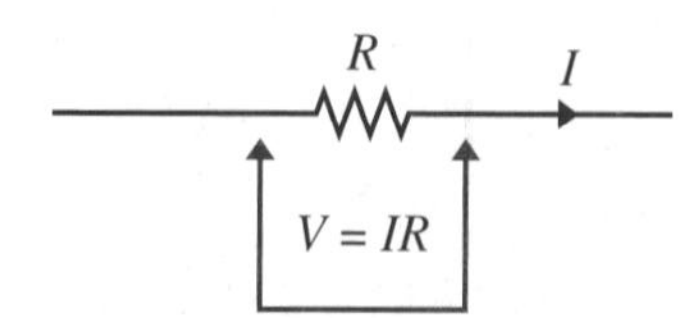

Figure 2.4.2 Ohm's Law: the voltage across the resistor is $V = IR$.

Kirchhoff's Laws

Kirchhoff's Current Law: At a node or junction where several currents enter, the signed sum of the currents entering the node is zero. (See Figure 2.4.3.)

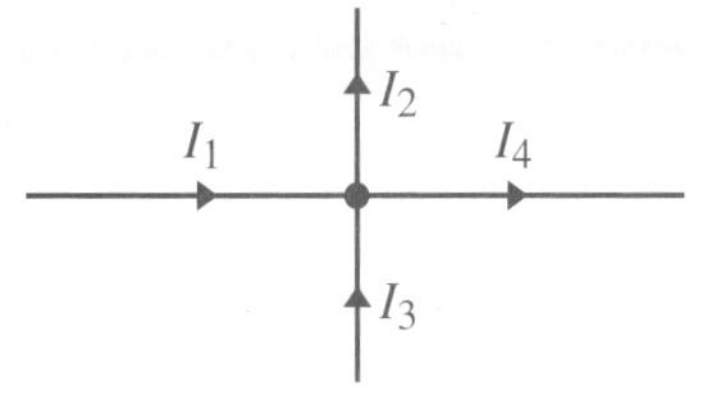

Figure 2.4.3 Kirchhoff's Current Law: $I_1 - I_2 + I_3 - I_4 = 0$.

Kirchhoff's Voltage Law: In a closed loop consisting of only resistors and an **electromotive force** E (for example, E might be due to a battery), the sum of the voltage drops across resistors is equal to E. (See Figure 2.4.4.)

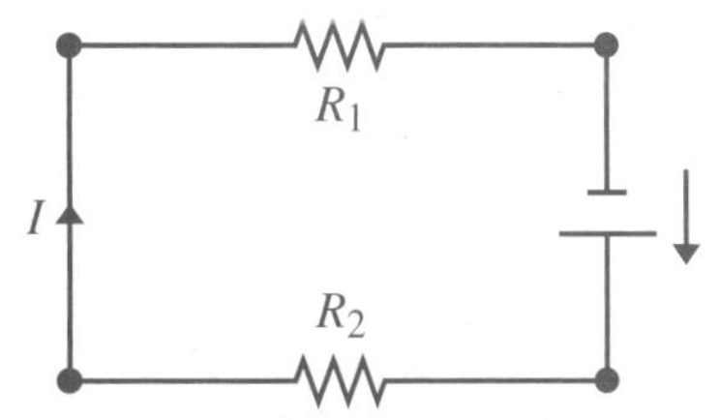

Figure 2.4.4 Kirchhoff's Voltage Law: $E = R_1 I + R_2 I$.

Note that we adopt the convention of drawing an arrow to show the direction of I or of E. These arrows can be assigned arbitrarily, and then the circuit laws will determine whether the quantity has a positive or negative sign. If the quantity has a negative sign, then it means the flow is in the opposite direction of the arrow. It is important to be consistent in using these assigned directions when you use Kirchhoff's Voltage Law.

Sometimes it is necessary to determine the current flowing in each of the loops of a network of loops as shown in Figure 2.4.5. (If the sources of electromotive force are distributed in various places, it will not be sufficient to deal with the problems as a collection of resistors "in parallel and/or in series.") In such problems, it is convenient to introduce the idea of the "current in the loop," which will be denoted i. The true current across any circuit element is given as the algebraic (signed) sum of the "loop currents" flowing through that circuit element. For example, in Figure 2.4.5, the circuit consists of four loops, and a loop current has been indicated in each loop. Across the resistor R_1 in the figure, the true current is simply the loop current i_1; however, across the resistor R_2, the true current (directed from top to bottom) is $i_1 - i_2$. Similarly, across R_4, the true current (from right to left) is $i_1 - i_3$.

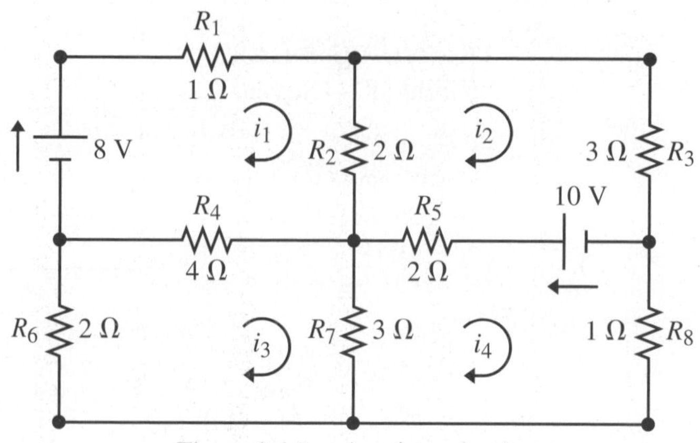

Figure 2.4.5 A resistor circuit.

EXAMPLE 2.4.2 Find all loop currents in the electric circuit in Figure 2.4.5.

Solution: We use Kirchhoff's Voltage Law on each of the four loops:

the top left loop: $i_1 + 2(i_1 - i_2) + 4(i_1 - i_3) = 8$

the top right loop: $3i_2 + 2(i_2 - i_4) + 2(i_2 - i_1) = 10$

the bottom left loop: $2i_3 + 4(i_3 - i_1) + 4(i_3 - i_4) = 0$

the bottom right loop: $2i_4 + 4(i_4 - i_3) + 2(i_4 - i_2) = -10$

Solving this system of 4 equations in the 4 unknowns I_1, I_2, I_3, I_4, we find that the currents are

$$\begin{bmatrix} I_1 \\ I_2 \\ I_3 \\ I_4 \end{bmatrix} = \frac{1}{1069} \begin{bmatrix} 2172 \\ 1922 \\ 702 \\ -790 \end{bmatrix}$$

In addition to showing you an application of linear algebra, the point of this example is to show that even for a fairly simple electrical circuit with the most basic elements (resistors), the analysis requires you to be competent in dealing with large systems of linear equations. Systematic, efficient methods of solution are essential.

Obviously, as the number of nodes and loops in the network increases, so does the number of variables and equations. For larger systems, it is important to know whether you have the correct number of equations to determine the unknowns. Thus, the theorems in Sections 2.1 and 2.2, the idea of rank, and the idea of linear independence are all important.

The Moral of This Example Linear algebra is an essential tool for dealing with large systems of linear equations that may arise in dealing with circuits; really interesting examples cannot be given without assuming greater knowledge of electrical circuits and their components.

Water Flow

Kirchhoff's Current Law can be applied to many other situations. Here we will look at water flowing through a system of pipes. However, the same principles and techniques can be applied to a variety of other problems involving networks; for example, in communication networks (including Internet connectivity), transportation networks, and economic networks.

EXAMPLE 2.4.3

The diagram below indicates a system of pipes and the water flow entering or leaving the system. An irrigation engineer may be interested in determining the amount of water flowing through each pipe and the effects of changing the amount of flow in a certain pipe by use of a ball valve.

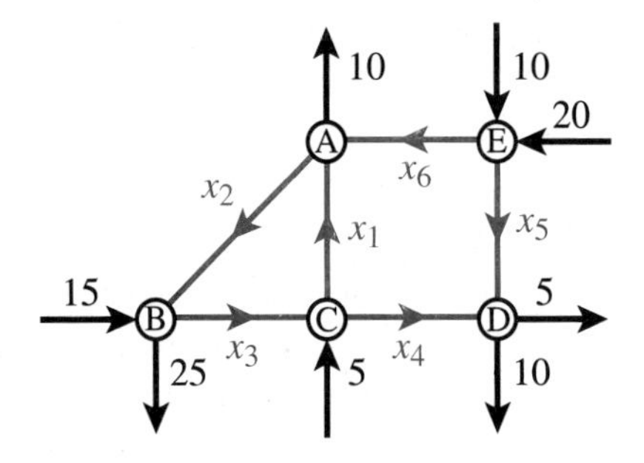

We set up a system of linear equations by equating the rate in and the rate out at each intersection. We get

Intersection	rate in	=	rate out
A	$x_1 + x_6$	=	$10 + x_2$
B	$x_2 + 15$	=	$25 + x_3$
C	$x_3 + 5$	=	$x_1 + x_4$
D	$x_4 + x_5$	=	$5 + 10$
E	$10 + 20$	=	$x_5 + x_6$

Rearranging this gives the system of linear equations

$$\begin{aligned} x_1 - x_2 + x_6 &= 10 \\ x_2 - x_3 &= 10 \\ -x_1 + x_3 - x_4 &= -5 \\ x_4 + x_5 &= 15 \\ -x_5 - x_6 &= -30 \end{aligned}$$

Row reducing the corresponding augmented matrix gives

$$\left[\begin{array}{cccccc|c} 1 & -1 & 0 & 0 & 0 & 1 & 10 \\ 0 & 1 & -1 & 0 & 0 & 0 & 10 \\ -1 & 0 & 1 & -1 & 0 & 0 & -5 \\ 0 & 0 & 0 & 1 & 1 & 0 & 15 \\ 0 & 0 & 0 & 0 & -1 & -1 & -30 \end{array}\right] \sim \left[\begin{array}{cccccc|c} 1 & 0 & -1 & 0 & 0 & 1 & 20 \\ 0 & 1 & -1 & 0 & 0 & 0 & 10 \\ 0 & 0 & 0 & 1 & 0 & -1 & -15 \\ 0 & 0 & 0 & 0 & 1 & 1 & 30 \\ 0 & 0 & 0 & 0 & 0 & 0 & 0 \end{array}\right]$$

EXAMPLE 2.4.3
(continued)

Hence, the general solution is

$$\begin{aligned} x_1 &= 20 + x_3 - x_6 \\ x_2 &= 10 + x_3 \\ x_4 &= -15 + x_6 \\ x_5 &= 30 - x_6 \end{aligned}$$

with $x_3, x_6 \in \mathbb{R}$.

We do not have enough information to determine the exact flow through all pipes. If we were to able to install water flow meters on pipes x_3 and x_6, then we would be able to deduce the exact amounts flowing through the remaining pipes.

For example, if we find that $x_3 = 20$ L/s and $x_6 = 10$ L/s, then we get that

$$x_1 = 30 \text{ L/s}, x_2 = 30 \text{ L/s}, x_4 = -5 \text{ L/s}, x_5 = 20 \text{ L/s}$$

Observe that in this situation water is flowing along pipe x_4 in the opposite direction that we indicated. If we want to ensure that water is flowing along pipe x_4 to the right, we need to increase the rate of flow in pipe x_6 (by widening the pipe for example) so that the flow rate of pipe x_6 is greater than 15 L/s. Alternately, if we can get the flow rate of pipe x_6 to exactly 15 L/s, then we can safely remove pipe x_4.

Partial Fraction Decomposition

Consider a rational function $\dfrac{N(x)}{D(x)}$, where the degree of the polynomial $N(x)$ is less than the degree of the polynomial $D(x)$. There are several cases in mathematics where it is very useful to rewrite such a rational function as a sum of rational functions whose denominators are linear or irreducible quadratic factors. Most students will first encounter this when learning techniques of integration. They are also used when working with Taylor or Laurent series and when working with the inverse Laplace transform.

For example, given a rational function of the form

$$\frac{cx+d}{(x-a)(x-b)}$$

we want to find constants A and B such that

$$\frac{cx+d}{(x-a)(x-b)} = \frac{A}{x-a} + \frac{B}{x-b}$$

Essentially, this procedure, called the **partial fraction decomposition**, is the opposite of adding fractions.

We demonstrate this with a couple of examples.

EXAMPLE 2.4.4

Find constants A and B such that $\dfrac{1}{(x+3)(x-2)} = \dfrac{A}{x+3} + \dfrac{B}{x-2}$.

Solution: If we multiply both sides of the equation by $(x+3)(x-2)$, we get

$$1 = A(x-2) + B(x+3)$$

We rewrite this as

$$0x + 1 = (A+B)x + (-2A + 3B)$$

Upon comparing coefficients of like powers of x, we get a system of linear equations in the unknowns A and B

$$\begin{aligned} A + B &= 0 \\ -2A + 3B &= 1 \end{aligned}$$

Row reducing the corresponding augmented matrix gives

$$\left[\begin{array}{cc|c} 1 & 1 & 0 \\ -2 & 3 & 1 \end{array}\right] \sim \left[\begin{array}{cc|c} 1 & 0 & -1/5 \\ 0 & 1 & 1/5 \end{array}\right]$$

Hence,

$$\frac{1}{(x+3)(x-2)} = \frac{-1/5}{x+3} + \frac{1/5}{x-2}$$

It is easy to perform the addition on the right to verify this is correct.

EXAMPLE 2.4.5

Find constants A, B, C, D, E, and F such that

$$\frac{3x^5 - 2x^4 - x + 1}{(x^2+1)^2(x^2+x+1)} = \frac{Ax+B}{x^2+1} + \frac{Cx+D}{(x^2+1)^2} + \frac{Ex+F}{x^2+x+1}$$

Solution: We first multiply both sides of the equation by $(x^2+1)^2(x^2+x+1)$ to get

$$\begin{aligned} 3x^5 - 2x^4 - x + 1 =& (Ax+B)(x^2+1)(x^2+x+1) + (Cx+D)(x^2+x+1) \\ &+ (Ex+F)(x^2+1)^2 \end{aligned}$$

Expanding the right and collecting coefficients of like powers of x gives

$$\begin{aligned} 3x^5 - 2x^4 - x + 1 =& (A+E)x^5 + (A+B+F)x^4 + (2A+B+C+2E)x^3 \\ &+ (A+2B+C+D+2F)x^2 + (A+B+C+D+E)x + B + D + F \end{aligned}$$

Comparing coefficients of like powers of x gives the system of linear equations

$$\begin{aligned} A + E &= 3 \\ A + B + F &= -2 \\ 2A + B + C + 2E &= 0 \\ A + 2B + C + D + 2F &= 0 \\ A + B + C + D + E &= -1 \\ B + D + F &= 1 \end{aligned}$$

EXAMPLE 2.4.5
(continued)

Row reducing the corresponding augmented matrix gives

$$\left[\begin{array}{cccccc|c} 1 & 0 & 0 & 0 & 1 & 0 & 3 \\ 1 & 1 & 0 & 0 & 0 & 1 & -2 \\ 2 & 1 & 1 & 0 & 2 & 0 & 0 \\ 1 & 2 & 1 & 1 & 0 & 2 & 0 \\ 1 & 1 & 1 & 1 & 1 & 0 & -1 \\ 0 & 1 & 0 & 1 & 0 & 1 & 1 \end{array}\right] \sim \left[\begin{array}{cccccc|c} 1 & 0 & 0 & 0 & 0 & 0 & -1 \\ 0 & 1 & 0 & 0 & 0 & 0 & -7 \\ 0 & 0 & 1 & 0 & 0 & 0 & 1 \\ 0 & 0 & 0 & 1 & 0 & 0 & 2 \\ 0 & 0 & 0 & 0 & 1 & 0 & 4 \\ 0 & 0 & 0 & 0 & 0 & 1 & 6 \end{array}\right]$$

Hence,

$$\frac{3x^5 - 2x^4 - x + 1}{(x^2+1)^2(x^2+x+1)} = \frac{-x-7}{x^2+1} + \frac{x+2}{(x^2+1)^2} + \frac{4x+6}{x^2+x+1}$$

Balancing Chemical Equations

When molecules are combined under the correct conditions, a chemical reaction occurs. For example, in Andy Weir's novel *The Martian*, the main character, Mark Watney, uses an iridium catalyst to first convert hydrazine into nitrogen gas and hydrogen gas according to the **chemical equation**

$$N_2H_4 \to N_2 + H_2$$

He then burns the hydrogen gas with the oxygen in the Hab to get the chemical reaction

$$H_2 + O_2 \to H_2O$$

Although these chemical equations indicate how the **reactants** (the molecules on the left of the arrows) are rearranged into the **products** (the molecules on the right of the arrows), we see that the equations are not really complete. In particular, in the first chemical equation, there are four hydrogen atoms on the left side, but only two on the right side. Similarly, in the second chemical equation, there are two oxygen atoms on the left, but only one on the right.

A chemical equation is said to be **balanced** if both sides of the equation have the same number of atoms of each type. So, for example, the chemical equation

$$N_2H_4 \to N_2 + 2H_2$$

is balanced since both the left and right side have two nitrogen atoms and four hydrogen atoms. Similarly, we can see that the chemical equation

$$2H_2 + O_2 \to 2H_2O$$

is balanced. Of course, technically, the chemical equation

$$4H_2 + 2O_2 \to 4H_2O$$

is also balanced, but we always try to use the smallest positive integers that balance the equation.

As usual, we will demonstrate this with a couple of examples.

EXAMPLE 2.4.6

Balance the chemical equation

$$H_3PO_4 + Mg(OH)_2 \rightarrow Mg_3(PO_4)_2 + H_2O$$

Solution: We want to find constants x_1, x_2, x_3, x_4 such that

$$x_1 H_3PO_4 + x_2 Mg(OH)_2 \rightarrow x_3 Mg_3(PO_4)_2 + x_4 H_2O$$

is balanced. To turn this into a vector equation, we represent the molecules in the equation with the vectors in $\mathbb{R}^4$:

$$\begin{bmatrix} \text{\# of hydrogen atoms} \\ \text{\# of phosphorus atoms} \\ \text{\# of oxygen atoms} \\ \text{\# of magnesium atoms} \end{bmatrix}$$

We get

$$x_1 \begin{bmatrix} 3 \\ 1 \\ 4 \\ 0 \end{bmatrix} + x_2 \begin{bmatrix} 2 \\ 0 \\ 2 \\ 1 \end{bmatrix} = x_3 \begin{bmatrix} 0 \\ 2 \\ 8 \\ 3 \end{bmatrix} + x_4 \begin{bmatrix} 2 \\ 0 \\ 1 \\ 0 \end{bmatrix}$$

Moving all the terms to the left side and performing the linear combination of vectors, we get the homogeneous system

$$\begin{aligned} 3x_1 + 2x_2 - 2x_4 &= 0 \\ x_1 - 2x_3 &= 0 \\ 4x_1 + 2x_2 - 8x_3 - x_4 &= 0 \\ x_2 - 3x_3 &= 0 \end{aligned}$$

Row reducing the corresponding coefficient matrix gives

$$\begin{bmatrix} 3 & 2 & 0 & -2 \\ 1 & 0 & -2 & 0 \\ 4 & 2 & -8 & -1 \\ 0 & 1 & -3 & 0 \end{bmatrix} \sim \begin{bmatrix} 1 & 0 & 0 & -1/3 \\ 0 & 1 & 0 & -1/2 \\ 0 & 0 & 1 & -1/6 \\ 0 & 0 & 0 & 0 \end{bmatrix}$$

We find that a vector equation for the solution space is

$$\begin{bmatrix} x_1 \\ x_2 \\ x_3 \\ x_4 \end{bmatrix} = t \begin{bmatrix} 1/3 \\ 1/2 \\ 1/6 \\ 1 \end{bmatrix}, \quad t \in \mathbb{R}$$

To get the smallest positive integer values, we take $t = 6$. This gives $x_1 = 2$, $x_2 = 3$, $x_3 = 1$, and $x_4 = 6$. Thus, a balanced chemical equation is

$$2H_3PO_4 + 3Mg(OH)_2 \rightarrow Mg_3(PO_4)_2 + 6H_2O$$

EXAMPLE 2.4.7

Balance the chemical equation

$$(NH_4)_3PO_4 + Pb(NO_3)_4 \to Pb_3(PO_4)_4 + NH_4NO_3$$

Solution: We want to find constants x_1, x_2, x_3, x_4 such that

$$x_1(NH_4)_3PO_4 + x_2Pb(NO_3)_4 \to x_3Pb_3(PO_4)_4 + x_4NH_4NO_3$$

is balanced. Define vectors in $\mathbb{R}^5$ by

$$\begin{bmatrix} \text{\# of nitrogen atoms} \\ \text{\# of hydrogen atoms} \\ \text{\# of phosphorus atoms} \\ \text{\# of oxygen atoms} \\ \text{\# of lead atoms} \end{bmatrix}$$

We get the vector equation

$$x_1\begin{bmatrix} 3 \\ 12 \\ 1 \\ 4 \\ 0 \end{bmatrix} + x_2\begin{bmatrix} 4 \\ 0 \\ 0 \\ 12 \\ 1 \end{bmatrix} = x_3\begin{bmatrix} 0 \\ 0 \\ 4 \\ 16 \\ 3 \end{bmatrix} + x_4\begin{bmatrix} 2 \\ 4 \\ 0 \\ 3 \\ 0 \end{bmatrix}$$

Rearranging gives the homogeneous system

$$\begin{aligned} 3x_1 + 4x_2 - 2x_4 &= 0 \\ 12x_1 - 4x_4 &= 0 \\ x_1 - 4x_3 &= 0 \\ 4x_1 + 12x_2 - 16x_3 - 3x_4 &= 0 \\ x_2 - 3x_3 &= 0 \end{aligned}$$

Row reducing the corresponding coefficient matrix gives

$$\begin{bmatrix} 3 & 4 & 0 & -2 \\ 12 & 0 & 0 & -4 \\ 1 & 0 & -4 & 0 \\ 4 & 12 & -16 & -3 \\ 0 & 1 & -3 & 0 \end{bmatrix} \sim \begin{bmatrix} 1 & 0 & 0 & -1/3 \\ 0 & 1 & 0 & -1/4 \\ 0 & 0 & 1 & -1/12 \\ 0 & 0 & 0 & 0 \\ 0 & 0 & 0 & 0 \end{bmatrix}$$

We find that a vector equation for the solution space is

$$\begin{bmatrix} x_1 \\ x_2 \\ x_3 \\ x_4 \end{bmatrix} = t\begin{bmatrix} 1/3 \\ 1/4 \\ 1/12 \\ 1 \end{bmatrix}, \quad t \in \mathbb{R}$$

To get the smallest positive integer values, we take $t = 12$. This gives $x_1 = 4$, $x_2 = 3$, $x_3 = 1$, and $x_4 = 12$. Thus, a balanced chemical equation is

$$4(NH_4)_3PO_4 + 3Pb(NO_3)_4 \to Pb_3(PO_4)_4 + 12NH_4NO_3$$

Planar Trusses

It is common to use trusses, such as the one shown in Figure 2.4.6, in construction. For example, many bridges employ some variation of this design. When designing such structures, it is necessary to determine the **axial forces** in each **member** of the structure (that is, the force along the long axis of the member). To keep this simple, only two-dimensional trusses with hinged joints will be considered; it will be assumed that any displacements of the joints under loading are small enough to be negligible.

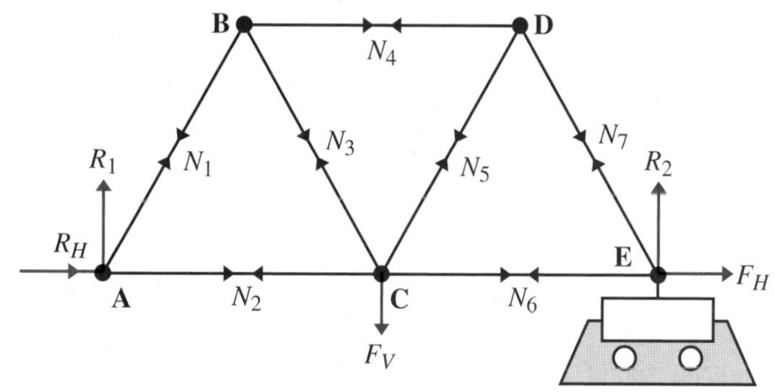

Figure 2.4.6 A planar truss. All triangles are equilateral with sides of length s.

The external loads (such as vehicles on a bridge, or wind or waves) are assumed to be given. The reaction forces at the supports (shown as R_1, R_2, and R_H in the figure) are also external forces; these forces must have values such that the total external force on the structure is zero. To get enough information to design a truss for a particular application, we must determine the forces in the members under various loadings. To illustrate the kinds of equations that arise, we shall consider only the very simple case of a vertical force F_V acting at C and a horizontal force F_H acting at E. Notice that in this figure, the right end of the truss is allowed to undergo small horizontal displacements; it turns out that if a reaction force were applied here as well, the equations would not uniquely determine all the unknown forces (the structure would be "statically indeterminate"), and other considerations would have to be introduced.

The geometry of the truss is assumed given: here it will be assumed that the triangles are equilateral with all sides equal to s metres.

First consider the equations that indicate that the total force on the structure is zero and that the total moment about some convenient point due to those forces is zero. Note that the axial force along the members does not appear in this first set of equations.

- Total horizontal force: $R_H + F_H = 0$
- Total vertical force: $R_1 + R_2 - F_V = 0$
- Moment about A: $-F_V(s) + R_2(2s) = 0$, so $R_2 = \frac{1}{2}F_V = R_1$

Next, we consider the system of equations obtained from the fact that the sum of the forces at each joint must be zero. The moments are automatically zero because the forces along the members act through the joints.

At a joint, each member at that joint exerts a force in the direction of the axis of the member. It will be assumed that each member is in tension, so it is "pulling" away from the joint; if it were compressed, it would be "pushing" at the joint. As indicated in the figure, the force exerted on this joint A by the upper-left-hand member has magnitude N_1; with the conventions that forces to the right are positive and forces up are positive, the force vector exerted by this member on the joint A is $\begin{bmatrix} N_1/2 \\ \sqrt{3}N_1/2 \end{bmatrix}$. On the joint B, the same member will exert a force $\begin{bmatrix} -N_1/2 \\ -\sqrt{3}N_1/2 \end{bmatrix}$. If N_1 is positive, the force is a tension force; if N_1 is negative, there is compression.

For each of the joints A, B, C, D, and E, there are two equations—the first for the sum of horizontal forces and the second for the sum of the vertical forces:

$$\begin{array}{lllllllll}
A1 & N_1/2 & +N_2 & & & & & +R_H & = 0 \\
A2 & \sqrt{3}N_1/2 & & & & & & +R_1 & = 0 \\
B1 & -N_1/2 & & +N_3/2 \; +N_4 & & & & & = 0 \\
B2 & -\sqrt{3}N_1/2 & & -\sqrt{3}N_3/2 & & & & & = 0 \\
C1 & & -N_2 & -N_3/2 & +N_5/2 & +N_6 & & & = 0 \\
C2 & & & \sqrt{3}N_3/2 & +\sqrt{3}N_5/2 & & & & = F_V \\
D1 & & & -N_4 & -N_5/2 & & +N_7/2 & & = 0 \\
D2 & & & & -\sqrt{3}N_5/2 & & -\sqrt{3}N_7/2 & & = 0 \\
E1 & & & & & -N_6 & -N_7/2 & & = -F_H \\
E2 & & & & & & \sqrt{3}N_7/2 & +R_2 & = 0
\end{array}$$

Notice that if the reaction forces are treated as unknowns, this is a system of 10 equations in 10 unknowns. The geometry of the truss and its supports determines the coefficient matrix of this system, and it could be shown that the system is necessarily consistent with a unique solution. Notice also that if the horizontal force equations (A1, B1, C1, D1, and E1) are added together, the sum is the total horizontal force equation, and similarly the sum of the vertical force equations is the total vertical force equation. A suitable combination of the equations would also produce the moment equation, so if those three equations are solved as above, then the 10 joint equations will still be a consistent system for the remaining 7 axial force variables.

For this particular truss, the system of equations is quite easy to solve, since some of the variables are already leading variables. For example, if $F_H = 0$, from A2 and E2 it follows that $N_1 = N_7 = -\frac{1}{\sqrt{3}}F_V$ and then B2, C2, and D2 give $N_3 = N_5 = \frac{1}{\sqrt{3}}F_V$; then A1 and E1 imply that $N_2 = N_6 = \frac{1}{2\sqrt{3}}F_V$, and B1 implies that $N_4 = -\frac{1}{\sqrt{3}}F_V$. Note that the members AC, BC, CD, and CE are under tension, and AB, BD, and DE experience compression, which makes intuitive sense.

This is a particularly simple truss. In the real world, trusses often involve many more members and use more complicated geometry; trusses may also be three-dimensional. Therefore, the systems of equations that arise may be considerably larger and more complicated. It is also sometimes essential to introduce considerations other than the equations of equilibrium of forces in statics. To study these questions, you need to know the basic facts of linear algebra.

It is worth noting that in the system of equations above, each of the quantities N_1, $N_2, \ldots, N_7$ appears with a non-zero coefficient in only some of the equations. Since each member touches only two joints, this sort of special structure will often occur in the equations that arise in the analysis of trusses. A deeper knowledge of linear algebra is important in understanding how such special features of linear equations may be exploited to produce efficient solution methods.

Linear Programming

Linear programming is a procedure for deciding the best way to allocate resources. "Best" may mean fastest, most profitable, least expensive, or best by whatever criterion is appropriate. For linear programming to be applicable, the problem must have some special features. These will be illustrated by an example.

In a primitive economy, a man decides to earn a living by making hinges and gate latches. He is able to obtain a supply of 25 kg per week of suitable metal at a price of 2 dollars per kg. His design requires 500 g to make a hinge and 250 g to make a gate latch. With his primitive tools, he finds that he can make a hinge in 1 hour, and it takes 3/4 hour to make a gate latch. He is willing to work 60 hours a week. The going price is 3 dollars for a hinge and 2 dollars for a gate latch. How many hinges and how many gate latches should he produce each week in order to maximize his net income?

To analyze the problem, let x be the number of hinges produced per week and let y be the number of gate latches. Then the amount of metal used is $(0.5x + 0.25y)$ kg. Clearly, this must be less than or equal to 25 kg:

$$0.5x + 0.25y \leq 25$$

Multiplying by 4 to clear the decimals gives

$$2x + y \leq 100$$

Such an inequality is called a **constraint** on x and y; it is a linear constraint because the corresponding equation is linear.

Our producer also has a time constraint: the time taken making hinges plus the time taken making gate latches cannot exceed 60 hours. Therefore,

$$1x + 0.75y \leq 60$$

which can be rewritten as

$$4x + 3y \leq 240$$

Obviously, also $x \geq 0$ and $y \geq 0$.

The producer's net revenue for selling x hinges and y gate latches is

$$R(x, y) = 3x + 2y - 2(25)$$

This is called the **objective function** for the problem. The mathematical problem can now be stated as follows:

Find the point (x, y) that maximizes the objective function

$$R(x, y) = 3x + 2y - 50$$

subject to the linear constraints

$$\begin{aligned} 2x + y &\leq 100 \\ 4x + 3y &\leq 240 \\ x &\geq 0 \\ y &\geq 0 \end{aligned}$$

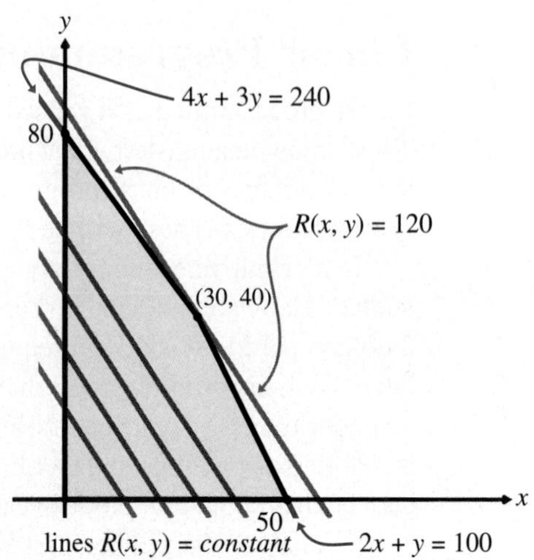

Figure 2.4.7 The feasible region for the linear programming example. The grey lines are level sets of the objective function R.

It is useful to introduce one piece of special vocabulary: the **feasible region** for the problem is the set of (x, y) satisfying all of the constraints. Our goal is to find which point in the feasible region gives us the maximum value of $R(x, y)$.

Consider sets of the form $R(x, y) = k$, where k is a constant. These sets form a family of parallel lines, called the **objective lines** as in Figure 2.4.7.

As we move further from the origin into the first quadrant, $R(x, y)$ increases. The biggest possible value for $R(x, y)$ will occur at a point where the set $R(x, y) = k$ (for some constant k to be determined) just touches the feasible region. For larger values of $R(x, y)$, the set $R(x, y) = k$ does not meet the feasible region at all, so there are no feasible points that give such bigger values of R. The touching must occur at a **vertex**—that is, at an intersection point of two of the boundary lines. (In general, the line $R(x, y) = k$ for the largest possible constant could touch the feasible region along a line segment that makes up part of the boundary. But such a line segment has two vertices as endpoints, so it is correct to say that the touching occurs at a vertex.)

For this particular problem, we see that the vertices of the feasible region are $(0, 0)$, $(50, 0)$, $(0, 80)$, and at the intersection of the two lines $2x + y = 100$ and $4x + 3y = 240$. We can find the intersection by solving

$$\begin{aligned} 2x + y &= 100 \\ 4x + 3y &= 240 \end{aligned}$$

The solution is $x = 30, y = 40$.

Now compare the values of $R(x, y)$ at all of these vertices:

$$\begin{aligned} R(0, 0) &= -50 \\ R(50, 0) &= 100 \\ R(0, 80) &= 110 \\ R(30, 40) &= 120 \end{aligned}$$

The vertex $(30, 40)$ gives the best net revenue, so the producer should make 30 hinges and 40 gate latches each week.

General Remarks Problems involving allocation of resources can be found in numerous areas such as management engineering, economics, politics, and biology. Problems such as scheduling ship transits through a canal can be analyzed this way. Oil companies must make choices about the grades of crude oil to use in their refineries, and about the amounts of various refined products to produce. Such problems often involve tens or even hundreds of variables—and similar numbers of constraints. The boundaries of the feasible region are hyperplanes in some $\mathbb{R}^n$, where n is large. Although the basic principles of the solution method remain the same as in this example (look for the best vertex), the problem is much more complicated because there are so many vertices. In fact, it is a challenge to find vertices; simply solving all possible combinations of systems of boundary equations is not good enough. Note in the simple two-dimensional example that the point $(60, 0)$ is the intersection point of two of the lines ($y = 0$ and $4x + 3y = 240$) that make up the boundary, but it is not a vertex of the feasible region because it fails to satisfy the constraint $2x + y \leq 100$. For higher-dimension problems, drawing pictures is not good enough, and an organized approach is called for.

The standard method for solving linear programming problems is the **simplex method**, which finds an initial vertex and then prescribes a method (very similar to row reduction) for moving to another vertex, improving the value of the objective function with each step.

PROBLEMS 2.4

Practice Problems

A1 For the spring-mass system below, assume that the spring constants are $k_1 = 2$ N/m, $k_2 = 1$ N/m, and $k_3 = 4$ N/m. Find the equilibrium displacements x_1, x_2, x_3 if forces $f_1 = 10$ N, $f_2 = 6$ N, and $f_3 = 8$ N are applied to masses m_1, m_2, and m_3 respectively.

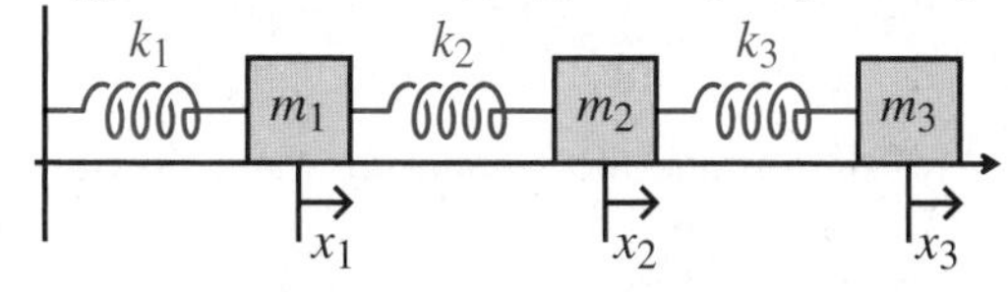

A2 Write the system of linear equations required to solve for the loop currents in the diagram.

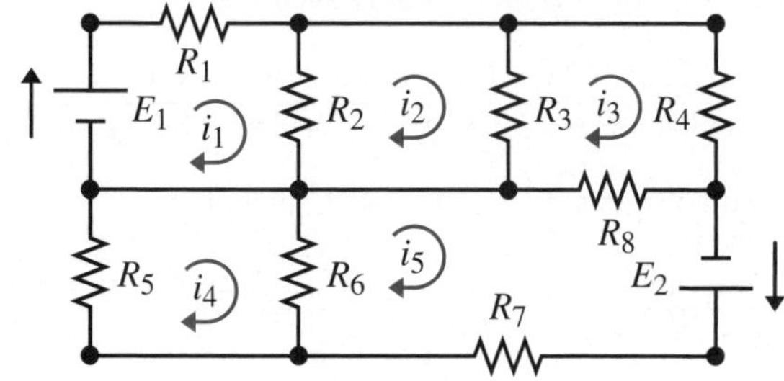

A3 Determine the amount of water flowing through each pipe.

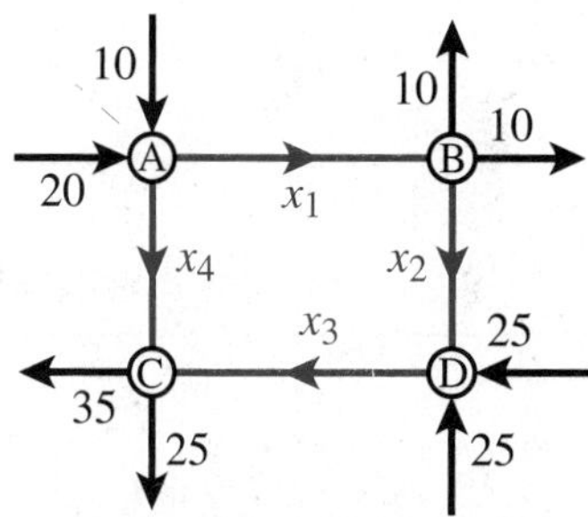

A4 Find constants A, B, C, D, and E such that

$$\frac{4x^4 + x^3 + x^2 + x + 1}{(x-1)(x^2+1)^2} = \frac{A}{x-1} + \frac{Bx+C}{x^2+1} + \frac{Dx+E}{(x^2+1)^2}$$

A5 Balance the chemical equation

$$Al(OH)_3 + H_2CO_3 \rightarrow Al_2(CO_3)_3 + H_2O$$

A6 Find the maximum value of the objective function $x+y$ subject to the constraints $0 \leq x \leq 100, 0 \leq y \leq 80$, and $4x + 5y \leq 600$. Sketch the feasible region.

A7 Determine the system of equations for the reaction forces and axial forces in members for the truss shown in the diagram to the right. Assume that all triangles are right-angled and isosceles, with side lengths s, s, and $\sqrt{2}s$.

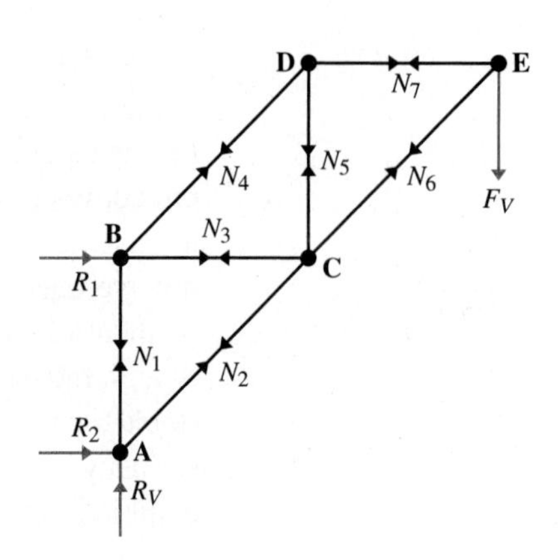

Homework Problems

B1 For the spring-mass system below, assume that the spring constants are $k_1 = 3$ N/m, $k_2 = 2$ N/m, $k_3 = 1$ N/m, and $k_4 = 4$ N/m. Find the equilibrium displacements x_1, x_2, x_3, x_4 if constant forces $f_1 = 2$ N, $f_2 = 3$ N, $f_3 = 3$ N, and $f_4 = 4$ N are applied to masses m_1, m_2, m_3, and m_4 respectively.

B2 For the spring-mass system below, assume that the spring constants are $k_1 = 2$ N/m, $k_2 = 3$ N/m, $k_3 = 1$ N/m, $k_4 = 2$ N/m, and $k_5 = 1$ N/m. Find the equilibrium displacements x_1, x_2, x_3 if a constant force of 1 N is applied to all three masses.

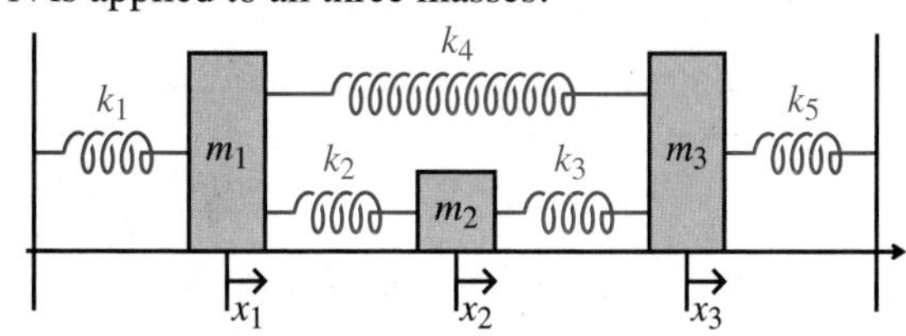

B3 Determine the currents I_1 and I_2 for the electric circuit below.

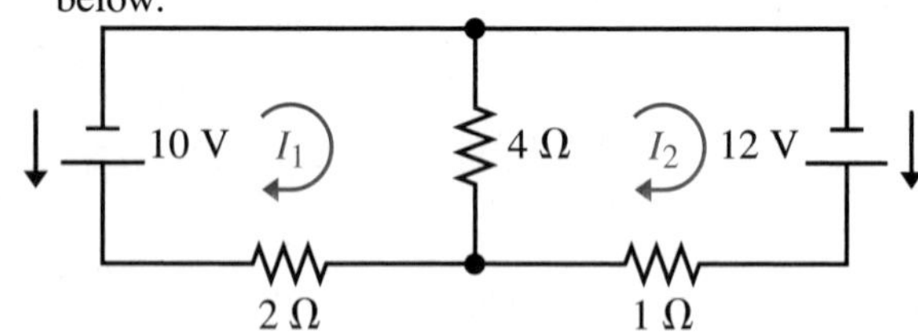

B4 Determine the currents I_1, I_2, and I_3 for the electric circuit below.

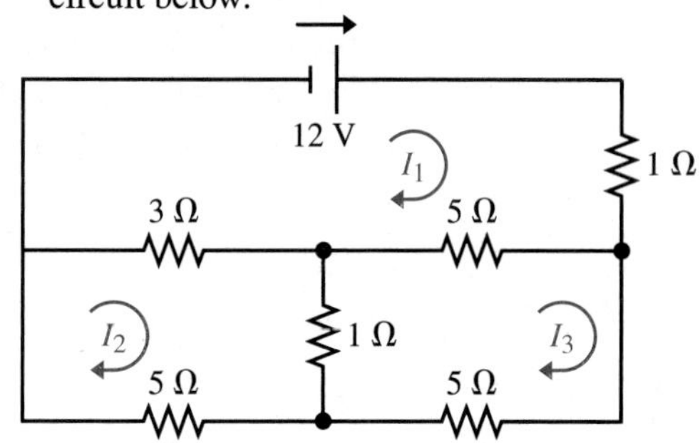

B5 Consider the pipe network below.

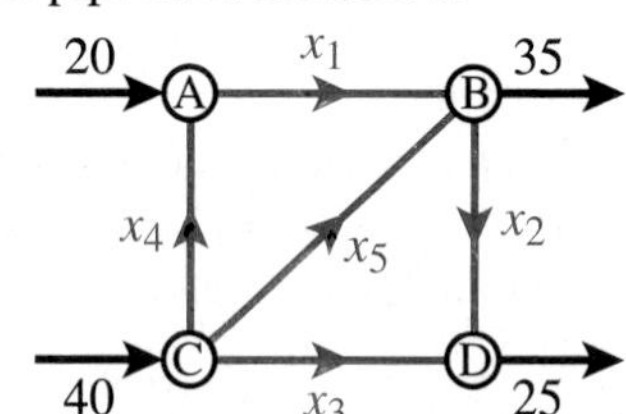

(a) Determine the amount of water flowing through each pipe.

(b) If x_4 and x_5 are both set to a flow rate of 20 L/s, what is the flow rate of x_1, x_2, and x_3?

B6 Find constants A, B, C, and D such that

$$\frac{2x^2 + 3x - 3}{(x^2 + 3)(x^2 + 3x + 3)} = \frac{Ax + B}{x^2 + 3} + \frac{Cx + D}{x^2 + 3x + 3}$$

B7 Find constants A, B, C, D, and E such that

$$\frac{x^3 + 2x^2 - 1}{(x - 1)(x^2 + 1)(x^2 + 3)} = \frac{A}{x - 1} + \frac{Bx + C}{x^2 + 1} + \frac{Dx + E}{x^2 + 3}$$

B8 Balance the chemical equation

$$C_3H_8 + O_2 \to CO_2 + H_2O$$

B9 Balance the chemical equation

$$Ca_3(PO)_4 + SiO_2 + C \to CaSiO_3 + CO + P$$

B10 Determine the system of equations for the reaction forces and axial forces in members for the truss shown in the diagram to the right. Assume that both triangles are right-angled and isosceles, with side lengths s, s, and $\sqrt{2}s$.

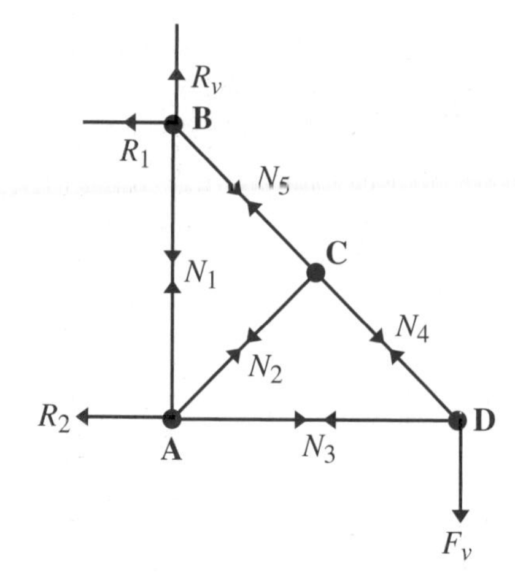

CHAPTER REVIEW

Suggestions for Student Review

Try to answer all of these questions before checking at the suggested locations. In particular, try to invent your own examples. These review suggestions are intended to help you carry out your review. They may not cover every idea you need to master. Working in small groups may improve your efficiency.

1 Describe the geometric interpretation of a system $\left[A \mid \vec{b}\right]$ of m linear equations in n variables and its solutions. What was the alternate view of a system and how does it relate to the content in Section 2.3? (Sections 2.1, 2.3)

2 When you row reduce an augmented matrix $\left[A \mid \vec{b}\right]$ to solve a system of linear equations, why can you stop when the matrix is in row echelon form? How do you use this form to decide if the system is consistent and if it has a unique solution? (Section 2.1)

3 How is reduced row echelon form different from row echelon form? (Section 2.2)

4 (a) Write the augmented matrix of a consistent non-homogeneous system of three linear equations in four variables, such that the coefficient matrix is in row echelon form (but not reduced row echelon form) and of rank 3.

(b) Determine the general solution of your system.

(c) Perform the following sequence of elementary row operations on your augmented matrix:

(i) Interchange the first and second rows.

(ii) Add the (new) second row to the first row.

(iii) Add twice the second row to the third row.

(iv) Add the third row to the second.

(d) Regard the result of (c) as the augmented matrix of a system and solve that system directly. (Don't just use the reverse operation in (c).) Check that your general solution agrees with (b).

5 State the definition of the rank of a matrix and the System-Rank Theorem. Explain how parts (2) and (3) of the System-Rank Theorem relate to both of our interpretations of a system of linear equations. (Sections 2.1, 2.2, 2.3)

6 For homogeneous systems, how can you use the row echelon form to determine whether there are non-trivial solutions and, if there are, how many parameters there are in the general solution? Is there any case where we know (by inspection) that a homogeneous system has non-trivial solutions? (Section 2.2)

7 Write a short explanation of how you use information about consistency of systems and uniqueness of solutions in testing for linear independence and in determining whether a vector $\vec{x}$ belongs to a given subspace of $\mathbb{R}^n$. (Section 2.3)

8 State the definition of a basis and the definition of the dimension of a subspace. What form must the reduced row echelon form of the coefficient matrix of the vector equation $t_1\vec{v}_1 + \cdots + t_k\vec{v}_k = \vec{b}$ have if the set is a basis for $\mathbb{R}^n$? (Section 2.3)

9 Consider a subspace $S = \text{Span}\{\vec{v}_1, \vec{v}_2\}$ of $\mathbb{R}^3$ where $\{\vec{v}_1, \vec{v}_2\}$ is linearly independent.

(a) Explain the procedure for finding a homogeneous system that represents the set.

(b) Explain the two ways of determining if a vector $\vec{x} \in \mathbb{R}^3$ is in S.

(c) Explain in general the advantages and disadvantages of having a system of linear equations representing a set of points rather than a vector equation.

10 Search online for some applications of systems of linear equations to your desired field of study. How do these relate with the applications covered so far in this text? (Section 2.4)

Chapter Quiz

In Problems E1 and E2:

(a) Write the augmented matrix.

(b) Row reduce the augmented matrix and determine the rank of the augmented matrix.

(c) Find the general solution of the system or explain why the system is inconsistent.

E1

$$\begin{aligned} x_2 - 2x_3 + x_4 &= 2 \\ 2x_1 - 2x_2 + 4x_3 - x_4 &= 10 \\ x_1 - x_2 + x_3 &= 2 \\ x_1 + x_3 &= 9 \end{aligned}$$

E2

$$\begin{aligned} 2x_1 + 4x_2 + x_3 - 6x_4 &= 7 \\ 4x_1 + 8x_2 - 3x_3 + 8x_4 &= -1 \\ -3x_1 - 6x_2 + 2x_3 - 5x_4 &= 0 \\ x_1 + 2x_2 + x_3 - 5x_4 &= 5 \end{aligned}$$

E3 Let $A = \begin{bmatrix} 0 & 3 & 3 & 0 & -1 \\ 1 & 1 & 3 & 3 & 1 \\ 2 & 4 & 9 & 6 & 1 \\ -2 & -4 & -6 & -3 & -1 \end{bmatrix}$.

(a) Find the reduced row echelon form of A and state the rank of A.

(b) Write the general solution of the homogeneous system $\left[A \mid \vec{0}\right]$.

(c) What is a basis and the dimension of the solution space of $\left[A \mid \vec{0}\right]$?

E4 Find a homogeneous system that defines Span $\left\{\begin{bmatrix} 1 \\ -1 \\ 2 \end{bmatrix}, \begin{bmatrix} 1 \\ 2 \\ 5 \end{bmatrix}\right\}$.

E5 The matrix $A = \left[\begin{array}{cccc|c} 1 & 2 & 3 & a & 2 \\ 0 & 2 & 1 & 0 & -3 \\ 0 & 0 & b+2 & 0 & b \\ 0 & 0 & 0 & c^2-1 & c+1 \end{array}\right]$ is the augmented matrix of a system of linear equations.

(a) Determine all values of (a, b, c) such that the system is consistent and all values of (a, b, c) such that the system is inconsistent.

(b) Determine all values of (a, b, c) such that the system has a unique solution.

E6 (a) Determine all vectors $\vec{x}$ in $\mathbb{R}^5$ that are orthogonal to

$$\vec{v}_1 = \begin{bmatrix} 1 \\ 1 \\ 3 \\ 1 \\ 4 \end{bmatrix}, \vec{v}_2 = \begin{bmatrix} 2 \\ 1 \\ 5 \\ 0 \\ 0 \end{bmatrix}, \vec{v}_3 = \begin{bmatrix} 3 \\ 2 \\ 8 \\ 5 \\ 9 \end{bmatrix}$$

(b) Let $\vec{u}$, $\vec{v}$, and $\vec{w}$ be three vectors in $\mathbb{R}^5$. Explain why there must be non-zero vectors orthogonal to all of $\vec{u}$, $\vec{v}$, and $\vec{w}$.

E7 Determine whether $\mathcal{B} = \left\{\begin{bmatrix} 1 \\ -3 \\ 2 \end{bmatrix}, \begin{bmatrix} 1 \\ 1 \\ 2 \end{bmatrix}, \begin{bmatrix} 2 \\ 1 \\ 1 \end{bmatrix}\right\}$ is linearly independent.

E8 Determine whether the set $\left\{\begin{bmatrix} 2 \\ 1 \\ 1 \end{bmatrix}, \begin{bmatrix} 3 \\ 2 \\ 1 \end{bmatrix}\right\}$ spans $\mathbb{R}^3$.

E9 Prove that $\mathcal{B} = \left\{\begin{bmatrix} 3 \\ 1 \\ 2 \end{bmatrix}, \begin{bmatrix} 1 \\ 1 \\ 6 \end{bmatrix}, \begin{bmatrix} 4 \\ 1 \\ 5 \end{bmatrix}\right\}$ is a basis for $\mathbb{R}^3$.

For Problems E10–E15, indicate whether the statement is true or false. Justify your answer with a brief explanation or counterexample.

E10 $x_1^2 + 2x_1x_2 + x_2 = 0$ is a linear equation.

E11 A consistent system must have a unique solution.

E12 If there are more equations than variables in a non-homogeneous system of linear equations, then the system must be inconsistent.

E13 If $\begin{bmatrix} 1 \\ 1 \\ 1 \end{bmatrix}$ is a solution of a system of linear equations, then $\begin{bmatrix} 2 \\ 2 \\ 2 \end{bmatrix}$ is also a solution of the system.

E14 Some homogeneous systems of linear equations have unique solutions.

E15 If there are more variables than equations in a system of linear equations, then the system cannot have a unique solution.

Further Problems

These problems are intended to be challenging.

F1 The purpose of this problem is to explore the relationship between the general solution of the system $\left[A \mid \vec{b}\,\right]$ and the general solution of the **corresponding homogeneous** system $\left[A \mid \vec{0}\,\right]$. This relation will be studied with different tools in Section 3.4. We begin by considering some examples where the coefficient matrix is in RREF.

(a) Let $R = \begin{bmatrix} 1 & 0 & r_{13} \\ 0 & 1 & r_{23} \end{bmatrix}$. Show that the general solution of the homogeneous system $\left[R \mid \vec{0}\,\right]$ is

$$\vec{x}_H = t\vec{v}, \quad t \in \mathbb{R}$$

where $\vec{v}$ is expressed in terms of r_{13} and r_{23}. Show that the general solution of the non-homogeneous system $[R \mid \vec{c}\,]$ is

$$\vec{x}_N = \vec{p} + \vec{x}_H$$

where $\vec{p}$ is expressed in terms of $\vec{c}$.

(b) Let $R = \begin{bmatrix} 1 & r_{12} & 0 & 0 & r_{15} \\ 0 & 0 & 1 & 0 & r_{25} \\ 0 & 0 & 0 & 1 & r_{35} \end{bmatrix}$. Show that the general solution of the homogeneous system $\left[R \mid \vec{0}\,\right]$ is

$$\vec{x}_H = t_1\vec{v}_1 + t_2\vec{v}_2, \quad t_i \in \mathbb{R}$$

where each of $\vec{v}_1$ and $\vec{v}_2$ can be expressed in terms of the entries r_{ij}. Express each $\vec{v}_i$ explicitly. Then show that the general solution of $[R \mid \vec{c}\,]$ can be written as

$$\vec{x}_N = \vec{p} + \vec{x}_H$$

where $\vec{p}$ is expressed in terms of the components $\vec{c}$, and $\vec{x}_H$ is the solution of the corresponding homogeneous system.

The pattern should now be apparent; if it is not, try again with another special case of R. In the next part of this exercise, create an effective labelling system so that you can clearly indicate what you want to say.

(c) Let R be a matrix in RREF, with m rows, n columns, and rank k. Show that the general solution of the homogeneous system $\left[R \mid \vec{0}\,\right]$ is

$$\vec{x}_H = t_1\vec{v}_1 + \cdots + t_{n-k}\vec{v}_{n-k}, \quad t_i \in \mathbb{R}$$

where each $\vec{v}_i$ is expressed in terms of the entries in R. Suppose that the system $[R \mid \vec{c}\,]$ is consistent and show that the general solution is

$$\vec{x}_N = \vec{p} + \vec{x}_H$$

where $\vec{p}$ is expressed in terms of the components of $\vec{c}$, and $\vec{x}_H$ is the solution of the corresponding homogeneous system.

(d) Use the result of (c) to discuss the relationship between the general solution of the consistent system $\left[A \mid \vec{b}\,\right]$ and the corresponding homogeneous system $\left[A \mid \vec{0}\,\right]$.

F2 *This problem involves comparing the efficiency of row reduction procedures.*

When we use a computer to solve large systems of linear equations, we want to keep the number of arithmetic operations as small as possible. This reduces the time taken for calculations, which is important in many industrial and commercial applications. It also tends to improve accuracy: every arithmetic operation is an opportunity to *lose* accuracy through truncation or round-off, subtraction of two nearly equal numbers, and so on.

We want to count the number of multiplications and/or divisions in solving a system by elimination. We focus on these operations because they are more time-consuming than addition or subtraction, and the number of additions is approximately the same as the number of multiplications. We make certain assumptions: the system $\left[A \mid \vec{b}\right]$ has n equations and n variables, and it is consistent with a unique solution. (Equivalently, A has n rows, n columns, and rank n.) We assume for simplicity that no row interchanges are required. (If row interchanges are required, they can be handled by renaming "addresses" in the computer.)

(a) How many multiplications and divisions are required to reduce $\left[A \mid \vec{b}\right]$ to a form $\left[C \mid \vec{d}\right]$ such that C is in row echelon form?

Hints

(1) To carry out the obvious first elementary row operation, compute $\frac{a_{21}}{a_{11}}$—one division. Since we know what will happen in the first column, we do not multiply a_{11} by $\frac{a_{21}}{a_{11}}$, but we must multiply every other element of the first row of $\left[A \mid \vec{b}\right]$ by this factor and subtract the product from the corresponding element of the second row—n multiplications.

(2) Obtain zeros in the remaining entries in the first column, then move to the $(n-1)$ by n blocks consisting of the reduced version of $\left[A \mid \vec{b}\right]$ with the first row and first column deleted.

(3) Note that $\sum_{i=1}^{n} i = \frac{n(n+1)}{2}$ and $\sum_{i=1}^{n} i^2 = \frac{n(n+1)(2n+1)}{6}$.

(4) The biggest term in your answer should be $n^3/3$. Note that n^3 is much greater than n^2 when n is large.

(b) Determine how many multiplications and divisions are required to solve the system with the augmented matrix $\left[C \mid \vec{d}\right]$ of part (a) by back-substitution.

(c) Show that the number of multiplications and divisions required to row reduce $[R \mid \vec{c}\,]$ to reduced row echelon form is the same as the number used in solving the system by back-substitution. Conclude that the Gauss-Jordan procedure is as efficient as Gaussian elimination with back-substitution. For large n, the number of multiplications and divisions is roughly $\frac{n^3}{3}$.

(d) Suppose that we do a "clumsy" Gauss-Jordan procedure. We do not first obtain row echelon form; instead we obtain zeros in all entries above and below a pivot before moving on to the next column. Show that the number of multiplications and divisions required in this procedure is roughly $\frac{n^3}{2}$, so that this procedure requires approximately 50% more operations than the more efficient procedures.

CHAPTER 3

Matrices, Linear Mappings, and Inverses

CHAPTER OUTLINE

In many applications of linear algebra, we use vectors in $\mathbb{R}^n$ to represent quantities, such as forces, and then use the tools of Chapters 1 and 2 to solve various problems. However, there are many times when it is useful to translate a problem into other linear algebra objects. In this chapter, we look at two of these fundamental objects: matrices and linear mappings. We now explore the properties of these objects and show how they are tied together with the material from Chapters 1 and 2.

3.1 Operations on Matrices

We used matrices essentially as bookkeeping devices in Chapter 2. Matrices also possess interesting algebraic properties, so they have wider and more powerful applications than is suggested by their use in solving systems of equations. We now look at some of these algebraic properties.

Definition
Matrix
$M_{m\times n}(\mathbb{R})$

An $m \times n$ **matrix** A is a rectangular array with m rows and n columns. We denote the entry in the i-th row and j-th column by a_{ij}. That is,

$$A = \begin{bmatrix} a_{11} & a_{12} & \cdots & a_{1j} & \cdots & a_{1n} \\ a_{21} & a_{22} & \cdots & a_{2j} & \cdots & a_{2n} \\ \vdots & \vdots & & \vdots & & \vdots \\ a_{i1} & a_{i2} & \cdots & a_{ij} & \cdots & a_{in} \\ \vdots & \vdots & & \vdots & & \vdots \\ a_{m1} & a_{m2} & \cdots & a_{mj} & \cdots & a_{mn} \end{bmatrix}$$

Two $m \times n$ matrices A and B are **equal** if $a_{ij} = b_{ij}$ for all $1 \leq i \leq m$, $1 \leq j \leq n$. The set of all $m \times n$ matrices with real entries is denoted by $M_{m\times n}(\mathbb{R})$.

Remarks

1. When working with multiple matrices we sometimes denote the ij-th entry of a matrix A by $(A)_{ij} = a_{ij}$.
2. In the notation for the set of all $m \times n$ matrices, the $(\mathbb{R})$ indicates that the entries of the matrices are real. In Chapter 9 we will look at the set $M_{m\times n}(\mathbb{C})$, the set of all $m \times n$ matrices with complex entries.

Several special types of matrices arise frequently in linear algebra.

Definition
Square Matrix

A matrix in $M_{n\times n}(\mathbb{R})$ (the number of rows of the matrix is equal to the number of columns) is called a **square matrix**.

Definition
Upper Triangular
Lower Triangular

A matrix $U \in M_{n\times n}(\mathbb{R})$ is said to be **upper triangular** if $u_{ij} = 0$ whenever $i > j$. A matrix $L \in M_{n\times n}(\mathbb{R})$ is said to be **lower triangular** if $l_{ij} = 0$ whenever $i < j$.

EXAMPLE 3.1.1

The matrices $\begin{bmatrix} 2 & 3 \\ 0 & 1 \end{bmatrix}$ and $\begin{bmatrix} 3 & 1 & 2 \\ 0 & 0 & 2 \\ 0 & 0 & 1 \end{bmatrix}$ are upper triangular.

The matrices $\begin{bmatrix} -3 & 0 \\ 1 & 2 \end{bmatrix}$ and $\begin{bmatrix} 0 & 0 & 0 \\ -2 & 3 & 0 \\ 0 & -2 & 1 \end{bmatrix}$ are lower triangular.

The matrix $\begin{bmatrix} 2 & 0 & 0 \\ 0 & 3 & 0 \\ 0 & 0 & 1 \end{bmatrix}$ is both upper and lower triangular.

Definition
Diagonal Matrix

A matrix $D \in M_{n\times n}(\mathbb{R})$ such that $d_{ij} = 0$ for all $i \neq j$ is called a **diagonal matrix** and is denoted by

$$D = \operatorname{diag}(d_{11}, d_{22}, \cdots, d_{nn})$$

EXAMPLE 3.1.2

We denote the diagonal matrix $D = \begin{bmatrix} \sqrt{3} & 0 \\ 0 & -2 \end{bmatrix}$ by $D = \operatorname{diag}(\sqrt{3}, -2)$.

The notation $\operatorname{diag}(0, 3, 1)$ denotes the diagonal matrix $\begin{bmatrix} 0 & 0 & 0 \\ 0 & 3 & 0 \\ 0 & 0 & 1 \end{bmatrix}$.

Observe that the columns of a diagonal matrix are just scalar multiples of the standard basis vectors for $\mathbb{R}^n$. That is,

$$\operatorname{diag}(d_{11}, d_{22}, \cdots, d_{nn}) = \begin{bmatrix} d_{11}\vec{e}_1 & \cdots & d_{nn}\vec{e}_n \end{bmatrix}$$

Matrices as Vectors

We have seen that matrices are useful in solving systems of linear equations. However, we shall see that matrices show up in different kinds of problems, and *it is important to be able to think of matrices as "things" that are worth studying and playing with—and these things may have no connection with a system of equations*. In particular, we now show that we can treat matrices in exactly the same way as we did vectors in $\mathbb{R}^n$ in Chapter 1.

Definition
Addition and Scalar Multiplication of Matrices

Let $A, B \in M_{m\times n}(\mathbb{R})$. We define **addition** of matrices by

$$(A+B)_{ij} = (A)_{ij} + (B)_{ij}$$

We define **scalar multiplication** of A by a scalar $t \in \mathbb{R}$ by

$$(tA)_{ij} = t(A)_{ij}$$

Remark

As with vectors in $\mathbb{R}^n$, $A - B$ is to be interpreted as $A + (-1)B$.

EXAMPLE 3.1.3

Perform the following operations.

(a) $\begin{bmatrix} 2 & 3 \\ 4 & 1 \end{bmatrix} + \begin{bmatrix} 5 & 1 \\ -2 & 7 \end{bmatrix}$

Solution: $\begin{bmatrix} 2 & 3 \\ 4 & 1 \end{bmatrix} + \begin{bmatrix} 5 & 1 \\ -2 & 7 \end{bmatrix} = \begin{bmatrix} 2+5 & 3+1 \\ 4+(-2) & 1+7 \end{bmatrix} = \begin{bmatrix} 7 & 4 \\ 2 & 8 \end{bmatrix}$

(b) $\begin{bmatrix} 3 & 0 \\ 1 & -5 \end{bmatrix} - \begin{bmatrix} 1 & -1 \\ -2 & 0 \end{bmatrix}$

Solution: $\begin{bmatrix} 3 & 0 \\ 1 & -5 \end{bmatrix} - \begin{bmatrix} 1 & -1 \\ -2 & 0 \end{bmatrix} = \begin{bmatrix} 3-1 & 0-(-1) \\ 1-(-2) & -5-0 \end{bmatrix} = \begin{bmatrix} 2 & 1 \\ 3 & -5 \end{bmatrix}$

(c) $5\begin{bmatrix} 2 & 3 \\ 4 & 1 \end{bmatrix}$

Solution: $5\begin{bmatrix} 2 & 3 \\ 4 & 1 \end{bmatrix} = \begin{bmatrix} 5(2) & 5(3) \\ 5(4) & 5(1) \end{bmatrix} = \begin{bmatrix} 10 & 15 \\ 20 & 5 \end{bmatrix}$

(d) $2\begin{bmatrix} 1 & 3 \\ 0 & -1 \end{bmatrix} + 3\begin{bmatrix} 4 & 0 \\ 1 & 2 \end{bmatrix}$

Solution: $2\begin{bmatrix} 1 & 3 \\ 0 & -1 \end{bmatrix} + 3\begin{bmatrix} 4 & 0 \\ 1 & 2 \end{bmatrix} = \begin{bmatrix} 2 & 6 \\ 0 & -2 \end{bmatrix} + \begin{bmatrix} 12 & 0 \\ 3 & 6 \end{bmatrix} = \begin{bmatrix} 2+12 & 6+0 \\ 0+3 & -2+6 \end{bmatrix} = \begin{bmatrix} 14 & 6 \\ 3 & 4 \end{bmatrix}$

Note that matrix addition is defined only if the matrices are the same size.

Properties of Matrix Addition and Scalar Multiplication

We now look at the properties of addition and scalar multiplication of matrices. It is very important to notice that these are the exact same ten properties discussed in Theorem 1.4.1 for addition and scalar multiplication of vectors in $\mathbb{R}^n$.

Theorem 3.1.1

For all $A, B, C \in M_{m\times n}(\mathbb{R})$ and $s, t \in \mathbb{R}$ we have

(1) $A + B \in M_{m\times n}(\mathbb{R})$ (closed under addition)
(2) $A + B = B + A$ (addition is commutative)
(3) $(A + B) + C = A + (B + C)$ (addition is associative)
(4) There exists a matrix $O_{m,n} \in M_{m\times n}(\mathbb{R})$, such that $A + O_{m,n} = A$ for all $A \in M_{m\times n}(\mathbb{R})$ (zero matrix)
(5) For each $A \in M_{m\times n}(\mathbb{R})$, there exists $(-A) \in M_{m\times n}(\mathbb{R})$ such that $A+(-A) = O_{m,n}$ (additive inverses)
(6) $sA \in M_{m\times n}(\mathbb{R})$ (closed under scalar multiplication)
(7) $s(tA) = (st)A$ (scalar multiplication is associative)
(8) $(s + t)A = sA + tA$ (a distributive law)
(9) $s(A + B) = sA + sB$ (another distributive law)
(10) $1A = A$ (scalar multiplicative identity)

These properties follow easily from the definitions of addition and multiplication by scalars. The proofs are left to the reader.

The matrix $O_{m,n}$, called the **zero matrix**, is the $m \times n$ matrix with all entries as zero. The additive inverse $(-A)$ of a matrix A is defined by $(-A)_{ij} = -(A)_{ij}$.

As before, we call a sum of linear combinations of matrices a **linear combination** of matrices.

Spanning and Linear Independence for Matrices

To stress the fact that matrices can be treated in the same way as vectors in $\mathbb{R}^n$, we now briefly look at the concepts of spanning and linear independence for sets of matrices.

Definition
Span

Let $\mathcal{B} = \{A_1, \dots, A_k\}$ be a set of $m \times n$ matrices. The **span** of $\mathcal{B}$ is defined as

$$\operatorname{Span}\mathcal{B} = \{t_1A_1 + \cdots + t_kA_k \mid t_1, \dots, t_k \in \mathbb{R}\}$$

Definition
Linearly Dependent
Linearly Independent

Let $\mathcal{B} = \{A_1, \dots, A_k\}$ be a set of $m \times n$ matrices. The set $\mathcal{B}$ is said to be **linearly dependent** if there exist real coefficients $c_1, \dots, c_k$ not all zero such that

$$c_1A_1 + \cdots + c_kA_k = O_{m,n}$$

The set $\mathcal{B}$ is said to be **linearly independent** if the only solution to

$$c_1A_1 + \cdots + c_kA_k = O_{m,n}$$

is the trivial solution $c_1 = \cdots = c_k = 0$.

EXAMPLE 3.1.4

Determine whether $\begin{bmatrix} 1 & 2 \\ 3 & 4 \end{bmatrix}$ is in the span of $\mathcal{B} = \left\{ \begin{bmatrix} 1 & 1 \\ 0 & 0 \end{bmatrix}, \begin{bmatrix} 1 & 0 \\ 0 & 1 \end{bmatrix}, \begin{bmatrix} 0 & 1 \\ 1 & 0 \end{bmatrix}, \begin{bmatrix} 0 & 1 \\ 0 & 1 \end{bmatrix} \right\}$.

Solution: We want to find if there exists $t_1, t_2, t_3, t_4 \in \mathbb{R}$ such that

$$\begin{bmatrix} 1 & 2 \\ 3 & 4 \end{bmatrix} = t_1 \begin{bmatrix} 1 & 1 \\ 0 & 0 \end{bmatrix} + t_2 \begin{bmatrix} 1 & 0 \\ 0 & 1 \end{bmatrix} + t_3 \begin{bmatrix} 0 & 1 \\ 1 & 0 \end{bmatrix} + t_4 \begin{bmatrix} 0 & 1 \\ 0 & 1 \end{bmatrix} = \begin{bmatrix} t_1 + t_2 & t_1 + t_3 + t_4 \\ t_3 & t_2 + t_4 \end{bmatrix}$$

Since two matrices are equal if and only if their corresponding entries are equal, this gives the system of linear equations

$$\begin{aligned} t_1 + t_2 &= 1 \\ t_1 + t_3 + t_4 &= 2 \\ t_3 &= 3 \\ t_2 + t_4 &= 4 \end{aligned}$$

Row reducing the corresponding augmented matrix gives

$$\left[\begin{array}{cccc|c} 1 & 1 & 0 & 0 & 1 \\ 1 & 0 & 1 & 1 & 2 \\ 0 & 0 & 1 & 0 & 3 \\ 0 & 1 & 0 & 1 & 4 \end{array}\right] \sim \left[\begin{array}{cccc|c} 1 & 0 & 0 & 0 & -2 \\ 0 & 1 & 0 & 0 & 3 \\ 0 & 0 & 1 & 0 & 3 \\ 0 & 0 & 0 & 1 & 1 \end{array}\right]$$

We see that the system is consistent. Therefore, $\begin{bmatrix} 1 & 2 \\ 3 & 4 \end{bmatrix}$ is in the span of $\mathcal{B}$. In particular, we have $t_1 = -2, t_2 = 3, t_3 = 3$, and $t_4 = 1$.

EXAMPLE 3.1.5

Determine whether the set $C = \left\{ \begin{bmatrix} 1 & 2 \\ 2 & -1 \end{bmatrix}, \begin{bmatrix} 3 & 2 \\ 1 & 1 \end{bmatrix}, \begin{bmatrix} 0 & 0 \\ 2 & 2 \end{bmatrix} \right\}$ is linearly independent.

Solution: We consider the equation

$$\begin{bmatrix} 0 & 0 \\ 0 & 0 \end{bmatrix} = t_1 \begin{bmatrix} 1 & 2 \\ 2 & -1 \end{bmatrix} + t_2 \begin{bmatrix} 3 & 2 \\ 1 & 1 \end{bmatrix} + t_3 \begin{bmatrix} 0 & 0 \\ 2 & 2 \end{bmatrix} = \begin{bmatrix} t_1 + 3t_2 & 2t_1 + 2t_2 \\ 2t_1 + t_2 + 2t_3 & -t_1 + t_2 + 2t_3 \end{bmatrix}$$

This gives the homogeneous system of equations

$$\begin{aligned} t_1 + 3t_2 &= 0 \\ 2t_1 + 2t_2 &= 0 \\ 2t_1 + t_2 + 2t_3 &= 0 \\ -t_1 + t_2 + 2t_3 &= 0 \end{aligned}$$

Row reducing the coefficient matrix of this system gives

$$\begin{bmatrix} 1 & 3 & 0 \\ 2 & 2 & 0 \\ 2 & 1 & 2 \\ -1 & 1 & 2 \end{bmatrix} \sim \begin{bmatrix} 1 & 0 & 0 \\ 0 & 1 & 0 \\ 0 & 0 & 1 \\ 0 & 0 & 0 \end{bmatrix}$$

The only solution is the trivial solution $t_1 = t_2 = t_3 = 0$, so C is linearly independent.

EXERCISE 3.1.1

Determine whether $\mathcal{B} = \left\{ \begin{bmatrix} 1 & 2 \\ 1 & 1 \end{bmatrix}, \begin{bmatrix} 1 & 1 \\ 3 & 1 \end{bmatrix}, \begin{bmatrix} 3 & 5 \\ 5 & 3 \end{bmatrix}, \begin{bmatrix} 0 & -1 \\ -2 & 0 \end{bmatrix} \right\}$ is linearly independent. Is $X = \begin{bmatrix} 1 & 5 \\ -5 & 1 \end{bmatrix}$ in the span of $\mathcal{B}$?

EXERCISE 3.1.2

Consider $\mathcal{B} = \left\{ \begin{bmatrix} 1 & 0 \\ 0 & 0 \end{bmatrix}, \begin{bmatrix} 0 & 1 \\ 0 & 0 \end{bmatrix}, \begin{bmatrix} 0 & 0 \\ 1 & 0 \end{bmatrix}, \begin{bmatrix} 0 & 0 \\ 0 & 1 \end{bmatrix} \right\}$. Prove that $\mathcal{B}$ is linearly independent and show that Span $\mathcal{B} = M_{2\times 2}(\mathbb{R})$. Compare $\mathcal{B}$ with the standard basis for $\mathbb{R}^4$.

Transpose

We will soon see that we sometimes wish to treat the rows of an $m \times n$ matrix as vectors in $\mathbb{R}^n$. To preserve our convention of writing vectors in $\mathbb{R}^n$ as column vectors, we invent some notation for turning columns into rows and vice versa.

Definition
Transpose (vector)

Let $\vec{x} = \begin{bmatrix} x_1 \\ \vdots \\ x_n \end{bmatrix} \in \mathbb{R}^n$. The **transpose** of $\vec{x}$, denoted $\vec{x}^T$, is the **row vector**

$$\vec{x}^T = \begin{bmatrix} x_1 & \cdots & x_n \end{bmatrix}$$

EXAMPLE 3.1.6

If $\vec{a}_1 = \begin{bmatrix} 1 \\ 3 \\ 1 \end{bmatrix}$ and $\vec{a}_2 = \begin{bmatrix} 2 \\ -1 \\ 4 \end{bmatrix}$, then what is the matrix $A = \begin{bmatrix} \vec{a}_1^T \\ \vec{a}_2^T \end{bmatrix}$?

Solution: Since $\vec{a}_1^T = \begin{bmatrix} 1 & 3 & 1 \end{bmatrix}$ and $\vec{a}_2^T = \begin{bmatrix} 2 & -1 & 4 \end{bmatrix}$, we get $A = \begin{bmatrix} 1 & 3 & 1 \\ 2 & -1 & 4 \end{bmatrix}$.

EXERCISE 3.1.3

If $\begin{bmatrix} \vec{a}_1^T \\ \vec{a}_2^T \\ \vec{a}_3^T \end{bmatrix} = \begin{bmatrix} 1 & 3 & 1 \\ -4 & 0 & 2 \\ 5 & 9 & -3 \end{bmatrix}$, then what are $\vec{a}_1, \vec{a}_2$, and $\vec{a}_3$?

We now extend this operation to matrices. We will see throughout the book that the transpose of a matrix is very useful in helping us solve a variety problems.

Definition
Transpose (matrix)

Let $A \in M_{m\times n}(\mathbb{R})$. The **transpose** of A is the $n \times m$ matrix, denoted A^T, whose ij-th entry is the ji-th entry of A. That is,

$$(A^T)_{ij} = (A)_{ji}$$

EXAMPLE 3.1.7

Determine the transpose of $A = \begin{bmatrix} -1 & 6 & -4 \\ 3 & 5 & 2 \end{bmatrix}$ and $B = \begin{bmatrix} 1 & 2 \\ 0 & 3 \\ -1 & 5 \end{bmatrix}$.

Solution: $A^T = \begin{bmatrix} -1 & 6 & -4 \\ 3 & 5 & 2 \end{bmatrix}^T = \begin{bmatrix} -1 & 3 \\ 6 & 5 \\ -4 & 2 \end{bmatrix}$ and $B^T = \begin{bmatrix} 1 & 2 \\ 0 & 3 \\ -1 & 5 \end{bmatrix}^T = \begin{bmatrix} 1 & 0 & -1 \\ 2 & 3 & 5 \end{bmatrix}$.

Theorem 3.1.2

If A and B are matrices, column vectors, or row vectors of the correct size so that the required operations are defined, and $s \in \mathbb{R}$, then

(1) $(A^T)^T = A$

(2) $(A + B)^T = A^T + B^T$

(3) $(sA)^T = sA^T$

Proof: We prove (2) and leave (1) and (3) as exercises. For (2) we have

$$((A + B)^T)_{ij} = (A + B)_{ji} = (A)_{ji} + (B)_{ji} = (A^T)_{ij} + (B^T)_{ij} = (A^T + B^T)_{ij}$$

■

EXERCISE 3.1.4

Let $A = \begin{bmatrix} 2 & 3 & 1 \\ -1 & 0 & 5 \end{bmatrix}$. Verify that $(A^T)^T = A$ and $(3A)^T = 3A^T$.

An Introduction to Matrix Multiplication

The whole purpose of algebra is to use symbols to make writing and manipulating mathematical equations easier. For example, rather than having to write

"The area of a circle is equal to the radius of the circle multiplied by itself and then multiplied by 3.141592653589...," we instead just write

$$A = \pi r^2$$

We could now use our rules of operations on real numbers to manipulate this equation as desired.

Our goal is to define matrix-vector multiplication and matrix multiplication to facilitate working with linear equations.

Matrix-Vector Multiplication Using Rows

We motivate our first definition of matrix-vector multiplication with an example.

EXAMPLE 3.1.8

According to the USDA, 1 kg of 1% milk contains 34 grams of protein and 50 grams of sugar, and 1 kg of apples contains 3 grams of protein and 100 grams of sugar. If one drinks m kg of milk and eats a kg of apples, then the amount of protein p and the amount of sugar s they will have consumed is given by the system of linear equations

$$\begin{aligned} 34m + 3a &= p \\ 50m + 100a &= s \end{aligned}$$

Observe that we can write the amount of protein being consumed as the dot product

$$\begin{bmatrix} 34 \\ 3 \end{bmatrix} \cdot \begin{bmatrix} m \\ a \end{bmatrix} = p$$

Similarly, the amount of sugar being consumed is

$$\begin{bmatrix} 50 \\ 100 \end{bmatrix} \cdot \begin{bmatrix} m \\ a \end{bmatrix} = s$$

If we let $\vec{a}_1 = \begin{bmatrix} 34 \\ 3 \end{bmatrix}$, $\vec{a}_2 = \begin{bmatrix} 50 \\ 100 \end{bmatrix}$, and $\vec{x} = \begin{bmatrix} m \\ a \end{bmatrix}$, then, since vectors in $\mathbb{R}^n$ are equal if and only if they have equal entries, we can represent the system of linear equations in the form

$$\begin{bmatrix} \vec{a}_1 \cdot \vec{x} \\ \vec{a}_2 \cdot \vec{x} \end{bmatrix} = \begin{bmatrix} p \\ s \end{bmatrix} \tag{3.1}$$

We now define **matrix-vector multiplication**, so that we can write equation (3.1) in the form of

$$A\vec{x} = \begin{bmatrix} p \\ s \end{bmatrix}$$

where $A = \begin{bmatrix} \vec{a}_1^T \\ \vec{a}_2^T \end{bmatrix}$ is the coefficient matrix of the original system of linear equations.

Definition
Matrix-Vector Multiplication

Let $A \in M_{m\times n}(\mathbb{R})$ whose rows are denoted $\vec{a}_i^T$ for $1 \leq i \leq m$. For any $\vec{x} \in \mathbb{R}^n$, we define $A\vec{x}$ by

$$A\vec{x} = \begin{bmatrix} \vec{a}_1 \cdot \vec{x} \\ \vdots \\ \vec{a}_m \cdot \vec{x} \end{bmatrix}$$

It is important to note that if A is an $m\times n$ matrix, then $A\vec{x}$ is defined only if $\vec{x} \in \mathbb{R}^n$. Moreover, if $\vec{x} \in \mathbb{R}^n$, then $A\vec{x} \in \mathbb{R}^m$.

EXAMPLE 3.1.9

Let $A = \begin{bmatrix} 3 & 4 & -5 \\ 1 & 0 & 2 \end{bmatrix}$, $\vec{x} = \begin{bmatrix} 2 \\ -1 \\ 6 \end{bmatrix}$, $\vec{y} = \begin{bmatrix} 1 \\ 0 \\ 0 \end{bmatrix}$, and $\vec{z} = \begin{bmatrix} 0 \\ 0 \\ 1 \end{bmatrix}$. Calculate $A\vec{x}$, $A\vec{y}$, and $A\vec{z}$.

Solution: Using the definition of matrix-vector multiplication gives

$$A\vec{x} = \begin{bmatrix} 3 & 4 & -5 \\ 1 & 0 & 2 \end{bmatrix}\begin{bmatrix} 2 \\ -1 \\ 6 \end{bmatrix} = \begin{bmatrix} 3(2) + 4(-1) + (-5)(6) \\ 1(2) + 0(-1) + (2)(6) \end{bmatrix} = \begin{bmatrix} -28 \\ 14 \end{bmatrix}$$

$$A\vec{y} = \begin{bmatrix} 3 & 4 & -5 \\ 1 & 0 & 2 \end{bmatrix}\begin{bmatrix} 1 \\ 0 \\ 0 \end{bmatrix} = \begin{bmatrix} 3(1) + 4(0) + (-5)(0) \\ 1(1) + 0(0) + (2)(0) \end{bmatrix} = \begin{bmatrix} 3 \\ 1 \end{bmatrix}$$

$$A\vec{z} = \begin{bmatrix} 3 & 4 & -5 \\ 1 & 0 & 2 \end{bmatrix}\begin{bmatrix} 0 \\ 0 \\ 1 \end{bmatrix} = \begin{bmatrix} 3(0) + 4(0) + (-5)(1) \\ 1(0) + 0(0) + (2)(1) \end{bmatrix} = \begin{bmatrix} -5 \\ 2 \end{bmatrix}$$

EXERCISE 3.1.5

Calculate the following matrix-vector products.

(a) $\begin{bmatrix} 1 & 3 & 2 \\ -1 & 4 & 5 \end{bmatrix}\begin{bmatrix} 2 \\ -1 \\ 6 \end{bmatrix}$ (b) $\begin{bmatrix} 6 & -1 & 1 \end{bmatrix}\begin{bmatrix} 2 \\ 3 \\ 1 \end{bmatrix}$ (c) $\begin{bmatrix} 1 & -4 \\ 3 & 1 \\ 1 & -2 \end{bmatrix}\begin{bmatrix} 0 \\ 1 \end{bmatrix}$

EXAMPLE 3.1.10

Let $A = \begin{bmatrix} 1 & 3 \\ 2 & -4 \\ 9 & -1 \end{bmatrix}$, $\vec{x} = \begin{bmatrix} x_1 \\ x_2 \end{bmatrix}$, and $\vec{b} = \begin{bmatrix} 5 \\ 0 \\ 8 \end{bmatrix}$. Write the system of linear equations represented by $A\vec{x} = \vec{b}$.

Solution: By definition, we have that

$$A\vec{x} = \begin{bmatrix} 1 & 3 \\ 2 & -4 \\ 9 & -1 \end{bmatrix}\begin{bmatrix} x_1 \\ x_2 \end{bmatrix} = \begin{bmatrix} x_1 + 3x_2 \\ 2x_1 - 4x_2 \\ 9x_1 - x_2 \end{bmatrix}$$

Hence, $A\vec{x} = \vec{b}$ gives

$$\begin{bmatrix} x_1 + 3x_2 \\ 2x_1 - 4x_2 \\ 9x_1 - x_2 \end{bmatrix} = \begin{bmatrix} 5 \\ 0 \\ 8 \end{bmatrix}$$

Since vectors are equal if and only if their corresponding entries are equal, this gives the system of linear equations

$$\begin{aligned} x_1 + 3x_2 &= 5 \\ 2x_1 - 4x_2 &= 0 \\ 9x_1 - x_2 &= 8 \end{aligned}$$

EXERCISE 3.1.6

Write the following system of linear equations in the form $A\vec{x} = \vec{b}$.

$$\begin{aligned} x_1 + \ \ x_2 &= 3 \\ 2x_1 \qquad &= 15 \\ -2x_1 - 3x_2 &= -5 \end{aligned}$$

EXAMPLE 3.1.11

Consider the spring-mass system depicted below. Rather than finding the equilibrium displacement as we did in Section 2.4, we will now determine how to find the forces f_1 and f_2 acting on the masses m_1 and m_2 given displacements x_1 and x_2. By Hooke's Law, we find that the forces f_1 and f_2 satisfy

$$\begin{aligned} f_1 &= (k_1 + k_2)x_1 - k_2x_2 \\ f_2 &= -k_2x_1 + (k_2 + k_3)x_2 \end{aligned}$$

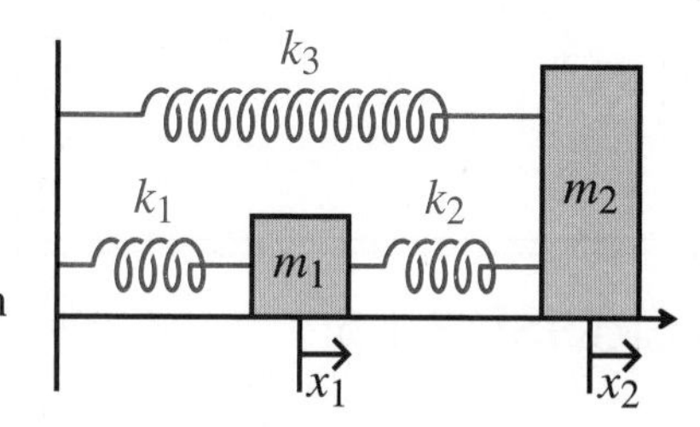

Using matrix-vector multiplication, we can rewrite this as

$$\begin{bmatrix} k_1 + k_2 & -k_2 \\ -k_2 & k_2 + k_3 \end{bmatrix} \begin{bmatrix} x_1 \\ x_2 \end{bmatrix} = \begin{bmatrix} f_1 \\ f_2 \end{bmatrix}$$

We call the matrix $K = \begin{bmatrix} k_1 + k_2 & -k_2 \\ -k_2 & k_2 + k_3 \end{bmatrix}$ the **stiffness matrix** of the system.

In most cases, we will now represent a system of linear equations $\left[A \mid \vec{b}\right]$ as $A\vec{x} = \vec{b}$. This will allow us to use properties of matrix multiplication (see Theorem 3.1.4 below) when dealing with systems of linear equations.

Matrix-Vector Multiplication Using Columns

In Example 3.1.8 we derived a definition of matrix-vector multiplication by focusing on each equation. However, in some cases it may make more sense to focus on the products themselves. In particular, observe that we instead could write the system

$$\begin{aligned} 34m + 3a &= p \\ 50m + 100a &= s \end{aligned}$$

in the form

$$\begin{bmatrix} 34m + 3a \\ 50m + 100a \end{bmatrix} = \begin{bmatrix} p \\ s \end{bmatrix}$$

Then, we could use operations on vectors to get

$$m\begin{bmatrix} 34 \\ 50 \end{bmatrix} + a\begin{bmatrix} 3 \\ 100 \end{bmatrix} = \begin{bmatrix} p \\ s \end{bmatrix}$$

We still want the left-hand side to be written in the form $A\vec{x}$. So, this gives us an alternative definition of matrix-vector multiplication.

Definition
Matrix-Vector Multiplication

Let $A = \begin{bmatrix} \vec{a}_1 & \cdots & \vec{a}_n \end{bmatrix} \in M_{m\times n}(\mathbb{R})$. For any $\vec{x} = \begin{bmatrix} x_1 \\ \vdots \\ x_n \end{bmatrix} \in \mathbb{R}^n$, we define $A\vec{x}$ by

$$A\vec{x} = x_1\vec{a}_1 + \cdots + x_n\vec{a}_n$$

Remarks

1. As before, if A is an $m\times n$ matrix, then $A\vec{x}$ only makes sense if $\vec{x} \in \mathbb{R}^n$. Moreover, if $\vec{x} \in \mathbb{R}^n$, then $A\vec{x} \in \mathbb{R}^m$.
2. By definition, for any $\vec{x} \in \mathbb{R}^n$, $A\vec{x}$ is a linear combination of the columns of A.

EXAMPLE 3.1.12

Let $A = \begin{bmatrix} 3 & 4 & -5 \\ 1 & 0 & 2 \end{bmatrix}$, $\vec{x} = \begin{bmatrix} 2 \\ -1 \\ 6 \end{bmatrix}$, and $\vec{y} = \begin{bmatrix} 0 \\ 1 \\ 0 \end{bmatrix}$. Calculate $A\vec{x}$ and $A\vec{y}$.

Solution: We have

$$A\vec{x} = \begin{bmatrix} 3 & 4 & -5 \\ 1 & 0 & 2 \end{bmatrix}\begin{bmatrix} 2 \\ -1 \\ 6 \end{bmatrix} = 2\begin{bmatrix} 3 \\ 1 \end{bmatrix} + (-1)\begin{bmatrix} 4 \\ 0 \end{bmatrix} + 6\begin{bmatrix} -5 \\ 2 \end{bmatrix} = \begin{bmatrix} -28 \\ 14 \end{bmatrix}$$

$$A\vec{y} = \begin{bmatrix} 3 & 4 & -5 \\ 1 & 0 & 2 \end{bmatrix}\begin{bmatrix} 0 \\ 1 \\ 0 \end{bmatrix} = 0\begin{bmatrix} 3 \\ 1 \end{bmatrix} + 1\begin{bmatrix} 4 \\ 0 \end{bmatrix} + 0\begin{bmatrix} -5 \\ 2 \end{bmatrix} = \begin{bmatrix} 4 \\ 0 \end{bmatrix}$$

EXERCISE 3.1.7

Calculate the following matrix-vector products by expanding them as a linear combination of the columns of the matrix.

(a) $\begin{bmatrix} 1 & 3 & 2 \\ -1 & 4 & 5 \end{bmatrix}\begin{bmatrix} 2 \\ -1 \\ 6 \end{bmatrix}$ (b) $\begin{bmatrix} 6 & -1 & 1 \end{bmatrix}\begin{bmatrix} 2 \\ 3 \\ 1 \end{bmatrix}$ (c) $\begin{bmatrix} 1 & -4 \\ 3 & 1 \\ 1 & -2 \end{bmatrix}\begin{bmatrix} 0 \\ 1 \end{bmatrix}$

Observe that this exercise contained the same problems as Exercise 3.1.5. This was to demonstrate that both definitions of matrix-vector multiplication indeed give the same answer. At this point you may wonder why we have two different definitions for the same thing. The reason is that they both have different uses. We use the first method for computing matrix-vector products when we want to work with the rows of a matrix. When we are working with the columns of a matrix or linear combinations of vectors, then we use the second method.

The following is a simple but useful theorem.

Theorem 3.1.3

If $A = \begin{bmatrix} \vec{a}_1 & \cdots & \vec{a}_n \end{bmatrix} \in M_{m\times n}(\mathbb{R})$ and $\vec{e}_i$ is the i-th standard basis vector of $\mathbb{R}^n$, then

$$A\vec{e}_i = \vec{a}_i$$

The proof is left as Problem A34.

Matrix Multiplication

It is important to observe that matrix-vector multiplication behaves like a function. It inputs a vector in $\mathbb{R}^n$ and outputs a vector in $\mathbb{R}^m$. For instance, in Example 3.1.8 we formed the function which inputs the amount of milk and apples and outputs the amount of protein and sugar. We now want to define matrix multiplication to represent a composition of functions. We again use an example to demonstrate this.

EXAMPLE 3.1.13

Suppose that apple rice pudding contains 300 g of milk and 240 g of apples, while apple milk drink contains 250 g of milk and 100 g of apples. If we let $B = \begin{bmatrix} .3 & .25 \\ .24 & .1 \end{bmatrix}$, then we can represent the amount of milk m and apples a consumed by having r kg of apple rice pudding and d kg of apple milk drink by

$$\begin{bmatrix} m \\ a \end{bmatrix} = \begin{bmatrix} .3r + .25d \\ .24r + .1d \end{bmatrix} = \begin{bmatrix} .3 & .25 \\ .24 & .1 \end{bmatrix} \begin{bmatrix} r \\ d \end{bmatrix} = B \begin{bmatrix} r \\ d \end{bmatrix}$$

Using the matrix $A = \begin{bmatrix} \vec{a}_1^T \\ \vec{a}_2^T \end{bmatrix} = \begin{bmatrix} 34 & 3 \\ 50 & 100 \end{bmatrix}$ from Example 3.1.8, we get that the amount of protein and sugar consumed by having these is

$$\begin{aligned}
\begin{bmatrix} p \\ s \end{bmatrix} &= A \begin{bmatrix} m \\ a \end{bmatrix} \\
&= \begin{bmatrix} 34 & 3 \\ 50 & 100 \end{bmatrix} \begin{bmatrix} .3r + .25d \\ .24r + .1d \end{bmatrix} \\
&= \begin{bmatrix} 34(.3r + .25d) + 3(.24r + .1d) \\ 50(.3r + .25d) + 100(.24r + .1d) \end{bmatrix} \\
&= \begin{bmatrix} \big(34(.3) + 3(.24)\big)r + \big(34(.25) + 3(.1)\big)d \\ \big(50(.3) + 100(.24)\big)r + \big(50(.25) + 100(.1)\big)d \end{bmatrix} \\
&= \begin{bmatrix} 34(.3) + 3(.24) & 34(.25) + 3(.1) \\ 50(.3) + 100(.24) & 50(.25) + 100(.1) \end{bmatrix} \begin{bmatrix} r \\ d \end{bmatrix}
\end{aligned}$$

Since we also have that

$$\begin{bmatrix} p \\ s \end{bmatrix} = A \begin{bmatrix} m \\ a \end{bmatrix} = AB \begin{bmatrix} r \\ d \end{bmatrix}$$

This shows us how to define the matrix product AB.

Careful inspection of the result above shows that the product AB must be defined by the following rules:

- $(AB)_{11}$ is the dot product of the first row of A and the first column of B.
- $(AB)_{12}$ is the dot product of the first row of A and the second column of B.
- $(AB)_{21}$ is the dot product of the second row of A and the first column of B.
- $(AB)_{22}$ is the dot product of the second row of A and the second column of B.

EXAMPLE 3.1.14

Calculate $\begin{bmatrix} 2 & 3 \\ 4 & 1 \end{bmatrix}\begin{bmatrix} 5 & 1 \\ -2 & 7 \end{bmatrix}$.

Solution: Taking dot products of the rows of the first matrix with columns of the second matrix gives

$$\begin{bmatrix} 2 & 3 \\ 4 & 1 \end{bmatrix}\begin{bmatrix} 5 & 1 \\ -2 & 7 \end{bmatrix} = \begin{bmatrix} 2(5)+3(-2) & 2(1)+3(7) \\ 4(5)+1(-2) & 4(1)+1(7) \end{bmatrix} = \begin{bmatrix} 4 & 23 \\ 18 & 11 \end{bmatrix}$$

To formalize the definition of **matrix multiplication**, it will be convenient to use $\vec{a}_i^T$ to represent the i-th row of A and $\vec{b}_j$ to represent the j-th column of B. Observe from our work above that we want the ij-th entry of AB to be the dot product of the i-th row of A and the j-th column of B. However, for this to be defined, $\vec{a}_i^T$ must have the same number of entries as $\vec{b}_j$. Hence, the number of entries in the rows of the matrix A (that is, the number of columns of A) must be equal to the number of entries in the columns of B (that is, the number of rows of B). We can now make a precise definition.

Definition
Matrix Multiplication

Let $A \in M_{m\times n}(\mathbb{R})$ with rows $\vec{a}_1^T, \dots, \vec{a}_m^T$ and let $B \in M_{n\times p}(\mathbb{R})$ with columns $\vec{b}_1, \dots, \vec{b}_p$. We define AB to be the $m \times p$ matrix whose ij-th entry is

$$(AB)_{ij} = \vec{a}_i \cdot \vec{b}_j$$

To emphasize the point, if A is an $m \times n$ matrix and B is a $q \times p$ matrix, then AB is defined only if $n = q$.

EXAMPLE 3.1.15

Calculate $\begin{bmatrix} 2 & 3 & 0 & 1 \\ 4 & 1 & 2 & 1 \\ 0 & 0 & 0 & 1 \end{bmatrix}\begin{bmatrix} 3 & 1 \\ 1 & 2 \\ 2 & 3 \\ 0 & 5 \end{bmatrix}$.

Solution: We have

$$\begin{bmatrix} 2 & 3 & 0 & 1 \\ 4 & 1 & 2 & 1 \\ 0 & 0 & 0 & 1 \end{bmatrix}\begin{bmatrix} 3 & 1 \\ 1 & 2 \\ 2 & 3 \\ 0 & 5 \end{bmatrix} = \begin{bmatrix} 2(3)+3(1)+0(2)+1(0) & 2(1)+3(2)+0(3)+1(5) \\ 4(3)+1(1)+2(2)+1(0) & 4(1)+1(2)+2(3)+1(5) \\ 0(3)+0(1)+0(2)+1(0) & 0(1)+0(2)+0(3)+1(5) \end{bmatrix}$$

$$= \begin{bmatrix} 9 & 13 \\ 17 & 17 \\ 0 & 5 \end{bmatrix}$$

EXAMPLE 3.1.16 Calculate the following or explain why they are not defined.

(a) $\begin{bmatrix} 1 & 1 & 2 \\ -3 & 1 & 3 \\ 0 & 0 & 1 \end{bmatrix}\begin{bmatrix} 5 & 6 \\ 4 & 7 \\ 2 & 5 \end{bmatrix}$ (b) $\begin{bmatrix} 2 & 3 \\ 1 & -3 \end{bmatrix}\begin{bmatrix} 2 & -3 \\ 4 & 1 \\ 5 & 7 \end{bmatrix}$

Solution: For (a) we have

$$\begin{bmatrix} 1 & 1 & 2 \\ -3 & 1 & 3 \\ 0 & 0 & 1 \end{bmatrix}\begin{bmatrix} 5 & 6 \\ 4 & 7 \\ 2 & 5 \end{bmatrix} = \begin{bmatrix} 1(5)+1(4)+2(2) & 1(6)+1(7)+2(5) \\ (-3)(5)+1(4)+3(2) & (-3)(6)+1(7)+3(5) \\ 0(5)+0(4)+1(2) & 0(6)+0(7)+1(5) \end{bmatrix} = \begin{bmatrix} 13 & 23 \\ -5 & 4 \\ 2 & 5 \end{bmatrix}$$

For (b) we see that the product is not defined because the first matrix has two columns but the second matrix has three rows.

EXERCISE 3.1.8 Let $A = \begin{bmatrix} 1 & 2 & -1 \\ 2 & 3 & 1 \end{bmatrix}$ and $B = \begin{bmatrix} 2 & 1 \\ 1 & 0 \end{bmatrix}$. Calculate the following or explain why they are not defined.

(a) AB (b) BA (c) A^TA (d) BB^T

EXAMPLE 3.1.17 Let $A = \begin{bmatrix} 5 & 3 & -1 \\ 4 & 2 & 1 \end{bmatrix}$ and $\vec{x} = \begin{bmatrix} -3 \\ 1 \end{bmatrix}$. Calculate $\vec{x}^TA$.

Solution: To compute this, we interpret the row vector $\vec{x}^T$ as a 1×2 matrix and use the definition of matrix multiplication. Since the number of columns of the first matrix equals the number of rows of the second matrix, the product is defined. We get

$$\begin{aligned} \vec{x}^TA &= \begin{bmatrix} -3 & 1 \end{bmatrix}\begin{bmatrix} 5 & 3 & -1 \\ 4 & 2 & 1 \end{bmatrix} = \begin{bmatrix} (-3)(5)+1(4) & (-3)(3)+1(2) & (-3)(-1)+1(1) \end{bmatrix} \\ &= \begin{bmatrix} -11 & -7 & 4 \end{bmatrix} \end{aligned}$$

EXAMPLE 3.1.18 Let $\vec{x} = \begin{bmatrix} 1 \\ 2 \\ 3 \end{bmatrix}, \vec{y} = \begin{bmatrix} 6 \\ 5 \\ 4 \end{bmatrix} \in \mathbb{R}^3$. Compute $\vec{x}^T\vec{y}$.

Solution: Using the definition of matrix multiplication, we get

$$\vec{x}^T\vec{y} = \begin{bmatrix} 1 & 2 & 3 \end{bmatrix}\begin{bmatrix} 6 \\ 5 \\ 4 \end{bmatrix} = \begin{bmatrix} 1(6)+2(5)+3(4) \end{bmatrix} = \begin{bmatrix} 28 \end{bmatrix}$$

Observe that the vectors $\vec{x}$ and $\vec{y}$ in Example 3.1.18 satisfy $\vec{x} \cdot \vec{y} = 28$ which matches the result in the example. This should not be surprising since we have defined matrix multiplication in terms of the dot product. More generally, for any $\vec{x}, \vec{y} \in \mathbb{R}^n$ we have

$$\vec{x}^T\vec{y} = \vec{x} \cdot \vec{y}$$

where we interpret the 1×1 matrix on the right-hand side as a scalar. This formula will be used frequently later in the book.

Defining matrix multiplication with the dot product fits our first view of matrix-vector multiplication. We now look at how we could define matrix multiplication by using our alternate view of matrix-vector multiplication.

EXAMPLE 3.1.19

In Example 3.1.13 we found that $\begin{bmatrix} p \\ s \end{bmatrix} = \begin{bmatrix} 34(.3)+3(.24) & 34(.25)+3(.1) \\ 50(.3)+100(.24) & 50(.25)+100(.1) \end{bmatrix}\begin{bmatrix} r \\ d \end{bmatrix}$. Observe that the first column of this matrix can be written as

$$\begin{bmatrix} 34(.3)+3(.24) \\ 50(.3)+100(.24) \end{bmatrix} = (.3)\begin{bmatrix} 34 \\ 50 \end{bmatrix} + (.24)\begin{bmatrix} 3 \\ 100 \end{bmatrix} = A\begin{bmatrix} .3 \\ .24 \end{bmatrix}$$

Similarly, the second column of the matrix is

$$\begin{bmatrix} 34(.25)+3(.1) \\ 50(.25)+100(.1) \end{bmatrix} = (.25)\begin{bmatrix} 34 \\ 50 \end{bmatrix} + (.1)\begin{bmatrix} 3 \\ 100 \end{bmatrix} = A\begin{bmatrix} .25 \\ .1 \end{bmatrix}$$

So, the i-th column of AB is the matrix-vector product of A and the i-th column of B.

Thus, we can alternatively define matrix multiplication in the following way.

Definition
Matrix Multiplication

For $A \in M_{m\times n}(\mathbb{R})$ and $B = \begin{bmatrix} \vec{b}_1 & \cdots & \vec{b}_p \end{bmatrix} \in M_{n\times p}(\mathbb{R})$ we define AB to be the $m \times p$ matrix

$$AB = A\begin{bmatrix} \vec{b}_1 & \cdots & \vec{b}_p \end{bmatrix} = \begin{bmatrix} A\vec{b}_1 & \cdots & A\vec{b}_p \end{bmatrix} \tag{3.2}$$

EXAMPLE 3.1.20

Calculate $\begin{bmatrix} 2 & 3 & 0 & 1 \\ 4 & 1 & 2 & 1 \\ 0 & 0 & 0 & 1 \end{bmatrix}\begin{bmatrix} 3 & 1 \\ 1 & 2 \\ 2 & 3 \\ 0 & 5 \end{bmatrix}$.

Solution: We have

$$\begin{bmatrix} 2 & 3 & 0 & 1 \\ 4 & 1 & 2 & 1 \\ 0 & 0 & 0 & 1 \end{bmatrix}\begin{bmatrix} 3 \\ 1 \\ 2 \\ 0 \end{bmatrix} = \begin{bmatrix} 9 \\ 17 \\ 0 \end{bmatrix}, \quad \begin{bmatrix} 2 & 3 & 0 & 1 \\ 4 & 1 & 2 & 1 \\ 0 & 0 & 0 & 1 \end{bmatrix}\begin{bmatrix} 1 \\ 2 \\ 3 \\ 5 \end{bmatrix} = \begin{bmatrix} 13 \\ 17 \\ 5 \end{bmatrix}$$

Hence,

$$\begin{bmatrix} 2 & 3 & 0 & 1 \\ 4 & 1 & 2 & 1 \\ 0 & 0 & 0 & 1 \end{bmatrix}\begin{bmatrix} 3 & 1 \\ 1 & 2 \\ 2 & 3 \\ 0 & 5 \end{bmatrix} = \begin{bmatrix} 9 & 13 \\ 17 & 17 \\ 0 & 5 \end{bmatrix}$$

Both interpretations of matrix multiplication will be very useful, so it is important to know and understand both of them.

CONNECTION

We now see that linear combinations of vectors (and hence concepts such as spanning and linear independence), solving systems of linear equations, and matrix multiplication are all closely tied together. We will continue to see these connections later in this chapter and throughout the book.

EXAMPLE 3.1.21

The diagram indicates how six webpages are connected by hyperlinks. For example, the single directional arrow from A_3 to A_6 indicates there is a hyperlink on page A_3 that takes the user to page A_6. The double directional arrow between A_4 and A_5 indicates that there is a hyperlink on page A_4 to page A_5 and a hyperlink on page A_5 to A_4.

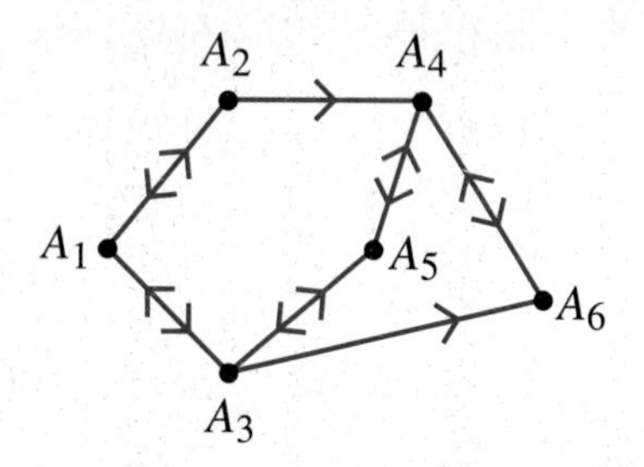

We create a matrix C to represent the network by putting a 1 in the i, j-th entry of C if webpage A_i has a hyperlink to page A_j. Thus, because of the connection from A_3 to A_6 we get $(C)_{3,6} = 1$. Since A_6 does not have a hyperlink to A_3, we have that $(C)_{6,3} = 0$. The bidirectional connection between A_4 and A_5 means that we have $(C)_{4,5} = 1$ and $(C)_{5,4} = 1$. Note, that in our example, no page links to itself (although that certainly happens), so we have $(C)_{ii} = 0$ for $1 \leq i \leq 6$. Completing the matrix we find that

$$C = \begin{bmatrix} 0 & 1 & 1 & 0 & 0 & 0 \\ 1 & 0 & 0 & 1 & 0 & 0 \\ 1 & 0 & 0 & 0 & 1 & 1 \\ 0 & 0 & 0 & 0 & 1 & 1 \\ 0 & 0 & 1 & 1 & 0 & 0 \\ 0 & 0 & 0 & 1 & 0 & 0 \end{bmatrix}$$

We call this the **connectivity matrix** of the network as it indicates all the direct communications (travel) between webpages.

Matrix multiplication has been defined to represent a composition of functions. So, if the matrix C indicates travel from one webpage to another by following a single hyperlink, then $C^2 = CC$ indicates travel from one webpage to another by following two hyperlinks. We find that

$$C^2 = \begin{bmatrix} 2 & 0 & 0 & 1 & 1 & 1 \\ 0 & 1 & 1 & 0 & 1 & 1 \\ 0 & 1 & 2 & 2 & 0 & 0 \\ 0 & 0 & 1 & 2 & 0 & 0 \\ 1 & 0 & 0 & 0 & 2 & 2 \\ 0 & 0 & 0 & 0 & 1 & 1 \end{bmatrix}$$

The fact that $(C^2)_{3,4} = 2$ indicates that there are two ways of clicking a sequence of two hyperlinks from A_3 to A_4, while $(C^2)_{1,2} = 0$ indicates that there is no sequence of two hyperlinks (even though there is a direct link) from A_1 to A_2. A sequence of n hyperlinks links moving from one page to another is called a **walk of length** n.

The ij-th entry of $C + C^2 + C^3$ is the number of walks of length at most 3 from A_i to A_j. For example, we can calculate that $(C + C^2 + C^3)_{1,6} = 2$. This indicates that there are two ways of getting from page A_1 to page A_6 by clicking at most 3 hyperlinks. The value $(C + C^2 + C^3)_{6,2} = 0$ indicates that it would take clicking more than 3 hyperlinks to move from page A_6 to page A_2.

Properties of Matrix Multiplication

Theorem 3.1.4

If A, B, and C are matrices, column vectors, or row vectors of the correct size so that the required operations are defined, and $s \in \mathbb{R}$, then

(1) $A(B + C) = AB + AC$
(2) $(A + B)C = AC + BC$
(3) $s(AB) = (sA)B = A(sB)$
(4) $A(BC) = (AB)C$
(5) $(AB)^T = B^T A^T$

These properties follows easily from the definition of matrix multiplication. However, the proofs are not particularly illuminating and so are omitted.

Three Important Facts:

1. Matrix multiplication is not commutative: That is, in general, $AB \neq BA$. In fact, if BA is defined, it is not necessarily true that AB is even defined. For example, if B is 2×2 and A is 2×3, then BA is defined, but AB is not. However, even if both AB and BA are defined, they are usually not equal. $AB = BA$ is true only in very special circumstances.

EXAMPLE 3.1.22

Show that if $A = \begin{bmatrix} 2 & 3 \\ 4 & -1 \end{bmatrix}$ and $B = \begin{bmatrix} 5 & 1 \\ -2 & 7 \end{bmatrix}$, then $AB \neq BA$.

Solution: $AB = \begin{bmatrix} 2 & 3 \\ 4 & -1 \end{bmatrix}\begin{bmatrix} 5 & 1 \\ -2 & 7 \end{bmatrix} = \begin{bmatrix} 4 & 23 \\ 22 & -3 \end{bmatrix}$,

but

$$BA = \begin{bmatrix} 5 & 1 \\ -2 & 7 \end{bmatrix}\begin{bmatrix} 2 & 3 \\ 4 & -1 \end{bmatrix} = \begin{bmatrix} 14 & 14 \\ 24 & -13 \end{bmatrix}$$

When multiplying both sides of a matrix equation by a matrix we must either multiply it on the right of both sides of the equation or on the left of both sides of the equation. That is, if we have the matrix equation

$$AX = B$$

then multiplying by the matrix C can give either

$$CAX = CB \quad \text{or} \quad AXC = BC$$

2. The cancellation law is almost never valid for matrix multiplication: That is, if $AB = AC$, then we cannot guarantee that $B = C$.

EXAMPLE 3.1.23

Let $A = \begin{bmatrix} 0 & 0 \\ 0 & 1 \end{bmatrix}$, $B = \begin{bmatrix} 5 & 6 \\ 7 & 8 \end{bmatrix}$, and $C = \begin{bmatrix} 2 & 3 \\ 7 & 8 \end{bmatrix}$. Then,

$$AB = \begin{bmatrix} 0 & 0 \\ 0 & 1 \end{bmatrix}\begin{bmatrix} 5 & 6 \\ 7 & 8 \end{bmatrix} = \begin{bmatrix} 0 & 0 \\ 7 & 8 \end{bmatrix} = \begin{bmatrix} 0 & 0 \\ 0 & 1 \end{bmatrix}\begin{bmatrix} 2 & 3 \\ 7 & 8 \end{bmatrix} = AC$$

so $AB = AC$ but $B \neq C$.

Remark

The fact that we do not have the cancellation law for matrix multiplication comes from the fact that **we do not have division for matrices**.

We must distinguish carefully between a general cancellation law and the following theorem, which we will use many times.

Theorem 3.1.5

Matrices Equal Theorem

If $A, B \in M_{m\times n}(\mathbb{R})$ such that $A\vec{x} = B\vec{x}$ for every $\vec{x} \in \mathbb{R}^n$, then $A = B$.

You are asked to prove Theorem 3.1.5, with hints, in Problem C1.

Note that it is the assumption that equality holds for *every* $\vec{x} \in \mathbb{R}^n$ that distinguishes this from a cancellation law.

3. $AB = O_{m,n}$ **does not imply that one of** A **or** B **is the zero matrix.**

EXAMPLE 3.1.24

If $A = \begin{bmatrix} 1 & -3 \\ 2 & -6 \end{bmatrix}$ and $B = \begin{bmatrix} 3 & 3 \\ 1 & 1 \end{bmatrix}$, then $AB = \begin{bmatrix} 1 & -3 \\ 2 & -6 \end{bmatrix}\begin{bmatrix} 3 & 3 \\ 1 & 1 \end{bmatrix} = \begin{bmatrix} 0 & 0 \\ 0 & 0 \end{bmatrix}$.

But neither A nor B is the zero matrix.

Identity Matrix

We have seen that the zero matrix $O_{m,n}$ is the additive identity for addition of $m \times n$ matrices. However, since we also have multiplication of matrices, it is important to determine whether we have a multiplicative identity. If we do, we need to determine what the multiplicative identity is. First, we observe that for there to exist a matrix A and a matrix I such that $AI = A = IA$, both A and I must be $n \times n$ matrices as otherwise either AI or IA is undefined. The multiplicative identity I is the $n \times n$ matrix that has this property for all $n \times n$ matrices A.

To find how to define I, we begin with a simple case. Let $A = \begin{bmatrix} a & b \\ c & d \end{bmatrix}$. We want to find a matrix $I = \begin{bmatrix} e & f \\ g & h \end{bmatrix}$ such that $AI = A$. By matrix multiplication, we get

$$\begin{bmatrix} a & b \\ c & d \end{bmatrix} = \begin{bmatrix} a & b \\ c & d \end{bmatrix}\begin{bmatrix} e & f \\ g & h \end{bmatrix} = \begin{bmatrix} ae + bg & af + bh \\ ce + dg & cf + dh \end{bmatrix}$$

Thus, we must have $a = ae + bg$, $b = af + bh$, $c = ce + dg$, and $d = cf + dh$. Although this system of equations is not linear, it is still easy to solve. We find that we must have $e = 1 = h$ and $f = g = 0$. Thus,

$$I = \begin{bmatrix} 1 & 0 \\ 0 & 1 \end{bmatrix} = \text{diag}(1, 1)$$

It is easy to verify that I also satisfies $IA = A$. Hence, I is the multiplicative identity for 2×2 matrices. We now extend this definition to the $n \times n$ case.

Definition
Identity Matrix

The $n \times n$ matrix $I = \text{diag}(1, 1, \dots, 1)$ is called the **identity matrix**.

EXAMPLE 3.1.25

The 3×3 identity matrix is $I = \text{diag}(1, 1, 1) = \begin{bmatrix} 1 & 0 & 0 \\ 0 & 1 & 0 \\ 0 & 0 & 1 \end{bmatrix}$.

The 4×4 identity matrix is $I = \text{diag}(1, 1, 1, 1) = \begin{bmatrix} 1 & 0 & 0 & 0 \\ 0 & 1 & 0 & 0 \\ 0 & 0 & 1 & 0 \\ 0 & 0 & 0 & 1 \end{bmatrix}$.

Remarks

1. In general, the size of I (the value of n) is clear from the given context. However, in some cases, we stress the size of the identity matrix by denoting it by I_n. For example, I_2 is the 2×2 identity matrix, and I_m is the $m \times m$ identity matrix.

2. The columns of the identity matrix should seem familiar. If $\{\vec{e}_1, \dots, \vec{e}_n\}$ is the standard basis for $\mathbb{R}^n$, then

$$I_n = \begin{bmatrix} \vec{e}_1 & \cdots & \vec{e}_n \end{bmatrix}$$

Theorem 3.1.6

If $A \in M_{m\times n}(\mathbb{R})$, then $I_m A = A = A I_n$.

You are asked to prove this theorem in Problem C2. It implies that I_n is the multiplicative identity for the set of $n \times n$ matrices.

Remark

Often the best way of understanding a theorem or definition is to write down some simple examples. So, to help us understand Theorem 3.1.6, we randomly choose a matrix A, say $A = \begin{bmatrix} 2 & 3 & 1 \\ -1 & 4 & 1 \end{bmatrix}$. Since A is 2×3, to multiply this by I_m on the left we must have $m = 2$ so that the matrix multiplication is valid. Similarly, to multiply A on the right by I_n we must have $n = 3$. So, Theorem 3.1.6 says that we have

$$\begin{bmatrix} 1 & 0 \\ 0 & 1 \end{bmatrix}\begin{bmatrix} 2 & 3 & 1 \\ -1 & 4 & 1 \end{bmatrix} = \begin{bmatrix} 2 & 3 & 1 \\ -1 & 4 & 1 \end{bmatrix} = \begin{bmatrix} 2 & 3 & 1 \\ -1 & 4 & 1 \end{bmatrix}\begin{bmatrix} 1 & 0 & 0 \\ 0 & 1 & 0 \\ 0 & 0 & 1 \end{bmatrix}$$

EXERCISE 3.1.9

Let $A \in M_{m\times n}(\mathbb{R})$ such that the system of linear equations $A\vec{x} = \vec{e}_i$ is consistent for all $1 \leq i \leq m$.

(a) Prove that the system of equations $A\vec{x} = \vec{y}$ is consistent for all $\vec{y} \in \mathbb{R}^m$.

(b) What can you conclude about the rank of A from the result in part (a)?

(c) Prove that there exists a matrix B such that $AB = I_m$.

(d) Let $A = \begin{bmatrix} 1 & 2 & 1 \\ 0 & 1 & 1 \end{bmatrix}$. Use your method in part (c) to find a matrix B such that $AB = I_2$.

Block Multiplication

Observe that in our second interpretation of matrix multiplication, equation (3.2), we calculated the product AB in **blocks**. That is, we computed the smaller matrix products $A\vec{b}_1, A\vec{b}_2, \dots, A\vec{b}_p$ and put these in the appropriate positions to create AB. This is a very simple example of **block multiplication**. Observe that we could also regard the rows of A as blocks and write

$$AB = \begin{bmatrix} \vec{a}_1^T B \\ \vdots \\ \vec{a}_p^T B \end{bmatrix}$$

There are more general statements about the products of two matrices, each of which have been partitioned into blocks. In addition to clarifying the meaning of some calculations, block multiplication is used in organizing calculations with very large matrices.

Roughly speaking, as long as the sizes of the blocks are chosen so that the products of the blocks are defined and fit together as required, block multiplication is defined by an extension of the usual rules of matrix multiplication. We demonstrate this with a couple of examples.

EXAMPLE 3.1.26

Suppose that $A \in M_{m\times n}(\mathbb{R})$ and $B \in M_{n\times p}(\mathbb{R})$ such that A and B are **partitioned** into blocks as indicated:

$$A = \begin{bmatrix} A_1 \\ A_2 \end{bmatrix}, \qquad B = \begin{bmatrix} B_1 & B_2 \end{bmatrix}$$

Say that A_1 is $r \times n$ so that A_2 is $(m - r) \times n$, while B_1 is $n \times q$ and B_2 is $n \times (p - q)$. Now, the product of a 2×1 matrix and a 1×2 matrix is given by

$$\begin{bmatrix} a_1 \\ a_2 \end{bmatrix} \begin{bmatrix} b_1 & b_2 \end{bmatrix} = \begin{bmatrix} a_1b_1 & a_1b_2 \\ a_2b_1 & a_2b_2 \end{bmatrix}$$

So, for the partitioned block matrices, we have

$$\begin{bmatrix} A_1 \\ A_2 \end{bmatrix} \begin{bmatrix} B_1 & B_2 \end{bmatrix} = \begin{bmatrix} A_1B_1 & A_1B_2 \\ A_2B_1 & A_2B_2 \end{bmatrix}$$

Observe that all the products are defined and the size of the resulting matrix is $m \times p$.

EXAMPLE 3.1.27

Let $A = \begin{bmatrix} 1 & 2 & -3 \\ 0 & 3 & 1 \\ 1 & 0 & 2 \end{bmatrix}$ and $B = \begin{bmatrix} 2 & 3 & 1 \\ 0 & 1 & -2 \\ 0 & 0 & 3 \end{bmatrix}$.

Let $A_{11} = \begin{bmatrix} 1 \end{bmatrix}$, $A_{12} = \begin{bmatrix} 2 & -3 \end{bmatrix}$, $A_{21} = \begin{bmatrix} 0 \\ 1 \end{bmatrix}$, and $A_{22} = \begin{bmatrix} 3 & 1 \\ 0 & 2 \end{bmatrix}$ so that $A = \begin{bmatrix} A_{11} & A_{12} \\ A_{21} & A_{22} \end{bmatrix}$.

Let $B_{11} = \begin{bmatrix} 2 \end{bmatrix}$, $B_{12} = \begin{bmatrix} 3 & 1 \end{bmatrix}$, $B_{21} = \begin{bmatrix} 0 \\ 0 \end{bmatrix}$, and $B_{22} = \begin{bmatrix} 1 & -2 \\ 0 & 3 \end{bmatrix}$, so that $B = \begin{bmatrix} B_{11} & B_{12} \\ B_{21} & B_{22} \end{bmatrix}$.

Use block multiplication to calculate AB.

Solution: According to normal matrix-matrix multiplication rules we have

$$AB = \begin{bmatrix} A_{11} & A_{12} \\ A_{21} & A_{22} \end{bmatrix} \begin{bmatrix} B_{11} & B_{12} \\ B_{21} & B_{22} \end{bmatrix} = \begin{bmatrix} A_{11}B_{11} + A_{12}B_{21} & A_{11}B_{12} + A_{12}B_{22} \\ A_{21}B_{11} + A_{22}B_{21} & A_{21}B_{12} + A_{22}B_{22} \end{bmatrix}$$

Computing each entry we get

$$A_{11}B_{11} + A_{12}B_{21} = \begin{bmatrix} 1 \end{bmatrix}\begin{bmatrix} 2 \end{bmatrix} + \begin{bmatrix} 2 & -3 \end{bmatrix}\begin{bmatrix} 0 \\ 0 \end{bmatrix} = \begin{bmatrix} 2 \end{bmatrix} + \begin{bmatrix} 0 \end{bmatrix} = \begin{bmatrix} 2 \end{bmatrix}$$

$$A_{11}B_{12} + A_{12}B_{22} = \begin{bmatrix} 1 \end{bmatrix}\begin{bmatrix} 3 & 1 \end{bmatrix} + \begin{bmatrix} 2 & -3 \end{bmatrix}\begin{bmatrix} 1 & -2 \\ 0 & 3 \end{bmatrix} = \begin{bmatrix} 3 & 1 \end{bmatrix} + \begin{bmatrix} 2 & -13 \end{bmatrix} = \begin{bmatrix} 5 & -12 \end{bmatrix}$$

$$A_{21}B_{11} + A_{22}B_{21} = \begin{bmatrix} 0 \\ 1 \end{bmatrix}\begin{bmatrix} 2 \end{bmatrix} + \begin{bmatrix} 3 & 1 \\ 0 & 2 \end{bmatrix}\begin{bmatrix} 0 \\ 0 \end{bmatrix} = \begin{bmatrix} 0 \\ 2 \end{bmatrix} + \begin{bmatrix} 0 \\ 0 \end{bmatrix} = \begin{bmatrix} 0 \\ 2 \end{bmatrix}$$

$$A_{21}B_{12} + A_{22}B_{22} = \begin{bmatrix} 0 \\ 1 \end{bmatrix}\begin{bmatrix} 3 & 1 \end{bmatrix} + \begin{bmatrix} 3 & 1 \\ 0 & 2 \end{bmatrix}\begin{bmatrix} 1 & -2 \\ 0 & 3 \end{bmatrix} = \begin{bmatrix} 0 & 0 \\ 3 & 1 \end{bmatrix} + \begin{bmatrix} 3 & -3 \\ 0 & 6 \end{bmatrix} = \begin{bmatrix} 3 & -3 \\ 3 & 7 \end{bmatrix}$$

Hence,

$$AB = \begin{bmatrix} 2 & 5 & -12 \\ 0 & 3 & -3 \\ 2 & 3 & 7 \end{bmatrix}$$

Remark

To understand why block multiplication works, try multiplying out AB without using block multiplication. Carefully compare your calculations for each entry with the calculations in Example 3.1.27.

PROBLEMS 3.1

Practice Problems

For Problems A1–A9, compute the expression or explain why it is not defined. Let

$$A = \begin{bmatrix} 2 & -2 & 3 \\ 4 & 1 & -1 \end{bmatrix}, B = \begin{bmatrix} -3 & -4 & 1 \\ 2 & -5 & 3 \end{bmatrix},$$

$$C = \begin{bmatrix} 1 & -2 \\ 2 & 1 \\ 4 & -2 \end{bmatrix}, D = \begin{bmatrix} 5 & 3 \\ -1/2 & 1/3 \end{bmatrix}, \text{ and } \vec{x} = \begin{bmatrix} 3 \\ 1 \\ 0 \end{bmatrix}.$$

A1 $A + B$ **A2** $3A - 2B + C^T$ **A3** $C\vec{x}$

A4 AB **A5** AC **A6** CB

A7 CD **A8** $D^T A\vec{x}$ **A9** $\vec{x}^T\vec{x}$

For Problems A10–A13, find a matrix A and vectors $\vec{x}$ and $\vec{b}$ such that $A\vec{x} = \vec{b}$ represents the given system.

A10
$$\begin{aligned} 3x_1 + 2x_2 - x_3 &= 4 \\ 2x_1 - x_2 + 5x_3 &= 5 \end{aligned}$$

A11
$$\begin{aligned} x_1 - 4x_2 + x_3 - 2x_4 &= 1 \\ x_1 - x_2 + 3x_3 \quad &= 0 \end{aligned}$$

A12
$$\begin{aligned} \frac{1}{3}x_1 + 3x_2 - \frac{1}{4}x_3 &= 1 \\ x_1 \quad + x_3 &= \frac{2}{3} \\ x_1 - x_2 \quad &= 3 \end{aligned}$$

A13
$$\begin{aligned} x_1 - x_2 &= 3 \\ 3x_1 + x_2 &= 4 \\ 5x_1 - 8x_2 &= 17 \end{aligned}$$

For Problems A14–A19, determine whether the statement is true or false. Justify your answer.

A14 If $A \in M_{3\times 2}(\mathbb{R})$ and $A\vec{x}$ is defined, then $\vec{x} \in \mathbb{R}^2$.

A15 If $A \in M_{2\times 4}(\mathbb{R})$ and $A\vec{x}$ is defined, then $A\vec{x} \in \mathbb{R}^4$.

A16 If $A \in M_{3\times 3}(\mathbb{R})$, then there is no matrix B such that $AB = BA$.

A17 If $A \in M_{m\times n}(\mathbb{R})$, then $A^T A$ is a square matrix.

A18 The only 2×2 matrix A such that $A^2 = O_{2,2}$ is $O_{2,2}$. (NOTE: As usual, by A^2 we mean AA.)

A19 If A and B are 2×2 matrices such that $AB = O_{2,2}$, then either $A = O_{2,2}$ or $B = O_{2,2}$.

For Problems A20 and A21, check whether $A + B$ and AB are defined. If so, check that $(A + B)^T = A^T + B^T$ and/or $(AB)^T = B^T A^T$.

A20 $A = \begin{bmatrix} 1 & 2 \\ 1 & 3 \\ -2 & 1 \end{bmatrix}, B = \begin{bmatrix} -4 & -3 \\ 1 & -1 \\ 3 & 2 \end{bmatrix}$

A21 $A = \begin{bmatrix} 2 & -4 & 5 \\ 4 & 1 & -3 \end{bmatrix}, B = \begin{bmatrix} -3 & -4 \\ 5 & -2 \\ 1 & 3 \end{bmatrix}$

For Problems A22–A30, compute the product or explain why it is not defined. Let

$$A = \begin{bmatrix} 2 & 5 \\ -1 & 3 \end{bmatrix}, B = \begin{bmatrix} -1 & 3 & -4 \\ 3 & 5 & 2 \end{bmatrix}, C = \begin{bmatrix} 1 & 4 \\ 1 & 3 \\ 4 & -3 \end{bmatrix},$$

$$D = \begin{bmatrix} 4 & 3 & 2 & 1 \\ -1 & 0 & 1 & 2 \\ 2 & 1 & 0 & 3 \end{bmatrix}, \vec{x} = \begin{bmatrix} x_1 \\ x_2 \\ x_3 \end{bmatrix}, \text{ and } \vec{y} = \begin{bmatrix} 1 \\ 2 \\ 1 \end{bmatrix}.$$

A22 AB **A23** BA **A24** DC

A25 $C^T D$ **A26** $\vec{x}^T\vec{y}$ **A27** $\vec{x}\vec{x}$

A28 $A(BC)$ **A29** $(AB)C$ **A30** $(AB)^T$

A31 Let $A = \begin{bmatrix} 2 & 3 & 1 \\ 3 & -1 & 4 \\ -1 & 0 & 1 \end{bmatrix}, \vec{x} = \begin{bmatrix} 1 \\ 2 \\ 4 \end{bmatrix}, \vec{y} = \begin{bmatrix} 3 \\ 1 \\ -1 \end{bmatrix}, \vec{z} = \begin{bmatrix} 0 \\ -1 \\ 1 \end{bmatrix}$.

(a) Determine $A\vec{x}$, $A\vec{y}$, and $A\vec{z}$ using both definitions of matrix-vector multiplication.

(b) Use the result of (a) to determine $A\begin{bmatrix} 1 & 3 & 0 \\ 2 & 1 & -1 \\ 4 & -1 & 1 \end{bmatrix}$.

A32 Verify the following case of block multiplication by calculating both sides of the equation and comparing.

$$\left[\begin{array}{cc|cc} 2 & 3 & -4 & 5 \\ -4 & 1 & 2 & 1 \end{array}\right]\left[\begin{array}{cc} 6 & 3 \\ -2 & 4 \\ \hline 1 & 3 \\ -3 & 2 \end{array}\right] = \begin{bmatrix} 2 & 3 \\ -4 & 1 \end{bmatrix}\begin{bmatrix} 6 & 3 \\ -2 & 4 \end{bmatrix} + \begin{bmatrix} -4 & 5 \\ 2 & 1 \end{bmatrix}\begin{bmatrix} 1 & 3 \\ -3 & 2 \end{bmatrix}$$

A33 Let $\mathcal{B} = \left\{ \begin{bmatrix} 1 & 2 \\ 1 & 0 \end{bmatrix}, \begin{bmatrix} 0 & 1 \\ -1 & 2 \end{bmatrix}, \begin{bmatrix} 1 & 1 \\ 3 & -1 \end{bmatrix} \right\}$.

(a) Determine whether $A = \begin{bmatrix} 2 & 3 \\ 2 & -3 \end{bmatrix}$ is in Span $\mathcal{B}$.

(b) Determine whether the set $\mathcal{B}$ is linearly independent.

A34 Prove if $A = \begin{bmatrix} \vec{a}_1 & \cdots & \vec{a}_n \end{bmatrix}$ is an $m \times n$ matrix and $\vec{e}_i$ is the i-th standard basis vector of $\mathbb{R}^n$, then $A\vec{e}_i = \vec{a}_i$.

Homework Problems

For Problems B1–B12, compute the expression or explain why it is not defined. Let $A = \begin{bmatrix} 8 & 1 \\ -2 & 1 \\ 0 & 4 \end{bmatrix}$, $B = \begin{bmatrix} 1 & 0 \\ 0 & 1 \\ 0 & 0 \end{bmatrix}$,

$C = \begin{bmatrix} 3 & -4 \\ 1 & -1 \\ 2 & -2 \end{bmatrix}$, $D = \begin{bmatrix} 2 & 3 \\ -1 & 4 \end{bmatrix}$, $E = \begin{bmatrix} 2 & 7 \\ 6 & -1 \end{bmatrix}$, and $\vec{x} = \begin{bmatrix} 1 \\ -1 \\ 2 \end{bmatrix}$.

B1 $A + C$ **B2** $2A + B - 2C$ **B3** $A + D$
B4 $C\vec{x}$ **B5** $C^T\vec{x}$ **B6** DE
B7 AE **B8** D^TB **B9** ED
B10 $E\vec{x}$ **B11** $AC^T\vec{x}$ **B12** $\vec{x}^T\vec{x}$

For Problems B13–B16, find a matrix A and vectors $\vec{x}$ and $\vec{b}$ such that $A\vec{x} = \vec{b}$ represents the given system.

B13 $\begin{aligned} x_1 - 3x_2 + x_3 - x_4 &= 1 \\ x_1 + x_2 + 3x_3 + 4x_4 &= 5 \end{aligned}$ **B14** $\begin{aligned} 2x_1 + x_2 + 5x_3 &= 0 \\ 3x_1 - x_2 - 2x_3 &= 0 \end{aligned}$

B15 $\begin{aligned} x_1 + x_2 - x_3 &= 1 \\ x_1 + 2x_2 + x_3 &= 9 \\ 2x_1 - 3x_2 \quad &= -3 \end{aligned}$ **B16** $\begin{aligned} 2x_1 + 3x_2 &= 1 \\ 8x_1 - x_2 &= 1 \\ 7x_1 - 4x_2 &= 1 \end{aligned}$

For Problems B17–B22, determine whether the statement is true or false. Justify your answer.

B17 If $A, B \in M_{2\times 2}(\mathbb{R})$, then $AB = BA$.

B18 If $A \in M_{4\times 3}(\mathbb{R})$ and $A\vec{x}$ is defined, then $A\vec{x} \in \mathbb{R}^4$.

B19 If $A \in M_{2\times 3}(\mathbb{R})$, then there is no matrix B such that $AB = BA$.

B20 If $A, B \in M_{2\times 2}(\mathbb{R})$, then $(A + B)(A - B) = A^2 - B^2$.

B21 If $A \in M_{m\times n}(\mathbb{R})$ and $A^T = A$, then A is a square matrix.

B22 If $A, B \in M_{2\times 2}(\mathbb{R})$ and $A\vec{x} = B\vec{x}$ for some $\vec{x} \in \mathbb{R}^2$, then $A = B$.

For Problems B23 and B24, check whether $A + B$ and AB are defined. If so, check that $(A + B)^T = A^T + B^T$ and/or $(AB)^T = B^TA^T$.

B23 $A = \begin{bmatrix} 5 & -2 \\ 3 & 4 \end{bmatrix}$, $B = \begin{bmatrix} 2 & 2 \\ -1 & 3 \end{bmatrix}$

B24 $A = \begin{bmatrix} 6 & 9 \\ -2 & 1 \\ 0 & 3 \end{bmatrix}$, $B = \begin{bmatrix} 1 & 4 & 1 \\ 2 & 2 & -5 \end{bmatrix}$

For Problems B25–B36, compute the product or explain why it is not defined. Let $A = \begin{bmatrix} 6 & 3 & 3 \\ -1 & 2 & 3 \end{bmatrix}$,

$B = \begin{bmatrix} 1 & 2 \\ 1 & -1 \\ 3 & 8 \end{bmatrix}$, $C = \begin{bmatrix} 2 & 0 \\ -1 & 3 \end{bmatrix}$, $\vec{x} = \begin{bmatrix} x_1 \\ x_2 \\ x_3 \end{bmatrix}$, $\vec{y} = \begin{bmatrix} 2 \\ 2 \\ 1 \end{bmatrix}$, and $\vec{z} = \begin{bmatrix} 2 \\ 1 \end{bmatrix}$.

B25 AB **B26** BA **B27** AC
B28 CA **B29** $A\vec{x}$ **B30** $B\vec{z}$
B31 $\vec{x}^T\vec{y}$ **B32** $\vec{y}^T\vec{z}$ **B33** $\vec{z}^TC\vec{z}$
B34 $A(BC)$ **B35** $(CA)^T$ **B36** A^TC^T

B37 Let $A = \begin{bmatrix} 2 & -8 & 5 \\ 0 & 1 & 5 \\ -7 & 3 & 9 \end{bmatrix}$, $\vec{x} = \begin{bmatrix} 4 \\ 1 \\ 3 \end{bmatrix}$, $\vec{y} = \begin{bmatrix} 1 \\ 4 \\ -2 \end{bmatrix}$, $\vec{z} = \begin{bmatrix} 1 \\ 0 \\ 1 \end{bmatrix}$.

(a) Determine $A\vec{x}$, $A\vec{y}$, and $A\vec{z}$ using both definitions of matrix-vector multiplication.

(b) Use the result of (a) to determine $A\begin{bmatrix} 4 & 1 & 1 \\ 1 & 4 & 0 \\ 3 & -2 & 1 \end{bmatrix}$.

B38 Let $A = \begin{bmatrix} 1 & -2 & 3 \\ 2 & -3 & 5 \\ 1 & -4 & 6 \end{bmatrix}$, $\vec{x} = \begin{bmatrix} 2 \\ -7 \\ -5 \end{bmatrix}$, $\vec{y} = \begin{bmatrix} 0 \\ 3 \\ 2 \end{bmatrix}$, $\vec{z} = \begin{bmatrix} -1 \\ 1 \\ 1 \end{bmatrix}$.

(a) Determine $A\vec{x}$, $A\vec{y}$, and $A\vec{z}$ using both definitions of matrix-vector multiplication.

(b) Use the result of (a) to determine $A\begin{bmatrix} 2 & 0 & -1 \\ -7 & 3 & 1 \\ -5 & 2 & 1 \end{bmatrix}$.

B39 Verify the following case of block multiplication by calculating both sides of the equation and comparing.

$$\left[\begin{array}{c|cc} 6 & 3 & -2 \\ 1 & 2 & -1 \end{array}\right]\left[\begin{array}{cc} -4 & 8 \\ \hline 2 & 1 \\ 0 & 6 \end{array}\right] = \begin{bmatrix} 6 \\ 1 \end{bmatrix}\begin{bmatrix} -4 & 8 \end{bmatrix} + \begin{bmatrix} 3 & -2 \\ 2 & -1 \end{bmatrix}\begin{bmatrix} 2 & 1 \\ 0 & 6 \end{bmatrix}$$

B40 Let $\mathcal{B} = \left\{\begin{bmatrix} 1 & 0 \\ -1 & 1 \end{bmatrix}, \begin{bmatrix} 2 & 1 \\ 2 & 1 \end{bmatrix}, \begin{bmatrix} 0 & -1 \\ -4 & -1 \end{bmatrix}\right\}$.

(a) Determine whether $A = \begin{bmatrix} 2 & 3 \\ 10 & -1 \end{bmatrix}$ is in Span $\mathcal{B}$.

(b) Determine whether the set $\mathcal{B}$ is linearly independent.

B41 Let $C = \left\{ \begin{bmatrix} 1 & 1 \\ 2 & 0 \end{bmatrix}, \begin{bmatrix} 2 & 3 \\ 5 & 1 \end{bmatrix}, \begin{bmatrix} 0 & 1 \\ 1 & 0 \end{bmatrix} \right\}$.

(a) Determine whether $A = \begin{bmatrix} 1 & 1 \\ 1 & 2 \end{bmatrix}$ is in Span C.

(b) Determine whether the set C is linearly independent.

B42 Let $\mathcal{B} = \left\{ \begin{bmatrix} 1 & 0 \\ 1 & 0 \end{bmatrix}, \begin{bmatrix} 1 & 1 \\ 0 & 0 \end{bmatrix}, \begin{bmatrix} 1 & 0 \\ 0 & 1 \end{bmatrix}, \begin{bmatrix} 0 & 1 \\ 0 & 1 \end{bmatrix} \right\}$.

(a) Determine whether $A = \begin{bmatrix} -2 & 3 \\ -1 & 1 \end{bmatrix}$ is in Span $\mathcal{B}$.

(b) Determine whether the set $\mathcal{B}$ is linearly independent.

B43 Denote the number of individuals in year t of two adjacent cities Neville and Maggiton by n_t and m_t respectively. Suppose that the only changes in these populations occur via individuals moving from one city to the other. Suppose that each year, $\frac{1}{10}$ of the population of Neville moves to Maggiton, while $\frac{1}{5}$ of the population of Maggiton migrates to Neville. We can express the year-to-year change in the populations as:

$$n_{t+1} = \frac{9}{10}x_t + \frac{1}{5}y_t$$
$$m_{t+1} = \frac{1}{10}x_t + \frac{4}{5}y_t$$

(a) Express the year-to-year change in the population as a matrix-vector product.

$$\begin{bmatrix} n_{t+1} \\ m_{t+1} \end{bmatrix} = A \begin{bmatrix} n_t \\ m_t \end{bmatrix}$$

(b) Suppose that in year $t = 0$ the populations are given by $n_0 = 1000$, $m_0 = 2000$. Use your formula from part (a) to calculate the populations in years $t = 1, 2$, and 3. Confirm that in each year, the total population $(n_t + m_t)$ is 3000. Why should this be expected?

(c) Calculate $A^3 = AAA$ and use it to calculate $A^3 \begin{bmatrix} 1000 \\ 2000 \end{bmatrix}$. What would $A^{10} \begin{bmatrix} 1000 \\ 2000 \end{bmatrix}$ represent?

For Problems B44 and B45:

(a) Find the connectivity matrix.

(b) Determine how many paths of length 3 there are from A_2 to A_6.

(c) Determine how many paths of length at most 3 there are from A_5 to A_1.

(Use a computer for the calculations.)

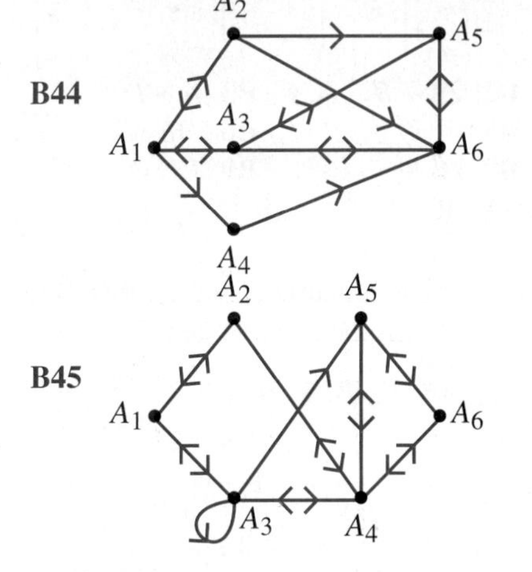

B46 The Fibonacci sequence f_n is defined by

$$f_1 = 1, \quad f_2 = 1, \quad f_{n+2} = f_n + f_{n+1}$$

We get the sequence $1, 1, 2, 3, 5, 8, 13, 21, 34, 55, \ldots$. This is called a *two-step recursion*, because each term depends on the two preceding terms. The sequence can also be constructed as a pair of one-step recursions as follows:

$$x_{n+1} = y_n$$
$$y_{n+1} = x_n + y_n$$

(a) Verify that using this rule and taking $x_0 = 1$, $y_0 = 1$, the x_n sequence is the Fibonacci sequence, while the y_n sequence is a shifted version of the Fibonacci sequence.

(b) An advantage of the one-step recursion form of the sequence definition is that it can be written as a matrix product. Find the matrix $\mathbf{A}$ so that the recursion takes the form

$$\begin{bmatrix} x_{n+1} \\ y_{n+1} \end{bmatrix} = A \begin{bmatrix} x_n \\ y_n \end{bmatrix}$$

Conceptual Problems

C1 Prove Theorem 3.1.5, using the following hints. To prove $A = B$, prove that $A - B = O_{m,n}$; note that $A\vec{x} = B\vec{x}$ for every $\vec{x} \in \mathbb{R}^n$ if and only if $(A - B)\vec{x} = \vec{0}$ for every $\vec{x} \in \mathbb{R}^n$. Now, suppose that $C\vec{x} = \vec{0}$ for every $\vec{x} \in \mathbb{R}^n$. Consider the case where $\vec{x} = \vec{e}_i$ and conclude that each column of C must be the zero vector.

C2 Prove Theorem 3.1.6.

C3 Let $A \in M_{m\times n}(\mathbb{R})$ such that the system of linear equations $A\vec{x} = \vec{e}_i$ is consistent for all $1 \leq i \leq m$.

(a) Prove that the system of equations $A\vec{x} = \vec{y}$ is consistent for all $\vec{y} \in \mathbb{R}^m$.

(b) What can you conclude about the rank of A from the result in part (a)?

(c) Prove there exists a matrix B such that $AB = I_m$.

(d) Let $A = \begin{bmatrix} 1 & 2 & 1 \\ 0 & 1 & 1 \end{bmatrix}$. Use your method in part (c) to find a matrix B such that $AB = I_2$.

C4 Find a formula to calculate the ij-th entry of AA^T and of A^TA. Explain why it follows that if AA^T or A^TA is the zero matrix, then A is the zero matrix.

C5 A square matrix A is called **symmetric** if $A^T = A$. Prove that for any $B \in M_{m\times n}(\mathbb{R})$ both BB^T and B^TB are symmetric.

C6 (a) Construct $A \in M_{2\times 2}(\mathbb{R})$ that is not the zero matrix yet satisfies $A^2 = O_{2,2}$.

(b) Find $A, B \in M_{2\times 2}(\mathbb{R})$ with $A \neq B$ and neither $A = O_{2,2}$ nor $B = O_{2,2}$, such that

$$A^2 - AB - BA + B^2 = O_{2,2}$$

C7 Find as many 2×2 matrices A as you can that satisfy $A^2 = I$.

C8 We define the **trace** of $A \in M_{n\times n}(\mathbb{R})$ by

$$\operatorname{tr} A = a_{11} + a_{22} + \cdots + a_{nn}$$

Let $A, B \in M_{n\times n}(\mathbb{R})$. Prove that

$$\operatorname{tr}(A + B) = \operatorname{tr} A + \operatorname{tr} B$$

C9 Let $D = \operatorname{diag}(\lambda_1, \dots, \lambda_n) \in M_{n\times n}(\mathbb{R})$.

(a) Show that $D^2 = \operatorname{diag}(\lambda_1^2, \dots, \lambda_n^2)$.

(b) Use induction to prove that $D^k = \operatorname{diag}(\lambda_1^k, \dots, \lambda_n^k)$.

C10 Let $A \in M_{m\times n}(\mathbb{R})$ such that $n > m$. Prove that if $\operatorname{rank} A = m$, then there are infinitely many matrices $B \in M_{n\times m}(\mathbb{R})$ such that $AB = I_m$.

C11 Let $A = \begin{bmatrix} \vec{v}_1 & \cdots & \vec{v}_k \end{bmatrix}$ and $B = \begin{bmatrix} \vec{u}_1 & \cdots & \vec{u}_k \end{bmatrix}$ be $n\times k$ matrices. Prove that if

$$\operatorname{Span}\{\vec{v}_1, \dots, \vec{v}_k\} \subseteq \operatorname{Span}\{\vec{u}_1, \dots, \vec{u}_k\}$$

then there exists a matrix C such that $A = BC$.

C12 Let $A \in M_{m\times n}(\mathbb{R})$ with one row of zeros and $B \in M_{n\times p}(\mathbb{R})$.

(a) Use block multiplication to prove that AB also has at least one row of zeros.

(b) Give an example where AB has more than one row of zeros.

3.2 Matrix Mappings and Linear Mappings

Functions are a fundamental concept in mathematics. Recall that a **function** f *is a rule that assigns to every element* x *of an initial set called the* **domain** *of the function a unique value* y *in another set called the* **codomain** *of* f. *We say that* f **maps** x *to* y *or that* y *is the* **image** *of* x *under* f, *and we write* f(x) = y. *If* f *is a function with domain* U *and codomain* V, *then we say that* f *maps* U *to* V *and denote this by* f:U → V. *In your earlier mathematics, you met functions* f : ℝ → ℝ *such as* f(x) = x² *and looked at their various properties. In this section, we will start by looking at more general functions* f : $\mathbb{R}^n \to \mathbb{R}^m$, *commonly called mappings or transformations. We will also look at a class of functions called linear mappings that are very important in linear algebra and its applications.*

Matrix Mappings

We saw in the preceding section that our rule for matrix-vector multiplication behaves like a function whose domain is $\mathbb{R}^n$ and whose codomain is $\mathbb{R}^m$. In particular, for any $m \times n$ matrix A and vector $\vec{x} \in \mathbb{R}^n$, the product $A\vec{x}$ is a vector in $\mathbb{R}^m$.

Definition
Matrix Mapping

For any $A \in M_{m\times n}(\mathbb{R})$, we define a function $f_A : \mathbb{R}^n \to \mathbb{R}^m$ called the **matrix mapping** corresponding to A by

$$f_A(\vec{x}) = A\vec{x}, \text{ for all } \vec{x} \in \mathbb{R}^n$$

Remark

Although a matrix mapping sends vectors to vectors, it is much more common to view functions as mapping points to points. Thus, when dealing with mappings in this text, we will often write

$$f(x_1, \dots, x_n) = (y_1, \dots, y_m), \quad \text{or} \quad f(x_1, \dots, x_n) = \begin{bmatrix} y_1 \\ \vdots \\ y_m \end{bmatrix}$$

EXAMPLE 3.2.1

Let $A = \begin{bmatrix} 2 & 3 \\ -1 & 4 \\ 0 & 1 \end{bmatrix}$. Find $f_A(1,2)$ and $f_A(-1,4)$.

Solution: We have

$$f_A(1,2) = \begin{bmatrix} 2 & 3 \\ -1 & 4 \\ 0 & 1 \end{bmatrix}\begin{bmatrix} 1 \\ 2 \end{bmatrix} = \begin{bmatrix} 8 \\ 7 \\ 2 \end{bmatrix}$$

$$f_A(-1,4) = \begin{bmatrix} 2 & 3 \\ -1 & 4 \\ 0 & 1 \end{bmatrix}\begin{bmatrix} -1 \\ 4 \end{bmatrix} = \begin{bmatrix} 10 \\ 17 \\ 4 \end{bmatrix}$$

EXERCISE 3.2.1

Let $A = \begin{bmatrix} 1 & 2 & -1 & 1 \\ 2 & 0 & 2 & 6 \\ 3 & 2 & 1 & 7 \end{bmatrix}$. Find $f_A(-1, 1, 1, 0)$ and $f_A(-3, 1, 0, 1)$.

EXERCISE 3.2.2

Let $A = \begin{bmatrix} 1 & 2 \\ 3 & -1 \end{bmatrix}$. Find $f_A(1, 0)$, $f_A(0, 1)$, and $f_A(2, 3)$.
What is the relationship between the value of $f_A(2, 3)$ and the values of $f_A(1, 0)$ and $f_A(0, 1)$?

Based on our results in Exercise 3.2.2, is such a relationship always true? A good way to explore this is to look at a more general example.

EXAMPLE 3.2.2

Let $A = \begin{bmatrix} a_{11} & a_{12} \\ a_{21} & a_{22} \end{bmatrix}$ and find the values of $f_A(1, 0)$, $f_A(0, 1)$, and $f_A(x_1, x_2)$.

Solution: We have

$$f_A(1, 0) = \begin{bmatrix} a_{11} & a_{12} \\ a_{21} & a_{22} \end{bmatrix} \begin{bmatrix} 1 \\ 0 \end{bmatrix} = \begin{bmatrix} a_{11} \\ a_{21} \end{bmatrix}$$

$$f_A(0, 1) = \begin{bmatrix} a_{11} & a_{12} \\ a_{21} & a_{22} \end{bmatrix} \begin{bmatrix} 0 \\ 1 \end{bmatrix} = \begin{bmatrix} a_{12} \\ a_{22} \end{bmatrix}$$

Then we get

$$f_A(x_1, x_2) = \begin{bmatrix} a_{11} & a_{12} \\ a_{21} & a_{22} \end{bmatrix} \begin{bmatrix} x_1 \\ x_2 \end{bmatrix} = \begin{bmatrix} a_{11}x_1 + a_{12}x_2 \\ a_{21}x_1 + a_{22}x_2 \end{bmatrix} = x_1 \begin{bmatrix} a_{11} \\ a_{21} \end{bmatrix} + x_2 \begin{bmatrix} a_{12} \\ a_{22} \end{bmatrix}$$

We can now clearly see the relationship between the image of the standard basis vectors in $\mathbb{R}^2$ and the image of any vector $\vec{x}$. We conjecture that this works for any $m \times n$ matrix A.

Theorem 3.2.1

Let $\vec{e}_1, \ldots, \vec{e}_n$ be the standard basis vectors of $\mathbb{R}^n$. If $A \in M_{m\times n}(\mathbb{R})$ and $f_A : \mathbb{R}^n \to \mathbb{R}^m$ is the corresponding matrix mapping, then for any vector $\vec{x} \in \mathbb{R}^n$ we have

$$f_A(\vec{x}) = x_1 f_A(\vec{e}_1) + x_2 f_A(\vec{e}_2) + \cdots + x_n f_A(\vec{e}_n)$$

Proof: Let $A = \begin{bmatrix} \vec{a}_1 & \vec{a}_2 & \cdots & \vec{a}_n \end{bmatrix}$. Then,

$$\begin{aligned} f_A(\vec{x}) = A\vec{x} &= x_1\vec{a}_1 + \cdots + x_n\vec{a}_n && \text{by definition of matrix-vector multiplication} \\ &= x_1 A\vec{e}_1 + \cdots + x_n A\vec{e}_n && \text{by Theorem 3.1.3} \\ &= x_1 f_A(\vec{e}_1) + \cdots x_n f_A(\vec{e}_n) \end{aligned}$$

■

Since the images of the standard basis vectors are just the columns of A, we see that the image of any vector $\vec{x} \in \mathbb{R}^n$ is a linear combination of the columns of A. This should not be surprising as this is one of our interpretations of matrix-vector multiplication. However, it does make us think about how a matrix mapping will affect a linear combination of vectors in $\mathbb{R}^n$.

Theorem 3.2.2

If $A \in M_{m\times n}(\mathbb{R})$ with corresponding matrix mapping $f_A : \mathbb{R}^n \to \mathbb{R}^m$, then for any $\vec{x}, \vec{y} \in \mathbb{R}^n$ and any $t \in \mathbb{R}$ we have

$$f_A(\vec{x} + \vec{y}) = f_A(\vec{x}) + f_A(\vec{y}) \tag{3.3}$$

$$f_A(t\vec{x}) = tf_A(\vec{x}) \tag{3.4}$$

Proof: Using properties of matrix multiplication, we get

$$f_A(\vec{x} + \vec{y}) = A(\vec{x} + \vec{y}) = A\vec{x} + A\vec{y} = f_A(\vec{x}) + f_A(\vec{y})$$

and

$$f_A(t\vec{x}) = A(t\vec{x}) = tA\vec{x} = tf_A(\vec{x})$$

■

A function that satisfies equation (3.3) is said to **preserve addition**. Similarly, a function satisfying equation (3.4) is said to **preserve scalar multiplication**. Notice that a function that satisfies both properties will in fact **preserve linear combinations**—that is,

$$f_A(t_1\vec{x}_1 + \cdots + t_n\vec{x}_n) = t_1 f_A(\vec{x}_1) + \cdots + t_n f_A(\vec{x}_n)$$

We call such functions **linear mappings**.

Linear Mappings

Definition
Linear Mapping
Linear Operator

A function $L : \mathbb{R}^n \to \mathbb{R}^m$ is called a **linear mapping** (or **linear transformation**) if for every $\vec{x}, \vec{y} \in \mathbb{R}^n$ and $s, t \in \mathbb{R}$ it satisfies

$$L(s\vec{x} + t\vec{y}) = sL(\vec{x}) + tL(\vec{y})$$

A linear mapping whose domain and codomain are the same is sometimes called a **linear operator**.

Remarks

1. *Linear transformation* and *linear mapping* mean exactly the same thing. Some people prefer one or the other, but we shall use both.
2. Since a linear operator L has the same domain and codomain, we often speak of a **linear operator** L **on** $\mathbb{R}^n$ to indicate that L is a linear mapping from $\mathbb{R}^n$ to $\mathbb{R}^n$.
3. For the time being, we have defined only linear mappings whose domain is $\mathbb{R}^n$ and whose codomain is $\mathbb{R}^m$. In Chapter 4, we will look at other sets that can be the domain and/or codomain of linear mappings.
4. Two linear mappings $L : \mathbb{R}^n \to \mathbb{R}^m$ and $M : \mathbb{R}^n \to \mathbb{R}^m$ are **equal** if $L(\vec{x}) = M(\vec{x})$ for all $\vec{x} \in \mathbb{R}^n$. In this case, we can write $L = M$.

EXAMPLE 3.2.3

Show that the mapping $f : \mathbb{R}^2 \to \mathbb{R}^2$ defined by $f(x_1, x_2) = (2x_1 + x_2, -3x_1 + 5x_2)$ is linear.

Solution: For any $\vec{y}, \vec{z} \in \mathbb{R}^2$, we have

$$\begin{aligned} f(s\vec{y} + t\vec{z}) &= f(sy_1 + tz_1, sy_2 + tz_2) \\ &= \begin{bmatrix} 2(sy_1 + tz_1) + (sy_2 + tz_2) \\ -3(sy_1 + tz_1) + 5(sy_2 + tz_2) \end{bmatrix} \\ &= s\begin{bmatrix} 2y_1 + y_2 \\ -3y_1 + 5y_2 \end{bmatrix} + t\begin{bmatrix} 2z_1 + z_2 \\ -3z_1 + 5z_2 \end{bmatrix} \\ &= sf(\vec{y}) + tf(\vec{z}) \end{aligned}$$

Thus, f is closed under linear combinations and therefore is a linear operator.

EXAMPLE 3.2.4

Determine whether the mapping $f : \mathbb{R}^3 \to \mathbb{R}$ defined by $f(\vec{x}) = \|\vec{x}\|$ is linear.

Solution: Let $\vec{x}, \vec{y} \in \mathbb{R}^3$ and consider

$$f(\vec{x} + \vec{y}) = \|\vec{x} + \vec{y}\| \quad \text{and} \quad f(\vec{x}) + f(\vec{y}) = \|\vec{x}\| + \|\vec{y}\|$$

Are these equal? By the triangle inequality, we get

$$\|\vec{x} + \vec{y}\| \leq \|\vec{x}\| + \|\vec{y}\|$$

and we expect equality only when one of $\vec{x}, \vec{y}$ is a multiple of the other. Therefore, we believe that these are not always equal, and consequently f is not closed under addition. To demonstrate this, we give a counterexample: if $\vec{x} = \begin{bmatrix} 1 \\ 0 \\ 0 \end{bmatrix}$ and $\vec{y} = \begin{bmatrix} 0 \\ 1 \\ 0 \end{bmatrix}$, then

$$f(\vec{x} + \vec{y}) = f(1, 1, 0) = \left\| \begin{bmatrix} 1 \\ 1 \\ 0 \end{bmatrix} \right\| = \sqrt{2}$$

but

$$f(\vec{x}) + f(\vec{y}) = \left\| \begin{bmatrix} 1 \\ 0 \\ 0 \end{bmatrix} \right\| + \left\| \begin{bmatrix} 0 \\ 1 \\ 0 \end{bmatrix} \right\| = 1 + 1 = 2$$

Thus, $f(\vec{x} + \vec{y}) \neq f(\vec{x}) + f(\vec{y})$ for all pairs of vectors $\vec{x}, \vec{y}$ in $\mathbb{R}^3$, hence f is not linear.

EXERCISE 3.2.3

Determine whether the following mappings are linear.

(a) $f : \mathbb{R}^2 \to \mathbb{R}^2$ defined by $f(x_1, x_2) = (x_1^2, x_1 + x_2)$

(b) $g : \mathbb{R}^2 \to \mathbb{R}^2$ defined by $g(x_1, x_2) = (x_2, x_1 - x_2)$

Is Every Linear Mapping a Matrix Mapping?

We saw that every matrix determines a corresponding linear mapping. It is natural to ask whether every linear mapping can be represented as a matrix mapping.

EXAMPLE 3.2.5

Can the linear mapping f defined by $f(x_1, x_2) = (2x_1 + x_2, -3x_1 + 5x_2)$, be represented as a matrix-mapping?

Solution: If $f(\vec{x}) = A\vec{x}$, then our work with matrix mappings suggests that the columns of A are the images of the standard basis vectors. We have

$$f(1,0) = \begin{bmatrix} 2 \\ -3 \end{bmatrix}, \qquad f(0,1) = \begin{bmatrix} 1 \\ 5 \end{bmatrix}$$

Thus, we define

$$A = \begin{bmatrix} f(1,0) & f(0,1) \end{bmatrix} = \begin{bmatrix} 2 & 1 \\ -3 & 5 \end{bmatrix}$$

This gives

$$f_A(x_1, x_2) = \begin{bmatrix} 2 & 1 \\ -3 & 5 \end{bmatrix} \begin{bmatrix} x_1 \\ x_2 \end{bmatrix} = \begin{bmatrix} 2x_1 + x_2 \\ -3x_1 + 5x_2 \end{bmatrix} = f(x_1, x_2)$$

Hence, f can be represented as a matrix mapping.

This example not only gives us a good reason to believe it is always true but indicates how we can find the matrix for a given linear mapping L.

Theorem 3.2.3

Let $L : \mathbb{R}^n \to \mathbb{R}^m$ be a linear mapping. If we define

$$[L] = \begin{bmatrix} L(\vec{e}_1) & L(\vec{e}_2) & \cdots & L(\vec{e}_n) \end{bmatrix}$$

then we have

$$L(\vec{x}) = [L]\vec{x}$$

Proof: Let $\vec{x} \in \mathbb{R}^n$. Writing $\vec{x}$ as a linear combination of the standard basis vectors $\vec{e}_1, \dots, \vec{e}_n$ gives

$$\begin{aligned} L(\vec{x}) &= L(x_1\vec{e}_1 + \cdots + x_n\vec{e}_n) \\ &= x_1 L(\vec{e}_1) + \cdots + x_n L(\vec{e}_n) && \text{since } L \text{ is linear} \\ &= \begin{bmatrix} L(\vec{e}_1) & \cdots & L(\vec{e}_n) \end{bmatrix} \begin{bmatrix} x_1 \\ \vdots \\ x_n \end{bmatrix} && \text{by definition of matrix-vector multiplication} \\ &= [L]\vec{x} \end{aligned}$$

■

Remarks

1. The matrix $[L]$ in Theorem 3.2.3 is called the **standard matrix** of L.
2. Combining Theorem 3.2.3 with Theorem 3.2.2 shows that a mapping is linear if and only if it is a matrix mapping.

EXAMPLE 3.2.6

Let $\vec{v} = \begin{bmatrix} 3 \\ 4 \end{bmatrix}$ and $\vec{u} = \begin{bmatrix} 1 \\ 2 \end{bmatrix}$. Find the standard matrix of the mapping $\text{proj}_{\vec{v}} : \mathbb{R}^2 \to \mathbb{R}^2$ and use it to find $\text{proj}_{\vec{v}}(\vec{u})$.

Solution: Since $\text{proj}_{\vec{v}}$ is linear (see Section 1.5), we can apply Theorem 3.2.3 to find $[\text{proj}_{\vec{v}}]$. The first column of the matrix is the image of the first standard basis vector under $\text{proj}_{\vec{v}}$:

$$\text{proj}_{\vec{v}}(\vec{e}_1) = \frac{\vec{e}_1 \cdot \vec{v}}{\|\vec{v}\|^2}\vec{v} = \frac{1(3) + 0(4)}{3^2 + 4^2}\begin{bmatrix} 3 \\ 4 \end{bmatrix} = \begin{bmatrix} 9/25 \\ 12/25 \end{bmatrix}$$

Similarly, the second column is the image of the second basis vector:

$$\text{proj}_{\vec{v}}(\vec{e}_2) = \frac{\vec{e}_2 \cdot \vec{v}}{\|\vec{v}\|^2}\vec{v} = \frac{0(3) + 1(4)}{25}\begin{bmatrix} 3 \\ 4 \end{bmatrix} = \begin{bmatrix} 12/25 \\ 16/25 \end{bmatrix}$$

Hence, the standard matrix of the linear mapping is

$$[\text{proj}_{\vec{v}}] = \begin{bmatrix} \text{proj}_{\vec{v}}(\vec{e}_1) & \text{proj}_{\vec{v}}(\vec{e}_2) \end{bmatrix} = \begin{bmatrix} 9/25 & 12/25 \\ 12/25 & 16/25 \end{bmatrix}$$

Therefore, we have

$$\text{proj}_{\vec{v}}(\vec{u}) = [\text{proj}_{\vec{v}}]\vec{u} = \begin{bmatrix} 9/25 & 12/25 \\ 12/25 & 16/25 \end{bmatrix}\begin{bmatrix} 1 \\ 2 \end{bmatrix} = \begin{bmatrix} 33/25 \\ 44/25 \end{bmatrix}$$

EXAMPLE 3.2.7

Let $G : \mathbb{R}^3 \to \mathbb{R}^2$ be defined by $G(x_1, x_2, x_3) = (x_1, x_2)$. Prove that G is linear and find the standard matrix of G.

Solution: We first prove that G is linear. For any $\vec{x}, \vec{y} \in \mathbb{R}^3$ and $s, t \in \mathbb{R}$, we have

$$\begin{aligned} G(s\vec{x} + t\vec{y}) &= G(sx_1 + ty_1, sx_2 + ty_2, sx_3 + ty_3) \\ &= \begin{bmatrix} sx_1 + ty_1 \\ sx_2 + ty_2 \end{bmatrix} \\ &= s\begin{bmatrix} x_1 \\ x_2 \end{bmatrix} + t\begin{bmatrix} y_1 \\ y_2 \end{bmatrix} \\ &= sG(\vec{x}) + tG(\vec{y}) \end{aligned}$$

Hence, G is linear. Thus, we can apply Theorem 3.2.3 to find its standard matrix. The images of the standard basis vectors are

$$G(\vec{e}_1) = \begin{bmatrix} 1 \\ 0 \end{bmatrix}, \qquad G(\vec{e}_2) = \begin{bmatrix} 0 \\ 1 \end{bmatrix}, \qquad G(\vec{e}_3) = \begin{bmatrix} 0 \\ 0 \end{bmatrix}$$

So, by definition, $[G] = \begin{bmatrix} G(\vec{e}_1) & G(\vec{e}_2) & G(\vec{e}_3) \end{bmatrix} = \begin{bmatrix} 1 & 0 & 0 \\ 0 & 1 & 0 \end{bmatrix}$.

Did we really need to prove that G was linear first? Could we have not just constructed $[G]$ using the image of the standard basis vectors and then said that G is linear because it is a matrix mapping? *No*! We must always check the hypotheses of a theorem before using it. Theorem 3.2.3 says that *if* f is linear, *then* $[f]$ can be constructed from the images of the standard basis vectors. The converse is not true!

For example, consider the mapping $f : \mathbb{R}^2 \to \mathbb{R}^2$ defined by $f(x_1, x_2) = \begin{bmatrix} x_1x_2 \\ 0 \end{bmatrix}$. The images of the standard basis vectors are $f(\vec{e}_1) = \begin{bmatrix} 0 \\ 0 \end{bmatrix}$ and $f(\vec{e}_2) = \begin{bmatrix} 0 \\ 0 \end{bmatrix}$, so we can construct the matrix $\begin{bmatrix} 0 & 0 \\ 0 & 0 \end{bmatrix}$. But this matrix does not represent the mapping! In particular, observe that $f(1, 1) = \begin{bmatrix} 1 \\ 0 \end{bmatrix}$ but $\begin{bmatrix} 0 & 0 \\ 0 & 0 \end{bmatrix}\begin{bmatrix} 1 \\ 1 \end{bmatrix} = \begin{bmatrix} 0 \\ 0 \end{bmatrix}$. Hence, even though we can create a matrix using the images of the standard basis vectors, it does not imply that the matrix will represent that mapping, unless we already know the mapping is linear.

EXERCISE 3.2.4

Let $H : \mathbb{R}^4 \to \mathbb{R}^2$ be defined by $H(x_1, x_2, x_3, x_4) = (x_3 + x_4, x_1)$. Prove that H is linear and find the standard matrix of H.

Linear Combinations and Compositions of Linear Mappings

We now look at the usual operations on functions and how these affect linear mappings.

Definition
Addition and Scalar Multiplication of Linear Mappings

Let $L : \mathbb{R}^n \to \mathbb{R}^m$ and $M : \mathbb{R}^n \to \mathbb{R}^m$ be linear mappings. We define $(L + M) : \mathbb{R}^n \to \mathbb{R}^m$ by

$$(L + M)(\vec{x}) = L(\vec{x}) + M(\vec{x}), \quad \text{for all } \vec{x} \in \mathbb{R}^n$$

For any $c \in \mathbb{R}$, we define $(cL) : \mathbb{R}^n \to \mathbb{R}^m$ by

$$(cL)(\vec{x}) = cL(\vec{x}), \quad \text{for all } \vec{x} \in \mathbb{R}^n$$

EXAMPLE 3.2.8

Let $L : \mathbb{R}^2 \to \mathbb{R}^3$ and $M : \mathbb{R}^2 \to \mathbb{R}^3$ be defined by $L(x_1, x_2) = (x_1, x_1 + x_2, -x_2)$ and $M(x_1, x_2) = (x_1, x_1, x_2)$. Calculate $L + M$ and $3L$.

Solution: $L + M$ is the mapping defined by

$$(L + M)(\vec{x}) = L(\vec{x}) + M(\vec{x}) = (x_1, x_1 + x_2, -x_2) + (x_1, x_1, x_2) = (2x_1, 2x_1 + x_2, 0)$$

and $3L$ is the mapping defined by

$$(3L)(\vec{x}) = 3L(\vec{x}) = 3(x_1, x_1 + x_2, -x_2) = (3x_1, 3x_1 + 3x_2, -3x_2)$$

By analyzing Example 3.2.8 we can learn more than just how to add linear mappings and how to multiply them by a scalar. We first observe that $L + M$ and $3L$ are both linear mappings. This makes us realize that we can rewrite the calculations in the example in terms of standard matrices. For example,

$$\begin{aligned}(L+M)(\vec{x}) &= L(\vec{x}) + M(\vec{x}) = [L]\vec{x} + [M]\vec{x} = ([L]+[M])\vec{x} \\ &= \left(\begin{bmatrix}1 & 0\\1 & 1\\0 & -1\end{bmatrix} + \begin{bmatrix}1 & 0\\1 & 0\\0 & 1\end{bmatrix}\right)\vec{x} = \begin{bmatrix}2 & 0\\2 & 1\\0 & 0\end{bmatrix}\begin{bmatrix}x_1\\x_2\end{bmatrix} = \begin{bmatrix}2x_1\\2x_1+x_2\\0\end{bmatrix}\end{aligned}$$

This matches our result above. Generalizing what we did here gives the following theorem.

Theorem 3.2.4

If $L, M : \mathbb{R}^n \to \mathbb{R}^m$ are linear mappings and $c \in \mathbb{R}$, then $L + M : \mathbb{R}^n \to \mathbb{R}^m$ and $cL : \mathbb{R}^n \to \mathbb{R}^m$ are linear mappings. Moreover, we have

$$[L+M] = [L] + [M] \quad \text{and} \quad [cL] = c[L]$$

Proof: We will prove the result for cL. The result for $L + M$ is left as Problem C3. Let $\vec{x}, \vec{y} \in \mathbb{R}^n$ and $s, t \in \mathbb{R}$. Then,

$$\begin{aligned}(cL)(s\vec{x} + t\vec{y}) &= cL(s\vec{x} + t\vec{y}) && \text{by definition of } cL\\ &= c\left(sL(\vec{x}) + tL(\vec{y})\right) && \text{since } L \text{ is linear}\\ &= csL(\vec{x}) + ctL(\vec{y}) && \text{by properties V9,V8 of Theorem 1.4.1}\\ &= s(cL)(\vec{x}) + t(cL)(\vec{y}) && \text{by definition of } cL\end{aligned}$$

Hence, (cL) is linear. Moreover, for any $\vec{x} \in \mathbb{R}^n$, we have

$$\begin{aligned}[cL]\vec{x} &= (cL)(\vec{x}) && \text{by definition of the standard matrix of } cL\\ &= c(L(\vec{x})) && \text{by definition of } cL\\ &= c[L]\vec{x} && \text{by definition of the standard matrix of } L\end{aligned}$$

Thus, by the Matrices Equal Theorem, $[cL] = c[L]$. ■

Definition
Composition of Linear Mappings

Let $L : \mathbb{R}^n \to \mathbb{R}^m$ and $M : \mathbb{R}^m \to \mathbb{R}^p$ be linear mappings. The **composition** $M \circ L : \mathbb{R}^n \to \mathbb{R}^p$ is defined by

$$(M \circ L)(\vec{x}) = M(L(\vec{x}))$$

for all $\vec{x} \in \mathbb{R}^n$.

Note that the definition makes sense only if the domain of the second map M contains the codomain of the first map L, as we are evaluating M at $L(\vec{x})$. Moreover, observe that the order of the mappings in the definition is important.

EXAMPLE 3.2.9

Let $L : \mathbb{R}^2 \to \mathbb{R}^3$ and $M : \mathbb{R}^3 \to \mathbb{R}^1$ be the linear mappings defined by $L(x_1, x_2) = (x_1 + x_2, 0, x_1 - 2x_2)$ and $M(x_1, x_2, x_3) = (x_1 + x_2 + x_3)$. Find $M \circ L$.

Solution: We have

$$(M \circ L)(x_1, x_2) = M(x_1 + x_2, 0, x_1 - 2x_2) = ((x_1 + x_2) + 0 + (x_1 - 2x_2)) = (2x_1 - x_2)$$

EXAMPLE 3.2.10

Let $L : \mathbb{R}^2 \to \mathbb{R}^2$ and $M : \mathbb{R}^2 \to \mathbb{R}^2$ be the linear mappings defined by $L(x_1, x_2) = (2x_1 + x_2, x_1 + x_2)$ and $M(x_1, x_2) = (x_1 - x_2, -x_1 + 2x_2)$. Then, $M \circ L$ is mapping defined by

$$\begin{aligned}(M \circ L)(x_1, x_2) &= M(2x_1 + x_2, x_1 + x_2) \\ &= \Big((2x_1 + x_2) - (x_1 + x_2), -(2x_1 + x_2) + 2(x_1 + x_2)\Big) \\ &= (x_1, x_2)\end{aligned}$$

As with addition and scalar multiplication, we observe that a composition of linear mappings is linear. Moreover, we recall that we defined matrix multiplication to represent a composition of functions. Hence, the next theorem is not surprising.

Theorem 3.2.5

If $L : \mathbb{R}^n \to \mathbb{R}^m$ and $M : \mathbb{R}^m \to \mathbb{R}^p$ are linear mappings, then $M \circ L : \mathbb{R}^n \to \mathbb{R}^p$ is a linear mapping and

$$[M \circ L] = [M][L]$$

We leave the proof as Problem C4.

EXAMPLE 3.2.11

Let $L : \mathbb{R}^2 \to \mathbb{R}^2$ and $M : \mathbb{R}^2 \to \mathbb{R}^2$ be the linear mappings defined by $L(x_1, x_2) = (2x_1 + x_2, x_1 + x_2)$ and $M(x_1, x_2) = (x_1 - x_2, -x_1 + 2x_2)$. Find $[M \circ L]$.

Solution: We have

$$[M \circ L] = [M][L] = \begin{bmatrix} 1 & -1 \\ -1 & 2 \end{bmatrix}\begin{bmatrix} 2 & 1 \\ 1 & 1 \end{bmatrix} = \begin{bmatrix} 1 & 0 \\ 0 & 1 \end{bmatrix}$$

In Example 3.2.10, $M \circ L$ is the mapping such that $(M \circ L)(\vec{x}) = \vec{x}$ for all $\vec{x} \in \mathbb{R}^2$. We see from Example 3.2.11 that the standard matrix of this mapping is the identity matrix. Thus, we make the following definition.

Definition
Identity Mapping

The **identity mapping** is the linear mapping $\mathrm{Id} : \mathbb{R}^n \to \mathbb{R}^n$ defined by

$$\mathrm{Id}(\vec{x}) = \vec{x}$$

CONNECTION

The mappings L and M in Example 3.2.10 also satisfy $L \circ M = \mathrm{Id}$ and hence are said to be **inverses** of each other, as are the matrices $[L]$ and $[M]$. We will look at inverse mappings and matrices in Section 3.5.

PROBLEMS 3.2

Practice Problems

A1 Let $A = \begin{bmatrix} -2 & 3 \\ 3 & 0 \\ 1 & 5 \\ 4 & -6 \end{bmatrix}$ and let $f_A(\vec{x}) = A\vec{x}$ be the corresponding matrix mapping.

(a) Determine the domain and codomain of f_A.

(b) Determine $f_A(2, -5)$ and $f_A(-3, 4)$.

(c) Find the images of the standard basis vectors for the domain under f_A.

(d) Determine $f_A(\vec{x})$.

(e) Check your answers in (c) and (d) by calculating $[f_A(\vec{x})]$ using Theorem 3.2.3.

A2 Let $A = \begin{bmatrix} 1 & 2 & -3 & 0 \\ 2 & -1 & 0 & 3 \\ 1 & 0 & 2 & -1 \end{bmatrix}$ and let $f_A(\vec{x}) = A\vec{x}$.

(a) Determine the domain and codomain of f_A.

(b) Determine $f_A(2, -2, 3, 1)$ and $f_A(-3, 1, 4, 2)$.

(c) Find the images of the standard basis vectors for the domain under f_A.

(d) Determine $f_A(\vec{x})$.

(e) Check your answers in (c) and (d) by calculating $[f_A(\vec{x})]$ using Theorem 3.2.3.

For Problems A3–A16, state the domain and codomain of the mapping. Either prove that the mapping is linear or give a counterexample to show why it cannot be linear.

A3 $f(x_1, x_2) = (\sin x_1, e^{x_2^2})$

A4 $f(x_1, x_2, x_3) = (0, x_1x_2x_3)$

A5 $g(x_1, x_2, x_3) = (1, 1, 1)$

A6 $g(x_1, x_2) = (2x_1 + 3x_2, x_1 - x_2)$

A7 $h(x_1, x_2) = (2x_1 + 3x_2, x_1 - x_2, x_1x_2)$

A8 $k(x_1, x_2, x_3) = (x_1 + x_2, 0, x_2 - x_3)$

A9 $\ell(x_1, x_2, x_3) = (x_2, |x_1|)$

A10 $m(x_1) = (x_1, 1, 0)$

A11 $L(x_1, x_2, x_3) = (2x_1, x_1 - x_2 + 3x_3)$

A12 $L(x_1, x_2) = (x_1^2 - x_2^2, x_1)$

A13 $M(x_1, x_2, x_3) = (x_1 + 2, x_2 + 2)$

A14 $M(x_1, x_2, x_3) = (x_1 + 3x_3, x_2 - 2x_3, x_1 + x_2)$

A15 $N(x_1, x_2) = (-x_1, 0, x_1)$

A16 $N(x_1, x_2, x_3) = (x_1x_2, x_1x_3, x_2x_3)$

For Problems A17–A19, determine $[\text{proj}_{\vec{v}}]$.

A17 $\vec{v} = \begin{bmatrix} -2 \\ 1 \end{bmatrix}$ **A18** $\vec{v} = \begin{bmatrix} 4 \\ 5 \end{bmatrix}$ **A19** $\vec{v} = \begin{bmatrix} 2 \\ 2 \\ -1 \end{bmatrix}$

For Problems A20–A22, determine $[\text{perp}_{\vec{v}}]$.

A20 $\vec{v} = \begin{bmatrix} -2 \\ 1 \end{bmatrix}$ **A21** $\vec{v} = \begin{bmatrix} 1 \\ 4 \end{bmatrix}$ **A22** $\vec{v} = \begin{bmatrix} 1 \\ 2 \\ 3 \end{bmatrix}$

For Problems A23–A31, determine the domain, the codomain, and the standard matrix of the linear mapping.

A23 $L(x_1, x_2) = (-3x_1 + 5x_2, -x_1 - 2x_2)$

A24 $L(x_1, x_2) = (x_1, x_2, x_1 + x_2)$

A25 $L(x_1) = (x_1, 0, 3x_1)$

A26 $M(x_1, x_2, x_3) = (x_1 - x_2 + \sqrt{2}x_3)$

A27 $M(x_1, x_2, x_3) = (2x_1 - x_3, 2x_1 - x_3)$

A28 $N(x_1, x_2, x_3) = (0, 0, 0, 0)$

A29 $L(x_1, x_2, x_3) = (2x_1 - 3x_2 + x_3, x_2 - 5x_3)$

A30 $K(x_1, x_2, x_3, x_4) = (5x_1 + 3x_3 - x_4, x_2 - 7x_3 + 3x_4)$

A31 $M(x_1, x_2, x_3, x_4) = (x_1 - x_3 + x_4, x_1 + 2x_2 - 3x_4, x_2 + x_3)$

For Problems A32–A41, find the standard matrix of the linear mapping with the given properties.

A32 $L : \mathbb{R}^2 \to \mathbb{R}^2, L(\vec{e}_1) = (3, 5), L(\vec{e}_2) = (1, -2)$.

A33 $L : \mathbb{R}^2 \to \mathbb{R}^2, L(\vec{e}_1) = (-1, 11), L(\vec{e}_2) = (13, -21)$.

A34 $L : \mathbb{R}^2 \to \mathbb{R}^3, L(\vec{e}_1) = (1, 0, 1), L(\vec{e}_2) = (1, 0, 1)$.

A35 $L : \mathbb{R}^3 \to \mathbb{R}^3, L(\vec{e}_1) = (1, 0, 1), L(\vec{e}_2) = (1, 0, 1), L(\vec{e}_3) = (1, 0, 1)$.

A36 $L : \mathbb{R}^3 \to \mathbb{R}^3, L(\vec{e}_1) = (5, 0, 0), L(\vec{e}_2) = (0, 3, 0), L(\vec{e}_3) = (0, 0, 2)$.

A37 $L : \mathbb{R}^3 \to \mathbb{R}^2, L(\vec{e}_1) = (-1, \sqrt{2}), L(\vec{e}_2) = (1/2, 0), L(\vec{e}_3) = (-1, -1)$.

A38 $L : \mathbb{R}^3 \to \mathbb{R}^3, L(\vec{e}_1) = (2, 1, 1), L(\vec{e}_2) = (1, -2, 1), L(\vec{e}_3) = (0, 1, -2)$.

A39 $L : \mathbb{R}^2 \to \mathbb{R}^2, L(1, 0) = (3, 5), L(1, 1) = (5, -2)$.

A40 $L : \mathbb{R}^2 \to \mathbb{R}^3, L(1, 1) = (3, 2, 0), L(1, -1) = (1, 0, -2)$.

A41 $L : \mathbb{R}^3 \to \mathbb{R}^2, L(1, 0, 1) = (2, 0), L(1, 1, 1) = (4, 5), L(1, 1, 0) = (5, 6)$.

A42 Suppose that S and T are linear mappings with standard matrices

$$[S] = \begin{bmatrix} 2 & 1 & 3 \\ -1 & 0 & 2 \end{bmatrix}, \quad [T] = \begin{bmatrix} 1 & 2 & -1 \\ 2 & 2 & 3 \end{bmatrix}$$

(a) Determine the domain and codomain of S and T.
(b) Determine $[S + T]$ and $[2S - 3T]$.

A43 Suppose that S and T are linear mappings with standard matrices

$$[S] = \begin{bmatrix} -3 & -3 & 0 & 1 \\ 0 & 2 & 4 & 2 \end{bmatrix}, \quad [T] = \begin{bmatrix} 1 & 4 \\ -2 & 1 \\ 2 & -1 \\ 3 & -4 \end{bmatrix}$$

(a) Determine the domain and codomain of S and T.
(b) Determine $[S \circ T]$ and $[T \circ S]$.

For Problems A44–A49, suppose that L, M, and N are linear mappings with standard matrices $[L] = \begin{bmatrix} 2 & 3 \\ -1 & 4 \\ 0 & 1 \end{bmatrix}$, $[M] = \begin{bmatrix} 1 & 1 & 2 \\ 3 & -2 & -1 \end{bmatrix}$, $[N] = \begin{bmatrix} 2 & 1 \\ 3 & 1 \\ -3 & 0 \\ 1 & 4 \end{bmatrix}$. Determine whether the given composition is defined. If so, calculate the standard matrix of the composition.

A44 $L \circ M$ **A45** $M \circ L$ **A46** $L \circ N$
A47 $N \circ L$ **A48** $M \circ N$ **A49** $N \circ M$

A50 (a) Invent a linear mapping $L : \mathbb{R}^2 \to \mathbb{R}^3$ such that $L(1, 0) = (3, -1, 4)$ and $L(0, 1) = (1, -5, 9)$.
(b) Invent a *non*-linear mapping $L : \mathbb{R}^2 \to \mathbb{R}^3$ such that $L(1, 0) = (3, -1, 4)$ and $L(0, 1) = (1, -5, 9)$.

Homework Problems

B1 Let $A = \begin{bmatrix} 6 & 2 & 3 \\ -1 & 6 & 4 \end{bmatrix}$ and let $f_A(\vec{x}) = A\vec{x}$ be the corresponding matrix mapping.

(a) Determine the domain and codomain of f_A.
(b) Determine $f_A(2, 1, -1)$ and $f_A(3, -8, 9)$.
(c) Find the images of the standard basis vectors for the domain under f_A.
(d) Determine $f_A(\vec{x})$.
(e) Check your answers in (c) and (d) by calculating $[f_A(\vec{x})]$ using Theorem 3.2.3.

B2 Let $A = \begin{bmatrix} 1 & 2 & -2 & 1 \\ 0 & 1 & 3 & 1 \\ 4 & -1 & 0 & 1 \end{bmatrix}$ and let $f_A(\vec{x}) = A\vec{x}$.

(a) Determine the domain and codomain of f_A.
(b) Determine $f_A(1, 1, 1, 1)$ and $f_A(3, 1, -2, 3)$.
(c) Find the images of the standard basis vectors for the domain under f_A.
(d) Determine $f_A(\vec{x})$.
(e) Check your answers in (c) and (d) by calculating $[f_A(\vec{x})]$ using Theorem 3.2.3.

For Problems B3–B9, state the domain and codomain of the mapping. Either prove that the mapping is linear or give a counterexample to show why it cannot be linear.

B3 $f(x_1, x_2, x_3) = (x_1^2, x_2^2, x_3^2)$

B4 $g(x_1, x_2) = (2x_1 - x_2, 3x_1 + x_2)$

B5 $h(x_1, x_2) = (0, 0)$

B6 $k(x_1, x_2, x_3) = (x_1, x_1 + x_2, x_2)$

B7 $\ell(x_1, x_2, x_3, x_4) = (1, 1)$

B8 $m(x_1, x_2, x_3, x_4) = (x_1 + x_2, x_3 + x_4)$

B9 $L(x_1, x_2, x_3) = (x_1 + x_2, x_2 + x_3, x_3 + 5)$

For Problems B10–B12, determine $[\text{proj}_{\vec{v}}]$.

B10 $\vec{v} = \begin{bmatrix} 2 \\ 3 \end{bmatrix}$ **B11** $\vec{v} = \begin{bmatrix} 5 \\ -1 \end{bmatrix}$ **B12** $\vec{v} = \begin{bmatrix} 2 \\ 0 \\ 1 \end{bmatrix}$

For Problems B13–B15, determine $[\text{perp}_{\vec{v}}]$.

B13 $\vec{v} = \begin{bmatrix} -1 \\ -1 \end{bmatrix}$ **B14** $\vec{v} = \begin{bmatrix} 3 \\ 6 \end{bmatrix}$ **B15** $\vec{v} = \begin{bmatrix} 2 \\ 0 \\ 1 \end{bmatrix}$

For Problems B16–B21, determine the domain, the codomain, and the standard matrix of the linear mapping.

B16 $L(x_1, x_2) = (7x_1 - 2x_2, \pi x_1)$

B17 $L(x_1, x_2, x_3) = (x_1, x_2)$

B18 $L(x_1, x_2, x_3) = (x_1 - x_3, x_2 + x_3, x_1 + x_2 + x_3)$

B19 $K(x_1, x_2) = (2x_1 + 4x_2, x_1 + 2x_2, x_1 - x_2, x_1 + 3x_2)$

B20 $M(x_1, x_2, x_3, x_4) = (x_1 + 2x_2, x_1 - x_3, x_1 + 2x_3)$

B21 $N(x_1, x_2, x_3) = (x_3, x_1, x_2)$

For Problems B22–B25, find the standard matrix of the linear mapping with the given properties.

B22 $L : \mathbb{R}^2 \to \mathbb{R}^2$, $L(1,0) = (8,3)$, $L(0,1) = (7,-1)$.

B23 $L : \mathbb{R}^3 \to \mathbb{R}^2$, $L(1,0,0) = (1,5)$, $L(0,1,0) = (1,2)$, $L(0,0,1) = (4,2)$.

B24 $L : \mathbb{R}^2 \to \mathbb{R}^2$, $L(1,1) = (7,2)$, $L(1,0) = (1,-5)$.

B25 $L : \mathbb{R}^2 \to \mathbb{R}^2$, $L(1,2) = (-1,6)$, $L(2,1) = (3,4)$.

B26 Suppose that S and T are linear mappings with standard matrices

$$[S] = \begin{bmatrix} 3 & 5 \\ 0 & 0 \\ 1 & 2 \end{bmatrix} \quad \text{and} \quad [T] = \begin{bmatrix} 1 & -2 \\ -3 & 9 \\ 5 & 5 \end{bmatrix}$$

(a) Determine the domain and codomain of S and T.
(b) Determine $[S + T]$ and $[-S + T]$.

B27 Suppose that S and T are linear mappings with standard matrices

$$[S] = \begin{bmatrix} 3 & 7 \\ 2 & 4 \\ -1 & 1 \end{bmatrix}, [T] = \begin{bmatrix} 4 & 0 & 2 \\ -2 & 1 & 3 \end{bmatrix}$$

(a) Determine the domain and codomain of S and T.
(b) Determine $[S \circ T]$ and $[T \circ S]$.

For Problems B28–B33, suppose that L, M, and N are linear mappings with standard matrices $[L] = \begin{bmatrix} 5 & -1 & 2 \\ 0 & 1 & 2 \end{bmatrix}$, $[M] = \begin{bmatrix} 2 & 2 \\ 1 & 2 \\ -1 & 6 \end{bmatrix}$, $[N] = \begin{bmatrix} 2 & 1 \\ 7 & -4 \end{bmatrix}$. Determine whether the given composition is defined. If so, calculate the standard matrix of the composition.

B28 $L \circ M$ **B29** $M \circ L$ **B30** $L \circ N$
B31 $N \circ L$ **B32** $M \circ N$ **B33** $N \circ M$

Conceptual Problems

C1 Let $L : \mathbb{R}^n \to \mathbb{R}^m$. Show that for any $\vec{x}, \vec{y} \in \mathbb{R}^n$ and $s, t \in \mathbb{R}$, L satisfies

$$L(\vec{x} + \vec{y}) = L(\vec{x}) + L(\vec{y}) \quad \text{and} \quad L(s\vec{x}) = sL(\vec{x})$$

if and only if

$$L(s\vec{x} + t\vec{y}) = sL(\vec{x}) + tL(\vec{y})$$

C2 Let $L : \mathbb{R}^n \to \mathbb{R}^m$ be a linear mapping.
(a) Prove that $L(\vec{0}) = \vec{0}$.
(b) Explain what (a) says about a mapping $M : \mathbb{R}^n \to \mathbb{R}^m$ such that $M(\vec{0}) \neq \vec{0}$.

C3 Let $L : \mathbb{R}^n \to \mathbb{R}^m$ and $M : \mathbb{R}^n \to \mathbb{R}^m$ be linear mappings. Prove that $(L + M)$ is linear and that

$$[L + M] = [L] + [M]$$

C4 Let $L : \mathbb{R}^n \to \mathbb{R}^m$ and $M : \mathbb{R}^m \to \mathbb{R}^p$ be linear mappings. Prove that $(M \circ L)$ is linear and that

$$[M \circ L] = [M][L]$$

C5 Let $\vec{v} \in \mathbb{R}^3$ be a fixed vector and define a mapping $\text{CROSS}_{\vec{v}}$ by

$$\text{CROSS}_{\vec{v}}(\vec{x}) = \vec{v} \times \vec{x}$$

Verify that $\text{CROSS}_{\vec{v}}$ is a linear mapping and determine its domain, codomain, and standard matrix.

C6 Let $\vec{v} \in \mathbb{R}^n$ be a fixed vector and define a mapping $\text{DOT}_{\vec{v}}$ by

$$\text{DOT}_{\vec{v}}(\vec{x}) = \vec{v} \cdot \vec{x}$$

Verify that $\text{DOT}_{\vec{v}}$ is a linear mapping. What is its domain and codomain? Verify that the matrix of this linear mapping can be written as $\vec{v}^T$.

C7 If $\vec{u}$ is a unit vector, show that $[\text{proj}_{\vec{u}}] = \vec{u}\vec{u}^T$.

C8 (a) Let $L : \mathbb{R}^n \to \mathbb{R}^m$ be a linear mapping. Prove that if $\{L(\vec{v}_1), \dots, L(\vec{v}_k)\}$ is a linearly independent set in $\mathbb{R}^m$, then $\{\vec{v}_1, \dots, \vec{v}_k\}$ is linearly independent.
(b) Give an example of a linear mapping $L : \mathbb{R}^n \to \mathbb{R}^m$ where $\{\vec{v}_1, \dots, \vec{v}_k\}$ is linearly independent in $\mathbb{R}^n$, but $\{L(\vec{v}_1), \dots, L(\vec{v}_k)\}$ is linearly dependent.

C9 Let $L : \mathbb{R}^n \to \mathbb{R}^m$ be a linear mapping and let $\mathbb{S}$ be a subspace of $\mathbb{R}^m$. Prove that the set

$$T = \{\vec{v} \in \mathbb{R}^n \mid L(\vec{v}) \in \mathbb{S}\}$$

is a subspace of $\mathbb{R}^n$. T is called the **pre-image** of $\mathbb{S}$.

3.3 Geometrical Transformations

Geometrical transformations have long been of great interest to mathematicians. They have many important applications. Physicists and engineers often rely on simple geometrical transformations to gain understanding of the properties of materials or structures they wish to examine. For example, structural engineers use stretches, shears, and rotations to understand the deformation of materials. Material scientists use rotations and reflections to analyze crystals and other fine structures. Many of these simple geometrical transformations in $\mathbb{R}^2$ and $\mathbb{R}^3$ are linear. The following is a brief partial catalogue of some of these transformations and their matrix representations. ($\text{proj}_{\vec{v}}$ and $\text{perp}_{\vec{v}}$ belong to the list of geometrical transformations, too, but they were discussed in Chapter 1 and so are not included here).

Rotations in the Plane

Let $R_\theta : \mathbb{R}^2 \to \mathbb{R}^2$ denote the function that rotates a vector $\vec{x} \in \mathbb{R}^2$ about the origin through an angle θ as depicted in Figure 3.3.1. Using trigonometric identities, we can show that

$$R_\theta(x_1, x_2) = (x_1 \cos\theta - x_2 \sin\theta, x_1 \sin\theta + x_2 \cos\theta)$$

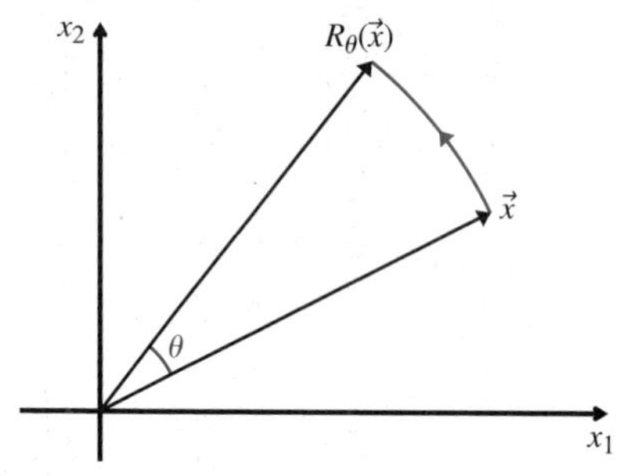

Figure 3.3.1 Counterclockwise rotation through angle θ in the plane.

From this, it is easy to prove that R_θ is linear. Hence, we can use the definition of the standard matrix to calculate $[R_\theta]$. We have

$$R_\theta(1,0) = \begin{bmatrix} \cos\theta \\ \sin\theta \end{bmatrix}$$

$$R_\theta(0,1) = \begin{bmatrix} -\sin\theta \\ \cos\theta \end{bmatrix}$$

Hence,

$$[R_\theta] = \begin{bmatrix} \cos\theta & -\sin\theta \\ \sin\theta & \cos\theta \end{bmatrix}$$

Definition
Rotation Matrix

A matrix $[R_\theta] = \begin{bmatrix} \cos\theta & -\sin\theta \\ \sin\theta & \cos\theta \end{bmatrix}$ is called a **rotation matrix**.

EXAMPLE 3.3.1

What is the matrix of rotation of $\mathbb{R}^2$ through angle $2\pi/3$?

Solution: We have

$$[R_{2\pi/3}] = \begin{bmatrix} \cos 2\pi/3 & -\sin 2\pi/3 \\ \sin 2\pi/3 & \cos 2\pi/3 \end{bmatrix} = \begin{bmatrix} -1/2 & -\sqrt{3}/2 \\ \sqrt{3}/2 & -1/2 \end{bmatrix}$$

EXERCISE 3.3.1

Determine $[R_{\pi/4}]$ and use it to calculate $R_{\pi/4}(1, 1)$. Illustrate with a sketch.

CONNECTION

Other than the application of using rotation matrices to rotate computer graphics, one might think that rotation matrices are not particularly important. However, since the columns of a rotation matrix are both unit vectors and are orthogonal to each other, they form an **orthonormal basis** for $\mathbb{R}^2$ (see Problem C13 in Section 1.5). Such bases, and hence rotation matrices, are very useful in solving many problems. We will explore this in Chapter 8.

Rotation Through Angle θ About the x_3-axis in $\mathbb{R}^3$

Figure 3.3.2 demonstrates a counterclockwise rotation with respect to the right-handed standard basis. This rotation leaves x_3 unchanged, so that if the transformation is denoted by R, then $R(0, 0, 1) = (0, 0, 1)$. Together with the previous case, this tells us that the standard matrix of this rotation is

$$[R] = \begin{bmatrix} \cos\theta & -\sin\theta & 0 \\ \sin\theta & \cos\theta & 0 \\ 0 & 0 & 1 \end{bmatrix}$$

These ideas can be adapted to give rotations about the other coordinate axes. We shall see how to determine the standard matrix of a rotation about an arbitrary axis in Chapter 7.

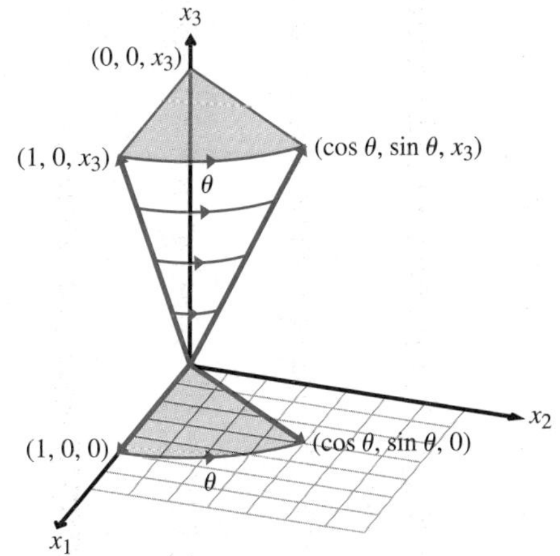

Figure 3.3.2 A right-handed counterclockwise rotation about the x_3-axis in $\mathbb{R}^3$.

Stretches

Let $S : \mathbb{R}^2 \to \mathbb{R}^2$ be denote the function that multiplies all lengths in the x_1-direction by a scalar factor $t > 0$, while lengths in the x_2-direction are left unchanged (Figure 3.3.3). This linear operator, called a "stretch by factor t in the x_1-direction," has standard matrix

$$[S] = \begin{bmatrix} t & 0 \\ 0 & 1 \end{bmatrix}$$

(If $t < 1$, you might prefer to call this a *shrink*.) Stretches can also be defined in the x_2-direction and in higher dimensions. Stretches are important in understanding the deformation of solids (see Section 8.4).

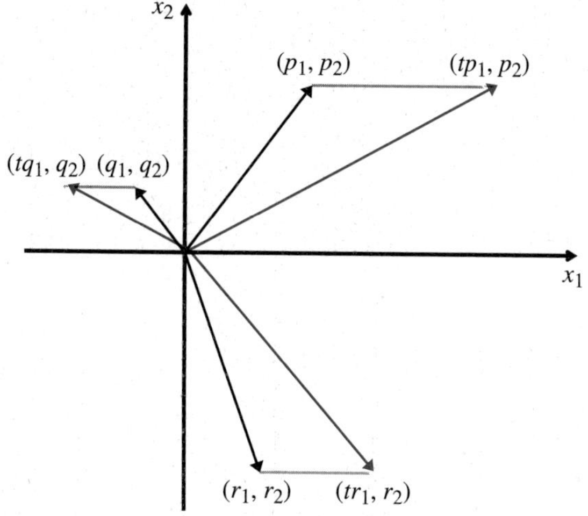

Figure 3.3.3 A stretch by a factor t in the x_1-direction.

EXERCISE 3.3.2

Let S be the stretch by factor 3 in the x_2-direction. Write the standard matrix $[S]$ of S and use it to calculate $S(\vec{x})$ where $\vec{x} = \begin{bmatrix} 2 \\ 1 \end{bmatrix}$. Illustrate with a sketch.

Contractions and Dilations

A linear operator $T : \mathbb{R}^2 \to \mathbb{R}^2$ with standard matrix

$$[T] = \begin{bmatrix} t & 0 \\ 0 & t \end{bmatrix}$$

transforms vectors in all directions by the same factor. Thus, for example, a circle of radius 1 centred at the origin is mapped to a circle of radius t at the origin. If $0 < t < 1$, such a transformation is called a **contraction**; if $t > 1$, it is a **dilation**.

EXERCISE 3.3.3

Let T be the dilation by factor 3. Write the standard matrix $[T]$ of T and use it to calculate $T(\vec{x})$ where $\vec{x} = \begin{bmatrix} 2 \\ 1 \end{bmatrix}$. Illustrate with a sketch.

Shears

Sometimes a force applied to a rectangle will cause it to deform into a parallelogram, as shown in Figure 3.3.4. The change can be described by the function $H : \mathbb{R}^2 \to \mathbb{R}^2$, such that $H(1,0) = (1,0)$ and $H(0,1) = (s,1)$. Although the deformation of a real solid may be more complicated, it is customary to assume that the transformation H is linear. Such a linear transformation is called a **horizontal shear** by amount s. Since the action of H on the standard basis vectors is known, we find that its standard matrix is

$$[H] = \begin{bmatrix} 1 & s \\ 0 & 1 \end{bmatrix}$$

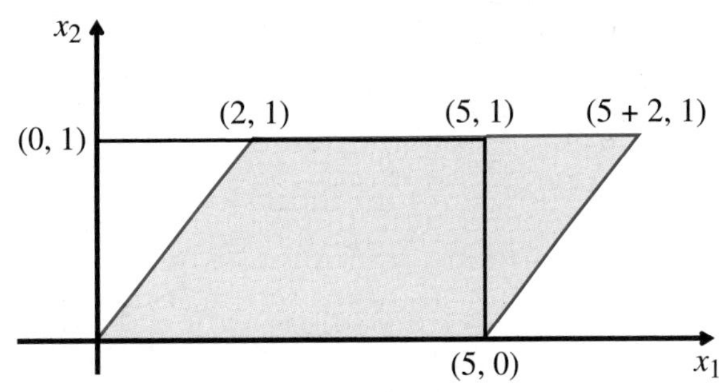

Figure 3.3.4 A horizontal shear by amount 2.

The standard matrix of a **vertical shear** $V : \mathbb{R}^2 \to \mathbb{R}^2$ by amount s will be

$$[V] = \begin{bmatrix} 1 & 0 \\ s & 1 \end{bmatrix}$$

EXERCISE 3.3.4

Let V be a vertical shear by amount 2. Write the standard matrix $[V]$ of V. Determine $V(1,0)$ and $V(0,1)$, and illustrate with a sketch how the unit square is deformed into a parallelogram by V.

CONNECTION

Recall that the 2×2 identity matrix is $I_2 = \begin{bmatrix} 1 & 0 \\ 0 & 1 \end{bmatrix}$. Since $I_2\vec{x} = \vec{x}$ for all $\vec{x} \in \mathbb{R}^2$, geometrically I_2 does nothing. If we perform the elementary row operation "multiply a row by a non-zero scalar t" on I_2, then we get the standard matrix of a stretch. Similarly, if we perform the elementary row operation "add s times row 2 to row 1" on I_2, then we get a horizontal shear. That is, elementary row operations have very simple geometrical interpretations. What is the geometry of "swapping row 1 and row 2"?

Reflections in Coordinate Axes in $\mathbb{R}^2$ or Coordinate Planes in $\mathbb{R}^3$

Let $F : \mathbb{R}^2 \to \mathbb{R}^2$ be a reflection over the x_1-axis. See Figure 3.3.5. Then each vector corresponding to a point above the axis is mapped by F to the mirror image vector below. Hence,

$$F(x_1, x_2) = (x_1, -x_2)$$

It follows that this transformation is linear with standard matrix

$$[F] = \begin{bmatrix} 1 & 0 \\ 0 & -1 \end{bmatrix}$$

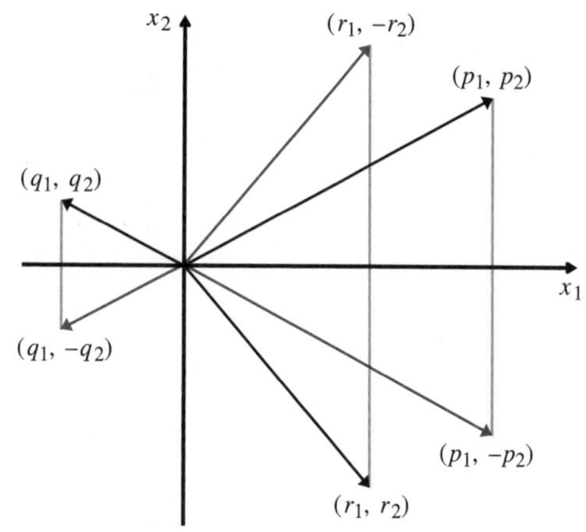

Figure 3.3.5 A reflection in $\mathbb{R}^2$ over the x_1-axis.

EXERCISE 3.3.5 Let F be a reflection in the x_2-axis. Write the standard matrix $[F]$ of F and use it to calculate $F(\vec{x})$ where $\vec{x} = \begin{bmatrix} 2 \\ 1 \end{bmatrix}$. Illustrate with a sketch.

Next, consider the reflection $F : \mathbb{R}^3 \to \mathbb{R}^3$ that reflects in the x_1x_2-plane (that is, the plane $x_3 = 0$). Points above the plane are reflected to points below the plane. The standard matrix of this reflection is

$$[F] = \begin{bmatrix} 1 & 0 & 0 \\ 0 & 1 & 0 \\ 0 & 0 & -1 \end{bmatrix}$$

EXERCISE 3.3.6 Write the matrices for the reflections in the other two coordinate planes in $\mathbb{R}^3$.

General Reflections

We consider only reflections in (or "across") lines in $\mathbb{R}^2$ or planes in $\mathbb{R}^3$ that pass through the origin. Reflections in lines or planes not containing the origin involve translations (which are not linear) as well as linear mappings.

Consider the plane in $\mathbb{R}^3$ with equation $\vec{n} \cdot \vec{x} = 0$. Since a reflection is related to $\text{proj}_{\vec{n}}$, a **reflection in the plane with normal vector** $\vec{n}$ will be denoted $\text{refl}_{\vec{n}}$. If a vector $\vec{p}$ corresponds to a point P that does not lie in the plane, its image under $\text{refl}_{\vec{n}}$ is the vector that corresponds to the point on the opposite side of the plane, lying on a line through P perpendicular to the plane of reflection, at the same distance from the plane as P. Figure 3.3.6 shows reflection in a line. From the figure, we see that

$$\text{refl}_{\vec{n}}(\vec{p}) = \vec{p} - 2\,\text{proj}_{\vec{n}}(\vec{p})$$

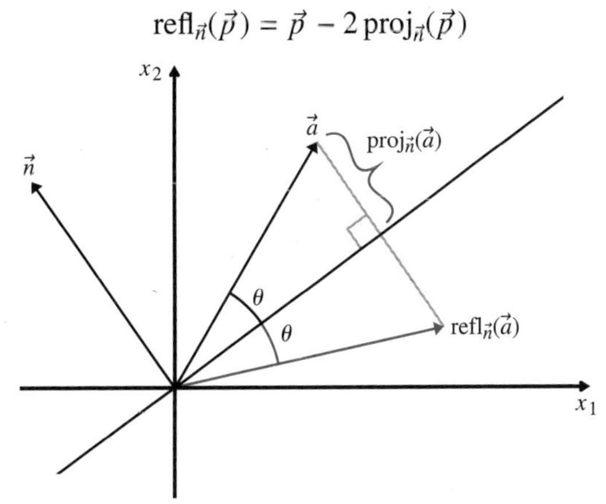

Figure 3.3.6 A reflection in $\mathbb{R}^2$ over the line with normal vector $\vec{n}$.

EXAMPLE 3.3.2

Prove that $\text{refl}_{\vec{n}} : \mathbb{R}^3 \to \mathbb{R}^3$ is a linear mapping.

Solution: Since $\text{proj}_{\vec{n}}$ is a linear mapping we get

$$\begin{aligned}
\text{refl}_{\vec{n}}(b\vec{x} + c\vec{y}) &= (b\vec{x} + c\vec{y}) - 2\,\text{proj}_{\vec{n}}(b\vec{x} + c\vec{y}) \\
&= b\vec{x} + c\vec{y} - 2(b\,\text{proj}_{\vec{n}}(\vec{x}) + c\,\text{proj}_{\vec{n}}(\vec{y})) \\
&= b(\vec{x} - 2\,\text{proj}_{\vec{n}}(\vec{x})) + c(\vec{y} - 2\,\text{proj}_{\vec{n}}(\vec{y})) \\
&= b\,\text{refl}_{\vec{n}}(\vec{x}) + c\,\text{refl}_{\vec{n}}(\vec{y})
\end{aligned}$$

for all $\vec{x}, \vec{y} \in \mathbb{R}^n$ and $b, c \in \mathbb{R}$ as required.

It is important to note that $\text{refl}_{\vec{n}}$ is a reflection over a line or plane passing through the origin with normal vector $\vec{n}$. The calculations for reflection in a line in $\mathbb{R}^2$ are similar to those for a plane, provided that the equation of the line is given in scalar form $\vec{n} \cdot \vec{x} = 0$. If the vector equation of the line is given as $\vec{x} = t\vec{d}$, then either we must find a normal vector $\vec{n}$ and proceed as above, or, in terms of the direction vector $\vec{d}$, the reflection will map $\vec{p}$ to $(\vec{p} - 2\,\text{perp}_{\vec{d}}(\vec{p}))$.

EXAMPLE 3.3.3

Consider a reflection $\text{refl}_{\vec{n}} : \mathbb{R}^3 \to \mathbb{R}^3$ over the plane with normal vector $\vec{n} = \begin{bmatrix} 1 \\ -1 \\ 2 \end{bmatrix}$.

Determine the standard matrix $[\text{refl}_{\vec{n}}]$.

Solution: We have

$$\text{refl}_{\vec{n}}(\vec{e}_1) = \vec{e}_1 - 2\frac{\vec{e}_1 \cdot \vec{n}}{\|\vec{n}\|^2}\vec{n} = \begin{bmatrix} 1 \\ 0 \\ 0 \end{bmatrix} - 2\left(\frac{1}{6}\right)\begin{bmatrix} 1 \\ -1 \\ 2 \end{bmatrix} = \begin{bmatrix} 2/3 \\ 1/3 \\ -2/3 \end{bmatrix}$$

$$\text{refl}_{\vec{n}}(\vec{e}_2) = \vec{e}_2 - 2\frac{\vec{e}_2 \cdot \vec{n}}{\|\vec{n}\|^2}\vec{n} = \begin{bmatrix} 0 \\ 1 \\ 0 \end{bmatrix} - 2\left(\frac{-1}{6}\right)\begin{bmatrix} 1 \\ -1 \\ 2 \end{bmatrix} = \begin{bmatrix} 1/3 \\ 2/3 \\ 2/3 \end{bmatrix}$$

$$\text{refl}_{\vec{n}}(\vec{e}_3) = \vec{e}_3 - 2\frac{\vec{e}_3 \cdot \vec{n}}{\|\vec{n}\|^2}\vec{n} = \begin{bmatrix} 0 \\ 0 \\ 1 \end{bmatrix} - 2\left(\frac{2}{6}\right)\begin{bmatrix} 1 \\ -1 \\ 2 \end{bmatrix} = \begin{bmatrix} -2/3 \\ 2/3 \\ -1/3 \end{bmatrix}$$

Hence,

$$[\text{refl}_{\vec{n}}] = \begin{bmatrix} 2/3 & 1/3 & -2/3 \\ 1/3 & 2/3 & 2/3 \\ -2/3 & 2/3 & -1/3 \end{bmatrix}$$

PROBLEMS 3.3

Practice Problems

For Problems A1–A4, determine the matrix of the rotation in the plane through the given angle.

A1 $\frac{\pi}{2}$ **A2** π **A3** $-\frac{\pi}{4}$ **A4** $\frac{2\pi}{5}$

For Problems A5–A9, in $\mathbb{R}^2$, let V be a vertical shear by amount 3 and S be a stretch by a factor 5 in the x_2-direction.

A5 Determine $[V]$ and $[S]$.

A6 Calculate the composition of S *followed by* V.

A7 Calculate the composition of S *following* V.

A8 Calculate the composition of S *followed by* a rotation through angle θ.

A9 Calculate the composition of S *following* a rotation through angle θ.

For Problems A10–A14, in $\mathbb{R}^2$, let H be a horizontal shear by amount 1 and V be a vertical shear by amount -2.

A10 Determine $[H]$ and $[V]$.

A11 Calculate the composition of H *followed by* V.

A12 Calculate the composition of H *following* V.

A13 Calculate the composition of H *followed by* a reflection over the x_1-axis.

A14 Calculate the composition of H *following* a reflection over the x_1-axis.

For Problems A15–A18, determine the matrix of the reflection over the given line in $\mathbb{R}^2$.

A15 $x_1 + 3x_2 = 0$ **A16** $2x_1 - x_2 = 0$

A17 $-4x_1 + x_2 = 0$ **A18** $3x_1 - 5x_2 = 0$

For Problems A19–A22, determine the matrix of the reflection over the given plane in $\mathbb{R}^3$.

A19 $x_1 + x_2 + x_3 = 0$ **A20** $2x_1 - 2x_2 - x_3 = 0$

A21 $x_1 - x_3 = 0$ **A22** $x_1 + 2x_2 - 3x_3 = 0$

A23 Let $D : \mathbb{R}^3 \to \mathbb{R}^3$ be the dilation with factor $t = 5$ and let $\text{inj} : \mathbb{R}^3 \to \mathbb{R}^4$ be the linear mapping defined by $\text{inj}(x_1, x_2, x_3) = (x_1, x_2, 0, x_3)$. Determine the matrix of $\text{inj} \circ D$.

A24 (a) Let $P : \mathbb{R}^3 \to \mathbb{R}^2$ be the linear mapping defined by $P(x_1, x_2, x_3) = (x_2, x_3)$ and let S be the shear in $\mathbb{R}^3$ such that $S(x_1, x_2, x_3) = (x_1, x_2, x_3 + 2x_1)$. Determine the matrix of $P \circ S$.

(b) Can you define a shear $T : \mathbb{R}^2 \to \mathbb{R}^2$ such that $T \circ P = P \circ S$, where P and S are as in part (a)?

(c) Let $Q : \mathbb{R}^3 \to \mathbb{R}^2$ be the linear mapping defined by $Q(x_1, x_2, x_3) = (x_1, x_2)$. Determine the matrix of $Q \circ S$, where S is the mapping in part (a).

Homework Problems

For Problems B1–B4, determine the matrix of the rotation in the plane through the given angle.

B1 $\frac{3\pi}{2}$ **B2** $\pi/3$ **B3** $-\frac{3\pi}{4}$ **B4** $\frac{5\pi}{6}$

For Problems B5–B9, in $\mathbb{R}^2$, let H be a horizontal shear by amount -2 and S be a stretch by a factor 3 in the x_1-direction.

B5 Determine $[H]$ and $[S]$.

B6 Calculate the composition of S *followed by* H.

B7 Calculate the composition of S *following* H.

B8 Calculate the composition of S *followed by* a rotation through angle $\frac{\pi}{3}$.

B9 Calculate the composition of S *following* a rotation through angle $\frac{\pi}{3}$.

For Problems B10–B14, in $\mathbb{R}^2$, let V be a vertical shear by amount -1 and T be a contraction by a factor $t = 1/2$.

B10 Determine $[V]$ and $[T]$.

B11 Calculate the composition of T *followed by* V.

B12 Calculate the composition of T *following* V.

B13 Find a linear mapping J such that $V \circ J = \mathrm{Id}$.

B14 Find a linear mapping K such that $K \circ T = \mathrm{Id}$.

For Problems B15–B18, determine the matrix of the reflection over the given line in $\mathbb{R}^2$.

B15 $2x_1 + 3x_2 = 0$ **B16** $5x_1 - x_2 = 0$

B17 $4x_1 - 3x_2 = 0$ **B18** $x_1 - 5x_2 = 0$

For Problems B19–B22, determine the matrix of the reflection over the given plane in $\mathbb{R}^3$.

B19 $x_1 + 2x_2 - 4x_3 = 0$ **B20** $x_1 + 2x_2 - 3x_3 = 0$

B21 $x_1 + 2x_3 = 0$ **B22** $2x_1 - 2x_2 + 3x_3 = 0$

B23 (a) Let $C : \mathbb{R}^3 \to \mathbb{R}^3$ be a contraction with factor $1/3$ and let $\mathrm{inj} : \mathbb{R}^3 \to \mathbb{R}^5$ be the linear mapping defined by $\mathrm{inj}(x_1, x_2, x_3) = (0, x_1, 0, x_2, x_3)$. Determine the matrix of $\mathrm{inj} \circ C$.

(b) Let $S : \mathbb{R}^3 \to \mathbb{R}^2$ be the shear defined by $S(x_1, x_2, x_3) = (x_1, x_2 - 2x_3, x_3)$. Determine the matrices $C \circ S$ and $S \circ C$, where C is the contraction in part (a).

(c) Let $T : \mathbb{R}^3 \to \mathbb{R}^3$ be the shear defined by $T(x_1, x_2, x_3) = (x_1 + 3x_2, x_2, x_3)$. Determine the matrix of $S \circ T$ and $T \circ S$, where S is the mapping in part (b).

Conceptual Problems

C1 Let R denote the reflection in Problem A15 and let S denote the reflection in Problem A16. Show that the composition of R and S can be identified as a rotation. Determine the angle of the rotation. Draw a picture illustrating how the composition of these reflections is a rotation.

C2 In $\mathbb{R}^3$, calculate the matrix of the composition of a reflection in the x_2x_3-plane followed by a reflection in the x_1x_2-plane and identify it as a rotation about some coordinate axis. What is the angle of the rotation?

C3 Consider rotations R_θ and R_α in $\mathbb{R}^2$.

(a) From geometrical considerations, we know that $R_\alpha \circ R_\theta = R_{\alpha+\theta}$. Verify the corresponding matrix equation.

(b) Prove that $[R_{-\theta}]^T = [R_\theta]$.

(c) Prove that the columns of R_θ form an orthonormal basis for $\mathbb{R}^2$.

C4 Let $\vec{u}$ be a unit vector in $\mathbb{R}^n$. A matrix A is said to be symmetric if $A^T = A$.

(a) Use the result of Section 3.2 Problem C7 to prove that $[\mathrm{proj}_{\vec{u}}]$ is symmetric.

(b) Prove that $[\mathrm{refl}_{\vec{u}}] = I - 2[\mathrm{proj}_{\vec{u}}]$.

(c) Prove that $[\mathrm{refl}_{\vec{u}}]$ is symmetric.

(d) From geometrical considerations, we know that $\mathrm{refl}_{\vec{n}} \circ \mathrm{refl}_{\vec{n}} = \mathrm{Id}$. Verify the corresponding matrix equation. (Hint: use part (b) and the fact that $\mathrm{proj}_{\vec{n}}$ satisfies the projection property (L2) from Section 1.5.)

C5 (a) Construct a 2×2 matrix $A \neq I$ such that $A^3 = I$. (Hint: think geometrically.)

(b) Construct a 2×2 matrix $A \neq I$ such that $A^5 = I$.

3.4 Special Subspaces

When working with a function we are often interested in the subset of the codomain which contains all possible images under the function. This is called the **range** *of the function. In this section we will look not only at the range of a linear mapping but also at a special subset of the domain. We will then use the connection between linear mappings and their standard matrices to derive four important subspaces of a matrix.*

The other main purpose of this section is to review and connect many of the concepts we have covered so far in the text. In addition to using what we learned about matrices and linear mappings in Section 3.1 and 3.2, we will be using:

- *subspaces and bases (Section 1.4)*
- *dot products and orthogonality (Section 1.5)*
- *rank of a matrix (Section 2.2)*
- *dimension of a subspace (Section 2.3)*

Special Subspaces of Linear Mappings

Range

Definition
Range

The **range** of a linear mapping $L : \mathbb{R}^n \to \mathbb{R}^m$ is defined to be the set

$$\text{Range}(L) = \{L(\vec{x}) \in \mathbb{R}^m \mid \vec{x} \in \mathbb{R}^n\}$$

EXAMPLE 3.4.1

Let $L : \mathbb{R}^2 \to \mathbb{R}^3$ be the linear mapping defined by $L(x_1, x_2) = (2x_1 - x_2, 0, x_1 + x_2)$. Find Range($L$).

Solution: By definition of the range, if $L(\vec{x})$ is any vector in the range, then

$$L(\vec{x}) = \begin{bmatrix} 2x_1 - x_2 \\ 0 \\ x_1 + x_2 \end{bmatrix} = x_1 \begin{bmatrix} 2 \\ 0 \\ 1 \end{bmatrix} + x_2 \begin{bmatrix} -1 \\ 0 \\ 1 \end{bmatrix}$$

This is valid for any $x_1, x_2 \in \mathbb{R}$. Thus, $\text{Range}(L) = \text{Span}\left\{\begin{bmatrix} 2 \\ 0 \\ 1 \end{bmatrix}, \begin{bmatrix} -1 \\ 0 \\ 1 \end{bmatrix}\right\}$.

EXAMPLE 3.4.2

Find the range of a rotation $R_\theta : \mathbb{R}^2 \to \mathbb{R}^2$ through an angle θ.

Solution: Geometrically, it is clear that the range of the rotation is all of $\mathbb{R}^2$. Indeed, if we pick any vector $\vec{y} \in \mathbb{R}^2$, then we can get $\vec{y} = R_\theta(\vec{x})$ by taking $\vec{x}$ to be the vector which we get by rotating $\vec{y}$ by an angle of $-\theta$.

EXAMPLE 3.4.3

Let $\vec{v} = \begin{bmatrix} 1 \\ 1 \\ 3 \end{bmatrix}$. Find the range of the linear mappings $\text{proj}_{\vec{v}} : \mathbb{R}^3 \to \mathbb{R}^3$.

Solution: Geometrically, the range of the projection should be the line spanned by $\vec{v}$. Indeed, by definition, every image $\text{proj}_{\vec{v}}(\vec{x})$ is a vector on the line. Moreover, for any vector $c\vec{v}$ on the line we have

$$\text{proj}_{\vec{v}}(c\vec{v}) = c\vec{v}$$

Hence, $\text{Range}(\text{proj}_{\vec{v}}) = \text{Span}\{\vec{v}\}$.

EXERCISE 3.4.1

Find a spanning set for the range of the linear mapping $L : \mathbb{R}^3 \to \mathbb{R}^2$ defined by $L(x_1, x_2, x_3) = (x_1 - x_2, -2x_1 + 2x_2 + x_3)$.

Not surprisingly, we see in the examples that the range of a linear mapping is a subspace of the codomain. To prove that in general, we need the following useful result.

Theorem 3.4.1

If $L : \mathbb{R}^n \to \mathbb{R}^m$ is a linear mapping, then $L(\vec{0}) = \vec{0}$.

Proof: Since $0\vec{x} = \vec{0}$ for any $\vec{x} \in \mathbb{R}^n$, we have

$$\vec{0} = 0L(\vec{x}) = L(0\vec{x}) = L(\vec{0})$$

■

Theorem 3.4.2

If $L : \mathbb{R}^n \to \mathbb{R}^m$ is a linear mapping, then $\text{Range}(L)$ is a subspace of $\mathbb{R}^m$.

Nullspace

Definition
Nullspace

The **nullspace** of a linear mapping L is the set of all vectors whose image under L is the zero vector $\vec{0}$. We write

$$\text{Null}(L) = \{\vec{x} \in \mathbb{R}^n \mid L(\vec{x}) = \vec{0}\}$$

Remark

The word **kernel**—and the notation $\ker(L) = \{\vec{x} \in \mathbb{R}^n \mid L(\vec{x}) = \vec{0}\}$—is often used in place of *nullspace*.

EXAMPLE 3.4.4

Let $\vec{v} = \begin{bmatrix} 3 \\ -1 \end{bmatrix}$. Find the nullspace of $\text{proj}_{\vec{v}} : \mathbb{R}^2 \to \mathbb{R}^2$.

Solution: Let $\vec{x} \in \text{Null}(\text{proj}_{\vec{v}})$. By definition of the nullspace, we have that

$$\begin{bmatrix} 0 \\ 0 \end{bmatrix} = \text{proj}_{\vec{v}}(\vec{x}) = \frac{\vec{x} \cdot \vec{v}}{\|\vec{v}\|^2}\vec{v} = \frac{3x_1 - x_2}{10}\begin{bmatrix} 3 \\ -1 \end{bmatrix}$$

This implies that $3x_1 - x_2 = 0$. Thus, every $\vec{x} \in \text{Null}(\text{proj}_{\vec{v}})$ satisfies

$$\vec{x} = \begin{bmatrix} x_1 \\ x_2 \end{bmatrix} = \begin{bmatrix} x_1 \\ 3x_1 \end{bmatrix} = x_1\begin{bmatrix} 1 \\ 3 \end{bmatrix}$$

Thus, $\text{Null}(\text{proj}_{\vec{v}}) = \text{Span}\left\{\begin{bmatrix} 1 \\ 3 \end{bmatrix}\right\}$.

EXAMPLE 3.4.5

Find the nullspace of the linear mapping $L : \mathbb{R}^2 \to \mathbb{R}^3$ defined by

$$L(x_1, x_2) = (2x_1 - x_2, 0, x_1 + x_2)$$

Solution: Let $\begin{bmatrix} x_1 \\ x_2 \end{bmatrix} \in \text{Null}(L)$. Then, we have

$$\begin{bmatrix} 0 \\ 0 \end{bmatrix} = L(x_1, x_2) = \begin{bmatrix} 2x_1 - x_2 \\ 0 \\ x_1 + x_2 \end{bmatrix}$$

This gives us the homogeneous system

$$\begin{aligned} 2x_1 - x_2 &= 0 \\ x_1 + x_2 &= 0 \end{aligned}$$

Row reducing the corresponding coefficient matrix gives

$$\begin{bmatrix} 2 & -1 \\ 1 & 1 \end{bmatrix} \sim \begin{bmatrix} 1 & 0 \\ 0 & 1 \end{bmatrix}$$

This implies that the only solution is $\vec{x} = \vec{0}$. Thus, $\text{Null}(L) = \{\vec{0}\}$.

EXERCISE 3.4.2

Find a spanning set for the nullspace of the linear mapping $L : \mathbb{R}^3 \to \mathbb{R}^2$ defined by

$$L(x_1, x_2, x_3) = (x_1 - x_2, -2x_1 + 2x_2 + x_3)$$

Theorem 3.4.3 If $L : \mathbb{R}^n \to \mathbb{R}^m$ is a linear mapping, then $\text{Null}(L)$ is a subspace of $\mathbb{R}^n$.

You are asked to prove this in Problem C4.

The Four Fundamental Subspaces of a Matrix

We now look at the relationship of the standard matrix of a linear mapping to its range and nullspace. In doing so, we will derive four important subspaces of a matrix.

Column Space

The connection between the range of a linear mapping and its standard matrix follows easily from our second interpretation of matrix-vector multiplication.

Theorem 3.4.4

If $L : \mathbb{R}^n \to \mathbb{R}^m$ is a linear mapping with standard matrix $[L] = \begin{bmatrix} \vec{a}_1 & \cdots & \vec{a}_n \end{bmatrix}$, then

$$\text{Range}(L) = \text{Span}\{\vec{a}_1, \ldots, \vec{a}_n\}$$

Proof: For any $\vec{x} \in \mathbb{R}^n$ we have

$$L(\vec{x}) = [L]\vec{x} = \begin{bmatrix} \vec{a}_1 & \cdots & \vec{a}_n \end{bmatrix}\begin{bmatrix} x_1 \\ \vdots \\ x_n \end{bmatrix} = x_1\vec{a}_1 + \cdots + x_n\vec{a}_n$$

Thus, a vector $\vec{y}$ is in Range(L) if and only if it is in Span$\{\vec{a}_1, \ldots, \vec{a}_n\}$. Hence, Range($L$) = Span$\{\vec{a}_1, \ldots, \vec{a}_n\}$ as required. ■

Therefore, the range of a linear mapping $L : \mathbb{R}^n \to \mathbb{R}^m$ is the subspace of $\mathbb{R}^m$ spanned by the columns of its standard matrix.

Definition
Column Space

Let $A = \begin{bmatrix} \vec{a}_1 & \cdots & \vec{a}_n \end{bmatrix} \in M_{m\times n}(\mathbb{R})$. The **column space** of A is the subspace of $\mathbb{R}^m$ defined by

$$\text{Col}(A) = \text{Span}\{\vec{a}_1, \ldots, \vec{a}_n\} = \{A\vec{x} \in \mathbb{R}^m \mid \vec{x} \in \mathbb{R}^n\}$$

EXAMPLE 3.4.6

Let $A = \begin{bmatrix} 1 & 1 \\ 2 & 1 \\ 1 & 3 \end{bmatrix}$. Determine whether $\vec{c} = \begin{bmatrix} 1 \\ 3 \\ -1 \end{bmatrix}$ and $\vec{d} = \begin{bmatrix} 2 \\ 1 \\ 9 \end{bmatrix}$ are in the Col(A).

Solution: By definition, $\vec{c} \in \text{Col}(A)$ if there exists $\vec{x} \in \mathbb{R}^2$ such that $A\vec{x} = \vec{c}$. Similarly, $\vec{d} \in \text{Col}(A)$ if there exists $\vec{y} \in \mathbb{R}^2$ such that $A\vec{y} = \vec{d}$. Since the coefficient matrix is the same for the two systems, we can answer both questions simultaneously by row reducing the doubly augmented matrix $\left[\begin{array}{c|c|c} A & \vec{c} & \vec{d} \end{array}\right]$:

$$\left[\begin{array}{cc|c|c} 1 & 1 & 1 & 2 \\ 2 & 1 & 3 & 1 \\ 1 & 3 & -1 & 9 \end{array}\right] \sim \left[\begin{array}{cc|c|c} 1 & 1 & 1 & 2 \\ 0 & 1 & -1 & 3 \\ 0 & 0 & 0 & 1 \end{array}\right]$$

If we ignore the second augmented column, then this corresponds to solving the system $\left[\begin{array}{c|c} A & \vec{c} \end{array}\right]$. Hence, we see that $A\vec{x} = \vec{c}$ is consistent, so $\vec{c} \in \text{Col}(A)$.
Similarly, if we ignore the first augmented column, then this corresponds to solving the system $\left[\begin{array}{c|c} A & \vec{d} \end{array}\right]$. From this we see that $A\vec{x} = \vec{d}$ is inconsistent and hence $\vec{d} \notin \text{Col}(A)$.

EXAMPLE 3.4.7

Find a basis for the column space of $A = \begin{bmatrix} 1 & 2 & 3 \\ 2 & 1 & -1 \end{bmatrix}$ and state its dimension.

Solution: By definition, we have that

$$\text{Col}(A) = \text{Span}\left\{\begin{bmatrix} 1 \\ 2 \end{bmatrix}, \begin{bmatrix} 2 \\ 1 \end{bmatrix}, \begin{bmatrix} 3 \\ -1 \end{bmatrix}\right\}$$

We now just need to determine if the spanning set is linearly independent. So, we consider

$$c_1 \begin{bmatrix} 1 \\ 2 \end{bmatrix} + c_2 \begin{bmatrix} 2 \\ 1 \end{bmatrix} + c_3 \begin{bmatrix} 3 \\ -1 \end{bmatrix} = \begin{bmatrix} 0 \\ 0 \end{bmatrix}$$

As we saw in Section 2.3, this gives us the homogeneous system with coefficient matrix A. Row reducing A to RREF gives

$$\begin{bmatrix} 1 & 2 & 3 \\ 2 & 1 & -1 \end{bmatrix} \sim \begin{bmatrix} 1 & 0 & -5/3 \\ 0 & 1 & 7/3 \end{bmatrix}$$

The general solution is $\begin{bmatrix} c_1 \\ c_2 \\ c_3 \end{bmatrix} = t \begin{bmatrix} 5/3 \\ -7/3 \\ 1 \end{bmatrix}, t \in \mathbb{R}$. Taking $t = 1$, gives $c_1 = 5/3, c_2 = -7/3$, and $c_3 = 1$. Hence,

$$\frac{5}{3}\begin{bmatrix} 1 \\ 2 \end{bmatrix} - \frac{7}{3}\begin{bmatrix} 2 \\ 1 \end{bmatrix} + \begin{bmatrix} 3 \\ -1 \end{bmatrix} = \begin{bmatrix} 0 \\ 0 \end{bmatrix}$$

Therefore, $\begin{bmatrix} 3 \\ -1 \end{bmatrix}$ is a linear combination of $\begin{bmatrix} 1 \\ 2 \end{bmatrix}$ and $\begin{bmatrix} 2 \\ 1 \end{bmatrix}$. Hence, by Theorem 1.4.3 we get that

$$\text{Col}(A) = \text{Span}\left\{\begin{bmatrix} 1 \\ 2 \end{bmatrix}, \begin{bmatrix} 2 \\ 1 \end{bmatrix}\right\}$$

Since $\mathcal{B} = \left\{\begin{bmatrix} 1 \\ 2 \end{bmatrix}, \begin{bmatrix} 2 \\ 1 \end{bmatrix}\right\}$ is also linearly independent, we have that $\mathcal{B}$ is a basis for Col(A).

By definition, the dimension of a subspace is the number of vectors in a basis. Therefore, since $\mathcal{B}$ contains 2 vectors, we get that

$$\dim \text{Col}(A) = 2$$

Example 3.4.7 shows us that all we need to do to find a basis for the column space of a matrix A is to determine which columns of A can be written as linear combinations of the others and remove them. Let R be the reduced row echelon form of A. As in the example, the columns of A that correspond to columns in R without leading ones will be linear combinations of the columns of A corresponding to the columns in R with leading ones. We get the following theorem.

Theorem 3.4.5

If R is the reduced row echelon form of a matrix A, then the columns of A that correspond to the columns of R with leading ones form a basis of the column space of A. Moreover,

$$\dim \operatorname{Col}(A) = \operatorname{rank}(A)$$

The idea of the proof of Theorem 3.4.5 is the same as the method outlined in the solution of Example 3.4.7.

EXAMPLE 3.4.8

Let $A = \begin{bmatrix} 1 & 2 & 1 & 1 \\ 1 & 2 & 2 & 1 \\ 2 & 4 & 2 & 3 \\ 3 & 6 & 4 & 3 \end{bmatrix}$. Find a basis for Col($A$) and state its dimension.

Solution: Row reducing A gives

$$\begin{bmatrix} 1 & 2 & 1 & 1 \\ 1 & 2 & 2 & 1 \\ 2 & 4 & 2 & 3 \\ 3 & 6 & 4 & 3 \end{bmatrix} \sim \begin{bmatrix} 1 & 2 & 0 & 0 \\ 0 & 0 & 1 & 0 \\ 0 & 0 & 0 & 1 \\ 0 & 0 & 0 & 0 \end{bmatrix} = R$$

The first, third, and fourth columns of R contain leading ones. Therefore, by Theorem 3.4.5, the first, third, and fourth columns of matrix A form a basis for Col(A). So, a basis for Col(A) is

$$\mathcal{B} = \left\{ \begin{bmatrix} 1 \\ 1 \\ 2 \\ 3 \end{bmatrix}, \begin{bmatrix} 1 \\ 2 \\ 2 \\ 4 \end{bmatrix}, \begin{bmatrix} 1 \\ 1 \\ 3 \\ 3 \end{bmatrix} \right\}$$

We have

$$\dim \operatorname{Col}(A) = 3 = \operatorname{rank}(A)$$

Notice in Example 3.4.8 that last entry of every vector in the reduced row echelon form R of A is 0. Hence, $\operatorname{Col}(A) \neq \operatorname{Col}(R)$. That is, the first, third, and fourth columns of R *do not* form a basis for Col(A).

EXERCISE 3.4.3

Let $A = \begin{bmatrix} 1 & 1 & 2 & 0 & 3 \\ 1 & -1 & 0 & 2 & -3 \\ -1 & 2 & 1 & -3 & -2 \end{bmatrix}$. Find a basis for Col($A$).

Nullspace

We now connect the standard matrix of a linear mapping L with Null(L).

Theorem 3.4.6 If $L : \mathbb{R}^n \to \mathbb{R}^m$ is a linear mapping with standard matrix $[L]$, then $\vec{x} \in \text{Null}(L)$ if and only if $[L]\vec{x} = \vec{0}$.

Proof: If $\vec{x} \in \text{Null}(L)$, then $\vec{0} = L(\vec{x}) = [L]\vec{x}$.

On the other hand, if $[L]\vec{x} = \vec{0}$, then $\vec{0} = [L]\vec{x} = L(\vec{x})$, so $\vec{x} \in \text{Null}(L)$. ■

This motivates the following definition.

Definition
Nullspace

Let $A \in M_{m\times n}(\mathbb{R})$. The **nullspace (kernel)** of A is defined by

$$\text{Null}(A) = \{\vec{x} \in \mathbb{R}^n \mid A\vec{x} = \vec{0}\}$$

Remark

Since the nullspace of A is just the solution space of a homogeneous system, we get by Theorem 2.2.3 that the nullspace of A is a subspace of $\mathbb{R}^n$.

EXAMPLE 3.4.9

Let $A = \begin{bmatrix} 1 & 2 \\ -1 & 0 \end{bmatrix}$. Determine whether $\vec{x} = \begin{bmatrix} -1 \\ 2 \end{bmatrix}$ is in Null(A).

Solution: We have $A\vec{x} = \begin{bmatrix} 3 \\ 1 \end{bmatrix}$. Since $A\vec{x} \neq \vec{0}$, $\vec{x} \notin \text{Null}(A)$.

EXAMPLE 3.4.10

Let $A = \begin{bmatrix} 1 & 3 & 1 \\ -1 & -2 & 2 \end{bmatrix}$. Find a basis for the nullspace of A.

Solution: To find a basis for Null(A), we just need to find a basis for the solution space of $A\vec{x} = \vec{0}$ using the methods of Chapter 2.

Row reducing A gives

$$\begin{bmatrix} 1 & 3 & 1 \\ -1 & -2 & 2 \end{bmatrix} \sim \begin{bmatrix} 1 & 0 & -8 \\ 0 & 1 & 3 \end{bmatrix}$$

Thus, the solution space of the system $A\vec{x} = \vec{0}$ is

$$\vec{x} = t\begin{bmatrix} 8 \\ -3 \\ 1 \end{bmatrix}, \quad t \in \mathbb{R}$$

Hence, $\mathcal{B} = \left\{\begin{bmatrix} 8 \\ -3 \\ 1 \end{bmatrix}\right\}$ spans Null(A). Since $\mathcal{B}$ contains only one non-zero vector, it is also linearly independent. Therefore, $\mathcal{B}$ is a basis for Null(A).

EXAMPLE 3.4.11

Let $A = \begin{bmatrix} 1 & 2 & -1 \\ 5 & 0 & -4 \\ -2 & 6 & 4 \end{bmatrix}$. Find a basis for the nullspace of A.

Solution: Row reducing A gives

$$\begin{bmatrix} 1 & 2 & -1 \\ 5 & 0 & -4 \\ -2 & 6 & 4 \end{bmatrix} \sim \begin{bmatrix} 1 & 0 & 0 \\ 0 & 1 & 0 \\ 0 & 0 & 1 \end{bmatrix}$$

Thus, the only solution of the system $A\vec{x} = \vec{0}$ is $\vec{x} = \vec{0}$. Therefore, $\text{Null}(A) = \{\vec{0}\}$. By definition, a basis for this subspace is the empty set.

As indicated in the examples, to find a basis for the nullspace of a matrix we just use the methods of Chapter 2 to solve the homogeneous system $A\vec{x} = \vec{0}$. You may have already observed that the vector equation of the solution space we get from this method always contains a set of linearly independent vectors. We demonstrate this more carefully with an example.

EXAMPLE 3.4.12

Let $A = \begin{bmatrix} 1 & 2 & 0 & 3 & 4 \\ 0 & 0 & 1 & 5 & 6 \end{bmatrix}$. Find a basis for Null($A$) and relate the dimension of Null(A) to rank(A).

Solution: Observe that A is already in reduced row echelon form. We find that the general solution to $A\vec{x} = \vec{0}$ is

$$\vec{x} = t_1 \begin{bmatrix} -2 \\ 1 \\ 0 \\ 0 \\ 0 \end{bmatrix} + t_2 \begin{bmatrix} -3 \\ 0 \\ -5 \\ 1 \\ 0 \end{bmatrix} + t_3 \begin{bmatrix} -4 \\ 0 \\ -6 \\ 0 \\ 1 \end{bmatrix}, \quad t_1, t_2, t_3 \in \mathbb{R}$$

To see that the set containing these three vectors is linearly independent, observe that the vector corresponding to the free variable x_2 has a 1 as its second entry while the other vectors have a 0. Similarly, the vector corresponding to x_4 has a 1 as its fourth entry while the other vectors have 0, and so forth for the vector corresponding to x_5.

Since the vectors we got from the free variables formed a linearly independent set, we have that the dimension of the nullspace is the number of free variables. Hence,

$$\dim \text{Null}(A) = 3 = (\text{ \# of columns}) - \text{rank}(A)$$

Following the method in the example, we get the following theorem.

Theorem 3.4.7

If $A \in M_{m\times n}(\mathbb{R})$ with $\text{rank}(A) = r$, then

$$\dim \text{Null}(A) = n - r$$

Row Space and Left Nullspace

In a variety of applications, for example analyzing incidence matrices for electric circuits or the stoichiometry matrix for chemical reactions, we find that the column space and nullspace of A^T are are also important.

Definition
Row Space
Left Nullspace

Let $A \in M_{m\times n}(\mathbb{R})$. The **row space** of A is the subspace of $\mathbb{R}^n$ defined by

$$\text{Row}(A) = \{A^T\vec{x} \in \mathbb{R}^n \mid \vec{x} \in \mathbb{R}^m\} = \text{Col}(A^T)$$

The **left nullspace** of A is the subspace of $\mathbb{R}^m$ defined by

$$\text{Null}(A^T) = \{\vec{x} \in \mathbb{R}^m \mid A^T\vec{x} = \vec{0}\}$$

To find a basis for the left nullspace, we can just use our procedure for finding a basis for the nullspace on A^T. That is, we solve $A^T\vec{x} = \vec{0}$. To find a basis for the row space, we use the following theorem.

Theorem 3.4.8

If R is the reduced row echelon form of a matrix A, then the non-zero rows of R form a basis of the row space of A. Moreover,

$$\dim \text{Row}(A) = \text{rank}(A)$$

EXAMPLE 3.4.13

Let $A = \begin{bmatrix} 1 & 2 & -3 \\ 0 & 0 & 2 \\ 1 & 2 & 1 \end{bmatrix}$. Find a basis for Row($A$) and a basis for Null($A^T$).

Solution: Row reducing A gives

$$\begin{bmatrix} 1 & 2 & -3 \\ 0 & 0 & 2 \\ 1 & 2 & 1 \end{bmatrix} \sim \begin{bmatrix} 1 & 2 & 0 \\ 0 & 0 & 1 \\ 0 & 0 & 0 \end{bmatrix} = R$$

Hence, by Theorem 3.4.8, a basis for Row(A) is formed by taking the non-zero rows of R. So, a basis for Row(A) is $\left\{\begin{bmatrix} 1 \\ 2 \\ 0 \end{bmatrix}, \begin{bmatrix} 0 \\ 0 \\ 1 \end{bmatrix}\right\}$.

To find a basis for Null(A^T) we row reduce A^T and solve $A^T\vec{x} = \vec{0}$. We get

$$A^T = \begin{bmatrix} 1 & 0 & 1 \\ 2 & 0 & 2 \\ -3 & 2 & 1 \end{bmatrix} \sim \begin{bmatrix} 1 & 0 & 1 \\ 0 & 1 & 2 \\ 0 & 0 & 0 \end{bmatrix}$$

Hence, a basis for Null(A^T) is $\left\{\begin{bmatrix} -1 \\ -2 \\ 1 \end{bmatrix}\right\}$.

These four special subspaces of matrix are extremely important so we make the following definition.

Definition
Four Fundamental Subspaces

Let $A \in M_{m\times n}(\mathbb{R})$. We call $\text{Col}(A)$, $\text{Null}(A)$, $\text{Row}(A)$, and $\text{Null}(A^T)$ the **Four Fundamental Subspaces** of A.

Rank-Nullity Theorem

Combining Theorem 3.4.5 and Theorem 3.4.7 gives the following theorem.

Theorem 3.4.9

Rank-Nullity Theorem
If $A \in M_{m\times n}(\mathbb{R})$, then

$$\text{rank}(A) + \dim(\text{Null}(A)) = n$$

EXAMPLE 3.4.14

Find a basis for each of the four fundamental subspaces of $A = \begin{bmatrix} 1 & 2 & 0 & -1 \\ 3 & 6 & 1 & -1 \\ -2 & -4 & 2 & 6 \end{bmatrix}$ and verify the Rank-Nullity Theorem for both A and A^T.

Solution: Row reducing A and A^T gives

$$A \sim \begin{bmatrix} 1 & 2 & 0 & -1 \\ 0 & 0 & 1 & 2 \\ 0 & 0 & 0 & 0 \end{bmatrix}, \qquad A^T \sim \begin{bmatrix} 1 & 0 & -8 \\ 0 & 1 & 2 \\ 0 & 0 & 0 \\ 0 & 0 & 0 \end{bmatrix}$$

Thus:

A basis for $\text{Col}(A)$ is $\left\{ \begin{bmatrix} 1 \\ 3 \\ -2 \end{bmatrix}, \begin{bmatrix} 0 \\ 1 \\ 2 \end{bmatrix} \right\}$, and a basis for $\text{Null}(A^T)$ is $\left\{ \begin{bmatrix} 8 \\ -2 \\ 1 \end{bmatrix} \right\}$.

A basis for $\text{Row}(A)$ is $\left\{ \begin{bmatrix} 1 \\ 2 \\ 0 \\ -1 \end{bmatrix}, \begin{bmatrix} 0 \\ 0 \\ 1 \\ 2 \end{bmatrix} \right\}$, and a basis for $\text{Null}(A)$ is $\left\{ \begin{bmatrix} -2 \\ 1 \\ 0 \\ 0 \end{bmatrix}, \begin{bmatrix} 1 \\ 0 \\ -2 \\ 1 \end{bmatrix} \right\}$.

For A, we have that $\text{rank}(A) = 2$ and $\dim \text{Null}(A) = 2$, so indeed

$$\text{rank}(A) + \dim \text{Null}(A) = 2 + 2 = 4 = \text{ \# of columns of } A$$

as predicted by the Rank-Nullity Theorem.

Similarly, for A^T, we have that $\text{rank}(A^T) = 2$ and $\dim \text{Null}(A^T) = 1$ so

$$\text{rank}(A^T) + \dim \text{Null}(A^T) = 2 + 1 = 3 = \text{ \# of columns of } A^T$$

We can observe something amazing about the bases for the fundamental subspaces of A in Example 3.4.14. First, each basis vector of $\text{Col}(A)$ is orthogonal to each basis vector of $\text{Null}(A^T)$. Similarly, each basis vector of $\text{Row}(A)$ is orthogonal to each basis vector of $\text{Null}(A)$. This implies that if we combine the basis vectors for $\text{Col}(A)$ and $\text{Null}(A^T)$ we will get a basis for $\mathbb{R}^3$, and if we combine the basis vectors for $\text{Row}(A)$ and $\text{Null}(A)$ we will get a basis for $\mathbb{R}^4$.

EXERCISE 3.4.4

Find a basis for each of the four fundamental subspaces of $A = \begin{bmatrix} 1 & 1 & -3 & 1 \\ 2 & 3 & -8 & 4 \\ 0 & 1 & -2 & 3 \end{bmatrix}$ and verify the Rank-Nullity Theorem for A and A^T.

Fundamental Theorem of Linear Algebra

We now prove that our observations about the fundamental subspaces of the matrix A in Example 3.4.14 hold for all $m \times n$ matrices.

Theorem 3.4.10 Fundamental Theorem of Linear Algebra (FTLA)

If $A \in M_{m\times n}(\mathbb{R})$ with $\text{rank}(A) = k$, then

(1) $\text{Null}(A) = \{\vec{n} \in \mathbb{R}^n \mid \vec{n} \cdot \vec{r} = 0 \text{ for all } \vec{r} \in \text{Row}(A)\}$.

(2) $\text{Null}(A^T) = \{\vec{\ell} \in \mathbb{R}^m \mid \vec{\ell} \cdot \vec{c} = 0 \text{ for all } \vec{c} \in \text{Col}(A)\}$.

(3) If $\{\vec{w}_1, \dots, \vec{w}_k\}$ is a basis for Row(A) and $\{\vec{w}_{k+1}, \dots, \vec{w}_n\}$ is a basis for Null(A), then $\{\vec{w}_1, \dots, \vec{w}_k, \vec{w}_{k+1}, \dots, \vec{w}_n\}$ is a basis for $\mathbb{R}^n$.

(4) If $\{\vec{v}_1, \dots, \vec{v}_k\}$ is a basis for Col(A) and $\{\vec{v}_{k+1}, \dots, \vec{v}_m\}$ is a basis for Null(A^T), then $\{\vec{v}_1, \dots, \vec{v}_k, \vec{v}_{k+1}, \dots, \vec{v}_m\}$ is a basis for $\mathbb{R}^m$.

Proof: We prove (3) and (4) and leave (1) and (2) as Problem C10.

(3): Consider

$$c_1\vec{w}_1 + \cdots + c_k\vec{w}_k + c_{k+1}\vec{w}_{k+1} + \cdots + c_n\vec{w}_n = \vec{0} \tag{3.5}$$

Rearranging we can get

$$c_1\vec{w}_1 + \cdots + c_k\vec{w}_k = -c_{k+1}\vec{w}_{k+1} - \cdots - c_n\vec{w}_n$$

Observe that $c_1\vec{w}_1 + \cdots + c_k\vec{w}_k \in \text{Row}(A)$ and $-c_{k+1}\vec{w}_{k+1} - \cdots - c_n\vec{w}_n \in \text{Null}(A)$. Hence, $\vec{x} = c_1\vec{w}_1 + \cdots + c_k\vec{w}_k$ is a vector in both Row(A) and Null(A). Hence, by (1) we have that $\vec{x} \cdot \vec{x} = 0$. By Theorem 1.5.1 (2), this implies that $\vec{x} = \vec{0}$. Thus, we have

$$c_1\vec{w}_1 + \cdots + c_k\vec{w}_k = \vec{0} = -c_{k+1}\vec{w}_{k+1} - \cdots - c_n\vec{w}_n$$

So, $c_1 = \cdots = c_k = 0$ since $\{\vec{w}_1, \dots, \vec{w}_k\}$ is linearly independent, and $c_{k+1} = \cdots = c_n = 0$ since $\{\vec{w}_{k+1}, \dots, \vec{w}_n\}$ is linearly independent. Consequently, the only solution to equation (3.5) is $c_1 = \cdots = c_n = 0$ and so $\{\vec{w}_1, \dots, \vec{w}_k, \vec{w}_{k+1}, \dots, \vec{w}_n\}$ is linearly independent.

Since equation (3.5) has a unique solution, the coefficient matrix of this system, $\begin{bmatrix} \vec{w}_1 & \cdots & \vec{w}_n \end{bmatrix}$ has rank n by the System-Rank Theorem (2). Hence, by the System-Rank Theorem (3), the system

$$c_1\vec{w}_1 + \cdots + c_n\vec{w}_n = \vec{b}$$

is consistent for all $\vec{b} \in \mathbb{R}^n$. Therefore, $\{\vec{w}_1, \dots, \vec{w}_n\}$ also spans $\mathbb{R}^n$ and hence is a basis for $\mathbb{R}^n$.

(4): Follows immediately from (3) by substituting A^T in for A. ■

Part (1) says the nullspace of A is exactly the set of all vectors in $\mathbb{R}^n$ that are orthogonal to every vector in the row space of A. We say that Null(A) and Row(A) are **orthogonal complements** of each other.

Part (2) says that the left nullspace of A and the column space of A are also orthogonal complements.

Part (3) can be interpreted as saying that every vector $\vec{x}_0 \in \mathbb{R}^n$ can be written as a sum of a vector $\vec{r}_0 \in \text{Row}(A)$ and a vector $\vec{n}_0 \in \text{Null}(A)$. Combining this with Part (1), we can think of $\mathbb{R}^n$ as being split up by a pair of orthogonal axes, where one axis is the nullspace and the other axis is the row space. We depict this in Figure 3.4.1.

Part (4) indicates that every vector $\vec{y}_0 \in \mathbb{R}^m$ can be written as a sum of a vector $\vec{c}_0 \in \text{Col}(A)$ and a vector $\vec{\ell}_0 \in \text{Null}(A^T)$. Hence, combining this with Part (2), we can think of $\mathbb{R}^m$ as having axes of the left nullspace and the column space as depicted in Figure 3.4.1.

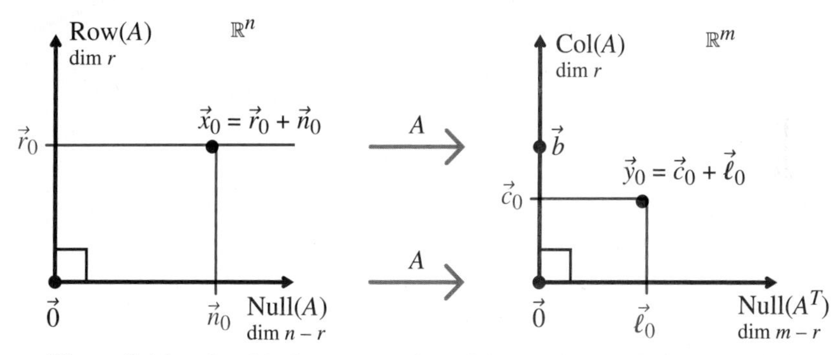

Figure 3.4.1 Graphical representation of the Fundamental Theorem of Linear Algebra.

The figure gives us a visualization of much of what we have done with solving systems of linear equations and more. For example, it shows that each horizontal line

$$\vec{x} = \vec{r}_0 + \vec{n}, \qquad \text{for all } \vec{n} \in \text{Null}(A)$$

in $\mathbb{R}^n$ is mapped by A to a unique vector $\vec{b} \in \text{Col}(A)$. That is, the matrix mapping $A\vec{x}$ is a one-to-one and onto mapping from Row(A) to Col(A).

This shows us that:

1. The general solution of $A\vec{x} = \vec{b}$ is $\vec{x} = \vec{r}_0 + \vec{n}$ for all $\vec{n} \in \text{Null}(A)$.
2. $A\vec{x} = \vec{b}$ has a unique solution if and only if $\text{Null}(A) = \{\vec{0}\}$.
3. $A\vec{x} = \vec{b}$ is consistent for all $\vec{b} \in \mathbb{R}^m$ if and only if $\text{Col}(A) = \mathbb{R}^m$ (i.e. $\text{rank}(A) = m$).
4. If $\vec{x}$ and $\vec{p}$ are both solutions of $A\vec{x} = \vec{b}$, then $\vec{x} - \vec{p} \in \text{Null}(A)$.

A Summary of Facts About Rank

For an $m \times n$ matrix A:

$$\begin{aligned}
\text{rank}(A) &= \text{the number of leading ones in the reduced row echelon form of } A \\
&= \text{the number of non-zero rows in any row echelon form of } A \\
&= \dim \text{Row}(A) \\
&= \dim \text{Col}(A) \\
&= n - \dim \text{Null}(A) \\
&= m - \dim \text{Null}(A^T) \\
&= \text{rank}(A^T)
\end{aligned}$$

PROBLEMS 3.4

Practice Problems

For Problems A1–A3, let L be the linear mapping with the given standard matrix.
(a) Is $\vec{y}_1 \in \text{Range}(L)$? If so, find $\vec{x}$ such that $L(\vec{x}) = \vec{y}_1$.
(b) Is $\vec{y}_2 \in \text{Range}(L)$? If so, find $\vec{x}$ such that $L(\vec{x}) = \vec{y}_2$.
(c) Is $\vec{v} \in \text{Null}(L)$?

A1 $[L] = \begin{bmatrix} 3 & 5 & -1 \\ 1 & 2 & 1 \end{bmatrix}, \vec{y}_1 = \begin{bmatrix} 12 \\ 3 \end{bmatrix}, \vec{y}_2 = \begin{bmatrix} 1 \\ 1 \end{bmatrix}, \vec{v} = \begin{bmatrix} -7 \\ 4 \\ -1 \end{bmatrix}$

A2 $[L] = \begin{bmatrix} 7 & -2 \\ 7 & -8 \\ 1 & 1 \end{bmatrix}, \vec{y}_1 = \begin{bmatrix} 3 \\ -2 \\ 5 \end{bmatrix}, \vec{y}_2 = \begin{bmatrix} 8 \\ -10 \\ 5 \end{bmatrix}, \vec{v} = \begin{bmatrix} 1 \\ 3 \end{bmatrix}$

A3 $[L] = \begin{bmatrix} 1 & 0 & -1 \\ 0 & 1 & 3 \\ 1 & 2 & 1 \\ 2 & 5 & 1 \end{bmatrix}, \vec{y}_1 = \begin{bmatrix} 3 \\ 1 \\ 6 \\ 1 \end{bmatrix}, \vec{y}_2 = \begin{bmatrix} 3 \\ -5 \\ 1 \\ 5 \end{bmatrix}, \vec{v} = \begin{bmatrix} 1 \\ 1 \\ 1 \\ 1 \end{bmatrix}$

For Problems A4–A14, find a basis for the range and a basis for the nullspace of the linear mapping.

A4 $L(x_1, x_2) = (3x_1 + 5x_2, x_1 - x_2, x_1 - x_2)$
A5 $L(x_1, x_2) = (2x_1 - x_2, 4x_1 - 2x_2)$
A6 $L(x_1, x_2) = (x_1 - 7x_2, x_1 + x_2)$
A7 $L(x_1, x_2) = (x_1, 2x_1, 3x_1)$
A8 $L(x_1, x_2, x_3) = (x_1 + x_2, 0)$
A9 $L(x_1, x_2, x_3) = (0, 0, 0)$
A10 $L(x_1, x_2, x_3) = (2x_1, -x_2 + 2x_3)$
A11 $L(x_1, x_2, x_3) = (x_1 + 7x_3, x_1 + x_2 + x_3, x_1 + x_2 + x_3)$
A12 $L(x_1, x_2, x_3) = (-x_2 + 2x_3, x_2 - 4x_3, -2x_2 + 4x_3)$
A13 $L(x_1, x_2, x_3, x_4) = (x_4, x_3, 0, x_2, x_1 + x_2 - x_3)$
A14 $L(x_1, x_2, x_3, x_4) = (x_1 - x_2 + 2x_3 + x_4, x_2 - 3x_3 + 3x_4)$

For Problems A15–A19, find a basis for each of the four fundamental subspaces of the standard matrix $[L]$ of L.

A15 The linear mapping in A4.
A16 The linear mapping in A8.
A17 The linear mapping in A9.
A18 The linear mapping in A10.
A19 The linear mapping in A13.

For Problems A20–A22, use a geometrical argument to give a basis for the nullspace and a basis for the range of the linear mapping.

A20 $\text{proj}_{\vec{v}} : \mathbb{R}^3 \to \mathbb{R}^3$, where $\vec{v} = \begin{bmatrix} 1 \\ -2 \\ 2 \end{bmatrix}$

A21 $\text{perp}_{\vec{v}} : \mathbb{R}^3 \to \mathbb{R}^3$, where $\vec{v} = \begin{bmatrix} 3 \\ 1 \\ 2 \end{bmatrix}$

A22 $\text{refl}_{\vec{v}} : \mathbb{R}^3 \to \mathbb{R}^3$, where $\vec{v} = \begin{bmatrix} 0 \\ 1 \\ 0 \end{bmatrix}$

For Problems A23–A26, determine the standard matrix of a linear mapping L with the given nullspace and range.

A23 $\text{Null}(L) = \left\{ \begin{bmatrix} 0 \\ 0 \end{bmatrix} \right\}, \text{Range}(L) = \text{Span}\left\{ \begin{bmatrix} 1 \\ -1 \\ 1 \end{bmatrix}, \begin{bmatrix} 2 \\ 5 \\ -3 \end{bmatrix} \right\}$

A24 $\text{Null}(L) = \text{Span}\left\{ \begin{bmatrix} 1 \\ 1 \end{bmatrix} \right\}, \text{Range}(L) = \text{Span}\left\{ \begin{bmatrix} 1 \\ 2 \\ 3 \end{bmatrix} \right\}$

A25 $\text{Null}(L) = \text{Span}\left\{ \begin{bmatrix} 1 \\ -2 \end{bmatrix} \right\}, \text{Range}(L) = \text{Span}\left\{ \begin{bmatrix} 1 \\ 1 \\ 1 \end{bmatrix} \right\}$

A26 $\text{Null}(L) = \text{Span}\left\{ \begin{bmatrix} 3 \\ -2 \\ 1 \end{bmatrix} \right\}, \text{Range}(L) = \text{Span}\left\{ \begin{bmatrix} 1 \\ 0 \\ 0 \end{bmatrix}, \begin{bmatrix} 0 \\ 1 \\ 1 \end{bmatrix} \right\}$

A27 Give a mathematical proof of the statement "If $\vec{r}_0$ is a solution of $A\vec{x} = \vec{b}$, then $\vec{r}_0 + \vec{n}$ is a solution of $A\vec{x} = \vec{b}$ for all $\vec{n} \in \text{Null}(A)$."

For Problems A28 and A29, suppose that the matrix is the coefficient matrix of a homogeneous system of equations. State the following:
(a) The number of variables in the system.
(b) The rank of the matrix.
(c) The dimension of the solution space.

A28 $\begin{bmatrix} 1 & 2 & 1 & 3 \\ 0 & 0 & 1 & -2 \end{bmatrix}$

A29 $\begin{bmatrix} 1 & -2 & 0 & 0 & 5 \\ 0 & 1 & 3 & 4 & -1 \\ 0 & 0 & 0 & 1 & 2 \end{bmatrix}$

For Problems A30–A42, find a basis for each of the four fundamental subspaces, and verify the Rank-Nullity Theorem for A and A^T.

A30 $\begin{bmatrix} 1 & 2 \\ 2 & 4 \end{bmatrix}$

A31 $\begin{bmatrix} -7 & 3 \\ -3 & 1 \end{bmatrix}$

A32 $\begin{bmatrix} 1 & 2 & 1 \\ 0 & 4 & 2 \end{bmatrix}$

A33 $\begin{bmatrix} 0 & -2 & 4 & 2 \\ 0 & 3 & -6 & -3 \end{bmatrix}$

A34 $\begin{bmatrix} 1 & 1 & -3 & 1 \\ 2 & 3 & -8 & 4 \\ 0 & 1 & -2 & 3 \end{bmatrix}$

A35 $\begin{bmatrix} 1 & 2 & 8 \\ 1 & 1 & 5 \\ 1 & 0 & -2 \end{bmatrix}$

A36 $\begin{bmatrix} 2 & 1 & 3 \\ 2 & -2 & 6 \\ 4 & 3 & 5 \end{bmatrix}$

A37 $\begin{bmatrix} 1 & 2 & 4 \\ 1 & 2 & 4 \\ 1 & 2 & 4 \end{bmatrix}$

A38 $\begin{bmatrix} 1 & 2 & 9 \\ 0 & 1 & 7 \\ -2 & -2 & -4 \\ 1 & 1 & 2 \end{bmatrix}$

A39 $\begin{bmatrix} 3 & -1 & 6 \\ 1 & 2 & 5 \\ 1 & 3 & 3 \end{bmatrix}$

A40 $\begin{bmatrix} 1 & -1 & 1 \\ 1 & 0 & 0 \\ 1 & 1 & 1 \\ 1 & 2 & 4 \end{bmatrix}$

A41 $\begin{bmatrix} 1 & 1 & 0 \\ 2 & 2 & 0 \\ 1 & 1 & 2 \\ 3 & 3 & 4 \end{bmatrix}$

A42 $\begin{bmatrix} 1 & 2 & 0 & 3 & 0 \\ 1 & 2 & 1 & 7 & 1 \\ 2 & 4 & 0 & 6 & 1 \\ 3 & 6 & 1 & 13 & 2 \end{bmatrix}$

For Problems A43–A48, let $A = \begin{bmatrix} 1 & 1 & 1 & 1 & 5 \\ 2 & 3 & 1 & 2 & 11 \\ 1 & 1 & 1 & 3 & 7 \\ 1 & 2 & 0 & -1 & 4 \end{bmatrix}$ which has RREF $R = \begin{bmatrix} 1 & 0 & 2 & 0 & 3 \\ 0 & 1 & -1 & 0 & 1 \\ 0 & 0 & 0 & 1 & 1 \\ 0 & 0 & 0 & 0 & 0 \end{bmatrix}$. Let

$\vec{u} = \begin{bmatrix} 0 \\ -1 \\ 1 \\ 1 \end{bmatrix}, \vec{v} = \begin{bmatrix} 2 \\ 3 \\ 3 \\ 1 \end{bmatrix}, \vec{w} = \begin{bmatrix} -2 \\ 1 \\ 1 \\ -1 \\ 1 \end{bmatrix}, \vec{z} = \begin{bmatrix} 3 \\ -4 \\ 10 \\ 2 \\ 7 \end{bmatrix}, \vec{b} = \begin{bmatrix} 1 \\ 2 \\ 1 \\ 1 \end{bmatrix}.$

A43 Is $\vec{u} \in \text{Null}(A^T)$?

A44 Is $\vec{v} \in \text{Col}(A)$?

A45 Is $\vec{w} \in \text{Null}(A)$?

A46 Is $\vec{z} \in \text{Row}(A)$?

A47 Find a basis for Null(A).

A48 Find all solutions of $A\vec{x} = \vec{b}$.

Homework Problems

For Problems B1 and B2, let L be the linear mapping with the given standard matrix.

(a) Is $\vec{y}_1 \in \text{Range}(L)$? If so, find $\vec{x}$ such that $L(\vec{x}) = \vec{y}_1$.

(b) Is $\vec{y}_2 \in \text{Range}(L)$? If so, find $\vec{x}$ such that $L(\vec{x}) = \vec{y}_2$.

(c) Is $\vec{v} \in \text{Null}(L)$?

B1 $[L] = \begin{bmatrix} 1 & 1 & 5 \\ 1 & 2 & 7 \\ 1 & -1 & 1 \end{bmatrix}, \vec{y}_1 = \begin{bmatrix} 1 \\ 6 \\ -9 \end{bmatrix}, \vec{y}_2 = \begin{bmatrix} 8 \\ 8 \\ 3 \end{bmatrix}, \vec{v} = \begin{bmatrix} -4 \\ -1 \\ 1 \end{bmatrix}$

B2 $[L] = \begin{bmatrix} 0 & -2 & 1 \\ 2 & 4 & 1 \\ 1 & 3 & 0 \end{bmatrix}, \vec{y}_1 = \begin{bmatrix} 5 \\ 5 \\ 3 \end{bmatrix}, \vec{y}_2 = \begin{bmatrix} 2 \\ 1 \\ -2 \end{bmatrix}, \vec{v} = \begin{bmatrix} 3 \\ -1 \\ -2 \end{bmatrix}$

For Problems B3–B11, find a basis for the range and a basis for the nullspace of the linear mapping.

B3 $L(x_1, x_2, x_3) = (x_1 - x_2, x_2 + x_3)$

B4 $L(x_1, x_2, x_3, x_4) = (x_1, x_2 + x_4)$

B5 $L(x_1, x_2) = (0, x_1 + x_2)$

B6 $L(x_1, x_2, x_3, x_4) = (-x_1 + x_4, 2x_2 + 3x_3, x_1 - x_4)$

B7 $L(x_1) = (x_1, 2x_1, -x_1)$

B8 $L(x_1, x_2, x_3) = (2x_1 + x_3, x_1 - x_3, x_1 + x_3)$

B9 $L(x_1, x_2, x_3) = (x_1 + 2x_2, 3x_1 + 6x_2 + x_3, x_1 + 2x_2 + x_3)$

B10 $L(x_1, x_2, x_3) = (x_1 - x_3, x_2 - x_3, x_1 - x_2)$

B11 $L(x_1, x_2, x_3, x_4) = (x_1 + x_3, 2x_1 + 2x_3 + x_4)$

For Problems B12–B17, find a basis for each of the four fundamental subspaces of the standard matrix $[L]$ of L.

B12 The linear mapping in B3.

B13 The linear mapping in B4.

B14 The linear mapping in B5.

B15 The linear mapping in B6.

B16 The linear mapping in B7.

B17 The linear mapping in B8.

For Problems B18–B20, use a geometrical argument to give a basis for the nullspace and a basis for the range of the linear mapping.

B18 $R_\theta : \mathbb{R}^2 \to \mathbb{R}^2$

B19 $\text{proj}_{\vec{v}} : \mathbb{R}^3 \to \mathbb{R}^3$, where $\vec{v} = \begin{bmatrix} -1 \\ 1 \\ 1 \end{bmatrix}$

B20 $\text{refl}_{\vec{v}} : \mathbb{R}^3 \to \mathbb{R}^3$, where $\vec{v} = \begin{bmatrix} 1 \\ 1 \\ 1 \end{bmatrix}$

For Problems B21–B24, determine the standard matrix of a linear mapping L with the given nullspace and range.

B21 $\text{Null}(L) = \text{Span}\left\{\begin{bmatrix} 1 \\ -2 \end{bmatrix}\right\}, \text{Range}(L) = \text{Span}\left\{\begin{bmatrix} 1 \\ 1 \end{bmatrix}\right\}$

B22 $\text{Null}(L) = \text{Span}\left\{\begin{bmatrix} 3 \\ 2 \end{bmatrix}\right\}, \text{Range}(L) = \text{Span}\left\{\begin{bmatrix} 0 \\ 1 \\ -1 \end{bmatrix}\right\}$

B23 $\text{Null}(L) = \text{Span}\left\{\begin{bmatrix} 1 \\ 0 \end{bmatrix}\right\}, \text{Range}(L) = \text{Span}\left\{\begin{bmatrix} 1 \\ 0 \end{bmatrix}\right\}$

B24 $\text{Null}(L) = \mathbb{R}^4, \text{Range}(L) = \left\{\begin{bmatrix} 0 \\ 0 \\ 0 \end{bmatrix}\right\}$

For Problems B25–B28, suppose that the matrix is the coefficient matrix of a homogeneous system of equations. State the following:

(a) The number of variables in the system.
(b) The rank of the matrix.
(c) The dimension of the solution space.

B25 $\begin{bmatrix} 1 & 0 & 2 & 4 \\ 0 & 1 & -1 & -1 \end{bmatrix}$ **B26** $\begin{bmatrix} 1 & 0 & 0 & 0 \\ 0 & 0 & 1 & 0 \end{bmatrix}$

B27 $\begin{bmatrix} 1 & 0 & -3 \\ 0 & 1 & 1 \\ 0 & 0 & 0 \end{bmatrix}$ **B28** $\begin{bmatrix} 0 & 1 & -1 & 0 & 0 \\ 0 & 0 & 0 & 1 & 0 \\ 0 & 0 & 0 & 0 & 1 \end{bmatrix}$

For Problems B29–B37, find a basis for each of the four fundamental subspaces, and verify the Rank-Nullity Theorem for A and A^T.

B29 $\begin{bmatrix} 1 & 2 \\ 3 & 4 \end{bmatrix}$ **B30** $\begin{bmatrix} 0 & 3 \\ 0 & -11 \end{bmatrix}$ **B31** $\begin{bmatrix} 1 & 1 & 1 \\ 0 & 1 & -1 \end{bmatrix}$

B32 $\begin{bmatrix} 2 & 2 & 1 \\ -2 & -2 & -1 \\ 1 & 4 & 2 \end{bmatrix}$ **B33** $\begin{bmatrix} 1 & 1 & 2 & 1 \\ -2 & -2 & -5 & 1 \\ -1 & -1 & -3 & 2 \end{bmatrix}$

B34 $\begin{bmatrix} 0 & 1 & -3 \\ 1 & 0 & 2 \\ 1 & 2 & -4 \end{bmatrix}$ **B35** $\begin{bmatrix} 3 & 2 & 0 & -1 \\ 3 & 2 & 0 & -1 \\ 3 & 2 & 0 & -1 \end{bmatrix}$

B36 $\begin{bmatrix} 0 & 1 & 4 \\ 0 & 2 & -4 \\ 0 & 1 & 5 \end{bmatrix}$ **B37** $\begin{bmatrix} 2 & 7 & -1 & 9 \\ 1 & 14 & 3 & 6 \\ 1 & 8 & 1 & -7 \end{bmatrix}$

For Problems B38–B43, let $A = \begin{bmatrix} 1 & 2 & 0 & 0 & 3 & 0 \\ 1 & 2 & 1 & 0 & 2 & 0 \\ 1 & 2 & 0 & 1 & 1 & 1 \\ 1 & 2 & 1 & 0 & 2 & 1 \\ 3 & 6 & 2 & 1 & 5 & 2 \end{bmatrix}$

which has RREF $R = \begin{bmatrix} 1 & 2 & 0 & 0 & 3 & 0 \\ 0 & 0 & 1 & 0 & -1 & 0 \\ 0 & 0 & 0 & 1 & -2 & 0 \\ 0 & 0 & 0 & 0 & 0 & 1 \\ 0 & 0 & 0 & 0 & 0 & 0 \end{bmatrix}$. Let

$\vec{u} = \begin{bmatrix} 0 \\ -1 \\ -1 \\ -1 \\ 1 \end{bmatrix}, \vec{v} = \begin{bmatrix} 0 \\ 1 \\ 0 \\ 0 \\ 1 \end{bmatrix}, \vec{w} = \begin{bmatrix} 1 \\ -2 \\ 1 \\ 2 \\ 1 \\ 0 \end{bmatrix}, \vec{z} = \begin{bmatrix} 2 \\ 2 \\ 0 \\ 1 \\ 0 \\ 0 \end{bmatrix}, \vec{b} = \begin{bmatrix} 0 \\ 0 \\ 1 \\ 0 \\ 1 \end{bmatrix}.$

B38 Is $\vec{u} \in \text{Null}(A^T)$?

B39 Is $\vec{v} \in \text{Col}(A)$?

B40 Is $\vec{w} \in \text{Null}(A)$?

B41 Is $\vec{z} \in \text{Row}(A)$?

B42 Find a basis for Null(A).

B43 Find all solution of $A\vec{x} = \vec{b}$.

Conceptual Problems

C1 Let $L : \mathbb{R}^n \to \mathbb{R}^m$ be a linear mapping. Prove that

$$\dim(\text{Range}(L)) + \dim(\text{Null}(L)) = n$$

C2 Suppose that $\{\vec{v}_1, \ldots, \vec{v}_k\}$ is a linearly independent set in $\mathbb{R}^n$ and that $L : \mathbb{R}^n \to \mathbb{R}^m$ is a linear mapping with $\text{Null}(L) = \{\vec{0}\}$. Prove that $\{L(\vec{v}_1), \ldots, L(\vec{v}_k)\}$ is linearly independent.

C3 Let $L : \mathbb{R}^n \to \mathbb{R}^n$ be a linear mapping. Prove that $\text{Null}(L) = \{\vec{0}\}$ if and only if $\text{Range}(L) = \mathbb{R}^n$.

C4 Prove if $L : \mathbb{R}^n \to \mathbb{R}^m$ is a linear mapping, then Null(L) is a subspace of $\mathbb{R}^n$.

C5 Suppose that $L : \mathbb{R}^n \to \mathbb{R}^m$ and $M : \mathbb{R}^m \to \mathbb{R}^p$ are linear mappings.

(a) Show that the range of $M \circ L$ is a subspace of the range of M.
(b) Give an example such that the range of $M \circ L$ is not equal to the range of M.
(c) Show that the nullspace of L is a subspace of the nullspace of $M \circ L$.

C6 If A is a 5×7 matrix and $\text{rank}(A) = 4$, then what is the nullity of A, and what is the dimension of the column space of A?

C7 If A is a 5×4 matrix, then what is the largest possible dimension of the nullspace of A? What is the largest possible rank of A?

C8 If A is a 4×5 matrix and $\text{nullity}(A) = 3$, then what is the dimension of the row space of A?

C9 Let $A \in M_{n\times n}(\mathbb{R})$ such that $A^2 = O_{n,n}$. Prove that the column space of A is a subset of the nullspace of A.

C10 Prove parts (1) and (2) of the Fundamental Theorem of Linear Algebra.

C11 Let $A \in M_{m\times n}(\mathbb{R})$. If $A\vec{x} = \vec{b}$ is inconsistent, then there exists $\vec{y} \in \mathbb{R}^m$ such that $A^T\vec{y} = \vec{0}$ with $\vec{y}^T\vec{b} \neq 0$.

C12 Assume $\vec{r}_0$ is a solution of $A\vec{x} = \vec{b}$. Prove that if $\vec{y}$ is also a solution of $A\vec{x} = \vec{b}$, then $\vec{y} = \vec{r}_0 + \vec{n}$ for some $\vec{n} \in \text{Null}(A)$.

3.5 Inverse Matrices and Inverse Mappings

In this section we will look at inverses of matrices and linear mappings. We will make many connections with the material we have covered so far and provide useful tools for the material contained in the rest of the book.

EXAMPLE 3.5.1

In Example 3.1.11 on page 156 we saw that stiffness matrix $K = \begin{bmatrix} k_1 + k_2 & -k_2 \\ -k_2 & k_2 + k_3 \end{bmatrix}$ of the spring-mass system below allowed us to calculate the constant forces f_1, f_2 required to achieve desired equilibrium displacements x_1 and x_2 via the formula

$$K\begin{bmatrix} x_1 \\ x_2 \end{bmatrix} = \begin{bmatrix} f_1 \\ f_2 \end{bmatrix} \qquad (3.6)$$

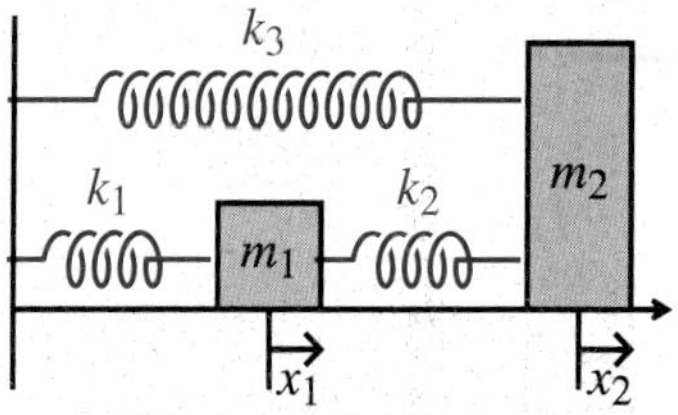

Say that we were instead interested in finding the equilibrium displacements x_1, x_2 for given forces f_1 and f_2. That is, we would like to find a matrix K' such that

$$K'\begin{bmatrix} f_1 \\ f_2 \end{bmatrix} = \begin{bmatrix} x_1 \\ x_2 \end{bmatrix} \qquad (3.7)$$

Substituting equation (3.7) into equation (3.6) gives

$$K\left(K'\begin{bmatrix} f_1 \\ f_2 \end{bmatrix}\right) = \begin{bmatrix} f_1 \\ f_2 \end{bmatrix}$$

which we can rewrite as

$$(KK')\begin{bmatrix} f_1 \\ f_2 \end{bmatrix} = I\begin{bmatrix} f_1 \\ f_2 \end{bmatrix}$$

where I is the 2×2 identity matrix (see Section 3.1). Since this is valid for all $\begin{bmatrix} f_1 \\ f_2 \end{bmatrix} \in \mathbb{R}^2$, the Matrices Equal Theorem gives

$$KK' = I$$

Similarly, if we instead substitute equation (3.6) into equation (3.7) we find that

$$K'K = I$$

Thus, the matrix K' is the multiplicative inverse of K. Since it computes the equilibrium displacements given the forces on the masses, it is called the **flexibility matrix** of the system.

Definition
Inverse

Let $A \in M_{n\times n}(\mathbb{R})$. If there exists $B \in M_{n\times n}(\mathbb{R})$ such that $AB = I = BA$, then A is said to be **invertible**, and B is called the **inverse** of A (and A is the inverse of B). The inverse of A is denoted A^{-1}.

EXAMPLE 3.5.2

The matrix $\begin{bmatrix} 2 & -1 \\ -1 & 1 \end{bmatrix}$ is the inverse of the matrix $\begin{bmatrix} 1 & 1 \\ 1 & 2 \end{bmatrix}$ because

$$\begin{bmatrix} 1 & 1 \\ 1 & 2 \end{bmatrix}\begin{bmatrix} 2 & -1 \\ -1 & 1 \end{bmatrix} = \begin{bmatrix} 1 & 0 \\ 0 & 1 \end{bmatrix} = I$$

and

$$\begin{bmatrix} 2 & -1 \\ -1 & 1 \end{bmatrix}\begin{bmatrix} 1 & 1 \\ 1 & 2 \end{bmatrix} = \begin{bmatrix} 1 & 0 \\ 0 & 1 \end{bmatrix} = I$$

Theorem 3.5.1 If A is invertible, then A^{-1} is unique.

Proof: If B and C are both inverses of A, then $AB = I = BA$ and $AC = I = CA$. Hence,

$$B = BI = B(AC) = (BA)C = IC = C$$

Therefore, the inverse is unique. ■

Notice in the proof that we actually only need to assume that $I = AC$ and that $BA = I$. In such a case, we call C a **right inverse** of A, and we call B a **left inverse** of A. Observe that the distinction is important since matrix multiplication is not commutative. The proof shows that for a square matrix, any left inverse must be equal to any right inverse. However, non-square matrices may have only a right inverse or a left inverse, but not both (see Problem C8). We will now show that for square matrices, a right inverse is automatically a left inverse.

Theorem 3.5.2 If $A, B \in M_{n\times n}(\mathbb{R})$ such that $AB = I$, then $BA = I$ and hence $B = A^{-1}$. Moreover, $\text{rank}(B) = n = \text{rank}(A)$.

Proof: Consider the homogeneous system $B\vec{x} = \vec{0}$. Since $AB = I$, we find that

$$\vec{x} = I\vec{x} = (AB)\vec{x} = A(B\vec{x}) = A(\vec{0}) = \vec{0}$$

Thus, $B\vec{x} = \vec{0}$ has a unique solution which, by the System-Rank Theorem (2), implies that rank $B = n$.
Let $\vec{y} \in \mathbb{R}^n$. Since $\text{rank}(B) = n$, by the System-Rank Theorem (3), we get that there exists a vector $\vec{x}$ such that $B\vec{x} = \vec{y}$. Hence,

$$BA\vec{y} = BA(B\vec{x}) = B(AB)\vec{x} = BI\vec{x} = B\vec{x} = \vec{y} = I\vec{y}$$

Therefore, $BA = I$ by the Matrices Equal Theorem. Therefore, $AB = I$ and $BA = I$, so that $B = A^{-1}$.
Since we have $BA = I$, we see that $\text{rank}(A) = n$, by the same argument we used to prove $\text{rank}(B) = n$. ■

Theorem 3.5.2 makes it very easy to prove some useful properties of the matrix inverse. In particular, to show that $A^{-1} = B$, we only need to show that $AB = I$.

Theorem 3.5.3 If A and B are invertible matrices and t is a non-zero real number, then
(1) $(tA)^{-1} = \frac{1}{t}A^{-1}$
(2) $(AB)^{-1} = B^{-1}A^{-1}$
(3) $(A^T)^{-1} = (A^{-1})^T$

Proof: We have

$$(tA)\left(\frac{1}{t}A^{-1}\right) = \left(\frac{t}{t}\right)AA^{-1} = 1I = I$$
$$(AB)(B^{-1}A^{-1}) = A(BB^{-1})A^{-1} = AIA^{-1} = AA^{-1} = I$$
$$(A^T)(A^{-1})^T = (A^{-1}A)^T = I^T = I$$

■

Finding the Inverse of a Matrix

For any given square matrix A, we would like to determine whether it has an inverse and, if so, what the inverse is. Fortunately, one procedure answers both questions. We begin by trying to solve the matrix equation $AX = I$ for the unknown square matrix X. If a solution X can be found, then $X = A^{-1}$ by Theorem 3.5.2. If no such matrix X can be found, then A is not invertible.

To keep it simple, the procedure is examined in the case where A is 3×3, but it should be clear that it can be applied to any square matrix. Write the matrix equation $AX = I$ in the form

$$\begin{bmatrix} A\vec{x}_1 & A\vec{x}_2 & A\vec{x}_3 \end{bmatrix} = \begin{bmatrix} \vec{e}_1 & \vec{e}_2 & \vec{e}_3 \end{bmatrix}$$

Hence, we have

$$A\vec{x}_1 = \vec{e}_1, \quad A\vec{x}_2 = \vec{e}_2, \quad A\vec{x}_3 = \vec{e}_3$$

So, it is necessary to solve three systems of equations, one for each column of X. Note that each system has a different standard basis vector as its right-hand side, but all have the same coefficient matrix. Since the procedure for solving systems of equations requires that we row reduce the coefficient matrix, we might as well write out a "triple-augmented matrix" and solve all three systems at once. Therefore, write

$$\left[\begin{array}{c|ccc} A & \vec{e}_1 & \vec{e}_2 & \vec{e}_3 \end{array}\right] = \left[\begin{array}{c|c} A & I \end{array}\right]$$

and row reduce to reduced row echelon form to solve.

Suppose that A is row equivalent to I, so that the reduction gives

$$\left[\begin{array}{c|c} A & I \end{array}\right] \sim \left[\begin{array}{c|ccc} I & \vec{b}_1 & \vec{b}_2 & \vec{b}_3 \end{array}\right]$$

This tells us that $\vec{b}_1$ is a solution of $A\vec{x} = \vec{e}_1$, $\vec{b}_2$ is a solution of $A\vec{x} = \vec{e}_2$, and $\vec{b}_3$ is a solution of $A\vec{x} = \vec{e}_3$. That is, we have

$$A\begin{bmatrix} \vec{b}_1 & \vec{b}_2 & \vec{b}_3 \end{bmatrix} = \begin{bmatrix} A\vec{b}_1 & A\vec{b}_2 & A\vec{b}_3 \end{bmatrix} = \begin{bmatrix} \vec{e}_1 & \vec{e}_2 & \vec{e}_3 \end{bmatrix} = I$$

Thus,

$$A^{-1} = \begin{bmatrix} \vec{b}_1 & \vec{b}_2 & \vec{b}_3 \end{bmatrix}$$

If the reduced row echelon form of A is not I, then $\operatorname{rank}(A) < n$. Hence, A is not invertible, since Theorem 3.5.2 tells us that if A is invertible, then $\operatorname{rank}(A) = n$.

First, we summarize the procedure and then give an example.

Algorithm 3.5.1

Finding A^{-1}

To find the inverse of a square matrix A,

(1) Row reduce the multi-augmented matrix $\left[\begin{array}{c|c} A & I \end{array}\right]$ so that the left block is in reduced row echelon form.
(2) If the reduced row echelon form is $\left[\begin{array}{c|c} I & B \end{array}\right]$, then $A^{-1} = B$.
(3) If the reduced row echelon form of A is not I, then A is not invertible.

EXAMPLE 3.5.3

Determine whether $A = \begin{bmatrix} 1 & 1 & 2 \\ 1 & 2 & 2 \\ 2 & 4 & 3 \end{bmatrix}$ is invertible, and if it is, determine its inverse.

Solution: Write the matrix $\left[\begin{array}{c|c} A & I \end{array}\right]$ and row reduce:

$$\left[\begin{array}{ccc|ccc} 1 & 1 & 2 & 1 & 0 & 0 \\ 1 & 2 & 2 & 0 & 1 & 0 \\ 2 & 4 & 3 & 0 & 0 & 1 \end{array}\right] \sim \left[\begin{array}{ccc|ccc} 1 & 1 & 2 & 1 & 0 & 0 \\ 0 & 1 & 0 & -1 & 1 & 0 \\ 0 & 2 & -1 & -2 & 0 & 1 \end{array}\right] \sim$$

$$\left[\begin{array}{ccc|ccc} 1 & 0 & 2 & 2 & -1 & 0 \\ 0 & 1 & 0 & -1 & 1 & 0 \\ 0 & 0 & -1 & 0 & -2 & 1 \end{array}\right] \sim \left[\begin{array}{ccc|ccc} 1 & 0 & 0 & 2 & -5 & 2 \\ 0 & 1 & 0 & -1 & 1 & 0 \\ 0 & 0 & 1 & 0 & 2 & -1 \end{array}\right]$$

Hence, A is invertible and $A^{-1} = \begin{bmatrix} 2 & -5 & 2 \\ -1 & 1 & 0 \\ 0 & 2 & -1 \end{bmatrix}$.

You should check that the inverse has been correctly calculated by verifying that $AA^{-1} = I$.

EXAMPLE 3.5.4

Let $A \in M_{2\times 2}(\mathbb{R})$. Find a condition on the entries of A that guarantees that A is invertible and then, assuming that A is invertible, find A^{-1}.

Solution: Let $A = \begin{bmatrix} a & b \\ c & d \end{bmatrix}$. For A to be invertible, we require that $\operatorname{rank} A = 2$. Thus, we cannot have $a = 0 = c$. Assume that $a \neq 0$. Row reducing $[A \mid I]$ we get

$$\left[\begin{array}{cc|cc} a & b & 1 & 0 \\ c & d & 0 & 1 \end{array}\right] \begin{array}{c} \\ R_2 - \frac{c}{a}R_1 \end{array} \sim \left[\begin{array}{cc|cc} a & b & 1 & 0 \\ 0 & d - \frac{bc}{a} & -\frac{c}{a} & 1 \end{array}\right] \begin{array}{c} \\ aR_2 \end{array} \sim \left[\begin{array}{cc|cc} a & b & 1 & 0 \\ 0 & ad - bc & -c & a \end{array}\right]$$

Since we need $\operatorname{rank} A = 2$, we now require that $ad - bc \neq 0$. Assuming this, we continue row reducing to get

$$\left[\begin{array}{cc|cc} a & b & 1 & 0 \\ 0 & ad - bc & -c & a \end{array}\right] \sim \left[\begin{array}{cc|cc} 1 & 0 & \frac{d}{ad-bc} & \frac{-b}{ad-bc} \\ 0 & 1 & \frac{-c}{ad-bc} & \frac{a}{ad-bc} \end{array}\right]$$

Thus, we get that A is invertible if and only if $ad - bc \neq 0$. Moreover, if A is invertible, then

$$A^{-1} = \frac{1}{ad - bc}\begin{bmatrix} d & -b \\ -c & a \end{bmatrix}$$

Repeating the argument in the case that $c \neq 0$ gives the same result.

CONNECTION

The quantity $ad - bc$ determines whether a 2×2 matrix is invertible or not. It is called the **determinant** of the matrix. We will examine this in more detail in Chapter 5.

EXAMPLE 3.5.5

Determine whether $A = \begin{bmatrix} 1 & 2 \\ 2 & 4 \end{bmatrix}$ is invertible, and if it is, determine its inverse.

Solution: Write the matrix $\left[\begin{array}{c|c} A & I \end{array}\right]$ and row reduce:

$$\left[\begin{array}{cc|cc} 1 & 2 & 1 & 0 \\ 2 & 4 & 0 & 1 \end{array}\right] \sim \left[\begin{array}{cc|cc} 1 & 2 & 1 & 0 \\ 0 & 0 & -2 & 1 \end{array}\right]$$

Hence, A is not invertible.

EXERCISE 3.5.1

Assume that the stiffness matrix for a spring-mass system is $K = \begin{bmatrix} 6 & -4 \\ -4 & 7 \end{bmatrix}$. Find K^{-1} and use it to find the equilibrium displacements x_1, x_2 when the constant forces are $f_1 = 10$ and $f_2 = 5$. Compare with Example 2.4.1.

Invertible Matrix Theorem

In Theorem 3.5.2 and in the description of the procedure for finding the inverse matrix, we used some facts about systems of equations with square matrices. It is worth stating them clearly as a theorem. This theorem is widely considered as one of the most important theorems in linear algebra since it shows us how a variety of concepts are tied together. Note that most of the conclusions follow from previous results.

Theorem 3.5.4

Invertible Matrix Theorem

If $A \in M_{n\times n}(\mathbb{R})$, then the following statements are equivalent (that is, one is true if and only if each of the others is true).

(1) A is invertible.
(2) $\text{rank}(A) = n$.
(3) The reduced row echelon form of A is I.
(4) For all $\vec{b} \in \mathbb{R}^n$, the system $A\vec{x} = \vec{b}$ is consistent and has a unique solution.
(5) $\text{Null}(A) = \{\vec{0}\}$
(6) $\text{Col}(A) = \mathbb{R}^n$
(7) $\text{Row}(A) = \mathbb{R}^n$
(8) $\text{Null}(A^T) = \{\vec{0}\}$
(9) A^T is invertible

You are asked to prove the Invertible Matrix Theorem in Problem C16.

Amongst many other things, the Invertible Matrix Theorem tells us that if a matrix A is invertible, then the system $A\vec{x} = \vec{b}$ is consistent with a unique solution. In particular, if we multiply both sides of the equation by A^{-1} on the left we get

$$\begin{aligned} A^{-1}A\vec{x} &= A^{-1}\vec{b} \\ I\vec{x} &= A^{-1}\vec{b} \\ \vec{x} &= A^{-1}\vec{b} \end{aligned}$$

EXAMPLE 3.5.6

Let $A = \begin{bmatrix} 1 & 1 \\ 1 & 2 \end{bmatrix}$. Find the solution of $A\vec{x} = \begin{bmatrix} 2 \\ 4 \end{bmatrix}$.

Solution: By the result of Example 3.5.4, we have that $A^{-1} = \begin{bmatrix} 2 & -1 \\ -1 & 1 \end{bmatrix}$.

Thus, the solution of $A\vec{x} = \begin{bmatrix} 2 \\ 4 \end{bmatrix}$ is $\vec{x} = A^{-1}\vec{b} = \begin{bmatrix} 2 & -1 \\ -1 & 1 \end{bmatrix}\begin{bmatrix} 2 \\ 4 \end{bmatrix} = \begin{bmatrix} 0 \\ 2 \end{bmatrix}$.

CONNECTION

It might seem that we solved the system of linear equations in Example 3.5.6 without performing any elementary row operations. However, with some thought, one realizes that the elementary row operations are "contained" inside the inverse of the matrix (we obtained the inverse by row reducing). In the next section, we will see more of the connection between matrix multiplication and elementary row operations.

EXERCISE 3.5.2

Let $A = \begin{bmatrix} 1 & 1 & 1 \\ 1 & -1 & 0 \\ 1 & 2 & 1 \end{bmatrix}$ and $\vec{b} = \begin{bmatrix} 4 \\ 5 \\ 1 \end{bmatrix}$. Determine A^{-1} and use it to solve $A\vec{x} = \vec{b}$.

It likely seems very inefficient to solve Exercise 3.5.2 by the method described. One would think that simply row reducing the augmented matrix of the system would make more sense. However, if we wanted to solve many systems of equations with the same coefficient matrix A, then we would only need to compute A^{-1} once and each system could then be solved by simple matrix-vector multiplication.

Inverse Linear Mappings

It is useful to introduce the **inverse of a linear mapping** here because many geometrical transformations provide nice examples of inverses. Note that just as the inverse matrix is defined only for square matrices, the inverse of a linear mapping is defined only for linear operators. Recall that the identity transformation $\mathrm{Id} : \mathbb{R}^n \to \mathbb{R}^n$ is the linear mapping defined by $\mathrm{Id}(\vec{x}) = \vec{x}$ for all $\vec{x} \in \mathbb{R}^n$.

Definition
Inverse of a Linear Mapping

If $L : \mathbb{R}^n \to \mathbb{R}^n$ is a linear mapping and there exists another linear mapping $M : \mathbb{R}^n \to \mathbb{R}^n$ such that $M \circ L = \mathrm{Id} = L \circ M$, then L is said to be **invertible**, and M is called the **inverse** of L, usually denoted L^{-1}.

Observe that if M is the inverse of L and $L(\vec{v}) = \vec{w}$, then

$$M(\vec{w}) = M(L(\vec{v})) = (M \circ L)(\vec{v}) = \mathrm{Id}(\vec{v}) = \vec{v}$$

Similarly, if $M(\vec{w}) = \vec{v}$, then

$$L(\vec{v}) = L(M(\vec{w})) = (L \circ M)(\vec{w}) = \mathrm{Id}(\vec{w}) = \vec{w}$$

So, we have $L(\vec{v}) = \vec{w}$ if and only if $M(\vec{w}) = \vec{v}$.

Theorem 3.5.5

If $L : \mathbb{R}^n \to \mathbb{R}^n$ is a linear mapping with standard matrix $[L] = A$ and $M : \mathbb{R}^n \to \mathbb{R}^n$ is a linear mapping with standard matrix $[M] = B$, then M is the inverse of L if and only if B is the inverse of A.

Proof: By Theorem 3.2.5, $[M \circ L] = [M][L]$. Hence, $L \circ M = \text{Id} = M \circ L$ if and only if $AB = I = BA$. ■

For many of the geometrical transformations of Section 3.3, an inverse transformation is easily found by geometrical arguments, and these provide many examples of inverse matrices.

EXAMPLE 3.5.7

For each of the following geometrical transformations, determine the inverse transformation. Verify that the product of the standard matrix of the transformation and its inverse is the identity matrix.

(a) The rotation R_θ of the plane

(b) In the plane, a stretch T by a factor of $t > 0$ in the x_1-direction

Solution: (a) The inverse transformation is a rotation by angle $-\theta$. That is, $(R_\theta)^{-1} = R_{-\theta}$. We have

$$[R_{-\theta}] = \begin{bmatrix} \cos(-\theta) & -\sin(-\theta) \\ \sin(-\theta) & \cos(-\theta) \end{bmatrix} = \begin{bmatrix} \cos\theta & \sin\theta \\ -\sin\theta & \cos\theta \end{bmatrix}$$

since $\sin(-\theta) = -\sin\theta$ and $\cos(-\theta) = \cos\theta$. Hence,

$$\begin{aligned} [R_\theta][R_{-\theta}] &= \begin{bmatrix} \cos\theta & -\sin\theta \\ \sin\theta & \cos\theta \end{bmatrix} \begin{bmatrix} \cos\theta & \sin\theta \\ -\sin\theta & \cos\theta \end{bmatrix} \\ &= \begin{bmatrix} \cos^2\theta + \sin^2\theta & \cos\theta\sin\theta - \cos\theta\sin\theta \\ -\sin\theta\cos\theta + \sin\theta\cos\theta & \cos^2\theta + \sin^2\theta \end{bmatrix} = \begin{bmatrix} 1 & 0 \\ 0 & 1 \end{bmatrix} \end{aligned}$$

(b) The inverse transformation T^{-1} is a stretch by a factor of $\frac{1}{t}$ in the x_1-direction:

$$[T][T^{-1}] = \begin{bmatrix} t & 0 \\ 0 & 1 \end{bmatrix} \begin{bmatrix} 1/t & 0 \\ 0 & 1 \end{bmatrix} = \begin{bmatrix} 1 & 0 \\ 0 & 1 \end{bmatrix}$$

CONNECTION

Observe that $[R_\theta]^{-1} = [R_\theta]^T$. This amazing fact is a result of the columns of R_θ being unit vectors that are orthogonal to each other. See Section 1.5 Problem C12. We shall look at other **orthonormal sets** in Chapter 7.

EXERCISE 3.5.3

For each of the following geometrical transformations, determine the inverse transformation. Verify that the product of the standard matrix of the transformation and its inverse is the identity matrix.

(a) A reflection over the line $x_2 = x_1$ in the plane

(b) A shear in the plane by a factor of t in the x_1-direction

---

Observe that if $\vec{y} \in \mathbb{R}^n$ is in the domain of the inverse M, then it must be in the range of the original L. Therefore, it follows that if L has an inverse, the range of L must be all of the codomain $\mathbb{R}^n$. Moreover, if $L(\vec{x}_1) = \vec{y} = L(\vec{x}_2)$, then by applying M to both sides, we have

$$M(L(\vec{x}_1)) = M(L(\vec{x}_2)) \Rightarrow x_1 = x_2$$

Hence, we have shown that for any $\vec{y} \in \mathbb{R}^n$, there exists a unique $\vec{x} \in \mathbb{R}^n$ such that $L(\vec{x}) = \vec{y}$. This property is the linear mapping version of statement (4) of Theorem 3.5.4 about square matrices.

EXAMPLE 3.5.8

Prove that the linear mapping $\text{proj}_{\vec{v}}$ is not invertible for any $\vec{v} \in \mathbb{R}^n$, $n \geq 2$.

Solution: By definition, $\text{proj}_{\vec{v}}(\vec{x}) = \dfrac{\vec{x} \cdot \vec{v}}{\|\vec{v}\|^2}\vec{v}$. So, any vector $\vec{y} \in \mathbb{R}^n$ that is not a scalar multiple of $\vec{v}$ cannot be in the range of $\text{proj}_{\vec{v}}$. Thus, $\text{Range}(\text{proj}_{\vec{v}}) \neq \mathbb{R}^n$, and hence $\text{proj}_{\vec{v}}$ is not invertible.

EXAMPLE 3.5.9

Prove that the linear mapping $L : \mathbb{R}^3 \to \mathbb{R}^3$ defined by

$$L(x_1, x_2, x_3) = (2x_1 + x_2, x_3, x_2 - 2x_3)$$

is invertible.

Solution: Assume that $\vec{x}$ is in the nullspace of L. Then $L(\vec{x}) = \vec{0}$, so by definition of L, we have

$$\begin{aligned} 2x_1 + x_2 &= 0 \\ x_3 &= 0 \\ x_2 - 2x_3 &= 0 \end{aligned}$$

The only solution to this system is $x_1 = x_2 = x_3 = 0$. Thus, $\text{Null}(L) = \{\vec{0}\}$. By Theorem 3.4.6, this implies that $\text{Null}([L]) = \{\vec{0}\}$. Thus, by the Invertible Matrix Theorem $[L]$ is invertible and hence by Theorem 3.5.5, L is also invertible.

Finally, recall that the matrix condition $AB = I = BA$ implies that the matrix inverse can be defined only for square matrices. Here is an example that illustrates for linear mappings that the domain and codomain of L must be the same if it is to have an inverse.

EXAMPLE 3.5.10

Consider the linear mappings $P : \mathbb{R}^4 \to \mathbb{R}^3$ defined by $P(x_1, x_2, x_3, x_4) = (x_1, x_2, x_3)$ and $\text{inj} : \mathbb{R}^3 \to \mathbb{R}^4$ defined by $\text{inj}(x_1, x_2, x_3) = (x_1, x_2, x_3, 0)$.

It is easy to see that $P \circ \text{inj} = \text{Id}$ but that $\text{inj} \circ P \neq \text{Id}$. Thus, P is not an inverse for inj. Notice that P satisfies the condition that its range is all of its codomain, but it fails the condition that its nullspace is trivial. On the other hand, inj satisfies the condition that its nullspace is trivial but fails the condition that its range is all of its codomain.

PROBLEMS 3.5

Practice Problems

For Problems A1–A19, either show that the matrix is not invertible or find its inverse. Check by multiplication.

A1 $\begin{bmatrix} 3 & 8 \\ -3 & -8 \end{bmatrix}$ **A2** $\begin{bmatrix} 4 & 6 \\ -5 & 1 \end{bmatrix}$ **A3** $\begin{bmatrix} 4 & 6 \\ -2 & 3 \end{bmatrix}$

A4 $\begin{bmatrix} 3 & -4 \\ 2 & 5 \end{bmatrix}$ **A5** $\begin{bmatrix} 1 & 0 & 1 \\ 2 & 1 & 3 \\ 1 & 0 & 2 \end{bmatrix}$ **A6** $\begin{bmatrix} 1 & 0 & 2 \\ 1 & 1 & 3 \\ 3 & 1 & 7 \end{bmatrix}$

A7 $\begin{bmatrix} 1 & 1 \\ 2 & 2 \end{bmatrix}$ **A8** $\begin{bmatrix} 2 & -1 & 3 \\ 0 & 2 & 1 \\ 0 & 0 & 1 \end{bmatrix}$ **A9** $\begin{bmatrix} 1 & -2 & 1 \\ 1 & -3 & 2 \\ 1 & 1 & -3 \end{bmatrix}$

A10 $\begin{bmatrix} 0 & 0 & 1 \\ 0 & 1 & 1 \\ 1 & 1 & 1 \end{bmatrix}$ **A11** $\begin{bmatrix} 0 & 2 & 3 \\ 1 & 2 & 3 \\ 1 & -5 & -6 \end{bmatrix}$ **A12** $\begin{bmatrix} 1 & 3 & -3 \\ 2 & 1 & 4 \\ 2 & -1 & 8 \end{bmatrix}$

A13 $\begin{bmatrix} 0 & 1 & 0 \\ 0 & 0 & 1 \\ 1 & 0 & 0 \end{bmatrix}$ **A14** $\begin{bmatrix} 1 & 2 & 3 \\ 4 & 5 & 6 \\ 7 & 8 & 9 \end{bmatrix}$ **A15** $\begin{bmatrix} 1 & 1 & -1 \\ 0 & 1 & 4 \\ 2 & 3 & 2 \end{bmatrix}$

A16 $\begin{bmatrix} 3 & 2 & 4 & 3 \\ 0 & 1 & 0 & 1 \\ 2 & 2 & 4 & 2 \\ 0 & -1 & 1 & -1 \end{bmatrix}$ **A17** $\begin{bmatrix} 1 & 1 & 3 & 1 \\ 0 & 2 & 1 & 0 \\ 2 & 2 & 7 & 1 \\ 0 & 6 & 3 & 1 \end{bmatrix}$

A18 $\begin{bmatrix} 1 & 2 & 1 & 2 \\ -1 & 1 & -7 & 3 \\ 0 & 1 & -2 & 2 \\ -1 & 2 & -9 & 2 \end{bmatrix}$ **A19** $\begin{bmatrix} 1 & 0 & 1 & 0 & 1 \\ 0 & 1 & 0 & 1 & 0 \\ 0 & 0 & 1 & 1 & 1 \\ 0 & 0 & 0 & 1 & 2 \\ 0 & 0 & 0 & 0 & 1 \end{bmatrix}$

A20 Let $B = \begin{bmatrix} 1 & 0 & 1 \\ 2 & 1 & 3 \\ 1 & 0 & 2 \end{bmatrix}$. Find B^{-1} and use it to find the solution of

(a) $B\vec{x} = \begin{bmatrix} 1 \\ 1 \\ 1 \end{bmatrix}$ (b) $B\vec{x} = \begin{bmatrix} -1 \\ 0 \\ 1 \end{bmatrix}$

(c) $B\vec{x} = \left(\begin{bmatrix} 1 \\ 1 \\ 1 \end{bmatrix} + \begin{bmatrix} -1 \\ 0 \\ 1 \end{bmatrix}\right)$ (d) $B\vec{x} = \begin{bmatrix} 3 \\ 1 \\ -2 \end{bmatrix}$

A21 Let $A = \begin{bmatrix} 2 & 1 \\ 3 & 2 \end{bmatrix}$ and $B = \begin{bmatrix} 1 & 2 \\ 3 & 5 \end{bmatrix}$.

(a) Find A^{-1} and B^{-1}.

(b) Calculate AB and $(AB)^{-1}$ and check that $(AB)^{-1} = B^{-1}A^{-1}$.

(c) Calculate $(3A)^{-1}$ and check that it equals $\frac{1}{3}A^{-1}$.

(d) Calculate $(A^T)^{-1}$ and check that $A^T(A^T)^{-1} = I$.

(e) Show that $(A + B)^{-1} \neq A^{-1} + B^{-1}$

A22 The mappings in this problem are from $\mathbb{R}^2 \to \mathbb{R}^2$.

(a) Determine the matrix of the rotation $R_{\pi/6}$ and the matrix of $R_{\pi/6}^{-1}$.

(b) Determine the matrix of a vertical shear S by amount 2 and the matrix of S^{-1}.

(c) Determine the matrix of the reflection R in the line $x_1 - x_2 = 0$ and the matrix of R^{-1}.

(d) Determine the matrix of $(R \circ S)^{-1}$ and the matrix of $(S \circ R)^{-1}$ (without determining the matrices of $R \circ S$ and $S \circ R$).

A23 Suppose that $L : \mathbb{R}^n \to \mathbb{R}^n$ is a linear mapping and that $M : \mathbb{R}^n \to \mathbb{R}^n$ is a function (not assumed to be linear) such that $\vec{x} = M(\vec{y})$ if and only if $\vec{y} = L(\vec{x})$. Show that M is also linear.

For Problems A24–A29, use a geometrical argument to determine the inverse of the matrix. What do you notice about the matrix and its inverse?

A24 $\begin{bmatrix} 1 & -3 \\ 0 & 1 \end{bmatrix}$ **A25** $\begin{bmatrix} 5 & 0 \\ 0 & 1 \end{bmatrix}$ **A26** $\begin{bmatrix} 1 & 0 & 0 \\ 0 & -1 & 0 \\ 0 & 0 & 1 \end{bmatrix}$

A27 $\begin{bmatrix} 1 & 0 \\ 4 & 1 \end{bmatrix}$ **A28** $\begin{bmatrix} 1 & 0 & 0 \\ 0 & 1 & 2 \\ 0 & 0 & 1 \end{bmatrix}$ **A29** $\begin{bmatrix} 1 & 0 & 0 \\ 0 & 0 & 1 \\ 0 & 1 & 0 \end{bmatrix}$

For Problems A30 and A31, find the inverse of the linear mapping.

A30 $L(x_1, x_2) = (2x_1 + 3x_2, x_1 + 5x_2)$

A31 $L(x_1, x_2, x_3) = (x_1 + x_2 + 3x_3, x_2 + x_3, 2x_1 + 3x_2 + 6x_3)$

For Problems A32 and A33, solve the matrix equation for the matrix X.

A32 $\begin{bmatrix} 5 & 7 \\ 2 & 3 \end{bmatrix} X = \begin{bmatrix} 1 & 2 & -1 \\ 3 & 3 & -1 \end{bmatrix}$ **A33** $X\begin{bmatrix} 5 & 7 \\ 2 & 3 \end{bmatrix} = \begin{bmatrix} 1 & 2 \\ 3 & 5 \\ -2 & 1 \end{bmatrix}$

For Problems A34–A36, let $A, B \in M_{n\times n}(\mathbb{R})$. Prove or disprove the statement.

A34 If A is invertible, then the columns of A^T form a linearly independent set.

A35 If A and B are invertible matrices, then $A + B$ is invertible.

A36 If $AX = XC$ and X is invertible, then $A = C$.

Homework Problems

For Problems B1–B16, either show that the matrix is not invertible or find its inverse. Check by multiplication.

B1 $\begin{bmatrix} 2 & 4 \\ 3 & -1 \end{bmatrix}$ **B2** $\begin{bmatrix} -3 & 2 \\ -1 & -1 \end{bmatrix}$ **B3** $\begin{bmatrix} 1 & 2 & 5 \\ 0 & 1 & 3 \\ 2 & 3 & 8 \end{bmatrix}$

B4 $\begin{bmatrix} 2 & 0 \\ 5 & 0 \end{bmatrix}$ **B5** $\begin{bmatrix} 7 & 8 \\ 2 & 0 \end{bmatrix}$ **B6** $\begin{bmatrix} 4 & 0 & -4 \\ 1 & 1 & -1 \\ 6 & 2 & -5 \end{bmatrix}$

B7 $\begin{bmatrix} 6 & -9 \\ -4 & 6 \end{bmatrix}$ **B8** $\begin{bmatrix} 3 & -3 & -3 \\ 0 & -1 & -1 \\ 0 & 1 & 2 \end{bmatrix}$ **B9** $\begin{bmatrix} 1 & 2 & -3 \\ -2 & -3 & 4 \\ 2 & 1 & 0 \end{bmatrix}$

B10 $\begin{bmatrix} 1 & 2 \\ -5 & -8 \end{bmatrix}$ **B11** $\begin{bmatrix} 1 & 0 & 2 \\ 0 & 2 & 0 \\ -2 & 0 & -2 \end{bmatrix}$ **B12** $\begin{bmatrix} 1 & 3 & 1 \\ 2 & 9 & 8 \\ -1 & -3 & -3 \end{bmatrix}$

B13 $\begin{bmatrix} 1 & 0 & 3 & -5 \\ 1 & 1 & 3 & -4 \\ 0 & 0 & 1 & -2 \\ 2 & 2 & 6 & -8 \end{bmatrix}$ **B14** $\begin{bmatrix} 1 & -1 & -1 & 4 \\ 0 & 1 & 1 & -2 \\ -2 & 2 & 3 & -11 \\ 0 & 1 & 3 & -7 \end{bmatrix}$

B15 $\begin{bmatrix} 1 & 1 & 2 & 1 \\ 1 & 2 & 2 & 4 \\ 0 & 1 & 1 & 2 \\ 1 & 1 & 3 & 1 \end{bmatrix}$ **B16** $\begin{bmatrix} 0 & 3 & -1 & 1 \\ 2 & 3 & -1 & 1 \\ 0 & 6 & -3 & 2 \\ 2 & -3 & 2 & 0 \end{bmatrix}$

B17 Let $B = \begin{bmatrix} 0 & 1 & 6 \\ 2 & -1 & -4 \\ 4 & -2 & -6 \end{bmatrix}$. Find B^{-1} and use it to find the solution of $B\vec{x} = \vec{b}$ where:

(a) $\vec{b} = \begin{bmatrix} 1 \\ 1 \\ 1 \end{bmatrix}$ (b) $\vec{b} = \begin{bmatrix} 2 \\ 1 \\ 3 \end{bmatrix}$ (c) $\vec{b} = \begin{bmatrix} -2 \\ -6 \\ 2 \end{bmatrix}$

B18 Let $B = \begin{bmatrix} 3 & 4 & -3 \\ 3 & 8 & -5 \\ -3 & -4 & 4 \end{bmatrix}$. Find B^{-1} and use it to find the solution of $B\vec{x} = \vec{b}$ where:

(a) $\vec{b} = \begin{bmatrix} 4 \\ 12 \\ 6 \end{bmatrix}$ (b) $\vec{b} = \begin{bmatrix} 4 \\ 6 \\ 3 \end{bmatrix}$ (c) $\vec{b} = \begin{bmatrix} -1 \\ 1 \\ 0 \end{bmatrix}$

B19 Let $A = \begin{bmatrix} 4 & 7 \\ 3 & 5 \end{bmatrix}$ and $B = \begin{bmatrix} 0 & 1 \\ 1 & 2 \end{bmatrix}$.

(a) Find A^{-1} and B^{-1}.

(b) Calculate AB and $(AB)^{-1}$ and check that $(AB)^{-1} = B^{-1}A^{-1}$.

(c) Calculate $(2A)^{-1}$ and check that it equals $\frac{1}{2}A^{-1}$.

(d) Calculate $(A^T)^{-1}$ and check that $A^T(A^T)^{-1} = I$.

B20 The mappings in this problem are from $\mathbb{R}^2 \to \mathbb{R}^2$.

(a) Determine the matrix of the rotation $R_{\pi/3}$ and the matrix of $R_{\pi/3}^{-1}$.

(b) Determine the matrix of a contraction T with $t = 1/2$ and the matrix of T^{-1}.

(c) Determine the matrix of the reflection R in the x_1-axis and the matrix of R^{-1}.

(d) Determine the matrix of $(R \circ T)^{-1}$ and the matrix of $(T \circ R)^{-1}$ (without determining the matrices of $R \circ T$ and $T \circ R$).

For Problems B21–B26, use a geometrical argument to determine the inverse of the matrix. What do you notice about the matrix and its inverse?

B21 $\begin{bmatrix} 1 & 0 \\ 0 & -2 \end{bmatrix}$ **B22** $\begin{bmatrix} 1 & 0 \\ 2 & 1 \end{bmatrix}$ **B23** $\begin{bmatrix} 0 & 0 & 1 \\ 0 & 1 & 0 \\ 1 & 0 & 0 \end{bmatrix}$

B24 $\begin{bmatrix} 1 & -3 \\ 0 & 1 \end{bmatrix}$ **B25** $\begin{bmatrix} 1/2 & 0 & 0 \\ 0 & 1 & 0 \\ 0 & 0 & 1 \end{bmatrix}$ **B26** $\begin{bmatrix} 1 & 0 & 0 \\ 0 & 1 & 0 \\ 1 & 0 & 1 \end{bmatrix}$

For Problems B27–B30, find the inverse of the linear mapping.

B27 $L(x_1, x_2) = (x_1 - 4x_2, 2x_1 + 2x_2)$

B28 $L(x_1, x_2) = (-2x_1 - x_2, 7x_1 + 4x_2)$

B29 $L(x_1, x_2, x_3) = (x_1 + 2x_2 + 3x_3, x_1 + 3x_2 + 3x_3, -2x_2 + x_3)$

B30 $L(x_1, x_2, x_3) = (x_2 + 2x_3, x_1 + 2x_2 + 5x_3, -x_1 - x_2 - 2x_3)$

For Problems B31–B38, solve the matrix equation for the matrix X.

B31 $\begin{bmatrix} 2 & 5 \\ 1 & 2 \end{bmatrix} X = \begin{bmatrix} 3 & 1 \\ 4 & 1 \end{bmatrix}$ **B32** $\begin{bmatrix} 3 & 1 \\ 5 & 2 \end{bmatrix} X = \begin{bmatrix} 1 & 0 \\ 1 & 1 \end{bmatrix}$

B33 $X \begin{bmatrix} 4 & 3 \\ 2 & 1 \end{bmatrix} = \begin{bmatrix} 6 & -2 \\ 2 & 4 \end{bmatrix}$ **B34** $X \begin{bmatrix} 6 & -1 \\ 4 & -3 \end{bmatrix} = \begin{bmatrix} 2 & 1 \\ 4 & 2 \end{bmatrix}$

B35 $\begin{bmatrix} 1 & -1 & 1 \\ 1 & 0 & 2 \\ -1 & 2 & 1 \end{bmatrix} X = \begin{bmatrix} 1 & 1 \\ 1 & 1 \\ 2 & 2 \end{bmatrix}$

B36 $X \begin{bmatrix} 2 & 1 & 4 \\ 4 & 3 & 12 \\ 4 & 4 & 18 \end{bmatrix} = \begin{bmatrix} 1 & 0 & 1 \\ -1 & -2 & -3 \end{bmatrix}$

B37 $\begin{bmatrix} 1 & -2 \\ 0 & 1 \end{bmatrix} X \begin{bmatrix} 1 & 1 \\ 1 & 2 \end{bmatrix} = \begin{bmatrix} 2 & 2 \\ 2 & 2 \end{bmatrix}$

B38 $\begin{bmatrix} 1 & 4 \\ 2 & 7 \end{bmatrix} X = X \begin{bmatrix} 0 & 2 \\ 1 & 1 \end{bmatrix}$

Conceptual Problems

C1 Determine an expression in terms of A^{-1} and B^{-1} for $((AB)^T)^{-1}$.

C2 Suppose that $A \in M_{n\times n}(\mathbb{R})$ such that $A^3 = I$. Find an expression for A^{-1} in terms of A. (Hint: find X such that $AX = I$.)

C3 Suppose that B satisfies $B^5 + B^3 + B = I$. Find an expression for B^{-1} in terms of B.

C4 (a) Prove that if A and B are square matrices such that AB is invertible, then A and B are invertible.
(b) Find non-invertible matrices C and D such that CD is invertible.

For Problems C5–C12, recall the following definitions. Let $A \in M_{m\times n}(\mathbb{R})$. If $B \in M_{n\times m}(\mathbb{R})$ such that $AB = I$, then A is called a **left inverse** of B and B is called a **right inverse** of A.

C5 Give a method for either finding a right inverse B of A or showing that A does not have a right inverse.

C6 Prove if A has a right inverse and $m < n$, then A has infinitely many right inverses.

C7 Prove if $m > n$, then A cannot have a right inverse.

C8 Show that a non-square matrix cannot have both a left and a right inverse. [Hint: use Problem C7.]

C9 Find all right inverses of $A = \begin{bmatrix} 3 & 1 & 0 \\ 1 & 2 & 0 \end{bmatrix}$.

C10 Find all right inverses of $A = \begin{bmatrix} 1 & -2 & 1 \\ 1 & -1 & 1 \end{bmatrix}$.

C11 Find all left inverses of $B = \begin{bmatrix} 1 & 1 \\ -2 & -1 \\ 1 & 1 \end{bmatrix}$.

C12 Find all left inverses of $B = \begin{bmatrix} 1 & 2 \\ 0 & -1 \\ 3 & 3 \end{bmatrix}$.

C13 Assume that $A \in M_{n\times n}(\mathbb{R})$ is invertible. Prove that if $\{\vec{v}_1, \dots, \vec{v}_k\}$ is a linearly independent set in $\mathbb{R}^n$, then $\{A\vec{v}_1, \dots, A\vec{v}_k\}$ is also linearly independent.

C14 Assume that $A \in M_{n\times n}(\mathbb{R})$ is invertible. Prove that if $\{\vec{w}_1, \dots, \vec{w}_k\}$ spans $\mathbb{R}^n$, the $\{A\vec{w}_1, \dots, A\vec{w}_k\}$ also spans $\mathbb{R}^n$.

C15 (a) Find $A, B \in M_{n\times n}(\mathbb{R})$ such that $(AB)^{-1} \neq A^{-1}B^{-1}$.
(b) What condition is required on AB so that we do have $(AB)^{-1} = A^{-1}B^{-1}$.

C16 Prove the Invertible Matrix Theorem by proving the following statements in the given order.
(a) (1) if and only if (2)
(b) (2) implies (3)
(c) (3) implies (4)
(d) (4) implies (5)
(e) (5) implies (6)
(f) (6) implies (7)
(g) (7) implies (8)
(h) (8) implies (9)
(i) (9) implies (1)

For Problems C17–C26, let $A, B \in M_{n\times n}(\mathbb{R})$. Prove or disprove the statement.

C17 If P is an invertible matrix such that $P^{-1}AP = B$, then $A = B$.

C18 If $AA = A$ and A is not the zero matrix, then A is invertible.

C19 If AA is invertible, then A is invertible.

C20 If A and B satisfy $AB = O_{n,n}$, then B is not invertible.

C21 There exists a linearly independent set $\{A_1, A_2, A_3, A_4\}$ of 2×2 matrices such that each A_i is invertible.

C22 If A has a column of zeroes, then A is not invertible.

C23 If $A\vec{x} = \vec{0}$ has a unique solution, then $\mathrm{Col}(A) = \mathbb{R}^n$.

C24 If A satisfies $A^2 - 2A + I = O_{n,n}$, then A is invertible.

C25 If A is not invertible, then the columns of A form a linearly dependent set.

C26 If the columns of A form a linearly dependent set, then A is not invertible.

3.6 Elementary Matrices

In Sections 3.1 and 3.5, we saw connections between matrix-vector multiplication and systems of linear equations. In Section 3.2, we observed the connection between linear mappings and matrix-vector multiplication. Since matrix multiplication is an extension of matrix-vector multiplication, it should not be surprising that there is a connection between matrix multiplication, systems of linear equations, and linear mappings. We examine this connection through the use of elementary matrices.

Definition
Elementary Matrix

A matrix that can be obtained from the identity matrix by performing a single elementary row operation is called an **elementary matrix**.

Note that it follows from the definition that an elementary matrix must be square.

EXAMPLE 3.6.1

Find the 2×2 elementary matrix corresponding to each row operation by performing the row operation on the 2×2 identity matrix.

(a) $R_2 - 3R_1$ (b) $3R_2$ (c) $R_1 \updownarrow R_2$

Solution: For (a) we have

$$\begin{bmatrix} 1 & 0 \\ 0 & 1 \end{bmatrix} \; R_2 - 3R_1 \quad \sim \begin{bmatrix} 1 & 0 \\ -3 & 1 \end{bmatrix}$$

Thus, the corresponding elementary matrix is $E_1 = \begin{bmatrix} 1 & 0 \\ -3 & 1 \end{bmatrix}$.

For (b) we have

$$\begin{bmatrix} 1 & 0 \\ 0 & 1 \end{bmatrix} \; 3R_2 \quad \sim \begin{bmatrix} 1 & 0 \\ 0 & 3 \end{bmatrix}$$

Thus, the corresponding elementary matrix is $E_2 = \begin{bmatrix} 1 & 0 \\ 0 & 3 \end{bmatrix}$.

For (c) we have

$$\begin{bmatrix} 1 & 0 \\ 0 & 1 \end{bmatrix} \; R_1 \updownarrow R_2 \quad \sim \begin{bmatrix} 0 & 1 \\ 1 & 0 \end{bmatrix}$$

Thus, the corresponding elementary matrix is $E_3 = \begin{bmatrix} 0 & 1 \\ 1 & 0 \end{bmatrix}$.

CONNECTION

Observe in Example 3.6.1 that the matrix E_1 is the standard matrix of a vertical shear by amount -3, E_2 is the standard matrix of a stretch in the x_2-direction by a factor 3, and E_3 is the standard matrix of a reflection over the line $x_1 = x_2$. It can be shown that every $n \times n$ elementary matrix is the standard matrix of a shear, a stretch, or a reflection.

EXAMPLE 3.6.2

Determine which of the following matrices are elementary. For each elementary matrix, indicate the associated elementary row operation.

(a) $\begin{bmatrix} 1 & 0 & 2 \\ 0 & 1 & 0 \\ 0 & 0 & 1 \end{bmatrix}$ (b) $\begin{bmatrix} 1 & 0 & 0 \\ 0 & 2 & 0 \\ 0 & 0 & 1 \end{bmatrix}$ (c) $\begin{bmatrix} 3 & 0 & 0 \\ 0 & 3 & 0 \\ 0 & 0 & 3 \end{bmatrix}$

Solution: For (a), the matrix is elementary as it can be obtained from the 3×3 identity matrix by performing the row operation $R_1 + 2R_3$.
For (b), the matrix is elementary as it can be obtained from the 3×3 identity matrix by performing the row operation $2R_2$.
For (c), the matrix is not elementary as it would require three elementary row operations to get this matrix from the 3×3 identity matrix.

It is clear that the RREF of every elementary matrix is I; we just need to perform the inverse (opposite) row operation to turn the elementary matrix E back to I. Thus, we not only have that every elementary matrix is invertible, but the inverse is the elementary matrix associated with the reverse row operation.

Theorem 3.6.1

If E is an elementary matrix, then E is invertible and E^{-1} is also an elementary matrix.

EXAMPLE 3.6.3

Find the inverse of each of the following elementary matrices. Check your answer by multiplying the matrices together.

(a) $E_1 = \begin{bmatrix} 1 & 2 \\ 0 & 1 \end{bmatrix}$ (b) $E_2 = \begin{bmatrix} 0 & 0 & 1 \\ 0 & 1 & 0 \\ 1 & 0 & 0 \end{bmatrix}$ (c) $E_3 = \begin{bmatrix} 1 & 0 & 0 \\ 0 & 1 & 0 \\ 0 & 0 & -3 \end{bmatrix}$

Solution: The inverse matrix E_1^{-1} is the elementary matrix associated with the row operation required to bring E_1 back to I. That is, $R_1 - 2R_2$. Therefore,

$$E_1^{-1} = \begin{bmatrix} 1 & -2 \\ 0 & 1 \end{bmatrix}$$

Checking, we get $E_1^{-1}E_1 = \begin{bmatrix} 1 & -2 \\ 0 & 1 \end{bmatrix}\begin{bmatrix} 1 & 2 \\ 0 & 1 \end{bmatrix} = \begin{bmatrix} 1 & 0 \\ 0 & 1 \end{bmatrix}$.

The inverse matrix E_2^{-1} is the elementary matrix associated with the row operation required to bring E_2 back to I. That is, $R_1 \updownarrow R_3$. Therefore,

$$E_2^{-1} = \begin{bmatrix} 0 & 0 & 1 \\ 0 & 1 & 0 \\ 1 & 0 & 0 \end{bmatrix}$$

Checking, we get $E_2^{-1}E_2 = \begin{bmatrix} 0 & 0 & 1 \\ 0 & 1 & 0 \\ 1 & 0 & 0 \end{bmatrix}\begin{bmatrix} 0 & 0 & 1 \\ 0 & 1 & 0 \\ 1 & 0 & 0 \end{bmatrix} = \begin{bmatrix} 1 & 0 & 0 \\ 0 & 1 & 0 \\ 0 & 0 & 1 \end{bmatrix}$.

EXAMPLE 3.6.3 (continued)

The inverse matrix E_3^{-1} is the elementary matrix associated with the row operation required to bring E_3 back to I. That is, $(-1/3)R_3$. Therefore,

$$E_3^{-1} = \begin{bmatrix} 1 & 0 & 0 \\ 0 & 1 & 0 \\ 0 & 0 & -1/3 \end{bmatrix}$$

Checking, we get $E_3^{-1}E_3 = \begin{bmatrix} 1 & 0 & 0 \\ 0 & 1 & 0 \\ 0 & 0 & -1/3 \end{bmatrix}\begin{bmatrix} 1 & 0 & 0 \\ 0 & 1 & 0 \\ 0 & 0 & -3 \end{bmatrix} = \begin{bmatrix} 1 & 0 & 0 \\ 0 & 1 & 0 \\ 0 & 0 & 1 \end{bmatrix}$.

If we look at our calculations in the example above, we see something interesting. Consider an elementary matrix E. We saw that the product EI is the matrix obtained from I by performing the row operations associated with E on I, and that $E^{-1}E$ is the matrix obtained from E by performing the row operation associated with E^{-1} on E. We demonstrate this further with another example.

EXAMPLE 3.6.4

Let $E_1 = \begin{bmatrix} 1 & 0 & 0 \\ 2 & 1 & 0 \\ 0 & 0 & 1 \end{bmatrix}$, $E_2 = \begin{bmatrix} 1 & 0 & 0 \\ 0 & 1 & 0 \\ 0 & 0 & 3 \end{bmatrix}$, and $A = \begin{bmatrix} 1 & 1 & 3 \\ 0 & -1 & 2 \\ 0 & 5 & 6 \end{bmatrix}$. Calculate E_1A and E_2E_1A. Describe the products in terms of matrices obtained from A by elementary row operations.

Solution: By matrix multiplication, we get

$$E_1A = \begin{bmatrix} 1 & 0 & 0 \\ 2 & 1 & 0 \\ 0 & 0 & 1 \end{bmatrix}\begin{bmatrix} 1 & 1 & 3 \\ 0 & -1 & 2 \\ 0 & 5 & 6 \end{bmatrix} = \begin{bmatrix} 1 & 1 & 3 \\ 2 & 1 & 8 \\ 0 & 5 & 6 \end{bmatrix}$$

Observe that E_1A is the matrix obtained from A by performing the row operation $R_2 + 2R_1$. That is, by performing the row operation associated with E_1.

$$E_2E_1A = \begin{bmatrix} 1 & 0 & 0 \\ 0 & 1 & 0 \\ 0 & 0 & 3 \end{bmatrix}\begin{bmatrix} 1 & 1 & 3 \\ 2 & 1 & 8 \\ 0 & 5 & 6 \end{bmatrix} = \begin{bmatrix} 1 & 1 & 3 \\ 2 & 1 & 8 \\ 0 & 15 & 18 \end{bmatrix}$$

Thus, E_2E_1A is the matrix obtained from A by first performing the row operation $R_2 + 2R_1$, and then performing the row operation associated with E_2, namely $3R_3$.

We now state the general theorem.

Theorem 3.6.2 If $A \in M_{m\times n}(\mathbb{R})$ and E is the $m \times m$ elementary matrix corresponding to a certain elementary row operation, then the product EA is the matrix obtained from A by performing the same elementary row operation.

It would be tedious to write the proof of Theorem 3.6.2 in the general $n \times n$ case, so the proof is omitted.

Matrix Decomposition into Elementary Matrices

Consider the system of linear equations $A\vec{x} = \vec{b}$ where A is invertible. We compare our two methods of solving this system. First, we can solve this system using the approach shown in Chapter 2 to row reduce the augmented matrix:

$$\left[A \mid \vec{b}\right] \sim [I \mid \vec{x}]$$

Alternatively, we find A^{-1} using the method in the previous section. We row reduce

$$[A \mid I] \sim \left[I \mid A^{-1}\right]$$

and then solve the system by computing $\vec{x} = A^{-1}\vec{b}$.

Observe that both methods use exactly the same row operations to row reduce A to I. In the first method, we are applying those row operations directly to $\vec{b}$ to determine $\vec{x}$. In the second method, the row operations are being "stored" inside of A^{-1} and the matrix-vector product $A^{-1}\vec{b}$ is "performing" those row operations on $\vec{b}$ so that we get the same answer we obtained in the first method. From our work above, we conjecture that these elementary row operations are stored as elementary matrices.

Theorem 3.6.3 If $A \in M_{m\times n}(\mathbb{R})$ with reduced row echelon form R, then there exists a sequence of elementary matrices, $E_1, E_2, \dots, E_k$, such that

$$E_k \cdots E_2E_1A = R$$

Proof: From our work in Chapter 2, we know that there is a sequence of elementary row operations to bring A into its reduced row echelon form. Call the elementary matrix corresponding to the first operation E_1, the elementary matrix corresponding to the second operation E_2, and so on, until the final elementary row operation corresponds to E_k. By Theorem 3.6.2 we get that E_1A is the matrix obtained by performing the first elementary row operation on A, E_2E_1A is the matrix obtained by performing the second elementary row operation on E_1A (that is, performing the first two elementary row operations on A), and $E_k \cdots E_2E_1A$ is the matrix obtained after performing all of the elementary row operations on A, in the specified order. ■

EXAMPLE 3.6.5 Let $A = \begin{bmatrix} 1 & 2 & 1 \\ 2 & 4 & 4 \end{bmatrix}$. Find a sequence of elementary matrices $E_1, \dots, E_k$ such that $E_k \cdots E_1A$ is the reduced row echelon form of A.

Solution: We row reduce A keeping track of our elementary row operations:

$$\begin{bmatrix} 1 & 2 & 1 \\ 2 & 4 & 4 \end{bmatrix} \begin{matrix} \\ R_2 - 2R_1 \end{matrix} \sim \begin{bmatrix} 1 & 2 & 1 \\ 0 & 0 & 2 \end{bmatrix} \begin{matrix} \\ \frac{1}{2}R_2 \end{matrix} \sim \begin{bmatrix} 1 & 2 & 1 \\ 0 & 0 & 1 \end{bmatrix} \begin{matrix} R_1 - R_2 \\ \\ \end{matrix} \sim \begin{bmatrix} 1 & 2 & 0 \\ 0 & 0 & 1 \end{bmatrix}$$

The first elementary row operation is $R_2 - 2R_1$, so $E_1 = \begin{bmatrix} 1 & 0 \\ -2 & 1 \end{bmatrix}$.

The second elementary row operation is $\frac{1}{2}R_2$, so $E_2 = \begin{bmatrix} 1 & 0 \\ 0 & 1/2 \end{bmatrix}$.

The third elementary row operation is $R_1 - R_2$, so $E_3 = \begin{bmatrix} 1 & -1 \\ 0 & 1 \end{bmatrix}$.

Thus, $E_3E_2E_1A = \begin{bmatrix} 1 & -1 \\ 0 & 1 \end{bmatrix}\begin{bmatrix} 1 & 0 \\ 0 & 1/2 \end{bmatrix}\begin{bmatrix} 1 & 0 \\ -2 & 1 \end{bmatrix}\begin{bmatrix} 1 & 2 & 1 \\ 2 & 4 & 4 \end{bmatrix} = \begin{bmatrix} 1 & 2 & 0 \\ 0 & 0 & 1 \end{bmatrix}$.

Remark

We know that the elementary matrices in Example 3.6.5 must be 2×2 for two reasons. First, we had only two rows in A to perform elementary row operations on, so this must be the same with the corresponding elementary matrices. Second, for the matrix multiplication $E_1 A$ to be defined, we know that the number of columns in E_1 must be equal to the number of rows in A. Also, E_1 is square since it is elementary.

EXERCISE 3.6.1

Let $A = \begin{bmatrix} 1 & 1 \\ 2 & 2 \\ 3 & 2 \end{bmatrix}$. Find a sequence of elementary matrices $E_1, \dots, E_k$ such that $E_k \cdots E_1 A$ is the reduced row echelon form of A.

In the special case where A is an invertible square matrix, the reduced row echelon form of A is I. Hence, by Theorem 3.6.3, there exists a sequence of elementary row operations such that

$$E_k \cdots E_1 A = I$$

Thus, the matrix

$$B = E_k \cdots E_1$$

satisfies $BA = I$, so B is the inverse of A. Observe that this result corresponds exactly to two facts we observed in Section 3.5. First, it demonstrates our procedure for finding the inverse of a matrix by row reducing $\left[\begin{array}{c|c} A & I \end{array}\right]$. Second, it shows that solving a system $A\vec{x} = \vec{b}$ by row reducing or by computing $\vec{x} = A^{-1}\vec{b}$ yields the same result.

Theorem 3.6.4 If A is invertible, then there exists a sequence of elementary matrices $E_1, \dots, E_k$ such that

$$A^{-1} = E_k \cdots E_1$$

and

$$A = E_1^{-1} E_2^{-2} \cdots E_k^{-1}$$

Proof: By Theorem 3.6.3, there exists a sequence of elementary row operations such that $E_k \cdots E_1 A = I$. Thus, by Theorem 3.5.2 we have that

$$A^{-1} = E_k \cdots E_1 \tag{3.8}$$

Taking the inverse of both sides of equation (3.8) gives

$$A = (E_K \cdots E_1)^{-1}$$

Since every elementary matrix is invertible, we can use Theorem 3.5.3 (2) to get

$$A = E_1^{-1} E_2^{-1} \cdots E_k^{-1}$$

as required. ■

CONNECTION

Observe that writing A as a product of simpler matrices is kind of like factoring a polynomial (although it is definitely not the same). This is an example of a **matrix decomposition**. There are many very important matrix decompositions in linear algebra. We will look at a useful matrix decomposition in the next section and a couple more of them later in the book.

EXAMPLE 3.6.6

Let $A = \begin{bmatrix} 1 & 0 & 1 \\ 1 & 0 & 3 \\ -1 & 1 & -1 \end{bmatrix}$. Write A and A^{-1} as products of elementary matrices.

Solution: Row reducing A to RREF gives

$$\begin{bmatrix} 1 & 0 & 1 \\ 1 & 0 & 3 \\ -1 & 1 & -1 \end{bmatrix} \begin{matrix} \\ R_2 - R_1 \\ R_3 + R_1 \end{matrix} \sim \begin{bmatrix} 1 & 0 & 1 \\ 0 & 0 & 2 \\ 0 & 1 & 0 \end{bmatrix} \begin{matrix} \\ \frac{1}{2}R_2 \\ \\ \end{matrix} \sim$$

$$\begin{bmatrix} 1 & 0 & 1 \\ 0 & 0 & 1 \\ 0 & 1 & 0 \end{bmatrix} \begin{matrix} \\ \\ R_2 \updownarrow R_3 \end{matrix} \sim \begin{bmatrix} 1 & 0 & 1 \\ 0 & 1 & 0 \\ 0 & 0 & 1 \end{bmatrix} \begin{matrix} R_1 - R_3 \\ \\ \\ \end{matrix} \sim \begin{bmatrix} 1 & 0 & 0 \\ 0 & 1 & 0 \\ 0 & 0 & 1 \end{bmatrix}$$

Hence,

$$E_1 = \begin{bmatrix} 1 & 0 & 0 \\ -1 & 1 & 0 \\ 0 & 0 & 1 \end{bmatrix} E_2 = \begin{bmatrix} 1 & 0 & 0 \\ 0 & 1 & 0 \\ 1 & 0 & 1 \end{bmatrix} E_3 = \begin{bmatrix} 1 & 0 & 0 \\ 0 & 1/2 & 0 \\ 0 & 0 & 1 \end{bmatrix} E_4 = \begin{bmatrix} 1 & 0 & 0 \\ 0 & 0 & 1 \\ 0 & 1 & 0 \end{bmatrix} E_5 = \begin{bmatrix} 1 & 0 & -1 \\ 0 & 1 & 0 \\ 0 & 0 & 1 \end{bmatrix}$$

and $E_5E_4E_3E_2E_1A = I$. Therefore,

$$A^{-1} = E_5E_4E_3E_2E_1 = \begin{bmatrix} 1 & 0 & -1 \\ 0 & 1 & 0 \\ 0 & 0 & 1 \end{bmatrix} \begin{bmatrix} 1 & 0 & 0 \\ 0 & 0 & 1 \\ 0 & 1 & 0 \end{bmatrix} \begin{bmatrix} 1 & 0 & 0 \\ 0 & 1/2 & 0 \\ 0 & 0 & 1 \end{bmatrix} \begin{bmatrix} 1 & 0 & 0 \\ 0 & 1 & 0 \\ 1 & 0 & 1 \end{bmatrix} \begin{bmatrix} 1 & 0 & 0 \\ -1 & 1 & 0 \\ 0 & 0 & 1 \end{bmatrix}$$

and

$$A = E_1^{-1}E_2^{-1}E_3^{-1}E_4^{-1}E_5^{-1} = \begin{bmatrix} 1 & 0 & 0 \\ 1 & 1 & 0 \\ 0 & 0 & 1 \end{bmatrix} \begin{bmatrix} 1 & 0 & 0 \\ 0 & 1 & 0 \\ -1 & 0 & 1 \end{bmatrix} \begin{bmatrix} 1 & 0 & 0 \\ 0 & 2 & 0 \\ 0 & 0 & 1 \end{bmatrix} \begin{bmatrix} 1 & 0 & 0 \\ 0 & 0 & 1 \\ 0 & 1 & 0 \end{bmatrix} \begin{bmatrix} 1 & 0 & 1 \\ 0 & 1 & 0 \\ 0 & 0 & 1 \end{bmatrix}$$

PROBLEMS 3.6

Practice Problems

For Problems A1–A5, write the 3×3 elementary matrix that corresponds to the elementary row operation. Let $A = \begin{bmatrix} 1 & 2 & 3 \\ -1 & 3 & 4 \\ 4 & 2 & 0 \end{bmatrix}$ and verify that the product EA is the matrix obtained from A by performing the elementary row operation on A.

A1 Add -5 times the second row to the first row.

A2 Swap the second and third rows.

A3 Multiply the third row by -1.

A4 Multiply the second row by 6.

A5 Add 4 times the first row to the third row.

For Problems A6–A13, write the 4×4 elementary matrix that corresponds to the elementary row operation, and write the corresponding inverse elementary matrix.

A6 Add -3 times the third row to the fourth row.

A7 Swap the second and fourth rows.

A8 Multiply the third row by -3.

A9 Add 2 times the first row to the third row.

A10 Multiply the first row by 3.

A11 Swap the first and third rows.

A12 Add 1 times the second row to the first row.

A13 Add -3 times the fourth row to the first row.

For Problems A14–A22, either state that the matrix is elementary and state the corresponding elementary row operation, or explain why the matrix is not elementary.

A14 $\begin{bmatrix} 5 & 0 \\ 0 & 1 \end{bmatrix}$ **A15** $\begin{bmatrix} 1 & 0 \\ 1 & 0 \end{bmatrix}$ **A16** $\begin{bmatrix} 1 & 2 \\ 0 & 1 \end{bmatrix}$

A17 $\begin{bmatrix} 1 & 0 & 0 \\ 0 & 1 & 0 \\ 0 & -4 & 1 \end{bmatrix}$ **A18** $\begin{bmatrix} -1 & 0 & 0 \\ 0 & 1 & 0 \\ 0 & 0 & -1 \end{bmatrix}$ **A19** $\begin{bmatrix} 3 & 0 & 1 \\ 0 & 1 & 0 \\ 0 & 0 & 1 \end{bmatrix}$

A20 $\begin{bmatrix} 0 & 0 & 1 \\ 0 & 1 & 0 \\ 1 & 0 & 0 \end{bmatrix}$ **A21** $\begin{bmatrix} 0 & 1 & 0 \\ 0 & 0 & 1 \\ 1 & 0 & 0 \end{bmatrix}$ **A22** $\begin{bmatrix} 1 & 0 & 0 \\ 0 & 1 & 0 \\ 0 & 0 & 1 \end{bmatrix}$

For Problems A23–A28:

(a) Find a sequence of elementary matrices $E_k, \dots, E_1$ such that $E_k \cdots E_1 A = I$.

(b) Determine A^{-1} by computing $E_k \cdots E_1$.

(c) Write A as a product of elementary matrices.

A23 $A = \begin{bmatrix} 1 & 3 & 4 \\ 0 & 0 & 2 \\ 0 & 1 & 0 \end{bmatrix}$ **A24** $A = \begin{bmatrix} 1 & 2 & 2 \\ 0 & 1 & 3 \\ 2 & 4 & 5 \end{bmatrix}$

A25 $A = \begin{bmatrix} 2 & 0 & -4 \\ -2 & 1 & 4 \\ 3 & -1 & -5 \end{bmatrix}$ **A26** $A = \begin{bmatrix} 0 & 3 & 1 \\ 1 & 6 & 2 \\ 1 & 3 & 2 \end{bmatrix}$

A27 $A = \begin{bmatrix} 1 & -2 & 4 \\ -1 & 3 & -4 \\ 0 & 1 & 2 \end{bmatrix}$ **A28** $A = \begin{bmatrix} 1 & 0 & -1 \\ -2 & 0 & -2 \\ -4 & 1 & 4 \end{bmatrix}$

Homework Problems

For Problems B1–B6, write the 3×3 elementary matrix that corresponds to the elementary row operation. Let $A = \begin{bmatrix} 1 & 0 & 1 \\ 0 & 2 & -1 \\ 1 & 3 & -1 \end{bmatrix}$ and verify that the product EA is the matrix obtained from A by performing the elementary row operation on A.

B1 Swap the first and third rows.

B2 Add -2 times the first row to the second row.

B3 Multiply the third row by $1/2$.

B4 Multiply the first row by -1.

B5 Add 6 times the third row to the first row.

B6 Add 1 times the third row to the second row.

For Problems B7–B16, write the 3×3 elementary matrix that corresponds to the elementary row operation, and write the corresponding inverse elementary matrix.

B7 Add 3 times the third row to the first row.

B8 Swap the first and second rows.

B9 Swap the first and third rows.

B10 Add -4 times the first row to the second row.

B11 Add 2 times the second row to the third row.

B12 Multiply the third row by 2.

B13 Multiply the first row by $-1/3$.

B14 Add -1 times the first row to the third row.

B15 Swap the second and third rows.

B16 Multiply the second row by -3.

For Problems B17–B22, either state that the matrix is elementary and state the corresponding elementary row operation, or explain why the matrix is not elementary.

B17 $\begin{bmatrix} 1 & 0 & -1/2 \\ 0 & 1 & 0 \\ 0 & 0 & 1 \end{bmatrix}$

B18 $\begin{bmatrix} 2 & 0 & 0 \\ 0 & 2 & 0 \\ 0 & 0 & 2 \end{bmatrix}$

B19 $\begin{bmatrix} 1 & 0 & 1 \\ 0 & 1 & 0 \\ 1 & 0 & 1 \end{bmatrix}$

B20 $\begin{bmatrix} 0 & 1 & 0 \\ 1 & 0 & 0 \\ 0 & 0 & 1 \end{bmatrix}$

B21 $\begin{bmatrix} 1 & 0 & 0 \\ 0 & 1 & 0 \\ -2 & 0 & 0 \end{bmatrix}$

B22 $\begin{bmatrix} 1 & 0 & 0 \\ 0 & 3 & 0 \\ 0 & 0 & 1 \end{bmatrix}$

B23 Without using matrix-matrix multiplication, calculate

$$\begin{bmatrix} 1 & 2 \\ 0 & 1 \end{bmatrix}\begin{bmatrix} 1/2 & 0 \\ 0 & 1 \end{bmatrix}\begin{bmatrix} 0 & 1 \\ 1 & 0 \end{bmatrix}\begin{bmatrix} 1 & 0 \\ -3 & 1 \end{bmatrix}\begin{bmatrix} 1 & 0 \\ 0 & 3 \end{bmatrix}\begin{bmatrix} 1 & 1 \\ 0 & 1 \end{bmatrix}$$

For Problems B24–B31:

(a) Find a sequence of elementary matrices $E_k, \dots, E_1$ such that $E_k \cdots E_1 A = I$.

(b) Determine A^{-1} by computing $E_k \cdots E_1$.

(c) Write A as a product of elementary matrices.

B24 $A = \begin{bmatrix} 2 & 1 \\ 3 & 3 \end{bmatrix}$

B25 $A = \begin{bmatrix} 0 & 3 \\ 1 & -2 \end{bmatrix}$

B26 $A = \begin{bmatrix} 4 & 2 \\ 2 & 2 \end{bmatrix}$

B27 $A = \begin{bmatrix} 1 & 3 \\ -3 & 1 \end{bmatrix}$

B28 $A = \begin{bmatrix} 1 & 1 & 3 \\ 1 & 2 & 2 \\ 1 & 1 & 1 \end{bmatrix}$

B29 $A = \begin{bmatrix} 4 & 0 & -4 \\ 1 & 1 & -1 \\ 6 & 2 & -5 \end{bmatrix}$

B30 $A = \begin{bmatrix} 1 & 2 & 5 \\ 0 & 1 & 3 \\ 2 & 3 & 8 \end{bmatrix}$

B31 $A = \begin{bmatrix} 0 & 2 & 6 \\ 1 & 4 & 4 \\ -1 & 2 & 8 \end{bmatrix}$

Conceptual Problems

C1 (a) Let $L : \mathbb{R}^2 \to \mathbb{R}^2$ be the invertible linear operator with standard matrix $A = \begin{bmatrix} 0 & -2 \\ 1 & -4 \end{bmatrix}$. By writing A as a product of elementary matrices, show that L can be written as a composition of shears, stretches, and reflections.

(b) Explain how we know that every invertible linear operator $L : \mathbb{R}^n \to \mathbb{R}^n$ can be written as a composition of shears, stretches, and reflections.

C2 Let $A = \begin{bmatrix} 1 & 2 \\ 2 & 6 \end{bmatrix}$ and $\vec{b} = \begin{bmatrix} 3 \\ 5 \end{bmatrix}$.

(a) Determine elementary matrices E_1, E_2, and E_3 such that $E_3E_2E_1A = I$.

(b) Since A is invertible, we know that the system $A\vec{x} = \vec{b}$ has unique solution

$$\vec{x} = A^{-1}\vec{b} = E_3E_2E_1\vec{b}$$

Calculate the solution $\vec{x}$ without using matrix-matrix multiplication.

(c) Solve the system $A\vec{x} = \vec{b}$ by row reducing $\left[A \mid \vec{b}\right]$. Compare the operations that you use on the augmented part of the system with the operations in part (b).

C3 Let $A \in M_{m\times n}(\mathbb{R})$ with reduced row echelon form R.

(a) Use elementary matrices to prove that there exists an invertible matrix E such that $EA = R$.

(b) Is the matrix E in part (a) unique?

C4 We have seen that multiplying an $m \times n$ matrix A on the left by an elementary matrix has the same effect as performing the elementary row operation corresponding to the elementary matrix on A. If E is an $n \times n$ elementary matrix, then what effect does multiplying by A on the right by E have? Justify your answer.

For Problems C5 and C6, observe that we can now say that A is **row equivalent** to B if there exists a sequence of elementary matrices $E_1, \dots, E_K$ such that $E_k \cdots E_1 A = B$.

C5 Prove that if A is row equivalent to a matrix B, then B is row equivalent to A.

C6 Prove that if A is row equivalent to a matrix B and C is also row equivalent to B, then A is row equivalent to C.

3.7 *LU*-Decomposition

One of the most basic and useful ideas in mathematics is the concept of a factorization of an object. You have already seen that it can be very useful to factor a number into primes or to factor a polynomial. Similarly, in many applications of linear algebra, we may want to decompose a matrix into factors that have certain properties.

In applied linear algebra, we often need to quickly solve multiple systems $A\vec{x} = \vec{b}$, where the coefficient matrix A remains the same but the vector $\vec{b}$ changes. The goal of this section is to derive a matrix factorization called the LU-decomposition, which is commonly used in computer algorithms to solve such problems.

We now start our look at the LU-decomposition by recalling the definition of upper triangular and lower triangular matrices.

Definition
Upper Triangular
Lower Triangular

A matrix $U \in M_{n\times n}(\mathbb{R})$ is said to be **upper triangular** if the entries beneath the main diagonal are all zero—that is, $(U)_{ij} = 0$ whenever $i > j$. A matrix $L \in M_{n\times n}(\mathbb{R})$ is said to be **lower triangular** if the entries above the main diagonal are all zero—in particular, $(L)_{ij} = 0$ whenever $i < j$.

Observe that for any system $A\vec{x} = \vec{b}$ with the same coefficient matrix A, we can use the same row operations to row reduce $\left[\begin{array}{c|c} A & \vec{b} \end{array}\right]$ to REF. Hence, the only difference between solving the systems $\left[\begin{array}{c|c} A & \vec{b}_1 \end{array}\right]$ and $\left[\begin{array}{c|c} A & \vec{b}_2 \end{array}\right]$ will then be the effect of the row operations on $\vec{b}_1$ and $\vec{b}_2$. In particular, we see that the two important pieces of information we require are an REF of A and the elementary row operations used.

For our purposes, we will assume that our $n\times n$ coefficient matrix A can be brought into REF using only elementary row operations of the form $R_i + sR_j$, where $i > j$. Since we can row reduce a matrix to a REF without multiplying a row by a non-zero constant, omitting this row operation is not a problem. However, omitting row interchanges may seem rather serious: without row interchanges, we cannot bring a matrix such as $\begin{bmatrix} 0 & 1 \\ 1 & 2 \end{bmatrix}$ into REF using only $R_i + sR_j$, where $i > j$. However, we omit row interchanges only because it is difficult to keep track of them by hand. A computer can keep track of row interchanges without physically moving entries from one location to another. At the end of the section, we will comment on the case where swapping rows is required.

Row operations of the form $R_i + sR_j$, where $i > j$, have a corresponding elementary matrix that is lower triangular and has ones along the main diagonal. So, under our assumption about A, there are elementary matrices $E_1, \dots, E_k$ that are all lower triangular such that

$$E_k \cdots E_1 A = U$$

where U is a REF of A. Since $E_k \cdots E_1$ is invertible, we can write $A = (E_k \cdots E_1)^{-1} U$ and define

$$L = (E_k \cdots E_1)^{-1} = E_1^{-1} \cdots E_k^{-1}$$

We get that L is lower triangular because the inverse of a lower triangular elementary matrix is lower triangular, and a product of lower triangular matrices is lower triangular (see Problem A1). Additionally, by definition of REF, U is upper triangular. Consequently, we have a matrix decomposition $A = LU$, where U is upper triangular and L is lower triangular. Moreover, L contains the information about the row operations used to bring A to U.

Theorem 3.7.1 If $A \in M_{n\times n}(\mathbb{R})$ can be row reduced to REF without swapping rows, then there exists an upper triangular matrix U and lower triangular matrix L such that $A = LU$.

Definition
LU-**Decomposition**

Writing $A \in M_{n\times n}(\mathbb{R})$ as a product LU, where L is lower triangular and U is upper triangular, is called an ***LU*-decomposition** of A.

Our derivation has given an algorithm for finding an LU-decomposition of a matrix that can be row reduced to REF using only operations of the form $R_i + sR_j$, where $i > j$.

EXAMPLE 3.7.1

Find an LU-decomposition of $A = \begin{bmatrix} 2 & -1 & 4 \\ 4 & -1 & 6 \\ -1 & -1 & 3 \end{bmatrix}$.

Solution: Row reducing and keeping track of our row operations gives

$$\begin{bmatrix} 2 & -1 & 4 \\ 4 & -1 & 6 \\ -1 & -1 & 3 \end{bmatrix} \begin{matrix} \\ R_2 - 2R_1 \\ R_3 + \frac{1}{2}R_1 \end{matrix} \sim \begin{bmatrix} 2 & -1 & 4 \\ 0 & 1 & -2 \\ 0 & -3/2 & 5 \end{bmatrix} \begin{matrix} \\ \\ R_3 + \frac{3}{2}R_2 \end{matrix} \sim \begin{bmatrix} 2 & -1 & 4 \\ 0 & 1 & -2 \\ 0 & 0 & 2 \end{bmatrix} = U$$

We have $E_3E_2E_1A = U$, so $A = E_1^{-1}E_2^{-1}E_3^{-1}U$, where

$$E_1 = \begin{bmatrix} 1 & 0 & 0 \\ -2 & 1 & 0 \\ 0 & 0 & 1 \end{bmatrix}, E_2 = \begin{bmatrix} 1 & 0 & 0 \\ 0 & 1 & 0 \\ 1/2 & 0 & 1 \end{bmatrix}, E_3 = \begin{bmatrix} 1 & 0 & 0 \\ 0 & 1 & 0 \\ 0 & 3/2 & 1 \end{bmatrix}$$

Hence, we let

$$L = E_1^{-1}E_2^{-1}E_3^{-1} = \begin{bmatrix} 1 & 0 & 0 \\ 2 & 1 & 0 \\ 0 & 0 & 1 \end{bmatrix}\begin{bmatrix} 1 & 0 & 0 \\ 0 & 1 & 0 \\ -1/2 & 0 & 1 \end{bmatrix}\begin{bmatrix} 1 & 0 & 0 \\ 0 & 1 & 0 \\ 0 & -3/2 & 1 \end{bmatrix}$$

$$= \begin{bmatrix} 1 & 0 & 0 \\ 2 & 1 & 0 \\ 0 & 0 & 1 \end{bmatrix}\begin{bmatrix} 1 & 0 & 0 \\ 0 & 1 & 0 \\ -1/2 & -3/2 & 1 \end{bmatrix} = \begin{bmatrix} 1 & 0 & 0 \\ 2 & 1 & 0 \\ -1/2 & -3/2 & 1 \end{bmatrix}$$

And we get $A = LU$.

Observe from this example that the entries in L that are beneath the main diagonal are just the negative of the multipliers used to put a zero in the corresponding entry of U. That is, if we use the operation $R_i + cR_j$ to put a 0 in the i, j-th entry of U, then the i, j-th entry of L is $-c$. To see why this is the case, observe that if $E_k \cdots E_1A = U$, then

$$(E_k \cdots E_1)L = (E_k \cdots E_1)(E_k \cdots E_1)^{-1} = I$$

Hence, the same row operations that reduce A to U will reduce L to I.

Also note that the diagonal entries of L will all be 1s, and, since L is to be lower triangular, all the entries above the main diagonal are all 0s. These observations makes the LU-decomposition extremely easy to find.

EXAMPLE 3.7.2

Find an LU-decomposition of $B = \begin{bmatrix} 2 & 1 & -1 \\ -4 & 3 & 3 \\ 6 & 8 & -3 \end{bmatrix}$.

Solution: By row reducing, we get

$$\begin{bmatrix} 2 & 1 & -1 \\ -4 & 3 & 3 \\ 6 & 8 & -3 \end{bmatrix} \begin{matrix} \\ R_2 + 2R_1 \\ R_3 - 3R_1 \end{matrix} \sim \begin{bmatrix} 2 & 1 & -1 \\ 0 & 5 & 1 \\ 0 & 5 & 0 \end{bmatrix} \begin{matrix} \\ \\ R_3 - R_2 \end{matrix} \sim \begin{bmatrix} 2 & 1 & -1 \\ 0 & 5 & 1 \\ 0 & 0 & -1 \end{bmatrix} = U$$

We used the multiplier 2 to get a 0 in the 2, 1-entry of U, so $(L)_{21} = -2$.
We used the multiplier -3 to get a 0 in the 3, 1-entry of U, so $(L)_{31} = 3$.
We used the multiplier -1 to get a 0 in the 3, 2-entry of U, we $(L)_{32} = 1$.
Thus,

$$L = \begin{bmatrix} 1 & 0 & 0 \\ -2 & 1 & 0 \\ 3 & 1 & 1 \end{bmatrix}$$

We can easily verify that $B = LU$.

EXAMPLE 3.7.3

Find an LU-decomposition of $C = \begin{bmatrix} 1 & 2 & -3 \\ 2 & 2 & 3 \\ -4 & -2 & 1 \end{bmatrix}$.

Solution: By row reducing, we get

$$\begin{bmatrix} 1 & 2 & -3 \\ 2 & 2 & 3 \\ -4 & -2 & 1 \end{bmatrix} \begin{matrix} \\ R_2 - 2R_1 \\ R_3 + 4R_1 \end{matrix} \sim \begin{bmatrix} 1 & 2 & -3 \\ 0 & -2 & 9 \\ 0 & 6 & -11 \end{bmatrix} \begin{matrix} \\ \\ R_3 + 3R_2 \end{matrix} \sim \begin{bmatrix} 1 & 2 & -3 \\ 0 & -2 & 9 \\ 0 & 0 & 16 \end{bmatrix} = U$$

From our elementary row operations, we get that

$$\begin{aligned} (L)_{21} &= 2 \\ (L)_{31} &= -4 \\ (L)_{32} &= -3 \end{aligned}$$

Hence,

$$L = \begin{bmatrix} 1 & 0 & 0 \\ 2 & 1 & 0 \\ -4 & -3 & 1 \end{bmatrix}$$

We then have $C = LU$.

It is important to note that the method we are using to find L only works when we row reduce the matrix by in order from the leftmost column to the rightmost, and from top to bottom within each column. We also must always be using the appropriate diagonal entry as the pivot.

EXERCISE 3.7.1

Find an LU-decomposition of $A = \begin{bmatrix} -1 & 1 & 2 \\ 4 & -1 & -3 \\ -3 & -3 & 1 \end{bmatrix}$.

Solving Systems with the *LU*-Decomposition

We now look at how to use the LU-decomposition to solve the system $A\vec{x} = \vec{b}$. If $A = LU$, the system can be written as

$$LU\vec{x} = \vec{b}$$

Letting $\vec{y} = U\vec{x}$, we can write $LU\vec{x} = \vec{b}$ as two systems:

$$L\vec{y} = \vec{b} \qquad \text{and} \qquad U\vec{x} = \vec{y}$$

which both have triangular coefficient matrices. This allows us to solve both systems immediately, using substitution. In particular, since L is lower triangular, we use forward-substitution to solve $\vec{y}$ and then solve $U\vec{x} = \vec{y}$ for $\vec{x}$ using back-substitution.

Remark

Observe that the first system is really calculating how performing the row operations on A would have affected $\vec{b}$.

EXAMPLE 3.7.4

Let $B = \begin{bmatrix} 2 & 1 & -1 \\ -4 & 3 & 3 \\ 6 & 8 & -3 \end{bmatrix}$ and $\vec{b} = \begin{bmatrix} 3 \\ -13 \\ 4 \end{bmatrix}$. Use an LU-decomposition of B to solve $B\vec{x} = \vec{b}$.

Solution: In Example 3.7.2 we found an LU-decomposition of B. We write $B\vec{x} = \vec{b}$ as $LU\vec{x} = \vec{b}$ and take $\vec{y} = U\vec{x}$. Writing out the system $L\vec{y} = \vec{b}$, we get

$$\begin{aligned} y_1 &= 3 \\ -2y_1 + y_2 &= -13 \\ 3y_1 + y_2 + y_3 &= 4 \end{aligned}$$

Using forward-substitution, we find that $y_1 = 3$, so $y_2 = -13 + 2(3) = -7$ and $y_3 = 4 - 3(3) - (-7) = 2$. Hence, $\vec{y} = \begin{bmatrix} 3 \\ -7 \\ 2 \end{bmatrix}$.

Thus, our system $U\vec{x} = \vec{y}$ is

$$\begin{aligned} 2x_1 + x_2 - x_3 &= 3 \\ 5x_2 + x_3 &= -7 \\ -x_3 &= 2 \end{aligned}$$

Using back-substitution, we get $x_3 = -2, 5x_2 = -7 - (-2) \Rightarrow x_2 = -1$ and $2x_1 = 3 - (-1) + (-2) \Rightarrow x_1 = 1$. Thus, the solution is $\vec{x} = \begin{bmatrix} 1 \\ -1 \\ -2 \end{bmatrix}$.

EXAMPLE 3.7.5

Let $A = \begin{bmatrix} 1 & 1 & 1 \\ -1 & -2 & 3 \\ -2 & -4 & 6 \end{bmatrix}$ and $\vec{b} = \begin{bmatrix} 1 \\ 6 \\ 12 \end{bmatrix}$. Use an LU-decomposition to solve $A\vec{x} = \vec{b}$.

Solution: We first find an LU-decomposition for A. Row reducing gives

$$\begin{bmatrix} 1 & 1 & 1 \\ -1 & -2 & 3 \\ -2 & -4 & 6 \end{bmatrix} \begin{matrix} \\ R_2 + R_1 \\ R_3 + 2R_1 \end{matrix} \sim \begin{bmatrix} 1 & 1 & 1 \\ 0 & -1 & 4 \\ 0 & -2 & 8 \end{bmatrix} \begin{matrix} \\ \\ R_3 - 2R_2 \end{matrix} \sim \begin{bmatrix} 1 & 1 & 1 \\ 0 & -1 & 4 \\ 0 & 0 & 0 \end{bmatrix} = U$$

From our elementary row operations, we find that $L = \begin{bmatrix} 1 & 0 & 0 \\ -1 & 1 & 0 \\ -2 & 2 & 1 \end{bmatrix}$.

We let $\vec{y} = U\vec{x}$ and solve $L\vec{y} = \vec{b}$. This gives

$$\begin{aligned} y_1 &= 1 \\ -y_1 + y_2 &= 6 \\ -2y_1 + 2y_2 + y_3 &= 12 \end{aligned}$$

Hence, $y_1 = 1$, $y_2 = 6 + 1 = 7$, and $y_3 = 12 + 2 - 14 = 0$. Next we solve $U\vec{x} = \begin{bmatrix} 1 \\ 7 \\ 0 \end{bmatrix}$.

$$\begin{aligned} x_1 + x_2 + x_3 &= 1 \\ -x_2 + 4x_3 &= 7 \\ 0x_3 &= 0 \end{aligned}$$

This gives $x_3 = t \in \mathbb{R}$, $x_2 = -7 + 4t$, and $x_1 = 1 + (7 - 4t) - t = 8 - 5t$. Thus,

$$\vec{x} = \begin{bmatrix} 8 - 5t \\ -7 + 4t \\ t \end{bmatrix} = \begin{bmatrix} 8 \\ -7 \\ 0 \end{bmatrix} + t \begin{bmatrix} -5 \\ 4 \\ 1 \end{bmatrix}, \quad t \in \mathbb{R}$$

EXERCISE 3.7.2

Let $A = \begin{bmatrix} -1 & 1 & 2 \\ 4 & -1 & -3 \\ -3 & -3 & 1 \end{bmatrix}$, $\vec{b}_1 = \begin{bmatrix} 3 \\ 2 \\ 6 \end{bmatrix}$ and $\vec{b}_2 = \begin{bmatrix} 8 \\ 2 \\ -9 \end{bmatrix}$. Use the LU-decomposition of A that you found in Exercise 3.7.1 to solve the system $A\vec{x}_i = \vec{b}_i$, for $i = 1, 2$.

A Comment About Swapping Rows

For any $A \in M_{n\times n}(\mathbb{R})$, we can first rearrange the rows of A to get a matrix that has an LU-factorization. In particular, for every matrix A, there exists a matrix P, called a permutation matrix, that can be obtained by only performing row swaps on the identity matrix such that

$$PA = LU$$

Then, we can solve $A\vec{x} = \vec{b}$ by finding the solutions of

$$PA\vec{x} = P\vec{b}$$

PROBLEMS 3.7

Practice Problems

A1 (a) Prove that the inverse of a lower triangular elementary matrix is lower triangular.
(b) Prove that a product of lower triangular matrices is lower triangular.

For Problems A2–A7, find an LU-decomposition for the matrix.

A2 $\begin{bmatrix} -2 & -1 & 5 \\ -4 & 0 & -2 \\ 2 & 1 & 3 \end{bmatrix}$

A3 $\begin{bmatrix} 1 & -2 & 4 \\ 3 & -2 & 4 \\ 2 & 2 & -5 \end{bmatrix}$

A4 $\begin{bmatrix} 2 & -4 & 5 \\ 2 & 5 & 2 \\ 2 & -1 & 5 \end{bmatrix}$

A5 $\begin{bmatrix} 1 & 5 & 3 & 4 \\ -2 & -6 & -1 & 3 \\ 0 & 2 & -1 & -1 \\ 0 & 0 & 0 & 0 \end{bmatrix}$

A6 $\begin{bmatrix} 1 & -2 & 1 & 1 \\ 0 & -3 & -2 & 1 \\ 3 & -3 & 2 & -1 \\ 0 & 4 & -3 & 0 \end{bmatrix}$

A7 $\begin{bmatrix} -2 & -1 & 2 & 0 \\ 4 & 3 & -2 & 2 \\ 3 & 3 & 4 & 3 \\ 2 & -1 & 2 & -4 \end{bmatrix}$

For Problems A8–A12, find an LU-decomposition for A and use it to solve $A\vec{x} = \vec{b}_i$, for $i = 1, 2$.

A8 $A = \begin{bmatrix} 3 & -2 \\ -1 & 5 \end{bmatrix}, \vec{b}_1 = \begin{bmatrix} 3 \\ -1 \end{bmatrix}, \vec{b}_2 = \begin{bmatrix} 4 \\ 9 \end{bmatrix}$

A9 $A = \begin{bmatrix} 1 & 0 & 3 \\ -2 & 1 & -3 \\ -1 & 4 & 5 \end{bmatrix}, \vec{b}_1 = \begin{bmatrix} 3 \\ -4 \\ -3 \end{bmatrix}, \vec{b}_2 = \begin{bmatrix} 2 \\ -5 \\ -2 \end{bmatrix}$

A10 $A = \begin{bmatrix} 1 & 0 & -2 \\ -1 & -4 & 4 \\ 3 & -4 & -1 \end{bmatrix}, \vec{b}_1 = \begin{bmatrix} -1 \\ -7 \\ -5 \end{bmatrix}, \vec{b}_2 = \begin{bmatrix} 2 \\ 0 \\ -1 \end{bmatrix}$

A11 $A = \begin{bmatrix} 1 & 0 & 1 \\ -3 & 2 & -1 \\ -3 & 4 & 2 \end{bmatrix}, \vec{b}_1 = \begin{bmatrix} 3 \\ -5 \\ -1 \end{bmatrix}, \vec{b}_2 = \begin{bmatrix} -4 \\ 4 \\ -5 \end{bmatrix}$

A12 $A = \begin{bmatrix} -1 & 2 & -3 & 0 \\ 0 & -1 & 3 & 1 \\ 3 & -8 & 3 & 2 \\ 1 & -2 & 3 & 1 \end{bmatrix}, \vec{b}_1 = \begin{bmatrix} -6 \\ 7 \\ -4 \\ 5 \end{bmatrix}, \vec{b}_2 = \begin{bmatrix} 5 \\ -3 \\ 3 \\ -5 \end{bmatrix}$

Homework Problems

For Problems B1–B12, find an LU-decomposition for the matrix.

B1 $\begin{bmatrix} 2 & 3 \\ 4 & -1 \end{bmatrix}$

B2 $\begin{bmatrix} 1 & 2 & 3 \\ -2 & 1 & 1 \\ 3 & 6 & 1 \end{bmatrix}$

B3 $\begin{bmatrix} -1 & 0 & -1 \\ 3 & 3 & 1 \\ -2 & 6 & 0 \end{bmatrix}$

B4 $\begin{bmatrix} 2 & 8 & 5 \\ 2 & 7 & 6 \\ -2 & -6 & 5 \end{bmatrix}$

B5 $\begin{bmatrix} 2 & 0 & 1 \\ 4 & 1 & 1 \\ -8 & 2 & -3 \end{bmatrix}$

B6 $\begin{bmatrix} 1 & 2 & 3 \\ 1 & 4 & 1 \\ 2 & 1 & 9 \end{bmatrix}$

B7 $\begin{bmatrix} 5 & 10 & 1 \\ 2 & 5 & 1 \\ -5 & -4 & 1 \end{bmatrix}$

B8 $\begin{bmatrix} 3 & 1 & 2 \\ 3 & 1 & 2 \\ 3 & 1 & 2 \end{bmatrix}$

B9 $\begin{bmatrix} -1 & 3 & 6 \\ 3 & 1 & 0 \\ 5 & -5 & -3 \end{bmatrix}$

B10 $\begin{bmatrix} 1 & 2 & 7 \\ 1 & -6 & 5 \\ 1 & -4 & -2 \end{bmatrix}$

B11 $\begin{bmatrix} 1 & -2 & 1 & 1 \\ -2 & 5 & 0 & 7 \\ 6 & -1 & 5 & -9 \\ 3 & 4 & 0 & 3 \end{bmatrix}$

B12 $\begin{bmatrix} 2 & 3 & -1 & 3 \\ 4 & 8 & 2 & 1 \\ -2 & 1 & 4 & -1 \\ 1 & 1 & 1 & -1 \end{bmatrix}$

For Problems B13–B18, find an LU-decomposition for A and use it to solve $A\vec{x} = \vec{b}_i$, for $i = 1, 2$.

B13 $A = \begin{bmatrix} 2 & 3 \\ 1 & 1 \end{bmatrix}, \vec{b}_1 = \begin{bmatrix} 3 \\ -1 \end{bmatrix}, \vec{b}_2 = \begin{bmatrix} 5 \\ 6 \end{bmatrix}$

B14 $A = \begin{bmatrix} 1 & 2 & 3 \\ -2 & 1 & 1 \\ 3 & 6 & 1 \end{bmatrix}, \vec{b}_1 = \begin{bmatrix} -9 \\ 2 \\ -3 \end{bmatrix}, \vec{b}_2 = \begin{bmatrix} -2 \\ -8 \\ 2 \end{bmatrix}$

B15 $A = \begin{bmatrix} 1 & 1 & 3 \\ 2 & 3 & 5 \\ 1 & 1 & 4 \end{bmatrix}, \vec{b}_1 = \begin{bmatrix} 5 \\ 3 \\ 7 \end{bmatrix}, \vec{b}_2 = \begin{bmatrix} 4 \\ 8 \\ 8 \end{bmatrix}$

B16 $A = \begin{bmatrix} 1 & 0 & 4 \\ 3 & 3 & 15 \\ -4 & 6 & -10 \end{bmatrix}, \vec{b}_1 = \begin{bmatrix} 2 \\ 15 \\ 10 \end{bmatrix}, \vec{b}_2 = \begin{bmatrix} 2 \\ 4 \\ -12 \end{bmatrix}$

B17 $A = \begin{bmatrix} -3 & 2 & -5 \\ 9 & -5 & 8 \\ -9 & 3 & 5 \end{bmatrix}, \vec{b}_1 = \begin{bmatrix} 9 \\ -12 \\ -16 \end{bmatrix}, \vec{b}_2 = \begin{bmatrix} -10 \\ 13 \\ 19 \end{bmatrix}$

B18 $A = \begin{bmatrix} 1 & -3 & 1 & 1 \\ -2 & 7 & -2 & -2 \\ 4 & -12 & 1 & 2 \\ 3 & -5 & 3 & -1 \end{bmatrix}, \vec{b}_1 = \begin{bmatrix} 1 \\ -1 \\ 1 \\ -5 \end{bmatrix}, \vec{b}_2 = \begin{bmatrix} 0 \\ 0 \\ 1 \\ -1 \end{bmatrix}$

CHAPTER REVIEW

Suggestions for Student Review

1 State the definition of addition and scalar multiplication for matrices in $M_{m\times n}(\mathbb{R})$ and list the ten properties in Theorem 3.1.1. (Section 3.1)

2 Give the definitions of the special types of matrices: square, upper triangular, lower triangular, and diagonal. List where we have used these so far and any special properties they have. (All Chapter 3)

3 State both definitions of matrix-vector multiplication, clearly stating the condition required for the product to be defined. How does each of these rules correspond to writing a system of linear equations? Why do we have two different definitions? List the properties of matrix-vector multiplication. (Section 3.1)

4 State the two methods for matrix-matrix multiplication. What condition must be satisfied by the sizes of A and B for AB to be defined? Create an example to show that matrix-matrix multiplication represents a composition of functions. List the properties of matrix-matrix multiplication. (Section 3.1)

5 State the definition and properties of the transpose of a vector, and the definition and properties of the transpose of a matrix. (Section 3.1)

6 State the definition of a linear mapping and explain the relationship between a linear mapping and the corresponding matrix mapping. Create some examples to show how to find and use a standard matrix of a linear mapping. (Section 3.2)

7 Give the standard matrix of a rotation, shear, stretch, and dilation in $\mathbb{R}^2$. (Section 3.3)

8 Give definitions of the two special subspaces of a linear mapping. Give a general procedure for finding a basis for each of these subspaces. (Section 3.4)

9 Give definitions of the four fundamental subspaces of a matrix and state the Fundamental Theorem of Linear Algebra. State the general procedure for finding a basis of each subspace, and state the dimension of each subspace. (Section 3.4)

10 (a) How many ways can you recognize the rank of a matrix? State them all.

(b) State the connection between the rank of a matrix A and the dimension of the solution space of $A\vec{x} = \vec{0}$.

(c) Illustrate your answers to (a) and (b) by constructing examples of 4×5 matrices in RREF of (i) rank 4; (ii) rank 3; and (iii) rank 2. In each case, actually determine the general solution of the system $A\vec{x} = \vec{0}$ and check that the solution space has the correct dimension. (Section 3.4)

11 (a) Outline the procedure for determining the inverse of a matrix. Indicate why it might not produce an inverse for a matrix A. Use the matrices of some geometric linear mappings to give two or three examples of matrices that have inverses and two examples of square matrices that do not have inverses. (Section 3.5)

(b) Pick a fairly simple 3×3 matrix (that does not contain too many zeros) and try to find its inverse. If it is not invertible, try another. When you have an inverse, check its correctness by multiplication. (Section 3.5)

12 State as many theorems and properties of invertible matrices as you can. (Section 3.5)

13 For 3×3 matrices, choose one elementary row operation of each of the three types; call these E_1, E_2, E_3. Choose an arbitrary 3×3 matrix A and check that E_iA is the matrix obtained from A by the appropriate elementary row operations. What is the relationship of elementary matrices with our geometrical mappings? (Section 3.6, 3.3)

14 Explain the algorithm for writing a matrix A as a product of elementary matrices. (Section 3.6)

15 Explain the procedure for finding an LU-decomposition of a matrix A, and how to use the LU-decomposition to solve a system $A\vec{x} = \vec{b}$. (Section 3.7)

Chapter Quiz

For Problems E1–E3, let $A = \begin{bmatrix} 2 & -5 & -3 \\ -3 & 4 & -7 \end{bmatrix}$ and $B = \begin{bmatrix} 2 & -1 & 4 \\ 3 & 0 & 2 \\ 1 & -1 & 5 \end{bmatrix}$. Compute the product or explain why it is not defined.

E1 AB **E2** BA **E3** BA^T

E4 (a) Let $A = \begin{bmatrix} -3 & 0 & 4 \\ 2 & 1 & -1 \end{bmatrix}$, $\vec{u} = \begin{bmatrix} 1 \\ 1 \\ -2 \end{bmatrix}$, $\vec{v} = \begin{bmatrix} 4 \\ -2 \\ -1 \end{bmatrix}$, and let f_A be the matrix mapping with matrix A. Determine $f_A(\vec{u})$ and $f_A(\vec{v})$.

(b) Use the result of part (a) to calculate $A\begin{bmatrix} 4 & 1 \\ -2 & 1 \\ -1 & -2 \end{bmatrix}$.

E5 Let R be the rotation through angle $\frac{\pi}{3}$ about the x_3-axis in $\mathbb{R}^3$. Let M be a reflection in $\mathbb{R}^3$ in the plane with equation $-x_1 - x_2 + 2x_3 = 0$. Determine

(a) The matrix of R
(b) The matrix of M
(c) The matrix of $[R \circ M]$

E6 Let $A = \begin{bmatrix} 1 & 0 & 2 & 1 & 0 \\ 2 & 1 & 3 & 2 & 0 \\ 1 & 1 & 1 & 1 & 1 \end{bmatrix}$ and $\vec{b} = \begin{bmatrix} 5 \\ 16 \\ 18 \end{bmatrix}$. Determine the solution set of $A\vec{x} = \vec{b}$ and the solution space of $A\vec{x} = \vec{0}$. Discuss the relationship between the two sets.

E7 Let $B = \begin{bmatrix} 1 & 2 & 0 \\ -1 & -1 & -1 \\ 1 & 3 & 0 \\ 0 & 2 & -1 \end{bmatrix}$, $\vec{u} = \begin{bmatrix} 4 \\ -3 \\ 5 \\ 3 \end{bmatrix}$, and $\vec{v} = \begin{bmatrix} -5 \\ 6 \\ -7 \\ -1 \end{bmatrix}$.

(a) Determine whether $\vec{u}$ or $\vec{v}$ is in Col(B).
(b) Determine from your calculation in part (a) a vector $\vec{x}$ such that $B\vec{x} = \vec{v}$.

E8 Find a basis for each of the four fundamental subspaces of $A = \begin{bmatrix} 1 & 0 & 1 & 1 & 1 \\ 2 & 1 & 1 & 2 & 5 \\ 0 & 2 & -2 & 1 & 8 \\ 3 & 3 & 0 & 4 & 14 \end{bmatrix}$.

E9 Determine the inverse of $A = \begin{bmatrix} 1 & 0 & 0 & -1 \\ 0 & 0 & 1 & 0 \\ 0 & 2 & 0 & 1 \\ 1 & 0 & 0 & 2 \end{bmatrix}$.

E10 Determine all values of p such that the matrix $\begin{bmatrix} 1 & 0 & p \\ 1 & 1 & 0 \\ 2 & 1 & 1 \end{bmatrix}$ is invertible and determine its inverse.

E11 Prove that the range of a linear mapping $L : \mathbb{R}^n \to \mathbb{R}^m$ is a subspace of the codomain.

E12 Let $\{\vec{v}_1, \dots, \vec{v}_k\}$ be a linearly independent set in $\mathbb{R}^n$ and let $L : \mathbb{R}^n \to \mathbb{R}^m$ be a linear mapping. Prove that if Null(L) $= \{\vec{0}\}$, then $\{L(\vec{v}_1), \dots, L(\vec{v}_k)\}$ is a linearly independent set in $\mathbb{R}^m$.

E13 Let $A = \begin{bmatrix} 1 & 0 & -2 \\ 0 & 2 & -3 \\ 0 & 0 & 4 \end{bmatrix}$.

(a) Determine a sequence of elementary matrices $E_1, \dots, E_k$, such that $E_k \cdots E_1 A = I$.
(b) By inverting the elementary matrices in part (a), write A as a product of elementary matrices.

For Problems E14–E19, either give an example or explain (in terms of theorems or definitions) why no such example can exist.

E14 A matrix K such that $KM = MK$ for all 3×3 matrices M.

E15 A matrix K such that $KM = MK$ for all 3×4 matrices M.

E16 The matrix of a linear mapping $L:\mathbb{R}^2 \to \mathbb{R}^3$ whose range is Span$\left\{\begin{bmatrix} 1 \\ 1 \end{bmatrix}\right\}$ and whose nullspace is Span$\left\{\begin{bmatrix} 2 \\ 3 \end{bmatrix}\right\}$.

E17 The matrix of a linear mapping $L:\mathbb{R}^2 \to \mathbb{R}^3$ whose range is Span$\left\{\begin{bmatrix} 1 \\ 1 \\ 2 \end{bmatrix}\right\}$ and whose nullspace is Span$\left\{\begin{bmatrix} 2 \\ 3 \end{bmatrix}\right\}$.

E18 A linear mapping $L : \mathbb{R}^3 \to \mathbb{R}^3$ such that the range of L is all of $\mathbb{R}^3$ and the nullspace of L is Span$\left\{\begin{bmatrix} 1 \\ -1 \\ 1 \end{bmatrix}\right\}$.

E19 An invertible 4×4 matrix of rank 3.

Further Problems

These problems are intended to be challenging.

F1 Let $A \in M_{m\times n}(\mathbb{R})$ with $A \neq O_{m,n}$. Assume that $\{\vec{v}_1, \dots, \vec{v}_k, \vec{v}_{k+1}, \dots, \vec{v}_n\}$ is a basis for $\mathbb{R}^n$ such that $\{\vec{v}_{k+1}, \dots, \vec{v}_n\}$ is a basis for Null(A).

(a) Prove that $\{A\vec{v}_1, \dots, A\vec{v}_k\}$ is a basis for Col(A).

(b) Use the result of part (a) to prove the Rank-Nullity Theorem.

F2 Let $A = \begin{bmatrix} \vec{v}_1 & \cdots & \vec{v}_k \end{bmatrix}$ and $B = \begin{bmatrix} \vec{u}_1 & \cdots & \vec{u}_k \end{bmatrix}$ be $n\times k$ matrices. Prove that if

$$\text{Span}\{\vec{v}_1, \dots, \vec{v}_k\} \subseteq \text{Span}\{\vec{u}_1, \dots, \vec{u}_k\}$$

then there exists a matrix C such that $A = BC$.

F3 We say that a matrix C **commutes** with a matrix D if $CD = DC$. Show that the set of matrices that commute with $A = \begin{bmatrix} 3 & 2 \\ 0 & 1 \end{bmatrix}$ is the set of matrices of the form $pI + qA$, where p and q are arbitrary scalars.

F4 Let A be some fixed $n \times n$ matrix. Show that the set $C(A)$ of matrices that commutes with A is closed under addition, scalar multiplication, and matrix multiplication.

F5 A square matrix A is said to be **nilpotent** if some power of A is equal to the zero matrix. Show that the matrix $\begin{bmatrix} 0 & a_{12} & a_{13} \\ 0 & 0 & a_{23} \\ 0 & 0 & 0 \end{bmatrix}$ is nilpotent. Generalize.

F6 (a) Suppose that ℓ is a line in $\mathbb{R}^2$ passing through the origin and making an angle θ with the positive x_1-axis. Let refl_θ denote a reflection in this line. Determine the matrix $[\text{refl}_\theta]$ in terms of functions of θ.

(b) Let refl_α denote a reflection in a second line, and by considering the matrix $[\text{refl}_\alpha \circ \text{refl}_\theta]$, show that the composition of two reflections in the plane is a rotation. Express the angle of the rotation in terms of α and θ.

F7 (Isometries) A linear transformation $L : \mathbb{R}^2 \to \mathbb{R}^2$ is an **isometry** of $\mathbb{R}^2$ if L preserves lengths (that is, if $\|L(\vec{x})\| = \|\vec{x}\|$ for every $\vec{x} \in \mathbb{R}^2$).

(a) Show that an isometry preserves the dot product (that is, $L(\vec{x}) \cdot L(\vec{y}) = \vec{x} \cdot \vec{y}$ for every $\vec{x}, \vec{y} \in \mathbb{R}^2$). (Hint: consider $L(\vec{x} + \vec{y})$.)

(b) Show that the columns of that matrix $[L]$ must be orthogonal to each other and of length 1. Deduce that any isometry of $\mathbb{R}^2$ must be the composition of a reflection and a rotation. (Hint: you may find it helpful to use the result of Problem F4 (a).)

F8 (a) Suppose that A and B are $n \times n$ matrices such that $A + B$ and $A - B$ are invertible and that C and D are arbitrary $n \times n$ matrices. Show that there are $n \times n$ matrices X and Y satisfying the system

$$AX + BY = C$$
$$BX + AY = D$$

(b) With the same assumptions as in part (a), give a careful explanation of why the matrix $\begin{bmatrix} A & B \\ B & A \end{bmatrix}$ must be invertible. Obtain an expression for its inverse in terms of $(A + B)^{-1}$ and $(A - B)^{-1}$.

CHAPTER 4

Vector Spaces

CHAPTER OUTLINE

In the first three chapters, we explored some of the most important concepts in linear algebra, including spanning, linear independence, and bases, in the context of vectors in $\mathbb{R}^n$ and matrices. In this chapter, we will examine these concepts, as well as other important ideas in linear algebra, in a more general setting. However, as we will see, although the setting is more abstract, we will use the same procedures and tools as we did with vectors in $\mathbb{R}^n$.

4.1 Spaces of Polynomials

We now compare sets of polynomials under standard addition and scalar multiplication to sets of vectors in $\mathbb{R}^n$ and sets of matrices.

Addition and Scalar Multiplication of Polynomials

Definition
$P_n(\mathbb{R})$
Addition of Polynomials
Scalar Multiplication of Polynomials

We let $P_n(\mathbb{R})$ denote the set of all polynomials $\mathbf{p}(x) = a_0 + a_1x + \cdots + a_nx^n$ of degree *at most n* where the coefficients a_i are real numbers.

The polynomials $\mathbf{p}(x) = a_0 + a_1x + \cdots + a_nx^n$, $\mathbf{q}(x) = b_0 + b_1x + \cdots + b_nx^n \in P_n(\mathbb{R})$ are said to be **equal** if $a_i = b_i$ for $0 \leq i \leq n$.

We define the **addition of** of $\mathbf{p}$ and $\mathbf{q}$, denoted $\mathbf{p} + \mathbf{q}$, by

$$(\mathbf{p} + \mathbf{q})(x) = (a_0 + b_0) + (a_1 + b_1)x + \cdots + (a_n + b_n)x^n$$

We define **scalar multiplication** of $\mathbf{p}$ by a scalar $t \in \mathbb{R}$ by

$$(t\mathbf{p})(x) = ta_0 + (ta_1)x + \cdots + (ta_n)x^n$$

As before, when we write $\mathbf{p} - \mathbf{q}$ we mean $\mathbf{p} + (-1)\mathbf{q}$.

EXAMPLE 4.1.1

Evaluate the following linear combinations of polynomials

(a) $(2 + 3x + 4x^2 + x^3) + (5 + x - 2x^2 + 7x^3)$

(b) $2(1 + 3x - x^3) + 3(4 + x^2 + 2x^3)$

Solution: For (a) we have

$$\begin{aligned}(2 + 3x + 4x^2 + x^3) + (5 + x - 2x^2 + 7x^3) &= 2 + 5 + (3 + 1)x + (4 - 2)x^2 + (1 + 7)x^3 \\ &= 7 + 4x + 2x^2 + 8x^3\end{aligned}$$

For (b) we have

$$\begin{aligned}2(1 + 3x - x^3) + 3(4 + x^2 + 2x^3) &= 2(1) + 2(3)x + 2(0)x^2 + 2(-1)x^3 \\ &\quad + 3(4) + 3(0)x + 3(1)x^2 + 3(2)x^3 \\ &= 2 + 12 + (6 + 0)x + (0 + 3)x^2 + (-2 + 6)x^3 \\ &= 14 + 6x + 3x^2 + 4x^3\end{aligned}$$

Properties of Polynomial Addition and Scalar Multiplication

Theorem 4.1.1

For all $\mathbf{p}, \mathbf{q}, \mathbf{r} \in P_n(\mathbb{R})$ and $s, t \in \mathbb{R}$ we have

(1) $\mathbf{p} + \mathbf{q} \in P_n(\mathbb{R})$

(2) $\mathbf{p} + \mathbf{q} = \mathbf{q} + \mathbf{p}$

(3) $(\mathbf{p} + \mathbf{q}) + \mathbf{r} = \mathbf{p} + (\mathbf{q} + \mathbf{r})$

(4) There exists a polynomial $\mathbf{0} \in P_n(\mathbb{R})$ such that $\mathbf{p} + \mathbf{0} = \mathbf{p}$ for all $\mathbf{p} \in P_n(\mathbb{R})$.

(5) For each polynomial $\mathbf{p} \in P_n(\mathbb{R})$, there exists a polynomial $(-\mathbf{p}) \in P_n(\mathbb{R})$ such that $\mathbf{p} + (-\mathbf{p}) = \mathbf{0}$

(6) $s\mathbf{p} \in P_n(\mathbb{R})$

(7) $s(t\mathbf{p}) = (st)\mathbf{p}$

(8) $(s + t)\mathbf{p} = s\mathbf{p} + t\mathbf{p}$

(9) $s(\mathbf{p} + \mathbf{q}) = s\mathbf{p} + s\mathbf{q}$

(10) $1\mathbf{p} = \mathbf{p}$

These properties follow easily from the definitions of addition and scalar multiplication and are very similar to those for vectors in $\mathbb{R}^n$. Thus, the proofs are left to the reader.

The polynomial $\mathbf{0}$, called the **zero polynomial**, is defined by

$$\mathbf{0}(x) = 0 = 0 + 0x + \cdots + 0x^n, \quad \text{for all } x \in \mathbb{R}$$

The additive inverse $(-\mathbf{p})$ of a polynomial $\mathbf{p}(x) = a_0 + a_1x + \cdots + a_nx^n$ is

$$(-\mathbf{p})(x) = -a_0 - a_1x - \cdots - a_nx^n, \quad \text{for all } x \in \mathbb{R}$$

It is important to recognize that these are the same ten properties we had for addition and scalar multiplication of vectors in $\mathbb{R}^n$ (Theorem 1.4.1) and of matrices (Theorem 3.1.1).

To match what we did with vectors in $\mathbb{R}^n$ and matrices in $M_{m\times n}(\mathbb{R})$, we now define the concepts of spanning and linear independence for polynomials.

Definition
Span

Let $\mathcal{B} = \{\mathbf{p}_1, \dots, \mathbf{p}_k\}$ be a set of polynomials in $P_n(\mathbb{R})$. The **span** of $\mathcal{B}$ is defined as

$$\operatorname{Span} \mathcal{B} = \{t_1\mathbf{p}_1 + \cdots + t_k\mathbf{p}_k \mid t_1, \dots, t_k \in \mathbb{R}\}$$

Definition
Linearly Dependent
Linearly Independent

Let $\mathcal{B} = \{\mathbf{p}_1, \dots, \mathbf{p}_k\}$ be a set of polynomials in $P_n(\mathbb{R})$. The set $\mathcal{B}$ is said to be **linearly dependent** if there exists real coefficients $t_1, \dots, t_k$ not all zero such that

$$t_1\mathbf{p}_1 + \cdots + t_k\mathbf{p}_k = \mathbf{0}$$

The set $\mathcal{B}$ is said to be **linearly independent** if the only solution to

$$t_1\mathbf{p}_1 + \cdots + t_k\mathbf{p}_k = \mathbf{0}$$

is the trivial solution $t_1 = \cdots = t_k = 0$.

EXAMPLE 4.1.2

Determine whether $\mathbf{p}(x) = 1 + 2x + 3x^2 + 4x^3$ is in the span of $\mathcal{B} = \{1 + x, 1 + x^3, x + x^2, x + x^3\}$.

Solution: We want to determine if there are t_1, t_2, t_3, t_4 such that

$$t_1(1 + x) + t_2(1 + x^3) + t_3(x + x^2) + t_4(x + x^3) = 1 + 2x + 3x^2 + 4x^3$$

Performing the linear combination on the left-hand side gives

$$(t_1 + t_2) + (t_1 + t_3 + t_4)x + t_3x^2 + (t_2 + t_4)x^3 = 1 + 2x + 3x^2 + 4x^3$$

Comparing the coefficients of powers of x on both sides of the equation, we get the system of linear equations

$$\begin{aligned} t_1 + t_2 &= 1 \\ t_1 + t_3 + t_4 &= 2 \\ t_3 &= 3 \\ t_2 + t_4 &= 4 \end{aligned}$$

Row reducing the corresponding augmented matrix gives

$$\left[\begin{array}{cccc|c} 1 & 1 & 0 & 0 & 1 \\ 1 & 0 & 1 & 1 & 2 \\ 0 & 0 & 1 & 0 & 3 \\ 0 & 1 & 0 & 1 & 4 \end{array}\right] \sim \left[\begin{array}{cccc|c} 1 & 0 & 0 & 0 & -2 \\ 0 & 1 & 0 & 0 & 3 \\ 0 & 0 & 1 & 0 & 3 \\ 0 & 0 & 0 & 1 & 1 \end{array}\right]$$

We see that the system is consistent; therefore, $1 + 2x + 3x^2 + 4x^3$ is in the span of $\mathcal{B}$. In particular, we have

$$(-2)(1 + x) + 3(1 + x^3) + 3(x + x^2) + (x + x^3) = 1 + 2x + 3x^2 + 4x^3$$

EXAMPLE 4.1.3 Determine whether the set

$$\mathcal{B} = \{1 + 2x + 2x^2 - x^3, 3 + 2x + x^2 + x^3, 2x^2 + 2x^3\}$$

is linearly independent in $P_3(\mathbb{R})$.

Solution: Consider

$$t_1(1 + 2x + 2x^2 - x^3) + t_2(3 + 2x + x^2 + x^3) + t_3(2x^2 + 2x^3) = 0$$

Performing the linear combination on the left-hand side gives

$$(t_1 + 3t_2) + (2t_1 + 2t_2)x + (2t_1 + t_2 + 2t_3)x^2 + (-t_1 + t_2 + 2t_3)x^3 = 0$$

Comparing coefficients of the powers of x, we get the homogeneous system of linear equations

$$\begin{aligned} t_1 + 3t_2 &= 0 \\ 2t_1 + 2t_2 &= 0 \\ 2t_1 + t_2 + 2t_3 &= 0 \\ -t_1 + t_2 + 2t_3 &= 0 \end{aligned}$$

Row reducing the associated coefficient matrix gives

$$\begin{bmatrix} 1 & 3 & 0 \\ 2 & 2 & 0 \\ 2 & 1 & 2 \\ -1 & 1 & 2 \end{bmatrix} \sim \begin{bmatrix} 1 & 0 & 0 \\ 0 & 1 & 0 \\ 0 & 0 & 1 \\ 0 & 0 & 0 \end{bmatrix}$$

The only solution is $t_1 = t_2 = t_3 = 0$, so $\mathcal{B}$ is linearly independent.

EXERCISE 4.1.1 Determine whether

$$\mathcal{B} = \{1 + 2x + x^2 + x^3, 1 + x + 3x^2 + x^3, 3 + 5x + 5x^2 - 3x^3, -x - 2x^2\}$$

is linearly dependent or linearly independent.

Is $\mathbf{p}(x) = 1 + 5x - 5x^2 + x^3$ in the span of $\mathcal{B}$?

EXERCISE 4.1.2 Consider $\mathcal{B} = \{1, x, x^2, x^3\}$. Prove that $\mathcal{B}$ is linearly independent and show that $\operatorname{Span}\mathcal{B} = P_3(\mathbb{R})$.

PROBLEMS 4.1

Practice Problems

For Problems A1–A7, evaluate the expression.

A1 $(2 - 2x + 3x^2 + 4x^3) + (-3 - 4x + x^2 + 2x^3)$

A2 $(-3)(1 - 2x + 2x^2 + x^3 + 4x^4)$

A3 $(2 + 3x + x^2 - 2x^3) - 3(1 - 2x + 4x^2 + 5x^3)$

A4 $(2 + 3x + 4x^2) - (5 + x - 2x^2)$

A5 $-2(-5 + x + x^2) + 3(-1 - x^2)$

A6 $2\left(\frac{2}{3} - \frac{1}{3}x + 2x^2\right) + \frac{1}{3}\left(3 - 2x + x^2\right)$

A7 $\sqrt{2}(1 + x + x^2) + \pi(-1 + x^2)$

For Problems A8–A13, either express the polynomial as a linear combination of the polynomials in $\mathcal{B} = \{1 + x^2 + x^3, 2 + x + x^3, -1 + x + 2x^2 + x^3\}$ or show that it is not in Span $\mathcal{B}$.

A8 $\mathbf{p}(x) = 0$

A9 $\mathbf{p}(x) = 2 + 4x + 3x^2 + 4x^3$

A10 $\mathbf{p}(x) = -x + 2x^2 + x^3$

A11 $\mathbf{p}(x) = -4 - x + 3x^2$

A12 $\mathbf{p}(x) = -1 + 7x + 5x^2 + 4x^3$

A13 $\mathbf{p}(x) = 2 + x + 5x^3$

For Problems A14–A17, determine whether the set is linearly independent. If a set is linearly dependent, find all linear combinations of the polynomials that equal the zero polynomial.

A14 $\left\{1 + 2x + x^2 - x^3, 5x + x^2, 1 - 3x + 2x^2 + x^3\right\}$

A15 $\left\{1 + x + x^2, x, x^2 + x^3, 3 + 2x + 2x^2 - x^3\right\}$

A16 $\left\{3 + x + x^2, 4 + x - x^2, 1 + 2x + x^2 + 2x^3, -1 + 5x^2 + x^3\right\}$

A17 $\left\{1 + x + x^3 + x^4, 2 + x - x^2 + x^3 + x^4, x + x^2 + x^3 + x^4\right\}$

A18 Prove that the set $\mathcal{B} = \{1, x - 1, (x - 1)^2\}$ is linearly independent and show that Span $\mathcal{B} = P_2(\mathbb{R})$.

Homework Problems

For Problems B1–B7, evaluate the expression.

B1 $(1 + 2x - 3x^2 + 4x^3) - (2 - x^2 + 2x^3)$

B2 $(-2)(1 + 2x - 3x^2 + 3x^3 + 5x^4)$

B3 $2(1 + 2x + 7x^2 - 3x^3) - 3(1 + 6x - 2x^2)$

B4 $(-4)(1 + x^2 - 2x^3) + (1 - x^2 + x^3)$

B5 $0(1 + 2x^2 + 3x^4)$

B6 $\frac{1}{2}\left(2 + \frac{1}{3}x + x^2\right) + \frac{1}{4}\left(4 - 2x + 2x^2\right)$

B7 $\left(1 - \sqrt{2}\right)\left(1 + \sqrt{2} + (\sqrt{2} + 1)x^2\right) - \frac{1}{2}\left(-2 + 2x^2\right)$

For Problems B8–B11, $\mathcal{B} = \{1 + x + x^3, 1 + 2x + x^2, x + x^3\}$. Either express the polynomial as a linear combination of the polynomials in $\mathcal{B}$ or show that it is not in Span $\mathcal{B}$.

B8 $\mathbf{p}(x) = 2$

B9 $\mathbf{p}(x) = 1 - x + x^3$

B10 $\mathbf{p}(x) = 2 + 2x - x^2 + 4x^3$

B11 $\mathbf{p}(x) = 6 + 4x + 2x^2$

For Problems B12–B18, determine whether the set is linearly independent. If a set is linearly dependent, find all linear combinations of the polynomials that equal the zero polynomial.

B12 $\left\{0, 1 + x^2, x + x^2 - x^3\right\}$

B13 $\left\{1 + x, 1 - x^2, x + x^3, 1 + 2x^2 + x^3\right\}$

B14 $\left\{1 + 2x + 5x^2, x + 3x^2, 2 + 3x + 8x^2\right\}$

B15 $\{4x - 3x^2, 1 - 3x + 2x^2, 2 - 2x + x^2\}$

B16 $\{1 + 3x + x^2, 8 + 9x + 2x^2, -3 - 3x - x^2\}$

B17 $\{4 + x - 2x^3, 5 + 2x + x^3, -2 + x + 8x^3\}$

B18 $\{1 - x + x^2, 1 + 2x^2, -1 + 2x + x^2\}$

B19 Prove that $\mathcal{B} = \{1, 1 - x, (1 - x)^2\}$ is linearly independent and show that Span $\mathcal{B} = P_2(\mathbb{R})$.

Conceptual Problems

For Problems C1–C3, let $\mathcal{B} = \{\mathbf{p}_1, \ldots, \mathbf{p}_k\}$ be a set of polynomials in $P_n(\mathbb{R})$.

C1 Prove if $k < n + 1$, then there exists a polynomial $\mathbf{q} \in P_n(\mathbb{R})$ such that $\mathbf{q} \notin$ Span $\mathcal{B}$.

C2 Prove if $k > n + 1$, then $\mathcal{B}$ must be linearly dependent.

C3 Prove if $k = n + 1$ and $\mathcal{B}$ is linearly independent, then Span $\mathcal{B} = P_n(\mathbb{R})$.

4.2 Vector Spaces

We have now seen that addition and scalar multiplication of vectors in $\mathbb{R}^n$, matrices in $M_{m\times n}(\mathbb{R})$, and polynomials in $P_n(\mathbb{R})$ satisfy the same ten properties. Moreover, we can show that addition and scalar multiplication of linear mappings also satisfy these same ten properties. In fact, many other mathematical objects also have these important properties. Instead of analyzing each of these objects separately, it is useful to define one abstract concept that encompasses all of them.

Vector Spaces

Definition
Vector Space over $\mathbb{R}$

A **vector space over** $\mathbb{R}$ is a set $\mathbb{V}$ together with an operation of **addition**, usually denoted $\mathbf{x} + \mathbf{y}$ for any $\mathbf{x}, \mathbf{y} \in \mathbb{V}$, and an operation of **scalar multiplication**, usually denoted $s\mathbf{x}$ for any $\mathbf{x} \in \mathbb{V}$ and $s \in \mathbb{R}$, such that for any $\mathbf{x}, \mathbf{y}, \mathbf{z} \in \mathbb{V}$ and $s, t \in \mathbb{R}$ we have all of the following properties:

V1 $\mathbf{x} + \mathbf{y} \in \mathbb{V}$
V2 $\mathbf{x} + \mathbf{y} = \mathbf{y} + \mathbf{x}$
V3 $(\mathbf{x} + \mathbf{y}) + \mathbf{z} = \mathbf{x} + (\mathbf{y} + \mathbf{z})$
V4 There exists $\mathbf{0} \in \mathbb{V}$, called the **zero vector**, such that $\mathbf{x} + \mathbf{0} = \mathbf{x}$ for all $\mathbf{x} \in \mathbb{V}$
V5 For each $\mathbf{x} \in \mathbb{V}$, there exists $(-\mathbf{x}) \in \mathbb{V}$ such that $\mathbf{x} + (-\mathbf{x}) = \mathbf{0}$
V6 $s\mathbf{x} \in \mathbb{V}$
V7 $s(t\mathbf{x}) = (st)\mathbf{x}$
V8 $(s + t)\mathbf{x} = s\mathbf{x} + t\mathbf{x}$
V9 $s(\mathbf{x} + \mathbf{y}) = s\mathbf{x} + s\mathbf{y}$
V10 $1\mathbf{x} = \mathbf{x}$

Remarks

1. The elements of a vector space are called **vectors**. Note that these can be very different objects than vectors in $\mathbb{R}^n$. Thus, we will always denote, as in the definition above, a vector in a general vector space in boldface (for example, $\mathbf{x}$). However, for vector spaces such as $\mathbb{R}^n$ or $M_{m\times n}(\mathbb{R})$ we will often use the notation we introduced earlier.

2. As usual, we call a sum of scalar multiples of vectors a **linear combination**, and when we write $\mathbf{x} - \mathbf{y}$ we mean $\mathbf{x} + (-1)\mathbf{y}$.

3. Some people denote the operations of addition and scalar multiplication in general vector spaces by $\oplus$ and $\odot$, respectively, to stress the fact that these do not need to be "standard" addition and scalar multiplication.

4. Since every vector space contains a zero vector by V4, the empty set cannot be a vector space.

5. When working with multiple vector spaces, we sometimes use a subscript to denote the vector space to which the zero vector belongs. For example, $\mathbf{0}_{\mathbb{V}}$ would represent the zero vector in the vector space $\mathbb{V}$.

6. Vector spaces can be defined using other number systems as the scalars. For example, the definition makes perfect sense if rational numbers are used instead of the real numbers. Vector spaces over the complex numbers are discussed in Chapter 9. Until Chapter 9, "vector space" means "vector space over $\mathbb{R}$."

7. We have defined vector spaces to have the same structure as $\mathbb{R}^n$. The study of vector spaces is the study of this common structure. However, it is possible that vectors in individual vector spaces have other aspects not common to all vector spaces, such as matrix multiplication or factorization of polynomials.

EXAMPLE 4.2.1

$\mathbb{R}^n$ is a vector space with addition and scalar multiplication defined in the usual way. We call these standard addition and scalar multiplication of vectors in $\mathbb{R}^n$.

EXAMPLE 4.2.2

$P_n(\mathbb{R})$, the set of all polynomials of degree at most n with real coefficients, is a vector space with standard addition and scalar multiplication of polynomials.

EXAMPLE 4.2.3

$M_{m\times n}(\mathbb{R})$, the set of all $m \times n$ matrices with real entries, is a vector space with standard addition and scalar multiplication of matrices.

EXAMPLE 4.2.4

Consider the set of polynomials of degree n with real coefficients. Is this a vector space with standard addition and scalar multiplication? No, it does not contain the zero polynomial since the zero polynomial is not of degree n. Note also that the sum of two polynomials of degree n may not be of degree n. For example, $(1 + x^n) + (1 - x^n) = 2$, which is of degree 0. Thus, the set is also not closed under addition.

EXAMPLE 4.2.5

Let $\mathcal{F}(a, b)$ denote the set of all functions $f : (a, b) \to \mathbb{R}$. If $f, g \in \mathcal{F}(a, b)$, then the sum is defined by $(f + g)(x) = f(x) + g(x)$, and multiplication by a scalar $t \in \mathbb{R}$ is defined by $(tf)(x) = tf(x)$. With these definitions, $\mathcal{F}(a, b)$ is a vector space.

EXAMPLE 4.2.6

Let $C(a, b)$ denote the set of all functions that are continuous on the interval (a, b). Since the sum of continuous functions is continuous and a scalar multiple of a continuous function is continuous, $C(a, b)$ is a vector space.

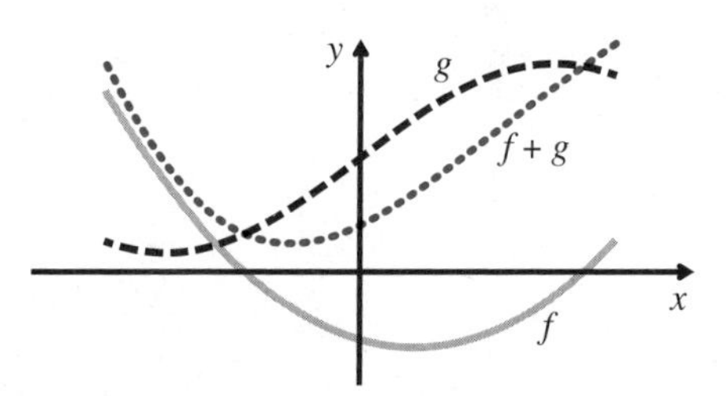

EXAMPLE 4.2.7

Let $\mathbb{T}$ be the set of all solutions to $x_1 + 2x_2 = 1, 2x_1 + 3x_2 = 0$. Is $\mathbb{T}$ a vector space with standard addition and scalar multiplication?

Solution: No. This set with these operations does not satisfy many of the vector space axioms. For example, V6 does not hold since $\begin{bmatrix} -3 \\ 2 \end{bmatrix} \in \mathbb{T}$ is a solution of this system, but $2\begin{bmatrix} -3 \\ 2 \end{bmatrix} = \begin{bmatrix} -6 \\ 4 \end{bmatrix}$ is not a solution of the system and hence is not in $\mathbb{T}$.

EXAMPLE 4.2.8

Consider $\mathbb{V} = \{(x, y) \mid x, y \in \mathbb{R}\}$ with addition and scalar multiplication defined by

$$(x_1, y_1) + (x_2, y_2) = (x_1 + x_2, y_1 + y_2)$$
$$k(x, y) = (ky, kx)$$

Is $\mathbb{V}$ a vector space?

Solution: No, since $1(2, 3) = (3, 2) \neq (2, 3)$, it does not satisfy V10. Note that it also does not satisfy V7.

EXAMPLE 4.2.9

Is $\mathbb{S} = \left\{ \begin{bmatrix} x_1 \\ x_2 \\ x_1 + x_2 \end{bmatrix} \mid x_1, x_2 \in \mathbb{R} \right\}$ a vector space with standard addition and scalar multiplication in $\mathbb{R}^3$?

Solution: Yes! To prove it we need to verify that all ten axioms hold. However, since axioms V2, V3, V7, V8, V9, and V10 refer only to the operations of addition and scalar multiplication, we know by Theorem 1.4.1 that these all hold. We now prove that the remaining axioms also hold.

Let $\vec{x} = \begin{bmatrix} x_1 \\ x_2 \\ x_1 + x_2 \end{bmatrix}$ and $\vec{y} = \begin{bmatrix} y_1 \\ y_2 \\ y_1 + y_2 \end{bmatrix}$ be vectors in $\mathbb{S}$.

V1: We have

$$\vec{x} + \vec{y} = \begin{bmatrix} x_1 \\ x_2 \\ x_1 + x_2 \end{bmatrix} + \begin{bmatrix} y_1 \\ y_2 \\ y_1 + y_2 \end{bmatrix} = \begin{bmatrix} x_1 + y_1 \\ x_2 + y_2 \\ x_1 + y_1 + x_2 + y_2 \end{bmatrix}$$

If we let $z_1 = x_1 + y_1$ and $z_2 = x_2 + y_2$, then $z_1 + z_2 = x_1 + y_1 + x_2 + y_2$ and hence

$$\vec{x} + \vec{y} = \begin{bmatrix} z_1 \\ z_2 \\ z_1 + z_2 \end{bmatrix} \in \mathbb{S}$$

V6: For any $t \in \mathbb{R}$, $t\vec{x} = t\begin{bmatrix} x_1 \\ x_2 \\ x_1 + x_2 \end{bmatrix} = \begin{bmatrix} tx_1 \\ tx_2 \\ tx_1 + tx_2 \end{bmatrix} \in \mathbb{S}$.

V4: Since V6 holds, we have that $\begin{bmatrix} 0 \\ 0 \\ 0 \end{bmatrix} = 0\vec{x} \in \mathbb{S}$.

V5: Since V6 holds, we have that $\begin{bmatrix} -x_1 \\ -x_2 \\ -x_1 + (-x_2) \end{bmatrix} = (-1)\vec{x} \in \mathbb{S}$.

Thus, $\mathbb{S}$ with these operators is a vector space as it satisfies all ten axioms.

EXERCISE 4.2.1 Prove that the set $\mathbb{S} = \left\{ \begin{bmatrix} x_1 \\ x_2 \end{bmatrix} \mid x_1, x_2 \in \mathbb{Z} \right\}$ is not a vector space using standard addition and scalar multiplication of vectors in $\mathbb{R}^2$.

EXERCISE 4.2.2 Let $\mathbb{S} = \left\{ \begin{bmatrix} a_1 & 0 \\ 0 & a_2 \end{bmatrix} \mid a_1, a_2 \in \mathbb{R} \right\}$. Prove that $\mathbb{S}$ is a vector space using standard addition and scalar multiplication of matrices. This is the vector space of 2×2 diagonal matrices.

EXERCISE 4.2.3 Let $\mathbb{T} = \{a_1 + a_2x^3 \mid a_1, a_2 \in \mathbb{R}\}$. Prove that $\mathbb{T}$ is a vector space using standard addition and scalar multiplication of polynomials in $P_3(\mathbb{R})$. How does this vector space compare with the vector space in Exercise 4.2.2?

Again, one advantage of having the abstract concept of a vector space is that when we prove a result about a general vector space, it instantly applies to all of the examples of vector spaces. To demonstrate this, we give three additional properties that follow from the vector space axioms.

Theorem 4.2.1 If $\mathbb{V}$ is a vector space, then

(1) $0\mathbf{x} = \mathbf{0}$ for all $\mathbf{x} \in \mathbb{V}$

(2) $(-1)\mathbf{x} = -\mathbf{x}$ for all $\mathbf{x} \in \mathbb{V}$

(3) $t\mathbf{0} = \mathbf{0}$ for all $t \in \mathbb{R}$

Proof: We will prove (1). You are asked to prove (2) in Problem C1 and to prove (3) in Problem C2.

For any $\mathbf{x} \in \mathbb{V}$ we have

$$\begin{aligned}
0\mathbf{x} &= 0\mathbf{x} + \mathbf{0} && \text{by V4} \\
&= 0\mathbf{x} + [\mathbf{x} + (-\mathbf{x})] && \text{by V5} \\
&= 0\mathbf{x} + [1\mathbf{x} + (-\mathbf{x})] && \text{by V10} \\
&= [0\mathbf{x} + 1\mathbf{x}] + (-\mathbf{x}) && \text{by V3} \\
&= (0 + 1)\mathbf{x} + (-\mathbf{x}) && \text{by V8} \\
&= 1\mathbf{x} + (-\mathbf{x}) && \text{operation of numbers in } \mathbb{R} \\
&= \mathbf{x} + (-\mathbf{x}) && \text{by V10} \\
&= \mathbf{0} && \text{by V5}
\end{aligned}$$

■

Thus, if we know that $\mathbb{V}$ is a vector space, we can determine the zero vector of $\mathbb{V}$ by finding $0\mathbf{x}$ for any $\mathbf{x} \in \mathbb{V}$. Similarly, we can determine the additive inverse of any vector $\mathbf{x} \in \mathbb{V}$ by computing $(-1)\mathbf{x}$.

EXAMPLE 4.2.10

Let $\mathbb{V} = \{(a, b) \mid a, b \in \mathbb{R}, b > 0\}$ and define addition by

$$(a, b) \oplus (c, d) = (ad + bc, bd)$$

and define scalar multiplication by

$$t \odot (a, b) = (tab^{t-1}, b^t)$$

Use Theorem 4.2.1 to show that axioms V4 and V5 hold for $\mathbb{V}$ with these operations. (Note that we are using $\oplus$ and $\odot$ to represent the operations of addition and scalar multiplication in the vector space to help distinguish the difference between these and the operations of addition and multiplication of real numbers.)

Solution: We do not know if $\mathbb{V}$ is a vector space, but if it is, then by Theorem 4.2.1 we must have

$$\mathbf{0} = 0 \odot (a, b) = (0ab^{-1}, b^0) = (0, 1)$$

Observe that $(0, 1) \in \mathbb{V}$ and for any $(a, b) \in \mathbb{V}$ we have

$$(a, b) \oplus (0, 1) = (a(1) + b(0), b(1)) = (a, b)$$

So, $\mathbb{V}$ satisfies V4 using $\mathbf{0} = (0, 1)$.

Similarly, if $\mathbb{V}$ is a vector space, then by Theorem 4.2.1 for any $\mathbf{x} = (a, b) \in \mathbb{V}$ we must have

$$(-\mathbf{x}) = (-1) \odot (a, b) = (-ab^{-2}, b^{-1})$$

Observe that for any $(a, b) \in \mathbb{V}$ we have $(-ab^{-2}, b^{-1}) \in \mathbb{V}$ since $b^{-1} > 0$ whenever $b > 0$. Also,

$$(a, b) \oplus (-ab^{-2}, b^{-1}) = (ab^{-1} + b(-ab^{-2}), bb^{-1}) = (ab^{-1} - ab^{-1}, 1) = (0, 1)$$

So, $\mathbb{V}$ satisfies V5 using $-(a, b) = (-ab^{-2}, b^{-1})$.

You are asked to complete the proof that $\mathbb{V}$ is indeed a vector space in Problem C5.

Subspaces

In Example 4.2.9 we showed that $\mathbb{S}$ is a vector space that is contained inside the vector space $\mathbb{R}^3$. Upon inspection, we see that our steps in the solution of the example also show that $\mathbb{S}$ is a subspace of $\mathbb{R}^3$.

Definition
Subspace

Suppose that $\mathbb{V}$ is a vector space. A non-empty subset $\mathbb{S}$ of $\mathbb{V}$ is called a **subspace** of $\mathbb{V}$ if for all $\mathbf{x}, \mathbf{y} \in \mathbb{S}$ and $s, t \in \mathbb{R}$ we have

$$s\mathbf{x} + t\mathbf{y} \in \mathbb{S} \tag{4.1}$$

Equivalently, if $\mathbb{S}$ is a subset of a vector space $\mathbb{V}$ and $\mathbb{S}$ is also a vector space using the same operations as $\mathbb{V}$, then $\mathbb{S}$ is a **subspace** of $\mathbb{V}$.

To prove that both definitions are equivalent, we first observe that if $\mathbb{S}$ is a vector space, then it satisfies vector space axioms V1 and V6. Combining V1 and V6 proves that $s\mathbf{x} + t\mathbf{y} \in \mathbb{S}$ for all $\mathbf{x}, \mathbf{y} \in \mathbb{S}$ and $s, t \in \mathbb{R}$.
On the other hand, as in Example 4.2.9, we know the operations must satisfy axioms V2, V3, V7, V8, V9, and V10 since $\mathbb{V}$ is a vector space. For the remaining axioms we have

V1: Taking $s = t = 1$ in equation (4.1) gives $\mathbf{x} + \mathbf{y} \in \mathbb{S}$.

V4: Taking $s = t = 0$ in equation (4.1) gives $0\mathbf{x} + 0\mathbf{y} \in \mathbb{S}$. But, $0\mathbf{x} + 0\mathbf{y} = \mathbf{0}$ by Theorem 4.2.1. So, $\mathbf{0} \in \mathbb{S}$.

V5: By Theorem 4.2.1 we know that $(-\mathbf{x}) = (-1)\mathbf{x}$. So, taking $s = -1$, $t = 0$ in equation (4.1) gives

$$(-\mathbf{x}) = (-1)\mathbf{x} + \mathbf{0} = (-1)\mathbf{x} + 0\mathbf{y} \in \mathbb{S}$$

V6: Taking $t = 0$ in equation (4.1) gives $s\mathbf{x} = s\mathbf{x} + \mathbf{0} = s\mathbf{x} + 0\mathbf{y} \in \mathbb{S}$.

Hence, all 10 axioms are satisfied. Therefore, $\mathbb{S}$ is also a vector space under the operations of $\mathbb{V}$.

Remarks

1. When proving that a set $\mathbb{S}$ is a subspace of a vector space $\mathbb{V}$, it is important not to forget to show that $\mathbb{S}$ is actually a subset of $\mathbb{V}$.

2. As with subspaces of $\mathbb{R}^n$ in Section 1.4, we typically show that the subset is non-empty by showing that it contains the zero vector of $\mathbb{V}$.

EXAMPLE 4.2.11

In Exercise 4.2.2 you proved that $\mathbb{S} = \left\{ \begin{bmatrix} a_1 & 0 \\ 0 & a_2 \end{bmatrix} \mid a_1, a_2 \in \mathbb{R} \right\}$ is a vector space. Thus, since $\mathbb{S}$ is a subset of $M_{2\times 2}(\mathbb{R})$, it is a subspace of $M_{2\times 2}(\mathbb{R})$.

EXAMPLE 4.2.12

Let $\mathbb{U} = \{\mathbf{p} \in P_3(\mathbb{R}) \mid \mathbf{p}(3) = 0\}$. Show that $\mathbb{U}$ is a subspace of $P_3(\mathbb{R})$.

Solution: By definition, $\mathbb{U}$ is a subset of $P_3(\mathbb{R})$. The zero vector $\mathbf{0} \in P_3(\mathbb{R})$ maps x to 0 for all x; hence it maps 3 to 0. Therefore, the zero vector of $P_3(\mathbb{R})$ is in $\mathbb{U}$, and hence $\mathbb{U}$ is non-empty.

Let $\mathbf{p}, \mathbf{q} \in \mathbb{U}$. Then, by definition of $\mathbb{U}$, polynomials $\mathbf{p}$ and $\mathbf{q}$ satisfy $\mathbf{p}(3) = 0$ and $\mathbf{q}(3) = 0$. Now, for any $s, t \in \mathbb{R}$, we need to show that the polynomial $s\mathbf{p} + t\mathbf{q}$ also satisfies the condition on $\mathbb{U}$. We have that

$$(s\mathbf{p} + t\mathbf{q})(3) = s\mathbf{p}(3) + t\mathbf{q}(3) = s(0) + t(0) = 0$$

Hence, $\mathbb{U}$ is a subspace of $P_3(\mathbb{R})$. This implies that $\mathbb{U}$ is itself a vector space using the operations of $P_3(\mathbb{R})$.

EXAMPLE 4.2.13

Define the **trace** of a 2×2 matrix by $\text{tr}\left(\begin{bmatrix} a_{11} & a_{12} \\ a_{21} & a_{22} \end{bmatrix}\right) = a_{11} + a_{22}$. Prove that $\mathbb{S} = \{A \in M_{2\times 2}(\mathbb{R}) \mid \text{tr}(A) = 0\}$ is a subspace of $M_{2\times 2}(\mathbb{R})$.

Solution: By definition, $\mathbb{S}$ is a subset of $M_{2\times 2}(\mathbb{R})$. The zero vector of $M_{2\times 2}(\mathbb{R})$ is $O_{2,2} = \begin{bmatrix} 0 & 0 \\ 0 & 0 \end{bmatrix}$. We have $\text{tr}(O_{2,2}) = 0 + 0 = 0$, so $O_{2,2} \in \mathbb{S}$.

Let $A, B \in \mathbb{S}$ and $s, t \in \mathbb{R}$. Then $a_{11} + a_{22} = \text{tr}(A) = 0$ and $b_{11} + b_{22} = \text{tr}(B) = 0$. Hence,

$$\begin{aligned} \text{tr}(sA + tB) &= \text{tr}\left(\begin{bmatrix} sa_{11} + tb_{11} & sa_{12} + tb_{12} \\ sa_{21} + tb_{21} & sa_{22} + tb_{22} \end{bmatrix}\right) \\ &= sa_{11} + tb_{11} + sa_{22} + tb_{22} \\ &= s(a_{11} + a_{22}) + t(b_{11} + b_{22}) \\ &= s(0) + t(0) = 0 \end{aligned}$$

Hence, $sA + tB \in \mathbb{S}$ and so $\mathbb{S}$ is a subspace of $M_{2\times 2}(\mathbb{R})$.

EXAMPLE 4.2.14

The vector space $\mathbb{R}^2$ is *not* a subspace of $\mathbb{R}^3$, since $\mathbb{R}^2$ is not a subset of $\mathbb{R}^3$. That is, if we take any vector $\vec{x} = \begin{bmatrix} x_1 \\ x_2 \end{bmatrix} \in \mathbb{R}^2$, this is not a vector in $\mathbb{R}^3$, since a vector in $\mathbb{R}^3$ has three components.

EXERCISE 4.2.4

Prove that $\mathbb{U} = \{a + bx + cx^2 \in P_2(\mathbb{R}) \mid b + c = a\}$ is a subspace of $P_2(\mathbb{R})$.

EXERCISE 4.2.5

Let $\mathbb{V}$ be a vector space. Prove that $\{\mathbf{0}\}$ is also a vector space, called the **trivial vector space**, under the same operations as $\mathbb{V}$ by proving it is a subspace of $\mathbb{V}$.

In Exercise 4.2.5 you proved that $\{\mathbf{0}\}$ is a subspace of any vector space $\mathbb{V}$. Furthermore, by definition, $\mathbb{V}$ is a subspace of itself. As in Chapter 1, the set of all possible linear combinations of a set of vectors in a vector space $\mathbb{V}$ is also a subspace.

Theorem 4.2.2

If $\{\mathbf{v}_1, \dots, \mathbf{v}_k\}$ is a set of vectors in a vector space $\mathbb{V}$ and $\mathbb{S}$ is the set of all possible linear combinations of these vectors,

$$\mathbb{S} = \{t_1\mathbf{v}_1 + \cdots + t_k\mathbf{v}_k \mid t_1, \dots, t_k \in \mathbb{R}\}$$

then $\mathbb{S}$ is a subspace of $\mathbb{V}$.

The proof of Theorem 4.2.2 is identical to the proof of Theorem 1.4.2 and hence is omitted.

PROBLEMS 4.2

Practice Problems

For Problems A1–A8, determine, with proof, whether the set is a subspace of the given vector space.

A1 $\left\{\begin{bmatrix} x_1 \\ x_2 \\ x_3 \\ x_4 \end{bmatrix} \in \mathbb{R}^4 \mid x_1 + 2x_2 = 0\right\}$ of $\mathbb{R}^4$

A2 $\left\{\begin{bmatrix} a_1 & a_2 \\ a_3 & a_4 \end{bmatrix} \in M_{2\times 2}(\mathbb{R}) \mid a_1 + 2a_2 = 0\right\}$ of $M_{2\times 2}(\mathbb{R})$

A3 $\{a_0 + a_1x + a_2x^2 + a_3x^3 \in P_3(\mathbb{R}) \mid a_0 + 2a_1 = 0\}$ of $P_3(\mathbb{R})$

A4 $\left\{\begin{bmatrix} a_1 & a_2 \\ a_3 & a_4 \end{bmatrix} \mid a_1, a_2, a_3, a_4 \in \mathbb{Z}\right\}$ of $M_{2\times 2}(\mathbb{R})$

A5 $\left\{\begin{bmatrix} a_1 & a_2 \\ a_3 & a_4 \end{bmatrix} \in M_{2\times 2}(\mathbb{R}) \mid a_1a_4 - a_2a_3 = 0\right\}$ of $M_{2\times 2}(\mathbb{R})$

A6 $\left\{\begin{bmatrix} a_1 & a_2 \\ 0 & 0 \end{bmatrix} \in M_{2\times 2}(\mathbb{R}) \mid a_1 = a_2\right\}$ of $M_{2\times 2}(\mathbb{R})$

A7 $\{\mathbf{p} \in P_3(\mathbb{R}) \mid \mathbf{p}(1) = 1\}$ of $P_3(\mathbb{R})$

A8 $\{\mathbf{p} \in P_3(\mathbb{R}) \mid \mathbf{p}(2) = 0 \in \mathbb{R}\}$ of $P_3(\mathbb{R})$

For Problems A9–A11, determine, with proof, whether the subset of $M_{n\times n}(\mathbb{R})$ is a subspace of $M_{n\times n}(\mathbb{R})$.

A9 The subset of matrices that are in row echelon form.

A10 The subset of upper triangular matrices.

A11 The subset of matrices such that $A^T = A$.

For Problems A12–A16, determine, with proof, whether the subset of $P_5(\mathbb{R})$ is a subspace of $P_5(\mathbb{R})$.

A12 $\{\mathbf{p} \in P_5(\mathbb{R}) \mid \mathbf{p}(-x) = \mathbf{p}(x) \text{ for all } x \in \mathbb{R}\}$

A13 $\{(1 + x^2)\mathbf{p} \mid \mathbf{p} \in P_3(\mathbb{R})\}$

A14 $\{a_0 + a_1x + \cdots + a_4x^4 \in P_5(\mathbb{R}) \mid a_0 = a_4, a_1 = a_3\}$

A15 $\{\mathbf{p} \in P_5(\mathbb{R}) \mid p(0) = 1\}$

A16 $\{a_0 + a_1x + a_2x^2 \mid a_0, a_1, a_2 \in \mathbb{R}\}$

For Problems A17–A20, let $\mathcal{F}$ be the vector space of all real-valued functions of a real variable. Determine, with proof, whether the subset of $\mathcal{F}$ is a subspace of $\mathcal{F}$.

A17 $\{f \in \mathcal{F} \mid f(3) = 0\}$

A18 $\{f \in \mathcal{F} \mid f(3) = 1\}$

A19 $\{f \in \mathcal{F} \mid f(-x) = f(x) \text{ for all } x \in \mathbb{R}\}$

A20 $\{f \in \mathcal{F} \mid f(x) \geq 0 \text{ for all } x \in \mathbb{R}\}$

Homework Problems

For Problems B1–B9, determine, with proof, whether the set is a subspace of the given vector space.

B1 $\left\{\begin{bmatrix} x_1 \\ x_2 \\ x_3 \end{bmatrix} \in \mathbb{R}^3 \mid x_1x_2 = x_3\right\}$ of $\mathbb{R}^3$

B2 $\left\{\begin{bmatrix} x_1 \\ x_2 \\ x_3 \end{bmatrix} \in \mathbb{R}^3 \mid x_1 = x_2 + 2x_3\right\}$ of $\mathbb{R}^3$

B3 $\left\{\begin{bmatrix} a_1 & a_2 \\ a_3 & a_4 \end{bmatrix} \in M_{2\times 2}(\mathbb{R}) \mid a_1 = a_2 = a_4\right\}$ of $M_{2\times 2}(\mathbb{R})$

B4 $\{a_0 + a_1x + a_2x^2 \in P_2(\mathbb{R}) \mid a_0 + a_1 = 1\}$ of $P_2(\mathbb{R})$

B5 $\left\{\begin{bmatrix} a_1 & a_2 \\ a_3 & a_4 \end{bmatrix} \mid a_2^2 = a_3^2\right\}$ of $M_{2\times 2}(\mathbb{R})$

B6 $\{\mathbf{p} \in P_2(\mathbb{R}) \mid \mathbf{p}(-1) = 0 \in \mathbb{R}\}$ of $P_2(\mathbb{R})$

B7 $\left\{\begin{bmatrix} 0 & a_1 \\ a_2 & 0 \end{bmatrix} \in M_{2\times 2}(\mathbb{R}) \mid a_1 = -a_2\right\}$ of $M_{2\times 2}(\mathbb{R})$

B8 $\left\{\begin{bmatrix} a_1 & a_2 \\ 1 & 1 \end{bmatrix} \in M_{2\times 2}(\mathbb{R}) \mid 2a_1 = 3a_2\right\}$ of $M_{2\times 2}(\mathbb{R})$

B9 $\{\mathbf{p} \in P_2(\mathbb{R}) \mid \mathbf{p}(1) = 0 \text{ and } \mathbf{p}(2) = 0\}$ of $P_2(\mathbb{R})$

For Problems B10–B17, determine, with proof, whether the subset of $M_{2\times 2}(\mathbb{R})$ is a subspace of $M_{2\times 2}(\mathbb{R})$.

B10 The subset of matrices A such that $A\begin{bmatrix} 1 \\ 1 \end{bmatrix} = \begin{bmatrix} 0 \\ 0 \end{bmatrix}$.

B11 The subset of matrices A such that $A\begin{bmatrix} 1 \\ 2 \end{bmatrix} = \begin{bmatrix} 1 \\ 2 \end{bmatrix}$.

B12 The subset of matrices A such that

$$A\begin{bmatrix} 1 & 2 \\ 1 & 2 \end{bmatrix} = \begin{bmatrix} 0 & 0 \\ 0 & 0 \end{bmatrix}$$

B13 The subset of matrices A such that

$$A\begin{bmatrix} 1 & 2 \\ 1 & 2 \end{bmatrix} = \begin{bmatrix} 1 & 2 \\ 1 & 2 \end{bmatrix}A$$

B14 The subset of elementary matrices.

B15 The subset of non-invertible matrices.

B16 The subset of matrices such that $A^T = -A$.

B17 The subset of matrices such that $A^2 = O_{2,2}$.

For Problems B18–B22, determine, with proof, whether the subset of $P_2(\mathbb{R})$ is a subspace of $P_2(\mathbb{R})$.

B18 $\{\mathbf{p} \mid \mathbf{p}(-x) = -\mathbf{p}(x) \text{ for all } x \in \mathbb{R}\}$

B19 $\{x\mathbf{p} \mid \mathbf{p} \in P_1(\mathbb{R})\}$

B20 The subset of polynomials $\mathbf{p}$ that have derivative equal to 0.

B21 The subset of polynomials $\mathbf{p}$ that have all real roots.

B22 The subset of polynomials $\mathbf{p}$ that have degree 1.

For Problems B23–B27, let $\mathcal{F}$ be the vector space of all real-valued functions of a real variable. Determine, with proof, whether the subset of $\mathcal{F}$ is a subspace of $\mathcal{F}$.

B23 $\{f \in \mathcal{F} \mid f(1) = f(-1)\}$

B24 $\{f \in \mathcal{F} \mid f(2) + f(3) = 0\}$

B25 $\{f \in \mathcal{F} \mid f(1) + f(2) = 1\}$

B26 $\{f \in \mathcal{F} \mid |f(x)| \leq 1\}$

B27 $\{f \in \mathcal{F} \mid f \text{ is non decreasing on } \mathbb{R}\}$

Conceptual Problems

For Problems C1–C4, let $\mathbb{V}$ be a vector space. Prove the statement, clearly justifying each step.

C1 $-\mathbf{x} = (-1)\mathbf{x}$ for every $\mathbf{x} \in \mathbb{V}$.

C2 $t\mathbf{0} = \mathbf{0}$ for every $t \in \mathbb{R}$.

C3 If $t\mathbf{x} = \mathbf{0}$, then either $t = 0$ or $\mathbf{x} = \mathbf{0}$.

C4 If $\mathbf{x} + \mathbf{y} = \mathbf{z} + \mathbf{x}$, then $\mathbf{y} = \mathbf{z}$.

C5 Let $\mathbb{V} = \{(a, b) \mid a, b \in \mathbb{R}, b > 0\}$ and define addition by $(a, b) \oplus (c, d) = (ad + bc, bd)$ and scalar multiplication by $t \odot (a, b) = (tab^{t-1}, b^t)$ for any $t \in \mathbb{R}$. Prove that $\mathbb{V}$ is a vector space with these operations.

C6 Let $\mathbb{V} = \{x \in \mathbb{R} \mid x > 0\}$ and define addition by $x \oplus y = xy$ and scalar multiplication by $t \odot x = x^t$ for any $t \in \mathbb{R}$. Prove that $\mathbb{V}$ is a vector space with these operations.

C7 Let $\mathbb{C} = \{a + bi \mid a, b \in \mathbb{R}\}$ be the set of complex numbers. Prove that $\mathbb{C}$ is a vector space with addition and scalar multiplication defined by

$$(a + bi) + (c + di) = (a + c) + (b + d)i$$
$$t(a + bi) = (ta) + (tb)i, \quad \text{for all } t \in \mathbb{R}$$

C8 Let $\mathbb{L}$ denote the set of all linear operators $L : \mathbb{R}^n \to \mathbb{R}^n$ with standard addition and scalar multiplication of linear mappings. Prove that $\mathbb{L}$ is a vector space under these operations.

C9 Categorize all subspaces of $\mathbb{R}^2$.

C10 Suppose that $\mathbb{U}$ and $\mathbb{V}$ are vector spaces over $\mathbb{R}$. The **Cartesian product** of $\mathbb{U}$ and $\mathbb{V}$ is defined to be

$$\mathbb{U} \times \mathbb{V} = \{(\mathbf{u}, \mathbf{v}) \mid \mathbf{u} \in \mathbb{U}, \mathbf{v} \in \mathbb{V}\}$$

(a) In $\mathbb{U} \times \mathbb{V}$ define addition and scalar multiplication by

$$(\mathbf{u}_1, \mathbf{v}_1) \oplus (\mathbf{u}_2, \mathbf{v}_2) = (\mathbf{u}_1 + \mathbf{u}_2, \mathbf{v}_1 + \mathbf{v}_2)$$
$$t \odot (\mathbf{u}_1, \mathbf{v}_1) = (t\mathbf{u}_1, t\mathbf{v}_1)$$

Verify that with these operations that $\mathbb{U} \times \mathbb{V}$ is a vector space.

(b) Verify that $\mathbb{U} \times \{\mathbf{0}_{\mathbb{V}}\}$ is a subspace of $\mathbb{U} \times \mathbb{V}$.

(c) Suppose instead that scalar multiplication is defined by $t \odot (\mathbf{u}, \mathbf{v}) = (t\mathbf{u}, \mathbf{v})$, while addition is defined as in part (a). Is $\mathbb{U} \times \mathbb{V}$ a vector space with these operations?

For Problems C11–C13, let $\mathbb{V} = \{(x, y) \mid x, y \in \mathbb{R}\}$. Show that $\mathbb{V}$ is not a vector space for the given operations of addition and scalar multiplication by listing all the properties that fail to hold.

C11
$$(x_1, y_1) + (x_2, y_2) = (x_1 + y_2, x_2 + y_1)$$
$$t(x_1, x_2) = (tx_2, tx_1)$$

C12
$$(x_1, y_1) + (x_2, y_2) = (x_1 + x_2, 0)$$
$$t(x_1, x_2) = (tx_1, 0)$$

C13
$$(x_1, y_1) + (x_2, y_2) = (x_1 + x_2 + 1, y_1 + y_2 + 1)$$
$$t(x_1, x_2) = (tx_1, tx_2)$$

4.3 Bases and Dimensions

We have seen the importance and usefulness of bases in $\mathbb{R}^n$. A basis for a subspace $\mathbb{S}$ of $\mathbb{R}^n$ provides an explicit formula for every vector in $\mathbb{S}$. Moreover, we can geometrically view this explicit formula as a coordinate system. That is, a basis of a subspace of $\mathbb{R}^n$ forms a set of coordinate axes for that subspace. Thus, we desire to extend the concept of bases to general vector spaces.

We begin by defining the concepts of spanning and linear independence in general vector spaces.

Definition
Span
Spanning Set

Let $\mathcal{B} = \{\mathbf{v}_1, \dots, \mathbf{v}_k\}$ be a set of vectors in a vector space $\mathbb{V}$. The **span** of $\mathcal{B}$ is defined as

$$\operatorname{Span}\mathcal{B} = \operatorname{Span}\{\mathbf{v}_1, \dots, \mathbf{v}_k\} = \{c_1\mathbf{v}_1 + \cdots + c_k\mathbf{v}_k \mid c_1, \dots, c_k \in \mathbb{R}\}$$

The subspace $\mathbb{S} = \operatorname{Span}\mathcal{B}$ of $\mathbb{V}$ is said to be **spanned** by $\mathcal{B}$, and $\mathcal{B}$ is called a **spanning set** for $\mathbb{S}$.

Definition
Linearly Dependent
Linearly Independent

Let $\mathcal{B} = \{\mathbf{v}_1, \dots, \mathbf{v}_k\}$ be a set of vectors in a vector space $\mathbb{V}$. The set $\mathcal{B}$ is said to be **linearly dependent** if there exists real coefficients $t_1, \dots, t_k$ not all zero such that

$$t_1\mathbf{v}_1 + \cdots + t_k\mathbf{v}_k = \mathbf{0}$$

The set $\mathcal{B}$ is said to be **linearly independent** if the only solution to

$$t_1\mathbf{v}_1 + \cdots + t_k\mathbf{v}_k = \mathbf{0}$$

is the trivial solution $t_1 = \cdots = t_k = 0$.

In a general vector space, the procedure for determining if a vector is in a span, if a set spans a subspace, or if a set is linearly independent is exactly the same as we saw for $\mathbb{R}^n$, $M_{m\times n}(\mathbb{R})$, and $P_n(\mathbb{R})$ in Sections 1.4, 2.3, 3.1, and 4.1.

Bases

Definition
Basis

A set $\mathcal{B}$ in a vector space $\mathbb{V}$ is called a **basis** for $\mathbb{V}$ if
(1) $\mathcal{B}$ is linearly independent, and
(2) $\operatorname{Span}\mathcal{B} = \mathbb{V}$.
A basis for the trivial vector space $\{\mathbf{0}\}$ is defined to be the empty set $\emptyset = \{\,\}$.

It is clear that we should want a basis $\mathcal{B}$ for a vector space $\mathbb{V}$ to be a spanning set so that every vector in $\mathbb{V}$ can be written as a linear combination of the vectors in $\mathcal{B}$. Why would it be important that the set $\mathcal{B}$ be linearly independent? The following theorem answers this question.

Theorem 4.3.1 Unique Representation Theorem

Let $\mathcal{B} = \{\mathbf{v}_1, \dots, \mathbf{v}_n\}$ be a spanning set for a vector space $\mathbb{V}$. Every vector in $\mathbb{V}$ can be expressed in a *unique* way as a linear combination of the vectors of $\mathcal{B}$ if and only if the set $\mathcal{B}$ is linearly independent.

Proof: Let $\mathbf{x}$ be any vector in $\mathbb{V}$. Since Span $\mathcal{B} = \mathbb{V}$, we have that $\mathbf{x}$ can be written as a linear combination of the vectors in $\mathcal{B}$. Assume that there are linear combinations

$$\mathbf{x} = a_1\mathbf{v}_1 + \cdots + a_n\mathbf{v}_n \quad \text{and} \quad \mathbf{x} = b_1\mathbf{v}_1 + \cdots + b_n\mathbf{v}_n$$

This gives

$$a_1\mathbf{v}_1 + \cdots + a_n\mathbf{v}_n = \mathbf{x} = b_1\mathbf{v}_1 + \cdots + b_n\mathbf{v}_n$$

which implies

$$\mathbf{0} = \mathbf{x} - \mathbf{x} = (a_1\mathbf{v}_1 + \cdots + a_n\mathbf{v}_n) - (b_1\mathbf{v}_1 + \cdots + b_n\mathbf{v}_n) = (a_1 - b_1)\mathbf{v}_1 + \cdots + (a_n - b_n)\mathbf{v}_n$$

If $\mathcal{B}$ is linearly independent, then we must have $a_i - b_i = 0$, so $a_i = b_i$ for $1 \leq i \leq n$. Hence, $\mathbf{x}$ has a unique representation.

On the other hand, if $\mathcal{B}$ is linearly dependent, then

$$\mathbf{0} = t_1\mathbf{v}_1 + \cdots + t_n\mathbf{v}_n$$

has a solution where at least one of the coefficients is non-zero. But

$$\mathbf{0} = 0\mathbf{v}_1 + \cdots + 0\mathbf{v}_n$$

Hence, $\mathbf{0}$ can be expressed as a linear combination of the vectors in $\mathcal{B}$ in multiple ways. ■

Thus, if $\mathcal{B}$ is a basis for a vector space $\mathbb{V}$, then every vector in $\mathbb{V}$ can be written as a unique linear combination of the vectors in $\mathcal{B}$.

EXAMPLE 4.3.1

The set of vectors $\left\{\begin{bmatrix}1 & 0\\0 & 0\end{bmatrix}, \begin{bmatrix}0 & 1\\0 & 0\end{bmatrix}, \begin{bmatrix}0 & 0\\1 & 0\end{bmatrix}, \begin{bmatrix}0 & 0\\0 & 1\end{bmatrix}\right\}$ in Exercise 3.1.2 is a basis for $M_{2\times2}(\mathbb{R})$. It is called the **standard basis** for $M_{2\times2}(\mathbb{R})$.

EXAMPLE 4.3.2

The set of vectors $\{1, x, x^2, x^3\}$ in Exercise 4.1.2 is a basis for $P_3(\mathbb{R})$. In particular, the set $\{1, x, \dots, x^n\}$ is called the **standard basis** for $P_n(\mathbb{R})$.

Not surprisingly, the method for proving a set $\mathcal{B}$ is a basis for a vector space $\mathbb{V}$ is the same as the procedure we used in Section 2.3 for subspaces of $\mathbb{R}^n$.

EXAMPLE 4.3.3

Is the set $C = \{3 + 2x + 2x^2, 1 + x^2, 1 + x + x^2\}$ a basis for $P_2(\mathbb{R})$?

Solution: We need to determine whether C is linearly independent and if it spans $P_2(\mathbb{R})$. Consider the equation

$$\begin{aligned} a_0 + a_1x + a_2x^2 &= t_1(3 + 2x + 2x^2) + t_2(1 + x^2) + t_3(1 + x + x^2) \\ &= (3t_1 + t_2 + t_3) + (2t_1 + t_3)x + (2t_1 + t_2 + t_3)x^2 \end{aligned}$$

Row reducing the coefficient matrix of the corresponding system gives

$$\begin{bmatrix} 3 & 1 & 1 \\ 2 & 0 & 1 \\ 2 & 1 & 1 \end{bmatrix} \sim \begin{bmatrix} 1 & 0 & 0 \\ 0 & 1 & 0 \\ 0 & 0 & 1 \end{bmatrix}$$

By the System-Rank Theorem (3), the system is consistent for all $a_0, a_1, a_2 \in \mathbb{R}$. Thus, C spans $P_2(\mathbb{R})$. By the System-Rank Theorem (2), the system has a unique solution for all $a_0, a_1, a_2 \in \mathbb{R}$ (including $a_0 = a_1 = a_2 = 0$). Thus, C is also linearly independent and hence is a basis for $P_2(\mathbb{R})$.

EXAMPLE 4.3.4

Determine whether the set $C = \left\{ \begin{bmatrix} 1 & 0 \\ 1 & 0 \end{bmatrix}, \begin{bmatrix} 1 & 2 \\ 1 & 1 \end{bmatrix}, \begin{bmatrix} 1 & 3 \\ 0 & 1 \end{bmatrix} \right\}$ is a basis for $M_{2\times 2}(\mathbb{R})$.

Solution: Consider

$$\begin{bmatrix} x_1 & x_2 \\ x_3 & x_4 \end{bmatrix} = t_1 \begin{bmatrix} 1 & 0 \\ 1 & 0 \end{bmatrix} + t_2 \begin{bmatrix} 1 & 2 \\ 1 & 1 \end{bmatrix} + t_3 \begin{bmatrix} 1 & 3 \\ 0 & 1 \end{bmatrix} = \begin{bmatrix} t_1 + t_2 + t_3 & 2t_2 + 3t_3 \\ t_1 + t_2 & t_2 + t_3 \end{bmatrix}$$

This gives a system of 4 equation in the 3 variables t_1, t_2, t_3. The maximum rank of the coefficient matrix is then 3. Therefore, by the System-Rank Theorem (3), the system is not consistent for all $x_1, x_2, x_3, x_4 \in \mathbb{R}$ and hence C cannot span $M_{2\times 2}(\mathbb{R})$.

EXAMPLE 4.3.5

Determine whether the set $\mathcal{B} = \left\{ \begin{bmatrix} 1 & 2 \\ -1 & 1 \end{bmatrix}, \begin{bmatrix} 0 & 1 \\ 3 & 1 \end{bmatrix}, \begin{bmatrix} 2 & 5 \\ 1 & 3 \end{bmatrix} \right\}$ is a basis for the subspace Span $\mathcal{B}$ of $M_{2\times 2}(\mathbb{R})$.

Solution: Since $\mathcal{B}$ is a spanning set for Span $\mathcal{B}$, we just need to check if the vectors in $\mathcal{B}$ are linearly independent. Consider the equation

$$\begin{bmatrix} 0 & 0 \\ 0 & 0 \end{bmatrix} = t_1 \begin{bmatrix} 1 & 2 \\ -1 & 1 \end{bmatrix} + t_2 \begin{bmatrix} 0 & 1 \\ 3 & 1 \end{bmatrix} + t_3 \begin{bmatrix} 2 & 5 \\ 1 & 3 \end{bmatrix} = \begin{bmatrix} t_1 + 2t_3 & 2t_1 + t_2 + 5t_3 \\ -t_1 + 3t_2 + t_3 & t_1 + t_2 + 3t_3 \end{bmatrix}$$

Row reducing the coefficient matrix of the corresponding system gives

$$\begin{bmatrix} 1 & 0 & 2 \\ 2 & 1 & 5 \\ -1 & 3 & 1 \\ 1 & 1 & 3 \end{bmatrix} \sim \begin{bmatrix} 1 & 0 & 2 \\ 0 & 1 & 1 \\ 0 & 0 & 0 \\ 0 & 0 & 0 \end{bmatrix}$$

By the System-Rank Theorem (2), the system has infinitely many solutions. Thus, $\mathcal{B}$ is linearly dependent and hence is not a basis.

Obtaining a Basis from a Finite Spanning Set

Many times throughout the rest of this book, we will need to determine a basis for a vector space. One standard way of doing this is to first determine a spanning set for the vector space and then to remove vectors from the spanning set until we have a basis. We will need the following theorem.

Theorem 4.3.2 Basis Reduction Theorem

If $\mathcal{T} = \{\mathbf{v}_1, \dots, \mathbf{v}_k\}$ is a spanning set for a non-trivial vector space $\mathbb{V}$, then some subset of $\mathcal{T}$ is a basis for $\mathbb{V}$.

Proof: If $\mathcal{T}$ is linearly independent, then $\mathcal{T}$ is a basis for $\mathbb{V}$, and we are done. If $\mathcal{T}$ is linearly dependent, then $t_1\mathbf{v}_1 + \cdots + t_k\mathbf{v}_k = \mathbf{0}$ has a solution where at least one of the coefficients is non-zero, say $t_i \neq 0$. Then, we can solve the equation for $\mathbf{v}_i$ to get

$$\mathbf{v}_i = -\frac{1}{t_i}(t_1\mathbf{v}_1 + \cdots + t_{i-1}\mathbf{v}_{i-1} + t_{i+1}\mathbf{v}_{i+1} + \cdots + t_k\mathbf{v}_k)$$

So, for any $\mathbf{x} \in \mathbb{V}$ we have

$$\begin{aligned}\mathbf{x} =& a_1\mathbf{v}_1 + \cdots + a_{i-1}\mathbf{v}_{i-1} + a_i\mathbf{v}_i + a_{i+1}\mathbf{v}_{i+1} + \cdots + a_k\mathbf{v}_k \\ =& a_1\mathbf{v}_1 + \cdots + a_{i-1}\mathbf{v}_{i-1} + a_i\left[-\frac{1}{t_i}(t_1\mathbf{v}_1 + \cdots + t_{i-1}\mathbf{v}_{i-1} + t_{i+1}\mathbf{v}_{i+1} + \cdots + t_k\mathbf{v}_k)\right] \\ &+ a_{i+1}\mathbf{v}_{i+1} + \cdots + a_k\mathbf{v}_k \\ =& (a_1 - \frac{a_it_1}{t_i})\mathbf{v}_1 + \cdots + (a_{i-1} - \frac{a_it_{i-1}}{t_i})\mathbf{v}_{i-1} + (a_{i+1} - \frac{a_it_{i+1}}{t_i})\mathbf{v}_{i+1} + \cdots + (a_n - \frac{a_it_n}{t_i})\mathbf{v}_n\end{aligned}$$

Thus, $\mathbf{x} \in \operatorname{Span}\{\mathbf{v}_1, \dots, \mathbf{v}_{i-1}, \mathbf{v}_{i+1}, \dots, \mathbf{v}_n\}$. That is, $\mathcal{T}_1 = \{\mathbf{v}_1, \dots, \mathbf{v}_{i-1}, \mathbf{v}_{i+1}, \dots, \mathbf{v}_n\}$ is a spanning set for $\mathbb{V}$. If $\mathcal{T}_1$ is linearly independent, it is a basis for $\mathbb{V}$, and the procedure is finished. Otherwise, we repeat the procedure to omit a second vector, say $\mathbf{v}_j$, and get another subset $\mathcal{T}_2$, which still spans $\mathbb{V}$. In this fashion, we must eventually get a linearly independent set. (Certainly, if there is only one non-zero vector left, it forms a linearly independent set.) Thus, we obtain a subset of $\mathcal{T}$ that is a basis for $\mathbb{V}$. ■

EXAMPLE 4.3.6 Determine a basis for the subspace $\mathbb{S} = \{\mathbf{p} \in P_2(\mathbb{R}) \mid \mathbf{p}(1) = 0\}$ of $P_2(\mathbb{R})$.

Solution: We first find a spanning set for $\mathbb{S}$. By the Factor Theorem, if $\mathbf{p}(1) = 0$, then $(x - 1)$ is a factor of $\mathbf{p}$. That is, every polynomial $\mathbf{p} \in \mathbb{S}$ can be written in the form

$$\mathbf{p}(x) = (x - 1)(ax + b) = a(x^2 - x) + b(x - 1)$$

Thus, we see that $\mathcal{T} = \{x^2 - x, x - 1\}$ spans $\mathbb{S}$.

Next, we need to determine whether $\mathcal{T}$ is linearly independent. Consider

$$0 = t_1(x^2 - x) + t_2(x - 1) = t_1x^2 + (-t_1 + t_2)x - t_2$$

The only solution is $t_1 = t_2 = 0$. Hence, $\mathcal{T}$ is linearly independent. Thus, $\mathcal{T}$ is a basis for $\mathbb{S}$.

EXAMPLE 4.3.7

Consider the subspace of $\mathbb{R}^3$ spanned by $\mathcal{T} = \left\{ \begin{bmatrix} 1 \\ 1 \\ -2 \end{bmatrix}, \begin{bmatrix} 2 \\ -1 \\ 1 \end{bmatrix}, \begin{bmatrix} 1 \\ -2 \\ 3 \end{bmatrix}, \begin{bmatrix} 1 \\ 5 \\ 3 \end{bmatrix} \right\}$. Determine a subset of $\mathcal{T}$ that is a basis for Span $\mathcal{T}$.

Solution: Consider

$$\begin{bmatrix} 0 \\ 0 \\ 0 \end{bmatrix} = t_1 \begin{bmatrix} 1 \\ 1 \\ -2 \end{bmatrix} + t_2 \begin{bmatrix} 2 \\ -1 \\ 1 \end{bmatrix} + t_3 \begin{bmatrix} 1 \\ -2 \\ 3 \end{bmatrix} + t_4 \begin{bmatrix} 1 \\ 5 \\ 3 \end{bmatrix} = \begin{bmatrix} t_1 + 2t_2 + t_3 + t_4 \\ t_1 - t_2 - 2t_3 + 5t_4 \\ -2t_1 + t_2 + 3t_3 + 3t_4 \end{bmatrix}$$

We row reduce the corresponding coefficent matrix

$$\begin{bmatrix} 1 & 2 & 1 & 1 \\ 1 & -1 & -2 & 5 \\ -2 & 1 & 3 & 3 \end{bmatrix} \sim \begin{bmatrix} 1 & 0 & -1 & 0 \\ 0 & 1 & 1 & 0 \\ 0 & 0 & 0 & 1 \end{bmatrix}$$

The general solution is $\begin{bmatrix} t_1 \\ t_2 \\ t_3 \\ t_4 \end{bmatrix} = s \begin{bmatrix} 1 \\ -1 \\ 1 \\ 0 \end{bmatrix}$, $s \in \mathbb{R}$. Taking $s = 1$, we get $\begin{bmatrix} t_1 \\ t_2 \\ t_3 \\ t_4 \end{bmatrix} = \begin{bmatrix} 1 \\ -1 \\ 1 \\ 0 \end{bmatrix}$, which gives

$$\begin{bmatrix} 1 \\ 1 \\ -2 \end{bmatrix} - \begin{bmatrix} 2 \\ -1 \\ 1 \end{bmatrix} + \begin{bmatrix} 1 \\ -2 \\ 3 \end{bmatrix} = \begin{bmatrix} 0 \\ 0 \\ 0 \end{bmatrix} \quad \text{or} \quad \begin{bmatrix} 1 \\ -2 \\ 3 \end{bmatrix} = -\begin{bmatrix} 1 \\ 1 \\ -2 \end{bmatrix} + \begin{bmatrix} 2 \\ -1 \\ 1 \end{bmatrix}$$

Thus, we can omit $\begin{bmatrix} 1 \\ -2 \\ 3 \end{bmatrix}$ from $\mathcal{T}$ and consider $\mathcal{T}_1 = \left\{ \begin{bmatrix} 1 \\ 1 \\ -2 \end{bmatrix}, \begin{bmatrix} 2 \\ -1 \\ 1 \end{bmatrix}, \begin{bmatrix} 1 \\ 5 \\ 3 \end{bmatrix} \right\}$.

Now consider

$$\begin{bmatrix} 0 \\ 0 \\ 0 \end{bmatrix} = t_1 \begin{bmatrix} 1 \\ 1 \\ -2 \end{bmatrix} + t_2 \begin{bmatrix} 2 \\ -1 \\ 1 \end{bmatrix} + t_3 \begin{bmatrix} 1 \\ 5 \\ 3 \end{bmatrix}$$

The coefficient matrix corresponding to this system is the same as above except that the third column is omitted, so the same row operations give

$$\begin{bmatrix} 1 & 2 & 1 \\ 1 & -1 & 5 \\ -2 & 1 & 3 \end{bmatrix} \sim \begin{bmatrix} 1 & 0 & 0 \\ 0 & 1 & 0 \\ 0 & 0 & 1 \end{bmatrix}$$

Hence, the only solution is $t_1 = t_2 = t_3 = 0$ and we conclude that $\left\{ \begin{bmatrix} 1 \\ 1 \\ -2 \end{bmatrix}, \begin{bmatrix} 2 \\ -1 \\ 1 \end{bmatrix}, \begin{bmatrix} 1 \\ 5 \\ 3 \end{bmatrix} \right\}$ is linearly independent and thus a basis for Span $\mathcal{T}$.

CONNECTION

Compare this to how we solved Example 3.4.7 on page 196 and our procedure for finding a basis for the column space of a matrix (Theorem 3.4.5).

EXERCISE 4.3.1 Consider the subspace of $P_2(\mathbb{R})$ spanned by $\mathcal{B} = \{1 - x, 2 + 2x + x^2, x + x^2, 1 + x^2\}$. Determine a subset of $\mathcal{B}$ that is a basis for Span $\mathcal{B}$.

Dimension

We saw in Section 2.3 that every basis for a subspace $\mathbb{S}$ of $\mathbb{R}^n$ contains the same number of vectors. We now prove that this result holds for general vector spaces. Observe that the proof is identical to that in Section 2.3.

Theorem 4.3.3 Suppose that $\mathcal{B} = \{\mathbf{v}_1, \dots, \mathbf{v}_\ell\}$ is a basis for a non-trivial vector space $\mathbb{V}$ and that $\{\mathbf{u}_1, \dots, \mathbf{u}_k\}$ is a set in $\mathbb{V}$. If $k > \ell$, then $\{\mathbf{u}_1, \dots, \mathbf{u}_k\}$ is linearly dependent.

Proof: Since each $\mathbf{u}_i$, $1 \leq i \leq k$, is a vector in $\mathbb{V}$ and $\mathcal{B}$ is a basis for $\mathbb{V}$, by the Unique Representation Theorem, each $\mathbf{u}_i$ can be written as a unique linear combination of the vectors in $\mathcal{B}$. We get

$$\begin{aligned}
\mathbf{u}_1 &= a_{11}\mathbf{v}_1 + a_{21}\mathbf{v}_2 + \cdots + a_{\ell 1}\mathbf{v}_\ell \\
\mathbf{u}_2 &= a_{12}\mathbf{v}_1 + a_{22}\mathbf{v}_2 + \cdots + a_{\ell 2}\mathbf{v}_\ell \\
&\vdots \\
\mathbf{u}_k &= a_{1k}\mathbf{v}_1 + a_{2k}\mathbf{v}_2 + \cdots + a_{\ell k}\mathbf{v}_\ell
\end{aligned}$$

Consider the equation

$$\begin{aligned}
\mathbf{0} &= t_1\mathbf{u}_1 + \cdots + t_k\mathbf{u}_k \qquad (4.2) \\
&= t_1(a_{11}\mathbf{v}_1 + a_{21}\mathbf{v}_2 + \cdots + a_{\ell 1}\mathbf{v}_\ell) + \cdots + t_k(a_{1k}\mathbf{v}_1 + a_{2k}\mathbf{v}_2 + \cdots + a_{\ell k}\mathbf{v}_\ell) \\
&= (a_{11}t_1 + \cdots + a_{1k}t_k)\mathbf{v}_1 + \cdots + (a_{\ell 1}t_1 + \cdots + a_{\ell k}t_k)\mathbf{v}_\ell
\end{aligned}$$

But, $\{\mathbf{v}_1, \dots, \mathbf{v}_\ell\}$ is linearly independent, so the only solution to this equation is

$$\begin{aligned}
a_{11}t_1 + \cdots + a_{1k}t_k &= 0 \\
\vdots \qquad\qquad & \quad \vdots \\
a_{\ell 1}t_1 + \cdots + a_{\ell k}t_k &= 0
\end{aligned}$$

The rank of the coefficient matrix of this homogeneous system is at most ℓ because $\ell < k$. Hence, by the System-Rank Theorem (2), the solution space has at least $k - \ell > 0$ parameters. Therefore, there are infinitely many possible $t_1, \dots, t_k$ and so $\{\mathbf{u}_1, \dots, \mathbf{u}_k\}$ is linearly dependent since equation (4.2) has infinitely many solutions. ■

Theorem 4.3.4

Dimension Theorem

If $\{\mathbf{v}_1, \ldots, \mathbf{v}_\ell\}$ and $\{\mathbf{u}_1, \ldots, \mathbf{u}_k\}$ are both bases of a vector space $\mathbb{V}$, then $k = \ell$.

Proof: Since $\{\mathbf{v}_1, \ldots, \mathbf{v}_\ell\}$ is a basis for S and $\{\mathbf{u}_1, \ldots, \mathbf{u}_k\}$ is linearly independent, by Theorem 4.3.3, we must have $k \leq \ell$. Similarly, since $\{\mathbf{u}_1, \ldots, \mathbf{u}_k\}$ is a basis and $\{\mathbf{v}_1, \ldots, \mathbf{v}_\ell\}$ is linearly independent, we must also have $\ell \leq k$. Therefore, $\ell = k$. ■

As in Section 2.3, this theorem justifies the following definition of the dimension of a vector space.

Definition
Dimension

If a vector space $\mathbb{V}$ has a basis with n vectors, then we say that the **dimension** of $\mathbb{V}$ is n and write

$$\dim \mathbb{V} = n$$

If a vector space $\mathbb{V}$ does not have a basis with finitely many elements, then $\mathbb{V}$ is called **infinite-dimensional**.

Remarks

1. By definition, the trivial vector space $\{\mathbf{0}\}$ is zero dimensional since we defined a basis for this vector space to be the empty set.
2. Properties of infinite-dimensional spaces are beyond the scope of this book.

EXAMPLE 4.3.8

(a) $\mathbb{R}^n$ is n-dimensional because the standard basis contains n vectors.

(b) The vector space $M_{m\times n}(\mathbb{R})$ is $(m \times n)$-dimensional since the standard basis has $m \times n$ vectors.

(c) The vector space $P_n(\mathbb{R})$ is $(n + 1)$-dimensional as it has the standard basis $\{1, x, x^2, \ldots, x^n\}$.

(d) The vector space $C(a, b)$ is infinite-dimensional as it contains all polynomials (along with many other types of functions). Most function spaces are infinite-dimensional.

EXAMPLE 4.3.9

Let $\mathbb{S} = \operatorname{Span}\left\{\begin{bmatrix}1\\3\\2\end{bmatrix}, \begin{bmatrix}-1\\-2\\-1\end{bmatrix}, \begin{bmatrix}-1\\2\\3\end{bmatrix}, \begin{bmatrix}-1\\3\\4\end{bmatrix}\right\}$. Show that $\dim \mathbb{S} = 2$.

Solution: Consider

$$\begin{bmatrix}0\\0\\0\end{bmatrix} = t_1\begin{bmatrix}1\\3\\2\end{bmatrix} + t_2\begin{bmatrix}-1\\-2\\-1\end{bmatrix} + t_3\begin{bmatrix}-1\\2\\3\end{bmatrix} + t_4\begin{bmatrix}-1\\3\\4\end{bmatrix} = \begin{bmatrix}t_1 - t_2 - t_3 - t_4\\3t_1 - 2t_2 + 2t_3 + 3t_4\\2t_1 - t_2 + 3t_3 + 4t_4\end{bmatrix}$$

We row reduce the corresponding coefficient matrix

$$\begin{bmatrix}1 & -1 & -1 & -1\\3 & -2 & 2 & 3\\2 & -1 & 3 & 4\end{bmatrix} \sim \begin{bmatrix}1 & 0 & 4 & 5\\0 & 1 & 5 & 6\\0 & 0 & 0 & 0\end{bmatrix}$$

This implies that $\begin{bmatrix}-1\\2\\3\end{bmatrix}$ and $\begin{bmatrix}-1\\3\\4\end{bmatrix}$ can be written as linear combinations of the first two vectors. Thus, $\mathbb{S} = \operatorname{Span}\left\{\begin{bmatrix}1\\3\\2\end{bmatrix}, \begin{bmatrix}-1\\-2\\-1\end{bmatrix}\right\}$. Moreover, $\mathcal{B} = \left\{\begin{bmatrix}1\\3\\2\end{bmatrix}, \begin{bmatrix}-1\\-2\\-1\end{bmatrix}\right\}$ is clearly linearly independent since neither vector is a scalar multiple of the other, hence $\mathcal{B}$ is a basis for $\mathbb{S}$. Thus, $\dim \mathbb{S} = 2$.

EXAMPLE 4.3.10

Let $\mathbb{S} = \left\{\begin{bmatrix}a & b\\c & d\end{bmatrix} \in M_{2\times 2}(\mathbb{R}) \mid a + b = d\right\}$. Determine the dimension of $\mathbb{S}$.

Solution: Since $d = a + b$, observe that every matrix in $\mathbb{S}$ has the form

$$\begin{bmatrix}a & b\\c & d\end{bmatrix} = \begin{bmatrix}a & b\\c & a+b\end{bmatrix} = a\begin{bmatrix}1 & 0\\0 & 1\end{bmatrix} + b\begin{bmatrix}0 & 1\\0 & 1\end{bmatrix} + c\begin{bmatrix}0 & 0\\1 & 0\end{bmatrix}$$

Thus,

$$\mathbb{S} = \operatorname{Span}\left\{\begin{bmatrix}1 & 0\\0 & 1\end{bmatrix}, \begin{bmatrix}0 & 1\\0 & 1\end{bmatrix}, \begin{bmatrix}0 & 0\\1 & 0\end{bmatrix}\right\}$$

It is easy to show that $\left\{\begin{bmatrix}1 & 0\\0 & 1\end{bmatrix}, \begin{bmatrix}0 & 1\\0 & 1\end{bmatrix}, \begin{bmatrix}0 & 0\\1 & 0\end{bmatrix}\right\}$ is also linearly independent and hence is a basis for $\mathbb{S}$. Thus, $\dim \mathbb{S} = 3$.

EXERCISE 4.3.2

Find the dimension of $\mathbb{S} = \{a + bx + cx^2 + dx^3 \in P_3(\mathbb{R}) \mid a + b + c + d = 0\}$.

Extending a Linearly Independent Subset to a Basis

Sometimes a linearly independent set $\mathcal{T} = \{\mathbf{v}_1, \ldots, \mathbf{v}_k\}$ is given in an n-dimensional vector space $\mathbb{V}$, and it is necessary to include the vectors in $\mathcal{T}$ in a basis for $\mathbb{V}$.

Theorem 4.3.5 Basis Extension Theorem

If $\mathbb{V}$ is an n-dimensional vector space and $\mathcal{T} = \{\mathbf{v}_1, \ldots, \mathbf{v}_k\}$ is a linearly independent set in $\mathbb{V}$ with $k < n$, then there exist vectors $\mathbf{w}_{k+1}, \ldots, \mathbf{w}_n$ such that $\{\mathbf{v}_1, \ldots, \mathbf{v}_k, \mathbf{w}_{k+1}, \ldots, \mathbf{w}_n\}$ is a basis for $\mathbb{V}$.

Proof: Since $k < n$, the set $\mathcal{T} = \{\mathbf{v}_1, \ldots, \mathbf{v}_k\}$ cannot be a basis for $\mathbb{V}$ as that would contradict the Dimension Theorem. Thus, since the set is linearly independent, it must not span $\mathbb{V}$. Therefore, there exists a vector $\mathbf{w}_{k+1}$ that is in $\mathbb{V}$ but not in Span $\mathcal{T}$. Now consider

$$t_1\mathbf{v}_1 + \cdots + t_k\mathbf{v}_k + t_{k+1}\mathbf{w}_{k+1} = \mathbf{0} \tag{4.3}$$

If $t_{k+1} \neq 0$, then we have

$$\mathbf{w}_{k+1} = -\frac{t_1}{t_{k+1}}\mathbf{v}_1 - \cdots - \frac{t_k}{t_{k+1}}\mathbf{v}_k$$

and so $\mathbf{w}_{k+1}$ can be written as a linear combination of the vectors in $\mathcal{T}$, which cannot be since $\mathbf{w}_{k+1} \notin \operatorname{Span}\mathcal{T}$. Therefore, we must have $t_{k+1} = 0$. In this case, equation (4.3) becomes

$$\mathbf{0} = t_1\mathbf{v}_1 + \cdots + t_k\mathbf{v}_k$$

But, $\mathcal{T}$ is linearly independent, which implies that $t_1 = \cdots = t_k = 0$. Thus, the only solution to equation (4.3) is $t_1 = \cdots = t_{k+1} = 0$, and hence $\mathcal{T}_1 = \{\mathbf{v}_1, \ldots, \mathbf{v}_k, \mathbf{w}_{k+1}\}$ is linearly independent.

Now, if $\operatorname{Span}\{\mathbf{v}_1, \ldots, \mathbf{v}_k, \mathbf{w}_{k+1}\} = \mathbb{V}$, then $\mathcal{T}_1$ is a basis for $\mathbb{V}$. If not, we repeat the procedure to add another vector $\mathbf{w}_{k+2}$ to get $\mathcal{T}_2 = \{\mathbf{v}_1, \ldots, \mathbf{v}_k, \mathbf{w}_{k+1}, \mathbf{w}_{k+2}\}$, which is linearly independent. In this fashion, we must eventually get a basis, since according to Theorem 4.3.3, there cannot be more than n linearly independent vectors in an n-dimensional vector space. ■

The Basis Extension Theorem proves that every n-dimensional vector space $\mathbb{V}$ has a basis. In particular, $\{\mathbf{0}\}$ has a basis by definition. If $n \geq 1$, then we can pick any non-zero vector $\mathbf{v} \in \mathbb{V}$ and then extend $\{\mathbf{v}\}$ to a basis for $\mathbb{V}$.

It also gives us the following useful result.

Theorem 4.3.6 If $\mathbb{S}$ is a subspace of an n-dimensional vector space $\mathbb{V}$, then

$$\dim \mathbb{S} \leq n$$

EXAMPLE 4.3.11 Let $\mathcal{T} = \left\{ \begin{bmatrix} 1 & 1 \\ 0 & 1 \end{bmatrix}, \begin{bmatrix} -2 & -1 \\ 1 & 1 \end{bmatrix} \right\}$. Extend $\mathcal{T}$ to a basis for $M_{2\times2}(\mathbb{R})$.

Solution: We first want to determine whether $\mathcal{T}$ is a spanning set for $M_{2\times2}(\mathbb{R})$. Consider

$$\begin{bmatrix} b_1 & b_2 \\ b_3 & b_4 \end{bmatrix} = t_1 \begin{bmatrix} 1 & 1 \\ 0 & 1 \end{bmatrix} + t_2 \begin{bmatrix} -2 & -1 \\ 1 & 1 \end{bmatrix} = \begin{bmatrix} t_1 - 2t_2 & t_1 - t_2 \\ t_2 & t_1 + t_2 \end{bmatrix}$$

Row reducing the augmented matrix of the associated system gives

$$\left[\begin{array}{cc|c} 1 & -2 & b_1 \\ 1 & -1 & b_2 \\ 0 & 1 & b_3 \\ 1 & 1 & b_4 \end{array}\right] \sim \left[\begin{array}{cc|c} 1 & -2 & b_1 \\ 0 & 1 & b_2 - b_1 \\ 0 & 0 & b_1 - b_2 + b_3 \\ 0 & 0 & 2b_1 - 3b_2 + b_4 \end{array}\right]$$

Observe that if $b_1 - b_2 + b_3 \neq 0$ (or $2b_1 - 3b_2 + b_4 \neq 0$), then the system is inconsistent. Therefore, $\mathcal{T}$ is not a spanning set of $M_{2\times2}(\mathbb{R})$ since any matrix $\begin{bmatrix} b_1 & b_2 \\ b_3 & b_4 \end{bmatrix}$ with $b_1 - b_2 + b_3 \neq 0$ is not in Span $\mathcal{T}$. In particular, $\begin{bmatrix} 0 & 0 \\ 1 & 0 \end{bmatrix}$ is not in the span of $\mathcal{T}$. Following the steps in the proof, we should add this matrix to $\mathcal{T}$. We let

$$\mathcal{T}_1 = \left\{ \begin{bmatrix} 1 & 1 \\ 0 & 1 \end{bmatrix}, \begin{bmatrix} -2 & -1 \\ 1 & 1 \end{bmatrix}, \begin{bmatrix} 0 & 0 \\ 1 & 0 \end{bmatrix} \right\}$$

and repeat the procedure. Consider

$$\begin{bmatrix} b_1 & b_2 \\ b_3 & b_4 \end{bmatrix} = t_1 \begin{bmatrix} 1 & 1 \\ 0 & 1 \end{bmatrix} + t_2 \begin{bmatrix} -2 & -1 \\ 1 & 1 \end{bmatrix} + t_3 \begin{bmatrix} 0 & 0 \\ 1 & 0 \end{bmatrix} = \begin{bmatrix} t_1 - 2t_2 & t_1 - t_2 \\ t_2 + t_3 & t_1 + t_2 \end{bmatrix}$$

Row reducing the augmented matrix of the associated system gives

$$\left[\begin{array}{ccc|c} 1 & -2 & 0 & b_1 \\ 1 & -1 & 0 & b_2 \\ 0 & 1 & 1 & b_3 \\ 1 & 1 & 0 & b_4 \end{array}\right] \sim \left[\begin{array}{ccc|c} 1 & -2 & 0 & b_1 \\ 0 & 1 & 0 & b_2 - b_1 \\ 0 & 0 & 1 & b_1 - b_2 + b_3 \\ 0 & 0 & 0 & 2b_1 - 3b_2 + b_4 \end{array}\right]$$

So, any matrix $\begin{bmatrix} b_1 & b_2 \\ b_3 & b_4 \end{bmatrix}$ with $2b_1 - 3b_2 + b_4 \neq 0$ is not in Span $\mathcal{T}_1$. For example, $\begin{bmatrix} 0 & 0 \\ 0 & 1 \end{bmatrix}$ is not in the span of $\mathcal{T}_1$ and thus $\mathcal{T}_1$ is not a basis for $M_{2\times2}(\mathbb{R})$. Adding $\begin{bmatrix} 0 & 0 \\ 0 & 1 \end{bmatrix}$ to $\mathcal{T}_1$ we get

$$\mathcal{T}_2 = \left\{ \begin{bmatrix} 1 & 1 \\ 0 & 1 \end{bmatrix}, \begin{bmatrix} -2 & -1 \\ 1 & 1 \end{bmatrix}, \begin{bmatrix} 0 & 0 \\ 1 & 0 \end{bmatrix}, \begin{bmatrix} 0 & 0 \\ 0 & 1 \end{bmatrix} \right\}$$

By construction $\mathcal{T}_2$ is linearly independent. Moreover, we can show that it spans $M_{2\times2}(\mathbb{R})$. Thus, it is a basis for $M_{2\times2}(\mathbb{R})$.

EXERCISE 4.3.3

Extend the set $\mathcal{T} = \left\{ \begin{bmatrix} 1 \\ 1 \\ 1 \end{bmatrix} \right\}$ to a basis for $\mathbb{R}^3$.

Knowing the dimension of a finite dimensional vector space $\mathbb{V}$ is very useful when trying to construct a basis for $\mathbb{V}$, as the next theorem demonstrates.

Theorem 4.3.7

Basis Theorem

If $\mathbb{V}$ is an n-dimensional vector space, then

(1) A set of more than n vectors in $\mathbb{V}$ must be linearly dependent.

(2) A set of fewer than n vectors cannot span $\mathbb{V}$.

(3) If $\mathcal{B}$ contains n elements of $\mathbb{V}$ and Span $\mathcal{B} = \mathbb{V}$, then $\mathcal{B}$ is a basis for $\mathbb{V}$.

(4) If $\mathcal{B}$ contains n elements of $\mathbb{V}$ and $\mathcal{B}$ is linearly independent, then $\mathcal{B}$ is a basis for $\mathbb{V}$.

Proof: (1) This is Theorem 4.3.3 above.

(2) Suppose that $\mathbb{V}$ can be spanned by a set $\mathcal{B} = \{\vec{v}_1, \dots, \vec{v}_k\}$ where $k < n$. Then, by the Basis Reduction Theorem, some subset of $\mathcal{B}$ is a basis for $\mathbb{V}$. Hence, there is a basis of $\mathbb{V}$ that contains less than n vectors which contradicts the Dimension Theorem.

(3) If $\mathcal{B}$ is a spanning set for $\mathbb{V}$ that is not linearly independent, then by the Basis Reduction Theorem, there is a proper subset of $\mathcal{B}$ that is a basis for $\mathbb{V}$. But, this would contradict the Dimension Theorem.

(4) If $\mathcal{B}$ is a linearly independent set of n vectors that does not span $\mathbb{V}$, then it can be extended to a basis for $\mathbb{V}$ by the Basis Extension Theorem. But, this would contradict the Dimension Theorem. ■

Remarks

1. The Basis Theorem categorizes a basis as a **maximally linearly independent set**. That is, if we have a linearly independent set $\mathcal{B}$ in an n-dimensional vector space $\mathbb{V}$ such that adding any other vector in $\mathbb{V}$ to $\mathcal{B}$ makes $\mathcal{B}$ linearly dependent, then $\mathcal{B}$ is a basis for $\mathbb{V}$.

2. The Basis Theorem also categorizes a basis as a **minimal spanning set**. That is, if we have a spanning set C in an n-dimensional vector space $\mathbb{V}$ such that removing any vector from C would result in a set that does not span $\mathbb{V}$, then C is a basis for $\mathbb{V}$.

EXAMPLE 4.3.12

Consider the plane $\mathcal{P}$ in $\mathbb{R}^3$ with equation $x_1 + 2x_2 - x_3 = 0$.

(a) Produce a basis $\mathcal{B}$ for $\mathcal{P}$.

(b) Extend the basis $\mathcal{B}$ to obtain a basis C for $\mathbb{R}^3$.

Solution: (a) By definition, a plane in $\mathbb{R}^3$ has dimension 2. So, by the Basis Theorem, we just need to pick two linearly independent vectors that lie in the plane.

Observe that $\vec{v}_1 = \begin{bmatrix} 1 \\ 0 \\ 1 \end{bmatrix}$ and $\vec{v}_2 = \begin{bmatrix} 0 \\ 1 \\ 2 \end{bmatrix}$ both satisfy the equation of the plane and hence $\vec{v}_1, \vec{v}_2 \in \mathcal{P}$.

Since neither $\vec{v}_1$ nor $\vec{v}_2$ is a scalar multiple of the other, we have that $\mathcal{B} = \{\vec{v}_1, \vec{v}_2\}$ is linearly independent.

Therefore, $\mathcal{B}$ is a linearly independent set of two vectors in a 2-dimensional vector space. Hence, by the Basis Theorem, it is a basis for the plane $\mathcal{P}$.

(b) From the proof of the Basis Extension Theorem, we just need to add a vector that is not in the span of $\{\vec{v}_1, \vec{v}_2\}$. But, Span$\{\vec{v}_1, \vec{v}_2\}$ is the plane, so we need to pick any vector not in the plane. We pick a normal vector $\vec{n} = \begin{bmatrix} 1 \\ 2 \\ -1 \end{bmatrix}$.

Thus, $\{\vec{v}_1, \vec{v}_2, \vec{n}\}$ is a linearly independent set of three vectors in $\mathbb{R}^3$ and therefore, by the Basis Theorem, is a basis for $\mathbb{R}^3$.

CONNECTION

To extend the basis $\mathcal{B}$ in Example 4.3.12, we just need to pick a vector that does not satisfy the equation of the plane. However, we will see in Section 4.6 that there can be a benefit to picking a normal vector for the plane over other vectors.

EXERCISE 4.3.4

Produce a basis for the hyperplane in $\mathbb{R}^4$ with equation $x_1 - x_2 + x_3 - 2x_4 = 0$ and extend the basis to obtain a basis for $\mathbb{R}^4$.

PROBLEMS 4.3

Practice Problems

For Problems A1–A4, determine whether the set is a basis for $\mathbb{R}^3$.

A1 $\left\{\begin{bmatrix}1\\1\\2\end{bmatrix},\begin{bmatrix}1\\-1\\-1\end{bmatrix},\begin{bmatrix}2\\1\\1\end{bmatrix}\right\}$

A2 $\left\{\begin{bmatrix}-2\\2\\1\end{bmatrix},\begin{bmatrix}3\\-1\\2\end{bmatrix}\right\}$

A3 $\left\{\begin{bmatrix}-1\\3\\5\end{bmatrix},\begin{bmatrix}2\\4\\0\end{bmatrix},\begin{bmatrix}1\\4\\2\end{bmatrix}\right\}$

A4 $\left\{\begin{bmatrix}1\\0\\1\end{bmatrix},\begin{bmatrix}-1\\2\\1\end{bmatrix},\begin{bmatrix}1\\3\\5\end{bmatrix},\begin{bmatrix}2\\-1\\-4\end{bmatrix}\right\}$

For Problems A5–A7, determine whether the set is a basis for $P_2(\mathbb{R})$.

A5 $\{1 + x + 2x^2, 1 - x - x^2, 2 + x + x^2\}$

A6 $\{-2 + 2x + x^2, 3 - x + 2x^2\}$

A7 $\{1 - x + x^2, 1 + 2x - x^2, 3 + x^2\}$

A8 Determine whether $\mathcal{B} = \left\{\begin{bmatrix}1 & 1\\0 & 1\end{bmatrix},\begin{bmatrix}1 & 2\\0 & -1\end{bmatrix},\begin{bmatrix}1 & 3\\0 & 2\end{bmatrix}\right\}$ is a basis of the subspace $\mathbb{U} = \left\{\begin{bmatrix}a & b\\0 & c\end{bmatrix} \mid a, b, c \in \mathbb{R}\right\}$ of $M_{2\times2}(\mathbb{R})$.

For Problems A9 and A10, determine the dimension of the subspace of $\mathbb{R}^3$ spanned by $\mathcal{B}$.

A9 $\mathcal{B} = \left\{\begin{bmatrix}1\\-2\\1\end{bmatrix},\begin{bmatrix}0\\1\\2\end{bmatrix},\begin{bmatrix}2\\0\\10\end{bmatrix},\begin{bmatrix}1\\1\\7\end{bmatrix}\right\}$

A10 $\mathcal{B} = \left\{\begin{bmatrix}1\\3\\2\end{bmatrix},\begin{bmatrix}-2\\-6\\-4\end{bmatrix},\begin{bmatrix}-1\\-1\\2\end{bmatrix},\begin{bmatrix}0\\4\\8\end{bmatrix},\begin{bmatrix}0\\1\\1\end{bmatrix}\right\}$

For Problems A11–A13, determine the dimension of the subspace of $M_{2\times2}(\mathbb{R})$ spanned by $\mathcal{B}$.

A11 $\mathcal{B} = \left\{\begin{bmatrix}1 & 1\\-1 & 1\end{bmatrix},\begin{bmatrix}0 & 1\\3 & -1\end{bmatrix},\begin{bmatrix}1 & -1\\2 & -3\end{bmatrix},\begin{bmatrix}2 & 1\\4 & -3\end{bmatrix}\right\}$

A12 $\mathcal{B} = \left\{\begin{bmatrix}1 & 1\\1 & 1\end{bmatrix},\begin{bmatrix}2 & 2\\2 & 2\end{bmatrix},\begin{bmatrix}0 & 2\\1 & 1\end{bmatrix},\begin{bmatrix}2 & 0\\1 & 1\end{bmatrix},\begin{bmatrix}3 & 1\\2 & 2\end{bmatrix}\right\}$

A13 $\mathcal{B} = \left\{\begin{bmatrix}1 & 0\\0 & 1\end{bmatrix},\begin{bmatrix}0 & 1\\1 & 0\end{bmatrix},\begin{bmatrix}0 & 1\\0 & -1\end{bmatrix},\begin{bmatrix}1 & 1\\1 & 0\end{bmatrix}\right\}$

For Problems A14–A16, determine the dimension of the subspace of $P_3(\mathbb{R})$ spanned by $\mathcal{B}$.

A14 $\mathcal{B} = \{1 + x, 1 + x + x^2, 1 + x^3\}$

A15 $\mathcal{B} = \{1 + x, 1 - x, 1 + x^3, 1 - x^3\}$

A16 $\mathcal{B} = \{1 + x + x^2, 1 - x^3, 1 - 2x + 2x^2 - x^3, 1 - x^2 + 2x^3, x^2 + x^3\}$

A17 (a) Using the method in Example 4.3.12, determine a basis for the plane in $\mathbb{R}^3$ with equation $2x_1 - x_2 - x_3 = 0$.

(b) Extend the basis of part (a) to obtain a basis for $\mathbb{R}^3$.

A18 (a) Using the method in Example 4.3.12, determine a basis for the hyperplane in $\mathbb{R}^4$ with equation $x_1 - x_2 + x_3 - x_4 = 0$.

(b) Extend the basis of part (a) to obtain a basis for $\mathbb{R}^4$.

A19 Find a basis for $P_3(\mathbb{R})$ that includes the vectors $\vec{v}_1 = 1 + x + x^3$ and $\vec{v}_2 = 1 + x^2$.

A20 Find a basis for $M_{2\times2}(\mathbb{R})$ that includes the vectors $\vec{v}_1 = \begin{bmatrix}1 & 2\\2 & 1\end{bmatrix}$ and $\vec{v}_2 = \begin{bmatrix}1 & 2\\3 & 4\end{bmatrix}$.

A21 State the standard basis for $M_{2\times3}(\mathbb{R})$.

For Problems A22–A27, obtain a basis for the vector space and determine its dimension.

A22 $\mathbb{S} = \{a + bx + cx^2 \in P_2(\mathbb{R}) \mid a = -c\}$

A23 $\mathbb{S} = \{a + bx + cx^2 + dx^3 \in P_3(\mathbb{R}) \mid a - 2b = d\}$

A24 $\mathbb{S} = \left\{\begin{bmatrix}a & b\\0 & c\end{bmatrix} \in M_{2\times2}(\mathbb{R}) \mid a, b, c \in \mathbb{R}\right\}$

A25 $\mathbb{S} = \left\{\begin{bmatrix}x_1\\x_2\\x_3\end{bmatrix} \in \mathbb{R}^3 \mid \begin{bmatrix}x_1\\x_2\\x_3\end{bmatrix} \cdot \begin{bmatrix}1\\1\\1\end{bmatrix} = 0\right\}$

A26 $\mathbb{S} = \{\mathbf{p} \in P_2(\mathbb{R}) \mid \mathbf{p}(2) = 0 \text{ and } \mathbf{p}(3) = 0\}$

A27 $\mathbb{S} = \left\{\begin{bmatrix}a & b\\c & d\end{bmatrix} \in M_{2\times2}(\mathbb{R}) \mid a = -c \text{ and } b = -c\right\}$

Homework Problems

For Problems B1–B6, determine whether the set is a basis for $\mathbb{R}^3$.

B1 $\left\{\begin{bmatrix}1\\3\\2\end{bmatrix}, \begin{bmatrix}1\\2\\1\end{bmatrix}, \begin{bmatrix}3\\3\\-2\end{bmatrix}\right\}$

B2 $\left\{\begin{bmatrix}3\\0\\3\end{bmatrix}, \begin{bmatrix}5\\2\\0\end{bmatrix}, \begin{bmatrix}4\\-1\\3\end{bmatrix}\right\}$

B3 $\left\{\begin{bmatrix}1\\1\\-2\end{bmatrix}, \begin{bmatrix}2\\-6\\3\end{bmatrix}, \begin{bmatrix}6\\-2\\-5\end{bmatrix}\right\}$

B4 $\left\{\begin{bmatrix}2\\-1\\-4\end{bmatrix}, \begin{bmatrix}5\\7\\-5\end{bmatrix}\right\}$

B5 $\left\{\begin{bmatrix}3\\2\\-2\end{bmatrix}, \begin{bmatrix}4\\-1\\1\end{bmatrix}, \begin{bmatrix}6\\7\\-4\end{bmatrix}\right\}$

B6 $\left\{\begin{bmatrix}3\\1\\4\end{bmatrix}, \begin{bmatrix}1\\5\\9\end{bmatrix}, \begin{bmatrix}2\\6\\5\end{bmatrix}, \begin{bmatrix}3\\5\\8\end{bmatrix}\right\}$

For Problems B7–B10, determine whether the set is a basis for $P_2(\mathbb{R})$.

B7 $\{1 + 2x, 3 + 5x + x^2, 4 + x + 2x^2\}$

B8 $\{2 + x + 3x^2, 3 + x + 5x^2, 1 + x + x^2\}$

B9 $\{1 - x, 1 + x + x^2, 1 + 2x^2, x + 3x^2\}$

B10 $\{1 + 2x + 3x^2, -1 - 4x + 3x^2\}$

For Problems B11–B13, determine whether the set is a basis for the subspace $\mathbb{S}$ of $M_{2\times 2}(\mathbb{R})$ defined by

$$\mathbb{S} = \left\{\begin{bmatrix}a & b\\c & d\end{bmatrix} \mid a = 2b + c + d \in \mathbb{R}\right\}$$

B11 $\mathcal{B}_1 = \left\{\begin{bmatrix}2 & 1\\0 & 0\end{bmatrix}, \begin{bmatrix}1 & 0\\1 & 0\end{bmatrix}, \begin{bmatrix}1 & 0\\0 & 1\end{bmatrix}\right\}$

B12 $\mathcal{B}_2 = \left\{\begin{bmatrix}4 & 1\\1 & 1\end{bmatrix}, \begin{bmatrix}5 & 2\\0 & 1\end{bmatrix}\right\}$

B13 $\mathcal{B}_3 = \left\{\begin{bmatrix}0 & 0\\1 & -1\end{bmatrix}, \begin{bmatrix}0 & 1\\0 & -2\end{bmatrix}, \begin{bmatrix}2 & 0\\1 & 1\end{bmatrix}\right\}$

For Problems B14–B17, determine the dimension of the subspace of $\mathbb{R}^3$ spanned by $\mathcal{B}$.

B14 $\mathcal{B} = \left\{\begin{bmatrix}1\\0\\2\end{bmatrix}, \begin{bmatrix}3\\1\\7\end{bmatrix}, \begin{bmatrix}0\\-1\\-1\end{bmatrix}, \begin{bmatrix}2\\2\\1\end{bmatrix}\right\}$

B15 $\mathcal{B} = \left\{\begin{bmatrix}1\\2\\0\end{bmatrix}, \begin{bmatrix}3\\5\\7\end{bmatrix}\right\}$

B16 $\mathcal{B} = \left\{\begin{bmatrix}1\\3\\2\end{bmatrix}, \begin{bmatrix}3\\9\\6\end{bmatrix}, \begin{bmatrix}4\\1\\2\end{bmatrix}, \begin{bmatrix}-5\\7\\2\end{bmatrix}, \begin{bmatrix}5\\4\\4\end{bmatrix}\right\}$

B17 $\mathcal{B} = \left\{\begin{bmatrix}0\\1\\1\end{bmatrix}, \begin{bmatrix}3\\1\\0\end{bmatrix}, \begin{bmatrix}9\\-1\\-4\end{bmatrix}, \begin{bmatrix}-8\\2\\4\end{bmatrix}\right\}$

For Problems B18–B20, determine the dimension of the subspace of $M_{2\times 2}(\mathbb{R})$ spanned by $\mathcal{B}$.

B18 $\mathcal{B} = \left\{\begin{bmatrix}1 & 1\\-1 & 1\end{bmatrix}, \begin{bmatrix}2 & 4\\1 & 2\end{bmatrix}, \begin{bmatrix}3 & 1\\2 & 1\end{bmatrix}, \begin{bmatrix}-1 & 5\\-6 & 3\end{bmatrix}\right\}$

B19 $\mathcal{B} = \left\{\begin{bmatrix}3 & 2\\1 & 2\end{bmatrix}, \begin{bmatrix}4 & 1\\1 & 1\end{bmatrix}, \begin{bmatrix}-6 & 2\\-2 & 3\end{bmatrix}, \begin{bmatrix}2 & 1\\-1 & 2\end{bmatrix}\right\}$

B20 $\mathcal{B} = \left\{\begin{bmatrix}5 & 3\\3 & 4\end{bmatrix}, \begin{bmatrix}4 & 6\\2 & 3\end{bmatrix}, \begin{bmatrix}-7 & 3\\-5 & -6\end{bmatrix}, \begin{bmatrix}3 & 9\\1 & 2\end{bmatrix}\right\}$

For Problems B21–B23, determine the dimension of the subspace of $P_2(\mathbb{R})$ spanned by $\mathcal{B}$.

B21 $\mathcal{B} = \{1, 1 + 2x, 1 + 4x + 4x^2\}$

B22 $\mathcal{B} = \{1 + 2x + 5x^2, -2 + 3x - 8x^2, 1 + 9x + 7x^2, 4 + x\}$

B23 $\mathcal{B} = \{2 + x^2, 1 + x + x^2, 8 + 2x + 5x^2, -2 + 4x + x^2\}$

For Problems B24–B27:

(a) Determine a basis $\mathcal{B}$ for the given plane in $\mathbb{R}^3$.

(b) Extend $\mathcal{B}$ to obtain a basis for $\mathbb{R}^3$.

B24 $x_1 - 2x_2 + x_3 = 0$

B25 $3x_1 + 5x_2 - x_3 = 0$

B26 $2x_1 + 4x_2 + 3x_3 = 0$

B27 $3x_1 - 2x_3 = 0$

For Problems B28 and B29:

(a) Determine a basis $\mathcal{B}$ for the given hyperplane in $\mathbb{R}^4$.

(b) Extend $\mathcal{B}$ to obtain a basis for $\mathbb{R}^4$.

B28 $x_1 + 3x_2 - x_3 + 2x_4 = 0$

B29 $-x_1 + 2x_2 + 3x_3 + x_4 = 0$

B30 Find a basis for $P_3(\mathbb{R})$ that includes the vectors $\vec{v}_1 = 1 - x^2 + x^3$ and $\vec{v}_2 = 1 + x + 2x^2 + x^3$.

B31 Find a basis for $M_{2\times 2}(\mathbb{R})$ that includes the vectors $\vec{v}_1 = \begin{bmatrix}1 & -1\\3 & 2\end{bmatrix}$ and $\vec{v}_2 = \begin{bmatrix}2 & -3\\3 & 1\end{bmatrix}$.

For Problems B32–B37, obtain a basis for the vector space and determine its dimension.

B32 $\mathbb{S} = \left\{\begin{bmatrix}x_1\\x_2\\x_3\end{bmatrix} \in \mathbb{R}^3 \mid 3x_2 = 5x_3 \text{ and } x_1 = -2x_3\right\}$

B33 $\mathbb{S} = \{a + bx + cx^2 \in P_2(\mathbb{R}) \mid a + b = c\}$

B34 $\mathbb{S} = \{a + bx + cx^2 + dx^3 \in P_3(\mathbb{R}) \mid a + b = d\}$

B35 $\mathbb{S} = \left\{\begin{bmatrix}a & b\\0 & c\end{bmatrix} \in M_{2\times 2}(\mathbb{R}) \mid a - 2c = 3b\right\}$

B36 $\mathbb{S} = \{\mathbf{p} \in P_2(\mathbb{R}) \mid \mathbf{p}(0) = 0 \text{ and } \mathbf{p}(1) = 0\}$

B37 $\mathbb{S} = \left\{\begin{bmatrix}a & b\\c & d\end{bmatrix} \in M_{2\times 2}(\mathbb{R}) \mid a + 2b = c - d\right\}$

Conceptual Problems

C1 Let $\mathbb{V}$ be an n-dimensional vector space. Prove that if $\mathbb{S}$ is a subspace of $\mathbb{V}$ and $\dim \mathbb{S} = n$, then $\mathbb{S} = \mathbb{V}$.

C2 Show that if $\{\mathbf{v}_1, \mathbf{v}_2\}$ is a basis for a vector space $\mathbb{V}$, then for any real number t, $\{\mathbf{v}_1, \mathbf{v}_2 + t\mathbf{v}_1\}$ is also a basis for $\mathbb{V}$.

C3 Show that if $\{\mathbf{v}_1, \mathbf{v}_2, \mathbf{v}_3\}$ is a basis for a vector space $\mathbb{V}$, then $\{\mathbf{v}_1, \mathbf{v}_2, \mathbf{v}_3 + t\mathbf{v}_1 + s\mathbf{v}_2\}$ is also a basis for $\mathbb{V}$ for any $s, t \in \mathbb{R}$.

For Problems C4–C11, prove or disprove the statement.

C4 If $\mathbb{V}$ is an n-dimensional vector space and $\{\vec{v}_1, \dots, \vec{v}_k\}$ is a linearly independent set in $\mathbb{V}$, then $k \leq n$.

C5 Every basis for $P_2(\mathbb{R})$ has exactly two vectors in it.

C6 If $\{\vec{v}_1, \vec{v}_2\}$ is a basis for a 2-dimensional vector space $\mathbb{V}$, then $\{a\vec{v}_1 + b\vec{v}_2, c\vec{v}_1 + d\vec{v}_2\}$ is also a basis for $\mathbb{V}$ for any non-zero real numbers a, b, c, d.

C7 If $\mathcal{B} = \{\vec{v}_1, \dots, \vec{v}_k\}$ spans a vector space $\mathbb{V}$, then some subset of $\mathcal{B}$ is a basis for $\mathbb{V}$.

C8 If $\{\vec{v}_1, \dots, \vec{v}_k\}$ is a linearly independent set in a vector space $\mathbb{V}$, then $\dim \mathbb{V} \geq k$.

C9 Every set of four non-zero matrices in $M_{2\times2}(\mathbb{R})$ is a basis for $M_{2\times2}(\mathbb{R})$.

C10 If $\mathbb{U}$ is a subspace of a finite dimensional vector space $\mathbb{V}$ and $\mathcal{B} = \{\vec{v}_1, \dots, \vec{v}_n\}$ is a basis for $\mathbb{V}$, then some subset of $\mathcal{B}$ is a basis for $\mathbb{U}$.

C11 If $\mathcal{B} = \{\vec{x}, \vec{y}, \vec{z}\}$ is a basis for a vector space $\mathbb{V}$, then $C = \{\vec{x} - \vec{y}, \vec{x} + \vec{y}, 2\vec{z}\}$ is also a basis for $\mathbb{V}$.

C12 Prove that if $\mathbb{S}$ and $\mathbb{T}$ are both 3-dimensional subspaces of a 5-dimensional vector space $\mathbb{V}$, then $\mathbb{S} \cap \mathbb{T} \neq \{\vec{0}\}$.

C13 In Section 4.2 Problem C5 you proved that

$$\mathbb{V} = \{(a, b) \mid a, b \in \mathbb{R}, b > 0\}$$

with addition and scalar multiplication defined by

$$\begin{aligned} (a, b) \oplus (c, d) &= (ad + bc, bd) \\ t \odot (a, b) &= (tab^{t-1}, b^t) \quad \text{for all } t \in \mathbb{R} \end{aligned}$$

is a vector space. Find, with justification, a basis for $\mathbb{V}$ and hence determine the dimension of $\mathbb{V}$.

C14 In Section 4.2 Problem C6 you proved that $\mathbb{V} = \{x \in \mathbb{R} \mid x > 0\}$ with addition and scalar multiplication defined by

$$\begin{aligned} x \oplus y &= xy \\ t \odot x &= x^t \quad \text{for all } t \in \mathbb{R} \end{aligned}$$

is a vector space. Find, with justification, a basis for $\mathbb{V}$ and hence determine the dimension of $\mathbb{V}$.

C15 In Section 4.2 Problem C7 you proved that $\mathbb{C} = \{a + bi \mid a, b \in \mathbb{R}\}$ with addition and scalar multiplication defined by

$$\begin{aligned} (a + bi) + (c + di) &= (a + c) + (b + d)i \\ t(a + bi) &= (ta) + (tb)i, \quad \text{for all } t \in \mathbb{R} \end{aligned}$$

is a vector space. Find, with justification, a basis for $\mathbb{V}$ and hence determine the dimension of $\mathbb{V}$.

C16 Consider the vector space $\mathbb{V} = \{(a, 1 + a) \mid a \in \mathbb{R}\}$ with addition and scalar multiplication defined by

$$\begin{aligned} (a, 1 + a) \oplus (b, 1 + b) &= (a + b, 1 + a + b) \\ k \odot (a, 1 + a) &= (ka, 1 + ka), \quad k \in \mathbb{R} \end{aligned}$$

Find, with justification, a basis for $\mathbb{V}$ and hence determine the dimension of $\mathbb{V}$.

4.4 Coordinates

Recall that when we write $\vec{x} = \begin{bmatrix} x_1 \\ \vdots \\ x_n \end{bmatrix}$ *in* $\mathbb{R}^n$ *we really mean* $\vec{x} = x_1\vec{e}_1 + \cdots + x_n\vec{e}_n$ *where* $\{\vec{e}_1, \dots, \vec{e}_n\}$ *is the standard basis for* $\mathbb{R}^n$. *For example, when you originally learned to plot the point* $(1, 2)$ *in the xy-plane, you were taught that this means you move 1 in the x-direction* ($\vec{e}_1$) *and 2 in the y-direction* ($\vec{e}_2$). *We can extend this idea to bases for general vector spaces.*

Definition
Coordinates
Coordinate Vector

Suppose that $\mathcal{B} = \{\mathbf{v}_1, \dots, \mathbf{v}_n\}$ is a basis for the vector space $\mathbb{V}$. If $\mathbf{x} \in \mathbb{V}$ with $\mathbf{x} = c_1\mathbf{v}_1 + c_2\mathbf{v}_2 + \cdots + c_n\mathbf{v}_n$, then $c_1, c_2, \dots, c_n$ are called the **coordinates** of $\mathbf{x}$ with respect to $\mathcal{B}$ (or the **$\mathcal{B}$-coordinates**) and

$$[\mathbf{x}]_{\mathcal{B}} = \begin{bmatrix} c_1 \\ \vdots \\ c_n \end{bmatrix}$$

is called the **coordinate vector** of $\mathbf{x}$ with respect to the basis $\mathcal{B}$ (or the **$\mathcal{B}$-coordinate vector**).

Remarks

1. This definition makes sense because of the Unique Representation Theorem.
2. It is important to observe that $[\mathbf{x}]_{\mathcal{B}}$ is a vector in $\mathbb{R}^n$.
3. Observe that the coordinate vector $[\mathbf{x}]_{\mathcal{B}}$ depends on the order in which the basis vectors appear. In this book, "basis" always means **ordered basis**; that is, it is always assumed that a basis is specified in the order in which the basis vectors are listed.

EXAMPLE 4.4.1

Find the coordinate vector of $\mathbf{p}(x) = 4 + x$ with respect to the basis $\mathcal{B} = \{-1 + 2x, 1 + x\}$ of $P_1(\mathbb{R})$.

Solution: By definition, we need to write $\mathbf{p}$ as a linear combination of the vectors in $\mathcal{B}$. We consider

$$4 + x = c_1(-1 + 2x) + c_2(1 + x) = (-c_1 + c_2) + (2c_1 + c_2)x$$

Row reducing the corresponding augmented matrix gives

$$\left[\begin{array}{cc|c} -1 & 1 & 4 \\ 2 & 1 & 1 \end{array}\right] \sim \left[\begin{array}{cc|c} 1 & 0 & -1 \\ 0 & 1 & 3 \end{array}\right]$$

Thus,

$$4 + x = (-1)(-1 + 2x) + 3(1 + x)$$

Hence,

$$[\mathbf{p}]_{\mathcal{B}} = \begin{bmatrix} -1 \\ 3 \end{bmatrix}$$

Compare Example 4.4.1 to Example 1.2.10 on page 25. The similarity is not surprising upon realizing that the coordinate vectors of $4 + x$, $-1 + 2x$, and $1 + x$ with respect to the standard basis $\{1, x\}$ for $P_1(\mathbb{R})$ are $\begin{bmatrix} 4 \\ 1 \end{bmatrix}$, $\begin{bmatrix} -1 \\ 2 \end{bmatrix}$, and $\begin{bmatrix} 1 \\ 1 \end{bmatrix}$ respectively. In fact, we can draw the same diagram where the x_1- and x_2-axes are now in terms of the standard basis $\{1, x\}$ for $P_1(\mathbb{R})$.

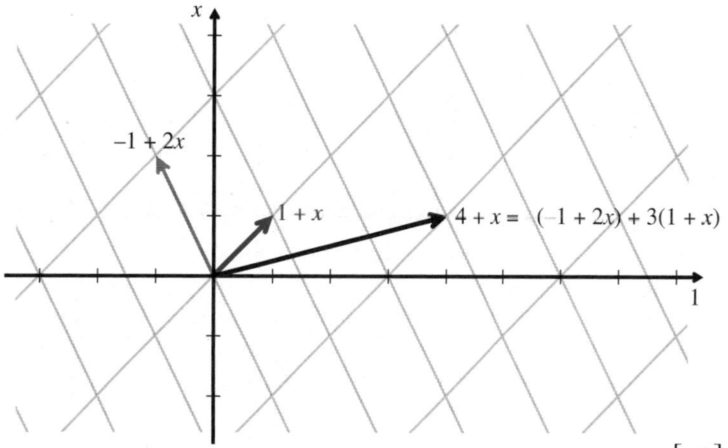

Figure 4.4.1 The basis $\mathcal{B} = \{-1 + 2x, 1 + x\}$ in $P_1(\mathbb{R})$; $[4 + x]_{\mathcal{B}} = \begin{bmatrix} -1 \\ 3 \end{bmatrix}$.

In particular, finding the coordinate vector of a vector $\vec{v}$ with respect to a basis $\mathcal{B}$ of an n-dimensional vector space $\mathbb{V}$ is a way of transforming the vector $\vec{v}$ into a vector in $\mathbb{R}^n$. Therefore, no matter how complicated a vector space looks, as long as we have a basis for the vector space, we can convert its vectors into vectors in $\mathbb{R}^n$. We will look more at this in Section 4.7.

EXAMPLE 4.4.2

Given that $\mathcal{B} = \left\{ \begin{bmatrix} 3 & 2 \\ 2 & 2 \end{bmatrix}, \begin{bmatrix} 1 & 0 \\ 1 & 1 \end{bmatrix}, \begin{bmatrix} 1 & 1 \\ 1 & 0 \end{bmatrix} \right\}$ is a basis for the subspace Span $\mathcal{B}$ of $M_{2\times 2}(\mathbb{R})$, find the coordinate vector of $A = \begin{bmatrix} 1 & -1 \\ 0 & 3 \end{bmatrix}$.

Solution: We are required to determine whether there are numbers c_1, c_2, c_3 such that

$$\begin{bmatrix} 1 & -1 \\ 0 & 3 \end{bmatrix} = c_1 \begin{bmatrix} 3 & 2 \\ 2 & 2 \end{bmatrix} + c_2 \begin{bmatrix} 1 & 0 \\ 1 & 1 \end{bmatrix} + c_3 \begin{bmatrix} 1 & 1 \\ 1 & 0 \end{bmatrix} = \begin{bmatrix} 3c_1 + c_2 + c_3 & 2c_1 + c_3 \\ 2c_1 + c_2 + c_3 & 2c_1 + c_2 \end{bmatrix}$$

Row reducing the augmented matrix gives

$$\left[\begin{array}{ccc|c} 3 & 1 & 1 & 1 \\ 2 & 0 & 1 & -1 \\ 2 & 1 & 1 & 0 \\ 2 & 1 & 0 & 3 \end{array}\right] \sim \left[\begin{array}{ccc|c} 1 & 0 & 0 & 1 \\ 0 & 1 & 0 & 1 \\ 0 & 0 & 1 & -3 \\ 0 & 0 & 0 & 0 \end{array}\right]$$

Thus,

$$[A]_{\mathcal{B}} = \begin{bmatrix} 1 \\ 1 \\ -3 \end{bmatrix}$$

Note that there are only three $\mathcal{B}$-coordinates because the basis $\mathcal{B}$ has only three vectors.

EXERCISE 4.4.1

Find the coordinate vector of $\begin{bmatrix} 0 \\ 2 \\ 2 \end{bmatrix}$ with respect to the basis $\mathcal{B} = \left\{ \begin{bmatrix} 2 \\ 1 \\ -1 \end{bmatrix}, \begin{bmatrix} -3 \\ -1 \\ 2 \end{bmatrix} \right\}$ of Span $\mathcal{B}$.

EXAMPLE 4.4.3

Suppose that you have written a computer program to perform certain operations with polynomials in $P_2(\mathbb{R})$. You need to include some method of inputting the polynomial you are considering. If you use the standard basis $\{1, x, x^2\}$ for $P_2(\mathbb{R})$ to input the polynomial $3 - 5x + 2x^2$, you would surely write your program in such a way that you would type "3, −5, 2," the standard coordinates, as the input.

On the other hand, for some problems in differential equations, you might prefer the basis $\mathcal{B} = \{1 - x^2, x, 1 + x^2\}$. To find the $\mathcal{B}$-coordinates of $3 - 5x + 2x^2$, we must find t_1, t_2 and t_3 such that

$$3 - 5x + 2x^2 = t_1(1 - x^2) + t_2x + t_3(1 + x^2) = (t_1 + t_3) + t_2x + (t_3 - t_1)x^2$$

Row reducing the corresponding augmented matrix gives

$$\left[\begin{array}{ccc|c} 1 & 0 & 1 & 3 \\ 0 & 1 & 0 & -5 \\ -1 & 0 & 1 & 2 \end{array}\right] \sim \left[\begin{array}{ccc|c} 1 & 0 & 0 & 1/2 \\ 0 & 1 & 0 & -5 \\ 0 & 0 & 1 & 5/2 \end{array}\right]$$

It follows that the coordinates of $3 - 5x + 2x^2$ with respect to $\mathcal{B}$ are $1/2, -5, 5/2$. Thus, if your computer program is written to work in the basis $\mathcal{B}$, then you would input "0.5, −5, 2.5."

EXERCISE 4.4.2

Prove that if $\mathcal{S} = \{\vec{e}_1, \dots, \vec{e}_n\}$ is the standard basis for $\mathbb{R}^n$, then $[\vec{x}]_{\mathcal{S}} = \vec{x}$ for any $\vec{x} \in \mathbb{R}^n$.

Change of Coordinates

We might need to input several polynomials into the computer program written to work in the basis $\mathcal{B}$. In this case, we would want a much faster way of converting standard coordinates to coordinates with respect to the basis $\mathcal{B}$. We now develop a method for doing this. We will require the following theorem.

Theorem 4.4.1

If $\mathcal{B}$ is a basis for a finite dimensional vector space $\mathbb{V}$, then for any $\mathbf{x}, \mathbf{y} \in \mathbb{V}$ and $s, t \in \mathbb{R}$ we have

$$[s\mathbf{x} + t\mathbf{y}]_{\mathcal{B}} = s[\mathbf{x}]_{\mathcal{B}} + t[\mathbf{y}]_{\mathcal{B}}$$

Proof: Let $\mathcal{B} = \{\mathbf{v}_1, \dots, \mathbf{v}_n\}$, $\mathbf{x} = x_1\mathbf{v}_1 + \cdots + x_n\mathbf{v}_n$, and $\mathbf{y} = y_1\mathbf{v}_1 + \cdots + y_n\mathbf{v}_n$ in $\mathbb{V}$. Then, we have

$$s\mathbf{x} + t\mathbf{y} = (sx_1 + ty_1)\mathbf{v}_1 + \cdots + (sx_n + ty_n)\mathbf{v}_n$$

Thus,

$$[s\mathbf{x} + t\mathbf{y}]_{\mathcal{B}} = \begin{bmatrix} sx_1 + ty_1 \\ \vdots \\ sx_n + ty_n \end{bmatrix} = s\begin{bmatrix} x_1 \\ \vdots \\ x_n \end{bmatrix} + t\begin{bmatrix} y_1 \\ \vdots \\ y_n \end{bmatrix} = s[\mathbf{x}]_{\mathcal{B}} + t[\mathbf{y}]_{\mathcal{B}}$$

■

Let $\mathcal{B}$ be a basis for an n-dimensional vector space $\mathbb{V}$ and let $C = \{\mathbf{w}_1, \dots, \mathbf{w}_n\}$ be another basis for $\mathbb{V}$. Consider $\mathbf{x} \in \mathbb{V}$. Writing $\mathbf{x}$ as a linear combination of the vectors in C gives

$$\mathbf{x} = x_1\mathbf{w}_1 + \cdots + x_n\mathbf{w}_n$$

Taking $\mathcal{B}$-coordinates gives

$$\begin{aligned}
[\mathbf{x}]_B &= [x_1\mathbf{w}_1 + \cdots + x_n\mathbf{w}_n]_\mathcal{B} \\
&= x_1[\mathbf{w}_1]_\mathcal{B} + \cdots + x_n[\mathbf{w}_n]_\mathcal{B} && \text{by Theorem 4.4.1} \\
&= \begin{bmatrix} [\mathbf{w}_1]_\mathcal{B} & \cdots & [\mathbf{w}_n]_\mathcal{B} \end{bmatrix} \begin{bmatrix} x_1 \\ \vdots \\ x_n \end{bmatrix} && \text{by definition of matrix-vector multiplication}
\end{aligned}$$

Since $\begin{bmatrix} x_1 \\ \vdots \\ x_n \end{bmatrix} = [\mathbf{x}]_C$, we see that this equation gives a formula for calculating the $\mathcal{B}$-coordinates of $\mathbf{x}$ from the C-coordinates of $\mathbf{x}$ simply using matrix-vector multiplication. We call this equation the **change of coordinates equation** and make the following definition.

Definition
Change of Coordinates Matrix

Let $\mathcal{B}$ and $C = \{\mathbf{w}_1, \dots, \mathbf{w}_n\}$ both be bases for a vector space $\mathbb{V}$. The matrix

$$P = \begin{bmatrix} [\mathbf{w}_1]_\mathcal{B} & \cdots & [\mathbf{w}_n]_\mathcal{B} \end{bmatrix}$$

is called the **change of coordinates matrix** from C-coordinates to $\mathcal{B}$-coordinates and satisfies

$$[\mathbf{x}]_\mathcal{B} = P[\mathbf{x}]_C$$

Of course, we could exchange the roles of $\mathcal{B}$ and C to find the change of coordinates matrix Q from $\mathcal{B}$-coordinates to C-coordinates.

Theorem 4.4.2

Let $\mathcal{B}$ and C both be bases for a finite-dimensional vector space $\mathbb{V}$. If P is the change of coordinates matrix from C-coordinates to $\mathcal{B}$-coordinates, then P is invertible and P^{-1} is the change of coordinates matrix from $\mathcal{B}$-coordinates to C-coordinates.

The proof of Theorem 4.4.2 is left to Problem C4.

EXAMPLE 4.4.4

Let $\mathcal{S} = \{\vec{e}_1, \vec{e}_2, \vec{e}_3\}$ be the standard basis for $\mathbb{R}^3$ and let $\mathcal{B} = \left\{ \begin{bmatrix} 1 \\ 3 \\ -1 \end{bmatrix}, \begin{bmatrix} 2 \\ 1 \\ 1 \end{bmatrix}, \begin{bmatrix} 3 \\ 4 \\ 1 \end{bmatrix} \right\}$. Find the change of coordinates matrix Q from $\mathcal{B}$-coordinates to $\mathcal{S}$-coordinates. Find the change of coordinates matrix P from $\mathcal{S}$-coordinates to $\mathcal{B}$-coordinates. Verify that $PQ = I$.

Solution: To find the change of coordinates matrix Q, we need to find the coordinates of the vectors in $\mathcal{B}$ with respect to the standard basis $\mathcal{S}$. We get

$$Q = \begin{bmatrix} \begin{bmatrix} 1 \\ 3 \\ -1 \end{bmatrix}_{\mathcal{S}} & \begin{bmatrix} 2 \\ 1 \\ 1 \end{bmatrix}_{\mathcal{S}} & \begin{bmatrix} 3 \\ 4 \\ 1 \end{bmatrix}_{\mathcal{S}} \end{bmatrix} = \begin{bmatrix} 1 & 2 & 3 \\ 3 & 1 & 4 \\ -1 & 1 & 1 \end{bmatrix}$$

To find the change of coordinates matrix P, we need to find the coordinates of the standard basis vectors with respect to the basis $\mathcal{B}$. To do this, we need to solve the systems

$$a_1 \begin{bmatrix} 1 \\ 3 \\ -1 \end{bmatrix} + a_2 \begin{bmatrix} 2 \\ 1 \\ 1 \end{bmatrix} + a_3 \begin{bmatrix} 3 \\ 4 \\ 1 \end{bmatrix} = \begin{bmatrix} 1 \\ 0 \\ 0 \end{bmatrix}$$

$$b_1 \begin{bmatrix} 1 \\ 3 \\ -1 \end{bmatrix} + b_2 \begin{bmatrix} 2 \\ 1 \\ 1 \end{bmatrix} + b_3 \begin{bmatrix} 3 \\ 4 \\ 1 \end{bmatrix} = \begin{bmatrix} 0 \\ 1 \\ 0 \end{bmatrix}$$

$$c_1 \begin{bmatrix} 1 \\ 3 \\ -1 \end{bmatrix} + c_2 \begin{bmatrix} 2 \\ 1 \\ 1 \end{bmatrix} + c_3 \begin{bmatrix} 3 \\ 4 \\ 1 \end{bmatrix} = \begin{bmatrix} 0 \\ 0 \\ 1 \end{bmatrix}$$

This gives three systems of linear equations with the same coefficient matrix. To make this easier, we row reduce the corresponding triple-augmented matrix.

$$\left[\begin{array}{ccc|ccc} 1 & 2 & 3 & 1 & 0 & 0 \\ 3 & 1 & 4 & 0 & 1 & 0 \\ -1 & 1 & 1 & 0 & 0 & 1 \end{array}\right] \sim \left[\begin{array}{ccc|ccc} 1 & 0 & 0 & 3/5 & -1/5 & -1 \\ 0 & 1 & 0 & 7/5 & -4/5 & -1 \\ 0 & 0 & 1 & -4/5 & 3/5 & 1 \end{array}\right]$$

Thus,

$$P = \begin{bmatrix} \begin{bmatrix} 1 \\ 0 \\ 0 \end{bmatrix}_{\mathcal{B}} & \begin{bmatrix} 0 \\ 1 \\ 0 \end{bmatrix}_{\mathcal{B}} & \begin{bmatrix} 0 \\ 0 \\ 1 \end{bmatrix}_{\mathcal{B}} \end{bmatrix} = \begin{bmatrix} 3/5 & -1/5 & -1 \\ 7/5 & -4/5 & -1 \\ -4/5 & 3/5 & 1 \end{bmatrix}$$

We now see that

$$\begin{bmatrix} 3/5 & -1/5 & -1 \\ 7/5 & -4/5 & -1 \\ -4/5 & 3/5 & 1 \end{bmatrix} \begin{bmatrix} 1 & 2 & 3 \\ 3 & 1 & 4 \\ -1 & 1 & 1 \end{bmatrix} = \begin{bmatrix} 1 & 0 & 0 \\ 0 & 1 & 0 \\ 0 & 0 & 1 \end{bmatrix}$$

EXAMPLE 4.4.5

Find the change of coordinates matrix P from the standard basis $\mathcal{S} = \{1, x, x^2\}$ to the basis $\mathcal{B} = \{1 - x^2, x, 1 + x^2\}$ in $P_2(\mathbb{R})$. Use P to find $[a + bx + cx^2]_{\mathcal{B}}$.

Solution: We need to find the coordinates of the standard basis vectors with respect to the basis $\mathcal{B}$. To do this, we solve the three systems of linear equations given by

$$\begin{aligned} t_{11}(1 - x^2) + t_{12}(x) + t_{13}(1 + x^2) &= 1 \\ t_{21}(1 - x^2) + t_{22}(x) + t_{23}(1 + x^2) &= x \\ t_{31}(1 - x^2) + t_{32}(x) + t_{33}(1 + x^2) &= x^2 \end{aligned}$$

We row reduce the triple-augmented matrix to get

$$\left[\begin{array}{ccc|ccc} 1 & 0 & 1 & 1 & 0 & 0 \\ 0 & 1 & 0 & 0 & 1 & 0 \\ -1 & 0 & 1 & 0 & 0 & 1 \end{array}\right] \sim \left[\begin{array}{ccc|ccc} 1 & 0 & 0 & 1/2 & 0 & -1/2 \\ 0 & 1 & 0 & 0 & 1 & 0 \\ 0 & 0 & 1 & 1/2 & 0 & 1/2 \end{array}\right]$$

Hence,

$$P = \begin{bmatrix} 1/2 & 0 & -1/2 \\ 0 & 1 & 0 \\ 1/2 & 0 & 1/2 \end{bmatrix}$$

To calculate $[a + bx + cx^2]_{\mathcal{B}}$, we first observe that

$$[a + bx + cx^2]_{\mathcal{S}} = \begin{bmatrix} a \\ b \\ c \end{bmatrix}$$

Thus,

$$\begin{aligned} [a + bx + cx^2]_{\mathcal{B}} &= P[a + bx + cx^2]_{\mathcal{S}} \\ &= \begin{bmatrix} 1/2 & 0 & -1/2 \\ 0 & 1 & 0 \\ 1/2 & 0 & 1/2 \end{bmatrix} \begin{bmatrix} a \\ b \\ c \end{bmatrix} \\ &= \begin{bmatrix} \frac{1}{2}a - \frac{1}{2}c \\ b \\ \frac{1}{2}a + \frac{1}{2}c \end{bmatrix} \end{aligned}$$

EXERCISE 4.4.3

Let $\mathcal{S} = \{\vec{e}_1, \vec{e}_2, \vec{e}_3\}$ be the standard basis for $\mathbb{R}^3$ and let $\mathcal{B} = \left\{ \begin{bmatrix} 1 \\ 1 \\ 2 \end{bmatrix}, \begin{bmatrix} 1 \\ 2 \\ 4 \end{bmatrix}, \begin{bmatrix} 2 \\ 2 \\ 3 \end{bmatrix} \right\}$. Find the change of coordinates matrix Q from $\mathcal{B}$-coordinates to $\mathcal{S}$-coordinates. Find the change of coordinates matrix P from $\mathcal{S}$-coordinates to $\mathcal{B}$-coordinates. Verify that $PQ = I$.

PROBLEMS 4.4
Practice Problems

A1 (a) Verify that $\mathcal{B} = \left\{\begin{bmatrix}1\\2\\0\end{bmatrix}, \begin{bmatrix}0\\2\\-1\end{bmatrix}\right\}$ is a basis for the plane with equation $2x_1 - x_2 - 2x_3 = 0$.

(b) Determine whether $\vec{x}$ lies in the plane of part (a). If it does, find $[\vec{x}]_{\mathcal{B}}$.

(i) $\vec{x} = \begin{bmatrix}3\\2\\1\end{bmatrix}$ (ii) $\vec{x} = \begin{bmatrix}3\\2\\2\end{bmatrix}$ (iii) $\vec{x} = \begin{bmatrix}5\\2\\3\end{bmatrix}$

A2 Consider the basis $\mathcal{B} = \{1+x^2, 1-x+2x^2, -1-x+x^2\}$ for $P_2(\mathbb{R})$.

(a) Find the $\mathcal{B}$-coordinates of each polynomials.

(i) $\mathbf{q}(x) = 4-2x+7x^2$ (ii) $\mathbf{r}(x) = -2-2x+3x^2$

(b) Determine $[2-4x+10x^2]_{\mathcal{B}}$ and use your answers to part (a) to check that

$$[4-2x+7x^2]_{\mathcal{B}} + [-2-2x+3x^2]_{\mathcal{B}} = [(4-2)+(-2-2)x+(7+3)x^2]_{\mathcal{B}}$$

For Problems A3–A13, determine the coordinates of the vectors with respect to the given basis $\mathcal{B}$.

A3 $\mathcal{B} = \left\{\begin{bmatrix}1\\3\end{bmatrix}, \begin{bmatrix}2\\-5\end{bmatrix}\right\}, \mathbf{x} = \begin{bmatrix}5\\-7\end{bmatrix}, \mathbf{y} = \begin{bmatrix}-7\\8\end{bmatrix}$

A4 $\mathcal{B} = \left\{\begin{bmatrix}2\\-5\end{bmatrix}, \begin{bmatrix}1\\3\end{bmatrix}\right\}, \mathbf{x} = \begin{bmatrix}5\\-7\end{bmatrix}, \mathbf{y} = \begin{bmatrix}9\\-5\end{bmatrix}$

A5 $\mathcal{B} = \left\{\begin{bmatrix}1\\0\\1\end{bmatrix}, \begin{bmatrix}0\\1\\1\end{bmatrix}, \begin{bmatrix}1\\1\\0\end{bmatrix}\right\}, \mathbf{x} = \begin{bmatrix}8\\-7\\3\end{bmatrix}, \mathbf{y} = \begin{bmatrix}3\\-5\\2\end{bmatrix}$

A6 $\mathcal{B} = \left\{\begin{bmatrix}1\\0\\1\\1\end{bmatrix}, \begin{bmatrix}-1\\1\\-1\\0\end{bmatrix}\right\}, \mathbf{x} = \begin{bmatrix}5\\-2\\5\\3\end{bmatrix}, \mathbf{y} = \begin{bmatrix}-1\\3\\-1\\2\end{bmatrix}$

A7 $\mathcal{B} = \{1+3x, 2-5x\}, \mathbf{p}(x) = 5-7x, \mathbf{q}(x) = 1$

A8 $\mathcal{B} = \{1+x+x^2, 1+3x+2x^2, 4+x^2\}$, $\mathbf{p}(x) = -2+8x+5x^2, \mathbf{q}(x) = -4+8x+4x^2$

A9 $\mathcal{B} = \{1+x^2, 1+x+2x^2+x^3, x-x^2+x^3\}$, $\mathbf{p}(x) = 2+x-5x^2+x^3, \mathbf{q}(x) = 1+x+4x^2+x^3$

A10 $\mathcal{B} = \left\{\begin{bmatrix}1 & 1\\1 & 0\end{bmatrix}, \begin{bmatrix}0 & 1\\1 & 1\end{bmatrix}, \begin{bmatrix}2 & 0\\0 & -1\end{bmatrix}\right\}$, $A = \begin{bmatrix}0 & 1\\1 & 2\end{bmatrix}, B = \begin{bmatrix}-4 & 1\\1 & 4\end{bmatrix}$

A11 $\mathcal{B} = \left\{\begin{bmatrix}1 & 1 & 0\\0 & 1 & 1\end{bmatrix}, \begin{bmatrix}0 & 2 & -1\\1 & 3 & -1\end{bmatrix}\right\}$, $A = \begin{bmatrix}1 & 3 & -1\\1 & 4 & 0\end{bmatrix}, B = \begin{bmatrix}3 & -1 & 2\\-2 & -3 & 5\end{bmatrix}$

A12 $\mathcal{B} = \left\{\begin{bmatrix}1 & 2\\1 & 3\end{bmatrix}, \begin{bmatrix}2 & 1\\-1 & 2\end{bmatrix}, \begin{bmatrix}-2 & 2\\4 & 10\end{bmatrix}\right\}$, $A = \begin{bmatrix}-1 & 1\\2 & 7\end{bmatrix}, B = \begin{bmatrix}6 & 3\\-3 & 2\end{bmatrix}$

A13 $\mathcal{B} = \left\{\begin{bmatrix}1 & 3\\2 & 3\end{bmatrix}, \begin{bmatrix}1 & -2\\1 & 2\end{bmatrix}, \begin{bmatrix}0 & 1\\-1 & 1\end{bmatrix}\right\}$, $A = \begin{bmatrix}3 & -5\\8 & 3\end{bmatrix}, B = \begin{bmatrix}0 & 3\\3 & -1\end{bmatrix}$

For Problems A14–A20, find the change of coordinates matrix to and from the basis $\mathcal{B}$ and the standard basis of the given vector space.

A14 $\mathcal{B} = \left\{\begin{bmatrix}1\\1\end{bmatrix}, \begin{bmatrix}0\\2\end{bmatrix}\right\}$ for $\mathbb{R}^2$.

A15 $\mathcal{B} = \left\{\begin{bmatrix}3\\4\\1\end{bmatrix}, \begin{bmatrix}0\\1\\0\end{bmatrix}, \begin{bmatrix}-2\\-3\\3\end{bmatrix}\right\}$ for $\mathbb{R}^3$

A16 $\mathcal{B} = \{1, 1-2x, 1-4x+4x^2\}$ for $P_2(\mathbb{R})$

A17 $\mathcal{B} = \{1+2x+x^2, x+x^2, 1+3x\}$ for $P_2(\mathbb{R})$

A18 $\mathcal{B} = \{1-2x+5x^2, 1-2x^2, x+x^2\}$ for $P_2(\mathbb{R})$

A19 $\mathcal{B} = \{x^2, x^3, x, 1\}$ for $P_3(\mathbb{R})$

A20 $\mathcal{B} = \left\{\begin{bmatrix}1 & -1\\0 & -1\end{bmatrix}, \begin{bmatrix}0 & -4\\0 & -1\end{bmatrix}, \begin{bmatrix}2 & 1\\0 & 1\end{bmatrix}\right\}$ for the subspace of $M_{2\times 2}(\mathbb{R})$ of upper-triangular matrices

For Problems A21–A23, find the change of coordinates matrix Q from $\mathcal{B}$-coordinates to $\mathcal{C}$-coordinates and the change of coordinates matrix P from $\mathcal{C}$-coordinates to $\mathcal{B}$-coordinates. Verify that $PQ = I$.

A21 In $\mathbb{R}^2$: $\mathcal{B} = \left\{\begin{bmatrix}3\\1\end{bmatrix}, \begin{bmatrix}5\\3\end{bmatrix}\right\}, \mathcal{C} = \left\{\begin{bmatrix}2\\1\end{bmatrix}, \begin{bmatrix}5\\2\end{bmatrix}\right\}$

A22 In $\mathbb{R}^2$: $\mathcal{B} = \left\{\begin{bmatrix}1\\3\end{bmatrix}, \begin{bmatrix}2\\1\end{bmatrix}\right\}, \mathcal{C} = \left\{\begin{bmatrix}-1\\1\end{bmatrix}, \begin{bmatrix}5\\-4\end{bmatrix}\right\}$

A23 In $P_2(\mathbb{R})$: $\mathcal{B} = \{1, -1+x, (-1+x)^2\}$, $\mathcal{C} = \{1+x+x^2, 1+3x-x^2, 1-x-x^2\}$

Homework Problems

B1 (a) Verify that $\mathcal{B} = \left\{ \begin{bmatrix} 1 \\ 0 \\ 3 \end{bmatrix}, \begin{bmatrix} 1 \\ 1 \\ 5 \end{bmatrix} \right\}$ is a basis for the plane with equation $3x_1 + 2x_2 - x_3 = 0$.

(b) For each of the following vectors, determine whether it lies in the plane of part (a). If it does, find the vector's $\mathcal{B}$-coordinates.

(i) $\vec{x}_1 = \begin{bmatrix} 1 \\ 3 \\ 9 \end{bmatrix}$ (ii) $\vec{x}_2 = \begin{bmatrix} 1 \\ 2 \\ 4 \end{bmatrix}$ (iii) $\vec{x}_3 = \begin{bmatrix} 1 \\ -1 \\ 1 \end{bmatrix}$

B2 (a) Verify that $\mathcal{B} = \left\{ \begin{bmatrix} 0 \\ 3 \\ 2 \end{bmatrix}, \begin{bmatrix} -2 \\ 1 \\ 1 \end{bmatrix} \right\}$ is a basis for the plane with equation $x_1 - 4x_2 + 6x_3 = 0$.

(b) For each of the following vectors, determine whether it lies in the plane of part (a). If it does, find the vector's $\mathcal{B}$-coordinates.

(i) $\vec{x}_1 = \begin{bmatrix} 2 \\ 8 \\ 5 \end{bmatrix}$ (ii) $\vec{x}_2 = \begin{bmatrix} -4 \\ 11 \\ 8 \end{bmatrix}$ (iii) $\vec{x}_3 = \begin{bmatrix} -2 \\ 0 \\ 1/3 \end{bmatrix}$

B3 Consider the basis $\mathcal{B} = \{1+2x+x^2, 1+x+2x^2, x-2x^2\}$ for $P_2(\mathbb{R})$.

(a) Determine the $\mathcal{B}$-coordinates of the following polynomials.

(i) $\mathbf{p}(x) = 1$
(ii) $\mathbf{q}(x) = 4 - 2x + 7x^2$
(iii) $\mathbf{r}(x) = -2 - 2x + 3x^2$

(b) Determine $[2 - 4x + 10x^2]_{\mathcal{B}}$ and use your answers to part (a) to check that

$$[4 - 2x + 7x^2]_{\mathcal{B}} + [-2 - 2x + 3x^2]_{\mathcal{B}} = [(4-2) + (-2-2)x + (7+3)x^2]_{\mathcal{B}}$$

For Problems B4–B16, determine the coordinates of the vectors with respect to the given basis $\mathcal{B}$.

B4 $\mathcal{B} = \left\{ \begin{bmatrix} 1 \\ 2 \\ 1 \end{bmatrix}, \begin{bmatrix} -1 \\ 3 \\ 1 \end{bmatrix} \right\}, \mathbf{x} = \begin{bmatrix} 1 \\ 7 \\ 3 \end{bmatrix}, \mathbf{y} = \begin{bmatrix} -8 \\ 9 \\ 2 \end{bmatrix}$

B5 $\mathcal{B} = \left\{ \begin{bmatrix} 3 \\ 2 \\ 2 \end{bmatrix}, \begin{bmatrix} 1 \\ 1 \\ 4 \end{bmatrix}, \begin{bmatrix} 5 \\ 5 \\ 2 \end{bmatrix} \right\}, \mathbf{x} = \begin{bmatrix} 2 \\ 0 \\ 6 \end{bmatrix}, \mathbf{y} = \begin{bmatrix} 3 \\ 0 \\ 18 \end{bmatrix}$

B6 $\mathcal{B} = \{1 + 2x^2, 2 + x + 3x^2, 4 - x - 2x^2\}$, $\mathbf{p}(x) = 7 - 5x - 3x^2$, $\mathbf{q}(x) = 7 - 3x - 5x^2$

B7 $\mathcal{B} = \{1 - x - x^2, 2 - x + 5x^2, 1 + x + 5x^2\}$, $\mathbf{p}(x) = 4 - 3x + 7x^2$, $\mathbf{q}(x) = 4 + 3x - 3x^2$

B8 $\mathcal{B} = \left\{ \begin{bmatrix} 0 & 1 \\ 2 & 1 \end{bmatrix}, \begin{bmatrix} 1 & 0 \\ 3 & 1 \end{bmatrix}, \begin{bmatrix} 2 & 1 \\ 2 & 1 \end{bmatrix} \right\}$, $A = \begin{bmatrix} 7 & -7 \\ -5 & -4 \end{bmatrix}, B = \begin{bmatrix} -1 & 6 \\ 3 & 3 \end{bmatrix}$

B9 $\mathcal{B} = \left\{ \begin{bmatrix} 1 & 0 \\ 1 & 0 \end{bmatrix}, \begin{bmatrix} 1 & 1 \\ 0 & 0 \end{bmatrix}, \begin{bmatrix} 0 & 1 \\ 0 & 1 \end{bmatrix}, \begin{bmatrix} 1 & 0 \\ 0 & 1 \end{bmatrix} \right\}$, $A = \begin{bmatrix} 5 & 4 \\ 2 & 5 \end{bmatrix}, B = \begin{bmatrix} 3 & 2 \\ 2 & 6 \end{bmatrix}$

B10 $\mathcal{B} = \left\{ \begin{bmatrix} 1 & 1 & 0 \\ 0 & 1 & 1 \end{bmatrix}, \begin{bmatrix} 0 & 2 & -1 \\ 1 & 3 & -1 \end{bmatrix} \right\}$, $A = \begin{bmatrix} -4 & 2 & -3 \\ 3 & 5 & -7 \end{bmatrix}, B = \begin{bmatrix} 3 & 5 & -1 \\ 1 & 6 & 2 \end{bmatrix}$

B11 $\mathcal{B} = \left\{ \begin{bmatrix} 1 & 0 \\ 2 & 3 \\ 1 & 4 \end{bmatrix}, \begin{bmatrix} 0 & 1 \\ 1 & 4 \\ 2 & 3 \end{bmatrix}, \begin{bmatrix} 0 & 0 \\ 1 & 1 \\ 1 & 1 \end{bmatrix} \right\}$, $A = \begin{bmatrix} 5 & -4 \\ 5 & -2 \\ -4 & 7 \end{bmatrix}, B = \begin{bmatrix} 3 & -5 \\ 7 & -5 \\ -1 & 3 \end{bmatrix}$

B12 $\mathcal{B} = \{1 + x^2 + x^3, 3 + 2x + x^3, 2x + x^2\}$, $\mathbf{p}(x) = 3 + 2x + 2x^2 + 2x^3$, $\mathbf{q}(x) = 7 + 8x + x^2 + 2x^3$

B13 $\mathcal{B} = \left\{ \begin{bmatrix} 2 & 3 \\ 2 & 3 \end{bmatrix}, \begin{bmatrix} 4 & 1 \\ -1 & -2 \end{bmatrix} \right\}$, $A = \begin{bmatrix} 14 & 6 \\ -1 & -3 \end{bmatrix}, B = \begin{bmatrix} 2 & -7 \\ -8 & -13 \end{bmatrix}$

B14 $\mathcal{B} = \left\{ \begin{bmatrix} 1 & 0 \\ 2 & 3 \end{bmatrix}, \begin{bmatrix} 1 & 1 \\ 4 & 3 \end{bmatrix}, \begin{bmatrix} 3 & 1 \\ 2 & 1 \end{bmatrix} \right\}$, $A = \begin{bmatrix} -6 & -1 \\ 4 & 6 \end{bmatrix}, B = \begin{bmatrix} 1 & 1 \\ 1 & -1 \end{bmatrix}$

B15 $\mathcal{B} = \left\{ \begin{bmatrix} 3 & 5 \\ 1 & 4 \end{bmatrix}, \begin{bmatrix} 7 & 16 \\ 3 & 9 \end{bmatrix}, \begin{bmatrix} -1 & 5 \\ 1 & 2 \end{bmatrix} \right\}$, $A = \begin{bmatrix} -4 & 9 \\ 2 & 5 \end{bmatrix}, B = \begin{bmatrix} 0 & 9 \\ 2 & 7 \end{bmatrix}$

B16 $\mathcal{B} = \left\{ \begin{bmatrix} 2 & 2 \\ 4 & 2 \end{bmatrix}, \begin{bmatrix} 1 & 3 \\ 1 & 2 \end{bmatrix}, \begin{bmatrix} 4 & 2 \\ 3 & 4 \end{bmatrix} \right\}$, $A = \begin{bmatrix} -4 & 6 \\ 5 & -2 \end{bmatrix}, B = \begin{bmatrix} 7 & 9 \\ 7 & 9 \end{bmatrix}$

For Problems B17–B22, find the change of coordinates matrix to and from the basis $\mathcal{B}$ and the standard basis of the given vector space.

B17 $\mathcal{B} = \left\{ \begin{bmatrix} 1 \\ 1 \end{bmatrix}, \begin{bmatrix} 1 \\ -1 \end{bmatrix} \right\}$ for $\mathbb{R}^2$

B18 $\mathcal{B} = \left\{\begin{bmatrix}2\\1\end{bmatrix}, \begin{bmatrix}5\\3\end{bmatrix}\right\}$ for $\mathbb{R}^2$

B19 $\mathcal{B} = \left\{\begin{bmatrix}3\\2\\1\end{bmatrix}, \begin{bmatrix}4\\2\\1\end{bmatrix}, \begin{bmatrix}5\\-1\\-1\end{bmatrix}\right\}$ for $\mathbb{R}^3$

B20 $\mathcal{B} = \{1, -2 + x, 4 - 4x + x^2\}$ for $P_2(\mathbb{R})$

B21 $\mathcal{B} = \{1 + x + x^2, 1 - 2x^2, 2 + 2x + x^2\}$ for $P_2(\mathbb{R})$

B22 $\mathcal{B} = \left\{\begin{bmatrix}2 & 0\\0 & 3\end{bmatrix}, \begin{bmatrix}-4 & 0\\0 & 5\end{bmatrix}\right\}$ for the subspace of $M_{2\times 2}(\mathbb{R})$ of diagonal matrices

For Problems B23–B25, find the change of coordinates matrix Q from $\mathcal{B}$-coordinates to C-coordinates and the change of coordinates matrix P from C-coordinates to $\mathcal{B}$-coordinates. Verify that $PQ = I$.

B23 In $\mathbb{R}^2$: $\mathcal{B} = \left\{\begin{bmatrix}1\\4\end{bmatrix}, \begin{bmatrix}2\\3\end{bmatrix}\right\}, C = \left\{\begin{bmatrix}3\\2\end{bmatrix}, \begin{bmatrix}-4\\-1\end{bmatrix}\right\}$

B24 In $\mathbb{R}^2$: $\mathcal{B} = \left\{\begin{bmatrix}4\\5\end{bmatrix}, \begin{bmatrix}1\\1\end{bmatrix}\right\}, C = \left\{\begin{bmatrix}5\\7\end{bmatrix}, \begin{bmatrix}1\\-1\end{bmatrix}\right\}$

B25 In $P_2(\mathbb{R})$: $\mathcal{B} = \{1, 3 + 2x, (3 + 2x)^2\}$,
$C = \{1 - 2x - 4x^2, 2 + 2x, -2x - 4x^2\}$

Conceptual Problems

C1 Suppose that $\mathcal{B} = \{\mathbf{v}_1, \dots, \mathbf{v}_k\}$ is a basis for a vector space $\mathbb{V}$ and that $C = \{\mathbf{w}_1, \dots, \mathbf{w}_k\}$ is another basis for $\mathbb{V}$ and that for every $\mathbf{x} \in \mathbb{V}$, $[\mathbf{x}]_{\mathcal{B}} = [\mathbf{x}]_C$. Must it be true that $\mathbf{v}_i = \mathbf{w}_i$ for each $1 \leq i \leq k$? Explain or prove your conclusion.

C2 Suppose $\mathbb{V}$ is a vector space with basis $\mathcal{B} = \{\mathbf{v}_1, \mathbf{v}_2, \mathbf{v}_3, \mathbf{v}_4\}$. Then $C = \{\mathbf{v}_3, \mathbf{v}_2, \mathbf{v}_4, \mathbf{v}_1\}$ is also a basis of $\mathbb{V}$. Find a matrix P such that $P[\vec{x}]_{\mathcal{B}} = [\vec{x}]_C$.

C3 Let $B = \left\{\begin{bmatrix}2\\3\end{bmatrix}, \begin{bmatrix}1\\2\end{bmatrix}\right\}$ and $C = \left\{\begin{bmatrix}2\\1\end{bmatrix}, \begin{bmatrix}1\\1\end{bmatrix}\right\}$ and let $L : \mathbb{R}^2 \to \mathbb{R}^2$ be the linear mapping such that

$$[\vec{x}]_B = [L(\vec{x})]_C$$

(a) Find $L\left(\begin{bmatrix}3\\5\end{bmatrix}\right)$.

(b) Find $L\left(\begin{bmatrix}x_1\\x_2\end{bmatrix}\right)$.

C4 Let $\mathbb{V}$ be a finite dimensional vector space, and let $\mathcal{B}$ and C both be bases for $\mathbb{V}$. Prove if P is the change of coordinates matrix from C-coordinates to $\mathcal{B}$-coordinates, then P is invertible and P^{-1} is the change of coordinates matrix from $\mathcal{B}$-coordinates to C-coordinates.

C5 If $\mathcal{B} = \{\mathbf{v}_1, \dots, \mathbf{v}_n\}$ is a basis for $\mathbb{V}$, show that $\{[\mathbf{v}_1]_{\mathcal{B}}, \dots, [\mathbf{v}_n]_{\mathcal{B}}\}$ is a basis for $\mathbb{R}^n$.

For Problems C6–C9, let $\mathbb{V}$ be an n-dimensional vector space and let $\mathcal{B}$ be a basis for $\mathbb{V}$. Determine whether the statement is true or false. Justify your answer.

C6 $[\mathbf{v}]_{\mathcal{B}} \in \mathbb{R}^n$ for any $\mathbf{v} \in \mathbb{V}$.

C7 If C is another basis for $\mathbb{V}$, then $[\mathbf{v}]_{\mathcal{B}} = [\mathbf{v}]_C$ for any $\mathbf{v} \in \mathbb{V}$.

C8 If $\mathbf{v}, \mathbf{w} \in \mathbb{V}$ such that $[\mathbf{v}]_{\mathcal{B}} = [\mathbf{w}]_{\mathcal{B}}$, then $\mathbf{v} = \mathbf{w}$.

C9 If $\vec{x} \in \mathbb{R}^n$, then there exists $\mathbf{v} \in \mathbb{V}$ such that $[\mathbf{v}]_{\mathcal{B}} = \vec{x}$.

C10 Let $\mathcal{B} = \{\vec{v}_1, \dots, \vec{v}_n\}$ be a basis for $\mathbb{R}^n$ and let $\mathcal{S} = \{\vec{e}_1, \dots, \vec{e}_n\}$ be the standard basis for $\mathbb{R}^n$.

(a) Find the change of coordinates matrix P from $\mathcal{B}$-coordinates to $\mathcal{S}$-coordinates.

(b) Show that $[\vec{x}]_{\mathcal{B}} = P^{-1}\vec{x}$.

(c) Let A be an $n \times n$ matrix and define

$$B = \begin{bmatrix}[A\vec{v}_1]_{\mathcal{B}} & \cdots & [A\vec{v}_n]_{\mathcal{B}}\end{bmatrix}$$

Prove that $B = P^{-1}AP$.

4.5 General Linear Mappings

In Chapter 3 we looked at linear mappings $L : \mathbb{R}^n \to \mathbb{R}^m$ *and found that they can be useful in solving some problems. Since vector spaces encompass the essential properties of* $\mathbb{R}^n$*, it makes sense that we can also define linear mappings whose domain and codomain are other vector spaces. This also turns out to be extremely useful and important in many real-world applications.*

Definition
Linear Mapping
Linear Transformation
Linear Operator

Let $\mathbb{V}$ and $\mathbb{W}$ be vector spaces. A function $L : \mathbb{V} \to \mathbb{W}$ is called a **linear mapping** (or **linear transformation**) if for every $\mathbf{x}, \mathbf{y} \in \mathbb{V}$ and $s, t \in \mathbb{R}$ it satisfies

$$L(s\mathbf{x} + t\mathbf{y}) = sL(\mathbf{x}) + tL(\mathbf{y})$$

If $\mathbb{W} = \mathbb{V}$, then L may be called a **linear operator**.

As before, two linear mappings $L : \mathbb{V} \to \mathbb{W}$ and $M : \mathbb{V} \to \mathbb{W}$ are said to be **equal** if

$$L(\mathbf{v}) = M(\mathbf{v})$$

for all $\mathbf{v} \in \mathbb{V}$.

EXAMPLE 4.5.1

Let $L : M_{2\times 2}(\mathbb{R}) \to P_2(\mathbb{R})$ be defined by $L\left(\begin{bmatrix} a & b \\ c & d \end{bmatrix}\right) = d + (b + d)x + ax^2$. Prove that L is a linear mapping.

Solution: For any $\begin{bmatrix} a_1 & b_1 \\ c_1 & d_1 \end{bmatrix}, \begin{bmatrix} a_2 & b_2 \\ c_2 & d_2 \end{bmatrix} \in M_{2\times 2}(\mathbb{R})$ and $s, t \in \mathbb{R}$, we have

$$\begin{aligned} L\left(s\begin{bmatrix} a_1 & b_1 \\ c_1 & d_1 \end{bmatrix} + t\begin{bmatrix} a_2 & b_2 \\ c_2 & d_2 \end{bmatrix}\right) &= L\left(\begin{bmatrix} sa_1 + ta_2 & sb_1 + tb_2 \\ sc_1 + tc_2 & sd_1 + td_2 \end{bmatrix}\right) \\ &= (sd_1 + td_2) + (sb_1 + tb_2 + sd_1 + td_2)x + (sa_1 + ta_2)x^2 \\ &= s(d_1 + (b_1 + d_1)x + a_1x^2) + t(d_2 + (b_2 + d_2)x + a_2x^2) \\ &= sL\left(\begin{bmatrix} a_1 & b_1 \\ c_1 & d_1 \end{bmatrix}\right) + tL\left(\begin{bmatrix} a_2 & b_2 \\ c_2 & d_2 \end{bmatrix}\right) \end{aligned}$$

So, L is linear.

EXAMPLE 4.5.2

Let $M : P_2(\mathbb{R}) \to P_2(\mathbb{R})$ be defined by $M(a_0 + a_1x + a_2x^2) = a_1 + 2a_2x$. Prove that M is a linear operator.

Solution: Let $\mathbf{p}(x) = a_0 + a_1x + a_2x^2$, $\mathbf{q}(x) = b_0 + b_1x + b_2x^2$, and $s, t \in \mathbb{R}$. Then,

$$\begin{aligned} M(s\mathbf{p} + t\mathbf{q}) &= M((sa_0 + tb_0) + (sa_1 + tb_1)x + (sa_2 + tb_2)x^2) \\ &= (sa_1 + tb_1) + 2(sa_2 + tb_2)x \\ &= s(a_1 + 2a_2x) + t(b_1 + 2b_2x) \\ &= sM(\mathbf{p}) + tM(\mathbf{q}) \end{aligned}$$

Hence, M is linear.

EXERCISE 4.5.1

Let $L : \mathbb{R}^3 \to M_{2\times 2}(\mathbb{R})$ be defined by $L\left(\begin{bmatrix} x_1 \\ x_2 \\ x_3 \end{bmatrix}\right) = \begin{bmatrix} x_1 & x_1 + x_2 + x_3 \\ 0 & x_2 \end{bmatrix}$.

Prove that L is linear.

EXERCISE 4.5.2

Let $\mathcal{D} : P_2(\mathbb{R}) \to P_1(\mathbb{R})$ be the differential operator defined by $\mathcal{D}(\mathbf{p}) = \mathbf{p}'$. Prove that $\mathcal{D}$ is linear.

Definition
Zero Mapping

The **zero mapping** is the linear mapping $Z : \mathbb{V} \to \mathbb{W}$ defined by

$$Z(\vec{v}) = \vec{0}_{\mathbb{W}}, \qquad \text{for all } \vec{v} \in \mathbb{V}$$

Definition
Identity Mapping

The **identity mapping** is the linear operator $\mathrm{Id} : \mathbb{V} \to \mathbb{V}$ defined by

$$\mathrm{Id}(\vec{v}) = \vec{v}, \qquad \text{for all } \vec{v} \in \mathbb{V}$$

Range and Nullspace

Definition
Range
Nullspace

The **range** of a linear mapping $L : \mathbb{V} \to \mathbb{W}$ is defined to be the set

$$\mathrm{Range}(L) = \{L(\mathbf{x}) \in \mathbb{W} \mid \mathbf{x} \in \mathbb{V}\}$$

The **nullspace** of L is the set of all vectors in $\mathbb{V}$ whose image under L is the zero vector $\mathbf{0}_{\mathbb{W}}$. We write

$$\mathrm{Null}(L) = \{\mathbf{x} \in \mathbb{V} \mid L(\mathbf{x}) = \mathbf{0}_{\mathbb{W}}\}$$

EXAMPLE 4.5.3

Let $A = \begin{bmatrix} 0 & -1 \\ 0 & 1 \end{bmatrix}$ and let $L : M_{2\times 2}(\mathbb{R}) \to P_2(\mathbb{R})$ be the linear mapping defined by $L\left(\begin{bmatrix} a & b \\ c & d \end{bmatrix}\right) = c + (b+d)x + ax^2$. Determine whether A is in the nullspace of L.

Solution: We have

$$L\left(\begin{bmatrix} 0 & -1 \\ 0 & 1 \end{bmatrix}\right) = 0 + (-1+1)x + 0x^2 = 0 = \mathbf{0}_{P_2(\mathbb{R})}$$

Hence, $A \in \mathrm{Null}(L)$.

EXAMPLE 4.5.4

Let $L : P_2(\mathbb{R}) \to \mathbb{R}^3$ be the linear mapping defined by $L(a + bx + cx^2) = \begin{bmatrix} a-b \\ b-c \\ c-a \end{bmatrix}$.

Determine whether $\vec{x} = \begin{bmatrix} 1 \\ 1 \\ 1 \end{bmatrix}$ is in the range of L.

Solution: We want to find a, b, and c such that

$$\begin{bmatrix} 1 \\ 1 \\ 1 \end{bmatrix} = L(a + bx + cx^2) = \begin{bmatrix} a-b \\ b-c \\ c-a \end{bmatrix}$$

This gives us the system of linear equations $a - b = 1, b - c = 1$, and $c - a = 1$. Row reducing the corresponding augmented matrix gives

$$\left[\begin{array}{ccc|c} 1 & -1 & 0 & 1 \\ 0 & 1 & -1 & 1 \\ -1 & 0 & 1 & 1 \end{array}\right] \sim \left[\begin{array}{ccc|c} 1 & -1 & 0 & 1 \\ 0 & 1 & -1 & 1 \\ 0 & 0 & 0 & 3 \end{array}\right]$$

Hence, the system is inconsistent, so $\vec{x}$ is not in the range of L.

Theorem 4.5.1

If $L : \mathbb{V} \to \mathbb{W}$ is a linear mapping, then

(1) $L(\mathbf{0}_{\mathbb{V}}) = \mathbf{0}_{\mathbb{W}}$

(2) Null(L) is a subspace of $\mathbb{V}$

(3) Range(L) is a subspace of $\mathbb{W}$

The proof of Theorem 4.5.1 is left as Problems C1, C2, and C3.

EXAMPLE 4.5.5

Determine a basis for the range and a basis for the nullspace of the linear mapping $L : P_1(\mathbb{R}) \to \mathbb{R}^3$ defined by $L(a + bx) = \begin{bmatrix} a \\ 0 \\ a-2b \end{bmatrix}$.

Solution: If $a + bx \in \text{Null}(L)$, then we have $\begin{bmatrix} a \\ 0 \\ a-2b \end{bmatrix} = L(a + bx) = \begin{bmatrix} 0 \\ 0 \\ 0 \end{bmatrix}$. Hence, $a = 0$ and $a - 2b = 0$, which implies that $b = 0$. Thus, the only polynomial in the nullspace of L is the zero polynomial. That is, Null(L) $= \{0\}$, and so a basis for Null(L) is the empty set. Any vector $\vec{y}$ in the range of L has the form

$$\vec{y} = \begin{bmatrix} a \\ 0 \\ a-2b \end{bmatrix} = a\begin{bmatrix} 1 \\ 0 \\ 1 \end{bmatrix} + b\begin{bmatrix} 0 \\ 0 \\ -2 \end{bmatrix}$$

Thus, Range(L) $=$ Span C, where $C = \left\{\begin{bmatrix} 1 \\ 0 \\ 1 \end{bmatrix}, \begin{bmatrix} 0 \\ 0 \\ -2 \end{bmatrix}\right\}$. Moreover, C is clearly linearly independent. Consequently, C is a basis for the range of L.

EXAMPLE 4.5.6

Determine a basis for the range and a basis for the nullspace of the linear mapping $L : M_{2\times 2}(\mathbb{R}) \to P_2(\mathbb{R})$ defined by $L\left(\begin{bmatrix} a & b \\ c & d \end{bmatrix}\right) = (b+c) + (c-d)x^2$.

Solution: If $\begin{bmatrix} a & b \\ c & d \end{bmatrix} \in \text{Null}(L)$, then

$$0 + 0x + 0x^2 = L\left(\begin{bmatrix} a & b \\ c & d \end{bmatrix}\right) = (b+c) + (c-d)x^2$$

So, $b + c = 0$ and $c - d = 0$. Thus, $b = -c$ and $d = c$, so every matrix in the nullspace of L has the form

$$\begin{bmatrix} a & b \\ c & d \end{bmatrix} = \begin{bmatrix} a & -c \\ c & c \end{bmatrix} = a\begin{bmatrix} 1 & 0 \\ 0 & 0 \end{bmatrix} + c\begin{bmatrix} 0 & -1 \\ 1 & 1 \end{bmatrix}$$

Thus, $\mathcal{B} = \left\{\begin{bmatrix} 1 & 0 \\ 0 & 0 \end{bmatrix}, \begin{bmatrix} 0 & -1 \\ 1 & 1 \end{bmatrix}\right\}$ spans Null(L) and is clearly linearly independent. Consequently, $\mathcal{B}$ is a basis for Null(L).

Any polynomial $\mathbf{p} \in \text{Range}(L)$ has the form

$$\mathbf{p}(x) = L\left(\begin{bmatrix} a & b \\ c & d \end{bmatrix}\right) = (b+c) + (c-d)x^2 = b(1) + c(1+x^2) - dx^2$$

Hence $\{1, 1+x^2, -x^2\}$ spans Range(L). But this is linearly dependent. Clearly, we have $1 + (-1)(-x^2) = 1 + x^2$. Thus, $C = \{1, -x^2\}$ also spans Range(L) and is linearly independent. Therefore, C is a basis for Range(L).

EXERCISE 4.5.3

Determine a basis for the range and a basis for the nullspace of the linear mapping $L : \mathbb{R}^3 \to M_{2\times 2}(\mathbb{R})$ defined by $L\left(\begin{bmatrix} x_1 \\ x_2 \\ x_3 \end{bmatrix}\right) = \begin{bmatrix} x_1 & x_2 + x_3 \\ x_2 + x_3 & x_1 \end{bmatrix}$.

Observe that in each of these examples, the dimension of the range of L plus the dimension of the nullspace of L equals the dimension of the domain of L. This matches the Rank-Nullity Theorem (Theorem 3.4.9). Before we extend this to general linear mappings, we make some definitions.

Definition
Rank of a Linear Mapping

The **rank of a linear mapping** $L : \mathbb{V} \to \mathbb{W}$ is the dimension of the range of L:

$$\text{rank}(L) = \dim(\text{Range}(L))$$

Definition
Nullity of a Linear Mapping

The **nullity of a linear mapping** $L : \mathbb{V} \to \mathbb{W}$ is the dimension of the nullspace of L:

$$\text{nullity}(L) = \dim(\text{Null}(L))$$

Theorem 4.5.2 Rank-Nullity Theorem

Let $\mathbb{V}$ and $\mathbb{W}$ be vector spaces with $\dim \mathbb{V} = n$. If $L : \mathbb{V} \to \mathbb{W}$ is a linear mapping, then

$$\text{rank}(L) + \text{nullity}(L) = n$$

Proof: The idea of the proof is to assume that a basis for the nullspace of L contains k vectors and show that we can then construct a basis for the range of L that contains $n - k$ vectors.

Let $\mathcal{B} = \{\mathbf{v}_1, \dots, \mathbf{v}_k\}$ be a basis for Null(L), so that nullity(L) $= k$. By the Basis Extension Theorem, there exist vectors $\mathbf{u}_{k+1}, \dots, \mathbf{u}_n$ such that $\{\mathbf{v}_1, \dots, \mathbf{v}_k, \mathbf{u}_{k+1}, \dots, \mathbf{u}_n\}$ is a basis for $\mathbb{V}$.

Now consider any vector $\mathbf{w}$ in the range of L. Then $\mathbf{w} = L(\mathbf{x})$ for some $\mathbf{x} \in \mathbb{V}$. But any $\mathbf{x} \in \mathbb{V}$ can be written as a linear combination of the vectors in the basis $\{\mathbf{v}_1, \dots, \mathbf{v}_k, \mathbf{u}_{k+1}, \dots, \mathbf{u}_n\}$, so there exists $t_1, \dots, t_n$ such that $\mathbf{x} = t_1\mathbf{v}_1 + \cdots + t_k\mathbf{v}_k + t_{k+1}\mathbf{u}_{k+1} + \cdots + t_n\mathbf{u}_n$. Then,

$$\begin{aligned} \mathbf{w} &= L(t_1\mathbf{v}_1 + \cdots + t_k\mathbf{v}_k + t_{k+1}\mathbf{u}_{k+1} + \cdots + t_n\mathbf{u}_n) \\ &= t_1L(\mathbf{v}_1) + \cdots + t_kL(\mathbf{v}_k) + t_{k+1}L(\mathbf{u}_{k+1}) + \cdots + t_nL(\mathbf{u}_n) \end{aligned}$$

But each $\mathbf{v}_i$ is in the nullspace of L, so $L(\mathbf{v}_i) = \mathbf{0}$, and thus we have

$$\mathbf{w} = t_{k+1}L(\mathbf{u}_{k+1}) + \cdots + t_nL(\mathbf{u}_n)$$

Therefore, any $\mathbf{w} \in$ Range(L) can be expressed as a linear combination of the vectors in the set $C = \{L(\mathbf{u}_{k+1}), \dots, L(\mathbf{u}_n)\}$. Thus, C is a spanning set for Range(L). Is it linearly independent? We consider

$$t_{k+1}L(\mathbf{u}_{k+1}) + \cdots + t_nL(\mathbf{u}_n) = \mathbf{0}_{\mathbb{W}}$$

By the linearity of L, this is equivalent to

$$L(t_{k+1}\mathbf{u}_{k+1} + \cdots + t_n\mathbf{u}_n) = \mathbf{0}_{\mathbb{W}}$$

If this is true, then $t_{k+1}\mathbf{u}_{k+1} + \cdots + t_n\mathbf{u}_n$ is a vector in the nullspace of L. Hence, for some $d_1, \dots, d_k$, we have

$$t_{k+1}\mathbf{u}_{k+1} + \cdots + t_n\mathbf{u}_n = d_1\mathbf{v}_1 + \cdots + d_k\mathbf{v}_k$$

But this is impossible unless all t_i and d_i are zero, because $\{\mathbf{v}_1, \dots, \mathbf{v}_k, \mathbf{u}_{k+1}, \dots, \mathbf{u}_n\}$ is a basis for $\mathbb{V}$ and hence linearly independent.

It follows that C is a linearly independent spanning set for Range(L). Hence, it is a basis for Range(L) containing $n - k$ vectors. Thus, rank(L) $= n - k$ and

$$\text{rank}(L) + \text{nullity}(L) = (n - k) + k = n$$

as required. ■

EXAMPLE 4.5.7

Determine the dimension of the range and the dimension of the kernel of the linear mapping $L : P_2(\mathbb{R}) \to M_{2\times 2}(\mathbb{R})$ defined by

$$L(a + bx + cx^2) = \begin{bmatrix} a+c & b+c \\ 0 & a-b \end{bmatrix}$$

Solution: If $a + bx + cx^2 \in \text{Null}(L)$, then

$$\begin{bmatrix} 0 & 0 \\ 0 & 0 \end{bmatrix} = L(a + bx + cx^2) = \begin{bmatrix} a+c & b+c \\ 0 & a-b \end{bmatrix}$$

This gives the system of equations $a + c = 0, b + c = 0, a - b = 0$. Row reducing the corresponding coefficient matrix gives

$$\begin{bmatrix} 1 & 0 & 1 \\ 0 & 1 & 1 \\ 1 & -1 & 0 \end{bmatrix} \sim \begin{bmatrix} 1 & 0 & 1 \\ 0 & 1 & 1 \\ 0 & 0 & 0 \end{bmatrix}$$

The solution is

$$\begin{bmatrix} a \\ b \\ c \end{bmatrix} = c \begin{bmatrix} -1 \\ -1 \\ 1 \end{bmatrix}$$

That is, $a = -c, b = -c$, for any $c \in \mathbb{R}$. Hence, every vector in Null(L) has the form

$$-c - cx + cx^2 = c(-1 - x + x^2)$$

Thus, $\{-1 - x + x^2\}$ is a basis for Null(L) and so nullity(L) = 1.

Every matrix $A \in \text{Range}(L)$ has the form

$$A = L(a + bx + cx^2) = \begin{bmatrix} a+c & b+c \\ 0 & a-b \end{bmatrix} = a\begin{bmatrix} 1 & 0 \\ 0 & 1 \end{bmatrix} + b\begin{bmatrix} 0 & 1 \\ 0 & -1 \end{bmatrix} + c\begin{bmatrix} 1 & 1 \\ 0 & 0 \end{bmatrix}$$

Thus, $\left\{\begin{bmatrix} 1 & 0 \\ 0 & 1 \end{bmatrix}, \begin{bmatrix} 0 & 1 \\ 0 & -1 \end{bmatrix}, \begin{bmatrix} 1 & 1 \\ 0 & 0 \end{bmatrix}\right\}$ spans Range(L). To determine if it is linearly independent, we consider

$$\begin{bmatrix} 0 & 0 \\ 0 & 0 \end{bmatrix} = c_1\begin{bmatrix} 1 & 0 \\ 0 & 1 \end{bmatrix} + c_2\begin{bmatrix} 0 & 1 \\ 0 & -1 \end{bmatrix} + c_3\begin{bmatrix} 1 & 1 \\ 0 & 0 \end{bmatrix}$$

We observe that this gives the same coefficient matrix as above. Our row reduction above shows us that the third matrix is the sum of the first two. Hence,

$$C = \left\{\begin{bmatrix} 1 & 0 \\ 0 & 1 \end{bmatrix}, \begin{bmatrix} 0 & 1 \\ 0 & -1 \end{bmatrix}\right\}$$

is a basis for Range(L) as it is linearly independent and spans Range(L). Therefore, rank(L) = 2.

Then, as predicted by the Rank-Nullity Theorem, we have

$$\text{rank}(L) + \text{nullity}(L) = 2 + 1 = 3 = \dim P_2(\mathbb{R})$$

Inverse Linear Mappings

Definition
Composition of Linear Mappings

Let $\mathbb{U}$, $\mathbb{V}$, and $\mathbb{W}$ be vector spaces. If $L : \mathbb{V} \to \mathbb{W}$ and $M : \mathbb{W} \to \mathbb{U}$ are linear mappings, then the **composition** $M \circ L : \mathbb{V} \to \mathbb{U}$ is defined by

$$(M \circ L)(\mathbf{x}) = M(L(\mathbf{x}))$$

for all $\mathbf{x} \in \mathbb{V}$.

EXAMPLE 4.5.8

If $L : P_2(\mathbb{R}) \to \mathbb{R}^3$ is defined by $L(a + bx + cx^2) = \begin{bmatrix} a+b \\ c \\ 0 \end{bmatrix}$ and $M : \mathbb{R}^3 \to M_{2\times2}(\mathbb{R})$ is defined by $M\left(\begin{bmatrix} x_1 \\ x_2 \\ x_3 \end{bmatrix}\right) = \begin{bmatrix} x_1 & 0 \\ x_2 + x_3 & 0 \end{bmatrix}$, then $M \circ L$ is the mapping defined by

$$(M \circ L)(a + bx + cx^2) = M(L(a + bx + cx^2)) = M\left(\begin{bmatrix} a+b \\ c \\ 0 \end{bmatrix}\right) = \begin{bmatrix} a+b & 0 \\ c & 0 \end{bmatrix}$$

Observe that $M \circ L$ is in fact a linear mapping from $P_2(\mathbb{R})$ to $M_{2\times2}(\mathbb{R})$.

Theorem 4.5.3

If $L : \mathbb{V} \to \mathbb{W}$ and $M : \mathbb{W} \to \mathbb{U}$ are linear mappings, then $M \circ L : \mathbb{V} \to \mathbb{U}$ is also a linear mapping.

The proof of Theorem 4.5.3 is left as Problem C4.

Definition
Inverse Mapping

If $L : \mathbb{V} \to \mathbb{W}$ is a linear mapping and there exists another linear mapping $M : \mathbb{W} \to \mathbb{V}$ such that

$$M \circ L = \text{Id}$$
$$L \circ M = \text{Id}$$

then L is said to be **invertible**, and M is called the **inverse** of L, denoted L^{-1}.

Remarks

1. By definition, if M is the inverse of L, then L is the inverse of M.
2. Note that the domain and codomain of an invertible linear mapping need not be the same vector space. It is important to ask what condition on $\mathbb{V}$ and $\mathbb{W}$ is required so that it is possible for $L : \mathbb{V} \to \mathbb{W}$ to be invertible.

EXAMPLE 4.5.9

Let $L : P_3(\mathbb{R}) \to M_{2\times 2}(\mathbb{R})$ be the linear mapping defined by

$$L(a + bx + cx^2 + dx^3) = \begin{bmatrix} a+b+c & -a+b+c+d \\ a+c-d & a+b+d \end{bmatrix}$$

and $M : M_{2\times 2}(\mathbb{R}) \to P_3(\mathbb{R})$ be the linear mapping defined by

$$M\left(\begin{bmatrix} a & b \\ c & d \end{bmatrix}\right) = (-a+c+d)+(4a-b-3c-2d)x+(-2a+b+2c+d)x^2+(-3a+b+2c+2d)x^3$$

Prove that L and M are inverses of each other.

Solution: Observe that

$$\begin{aligned}(M \circ L)(a + bx + cx^2 + dx^3) &= M\left(\begin{bmatrix} a+b+c & -a+b+c+d \\ a+c-d & a+b+d \end{bmatrix}\right) \\ &= a + bx + cx^2 + dx^3 \\ (L \circ M)\left(\begin{bmatrix} a & b \\ c & d \end{bmatrix}\right) &= L((-a+c+d) + (4a-b-3c-2d)x \\ &\quad + (-2a+b+2c+d)x^2 + (-3a+b+2c+2d)x^3) \\ &= \begin{bmatrix} a & b \\ c & d \end{bmatrix}\end{aligned}$$

Thus, L and M are inverses of each other.

Theorem 4.5.4 If $L : \mathbb{V} \to \mathbb{W}$ is an invertible linear mapping, then L^{-1} is unique.

The proof of Theorem 4.5.4 is left as Problem C5.

Theorem 4.5.5 If $L : \mathbb{V} \to \mathbb{W}$ is invertible, then $\text{Null}(L) = \{\mathbf{0}_{\mathbb{V}}\}$ and $\text{Range}(L) = \mathbb{W}$.

Proof: Let $\mathbf{v} \in \text{Null}(L)$. Using Theorem 4.5.1 (1), we get

$$\mathbf{0} = L^{-1}(\mathbf{0}) = L^{-1}(L(\mathbf{v})) = \text{Id}(\mathbf{v}) = \mathbf{v}$$

Hence, $\text{Null}(L) = \{\mathbf{0}_{\mathbb{V}}\}$.

Let $\mathbf{y} \in \mathbb{W}$. Let $\mathbf{x} \in \mathbb{V}$ be the vector such that $L^{-1}(\mathbf{y}) = \mathbf{x}$. Then,

$$L(\mathbf{x}) = L\left(L^{-1}(\mathbf{y})\right) = \text{Id}(\mathbf{y}) = \mathbf{y}$$

Hence, $\mathbf{y} \in \text{Range}(L)$ and so $\text{Range}(L) = \mathbb{W}$. ■

PROBLEMS 4.5

Practice Problems

For Problems A1–A4, prove that the mapping is linear.

A1 $L : \mathbb{R}^3 \to \mathbb{R}^2$ defined by $L(x_1, x_2, x_3) = (x_1 + x_2, x_1 + x_2 + x_3)$

A2 $L : \mathbb{R}^3 \to P_1(\mathbb{R})$ defined by $L(a, b, c) = (a + b) + (a + b + c)x$

A3 $\text{tr} : M_{2\times 2}(\mathbb{R}) \to \mathbb{R}$ defined by $\text{tr}\left(\begin{bmatrix} a & b \\ c & d \end{bmatrix}\right) = a + d$

A4 $T : P_3(\mathbb{R}) \to M_{2\times 2}(\mathbb{R})$ defined by $T(a + bx + cx^2 + dx^3) = \begin{bmatrix} a & b \\ c & d \end{bmatrix}$

For Problems A5–A8, determine whether the mapping is linear.

A5 $D : M_{2\times 2}(\mathbb{R}) \to \mathbb{R}$ defined by $D(A) = \det A$

A6 $L : P_2(\mathbb{R}) \to P_2(\mathbb{R})$ defined by $L(a + bx + cx^2) = (a - b) + (b + c)x^2$

A7 $T : \mathbb{R}^2 \to M_{2\times 2}(\mathbb{R})$ defined by $T(x_1, x_2) = \begin{bmatrix} x_1 & 1 \\ 1 & x_2 \end{bmatrix}$

A8 $M : M_{2\times 2}(\mathbb{R}) \to M_{2\times 2}(\mathbb{R})$ defined by $M\left(\begin{bmatrix} a & b \\ c & d \end{bmatrix}\right) = \begin{bmatrix} 0 & 0 \\ 0 & 0 \end{bmatrix}$

For Problems A9–A12, determine whether the given vector $\mathbf{y}$ is in the range of the given linear mapping $L : \mathbb{V} \to \mathbb{W}$. If it is, find a vector $\mathbf{x} \in \mathbb{V}$ such that $L(\mathbf{x}) = \mathbf{y}$.

A9 $L : \mathbb{R}^3 \to \mathbb{R}^3$ defined by $L(x_1, x_2, x_3) = \begin{bmatrix} x_1 + x_3 \\ 0 \\ x_2 + x_3 \end{bmatrix}, \mathbf{y} = \begin{bmatrix} 2 \\ 0 \\ 3 \end{bmatrix}$

A10 $L : P_2(\mathbb{R}) \to M_{2\times 2}(\mathbb{R})$ defined by $L(a + bx + cx^2) = \begin{bmatrix} a + c & 0 \\ 0 & b + c \end{bmatrix}, \mathbf{y} = \begin{bmatrix} 2 & 0 \\ 0 & 3 \end{bmatrix}$

A11 $L : P_2(\mathbb{R}) \to P_1(\mathbb{R})$ defined by $L(a + bx + cx^2) = (b + c) + (-b - c)x, \mathbf{y} = 1 + x$

A12 $L : \mathbb{R}^4 \to M_{2\times 2}(\mathbb{R})$ defined by $L(\vec{x}) = \begin{bmatrix} -2x_2 - 2x_3 - 2x_4 & x_1 + x_4 \\ -2x_1 - x_2 - x_4 & 2x_1 - 2x_2 - x_3 + 2x_4 \end{bmatrix}$, $\mathbf{y} = \begin{bmatrix} -1 & -1 \\ -2 & 2 \end{bmatrix}$

For Problems A13–A18, find a basis for the range and a basis for the nullspace of the linear mapping and verify the Rank-Nullity Theorem.

A13 $L : \mathbb{R}^3 \to \mathbb{R}^2$ defined by $L(x_1, x_2, x_3) = (x_1 + x_2, x_1 + x_2 + x_3)$

A14 $L : \mathbb{R}^3 \to P_1(\mathbb{R})$ defined by $L(a, b, c) = (a + b) + (a + b + c)x$

A15 $L : P_1(\mathbb{R}) \to P_2(\mathbb{R})$ defined by $L(a + bx) = a - 2b + (a - 2b)x^2$

A16 $L : P_2(\mathbb{R}) \to \mathbb{R}^3$ defined by $L\left(a + bx + cx^2\right) = (a - b, 2a + b, a - b + c)$

A17 $\text{tr} : M_{2\times 2}(\mathbb{R}) \to \mathbb{R}$ defined by $\text{tr}\left(\begin{bmatrix} a & b \\ c & d \end{bmatrix}\right) = a + d$

A18 $T : P_3(\mathbb{R}) \to M_{2\times 2}(\mathbb{R})$ defined by $T(a + bx + cx^2 + dx^3) = \begin{bmatrix} a & b \\ c & d \end{bmatrix}$

For Problems A19 and A20, determine whether L and M are inverses.

A19 $L : \mathbb{R}^2 \to P_1(\mathbb{R})$ defined by $L(y_1, y_2) = (y_1 + y_2) + (2y_1 + 3y_2)x$
$M : P_1(\mathbb{R}) \to \mathbb{R}^2$ defined by $M(a + bx) = (3a - b, -a + b)$

A20 $L : P_2(\mathbb{R}) \to \mathbb{R}^3$ defined by $L(a + bx + cx^2) = (a - b + c, b + 2c, c)$
$M : \mathbb{R}^3 \to P_2(\mathbb{R})$ defined by $M(y_1, y_2, y_3) = (y_1 + y_2 - 3y_3) + (y_2 - 2y_3)x + y_3x^2$

For Problems A21–A23, construct a linear mapping $L : \mathbb{V} \to \mathbb{W}$ that satisfies the given properties.

A21 $\mathbb{V} = \mathbb{R}^3, \mathbb{W} = P_2(\mathbb{R})$; $L(1, 0, 0) = x^2$, $L(0, 1, 0) = 2x$, $L(0, 0, 1) = 1 + x + x^2$

A22 $\mathbb{V} = P_2(\mathbb{R})$, $\mathbb{W} = M_{2\times 2}(\mathbb{R})$; $\text{Null}(L) = \{0\}$ and $\text{Range}(L) = \text{Span}\left\{\begin{bmatrix} 1 & 0 \\ 0 & 0 \end{bmatrix}, \begin{bmatrix} 0 & 1 \\ 0 & 0 \end{bmatrix}, \begin{bmatrix} 0 & 0 \\ 0 & 1 \end{bmatrix}\right\}$

A23 $\mathbb{V} = M_{2\times 2}(\mathbb{R}), \mathbb{W} = \mathbb{R}^4$; $\text{nullity}(L) = 2$, $\text{rank}(L) = 2$, and $L\left(\begin{bmatrix} 1 & 0 \\ 0 & 1 \end{bmatrix}\right) = \begin{bmatrix} 0 \\ 1 \\ 1 \\ 0 \end{bmatrix}$

Homework Problems

For Problems B1–B7, prove that the mapping is linear.

B1 $L : \mathbb{R}^3 \to P_2(\mathbb{R})$ defined by $L(a, b, c) = a + bx + cx^2$

B2 $L : P_1(\mathbb{R}) \to P_1(\mathbb{R})$ defined by $L(a + bx) = a + b - 2ax$

B3 $L : M_{2\times 2}(\mathbb{R}) \to \mathbb{R}^2$ defined by $L\left(\begin{bmatrix} a & b \\ c & d \end{bmatrix}\right) = \begin{bmatrix} 0 \\ 0 \end{bmatrix}$

B4 Let $\mathbb{D}$ be the subspace of $M_{2\times 2}(\mathbb{R})$ of diagonal matrices; $L : P_1(\mathbb{R}) \to \mathbb{D}$ defined by $L(a + bx) = \begin{bmatrix} a - 2b & 0 \\ 0 & a - 2b \end{bmatrix}$

B5 $L : M_{2\times 2}(\mathbb{R}) \to M_{2\times 2}(\mathbb{R})$ defined by $L\left(\begin{bmatrix} a & b \\ c & d \end{bmatrix}\right) = \begin{bmatrix} a - b & 0 \\ a - d & -a + d \end{bmatrix}$

B6 $L : M_{2\times 2}(\mathbb{R}) \to \mathbb{R}^3$ defined by $L\left(\begin{bmatrix} a & b \\ c & d \end{bmatrix}\right) = \begin{bmatrix} a + b \\ 0 \\ c + d \end{bmatrix}$

B7 $T : M_{2\times 2}(\mathbb{R}) \to M_{2\times 2}(\mathbb{R})$ defined by $T(A) = A^T$

For Problems B8–B12, determine whether the mapping is linear.

B8 $L : P_2(\mathbb{R}) \to \mathbb{R}$ defined by $L(a + bx + cx^2) = \left\| \begin{bmatrix} a \\ b \\ c \end{bmatrix} \right\|$

B9 $M : P_1(\mathbb{R}) \to \mathbb{R}$ defined by $M(a + bx) = b - a$

B10 $N : \mathbb{R}^3 \to M_{2\times 2}(\mathbb{R})$ defined by $N(x_1, x_2, x_3) = \begin{bmatrix} x_1 - x_3 & 1 \\ x_3 & x_1 \end{bmatrix}$

B11 $L : M_{2\times 2}(\mathbb{R}) \to M_{2\times 2}(\mathbb{R})$ defined by $L(A) = A\begin{bmatrix} 1 & -1 \\ 1 & -1 \end{bmatrix}$

B12 $T : M_{2\times 2}(\mathbb{R}) \to P_2(\mathbb{R})$ defined by $T\left(\begin{bmatrix} a & b \\ c & d \end{bmatrix}\right) = a + (a + d)x + (b + c)x^2$

For Problems B13–B18, determine whether the given vector $\mathbf{y}$ is in the range of the given linear mapping $L : \mathbb{V} \to \mathbb{W}$. If it is, find a vector $\mathbf{x} \in \mathbb{V}$ such that $L(\mathbf{x}) = \mathbf{y}$.

B13 $L : \mathbb{R}^3 \to \mathbb{R}^3$ defined by $L\left(\begin{bmatrix} x_1 \\ x_2 \\ x_3 \end{bmatrix}\right) = \begin{bmatrix} -x_1 - 2x_2 \\ 2x_1 + x_3 \\ -2x_1 + x_2 - 2x_3 \end{bmatrix}, \mathbf{y} = \begin{bmatrix} -1 \\ 1 \\ -1 \end{bmatrix}$

B14 $L : P_1(\mathbb{R}) \to P_1(\mathbb{R})$ defined by $L(a + bx) = (a + 2b) + (2a + 3b)x, \mathbf{y} = 3 - 4x$

B15 $L : P_2(\mathbb{R}) \to \mathbb{R}^3$ defined by $L(a + bx + cx^2) = \begin{bmatrix} a + b \\ b + c \\ a - c \end{bmatrix}, \mathbf{y} = \begin{bmatrix} 2 \\ -1 \\ 2 \end{bmatrix}$

B16 $L : \mathbb{R}^2 \to M_{2\times 2}(\mathbb{R})$ defined by $L\left(\begin{bmatrix} x_1 \\ x_2 \end{bmatrix}\right) = \begin{bmatrix} x_1 + 2x_2 & x_1 - x_2 \\ 0 & 2x_1 + x_2 \end{bmatrix}, \mathbf{y} = \begin{bmatrix} 3 & 1 \\ 0 & 4 \end{bmatrix}$

B17 Let $\mathbb{T}$ denote the subspace of 2×2 upper-triangular matrices; $L : \mathbb{T} \to P_2(\mathbb{R})$ defined by $L\left(\begin{bmatrix} a & b \\ 0 & c \end{bmatrix}\right) = (-a - c) + (a - 2b)x^2, \mathbf{y} = 2 + x^2$

B18 $L : P_1(\mathbb{R}) \to M_{2\times 2}(\mathbb{R})$ defined by $L(a + bx) = \begin{bmatrix} -a & 2b \\ a - b & -2b \end{bmatrix}, \mathbf{y} = \begin{bmatrix} -2 & 2 \\ 0 & -2 \end{bmatrix}$

For Problems B19–B28, find a basis for the range and a basis for the nullspace of the linear mapping and verify the Rank-Nullity Theorem.

B19 $L : \mathbb{R}^3 \to P_2(\mathbb{R})$ defined by $L(a, b, c) = a + bx + cx^2$

B20 $L : P_1(\mathbb{R}) \to \mathbb{R}^3$ defined by $L(a + bx) = \begin{bmatrix} a - 2b \\ a + b \\ a - b \end{bmatrix}$

B21 $L : P_2(\mathbb{R}) \to P_1(\mathbb{R})$ defined by $L\left(a + bx + cx^2\right) = b + 2ax$

B22 $L : M_{2\times 2}(\mathbb{R}) \to \mathbb{R}^2$ defined by $L\left(\begin{bmatrix} a & b \\ c & d \end{bmatrix}\right) = \begin{bmatrix} 0 \\ 0 \end{bmatrix}$

B23 $L : \mathbb{R}^3 \to M_{2\times 2}(\mathbb{R})$ defined by $L(x_1, x_2, x_3) = \begin{bmatrix} x_1 + 2x_2 & -x_1 + 2x_3 \\ x_2 + x_3 & x_1 + 2x_2 \end{bmatrix}$

B24 Let $\mathbb{D}$ be the subspace of $M_{2\times 2}(\mathbb{R})$ of diagonal matrices; $L : \mathbb{D} \to P_2(\mathbb{R})$ defined by $L\left(\begin{bmatrix} a & 0 \\ 0 & b \end{bmatrix}\right) = a + (a + b)x + bx^2$

B25 $L : M_{2\times 2}(\mathbb{R}) \to M_{2\times 2}(\mathbb{R})$ defined by $L\left(\begin{bmatrix} a & b \\ c & d \end{bmatrix}\right) = \begin{bmatrix} a - b & b - c \\ c - d & d - a \end{bmatrix}$

B26 $L : M_{2\times 2}(\mathbb{R}) \to P_2(\mathbb{R})$ defined by $L\left(\begin{bmatrix} a & b \\ c & d \end{bmatrix}\right) = (a + b) + (c + d)x^2$

B27 $T : M_{2\times2}(\mathbb{R}) \to M_{2\times2}(\mathbb{R})$ defined by $T(A) = A^T$

B28 $L : P_2(\mathbb{R}) \to M_{2\times2}(\mathbb{R})$ defined by
$L(a + bx + cx^2) = \begin{bmatrix} -a - 2c & 2b - c \\ -2a + 2c & -2b - c \end{bmatrix}$

For Problems B29–B31, determine whether L and M are inverses.

B29 $L : \mathbb{R}^2 \to \mathbb{R}^2$ defined by
$L(x_1, x_2) = (2x_1 + 3x_2, x_1 + 3x_2)$
$M : \mathbb{R}^2 \to \mathbb{R}^2$ defined by
$M(x_1, x_2) = (x_1 - x_2, -x_1 + 2x_2)$

B30 $L : P_1(\mathbb{R}) \to \mathbb{R}^2$ defined by
$L(a + bx) = (4a - 3b, -7a + 5b)$
$M : \mathbb{R}^2 \to P_1(\mathbb{R})$ defined by
$M(y_1, y_2) = (-5y_1 - 3y_2) + (-7y_1 - 4y_2)x$

B31 $L : \mathbb{R}^3(\mathbb{R}) \to P_2(\mathbb{R})$ defined by
$L(a, b, c) = (a + b + 2c) + (a + 2c)x + (a + 2b + c)x^2$
$M : P_2(\mathbb{R}) \to \mathbb{R}^3$ defined by
$M(a + bx + cx^2) = (-4a + 3b + 2c, a - b, 2a - b - c)$

For Problems B32–B34, construct a linear mapping $L : \mathbb{V} \to \mathbb{W}$ that satisfies the given properties.

B32 $\mathbb{V} = P_2(\mathbb{R}), \mathbb{W} = \mathbb{R}^2; L(1) = \begin{bmatrix} 1 \\ -1 \end{bmatrix}$,
$L(x) = \begin{bmatrix} -1 \\ 1 \end{bmatrix}, L(x^2) = \begin{bmatrix} 3 \\ 2 \end{bmatrix}$.

B33 $\mathbb{V} = M_{2\times2}(\mathbb{R}), \mathbb{W} = \mathbb{R}^3$; nullity$(L) = 3$,
rank$(L) = 1$, and $L\left(\begin{bmatrix} 1 & 0 \\ 0 & 1 \end{bmatrix}\right) = \begin{bmatrix} 0 \\ 1 \\ 2 \end{bmatrix}$.

B34 $\mathbb{V} = P_2(\mathbb{R}), \mathbb{W} = P_2(\mathbb{R})$; Range$(L) = P_2(\mathbb{R})$,
$L(1 + x^2) = x, L(1 + x) = x + x^2$.

Conceptual Problems

C1 Prove if $L : \mathbb{V} \to \mathbb{W}$ is a linear mapping, then $L(\mathbf{0}_\mathbb{V}) = \mathbf{0}_\mathbb{W}$.

C2 Prove if $L : \mathbb{V} \to \mathbb{W}$ is a linear mapping, then Null(L) is a subspace of $\mathbb{V}$.

C3 Prove if $L : \mathbb{V} \to \mathbb{W}$ is a linear mapping, then Range(L) is a subspace of $\mathbb{W}$.

C4 Prove if $L : \mathbb{V} \to \mathbb{W}$ and $M : \mathbb{W} \to \mathbb{U}$ are linear mappings, then $M \circ L : \mathbb{V} \to \mathbb{U}$ is also a linear mapping.

C5 Prove if $L : \mathbb{V} \to \mathbb{W}$ is an invertible linear mapping, then L^{-1} is unique.

C6 (a) Let $\mathbb{V}$ and $\mathbb{W}$ be vector spaces and $L : \mathbb{V} \to \mathbb{W}$ be a linear mapping. Prove that if $\{L(\mathbf{v}_1), \dots, L(\mathbf{v}_k)\}$ is a linearly independent set in $\mathbb{W}$, then $\{\mathbf{v}_1, \dots, \mathbf{v}_k\}$ is a linearly independent set in $\mathbb{V}$.
(b) Give an example of a linear mapping $L : \mathbb{V} \to \mathbb{W}$, where $\{\mathbf{v}_1, \dots, \mathbf{v}_k\}$ is linearly independent in $\mathbb{V}$ but $\{L(\mathbf{v}_1), \dots, L(\mathbf{v}_k)\}$ is linearly dependent in $\mathbb{W}$.

C7 Let $\mathbb{V}$ and $\mathbb{W}$ be n-dimensional vector spaces and let $L : \mathbb{V} \to \mathbb{W}$ be a linear mapping. Prove that Range$(L) = \mathbb{W}$ if and only if Null$(L) = \{\mathbf{0}\}$.

C8 Let $\mathbb{U}, \mathbb{V}$, and $\mathbb{W}$ be finite-dimensional vector spaces over $\mathbb{R}$ and let $L : \mathbb{V} \to \mathbb{U}$ and $M : \mathbb{U} \to \mathbb{W}$ be linear mappings.
(a) Prove that rank$(M \circ L) \leq$ rank(M).
(b) Prove that rank$(M \circ L) \leq$ rank(L).
(c) Construct an example such that the rank of the composition is strictly less than the maximum of the ranks.

C9 Let $\mathbb{U}$ and $\mathbb{V}$ be finite-dimensional vector spaces and let $L : \mathbb{V} \to \mathbb{U}$ be a linear mapping and $M : \mathbb{U} \to \mathbb{U}$ be a linear operator such that Null$(M) = \{\mathbf{0}_\mathbb{U}\}$. Prove that rank$(M \circ L) =$ rank(L).

C10 Let S denote the set of all infinite sequences of real numbers. A typical element of S is $\mathbf{x} = (x_1, x_2, \dots, x_n, \dots)$. Define addition $\mathbf{x} + \mathbf{y}$ and scalar multiplication $t\mathbf{x}$ in the obvious way. Then S is a vector space. Define the left shift $L : S \to S$ by $L(x_1, x_2, x_3, \dots) = (x_2, x_3, x_4, \dots)$ and the right shift $R : S \to S$ by $R(x_1, x_2, x_3, \dots) = (0, x_1, x_2, x_3, \dots)$. Then it is easy to verify that L and R are linear. Check that $(L \circ R)(\mathbf{x}) = \mathbf{x}$ but that $(R \circ L)(\mathbf{x}) \neq \mathbf{x}$. L has a right inverse, but it does not have a left inverse. It is important in this example that S is infinite-dimensional.

4.6 Matrix of a Linear Mapping

In Section 3.2, we defined the standard matrix of a linear mapping $L : \mathbb{R}^n \to \mathbb{R}^m$ so that we could represent any such linear mapping as a matrix mapping. We now generalize this to finding a matrix representation of any linear mapping $L : \mathbb{V} \to \mathbb{V}$ with respect to any basis $\mathcal{B}$ of the vector space $\mathbb{V}$.

The Matrix of a Linear Mapping $L : \mathbb{R}^n \to \mathbb{R}^n$

Let $L : \mathbb{R}^n \to \mathbb{R}^n$ be a linear mapping and let $\mathcal{S} = \{\vec{e}_1, \dots, \vec{e}_n\}$ denote the standard basis for $\mathbb{R}^n$. Using the result of Exercise 4.4.2 on page 266, we can write the equations in Theorem 3.2.3 on page 176 regarding the standard matrix of L as

$$[L] = \begin{bmatrix} [L(\vec{e}_1)]_{\mathcal{S}} & \cdots & [L(\vec{e}_n)]_{\mathcal{S}} \end{bmatrix}$$

and

$$[L(\vec{x})]_{\mathcal{S}} = [L][\vec{x}]_{\mathcal{S}} \tag{4.4}$$

Our goal is to mimic these equations for any linear operator $L : \mathbb{R}^n \to \mathbb{R}^n$ with respect to any basis $\mathcal{B} = \{\vec{v}_1, \dots, \vec{v}_n\}$ for $\mathbb{R}^n$. We follow the same method as in the proof of Theorem 3.2.3. For any $\vec{x} \in \mathbb{R}^n$, we can write $\vec{x} = b_1\vec{v}_1 + \cdots + b_n\vec{v}_n$. Therefore,

$$L(\vec{x}) = L(b_1\vec{v}_1 + \cdots + b_n\vec{v}_n) = b_1L(\vec{v}_1) + \cdots + b_nL(\vec{v}_n)$$

Taking $\mathcal{B}$-coordinates of both sides gives

$$\begin{aligned} [L(\vec{x})]_{\mathcal{B}} &= [b_1L(\vec{v}_1) + \cdots + b_nL(\vec{v}_n)]_{\mathcal{B}} \\ &= b_1[L(\vec{v}_1)]_{\mathcal{B}} + \cdots + b_n[L(\vec{v}_n)]_{\mathcal{B}} \\ &= \begin{bmatrix} [L(\vec{v}_1)]_{\mathcal{B}} & \cdots & [L(\vec{v}_n)]_{\mathcal{B}} \end{bmatrix} \begin{bmatrix} b_1 \\ \vdots \\ b_n \end{bmatrix} \end{aligned}$$

Observe that $\begin{bmatrix} b_1 \\ \vdots \\ b_n \end{bmatrix} = [\vec{x}]_{\mathcal{B}}$, so this equation is the $\mathcal{B}$-coordinates version of equation (4.4) where the matrix $\begin{bmatrix} [L(\vec{v}_1)]_{\mathcal{B}} & \cdots & [L(\vec{v}_n)]_{\mathcal{B}} \end{bmatrix}$ is taking the place of the standard matrix of L.

Definition
Matrix of a Linear Mapping

Suppose that $\mathcal{B} = \{\vec{v}_1, \dots, \vec{v}_n\}$ is any basis for $\mathbb{R}^n$ and that $L : \mathbb{R}^n \to \mathbb{R}^n$ is a linear mapping. Define the **matrix of the linear mapping** L with respect to the basis $\mathcal{B}$ to be the matrix

$$[L]_{\mathcal{B}} = \begin{bmatrix} [L(\vec{v}_1)]_{\mathcal{B}} & \cdots & [L(\vec{v}_n)]_{\mathcal{B}} \end{bmatrix}$$

It satisfies

$$[L(\vec{x})]_{\mathcal{B}} = [L]_{\mathcal{B}}[\vec{x}]_{\mathcal{B}}$$

Note that the columns of $[L]_{\mathcal{B}}$ are the $\mathcal{B}$-coordinate vectors of the images of the $\mathcal{B}$-basis vectors under L. The pattern is exactly the same as before, except that everything is done in terms of the basis $\mathcal{B}$. It is important to emphasize again that by "basis," we always mean *ordered basis*; the order of the vectors in the basis determines the order of the columns of the matrix $[L]_{\mathcal{B}}$.

Remark

It is important to always distinguish between the two very similar-looking notations: $[L]_{\mathcal{B}}$ is the notation for the matrix of the linear mapping with respect to the basis $\mathcal{B}$. $[L(\vec{x})]_{\mathcal{B}}$ is the notation for the $\mathcal{B}$-coordinates of $L(\vec{x})$ with respect to the basis $\mathcal{B}$.

EXAMPLE 4.6.1

Let $L : \mathbb{R}^3 \to \mathbb{R}^3$ be the linear mapping defined by

$$L(x_1, x_2, x_3) = (x_1 + 2x_2 - 2x_3, -x_2 + 2x_3, x_1 + 2x_2)$$

Let $\mathcal{B} = \left\{ \begin{bmatrix} 2 \\ -1 \\ -1 \end{bmatrix}, \begin{bmatrix} 1 \\ 1 \\ 1 \end{bmatrix}, \begin{bmatrix} 0 \\ 0 \\ -1 \end{bmatrix} \right\}$ and let $\vec{x} \in \mathbb{R}^3$ such that $[\vec{x}]_{\mathcal{B}} = \begin{bmatrix} 1 \\ 2 \\ 3 \end{bmatrix}$. Find the matrix of L with respect to $\mathcal{B}$ and use it to determine $[L(\vec{x})]_{\mathcal{B}}$.

Solution: By definition, the columns of $[L]_{\mathcal{B}}$ are the $\mathcal{B}$-coordinates of the images of the vectors in $\mathcal{B}$ under L. So, we find these images and write them as a linear combination of the vectors in $\mathcal{B}$:

$$L(2, -1, -1) = \begin{bmatrix} 2 \\ -1 \\ 0 \end{bmatrix} = (1) \begin{bmatrix} 2 \\ -1 \\ -1 \end{bmatrix} + (0) \begin{bmatrix} 1 \\ 1 \\ 1 \end{bmatrix} + (-1) \begin{bmatrix} 0 \\ 0 \\ -1 \end{bmatrix}$$

$$L(1, 1, 1) = \begin{bmatrix} 1 \\ 1 \\ 3 \end{bmatrix} = (0) \begin{bmatrix} 2 \\ -1 \\ -1 \end{bmatrix} + (1) \begin{bmatrix} 1 \\ 1 \\ 1 \end{bmatrix} + (-2) \begin{bmatrix} 0 \\ 0 \\ -1 \end{bmatrix}$$

$$L(0, 0, -1) = \begin{bmatrix} 2 \\ -2 \\ 0 \end{bmatrix} = (4/3) \begin{bmatrix} 2 \\ -1 \\ -1 \end{bmatrix} + (-2/3) \begin{bmatrix} 1 \\ 1 \\ 1 \end{bmatrix} + (-2) \begin{bmatrix} 0 \\ 0 \\ -1 \end{bmatrix}$$

Hence,

$$[L]_{\mathcal{B}} = \Big[[L(2, -1, -1)]_{\mathcal{B}} \quad [L(1, 1, 1)]_{\mathcal{B}} \quad [L(0, 0, -1)]_{\mathcal{B}} \Big] = \begin{bmatrix} 1 & 0 & 4/3 \\ 0 & 1 & -2/3 \\ -1 & -2 & -2 \end{bmatrix}$$

Thus, $[L(\vec{x})]_{\mathcal{B}} = [L]_{\mathcal{B}}[\vec{x}]_{\mathcal{B}} = \begin{bmatrix} 1 & 0 & 4/3 \\ 0 & 1 & -2/3 \\ -1 & -2 & -2 \end{bmatrix} \begin{bmatrix} 1 \\ 2 \\ 3 \end{bmatrix} = \begin{bmatrix} 5 \\ 0 \\ -11 \end{bmatrix}$.

It is instructive to verify the answer in Example 4.6.1 by calculating $L(\vec{x})$ in two ways. First, if $[\vec{x}]_{\mathcal{B}} = \begin{bmatrix} 1 \\ 2 \\ 3 \end{bmatrix}$, then $\vec{x} = 1 \begin{bmatrix} 2 \\ -1 \\ -1 \end{bmatrix} + 2 \begin{bmatrix} 1 \\ 1 \\ 1 \end{bmatrix} + 3 \begin{bmatrix} 0 \\ 0 \\ -1 \end{bmatrix} = \begin{bmatrix} 4 \\ 1 \\ -2 \end{bmatrix}$, and, by definition of the mapping, we have $L(\vec{x}) = L(4, 1, -2) = (10, -5, 6)$.

Second, if $[L(\vec{x})]_{\mathcal{B}} = \begin{bmatrix} 5 \\ 0 \\ -11 \end{bmatrix}$, then $L(\vec{x}) = 5 \begin{bmatrix} 2 \\ -1 \\ -1 \end{bmatrix} + 0 \begin{bmatrix} 1 \\ 1 \\ 1 \end{bmatrix} - 11 \begin{bmatrix} 0 \\ 0 \\ -1 \end{bmatrix} = \begin{bmatrix} 10 \\ -5 \\ 6 \end{bmatrix}$.

It is natural to wonder why we might want to represent a linear mapping with respect to a basis other than the standard basis. However, in some problems it is much more convenient to use a different basis than the standard basis. For example, a stretch by a factor of 3 in $\mathbb{R}^2$ in the direction of the vector $\begin{bmatrix} 1 \\ 2 \end{bmatrix}$ is geometrically easy to understand. However, it would be awkward to determine the standard matrix of this stretch and then determine its effect on any other vector. It would be much better to have a basis that takes advantage of the vector $\begin{bmatrix} 1 \\ 2 \end{bmatrix}$ and of the orthogonal vector $\begin{bmatrix} -2 \\ 1 \end{bmatrix}$, which remain unchanged under the stretch.

Consider the reflection $\text{refl}_{\vec{n}} : \mathbb{R}^3 \to \mathbb{R}^3$ in the plane $x_1 + 2x_2 - 3x_3 = 0$. It is easy to describe this by saying that it reverses the normal vector $\vec{n} = \begin{bmatrix} 1 \\ 2 \\ -3 \end{bmatrix}$ to $\begin{bmatrix} -1 \\ -2 \\ 3 \end{bmatrix}$ and leaves unchanged any vectors lying in the plane (such as $\begin{bmatrix} 2 \\ -1 \\ 0 \end{bmatrix}$ and $\begin{bmatrix} 3 \\ 0 \\ 1 \end{bmatrix}$) as in Figure 4.6.1. Describing this reflection in terms of these vectors gives more geometrical information than describing it in terms of the standard basis vectors.

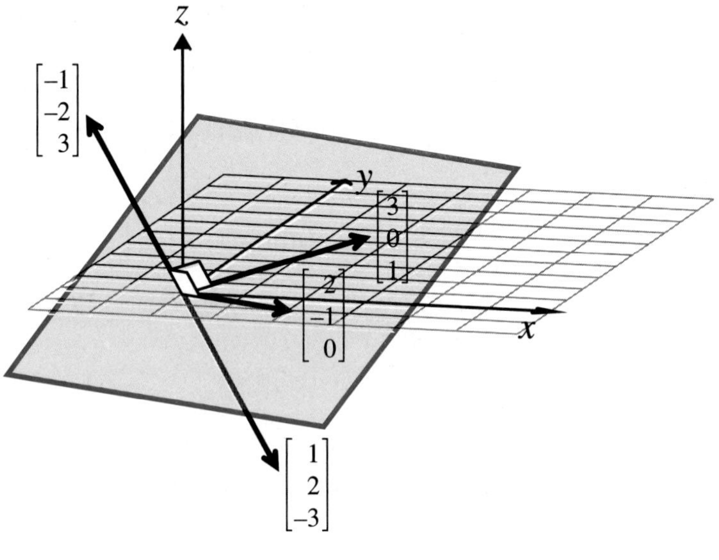

Figure 4.6.1 Reflection in the plane $x_1 + 2x_2 - 3x_3 = 0$.

EXERCISE 4.6.1

Let $\mathcal{B} = \left\{ \begin{bmatrix} 1 \\ 2 \\ -3 \end{bmatrix}, \begin{bmatrix} 2 \\ -1 \\ 0 \end{bmatrix}, \begin{bmatrix} 3 \\ 0 \\ 1 \end{bmatrix} \right\}$. Find $[\text{refl}_{\vec{n}}]_{\mathcal{B}}$ where $\text{refl}_{\vec{n}} : \mathbb{R}^3 \to \mathbb{R}^3$ is the reflection over the plane $x_1 + 2x_2 - 3x_3 = 0$.

Notice that in these examples, the geometry itself provides us with a preferred basis for the appropriate space.

EXAMPLE 4.6.2

Let $\vec{v} = \begin{bmatrix} 3 \\ 4 \end{bmatrix}$. In Example 3.2.6, we found the standard matrix of the linear operator $\text{proj}_{\vec{v}} : \mathbb{R}^2 \to \mathbb{R}^2$ to be

$$[\text{proj}_{\vec{v}}]_S = \begin{bmatrix} 9/25 & 12/25 \\ 12/25 & 16/25 \end{bmatrix}$$

Find the matrix of $\text{proj}_{\vec{v}}$ with respect to a basis that shows the geometry of the transformation more clearly.

Solution: For this linear transformation, it is natural to use a basis for $\mathbb{R}^2$ consisting of the vector $\vec{v}$, which is the direction vector for the projection, and a second vector orthogonal to $\vec{v}$, say $\vec{w} = \begin{bmatrix} -4 \\ 3 \end{bmatrix}$. Taking $\mathcal{B} = \{\vec{v}, \vec{w}\}$ we find that

$$\text{proj}_{\vec{v}}(\vec{v}) = \begin{bmatrix} 3 \\ 4 \end{bmatrix} = 1\vec{v} + 0\vec{w}$$

$$\text{proj}_{\vec{v}}(\vec{w}) = \begin{bmatrix} 0 \\ 0 \end{bmatrix} = 0\vec{v} + 0\vec{w}$$

Hence, $[\text{proj}_{\vec{v}}(\vec{v})]_{\mathcal{B}} = \begin{bmatrix} 1 \\ 0 \end{bmatrix}$ and $[\text{proj}_{\vec{v}}(\vec{w})]_{\mathcal{B}} = \begin{bmatrix} 0 \\ 0 \end{bmatrix}$. Thus,

$$[\text{proj}_{\vec{v}}]_{\mathcal{B}} = \begin{bmatrix} [\text{proj}_{\vec{v}}(\vec{v})]_{\mathcal{B}} & [\text{proj}_{\vec{v}}(\vec{w})]_{\mathcal{B}} \end{bmatrix} = \begin{bmatrix} 1 & 0 \\ 0 & 0 \end{bmatrix}$$

We can use $[\text{proj}_{\vec{v}}]_{\mathcal{B}}$ to get a simple geometrical description of $\text{proj}_{\vec{v}}$. Consider $\text{proj}_{\vec{v}}(\vec{x})$ for any $\vec{x} = b_1\vec{v} + b_2\vec{w} \in \mathbb{R}^2$. We have

$$[\text{proj}_{\vec{v}}(\vec{x})]_{\mathcal{B}} = [\text{proj}_{\vec{v}}]_{\mathcal{B}}[\vec{x}]_{\mathcal{B}} = \begin{bmatrix} 1 & 0 \\ 0 & 0 \end{bmatrix}\begin{bmatrix} b_1 \\ b_2 \end{bmatrix} = \begin{bmatrix} b_1 \\ 0 \end{bmatrix}$$

Consequently, $\text{proj}_{\vec{v}}$ is described as the linear mapping that sends $\begin{bmatrix} b_1 \\ b_2 \end{bmatrix}$ to $\begin{bmatrix} b_1 \\ 0 \end{bmatrix}$ as shown in Figure 4.6.2.

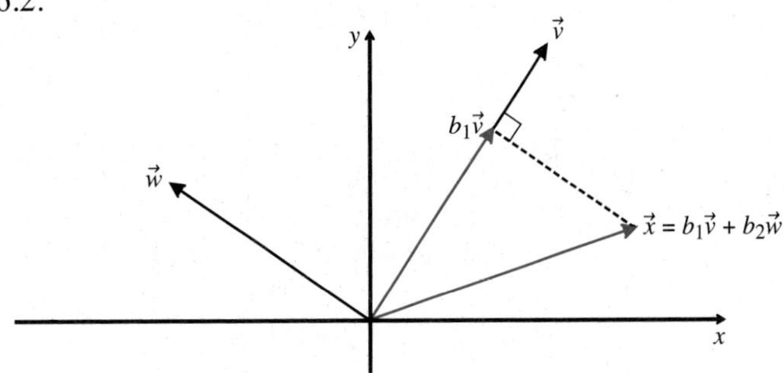

Figure 4.6.2 The geometry of $\text{proj}_{\vec{v}}$.

The Matrix of a General Linear Mapping

We now want to generalize our work above to find the matrix of any linear operator L on a vector space $\mathbb{V}$ with respect to any basis $\mathcal{B}$ for $\mathbb{V}$. To do this, we could repeat our argument from the beginning of this section. This is left as Problem C5.

We make the following definition.

Definition
Matrix of a General Linear Mapping

Suppose that $\mathcal{B} = \{\mathbf{v}_1, \dots, \mathbf{v}_n\}$ is any basis for a vector space $\mathbb{V}$ and that $L : \mathbb{V} \to \mathbb{V}$ is a linear mapping. Define the **matrix of the linear mapping** L with respect to the basis $\mathcal{B}$ to be the matrix

$$[L]_{\mathcal{B}} = \begin{bmatrix} [L(\mathbf{v}_1)]_{\mathcal{B}} & \cdots & [L(\mathbf{v}_n)]_{\mathcal{B}} \end{bmatrix}$$

It satisfies

$$[L(\mathbf{x})]_{\mathcal{B}} = [L]_{\mathcal{B}}[\mathbf{x}]_{\mathcal{B}}$$

EXAMPLE 4.6.3

Let $L : P_2(\mathbb{R}) \to P_2(\mathbb{R})$ be defined by $L(a + bx + cx^2) = (a + b) + bx + (a + b + c)x^2$. Find the matrix of L with respect to the basis $\mathcal{B} = \{1, x, x^2\}$ and use it to calculate $[L(1 + 2x + 3x^2)]_{\mathcal{B}}$.

Solution: We have

$$\begin{aligned} L(1) &= 1 + x^2 \\ L(x) &= 1 + x + x^2 \\ L(x^2) &= x^2 \end{aligned}$$

Hence,

$$[L]_{\mathcal{B}} = \begin{bmatrix} [L(1)]_{\mathcal{B}} & [L(x)]_{\mathcal{B}} & [L(x^2)]_{\mathcal{B}} \end{bmatrix} = \begin{bmatrix} 1 & 1 & 0 \\ 0 & 1 & 0 \\ 1 & 1 & 1 \end{bmatrix}$$

Since $[1 + 2x + 3x^2]_{\mathcal{B}} = \begin{bmatrix} 1 \\ 2 \\ 3 \end{bmatrix}$, we get that

$$[L(1 + 2x + 3x^2)]_{\mathcal{B}} = [L]_{\mathcal{B}}[1 + 2x + 3x^2]_{\mathcal{B}} = \begin{bmatrix} 1 & 1 & 0 \\ 0 & 1 & 0 \\ 1 & 1 & 1 \end{bmatrix}\begin{bmatrix} 1 \\ 2 \\ 3 \end{bmatrix} = \begin{bmatrix} 3 \\ 2 \\ 6 \end{bmatrix}$$

We can check the the matrix $[L]_{\mathcal{B}}$ in Example 4.6.3 is correct by observing that $[L(a + bx + cx^2)]_{\mathcal{B}} = \begin{bmatrix} a+b \\ b \\ a+b+c \end{bmatrix}$ and

$$[L(a + bx + cx^2)]_{\mathcal{B}} = [L]_{\mathcal{B}}[a + bx + cx^2]_{\mathcal{B}} = \begin{bmatrix} 1 & 1 & 0 \\ 0 & 1 & 0 \\ 1 & 1 & 1 \end{bmatrix}\begin{bmatrix} a \\ b \\ c \end{bmatrix} = \begin{bmatrix} a+b \\ b \\ a+b+c \end{bmatrix}$$

EXAMPLE 4.6.4

Let $L : P_2(\mathbb{R}) \to P_2(\mathbb{R})$ be defined by $L(a + bx + cx^2) = (a + b) + bx + (a + b + c)x^2$. Find the matrix of L with respect to the basis $\mathcal{B} = \{1 + x + x^2, 1 - x, 2\}$.

Solution: We have

$$\begin{aligned} L(1 + x + x^2) &= 2 + x + 3x^2 \\ L(1 - x) &= -x \\ L(2) &= 2 + 2x^2 \end{aligned}$$

In this case, we are not able to write the images as linear combinations of the vectors in $\mathcal{B}$ by inspection. So, we use the method of Section 4.4.

Consider

$$\begin{aligned} 2 + x + 3x^2 &= a_1(1 + x + x^2) + a_2(1 - x) + a_3(2) \\ 0 + (-1)x + 0x^2 &= b_1(1 + x + x^2) + b_2(1 - x) + b_3(2) \\ 2 + 2x^2 &= c_1(1 + x + x^2) + c_2(1 - x) + c_3(2) \end{aligned}$$

This gives us three systems of linear equations with the same coefficient matrix. Row reducing the triple augmented matrix gives

$$\left[\begin{array}{ccc|ccc} 1 & 1 & 2 & 2 & 0 & 2 \\ 1 & -1 & 0 & 1 & -1 & 0 \\ 1 & 0 & 0 & 3 & 0 & 2 \end{array}\right] \sim \left[\begin{array}{ccc|ccc} 1 & 0 & 0 & 3 & 0 & 2 \\ 0 & 1 & 0 & 2 & 1 & 2 \\ 0 & 0 & 1 & -3/2 & -1/2 & -1 \end{array}\right]$$

Hence,

$$[L]_{\mathcal{B}} = \Big[[L(1 + x + x^2)]_{\mathcal{B}} \quad [L(1 - x)]_{\mathcal{B}} \quad [L(2)]_{\mathcal{B}}\Big] = \begin{bmatrix} 3 & 0 & 2 \\ 2 & 1 & 2 \\ -3/2 & -1/2 & -1 \end{bmatrix}$$

EXERCISE 4.6.2

Let $L : M_{2\times 2}(\mathbb{R}) \to M_{2\times 2}(\mathbb{R})$ be defined by $L\left(\begin{bmatrix} a & b \\ c & d \end{bmatrix}\right) = \begin{bmatrix} a + b & a - b \\ c & a + b + d \end{bmatrix}$. Find the matrix of L with respect to the basis $\mathcal{B} = \left\{\begin{bmatrix} 1 & 1 \\ 1 & 1 \end{bmatrix}, \begin{bmatrix} 0 & 1 \\ 1 & 1 \end{bmatrix}, \begin{bmatrix} 0 & 0 \\ 1 & 1 \end{bmatrix}, \begin{bmatrix} 0 & 0 \\ 0 & 1 \end{bmatrix}\right\}$.

Observe that in each of the cases above, we have used the same basis for the domain and codomain of the linear mapping L. To make this as general as possible, we would like to define the matrix ${}_{\mathcal{C}}[L]_{\mathcal{B}}$ of a linear mapping $L : \mathbb{V} \to \mathbb{W}$, where $\mathcal{B}$ is a basis for the vector space $\mathbb{V}$ and $\mathcal{C}$ is a basis for the vector space $\mathbb{W}$. This is left as Problem C6.

Change of Coordinates and Linear Mappings

In Example 4.6.2, we used special geometrical properties of the linear transformation and of the chosen basis $\mathcal{B}$ to determine the $\mathcal{B}$-coordinate vectors that make up the $\mathcal{B}$-matrix $[L]_{\mathcal{B}}$ of the linear transformation $L : \mathbb{R}^2 \to \mathbb{R}^2$. In some applications of these ideas, the geometry does not provide such a simple way of determining $[L]_{\mathcal{B}}$. We need a general method for determining $[L]_{\mathcal{B}}$, given the standard matrix $[L]_{\mathcal{S}}$ of a linear operator $L : \mathbb{R}^n \to \mathbb{R}^n$ and a new basis $\mathcal{B}$.

Theorem 4.6.1

Let $L : \mathbb{R}^n \to \mathbb{R}^n$ be a linear operator, let $\mathcal{S}$ denote the standard basis for $\mathbb{R}^n$, and let $\mathcal{B}$ be any other basis for $\mathbb{R}^n$. If P is the change of coordinates matrix from $\mathcal{B}$-coordinates to $\mathcal{S}$-coordinates, then

$$[L]_{\mathcal{B}} = P^{-1}[L]_{\mathcal{S}}P$$

Proof: By definition of the change of coordinates matrix and Theorem 4.4.2, we have

$$P[\vec{x}]_{\mathcal{B}} = [\vec{x}]_{\mathcal{S}} \quad \text{and} \quad [\vec{x}]_{\mathcal{B}} = P^{-1}[\vec{x}]_{\mathcal{S}}$$

If we apply the change of coordinates equation to the vector $L(\vec{x})$ (which is in $\mathbb{R}^n$), we get

$$[L(\vec{x})]_{\mathcal{B}} = P^{-1}[L(\vec{x})]_{\mathcal{S}}$$

Substitute for $[L(\vec{x})]_{\mathcal{B}}$ and $[L(\vec{x})]_{\mathcal{S}}$ to get

$$[L]_{\mathcal{B}}[\vec{x}]_{\mathcal{B}} = P^{-1}[L]_{\mathcal{S}}[\vec{x}]_{\mathcal{S}}$$

But, $P[\vec{x}]_{\mathcal{B}} = [\vec{x}]_{\mathcal{S}}$, so we have

$$[L]_{\mathcal{B}}[\vec{x}]_{\mathcal{B}} = P^{-1}[L]_{\mathcal{S}}P[\vec{x}]_{\mathcal{B}}$$

Since this is true for every $[\vec{x}]_{\mathcal{B}} \in \mathbb{R}^n$, we get, by the Matrices Equal Theorem, that

$$[L]_{\mathcal{B}} = P^{-1}[L]_{\mathcal{S}}P$$

■

We analyze Theorem 4.6.1 by comparing it to Example 4.6.2.

EXAMPLE 4.6.5

Let $\vec{v} = \begin{bmatrix} 3 \\ 4 \end{bmatrix}$. In Example 4.6.2, we determined the matrix of $\text{proj}_{\vec{v}}$ with respect to a geometrically adapted basis $\mathcal{B}$. Let us verify that the change of basis method just described does transform the standard matrix $[\text{proj}_{\vec{v}}]_S$ to the $\mathcal{B}$-matrix $[\text{proj}_{\vec{v}}]_{\mathcal{B}}$.

The matrix $[\text{proj}_{\vec{v}}]_S = \begin{bmatrix} 9/25 & 12/25 \\ 12/25 & 16/25 \end{bmatrix}$. The basis $\mathcal{B} = \left\{ \begin{bmatrix} 3 \\ 4 \end{bmatrix}, \begin{bmatrix} -4 \\ 3 \end{bmatrix} \right\}$, so the change of coordinates matrix from $\mathcal{B}$ to S is $P = \begin{bmatrix} 3 & -4 \\ 4 & 3 \end{bmatrix}$. The inverse is found to be $P^{-1} = \begin{bmatrix} 3/25 & 4/25 \\ -4/25 & 3/25 \end{bmatrix}$. Hence, the $\mathcal{B}$-matrix of $\text{proj}_{\vec{v}}$ is given by

$$P^{-1}[\text{proj}_{\vec{v}}]_S P = \begin{bmatrix} 3/25 & 4/25 \\ -4/25 & 3/25 \end{bmatrix} \begin{bmatrix} 9/25 & 12/25 \\ 12/25 & 16/25 \end{bmatrix} \begin{bmatrix} 3 & -4 \\ 4 & 3 \end{bmatrix} = \begin{bmatrix} 1 & 0 \\ 0 & 0 \end{bmatrix}$$

Thus, we get exactly the same $\mathcal{B}$-matrix $[\text{proj}_{\vec{v}}]_{\mathcal{B}}$ as we obtained by the earlier geometric argument.

To make sure we understand precisely what this means, let us calculate the $\mathcal{B}$-coordinates of the image of the vector $\vec{x} = \begin{bmatrix} 5 \\ 2 \end{bmatrix}$ under $\text{proj}_{\vec{v}}$. We can do this in two ways.

Method 1. Use the fact that $[\text{proj}_{\vec{v}}\, \vec{x}]_{\mathcal{B}} = [\text{proj}_{\vec{v}}]_{\mathcal{B}}[\vec{x}]_{\mathcal{B}}$. We need the $\mathcal{B}$-coordinates of $\vec{x}$:

$$\begin{bmatrix} 5 \\ 2 \end{bmatrix}_{\mathcal{B}} = P^{-1} \begin{bmatrix} 5 \\ 2 \end{bmatrix}_S = \begin{bmatrix} 3/25 & 4/25 \\ -4/25 & 3/25 \end{bmatrix} \begin{bmatrix} 5 \\ 2 \end{bmatrix} = \begin{bmatrix} 23/25 \\ -14/25 \end{bmatrix}$$

Hence,

$$[\text{proj}_{\vec{v}}(\vec{x})]_{\mathcal{B}} = [\text{proj}_{\vec{v}}]_{\mathcal{B}}[\vec{x}]_{\mathcal{B}} = \begin{bmatrix} 1 & 0 \\ 0 & 0 \end{bmatrix} \begin{bmatrix} 23/25 \\ -14/25 \end{bmatrix} = \begin{bmatrix} 23/25 \\ 0 \end{bmatrix}$$

Method 2. Use the fact that $[\text{proj}_{\vec{v}}\, \vec{x}]_{\mathcal{B}} = P^{-1}[\text{proj}_{\vec{v}}\, \vec{x}]_S$:

$$[\text{proj}_{\vec{v}}(\vec{x})]_S = [\text{proj}_{\vec{v}}]_S \begin{bmatrix} 5 \\ 2 \end{bmatrix} = \begin{bmatrix} 9/25 & 12/25 \\ 12/25 & 16/25 \end{bmatrix} \begin{bmatrix} 5 \\ 2 \end{bmatrix} = \begin{bmatrix} 69/25 \\ 92/25 \end{bmatrix}$$

Therefore,

$$[\text{proj}_{\vec{v}}\, \vec{x}]_{\mathcal{B}} = \begin{bmatrix} 3/25 & 4/25 \\ -4/25 & 3/25 \end{bmatrix} \begin{bmatrix} 69/25 \\ 92/25 \end{bmatrix} = \begin{bmatrix} 23/25 \\ 0 \end{bmatrix}$$

The calculation is probably slightly easier if we use the first method, but that really is not the point. What is extremely important is that it is easy to get a geometrical understanding of what happens to vectors if you multiply by $\begin{bmatrix} 1 & 0 \\ 0 & 0 \end{bmatrix}$ (the $\mathcal{B}$-matrix); it is much more difficult to understand what happens if you multiply by $\begin{bmatrix} 9/25 & 12/25 \\ 12/25 & 16/25 \end{bmatrix}$ (the standard matrix). Using a non-standard basis may make it much easier to understand the geometry of a linear transformation.

EXAMPLE 4.6.6

Let L be the linear mapping with standard matrix $A = \begin{bmatrix} 2 & 3 \\ 4 & 5 \end{bmatrix}$. Let $\mathcal{B} = \left\{ \begin{bmatrix} 3 \\ 1 \end{bmatrix}, \begin{bmatrix} -1 \\ 1 \end{bmatrix} \right\}$ be a basis for $\mathbb{R}^2$. Find the matrix of L with respect to the basis $\mathcal{B}$.

Solution: The change of coordinates matrix from $\mathcal{B}$ to $\mathcal{S}$ is $P = \begin{bmatrix} 3 & -1 \\ 1 & 1 \end{bmatrix}$, and we have $P^{-1} = \frac{1}{4}\begin{bmatrix} 1 & 1 \\ -1 & 3 \end{bmatrix}$. It follows that the $\mathcal{B}$-matrix of L is

$$[L]_{\mathcal{B}} = P^{-1}AP = \frac{1}{4}\begin{bmatrix} 1 & 1 \\ -1 & 3 \end{bmatrix}\begin{bmatrix} 2 & 3 \\ 4 & 5 \end{bmatrix}\begin{bmatrix} 3 & -1 \\ 1 & 1 \end{bmatrix} = \begin{bmatrix} 13/2 & 1/2 \\ 21/2 & 1/2 \end{bmatrix}$$

EXAMPLE 4.6.7

Let $L(\vec{x}) = A\vec{x}$ where $A = \begin{bmatrix} -3 & 5 & -5 \\ -7 & 9 & -5 \\ -7 & 7 & -3 \end{bmatrix}$. Let $\mathcal{B}$ be the basis $\mathcal{B} = \left\{ \begin{bmatrix} 1 \\ 1 \\ 0 \end{bmatrix}, \begin{bmatrix} 1 \\ 1 \\ 1 \end{bmatrix}, \begin{bmatrix} 0 \\ 1 \\ 1 \end{bmatrix} \right\}$. Determine the matrix of L with respect to the basis $\mathcal{B}$ and use it to determine the geometry of L.

Solution: The change of coordinates matrix from $\mathcal{B}$ to $\mathcal{S}$ is $P = \begin{bmatrix} 1 & 1 & 0 \\ 1 & 1 & 1 \\ 0 & 1 & 1 \end{bmatrix}$, and we have $P^{-1} = \begin{bmatrix} 0 & 1 & -1 \\ 1 & -1 & 1 \\ -1 & 1 & 0 \end{bmatrix}$. Thus, the $\mathcal{B}$-matrix of the mapping L is

$$[L]_{\mathcal{B}} = P^{-1}AP = \begin{bmatrix} 2 & 0 & 0 \\ 0 & -3 & 0 \\ 0 & 0 & 4 \end{bmatrix}$$

Observe that for any $\vec{x} \in \mathbb{R}^3$ such that $[\vec{x}]_{\mathcal{B}} = \begin{bmatrix} b_1 \\ b_2 \\ b_3 \end{bmatrix}$ we have

$$[L(\vec{x})]_{\mathcal{B}} = [L]_{\mathcal{B}}[\vec{x}]_{\mathcal{B}} = \begin{bmatrix} 2 & 0 & 0 \\ 0 & -3 & 0 \\ 0 & 0 & 4 \end{bmatrix}\begin{bmatrix} b_1 \\ b_2 \\ b_3 \end{bmatrix} = \begin{bmatrix} 2b_1 \\ -3b_2 \\ 4b_3 \end{bmatrix}$$

Thus, L is a stretch by a factor of 2 in the direction of the first basis vector $\begin{bmatrix} 1 \\ 1 \\ 0 \end{bmatrix}$, a reflection (because of the minus sign) and a stretch by a factor of 3 in the second basis vector $\begin{bmatrix} 1 \\ 1 \\ 1 \end{bmatrix}$, and a stretch by a factor of 4 in the direction of the third basis vector $\begin{bmatrix} 0 \\ 1 \\ 1 \end{bmatrix}$. This gives a very clear geometrical picture of how the linear transformation maps vectors, which would not be obvious from looking at the standard matrix A.

CONNECTION

At this point it is natural to ask whether for any linear mapping $L : \mathbb{R}^n \to \mathbb{R}^n$ there exists a basis $\mathcal{B}$ of $\mathbb{R}^n$ such that the $\mathcal{B}$-matrix of L is in diagonal form, and how can we find such a basis if it exists? The answers to these questions are found in Chapter 6. However, in order to deal with these questions, one more computational tool is needed, the determinant, which is discussed in Chapter 5.

PROBLEMS 4.6

Practice Problems

A1 Let $\mathcal{B} = \{\vec{v}_1, \vec{v}_2\}$ be a basis for $\mathbb{R}^2$ and let $L : \mathbb{R}^2 \to \mathbb{R}^2$ be a linear mapping such that $L(\vec{v}_1) = \vec{v}_2$ and $L(\vec{v}_2) = 2\vec{v}_1 - \vec{v}_2$. Determine $[L]_{\mathcal{B}}$ and use it to calculate $[L(4\vec{v}_1 + 3\vec{v}_2)]_{\mathcal{B}}$.

A2 Let $\mathcal{B} = \{\vec{v}_1, \vec{v}_2, \vec{v}_3\}$ be a basis for $\mathbb{R}^3$ and let $L : \mathbb{R}^3 \to \mathbb{R}^3$ be a linear mapping such that $L(\vec{v}_1) = 2\vec{v}_1 - \vec{v}_3$, $L(\vec{v}_2) = 2\vec{v}_1 - \vec{v}_3$, and $L(\vec{v}_3) = 4\vec{v}_2 + 5\vec{v}_3$. Determine $[L]_{\mathcal{B}}$ and use it to calculate $[L(3\vec{v}_1 + 3\vec{v}_2 - \vec{v}_3)]_{\mathcal{B}}$.

For Problems A3 and A4, let $\mathcal{B} = \left\{\begin{bmatrix}1\\1\end{bmatrix}, \begin{bmatrix}-1\\2\end{bmatrix}\right\}$ be a basis of $\mathbb{R}^2$. Assume that L is a linear mapping. Compute $[L]_{\mathcal{B}}$.

A3 $L(1,1) = (-3,-3)$, $L(-1,2) = (-4,8)$

A4 $L(1,1) = (-1,2)$, $L(-1,2) = (2,2)$

For Problems A5–A9, compute $[L]_{\mathcal{B}}$ and use it to find $[L(\vec{x})]_{\mathcal{B}}$.

A5 $L(x_1, x_2) = (2x_1 + x_2, x_1 + 2x_2)$, $\mathcal{B} = \left\{\begin{bmatrix}1\\1\end{bmatrix}, \begin{bmatrix}1\\-1\end{bmatrix}\right\}, \vec{x} = \begin{bmatrix}2\\0\end{bmatrix}$

A6 $L(x_1, x_2) = (2x_1 + 3x_2, 2x_1 + 3x_2)$, $\mathcal{B} = \left\{\begin{bmatrix}1\\-3\end{bmatrix}, \begin{bmatrix}1\\1\end{bmatrix}\right\}, \vec{x} = \begin{bmatrix}0\\2\end{bmatrix}$

A7 $L(x_1, x_2) = (x_1 + 3x_2, -8x_1 + 7x_2)$, $\mathcal{B} = \left\{\begin{bmatrix}1\\2\end{bmatrix}, \begin{bmatrix}1\\4\end{bmatrix}\right\}, \vec{x} = \begin{bmatrix}0\\2\end{bmatrix}$

A8 $L(x_1, x_2, x_3) = (x_1 + x_2 + x_3, x_2, x_2 + x_3)$, $\mathcal{B} = \left\{\begin{bmatrix}-1\\1\\0\end{bmatrix}, \begin{bmatrix}0\\2\\1\end{bmatrix}, \begin{bmatrix}1\\0\\1\end{bmatrix}\right\}, \vec{x} = \begin{bmatrix}0\\3\\2\end{bmatrix}$

A9 $L(x_1, x_2, x_3) = (3x_1 - 2x_3, x_1 + x_2 + x_3, 2x_1 + x_2)$, $\mathcal{B} = \left\{\begin{bmatrix}1\\0\\0\end{bmatrix}, \begin{bmatrix}1\\1\\0\end{bmatrix}, \begin{bmatrix}1\\2\\1\end{bmatrix}\right\}, \vec{x} = \begin{bmatrix}3\\-2\\9\end{bmatrix}$

For Problems A10–A13, determine a geometrically natural basis $\mathcal{B}$ (as in Example 4.6.2) and determine the $\mathcal{B}$-matrix of the transformation.

A10 $\text{refl}_{(1,-2)} : \mathbb{R}^2 \to \mathbb{R}^2$

A11 $\text{perp}_{(1,-2)} : \mathbb{R}^2 \to \mathbb{R}^2$

A12 $\text{proj}_{(2,1,-1)} : \mathbb{R}^3 \to \mathbb{R}^3$

A13 $\text{refl}_{(-1,-1,1)} : \mathbb{R}^3 \to \mathbb{R}^3$

A14 Let $\mathcal{B} = \left\{\begin{bmatrix}1\\0\\1\end{bmatrix}, \begin{bmatrix}1\\-1\\0\end{bmatrix}, \begin{bmatrix}0\\1\\2\end{bmatrix}\right\}$ and let $L : \mathbb{R}^3 \to \mathbb{R}^3$ be the linear mapping such that

$$L(1,0,1) = (1,2,4)$$
$$L(1,-1,0) = (0,1,2)$$
$$L(0,1,2) = (2,-2,0)$$

(a) Let $\vec{x} = \begin{bmatrix}1\\2\\4\end{bmatrix}$. Find $[\vec{x}]_{\mathcal{B}}$.
(b) Find $[L]_{\mathcal{B}}$.
(c) Use parts (a) and (b) to determine $L(\vec{x})$.

A15 Let $\mathcal{B} = \left\{\begin{bmatrix}1\\0\\-1\end{bmatrix}, \begin{bmatrix}1\\2\\0\end{bmatrix}, \begin{bmatrix}0\\1\\1\end{bmatrix}\right\}$ and let $L : \mathbb{R}^3 \to \mathbb{R}^3$ be the linear mapping such that

$$L(1,0,-1) = (0,1,1)$$
$$L(1,2,0) = (-2,0,2)$$
$$L(0,1,1) = (5,3,-5)$$

(a) Let $\vec{x} = \begin{bmatrix}-1\\7\\6\end{bmatrix}$. Find $[\vec{x}]_{\mathcal{B}}$.
(b) Find $[L]_{\mathcal{B}}$.
(c) Use parts (a) and (b) to determine $L(\vec{x})$.

For Problems A16–A21, assume that the matrix A is the standard matrix of a linear mapping L. Determine the matrix of L with respect to the given basis $\mathcal{B}$ using Theorem 4.6.1. You may find it helpful to use a computer to find inverses and to multiply matrices.

A16 $A = \begin{bmatrix} 1 & 3 \\ -8 & 7 \end{bmatrix}, \mathcal{B} = \left\{ \begin{bmatrix} 1 \\ 2 \end{bmatrix}, \begin{bmatrix} 1 \\ 4 \end{bmatrix} \right\}$

A17 $A = \begin{bmatrix} 1 & -6 \\ -4 & -1 \end{bmatrix}, \mathcal{B} = \left\{ \begin{bmatrix} 3 \\ -2 \end{bmatrix}, \begin{bmatrix} 1 \\ 1 \end{bmatrix} \right\}$

A18 $A = \begin{bmatrix} 4 & -6 \\ 2 & 8 \end{bmatrix}, \mathcal{B} = \left\{ \begin{bmatrix} 3 \\ 1 \end{bmatrix}, \begin{bmatrix} 7 \\ 3 \end{bmatrix} \right\}$

A19 $A = \begin{bmatrix} 16 & -20 \\ 6 & -6 \end{bmatrix}, \mathcal{B} = \left\{ \begin{bmatrix} 5 \\ 3 \end{bmatrix}, \begin{bmatrix} 4 \\ 2 \end{bmatrix} \right\}$

A20 $A = \begin{bmatrix} 3 & 1 & 1 \\ 0 & 4 & 2 \\ 1 & -1 & 5 \end{bmatrix}, \mathcal{B} = \left\{ \begin{bmatrix} 1 \\ 1 \\ 0 \end{bmatrix}, \begin{bmatrix} 0 \\ 1 \\ 1 \end{bmatrix}, \begin{bmatrix} 1 \\ 0 \\ 1 \end{bmatrix} \right\}$

A21 $A = \begin{bmatrix} 4 & 1 & -3 \\ 16 & 4 & -18 \\ 6 & 1 & -5 \end{bmatrix}, \mathcal{B} = \left\{ \begin{bmatrix} 1 \\ 1 \\ 1 \end{bmatrix}, \begin{bmatrix} 0 \\ 3 \\ 1 \end{bmatrix}, \begin{bmatrix} 1 \\ 2 \\ 1 \end{bmatrix} \right\}$

For Problems A22–A25, find the matrix of each of the linear mapping with respect to the given basis $\mathcal{B}$.

A22 $L : \mathbb{R}^3 \to \mathbb{R}^3$ defined by
$L(x_1, x_2, x_3) = (x_1 + x_2, x_2 + x_3, x_1 - x_3)$,
$\mathcal{B} = \left\{ \begin{bmatrix} 1 \\ 1 \\ 1 \end{bmatrix}, \begin{bmatrix} 0 \\ 1 \\ 1 \end{bmatrix}, \begin{bmatrix} 0 \\ 0 \\ 1 \end{bmatrix} \right\}$

A23 $L : P_2(\mathbb{R}) \to P_2(\mathbb{R})$ defined by
$L(a + bx + cx^2) = a + (b + c)x^2$,
$\mathcal{B} = \{1 + x^2, -1 + x, 1 - x + x^2\}$

A24 $D : P_2(\mathbb{R}) \to P_2(\mathbb{R})$ defined by
$D(a + bx + cx^2) = b + 2cx, \mathcal{B} = \{1, x, x^2\}$

A25 $T : \mathbb{U} \to \mathbb{U}$, where $\mathbb{U}$ is the subspace of upper-triangular matrices in $M_{2\times2}(\mathbb{R})$, defined by
$T\left(\begin{bmatrix} a & b \\ 0 & c \end{bmatrix}\right) = \begin{bmatrix} a & b+c \\ 0 & a+b+c \end{bmatrix}$,
$\mathcal{B} = \left\{ \begin{bmatrix} 1 & 1 \\ 0 & 0 \end{bmatrix}, \begin{bmatrix} 1 & 0 \\ 0 & 1 \end{bmatrix}, \begin{bmatrix} 1 & 1 \\ 0 & 1 \end{bmatrix} \right\}$

Homework Problems

B1 Let $\mathcal{B} = \{\vec{v}_1, \vec{v}_2\}$ be a basis for $\mathbb{R}^2$ and let $L : \mathbb{R}^2 \to \mathbb{R}^2$ be a linear mapping such that $L(\vec{v}_1) = 2\vec{v}_1 + \vec{v}_2$, and $L(\vec{v}_2) = 3\vec{v}_1 + 4\vec{v}_2$. Determine $[L]_\mathcal{B}$ and use it to calculate $[L(-\vec{v}_1 + 3\vec{v}_2)]_\mathcal{B}$.

For Problems B2 and B3, let $\mathcal{B} = \{\vec{v}_1, \vec{v}_2, \vec{v}_3\}$ be a basis for $\mathbb{R}^3$, and let $L : \mathbb{R}^3 \to \mathbb{R}^3$ be a linear mapping. Determine $[L(\vec{x})]_\mathcal{B}$ for the given $[\vec{x}]_\mathcal{B}$.

B2 $L(\vec{v}_1) = 2\vec{v}_1 + \vec{v}_2 + 2\vec{v}_3, L(\vec{v}_2) = 2\vec{v}_2 + \vec{v}_3$,
$L(\vec{v}_3) = 4\vec{v}_1 + 3\vec{v}_3$; $[\vec{x}]_\mathcal{B} = \begin{bmatrix} -2 \\ 1 \\ 1 \end{bmatrix}$

B3 $L(\vec{v}_1) = \vec{v}_1 + 2\vec{v}_2 - \vec{v}_3, L(\vec{v}_2) = 2\vec{v}_2 + \vec{v}_3$,
$L(\vec{v}_3) = \vec{v}_1 - 2\vec{v}_3$; $[\vec{x}]_\mathcal{B} = \begin{bmatrix} 1 \\ 2 \\ 0 \end{bmatrix}$

For Problems B4–B6, let $\mathcal{B} = \left\{ \begin{bmatrix} 1 \\ 2 \end{bmatrix}, \begin{bmatrix} -2 \\ 1 \end{bmatrix} \right\}$ be a basis of $\mathbb{R}^2$ and assume that L is a linear mapping. Compute $[L]_\mathcal{B}$.

B4 $L(1, 2) = (0, 0), L(-2, 1) = (-1, 3)$

B5 $L(1, 2) = (4, -2), L(-2, 1) = (3, 6)$

B6 $L(1, 2) = (1, 2), L(-2, 1) = (1, 2)$

For Problems B7–B11, compute $[L]_\mathcal{B}$ and use it to find $[L(\vec{x})]_\mathcal{B}$.

B7 $L(x_1, x_2) = (-2x_1 + 3x_2, 4x_1 - 3x_2)$,
$\mathcal{B} = \left\{ \begin{bmatrix} 1 \\ 1 \end{bmatrix}, \begin{bmatrix} 3 \\ -4 \end{bmatrix} \right\}, \vec{x} = \begin{bmatrix} -2 \\ 5 \end{bmatrix}$

B8 $L(x_1, x_2) = (3x_1 + 2x_2, 6x_1 - x_2)$,
$\mathcal{B} = \left\{ \begin{bmatrix} 1 \\ 0 \end{bmatrix}, \begin{bmatrix} 3 \\ 2 \end{bmatrix} \right\}, \vec{x} = \begin{bmatrix} 1 \\ 1 \end{bmatrix}$

B9 $L(x_1, x_2) = (2x_1 + 5x_2, 3x_1 - 2x_2)$,
$\mathcal{B} = \left\{ \begin{bmatrix} 1 \\ -2 \end{bmatrix}, \begin{bmatrix} -2 \\ 1 \end{bmatrix} \right\}, \vec{x} = \begin{bmatrix} 1 \\ 1 \end{bmatrix}$

B10 $L(x_1, x_2, x_3) = (x_1 + 6x_2 + 3x_3, -2x_2, 3x_1 + 6x_2 + x_3)$,
$\mathcal{B} = \left\{ \begin{bmatrix} -2 \\ 1 \\ 0 \end{bmatrix}, \begin{bmatrix} -1 \\ 0 \\ 1 \end{bmatrix}, \begin{bmatrix} 1 \\ 0 \\ 1 \end{bmatrix} \right\}, \vec{x} = \begin{bmatrix} 1 \\ 1 \\ -1 \end{bmatrix}$

B11 $L(x_1, x_2, x_3) = (2x_1 - 7x_2 - 3x_3, x_1 - 2x_2 - x_3, -2x_2 - x_3)$,
$\mathcal{B} = \left\{ \begin{bmatrix} 1 \\ 0 \\ 1 \end{bmatrix}, \begin{bmatrix} 0 \\ 1 \\ 1 \end{bmatrix}, \begin{bmatrix} 1 \\ 1 \\ 1 \end{bmatrix} \right\}, \vec{x} = \begin{bmatrix} 1 \\ -2 \\ 3 \end{bmatrix}$

For Problems B12–B15, determine a geometrically natural basis $\mathcal{B}$ and determine the $\mathcal{B}$-matrix of the transformation.

B12 $\text{perp}_{(3,2)}$

B13 $\text{proj}_{(-1,-2)}$

B14 $\text{perp}_{(2,1,-2)}$

B15 $\text{refl}_{(1,2,3)}$

B16 Let $\mathcal{B} = \left\{ \begin{bmatrix} 1 \\ 1 \\ 0 \end{bmatrix}, \begin{bmatrix} 0 \\ 1 \\ 1 \end{bmatrix}, \begin{bmatrix} 1 \\ 0 \\ 1 \end{bmatrix} \right\}$ and let $L : \mathbb{R}^3 \to \mathbb{R}^3$ be the linear mapping such that

$$L(1,1,0) = (2,0,2)$$
$$L(0,1,1) = (1,1,2)$$
$$L(1,0,1) = (1,0,1)$$

(a) Let $\vec{x} = \begin{bmatrix} 5 \\ 4 \\ 5 \end{bmatrix}$. Find $[\vec{x}]_{\mathcal{B}}$.

(b) Find $[L]_{\mathcal{B}}$.

(c) Use parts (a) and (b) to determine $L(\vec{x})$.

B17 Let $\mathcal{B} = \{2 + x, -1 + x^2, 1 + x\}$ and let $L : P_2(\mathbb{R}) \to P_2(\mathbb{R})$ be the linear mapping such that

$$L(2 + x) = 4 + 2x$$
$$L(-1 + x^2) = 1 + x$$
$$L(1 + x) = 1 + x + x^2$$

(a) Let $\mathbf{p}(x) = 1 + 4x + 4x^2$. Find $[\vec{x}]_{\mathcal{B}}$.

(b) Find $[L]_{\mathcal{B}}$.

(c) Use parts (a) and (b) to determine $L(\mathbf{p})$.

For Problems B18–B21, assume that the matrix A is the standard matrix of a linear mapping $L : \mathbb{R}^2 \to \mathbb{R}^2$. Determine the matrix of L with respect to the given basis $\mathcal{B}$ using Theorem 4.6.1. Use $[L]_{\mathcal{B}}$ to determine $[L(\vec{x})]_{\mathcal{B}}$ for the given $\vec{x}$. You may find it helpful to use a computer to find inverses and to multiply matrices.

B18 $A = \begin{bmatrix} 3 & 1 \\ 4 & 2 \end{bmatrix}$, $\mathcal{B} = \left\{ \begin{bmatrix} 2 \\ 5 \end{bmatrix}, \begin{bmatrix} 1 \\ 3 \end{bmatrix} \right\}$, $[\vec{x}]_{\mathcal{B}} = \begin{bmatrix} 2 \\ 1 \end{bmatrix}$

B19 $A = \begin{bmatrix} 3 & 1 \\ 4 & 3 \end{bmatrix}$, $\mathcal{B} = \left\{ \begin{bmatrix} 1 \\ 2 \end{bmatrix}, \begin{bmatrix} -1 \\ 2 \end{bmatrix} \right\}$, $[\vec{x}]_{\mathcal{B}} = \begin{bmatrix} 5 \\ 3 \end{bmatrix}$

B20 $A = \begin{bmatrix} 1 & 3 \\ 3 & 9 \end{bmatrix}$, $\mathcal{B} = \left\{ \begin{bmatrix} 3 \\ -1 \end{bmatrix}, \begin{bmatrix} 2 \\ 1 \end{bmatrix} \right\}$, $[\vec{x}]_{\mathcal{B}} = \begin{bmatrix} -9 \\ 4 \end{bmatrix}$

B21 $A = \begin{bmatrix} 3 & -2 \\ -3 & 2 \end{bmatrix}$, $\mathcal{B} = \left\{ \begin{bmatrix} 2 \\ -4 \end{bmatrix}, \begin{bmatrix} -3 \\ 7 \end{bmatrix} \right\}$, $[\vec{x}]_{\mathcal{B}} = \begin{bmatrix} 2 \\ 1 \end{bmatrix}$

For Problems B22–B25, assume that the matrix A is the standard matrix of a linear mapping $L : \mathbb{R}^3 \to \mathbb{R}^3$. Determine the matrix of L with respect to the given basis $\mathcal{B}$ using Theorem 4.6.1. You may find it helpful to use a computer to find inverses and to multiply matrices.

B22 $\begin{bmatrix} 1 & 2 & 2 \\ 2 & 1 & 2 \\ 2 & 2 & 1 \end{bmatrix}$, $\mathcal{B} = \left\{ \begin{bmatrix} -1 \\ 0 \\ 1 \end{bmatrix}, \begin{bmatrix} 1 \\ 1 \\ 1 \end{bmatrix}, \begin{bmatrix} -1 \\ 1 \\ 0 \end{bmatrix} \right\}$

B23 $\begin{bmatrix} 3 & -1 & 3 \\ -1 & 9 & -3 \\ 3 & -3 & 11 \end{bmatrix}$, $\mathcal{B} = \left\{ \begin{bmatrix} 1 \\ 5 \\ 3 \end{bmatrix}, \begin{bmatrix} -3 \\ 0 \\ 1 \end{bmatrix}. \begin{bmatrix} 0 \\ 1 \\ 0 \end{bmatrix} \right\}$

B24 $\begin{bmatrix} 3 & 6 & 1 \\ 5 & -4 & 5 \\ 3 & -6 & 5 \end{bmatrix}$, $\mathcal{B} = \left\{ \begin{bmatrix} 1 \\ 0 \\ -1 \end{bmatrix}, \begin{bmatrix} 1 \\ 1 \\ 1 \end{bmatrix}, \begin{bmatrix} 2 \\ 1 \\ 0 \end{bmatrix} \right\}$

B25 $\begin{bmatrix} 1 & 1 & 1 \\ 1 & -1 & 3 \\ 0 & 2 & 1 \end{bmatrix}$, $\mathcal{B} = \left\{ \begin{bmatrix} 1 \\ 1 \\ 0 \end{bmatrix}, \begin{bmatrix} 1 \\ 2 \\ 1 \end{bmatrix}, \begin{bmatrix} -1 \\ 0 \\ 2 \end{bmatrix} \right\}$

For Problems B26–B31, find the matrix of each of the linear mapping with respect to the given basis $\mathcal{B}$.

B26 $L : P_1(\mathbb{R}) \to P_1(\mathbb{R})$ defined by
$L(a + bx) = (a + b) + (2a + 3b)x$, $\mathcal{B} = \{1 + x, 2 - x\}$

B27 $L : \mathbb{R}^3 \to \mathbb{R}^3$ defined by
$L(x_1, x_2, x_3) = (x_1 + x_2 + x_3, x_1 + 2x_2, x_1 + x_2 + x_3)$,
$\mathcal{B} = \left\{ \begin{bmatrix} 1 \\ 1 \\ 0 \end{bmatrix}, \begin{bmatrix} 0 \\ 1 \\ 1 \end{bmatrix}, \begin{bmatrix} 1 \\ 0 \\ 1 \end{bmatrix} \right\}$

B28 $L : P_2(\mathbb{R}) \to P_2(\mathbb{R})$ defined by
$L(a + bx + cx^2) = (a + b + c) + (a + 2b)x + (a + c)x^2$,
$\mathcal{B} = \{1 + x, x^2, 1 + x^2\}$

B29 $L : M_{2\times2}(\mathbb{R}) \to M_{2\times2}(\mathbb{R})$ defined by
$L\left(\begin{bmatrix} a & b \\ c & d \end{bmatrix}\right) = \begin{bmatrix} a + d & b + c \\ 0 & d \end{bmatrix}$,
$\mathcal{B} = \left\{ \begin{bmatrix} 1 & 1 \\ 0 & 0 \end{bmatrix}, \begin{bmatrix} 1 & 0 \\ 0 & 1 \end{bmatrix}, \begin{bmatrix} 1 & 1 \\ 0 & 1 \end{bmatrix}, \begin{bmatrix} 0 & 0 \\ 1 & 0 \end{bmatrix} \right\}$

B30 $D : P_2(\mathbb{R}) \to P_2(\mathbb{R})$ defined by
$D(a + bx + cx^2) = b + 2cx$,
$\mathcal{B} = \{1 + 2x + 3x^2, -2x + x^2, 1 + x + 3x^2\}$

B31 $T : \mathbb{D} \to \mathbb{D}$, where $\mathbb{D}$ is the subspace of diagonal matrices in $M_{2\times2}(\mathbb{R})$, defined by
$T\left(\begin{bmatrix} a & 0 \\ 0 & b \end{bmatrix}\right) = \begin{bmatrix} a + b & 0 \\ 0 & 2a + b \end{bmatrix}$, $\mathcal{B} = \left\{ \begin{bmatrix} 1 & 0 \\ 0 & 2 \end{bmatrix}, \begin{bmatrix} 2 & 0 \\ 0 & 3 \end{bmatrix} \right\}$

Conceptual Problems

C1 Suppose that $\mathcal{B}$ and $\mathcal{C}$ are bases for $\mathbb{R}^n$ and that $\mathcal{S}$ is the standard basis of $\mathbb{R}^n$. Suppose that P is the change of coordinates matrix from $\mathcal{B}$ to $\mathcal{S}$ and that Q is the change of coordinates matrix from $\mathcal{C}$ to $\mathcal{S}$. Let $L : \mathbb{R}^n \to \mathbb{R}^n$ be a linear mapping. Express the matrix $[L]_\mathcal{C}$ in terms of $[L]_\mathcal{B}$, P, and Q.

C2 Suppose that a 2×2 matrix A is the standard matrix of a linear mapping $L : \mathbb{R}^2 \to \mathbb{R}^2$. Let $\mathcal{B} = \{\vec{v}_1, \vec{v}_2\}$ be a basis for $\mathbb{R}^2$ and let P denote the change of coordinates matrix from $\mathcal{B}$ to the standard basis. What conditions will have to be satisfied by the vectors $\vec{v}_1$ and $\vec{v}_2$ in order for

$$P^{-1}AP = \begin{bmatrix} d_1 & 0 \\ 0 & d_2 \end{bmatrix} = D$$

for some $d_1, d_2 \in \mathbb{R}$? (Hint: consider the equation $AP = PD$, or $A\begin{bmatrix} \vec{v}_1 & \vec{v}_2 \end{bmatrix} = \begin{bmatrix} \vec{v}_1 & \vec{v}_2 \end{bmatrix} D$.)

C3 Let $\mathbb{V}$ be a finite dimensional vector space. Prove if $L : \mathbb{V} \to \mathbb{V}$ is an invertible linear mapping, then $[L]_\mathcal{B}$ is invertible for all bases $\mathcal{B}$ of $\mathbb{V}$.

C4 Let $\mathbb{V}$ be a finite dimensional vector space. Let $L : \mathbb{V} \to \mathbb{V}$ be a linear mapping and let $\mathcal{B}$ be any basis for $\mathbb{V}$.

(a) Prove if $\{\mathbf{v}_1, \dots, \mathbf{v}_k\}$ is a basis for Null(L), then $\{[\mathbf{v}_1]_\mathcal{B}, \dots, [\mathbf{v}_k]_\mathcal{B}\}$ is a basis for Null($[L]_\mathcal{B}$).

(b) Prove that rank(L) = rank($[L]_\mathcal{B}$).

C5 Let $\mathbb{V}$ be a vector space with basis $\mathcal{B} = \{\mathbf{v}_1, \dots, \mathbf{v}_n\}$ and let $L : \mathbb{V} \to \mathbb{V}$ be a linear mapping. Prove that the matrix $[L]_\mathcal{B}$ defined by

$$[L]_\mathcal{B} = \begin{bmatrix} [L(\mathbf{v}_1)]_\mathcal{B} & \cdots & [L(\mathbf{v}_n)]_\mathcal{B} \end{bmatrix}$$

satisfies $[L(\mathbf{x})]_\mathcal{B} = [L]_\mathcal{B}[\mathbf{x}]_\mathcal{B}$.

C6 Let $\mathbb{V}$ be a vector space with basis $\mathcal{B} = \{\mathbf{v}_1, \dots, \mathbf{v}_n\}$, let $\mathbb{W}$ be a vector space with basis $\mathcal{C}$, and let $L : \mathbb{V} \to \mathbb{W}$ be a linear mapping. Prove that the matrix ${}_\mathcal{C}[L]_\mathcal{B}$ defined by

$${}_\mathcal{C}[L]_\mathcal{B} = \begin{bmatrix} [L(\mathbf{v}_1)]_\mathcal{C} & \cdots & [L(\mathbf{v}_n)]_\mathcal{C} \end{bmatrix}$$

satisfies $[L(\mathbf{x})]_\mathcal{C} = {}_\mathcal{C}[L]_\mathcal{B}[\mathbf{x}]_\mathcal{B}$ and hence is the **matrix of** L with respect to basis $\mathcal{B}$ and $\mathcal{C}$.

For Problems C7–C10, using the definition given in Problem C6, determine the matrix of the linear mapping with respect to the given bases $\mathcal{B}$ and $\mathcal{C}$.

C7 $D : P_2(\mathbb{R}) \to P_1(\mathbb{R})$ defined by
$D(a + bx + cx^2) = b + 2cx$, $\mathcal{B} = \{1, x, x^2\}$, $\mathcal{C} = \{1, x\}$

C8 $L : \mathbb{R}^2 \to P_2(\mathbb{R})$ defined by
$L(a_1, a_2) = (a_1 + a_2) + a_1x^2$, $\mathcal{B} = \left\{\begin{bmatrix} 1 \\ -1 \end{bmatrix}, \begin{bmatrix} 1 \\ 2 \end{bmatrix}\right\}$,
$\mathcal{C} = \{1 + x^2, 1 + x, -1 - x + x^2\}$

C9 $T : \mathbb{R}^2 \to M_{2\times2}(\mathbb{R})$ defined by
$T\left(\begin{bmatrix} a \\ b \end{bmatrix}\right) = \begin{bmatrix} a+b & 0 \\ 0 & a-b \end{bmatrix}$, $\mathcal{B} = \left\{\begin{bmatrix} 2 \\ -1 \end{bmatrix}, \begin{bmatrix} 1 \\ 2 \end{bmatrix}\right\}$,
$\mathcal{C} = \left\{\begin{bmatrix} 1 & 1 \\ 0 & 0 \end{bmatrix}, \begin{bmatrix} 1 & 0 \\ 0 & 1 \end{bmatrix}, \begin{bmatrix} 1 & 1 \\ 0 & 1 \end{bmatrix}, \begin{bmatrix} 0 & 0 \\ 1 & 0 \end{bmatrix}\right\}$

C10 $L : P_2(\mathbb{R}) \to \mathbb{R}^2$ defined by
$L(a + bx + cx^2) = \begin{bmatrix} a+c \\ b-a \end{bmatrix}$,
$\mathcal{B} = \{1 + x^2, 1 + x, -1 + x + x^2\}$,
$\mathcal{C} = \left\{\begin{bmatrix} 1 \\ 0 \end{bmatrix}, \begin{bmatrix} 1 \\ 1 \end{bmatrix}\right\}$

4.7 Isomorphisms of Vector Spaces

Some of the ideas discussed in Chapters 3 and 4 lead to generalizations that are important in the further development of linear algebra (and also in abstract algebra). Some of these generalizations are outlined in this section. Most of the proofs are easy or simple variations on proofs given earlier, so they will be left as exercises.

One-to-One and Onto

Definition
One-to-one

A linear mapping $L : \mathbb{U} \rightarrow \mathbb{V}$ is said to be **one-to-one** if $L(\mathbf{u}_1) = L(\mathbf{u}_2)$ implies that $\mathbf{u}_1 = \mathbf{u}_2$.

That is, L is one-to-one if for each $\mathbf{w} \in \text{Range}(L)$ there is only one $\mathbf{v} \in \mathbb{V}$ such that $L(\mathbf{v}) = \mathbf{w}$. Figure 4.7.1 represents a one-to-one function, while Figure 4.7.2 represents a function that is not one-to-one (since there is an element in $\mathbb{W}$ that is mapped to by two vectors in $\mathbb{V}$).

Definition
Onto

A linear mapping $L : \mathbb{U} \rightarrow \mathbb{V}$ is said to be **onto** if for every $\mathbf{v} \in \mathbb{V}$ there exists some $\mathbf{u} \in \mathbb{U}$ such that $L(\mathbf{u}) = \mathbf{v}$.

By definition, L is onto if and only if $\text{Range}(L) = \mathbb{W}$. Thus, Figure 4.7.1 represents a function that is not onto (there is a vector in $\mathbb{W}$ that is not mapped to by a vector in $\mathbb{V}$), while Figure 4.7.2 represents an onto function.

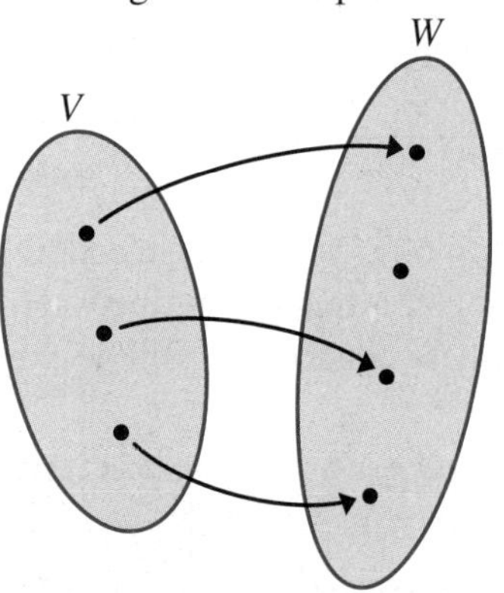

Figure 4.7.1 One-to-one, not onto.

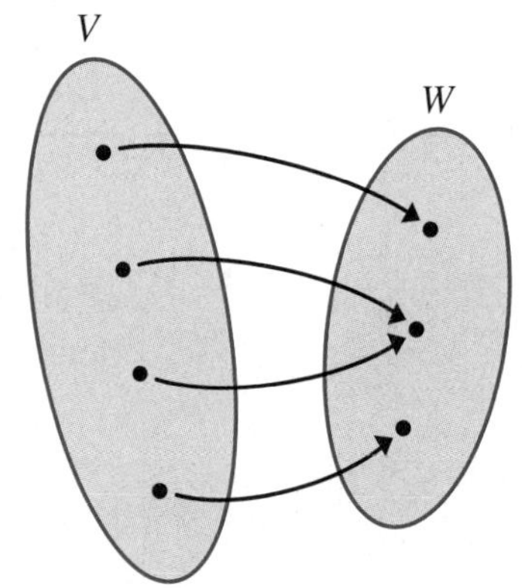

Figure 4.7.2 Onto, not one-to-one.

EXAMPLE 4.7.1

Prove that the linear mapping $L : \mathbb{R}^2 \rightarrow P_2(\mathbb{R})$ defined by $L(a_1, a_2) = a_1 + a_2x^2$ is one-to-one, but not onto.

Solution: If $L(a_1, a_2) = L(b_1, b_2)$, then $a_1 + a_2x^2 = b_1 + b_2x^2$ and hence $a_1 = b_1$ and $a_2 = b_2$. Thus, by definition, L is one-to-one.

L is not onto, since there is no $\vec{a} \in \mathbb{R}^2$ such that $L(\vec{a}) = x$.

EXAMPLE 4.7.2

Prove that the linear mapping $L : \mathbb{R}^3 \to P_1(\mathbb{R})$ defined by $L(y_1, y_2, y_3) = y_1 + (y_2 + y_3)x$ is onto, but not one-to-one.

Solution: Let $a + bx$ be any polynomial in $P_1(\mathbb{R})$. We need to find a vector $\vec{y} = \begin{bmatrix} y_1 \\ y_2 \\ y_3 \end{bmatrix} \in \mathbb{R}^3$, such that $L(\vec{y}) = a + bx$. We observe that we have

$$L(a, b, 0) = a + (b + 0)x = a + bx$$

So, L is onto. Notice that we also have

$$L(a, 0, b) = a + (0 + b)x = a + bx$$

Hence, L is not one-to-one.

EXAMPLE 4.7.3

Prove that the linear mapping $L : \mathbb{R}^2 \to \mathbb{R}^2$ defined by $L(x_1, x_2) = (x_1, x_1 + x_2)$ is one-to-one and onto.

Solution: Assume that $L(x_1, x_2) = L(y_1, y_2)$. Then we have $(x_1, x_1 + x_2) = (y_1, y_1 + y_2)$, and so $x_1 = y_1$ and $x_2 = y_2$. Thus, L is one-to-one.

Take any vector $(a, b) \in \mathbb{R}^2$. Observe that

$$L(a, b - a) = (a, a + (b - a)) = (a, b)$$

Thus, L is onto.

We have already observed the connection between L being onto and the range of L. We now establish a relationship between a mapping being one-to-one and its nullspace. Both of these connections will be exploited in some proofs below.

Theorem 4.7.1 A linear mapping $L : \mathbb{U} \to \mathbb{V}$ is one-to-one if and only if $\text{Null}(L) = \{\mathbf{0}\}$.

You are asked to prove Theorem 4.7.1 as Problem C1.

EXERCISE 4.7.1

Suppose that $\{\mathbf{u}_1, \dots, \mathbf{u}_k\}$ is a linearly independent set in $\mathbb{U}$ and $L : \mathbb{U} \to \mathbb{V}$ is one-to-one. Prove that $\{L(\mathbf{u}_1), \dots, L(\mathbf{u}_k)\}$ is linearly independent.

EXERCISE 4.7.2

Suppose that $\{\mathbf{u}_1, \dots, \mathbf{u}_k\}$ is a spanning set for $\mathbb{U}$ and $L : \mathbb{U} \to \mathbb{V}$ is onto. Prove that a spanning set for $\mathbb{V}$ is $\{L(\mathbf{u}_1), \dots, L(\mathbf{u}_k)\}$.

Theorem 4.7.2 The linear mapping $L : \mathbb{U} \to \mathbb{V}$ has an inverse linear mapping $L^{-1} : \mathbb{V} \to \mathbb{U}$ if and only if L is one-to-one and onto.

You are asked to prove Theorem 4.7.2 as Problem C4.

Isomorphisms

Our goal now is to identify the condition under which two vector spaces are "equal", and to create a mechanism for changing between two such vector spaces.

For two vector spaces $\mathbb{V}$ and $\mathbb{W}$ to be "equal" we need each vector $\mathbf{v} \in \mathbb{V}$ to correspond to a unique vector $\mathbf{w} \in \mathbb{W}$. However, to ensure that $\mathbb{V}$ and $\mathbb{W}$ have the same "structure", we also need to ensure that a linear combination of corresponding vectors gives corresponding results. That is, if $\mathbf{v}_i$ corresponds to $\mathbf{w}_i$ for $1 \leq i \leq n$, then $c_1\mathbf{v}_1 + \cdots + c_n\mathbf{v}_n$ corresponds to $c_1\mathbf{w}_1 + \cdots + c_n\mathbf{w}_n$.

To have a unique correspondence between vectors in $\mathbb{V}$ and $\mathbb{W}$, we need a mapping between $\mathbb{V}$ and $\mathbb{W}$ that is one-to-one and onto. To ensure that linear combinations of these vectors also correspond with one another, we need the mapping to be linear.

Definition
Isomorphism
Isomorphic

If $\mathbb{V}$ and $\mathbb{W}$ are vector spaces over $\mathbb{R}$, and if $L : \mathbb{V} \to \mathbb{W}$ is a linear, one-to-one, and onto mapping, then L is called an **isomorphism** (or a vector space isomorphism), and $\mathbb{V}$ and $\mathbb{W}$ are said to be **isomorphic**.

The word *isomorphism* comes from Greek words meaning "same form." The concept of an isomorphism is a very powerful and important one. It implies that the essential structure of the isomorphic vector spaces is the same. So, a vector space statement that is true in one space is immediately true in any isomorphic space. Of course, some vector spaces such as $M_{m\times n}(\mathbb{R})$ or $P_n(\mathbb{R})$ have some features that are not purely vector space properties (such as matrix multiplication and polynomial factorization), and these particular features cannot automatically be transferred.

EXAMPLE 4.7.4

Prove that $P_2(\mathbb{R})$ and $\mathbb{R}^3$ are isomorphic by constructing an explicit isomorphism L.

Solution: We define $L : P_2(\mathbb{R}) \to \mathbb{R}^3$ by $L(a_0 + a_1x + a_2x^2) = \begin{bmatrix} a_0 \\ a_1 \\ a_2 \end{bmatrix}$.

We need to prove that L is linear, one-to-one, and onto.

Linear: Let $\mathbf{p}(x) = a_0 + a_1x + a_2x^2, \mathbf{q}(x) = b_0 + b_1x + b_2x^2 \in P_2(\mathbb{R})$, and let $s, t \in \mathbb{R}$. Then, L is linear since

$$\begin{aligned} L(s\mathbf{p} + t\mathbf{q}) &= L(sa_0 + tb_0 + (sa_1 + tb_1)x + (sa_2 + tb_2)x^2) \\ &= \begin{bmatrix} sa_0 + tb_0 \\ sa_1 + tb_1 \\ sa_2 + tb_2 \end{bmatrix} \\ &= s\begin{bmatrix} a_0 \\ a_1 \\ a_2 \end{bmatrix} + t\begin{bmatrix} b_0 \\ b_1 \\ b_2 \end{bmatrix} = sL(\mathbf{p}) + tL(\mathbf{q}) \end{aligned}$$

One-to-one: Let $a_0 + a_1x + a_2x^2 \in \text{Null}(L)$. Then, $\begin{bmatrix} 0 \\ 0 \\ 0 \end{bmatrix} = L(a_0 + a_1x + a_2x^2) = \begin{bmatrix} a_0 \\ a_1 \\ a_2 \end{bmatrix}$.

Hence, $a_0 = a_1 = a_2 = 0$, so $\text{Null}(L) = \{\mathbf{0}\}$ and thus L is one-to-one by Theorem 4.7.1.

Onto: For any $\begin{bmatrix} a_0 \\ a_1 \\ a_2 \end{bmatrix} \in \mathbb{R}^3$, we have $L(a_0 + a_1x + a_2x^2) = \begin{bmatrix} a_0 \\ a_1 \\ a_2 \end{bmatrix}$. Hence, L is onto.

Thus, L is an isomorphism from $P_2(\mathbb{R})$ to $\mathbb{R}^3$. Hence $P_2(\mathbb{R})$ and $\mathbb{R}^3$ are isomorphic.

It is instructive to think carefully about the isomorphism in the example above. Observe that the images of the standard basis vectors for $P_2(\mathbb{R})$ are the standard basis vectors for $\mathbb{R}^3$. That is, the isomorphism is mapping basis vectors to basis vectors. We will keep this in mind when constructing an isomorphism in the next example.

EXAMPLE 4.7.5

Let $\mathbb{T} = \{\mathbf{p} \in P_2(\mathbb{R}) \mid \mathbf{p}(1) = 0\}$ and $\mathbb{D} = \left\{\begin{bmatrix} a & 0 \\ 0 & b \end{bmatrix} \in M_{2\times 2}(\mathbb{R}) \mid a, b \in \mathbb{R}\right\}$. Prove that $\mathbb{T}$ and $\mathbb{S}$ are isomorphic.

Solution: By the Factor Theorem, every vector $\mathbf{p} \in \mathbb{T}$ has the form $(1-x)(a+bx)$. Consequently, a basis for $\mathbb{T}$ is $\mathcal{B} = \{1-x, x-x^2\}$. We see that a basis for $\mathbb{D}$ is $C = \left\{\begin{bmatrix} 1 & 0 \\ 0 & 0 \end{bmatrix}, \begin{bmatrix} 0 & 0 \\ 0 & 1 \end{bmatrix}\right\}$. To define an isomorphism, we can map the vectors in $\mathcal{B}$ to the vectors in C. In particular, we define $L : \mathbb{T} \to \mathbb{S}$ by

$$L\Big(a(1-x) + b(x-x^2)\Big) = a\begin{bmatrix} 1 & 0 \\ 0 & 0 \end{bmatrix} + b\begin{bmatrix} 0 & 0 \\ 0 & 1 \end{bmatrix}$$

Linear: Let $\mathbf{p}(x) = a_1(1-x) + b_1(x-x^2), \mathbf{q}(x) = a_2(1-x) + b_2(x-x^2) \in \mathbb{T}$. For any $s, t \in \mathbb{R}$ we get

$$\begin{aligned} L(s\mathbf{p} + t\mathbf{q}) &= L\Big((sa_1 + ta_2)(1-x) + (sb_1 + tb_2)(x-x^2)\Big) \\ &= (sa_1 + ta_2)\begin{bmatrix} 1 & 0 \\ 0 & 0 \end{bmatrix} + (sb_1 + tb_2)\begin{bmatrix} 0 & 0 \\ 0 & 1 \end{bmatrix} \\ &= s\left(a_1\begin{bmatrix} 1 & 0 \\ 0 & 0 \end{bmatrix} + b_1\begin{bmatrix} 0 & 0 \\ 0 & 1 \end{bmatrix}\right) + t\left(a_2\begin{bmatrix} 1 & 0 \\ 0 & 0 \end{bmatrix} + b_2\begin{bmatrix} 0 & 0 \\ 0 & 1 \end{bmatrix}\right) \\ &= sL(\mathbf{p}) + tL(\mathbf{q}) \end{aligned}$$

Thus, L is linear.

One-to-one: Let $a(1-x) + b(x-x^2) \in \ker(L)$. Then,

$$\begin{bmatrix} 0 & 0 \\ 0 & 0 \end{bmatrix} = L\Big(a(1-x) + b(x-x^2)\Big) = a\begin{bmatrix} 1 & 0 \\ 0 & 0 \end{bmatrix} + b\begin{bmatrix} 0 & 0 \\ 0 & 1 \end{bmatrix} = \begin{bmatrix} a & 0 \\ 0 & b \end{bmatrix}$$

Hence, $a = b = 0$, so $\text{Null}(L) = \{0\}$. Therefore, by Theorem 4.7.1, L is one-to-one.

Onto: Let $\begin{bmatrix} a & 0 \\ 0 & b \end{bmatrix} \in \mathbb{S}$. We have

$$L\Big(a(1-x) + b(x-x^2)\Big) = a\begin{bmatrix} 1 & 0 \\ 0 & 0 \end{bmatrix} + b\begin{bmatrix} 0 & 0 \\ 0 & 1 \end{bmatrix} = \begin{bmatrix} a & 0 \\ 0 & b \end{bmatrix}$$

So, L is onto.

Thus, L is an isomorphism, and hence $\mathbb{T}$ and $\mathbb{S}$ are isomorphic.

Observe in the examples above that the isomorphic vector spaces have the same dimension. This makes sense since if we are mapping basis vectors to basis vectors, then both vector spaces need to have the same number of basis vectors.

Theorem 4.7.3 Suppose that $\mathbb{U}$ and $\mathbb{V}$ are finite-dimensional vector spaces over $\mathbb{R}$. Then $\mathbb{U}$ and $\mathbb{V}$ are isomorphic if and only if they are of the same dimension.

You are asked to prove Theorem 4.7.3 as Problem C5.

EXAMPLE 4.7.6

1. The vector space $M_{m\times n}(\mathbb{R})$ is isomorphic to $\mathbb{R}^{mn}$.
2. The vector space $P_n(\mathbb{R})$ is isomorphic to $\mathbb{R}^{n+1}$.
3. Every k-dimensional subspace of $\mathbb{R}^n$ is isomorphic to every k-dimensional subspace of $M_{m\times n}(\mathbb{R})$.

If we know that two vector spaces over $\mathbb{R}$ have the same dimension, then Theorem 4.7.3 says that they are isomorphic. However, even if we already know that two vector spaces are isomorphic, we may need to construct an explicit isomorphism between the two vector spaces. The following theorem shows that if we have two isomorphic vector spaces $\mathbb{U}$ and $\mathbb{V}$, then we only have to check if a linear mapping $L : \mathbb{U} \to \mathbb{V}$ is one-to-one or onto to prove that it is an isomorphism between these two spaces.

Theorem 4.7.4 If $\mathbb{U}$ and $\mathbb{V}$ are n-dimensional vector spaces over $\mathbb{R}$, then a linear mapping $L : \mathbb{U} \to \mathbb{V}$ is one-to-one if and only if it is onto.

You are asked to prove Theorem 4.7.4 as Problem C6.

Remark

If $\mathbb{V}$ and $\mathbb{W}$ are both n-dimensional, then it does **not** mean that every linear mapping $L : \mathbb{V} \to \mathbb{W}$ must be an isomorphism. For example, $L : \mathbb{R}^4 \to M_{2\times 2}(\mathbb{R})$ defined by $L(x_1, x_2, x_3, x_4) = \begin{bmatrix} 0 & 0 \\ 0 & 0 \end{bmatrix}$ is definitely not one-to-one nor onto!

We saw that we can define an isomorphism by mapping basis vectors of one vector space to the basis vectors of the other vector space. The following theorem shows that this property actually characterizes isomorphisms.

Theorem 4.7.5 Let $\{\vec{v}_1, \dots, \vec{v}_n\}$ be a basis for $\mathbb{V}$ and let $\mathbb{V}$ and $\mathbb{W}$ be isomorphic vector spaces. A linear mapping $L : \mathbb{V} \to \mathbb{W}$ is an isomorphism if and only if $\{L(\vec{v}_1), \dots, L(\vec{v}_n)\}$ is a basis for $\mathbb{W}$.

You are asked to prove Theorem 4.7.5 as Problem C7.

PROBLEMS 4.7

Practice Problems

For Problems A1_A12, determine whether the linear mapping is one-to-one and/or onto.

A1 $L : \mathbb{R}^2 \to \mathbb{R}^2$ defined by
$L(x_1, x_2) = (2x_1 - x_2, 3x_1 - 5x_2)$

A2 $L : \mathbb{R}^2 \to \mathbb{R}^3$ defined by
$L(x_1, x_2) = (x_1 + 2x_2, x_1 - x_2, x_1 + x_2)$

A3 $L : \mathbb{R}^3 \to \mathbb{R}^2$ defined by $L(x_1, x_2, x_3) = \begin{bmatrix} x_1 + x_2 \\ x_2 - x_3 \end{bmatrix}$

A4 $L : \mathbb{R}^3 \to \mathbb{R}^3$ defined by
$$L(x_1, x_2, x_3) = \begin{bmatrix} x_1 + 3x_3 \\ x_1 + x_2 + x_3 \\ -x_1 - 2x_2 + x_3 \end{bmatrix}$$

A5 $L : M_{2\times2}(\mathbb{R}) \to M_{2\times2}(\mathbb{R})$ defined by
$$L\left(\begin{bmatrix} a & b \\ c & d \end{bmatrix}\right) = \begin{bmatrix} a + c & b + c \\ a + d & b \end{bmatrix}$$

A6 $L : M_{2\times2}(\mathbb{R}) \to \mathbb{R}^3$ defined by $L\left(\begin{bmatrix} a & b \\ c & d \end{bmatrix}\right) = \begin{bmatrix} a + c \\ b - c \\ a + b \end{bmatrix}$

A7 $L : P_1(\mathbb{R}) \to P_1(\mathbb{R})$ defined by
$L(a + bx) = (3a - b) + (-3a + b)x$

A8 $L : P_2(\mathbb{R}) \to P_2(\mathbb{R})$ defined by
$L(a + bx + cx^2) = (a + 4b) + (2b + c)x + (a - 2c)x^2$

A9 $L : P_1(\mathbb{R}) \to \mathbb{R}_2$ defined by $L(\mathbf{p}) = \begin{bmatrix} \mathbf{p}(1) \\ \mathbf{p}(2) \end{bmatrix}$

A10 $L : P_1(\mathbb{R}) \to \mathbb{R}_3$ defined by $L(a + bx) = \begin{bmatrix} a + b \\ a - b \\ a + b \end{bmatrix}$

A11 $L : P_2(\mathbb{R}) \to M_{2\times2}(\mathbb{R})$ defined by
$$L(a + bx + cx^2) = \begin{bmatrix} a - b & b - c \\ a - c & 2a + b - 3c \end{bmatrix}$$

A12 $L : \mathbb{R}^3 \to P_2(\mathbb{R})$ defined by
$L(a, b, c) = (-b + 2c) + (a + 3b - c)x + (2a + 2b)x^2$

For Problems, A13_A21, determine whether the given vectors spaces are isomorphic. If so, then give, with proof, an explicit isomorphism.

A13 $P_2(\mathbb{R})$ and $\mathbb{R}^2$

A14 $P_3(\mathbb{R})$ and $\mathbb{R}^4$

A15 $M_{2\times2}(\mathbb{R})$ and $\mathbb{R}^4$

A16 $P_3(\mathbb{R})$ and $M_{2\times2}(\mathbb{R})$

A17 Any plane through the origin in $\mathbb{R}^3$ and $\mathbb{R}^2$

A18 The subspace $\mathbb{D}$ of diagonal matrices in $M_{2\times2}(\mathbb{R})$ and $P_1(\mathbb{R})$

A19 The subspace $\mathbb{T} = \{\vec{x} \in \mathbb{R}^4 \mid x_1 = x_4\}$ and $M_{2\times2}(\mathbb{R})$

A20 The subspace $\mathbb{S}$ of matrices in $M_{2\times2}(\mathbb{R})$ that satisfy $A^T = A$ and the subspace $\mathbb{V} = \{\vec{x} \in \mathbb{R}^3 \mid x_1 - x_2 = x_3\}$ of $\mathbb{R}^3$

A21 The subspace $\mathbb{U}$ of upper triangular matrices in $M_{2\times2}(\mathbb{R})$ and the subspace $\mathbb{P} = \{\mathbf{p} \in P_3(\mathbb{R}) \mid \mathbf{p}(1) = 0\}$ of $P_3(\mathbb{R})$

Homework Problems

For Problems B1_B7, determine whether the linear mapping is one-to-one and/or onto.

B1 $L : \mathbb{R}^2 \to P_1(\mathbb{R})$ defined by
$L(a, b) = (a + b) + (a + b)x$

B2 $L : \mathbb{R}^2 \to \mathbb{R}^3$ defined by $L(x_1, x_2) = \begin{bmatrix} x_1 + x_2 \\ x_1 - x_2 \\ x_2 \end{bmatrix}$

B3 $L : P_3(\mathbb{R}) \to P_2(\mathbb{R})$ defined by
$L(a + bx + cx^2 + dx^3) = (a + b) + (b + c)x + (c + d)x^2$

B4 $L : P_2(\mathbb{R}) \to P_2(\mathbb{R})$ defined by
$L(a + bx + cx^2) = (-2b - 3c) + (2a + c)x + (-3a + b)x^2$

B5 $L : P_1(\mathbb{R}) \to P_1(\mathbb{R})$ defined by
$L(a + bx) = (3a - 2b) + (2a + b)x$

B6 $L : \mathbb{R}^2 \to M_{2\times2}(\mathbb{R})$ defined by
$$L(x_1, x_2) = \begin{bmatrix} x_1 & x_1 \\ x_1 & x_2 \end{bmatrix}$$

B7 Let $\mathbb{U}$ be the subspace of $M_{2\times2}(\mathbb{R})$ of upper triangular matrices. $L : \mathbb{U} \to \mathbb{R}^3$ defined by
$$L\left(\begin{bmatrix} x_1 & x_2 \\ 0 & x_3 \end{bmatrix}\right) = \begin{bmatrix} x_1 + 2x_2 + x_3 \\ 3x_2 + 2x_3 \\ -x_1 + x_2 \end{bmatrix}$$

For Problems, B8–B14, determine whether the given vectors spaces are isomorphic. If so, then give, with proof, an explicit isomorphism.

B8 $P_1(\mathbb{R})$ and $\mathbb{R}^2$

B9 $P_2(\mathbb{R})$ and $M_{2\times 2}(\mathbb{R})$

B10 $M_{2\times 2}(\mathbb{R})$ and $P_3(\mathbb{R})$

B11 $\mathbb{R}^2$ and the subspace $\mathcal{S} = \operatorname{Span}\left\{\begin{bmatrix}1\\0\\1\end{bmatrix}, \begin{bmatrix}1\\2\\1\end{bmatrix}\right\}$ of $\mathbb{R}^3$

B12 The subspace $\mathbb{S} = \left\{\begin{bmatrix}x_1\\x_2\\0\\x_3\end{bmatrix} \mid x_1 - x_2 + x_3 = 0\right\}$ of $\mathbb{R}^4$ and the subspace $\mathbb{U} = \{a + bx + cx^2 \in P_2(\mathbb{R}) \mid a = b\}$ of $P_2(\mathbb{R})$

B13 The subspace $\mathbb{S}$ of matrices in $M_{2\times 2}(\mathbb{R})$ that satisfy $A^T = -A$ and the subspace $\mathbb{V} = \{\vec{x} \in \mathbb{R}^3 \mid x_1 = x_3\}$ of $\mathbb{R}^3$

B14 Any n-dimensional vector space $\mathbb{V}$ and $\mathbb{R}^n$

Conceptual Problems

C1 Prove L is one-to-one if and only if $\text{Null}(L) = \{\mathbf{0}\}$.

C2 (a) Prove that if L and M are one-to-one, then $M \circ L$ is one-to-one.

(b) Give an example where M is not one-to-one but $M \circ L$ is one-to-one.

(c) Is it possible to give an example where L is not one-to-one but $M \circ L$ is one-to-one? Explain.

C3 Prove that if L and M are onto, then $M \circ L$ is onto.

C4 Prove that the linear mapping $L : \mathbb{U} \to \mathbb{V}$ has an inverse linear mapping $L^{-1} : \mathbb{V} \to \mathbb{U}$ if and only if L is one-to-one and onto.

C5 Suppose $\mathbb{U}$ and $\mathbb{V}$ are finite-dimensional vector spaces over $\mathbb{R}$ and that $\mathcal{B} = \{\mathbf{u}_1, \dots, \mathbf{u}_n\}$ is a basis for $\mathbb{U}$.

(a) Prove if $\mathbb{U}$ and $\mathbb{V}$ are isomorphic, then $\dim \mathbb{U} = \dim \mathbb{V}$ by using an isomorphism $L : \mathbb{U} \to \mathbb{V}$ to construct a basis for $\mathbb{V}$. Make sure to prove that your basis is linearly independent and spans $\mathbb{V}$).

(b) Prove if $\dim \mathbb{U} = \dim \mathbb{V}$, then $\mathbb{U}$ and $\mathbb{V}$ are isomorphic by defining an explicit isomorphism between them . Make sure to prove that this is an isomorphism.

C6 Prove if $\mathbb{U}$ and $\mathbb{V}$ are n-dimensional vector spaces over $\mathbb{R}$, then a linear mapping $L : \mathbb{U} \to \mathbb{V}$ is one-to-one if and only if it is onto.

C7 Let $\{\vec{v}_1, \dots, \vec{v}_n\}$ be a basis for $\mathbb{V}$, let $\mathbb{V}$ and $\mathbb{W}$ be isomorphic vector spaces, and let $L : \mathbb{V} \to \mathbb{W}$ be a linear mapping. Prove $\{L(\vec{v}_1), \dots, L(\vec{v}_n)\}$ is a basis for $\mathbb{W}$ if and only if L is an isomorphism.

For Problems C8–C10, recall the definition of the Cartesian product from Section 4.2 Problem C7.

C8 Prove that $\mathbb{U} \times \{\mathbf{0}_\mathbb{V}\}$ is a subspace of $\mathbb{U} \times \mathbb{V}$ that is isomorphic to $\mathbb{U}$.

C9 Prove that $\mathbb{R}^2 \times \mathbb{R}$ is isomorphic to $\mathbb{R}^3$.

C10 Prove that $\mathbb{R}^n \times \mathbb{R}^m$ is isomorphic to $\mathbb{R}^{n+m}$.

C11 Suppose that $L : \mathbb{U} \to \mathbb{V}$ is a vector space isomorphism and that $M : \mathbb{V} \to \mathbb{V}$ is a linear mapping. Prove that $L^{-1} \circ M \circ L$ is a linear mapping from $\mathbb{U}$ to $\mathbb{U}$. Describe the nullspace and range of $L^{-1} \circ M \circ L$ in terms of the nullspace and range of M.

For Problems C12–C15, prove or disprove the statement. Assume that $\mathbb{U}, \mathbb{V}, \mathbb{W}$ are all finite dimensional vector spaces.

C12 If $\dim \mathbb{V} = \dim \mathbb{U}$ and $L : \mathbb{U} \to \mathbb{V}$ is linear, then L is an isomorphism.

C13 If $L : \mathbb{V} \to \mathbb{W}$ is a linear mapping and $\dim \mathbb{V} < \dim \mathbb{W}$, then L cannot be onto.

C14 If $L : \mathbb{V} \to \mathbb{W}$ is a linear mapping and $\dim \mathbb{V} > \dim \mathbb{W}$, then L is onto.

C15 If $L : \mathbb{V} \to \mathbb{W}$ is a linear mapping and $\dim \mathbb{V} > \dim \mathbb{W}$, then L cannot be one-to-one.

CHAPTER REVIEW

Suggestions for Student Review

1 State the ten properties of a vector space over $\mathbb{R}$. Why is the empty set not a vector space? Describe two or three examples of vector spaces that are not subspaces of $\mathbb{R}^n$. (Section 4.2)

2 State the definition of a subspace of a vector space $\mathbb{V}$. (Section 4.2)

3 State the definitions of spanning, linear independence, and bases. State the two ways we have of finding a basis. (Section 4.3)

4 (a) Explain the concept of dimension. What theorem is required to ensure that the concept of dimension is well defined? (Section 4.3)
  (b) Explain how knowing the dimension of a vector space is helpful when you have to find a basis for the vector space. (Section 4.3)

5 State the definition of $\mathcal{B}$-coordinates and the $\mathcal{B}$-coordinate vector of a vector $\mathbf{v}$ in a vector space $\mathbb{V}$. What theorem is required to ensure that the concept of coordinates is well defined? (Section 4.4)

6 Create and analyze an example as follows.
  (a) Give a basis $\mathcal{B}$ for a three-dimensional subspace in $\mathbb{R}^5$. (Do not make it too easy by choosing any standard basis vectors, but do not make it too hard by choosing completely random components.) (Section 4.3)
  (b) Determine the standard coordinates in $\mathbb{R}^5$ of the vector that has coordinate vector $\begin{bmatrix} 2 \\ -3 \\ 4 \end{bmatrix}$ with respect to your basis $\mathcal{B}$. (Section 4.4)
  (c) Take the vector you found in (b) and carry out the standard procedure to determine its coordinates with respect to $\mathcal{B}$. Did you get the right answer, $\begin{bmatrix} 2 \\ -3 \\ 4 \end{bmatrix}$? (Section 4.4)
  (d) Pick any two vectors in $\mathbb{R}^5$ and determine whether they lie in your subspace. Determine the coordinates of any vector that is in the subspace. (Section 4.4)

7 Write a short explanation of how you use information about consistency of systems and uniqueness of solutions in testing for linear independence and in determining whether a vector belongs to a given subspace. (Sections 4.3 and 4.4)

8 Define the change of coordinates matrix. Create two different bases $\mathcal{B}$ and C for $P_1(\mathbb{R})$. Find the change of coordinates matrix P from C-coordinates to $\mathcal{B}$-coordinates and its inverse. Check your answer in as many ways as you can. (Section 4.4)

9 Give the definition of a linear mapping $L : \mathbb{V} \to \mathbb{W}$. Explain the procedure for determining if a vector $\mathbf{y}$ is in the range of L. Describe how to find a basis for the nullspace and a basis for the range of L. (Section 4.5)

10 State the definition of the zero mapping and the identity mapping. (Section 4.5)

11 State the Rank-Nullity Theorem and demonstrate it with an example. In what kinds of problems have we used the Rank-Nullity Theorem? (Section 4.5 and 4.7)

12 State the definition of the inverse of a linear mapping and explain how it is related to the content in Sections 4.6 and 4.7. (Section 4.5, 4.6, and 4.7)

13 State how to determine the standard matrix and the $\mathcal{B}$-matrix of a linear mapping $L : \mathbb{V} \to \mathbb{V}$. Explain how $[L(\vec{x})]_{\mathcal{B}}$ is determined in terms of $[L]_{\mathcal{B}}$. (Section 4.6)

14 State the relationship between the matrix of a linear mapping and the change of coordinates matrix. Demonstrate with an example. (Section 4.6)

15 State and explain the concepts of one-to-one and onto. (Section 4.7)

16 State the definition of an isomorphism of vector spaces and give some examples. Explain why a finite-dimensional vector space cannot be isomorphic to a proper subspace of itself. (Section 4.7)

Chapter Quiz

For Problems E1–E4, determine whether the set is a vector space. Explain briefly.

E1 The set of 4×3 matrices such that the sum of the entries in the first row is zero ($a_{11} + a_{12} + a_{13} = 0$) under standard addition and scalar multiplication of matrices.

E2 The set of polynomials $\mathbf{p}(x) \in P_3(\mathbb{R})$ such that $\mathbf{p}(1) = 0$ and $\mathbf{p}(2) = 0$ under standard addition and scalar multiplication of polynomials.

E3 The set of 2×2 matrices such that all entries are integers under standard addition and scalar multiplication of matrices.

E4 The set of all $\begin{bmatrix} x_1 \\ x_2 \\ x_3 \end{bmatrix} \in \mathbb{R}^3$ such that $x_1 + x_2 + x_3 = 0$ under standard addition and scalar multiplication of vectors in $\mathbb{R}^3$.

For Problems E5–E8, determine with proof whether the given set is a subspace of the given vector space. If the set is a subspace, find a basis for the subspace.

E5 $\mathbb{S} = \{\mathbf{p} \in P_2(\mathbb{R}) \mid \mathbf{p}(0) = 1\}$ of $P_2(\mathbb{R})$

E6 $\mathbb{S} = \left\{ \begin{bmatrix} a & 0 \\ b & c \end{bmatrix} \in M_{2\times 2}(\mathbb{R}) \mid a + b = -2c \right\}$ of $M_{2\times 2}(\mathbb{R})$

E7 $\mathbb{S} = \{a + cx^2 \in P_2(\mathbb{R}) \mid a = c\}$ of $P_2(\mathbb{R})$

E8 $\mathbb{S} = \operatorname{Span}\{1 + x^2, 2 + x, x - 2x^2\}$ of $P_2(\mathbb{R})$

For Problems E9–E11, determine whether the given set is a basis for $M_{2\times 2}(\mathbb{R})$.

E9 $\left\{ \begin{bmatrix} 1 & 1 \\ 2 & 1 \end{bmatrix}, \begin{bmatrix} 0 & 1 \\ 1 & -1 \end{bmatrix}, \begin{bmatrix} 0 & 1 \\ 1 & 3 \end{bmatrix}, \begin{bmatrix} 2 & 2 \\ 4 & -2 \end{bmatrix}, \begin{bmatrix} 0 & 2 \\ 3 & 0 \end{bmatrix} \right\}$

E10 $\left\{ \begin{bmatrix} 1 & 1 \\ 2 & 1 \end{bmatrix}, \begin{bmatrix} 0 & 1 \\ 1 & -1 \end{bmatrix}, \begin{bmatrix} 0 & 1 \\ 1 & 3 \end{bmatrix}, \begin{bmatrix} 2 & 2 \\ 4 & -2 \end{bmatrix} \right\}$

E11 $\left\{ \begin{bmatrix} 1 & 3 \\ -1 & 4 \end{bmatrix}, \begin{bmatrix} 2 & 2 \\ 0 & 3 \end{bmatrix}, \begin{bmatrix} 1 & 0 \\ 2 & 0 \end{bmatrix} \right\}$

E12 (a) Let $\mathbb{S} = \operatorname{Span}\left\{ \begin{bmatrix} 1 \\ 0 \\ 1 \\ 1 \end{bmatrix}, \begin{bmatrix} 1 \\ 1 \\ 0 \\ 1 \end{bmatrix}, \begin{bmatrix} 3 \\ 3 \\ 1 \\ 0 \end{bmatrix}, \begin{bmatrix} 1 \\ 1 \\ 1 \\ -2 \end{bmatrix} \right\}$. Find a basis $\mathcal{B}$ for $\mathbb{S}$ and hence determine the dimension of $\mathbb{S}$.

(b) Determine the $\mathcal{B}$-coordinates of $\vec{x} = \begin{bmatrix} 0 \\ 2 \\ -1 \\ -3 \end{bmatrix}$.

E13 Let $\mathcal{B} = \{1 + x, 2 - x\}$ and $C = \{1 - 3x, 1 + 2x\}$ both be basis for $P_1(\mathbb{R})$. Find the change of coordinates matrix Q from C-coordinates to $\mathcal{B}$-coordinates and the change of coordinates matrix P from $\mathcal{B}$-coordinates to C-coordinates.

E14 (a) Find a basis for the plane in $\mathbb{R}^3$ with equation $x_1 - x_3 = 0$.

(b) Extend the basis you found in (a) to a basis $\mathcal{B}$ for $\mathbb{R}^3$.

(c) Let $L : \mathbb{R}^3 \to \mathbb{R}^3$ be a reflection in the plane from part (a). Determine $[L]_{\mathcal{B}}$.

(d) Using your result from part (c), determine the standard matrix $[L]_S$ of the reflection.

E15 Let $L : \mathbb{R}^3 \to \mathbb{R}^3$ be a linear mapping with standard matrix $\begin{bmatrix} 1 & -1 & 2 \\ -1 & 0 & 1 \\ -2 & 1 & 0 \end{bmatrix}$ and let $\mathcal{B} = \left\{ \begin{bmatrix} 1 \\ 1 \\ 0 \end{bmatrix}, \begin{bmatrix} 0 \\ 1 \\ 1 \end{bmatrix}, \begin{bmatrix} 1 \\ -1 \\ 1 \end{bmatrix} \right\}$. Determine the matrix $[L]_{\mathcal{B}}$.

E16 Suppose that $L : \mathbb{V} \to \mathbb{W}$ is a linear mapping with $\operatorname{Null}(L) = \{\mathbf{0}\}$. Suppose that $\{\mathbf{v}_1, \dots, \mathbf{v}_k\}$ is a linearly independent set in $\mathbb{V}$. Prove that $\{L(\mathbf{v}_1), \dots, L(\mathbf{v}_k)\}$ is a linearly independent set in $\mathbb{W}$.

For Problems E17–E22, decide whether each of the following statements is true or false. If it is true, explain briefly; if it is false, give an example to show that it is false.

E17 A subspace of $\mathbb{R}^n$ must have dimension less than n.

E18 A set of four polynomials in $P_2(\mathbb{R})$ cannot be a basis for $P_2(\mathbb{R})$.

E19 If $\mathcal{B}$ is a basis for a subspace of $\mathbb{R}^5$, then the $\mathcal{B}$-coordinate vector of some vector $\vec{x} \in \mathbb{R}^5$ has five components.

E20 For any linear mapping $L : \mathbb{R}^n \to \mathbb{R}^n$ and any basis $\mathcal{B}$ of $\mathbb{R}^n$, the rank of the matrix $[L]_{\mathcal{B}}$ is the same as the rank of the matrix $[L]_S$.

E21 For any linear mapping $L : \mathbb{V} \to \mathbb{V}$ and any basis $\mathcal{B}$ of $\mathbb{V}$, the column space of $[L]_{\mathcal{B}}$ equals the range of L.

E22 If $L : \mathbb{V} \to \mathbb{W}$ is one-to-one, then $\dim \mathbb{V} = \dim \mathbb{W}$.

Further Problems

These problems are intended to be challenging.

F1 Let $\mathbb{S}$ be a subspace of an n-dimensional vector space $\mathbb{V}$. Prove that there exists a linear operator $L : \mathbb{V} \to \mathbb{V}$ such that $\text{Null}(L) = \mathbb{S}$.

F2 Use the ideas of this chapter to prove the uniqueness of the reduced row echelon form for a given matrix A. (Hint: begin by assuming that there are two reduced row echelon forms R and S. What can you say about the columns with leading 1s in the two matrices?)

F3 **Magic Squares—An Exploration of Their Vector Space Properties**

We say that any matrix $A \in M_{3\times 3}(\mathbb{R})$ is a 3×3 **magic square** if the three **row sums** (where each row sum is the sum of the entries in one row of A) of A, the three **column sums** of A, and the two **diagonal sums** of A ($a_{11} + a_{22} + a_{33}$ and $a_{13} + a_{22} + a_{31}$) all have the same value k. The common sum k is called the **weight** of the magic square A and is denoted by $wt(A) = k$.

For example, $A = \begin{bmatrix} 2 & 2 & -1 \\ -2 & 1 & 4 \\ 3 & 0 & 0 \end{bmatrix}$ is a magic square with $wt(A) = 3$.

The aim of this exploration is to find all 3×3 magic squares. The subset of $M_{3\times 3}(\mathbb{R})$ consisting of magic squares is denoted MS_3.

(a) Show that MS_3 is a subspace of $M_{3\times 3}(\mathbb{R})$.

(b) Observe that weight determines a map $wt : MS_3 \to \mathbb{R}$. Show that wt is linear.

(c) Compute the nullspace of wt. Suppose that

$$\underline{X}_1 = \begin{bmatrix} 1 & 0 & a \\ b & c & d \\ e & f & g \end{bmatrix}, \quad \underline{X}_2 = \begin{bmatrix} 0 & 1 & h \\ i & j & k \\ l & m & n \end{bmatrix}$$

and

$$\underline{0} = \begin{bmatrix} 0 & 0 & p \\ q & r & s \\ t & u & v \end{bmatrix}$$

are all in the nullspace, where $a, b, c, \ldots, v$ denote unknown entries. Determine these unknown entries and prove that $\underline{X}_1$ and $\underline{X}_2$ form a basis for $\text{Null}(wt)$. (Hint: if $A \in \text{Null}(wt)$, consider $A - a_{11}\underline{X}_1 - a_{12}\underline{X}_2$.)

(d) Let $\underline{J} = \begin{bmatrix} 1 & 1 & 1 \\ 1 & 1 & 1 \\ 1 & 1 & 1 \end{bmatrix}$. Observe that $\underline{J}$ is a magic square with $wt(\underline{J}) = 3$. Show that all A in MS_3 that have weight k are of the form

$$(k/3)\underline{J} + p\underline{X}_1 + q\underline{X}_2, \quad \text{for some } p, q \in \mathbb{R}$$

(e) Show that $\mathcal{B} = \{\underline{J}, \underline{X}_1, \underline{X}_2\}$ is a basis for MS_3.

(f) Find the coordinates of $A = \begin{bmatrix} 3 & 1 & 2 \\ 1 & 2 & 3 \\ 2 & 3 & 1 \end{bmatrix}$ with respect to the basis $\mathcal{B}$.

Exercises F4–F8 require the following definitions.

If $\mathbb{S}$ and $\mathbb{T}$ are subspaces of the vector space $\mathbb{V}$, we define

$$\mathbb{S} + \mathbb{T} = \{\mathbf{s} + \mathbf{t} \mid \mathbf{s} \in \mathbb{S}, \mathbf{t} \in \mathbb{T}\}$$

If $\mathbb{S}$ and $\mathbb{T}$ are subspaces of $\mathbb{V}$ such that $\mathbb{S} + \mathbb{T} = \mathbb{V}$ and $\mathbb{S} \cap \mathbb{T} = \{\mathbf{0}\}$, then we say that $\mathbb{S}$ is a **complement** of $\mathbb{T}$ (and $\mathbb{T}$ is a complement of $\mathbb{S}$).

F4 Show that, in general, the complement of a subspace $\mathbb{T}$ is not unique.

F5 In the vector space of continuous real-valued functions of a real variable, show that the even functions and the odd functions form subspaces such that each is the complement of the other.

F6 (a) If $\mathbb{S}$ is a k-dimensional subspace of $\mathbb{R}^n$, show that any complement of $\mathbb{S}$ must be of dimension $n - k$.

(b) Suppose that $\mathbb{S}$ is a subspace of $\mathbb{R}^n$ that has a unique complement. Must it be true that $\mathbb{S}$ is either $\{\mathbf{0}\}$ or $\mathbb{R}^n$?

F7 Suppose that $\mathbf{v}$ and $\mathbf{w}$ are vectors in a vector space $\mathbb{V}$. Suppose also that $\mathbb{S}$ is a subspace of $\mathbb{V}$. Let $\mathbb{T}$ be the subspace spanned by $\mathbf{v}$ and $\mathbb{S}$. Let $\mathbb{U}$ be the subspace spanned by $\mathbf{w}$ and $\mathbb{S}$. Prove that if $\mathbf{w}$ is in $\mathbb{T}$ but not in $\mathbb{S}$, then $\mathbf{v}$ is in $\mathbb{U}$.

F8 Show that if $\mathbb{S}$ and $\mathbb{T}$ are finite-dimensional subspaces of $\mathbb{V}$, then

$$\dim \mathbb{S} + \dim \mathbb{T} = \dim(\mathbb{S} + \mathbb{T}) + \dim(\mathbb{S} \cap \mathbb{T})$$

CHAPTER 5

Determinants

CHAPTER OUTLINE

In Chapter 3, we saw that a 2×2 *matrix* $A = \begin{bmatrix} a & b \\ c & d \end{bmatrix}$ *is invertible if and only if* $ad - bc = 0$. *That is, the value* $ad - bc$ *determines if A is invertible or not. What is surprising is that the area of the parallelogram formed by vectors* $\begin{bmatrix} a \\ c \end{bmatrix}$ *and* $\begin{bmatrix} b \\ d \end{bmatrix}$ *according to the parallelogram rule for addition is also* $ad - bc$ *(see Section 5.4). This quantity* $ad - bc$*, which we call the* **determinant** *of the matrix, turns out to be extremely useful. For example, we will use it to find eigenvalues in Chapter 6, it can help us solve a variety of problems in geometry (Section 5.4), and it is even required in multivariable calculus.*

5.1 Determinants in Terms of Cofactors

Definition
Determinant of a 2×2 Matrix

The **determinant of a 2 × 2 matrix** $A = \begin{bmatrix} a_{11} & a_{12} \\ a_{21} & a_{22} \end{bmatrix}$ is defined by

$$\det A = \det \begin{bmatrix} a_{11} & a_{12} \\ a_{21} & a_{22} \end{bmatrix} = a_{11}a_{22} - a_{12}a_{21}$$

An Alternate Notation: The determinant is often denoted by vertical straight lines:

$$\begin{vmatrix} a_{11} & a_{12} \\ a_{21} & a_{22} \end{vmatrix} = \det \begin{bmatrix} a_{11} & a_{12} \\ a_{21} & a_{22} \end{bmatrix} = a_{11}a_{22} - a_{12}a_{21}$$

One risk with this notation is that one may fail to distinguish between a matrix and the determinant of the matrix. This is a rather gross error.

EXAMPLE 5.1.1

Find the determinant of $\begin{bmatrix} 1 & 3 \\ 2 & 4 \end{bmatrix}$ and the determinant of $\begin{bmatrix} 2 & 2 \\ 4 & 4 \end{bmatrix}$.

Solution: We have

$$\det \begin{bmatrix} 1 & 3 \\ 2 & 4 \end{bmatrix} = 1(4) - 3(2) = -2$$

$$\det \begin{bmatrix} 2 & 2 \\ 4 & 4 \end{bmatrix} = 2(4) - 2(4) = 0$$

EXERCISE 5.1.1

Calculate the following determinants.

(a) $\begin{vmatrix} 3 & 2 \\ 2 & 1 \end{vmatrix}$ (b) $\begin{vmatrix} 1 & 3 \\ 0 & -2 \end{vmatrix}$ (c) $\begin{vmatrix} 2 & 4 \\ 1 & 2 \end{vmatrix}$

The 3×3 *Case*

Let $A = \begin{bmatrix} a_{11} & a_{12} & a_{13} \\ a_{21} & a_{22} & a_{23} \\ a_{31} & a_{32} & a_{33} \end{bmatrix}$. We can show through elimination (with some effort) that A is invertible if and only if

$$D = a_{11}a_{22}a_{33} - a_{11}a_{23}a_{32} - a_{12}a_{21}a_{33} + a_{12}a_{23}a_{31} + a_{13}a_{21}a_{32} - a_{13}a_{22}a_{31} \neq 0$$

We would like to reorganize this expression so that we can remember it more easily, and so that we can determine how to generalize it to the $n \times n$ case. Notice that a_{11} is a common factor in the first pair of terms in D, a_{12} is a common factor in the second pair, and a_{13} is a common factor in the third pair. Thus, D can be rewritten as

$$\begin{aligned} D &= a_{11}(a_{22}a_{33} - a_{23}a_{32}) - a_{12}(a_{21}a_{33} - a_{23}a_{31}) + a_{13}(a_{21}a_{32} - a_{22}a_{31}) \\ &= a_{11}\begin{vmatrix} a_{22} & a_{23} \\ a_{32} & a_{33} \end{vmatrix} + a_{12}(-1)\begin{vmatrix} a_{21} & a_{23} \\ a_{31} & a_{33} \end{vmatrix} + a_{13}\begin{vmatrix} a_{21} & a_{22} \\ a_{31} & a_{32} \end{vmatrix} \qquad (5.1) \end{aligned}$$

Observe that the determinant being multiplied by a_{11} in equation (5.1) is the determinant of the 2×2 matrix formed by removing the first row and first column of A. Similarly, a_{12} is being multiplied by (-1) times the determinant of the matrix formed by removing the first row and second column of A, and a_{13} is being multiplied by the determinant of the matrix formed by removing the first row and third column of A. Hence, we make the following definitions.

Definition
Cofactors of a 3×3 Matrix

Let $A \in M_{3\times3}(\mathbb{R})$ and let $A(i, j)$ denote the 2×2 submatrix obtained from A by deleting the i-th row and j-th column. Define the (i, j)-**cofactor** of A to be

$$C_{ij} = (-1)^{(i+j)} \det A(i, j)$$

EXAMPLE 5.1.2

Find all nine of the cofactors of $A = \begin{bmatrix} 1 & 0 & 2 \\ 0 & -1 & 3 \\ -2 & -3 & 4 \end{bmatrix}$.

Solution: The $(1,1)$-cofactor C_{11} of A is defined by $C_{11} = (-1)^{1+1} \det A(1,1)$, where $A(1,1)$ is the matrix obtained from A by deleting the first row and first column. That is

$$A(1,1) = \begin{bmatrix} 1 & 0 & 2 \\ 0 & -1 & 3 \\ -2 & -3 & 4 \end{bmatrix} = \begin{bmatrix} -1 & 3 \\ -3 & 4 \end{bmatrix}$$

Hence, $C_{11} = (-1)^{1+1} \begin{vmatrix} -1 & 3 \\ -3 & 4 \end{vmatrix} = (1)[(-1)(4) - 3(-3)] = 5$.

The $(1,2)$-cofactor C_{12} is defined by $C_{12} = (-1)^{1+2} \det A(1,2)$, where $A(1,2)$ is the matrix obtained from A by deleting the first row and second column. We have

$$A(1,2) = \begin{bmatrix} 1 & 0 & 2 \\ 0 & -1 & 3 \\ -2 & -3 & 4 \end{bmatrix} = \begin{bmatrix} 0 & 3 \\ -2 & 4 \end{bmatrix}$$

Thus, $C_{12} = (-1)^{1+2} \begin{vmatrix} 0 & 3 \\ -2 & 4 \end{vmatrix} = (-1)[0(4) - 3(-2)] = -6$.

Similarly, the $(1,3)$-cofactor C_{13} is defined by $C_{13} = (-1)^{1+3} \det A(1,3)$, where $A(1,3)$ is the matrix obtained from A by deleting the first row and third column. We have

$$A(1,3) = \begin{bmatrix} 1 & 0 & 2 \\ 0 & -1 & 3 \\ -2 & -3 & 4 \end{bmatrix} = \begin{bmatrix} 0 & -1 \\ -2 & -3 \end{bmatrix}$$

Thus, $C_{13} = (-1)^{1+3} \begin{vmatrix} 0 & -1 \\ -2 & -3 \end{vmatrix} = (1)[0(-3) - (-1)(-2)] = -2$.

Continuing in this way, we can find the other cofactors are:

$$C_{21} = (-1)^{2+1} \begin{vmatrix} 0 & 2 \\ -3 & 4 \end{vmatrix} = (-1)[0(4) - 2(-3)] = -6$$

$$C_{22} = (-1)^{2+2} \begin{vmatrix} 1 & 2 \\ -2 & 4 \end{vmatrix} = (1)[1(4) - 2(-2)] = 8$$

$$C_{23} = (-1)^{2+3} \begin{vmatrix} 1 & 0 \\ -2 & -3 \end{vmatrix} = (-1)[1(-3) - 0(-2)] = 3$$

$$C_{31} = (-1)^{3+1} \begin{vmatrix} 0 & 2 \\ -1 & 3 \end{vmatrix} = (1)[0(3) - 2(-1)] = 2$$

$$C_{32} = (-1)^{3+2} \begin{vmatrix} 1 & 2 \\ 0 & 3 \end{vmatrix} = (-1)[1(3) - 2(0)] = -3$$

$$C_{33} = (-1)^{3+3} \begin{vmatrix} 1 & 0 \\ 0 & -1 \end{vmatrix} = (1)[1(-1) - 0(0)] = -1$$

CONNECTION

Comparing the cross product of the second and third column of the matrix in Example 5.1.2 with the cofactors from the first column of the matrix shows us that

$$\begin{bmatrix} 0 \\ -1 \\ -3 \end{bmatrix} \times \begin{bmatrix} 2 \\ 3 \\ 4 \end{bmatrix} = \begin{bmatrix} 5 \\ -6 \\ 2 \end{bmatrix} = \begin{bmatrix} C_{11} \\ C_{21} \\ C_{31} \end{bmatrix}$$

That is, the cross product is defined in terms of 2×2 determinants. We will explore this relationship a little further in Section 5.3.

Definition
Determinant of a 3×3 Matrix

The **determinant of a 3×3 matrix** A is defined by

$$\det A = a_{11}C_{11} + a_{12}C_{12} + a_{13}C_{13}$$

EXAMPLE 5.1.3

Let $A = \begin{bmatrix} 4 & -1 & 1 \\ 2 & 3 & 5 \\ 1 & 0 & 6 \end{bmatrix}$. Calculate the cofactors of the first row of A and use them to find the determinant of A.

Solution: By definition, the $(1,1)$-cofactor C_{11} is $(-1)^{1+1}$ times the determinant of the matrix obtained from A by deleting the first row and first column. Thus,

$$C_{11} = (-1)^{1+1}\det\begin{bmatrix} 3 & 5 \\ 0 & 6 \end{bmatrix} = 3(6) - 5(0) = 18$$

The $(1,2)$-cofactor C_{12} is $(-1)^{1+2}$ times the determinant of the matrix obtained from A by deleting the first row and second column. So,

$$C_{12} = (-1)^{1+2}\det\begin{bmatrix} 2 & 5 \\ 1 & 6 \end{bmatrix} = -[2(6) - 5(1)] = -7$$

Finally, the $(1,3)$-cofactor C_{13} is

$$C_{13} = (-1)^{1+3}\det\begin{bmatrix} 2 & 3 \\ 1 & 0 \end{bmatrix} = 2(0) - 3(1) = -3$$

Hence,

$$\det A = a_{11}C_{11} + a_{12}C_{12} + a_{13}C_{13} = 4(18) + (-1)(-7) + 1(-3) = 76$$

EXERCISE 5.1.2

Let $A = \begin{bmatrix} 1 & 2 & 3 \\ 0 & -1 & -2 \\ 4 & 0 & -3 \end{bmatrix}$. Calculate the cofactors of the first row of A and use them to find the determinant of A.

Generally, when expanding a determinant, we write the steps more compactly, as in the next example.

EXAMPLE 5.1.4

Calculate $\det\begin{bmatrix} 1 & 2 & 3 \\ -2 & 2 & 1 \\ 5 & 0 & -1 \end{bmatrix}$.

Solution: By definition, we have

$$\begin{aligned}\det\begin{bmatrix} 1 & 2 & 3 \\ -2 & 2 & 1 \\ 5 & 0 & -1 \end{bmatrix} &= a_{11}C_{11} + a_{12}C_{12} + a_{13}C_{13} \\ &= 1(-1)^{1+1}\begin{vmatrix} 2 & 1 \\ 0 & -1 \end{vmatrix} + 2(-1)^{1+2}\begin{vmatrix} -2 & 1 \\ 5 & -1 \end{vmatrix} + 3(-1)^{1+3}\begin{vmatrix} -2 & 2 \\ 5 & 0 \end{vmatrix} \\ &= 1[2(-1) - 1(0)] - 2[(-2)(-1) - 1(5)] + 3[(-2)(0) - 2(5)] \\ &= -2 + 6 - 30 = -26\end{aligned}$$

We now define the determinant of an $n \times n$ matrix by following the pattern of the definition for the 3×3 case.

Definition

Determinant of a $n \times n$ Matrix
Cofactors of an $n \times n$ Matrix

Let $A \in M_{n\times n}(\mathbb{R})$ with $n > 2$. Let $A(i, j)$ denote the $(n - 1) \times (n - 1)$ submatrix obtained from A by deleting the i-th row and j-th column.
The **determinant** of $A \in M_{n\times n}(\mathbb{R})$ is defined by

$$\det A = a_{11}C_{11} + a_{12}C_{12} + \cdots + a_{1n}C_{1n}$$

where the (i, j)-**cofactor** of A is defined to be

$$C_{ij} = (-1)^{(i+j)} \det A(i, j)$$

Remarks

1. This definition of the determinant is called the **Cofactor (Laplace) expansion of the determinant along the first row**. As we shall see in Theorem 5.1.1 below, a determinant can be expanded along any row or column.

2. The signs attached to cofactors can cause trouble if you are not careful. One helpful way to remember which sign to attach to which cofactor is to take a blank matrix and put a + in the top left corner and then alternate − and + both across and down: $\begin{bmatrix} + & - & + \\ - & + & - \\ + & - & + \end{bmatrix}$. This is shown for a 3×3 matrix, but it works for a square matrix of any size.

3. This is a recursive definition. The result for the $n \times n$ case is defined in terms of the $(n - 1) \times (n - 1)$ case, which in turn must be calculated in terms of the $(n - 2) \times (n - 2)$ case, and so on, until we get back to the 2×2 case, for which the result is given explicitly. Note that this also works with the 1×1 case since we define $\det\begin{bmatrix} a \end{bmatrix} = a$.

EXAMPLE 5.1.5 We calculate the following determinant by using the definition of the determinant. Note that * and ** represent cofactors whose values are irrelevant because they are multiplied by 0.

$$\begin{aligned}
\begin{vmatrix} 0 & 2 & 3 & 0 \\ 1 & 5 & 6 & 7 \\ -2 & 3 & 0 & 4 \\ -5 & 1 & 2 & 3 \end{vmatrix} &= a_{11}C_{11} + a_{12}C_{12} + a_{13}C_{13} + a_{14}C_{14} \\
&= 0(*) + 2(-1)^{1+2}\begin{vmatrix} 1 & 6 & 7 \\ -2 & 0 & 4 \\ -5 & 2 & 3 \end{vmatrix} + 3(-1)^{1+3}\begin{vmatrix} 1 & 5 & 7 \\ -2 & 3 & 4 \\ -5 & 1 & 3 \end{vmatrix} + 0(**) \\
&= -2\left(1(-1)^{1+1}\begin{vmatrix} 0 & 4 \\ 2 & 3 \end{vmatrix} + 6(-1)^{1+2}\begin{vmatrix} -2 & 4 \\ -5 & 3 \end{vmatrix} + 7(-1)^{1+3}\begin{vmatrix} -2 & 0 \\ -5 & 2 \end{vmatrix}\right) \\
&\quad +3\left(1(-1)^{1+1}\begin{vmatrix} 3 & 4 \\ 1 & 3 \end{vmatrix} + 5(-1)^{1+2}\begin{vmatrix} -2 & 4 \\ -5 & 3 \end{vmatrix} + 7(-1)^{1+3}\begin{vmatrix} -2 & 3 \\ -5 & 1 \end{vmatrix}\right) \\
&= -2((0-8) - 6(-6+20) + 7(-4-0)) \\
&\quad +3((9-4) - 5(-6+20) + 7(-2+15)) \\
&= -2(-8-84-28) + 3(5-70+91) \\
&= -2(-120) + 3(26) = 318
\end{aligned}$$

It is apparent that evaluating the determinant of a 4×4 matrix is a fairly lengthy calculation and will get worse for larger matrices. In applications it is not uncommon to have a matrix with thousands (or even millions) of columns. Thus, in this section and the next section, we will examine some theorems which help us evaluate determinants more efficiently.

Theorem 5.1.1 The determinant of $A \in M_{n\times n}(\mathbb{R})$ may be obtained by a **cofactor expansion** along any row or any column. In particular, the expansion of the determinant along the i-th row of A is

$$\det A = a_{i1}C_{i1} + a_{i2}C_{i2} + \cdots + a_{in}C_{in}$$

The expansion of the determinant along the j-th column of A is

$$\det A = a_{1j}C_{1j} + a_{2j}C_{2j} + \cdots + a_{nj}C_{nj}$$

We omit a proof here since there is no conceptually helpful proof, and it would be a bit grim to verify the result in the general case.

Theorem 5.1.1 is a very practical result. It allows us to *choose* from A the row or column along which we are going to expand. If one row or column has many zeros, it is sensible to expand along it since we do not have to evaluate the cofactors of the zero entries. This was demonstrated in Example 5.1.5, where we had to compute only two cofactors in the first step.

EXAMPLE 5.1.6

Calculate the determinant of $A = \begin{bmatrix} 1 & 2 & -1 \\ 3 & 1 & 0 \\ -1 & 5 & 0 \end{bmatrix}$, $B = \begin{bmatrix} 3 & 2 & 0 & -1 \\ 0 & 6 & 0 & 0 \\ 4 & 1 & 2 & 1 \\ 3 & -1 & 0 & 1 \end{bmatrix}$, and

$$C = \begin{bmatrix} 4 & 2 & 1 & -1 \\ 0 & 2 & 2 & 2 \\ 0 & 0 & -1 & 3 \\ 0 & 0 & 0 & 4 \end{bmatrix}.$$

Solution: For A, we expand along the third column to get

$$\begin{aligned} \det A &= a_{13}C_{13} + a_{23}C_{23} + a_{33}C_{33} \\ &= (-1)(-1)^{1+3}\begin{vmatrix} 3 & 1 \\ -1 & 5 \end{vmatrix} + 0 + 0 \\ &= -1(15 - (-1)) \\ &= -16 \end{aligned}$$

For B, we expand along the second row to get

$$\begin{aligned} \det B &= b_{21}C_{21} + b_{22}C_{22} + b_{23}C_{23} + b_{24}C_{24} \\ &= 0 + 6(-1)^{2+2}\begin{vmatrix} 3 & 0 & -1 \\ 4 & 2 & 1 \\ 3 & 0 & 1 \end{vmatrix} + 0 + 0 \end{aligned}$$

We now expand the 3×3 determinant along the second column to get

$$\begin{aligned} \det B &= 0 + 6\left(2(-1)^{2+2}\begin{vmatrix} 3 & -1 \\ 3 & 1 \end{vmatrix}\right) + 0 \\ &= 6(2)(3 - (-3)) \\ &= 72 \end{aligned}$$

For C we continuously expand along the bottom row to get

$$\begin{aligned} \det C &= c_{41}C_{41} + c_{42}C_{42} + c_{43}C_{43} + c_{44}C_{44} \\ &= 0 + 0 + 0 + 4(-1)^{4+4}\begin{vmatrix} 4 & 2 & 1 \\ 0 & 2 & 2 \\ 0 & 0 & -1 \end{vmatrix} \\ &= 4\left(0 + 0 + (-1)(-1)^{3+3}\begin{vmatrix} 4 & 2 \\ 0 & 2 \end{vmatrix}\right) \\ &= 4(-1)(4(2) - 0) = 4(-1)(4)(2) \\ &= -32 \end{aligned}$$

EXERCISE 5.1.3

Calculate the determinant of $A = \begin{bmatrix} 1 & 3 & 2 & 0 \\ 0 & 0 & -1 & 2 \\ 3 & 5 & -1 & 0 \\ -2 & 2 & -4 & 0 \end{bmatrix}$ by

(a) Expanding along the first column
(b) Expanding along the second row
(c) Expanding along the fourth column

Exercise 5.1.3 demonstrates the usefulness of doing a cofactor expansion along the row or column with the most zeros. Of course, if one row or column contains only zeros, then it is really easy.

Theorem 5.1.2 If one row (or column) of $A \in M_{n\times n}(\mathbb{R})$ contains only zeros, then $\det A = 0$.

Proof: If the i-th row of A contains only zeros, then expanding the determinant along the i-th row of A gives

$$\det A = a_{i1}C_{i1} + a_{i2}C_{i2} + \cdots + a_{in}C_{in} = 0 + 0 + \cdots + 0 = 0$$ ■

As we saw with the matrix C in Example 5.1.6, another useful special case is when the matrix is upper or lower triangular.

Theorem 5.1.3 If $A \in M_{n\times n}(\mathbb{R})$ is an upper or lower triangular matrix, then the determinant of A is the product of the diagonal entries of A. That is,

$$\det A = a_{11}a_{22}\cdots a_{nn}$$

The proof is left as Problem C1.

Finally, recall that taking the transpose of a matrix A turns rows into columns and vice versa. That is, the columns of A^T are identical to the rows of A. Thus, if we expand the determinant of A^T along its first column, we will get the same cofactors and coefficients we would get by expanding the determinant of A along its first row. We get the following theorem.

Theorem 5.1.4 If $A \in M_{n\times n}(\mathbb{R})$, then $\det A = \det A^T$.

With the tools we have so far, evaluation of determinants is still a very tedious business. Properties of the determinant with respect to elementary row operations make the evaluation much easier. These properties are discussed in the next section.

PROBLEMS 5.1

Practice Problems

For Problems A1–A9, evaluate the determinant.

A1 $\begin{vmatrix} 2 & -4 \\ 7 & 5 \end{vmatrix}$ **A2** $\begin{vmatrix} -3 & 1 \\ 2 & 1 \end{vmatrix}$ **A3** $\begin{vmatrix} 1 & 1 \\ 1 & 1 \end{vmatrix}$

A4 $\begin{vmatrix} 2 & 4 \\ 3 & 6 \end{vmatrix}$ **A5** $\begin{vmatrix} 7 & 5 \\ -3 & 7 \end{vmatrix}$ **A6** $\begin{vmatrix} -2 & -1 \\ -5 & -11 \end{vmatrix}$

A7 $\begin{vmatrix} 1 & 3 & -4 \\ 0 & 0 & 2 \\ 0 & 0 & 3 \end{vmatrix}$ **A8** $\begin{vmatrix} 3 & 4 & 0 & 7 \\ 3 & 4 & 0 & 2 \\ 1 & 5 & 0 & 5 \\ 1 & 2 & 0 & 0 \end{vmatrix}$ **A9** $\begin{vmatrix} 5 & 0 & 0 & 0 \\ 3 & -4 & 0 & 0 \\ 1 & 2 & 1 & 0 \\ 1 & 2 & 0 & 1 \end{vmatrix}$

For Problems A10–A15, evaluate the determinant by expanding along the first row.

A10 $\begin{vmatrix} 3 & 4 & 0 \\ 2 & 1 & -1 \\ -4 & -1 & 2 \end{vmatrix}$ **A11** $\begin{vmatrix} 3 & 2 & 1 \\ -1 & 4 & 5 \\ 3 & 2 & 1 \end{vmatrix}$

A12 $\begin{vmatrix} 0 & 5 & 0 \\ 1 & 8 & -9 \\ 0 & \sqrt{2} & 1 \end{vmatrix}$ **A13** $\begin{vmatrix} 5 & 0 & 0 \\ 8 & 1 & -9 \\ \sqrt{2} & 0 & 1 \end{vmatrix}$

A14 $\begin{vmatrix} 2 & 1 & 0 & -1 \\ 0 & 3 & 2 & 1 \\ -4 & 0 & 2 & -2 \\ 3 & -5 & 2 & 1 \end{vmatrix}$ **A15** $\begin{vmatrix} 1 & 0 & 4 & 0 \\ 2 & -3 & 4 & 1 \\ -1 & 3 & 2 & 4 \\ 1 & 1 & -2 & 4 \end{vmatrix}$

For Problems A16 and A17, show that $\det A$ is equal to $\det A^T$ by expanding along the second column of A and the second row of A^T.

A16 $A = \begin{bmatrix} 1 & 3 & -1 \\ 2 & 1 & 0 \\ -1 & 0 & 5 \end{bmatrix}$ **A17** $A = \begin{bmatrix} 1 & 2 & 3 & 4 \\ -2 & 0 & 2 & 5 \\ 3 & 0 & 1 & 4 \\ 4 & 5 & 1 & -2 \end{bmatrix}$

For Problems A18–A27, evaluate the determinant by expanding along the row or column of your choice.

A18 $\begin{vmatrix} 3 & 5 & 0 \\ -2 & 6 & 0 \\ 4 & 1 & 0 \end{vmatrix}$ **A19** $\begin{vmatrix} -5 & 2 & -4 \\ 2 & -4 & 6 \\ -6 & 2 & -3 \end{vmatrix}$

A20 $\begin{vmatrix} 1 & -3 & 4 \\ 9 & 5 & 0 \\ 0 & -2 & 0 \end{vmatrix}$ **A21** $\begin{vmatrix} 1 & -3 & 4 \\ 0 & -2 & 0 \\ 9 & 5 & 0 \end{vmatrix}$

A22 $\begin{vmatrix} 2 & 1 & 5 \\ 4 & 3 & -1 \\ 0 & 1 & -2 \end{vmatrix}$ **A23** $\begin{vmatrix} -3 & 4 & 0 & 1 \\ 4 & -1 & 0 & -6 \\ 1 & -1 & 0 & -3 \\ 4 & -2 & 3 & 6 \end{vmatrix}$

A24 $\begin{vmatrix} 1 & 5 & -7 & 8 \\ 2 & -1 & 3 & 0 \\ -4 & 2 & 0 & 0 \\ 1 & 0 & 0 & 0 \end{vmatrix}$ **A25** $\begin{vmatrix} 8 & 0 & 2 & 1 \\ 0 & 0 & 2 & 0 \\ 3 & 1 & 1 & 1 \\ -1 & 2 & 1 & 0 \end{vmatrix}$

A26 $\begin{vmatrix} 0 & 6 & 1 & 2 \\ 0 & 5 & -1 & 1 \\ 3 & -5 & -3 & -5 \\ 5 & 6 & -3 & -6 \end{vmatrix}$ **A27** $\begin{vmatrix} 1 & 3 & 4 & -5 & 7 \\ 0 & -3 & 1 & 2 & 3 \\ 0 & 0 & 4 & 1 & 0 \\ 0 & 0 & 0 & -1 & 8 \\ 0 & 0 & 0 & 4 & 3 \end{vmatrix}$

For Problems A28–A31, calculate the determinant of the elementary matrix.

A28 $E_1 = \begin{bmatrix} 1 & 0 & 0 \\ 0 & 0 & 1 \\ 0 & 1 & 0 \end{bmatrix}$ **A29** $E_2 = \begin{bmatrix} 1 & 0 & 3 \\ 0 & 1 & 0 \\ 0 & 0 & 1 \end{bmatrix}$

A30 $E_3 = \begin{bmatrix} -3 & 0 & 0 \\ 0 & 1 & 0 \\ 0 & 0 & 1 \end{bmatrix}$ **A31** $E_4 = \begin{bmatrix} 1 & 0 & 0 \\ -2 & 1 & 0 \\ 0 & 0 & 1 \end{bmatrix}$

Homework Problems

For Problems B1–B9, evaluate the determinant.

B1 $\begin{vmatrix} 5 & 3 \\ -1 & 7 \end{vmatrix}$ **B2** $\begin{vmatrix} 2 & 1 \\ -1 & 1/2 \end{vmatrix}$ **B3** $\begin{vmatrix} 3 & -4 \\ -6 & 8 \end{vmatrix}$

B4 $\begin{vmatrix} 3 & 1 \\ 2 & 0 \end{vmatrix}$ **B5** $\begin{vmatrix} 5 & 7 \\ -2 & 4 \end{vmatrix}$ **B6** $\begin{vmatrix} 2 & 0 & 0 \\ \sqrt{2} & 5 & 0 \\ -1 & 2 & 3 \end{vmatrix}$

B7 $\begin{vmatrix} 1 & 2 & 3 \\ 0 & 1 & 0 \\ 0 & 0 & 5 \end{vmatrix}$ **B8** $\begin{vmatrix} 3 & 5 & 2 & 1 \\ 3 & 3 & 9 & 1 \\ 0 & 0 & 0 & 0 \\ 1 & 1 & 1 & 1 \end{vmatrix}$ **B9** $\begin{vmatrix} 1 & 0 & 0 & 0 \\ 2 & 3 & 0 & 0 \\ 4 & 1 & 5 & 0 \\ 3 & 1 & 2 & 1 \end{vmatrix}$

For Problems B10–B13, evaluate the determinant by expanding along the first row.

B10 $\begin{vmatrix} 2 & 0 & 1 \\ 3 & 6 & 1 \\ 4 & 1 & 7 \end{vmatrix}$ **B11** $\begin{vmatrix} 2 & 3 & 0 \\ 1 & -1 & 1 \\ 0 & 1 & 2 \end{vmatrix}$

B12 $\begin{vmatrix} 0 & 0 & 1 \\ 0 & 3 & 1 \\ 2 & 2 & 1 \end{vmatrix}$ **B13** $\begin{vmatrix} 3 & 0 & 1 & 0 \\ 1 & 1 & 0 & 2 \\ 0 & 1 & 0 & 2 \\ 1 & 0 & 1 & x \end{vmatrix}$

For Problems B14 and B15, show that $\det A$ is equal to $\det A^T$ by expanding along the third row of A and the third column of A^T.

B14 $A = \begin{bmatrix} 3 & 1 & 2 \\ 5 & 6 & 4 \\ -2 & 7 & 0 \end{bmatrix}$ **B15** $A = \begin{bmatrix} 1 & 8 & -1 & 1 \\ 4 & 2 & 2 & 2 \\ 0 & 1 & 0 & 1 \\ 1 & 8 & -1 & -1 \end{bmatrix}$

For Problems B16–B23, evaluate the determinant by expanding along the row or column of your choice.

B16 $\begin{vmatrix} 2 & 0 & 3 \\ 4 & 1 & -1 \\ 5 & 0 & 4 \end{vmatrix}$ **B17** $\begin{vmatrix} 0 & -1 & 2 \\ 3 & 0 & 4 \\ 6 & 9 & 0 \end{vmatrix}$

B18 $\begin{vmatrix} 2 & 2 & 3 \\ -1 & 4 & 1 \\ 1 & 6 & 4 \end{vmatrix}$ **B19** $\begin{vmatrix} 1 & 0 & -5 \\ 1/2 & 1 & 0 \\ 0 & 8 & 1/2 \end{vmatrix}$

B20 $\begin{vmatrix} 0 & 3 & \pi \\ 1 & 1 & 1 \\ 2 & 1 & 3 \end{vmatrix}$ **B21** $\begin{vmatrix} 4 & 1 & -2 & 5 \\ 0 & 0 & -2 & 0 \\ 3 & 6 & 9 & 0 \\ 2 & 1 & 7 & 1 \end{vmatrix}$

B22 $\begin{vmatrix} 6 & 2 & 0 & 3 \\ 0 & 1 & 4 & 1 \\ 2 & 1 & 0 & 3 \\ 0 & 1 & 1 & 1 \end{vmatrix}$ **B23** $\begin{vmatrix} 4 & 8 & 6 & 5 & 3 \\ 2 & 4 & 3 & 5 & 0 \\ 1 & 4 & 3 & 8 & 0 \\ 2 & 1 & 0 & 0 & 0 \\ 3 & 0 & 0 & 0 & 0 \end{vmatrix}$

For Problems B24–B27, calculate the determinant of the elementary matrix.

B24 $E_1 = \begin{bmatrix} 1 & 0 & 0 \\ 0 & 1 & 4 \\ 0 & 0 & 1 \end{bmatrix}$ **B25** $E_2 = \begin{bmatrix} 1 & 0 & 0 \\ 0 & 1 & 0 \\ 0 & 0 & -2 \end{bmatrix}$

B26 $E_3 = \begin{bmatrix} 1 & 5 & 0 \\ 0 & 1 & 0 \\ 0 & 0 & 1 \end{bmatrix}$ **B27** $E_4 = \begin{bmatrix} 0 & 1 & 0 \\ 1 & 0 & 0 \\ 0 & 0 & 1 \end{bmatrix}$

For Problems B28–B31, find all the cofactors of the matrix.

B28 $\begin{bmatrix} 2 & 1 \\ 3 & 4 \end{bmatrix}$ **B29** $\begin{bmatrix} 1 & 0 & 0 \\ 3 & 5 & 0 \\ 1 & 2 & 1 \end{bmatrix}$

B30 $\begin{bmatrix} 1 & 1 & 1 \\ 1 & 0 & -2 \\ 1 & -1 & 1 \end{bmatrix}$ **B31** $\begin{bmatrix} 1 & 2 & 4 \\ 3 & -1 & 1 \\ 1 & 1 & -7 \end{bmatrix}$

For Problems B32–B35, let $A = \begin{bmatrix} 2 & -1 \\ 3 & 5 \end{bmatrix}$ and $B = \begin{bmatrix} 1 & 2 \\ 1 & 1 \end{bmatrix}$. Compute all the expressions and compare.

B32 $\det(A + B)$, $\det A + \det B$

B33 $(\det A)(\det B)$, $\det(AB)$, $\det(BA)$

B34 $\det(3A)$, $3 \det A$

B35 $\det A$, $\det A^{-1}$, $\det A^T$

Conceptual Problems

C1 Prove Theorem 5.1.3.

C2 Prove that if $A \in M_{n\times n}(\mathbb{R})$ is a diagonal matrix, then

$$\det A = a_{11}a_{22}\cdots a_{nn}$$

For Problems C3–C5, find $A, B \in M_{2\times 2}(\mathbb{R})$ such that the statement holds, and find another pair $A, B \in M_{2\times 2}(\mathbb{R})$ such that the statement does not hold.

C3 $\det(A + B) = \det A + \det B$

C4 $\det(cA) = c \det A$

C5 $\det A^{-1} = \det A$

C6 Use induction to prove that if $A \in M_{n\times n}(\mathbb{R})$ with $n \geq 2$ that has two identical rows, then $\det A = 0$.

C7 Prove if $E_1 \in M_{n\times n}(\mathbb{R})$ with $n \geq 2$ is an elementary matrix corresponding to cR_i, $c \neq 0$, then $\det E_1 = c$.

C8 Prove if $E_2 \in M_{n\times n}(\mathbb{R})$ with $n \geq 2$ is an elementary matrix corresponding to $R_i + cR_j$, then $\det E_2 = 1$.

C9 Prove if $E_3 \in M_{n\times n}(\mathbb{R})$ with $n \geq 2$ is an elementary matrix corresponding to $R_i \updownarrow R_j$, then $\det E_3 = -1$.

C10 (a) Consider the points (a_1, a_2) and (b_1, b_2) in $\mathbb{R}^2$. Show that the equation $\det \begin{bmatrix} x_1 & x_2 & 1 \\ a_1 & a_2 & 1 \\ b_1 & b_2 & 1 \end{bmatrix} = 0$ is the equation of the line containing the two points.

(b) Write an equation for the plane in $\mathbb{R}^3$ that contains the non-collinear points (a_1, a_2, a_3), (b_1, b_2, b_3), and (c_1, c_2, c_3).

5.2 Properties of the Determinant

Calculating the determinant of a large matrix with few zeros can be very lengthy. Theorem 5.1.3 suggests that an effective strategy for evaluating a determinant of a matrix is to row reduce the matrix to upper triangular form. We now look at how applying elementary row operations changes the determinant. This will lead us to some important properties of the determinant.

Elementary Row Operations and the Determinant

To see what happens to the determinant of a matrix A when we multiply a row of A by a constant, we first consider a 3×3 example. Following the example, we state and prove the general result.

EXAMPLE 5.2.1

Let $A = \begin{bmatrix} 0 & 1 & 3 \\ 1 & 4 & -1 \\ 5 & 2 & 6 \end{bmatrix}$ and let B be the matrix obtained from A by multiplying the third row of A by 3. Show that $\det B = 3 \det A$.

Solution: We have $B = \begin{bmatrix} 0 & 1 & 3 \\ 1 & 4 & -1 \\ 15 & 6 & 18 \end{bmatrix}$. Expand the determinant of B along its third row. The cofactors for this row are

$$C_{31} = (-1)^{3+1} \begin{vmatrix} 1 & 3 \\ 4 & -1 \end{vmatrix}, \quad C_{32} = (-1)^{3+2} \begin{vmatrix} 0 & 3 \\ 1 & -1 \end{vmatrix}, \quad C_{33} = (-1)^{3+3} \begin{vmatrix} 0 & 1 \\ 1 & 4 \end{vmatrix}$$

Observe that these are also the cofactors for the third row of A. Hence,

$$\begin{aligned} \det B &= 15C_{31} + 6C_{32} + 18C_{33} \\ &= 3(5C_{31} + 2C_{32} + 6C_{33}) \\ &= 3 \det A \end{aligned}$$

Theorem 5.2.1 Let $A \in M_{n \times n}(\mathbb{R})$. If B is the matrix obtained from A by multiplying the i-th row of A by the real number r, then $\det B = r \det A$.

Proof: As in the example, we expand the determinant of B along the i-th row. Notice that the cofactors of the elements in this row are exactly the cofactors of the i-th row of A since all the other rows of B are identical to the corresponding rows in A. Therefore,

$$\det B = ra_{i1}C_{i1} + \cdots + ra_{in}C_{in} = r(a_{i1}C_{i1} + \cdots + a_{in}C_{in}) = r \det A$$

■

Remark

It is important to be careful when using this theorem. It is not uncommon to incorrectly use the reciprocal $(1/r)$ of the factor. One way to counter this error is to think of factoring out the value of r from a row of the matrix. Keep this in mind when reading the following example.

EXAMPLE 5.2.2

Given that $\det\begin{bmatrix}1 & 1 & -2\\ 1 & 2 & 1\\ 0 & 3 & 1\end{bmatrix} = -8$, find $\det\begin{bmatrix}-2 & -2 & 4\\ 1 & 2 & 1\\ 0 & 3 & 1\end{bmatrix}$.

Solution: By Theorem 5.2.1,

$$\det\begin{bmatrix}-2 & -2 & 4\\ 1 & 2 & 1\\ 0 & 3 & 1\end{bmatrix} = (-2)\det\begin{bmatrix}1 & 1 & -2\\ 1 & 2 & 1\\ 0 & 3 & 1\end{bmatrix} = (-2)(-8) = 16$$

EXERCISE 5.2.1

Let $A \in M_{3\times 3}(\mathbb{R})$ and let $r \in \mathbb{R}$. Use Theorem 5.2.1 to show that $\det(rA) = r^3 \det A$.

Next, we consider the effect of swapping two rows. For a general 2×2 matrix we get

$$\det\begin{bmatrix}c & d\\ a & b\end{bmatrix} = cb - da = -(ad - bc) = -\det\begin{bmatrix}a & b\\ c & d\end{bmatrix}$$

Hence, it seems that swapping two rows of a matrix multiplies the determinant by -1.

EXAMPLE 5.2.3

Let $A = \begin{bmatrix}0 & 1 & 3\\ 1 & 4 & -1\\ 5 & 2 & 6\end{bmatrix}$ and let B be the matrix obtained from A by swapping the first and third row. Show that $\det B = -\det A$.

Solution: Expand the determinant of $B = \begin{bmatrix}5 & 2 & 6\\ 1 & 4 & -1\\ 0 & 1 & 3\end{bmatrix}$ along the row that was not swapped. The cofactors for this row are

$$C_{21} = (-1)^{2+1}\begin{vmatrix}2 & 6\\ 1 & 3\end{vmatrix}, \quad C_{22} = (-1)^{2+2}\begin{vmatrix}5 & 6\\ 0 & 3\end{vmatrix}, \quad C_{23} = (-1)^{2+3}\begin{vmatrix}5 & 2\\ 0 & 1\end{vmatrix}$$

Observe that these are just the cofactors of the second row of A with their rows swapped. Since these cofactors are determinants of 2×2 matrices, we know from our work above that these will just be negatives of the cofactors of A. That is, if we let C^*_{2j} denote the cofactors of A, then we have

$$C_{2j} = -C^*_{2j}, \quad \text{for } j = 1, 2, 3$$

Hence,

$$\begin{aligned}\det B &= 1C_{21} + 4C_{22} + (-1)C_{23} && \text{by the cofactor expansion along the second row}\\ &= 1(-C^*_{21}) + 4(-C^*_{22}) + (-1)(-C^*_{23}) && \text{since } C_{2j} = -C^*_{2j}\\ &= -\left(1C^*_{21} + 4C^*_{22} + (-1)C^*_{23}\right)\\ &= -\det A\end{aligned}$$

as this is the cofactor expansion of A along its second row.

Indeed, this result holds in general.

Theorem 5.2.2 Let $A \in M_{n\times n}(\mathbb{R})$. If B is the matrix obtained from A by swapping two rows, then $\det B = -\det A$.

Example 5.2.3 shows how the proof of the theorem will work. In particular, we see that in the 3×3 case we needed to refer back to the 2×2 case. This indicates that a proof by induction would be appropriate. We leave this as Problem C1.

EXAMPLE 5.2.4

Given that $\det \begin{bmatrix} 1 & 1 & -2 \\ 1 & 2 & 1 \\ 0 & 3 & 1 \end{bmatrix} = -8$, find $\det \begin{bmatrix} 1 & 2 & 1 \\ 1 & 1 & -2 \\ 0 & 3 & 1 \end{bmatrix}$.

Solution: Since we can obtain $\begin{bmatrix} 1 & 2 & 1 \\ 1 & 1 & -2 \\ 0 & 3 & 1 \end{bmatrix}$ from the original matrix by swapping the first and second rows, Theorem 5.2.2 tells us that

$$\det \begin{bmatrix} 1 & 2 & 1 \\ 1 & 1 & -2 \\ 0 & 3 & 1 \end{bmatrix} = (-1) \det \begin{bmatrix} 1 & 1 & -2 \\ 1 & 2 & 1 \\ 0 & 3 & 1 \end{bmatrix} = (-1)(-8) = 8$$

Theorem 5.2.3 If two rows of $A \in M_{n\times n}(\mathbb{R})$ are equal, then $\det A = 0$.

Proof: Let B be the matrix obtained from A by interchanging the two equal rows. Obviously $B = A$, so $\det B = \det A$. But, by Theorem 5.2.2, $\det B = -\det A$, so $\det A = -\det A$. This implies that $\det A = 0$. ■

Finally, we show that the third type of elementary row operation is particularly useful as it does not change the determinant. We again begin by considering the 2×2 case.

Let $A = \begin{bmatrix} a & b \\ c & d \end{bmatrix}$ and let B be the matrix obtained from A by adding r times the first row to the second row. We get

$$\begin{aligned} \det B &= \begin{bmatrix} a & b \\ c + ra & d + rb \end{bmatrix} \\ &= a(d + rb) - b(c + ra) \\ &= ad + arb - bc - arb \\ &= ad - bc \\ &= \det A \end{aligned}$$

Hence, in the 2×2 case, adding a multiple of one row to another does not change the determinant.

EXAMPLE 5.2.5

Let $A = \begin{bmatrix} 0 & 1 & 3 \\ 1 & 4 & -1 \\ 5 & 2 & 6 \end{bmatrix}$ and let B be the matrix obtained from A by adding r times the first row to the second row.

Solution: Expand the determinant of $B = \begin{bmatrix} 0 & 1 & 3 \\ 1+0r & 4+r & -1+3r \\ 5 & 2 & 6 \end{bmatrix}$ along the third row . The cofactors for this row are

$$C_{31} = (-1)^{3+1} \begin{vmatrix} 1 & 3 \\ 4+r & -1+3r \end{vmatrix}$$

$$C_{32} = (-1)^{3+2} \begin{vmatrix} 0 & 3 \\ 1+0r & -1+3r \end{vmatrix}$$

$$C_{33} = (-1)^{3+3} \begin{vmatrix} 0 & 1 \\ 1+0r & 4+r \end{vmatrix}$$

Observe that these are the 2×2 cofactors C^*_{3j} of A with the operation $R_2 + rR_1$ applied to them. Hence, our work above shows us that

$$C_{3j} = C^*_{3j}, \qquad \text{for } j = 1, 2, 3$$

Hence,
$$\det B = 5C_{31} + 2C_{32} + 6C_{33} = 5C^*_{31} + 2C^*_{32} + 6C^*_{33} = \det A$$

Theorem 5.2.4 Let $A \in M_{n \times n}(\mathbb{R})$. If B is the matrix obtained from A by adding r times the i-th row of A to the k-th row, then $\det B = \det A$.

Once again the example outlines how one can do the proof. The proof is left as Problem C2.

EXAMPLE 5.2.6

Given that $\det \begin{bmatrix} 1 & 1 & -2 \\ 1 & 2 & 1 \\ 0 & 3 & 1 \end{bmatrix} = -8$, find $\det \begin{bmatrix} 1 & 1 & -2 \\ 1 & 2 & 1 \\ 2 & 7 & 3 \end{bmatrix}$.

Solution: Since we can obtain the new matrix from the original matrix by performing $R_3 + 2R_2$, Theorem 5.2.4 tells us that

$$\det \begin{bmatrix} 1 & 1 & -2 \\ 1 & 2 & 1 \\ 2 & 7 & 3 \end{bmatrix} = \det \begin{bmatrix} 1 & 1 & -2 \\ 1 & 2 & 1 \\ 0 & 3 & 1 \end{bmatrix} = -8$$

Theorems 5.2.1, 5.2.2, and 5.2.4 confirm that an effective strategy for evaluating the determinant of a matrix is to row reduce the matrix to upper triangular form keeping track of how the elementary row operations change the determinant. For $n > 3$, it can be shown that in general, this strategy will require fewer arithmetic operations than using only cofactor expansions. The following example illustrates this strategy.

EXAMPLE 5.2.7

Let $A = \begin{bmatrix} 1 & 3 & 1 & 5 \\ 1 & 3 & -3 & -3 \\ 0 & 3 & 1 & 0 \\ 1 & 6 & 2 & 11 \end{bmatrix}$. Find $\det A$.

Solution: By Theorem 5.2.4, performing the row operations $R_2 - R_1$ and $R_4 - R_1$ do not change the determinant, so

$$\det A = \begin{vmatrix} 1 & 3 & 1 & 5 \\ 0 & 0 & -4 & -8 \\ 0 & 3 & 1 & 0 \\ 0 & 3 & 1 & 6 \end{vmatrix}$$

By Theorem 5.2.2, performing $R_2 \updownarrow R_3$ gives

$$\det A = (-1)\begin{vmatrix} 1 & 3 & 1 & 5 \\ 0 & 3 & 1 & 0 \\ 0 & 0 & -4 & -8 \\ 0 & 3 & 1 & 6 \end{vmatrix}$$

By Theorem 5.2.1, performing $(-1/4)R_3$ gives

$$\det A = (-1)(-4)\begin{vmatrix} 1 & 3 & 1 & 5 \\ 0 & 3 & 1 & 0 \\ 0 & 0 & 1 & 2 \\ 0 & 3 & 1 & 6 \end{vmatrix}$$

Finally, by Theorem 5.2.4, performing the row operation $R_4 - R_2$ gives

$$\det A = 4\begin{vmatrix} 1 & 3 & 1 & 5 \\ 0 & 3 & 1 & 0 \\ 0 & 0 & 1 & 2 \\ 0 & 0 & 0 & 6 \end{vmatrix} = 4\,(1)(3)(1)(6) = 72$$

EXERCISE 5.2.2

Let $A = \begin{bmatrix} 2 & 4 & -2 & 6 \\ -6 & -6 & -2 & 5 \\ 1 & 1 & 3 & -1 \\ 4 & 6 & -2 & 5 \end{bmatrix}$. Find $\det A$.

In some cases, it may be appropriate to use some combination of row operations and cofactor expansions. We demonstrate this in the following example.

EXAMPLE 5.2.8

Find the determinant of $A = \begin{bmatrix} 1 & 5 & 6 & 7 \\ 1 & 8 & 7 & 9 \\ 1 & 5 & 6 & 10 \\ 0 & 1 & 4 & -2 \end{bmatrix}$.

Solution: By Theorem 5.2.4, performing $R_2 - R_1$ and $R_3 - R_1$ gives

$$\det A = \begin{vmatrix} 1 & 5 & 6 & 7 \\ 0 & 3 & 1 & 2 \\ 0 & 0 & 0 & 3 \\ 0 & 1 & 4 & -2 \end{vmatrix}$$

Expanding along the first column gives

$$\det A = (-1)^{1+1} \begin{vmatrix} 3 & 1 & 2 \\ 0 & 0 & 3 \\ 1 & 4 & -2 \end{vmatrix}$$

Expanding along the second row gives

$$\det A = 1(3)(-1)^{2+3} \begin{vmatrix} 3 & 1 \\ 1 & 4 \end{vmatrix} = (-3)(12 - 1) = -33$$

In the next example we show how the fact that $\det A = \det A^T$ can be used to give us even more options for simplifying a determinant.

EXAMPLE 5.2.9

Let $A = \begin{bmatrix} 1 & 2 & 3 \\ 0 & 1 & 1 \\ -2 & 3 & 1 \end{bmatrix}$. Find $\det A$.

Solution: Observe that the sum of the first and second columns equals the third column. If we could just add the first column to the second column, we would have two identical columns which should give us a determinant of 0 (as we know that a matrix with two identical rows has determinant 0).

We can simulate a column operation by taking the transpose and using a row operation. In particular, taking the transpose and then using $R_2 + R_1$ gives

$$\det A = \det A^T = \begin{vmatrix} 1 & 0 & -2 \\ 2 & 1 & 3 \\ 3 & 1 & 1 \end{vmatrix} = \begin{vmatrix} 1 & 0 & -2 \\ 3 & 1 & 1 \\ 3 & 1 & 1 \end{vmatrix} = 0$$

Rather that having to take the transpose of A, we just allow the use of column operations when simplifying a determinant. We get

1. Adding a multiple of one column to another does not change the determinant.
2. Swapping two columns multiplies the determinant by -1.
3. Multiplying a column by a non-zero scalar c multiplies the determinant by c.

EXAMPLE 5.2.10

Evaluate the determinant of $A = \begin{bmatrix} 1 & 3 & -1 & 1 \\ -3 & 2 & 1 & 2 \\ 2 & -1 & 1 & 1 \\ 2 & -3 & 2 & -3 \end{bmatrix}$.

Solution: Using the row operation $R_3 + R_1$ gives

$$\det A = \begin{vmatrix} 1 & 3 & -1 & 1 \\ -3 & 2 & 1 & 2 \\ 3 & 2 & 0 & 2 \\ 2 & -3 & 2 & -3 \end{vmatrix}$$

We now use the column operation $C_2 - C_4$ to get

$$\det A = \begin{vmatrix} 1 & 2 & -1 & 1 \\ -3 & 0 & 1 & 2 \\ 3 & 0 & 0 & 2 \\ 2 & 0 & 2 & -3 \end{vmatrix}$$

Using the cofactor expansion along the second column gives

$$\det A = 2(-1)^{1+2} \begin{vmatrix} -3 & 1 & 2 \\ 3 & 0 & 2 \\ 2 & 2 & -3 \end{vmatrix}$$

Applying $R_3 - 2R_1$ and then using the cofactor expansion along the second column we get

$$\det A = -2 \begin{vmatrix} -3 & 1 & 2 \\ 3 & 0 & 2 \\ 8 & 0 & -7 \end{vmatrix} = (-2)(1)(-1)^{1+2} \begin{vmatrix} 3 & 2 \\ 8 & -7 \end{vmatrix} = -74$$

EXERCISE 5.2.3

Find the determinant of $A = \begin{bmatrix} -6 & -2 & 4 & -5 \\ 3 & 2 & -4 & 3 \\ -6 & 4 & 0 & 0 \\ -3 & 2 & -3 & -4 \end{bmatrix}$.

CONNECTION

We will see in Section 5.4 that there are very simple geometric reasons why row and column operations affect the determinant the way that they do.

At this time, we will also stress that determinants play a very important role in Chapter 6 (and hence in Chapters 8 and 9, which both use the content in Chapter 6). Being able to simplify a determinant effectively will be of great assistance in those chapters.

Properties of Determinants

It follows from Theorem 5.2.1, Theorem 5.2.2, and Theorem 5.2.4 that there is a connection between the determinant of a square matrix, its rank, and whether it is invertible. Thus, we extend the Invertible Matrix Theorem from Section 3.5.

Theorem 5.2.5 Invertible Matrix Theorem continued

If $A \in M_{n\times n}(\mathbb{R})$, then the following are equivalent:

(2) $\text{rank}(A) = n$

(10) $\det A \neq 0$

Proof: Theorem 5.2.1, Theorem 5.2.2, and Theorem 5.2.4 indicate that applying an elementary row operation can only multiply the determinant of a matrix by $c \neq 0$, -1, or 1. Thus, $\det A \neq 0$ if and only if its reduced row echelon form has a non-zero determinant. But, since A is $n \times n$, the reduced row echelon form has no zero rows if and only if there is a leading one in every row. That is, $\det A \neq 0$ if and only if $\text{rank}(A) = n$. ■

We shall see how to use the determinant in calculating the inverse in the next section. It is worth noting that Theorem 5.2.5 implies that "almost all" square matrices are invertible; a square matrix fails to be invertible only if it satisfies the special condition $\det A = 0$.

Determinant of a Product

Often it is necessary to calculate the determinant of the product of two square matrices A and B. When you remember that each entry of AB is the dot product of a row from A and a column from B, and that the rule for calculating determinants is quite complicated, you might expect a very complicated rule. So, Theorem 5.2.7 should be a welcome surprise. The next result will help us prove Theorem 5.2.7.

Theorem 5.2.6 If $E \in M_{n\times n}(\mathbb{R})$ is an elementary matrix and $B \in M_{n\times n}(\mathbb{R})$, then

$$\det(EB) = (\det E)(\det B)$$

Proof: If E is an elementary matrix, then, since E is obtained by performing a single row operation on the identity matrix, we get by Theorem 5.2.1, Theorem 5.2.2, or Theorem 5.2.4 that there exists $c \in \mathbb{R}$ such that $\det E = c$ depending on the elementary row operation used. Moreover, since EB is the matrix obtained by performing that row operation on B, we get by Theorem 5.2.1, Theorem 5.2.2, or Theorem 5.2.4 that

$$\det(EB) = c \det B = \det E \det B$$

■

Theorem 5.2.7 If $A, B \in M_{n\times n}(\mathbb{R})$, then $\det(AB) = (\det A)(\det B)$.

Proof: If $\det A = 0$, then A is not invertible by the Invertible Matrix Theorem. Assume that $\det(AB) \neq 0$. Then, AB is invertible by the Invertible Matrix Theorem. Hence, there exists $C \in M_{n\times n}(\mathbb{R})$ such that

$$I = (AB)C = A(BC)$$

But, then Theorem 3.5.2 implies that A is invertible, which is a contradiction. Thus, we must have

$$\det(AB) = 0 = 0 \det B = \det A \det B$$

If $\det A \neq 0$, then A is invertible by the Invertible Matrix Theorem. Thus, by Theorem 3.6.3, there exists a sequence of elementary matrices $E_1, \dots, E_k$ such that

$$A = E_1^{-1} \cdots E_k^{-1}$$

Hence, by repeated use of Theorem 5.2.6, we get

$$\det(AB) = \det(E_1^{-1} \cdots E_k^{-1} B) = (\det E_1^{-1}) \cdots (\det E_k^{-1}) \det B = (\det A)(\det B)$$

■

EXAMPLE 5.2.11 Verify Theorem 5.2.7 for $A = \begin{bmatrix} 3 & 0 & 1 \\ 2 & -1 & 4 \\ 5 & 2 & 0 \end{bmatrix}$ and $B = \begin{bmatrix} -1 & 2 & 4 \\ 7 & 1 & 0 \\ 1 & -2 & 3 \end{bmatrix}$.

Solution: Using $R_2 - 4R_1$ and a cofactor expansion along the third column gives

$$\det A = \begin{vmatrix} 3 & 0 & 1 \\ -10 & -1 & 0 \\ 5 & 2 & 0 \end{vmatrix} = 1(-1)^{1+3} \begin{vmatrix} -10 & -1 \\ 5 & 2 \end{vmatrix} = -15$$

Using $R_2 + 7R_1$ and $R_3 + R_1$ gives

$$\det B = \begin{vmatrix} -1 & 2 & 4 \\ 0 & 15 & 28 \\ 0 & 0 & 7 \end{vmatrix} = -105$$

So, $(\det A)(\det B) = (-15)(-105) = 1575$. Similarly, using $-\frac{1}{5}R_2$, $R_1 + 2R_2, R_3 - 9R_2$, and a cofactor expansion along the first column gives

$$\det(AB) = \det \begin{bmatrix} -2 & 4 & 15 \\ -5 & -5 & 20 \\ 9 & 12 & 20 \end{bmatrix} = (-5) \begin{vmatrix} -2 & 4 & 15 \\ 1 & 1 & -4 \\ 9 & 12 & 20 \end{vmatrix}$$

$$= (-5) \begin{vmatrix} 0 & 6 & 7 \\ 1 & 1 & -4 \\ 0 & 3 & 56 \end{vmatrix} = (-5)(-1)^{2+1} \begin{vmatrix} 6 & 7 \\ 3 & 56 \end{vmatrix}$$

$$= (5)(315) = 1575$$

PROBLEMS 5.2

Practice Problems

For Problems A1–A6, use row operations and triangular form to compute the determinants of the matrix. Show your work clearly. Decide whether the matrix is invertible.

A1 $\begin{bmatrix} 1 & 2 & 4 \\ 3 & 1 & 0 \\ -1 & 3 & 2 \end{bmatrix}$ **A2** $\begin{bmatrix} 3 & 2 & 2 \\ 2 & 2 & 1 \\ 1 & 1 & 1 \end{bmatrix}$

A3 $\begin{bmatrix} 5 & 2 & -1 & 1 \\ 1 & 2 & -1 & 1 \\ 3 & 2 & 1 & 4 \\ -2 & 0 & 3 & 5 \end{bmatrix}$ **A4** $\begin{bmatrix} 1 & 1 & 3 & 1 \\ -2 & -2 & -4 & -1 \\ 2 & 2 & 8 & 3 \\ 1 & 1 & 7 & 3 \end{bmatrix}$

A5 $\begin{bmatrix} 5 & 10 & 5 & -5 \\ 1 & 3 & 5 & 7 \\ 1 & 2 & 6 & 3 \\ -1 & 7 & 1 & 1 \end{bmatrix}$ **A6** $\begin{bmatrix} 1 & 2 & 1 & 2 \\ 1 & 5 & 0 & 5 \\ 2 & 4 & 4 & 6 \\ 1 & -1 & -4 & -5 \end{bmatrix}$

For Problems A7–A20, evaluate the determinant.

A7 $\begin{vmatrix} 1 & 2 & 4 \\ 1 & 2 & 4 \\ 1 & 2 & 4 \end{vmatrix}$ **A8** $\begin{vmatrix} 2 & 3 & 5 \\ -1 & 1 & 0 \\ 7 & -6 & 1 \end{vmatrix}$

A9 $\begin{vmatrix} 3 & 3 & 3 \\ 1 & 2 & -2 \\ -1 & 5 & -7 \end{vmatrix}$ **A10** $\begin{vmatrix} 2 & 4 & 5 \\ 1 & 1 & 3 \\ 3 & 5 & 5 \end{vmatrix}$

A11 $\begin{vmatrix} 1 & -1 & 2 \\ 1 & 1 & -2 \\ 1 & 2 & 3 \end{vmatrix}$ **A12** $\begin{vmatrix} 2 & 4 & 2 \\ 4 & 2 & 1 \\ -2 & 2 & 2 \end{vmatrix}$

A13 $\begin{vmatrix} 1 & 2 & 1 & 2 \\ 2 & 4 & 1 & 5 \\ 3 & 6 & 5 & 9 \\ 1 & 3 & 4 & 3 \end{vmatrix}$ **A14** $\begin{vmatrix} 2 & 0 & -2 & -6 \\ 2 & -6 & -4 & -1 \\ -3 & -4 & 5 & 3 \\ -2 & -1 & -3 & 2 \end{vmatrix}$

A15 $\begin{vmatrix} 1 & 2 & 3 \\ -1 & 3 & -8 \\ 2 & -5 & 2 \end{vmatrix}$ **A16** $\begin{vmatrix} 2 & 3 & 1 \\ -2 & 2 & 0 \\ 1 & 3 & 4 \end{vmatrix}$

A17 $\begin{vmatrix} 6 & 8 & -8 \\ 7 & 5 & 8 \\ -2 & -4 & 8 \end{vmatrix}$ **A18** $\begin{vmatrix} 1 & 10 & 7 & -9 \\ 7 & -7 & 7 & 7 \\ 2 & -2 & 6 & 2 \\ -3 & -3 & 4 & 1 \end{vmatrix}$

A19 $\begin{vmatrix} -1 & 2 & 6 & 4 \\ 0 & 3 & 5 & 6 \\ 1 & -1 & 4 & -2 \\ 1 & 2 & 1 & 2 \end{vmatrix}$ **A20** $\begin{vmatrix} 1 & a & a^2 & a^3 \\ 1 & b & b^2 & b^3 \\ 1 & c & c^2 & c^3 \\ 1 & d & d^2 & d^3 \end{vmatrix}$

For Problems A21–A28, determine all values of p such that the matrix is invertible.

A21 $\begin{bmatrix} 1 & p \\ 2 & 4 \end{bmatrix}$ **A22** $\begin{bmatrix} p & -2 \\ 1 & p \end{bmatrix}$

A23 $\begin{bmatrix} 1 & 0 & 0 \\ p & p & 0 \\ -1 & 3 & 1 \end{bmatrix}$ **A24** $\begin{bmatrix} 3 & 2 & 2 \\ -8 & p & p \\ 5 & -3 & -1 \end{bmatrix}$

A25 $\begin{bmatrix} 2 & 3 & 1 \\ 1 & 1 & -1 \\ 4 & p & -2 \end{bmatrix}$ **A26** $\begin{bmatrix} 2 & 3 & 1 & p \\ 0 & 1 & 2 & 1 \\ 0 & 1 & 7 & 6 \\ 1 & 0 & 1 & 0 \end{bmatrix}$

A27 $\begin{bmatrix} 1 & 1 & 1 & 1 \\ 1 & 2 & 3 & 4 \\ 1 & 4 & 9 & 16 \\ 1 & 8 & 27 & p \end{bmatrix}$ **A28** $\begin{bmatrix} 2 & p & 1 & 1 \\ 2 & 2 & p & 2 \\ 0 & -2 & -p & 0 \\ 3 & 2 & p & 3 \end{bmatrix}$

For Problems A29 and A30, find $\det A$, $\det B$, and $\det(AB)$. Verify that Theorem 5.2.7 holds.

A29 $A = \begin{bmatrix} 2 & -1 \\ 3 & 5 \end{bmatrix}, B = \begin{bmatrix} 3 & 2 \\ -1 & 4 \end{bmatrix}$

A30 $A = \begin{bmatrix} 2 & 4 & 1 \\ -1 & 2 & -1 \\ 3 & 0 & 2 \end{bmatrix}, B = \begin{bmatrix} 1 & 3 & 2 \\ -1 & 0 & 5 \\ 4 & 1 & 1 \end{bmatrix}$

For Problems A31–A36, determine all values of λ such that the determinant is 0. [Hint: for the 3×3 determinants, use row and column operations to simplify the determinant so that you do not have to factor a cubic polynomial.]

A31 $\begin{vmatrix} 3-\lambda & 2 \\ 4 & 5-\lambda \end{vmatrix}$ **A32** $\begin{vmatrix} 4-\lambda & -3 \\ 3 & -4-\lambda \end{vmatrix}$

A33 $\begin{vmatrix} 1-\lambda & 1 & 1 \\ -1 & -1-\lambda & 1 \\ 1 & 1 & 1-\lambda \end{vmatrix}$ **A34** $\begin{vmatrix} 2-\lambda & 2 & 2 \\ 2 & 3-\lambda & 1 \\ 2 & 1 & 3-\lambda \end{vmatrix}$

A35 $\begin{vmatrix} -3-\lambda & 6 & -2 \\ -1 & 2-\lambda & -1 \\ 1 & -3 & -\lambda \end{vmatrix}$ **A36** $\begin{vmatrix} 4-\lambda & 2 & 2 \\ 2 & 4-\lambda & 2 \\ 2 & 2 & 4-\lambda \end{vmatrix}$

A37 Suppose that A is an $n \times n$ matrix and $r \in \mathbb{R}$. Determine $\det(rA)$.

A38 If A is invertible, prove that $\det A^{-1} = \dfrac{1}{\det A}$.

A39 If $A^3 = I$, prove that A is invertible.

Homework Problems

For Problems B1–B8, use row operations and triangular form to compute the determinants of the matrix. Show your work clearly. Decide whether the matrix is invertible.

B1 $\begin{bmatrix} 1 & 3 & 5 \\ 2 & 7 & 8 \\ 1 & 5 & 2 \end{bmatrix}$

B2 $\begin{bmatrix} 3 & -2 & 1 \\ -6 & 4 & 1 \\ 9 & -6 & -7 \end{bmatrix}$

B3 $\begin{bmatrix} 3 & -1 & 0 \\ -3 & 7 & 2 \\ 6 & 1 & 1 \end{bmatrix}$

B4 $\begin{bmatrix} -2 & -4 & -2 \\ 1 & 6 & 3 \\ 2 & 8 & 2 \end{bmatrix}$

B5 $\begin{bmatrix} 1 & 1 & 1 & 3 \\ 2 & 0 & 3 & 7 \\ 3 & 1 & 2 & 8 \\ 1 & -3 & 3 & 6 \end{bmatrix}$

B6 $\begin{bmatrix} 2 & 2 & 1 & 2 \\ -2 & -5 & 0 & -3 \\ 0 & -6 & 3 & -1 \\ 1 & 4 & 1 & 4 \end{bmatrix}$

B7 $\begin{bmatrix} 1 & -2 & -1 & -2 \\ -1 & 3 & 2 & 3 \\ 2 & -2 & 1 & 1 \\ 1 & 1 & 5 & 10 \end{bmatrix}$

B8 $\begin{bmatrix} 1 & 5 & -4 & 2 \\ 1 & -3 & 8 & 6 \\ 1 & -1 & 5 & 7 \\ 1 & 1 & 2 & 1 \end{bmatrix}$

For Problems B9–B20, evaluate the determinant.

B9 $\begin{vmatrix} 2 & -3 & 4 \\ 1 & 1 & 4 \\ 5 & 2 & 4 \end{vmatrix}$

B10 $\begin{vmatrix} 3 & 3 & 6 \\ 3 & 2 & 1 \\ -2 & 4 & 2 \end{vmatrix}$

B11 $\begin{vmatrix} 2 & 3 & -6 \\ 5 & 5 & -5 \\ -4 & -3 & 0 \end{vmatrix}$

B12 $\begin{vmatrix} 5 & 4 & 1 \\ -2 & 4 & -6 \\ 3 & -2 & 5 \end{vmatrix}$

B13 $\begin{vmatrix} 0 & 2 & 3 & 3 \\ 2 & 0 & 1 & 1 \\ 1 & 1 & 0 & 2 \\ 3 & 4 & 5 & 5 \end{vmatrix}$

B14 $\begin{vmatrix} 1 & 2 & 4 & -2 \\ 2 & 5 & 1 & -2 \\ 1 & 3 & -2 & 1 \\ -1 & 3 & 1 & 7 \end{vmatrix}$

B15 $\begin{vmatrix} 3 & 2 & 2 & 1 \\ 4 & 2 & 3 & 2 \\ 1 & 4 & -1 & 1 \\ 5 & 2 & 5 & 2 \end{vmatrix}$

B16 $\begin{vmatrix} 0 & 3 & 1 & 3 \\ 1 & 4 & 2 & 5 \\ 2 & 3 & -4 & -6 \\ 2 & 6 & -3 & -2 \end{vmatrix}$

B17 $\begin{vmatrix} 3 & 2 & 3 & 3 \\ 6 & 7 & 5 & 7 \\ -3 & 9 & 7 & 9 \\ -6 & 10 & 10 & 12 \end{vmatrix}$

B18 $\begin{vmatrix} 11 & 3 & 9 & 5 \\ 3 & -2 & 3 & 1 \\ 8 & -4 & 8 & 4 \\ 5 & 0 & 5 & 5 \end{vmatrix}$

B19 $\begin{vmatrix} 1 & 2 & 4 & 8 \\ 1 & 3 & 9 & 27 \\ 1 & -1 & 1 & -1 \\ 1 & -2 & 4 & -8 \end{vmatrix}$

B20 $\begin{vmatrix} 0 & 1 & 2 & 3 \\ 1 & 0 & 1 & 2 \\ 2 & 1 & 0 & 1 \\ 3 & 2 & 1 & 0 \end{vmatrix}$

For Problems B21–B28, determine all values of p such that the matrix is invertible.

B21 $\begin{bmatrix} p & 2 & 1 \\ p & 3 & 4 \\ 1 & 1 & -2 \end{bmatrix}$

B22 $\begin{bmatrix} 4 & 2 & 3 \\ p & 5 & 4 \\ 3 & 2 & 3 \end{bmatrix}$

B23 $\begin{bmatrix} 3 & 4 & p \\ 2 & 1 & -1 \\ 2 & 2 & 2 \end{bmatrix}$

B24 $\begin{bmatrix} 4 & 8 & 3 \\ 1 & 1 & 1 \\ 2p & 4p & 2p \end{bmatrix}$

B25 $\begin{bmatrix} 1 & 3 & 4 & 3 \\ 2 & 1 & -1 & p \\ 1 & 2 & 1 & 1 \\ 1 & 1 & 2 & 2 \end{bmatrix}$

B26 $\begin{bmatrix} 1 & 8 & 6 & 6 \\ p & p & 1 & 2 \\ 2 & 12 & 2 & 2 \\ 3 & 18 & 3 & 3 \end{bmatrix}$

B27 $\begin{bmatrix} p & 2 & 3 & 0 \\ 5 & -4 & -3 & 5 \\ 6 & -6 & p & 6 \\ -9 & 7 & 6 & -9 \end{bmatrix}$

B28 $\begin{bmatrix} 0 & 2 & 3 & 2 \\ 3 & 5 & 4 & 2 \\ 4 & p & 5 & 1 \\ 1 & 1 & 1 & 0 \end{bmatrix}$

For Problems B29–B32, find $\det A$, $\det B$, and $\det(AB)$. Verify that Theorem 5.2.7 holds.

B29 $A = \begin{bmatrix} 3 & 1 \\ -4 & 5 \end{bmatrix}, B = \begin{bmatrix} 1 & 0 \\ 2 & -3 \end{bmatrix}$

B30 $A = \begin{bmatrix} 2 & 3 \\ 7 & 5 \end{bmatrix}, B = \begin{bmatrix} 5 & -3 \\ -7 & 2 \end{bmatrix}$

B31 $A = \begin{bmatrix} 2 & 1 & 3 \\ 3 & -1 & 2 \\ 2 & 4 & 6 \end{bmatrix}, B = \begin{bmatrix} 1 & 3 & -1 \\ 2 & 1 & 2 \\ 1 & 1 & 1 \end{bmatrix}$

B32 $A = \begin{bmatrix} 4 & 3 & 5 \\ 2 & 1 & 2 \\ 1 & 3 & 4 \end{bmatrix}, B = \begin{bmatrix} 2 & -3 & -1 \\ 6 & -11 & -2 \\ -5 & 9 & 2 \end{bmatrix}$

For Problems B33–B36, determine all values of λ such that the determinant is 0. [Hint: for the 3×3 determinants, use row and column operations to simplify the determinant so that you do not have to factor a cubic polynomial.]

B33 $\begin{vmatrix} 6-\lambda & 1 \\ 5 & 2-\lambda \end{vmatrix}$

B34 $\begin{vmatrix} 1-\lambda & 1 \\ 1 & 1-\lambda \end{vmatrix}$

B35 $\begin{vmatrix} 3-\lambda & -1 & 3 \\ -1 & 9-\lambda & -3 \\ 3 & -3 & 11-\lambda \end{vmatrix}$

B36 $\begin{vmatrix} 1-\lambda & 2 & -2 \\ 2 & 1-\lambda & 2 \\ -2 & 2 & -3-\lambda \end{vmatrix}$

Conceptual Problems

C1 Let $A \in M_{n\times n}(\mathbb{R})$. Prove that if B is the matrix obtained from A by swapping two rows, then

$$\det B = -\det A$$

C2 Let $A \in M_{n\times n}(\mathbb{R})$. Prove that if B is the matrix obtained from A by adding r times the j-th row of A to the k-th row, then

$$\det B = \det A$$

C3 A square matrix A is a **skew-symmetric** if $A^T = -A$. If $A \in M_{n\times n}(\mathbb{R})$ is skew-symmetric matrix, with n odd, prove that $\det A = 0$.

C4 A matrix A is called **orthogonal** if $A^T = A^{-1}$.
(a) If A is orthogonal, prove that $\det A = \pm 1$.
(b) Give an example of an orthogonal matrix for which $\det A = -1$.

C5 Two matrices $A, B \in M_{n\times n}(\mathbb{R})$ are said to be **similar** if there exists an invertible matrix P such that $P^{-1}AP = B$. Prove that if A and B are similar, then

$$\det A = \det B$$

C6 Assume that $A, B \in M_{n\times n}(\mathbb{R})$ are invertible. Prove that $\det A = \det B$ if and only if $A = UB$, where U is a matrix with $\det U = 1$.

C7 Use determinants to prove that if $A, B \in M_{n\times n}(R)$ such that $AB = -BA$, A is invertible and n is odd, then B is not invertible.

For Problems C8–C13, let $A, B \in M_{n\times n}(\mathbb{R})$. Determine whether the statement is true or false. Justify your answer.

C8 If the columns of A are linearly independent, then $\det A \neq 0$.

C9 $\det(A + B) = \det A + \det B$.

C10 $\det(A + B^T) = \det(A^T + B)$.

C11 The system of equations $A\vec{x} = \vec{b}$ is consistent only if $\det A \neq 0$.

C12 If $\det(AB) = 0$, then $\det A = 0$ or $\det B = 0$.

C13 If $A^2 = I$, then $\det A = 1$.

C14 (a) Prove that $\det \begin{bmatrix} a+p & b+q & c+r \\ d & e & f \\ g & h & k \end{bmatrix}$
$= \det \begin{bmatrix} a & b & c \\ d & e & f \\ g & h & k \end{bmatrix} + \det \begin{bmatrix} p & q & r \\ d & e & f \\ g & h & k \end{bmatrix}$.

(b) Use part (a) to express $\det \begin{bmatrix} a+p & b+q & c+r \\ d+x & e+y & f+z \\ g & h & k \end{bmatrix}$ as the sum of determinants of matrices whose entries are not sums.

C15 Prove that

$$\det \begin{bmatrix} a+b & p+q & u+v \\ b+c & q+r & v+w \\ c+a & r+p & w+u \end{bmatrix} = 2\det \begin{bmatrix} a & p & u \\ b & q & v \\ c & r & w \end{bmatrix}.$$

C16 Prove that $\det \begin{bmatrix} 1 & 1 & 1 \\ 1 & 1+a & 1+2a \\ 1 & (1+a)^2 & (1+2a)^2 \end{bmatrix} = 2a^3$.

5.3 Inverse by Cofactors, Cramer's Rule

The Invertible Matrix Theorem shows us that there is a close connection between whether a square matrix A is invertible, the number of solutions of the system of linear equations $A\vec{x} = \vec{b}$, and the determinant of A. We now examine this relationship a little further by looking at how to use determinants to find the inverse of a matrix and to solve systems of linear equations.

Inverse by Cofactors

Observe that we can write the inverse of a matrix $A = \begin{bmatrix} a_{11} & a_{12} \\ a_{21} & a_{22} \end{bmatrix}$ in the form

$$A^{-1} = \frac{1}{\det A}\begin{bmatrix} a_{22} & -a_{12} \\ -a_{21} & a_{11} \end{bmatrix} = \frac{1}{\det A}\begin{bmatrix} C_{11} & C_{21} \\ C_{12} & C_{22} \end{bmatrix}$$

That is, the entries of A^{-1} are scalar multiples of the cofactors of A. In particular,

$$(A^{-1})_{ij} = \frac{1}{\det A} C_{ji} \tag{5.2}$$

Take careful notice of the change of order in the subscripts in the line above.
It is instructive to verify this result by multiplying out A and A^{-1}.

$$\begin{aligned} AA^{-1} &= \begin{bmatrix} a_{11} & a_{12} \\ a_{21} & a_{22} \end{bmatrix}\left(\frac{1}{\det A}\begin{bmatrix} C_{11} & C_{21} \\ C_{12} & C_{22} \end{bmatrix}\right) \\ &= \frac{1}{\det A}\begin{bmatrix} a_{11}C_{11} + a_{12}C_{12} & a_{11}C_{21} + a_{12}C_{22} \\ a_{21}C_{11} + a_{22}C_{12} & a_{21}C_{21} + a_{22}C_{22} \end{bmatrix} \\ &= \frac{1}{\det A}\begin{bmatrix} \det A & a_{11}(-a_{12}) + a_{12}a_{11} \\ a_{21}a_{22} + a_{22}(-a_{21}) & \det A \end{bmatrix} \\ &= \begin{bmatrix} 1 & 0 \\ 0 & 1 \end{bmatrix} \end{aligned}$$

Our goal now is to prove that equation (5.2) holds in the $n \times n$ case. We begin with a useful theorem.

Theorem 5.3.1 False Expansion Theorem
If $A \in M_{n\times n}(\mathbb{R})$ and $i \neq k$, then

$$a_{i1}C_{k1} + \cdots + a_{in}C_{kn} = 0$$

Proof: Let B be the matrix obtained from A by replacing (not swapping) the k-th row of A by the i-th row of A. Then the i-th row of B is identical to the k-th row of B, hence $\det B = 0$ by Theorem 5.2.3. Since the cofactors C^*_{kj} of B are equal to the cofactors C_{kj} of A, and the coefficients b_{kj} of the k-th row of B are equal to the coefficients a_{ij} of the i-th row of A, we get

$$0 = \det(B) = b_{k1}C^*_{k1} + \cdots + b_{kn}C^*_{kn} = a_{i1}C_{k1} + \cdots + a_{in}C_{kn}$$

as required. ■

The False Expansion Theorem says that if we try to do a cofactor expansion of the determinant, but use the coefficients from one row and the cofactors from another row, then we will always get 0. Of course, this also applies to a false expansion along any column of the matrix. That is, if $j \neq k$, then

$$a_{1j}C_{1k} + \cdots + a_{nj}C_{nk} = 0 \tag{5.3}$$

EXAMPLE 5.3.1

Find a non-zero vector $\vec{n} \in \mathbb{R}^4$ that is orthogonal to $\vec{v}_1 = \begin{bmatrix}1\\1\\1\\1\end{bmatrix}$, $\vec{v}_2 = \begin{bmatrix}2\\1\\0\\1\end{bmatrix}$, and $\vec{v}_3 = \begin{bmatrix}-1\\3\\1\\2\end{bmatrix}$.

Solution: Rather than setting up a system of linear equations, we will solve this by using the False Expansion Theorem. Consider the matrix

$$A = \begin{bmatrix}\vec{0} & \vec{v}_1 & \vec{v}_2 & \vec{v}_3\end{bmatrix} = \begin{bmatrix}0 & 1 & 2 & -1\\0 & 1 & 1 & 3\\0 & 1 & 0 & 1\\0 & 1 & 1 & 2\end{bmatrix}$$

Let $\vec{n} = \begin{bmatrix}C_{11}\\C_{21}\\C_{31}\\C_{41}\end{bmatrix}$. Using equation (5.3) with $j = 2$ and $k = 1$ we get

$$0 = 1C_{11} + 1C_{21} + 1C_{31} + 1C_{41} = \vec{v}_1 \cdot \vec{n}$$

Using equation (5.3) with $j = 3$ and $k = 1$ we get

$$0 = 2C_{11} + 1C_{21} + 0C_{31} + 1C_{41} = \vec{v}_2 \cdot \vec{n}$$

Using equation (5.3) with $j = 4$ and $k = 1$ we get

$$0 = -1C_{11} + 3C_{21} + 1C_{31} + 2C_{41} = \vec{v}_3 \cdot \vec{n}$$

That is, $\vec{n}$ is orthogonal to $\vec{v}_1$, $\vec{v}_2$, and $\vec{v}_3$. Calculating the cofactors, we find that

$$\vec{n} = \begin{bmatrix}1\\4\\1\\-6\end{bmatrix}$$

This can be generalized to find a vector $\vec{x} \in \mathbb{R}^n$ that is orthogonal to $n-1$ vectors $\vec{v}_1, \dots, \vec{v}_{n-1} \in \mathbb{R}^n$. See Problem C1. The case where $n = 3$ is the cross product.

Theorem 5.3.2 If $A \in M_{n\times n}(\mathbb{R})$ is invertible, then $(A^{-1})_{ij} = \frac{1}{\det A} C_{ji}$.

Proof: Let B be the $n \times n$ matrix such that

$$b_{ij} = C_{ji} \quad \text{(the } (j, i)\text{-cofactor of } A\text{)}$$

Observe that the cofactors of the i-th row of A form the i-th column of B. Thus, the dot product of the i-th row of A and the i-th column of B is

$$a_{i1}C_{i1} + \cdots + a_{in}C_{in} = \det A$$

For $i \neq j$, the dot product of the i-th row of A and the j-th column of B is

$$a_{i1}C_{j1} + \cdots + a_{in}C_{jn} = 0$$

by the False Expansion Theorem. Hence, if $\vec{a}_i^T$ represents the i-th row of A and $\vec{b}_j$ represents the j-th column of B, then

$$\begin{aligned} AB &= \begin{bmatrix} \vec{a}_1^T \\ \vdots \\ \vec{a}_n^T \end{bmatrix} \begin{bmatrix} \vec{b}_1 & \cdots & \vec{b}_n \end{bmatrix} \\ &= \begin{bmatrix} \vec{a}_1 \cdot \vec{b}_1 & \cdots & \vec{a}_1 \cdot \vec{b}_n \\ \vdots & \ddots & \vdots \\ \vec{a}_n \cdot \vec{b}_1 & \cdots & \vec{a}_n \cdot \vec{b}_n \end{bmatrix} \\ &= \begin{bmatrix} \det A & 0 & \cdots & 0 \\ 0 & \det A & \ddots & \vdots \\ \vdots & \ddots & \ddots & 0 \\ 0 & \cdots & 0 & \det A \end{bmatrix} = (\det A)I \end{aligned}$$

Thus, by Theorem 3.5.2, $A^{-1} = \frac{1}{\det A} B$ as required. ■

Because of this result, we make the following definition.

Definition
Adjugate

Let $A \in M_{n\times n}(\mathbb{R})$. The **adjugate** of A is the matrix $\text{adj}(A)$ defined by

$$(\text{adj}(A))_{ij} = C_{ji}$$

From the proof of Theorem 5.3.2, we have that

$$A\,\text{adj}(A) = (\det A)I$$

Or, if A is invertible, then

$$A^{-1} = \frac{1}{\det A}\,\text{adj}(A)$$

EXAMPLE 5.3.2

Determine the adjugate of $A = \begin{bmatrix} 2 & 4 & -1 \\ 0 & 3 & 1 \\ 6 & -2 & 5 \end{bmatrix}$ and verify that $A^{-1} = \dfrac{1}{\det A}\operatorname{adj}(A)$.

Solution: The nine cofactors of A are

$$C_{11} = (1)\begin{vmatrix} 3 & 1 \\ -2 & 5 \end{vmatrix} = 17 \qquad C_{12} = (-1)\begin{vmatrix} 0 & 1 \\ 6 & 5 \end{vmatrix} = 6 \qquad C_{13} = (1)\begin{vmatrix} 0 & 3 \\ 6 & -2 \end{vmatrix} = -18$$

$$C_{21} = (-1)\begin{vmatrix} 4 & -1 \\ -2 & 5 \end{vmatrix} = -18 \qquad C_{22} = (1)\begin{vmatrix} 2 & -1 \\ 6 & 5 \end{vmatrix} = 16 \qquad C_{23} = (-1)\begin{vmatrix} 2 & 4 \\ 6 & -2 \end{vmatrix} = 28$$

$$C_{31} = (1)\begin{vmatrix} 4 & -1 \\ 3 & 1 \end{vmatrix} = 7 \qquad C_{32} = (-1)\begin{vmatrix} 2 & -1 \\ 0 & 1 \end{vmatrix} = -2 \qquad C_{33} = (1)\begin{vmatrix} 2 & 4 \\ 0 & 3 \end{vmatrix} = 6$$

Hence,

$$\operatorname{adj}(A) = \begin{bmatrix} C_{11} & C_{21} & C_{31} \\ C_{12} & C_{22} & C_{32} \\ C_{13} & C_{23} & C_{33} \end{bmatrix} = \begin{bmatrix} 17 & -18 & 7 \\ 6 & 16 & -2 \\ -18 & 28 & 6 \end{bmatrix}$$

Multiplying we find

$$A\operatorname{adj}(A) = \begin{bmatrix} 2 & 4 & -1 \\ 0 & 3 & 1 \\ 6 & -2 & 5 \end{bmatrix}\begin{bmatrix} 17 & -18 & 7 \\ 6 & 16 & -2 \\ -18 & 28 & 6 \end{bmatrix} = \begin{bmatrix} 76 & 0 & 0 \\ 0 & 76 & 0 \\ 0 & 0 & 76 \end{bmatrix}$$

Hence, $\det A = 76$ and

$$A^{-1} = \frac{1}{\det A}\operatorname{adj}(A) = \frac{1}{76}\begin{bmatrix} 17 & -18 & 7 \\ 6 & 16 & -2 \\ -18 & 28 & 6 \end{bmatrix}$$

EXERCISE 5.3.1

Use $\operatorname{adj}(A)$ to find the inverse of $A = \begin{bmatrix} 1 & 2 & 0 \\ -1 & 0 & 1 \\ 0 & 3 & 1 \end{bmatrix}$.

For 3×3 matrices, finding $\operatorname{adj}(A)$ requires the evaluation of nine 2×2 determinants. This is manageable, but it is more work than would be required by row reducing $\left[\begin{array}{c|c} A & I \end{array}\right]$ to reduced row echelon form. Finding the inverse of a 4×4 matrix by finding $\operatorname{adj}(A)$ would require the evaluation of sixteen 3×3 determinants; this method becomes extremely unattractive. However, it can be useful in some theoretic applications as it gives a formula for the entries of the inverse.

Cramer's Rule

Consider the system of n linear equations in n variables, $A\vec{x} = \vec{b}$. If $\det A \neq 0$ so that A is invertible, then the solution may be written in the form

$$\vec{x} = A^{-1}\vec{b} = \left(\frac{1}{\det A}\operatorname{adj}(A)\right)\vec{b}$$

$$\begin{bmatrix} x_1 \\ \vdots \\ x_i \\ \vdots \\ x_n \end{bmatrix} = \frac{1}{\det A}\begin{bmatrix} C_{11} & C_{21} & \cdots & C_{n1} \\ \vdots & \vdots & \vdots & \vdots \\ C_{1i} & C_{2i} & \cdots & C_{ni} \\ \vdots & \vdots & \ddots & \vdots \\ C_{1n} & C_{2n} & \cdots & C_{nn} \end{bmatrix}\begin{bmatrix} b_1 \\ \vdots \\ b_i \\ \vdots \\ b_n \end{bmatrix}$$

By our first definition of matrix-vector multiplication, the value of the i-th component of $\vec{x}$ is

$$x_i = \frac{1}{\det A}(b_1C_{1i} + b_2C_{2i} + \cdots + b_nC_{ni})$$

Now, let N_i be the matrix obtained from A by replacing the i-th column of A by $\vec{b}$. Then the cofactors of the i-th column of N_i will equal the cofactors of the i-th column of A, and hence we get

$$\det N_i = b_1C_{1i} + b_2C_{2i} + \cdots + b_nC_{ni}$$

Therefore, the i-th component of $\vec{x}$ in the solution of $A\vec{x} = \vec{b}$ is

$$x_i = \frac{\det N_i}{\det A}$$

This is called Cramer's Rule (or Method). We now demonstrate Cramer's Rule with a couple of examples.

EXAMPLE 5.3.3

Use Cramer's Rule to solve the system of equations.

$$\begin{aligned} x_1 + x_2 - x_3 &= b_1 \\ 2x_1 + 4x_2 + 5x_3 &= b_2 \\ x_1 + x_2 + 2x_3 &= b_3 \end{aligned}$$

Solution: The coefficient matrix is $A = \begin{bmatrix} 1 & 1 & -1 \\ 2 & 4 & 5 \\ 1 & 1 & 2 \end{bmatrix}$, so

$$\det A = \begin{vmatrix} 1 & 1 & -1 \\ 2 & 4 & 5 \\ 1 & 1 & 2 \end{vmatrix} = \begin{vmatrix} 1 & 1 & -1 \\ 0 & 2 & 7 \\ 0 & 0 & 3 \end{vmatrix} = (1)(2)(3) = 6$$

EXAMPLE 5.3.3
(continued)

Hence,

$$x_1 = \frac{\det N_1}{\det A} = \frac{1}{6}\begin{vmatrix} b_1 & 1 & -1 \\ b_2 & 4 & 5 \\ b_3 & 1 & 2 \end{vmatrix} = \frac{1}{6}\begin{vmatrix} b_1 & 1 & -1 \\ b_2 + 5b_1 & 9 & 0 \\ b_3 + 2b_1 & 3 & 0 \end{vmatrix} = \frac{3b_1 - 3b_2 + 9b_3}{6}$$

$$x_2 = \frac{\det N_2}{\det A} = \frac{1}{6}\begin{vmatrix} 1 & b_1 & -1 \\ 2 & b_2 & 5 \\ 1 & b_3 & 2 \end{vmatrix} = \frac{1}{6}\begin{vmatrix} 1 & b_1 & -1 \\ 7 & b_2 + 5b_1 & 0 \\ 3 & b_3 + 2b_1 & 0 \end{vmatrix} = \frac{b_1 + 3b_2 - 7b_3}{6}$$

$$x_3 = \frac{\det N_3}{\det A} = \frac{1}{6}\begin{vmatrix} 1 & 1 & b_1 \\ 2 & 4 & b_2 \\ 1 & 1 & b_3 \end{vmatrix} = \frac{1}{6}\begin{vmatrix} 1 & 1 & b_1 \\ 0 & 2 & b_2 - 2b_1 \\ 0 & 0 & b_3 - b_1 \end{vmatrix} = \frac{-2b_1 + 2b_3}{6}$$

To solve a system of n equations in n variables by using Cramer's Rule would require the evaluation of the determinant of $n+1$ matrices where each matrix is $n \times n$. Thus, solving a system by using Cramer's Rule requires far more calculation than elimination. However, Cramer's Rule is sometimes used to write a formula for the solution of a problem. This is used, for example, in electrical engineering, control engineering, and economics. It is particularly useful when the system contains variables.

EXAMPLE 5.3.4

Let $A = \begin{bmatrix} a & 1 & 2 \\ 0 & b & -1 \\ c & 1 & d \end{bmatrix}$. Assuming that $\det A \neq 0$, solve $A\vec{x} = \begin{bmatrix} 1 \\ -1 \\ 1 \end{bmatrix}$.

Solution: We have

$$\det A = a(bd+1) + c(-1-2b) = abd + a - c - 2cb$$

Thus,

$$x_1 = \frac{1}{\det A}\begin{vmatrix} 1 & 1 & 2 \\ -1 & b & -1 \\ 1 & 1 & d \end{vmatrix} = \frac{1}{\det A}\begin{vmatrix} 1 & 1 & 2 \\ 0 & b+1 & 1 \\ 0 & 0 & d-2 \end{vmatrix} = \frac{(b+1)(d-2)}{abd + a - c - 2cb}$$

$$x_2 = \frac{1}{\det A}\begin{vmatrix} a & 1 & 2 \\ 0 & -1 & -1 \\ c & 1 & d \end{vmatrix} = \frac{1}{\det A}\begin{vmatrix} a & 0 & 1 \\ 0 & -1 & -1 \\ c & 0 & d-1 \end{vmatrix} = \frac{-ad + a + c}{abd + a - c - 2cb}$$

$$x_3 = \frac{1}{\det A}\begin{vmatrix} a & 1 & 1 \\ 0 & b & -1 \\ c & 1 & 1 \end{vmatrix} = \frac{1}{\det A}\begin{vmatrix} a & 0 & 1 \\ 0 & b+1 & -1 \\ c & 0 & 1 \end{vmatrix} = \frac{(b+1)(a-c)}{abd + a - c - 2cb}$$

That is, the solution is $\vec{x} = \dfrac{1}{abd + a - c - 2cb}\begin{bmatrix} (b+1)(d-2) \\ -ad + a + c \\ (b+1)(a-c) \end{bmatrix}$.

PROBLEMS 5.3

Practice Problems

For Problems A1–A4, find a non-zero vector $\vec{x} \in \mathbb{R}^4$ that is orthogonal to the given vectors.

A1 $\begin{bmatrix} 2 \\ 2 \\ 1 \\ -1 \end{bmatrix}, \begin{bmatrix} 1 \\ 2 \\ 2 \\ 2 \end{bmatrix}, \begin{bmatrix} 2 \\ 1 \\ -1 \\ -1 \end{bmatrix}$

A2 $\begin{bmatrix} 2 \\ 2 \\ 2 \\ 1 \end{bmatrix}, \begin{bmatrix} 2 \\ 1 \\ 1 \\ 2 \end{bmatrix}, \begin{bmatrix} 2 \\ 2 \\ -1 \\ -1 \end{bmatrix}$

A3 $\begin{bmatrix} 1 \\ 0 \\ -1 \\ 1 \end{bmatrix}, \begin{bmatrix} 1 \\ 2 \\ 0 \\ 0 \end{bmatrix}, \begin{bmatrix} 2 \\ 4 \\ 1 \\ 1 \end{bmatrix}$

A4 $\begin{bmatrix} 3 \\ 1 \\ 2 \\ -3 \end{bmatrix}, \begin{bmatrix} 4 \\ 1 \\ 2 \\ 1 \end{bmatrix}, \begin{bmatrix} -4 \\ -1 \\ -3 \\ 2 \end{bmatrix}$

For Problems A5–A10:

(a) Determine adj(A).

(b) Calculate $A\,\text{adj}(A)$. Determine $\det A$, and find A^{-1}.

A5 $A = \begin{bmatrix} 1 & 3 & 0 \\ 2 & 1 & -1 \\ 1 & 1 & 0 \end{bmatrix}$

A6 $A = \begin{bmatrix} 1 & 2 & 3 \\ 0 & 1 & -5 \\ 0 & 0 & 7 \end{bmatrix}$

A7 $A = \begin{bmatrix} a & 0 & d \\ 0 & c & 0 \\ b & 0 & e \end{bmatrix}$

A8 $A = \begin{bmatrix} a & 2 & 3 \\ 0 & 1 & -5 \\ 0 & 0 & b \end{bmatrix}$

A9 $A = \begin{bmatrix} 2 & -3 & t \\ -2 & 0 & 1 \\ 3 & 1 & 1 \end{bmatrix}$

A10 $A = \begin{bmatrix} 3 & 2 & t \\ 1 & t & 1 \\ 1 & 1 & 0 \end{bmatrix}$

For Problems A11–A17, determine the inverse of the matrix by finding adj(A). Verify your answer by using multiplication.

A11 $\begin{bmatrix} 1 & 3 \\ 4 & 10 \end{bmatrix}$

A12 $\begin{bmatrix} 3 & -5 \\ 2 & -1 \end{bmatrix}$

A13 $\begin{bmatrix} 4 & 1 & 7 \\ 2 & -3 & 1 \\ -2 & 6 & 0 \end{bmatrix}$

A14 $\begin{bmatrix} 4 & 0 & -4 \\ 0 & -1 & 1 \\ -2 & 2 & -1 \end{bmatrix}$

A15 $\begin{bmatrix} 2 & 1 & 1 \\ 1 & 2 & 1 \\ 4 & 1 & 2 \end{bmatrix}$

A16 $\begin{bmatrix} 4 & 0 & -2 \\ 0 & -1 & 2 \\ -4 & 1 & -1 \end{bmatrix}$

A17 $\begin{bmatrix} 1 & 5 & 3 \\ 3 & 1 & 1 \\ -6 & -2 & 2 \end{bmatrix}$

For Problems A18–A23, use Cramer's Rule to solve the system.

A18
$$\begin{aligned} 2x_1 - 3x_2 &= 6 \\ 3x_1 + 5x_2 &= 7 \end{aligned}$$

A19
$$\begin{aligned} 3x_1 + 3x_2 &= 2 \\ 2x_1 - 3x_2 &= 5 \end{aligned}$$

A20
$$\begin{aligned} 7x_1 + x_2 - 4x_3 &= 3 \\ -6x_1 - 4x_2 + x_3 &= 0 \\ 4x_1 - x_2 - 2x_3 &= 6 \end{aligned}$$

A21
$$\begin{aligned} 2x_1 + 3x_2 - 5x_3 &= 2 \\ 3x_1 - x_2 + 2x_3 &= 1 \\ 5x_1 + 4x_2 - 6x_3 &= 3 \end{aligned}$$

A22
$$\begin{aligned} 5x_1 + 3x_2 + 5x_3 &= 2 \\ 2x_1 + 4x_2 + 5x_3 &= 1 \\ 7x_1 + 2x_2 + 4x_3 &= 1 \end{aligned}$$

A23
$$\begin{aligned} 2x_1 + 9x_2 + 3x_3 &= 1 \\ 2x_1 - 2x_2 + 3x_3 &= 1 \\ 3x_2 + 3x_3 &= -5 \end{aligned}$$

Homework Problems

For Problems B1–B6, find a non-zero vector $\vec{x} \in \mathbb{R}^4$ that is orthogonal to the given vectors.

B1 $\begin{bmatrix} 3 \\ 1 \\ -1 \\ 0 \end{bmatrix}, \begin{bmatrix} 3 \\ 3 \\ 1 \\ 1 \end{bmatrix}, \begin{bmatrix} 2 \\ 1 \\ 0 \\ 2 \end{bmatrix}$

B2 $\begin{bmatrix} 4 \\ 1 \\ 2 \\ 1 \end{bmatrix}, \begin{bmatrix} 3 \\ 2 \\ 2 \\ 1 \end{bmatrix}, \begin{bmatrix} 1 \\ 1 \\ -1 \\ -1 \end{bmatrix}$

B3 $\begin{bmatrix} 2 \\ 3 \\ 4 \\ 3 \end{bmatrix}, \begin{bmatrix} -1 \\ 2 \\ -2 \\ 1 \end{bmatrix}, \begin{bmatrix} 1 \\ 0 \\ 3 \\ 1 \end{bmatrix}$

B4 $\begin{bmatrix} 0 \\ 1 \\ 3 \\ 5 \end{bmatrix}, \begin{bmatrix} -1 \\ 2 \\ 1 \\ 3 \end{bmatrix}, \begin{bmatrix} 2 \\ 1 \\ 3 \\ -1 \end{bmatrix}$

B5 $\begin{bmatrix} 2 \\ 1 \\ -4 \\ 3 \end{bmatrix}, \begin{bmatrix} 2 \\ 1 \\ -4 \\ 4 \end{bmatrix}, \begin{bmatrix} 2 \\ 1 \\ -5 \\ 5 \end{bmatrix}$

B6 $\begin{bmatrix} 4 \\ -3 \\ 6 \\ 2 \end{bmatrix}, \begin{bmatrix} 4 \\ -3 \\ 2 \\ 3 \end{bmatrix}, \begin{bmatrix} -4 \\ 2 \\ -7 \\ 3 \end{bmatrix}$

For Problems B7–B14:

(a) Determine adj(A).

(b) Calculate $A\,\text{adj}(A)$ and determine $\det A$ and A^{-1}.

B7 $A = \begin{bmatrix} a & b \\ 0 & c \end{bmatrix}$

B8 $A = \begin{bmatrix} 5 & 0 & 0 \\ 3 & 10 & 0 \\ -9 & -3 & 1 \end{bmatrix}$

B9 $A = \begin{bmatrix} 1 & 1 & 3 \\ 0 & 2 & 1 \\ 1 & 1 & -1 \end{bmatrix}$

B10 $A = \begin{bmatrix} 2 & 5 & t \\ 1 & 1 & 3 \\ -1 & 1 & 3 \end{bmatrix}$

B11 $A = \begin{bmatrix} 1 & 4 & 3 \\ 3 & t & 0 \\ -2 & 1 & 1 \end{bmatrix}$

B12 $A = \begin{bmatrix} t & -2 & 3 \\ 1 & 1 & 2 \\ 1 & 1 & 3 \end{bmatrix}$

B13 $A = \begin{bmatrix} 2 & t & 3 \\ 0 & -1 & 1 \\ 1 & 1 & t \end{bmatrix}$

B14 $A = \begin{bmatrix} t & 1 & 1 \\ 1 & t & 1 \\ 1 & 1 & t \end{bmatrix}$

For Problems B15–B22, determine the inverse of the matrix by finding adj(A). Verify your answer by using multiplication.

B15 $\begin{bmatrix} 7 & -5 \\ 3 & 9 \end{bmatrix}$

B16 $\begin{bmatrix} 6 & 5 \\ 8 & 5 \end{bmatrix}$

B17 $\begin{bmatrix} 6 & 3 & 1 \\ 4 & 5 & 4 \\ 2 & 1 & 2 \end{bmatrix}$

B18 $\begin{bmatrix} 2 & 5 & 4 \\ 1 & 4 & 4 \\ -1 & 4 & 3 \end{bmatrix}$

B19 $\begin{bmatrix} 1 & 2 & -2 \\ 0 & 5 & -4 \\ 0 & 0 & 11 \end{bmatrix}$

B20 $\begin{bmatrix} 3 & -2 & 7 \\ 3 & -3 & 9 \\ 1 & 1 & 7 \end{bmatrix}$

B21 $\begin{bmatrix} 0 & 3 & 7 \\ 2 & 0 & -8 \\ 3 & 1 & 0 \end{bmatrix}$

B22 $\begin{bmatrix} 3 & 2 & 4 & 5 \\ 1 & 1 & -2 & 0 \\ 2 & -1 & 0 & 0 \\ 1 & 0 & 0 & 0 \end{bmatrix}$

For Problems B23–B30, use Cramer's Rule to solve the system.

B23 $\begin{aligned} x_1 + 2x_2 &= -5 \\ 4x_1 + 9x_2 &= -24 \end{aligned}$

B24 $\begin{aligned} 2x_1 - 3x_2 &= 4 \\ -7x_1 + 3x_2 &= 9 \end{aligned}$

B25 $\begin{aligned} 3x_1 + 5x_2 &= 7 \\ 11x_1 - 6x_2 &= 13 \end{aligned}$

B26 $\begin{aligned} 8x_1 - 3x_2 &= 15 \\ 6x_1 + 8x_2 &= 14 \end{aligned}$

B27 $\begin{aligned} x_1 + 3x_2 - x_3 &= 4 \\ -x_1 + x_2 + 2x_3 &= 1 \\ x_1 + 3x_2 + 4x_3 &= 4 \end{aligned}$

B28 $\begin{aligned} 2x_1 + x_2 + 3x_3 &= 1 \\ 4x_1 + 3x_2 + 6x_3 &= 1 \\ -2x_1 + x_2 + 5x_3 &= 1 \end{aligned}$

B29 $\begin{aligned} 2x_1 - 3x_2 + 4x_3 &= 1 \\ 2x_1 - 6x_2 + 8x_3 &= 2 \\ 4x_1 + 4x_2 + 4x_3 &= 9 \end{aligned}$

B30 $\begin{aligned} 5x_1 + x_2 + 4x_3 &= 1 \\ 3x_1 + 3x_3 &= 3 \\ x_1 - 2x_2 + 2x_3 &= -2 \end{aligned}$

Conceptual Problems

C1 Let $\vec{v}_1, \dots, \vec{v}_{n-1} \in \mathbb{R}^n$, let $A = \begin{bmatrix} \vec{0} & \vec{v}_1 & \cdots & \vec{v}_{n-1} \end{bmatrix}$, and let $\vec{n} = \begin{bmatrix} C_{11} \\ C_{21} \\ \vdots \\ C_{n1} \end{bmatrix}$ where C_{ij} is the (i, j)-cofactor of A.

(a) Prove that $\vec{n}$ is orthogonal to $\vec{v}_i$ for $1 \leq i \leq n-1$.

(b) Assume that $\{\vec{v}_1, \dots, \vec{v}_{n-1}\}$ is linearly independent. Find an equation for the hyperplane in $\mathbb{R}^n$ spanned by $\{\vec{v}_1, \dots, \vec{v}_{n-1}\}$.

C2 Let $A = \begin{bmatrix} 2 & -1 & 0 & 1 \\ 0 & -1 & 3 & 2 \\ 0 & 1 & 0 & 0 \\ 0 & 2 & 0 & 3 \end{bmatrix}$. Calculate $(A^{-1})_{23}$ and $(A^{-1})_{42}$. (If you calculate more than these two entries of A^{-1}, you have missed the point.)

C3 Let L be an invertible 3×3 upper triangular matrix. Prove that L^{-1} is also upper triangular.

C4 Suppose that $A = \begin{bmatrix} \vec{a}_1 & \cdots & \vec{a}_n \end{bmatrix}$ is an invertible $n \times n$ matrix.

(a) Verify by Cramer's Rule that the system of equations $A\vec{x} = \vec{a}_j$ has the unique solution $\vec{x} = \vec{e}_j$ (the j-th standard basis vector).

(b) Explain the result of part (a) in terms of linear transformations and/or matrix multiplication.

C5 Prove if $A \in M_{n\times n}(\mathbb{R})$ is not invertible, then adj A is not invertible.

For Problems C6–C10, let $A, B \in M_{n\times n}(\mathbb{R})$ with $n \geq 3$ both be invertible.

C6 Prove that $\operatorname{adj}(A^{-1}) = \dfrac{1}{\det A} A = (\operatorname{adj}(A))^{-1}$.

C7 Prove that $\det(\operatorname{adj}(A)) = (\det A)^{n-1}$.

C8 Prove that $\operatorname{adj}(\operatorname{adj}(A)) = \det(A)^{n-2} A$.

C9 Prove that $\operatorname{adj}(AB) = \operatorname{adj}(B)\operatorname{adj}(A)$.

C10 Prove that $\operatorname{adj}(A^T) = (\operatorname{adj}(A))^T$.

5.4 Area, Volume, and the Determinant

As mentioned at the beginning of this chapter, the value of a determinant has a very nice geometric interpretation. As a result, we can use determinants to solve a variety of geometric problems.

Area and the Determinant

In Chapter 1, we saw that we could construct a parallelogram from two vectors $\vec{u} = \begin{bmatrix} u_1 \\ u_2 \end{bmatrix}, \vec{v} = \begin{bmatrix} v_1 \\ v_2 \end{bmatrix} \in \mathbb{R}^2$ by making the vectors $\vec{u}$ and $\vec{v}$ as adjacent sides and having $\vec{u} + \vec{v}$ as the vertex of the parallelogram, opposite the origin. This is called the **parallelogram induced by $\vec{u}$ and $\vec{v}$**.

Theorem 5.4.1

The area of the parallelogram induced by $\vec{u} = \begin{bmatrix} u_1 \\ u_2 \end{bmatrix}, \vec{v} = \begin{bmatrix} v_1 \\ v_2 \end{bmatrix} \in \mathbb{R}^2$ is

$$\text{Area} = \left|\det\begin{bmatrix} \vec{u} & \vec{v} \end{bmatrix}\right|$$

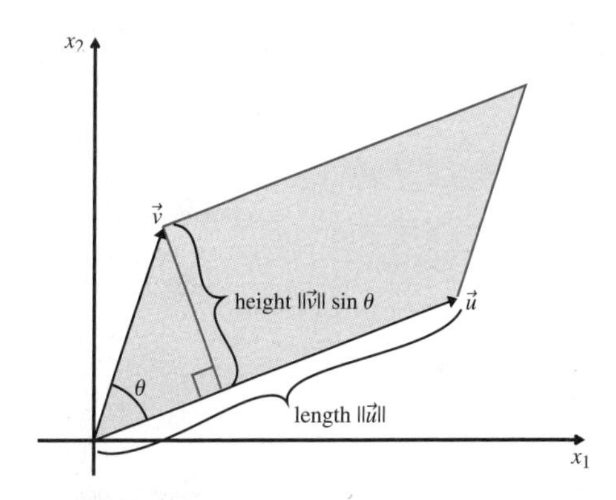

Proof: We have that

$$\text{Area} = \|\vec{u}\|\,\|\vec{v}\|\,|\sin\theta|$$

Recall from Section 1.3 that

$$\cos\theta = \frac{\vec{u} \cdot \vec{v}}{\|\vec{u}\|\,\|\vec{v}\|}$$

Using this gives

$$\begin{aligned}
(\text{Area})^2 &= \|\vec{u}\|^2\|\vec{v}\|^2 \sin^2\theta \\
&= \|\vec{u}\|^2\|\vec{v}\|^2(1 - \cos^2\theta) \\
&= \|\vec{u}\|^2\|\vec{v}\|^2 - \|\vec{u}\|^2\|\vec{v}\|^2 \cos^2\theta \\
&= \|\vec{u}\|^2\|\vec{v}\|^2 - (\vec{u} \cdot \vec{v})^2 \\
&= (u_1^2 + u_2^2)(v_1^2 + v_2^2) - (u_1v_1 + u_2v_2)^2 \\
&= u_1^2v_2^2 + u_2^2v_1^2 - 2(u_1v_2u_2v_1) \\
&= (u_1v_2 - u_2v_1)^2 \\
&= \begin{vmatrix} u_1 & v_1 \\ u_2 & v_2 \end{vmatrix}^2
\end{aligned}$$

Taking the square root of both sides gives Area $= \left|\det\begin{bmatrix} \vec{u} & \vec{v} \end{bmatrix}\right|$. ■

EXAMPLE 5.4.1

Draw the parallelogram induced by the following vectors and determine its area.

(a) $\vec{u} = \begin{bmatrix} 2 \\ 3 \end{bmatrix}, \vec{v} = \begin{bmatrix} -2 \\ 2 \end{bmatrix}$ (b) $\vec{u} = \begin{bmatrix} 1 \\ 1 \end{bmatrix}, \vec{v} = \begin{bmatrix} 1 \\ -2 \end{bmatrix}$

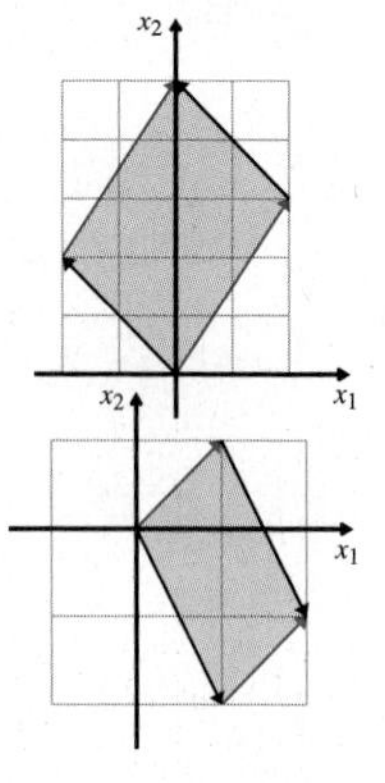

Solution: For (a), we have

$$\text{Area}\,(\vec{u}, \vec{v}) = \left|\det \begin{bmatrix} 2 & -2 \\ 3 & 2 \end{bmatrix}\right| = |2(2) - (-2)(3)| = 10$$

For (b), we have

$$\text{Area}(\vec{u}, \vec{v}) = \left|\det \begin{bmatrix} 1 & 1 \\ 1 & -2 \end{bmatrix}\right| = |1(-2) - 1(1)| = 3$$

Theorem 5.4.2

The area of a triangle with vertices (x_1, y_1), (x_2, y_2), and (x_3, y_3) is

$$\text{Area} = \frac{1}{2}\left|\det \begin{bmatrix} x_1 & y_1 & 1 \\ x_2 & y_2 & 1 \\ x_3 & y_3 & 1 \end{bmatrix}\right|$$

Proof: If we perform a translation by $-x_1$ in the x-direction and by $-y_1$ in the y-direction so that (x_1, y_1) is relocated to the origin, then the area of the resulting triangle will be half the area of the parallelogram induced by the vectors $\vec{u} = \begin{bmatrix} x_2 - x_1 \\ y_2 - y_1 \end{bmatrix}$ and $\vec{v} = \begin{bmatrix} x_3 - x_1 \\ y_3 - y_1 \end{bmatrix}$ as in the diagram below. That is,

$$\text{Area} = \frac{1}{2}\left|\det \begin{bmatrix} x_2 - x_1 & x_3 - x_1 \\ y_2 - y_1 & y_3 - y_1 \end{bmatrix}\right|$$

Rather than leaving it in this form, we expand the determinant to get

$$\text{Area} = \frac{1}{2}|(x_2y_3 - x_3y_2) - (x_1y_3 - x_3y_1) + (x_1y_2 - x_2y_1)|$$

which can be obtained from

$$\text{Area} = \frac{1}{2}\left|\det \begin{bmatrix} x_1 & y_1 & 1 \\ x_2 & y_2 & 1 \\ x_3 & y_3 & 1 \end{bmatrix}\right|$$

by performing a cofactor expansion along the third column.

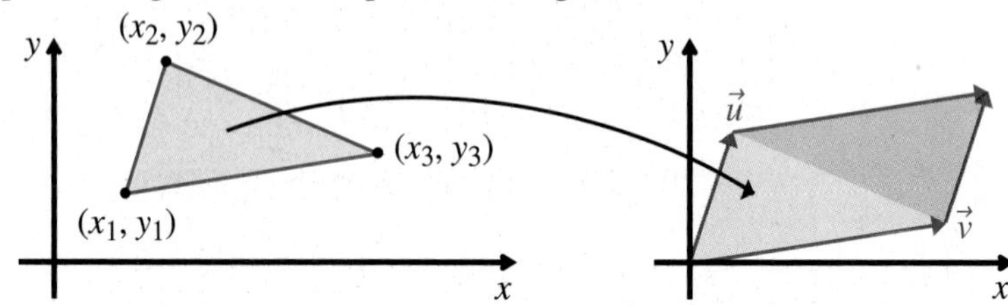

■

EXAMPLE 5.4.2

Find the area of the triangle with vertices $(-1,-1)$, $(2,3)$, and $(4,2)$.

Solution: We have

$$\text{Area} = \frac{1}{2}\left|\det\begin{bmatrix}-1 & -1 & 1\\ 2 & 3 & 1\\ 4 & 2 & 1\end{bmatrix}\right| = \frac{11}{2}$$

Notice that the area of the triangle will be 0 if and only if the three points are colinear. Hence, we get the following theorem.

Theorem 5.4.3

Three points (x_1, y_1), (x_2, y_2), and (x_3, y_3) are collinear if and only if

$$\det\begin{bmatrix}x_1 & y_1 & 1\\ x_2 & y_2 & 1\\ x_3 & y_3 & 1\end{bmatrix} = 0$$

If A is the standard matrix of a linear transformation $L : \mathbb{R}^2 \to \mathbb{R}^2$, then the images of $\vec{u}$ and $\vec{v}$ under L are $L(\vec{u}) = A\vec{u}$ and $L(\vec{v}) = A\vec{v}$. Moreover, the area of the **image parallelogram** is

$$\begin{aligned}
\text{Area}\,(A\vec{u}, A\vec{v}) &= \left|\det\begin{bmatrix}A\vec{u} & A\vec{v}\end{bmatrix}\right| && \text{by Theorem 5.4.1}\\
&= \left|\det\left(A\begin{bmatrix}\vec{u} & \vec{v}\end{bmatrix}\right)\right| && \text{by definition of matrix multiplication}\\
&= |\det A|\left|\det\begin{bmatrix}\vec{u} & \vec{v}\end{bmatrix}\right| && \text{since } \det(AB) = \det A \det B\\
&= |\det A|\,\text{Area}(\vec{u}, \vec{v}) && \text{by Theorem 5.4.1} \qquad (5.4)
\end{aligned}$$

In words: the absolute value of the determinant of the standard matrix A of a linear transformation is the factor by which area is changed under the linear transformation L. The result is illustrated in Figure 5.4.1.

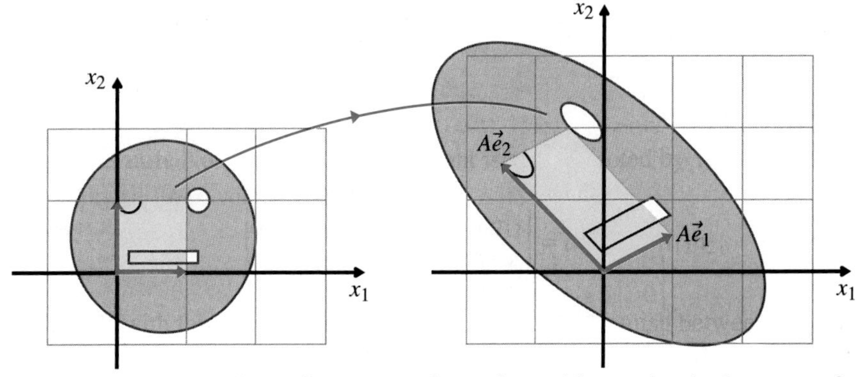

Figure 5.4.1 Under a linear transformation with matrix A, the area of a figure is changed by factor $|\det A|$.

EXAMPLE 5.4.3

Let $A = \begin{bmatrix} 1 & 3 \\ 0 & 2 \end{bmatrix}$ and L be the linear mapping $L(\vec{x}) = A\vec{x}$. Determine the image of $\vec{u} = \begin{bmatrix} 1 \\ 1 \end{bmatrix}$ and $\vec{v} = \begin{bmatrix} -1 \\ 1 \end{bmatrix}$ under L and compute the area determined by the image vectors in two ways.

Solution: The image of each vector under L is

$$L(\vec{u}) = \begin{bmatrix} 1 & 3 \\ 0 & 2 \end{bmatrix}\begin{bmatrix} 1 \\ 1 \end{bmatrix} = \begin{bmatrix} 4 \\ 2 \end{bmatrix}, \qquad L(\vec{v}) = \begin{bmatrix} 1 & 3 \\ 0 & 2 \end{bmatrix}\begin{bmatrix} -1 \\ 1 \end{bmatrix} = \begin{bmatrix} 2 \\ 2 \end{bmatrix}$$

Hence, the area determined by the image vectors is

$$\text{Area}\,(L(\vec{u}), L(\vec{v})) = \left|\det\begin{bmatrix} 4 & 2 \\ 2 & 2 \end{bmatrix}\right| = |8 - 4| = 4$$

Or, using equation (5.4) gives

$$\text{Area}\,(L(\vec{u}), L(\vec{v})) = |\det A|\,\text{Area}\,(\vec{u}, \vec{v}) = \left|\det\begin{bmatrix} 1 & 3 \\ 0 & 2 \end{bmatrix}\right|\left|\det\begin{bmatrix} 1 & -1 \\ 1 & 1 \end{bmatrix}\right| = 2(2) = 4$$

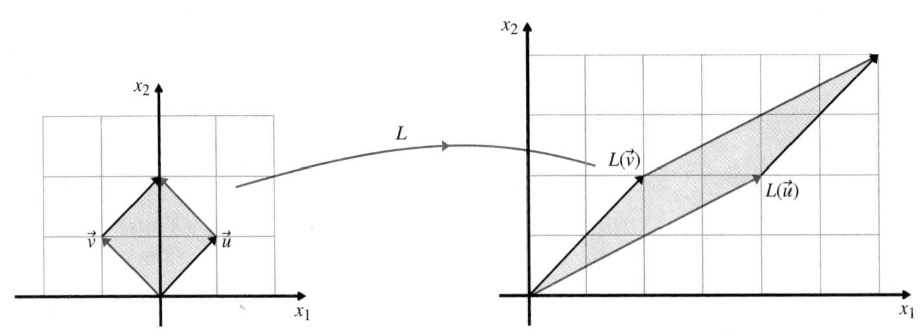

Figure 5.4.2 Illustration of Example 5.4.3.

EXERCISE 5.4.1

Let $A = \begin{bmatrix} t & 0 \\ 0 & 1 \end{bmatrix}$ be the standard matrix of the stretch $S : \mathbb{R}^2 \to \mathbb{R}^2$ in the x_1 direction by a factor of $t > 0$. Determine the image of the standard basis vectors $\vec{e}_1$ and $\vec{e}_2$ under S and compute the area determined by $A\vec{e}_1$ and $A\vec{e}_2$ by computing $\left|\det\begin{bmatrix} A\vec{e}_1 & A\vec{e}_2 \end{bmatrix}\right|$ and by using equation (5.4). Illustrate with a sketch.

The Determinant and Volume

Recall from Chapter 1 that if $\vec{u}$, $\vec{v}$, and $\vec{w}$ are vectors in $\mathbb{R}^3$, then the volume of the parallelepiped induced by $\vec{u}$, $\vec{v}$, and $\vec{w}$ is

$$\text{Volume}(\vec{u}, \vec{v}, \vec{w}) = |\vec{w} \cdot (\vec{u} \times \vec{v})|$$

Now observe that if $\vec{u} = \begin{bmatrix} u_1 \\ u_2 \\ u_3 \end{bmatrix}$, $\vec{v} = \begin{bmatrix} v_1 \\ v_2 \\ v_3 \end{bmatrix}$, and $\vec{w} = \begin{bmatrix} w_1 \\ w_2 \\ w_3 \end{bmatrix}$, then

$$|\vec{w} \cdot (\vec{u} \times \vec{v})| = |w_1(u_2v_3 - u_3v_2) - w_2(u_1v_3 - u_3v_1) + w_3(u_1v_2 - u_2v_1)| = \left|\det \begin{bmatrix} u_1 & v_1 & w_1 \\ u_2 & v_2 & w_2 \\ u_3 & v_3 & w_3 \end{bmatrix}\right|$$

Hence, the volume of the parallelepiped induced by $\vec{u}$, $\vec{v}$, and $\vec{w}$ is

$$\text{Volume}(\vec{u}, \vec{v}, \vec{w}) = \left|\det \begin{bmatrix} \vec{u} & \vec{v} & \vec{w} \end{bmatrix}\right|$$

EXAMPLE 5.4.4

Let $A = \begin{bmatrix} 4 & 1 & -1 \\ 2 & 4 & 1 \\ 1 & 1 & 4 \end{bmatrix}$, $\vec{u} = \begin{bmatrix} 1 \\ -1 \\ 0 \end{bmatrix}$, $\vec{v} = \begin{bmatrix} 0 \\ 1 \\ 2 \end{bmatrix}$, and $\vec{w} = \begin{bmatrix} -1 \\ 5 \\ 1 \end{bmatrix}$. Calculate the volume of the parallelepiped induced by $\vec{u}$, $\vec{v}$, and $\vec{w}$ and the volume of the parallelepiped induced by $A\vec{u}$, $A\vec{v}$, and $A\vec{w}$.

Solution: The volume determined by $\vec{u}$, $\vec{v}$, and $\vec{w}$ is

$$\text{Volume}(\vec{u}, \vec{v}, \vec{w}) = \left|\det \begin{bmatrix} 1 & 0 & -1 \\ -1 & 1 & 5 \\ 0 & 2 & 1 \end{bmatrix}\right| = |-7| = 7$$

The volume determined by $A\vec{u}$, $A\vec{v}$, and $A\vec{w}$ is

$$\begin{aligned} \text{Volume}(A\vec{u}, A\vec{v}, A\vec{w}) &= \left|\det \begin{bmatrix} A\vec{u} & A\vec{v} & A\vec{w} \end{bmatrix}\right| \\ &= \left|\det \begin{bmatrix} 3 & -1 & 0 \\ -2 & 6 & 19 \\ 0 & 9 & 8 \end{bmatrix}\right| = |-385| = 385 \end{aligned}$$

Moreover, $\det A = 55$, so

$$\text{Volume}\,(A\vec{u}, A\vec{v}, A\vec{w}) = |\det A| \text{Volume}\,(\vec{u}, \vec{v}, \vec{w})$$

which coincides with the result for the 2×2 case.

In general, if $\vec{v}_1, \dots, \vec{v}_n$ are n vectors in $\mathbb{R}^n$, then we say that they induce an **n-dimensional parallelotope** (the n-dimensional version of a parallelogram or parallelepiped). The **n-volume** of the parallelotope is

$$n\text{-Volume}(\vec{v}_1, \dots, \vec{v}_n) = \left|\det \begin{bmatrix} \vec{v}_1 & \cdots & \vec{v}_n \end{bmatrix}\right|$$

and if A is the standard matrix of a linear mapping $L : \mathbb{R}^n \to \mathbb{R}^n$, then

$$n\text{-Volume}\,(A\vec{v}_1, \dots, A\vec{v}_n) = |\det A| n\text{-Volume}\,(\vec{v}_1, \dots, \vec{v}_n)$$

PROBLEMS 5.4

Practice Problems

For Problems A1–A3:

(a) Calculate the area of the parallelogram induced by $\vec{u}$ and $\vec{v}$.

(b) Determine the area of the parallelogram induced by $A\vec{u}$ and $A\vec{v}$ by computing $\left|\det\begin{bmatrix} A\vec{u} & A\vec{v}\end{bmatrix}\right|$ and by using equation (5.4).

A1 $\vec{u} = \begin{bmatrix} 2 \\ 1 \end{bmatrix}, \vec{v} = \begin{bmatrix} -1 \\ 4 \end{bmatrix}, A = \begin{bmatrix} 3 & 2 \\ 1 & 2 \end{bmatrix}$

A2 $\vec{u} = \begin{bmatrix} 3 \\ 0 \end{bmatrix}, \vec{v} = \begin{bmatrix} 1 \\ 6 \end{bmatrix}, A = \begin{bmatrix} 1 & 2 \\ 1 & 2 \end{bmatrix}$

A3 $\vec{u} = \begin{bmatrix} 3 \\ 5 \end{bmatrix}, \vec{v} = \begin{bmatrix} 2 \\ 7 \end{bmatrix}, A = \begin{bmatrix} -4 & 6 \\ 3 & 2 \end{bmatrix}$

A4 Given that $A = \begin{bmatrix} 0 & 1 \\ 1 & 0 \end{bmatrix}$ is the standard matrix of the reflection $R : \mathbb{R}^2 \to \mathbb{R}^2$ over the line $x_2 = x_1$, find the image of $\vec{u} = \begin{bmatrix} 1 \\ 3 \end{bmatrix}$ and $\vec{v} = \begin{bmatrix} -2 \\ 2 \end{bmatrix}$ under R, and compute the area determined by $A\vec{u}$ and $A\vec{v}$.

For Problems A5 and A6:

(a) Calculate the area of the parallelogram induced by $\vec{u}$, $\vec{v}$, and $\vec{w}$.

(b) Determine the volume of the parallelepiped induced by $A\vec{u}$, $A\vec{v}$, and $A\vec{w}$.

A5 $\vec{u} = \begin{bmatrix} 2 \\ 3 \\ 4 \end{bmatrix}, \vec{v} = \begin{bmatrix} 2 \\ -1 \\ -5 \end{bmatrix}, \vec{w} = \begin{bmatrix} 1 \\ 5 \\ 2 \end{bmatrix}, A = \begin{bmatrix} 1 & -1 & 3 \\ 4 & 0 & 1 \\ 0 & 2 & 5 \end{bmatrix}$

A6 $\vec{u} = \begin{bmatrix} 0 \\ 2 \\ 3 \end{bmatrix}, \vec{v} = \begin{bmatrix} 1 \\ -5 \\ 5 \end{bmatrix}, \vec{w} = \begin{bmatrix} 2 \\ 1 \\ 6 \end{bmatrix}, A = \begin{bmatrix} 3 & -1 & 2 \\ 3 & 2 & -1 \\ 1 & 4 & 5 \end{bmatrix}$

For Problems A7–A11, determine whether the given points are collinear. If they are not collinear, find the area of the triangle which has those points as its vertices.

A7 $(-1, 3), (2, 5), (3, 7)$

A8 $(-4, -1), (2, 3), (5, 5)$

A9 $(0, 2), (3, 3), (4, 5)$

A10 $(1, 4), (6, 7), (9, 0)$

A11 $(-3, -26), (1, 2), (3, 16)$

A12 (a) Calculate the 4-volume of the 4-dimensional parallelotope determined by
$\vec{v}_1 = \begin{bmatrix} 1 \\ 0 \\ 2 \\ 0 \end{bmatrix}, \vec{v}_2 = \begin{bmatrix} 0 \\ 1 \\ 1 \\ 3 \end{bmatrix}, \vec{v}_3 = \begin{bmatrix} 0 \\ 2 \\ 3 \\ 0 \end{bmatrix}$, and $\vec{v}_4 = \begin{bmatrix} 1 \\ 0 \\ 2 \\ 5 \end{bmatrix}$.

(b) Calculate the 4-volume of the image of this parallelotope under the linear mapping with standard matrix $A = \begin{bmatrix} 2 & 3 & 1 & 1 \\ 5 & 4 & 3 & 0 \\ 0 & 0 & 7 & 3 \\ 0 & 0 & 0 & 1 \end{bmatrix}$.

A13 Let $\{\vec{v}_1, \dots, \vec{v}_n\}$ be vectors in $\mathbb{R}^n$. Prove that the n-volume of the parallelotope induced by $\vec{v}_1, \dots, \vec{v}_n$ is the same as the volume of the parallelotope induced by $\vec{v}_1, \dots, \vec{v}_{n-1}, \vec{v}_n + t\vec{v}_1$.

Homework Problems

For Problems B1–B3:

(a) Calculate the area of the parallelogram induced by $\vec{u}$ and $\vec{v}$.

(b) Determine the area of the parallelogram induced by $A\vec{u}$ and $A\vec{v}$ by computing $\left|\det\begin{bmatrix} A\vec{u} & A\vec{v}\end{bmatrix}\right|$ and by using equation (5.4).

B1 $\vec{u} = \begin{bmatrix} 5 \\ 6 \end{bmatrix}, \vec{v} = \begin{bmatrix} 8 \\ 9 \end{bmatrix}, A = \begin{bmatrix} 3 & -4 \\ -2 & 1 \end{bmatrix}$

B2 $\vec{u} = \begin{bmatrix} -3 \\ 5 \end{bmatrix}, \vec{v} = \begin{bmatrix} -2 \\ 1 \end{bmatrix}, A = \begin{bmatrix} 2 & 3 \\ 5 & 3 \end{bmatrix}$

B3 $\vec{u} = \begin{bmatrix} 4 \\ 1 \end{bmatrix}, \vec{v} = \begin{bmatrix} 4 \\ 2 \end{bmatrix}, A = \begin{bmatrix} -2 & 1 \\ 3 & -2 \end{bmatrix}$

For Problems B4–B10, determine the area of the parallelogram induced by the images of the standard basis vectors $\vec{e}_1$ and $\vec{e}_2$ of $\mathbb{R}^2$ under the given transformation.

B4 A rotation through angle $\theta = \pi/4$.

B5 A stretch in the x_2-direction by a factor of 3.

B6 A dilation by a factor of 2.

B7 A horizontal shear by an amount of 5.

B8 A vertical shear by an amount of -2.

B9 A stretch in the x_1-direction by a factor of 4.

B10 A horizontal shear by an amount of 3 followed by a stretch in the x_2-direction by a factor of 2.

For Problems B11 and B12:

(a) Calculate the volume of the parallelepiped induced by $\vec{u}$, $\vec{v}$, and $\vec{w}$.

(b) Determine the volume of the parallelepiped induced by $A\vec{u}$, $A\vec{v}$, and $A\vec{w}$.

B11 $\vec{u} = \begin{bmatrix} -3 \\ -3 \\ 1 \end{bmatrix}, \vec{v} = \begin{bmatrix} 2 \\ 2 \\ -1 \end{bmatrix}, \vec{w} = \begin{bmatrix} 1 \\ 3 \\ 2 \end{bmatrix}, A = \begin{bmatrix} 1 & 3 & 3 \\ 2 & 1 & 0 \\ -2 & 1 & 0 \end{bmatrix}$

B12 $\vec{u} = \begin{bmatrix} 1 \\ 1 \\ 1 \end{bmatrix}, \vec{v} = \begin{bmatrix} 2 \\ 1 \\ 0 \end{bmatrix}, \vec{w} = \begin{bmatrix} -1 \\ 2 \\ -1 \end{bmatrix}, A = \begin{bmatrix} 2 & 4 & 4 \\ 1 & 3 & 3 \\ 1 & -1 & 4 \end{bmatrix}$

For Problems B13–B19, determine whether the given points are collinear. If they are not collinear, find the area of the triangle which has those points as its vertices.

B13 $(-3, 1), (-2, 3), (1, -1)$

B14 $(0, -3), (1, 2), (3, 5)$

B15 $(-2, 2), (2, -1), (6, -4)$

B16 $(-3, -7), (1, -1), (4, 4)$

B17 $(-1, 2), (2, -1), (4, 3)$

B18 $(-5, 7), (1, -1), (4, -5)$

B19 $(-4, -2), (-1, 0), (3, 1)$

For Problems B20 and B21:

(a) Calculate the 4-volume of the 4-dimensional parallelotope determined by $\vec{v}_1, \vec{v}_2, \vec{v}_3, \vec{v}_4$.

(b) Calculate the 4-volume of the image of this parallelotope under the linear mapping with standard matrix $A = \begin{bmatrix} 1 & 3 & 4 & 3 \\ 2 & 2 & 4 & 2 \\ 0 & 1 & 1 & 2 \\ -1 & 1 & 1 & 3 \end{bmatrix}$.

B20 $\vec{v}_1 = \begin{bmatrix} 1 \\ 1 \\ -1 \\ 1 \end{bmatrix}, \vec{v}_2 = \begin{bmatrix} 1 \\ 2 \\ 3 \\ 3 \end{bmatrix}, \vec{v}_3 = \begin{bmatrix} 0 \\ 2 \\ 2 \\ 1 \end{bmatrix}$, and $\vec{v}_4 = \begin{bmatrix} 1 \\ 3 \\ 1 \\ 3 \end{bmatrix}$

B21 $\vec{v}_1 = \begin{bmatrix} 0 \\ 2 \\ 3 \\ 2 \end{bmatrix}, \vec{v}_2 = \begin{bmatrix} 1 \\ -1 \\ 2 \\ -1 \end{bmatrix}, \vec{v}_3 = \begin{bmatrix} 0 \\ 1 \\ 2 \\ 3 \end{bmatrix}$, and $\vec{v}_4 = \begin{bmatrix} 2 \\ 2 \\ -2 \\ -2 \end{bmatrix}$

Conceptual Problems

C1 Let $\{\vec{v}_1, \dots, \vec{v}_n\}$ be vectors in $\mathbb{R}^n$. Prove that the n-volume of the parallelotope induced by $\vec{v}_1, \dots, \vec{v}_n$ is half the volume of the parallelotope induced by $2\vec{v}_1, \vec{v}_2, \dots, \vec{v}_n$.

C2 Suppose that $L, M : \mathbb{R}^3 \to \mathbb{R}^3$ are linear mappings with standard matrices A and B, respectively. Prove that the factor by which a volume is multiplied under the composite map $M \circ L$ is $|\det BA|$.

CHAPTER REVIEW

Suggestions for Student Review

1 Define *cofactor* and explain cofactor expansion. Be especially careful about signs. (Section 5.1)

2 State as many facts as you can that simplify the evaluation of determinants. For each fact, explain why it is true. (Sections 5.1, 5.2)

3 List the properties of determinants, and list everything you can about a matrix A that is equivalent to $\det A \neq 0$. (Sections 5.2, 5.3)

4 Explain and justify the cofactor method for finding a matrix inverse. Write down a 3×3 matrix A and calculate $A(\text{cof } A)^T$. (Section 5.3)

5 State Cramer's Rule. Create a system of 3 linear equations in 3 variables that contains at least two unknown values a and b, and use Cramer's Rule to solve the system. (Section 5.3)

6 How are determinants connected to areas and volumes? (Section 5.4)

Chapter Quiz

For Problems E1–E6, find the determinant of the matrix. Decide whether the matrix is invertible.

E1 $\begin{bmatrix} 3 & 5 \\ 2 & -1 \end{bmatrix}$

E2 $\begin{bmatrix} 3 & 5 & -2 \\ 0 & 2 & \sqrt{2} \\ 0 & 0 & 3 \end{bmatrix}$

E3 $\begin{bmatrix} 3 & 2 & 1 \\ 2 & -2 & 2 \\ 1 & 4 & -1 \end{bmatrix}$

E4 $\begin{bmatrix} -2 & 4 & 0 & 0 \\ 1 & -2 & 2 & 9 \\ -3 & 6 & 0 & 3 \\ 1 & -1 & 0 & 0 \end{bmatrix}$

E5 $\begin{bmatrix} 3 & 2 & 7 & -8 \\ -6 & -1 & -9 & 20 \\ 3 & 8 & 21 & -17 \\ 3 & 5 & 12 & 1 \end{bmatrix}$

E6 $\begin{bmatrix} 0 & 2 & 0 & 0 & 0 \\ 0 & 0 & 0 & 3 & 0 \\ 0 & 0 & 0 & 0 & 1 \\ 0 & 0 & 4 & 0 & 0 \\ 5 & 0 & 0 & 0 & 6 \end{bmatrix}$

E7 Determine all values of k such that the matrix $\begin{bmatrix} k & 2 & 1 \\ 0 & 3 & k \\ 2 & -4 & 1 \end{bmatrix}$ is invertible.

For Problems E8–E12, suppose that $A \in M_{5\times 5}(\mathbb{R})$ and $\det A = 7$.

E8 If B is obtained from A by multiplying the fourth row of A by 3, what is $\det B$?

E9 If C is obtained from A by moving the first row to the bottom and moving all other rows up, what is $\det C$? Justify.

E10 What is $\det(2A)$?

E11 What is $\det(A^{-1})$?

E12 What is $\det(A^T A)$?

E13 Let $A = \begin{bmatrix} 2-\lambda & 2 & 1 \\ 2 & 1-\lambda & 2 \\ 2 & 2 & 1-\lambda \end{bmatrix}$. Determine all values of λ such that the determinant of A is 0.

E14 Let $A = \begin{bmatrix} 2 & 3 & 1 \\ 1 & 1 & 1 \\ -2 & 0 & 2 \end{bmatrix}$.

(a) Determine $\operatorname{adj}(A)$.

(b) Calculate $A \operatorname{adj}(A)$ and determine $\det A$.

(c) Determine $(A^{-1})_{31}$.

E15 Determine x_2 by using Cramer's Rule if

$$\begin{aligned} 2x_1 + 3x_2 + x_3 &= 1 \\ x_1 + x_2 - x_3 &= -1 \\ -2x_1 \quad\quad + 2x_3 &= 1 \end{aligned}$$

E16 (a) What is the volume of the parallelepiped induced by $\vec{u} = \begin{bmatrix} 1 \\ 1 \\ -2 \end{bmatrix}$, $\vec{v} = \begin{bmatrix} 2 \\ -1 \\ 3 \end{bmatrix}$, and $\vec{w} = \begin{bmatrix} 0 \\ 3 \\ 4 \end{bmatrix}$?

(b) If $A = \begin{bmatrix} 2 & 1 & 5 \\ 0 & 3 & -2 \\ 0 & 0 & -4 \end{bmatrix}$, what is the volume of the parallelepiped induced by $A\vec{u}$, $A\vec{v}$, and $A\vec{w}$?

E17 Find a non-zero vector $\vec{x} \in \mathbb{R}^4$ that is orthogonal to

$$\vec{v}_1 = \begin{bmatrix} 3 \\ 3 \\ -1 \\ -1 \end{bmatrix}, \quad \vec{v}_2 = \begin{bmatrix} 1 \\ 3 \\ 5 \\ 4 \end{bmatrix}, \quad \vec{v}_3 = \begin{bmatrix} 3 \\ 4 \\ 2 \\ 1 \end{bmatrix}$$

For Problems E18–E20, determine if the given points are collinear. If they are not collinear, find the area of the triangle which has those points as its vertices.

E18 $(-2, -1), (1, 8), (2, 11)$

E19 $(-1, 5), (2, 7), (3, 3)$

E20 $(-3, 4), (0, 3), (2, 2)$

Further Problems

These exercises are intended to be challenging.

F1 Suppose that $A \in M_{n\times n}(\mathbb{R})$ with all row sums equal to zero. (That is, $\sum_{j=1}^{n} a_{ij} = 0$ for $1 \leq i \leq n$.) Prove that $\det A = 0$.

F2 Suppose that A and A^{-1} both have all integer entries. Prove that $\det A = \pm 1$.

F3 Consider a triangle in the plane with side lengths a, b, and c. Let the angles opposite the sides with lengths a, b, and c be denoted by A, B, and C, respectively. By using trigonometry, show that

$$c = b\cos A + a\cos B$$

Write similar equations for the other two sides. Use Cramer's Rule to show that

$$\cos A = \frac{b^2 + c^2 - a^2}{2bc}$$

F4 Suppose that $A \in M_{3\times 3}(\mathbb{R})$ and $B \in M_{2\times 2}(\mathbb{R})$.

(a) Show that $\det\left[\begin{array}{c|c} A & O_{3,2} \\ \hline O_{2,3} & B \end{array}\right] = \det A \det B$.

(b) What is $\det\left[\begin{array}{c|c} O_{2,3} & B \\ \hline A & O_{3,2} \end{array}\right]$?

F5 (a) Let $V_3(a,b,c) = \det\begin{bmatrix} 1 & a & a^2 \\ 1 & b & b^2 \\ 1 & c & c^2 \end{bmatrix}$. Without expanding, argue that $(a-b)$, $(b-c)$, and $(c-a)$ are all factors of $V_3(a,b,c)$. By considering the cofactor of c^2, argue that

$$V_3(a,b,c) = (c-a)(c-b)(b-a)$$

(b) Let $V_4(a,b,c,d) = \det\begin{bmatrix} 1 & a & a^2 & a^3 \\ 1 & b & b^2 & b^3 \\ 1 & c & c^2 & c^3 \\ 1 & d & d^2 & d^3 \end{bmatrix}$. By using arguments similar to those in part (a) (and without expanding the determinant), argue that

$$V_4(a,b,c,d) = (d-a)(d-b)(d-c)V_3(a,b,c)$$

The determinant $V_n(x_1, \dots, x_n)$ is called the **Vandermonde determinant**. We will use the related Vandermonde matrix when we do the Method of Least Squares in Chapter 7.

F6 Suppose that $A \in M_{4\times 4}(\mathbb{R})$ is partitioned into 2×2 blocks:

$$A = \left[\begin{array}{c|c} A_1 & A_2 \\ \hline A_3 & A_4 \end{array}\right]$$

(a) If $A_3 = O_{2,2}$ (the 2×2 zero matrix), show that $\det A = \det A_1 \det A_4$.

(b) Give an example to show that, in general,

$$\det A \neq \det A_1 \det A_4 - \det A_2 \det A_3$$

CHAPTER 6

Eigenvectors and Diagonalization

CHAPTER OUTLINE

An eigenvector *is a special vector of a matrix that is mapped by the matrix to a scalar multiple of itself. Eigenvectors play an important role in many applications in the natural and physical sciences. In this chapter we will synthesize many of the concepts from the previous five chapters to develop this powerful tool.*

6.1 Eigenvalues and Eigenvectors

EXAMPLE 6.1.1

Consider the spring-mass system depicted below. Assume that the masses are identical and are attached to identical springs with spring constant k. As we did in Chapter 2, we denote the equilibrium displacements by x_1 and x_2 with a positive displacement being to the right. We would like to determine how to describe the motion of the masses that would result from the masses being displaced and then released.
According to Hooke's Law, the net force acting on the first mass is

$$-kx_1 + k(x_2 - x_1)$$

and the net force acting on the second mass is

$$-k(x_2 - x_1) - kx_2$$

By Newton's second law, the net force acting on a mass is also equal to

$$F = ma = m\frac{d^2x}{dt^2}$$

Thus, we have

$$-kx_1 + k(x_2 - x_1) = m\frac{d^2x_1}{dt^2}$$
$$-k(x_2 - x_1) - kx_2 = m\frac{d^2x_2}{dt^2}$$

EXAMPLE 6.1.1
(continued)

Experience tells us that the displacements x_1 and x_2 should undergo exponential decay. Thus, we take $x_1 = ae^{pt}$ and $x_2 = be^{pt}$. Substituting these into the equations above and rearranging, we can get

$$-2a + b = \frac{mp^2}{k}a$$
$$a - 2b = \frac{mp^2}{k}b$$

Defining $\lambda = \frac{mp^2}{k}$ and rewriting the system in matrix form gives

$$\begin{bmatrix} -2 & 1 \\ 1 & -2 \end{bmatrix}\begin{bmatrix} a \\ b \end{bmatrix} = \lambda \begin{bmatrix} a \\ b \end{bmatrix}$$

Solving this for λ gives us the value of p, which tells us the frequencies of oscillation.

Many problems in science and engineering require us to solve a system of linear equations of the form

$$A\vec{x} = \lambda\vec{x}$$

This is often referred to as the **eigenvalue problem**.

Definition

Eigenvector
Eigenvalue

Suppose that $A \in M_{n\times n}(\mathbb{R})$. If there exists a non-zero vector $\vec{v} \in \mathbb{R}^n$ such that

$$A\vec{v} = \lambda\vec{v}$$

then the scalar λ is called an **eigenvalue** of A and $\vec{v}$ is called an **eigenvector** of A corresponding to λ.

Remarks

1. The pairing of eigenvalues and eigenvectors is not one-to-one. In particular, we will see that each eigenvector of A will correspond to a distinct eigenvalue, while each eigenvalue will have infinitely many eigenvectors.

2. We have restricted our definition of eigenvectors (and hence eigenvalues) to be real. In Chapter 9 we will consider the case where we allow eigenvalues and eigenvectors to be complex.

3. The restriction that an eigenvector $\vec{v}$ be non-zero is natural and important. It is natural because $A\vec{0} = \vec{0}$ for any matrix A, so it is uninteresting to consider $\vec{0}$ as an eigenvector. It is important because most of the applications of eigenvectors make sense only for non-zero vectors.

Finding Eigenvalues and Eigenvectors

We now look at how to find eigenvalues and eigenvectors of a matrix. We begin by observing that it is easy to verify if a given vector is an eigenvector of a matrix.

EXAMPLE 6.1.2

Let $A = \begin{bmatrix} 0 & 2 \\ 1 & 1 \end{bmatrix}$. Determine whether $\vec{v}_1 = \begin{bmatrix} 1 \\ 2 \end{bmatrix}$ or $\vec{v}_2 = \begin{bmatrix} -2 \\ 1 \end{bmatrix}$ is an eigenvector of A. If so, give the corresponding eigenvalue.

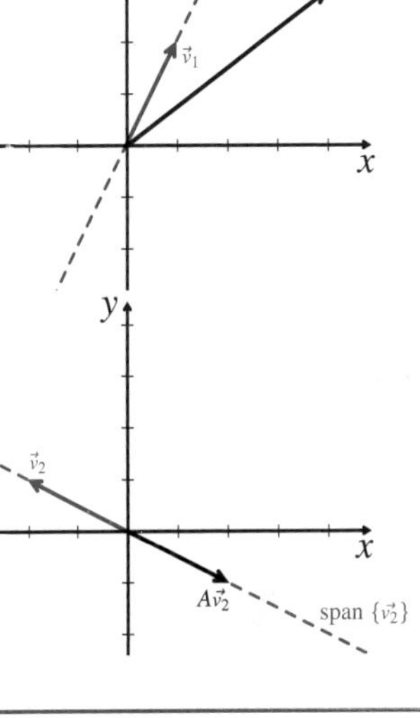

Solution: For $\vec{v}_1$, we find that

$$A\vec{v}_1 = \begin{bmatrix} 0 & 2 \\ 1 & 1 \end{bmatrix}\begin{bmatrix} 1 \\ 2 \end{bmatrix} = \begin{bmatrix} 4 \\ 3 \end{bmatrix}$$

Since, $A\vec{v}_1$ is not a scalar multiple of $\vec{v}_1$, $\vec{v}_1$ is not an eigenvector of A.
For $\vec{v}_2$, we find that

$$A\vec{v}_2 = \begin{bmatrix} 0 & 2 \\ 1 & 1 \end{bmatrix}\begin{bmatrix} -2 \\ 1 \end{bmatrix} = \begin{bmatrix} 2 \\ -1 \end{bmatrix} = (-1)\begin{bmatrix} -2 \\ 1 \end{bmatrix}$$

Thus, $A\vec{v}_2$ is scalar multiple of $\vec{v}_2$ (it is -1 times $\vec{v}_2$), so $\vec{v}_2$ is an eigenvector of A corresponding to the eigenvalue $\lambda = -1$.

If eigenvalues are going to be of any use, we need a systematic method for finding them. We first look at how to do this in a simple example before exploring the general procedure.

EXAMPLE 6.1.3

Let $A = \begin{bmatrix} 2 & 2 \\ 1 & 3 \end{bmatrix}$. Find all eigenvalues of A.

Solution: For $\lambda \in \mathbb{R}$ to be an eigenvalue of A there must exist $\vec{v} = \begin{bmatrix} v_1 \\ v_2 \end{bmatrix} \neq \vec{0}$ such that

$$\begin{bmatrix} 2 & 2 \\ 1 & 3 \end{bmatrix}\begin{bmatrix} v_1 \\ v_2 \end{bmatrix} = \lambda\begin{bmatrix} v_1 \\ v_2 \end{bmatrix}$$

We can rewrite this as the system of linear equations

$$\begin{aligned} 2v_1 + 2v_2 &= \lambda v_1 \\ 1v_1 + 3v_2 &= \lambda v_2 \end{aligned}$$

We put this into standard form of a system of linear equations by collecting the variables v_1 and v_2 on the left side. We get

$$\begin{aligned} (2-\lambda)v_1 + 2v_2 &= 0 \\ 1v_1 + (3-\lambda)v_2 &= 0 \end{aligned}$$

EXAMPLE 6.1.3
(continued)

For λ to be an eigenvalue, we need this system to have non-zero solutions so that $\vec{v}$ is a non-zero vector. By the Invertible Matrix Theorem, we know that this will only occur when the coefficient matrix $\begin{bmatrix} 2-\lambda & 2 \\ 1 & 3-\lambda \end{bmatrix}$ is not invertible. Hence, we can just determine when the determinant of this matrix equals 0. That is, we need to solve

$$0 = \begin{vmatrix} 2-\lambda & 2 \\ 1 & 3-\lambda \end{vmatrix} = (2-\lambda)(3-\lambda) - 2(1) = \lambda^2 - 5\lambda + 4 = (\lambda - 4)(\lambda - 1)$$

Hence, $A\vec{v} = \lambda\vec{v}$ has non-zero solutions if and only if $\lambda = 4$ or $\lambda = 1$. Therefore, the eigenvalues of A are $\lambda_1 = 4$ and $\lambda_2 = 1$.

Our work in Example 6.1.3 also shows us how to find the eigenvectors corresponding to an eigenvalue. For example, to find all infinitely many eigenvectors of the matrix A in Example 6.1.3 corresponding to $\lambda_1 = 4$, we just need to find the solution space of the homogeneous system with coefficient matrix

$$\begin{bmatrix} 2-\lambda_1 & 2 \\ 1 & 3-\lambda_1 \end{bmatrix}$$

In general, for any $A \in M_{n\times n}(\mathbb{R})$ a non-zero vector $\vec{v} \in \mathbb{R}^n$ is an eigenvector of A if and only if $A\vec{v} = \lambda\vec{v}$. So, as in Example 6.1.3, this condition can be rewritten as

$$(A - \lambda I)\vec{v} = \vec{0}$$

Thus, the eigenvector $\vec{v}$ is any non-trivial solution (since it cannot be the zero vector) of the homogeneous system of linear equations with coefficient matrix $(A - \lambda I)$. By the Invertible Matrix Theorem, the system $(A - \lambda I)\vec{v} = \vec{0}$ has non-trivial solutions if and only if the coefficient matrix $(A - \lambda I)$ has determinant equal to 0. Hence, for λ to be an eigenvalue, we must have $\det(A - \lambda I) = 0$. This is the key result in the procedure for finding the eigenvalues and eigenvectors, so it is worth summarizing as a theorem.

Theorem 6.1.1

Suppose that $A \in M_{n\times n}(\mathbb{R})$. A real number λ is an eigenvalue of A if and only if λ satisfies the equation

$$\det(A - \lambda I) = 0$$

If λ is an eigenvalue of A, then all non-zero solutions of the homogeneous system

$$(A - \lambda I)\vec{v} = \vec{0}$$

are all the eigenvectors of A that correspond to λ.

Observe that the set of all eigenvectors corresponding to an eigenvalue λ is just the nullspace of $A - \lambda I$, excluding the zero vector. In particular, the set containing all eigenvectors corresponding to λ and the zero vector is a subspace of $\mathbb{R}^n$. We make the following definition.

Definition
Eigenspace

Let λ be an eigenvalue of a matrix A. The set containing the zero vector and all eigenvectors of A corresponding to λ is called the **eigenspace** of λ and is denoted E_λ. In particular, we have

$$E_\lambda = \operatorname{Null}(A - \lambda I)$$

Remark

From our work preceding the theorem, we see that the eigenspace of any eigenvalue λ must contain at least one non-zero vector. Hence, the dimension of the eigenspace must be at least 1.

EXAMPLE 6.1.4

Find the eigenvalues and eigenvectors of the matrix $A = \begin{bmatrix} 17 & -15 \\ 20 & -18 \end{bmatrix}$.

Solution: We have

$$A - \lambda I = \begin{bmatrix} 17 & -15 \\ 20 & -18 \end{bmatrix} - \lambda \begin{bmatrix} 1 & 0 \\ 0 & 1 \end{bmatrix} = \begin{bmatrix} 17-\lambda & -15 \\ 20 & -18-\lambda \end{bmatrix}$$

(You should set up your calculations like this: you will need $A - \lambda I$ later when you find the eigenvectors.) Then

$$\det(A - \lambda I) = \begin{vmatrix} 17-\lambda & -15 \\ 20 & -18-\lambda \end{vmatrix} = \lambda^2 + \lambda - 6 = (\lambda + 3)(\lambda - 2)$$

Since $\det(A - \lambda I) = 0$ when $\lambda = -3, 2$, the eigenvalues of A are $\lambda_1 = -3$ and $\lambda_2 = 2$.

To find all the eigenvectors corresponding to $\lambda_1 = -3$, we solve $(A - \lambda_1 I)\vec{v} = \vec{0}$. Writing $A - \lambda_1 I$ and row reducing gives

$$A - (-3)I = \begin{bmatrix} 20 & -15 \\ 20 & -15 \end{bmatrix} \sim \begin{bmatrix} 1 & -3/4 \\ 0 & 0 \end{bmatrix}$$

The general solution of $(A - \lambda_1 I)\vec{v} = \vec{0}$ is $\vec{v} = t\begin{bmatrix} 3/4 \\ 1 \end{bmatrix}$, $t \in \mathbb{R}$. Thus, all eigenvectors of A corresponding to $\lambda_1 = -3$ are $\vec{v} = t\begin{bmatrix} 3/4 \\ 1 \end{bmatrix}$ for any *non-zero* value of t, and the eigenspace for $\lambda_1 = -3$ is $E_{\lambda_1} = \text{Span}\left\{\begin{bmatrix} 3/4 \\ 1 \end{bmatrix}\right\}$.

We repeat the process for the eigenvalue $\lambda_2 = 2$:

$$A - 2I = \begin{bmatrix} 15 & -15 \\ 20 & -20 \end{bmatrix} \sim \begin{bmatrix} 1 & -1 \\ 0 & 0 \end{bmatrix}$$

The general solution of $(A - \lambda_2 I)\vec{v} = \vec{0}$ is $\vec{v} = t\begin{bmatrix} 1 \\ 1 \end{bmatrix}$, $t \in \mathbb{R}$, so the eigenspace for $\lambda_2 = 2$ is $E_{\lambda_2} = \text{Span}\left\{\begin{bmatrix} 1 \\ 1 \end{bmatrix}\right\}$. In particular, all eigenvectors of A corresponding to $\lambda_2 = 2$ are all non-zero multiples of $\begin{bmatrix} 1 \\ 1 \end{bmatrix}$.

Observe in Example 6.1.4 that $\det(A - \lambda I)$ gave us a degree 2 polynomial. This motivates the following definition.

Definition
Characteristic Polynomial

Let $A \in M_{n\times n}(\mathbb{R})$. We call $C(\lambda) = \det(A - \lambda I)$ the **characteristic polynomial** of A.

For $A \in M_{n\times n}(\mathbb{R})$, the characteristic polynomial $C(\lambda)$ is of degree n, and the roots of $C(\lambda)$ are the eigenvalues of A. Note that the term of highest degree λ^n has coefficient $(-1)^n$; some other books prefer to work with the polynomial $\det(\lambda I - A)$ so that the coefficient of λ^n is always 1. In our notation, the constant term in the characteristic polynomial is $\det A$. See Section 6.2 Problem C7.

It is relevant here to recall some facts about the roots of an n-th degree polynomial with real coefficients:

(1) λ_1 is a root of $C(\lambda)$ if and only if $(\lambda - \lambda_1)$ is a factor of $C(\lambda)$.

(2) The total number of roots (real and complex, counting repetitions) is n.

(3) Complex roots of the equation occur in "conjugate pairs," so that the total number of complex roots must be even.

(4) If n is odd, there must be at least one real root.

(5) If the entries of A are integers, then, since the leading coefficient of the characteristic polynomial is ± 1, any rational root must be an integer.

EXAMPLE 6.1.5

Find the eigenvalues and a basis for each eigenspace of $A = \begin{bmatrix} 1 & 1 \\ 0 & 1 \end{bmatrix}$.

Solution: The characteristic polynomial is

$$C(\lambda) = \det(A - \lambda I) = \begin{vmatrix} 1-\lambda & 1 \\ 0 & 1-\lambda \end{vmatrix} = (1-\lambda)(1-\lambda)$$

So, $\lambda = 1$ is a double root (that is, $(\lambda - 1)$ appears as a factor of $C(\lambda)$ twice). Thus, $\lambda_1 = 1$ is the only distinct eigenvalue of A.
For $\lambda_1 = 1$, we have

$$A - \lambda_1 I = \begin{bmatrix} 0 & 1 \\ 0 & 0 \end{bmatrix}$$

which has the general solution $\vec{v} = t\begin{bmatrix} 1 \\ 0 \end{bmatrix}$, $t \in \mathbb{R}$. Thus, a basis for E_{λ_1} is $\left\{\begin{bmatrix} 1 \\ 0 \end{bmatrix}\right\}$.

EXERCISE 6.1.1

Find the eigenvalues and a basis for each eigenspace of $A = \begin{bmatrix} 1 & 2 \\ 2 & 4 \end{bmatrix}$.

EXAMPLE 6.1.6

Find the eigenvalues and a basis for each eigenspace of $A = \begin{bmatrix} -3 & 5 & -5 \\ -7 & 9 & -5 \\ -7 & 7 & -3 \end{bmatrix}$.

Solution: We have

$$C(\lambda) = \det(A - \lambda I) = \begin{vmatrix} -3-\lambda & 5 & -5 \\ -7 & 9-\lambda & -5 \\ -7 & 7 & -3-\lambda \end{vmatrix}$$

Expanding this determinant along some row or column will involve a fair number of calculations. Also, we will end up with a degree 3 polynomial, which may not be easy to factor. So, we use properties of determinants to make it easier. Using $R_2 - R_1$ and then $C_1 + C_2$ gives

$$C(\lambda) = \begin{vmatrix} -3-\lambda & 5 & -5 \\ -4+\lambda & 4-\lambda & 0 \\ -7 & 7 & -3-\lambda \end{vmatrix} = \begin{vmatrix} 2-\lambda & 5 & -5 \\ 0 & 4-\lambda & 0 \\ 0 & 7 & -3-\lambda \end{vmatrix}$$

Now, expanding this determinant along the first column gives

$$\begin{aligned} C(\lambda) &= (2-\lambda)[(4-\lambda)(-3-\lambda) - 0(7)] \\ &= (2-\lambda)(4-\lambda)(-3-\lambda) \end{aligned}$$

Hence, the eigenvalues of A are $\lambda_1 = 2$, $\lambda_2 = 4$, and $\lambda_3 = -3$.
For $\lambda_1 = 2$,

$$A - \lambda_1 I = \begin{bmatrix} -5 & 5 & -5 \\ -7 & 7 & -5 \\ -7 & 7 & -5 \end{bmatrix} \sim \begin{bmatrix} 1 & -1 & 0 \\ 0 & 0 & 1 \\ 0 & 0 & 0 \end{bmatrix}$$

The general solution of $(A - \lambda_1 I)\vec{v} = \vec{0}$ is $\vec{v} = t\begin{bmatrix} 1 \\ 1 \\ 0 \end{bmatrix}$, $t \in \mathbb{R}$. So, $\left\{\begin{bmatrix} 1 \\ 1 \\ 0 \end{bmatrix}\right\}$ is a basis for E_{λ_1}.

For $\lambda_2 = 4$,

$$A - \lambda_2 I = \begin{bmatrix} -7 & 5 & -5 \\ -7 & 5 & -5 \\ -7 & 7 & -7 \end{bmatrix} \sim \begin{bmatrix} 1 & 0 & 0 \\ 0 & 1 & -1 \\ 0 & 0 & 0 \end{bmatrix}$$

The general solution of $(A - \lambda_2 I)\vec{v} = \vec{0}$ is $\vec{v} = t\begin{bmatrix} 0 \\ 1 \\ 1 \end{bmatrix}$, $t \in \mathbb{R}$. So, $\left\{\begin{bmatrix} 0 \\ 1 \\ 1 \end{bmatrix}\right\}$ is a basis for E_{λ_2}.

For $\lambda_3 = -3$,

$$A - \lambda_3 I = \begin{bmatrix} 0 & 5 & -5 \\ -7 & 12 & -5 \\ -7 & 7 & 0 \end{bmatrix} \sim \begin{bmatrix} 1 & 0 & -1 \\ 0 & 1 & -1 \\ 0 & 0 & 0 \end{bmatrix}$$

The general solution of $(A - \lambda_3 I)\vec{v} = \vec{0}$ is $\vec{v} = t\begin{bmatrix} 1 \\ 1 \\ 1 \end{bmatrix}$, $t \in \mathbb{R}$. So, $\left\{\begin{bmatrix} 1 \\ 1 \\ 1 \end{bmatrix}\right\}$ is a basis for E_{λ_3}.

EXAMPLE 6.1.7

Find the eigenvalues and a basis for each eigenspace of $A = \begin{bmatrix} 1 & 1 & 1 \\ 1 & 1 & 1 \\ 1 & 1 & 1 \end{bmatrix}$.

Solution: Using $R_3 - R_2$ and then $C_2 + C_3$ gives

$$C(\lambda) = \begin{vmatrix} 1-\lambda & 1 & 1 \\ 1 & 1-\lambda & 1 \\ 1 & 1 & 1-\lambda \end{vmatrix} = \begin{vmatrix} 1-\lambda & 1 & 1 \\ 1 & 1-\lambda & 1 \\ 0 & \lambda & -\lambda \end{vmatrix}$$

$$= \begin{vmatrix} 1-\lambda & 2 & 1 \\ 1 & 2-\lambda & 1 \\ 0 & 0 & -\lambda \end{vmatrix} = -\lambda(\lambda^2 - 3\lambda) = -\lambda^2(\lambda - 3)$$

Therefore, the eigenvalues of A are $\lambda_1 = 0$ (which occurs twice) and $\lambda_2 = 3$. For $\lambda_1 = 0$,

$$A - \lambda_1 I = \begin{bmatrix} 1 & 1 & 1 \\ 1 & 1 & 1 \\ 1 & 1 & 1 \end{bmatrix} \sim \begin{bmatrix} 1 & 1 & 1 \\ 0 & 0 & 0 \\ 0 & 0 & 0 \end{bmatrix}$$

Hence, a basis for the eigenspace of $\lambda_1 = 0$ is $\left\{ \begin{bmatrix} -1 \\ 1 \\ 0 \end{bmatrix}, \begin{bmatrix} -1 \\ 0 \\ 1 \end{bmatrix} \right\}$.

For $\lambda_2 = 3$,

$$A - \lambda_2 I = \begin{bmatrix} -2 & 1 & 1 \\ 1 & -2 & 1 \\ 1 & 1 & -2 \end{bmatrix} \sim \begin{bmatrix} 1 & 0 & -1 \\ 0 & 1 & -1 \\ 0 & 0 & 0 \end{bmatrix}$$

Thus, a basis for the eigenspace of $\lambda_2 = 3$ is $\left\{ \begin{bmatrix} 1 \\ 1 \\ 1 \end{bmatrix} \right\}$.

These examples motivate the following definitions.

Definition
Algebraic Multiplicity
Geometric Multiplicity

Let $A \in M_{n\times n}(\mathbb{R})$ with eigenvalue λ. The **algebraic multiplicity** of λ is the number of times λ is repeated as a root of the characteristic polynomial. The **geometric multiplicity** of λ is the dimension of the eigenspace of λ.

EXAMPLE 6.1.8

In Example 6.1.5, the eigenvalue $\lambda_1 = 1$ has algebraic multiplicity 2 since the characteristic polynomial is $(\lambda - 1)(\lambda - 1)$, and it has geometric multiplicity 1 since a basis for its eigenspace is $\left\{ \begin{bmatrix} 1 \\ 0 \end{bmatrix} \right\}$.

In Example 6.1.6, all three eigenvalues have algebraic multiplicity 1 and they all have geometric multiplicity 1.

In Example 6.1.7, the eigenvalue $\lambda_1 = 0$ has algebraic and geometric multiplicity 2, and the eigenvalue $\lambda_2 = 3$ has algebraic and geometric multiplicity 1.

EXERCISE 6.1.2

Let $A = \begin{bmatrix} 5 & -3 & 2 \\ 0 & 0 & 2 \\ 0 & -2 & -4 \end{bmatrix}$. Show that $\lambda_1 = 5$ and $\lambda_2 = -2$ are both eigenvalues of A and determine the algebraic and geometric multiplicity of both of these eigenvalues.

These definitions lead to some theorems that will be very important in the next section.

Theorem 6.1.2

If λ is an eigenvalue of a matrix $A \in M_{n\times n}(\mathbb{R})$, then

$$1 \leq \text{geometric multiplicity} \leq \text{algebraic multiplicity}$$

The proof of Theorem 6.1.2 is beyond the scope of this course.

If the geometric multiplicity of an eigenvalue is less than its algebraic multiplicity, then we say that the eigenvalue is **deficient**. However, if $A \in M_{n\times n}(\mathbb{R})$ with distinct eigenvalues $\lambda_1, \dots, \lambda_k$, which all have the property that their geometric multiplicity equals their algebraic multiplicity, then the sum of the geometric multiplicities of all eigenvalues equals the sum of the algebraic multiplicities, which equals n (since an n-th degree polynomial has exactly n roots). Hence, if we collect the basis vectors from the eigenspaces of all k eigenvalues, we will end up with n vectors in $\mathbb{R}^n$. The next theorem states that eigenvectors from eigenspaces of different eigenvalues are necessarily linearly independent, and hence this collection of n eigenvectors will form a basis for $\mathbb{R}^n$.

Theorem 6.1.3

If $\lambda_1, \dots, \lambda_k$ are distinct eigenvalues of a matrix $A \in M_{n\times n}(\mathbb{R})$, with corresponding eigenvectors $\vec{v}_1, \dots, \vec{v}_k$, respectively, then $\{\vec{v}_1, \dots, \vec{v}_k\}$ is linearly independent.

The proof of Theorem 6.1.3 is left as Problem F4 at the end of the chapter.

Remark

In this book, most eigenvalues turn out to be integers. This is very unrealistic; in real world applications, eigenvalues are often not rational numbers. Effective computer methods for finding eigenvalues depend on the theory of eigenvectors and eigenvalues.

Theorem 6.1.4

Invertible Matrix Theorem continued

If $A \in M_{n\times n}(\mathbb{R})$, then the following are equivalent:

(1) A is invertible.

(11) $\lambda = 0$ is not an eigenvalue of A.

The proof of Theorem 6.1.4 is left as Problem C6.

Eigenvalues and Eigenvectors of Linear Mappings

The geometric meaning of eigenvectors and eigenvalues becomes much clearer when we think of them as belonging to linear transformations.

Definition
Eigenvector
Eigenvalue

Suppose that $L : \mathbb{R}^n \to \mathbb{R}^n$ is a linear mapping. If there exists a non-zero vector $\vec{v} \in \mathbb{R}^n$ such that $L(\vec{v}) = \lambda\vec{v}$, then λ is called an **eigenvalue** of L and $\vec{v}$ is called an **eigenvector** of L corresponding λ.

EXAMPLE 6.1.9

Eigenvectors and Eigenvalues of Projections and Reflections in $\mathbb{R}^3$

1. Because $\text{proj}_{\vec{n}}(\vec{n}) = 1\vec{n}$, $\vec{n}$ is an eigenvector of $\text{proj}_{\vec{n}}$ with corresponding eigenvalue 1. If $\vec{v} \neq \vec{0}$ is orthogonal to $\vec{n}$, then $\text{proj}_{\vec{n}}(\vec{v}) = \vec{0} = 0\vec{v}$, so $\vec{v}$ is an eigenvector of $\text{proj}_{\vec{n}}$ with corresponding eigenvalue 0. Observe that this means there is a whole plane of eigenvectors corresponding to the eigenvalue 0 as the set of vectors orthogonal to $\vec{n}$ is a plane in $\mathbb{R}^3$. For an arbitrary vector $\vec{u} \neq \vec{0}$, $\text{proj}_{\vec{n}}(\vec{u})$ is a multiple of $\vec{n}$, so that $\vec{u}$ is definitely *not* an eigenvector of $\text{proj}_{\vec{n}}$ unless it is a multiple of $\vec{n}$ or orthogonal to $\vec{n}$.

2. On the other hand, $\text{perp}_{\vec{n}}(\vec{n}) = \vec{0} = 0\vec{n}$, so $\vec{n}$ is an eigenvector of $\text{perp}_{\vec{n}}$ with eigenvalue 0. For $\vec{v} \neq \vec{0}$ orthogonal to $\vec{n}$, $\text{perp}_{\vec{n}}(\vec{v}) = 1\vec{v}$, so such a $\vec{v}$ is an eigenvector of $\text{perp}_{\vec{n}}$ with eigenvalue 1.

3. We have $\text{refl}_{\vec{n}}(\vec{n}) = -1\vec{n}$, so $\vec{n}$ is an eigenvector of $\text{refl}_{\vec{n}}$ with eigenvalue -1. For $\vec{v} \neq \vec{0}$ orthogonal to $\vec{n}$, $\text{refl}_{\vec{n}}(\vec{v}) = 1\vec{v}$. Hence, such a $\vec{v}$ is an eigenvector of $\text{refl}_{\vec{n}}$ with eigenvalue 1.

EXAMPLE 6.1.10

Eigenvectors and Eigenvalues of Rotations in $\mathbb{R}^2$

Consider the rotation $R_\theta : \mathbb{R}^2 \to \mathbb{R}^2$ with matrix $\begin{bmatrix} \cos\theta & -\sin\theta \\ \sin\theta & \cos\theta \end{bmatrix}$, where θ is not an integer multiple of π. Thinking geometrically, it is clear that there is no non-zero vector $\vec{v}$ in $\mathbb{R}^2$ such that $R_\theta(\vec{v}) = \lambda\vec{v}$ for some real number λ. This linear transformation has no real eigenvalues or real eigenvectors. In Chapter 9 we will see that it does have complex eigenvalues and complex eigenvectors.

EXERCISE 6.1.3

Let $R_\theta : \mathbb{R}^3 \to \mathbb{R}^3$ denote the rotation in $\mathbb{R}^3$ with matrix $\begin{bmatrix} \cos\theta & -\sin\theta & 0 \\ \sin\theta & \cos\theta & 0 \\ 0 & 0 & 1 \end{bmatrix}$ where θ is not an integer multiple of π. Determine any real eigenvectors of R_θ and the corresponding eigenvalues.

The Power Method of Determining Eigenvalues

Practical applications of eigenvalues often involve larger matrices with non-integer entries. Such problems often require efficient computer methods for determining eigenvalues. A thorough discussion of such methods is beyond the scope of this book, but we can indicate how powers of matrices provide one tool for finding eigenvalues.

Let $A \in M_{n\times n}(\mathbb{R})$. To simplify the discussion, we suppose that A has n distinct real eigenvalues $\lambda_1, \dots, \lambda_n$, with corresponding eigenvectors $\vec{v}_1, \dots, \vec{v}_n$. We suppose that $|\lambda_1| > |\lambda_i|$ for $2 \leq i \leq n$. We call λ_1 the **dominant** eigenvalue. By Theorem 6.1.3, the set $\{\vec{v}_1, \dots, \vec{v}_n\}$ is linearly independent and hence will form a basis for $\mathbb{R}^n$ by Theorem 2.3.6. Thus, any vector $\vec{x} \in \mathbb{R}^n$ can be written

$$\vec{x} = c_1\vec{v}_1 + \cdots + c_n\vec{v}_n$$

Multiplying both sides by A on the left gives

$$\begin{aligned} A\vec{x} &= A(c_1\vec{v}_1 + \cdots + c_n\vec{v}_n) \\ &= c_1A\vec{v}_1 + \cdots + c_nA\vec{v}_n \\ &= c_1\lambda_1\vec{v}_1 + \cdots + c_n\lambda_n\vec{v}_n \end{aligned}$$

If we multiply again by A on the left, we get

$$\begin{aligned} A^2\vec{x} &= A(c_1\lambda_1\vec{v}_1 + \cdots + c_n\lambda_n\vec{v}_n) \\ &= c_1\lambda_1A\vec{v}_1 + \cdots + c_n\lambda_nA\vec{v}_n \\ &= c_1\lambda_1^2\vec{v}_1 + \cdots + c_n\lambda_n^2\vec{v}_n \end{aligned}$$

Continuing this, we get

$$A^m\vec{x} = c_1\lambda_1^m\vec{v}_1 + \cdots + c_n\lambda_n^m\vec{v}_n$$

For m large, $|\lambda_1^m|$ is much greater than all other terms. Assuming $c_1 \neq 0$, if we divide by $c_1\lambda_1^m$, then all terms on the right-hand side will be negligibly small except for $\vec{v}_1$, so we will be able to identify $\vec{v}_1$. By calculating $A\vec{v}_1$, we determine λ_1.

To make this into an effective procedure, we must control the size of the vectors: if $\lambda_1 > 1$, then $\lambda_1^m \to \infty$ as m gets large, and the procedure would break down. Similarly, if all eigenvalues are between 0 and 1, then $A^m\vec{x} \to \vec{0}$, and the procedure would fail. To avoid these problems, we normalize the vector at each step (that is, convert it to a vector of length 1).

Algorithm 6.1.1 Power Method for Approximating Eigenvalues

For $A \in M_{n\times n}(\mathbb{R})$, pick an initial vector $\vec{x}_0 \in \mathbb{R}^n$ and calculate $\vec{y}_0 = \frac{1}{\|\vec{x}_0\|}\vec{x}_0$.

Let $\vec{x}_1 = A\vec{y}_0$ and then calculate $\vec{y}_1 = \frac{1}{\|\vec{x}_1\|}\vec{x}_1$.

Let $\vec{x}_2 = A\vec{y}_1$ and then calculate $\vec{y}_2 = \frac{1}{\|\vec{x}_2\|}\vec{x}_2$.

and so on.

We seek convergence of $\vec{y}_m$ to some limiting vector; if such a vector exists, it must be $\vec{v}_1$, a unit eigenvector for the largest eigenvalue λ_1. We can then calculate $A\vec{v}_1$ to determine λ_1.

This procedure is illustrated in the following example, which is simple enough that you can check the calculations.

EXAMPLE 6.1.11

Determine the eigenvalue of largest absolute value for the matrix $A = \begin{bmatrix} 13 & 6 \\ -12 & -5 \end{bmatrix}$ by using the power method.

Solution: We first choose any starting vector $\vec{x}_0$. We choose $\vec{x}_0 = \begin{bmatrix} 1 \\ 1 \end{bmatrix}$.

We then calculate

$$\vec{y}_0 = \frac{1}{\|\vec{x}_0\|}\vec{x}_0 = \frac{1}{\sqrt{2}}\begin{bmatrix} 1 \\ 1 \end{bmatrix} \approx \begin{bmatrix} 0.707 \\ 0.707 \end{bmatrix}$$

For our next iteration, we first define $\vec{x}_1$ by

$$\vec{x}_1 = A\vec{y}_0 \approx \begin{bmatrix} 13 & 6 \\ -12 & -5 \end{bmatrix}\begin{bmatrix} 0.707 \\ 0.707 \end{bmatrix} \approx \begin{bmatrix} 13.44 \\ -12.02 \end{bmatrix}$$

and then calculate

$$\vec{y}_1 = \frac{1}{\|\vec{x}_1\|}\vec{x}_1 \approx \begin{bmatrix} 0.745 \\ -0.667 \end{bmatrix}$$

Continuing, we get

$$\vec{x}_2 = A\vec{y}_1 \approx \begin{bmatrix} 5.683 \\ -5.605 \end{bmatrix}, \qquad \vec{y}_2 \approx \begin{bmatrix} 0.712 \\ -0.702 \end{bmatrix}$$

$$\vec{x}_3 = A\vec{y}_2 \approx \begin{bmatrix} 5.044 \\ -5.034 \end{bmatrix}, \qquad \vec{y}_3 \approx \begin{bmatrix} 0.7078 \\ -0.7063 \end{bmatrix}$$

$$\vec{x}_4 = A\vec{y}_3 \approx \begin{bmatrix} 4.9636 \\ -4.9621 \end{bmatrix}, \qquad \vec{y}_4 \approx \begin{bmatrix} 0.7072 \\ -0.7070 \end{bmatrix}$$

At this point, we judge that $\vec{y}_m \to \begin{bmatrix} 0.707 \\ -0.707 \end{bmatrix}$, so it seems that $\vec{v}_1 = \begin{bmatrix} 1 \\ -1 \end{bmatrix}$ is an eigenvector of A. We then find that

$$A\vec{v}_1 = \begin{bmatrix} 7 \\ -7 \end{bmatrix} = 7\vec{v}_1$$

Hence, $\vec{v}_1$ is an eigenvector and we get that the dominant eigenvalue is $\lambda_1 = 7$.

Many questions arise with the power method. What if we poorly choose the initial vector? If we choose $\vec{x}_0$ in the subspace spanned by all eigenvectors of A *except* $\vec{v}_1$, the method will fail to give $\vec{v}_1$. Can we quickly calculate large powers of the matrix? How do we decide when to stop repeating the steps of the procedure? For a computer version of the algorithm, it would be important to have tests to decide that the procedure has converged—or that it will never converge.

Once we have determined the dominant eigenvalue of A, how can we determine other eigenvalues? If A is invertible, the dominant eigenvalue of A^{-1} would give the reciprocal of the eigenvalue of A with the smallest absolute value. Another approach is to observe that if one eigenvalue λ_1 is known, then eigenvalues of $A - \lambda_1 I$ will give us information about other eigenvalues of A. (See Section 6.2 Problem C9.)

PROBLEMS 6.1

Practice Problems

For Problems A1–A5, determine whether the given vectors are eigenvectors of A. If a vector is an eigenvector, then determine the corresponding eigenvalue. Answer without calculating the characteristic polynomial.

A1 $A = \begin{bmatrix} 3 & 4 \\ 2 & 5 \end{bmatrix}, \vec{v}_1 = \begin{bmatrix} 1 \\ 1 \end{bmatrix}, \vec{v}_2 = \begin{bmatrix} -2 \\ 1 \end{bmatrix}$

A2 $A = \begin{bmatrix} 2 & 12 \\ -2 & -8 \end{bmatrix}, \vec{v}_1 = \begin{bmatrix} 1 \\ 1 \end{bmatrix}, \vec{v}_2 = \begin{bmatrix} -2 \\ 1 \end{bmatrix}$

A3 $A = \begin{bmatrix} -10 & 9 & 5 \\ -10 & 9 & 5 \\ 2 & 3 & -1 \end{bmatrix}, \vec{v}_1 = \begin{bmatrix} 1 \\ 0 \\ 1 \end{bmatrix}, \vec{v}_2 = \begin{bmatrix} 1 \\ 0 \\ 2 \end{bmatrix}$

A4 $A = \begin{bmatrix} -10 & 9 & 5 \\ -10 & 9 & 5 \\ 2 & 3 & -1 \end{bmatrix}, \vec{v}_1 = \begin{bmatrix} 1 \\ 1 \\ -1 \end{bmatrix}, \vec{v}_2 = \begin{bmatrix} 1 \\ -1 \\ 1 \end{bmatrix}$

A5 $A = \begin{bmatrix} 4 & 1 & 3 \\ 8 & 6 & 1 \\ -2 & -2 & 3 \end{bmatrix}, \vec{v}_1 = \begin{bmatrix} 1 \\ 3 \\ 3 \end{bmatrix}, \vec{v}_2 = \begin{bmatrix} -2 \\ 3 \\ 1 \end{bmatrix}$

For Problems A6–A15, find all eigenvalues of the matrix and a basis for the eigenspace of each eigenvalue.

A6 $\begin{bmatrix} 0 & 1 \\ -6 & 5 \end{bmatrix}$ **A7** $\begin{bmatrix} 1 & 3 \\ 0 & 1 \end{bmatrix}$ **A8** $\begin{bmatrix} 2 & 0 \\ 0 & 3 \end{bmatrix}$

A9 $\begin{bmatrix} -26 & 10 \\ -75 & 29 \end{bmatrix}$ **A10** $\begin{bmatrix} 1 & 3 \\ 4 & 2 \end{bmatrix}$ **A11** $\begin{bmatrix} 3 & -3 \\ 6 & -6 \end{bmatrix}$

A12 $\begin{bmatrix} 1 & 3 & 5 \\ 0 & 2 & 7 \\ 0 & 0 & 3 \end{bmatrix}$ **A13** $\begin{bmatrix} -4 & 6 & 6 \\ -2 & 2 & 4 \\ -1 & 3 & 1 \end{bmatrix}$

A14 $\begin{bmatrix} 3 & -1 & 2 \\ -1 & 3 & -2 \\ 0 & 2 & 0 \end{bmatrix}$ **A15** $\begin{bmatrix} 0 & 1 & 2 \\ 0 & 2 & 3 \\ 1 & -1 & -1 \end{bmatrix}$

For Problems A16–A22, determine the algebraic and geometric multiplicity of each eigenvalue of the matrix.

A16 $\begin{bmatrix} 2 & -2 \\ 0 & 3 \end{bmatrix}$ **A17** $\begin{bmatrix} 2 & -2 \\ 0 & 2 \end{bmatrix}$ **A18** $\begin{bmatrix} 1 & 1 \\ -1 & 3 \end{bmatrix}$

A19 $\begin{bmatrix} 0 & -5 & 3 \\ -2 & -6 & 6 \\ -2 & -7 & 7 \end{bmatrix}$ **A20** $\begin{bmatrix} 2 & 2 & 2 \\ 2 & 2 & 2 \\ 2 & 2 & 2 \end{bmatrix}$

A21 $\begin{bmatrix} 3 & 1 & 1 \\ 1 & 3 & 1 \\ 1 & 1 & 3 \end{bmatrix}$ **A22** $\begin{bmatrix} 1 & 3 & 1 \\ 0 & 1 & 2 \\ 0 & 2 & 1 \end{bmatrix}$

For Problems A23–A26, find the eigenvalue of the largest absolute value for the matrix by starting with the given $\vec{x}_0$ and using three iterations of the power method (up to y_3).

A23 $\begin{bmatrix} 27 & 84 \\ -7 & -22 \end{bmatrix}, \vec{x}_0 = \begin{bmatrix} 1 \\ 0 \end{bmatrix}$ **A24** $\begin{bmatrix} 5 & 0 \\ 0 & -2 \end{bmatrix}, \vec{x}_0 = \begin{bmatrix} 1 \\ 1 \end{bmatrix}$

A25 $\begin{bmatrix} 3.5 & 4.5 \\ 4.5 & 3.5 \end{bmatrix}, \vec{x}_0 = \begin{bmatrix} 1 \\ 0 \end{bmatrix}$ **A26** $\begin{bmatrix} 4 & 3 \\ 6 & 7 \end{bmatrix}, \vec{x}_0 = \begin{bmatrix} 1 \\ 1 \end{bmatrix}$

Homework Problems

For Problems B1–B5, determine whether the given vectors are eigenvectors of A. If a vector is an eigenvector, then determine the corresponding eigenvalue. Answer without calculating the characteristic polynomial.

B1 $A = \begin{bmatrix} 8 & 4 \\ -9 & -4 \end{bmatrix}, \vec{v}_1 = \begin{bmatrix} 1 \\ 2 \end{bmatrix}, \vec{v}_2 = \begin{bmatrix} -2 \\ 3 \end{bmatrix}$

B2 $A = \begin{bmatrix} -1 & 9 \\ -1 & 5 \end{bmatrix}, \vec{v}_1 = \begin{bmatrix} 1 \\ 1 \end{bmatrix}, \vec{v}_2 = \begin{bmatrix} 3 \\ 1 \end{bmatrix}$

B3 $A = \begin{bmatrix} 3 & -1 & 3 \\ -1 & 9 & -3 \\ 3 & -3 & 11 \end{bmatrix}, \vec{v}_1 = \begin{bmatrix} -3 \\ 0 \\ 1 \end{bmatrix}, \vec{v}_2 = \begin{bmatrix} -1 \\ 1 \\ 1 \end{bmatrix}$

B4 $A = \begin{bmatrix} 3 & -1 & 3 \\ -1 & 9 & -3 \\ 3 & -3 & 11 \end{bmatrix}, \vec{v}_1 = \begin{bmatrix} 1 \\ 5 \\ 3 \end{bmatrix}, \vec{v}_2 = \begin{bmatrix} 1 \\ -2 \\ 3 \end{bmatrix}$

B5 $A = \begin{bmatrix} 1 & 2 & 6 \\ 3 & -9 & 8 \\ -2 & 1 & 8 \end{bmatrix}, \vec{v}_1 = \begin{bmatrix} 2 \\ 1 \\ 1 \end{bmatrix}, \vec{v}_2 = \begin{bmatrix} 1 \\ 3 \\ 0 \end{bmatrix}$

For Problems B6–B17, find all eigenvalues of the matrix and a basis for the eigenspace of each eigenvalue.

B6 $\begin{bmatrix} 2 & 2 \\ 2 & 5 \end{bmatrix}$ **B7** $\begin{bmatrix} 3 & 3 \\ 2 & 4 \end{bmatrix}$ **B8** $\begin{bmatrix} -2 & 6 \\ 1 & 3 \end{bmatrix}$

B9 $\begin{bmatrix} 5 & 0 \\ -2 & -3 \end{bmatrix}$ **B10** $\begin{bmatrix} 5 & 1 \\ -9 & -1 \end{bmatrix}$ **B11** $\begin{bmatrix} -1 & 4 \\ 1 & -1 \end{bmatrix}$

B12 $\begin{bmatrix} 2 & 0 & 0 \\ -8 & 4 & 0 \\ 5 & -3 & 1 \end{bmatrix}$ **B13** $\begin{bmatrix} 1 & 2 & -2 \\ 2 & 1 & 2 \\ -2 & 2 & 1 \end{bmatrix}$

B14 $\begin{bmatrix} 1 & -1 & 1 \\ 1 & 3 & -1 \\ 1 & 1 & 1 \end{bmatrix}$ **B15** $\begin{bmatrix} 1 & -1 & 1 \\ 0 & 2 & -1 \\ -2 & -1 & 0 \end{bmatrix}$

B16 $\begin{bmatrix} 0 & 2 & 2 \\ 2 & 0 & 2 \\ 2 & 2 & 0 \end{bmatrix}$ **B17** $\begin{bmatrix} 7 & -1 & -1 \\ 2 & 4 & -2 \\ -1 & 1 & 7 \end{bmatrix}$

For Problems B18–B26, determine the algebraic and geometric multiplicity of each eigenvalue of the matrix.

B18 $\begin{bmatrix} 2 & 0 \\ 1 & 2 \end{bmatrix}$ **B19** $\begin{bmatrix} 5 & 4 \\ 2 & 3 \end{bmatrix}$ **B20** $\begin{bmatrix} 7 & 2 \\ -8 & -1 \end{bmatrix}$

B21 $\begin{bmatrix} -5 & 1 & -1 \\ 0 & 2 & 5 \\ 0 & 1 & -2 \end{bmatrix}$ **B22** $\begin{bmatrix} 3 & 8 & 10 \\ 1 & 4 & 4 \\ -1 & -5 & -5 \end{bmatrix}$

B23 $\begin{bmatrix} 2 & 3 & 3 \\ 3 & 2 & 3 \\ 3 & 3 & 2 \end{bmatrix}$ **B24** $\begin{bmatrix} -2 & -1 & -2 \\ 2 & 1 & 4 \\ 1 & 1 & 1 \end{bmatrix}$

B25 $\begin{bmatrix} 0 & 1 & 1 \\ -1 & -1 & -1 \\ 2 & 1 & 1 \end{bmatrix}$ **B26** $\begin{bmatrix} 3 & 4 & 1 \\ 1 & 2 & 1 \\ -3 & -4 & -1 \end{bmatrix}$

For Problems B27–B30, find the eigenvalue of the largest absolute value for the matrix by starting with the given $\vec{x}_0$ and using three iterations of the power method (up to y_3).

B27 $\begin{bmatrix} 5 & 3 \\ 15 & 17 \end{bmatrix}, \vec{x}_0 = \begin{bmatrix} 1 \\ 1 \end{bmatrix}$ **B28** $\begin{bmatrix} 5 & 8 \\ 2 & 5 \end{bmatrix}, \vec{x}_0 = \begin{bmatrix} 1 \\ 0 \end{bmatrix}$

B29 $\begin{bmatrix} 14 & -9 \\ 4 & -1 \end{bmatrix}, \vec{x}_0 = \begin{bmatrix} 1 \\ 0 \end{bmatrix}$ **B30** $\begin{bmatrix} 10 & -6 \\ 6 & -3 \end{bmatrix}, \vec{x}_0 = \begin{bmatrix} 2 \\ 1 \end{bmatrix}$

Conceptual Problems

C1 Invent $A \in M_{2\times 2}(\mathbb{R})$ that has eigenvalues 2 and 3 with corresponding eigenvectors $\begin{bmatrix} 1 \\ 2 \end{bmatrix}, \begin{bmatrix} 1 \\ 3 \end{bmatrix}$ respectively.

C2 Invent $A \in M_{3\times 3}(\mathbb{R})$ that has an eigenvalue λ such that the geometric multiplicity is less than its algebraic multiplicity. Justify your answer.

C3 Invent $A \in M_{3\times 3}(\mathbb{R})$ that has 3 distinct eigenvalues. Justify your answer.

C4 Suppose $A, B \in M_{n\times n}(\mathbb{R})$ and that $\vec{u}$ and $\vec{v}$ are eigenvectors of both A and B where $A\vec{u} = 6\vec{u}$, $B\vec{u} = 5\vec{u}$, $A\vec{v} = 10\vec{v}$, and $B\vec{v} = 3\vec{v}$.

(a) If $C = AB$, show that $5\vec{u} + 3\vec{v}$ is an eigenvector of C. Find the corresponding eigenvalue.

(b) If $n = 2$ and $\vec{w} = \begin{bmatrix} 0.2 \\ 1.4 \end{bmatrix}$, compute $AB\vec{w}$.

C5 Let $A, B \in M_{n\times n}(\mathbb{R})$ with eigenvalues λ and μ respectively. Does this imply that $\lambda\mu$ is an eigenvalue of AB? Justify your answer.

C6 Prove the following addition to the Invertible Matrix Theorem: A matrix A is invertible if and only if $\lambda = 0$ is not an eigenvalue of A.

C7 Show that if A is invertible and $\vec{v}$ is an eigenvector of A, then $\vec{v}$ is also an eigenvector of A^{-1}. How are the corresponding eigenvalues related?

C8 Suppose that $\vec{v}$ is an eigenvector of both the matrix A and the matrix B, with corresponding eigenvalue λ for A and corresponding eigenvalue μ for B. Show that $\vec{v}$ is an eigenvector of $(A + B)$ and of AB. Determine the corresponding eigenvalues.

C9 (a) Show that if λ is an eigenvalue of a matrix A, then λ^m is an eigenvalue of A^m. How are the corresponding eigenvectors related?

(b) Give an example of $A \in M_{2\times 2}(\mathbb{R})$ such that A has no real eigenvalues, but A^3 does have real eigenvalues. (Hint: see Section 3.3 Problem C5.)

C10 (a) Let $A \in M_{n\times n}(\mathbb{R})$ with $\text{rank}(A) = r < n$. Prove that 0 is an eigenvalue of A and determine its geometric multiplicity.

(b) Give an example of $A \in M_{3\times 3}(\mathbb{R})$ with $\text{rank}(A) = r < n$ such that the algebraic multiplicity of the eigenvalue 0 is greater than its geometric multiplicity.

C11 Suppose that $A \in M_{n\times n}(\mathbb{R})$ such that the sum of the entries in each row is the same. That is,

$$a_{i1} + a_{i2} + \cdots + a_{in} = c$$

for all $1 \leq i \leq n$. Show that $\vec{v} = \begin{bmatrix} 1 \\ \vdots \\ 1 \end{bmatrix}$ is an eigenvector of A. (Such matrices arise in probability theory.)

6.2 Diagonalization

In the last section, we found that if the k distinct eigenvalues $\lambda_1, \dots, \lambda_k$ of $A \in M_{n\times n}(\mathbb{R})$ all have the property that their geometric multiplicity is equal to their algebraic multiplicity, then, by combining the basis vectors from all k eigenspaces, we have a basis for $\mathbb{R}^n$ of eigenvectors of A. Such bases of eigenvectors are extremely useful in a wide variety of applications.

Definition
Diagonalizable

A matrix A is said to be **diagonalizable** if there exists an invertible matrix P and diagonal matrix D such that $P^{-1}AP = D$. In this case, we say that the matrix P **diagonalizes** A to its **diagonal form** D.

It may be tempting to think that $P^{-1}AP = D$ implies that $A = D$ since P and P^{-1} are inverses. However, this is *not* true in general since matrix multiplication is not commutative. Not surprisingly though, if A and B are matrices such that $P^{-1}AP = B$ for some invertible matrix P, then A and B have many similarities.

Theorem 6.2.1

If $A, B \in M_{n\times n}(\mathbb{R})$ such that $P^{-1}AP = B$ for some invertible matrix P, then A and B have

(1) the same determinant,
(2) the same eigenvalues,
(3) the same rank,
(4) the same **trace**, where the trace of A is defined by

$$\operatorname{tr}(A) = a_{11} + a_{22} + \cdots + a_{nn}$$

Section 5.2 Problem C5 proved (1). The proofs of (2), (3), and (4) are left as Problems C1, C2, and C3, respectively.

This theorem motivates the following definition.

Definition
Similar Matrices

If $A, B \in M_{n\times n}(\mathbb{R})$ such that $P^{-1}AP = B$ for some invertible matrix P, then A and B are said to be **similar**.

We could now restate the definition of a matrix being diagonalizable as the matrix being similar to a diagonal matrix.

CONNECTION

If you have previously covered Section 4.6, then the equation $P^{-1}AP = B$ should seem familiar to you. In particular, Theorem 4.6.1 says that $[L]_{\mathcal{B}} = P^{-1}[L]_S P$. That is, similar matrices, and hence diagonalization, have a very nice geometric interpretation when viewed in terms of finding a natural basis for a linear mapping.

The following theorem tells us when a matrix is diagonalizable.

Theorem 6.2.2 Diagonalization Theorem

A matrix $A \in M_{n\times n}(\mathbb{R})$ is diagonalizable if and only if there exists a basis for $\mathbb{R}^n$ which consists of eigenvectors of A.

Proof: First, suppose that A is diagonalizble. Then, by definition, there exists an invertible matrix $P = \begin{bmatrix} \vec{v}_1 & \cdots & \vec{v}_n \end{bmatrix}$ such that

$$P^{-1}AP = \operatorname{diag}(\lambda_1, \dots, \lambda_n)$$

Multiplying both sides on the left by P gives

$$\begin{aligned} AP &= P\operatorname{diag}(\lambda_1, \dots, \lambda_n) \\ A\begin{bmatrix} \vec{v}_1 & \cdots & \vec{v}_n \end{bmatrix} &= P\begin{bmatrix} \lambda_1\vec{e}_1 & \cdots & \lambda_n\vec{e}_n \end{bmatrix} \\ \begin{bmatrix} A\vec{v}_1 & \cdots & A\vec{v}_n \end{bmatrix} &= \begin{bmatrix} \lambda_1 P\vec{e}_1 & \cdots & \lambda_n P\vec{e}_n \end{bmatrix} \\ \begin{bmatrix} A\vec{v}_1 & \cdots & A\vec{v}_n \end{bmatrix} &= \begin{bmatrix} \lambda_1\vec{v}_1 & \cdots & \lambda_n\vec{v}_n \end{bmatrix} \end{aligned}$$

by Theorem 3.1.3. Thus, $A\vec{v}_i = \lambda_i\vec{v}_i$ for $1 \leq i \leq n$. Moreover, since P is invertible, $\{\vec{v}_1, \dots, \vec{v}_n\}$ is a basis for $\mathbb{R}^n$ by the Invertible Matrix Theorem which also implies that $\vec{v}_i \neq \vec{0}$ for $1 \leq i \leq n$. Hence, $\{\vec{v}_1, \dots, \vec{v}_n\}$ is a basis for $\mathbb{R}^n$ of eigenvectors of A.

On the other hand, if $\{\vec{v}_1, \dots, \vec{v}_n\}$ is a basis for $\mathbb{R}^n$ of eigenvectors, then the matrix $P = \begin{bmatrix} \vec{v}_1 & \cdots & \vec{v}_n \end{bmatrix}$ is invertible by the Invertible Matrix Theorem, and we have

$$\begin{aligned} P^{-1}AP &= P^{-1}\begin{bmatrix} A\vec{v}_1 & \cdots & A\vec{v}_n \end{bmatrix} \\ &= P^{-1}\begin{bmatrix} \lambda_1\vec{v}_1 & \cdots & \lambda_n\vec{v}_n \end{bmatrix} && \text{since } A\vec{v}_i = \lambda_i\vec{v}_i \\ &= P^{-1}\begin{bmatrix} \lambda_1 P\vec{e}_1 & \cdots & \lambda_n P\vec{e}_n \end{bmatrix} && \text{by Theorem 3.1.3} \\ &= P^{-1}P\begin{bmatrix} \lambda_1\vec{e}_1 & \cdots & \lambda_n\vec{e}_n \end{bmatrix} \\ &= \operatorname{diag}(\lambda_1, \dots, \lambda_n) \end{aligned}$$

is diagonal. Therefore, A is diagonalizable. ■

Remark

It is important to observe that the proof of the Diagonalization Theorem tells us that if $\{\vec{v}_1, \dots, \vec{v}_n\}$ is a basis of eigenvectors of A, then the matrix $P = \begin{bmatrix} \vec{v}_1 & \cdots & \vec{v}_n \end{bmatrix}$ diagonalizes A to a diagonal matrix $D = \operatorname{diag}(\lambda_1, \dots, \lambda_n)$, where λ_i is an eigenvalue of A corresponding to $\vec{v}_i$ for $1 \leq i \leq n$. In particular, we never actually need to multiply out $P^{-1}AP$ by hand (except to check our answer). Moreover, it shows us that we can put the eigenvectors $\vec{v}_1, \dots, \vec{v}_n$ in any order in P as long as we order the eigenvalues in D to match.

Combining the Diagonalization Theorem, Theorem 6.1.3, and the fact that the sum of algebraic multiplicities of an $n \times n$ matrix must be n by the Fundamental Theorem of Algebra, we get the following useful corollaries.

Theorem 6.2.3 A matrix is diagonalizable if and only if every eigenvalue of the matrix has its geometric multiplicity equal to its algebraic multiplicity.

Theorem 6.2.4 If $A \in M_{n\times n}(\mathbb{R})$ has n distinct eigenvalues, then A is diagonalizable.

CONNECTION

Observe that it is possible for a matrix A with real entries to have non-real eigenvalues, which will lead to non-real eigenvectors. In this case, there cannot exist a basis for $\mathbb{R}^n$ of eigenvectors of A, and so we will say that A is not diagonalizable over $\mathbb{R}$. In Chapter 9, we will examine the case where complex eigenvalues and eigenvectors are allowed.

EXAMPLE 6.2.1 Find an invertible matrix P and a diagonal matrix D such that $P^{-1}AP = D$, where $A = \begin{bmatrix} 2 & 3 \\ 3 & 2 \end{bmatrix}$.

Solution: We need to find a basis for $\mathbb{R}^2$ of eigenvectors of A. Hence, we need to find a basis for the eigenspace of each eigenvalue of A. The characteristic polynomial of A is

$$C(\lambda) = \det(A - \lambda I) = \begin{vmatrix} 2-\lambda & 3 \\ 3 & 2-\lambda \end{vmatrix} = \lambda^2 - 4\lambda - 5 = (\lambda - 5)(\lambda + 1)$$

Hence, the eigenvalues of A are $\lambda_1 = 5$ and $\lambda_2 = -1$.

For $\lambda_1 = 5$, we get

$$A - \lambda_1 I = \begin{bmatrix} -3 & 3 \\ 3 & -3 \end{bmatrix} \sim \begin{bmatrix} 1 & -1 \\ 0 & 0 \end{bmatrix}$$

So, $\vec{v}_1 = \begin{bmatrix} 1 \\ 1 \end{bmatrix}$ is an eigenvector for $\lambda_1 = 5$ and $\{\vec{v}_1\}$ is a basis for its eigenspace.

For $\lambda_2 = -1$, we get

$$A - \lambda_2 I = \begin{bmatrix} 3 & 3 \\ 3 & 3 \end{bmatrix} \sim \begin{bmatrix} 1 & 1 \\ 0 & 0 \end{bmatrix}$$

So, $\vec{v}_2 = \begin{bmatrix} -1 \\ 1 \end{bmatrix}$ is an eigenvector for $\lambda_2 = -1$ and $\{\vec{v}_2\}$ is a basis for its eigenspace.

Thus, $\{\vec{v}_1, \vec{v}_2\}$ is a basis for $\mathbb{R}^2$, and so if we let $P = \begin{bmatrix} \vec{v}_1 & \vec{v}_2 \end{bmatrix} = \begin{bmatrix} 1 & -1 \\ 1 & 1 \end{bmatrix}$, we get

$$P^{-1}AP = \operatorname{diag}(\lambda_1, \lambda_2) = \begin{bmatrix} 5 & 0 \\ 0 & -1 \end{bmatrix} = D$$

Note that we could have instead taken $P = \begin{bmatrix} \vec{v}_2 & \vec{v}_1 \end{bmatrix} = \begin{bmatrix} -1 & 1 \\ 1 & 1 \end{bmatrix}$, which would have given

$$P^{-1}AP = \operatorname{diag}(\lambda_2, \lambda_1) = \begin{bmatrix} -1 & 0 \\ 0 & 5 \end{bmatrix}$$

EXAMPLE 6.2.2

Determine whether $A = \begin{bmatrix} 0 & 3 & -2 \\ -2 & 5 & -2 \\ -2 & 3 & 0 \end{bmatrix}$ is diagonalizable. If it is, find an invertible matrix P and a diagonal matrix D such that $P^{-1}AP = D$.

Solution: To evaluate $\det(A - \lambda I)$, we use $C_1 - C_3$ and then $R_3 + R_1$ to get

$$\begin{aligned} C(\lambda) &= \begin{vmatrix} 0-\lambda & 3 & -2 \\ -2 & 5-\lambda & -2 \\ -2 & 3 & 0-\lambda \end{vmatrix} = \begin{vmatrix} 2-\lambda & 3 & -2 \\ 0 & 5-\lambda & -2 \\ -2+\lambda & 3 & 0-\lambda \end{vmatrix} \\ &= \begin{vmatrix} 2-\lambda & 3 & -2 \\ 0 & 5-\lambda & -2 \\ 0 & 6 & -2-\lambda \end{vmatrix} = (2-\lambda)(\lambda^2 - 3\lambda + 2) \\ &= -(\lambda-2)(\lambda-2)(\lambda-1) \end{aligned}$$

Hence, $\lambda_1 = 2$ is an eigenvalue with algebraic multiplicity 2, and $\lambda_2 = 1$ is an eigenvalue with algebraic multiplicity 1. By Theorem 6.1.2, the geometric multiplicity of $\lambda_2 = 1$ must equal 1. Thus, A is diagonalizable if and only if the geometric multiplicity of $\lambda_1 = 2$ is 2.

For $\lambda_1 = 2$, we get

$$A - \lambda_1 I = \begin{bmatrix} -2 & 3 & -2 \\ -2 & 3 & -2 \\ -2 & 3 & -2 \end{bmatrix} \sim \begin{bmatrix} 1 & -3/2 & 1 \\ 0 & 0 & 0 \\ 0 & 0 & 0 \end{bmatrix}$$

Thus, a basis for the eigenspace of λ_1 is $\left\{ \begin{bmatrix} 3/2 \\ 1 \\ 0 \end{bmatrix}, \begin{bmatrix} -1 \\ 0 \\ 1 \end{bmatrix} \right\}$. Hence, the geometric multiplicity of λ_1 equals its algebraic multiplicity.

Therefore, by Theorem 6.2.3, A is diagonalizable. To diagonalize A we still need to find a basis for the eigenspace of λ_2.

For $\lambda_2 = 1$, we get

$$A - \lambda_2 I = \begin{bmatrix} -1 & 3 & -2 \\ -2 & 4 & -2 \\ -2 & 3 & -1 \end{bmatrix} \sim \begin{bmatrix} 1 & 0 & -1 \\ 0 & 1 & -1 \\ 0 & 0 & 0 \end{bmatrix}$$

Therefore, $\left\{ \begin{bmatrix} 1 \\ 1 \\ 1 \end{bmatrix} \right\}$ is a basis for the eigenspace of λ_2.

So, we can take $P = \begin{bmatrix} 3/2 & -1 & 1 \\ 1 & 0 & 1 \\ 0 & 1 & 1 \end{bmatrix}$ and get $P^{-1}AP = \begin{bmatrix} 2 & 0 & 0 \\ 0 & 2 & 0 \\ 0 & 0 & 1 \end{bmatrix}$.

EXERCISE 6.2.1

Diagonalize $A = \begin{bmatrix} 1 & 0 & -1 \\ 11 & -4 & -7 \\ -7 & 3 & 4 \end{bmatrix}$.

EXAMPLE 6.2.3

Is the matrix $A = \begin{bmatrix} -1 & 7 & -5 \\ -4 & 11 & -6 \\ -4 & 8 & -3 \end{bmatrix}$ diagonalizable?

Solution: To evaluate $\det(A - \lambda I)$, we first use $R_3 - R_2$ and then $C_2 + C_3$ to get

$$C(\lambda) = \begin{vmatrix} -1-\lambda & 7 & -5 \\ -4 & 11-\lambda & -6 \\ -4 & 8 & -3-\lambda \end{vmatrix} = \begin{vmatrix} -1-\lambda & 7 & -5 \\ -4 & 11-\lambda & -6 \\ 0 & -3+\lambda & 3-\lambda \end{vmatrix}$$
$$= \begin{vmatrix} -1-\lambda & 2 & -5 \\ -4 & 5-\lambda & -6 \\ 0 & 0 & 3-\lambda \end{vmatrix} = (3-\lambda)(\lambda^2 - 4\lambda + 3)$$
$$= -(\lambda - 3)(\lambda - 3)(\lambda - 1)$$

Thus, $\lambda_1 = 3$ is an eigenvalue with algebraic multiplicity 2, and $\lambda_2 = 1$ is an eigenvalue with algebraic multiplicity 1.
For $\lambda_1 = 3$, we get

$$A - \lambda_1 I = \begin{bmatrix} -4 & 7 & -5 \\ -4 & 8 & -6 \\ -4 & 8 & -6 \end{bmatrix} \sim \begin{bmatrix} 1 & 0 & -1/2 \\ 0 & 1 & -1 \\ 0 & 0 & 0 \end{bmatrix}$$

Thus, a basis for the eigenspace of λ_1 is $\left\{ \begin{bmatrix} 1/2 \\ 1 \\ 1 \end{bmatrix} \right\}$. Hence, the geometric multiplicity of λ_1 is 1, which is less than its algebraic multiplicity. So, A is not diagonalizable by Theorem 6.2.3.

EXERCISE 6.2.2

Show that $A = \begin{bmatrix} 2 & 1 \\ 0 & 2 \end{bmatrix}$ is not diagonalizable.

Applications

Graphing Quadratic Forms A geometrical application of diagonalization occurs when we try to picture the graph of a quadratic equation in two variables, such as $ax^2 + 2bxy + cy^2 = d$. It turns out that we should consider the associated matrix $\begin{bmatrix} a & b \\ b & c \end{bmatrix}$. By diagonalizing this matrix, we can easily recognize the graph as an ellipse, a hyperbola, or perhaps some degenerate case. This application will be discussed in Section 8.3.

Deformation of Solids Imagine, for example, a small steel block that experiences a small deformation when some forces are applied. The change of shape in the block can be described in terms of a 3×3 *strain matrix*. This matrix can always be diagonalized, so it turns out that we can identify the change of shape as the composition of three stretches along mutually orthogonal directions. This application is discussed in Section 8.4.

Finding a Geometrically Natural Basis In Section 4.6 we saw that if $L : \mathbb{R}^n \to \mathbb{R}^n$ is a linear transformation, then its matrix with respect to the basis $\mathcal{B}$ is determined from its standard matrix by the equation

$$[L]_{\mathcal{B}} = P^{-1}[L]_S P$$

where $P = \begin{bmatrix} \vec{v}_1 & \cdots & \vec{v}_n \end{bmatrix}$ is the change of coordinates matrix. Example 4.6.5 and Example 4.6.7 show that we can more easily give a geometrical interpretation of a linear mapping L if there is a basis $\mathcal{B}$ such that $[L]_{\mathcal{B}}$ is in diagonal form. Hence, our diagonalization process is a method for finding such a geometrically natural basis. In particular, if the standard matrix of L is diagonalizable, then the basis for $\mathbb{R}^n$ of eigenvectors forms the geometrically natural basis.

Systems of Linear Difference Equations If $A \in M_{n\times n}(\mathbb{R})$ and $\vec{s}(m)$ is a vector for each positive integer m, then the matrix vector equation

$$\vec{s}(m+1) = A\vec{s}(m)$$

may be regarded as a system of n linear first-order difference equations, describing the coordinates $s_1, s_2, \dots, s_n$ at times $m+1$ in terms of those at time m. They are "first-order difference" equations because they involve only one time difference from m to $m+1$; the Fibonacci equation $s(m+1) = s(m) + s(m-1)$ is a second-order difference equation.

Linear difference equations arise in many settings. Consider, for example, a population that is divided into two groups; we count these two groups at regular intervals (say, once a month) so that at every time n, we have a vector $\vec{p} = \begin{bmatrix} p_1(n) \\ p_2(n) \end{bmatrix}$ that tells us how many are in each group. For some situations, the change from month to month can be described by saying that the vector $\vec{p}$ changes according to the rule

$$\vec{p}(n+1) = A\vec{p}(n)$$

where A is some known 2×2 matrix. It follows that $p(n) = A^n p(0)$. We are often interested in understanding what happens to the population "in the long run." This requires us to calculate A^n for n large.

Markov processes form a special class of this large class of systems of linear difference equations, but there are applications that do not fit the Markov assumptions. For example, in population models, we might wish to consider deaths (so that some column sums of A would be less than 1) or births (so that some entries in A would be greater than 1). Similar considerations apply to some economic models, which are represented by matrix models. A proper discussion of such models requires more theory than is discussed in this book.

PROBLEMS 6.2

Practice Problems

For Problems A1–A6, determine whether P diagonalizes A by checking whether the columns of P are eigenvectors of A. If so, determine P^{-1} and check that $P^{-1}AP$ is diagonal.

A1 $A = \begin{bmatrix} 11 & 6 \\ 9 & -4 \end{bmatrix}$, $P = \begin{bmatrix} 2 & -1 \\ 1 & 3 \end{bmatrix}$

A2 $A = \begin{bmatrix} 6 & 5 \\ 3 & -7 \end{bmatrix}$, $P = \begin{bmatrix} 1 & 2 \\ 1 & 1 \end{bmatrix}$

A3 $A = \begin{bmatrix} -2 & 2 \\ 2 & 0 \end{bmatrix}$, $P = \begin{bmatrix} 1 & -1 \\ 0 & 1 \end{bmatrix}$

A4 $A = \begin{bmatrix} 5 & -8 \\ 4 & -7 \end{bmatrix}$, $P = \begin{bmatrix} 2 & 1 \\ 1 & 1 \end{bmatrix}$

A5 $A = \begin{bmatrix} 2 & 4 & 4 \\ 4 & 2 & 4 \\ 4 & 4 & 2 \end{bmatrix}$, $P = \begin{bmatrix} -1 & -1 & 1 \\ 1 & 0 & 1 \\ 0 & 1 & 1 \end{bmatrix}$

A6 $A = \begin{bmatrix} 0 & 1 & 1 \\ 1 & 1 & 1 \\ 2 & 1 & 1 \end{bmatrix}$, $P = \begin{bmatrix} 3 & 0 & -1 \\ 4 & -1 & 0 \\ 5 & 1 & 1 \end{bmatrix}$

For Problems A7–A23, either diagonalize the matrix or show that the matrix is not diagonalizable.

A7 $\begin{bmatrix} 7 & 3 \\ 0 & -8 \end{bmatrix}$ **A8** $\begin{bmatrix} 5 & 2 \\ 0 & 5 \end{bmatrix}$ **A9** $\begin{bmatrix} 1 & 9 \\ 4 & -4 \end{bmatrix}$

A10 $\begin{bmatrix} 3 & 2 \\ 5 & 6 \end{bmatrix}$ **A11** $\begin{bmatrix} -2 & 3 \\ 4 & -3 \end{bmatrix}$ **A12** $\begin{bmatrix} 3 & 6 \\ -5 & -3 \end{bmatrix}$

A13 $\begin{bmatrix} 3 & 0 \\ -3 & 3 \end{bmatrix}$ **A14** $\begin{bmatrix} 4 & 4 \\ 4 & 4 \end{bmatrix}$ **A15** $\begin{bmatrix} -2 & 5 \\ 5 & -2 \end{bmatrix}$

A16 $\begin{bmatrix} 0 & 1 & 0 \\ 1 & 0 & 1 \\ 1 & 1 & 1 \end{bmatrix}$ **A17** $\begin{bmatrix} 6 & -9 & -5 \\ -4 & 9 & 4 \\ 9 & -17 & -8 \end{bmatrix}$

A18 $\begin{bmatrix} -2 & 7 & 3 \\ -1 & 2 & 1 \\ 0 & 2 & 1 \end{bmatrix}$ **A19** $\begin{bmatrix} -1 & 6 & 3 \\ 3 & -4 & -3 \\ -6 & 12 & 8 \end{bmatrix}$

A20 $\begin{bmatrix} 0 & 6 & -8 \\ -2 & 4 & -4 \\ -2 & 2 & -2 \end{bmatrix}$ **A21** $\begin{bmatrix} 2 & 2 & 2 \\ 2 & 2 & 2 \\ 3 & 2 & 1 \end{bmatrix}$

A22 $\begin{bmatrix} 2 & 0 & 0 \\ -1 & 0 & 1 \\ -1 & -2 & 3 \end{bmatrix}$ **A23** $\begin{bmatrix} -3 & -3 & 5 \\ 13 & 10 & -13 \\ 3 & 2 & -1 \end{bmatrix}$

Homework Problems

For Problems B1–B7, determine whether P diagonalizes A by checking whether the columns of P are eigenvectors of A. If so, determine P^{-1} and check that $P^{-1}AP$ is diagonal.

B1 $A = \begin{bmatrix} 4 & 2 \\ 6 & 3 \end{bmatrix}$, $P = \begin{bmatrix} 2 & -1 \\ 3 & 2 \end{bmatrix}$

B2 $A = \begin{bmatrix} 3 & 2 \\ 3 & -2 \end{bmatrix}$, $P = \begin{bmatrix} 2 & 1 \\ 1 & 1 \end{bmatrix}$

B3 $A = \begin{bmatrix} 0 & 1 \\ 2 & 3 \end{bmatrix}$, $P = \begin{bmatrix} 2 & 1 \\ 3 & -1 \end{bmatrix}$

B4 $A = \begin{bmatrix} 6 & 2 \\ -2 & 1 \end{bmatrix}$, $P = \begin{bmatrix} -1 & 2 \\ 2 & -1 \end{bmatrix}$

B5 $A = \begin{bmatrix} -5 & 8 & 18 \\ 2 & 1 & -6 \\ -4 & 4 & 13 \end{bmatrix}$, $P = \begin{bmatrix} 1 & 1 & 3 \\ -1 & 1 & 0 \\ 1 & 0 & 1 \end{bmatrix}$

B6 $A = \begin{bmatrix} 2 & 3 & 0 \\ 1 & 1 & 1 \\ -1 & 0 & 1 \end{bmatrix}$, $P = \begin{bmatrix} 1 & 1 & 1 \\ 0 & -2 & 1 \\ -1 & 1 & -2 \end{bmatrix}$

B7 $A = \begin{bmatrix} 7 & -4 & -4 \\ 4 & -1 & -4 \\ 4 & -4 & -1 \end{bmatrix}$, $P = \begin{bmatrix} 1 & 1 & 1 \\ 1 & 0 & 1 \\ 1 & 1 & 0 \end{bmatrix}$

For Problems B8–B23, either diagonalize the matrix or show that the matrix is not diagonalizable.

B8 $\begin{bmatrix} 3 & 0 \\ 1 & 1 \end{bmatrix}$ **B9** $\begin{bmatrix} 2 & 3 \\ 0 & 2 \end{bmatrix}$ **B10** $\begin{bmatrix} -2 & 8 \\ 1 & -4 \end{bmatrix}$

B11 $\begin{bmatrix} 3 & 2 \\ 5 & 6 \end{bmatrix}$ **B12** $\begin{bmatrix} 4 & -5 \\ 1 & -2 \end{bmatrix}$ **B13** $\begin{bmatrix} 8 & 4 \\ -1 & 4 \end{bmatrix}$

B14 $\begin{bmatrix} 2 & -3 & 1 \\ 0 & 3 & 1 \\ 0 & 0 & 2 \end{bmatrix}$ **B15** $\begin{bmatrix} 2 & -2 & -5 \\ -2 & -5 & -2 \\ -5 & 2 & 2 \end{bmatrix}$

B16 $\begin{bmatrix} 0 & 0 & 0 \\ 4 & 2 & 0 \\ -1 & 1 & 1 \end{bmatrix}$ **B17** $\begin{bmatrix} 1 & 2 & 1 \\ -2 & 5 & 1 \\ 2 & -2 & 2 \end{bmatrix}$

B18 $\begin{bmatrix} -4 & -3 & 7 \\ -2 & -9 & 14 \\ -1 & -3 & 4 \end{bmatrix}$ **B19** $\begin{bmatrix} 1 & 2 & -4 \\ 2 & -2 & -2 \\ -4 & -2 & 1 \end{bmatrix}$

B20 $\begin{bmatrix} 5 & 4 & -12 \\ 3 & 4 & -9 \\ 2 & 2 & -5 \end{bmatrix}$ **B21** $\begin{bmatrix} 3 & 1 & -4 \\ 3 & 2 & -5 \\ 2 & 1 & -3 \end{bmatrix}$

B22 $\begin{bmatrix} -5 & 6 & 5 \\ -2 & 0 & 2 \\ 1 & 2 & -1 \end{bmatrix}$ **B23** $\begin{bmatrix} 1 & 0 & -1 \\ 0 & 1 & 1 \\ -1 & 1 & 2 \end{bmatrix}$

Conceptual Problems

C1 Prove that if A and B are similar, then A and B have the same eigenvalues.

C2 Prove that if A and B are similar, then A and B have the same rank.

C3 (a) Let $A, B \in M_{n\times n}(\mathbb{R})$. Prove that $\text{tr}(AB) = \text{tr}(BA)$.

(b) Use the result of part (a) to prove that if A and B are similar, then $\text{tr}(A) = \text{tr}(B)$.

C4 (a) Suppose that P diagonalizes A and that the diagonal form is D. Show that $A = PDP^{-1}$.

(b) Use the result of part (a) and properties of eigenvectors to calculate a matrix that has eigenvalues 2 and 3 with corresponding eigenvectors $\begin{bmatrix}1\\2\end{bmatrix}$ and $\begin{bmatrix}1\\3\end{bmatrix}$, respectively.

(c) Determine a matrix that has eigenvalues 2, −2, and 3, with corresponding eigenvectors $\begin{bmatrix}1\\0\\1\end{bmatrix}$, $\begin{bmatrix}1\\1\\-1\end{bmatrix}$, and $\begin{bmatrix}1\\-1\\2\end{bmatrix}$, respectively.

C5 (a) Suppose that P diagonalizes A and that the diagonal form is D. Show that $A^k = PD^kP^{-1}$.

(b) Use the result of part (a) to calculate A^5, where $A = \begin{bmatrix}-1 & 6 & 3\\3 & -4 & -3\\-6 & 12 & 8\end{bmatrix}$ is the matrix from Problem A19.

C6 (a) Suppose that A is diagonalizable. Prove that $\text{tr}(A)$ is equal to the sum of the eigenvalues of A (including repeated eigenvalues) by using Theorem 6.2.1.

(b) Use the result of part (a) to determine, by inspection, the algebraic and geometric multiplicities of all of the eigenvalues of

$$A = \begin{bmatrix}a+b & a & a\\a & a+b & a\\a & a & a+b\end{bmatrix}.$$

C7 (a) Suppose that A is diagonalizable. Prove that $\det A$ is equal to the product of the eigenvalues of A (repeated according to their multiplicity) by considering $P^{-1}AP$.

(b) Show that the constant term in the characteristic polynomial is $\det A$. (Hint: how do you find the constant term in any polynomial $p(\lambda)$?)

(c) Without assuming that A is diagonalizable, show that $\det A$ is equal to the product of the roots of the characteristic equation of A (including any repeated roots and complex roots). (Hint: consider the constant term in the characteristic equation and the factored version of that equation.)

C8 Let $A \in M_{n\times n}(\mathbb{R})$. Prove that A is invertible if and only if A does not have 0 as an eigenvalue. (Hint: see Problem C7.)

C9 Suppose that A is diagonalized by the matrix P and that the eigenvalues of A are $\lambda_1, \dots, \lambda_n$. Show that the eigenvalues of $(A - \lambda_1 I)$ are $0, \lambda_2 - \lambda_1, \dots, \lambda_n - \lambda_1$. (Hint: $A - \lambda_1 I$ is diagonalized by P.)

6.3 Applications of Diagonalization

Powers of Matrices

In some applications of linear algebra, it is necessary to calculate large powers of a matrix. For example, say we wanted to calculate A^{1000} where $A = \begin{bmatrix} 1 & 2 \\ -1 & 4 \end{bmatrix}$. We certainly would not want to multiply A by itself 1000 times (especially if we were doing this by hand)! Instead, we use the benefits of diagonalization.

Observe that multiplying a diagonal matrix D by itself is easy. In particular, we have if $D = \text{diag}(d_1, \dots, d_n)$, then $D^k = \text{diag}(d_1^k, \dots, d_n^k)$. See Section 3.1 Problem C9. The following theorem, shows us how we can use this fact with diagonalization to quickly take powers of a diagonalizable matrix A.

Theorem 6.3.1 Let $A \in M_{n\times n}(\mathbb{R})$. If there exists an invertible matrix P and diagonal matrix D such that $P^{-1}AP = D$, then

$$A^k = PD^kP^{-1}$$

Proof: We prove the result by induction on k. If $k = 1$, then $P^{-1}AP = D$ implies $A = PDP^{-1}$ and so the result holds. Assume the result is true for some $k \geq 1$. We then have

$$A^{k+1} = A^kA = (PD^kP^{-1})(PDP^{-1}) = PD^kP^{-1}PDP^{-1} = PD^kIDP^{-1} = PD^{k+1}P^{-1}$$

as required. ■

We now demonstrate finding large powers of matrices with a couple of examples.

EXAMPLE 6.3.1 Let $A = \begin{bmatrix} 1 & 2 \\ -1 & 4 \end{bmatrix}$. Show that $A^{1000} = \begin{bmatrix} 2^{1001} - 3^{1000} & -2^{1001} + 2 \cdot 3^{1000} \\ 2^{1000} - 3^{1000} & -2^{1000} + 2 \cdot 3^{1000} \end{bmatrix}$.

Solution: We first diagonalize A. We have

$$0 = \det(A - \lambda I) = \begin{vmatrix} 1-\lambda & 2 \\ -1 & 4-\lambda \end{vmatrix} = \lambda^2 - 5\lambda + 6 = (\lambda - 2)(\lambda - 3)$$

Thus, the eigenvalues of A are $\lambda_1 = 2$ and $\lambda_2 = 3$.
For $\lambda_1 = 2$ we have

$$A - \lambda_1 I = \begin{bmatrix} -1 & 2 \\ -1 & 2 \end{bmatrix} \sim \begin{bmatrix} 1 & -2 \\ 0 & 0 \end{bmatrix}$$

So, a basis for E_{λ_1} is $\left\{\begin{bmatrix} 2 \\ 1 \end{bmatrix}\right\}$.

EXAMPLE 6.3.1
(continued)

For $\lambda_2 = 3$ we have

$$A - \lambda_2 I = \begin{bmatrix} -2 & 2 \\ -1 & 1 \end{bmatrix} \sim \begin{bmatrix} 1 & -1 \\ 0 & 0 \end{bmatrix}$$

So, a basis for E_{λ_2} is $\left\{ \begin{bmatrix} 1 \\ 1 \end{bmatrix} \right\}$.

Therefore, $P = \begin{bmatrix} 2 & 1 \\ 1 & 1 \end{bmatrix}$ and $D = \begin{bmatrix} 2 & 0 \\ 0 & 3 \end{bmatrix}$. We find that $P^{-1} = \begin{bmatrix} 1 & -1 \\ -1 & 2 \end{bmatrix}$. Hence,

$$\begin{aligned} A^{1000} &= PD^{1000}P^{-1} \\ &= \begin{bmatrix} 2 & 1 \\ 1 & 1 \end{bmatrix} \begin{bmatrix} 2^{1000} & 0 \\ 0 & 3^{1000} \end{bmatrix} \begin{bmatrix} 1 & -1 \\ -1 & 2 \end{bmatrix} \\ &= \begin{bmatrix} 2^{1001} & 3^{1000} \\ 2^{1000} & 3^{1000} \end{bmatrix} \begin{bmatrix} 1 & -1 \\ -1 & 2 \end{bmatrix} \\ &= \begin{bmatrix} 2^{1001} - 3^{1000} & -2^{1001} + 2 \cdot 3^{1000} \\ 2^{1000} - 3^{1000} & -2^{1000} + 2 \cdot 3^{1000} \end{bmatrix} \end{aligned}$$

EXAMPLE 6.3.2

Let $A = \begin{bmatrix} -2 & 2 \\ -3 & 5 \end{bmatrix}$. Calculate A^{200}.

Solution: We have

$$0 = \det(A - \lambda I) = \begin{vmatrix} -2-\lambda & 2 \\ -3 & 5-\lambda \end{vmatrix} = \lambda^2 - 3\lambda - 4 = (\lambda + 1)(\lambda - 4).$$

Thus, the eigenvalues of A are $\lambda_1 = -1$ and $\lambda_2 = 4$.

For $\lambda_1 = -1$ we have $A - \lambda_1 I = \begin{bmatrix} -1 & 2 \\ -3 & 6 \end{bmatrix} \sim \begin{bmatrix} 1 & -2 \\ 0 & 0 \end{bmatrix}$. So, a basis for E_{λ_1} is $\left\{ \begin{bmatrix} 2 \\ 1 \end{bmatrix} \right\}$.

For $\lambda_2 = 4$ we have $A - \lambda_2 I = \begin{bmatrix} -6 & 2 \\ -3 & 1 \end{bmatrix} \sim \begin{bmatrix} 1 & -1/3 \\ 0 & 0 \end{bmatrix}$. So, a basis for E_{λ_2} is $\left\{ \begin{bmatrix} 1 \\ 3 \end{bmatrix} \right\}$.

Therefore, $P = \begin{bmatrix} 2 & 1 \\ 1 & 3 \end{bmatrix}$ and $D = \begin{bmatrix} -1 & 0 \\ 0 & 4 \end{bmatrix}$. We find that $P^{-1} = \frac{1}{5}\begin{bmatrix} 3 & -1 \\ -1 & 2 \end{bmatrix}$. Hence,

$$\begin{aligned} A^{200} = PD^{200}P^{-1} &= \begin{bmatrix} 2 & 1 \\ 1 & 3 \end{bmatrix} \begin{bmatrix} 1 & 0 \\ 0 & 4^{200} \end{bmatrix} \left(\frac{1}{5} \begin{bmatrix} 3 & -1 \\ -1 & 2 \end{bmatrix} \right) \\ &= \frac{1}{5} \begin{bmatrix} 6 - 4^{200} & -2 + 2 \cdot 4^{200} \\ 3 - 3 \cdot 4^{200} & -1 + 6 \cdot 4^{200} \end{bmatrix} \end{aligned}$$

EXERCISE 6.3.1

Let $A = \begin{bmatrix} 0 & 1 \\ -2 & 3 \end{bmatrix}$. Calculate A^{100}.

Markov Processes

A Markov process, also commonly called a **Markov chain**, is a mathematical model of a system in which the system at any time occupies one of a finite number of states, and how the system transitions to the next state is dependent only on its current state.

Here is just a small sample of how Markov processes are used.

Chemical Reactions A chemical reaction network can be modelled as a Markov process where the states are the number of molecules of each species and the transitions are calculated from the rate of the chemical reactions.

Population Genetics Physical characteristics of animals are inherited from their parents based on their parents' genes. We can model the transition of characteristics from parent to child by a Markov process. For example, our possible states could be blue eyes or brown eyes, and the transition would be determined by considering the dominant and recessive genes.

Random Walks Many real world situations can be represented as randomly moving through a graph. For example, current passing through an electric network. These can be modelled by a Markov process where the states are the nodes of the graph and the transitions are based on the probability of moving from one node to the next.

We begin our look at Markov processes with an example.

EXAMPLE 6.3.3

Smith and Jones are the only competing suppliers of communication services in their community. At present, they each have a 50% share of the market. However, Smith has recently upgraded his service, and a survey indicates that from one month to the next, 90% of Smith's customers remain loyal, while 10% switch to Jones. On the other hand, 70% of Jones's customers remain loyal and 30% switch to Smith. If this goes on for six months, how large are their market shares? If this goes on for a long time, how big will Smith's share become?

Solution: Let S_m be Smith's market share (as a decimal) at the end of the m-th month and let J_m be Jones's share. Then $S_m + J_m = 1$, since between them they have 100% of the market. At the end of the $(m+1)$-st month, Smith has 90% of his previous customers and 30% of Jones's previous customers, so

$$S_{m+1} = 0.9S_m + 0.3J_m$$

Similarly,

$$J_{m+1} = 0.1S_m + 0.7J_m$$

We can rewrite these equations in matrix-vector form:

$$\begin{bmatrix} S_{m+1} \\ J_{m+1} \end{bmatrix} = \begin{bmatrix} 0.9 & 0.3 \\ 0.1 & 0.7 \end{bmatrix} \begin{bmatrix} S_m \\ J_m \end{bmatrix}$$

The matrix $T = \begin{bmatrix} 0.9 & 0.3 \\ 0.1 & 0.7 \end{bmatrix}$ is called the **transition matrix** for this problem: it describes the transition (change) from the **state** $\begin{bmatrix} S_m \\ J_m \end{bmatrix}$ at time m to the state $\begin{bmatrix} S_{m+1} \\ J_{m+1} \end{bmatrix}$ at time $m+1$.

EXAMPLE 6.3.3
(continued)

To answer the questions, we need to determine $T^6\begin{bmatrix}0.5\\0.5\end{bmatrix}$ and $T^m\begin{bmatrix}0.5\\0.5\end{bmatrix}$ for m large.

We could compute T^6 directly, but this approach is not reasonable for calculating T^m for large values of m. So, we diagonalize T. We find that $\lambda_1 = 1$ is an eigenvalue of T with eigenvector $\vec{v}_1 = \begin{bmatrix}3\\1\end{bmatrix}$, and $\lambda_2 = 0.6$ is the other eigenvalue, with eigenvector $\vec{v}_2 = \begin{bmatrix}1\\-1\end{bmatrix}$. Thus,

$$P = \begin{bmatrix}3 & 1\\1 & -1\end{bmatrix} \quad \text{and} \quad P^{-1} = \frac{1}{4}\begin{bmatrix}1 & 1\\1 & -3\end{bmatrix}$$

It follows that

$$T^m = PD^mP^{-1} = \begin{bmatrix}3 & 1\\1 & -1\end{bmatrix}\begin{bmatrix}1^m & 0\\0 & (0.6)^m\end{bmatrix}\frac{1}{4}\begin{bmatrix}1 & 1\\1 & -3\end{bmatrix}$$

We could now answer our question directly, but we get a simpler calculation if we observe that the eigenvectors form a basis, so we can write

$$\begin{bmatrix}S_0\\J_0\end{bmatrix} = c_1\begin{bmatrix}3\\1\end{bmatrix} + c_2\begin{bmatrix}1\\-1\end{bmatrix} = P\begin{bmatrix}c_1\\c_2\end{bmatrix}$$

Then,

$$\begin{bmatrix}c_1\\c_2\end{bmatrix} = P^{-1}\begin{bmatrix}S_0\\J_0\end{bmatrix} = \frac{1}{4}\begin{bmatrix}S_0 + J_0\\S_0 - 3J_0\end{bmatrix}$$

Hence,

$$\begin{aligned}
T^m\begin{bmatrix}S_0\\J_0\end{bmatrix} &= T^m\left(c_1\begin{bmatrix}3\\1\end{bmatrix} + c_2\begin{bmatrix}1\\-1\end{bmatrix}\right)\\
&= c_1T^m\begin{bmatrix}3\\1\end{bmatrix} + c_2T^m\begin{bmatrix}1\\-1\end{bmatrix}\\
&= c_1\lambda_1^m\begin{bmatrix}3\\1\end{bmatrix} + c_2\lambda_2^m\begin{bmatrix}1\\-1\end{bmatrix} \qquad \text{by Section 6.1 Problem C9 (a)}\\
&= \frac{1}{4}(S_0 + J_0)\begin{bmatrix}3\\1\end{bmatrix} + \frac{1}{4}(S_0 - 3J_0)(0.6)^m\begin{bmatrix}1\\-1\end{bmatrix}
\end{aligned}$$

Now $S_0 = J_0 = 0.5$. When $m = 6$,

$$\begin{aligned}
\begin{bmatrix}S_6\\J_6\end{bmatrix} &= \frac{1}{4}\begin{bmatrix}3\\1\end{bmatrix} - \frac{1}{4}(0.6)^6\begin{bmatrix}1\\-1\end{bmatrix}\\
&\approx \frac{1}{4}\begin{bmatrix}3 - 0.0467\\1 + 0.0467\end{bmatrix}\\
&\approx \begin{bmatrix}0.738\\0.262\end{bmatrix}
\end{aligned}$$

Thus, after six months, Smith has approximately 73.8% of the market.

When m is very large, $(0.6)^m$ is nearly zero, so for m large enough ($m \to \infty$), we have $S_\infty = 0.75$ and $J_\infty = 0.25$.

Thus, in this problem, Smith's share approaches 75% as m gets large, but it never gets larger than 75%. Now look carefully: we get the same answer in the long run, no matter what the initial value of S_0 and J_0 are because $(0.6)^m \to 0$ and $S_0 + J_0 = 1$.

By emphasizing some features of Example 6.3.3, we will be led to an important definition and several general properties:

(1) Each column of T has sum 1. This means that all of Smith's customers show up a month later as customers of Smith or Jones; the same is true for Jones's customers. No customers are lost from the system and none are added after the process begins.

(2) It is natural to interpret the entries t_{ij} as **probabilities**. For example, $t_{11} = 0.9$ is the probability that a Smith customer remains a Smith customer, with $t_{21} = 0.1$ as the probability that a Smith customer becomes a Jones customer. If we consider "Smith customer" as "state 1" and "Jones customer" as "state 2," then t_{ij} is the probability of **transition** from state j to state i between time m and time $m + 1$.

(3) The "initial state vector" is $\begin{bmatrix} S_0 \\ J_0 \end{bmatrix}$. The state vector at time m is $\begin{bmatrix} S_m \\ J_m \end{bmatrix} = T^m \begin{bmatrix} S_0 \\ J_0 \end{bmatrix}$.

(4) Note that

$$\begin{bmatrix} S_1 \\ J_1 \end{bmatrix} = T \begin{bmatrix} S_0 \\ J_0 \end{bmatrix} = S_0 \begin{bmatrix} t_{11} \\ t_{21} \end{bmatrix} + J_0 \begin{bmatrix} t_{12} \\ t_{22} \end{bmatrix}$$

Since $t_{11} + t_{21} = 1$ and $t_{12} + t_{22} = 1$, it follows that

$$S_1 + J_1 = S_0 + J_0$$

Thus, it follows from (1) that each state vector has the same column sum. In our example, S_0 and J_0 are decimal fractions, so $S_0 + J_0 = 1$, but we could consider a process whose states have some other constant column sum.

(5) Note that 1 is an eigenvalue of T with eigenvector $\begin{bmatrix} 3 \\ 1 \end{bmatrix}$. To get a state vector with the appropriate sum, we take the eigenvector to be $\begin{bmatrix} 3/4 \\ 1/4 \end{bmatrix}$. Thus,

$$T \begin{bmatrix} 3/4 \\ 1/4 \end{bmatrix} = \begin{bmatrix} 3/4 \\ 1/4 \end{bmatrix}$$

and the state vector $\begin{bmatrix} 3/4 \\ 1/4 \end{bmatrix}$ is **fixed** or **invariant** under the transformation with matrix T. Moreover, this fixed vector is the limiting state approached by $T^m \begin{bmatrix} S_0 \\ J_0 \end{bmatrix}$ for any $\begin{bmatrix} S_0 \\ J_0 \end{bmatrix}$.

The following definition captures the essential properties of this example.

Definition
Markov Matrix
Markov Process

A matrix $T \in M_{n\times n}(\mathbb{R})$ is the **Markov matrix** (or transition matrix) of an n-state **Markov process** if

(1) $t_{ij} \geq 0$, for each i and j.
(2) Each column sum is 1. That is, $t_{1j} + \cdots + t_{nj} = 1$ for each j.

We take possible states of the process to be the vectors $S = \begin{bmatrix} s_1 \\ \vdots \\ s_n \end{bmatrix}$ such that $s_i \geq 0$ for each i, and $s_1 + \cdots + s_n = 1$.

Remark

With minor changes, we could develop the theory with $s_1 + \cdots + s_n =$ constant.

EXAMPLE 6.3.4

The matrix $\begin{bmatrix} 0.1 & 0.3 \\ 0.9 & 0.8 \end{bmatrix}$ is not a Markov matrix since the sum of the entries in the second column does not equal 1.

EXAMPLE 6.3.5

Find the fixed-state vector for the Markov matrix $A = \begin{bmatrix} 0.1 & 0.3 \\ 0.9 & 0.7 \end{bmatrix}$.

Solution: We know the fixed-state vector is an eigenvector for the eigenvalue $\lambda = 1$. We have

$$A - I = \begin{bmatrix} -0.9 & 0.3 \\ 0.9 & -0.3 \end{bmatrix} \sim \begin{bmatrix} 1 & -1/3 \\ 0 & 0 \end{bmatrix}$$

Therefore, an eigenvector corresponding to $\lambda = 1$ is $\begin{bmatrix} 1 \\ 3 \end{bmatrix}$. The components in the state vector must sum to 1, so the invariant state is $\begin{bmatrix} 1/4 \\ 3/4 \end{bmatrix}$.

It is easy to verify that

$$\begin{bmatrix} 0.1 & 0.3 \\ 0.9 & 0.7 \end{bmatrix} \begin{bmatrix} 1/4 \\ 3/4 \end{bmatrix} = \begin{bmatrix} 1/4 \\ 3/4 \end{bmatrix}$$

EXERCISE 6.3.2

Determine which of the following matrices is a Markov matrix. Find the fixed-state vector of the Markov matrix.

(a) $A = \begin{bmatrix} 0.4 & 0.6 \\ 0.5 & 0.5 \end{bmatrix}$ (b) $B = \begin{bmatrix} 0.4 & 0.6 \\ 0.6 & 0.4 \end{bmatrix}$

The goal with the Markov process is to establish the behaviour of a sequence with states $\vec{s}, T\vec{s}, T^2\vec{s}, \ldots, T^m\vec{s}$. If possible, we want to say something about the limit of $T^m\vec{s}$ as $m \to \infty$. As we saw in Example 1, diagonalization of T is a key to solving the problem. It is beyond the scope of this book to establish all the properties of the Markov process, but some of the properties are easy to prove, and others are easy to illustrate if we make extra assumptions.

PROPERTY 1. One eigenvalue of a Markov matrix is $\lambda_1 = 1$.

Proof: Since each column of T has sum 1, each column of $(T - 1I)$ has sum 0. Hence, the sum of the rows of $(T - 1I)$ is the zero vector. Thus the rows are linearly dependent, and $(T - 1I)$ has rank less than n, so $\det(T - 1I) = 0$. Therefore, 1 is an eigenvalue of T. ■

PROPERTY 2. The eigenvector $\vec{s}^*$ for $\lambda_1 = 1$ has $s_j^* \geq 0$ for $1 \leq j \leq n$.
This property is important because it means that the eigenvector $\vec{s}^*$ is a real state of the process. In fact, it is a fixed, or invariant, state:

$$T\vec{s}^* = \lambda_1 \vec{s}^* = \vec{s}^*$$

PROPERTY 3. All other eigenvalues satisfy $|\lambda_i| \leq 1$.
To see why we expect this, let us assume that T is diagonalizable, with distinct eigenvalues $1, \lambda_2, \dots, \lambda_n$ and corresponding eigenvectors $\vec{s}^*, \vec{s}_2, \dots, \vec{s}_n$. Then any initial state $\vec{s}$ can be written

$$\vec{s} = c_1\vec{s}^* + c_2\vec{s}_2 + \cdots + c_n\vec{s}_n$$

It follows that

$$T^m\vec{s} = c_1 1^m\vec{s}^* + c_2\lambda_2^m\vec{s}_2 + \cdots + c_n\lambda_n^m\vec{s}_n$$

If any $|\lambda_i| > 1$, then the term $|\lambda_i^m|$ would become much larger than the other terms when m is large; it would follow that $T^m\vec{s}$ has some coordinates with magnitude greater than 1. This is impossible because state coordinates satisfy $0 \leq s_i \leq 1$, so we must have $|\lambda_i| \leq 1$.

PROPERTY 4. Suppose that for some m all the entries in T^m are not zero. Then all the eigenvalues of T except for $\lambda_1 = 1$ satisfy $|\lambda_i| < 1$. In this case, for any initial state $\vec{s}$, $T^m\vec{s} \to \vec{s}^*$ as $m \to \infty$: all states tend to the invariant state $\vec{s}^*$ under the process.

The proof of Property 4 is omitted. Notice that in the diagonalizable case, the fact that $T^m\vec{s} \to \vec{s}^*$ follows from the expression for $T^m\vec{s}$ given under Property 3.

EXERCISE 6.3.3

The Markov matrix $T = \begin{bmatrix} 0 & 1 \\ 1 & 0 \end{bmatrix}$ has eigenvalues 1 and -1; it does not satisfy the conclusion of Property 4. However, it also does not satisfy the extra assumption of Property 4. It is worthwhile to explore this "bad" case.

Let $\vec{s} = \begin{bmatrix} s_1 \\ s_2 \end{bmatrix}$. Determine the behaviour of the sequence $\vec{s}, T\vec{s}, T^2\vec{s}, \dots$. What is the fixed-state vector for T?

Differential Equations

We demonstrate how diagonalization can be used to help us solve systems of differential equations with an example.

EXAMPLE 6.3.6

Consider two tanks, Y and Z, each containing 1000 litres of a salt solution. At an initial time, $t = 0$ (in hours), the concentration of salt in tank Y is different from the concentration in tank Z. In each tank the solution is well stirred, so that the concentration is constant throughout the tank. The two tanks are joined by pipes; through one pipe, solution is pumped from Y to Z at a rate of 20 L/h; through the other, solution is pumped from Z to Y at the same rate. Determine the amount of salt in each tank at time t.

Solution: Let $y(t)$ be the amount of salt (in kilograms) in tank Y at time t, and let $z(t)$ be the amount of salt (in kilograms) in the tank Z at time t. Then the concentration in Y at time t is $(y(t)/1000)$ kg/L. Similarly, $(z(t)/1000)$ kg/L is the concentration in Z. Then for tank Y, salt is flowing out through one pipe at a rate of $(20)(y/1000)$ kg/h and in through the other pipe at a rate of $(20)(z/1000)$ kg/h. Since the rate of change is measured by the derivative, we have $\frac{dy}{dt} = -0.02y + 0.02z$. By consideration of Z, we get a second differential equation, so y and z are the solutions of the **system of linear ordinary differential equations**:

$$\frac{dy}{dt} = -0.02y + 0.02z$$
$$\frac{dz}{dt} = 0.02y - 0.02z$$

It is convenient to rewrite this system in the form $\frac{d}{dt}\begin{bmatrix} y \\ z \end{bmatrix} = \begin{bmatrix} -0.02 & 0.02 \\ 0.02 & -0.02 \end{bmatrix}\begin{bmatrix} y \\ z \end{bmatrix}$.

How can we solve this system? Well, it might be easier if we could change variables so that the 2×2 matrix is diagonalized. By standard methods, one eigenvalue of $A = \begin{bmatrix} -0.02 & 0.02 \\ 0.02 & -0.02 \end{bmatrix}$ is $\lambda_1 = 0$, with corresponding eigenvector $\begin{bmatrix} 1 \\ 1 \end{bmatrix}$. The other eigenvalue is $\lambda_2 = -0.04$, with corresponding eigenvector $\begin{bmatrix} -1 \\ 1 \end{bmatrix}$. Hence, A is diagonalized by $P = \begin{bmatrix} 1 & -1 \\ 1 & 1 \end{bmatrix}$ to $D = \begin{bmatrix} 0 & 0 \\ 0 & -0.04 \end{bmatrix}$.

Introduce new coordinates $\begin{bmatrix} y^* \\ z^* \end{bmatrix}$ by the change of coordinates equation $\begin{bmatrix} y \\ z \end{bmatrix} = P\begin{bmatrix} y^* \\ z^* \end{bmatrix}$. See Section 4.4. Substitute this for $\begin{bmatrix} y \\ z \end{bmatrix}$ on both sides of the system to obtain

$$\frac{d}{dt}P\begin{bmatrix} y^* \\ z^* \end{bmatrix} = AP\begin{bmatrix} y^* \\ z^* \end{bmatrix}$$

Since the entries in P are constants, it is easy to check that

$$\frac{d}{dt}P\begin{bmatrix} y^* \\ z^* \end{bmatrix} = P\frac{d}{dt}\begin{bmatrix} y^* \\ z^* \end{bmatrix}$$

EXAMPLE 6.3.6
(continued)

Multiply both sides of the system of equations (on the left) by P^{-1}. Since P diagonalizes A, we get

$$\frac{d}{dt}\begin{bmatrix} y^* \\ z^* \end{bmatrix} = P^{-1}AP\begin{bmatrix} y^* \\ z^* \end{bmatrix} = \begin{bmatrix} 0 & 0 \\ 0 & -0.04 \end{bmatrix}\begin{bmatrix} y^* \\ z^* \end{bmatrix}$$

Now write the pair of equations:

$$\frac{dy^*}{dt} = 0 \text{ and } \frac{dz^*}{dt} = -0.04z^*$$

These equations are "decoupled," and we can easily solve each of them by using simple one-variable calculus.

The only functions satisfying $\frac{dy^*}{dt} = 0$ are constants: we write $y^*(t) = a$. The only functions satisfying an equation of the form $\frac{dx}{dt} = kx$ are exponentials of the form $x(t) = ce^{kt}$ for a constant c. So, from $\frac{dz^*}{dt} = -0.04z^*$, we obtain $z^*(t) = be^{-0.04t}$, where b is a constant.

Now we need to express the solution in terms of the original variables y and z:

$$\begin{bmatrix} y \\ z \end{bmatrix} = P\begin{bmatrix} y^* \\ z^* \end{bmatrix} = \begin{bmatrix} 1 & -1 \\ 1 & 1 \end{bmatrix}\begin{bmatrix} y^* \\ z^* \end{bmatrix} = \begin{bmatrix} y^* - z^* \\ y^* + z^* \end{bmatrix} = \begin{bmatrix} a - be^{-0.04t} \\ a + be^{-0.04t} \end{bmatrix}$$

For later use, it is helpful to rewrite this as $\begin{bmatrix} y \\ z \end{bmatrix} = a\begin{bmatrix} 1 \\ 1 \end{bmatrix} + be^{-0.04t}\begin{bmatrix} -1 \\ 1 \end{bmatrix}$. This is the general solution of the problem. To determine the constants a and b, we would need to know the amounts $y(0)$ and $z(0)$ at the initial time $t = 0$. Then we would know y and z for all t. Note that as $t \to \infty$, y and z tend to a common value a, as we might expect.

The usual solution procedure takes advantage of the understanding obtained from this diagonalization argument, but it takes a major shortcut. Now that the expected form of the solution is known, we simply look for a solution of the form $\begin{bmatrix} y \\ z \end{bmatrix} = ce^{\lambda t}\vec{v}$. Substitute this into the original system and use the fact that $\frac{d}{dt}ce^{\lambda t}\vec{v} = \lambda ce^{\lambda t}\vec{v}$ to get

$$\lambda ce^{\lambda t}\vec{v} = Ace^{\lambda t}\vec{v}$$

After the common factor $ce^{\lambda t}$ is cancelled, this tells us that $\vec{v}$ is an eigenvector of A, with eigenvalue λ. We find the two eigenvalues λ_1 and λ_2 and the corresponding eigenvectors $\vec{v}_1$ and $\vec{v}_2$, as above. Observe that since our problem is a linear homogeneous problem, the general solution will be of the form

$$\begin{bmatrix} y \\ z \end{bmatrix} = ae^{\lambda_1 t}\vec{v}_1 + be^{\lambda_2 t}\vec{v}_2$$

General Discussion There are many other problems that give rise to systems of linear homogeneous ordinary differential equations (for example, electrical circuits or a mechanical system consisting of springs). Many of these systems are much larger than the example we considered. Methods for solving these systems make extensive use of eigenvectors and eigenvalues, and they require methods for dealing with cases where the characteristic equation has complex roots.

PROBLEMS 6.3

Practice Problems

A1 Let $A = \begin{bmatrix} 4 & -1 \\ -2 & 5 \end{bmatrix}$. Use diagonalization to calculate A^3. Verify your answer by computing A^3 directly.

A2 Calculate A^{100} where $A = \begin{bmatrix} -6 & -10 \\ 4 & 7 \end{bmatrix}$.

A3 Calculate A^{100} where $A = \begin{bmatrix} 2 & 2 \\ -3 & -5 \end{bmatrix}$.

A4 Calculate A^{200} where $A = \begin{bmatrix} -2 & 2 \\ -3 & 5 \end{bmatrix}$.

A5 Calculate A^{200} where $A = \begin{bmatrix} 6 & -6 \\ 2 & -1 \end{bmatrix}$.

A6 Calculate A^{100} where $A = \begin{bmatrix} -2 & 1 & 1 \\ -1 & 0 & 1 \\ -2 & 2 & 1 \end{bmatrix}$.

A7 Calculate A^{100} where $A = \begin{bmatrix} 7 & -3 & 2 \\ 8 & -4 & 2 \\ -10 & 4 & -3 \end{bmatrix}$.

A8 Calculate A^{100} where $A = \begin{bmatrix} 3 & -1 & 0 \\ 2 & 0 & 0 \\ -2 & 1 & 1 \end{bmatrix}$.

For Problems A9–A14, determine whether the matrix is a Markov matrix. If so, determine the invariant or fixed state (corresponding to the eigenvalue $\lambda = 1$).

A9 $\begin{bmatrix} 0.2 & 0.6 \\ 0.8 & 0.3 \end{bmatrix}$

A10 $\begin{bmatrix} 0.3 & 0.6 \\ 0.7 & 0.4 \end{bmatrix}$

A11 $\begin{bmatrix} 0.5 & 0.4 \\ 0.5 & 0.6 \end{bmatrix}$

A12 $\begin{bmatrix} 0.8 & 0.5 \\ 0.2 & 0.5 \end{bmatrix}$

A13 $\begin{bmatrix} 0.7 & 0.3 & 0.0 \\ 0.1 & 0.6 & 0.1 \\ 0.2 & 0.2 & 0.9 \end{bmatrix}$

A14 $\begin{bmatrix} 0.9 & 0.1 & 0.0 \\ 0.0 & 0.9 & 0.1 \\ 0.1 & 0.0 & 0.9 \end{bmatrix}$

A15 Suppose that census data show that every decade, 15% of people dwelling in rural areas move into towns and cities, while 5% of urban dwellers move into rural areas.

(a) What would be the eventual steady-state population distribution?

(b) If the population were 50% urban, 50% rural at some census, what would be the distribution after 50 years?

A16 A car rental company serving one city has three locations: the airport, the train station, and the city centre. Of the cars rented at the airport, 8/10 are returned to the airport, 1/10 are left at the train station, and 1/10 are left at the city centre. Of the cars rented at the train station, 3/10 are left at the airport, 6/10 are returned to the train station, and 1/10 are left at the city centre. Of the cars rented at the city centre, 3/10 go to the airport, 1/10 go to the train station, and 6/10 are returned to the city centre. Model this as a Markov process and determine the steady-state distribution for the cars.

A17 The town of Markov Centre has only two suppliers of widgets—Johnson and Thomson. All inhabitants buy their supply on the first day of each month. Neither supplier is very successful at keeping customers. 70% of the customers who deal with Johnson decide that they will "try the other guy" next time. Thomson does even worse: only 20% of his customers come back the next month, and the rest go to Johnson.

(a) Model this as a Markov process and determine the steady-state distribution of customers.

(b) Determine a general expression for Johnson and Thomson's shares of the customers, given an initial state where Johnson has 25% and Thomson has 75%.

For Problems A18–A27, find the general solution of the system of linear differential equations.

A18 $\frac{d}{dt}\begin{bmatrix} y \\ z \end{bmatrix} = \begin{bmatrix} 3 & 2 \\ 4 & -4 \end{bmatrix}\begin{bmatrix} y \\ z \end{bmatrix}$

A19 $\frac{d}{dt}\begin{bmatrix} y \\ z \end{bmatrix} = \begin{bmatrix} 0.2 & 0.7 \\ 0.1 & -0.4 \end{bmatrix}\begin{bmatrix} y \\ z \end{bmatrix}$

A20 $\frac{d}{dt}\begin{bmatrix} y \\ z \end{bmatrix} = \begin{bmatrix} -1 & 4 \\ 8 & -5 \end{bmatrix}\begin{bmatrix} y \\ z \end{bmatrix}$

A21 $\frac{d}{dt}\begin{bmatrix} y \\ z \end{bmatrix} = \begin{bmatrix} 0.5 & 0.3 \\ 0.1 & 0.3 \end{bmatrix}\begin{bmatrix} y \\ z \end{bmatrix}$

A22 $\frac{d}{dt}\begin{bmatrix} y \\ z \end{bmatrix} = \begin{bmatrix} 4 & 2 \\ 2 & 1 \end{bmatrix}\begin{bmatrix} y \\ z \end{bmatrix}$

A23 $\frac{d}{dt}\begin{bmatrix} y \\ z \end{bmatrix} = \begin{bmatrix} 3 & 6 \\ 5 & 2 \end{bmatrix}\begin{bmatrix} y \\ z \end{bmatrix}$

A24 $\frac{d}{dt}\begin{bmatrix} y \\ z \end{bmatrix} = \begin{bmatrix} 7 & 4 \\ 2 & 9 \end{bmatrix}\begin{bmatrix} y \\ z \end{bmatrix}$

A25 $\frac{d}{dt}\begin{bmatrix} y \\ z \end{bmatrix} = \begin{bmatrix} 0.4 & 0.4 \\ 0.6 & 0.6 \end{bmatrix}\begin{bmatrix} y \\ z \end{bmatrix}$

A26 $\frac{d}{dt}\begin{bmatrix} x \\ y \\ z \end{bmatrix} = \begin{bmatrix} -1 & -1 & 0 \\ -13 & 3 & 8 \\ 11 & -5 & -8 \end{bmatrix}\begin{bmatrix} x \\ y \\ z \end{bmatrix}$

A27 $\frac{d}{dt}\begin{bmatrix} x \\ y \\ z \end{bmatrix} = \begin{bmatrix} 1 & 5 & -5 \\ 6 & -3 & 6 \\ -4 & 1 & 2 \end{bmatrix}\begin{bmatrix} x \\ y \\ z \end{bmatrix}$

Homework Problems

B1 Calculate A^{100} where $A = \begin{bmatrix} 1 & 2 \\ 1 & 2 \end{bmatrix}$.

B2 Calculate A^{100} where $A = \begin{bmatrix} 3 & -2 \\ 1 & 0 \end{bmatrix}$.

B3 Calculate A^{100} where $A = \begin{bmatrix} 0 & -2 \\ -1 & 1 \end{bmatrix}$.

B4 Calculate A^{100} where $A = \begin{bmatrix} -8 & 4 \\ -15 & 8 \end{bmatrix}$.

B5 Calculate A^{100} where $A = \begin{bmatrix} 2 & -6 & -6 \\ 4 & -12 & -12 \\ -4 & 12 & 12 \end{bmatrix}$.

B6 Calculate A^{100} where $A = \begin{bmatrix} 0 & -1 & 1 \\ 0 & 2 & 0 \\ -2 & -1 & 3 \end{bmatrix}$.

For Problems B7–B12, determine whether the matrix is a Markov matrix. If so, determine the invariant or fixed state (corresponding to the eigenvalue $\lambda = 1$).

B7 $\begin{bmatrix} 0.4 & 0.7 \\ 0.6 & 0.3 \end{bmatrix}$

B8 $\begin{bmatrix} 0.6 & 0.5 \\ 0.5 & 0.4 \end{bmatrix}$

B9 $\begin{bmatrix} 0.9 & 0.2 \\ 0.1 & 0.8 \end{bmatrix}$

B10 $\begin{bmatrix} 0.3 & 0.7 \\ 0.2 & 0.8 \end{bmatrix}$

B11 $\begin{bmatrix} 0.8 & 0.3 & 0.2 \\ 0.0 & 0.6 & 0.2 \\ 0.2 & 0.1 & 0.6 \end{bmatrix}$

B12 $\begin{bmatrix} 0.8 & 0.1 & 0.2 \\ 0.1 & 0.8 & 0.6 \\ 0.1 & 0.1 & 0.2 \end{bmatrix}$

B13 Suppose that the only competing suppliers of Internet services are NewServices and RealWest. At present, they each have 50% of the market share in their community. However, NewServices is upgrading their connection speeds but at a higher cost. Because of this, each month 30% of RealWest's customers will switch to NewServices, while 10% of NewServices customers will switch to RealWest.

(a) Model this situation as a Markov process.

(b) Calculate the market share of each company for the first two months.

(c) Determine the steady-state distribution.

B14 Consider the population-migration system described in Section 3.1 Problem B43 on page 170.

(a) Given initial populations of n_0 and m_0, use diagonalization to verify that the populations in year t are given by

$$\begin{bmatrix} n_t \\ m_t \end{bmatrix} = A^t \begin{bmatrix} n_0 \\ m_0 \end{bmatrix} = \frac{1}{3} \begin{bmatrix} 2 + \left(\frac{7}{10}\right)^t & 2 - 2\left(\frac{7}{10}\right)^t \\ 1 - \left(\frac{7}{10}\right)^t & 1 + 2\left(\frac{7}{10}\right)^t \end{bmatrix} \begin{bmatrix} n_0 \\ m_0 \end{bmatrix}$$

(b) Determine the steady-state distribution for the population.

B15 A student society at a university campus decides to create a pool of bicycles that can be used by its members. Borrowed bicycles can be returned to the residence, the library, or the athletic centre. On the first day, 200 marked bicycles are left at each location. At the end of the day, at the residence, there are 160 bicycles that started at the residence, 40 that started at the library, and 60 that started at the athletic centre. At the library, there are 20 that started at the residence, 140 that started at the library, and 40 that started at the athletic centre. At the athletic centre, there are 20 that started at the residence, 20 that started at the library, and 100 that started at the athletic centre. If this pattern is repeated every day, what is the steady-state distribution of bicycles?

For Problems B16–B24, find the general solution of the system of linear differential equations.

B16 $\frac{d}{dt}\begin{bmatrix} y \\ z \end{bmatrix} = \begin{bmatrix} 2 & 1 \\ 2 & 3 \end{bmatrix}\begin{bmatrix} y \\ z \end{bmatrix}$

B17 $\frac{d}{dt}\begin{bmatrix} y \\ z \end{bmatrix} = \begin{bmatrix} 3 & 6 \\ 2 & -1 \end{bmatrix}\begin{bmatrix} y \\ z \end{bmatrix}$

B18 $\frac{d}{dt}\begin{bmatrix} y \\ z \end{bmatrix} = \begin{bmatrix} 3 & 8 \\ 2 & 3 \end{bmatrix}\begin{bmatrix} y \\ z \end{bmatrix}$

B19 $\frac{d}{dt}\begin{bmatrix} y \\ z \end{bmatrix} = \begin{bmatrix} 0 & 1 \\ -6 & 5 \end{bmatrix}\begin{bmatrix} y \\ z \end{bmatrix}$

B20 $\frac{d}{dt}\begin{bmatrix} y \\ z \end{bmatrix} = \begin{bmatrix} 1 & 2 \\ 3 & 6 \end{bmatrix}\begin{bmatrix} y \\ z \end{bmatrix}$

B21 $\frac{d}{dt}\begin{bmatrix} y \\ z \end{bmatrix} = \begin{bmatrix} 5 & -9 \\ 6 & -10 \end{bmatrix}\begin{bmatrix} y \\ z \end{bmatrix}$

B22 $\frac{d}{dt}\begin{bmatrix} x \\ y \\ z \end{bmatrix} = \begin{bmatrix} 1 & 1 & -1 \\ 4 & 1 & -4 \\ -2 & 1 & 2 \end{bmatrix}\begin{bmatrix} x \\ y \\ z \end{bmatrix}$

B23 $\frac{d}{dt}\begin{bmatrix} x \\ y \\ z \end{bmatrix} = \begin{bmatrix} -7 & 5 & 4 \\ -4 & 2 & 4 \\ -5 & 13 & 2 \end{bmatrix}\begin{bmatrix} x \\ y \\ z \end{bmatrix}$

B24 $\frac{d}{dt}\begin{bmatrix} x \\ y \\ z \end{bmatrix} = \begin{bmatrix} 3 & 0 & 3 \\ -2 & 10 & -6 \\ -1 & 4 & -1 \end{bmatrix}\begin{bmatrix} x \\ y \\ z \end{bmatrix}$

Conceptual Problems

C1 Assume that there exists an invertible matrix P such that $P^{-1}AP = B$. Show that $A^3 = PB^3P^{-1}$.

C2 (a) Let T be the transition matrix for a two-state Markov process. Show that the eigenvalue that is not 1 is $\lambda_2 = t_{11} + t_{22} - 1$.

(b) For a two-state Markov process with $t_{21} = a$ and $t_{12} = b$, show that the fixed state is $\frac{1}{a+b}\begin{bmatrix} b \\ a \end{bmatrix}$.

C3 Suppose that $T \in M_{n\times n}(\mathbb{R})$ is a Markov matrix.

(a) Show that for any state $\vec{x}$,

$$(T\vec{x})_1 + \cdots + (T\vec{x})_n = x_1 + \cdots + x_n$$

(b) Show that if $\vec{v}$ is an eigenvector of T with eigenvalue $\lambda \neq 1$, then $v_1 + \cdots + v_n = 0$.

CHAPTER REVIEW

Suggestions for Student Review

1 At the beginning of this chapter, we mentioned that we are synthesizing many of the concepts from Chapters 1 – 5. Make a list of these concepts and indicate which section they are from. (Sections 6.1 – 6.3)

2 Define eigenvalues and eigenvectors of a matrix A. Explain why A must be a square matrix to have eigenvectors. Also explain the connection between the statement that λ is an eigenvalue of A with eigenvector $\vec{v}$ and the condition $\det(A - \lambda I) = 0$. (Section 6.1)

3 Define the algebraic and geometric multiplicity of an eigenvalue of a matrix A. What is the relationship between these quantities? Give some examples showing the different possibilities. (Section 6.1)

4 Describe the geometric meaning of eigenvalues and eigenvectors. Create a few examples of eigenvalues and eigenvectors of linear mappings to support your description. (Section 6.1)

5 Is there a relationship between a matrix being diagonalizable and the matrix being invertible? (Section 6.1)

6 What does it mean to say that matrices A and B are similar? What properties do similar matrices have in common? (Section 6.2)

7 Suppose that $A \in M_{n\times n}(\mathbb{R})$ has eigenvalues $\lambda_1, \dots, \lambda_n$ (repeated according to multiplicity). (Section 6.2)

(a) What conditions on these eigenvalues guarantee that A is diagonalizable over $\mathbb{R}$?

(b) Is there any case where you can tell just from the eigenvalues that A is diagonalizable over $\mathbb{R}$?

(c) Assume A is diagonalizable. Can we choose the eigenvalues in any order? If so, how does this affect the matrix P which diagonalizes A? Demonstrate with some examples.

8 Use the idea suggested in Section 6.2 Problem C4 to create matrices for your classmates to diagonalize. (Section 6.2)

9 Explain the procedure for calculating large powers of a diagonalizable matrix. Will the procedure work if the matrix is not diagonalizable? (Section 6.3)

10 Describe what a Markov process is. Search the Internet for applications of Markov processes related to your field(s) of interest. (Section 6.3)

11 Suppose that $P^{-1}AP = D$, where D is a diagonal matrix with distinct diagonal entries $\lambda_1, \dots, \lambda_n$. How can we use this information to solve the system of linear differential equations $\frac{d}{dt}\vec{x} = A\vec{x}$? (Section 6.3)

Chapter Quiz

E1 Let $A = \begin{bmatrix} 5 & -16 & -4 \\ 2 & -7 & -2 \\ -2 & 8 & 3 \end{bmatrix}$.

(a) Determine whether the following are eigenvectors of A. State the corresponding eigenvalue of any eigenvector.

(i) $\begin{bmatrix} 3 \\ 1 \\ 0 \end{bmatrix}$ (ii) $\begin{bmatrix} 1 \\ 0 \\ 1 \end{bmatrix}$ (iii) $\begin{bmatrix} 4 \\ 1 \\ 0 \end{bmatrix}$ (iv) $\begin{bmatrix} 2 \\ 1 \\ -1 \end{bmatrix}$

(b) Use the results of part (a) to diagonalize A.

For Problems E2 and E3, determine the algebraic and geometric multiplicity of each eigenvalue.

E2 $\begin{bmatrix} 4 & 2 & 2 \\ -1 & 1 & -1 \\ 1 & 1 & 3 \end{bmatrix}$ **E3** $\begin{bmatrix} 2 & 4 & -5 \\ 1 & 4 & -4 \\ 1 & 3 & -3 \end{bmatrix}$

For Problems E4–E7, either diagonalize the matrix or show that the matrix is not diagonalizable.

E4 $\begin{bmatrix} -1 & 4 \\ -2 & 5 \end{bmatrix}$ **E5** $\begin{bmatrix} 4 & -1 \\ 4 & 0 \end{bmatrix}$

E6 $\begin{bmatrix} 5 & -4 & -2 \\ -4 & 5 & -2 \\ -2 & -2 & 8 \end{bmatrix}$ **E7** $\begin{bmatrix} -3 & 1 & 0 \\ 13 & -7 & -8 \\ -11 & 5 & 4 \end{bmatrix}$

E8 Use the power method to determine the eigenvalue of $A = \begin{bmatrix} 2 & 9 \\ -4 & 22 \end{bmatrix}$ of the largest absolute value.

E9 Let $A = \begin{bmatrix} 5 & 6 \\ -2 & -2 \end{bmatrix}$. Calculate A^{100}.

E10 Suppose that $A \in M_{3\times 3}(\mathbb{R})$ such that

$$\det A = 0, \quad \det(A + 2I) = 0, \quad \det(A - 3I) = 0$$

Answer the following questions and give a brief explanation in each case.

(a) What is the dimension of the solution space of $A\vec{x} = \vec{0}$?

(b) What is the dimension of the nullspace of the matrix $B = A - 2I$?

(c) What is the rank of A?

E11 Let $A = \begin{bmatrix} 0.9 & 0.1 & 0.0 \\ 0.0 & 0.8 & 0.1 \\ 0.1 & 0.1 & 0.9 \end{bmatrix}$. Verify that A is a Markov matrix and determine its invariant state $\vec{x}$ such that $x_1 + x_2 + x_3 = 1$.

E12 Find the general solution of the system of differential equations

$$\frac{d}{dt}\begin{bmatrix} y \\ z \end{bmatrix} = \begin{bmatrix} 0.1 & 0.2 \\ 0.3 & 0.2 \end{bmatrix}\begin{bmatrix} y \\ z \end{bmatrix}$$

E13 If λ is an eigenvalue of the invertible matrix A, prove that $\frac{1}{\lambda}$ is an eigenvalue of A^{-1}.

E14 Let $A \in M_{n\times n}(\mathbb{R})$ be a diagonalizable matrix such that λ_1 is an eigenvalue with algebraic multiplicity n. Prove that $A = \lambda_1 I$.

E15 Prove that if $\vec{v}$ is an eigenvector of a matrix A, then any non-zero scalar multiple of $\vec{v}$ is also an eigenvector of A.

E16 Prove if A is diagonalizable and B is similar to A, then B is also diagonalizable.

For Problems E17–E22, determine whether the statement is true or false. Justify your answer.

E17 Every invertible matrix is diagonalizable.

E18 Every diagonalizable matrix is invertible.

E19 If $A \in M_{n\times n}(\mathbb{R})$ has n distinct eigenvalues, then A is diagonalizable.

E20 If $A\vec{v} = \lambda\vec{v}$, then $\vec{v}$ is an eigenvector of A.

E21 If the reduced row echelon form of $A - \lambda I$ is I, then λ is not an eigenvalue of A.

E22 If $\vec{v}$ is an eigenvector of a linear mapping L, then $\vec{v}$ is an eigenvector of $[L]$.

Further Problems

These exercises are intended to be challenging.

F1 Diagonalize $A = \begin{bmatrix} a & b \\ b & c \end{bmatrix}$. Such a matrix A is called **symmetric** since $A^T = A$. It is important to observe that the columns of the diagonalizing matrix P are orthogonal to each other. We will look at the diagonalization of symmetric matrices in Chapter 8.

F2 (a) Suppose that A and B are square matrices such that $AB = BA$. Suppose that the eigenvalues of A all have algebraic multiplicity 1. Prove that any eigenvector of A is also an eigenvector of B.

(b) Give an example to illustrate that the result in part (a) may not be true if A has eigenvalues with algebraic multiplicity greater than 1.

F3 Let $A, B \in M_{n\times n}(\mathbb{R})$. If $\det B \neq 0$, prove that AB and BA have the same eigenvalues.

F4 Let $A \in M_{n\times n}(\mathbb{R})$. Let $\lambda_1, \dots, \lambda_k$ be distinct eigenvalues of A with corresponding eigenvectors $\vec{v}_1, \dots, \vec{v}_k$ respectively. Use the following steps to prove that $\{\vec{v}_1, \dots, \vec{v}_k\}$ is linearly independent.

(a) Show that for any i and j we have

$$(A - \lambda_i I)\vec{v}_j = (\lambda_j - \lambda_i)\vec{v}_j$$

(b) Use a proof by induction on k and the result of (a) to prove that $\{\vec{v}_1, \dots, \vec{v}_k\}$ is linearly independent.

F5 If $A \in M_{n\times n}(\mathbb{R})$ has n distinct eigenvalues $\lambda_1, \dots, \lambda_n$ with corresponding eigenvectors $\vec{v}_1, \dots, \vec{v}_n$ respectively, then, by representing $\vec{x}$ with respect to the basis of eigenvectors, show that

$$(A - \lambda_1 I)(A - \lambda_2 I) \cdots (A - \lambda_n I)\vec{x} = \vec{0}$$

for every $\vec{x} \in \mathbb{R}^n$, and hence conclude that "A is a root of its characteristic polynomial." That is, if the characteristic polynomial is

$$C(\lambda) = (-1)^n\lambda^n + c_{n-1}\lambda^{n-1} + \cdots + c_1\lambda + c_0$$

then

$$(-1)^n A^n + c_{n-1}A^{n-1} + \cdots + c_1 A + c_0 I = O_{n,n}$$

(Hint: write the characteristic polynomial in factored form.) This result, called the Cayley-Hamilton Theorem, is true for any square matrix A.

F6 If $A \in M_{n\times n}(\mathbb{R})$ is invertible, then use the Cayley-Hamilton Theorem to show that A^{-1} can be written as a polynomial of degree less than or equal to $n - 1$ in A (that is, a linear combination of $\{A^{n-1}, \dots, A^2, A, I\}$).

CHAPTER 7

Inner Products and Projections

CHAPTER OUTLINE

In Section 1.5 we saw that we can use a projection to find a point P *in a plane that is closest to some other point* Q. *We can view this as finding the point in the plane that best approximates* Q. *Since there are many other situations in which we want to find a best approximation, it is very useful to generalize our work with projections from Section 1.5 not only to subspaces of* $\mathbb{R}^n$, *but also to general vector spaces. You may find it helpful to review Section 1.5 carefully before proceeding with this chapter.*

7.1 Orthogonal Bases in $\mathbb{R}^n$

Most of our intuition about coordinate geometry is based on our experience with the standard basis for $\mathbb{R}^n$. *It is therefore a little uncomfortable for many beginners to deal with the arbitrary bases that arise in Chapter 1. Fortunately, for many problems, it is possible to work with bases that have the most essential property of the standard basis: the basis vectors are mutually orthogonal.*

Recall from Section 1.5 that two vectors $\vec{x}, \vec{y} \in \mathbb{R}^n$ are said to be orthogonal if $\vec{x} \cdot \vec{y} = 0$. We now extend the definition of orthogonality to sets.

Definition
Orthogonal Set

A set of vectors $\{\vec{v}_1, \dots, \vec{v}_k\}$ in $\mathbb{R}^n$ is called an **orthogonal set** if $\vec{v}_i \cdot \vec{v}_j = 0$ whenever $i \neq j$.

EXAMPLE 7.1.1

The set $\left\{\begin{bmatrix}1\\1\\1\\1\end{bmatrix}, \begin{bmatrix}1\\-1\\1\\-1\end{bmatrix}, \begin{bmatrix}-1\\0\\1\\0\end{bmatrix}\right\}$ is an orthogonal set of vectors in $\mathbb{R}^4$.

The set $\left\{\begin{bmatrix}1\\1\\1\\1\end{bmatrix}, \begin{bmatrix}1\\-1\\-1\\1\end{bmatrix}, \begin{bmatrix}0\\0\\0\\0\end{bmatrix}\right\}$ is also an orthogonal set of vectors in $\mathbb{R}^4$.

Geometrically, it seems clear that an orthogonal set will be linearly independent. However, since the zero vector is orthogonal to all other vectors, we must ensure that the zero vector is excluded from the set to ensure that this is true.

Theorem 7.1.1

If $\{\vec{v}_1, \dots, \vec{v}_k\}$ is an orthogonal set of non-zero vectors in $\mathbb{R}^n$, then it is linearly independent.

Proof: Consider

$$c_1\vec{v}_1 + \cdots + c_k\vec{v}_k = \vec{0}$$

Take the dot product of $\vec{v}_i$ with each side to get

$$\begin{aligned} \vec{v}_i \cdot (c_1\vec{v}_1 + \cdots + c_k\vec{v}_k) &= \vec{v}_i \cdot \vec{0} \\ c_1(\vec{v}_i \cdot \vec{v}_1) + \cdots + c_i(\vec{v}_i \cdot \vec{v}_i) + \cdots + c_k(\vec{v}_i \cdot \vec{v}_k) &= 0 \\ 0 + \cdots + 0 + c_i\|\vec{v}_i\|^2 + 0 + \cdots + 0 &= 0 \end{aligned}$$

since $\vec{v}_i \cdot \vec{v}_j = 0$ unless $i = j$. Moreover, $\vec{v}_i \neq \vec{0}$, so $\|\vec{v}_i\| \neq 0$ and hence $c_i = 0$. Since this works for all $1 \leq i \leq k$, it follows that $\{\vec{v}_1, \dots, \vec{v}_k\}$ is linearly independent. ■

Theorem 7.1.1 implies that an orthogonal set of non-zero vectors is a basis for the subspace it spans.

Definition
Orthogonal Basis

If $\mathcal{B}$ is an orthogonal set and $\mathcal{B}$ is a basis for a subspace $\mathbb{S}$ of $\mathbb{R}^n$, then $\mathcal{B}$ is called an **orthogonal basis** for $\mathbb{S}$.

Because the vectors in the basis are all orthogonal to each other, our geometric intuition tells us that it should be quite easy to determine how to write any vector in $\mathbb{S}$ as a linear combination of these basis vectors. The following theorem demonstrates this.

Theorem 7.1.2

If $\mathcal{B} = \{\vec{v}_1, \dots, \vec{v}_k\}$ is an orthogonal basis for a subspace $\mathbb{S}$ of $\mathbb{R}^n$, then the coefficients c_i, $1 \leq i \leq k$, when any vector $\vec{x} \in \mathbb{S}$ is written as a linear combination of the vectors in $\mathcal{B}$ are given by

$$c_i = \frac{\vec{v}_i \cdot \vec{x}}{\|\vec{v}_i\|^2}$$

Proof: Since $\mathcal{B}$ is a basis, there exists unique coefficients $c_1, \dots, c_k$ such that

$$\vec{x} = c_1\vec{v}_1 + \cdots + c_k\vec{v}_k$$

Take the dot product of $\vec{v}_i$ with each side to get

$$\begin{aligned} \vec{v}_i \cdot \vec{x} &= \vec{v}_i \cdot (c_1\vec{v}_1 + \cdots + c_k\vec{v}_k) \\ \vec{v}_i \cdot \vec{x} &= c_1(\vec{v}_i \cdot \vec{v}_1) + \cdots + c_k(\vec{v}_i \cdot \vec{v}_k) \\ \vec{v}_i \cdot \vec{x} &= 0 + \cdots + 0 + c_i\|\vec{v}_i\|^2 + 0 + \cdots + 0 \end{aligned}$$

since $\vec{v}_i \cdot \vec{v}_j = 0$ whenever $i \neq j$. Because $\mathcal{B}$ is linearly independent, $\vec{v}_i$ cannot be the zero vector and hence $\|\vec{v}_i\| \neq 0$. Therefore, we can solve for c_i to get

$$c_i = \frac{\vec{v}_i \cdot \vec{x}}{\|\vec{v}_i\|^2}$$

■

EXAMPLE 7.1.2

Given that $\mathcal{B} = \left\{ \begin{bmatrix} 1 \\ 2 \end{bmatrix}, \begin{bmatrix} -2 \\ 1 \end{bmatrix} \right\}$ is an orthogonal basis for $\mathbb{R}^2$, write $\vec{x} = \begin{bmatrix} 4 \\ 3 \end{bmatrix}$ as a linear combination of the vectors in $\mathcal{B}$ and illustrate with a sketch.

Solution: To simplify the notation, we denote the vectors in $\mathcal{B}$ by $\vec{v}_1$ and $\vec{v}_2$ respectively. By Theorem 7.1.2, we have

$$c_1 = \frac{\vec{v}_1 \cdot \vec{x}}{\|\vec{v}_1\|^2} = \frac{10}{5} = 2$$

$$c_2 = \frac{\vec{v}_2 \cdot \vec{x}}{\|\vec{v}_2\|^2} = \frac{-5}{5} = -1$$

Thus,

$$\begin{bmatrix} 4 \\ 3 \end{bmatrix} = 2\begin{bmatrix} 1 \\ 2 \end{bmatrix} + (-1)\begin{bmatrix} -2 \\ 1 \end{bmatrix}$$

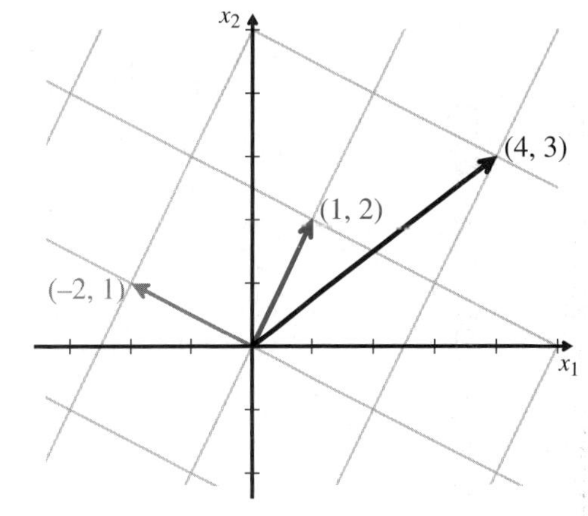

Another useful property of orthogonal sets is that they satisfy the multidimensional Pythagorean Theorem.

Theorem 7.1.3 If $\{\vec{v}_1, \dots, \vec{v}_k\}$ is an orthogonal set in $\mathbb{R}^n$, then

$$\|\vec{v}_1 + \cdots + \vec{v}_k\|^2 = \|\vec{v}_1\|^2 + \cdots + \|\vec{v}_k\|^2$$

Orthonormal Bases

Observe that the formula in Theorem 7.1.2 would be even simpler if all the basis vectors had length 1.

Definition
Orthonormal Set

A set $\{\vec{v}_1, \dots, \vec{v}_k\}$ of vectors in $\mathbb{R}^n$ is called an **orthonormal set** if it is an orthogonal set and each vector $\vec{v}_i$ is a unit vector (that is, $\|\vec{v}_i\| = 1$ for all $1 \leq i \leq k$).

Remark

Recall from Section 1.5 that we find a unit vector $\hat{x}$ in the direction of a vector $\vec{x}$ with the formula

$$\hat{x} = \frac{1}{\|\vec{x}\|}\vec{x}$$

This process is called **normalizing the vector**. Thus, the term “orthonormal” indicates an orthogonal set of non-zero vectors where all the vectors have been normalized.

EXAMPLE 7.1.3

Any subset of the standard basis vectors in $\mathbb{R}^n$ is an orthonormal set. For example, in $\mathbb{R}^6$, $\{\vec{e}_1, \vec{e}_2, \vec{e}_5, \vec{e}_6\}$ is an orthonormal set of four vectors (where, as usual, $\vec{e}_i$ is the i-th standard basis vector).

EXAMPLE 7.1.4

The set $\left\{ \frac{1}{2}\begin{bmatrix}1\\1\\1\\1\end{bmatrix}, \frac{1}{2}\begin{bmatrix}1\\-1\\1\\-1\end{bmatrix}, \frac{1}{\sqrt{2}}\begin{bmatrix}-1\\0\\1\\0\end{bmatrix} \right\}$ is an orthonormal set in $\mathbb{R}^4$. The vectors are multiples of the vectors in the first set in Example 7.1.1, so they are certainly mutually orthogonal. They have been normalized so that each vector has length 1.

EXERCISE 7.1.1

Verify that $\left\{ \begin{bmatrix}1\\1\\2\end{bmatrix}, \begin{bmatrix}-2\\0\\1\end{bmatrix}, \begin{bmatrix}1\\-5\\2\end{bmatrix} \right\}$ is an orthogonal set and then normalize the vectors to produce the corresponding orthonormal set.

By Theorem 7.1.1 an orthonormal set is necessarily linearly independent. Hence, an orthonormal set is a basis for the subspace it spans.

Definition
Orthonormal Basis

If $\mathcal{B}$ is an orthonormal set and $\mathcal{B}$ is a basis for a subspace $\mathbb{S}$ of $\mathbb{R}^n$, then $\mathcal{B}$ is called an **orthonormal basis** for $\mathbb{S}$.

EXAMPLE 7.1.5

Given that $\mathcal{B} = \left\{ \frac{1}{2}\begin{bmatrix}1\\1\\1\\1\end{bmatrix}, \frac{1}{2}\begin{bmatrix}1\\-1\\1\\-1\end{bmatrix}, \frac{1}{\sqrt{2}}\begin{bmatrix}-1\\0\\1\\0\end{bmatrix}, \frac{1}{\sqrt{2}}\begin{bmatrix}0\\1\\0\\-1\end{bmatrix} \right\}$ is an orthonormal basis for $\mathbb{R}^4$, write $\vec{x} = \begin{bmatrix}1\\2\\3\\4\end{bmatrix}$ as a linear combination of the vectors in $\mathcal{B}$.

Solution: Denote the vectors in $\mathcal{B}$ by $\vec{v}_1, \vec{v}_2, \vec{v}_3, \vec{v}_4$ respectively. Since each of these vectors has length 1, by Theorem 7.1.2, the coefficients when $\vec{x}$ is written as a linear combination of these vectors are

$$c_1 = \vec{v}_1 \cdot \vec{x} = \frac{1}{2}(1 + 2 + 3 + 4) = 5$$

$$c_2 = \vec{v}_2 \cdot \vec{x} = \frac{1}{2}(1 - 2 + 3 - 4) = -1$$

$$c_3 = \vec{v}_3 \cdot \vec{x} = \frac{1}{\sqrt{2}}(-1 + 0 + 3 + 0) = \sqrt{2}$$

$$c_4 = \vec{v}_4 \cdot \vec{x} = \frac{1}{\sqrt{2}}(0 + 2 + 0 - 4) = -\sqrt{2}$$

Hence,

$$\vec{x} = 5\vec{v}_1 - 1\vec{v}_2 + \sqrt{2}\vec{v}_3 - \sqrt{2}\vec{v}_4$$

EXERCISE 7.1.2

Let $\mathcal{B} = \left\{ \frac{1}{\sqrt{3}}\begin{bmatrix}1\\1\\1\end{bmatrix}, \frac{1}{\sqrt{2}}\begin{bmatrix}1\\0\\-1\end{bmatrix}, \frac{1}{\sqrt{6}}\begin{bmatrix}1\\-2\\1\end{bmatrix} \right\}$. Verify that $\mathcal{B}$ is an orthonormal basis for $\mathbb{R}^3$ and then write $\vec{x} = \begin{bmatrix}3\\1\\4\end{bmatrix}$ as a linear combination of the vectors in $\mathcal{B}$.

CONNECTION

Dot products and orthonormal bases in $\mathbb{R}^n$ have important generalizations to **inner products** and orthonormal bases in general vector spaces. These will be considered in Section 7.4.

Orthogonal Matrices

In Section 3.3 we saw that the standard matrix of a rotation $R_\theta : \mathbb{R}^2 \to \mathbb{R}^2$ about the origin through an angle θ is

$$[R_\theta] = \begin{bmatrix}\cos\theta & -\sin\theta\\ \sin\theta & \cos\theta\end{bmatrix}$$

We also saw in Section 3.5 that it satisfies

$$[R_\theta]^{-1} = [R_\theta]^T$$

We now show that this remarkable property, that the inverse of the matrix equals its transpose, holds whenever the columns or rows of the matrix form an orthonormal basis for $\mathbb{R}^n$.

Theorem 7.1.4

For $P \in M_{n\times n}(\mathbb{R})$, the following are equivalent:

(1) The columns of P form an orthonormal basis for $\mathbb{R}^n$
(2) $P^T = P^{-1}$
(3) The rows of P form an orthonormal basis for $\mathbb{R}^n$

Proof: Let $P = \begin{bmatrix}\vec{v}_1 & \cdots & \vec{v}_n\end{bmatrix}$. By definition of matrix-matrix multiplication we have that

$$(P^TP)_{ij} = \vec{v}_i \cdot \vec{v}_j$$

Hence, $P^TP = I$ if and only if $\vec{v}_i \cdot \vec{v}_i = 1$ for all i, and $\vec{v}_i \cdot \vec{v}_j = 0$ for all $i \neq j$. This is true if and only if the columns of P form an orthonormal set of n vectors in $\mathbb{R}^n$. But, by Theorem 7.1.1 and Theorem 2.3.6, such a set of n vectors in $\mathbb{R}^n$ is a basis for $\mathbb{R}^n$.

The result for the rows of P follows from consideration of the product $PP^T = I$. You are asked to show this in Problem C5. ■

Since such matrices are extremely useful, we make the following definition.

Definition
Orthogonal Matrix

A matrix $P \in M_{n\times n}(\mathbb{R})$ such that $P^{-1} = P^T$ is called an **orthogonal matrix**.

Remark

Observe that such matrices should probably be called orthonormal matrices, but the name *orthogonal matrix* is the generally accepted name. Be sure that you remember that an orthogonal matrix has *orthonormal* columns and rows.

EXAMPLE 7.1.6

The set $\left\{\begin{bmatrix}1\\1\\0\end{bmatrix}, \begin{bmatrix}1\\-1\\-1\end{bmatrix}, \begin{bmatrix}1\\-1\\2\end{bmatrix}\right\}$ is an orthogonal set (verify this). If the vectors are normalized, the resulting set is orthonormal, so the following matrix P is orthogonal:

$$P = \begin{bmatrix}1/\sqrt{2} & 1/\sqrt{3} & 1/\sqrt{6}\\ 1/\sqrt{2} & -1/\sqrt{3} & -1/\sqrt{6}\\ 0 & -1/\sqrt{3} & 2/\sqrt{6}\end{bmatrix}$$

Thus, P is invertible and

$$P^{-1} = P^T = \begin{bmatrix}1/\sqrt{2} & 1/\sqrt{2} & 0\\ 1/\sqrt{3} & -1/\sqrt{3} & -1/\sqrt{3}\\ 1/\sqrt{6} & -1/\sqrt{6} & 2/\sqrt{6}\end{bmatrix}$$

Moreover, observe that P^{-1} is also orthogonal.

EXERCISE 7.1.3

Verify that $P = \begin{bmatrix}1/\sqrt{2} & 1/\sqrt{2} & 0\\ 1/\sqrt{3} & -1/\sqrt{3} & 1/\sqrt{3}\\ -1/\sqrt{6} & 1/\sqrt{6} & 2/\sqrt{6}\end{bmatrix}$ is orthogonal by showing that $PP^T = I$.

Orthogonal matrices have the following useful properties.

Theorem 7.1.5 If $P, Q \in M_{n\times n}(\mathbb{R})$ are orthogonal matrices and $\vec{x}, \vec{y} \in \mathbb{R}^n$, then

(1) $(P\vec{x}) \cdot (P\vec{y}) = \vec{x} \cdot \vec{y}$
(2) $\|P\vec{x}\| = \|\vec{x}\|$
(3) $\det P = \pm 1$
(4) All real eigenvalues of P are 1 or -1
(5) PQ is an orthogonal matrix

Proof: We will prove (1) and leave the others as Problems C1, C2, C3, and C4 respectively.

Recall from page 160 the formula $\vec{x}^T\vec{y} = \vec{x} \cdot \vec{y}$. Using this formula we get

$$(P\vec{x}) \cdot (P\vec{y}) = (P\vec{x})^T(P\vec{y}) = \vec{x}^T P^T P\vec{y} = \vec{x}^T I\vec{y} = \vec{x}^T\vec{y} = \vec{x} \cdot \vec{y}$$

■

The most important applications of orthogonal matrices considered in this book are the diagonalization of symmetric matrices and the singular value decomposition in Chapter 8. Of course, there are many other applications of orthogonal matrices, such as for 3D computer graphics and Lorenz transformations in physics.

PROBLEMS 7.1

Practice Problems

For Problems A1–A8, determine whether the given set is an orthogonal set. If so, normalize the vectors to produce the corresponding orthonormal set.

A1 $\left\{\begin{bmatrix}1\\2\end{bmatrix}, \begin{bmatrix}2\\-1\end{bmatrix}\right\}$

A2 $\left\{\begin{bmatrix}1\\-3\end{bmatrix}, \begin{bmatrix}1\\3\end{bmatrix}\right\}$

A3 $\left\{\begin{bmatrix}4\\-1\\1\end{bmatrix}, \begin{bmatrix}-1\\0\\4\end{bmatrix}, \begin{bmatrix}4\\17\\1\end{bmatrix}\right\}$

A4 $\left\{\begin{bmatrix}1\\-1\\1\end{bmatrix}, \begin{bmatrix}1\\2\\1\end{bmatrix}, \begin{bmatrix}2\\-1\\1\end{bmatrix}\right\}$

A5 $\left\{\begin{bmatrix}1\\3\\-1\end{bmatrix}, \begin{bmatrix}2\\0\\2\end{bmatrix}, \begin{bmatrix}3\\-2\\-3\end{bmatrix}\right\}$

A6 $\left\{\begin{bmatrix}1\\-2\\2\end{bmatrix}, \begin{bmatrix}-2\\1\\2\end{bmatrix}, \begin{bmatrix}2\\2\\1\end{bmatrix}\right\}$

A7 $\left\{\begin{bmatrix}1\\1\\1\\1\end{bmatrix}, \begin{bmatrix}-2\\0\\1\\1\end{bmatrix}, \begin{bmatrix}0\\0\\1\\-1\end{bmatrix}\right\}$

A8 $\left\{\begin{bmatrix}1\\0\\1\\1\end{bmatrix}, \begin{bmatrix}2\\1\\1\\0\end{bmatrix}, \begin{bmatrix}1\\-3\\-1\\0\end{bmatrix}\right\}$

A9 Given that $\mathcal{B} = \left\{\begin{bmatrix}1\\-2\\2\end{bmatrix}, \begin{bmatrix}2\\2\\1\end{bmatrix}, \begin{bmatrix}-2\\1\\2\end{bmatrix}\right\}$ is an orthogonal basis for $\mathbb{R}^3$, use Theorem 7.1.2 to write the following vectors as a linear combination of the vectors in $\mathcal{B}$.

(a) $\vec{w} = \begin{bmatrix}4\\3\\5\end{bmatrix}$ (b) $\vec{x} = \begin{bmatrix}1\\0\\0\end{bmatrix}$ (c) $\vec{y} = \begin{bmatrix}-2\\2\\-1\end{bmatrix}$

A10 Given that $\mathcal{B} = \left\{\begin{bmatrix}1/3\\-2/3\\2/3\end{bmatrix}, \begin{bmatrix}2/3\\2/3\\1/3\end{bmatrix}, \begin{bmatrix}-2/3\\1/3\\2/3\end{bmatrix}\right\}$ is an orthonormal basis for $\mathbb{R}^3$, use Theorem 7.1.2 to write the following vectors as a linear combination of the vectors in $\mathcal{B}$.

(a) $\vec{w} = \begin{bmatrix}4\\3\\5\end{bmatrix}$ (b) $\vec{x} = \begin{bmatrix}3\\-7\\2\end{bmatrix}$ (c) $\vec{y} = \begin{bmatrix}2\\-4\\6\end{bmatrix}$

A11 Given that $\mathcal{B} = \left\{\begin{bmatrix}1\\1\\1\\1\end{bmatrix}, \begin{bmatrix}1\\-1\\1\\-1\end{bmatrix}, \begin{bmatrix}-1\\0\\1\\0\end{bmatrix}, \begin{bmatrix}0\\1\\0\\-1\end{bmatrix}\right\}$ is an orthogonal basis for $\mathbb{R}^4$, use Theorem 7.1.2 to write the following vectors as a linear combination of the vectors in $\mathcal{B}$.

(a) $\vec{x} = \begin{bmatrix}2\\4\\-3\\5\end{bmatrix}$ (b) $\vec{y} = \begin{bmatrix}-4\\1\\3\\-5\end{bmatrix}$ (c) $\vec{w} = \begin{bmatrix}3\\1\\0\\1\end{bmatrix}$

For Problems A12–A17, decide whether the matrix is orthogonal. If not, indicate how the columns fail to form an orthonormal set (for example, "the second and third columns are not orthogonal").

A12 $\begin{bmatrix}5/13 & 12/13\\12/13 & -5/13\end{bmatrix}$

A13 $\begin{bmatrix}3/5 & 4/5\\-4/5 & -3/5\end{bmatrix}$

A14 $\begin{bmatrix}2/5 & -1/5\\1/5 & 2/5\end{bmatrix}$

A15 $\begin{bmatrix}1/\sqrt{2} & -1/\sqrt{2}\\1/\sqrt{2} & 1/\sqrt{2}\end{bmatrix}$

A16 $\begin{bmatrix}1/3 & 2/3 & -2/3\\2/3 & -2/3 & 1/3\\2/3 & 1/3 & 2/3\end{bmatrix}$

A17 $\begin{bmatrix}1/3 & 2/3 & 2/3\\2/3 & -2/3 & 1/3\\2/3 & 1/3 & -2/3\end{bmatrix}$

A18 Given that $\mathcal{B} = \left\{\begin{bmatrix}1/\sqrt{6}\\-2/\sqrt{6}\\1/\sqrt{6}\end{bmatrix}, \begin{bmatrix}1/\sqrt{3}\\1/\sqrt{3}\\1/\sqrt{3}\end{bmatrix}, \begin{bmatrix}1/\sqrt{2}\\0\\-1/\sqrt{2}\end{bmatrix}\right\}$ is an orthonormal basis for $\mathbb{R}^3$. Determine another orthonormal basis for $\mathbb{R}^3$ that includes the vector $\begin{bmatrix}1/\sqrt{6}\\1/\sqrt{3}\\1/\sqrt{2}\end{bmatrix}$ and briefly explain why your basis is orthonormal.

A19 Prove that an orthonormal set $\mathcal{B}$ of n vectors in $\mathbb{R}^n$ is a basis for $\mathbb{R}^n$.

Homework Problems

For Problems B1–B6, determine whether the given set is an orthogonal set. If so, normalize the vectors to produce the corresponding orthonormal set.

B1 $\left\{\begin{bmatrix}2\\3\end{bmatrix}, \begin{bmatrix}3\\-2\end{bmatrix}\right\}$

B2 $\left\{\begin{bmatrix}2\\-1\end{bmatrix}, \begin{bmatrix}3\\6\end{bmatrix}\right\}$

B3 $\left\{\begin{bmatrix}1\\2\\-1\end{bmatrix}, \begin{bmatrix}-1\\1\\1\end{bmatrix}, \begin{bmatrix}1\\0\\1\end{bmatrix}\right\}$

B4 $\left\{\begin{bmatrix}2\\1\\1\end{bmatrix}, \begin{bmatrix}0\\-1\\1\end{bmatrix}, \begin{bmatrix}1\\3\\-1\end{bmatrix}\right\}$

B5 $\left\{\begin{bmatrix}1\\2\\-1\\1\end{bmatrix}, \begin{bmatrix}2\\1\\3\\-1\end{bmatrix}, \begin{bmatrix}3\\1\\0\\-2\end{bmatrix}\right\}$

B6 $\left\{\begin{bmatrix}0\\1\\-1\\2\end{bmatrix}, \begin{bmatrix}1\\-1\\1\\1\end{bmatrix}, \begin{bmatrix}-3\\0\\2\\1\end{bmatrix}\right\}$

B7 Given that $\mathcal{B} = \left\{\begin{bmatrix}1\\3\end{bmatrix}, \begin{bmatrix}-3\\1\end{bmatrix}\right\}$ is an orthogonal basis for $\mathbb{R}^2$, use Theorem 7.1.2 to write the following vectors as a linear combination of the vectors in $\mathcal{B}$.

(a) $\vec{w} = \begin{bmatrix}1\\0\end{bmatrix}$ (b) $\vec{x} = \begin{bmatrix}2\\7\end{bmatrix}$ (c) $\vec{y} = \begin{bmatrix}-3\\4\end{bmatrix}$

B8 Given that $\mathcal{B} = \left\{\begin{bmatrix}3\\-4\end{bmatrix}, \begin{bmatrix}4\\3\end{bmatrix}\right\}$ is an orthogonal basis for $\mathbb{R}^2$, use Theorem 7.1.2 to write the following vectors as a linear combination of the vectors in $\mathcal{B}$.

(a) $\vec{w} = \begin{bmatrix}1\\1\end{bmatrix}$ (b) $\vec{x} = \begin{bmatrix}-2\\3\end{bmatrix}$ (c) $\vec{y} = \begin{bmatrix}2\\3\end{bmatrix}$

B9 Given that $\mathcal{B} = \left\{\begin{bmatrix}2/\sqrt{5}\\-1/\sqrt{5}\end{bmatrix}, \begin{bmatrix}1/\sqrt{5}\\2/\sqrt{5}\end{bmatrix}\right\}$ is an orthonormal basis for $\mathbb{R}^2$, use Theorem 7.1.2 to write the following vectors as a linear combination of the vectors in $\mathcal{B}$.

(a) $\vec{w} = \begin{bmatrix}0\\1\end{bmatrix}$ (b) $\vec{x} = \begin{bmatrix}3\\-3\end{bmatrix}$ (c) $\vec{y} = \begin{bmatrix}-2\\-3\end{bmatrix}$

B10 Given that $\mathcal{B} = \left\{\begin{bmatrix}1\\2\\1\end{bmatrix}, \begin{bmatrix}-3\\1\\1\end{bmatrix}, \begin{bmatrix}1\\-4\\7\end{bmatrix}\right\}$ is an orthogonal basis for $\mathbb{R}^3$, use Theorem 7.1.2 to write the following vectors as a linear combination of the vectors in $\mathcal{B}$.

(a) $\vec{w} = \begin{bmatrix}2\\1\\1\end{bmatrix}$ (b) $\vec{x} = \begin{bmatrix}1\\-1\\0\end{bmatrix}$ (c) $\vec{y} = \begin{bmatrix}1\\-1\\-3\end{bmatrix}$

B11 Given that $\mathcal{B} = \left\{\begin{bmatrix}1\\2\\1\\1\end{bmatrix}, \begin{bmatrix}-2\\1\\-1\\1\end{bmatrix}, \begin{bmatrix}-2\\-4\\5\\5\end{bmatrix}\right\}$ is an orthogonal basis for the subspace of $\mathbb{R}^4$ it spans, use Theorem 7.1.2 to write the following vectors as a linear combination of the vectors in $\mathcal{B}$.

(a) $\vec{x} = \begin{bmatrix}7\\9\\-1\\-3\end{bmatrix}$ (b) $\vec{y} = \begin{bmatrix}4\\-2\\9\\5\end{bmatrix}$ (c) $\vec{w} = \begin{bmatrix}5\\10\\4\\4\end{bmatrix}$

For Problems B12–B16, decide whether the matrix is orthogonal. If not, indicate how the columns fail to form an orthonormal set.

B12 $\begin{bmatrix}1/\sqrt{5} & 0 & 1/\sqrt{3}\\ 2/\sqrt{5} & 1/\sqrt{2} & 1/\sqrt{3}\\ 0 & 1/\sqrt{2} & 1/\sqrt{3}\end{bmatrix}$

B13 $\begin{bmatrix}8/17 & -15/17\\ 15/17 & 8/17\end{bmatrix}$

B14 $\begin{bmatrix}1/3 & -4/\sqrt{18} & 0\\ 2/3 & 1/\sqrt{18} & 1/\sqrt{2}\\ 2/3 & 1/\sqrt{18} & -1/\sqrt{2}\end{bmatrix}$

B15 $\begin{bmatrix}3/\sqrt{24} & 5/\sqrt{24}\\ -5/\sqrt{24} & 3/\sqrt{24}\end{bmatrix}$

B16 $\begin{bmatrix}1/\sqrt{5} & 2/\sqrt{5} & 0\\ -2/\sqrt{6} & 1/\sqrt{6} & 1/\sqrt{6}\\ 2/\sqrt{30} & -1/\sqrt{30} & 5/\sqrt{30}\end{bmatrix}$

Conceptual Problems

C1 Prove Theorem 7.1.5 (2).

C2 (a) Prove Theorem 7.1.5 (3).
(b) Give an example of a 2×2 matrix A such that $\det A = 1$, but A is not orthogonal.

C3 Prove Theorem 7.1.5 (4).

C4 Prove Theorem 7.1.5 (5).

C5 Prove that $P \in M_{n\times n}(\mathbb{R})$ is orthogonal if and only if the rows of P form an orthonormal set.

7.2 Projections and the Gram-Schmidt Procedure

Projections onto a Subspace

The projection of a vector $\vec{y}$ onto another vector $\vec{x}$ was defined in Chapter 1 by finding a scalar multiple of $\vec{x}$, denoted $\text{proj}_{\vec{x}}(\vec{y})$, *and a vector perpendicular to $\vec{x}$, denoted* $\text{perp}_{\vec{x}}(\vec{y})$, *such that*

$$\vec{y} = \text{proj}_{\vec{x}}(\vec{y}) + \text{perp}_{\vec{x}}(\vec{y})$$

Since we were just trying to find a scalar multiple of $\vec{x}$, the projection of $\vec{y}$ onto $\vec{x}$ can be viewed as projecting $\vec{y}$ onto the subspace spanned by $\vec{x}$. Similarly, we saw how to find the projection of $\vec{y}$ onto a plane, which is just a 2-dimensional subspace. It is natural and useful to define the projection of vectors onto more general subspaces.

Let $\vec{y} \in \mathbb{R}^n$ and let $\mathbb{S}$ be a subspace of $\mathbb{R}^n$. To match what we did in Chapter 1, we want to write $\vec{y}$ as

$$\vec{y} = \text{proj}_{\mathbb{S}}(\vec{y}) + \text{perp}_{\mathbb{S}}(\vec{y})$$

where $\text{proj}_{\mathbb{S}}(\vec{y})$ is a vector in $\mathbb{S}$ and $\text{perp}_{\mathbb{S}}(\vec{y})$ is a vector orthogonal to $\mathbb{S}$. To do this, we first observe that we need to define precisely what we mean by a vector orthogonal to a subspace.

Definition
Orthogonal
Orthogonal Complement

Let $\mathbb{S}$ be a subspace of $\mathbb{R}^n$. We say that a vector $\vec{x} \in \mathbb{R}^n$ is **orthogonal** to $\mathbb{S}$ if

$$\vec{x} \cdot \vec{s} = 0 \qquad \text{for all } \vec{s} \in \mathbb{S}$$

We call the set of all vectors orthogonal to $\mathbb{S}$ the **orthogonal complement** of $\mathbb{S}$ and denote it by $\mathbb{S}^{\perp}$. That is,

$$\mathbb{S}^{\perp} = \{\vec{x} \in \mathbb{R}^n \mid \vec{x} \cdot \vec{s} = 0 \text{ for all } \vec{s} \in \mathbb{S}\}$$

EXAMPLE 7.2.1

Let $\mathbb{S}$ be a plane passing through the origin in $\mathbb{R}^3$ with normal vector $\vec{n}$. By definition, $\vec{n}$ is orthogonal to every vector in the plane. So, we say that $\vec{n}$ is orthogonal to the plane. Moreover, we know that any scalar multiple of $\vec{n}$ is also orthogonal to $\mathbb{S}$, so $\mathbb{S}^{\perp} = \text{Span}\{\vec{n}\}$.

On the other hand, we saw in Chapter 1 that the plane is the set of all vectors orthogonal to $\vec{n}$ (or any scalar multiple of $\vec{n}$), so the orthogonal complement of the subspace $\text{Span}\{\vec{n}\}$ is $\mathbb{S}$.

Instead of having to show that a vector $\vec{x} \in \mathbb{R}^n$ is orthogonal to every vector $\vec{s} \in \mathbb{S}$ to prove that $\vec{x} \in \mathbb{S}^{\perp}$, the following theorem tells us that we only need to show that $\vec{x}$ is orthogonal to every vector in a spanning set for $\mathbb{S}$.

Theorem 7.2.1

Let $\{\vec{v}_1, \dots, \vec{v}_k\}$ be a spanning set for a subspace $\mathbb{S}$ of $\mathbb{R}^n$, and let $\vec{x} \in \mathbb{R}^n$. We have that $\vec{x} \in \mathbb{S}^{\perp}$ if and only if $\vec{x} \cdot \vec{v}_i = 0$ for all $1 \leq i \leq k$.

EXAMPLE 7.2.2

Let $\mathbb{W} = \text{Span}\left\{\begin{bmatrix}1\\0\\0\\1\end{bmatrix}, \begin{bmatrix}1\\0\\1\\0\end{bmatrix}\right\}$. Find $\mathbb{W}^{\perp}$.

Solution: By Theorem 7.2.1, we just need to find all $\vec{x} = \begin{bmatrix}x_1\\x_2\\x_3\\x_4\end{bmatrix} \in \mathbb{R}^4$ such that

$$0 = \begin{bmatrix}x_1\\x_2\\x_3\\x_4\end{bmatrix} \cdot \begin{bmatrix}1\\0\\0\\1\end{bmatrix} = x_1 + x_4$$

$$0 = \begin{bmatrix}x_1\\x_2\\x_3\\x_4\end{bmatrix} \cdot \begin{bmatrix}1\\0\\1\\0\end{bmatrix} = x_1 + x_3$$

The solution space of this homogeneous system is $\text{Span}\left\{\begin{bmatrix}0\\1\\0\\0\end{bmatrix}, \begin{bmatrix}-1\\0\\1\\1\end{bmatrix}\right\}$. Hence,

$$\mathbb{W}^{\perp} = \text{Span}\left\{\begin{bmatrix}0\\1\\0\\0\end{bmatrix}, \begin{bmatrix}-1\\0\\1\\1\end{bmatrix}\right\}$$

EXERCISE 7.2.1

Let $\mathbb{S} = \text{Span}\left\{\begin{bmatrix}1\\1\\1\\0\end{bmatrix}, \begin{bmatrix}-1\\0\\1\\1\end{bmatrix}, \begin{bmatrix}0\\1\\-1\\1\end{bmatrix}\right\}$. Find $\mathbb{S}^{\perp}$.

We get the following important facts about $\mathbb{S}$ and $\mathbb{S}^{\perp}$.

Theorem 7.2.2

If $\mathbb{S}$ is a k-dimensional subspace of $\mathbb{R}^n$, then

(1) $\mathbb{S}^{\perp}$ is a subspace of $\mathbb{R}^n$.
(2) $\dim(\mathbb{S}^{\perp}) = n - k$
(3) $\mathbb{S} \cap \mathbb{S}^{\perp} = \{\vec{0}\}$
(4) If $\{\vec{v}_1, \dots, \vec{v}_k\}$ is an orthogonal basis for $\mathbb{S}$ and $\{\vec{v}_{k+1}, \dots, \vec{v}_n\}$ is an orthogonal basis for $\mathbb{S}^{\perp}$, then $\{\vec{v}_1, \dots, \vec{v}_k, \vec{v}_{k+1}, \dots, \vec{v}_n\}$ is an orthogonal basis for $\mathbb{R}^n$.

You are asked to prove (2), (3), and (4) in Problems C1, C2, and C3.

We are now able to return to our goal of defining the projection of a vector $\vec{x} \in \mathbb{R}^n$ onto a subspace $\mathbb{S}$ of $\mathbb{R}^n$. Assume that we have an orthogonal basis $\{\vec{v}_1, \dots, \vec{v}_k\}$ for $\mathbb{S}$ and an orthogonal basis $\{\vec{v}_{k+1}, \dots, \vec{v}_n\}$ for $\mathbb{S}^\perp$. Then, by Theorem 7.2.2 (4), we know that $\{\vec{v}_1, \dots, \vec{v}_n\}$ is an orthogonal basis for $\mathbb{R}^n$. Therefore, Theorem 7.1.2 gives

$$\vec{x} = \frac{\vec{v}_1 \cdot \vec{x}}{\|\vec{v}_1\|^2}\vec{v}_1 + \cdots + \frac{\vec{v}_k \cdot \vec{x}}{\|\vec{v}_k\|^2}\vec{v}_k + \frac{\vec{v}_{k+1} \cdot \vec{x}}{\|\vec{v}_{k+1}\|^2}\vec{v}_{k+1} + \cdots + \frac{\vec{v}_n \cdot \vec{x}}{\|\vec{v}_n\|^2}\vec{v}_n$$

But this is exactly what we have been looking for! In particular, we have written $\vec{x}$ as a sum of

$$\frac{\vec{v}_1 \cdot \vec{x}}{\|\vec{v}_1\|^2}\vec{v}_1 + \cdots + \frac{\vec{v}_k \cdot \vec{x}}{\|\vec{v}_k\|^2}\vec{v}_k \in \mathbb{S}$$

and

$$\frac{\vec{v}_{k+1} \cdot \vec{x}}{\|\vec{v}_{k+1}\|^2}\vec{v}_{k+1} + \cdots + \frac{\vec{v}_n \cdot \vec{x}}{\|\vec{v}_n\|^2}\vec{v}_n \in \mathbb{S}^\perp$$

Thus, we can make the following definition.

Definition

Projection onto a Subspace

Perpendicular of a Projection onto a Subspace

Let $\mathcal{B} = \{\vec{v}_1, \dots, \vec{v}_k\}$ be an *orthogonal basis* of a k-dimensional subspace $\mathbb{S}$ of $\mathbb{R}^n$. For any $\vec{x} \in \mathbb{R}^n$, the **projection** of $\vec{x}$ onto $\mathbb{S}$ is defined to be

$$\text{proj}_{\mathbb{S}}(\vec{x}) = \frac{\vec{v}_1 \cdot \vec{x}}{\|\vec{v}_1\|^2}\vec{v}_1 + \cdots + \frac{\vec{v}_k \cdot \vec{x}}{\|\vec{v}_k\|^2}\vec{v}_k$$

The **projection of $\vec{x}$ perpendicular to** $\mathbb{S}$ is defined to be

$$\text{perp}_{\mathbb{S}}(\vec{x}) = \vec{x} - \text{proj}_{\mathbb{S}}(\vec{x})$$

Remark

A key component for this definition is that we must have an orthogonal basis for the subspace $\mathbb{S}$. If you try to use the given formula on a basis that is not an orthogonal basis, you generally will not get the correct answer.

We have defined $\text{perp}_{\mathbb{S}}(\vec{x})$ so that we do not require an orthogonal basis for $\mathbb{S}^\perp$. To ensure that this is valid, we do need to verify that $\text{perp}_{\mathbb{S}}(\vec{x}) \in \mathbb{S}^\perp$. For any $1 \leq i \leq k$, we have

$$\begin{aligned}
\vec{v}_i \cdot \text{perp}_{\mathbb{S}}(\vec{x}) &= \vec{v}_i \cdot \left[\vec{x} - \left(\frac{\vec{v}_1 \cdot \vec{x}}{\|\vec{v}_1\|^2}\vec{v}_1 + \cdots + \frac{\vec{v}_k \cdot \vec{x}}{\|\vec{v}_k\|^2}\vec{v}_k\right)\right] \\
&= \vec{v}_i \cdot \vec{x} - \vec{v}_i \cdot \left(\frac{\vec{v}_1 \cdot \vec{x}}{\|\vec{v}_1\|^2}\vec{v}_1 + \cdots + \frac{\vec{v}_k \cdot \vec{x}}{\|\vec{v}_k\|^2}\vec{v}_k\right) \\
&= \vec{v}_i \cdot \vec{x} - \left(0 + \cdots + 0 + \frac{\vec{v}_i \cdot \vec{x}}{\|\vec{v}_i\|^2}(\vec{v}_i \cdot \vec{v}_i) + 0 + \cdots + 0\right) \\
&= \vec{v}_i \cdot \vec{x} - \vec{v}_i \cdot \vec{x} \\
&= 0
\end{aligned}$$

since $\mathcal{B}$ is an orthogonal basis. Hence, $\text{perp}_{\mathbb{S}}(\vec{x})$ is orthogonal to every vector in the orthogonal basis $\{\vec{v}_1, \dots, \vec{v}_k\}$ of $\mathbb{S}$. Consequently, $\text{perp}_{\mathbb{S}}(\vec{x}) \in \mathbb{S}^\perp$ by Theorem 7.2.1.

EXAMPLE 7.2.3

Let $\mathbb{S} = \text{Span}\left\{\begin{bmatrix}1\\1\\1\\1\end{bmatrix}, \begin{bmatrix}1\\-1\\1\\-1\end{bmatrix}\right\}$ and let $\vec{x} = \begin{bmatrix}2\\5\\-7\\3\end{bmatrix}$. Determine $\text{proj}_{\mathbb{S}}(\vec{x})$ and $\text{perp}_{\mathbb{S}}(\vec{x})$.

Solution: Observe that $\mathcal{B} = \{\vec{v}_1, \vec{v}_2\} = \left\{\begin{bmatrix}1\\1\\1\\1\end{bmatrix}, \begin{bmatrix}1\\-1\\1\\-1\end{bmatrix}\right\}$ is an orthogonal basis for $\mathbb{S}$. Thus,

$$\text{proj}_{\mathbb{S}}(\vec{x}) = \frac{\vec{v}_1 \cdot \vec{x}}{\|\vec{v}_1\|^2}\vec{v}_1 + \frac{\vec{v}_2 \cdot \vec{x}}{\|\vec{v}_2\|^2}\vec{v}_2 = \frac{3}{4}\begin{bmatrix}1\\1\\1\\1\end{bmatrix} + \frac{-13}{4}\begin{bmatrix}1\\-1\\1\\-1\end{bmatrix} = \begin{bmatrix}-5/2\\4\\-5/2\\4\end{bmatrix}$$

$$\text{perp}_{\mathbb{S}}(\vec{x}) = \vec{x} - \text{proj}_{\mathbb{S}}(\vec{x}) = \begin{bmatrix}2\\5\\-7\\3\end{bmatrix} - \begin{bmatrix}-5/2\\4\\-5/2\\4\end{bmatrix} = \begin{bmatrix}9/2\\1\\-9/2\\-1\end{bmatrix}$$

EXERCISE 7.2.2

Let $\mathcal{B} = \left\{\begin{bmatrix}1\\2\\1\end{bmatrix}, \begin{bmatrix}-1\\1\\-1\end{bmatrix}\right\}$ and let $\vec{x} = \begin{bmatrix}2\\1\\3\end{bmatrix}$. Show that $\mathcal{B}$ is an orthogonal basis for $\mathbb{S} = \text{Span}\,\mathcal{B}$, and determine $\text{proj}_{\mathbb{S}}(\vec{x})$ and $\text{perp}_{\mathbb{S}}(\vec{x})$.

Recall that we showed in Chapter 1 that the projection of a vector $\vec{x} \in \mathbb{R}^3$ onto a plane in $\mathbb{R}^3$ is the vector in the plane that is closest to $\vec{x}$. We now prove that the projection of $\vec{x} \in \mathbb{R}^n$ onto a subspace $\mathbb{S}$ of $\mathbb{R}^n$ is the vector in $\mathbb{S}$ that is closest to $\vec{x}$.

Theorem 7.2.3 Approximation Theorem

Let $\mathbb{S}$ be a subspace of $\mathbb{R}^n$. If $\vec{x} \in \mathbb{R}^n$, then the vector in $\mathbb{S}$ that is closest to $\vec{x}$ is $\text{proj}_{\mathbb{S}}(\vec{x})$. That is,

$$\|\vec{x} - \text{proj}_{\mathbb{S}}(\vec{x})\| < \|\vec{x} - \vec{v}\|$$

for all $\vec{v} \in \mathbb{S}$, $\vec{v} \neq \text{proj}_{\mathbb{S}}(\vec{x})$.

Proof: Consider $\vec{x} - \vec{v} = (\vec{x} - \text{proj}_{\mathbb{S}}(\vec{x})) + (\text{proj}_{\mathbb{S}}(\vec{x}) - \vec{v})$. Now, observe that $\{\vec{x} - \text{proj}_{\mathbb{S}}(\vec{x}), \text{proj}_{\mathbb{S}}(\vec{x}) - \vec{v}\}$ is an orthogonal set since $\vec{x} - \text{proj}_{\mathbb{S}}(\vec{x}) = \text{perp}_{\mathbb{S}}(\vec{x}) \in \mathbb{S}^\perp$ and $\text{proj}_{\mathbb{S}}(\vec{x}) - \vec{v} \in \mathbb{S}$. Therefore, by Theorem 7.1.3, we get

$$\begin{aligned}\|\vec{x} - \vec{v}\|^2 &= \|(\vec{x} - \text{proj}_{\mathbb{S}}(\vec{x})) + (\text{proj}_{\mathbb{S}}(\vec{x}) - \vec{v})\|^2 \\ &= \|\vec{x} - \text{proj}_{\mathbb{S}}(\vec{x})\|^2 + \|\text{proj}_{\mathbb{S}}(\vec{x}) - \vec{v}\|^2 \\ &> \|\vec{x} - \text{proj}_{\mathbb{S}}(\vec{x})\|^2\end{aligned}$$

since $\|\text{proj}_{\mathbb{S}}(\vec{x}) - \vec{v}\|^2 > 0$ if $\vec{v} \neq \text{proj}_{\mathbb{S}}(\vec{x})$. The result follows. ∎

The Gram-Schmidt Procedure

For many of the calculations in this chapter, we need an orthogonal (or orthonormal) basis for a subspace $\mathbb{S}$ of $\mathbb{R}^n$. If $\mathbb{S}$ is a k-dimensional subspace of $\mathbb{R}^n$, it is certainly possible to use the methods of Section 4.3 to produce some basis $\{\vec{w}_1, \ldots, \vec{w}_k\}$ for $\mathbb{S}$. We will now show that we can convert any such basis for $\mathbb{S}$ into an orthonormal basis $\{\vec{v}_1, \ldots, \vec{v}_k\}$ for $\mathbb{S}$.

The construction is recursive. We first take $\vec{v}_1 = \vec{w}_1$ and see that $\{\vec{v}_1\}$ is an orthogonal basis for Span$\{\vec{w}_1\}$. Then, whenever we have an orthogonal basis $\{\vec{v}_1, \ldots, \vec{v}_{i-1}\}$ for Span$\{\vec{w}_1, \ldots, \vec{w}_{i-1}\}$, we want to find a vector $\vec{v}_i$ such that $\{\vec{v}_1, \ldots, \vec{v}_{i-1}, \vec{v}_i\}$ is an orthogonal basis for Span$\{\vec{w}_1, \ldots, \vec{w}_i\}$. We will repeat this procedure, called the **Gram-Schmidt Procedure**, until we have the desired orthogonal basis $\{\vec{v}_1, \ldots, \vec{v}_k\}$ for Span$\{\vec{w}_1, \ldots, \vec{w}_k\}$. We will use the following theorem.

Theorem 7.2.4 If $\vec{v}_1, \ldots, \vec{v}_k \in \mathbb{R}^n$, then for any $t_1, \ldots, t_{k-1} \in \mathbb{R}$, we have

$$\text{Span}\{\vec{v}_1, \ldots, \vec{v}_k\} = \text{Span}\{\vec{v}_1, \ldots, \vec{v}_{k-1}, \vec{v}_k + t_1\vec{v}_1 + \cdots + t_{k-1}\vec{v}_{k-1}\}$$

We first demonstrate the **Gram-Schmidt Procedure** with an example.

EXAMPLE 7.2.4

Let $\vec{w}_1 = \begin{bmatrix} 1 \\ 1 \\ 0 \\ 1 \end{bmatrix}, \vec{w}_2 = \begin{bmatrix} 1 \\ 0 \\ 0 \\ 2 \end{bmatrix}, \vec{w}_3 = \begin{bmatrix} 1 \\ 2 \\ -1 \\ 3 \end{bmatrix}$. Find an orthogonal basis for $\mathbb{S} = \text{Span}\{\vec{w}_1, \vec{w}_2, \vec{w}_3\}$.

Solution: First step: Let $\vec{v}_1 = \vec{w}_1$ and let $\mathbb{S}_1 = \text{Span}\{\vec{v}_1\}$.

Second step: We want to find a vector $\vec{v}_2$ that is orthogonal to $\vec{v}_1$, and such that Span$\{\vec{v}_1, \vec{v}_2\}$ = Span$\{\vec{w}_1, \vec{w}_2\}$. We know that

$$\text{perp}_{\mathbb{S}_1}(\vec{w}_2) = \vec{w}_2 - \frac{\vec{v}_1 \cdot \vec{w}_2}{\|\vec{v}_1\|^2}\vec{v}_1 = \begin{bmatrix} 0 \\ -1 \\ 0 \\ 1 \end{bmatrix}$$

is orthogonal to $\vec{v}_1$. We also get that Span$\{\vec{v}_1, \vec{v}_2\}$ = Span$\{\vec{w}_1, \vec{w}_2\}$ by Theorem 7.2.4. Hence, we take $\vec{v}_2 = \text{perp}_{\mathbb{S}_1}(\vec{w}_2)$ and define $\mathbb{S}_2 = \text{Span}\{\vec{v}_1, \vec{v}_2\}$.

Third step: We want to find a vector $\vec{v}_3$ that it is orthogonal to both $\vec{v}_1$ and $\vec{v}_2$, and such that Span$\{\vec{v}_1, \vec{v}_2, \vec{v}_3\}$ = Span$\{\vec{w}_1, \vec{w}_2, \vec{w}_3\}$. Again, we see that we can use $\text{perp}_{\mathbb{S}_2}(\vec{w}_3)$. We let

$$\vec{v}_3 = \text{perp}_{\mathbb{S}_2}(\vec{w}_3) = \vec{w}_3 - \frac{\vec{v}_1 \cdot \vec{w}_3}{\|\vec{v}_1\|^2}\vec{v}_1 - \frac{\vec{v}_2 \cdot \vec{w}_3}{\|\vec{v}_2\|^2}\vec{v}_2 = \begin{bmatrix} -1 \\ 1/2 \\ -1 \\ 1/2 \end{bmatrix}$$

We now have that $\{\vec{v}_1, \vec{v}_2, \vec{v}_3\}$ is an orthogonal basis for $\mathbb{S}$.

Algorithm 7.2.1

The Gram-Schmidt Procedure:
Let $\{\vec{w}_1, \dots, \vec{w}_k\}$ be a basis for a subspace $\mathbb{S}$ of $\mathbb{R}^n$.
First step: Let $\vec{v}_1 = \vec{w}_1$ and let $\mathbb{S}_1 = \operatorname{Span}\{\vec{v}_1\}$.
Second step: Let

$$\vec{v}_2 = \operatorname{perp}_{\mathbb{S}_1}(\vec{w}_2) = \vec{w}_2 - \frac{\vec{v}_1 \cdot \vec{w}_2}{\|\vec{v}_1\|^2}\vec{v}_1$$

and define $\mathbb{S}_2 = \operatorname{Span}\{\vec{v}_1, \vec{v}_2\}$.
***i*-th step:** Suppose that $i-1$ steps have been carried out so that $\{\vec{v}_1, \dots, \vec{v}_{i-1}\}$ is an orthogonal set, and $\mathbb{S}_{i-1} = \operatorname{Span}\{\vec{v}_1, \dots, \vec{v}_{i-1}\} = \operatorname{Span}\{\vec{w}_1, \dots, \vec{w}_{i-1}\}$. Let

$$\vec{v}_i = \operatorname{perp}_{\mathbb{S}_{i-1}}(\vec{w}_i) = \vec{w}_i - \frac{\vec{v}_1 \cdot \vec{w}_i}{\|\vec{v}_1\|^2}\vec{v}_1 - \frac{\vec{v}_2 \cdot \vec{w}_i}{\|\vec{v}_2\|^2}\vec{v}_2 - \cdots - \frac{\vec{v}_{i-1} \cdot \vec{w}_i}{\|\vec{v}_{i-1}\|^2}\vec{v}_{i-1}$$

After the k-th step is completed, we will have an orthogonal basis $\{\vec{v}_1, \dots, \vec{v}_k\}$ for $\mathbb{S}$.

Remarks

1. It is an important feature of the construction that $\mathbb{S}_{i-1}$ is a subspace of the next $\mathbb{S}_i$ and that $\{\vec{v}_1, \dots, \vec{v}_i\}$ is an orthogonal basis for $\mathbb{S}_i$.
2. Since it is really only the direction of $\vec{v}_i$ that is important in this procedure, we can rescale each $\vec{v}_i$ in any convenient fashion to simplify the calculations.
3. The order of the vectors in the original basis has an effect on the calculations because each step takes the perpendicular part of the next vector. That is, if the original vectors were given in a different order, the procedure might produce a different orthogonal basis.
4. The Gram-Schmidt Procedure does not require that we start with a basis $\mathbb{S}$; only a spanning set is required. In particular, if we find that $\operatorname{perp}_{\mathbb{S}_{i-1}}(\vec{w}_i) = \vec{0}$, then $\vec{w}_i$ is a linear combination of $\{\vec{w}_1, \dots, \vec{w}_{i-1}\}$. In this case, we omit $\vec{w}_i$ from the spanning set and continue the procedure with the next vector. This is demonstrated in Example 7.2.6.

EXAMPLE 7.2.5

Use the Gram-Schmidt Procedure on the set $\left\{\begin{bmatrix}1\\1\\0\end{bmatrix}, \begin{bmatrix}-1\\2\\1\end{bmatrix}, \begin{bmatrix}0\\1\\1\end{bmatrix}\right\}$ to find an orthonormal basis for $\mathbb{R}^3$.

Solution: Call the vectors in the basis $\vec{w}_1$, $\vec{w}_2$, and $\vec{w}_3$, respectively.
First step: Let $\vec{v}_1 = \vec{w}_1$ and $\mathbb{S}_1 = \operatorname{Span}\{\vec{v}_1\}$.
Second step: Determine $\operatorname{perp}_{\mathbb{S}_1}(\vec{w}_2)$:

$$\operatorname{perp}_{\mathbb{S}_1}(\vec{w}_2) = \vec{w}_2 - \frac{\vec{v}_1 \cdot \vec{w}_2}{\|\vec{v}_1\|^2}\vec{v}_1 = \begin{bmatrix}-3/2\\3/2\\1\end{bmatrix}$$

(It is wise to check your arithmetic by verifying that $\vec{v}_1 \cdot \operatorname{perp}_{\mathbb{S}_1}(\vec{w}_2) = 0$.*)*
As mentioned above, we can take any non-zero scalar multiple of $\operatorname{perp}_{\mathbb{S}_1}(\vec{w}_2)$, so we take $\vec{v}_2 = \begin{bmatrix}-3\\3\\2\end{bmatrix}$ and define $\mathbb{S}_2 = \operatorname{Span}\{\vec{v}_1, \vec{v}_2\}$.

EXAMPLE 7.2.5
(continued)

Third step: Determine $\text{perp}_{\mathbb{S}_2}(\vec{w}_3)$:

$$\text{perp}_{\mathbb{S}_2}(\vec{w}_3) = \vec{w}_3 - \frac{\vec{v}_1 \cdot \vec{w}_3}{\|\vec{v}_1\|^2}\vec{v}_1 - \frac{\vec{v}_2 \cdot \vec{w}_3}{\|\vec{v}_2\|^2}\vec{v}_2 = \begin{bmatrix} 2/11 \\ -2/11 \\ 6/11 \end{bmatrix}$$

(Again, it is wise to check that $\text{perp}_{\mathbb{S}_2}(\vec{w}_3)$ *is orthogonal to both* $\vec{v}_1$ *and* $\vec{v}_2$*.)*

We now see that $\{\vec{v}_1, \vec{v}_2, \vec{v}_3\}$ is an orthogonal basis for $\mathbb{S}$. To obtain an orthonormal basis for $\mathbb{S}$, we divide each vector in this basis by its length. Thus, an orthonormal basis for $\mathbb{S}$ is

$$\left\{ \begin{bmatrix} 1/\sqrt{2} \\ 1/\sqrt{2} \\ 0 \end{bmatrix}, \begin{bmatrix} -3/\sqrt{22} \\ 3/\sqrt{22} \\ 2/\sqrt{22} \end{bmatrix}, \begin{bmatrix} 1/\sqrt{11} \\ -1/\sqrt{11} \\ 3/\sqrt{11} \end{bmatrix} \right\}$$

EXAMPLE 7.2.6

Use the Gram-Schmidt Procedure to find an orthogonal basis for the subspace

$$\mathbb{S} = \text{Span}\left\{ \begin{bmatrix} 1 \\ 1 \\ 0 \\ 1 \end{bmatrix}, \begin{bmatrix} 1 \\ 0 \\ 1 \\ 1 \end{bmatrix}, \begin{bmatrix} 2 \\ 1 \\ 1 \\ 2 \end{bmatrix}, \begin{bmatrix} 1 \\ 0 \\ 0 \\ 1 \end{bmatrix} \right\} = \text{Span}\{\vec{w}_1, \vec{w}_2, \vec{w}_3, \vec{w}_4\} \text{ of } \mathbb{R}^4.$$

Solution: First step: Let $\vec{v}_1 = \vec{w}_1$ and $\mathbb{S}_1 = \text{Span}\{\vec{v}_1\}$.

Second step: Determine $\text{perp}_{\mathbb{S}_1}(\vec{w}_2)$:

$$\text{perp}_{\mathbb{S}_1}(\vec{w}_2) = \vec{w}_2 - \frac{\vec{v}_1 \cdot \vec{w}_2}{\|\vec{v}_1\|^2}\vec{v}_1 = \begin{bmatrix} 1/3 \\ -2/3 \\ 1 \\ 1/3 \end{bmatrix}$$

We take $\vec{v}_2 = \begin{bmatrix} 1 \\ -2 \\ 3 \\ 1 \end{bmatrix}$ and $\mathbb{S}_2 = \text{Span}\{\vec{v}_1, \vec{v}_2\}$.

Third step: Determine $\text{perp}_{\mathbb{S}_2}(\vec{w}_3)$:

$$\text{perp}_{\mathbb{S}_2}(\vec{w}_3) = \vec{w}_3 - \frac{\vec{v}_1 \cdot \vec{w}_3}{\|\vec{v}_1\|^2}\vec{v}_1 - \frac{\vec{v}_2 \cdot \vec{w}_3}{\|\vec{v}_2\|^2}\vec{v}_2 = \begin{bmatrix} 0 \\ 0 \\ 0 \\ 0 \end{bmatrix}$$

Hence, $\vec{w}_3 \in \text{Span}\{\vec{w}_1, \vec{w}_2\}$. Therefore, we ignore $\vec{w}_3$ and instead calculate $\text{perp}_{\mathbb{S}_2}(\vec{w}_4)$.

$$\vec{v}_3 = \text{perp}_{\mathbb{S}_2}(\vec{w}_4) = \vec{w}_4 - \frac{\vec{v}_1 \cdot \vec{w}_4}{\|\vec{v}_1\|^2}\vec{v}_1 - \frac{\vec{v}_2 \cdot \vec{w}_4}{\|\vec{v}_2\|^2}\vec{v}_2 = \begin{bmatrix} 1/5 \\ -2/5 \\ -2/5 \\ 1/5 \end{bmatrix}$$

We get that $\{\vec{v}_1, \vec{v}_2, \vec{v}_3\}$ is an orthogonal basis for $\mathbb{S}$.

EXAMPLE 7.2.7

Let $\mathcal{B} = \left\{ \begin{bmatrix} 1 \\ 0 \\ -1 \\ 1 \end{bmatrix}, \begin{bmatrix} 1 \\ 1 \\ 0 \\ 1 \end{bmatrix}, \begin{bmatrix} -1 \\ 1 \\ 0 \\ -1 \end{bmatrix} \right\} = \{\vec{b}_1, \vec{b}_2, \vec{b}_3\}$ and let $\vec{x} = \begin{bmatrix} 4 \\ 3 \\ -2 \\ 5 \end{bmatrix}$. Find the projection of $\vec{x}$ onto the subspace $\mathbb{S} = \operatorname{Span} \mathcal{B}$ of $\mathbb{R}^4$.

Solution: Observe that $\mathcal{B}$ is not orthogonal. Therefore, our first step must be to perform the Gram-Schmidt Procedure on $\mathcal{B}$ to create an orthogonal basis for $\mathbb{S}$.

First step: Let $\vec{v}_1 = \begin{bmatrix} 1 \\ 0 \\ -1 \\ 1 \end{bmatrix}$ and $\mathbb{S}_1 = \operatorname{Span}\{\vec{v}_1\}$.

Second step: Determine $\operatorname{perp}_{\mathbb{S}_1}(\vec{b}_2)$:

$$\operatorname{perp}_{\mathbb{S}_1}(\vec{b}_2) = \vec{b}_2 - \frac{\vec{v}_1 \cdot \vec{b}_2}{\|\vec{v}_1\|^2}\vec{v}_1 = \begin{bmatrix} 1 \\ 1 \\ 0 \\ 1 \end{bmatrix} - \frac{2}{3}\begin{bmatrix} 1 \\ 0 \\ -1 \\ 1 \end{bmatrix} = \frac{1}{3}\begin{bmatrix} 1 \\ 3 \\ 2 \\ 1 \end{bmatrix}$$

To simplify calculations we use $\vec{v}_2 = \begin{bmatrix} 1 \\ 3 \\ 2 \\ 1 \end{bmatrix}$. Let $\mathbb{S}_2 = \operatorname{Span}\{\vec{v}_1, \vec{v}_2\}$.

Third step: Determine $\operatorname{perp}_{\mathbb{S}_2}(\vec{b}_3)$:

$$\operatorname{perp}_{\mathbb{S}_2}(\vec{b}_3) = \vec{b}_3 - \frac{\vec{v}_1 \cdot \vec{b}_3}{\|\vec{v}_1\|^2}\vec{v}_1 - \frac{\vec{v}_2 \cdot \vec{b}_3}{\|\vec{v}_2\|^2}\vec{v}_2 = \begin{bmatrix} -1 \\ 1 \\ 0 \\ -1 \end{bmatrix} + \frac{2}{3}\begin{bmatrix} 1 \\ 0 \\ -1 \\ 1 \end{bmatrix} - \frac{1}{15}\begin{bmatrix} 1 \\ 3 \\ 2 \\ 1 \end{bmatrix} = \frac{2}{5}\begin{bmatrix} -1 \\ 2 \\ -2 \\ -1 \end{bmatrix}$$

We take $\vec{v}_3 = \begin{bmatrix} -1 \\ 2 \\ -2 \\ -1 \end{bmatrix}$. Thus, the set $\{\vec{v}_1, \vec{v}_2, \vec{v}_3\}$ is an orthogonal basis for $\mathbb{S}$.

We can now determine the projection.

$$\operatorname{proj}_{\mathbb{S}}(\vec{x}) = \frac{\vec{v}_1 \cdot \vec{x}}{\|\vec{v}_1\|^2}\vec{v}_1 + \frac{\vec{v}_2 \cdot \vec{x}}{\|\vec{v}_2\|^2}\vec{v}_2 + \frac{\vec{v}_3 \cdot \vec{x}}{\|\vec{v}_3\|^2}\vec{v}_3 = \frac{11}{3}\vec{v}_1 + \frac{14}{15}\vec{v}_2 + \frac{1}{10}\vec{v}_3 = \begin{bmatrix} 9/2 \\ 3 \\ -2 \\ 9/2 \end{bmatrix}$$

PROBLEMS 7.2

Practice Problems

For Problems A1–A8, let $\mathbb{S}$ be the subspace spanned by the given orthogonal set. Determine $\text{proj}_{\mathbb{S}}(\vec{x})$ and $\text{perp}_{\mathbb{S}}(\vec{x})$.

A1 $\left\{\begin{bmatrix}2\\3\end{bmatrix},\begin{bmatrix}-3\\2\end{bmatrix}\right\}, \vec{x}=\begin{bmatrix}-1\\2\end{bmatrix}$

A2 $\left\{\begin{bmatrix}1\\0\\1\end{bmatrix},\begin{bmatrix}0\\1\\0\end{bmatrix}\right\}, \vec{x}=\begin{bmatrix}6\\9\\5\end{bmatrix}$

A3 $\left\{\begin{bmatrix}1\\0\\3\end{bmatrix},\begin{bmatrix}-3\\0\\1\end{bmatrix}\right\}, \vec{x}=\begin{bmatrix}2\\\sqrt{5}\\3\end{bmatrix}$

A4 $\left\{\begin{bmatrix}1\\-2\\1\end{bmatrix},\begin{bmatrix}1\\1\\1\end{bmatrix}\right\}, \vec{x}=\begin{bmatrix}7\\4\\1\end{bmatrix}$

A5 $\left\{\begin{bmatrix}1\\1\\2\end{bmatrix},\begin{bmatrix}1\\-1\\0\end{bmatrix},\begin{bmatrix}1\\1\\-1\end{bmatrix}\right\}, \vec{x}=\begin{bmatrix}1\\2\\3\end{bmatrix}$

A6 $\left\{\begin{bmatrix}1\\3\\2\end{bmatrix},\begin{bmatrix}1\\1\\-2\end{bmatrix}\right\}, \vec{x}=\begin{bmatrix}1\\0\\5\end{bmatrix}$

A7 $\left\{\begin{bmatrix}1\\0\\1\\0\end{bmatrix},\begin{bmatrix}0\\1\\0\\1\end{bmatrix},\begin{bmatrix}1\\0\\-1\\0\end{bmatrix}\right\}, \vec{x}=\begin{bmatrix}2\\3\\5\\6\end{bmatrix}$

A8 $\left\{\begin{bmatrix}1\\-1\\-1\\1\end{bmatrix},\begin{bmatrix}1\\2\\1\\2\end{bmatrix}\right\}, \vec{x}=\begin{bmatrix}2\\3\\5\\6\end{bmatrix}$

For Problems A9–13, find a basis for the orthogonal complement of the subspace spanned by the given set.

A9 $\left\{\begin{bmatrix}1\\2\\-1\end{bmatrix}\right\}$

A10 $\left\{\begin{bmatrix}1\\1\\1\end{bmatrix},\begin{bmatrix}-1\\1\\3\end{bmatrix}\right\}$

A11 $\left\{\begin{bmatrix}1\\0\\1\end{bmatrix},\begin{bmatrix}2\\-1\\1\end{bmatrix}\right\}$

A12 $\left\{\begin{bmatrix}2\\1\\-1\\0\end{bmatrix},\begin{bmatrix}1\\2\\1\\1\end{bmatrix},\begin{bmatrix}0\\-1\\3\\1\end{bmatrix}\right\}$

A13 $\left\{\begin{bmatrix}1\\1\\1\\2\end{bmatrix},\begin{bmatrix}2\\1\\2\\1\end{bmatrix},\begin{bmatrix}-1\\1\\-1\\4\end{bmatrix}\right\}$

For Problems A14–A20, use the Gram-Schmidt Procedure to produce an orthogonal basis for the subspace spanned by the given set.

A14 $\left\{\begin{bmatrix}1\\0\\0\end{bmatrix},\begin{bmatrix}1\\1\\0\end{bmatrix}\right\}$

A15 $\left\{\begin{bmatrix}1\\1\\0\end{bmatrix},\begin{bmatrix}1\\0\\0\end{bmatrix}\right\}$

A16 $\left\{\begin{bmatrix}1\\3\\2\end{bmatrix},\begin{bmatrix}1\\2\\-1\end{bmatrix}\right\}$

A17 $\left\{\begin{bmatrix}2\\1\\2\end{bmatrix},\begin{bmatrix}3\\1\\1\end{bmatrix},\begin{bmatrix}1\\1\\1\end{bmatrix}\right\}$

A18 $\left\{\begin{bmatrix}4\\1\\2\end{bmatrix},\begin{bmatrix}2\\3\\-4\end{bmatrix},\begin{bmatrix}1\\1\\-1\end{bmatrix}\right\}$

A19 $\left\{\begin{bmatrix}1\\0\\0\\1\end{bmatrix},\begin{bmatrix}2\\-1\\1\\2\end{bmatrix},\begin{bmatrix}-1\\1\\1\\1\end{bmatrix}\right\}$

A20 $\left\{\begin{bmatrix}1\\1\\0\\1\end{bmatrix},\begin{bmatrix}1\\0\\1\\1\end{bmatrix},\begin{bmatrix}1\\3\\-2\\1\end{bmatrix},\begin{bmatrix}1\\0\\0\\1\end{bmatrix}\right\}$

For Problems A21–A24, use the Gram-Schmidt Procedure to produce an orthonormal basis for the subspace spanned by the given set.

A21 $\left\{\begin{bmatrix}1\\1\\0\end{bmatrix},\begin{bmatrix}1\\1\\1\end{bmatrix}\right\}$

A22 $\left\{\begin{bmatrix}2\\1\\-2\end{bmatrix},\begin{bmatrix}2\\1\\1\end{bmatrix},\begin{bmatrix}-1\\-1\\1\end{bmatrix}\right\}$

A23 $\left\{\begin{bmatrix}1\\1\\0\\1\end{bmatrix},\begin{bmatrix}0\\1\\1\\1\end{bmatrix},\begin{bmatrix}1\\-1\\-1\\-1\end{bmatrix}\right\}$

A24 $\left\{\begin{bmatrix}1\\0\\1\\0\\1\end{bmatrix},\begin{bmatrix}1\\0\\-1\\1\\0\end{bmatrix},\begin{bmatrix}1\\1\\1\\1\\1\end{bmatrix}\right\}$

For Problems A25–A30, let $\mathbb{S}$ be the subspace spanned by the given set. Find the vector $\vec{y}$ in $\mathbb{S}$ that is closest to $\vec{x}$.

A25 $\left\{\begin{bmatrix}2\\3\end{bmatrix}\right\}, \vec{x}=\begin{bmatrix}-1\\2\end{bmatrix}$

A26 $\left\{\begin{bmatrix}3\\-4\\-1\end{bmatrix},\begin{bmatrix}1\\1\\-1\end{bmatrix}\right\}, \vec{x}=\begin{bmatrix}5\\0\\2\end{bmatrix}$

A27 $\left\{\begin{bmatrix}1\\1\end{bmatrix},\begin{bmatrix}2\\1\end{bmatrix}\right\}, \vec{x}=\begin{bmatrix}-1\\2\end{bmatrix}$

A28 $\left\{\begin{bmatrix}1\\1\\2\end{bmatrix},\begin{bmatrix}1\\-1\\-1\end{bmatrix},\begin{bmatrix}6\\0\\3\end{bmatrix}\right\}, \vec{x}=\begin{bmatrix}5\\4\\5\end{bmatrix}$

A29 $\left\{\begin{bmatrix}1\\1\\1\\0\end{bmatrix},\begin{bmatrix}1\\1\\2\\1\end{bmatrix}\right\}, \vec{x}=\begin{bmatrix}-1\\3\\-2\\3\end{bmatrix}$

A30 $\left\{\begin{bmatrix}1\\0\\-1\\1\end{bmatrix},\begin{bmatrix}1\\1\\-1\\-2\end{bmatrix}\right\}, \vec{x}=\begin{bmatrix}4\\3\\2\\1\end{bmatrix}$

Homework Problems

For Problems B1–B8, let $\mathbb{S}$ be the subspace spanned by the given orthogonal set. Determine $\text{proj}_{\mathbb{S}}(\vec{x})$ and $\text{perp}_{\mathbb{S}}(\vec{x})$.

B1 $\left\{\begin{bmatrix}1\\1\\2\end{bmatrix},\begin{bmatrix}1\\1\\-1\end{bmatrix}\right\}, \vec{x}=\begin{bmatrix}3\\1\\2\end{bmatrix}$

B2 $\left\{\begin{bmatrix}3\\1\\1\end{bmatrix},\begin{bmatrix}1\\-5\\2\end{bmatrix}\right\}, \vec{x}=\begin{bmatrix}1\\5\\3\end{bmatrix}$

B3 $\left\{\begin{bmatrix}2\\1\\0\end{bmatrix},\begin{bmatrix}0\\0\\1\end{bmatrix}\right\}, \vec{x}=\begin{bmatrix}2\\-3\\6\end{bmatrix}$

B4 $\left\{\begin{bmatrix}2\\2\\1\end{bmatrix},\begin{bmatrix}1\\1\\-4\end{bmatrix}\right\}, \vec{x}=\begin{bmatrix}3\\0\\3\end{bmatrix}$

B5 $\left\{\begin{bmatrix}3\\1\\2\end{bmatrix},\begin{bmatrix}1\\-1\\-1\end{bmatrix}\right\}, \vec{x}=\begin{bmatrix}3\\5\\7\end{bmatrix}$

B6 $\left\{\begin{bmatrix}1\\0\\1\\1\end{bmatrix},\begin{bmatrix}1\\2\\-2\\1\end{bmatrix},\begin{bmatrix}1\\-3\\-2\\1\end{bmatrix}\right\}, \vec{x}=\begin{bmatrix}1\\1\\2\\0\end{bmatrix}$

B7 $\left\{\begin{bmatrix}1\\-1\\-1\\1\end{bmatrix},\begin{bmatrix}1\\1\\1\\1\end{bmatrix}\right\}, \vec{x}=\begin{bmatrix}2\\1\\2\\3\end{bmatrix}$

B8 $\left\{\begin{bmatrix}1\\1\\0\\1\end{bmatrix},\begin{bmatrix}0\\0\\1\\0\end{bmatrix},\begin{bmatrix}-1\\-1\\0\\2\end{bmatrix}\right\}, \vec{x}=\begin{bmatrix}3\\1\\4\\7\end{bmatrix}$

For Problems B9–B16, find a basis for the orthogonal complement of the subspace spanned by the given set.

B9 $\left\{\begin{bmatrix}1\\3\\3\end{bmatrix}\right\}$

B10 $\left\{\begin{bmatrix}1\\2\\4\end{bmatrix},\begin{bmatrix}2\\1\\3\end{bmatrix}\right\}$

B11 $\left\{\begin{bmatrix}3\\1\\-1\end{bmatrix},\begin{bmatrix}4\\4\\1\end{bmatrix}\right\}$

B12 $\left\{\begin{bmatrix}1\\-1\\1\\2\end{bmatrix},\begin{bmatrix}2\\1\\0\\-1\end{bmatrix}\right\}$

B13 $\left\{\begin{bmatrix}2\\3\\1\\-3\end{bmatrix},\begin{bmatrix}1\\2\\3\\2\end{bmatrix}\right\}$

B14 $\left\{\begin{bmatrix}1\\2\\1\\2\end{bmatrix},\begin{bmatrix}2\\3\\1\\0\end{bmatrix},\begin{bmatrix}3\\3\\1\\2\end{bmatrix}\right\}$

B15 $\left\{\begin{bmatrix}3\\6\\-2\\3\end{bmatrix},\begin{bmatrix}6\\8\\-3\\1\end{bmatrix},\begin{bmatrix}-3\\2\\0\\7\end{bmatrix}\right\}$

B16 $\left\{\begin{bmatrix}1\\2\\3\\1\end{bmatrix},\begin{bmatrix}2\\4\\6\\2\end{bmatrix},\begin{bmatrix}-2\\-4\\1\\1\end{bmatrix},\begin{bmatrix}-3\\-6\\-2\\0\end{bmatrix}\right\}$

For Problems B17–B23, use the Gram-Schmidt Procedure to produce an orthogonal basis for the subspace spanned by the given set.

B17 $\left\{\begin{bmatrix}2\\1\\-1\end{bmatrix},\begin{bmatrix}1\\2\\7\end{bmatrix}\right\}$

B18 $\left\{\begin{bmatrix}1\\1\\1\end{bmatrix},\begin{bmatrix}3\\1\\2\end{bmatrix}\right\}$

B19 $\left\{\begin{bmatrix}2\\1\\2\end{bmatrix},\begin{bmatrix}5\\-2\\-4\end{bmatrix}\right\}$

B20 $\left\{\begin{bmatrix}1\\0\\-1\end{bmatrix},\begin{bmatrix}2\\2\\3\end{bmatrix},\begin{bmatrix}0\\-2\\-5\end{bmatrix}\right\}$

B21 $\left\{\begin{bmatrix}1\\2\\1\\-1\end{bmatrix},\begin{bmatrix}2\\-2\\1\\6\end{bmatrix},\begin{bmatrix}1\\-5\\6\\-3\end{bmatrix}\right\}$

B22 $\left\{\begin{bmatrix}-1\\2\\-1\\1\end{bmatrix},\begin{bmatrix}3\\-6\\3\\-3\end{bmatrix},\begin{bmatrix}1\\0\\-1\\1\end{bmatrix},\begin{bmatrix}1\\0\\0\\0\end{bmatrix}\right\}$

B23 $\left\{\begin{bmatrix}1\\2\\0\\2\end{bmatrix},\begin{bmatrix}1\\3\\-3\\1\end{bmatrix},\begin{bmatrix}2\\3\\5\\-1\end{bmatrix}\right\}$

For Problems B24–B30, use the Gram-Schmidt Procedure to produce an orthonormal basis for the subspace spanned by the given set.

B24 $\left\{\begin{bmatrix}2\\2\\1\end{bmatrix},\begin{bmatrix}1\\3\\1\end{bmatrix}\right\}$

B25 $\left\{\begin{bmatrix}2\\2\\4\end{bmatrix},\begin{bmatrix}1\\3\\3\end{bmatrix}\right\}$

B26 $\left\{\begin{bmatrix}1\\1\\1\end{bmatrix},\begin{bmatrix}2\\-2\\3\end{bmatrix}\right\}$

B27 $\left\{\begin{bmatrix}1\\-1\\1\\2\end{bmatrix},\begin{bmatrix}-1\\1\\-1\\1\end{bmatrix}\right\}$

B28 $\left\{\begin{bmatrix}1\\1\\1\\1\end{bmatrix},\begin{bmatrix}-1\\1\\-1\\1\end{bmatrix},\begin{bmatrix}1\\0\\1\\0\end{bmatrix},\begin{bmatrix}2\\1\\2\\2\end{bmatrix}\right\}$

B29 $\left\{\begin{bmatrix}0\\1\\2\\2\end{bmatrix},\begin{bmatrix}1\\1\\-1\\-1\end{bmatrix},\begin{bmatrix}4\\5\\1\\1\end{bmatrix}\right\}$

B30 $\left\{\begin{bmatrix}1\\1\\0\\1\end{bmatrix},\begin{bmatrix}1\\2\\1\\1\end{bmatrix},\begin{bmatrix}1\\3\\1\\2\end{bmatrix},\begin{bmatrix}1\\2\\3\\4\end{bmatrix}\right\}$

For Problems B31–B36, let $\mathbb{S}$ be the subspace spanned by the given set. Find the vector $\vec{y}$ in $\mathbb{S}$ that is closest to $\vec{x}$.

B31 $\left\{\begin{bmatrix}1\\1\\-2\end{bmatrix},\begin{bmatrix}0\\2\\1\end{bmatrix}\right\}, \vec{x}=\begin{bmatrix}3\\1\\3\end{bmatrix}$

B32 $\left\{\begin{bmatrix}1\\2\end{bmatrix},\begin{bmatrix}2\\4\end{bmatrix}\right\}, \vec{x}=\begin{bmatrix}4\\3\end{bmatrix}$

B33 $\left\{\begin{bmatrix}1\\0\\1\end{bmatrix},\begin{bmatrix}1\\1\\2\end{bmatrix}\right\}, \vec{x}=\begin{bmatrix}-4\\2\\1\end{bmatrix}$

B34 $\left\{\begin{bmatrix}-1\\1\\-1\end{bmatrix},\begin{bmatrix}3\\1\\1\end{bmatrix}\right\}, \vec{x}=\begin{bmatrix}0\\-2\\-5\end{bmatrix}$

B35 $\left\{\begin{bmatrix}1\\2\\2\end{bmatrix},\begin{bmatrix}2\\3\\2\end{bmatrix},\begin{bmatrix}4\\5\\2\end{bmatrix}\right\}, \vec{x}=\begin{bmatrix}1\\3\\1\end{bmatrix}$

B36 $\left\{\begin{bmatrix}1\\2\\1\\0\end{bmatrix},\begin{bmatrix}2\\-3\\-2\\1\end{bmatrix}\right\}, \vec{x}=\begin{bmatrix}4\\4\\0\\4\end{bmatrix}$

Conceptual Problems

C1 Prove that if $\mathbb{S}$ is a k-dimensional subspace of $\mathbb{R}^n$, then $\mathbb{S}^\perp$ is an $(n-k)$-dimensional subspace.

C2 Prove that if $\mathbb{S}$ is a k-dimensional subspace of $\mathbb{R}^n$, then $\mathbb{S}\cap\mathbb{S}^\perp=\{\vec{0}\}$.

C3 Prove that if $\{\vec{v}_1,\dots,\vec{v}_k\}$ is an orthonormal basis for $\mathbb{S}$ and $\{\vec{v}_{k+1},\dots,\vec{v}_n\}$ is an orthonormal basis for $\mathbb{S}^\perp$, then $\{\vec{v}_1,\dots,\vec{v}_n\}$ is an orthonormal basis for $\mathbb{R}^n$.

C4 Prove that if $\mathbb{S}$ is a k-dimensional subspace of $\mathbb{R}^n$, then $(\mathbb{S}^\perp)^\perp=\mathbb{S}$.

C5 Suppose that $\mathbb{S}$ is a k-dimensional subspace of $\mathbb{R}^n$. Prove that for any $\vec{x}\in\mathbb{R}^n$ we have

$$\text{proj}_{\mathbb{S}^\perp}(\vec{x})=\text{perp}_{\mathbb{S}}(\vec{x})$$

C6 Assume $\mathcal{B}=\{\vec{a}_1,\vec{a}_2\}$ is a linearly independent set in $\mathbb{R}^2$. Let $\vec{q}_1,\vec{q}_2$ denote the vectors that result from applying the Gram-Schmidt Procedure to $\{\vec{a}_1,\vec{a}_2\}$ (in order) and then normalizing. Prove that the matrix $R=\begin{bmatrix}\vec{a}_1\cdot\vec{q}_1 & \vec{a}_2\cdot\vec{q}_1\\ 0 & \vec{a}_2\cdot\vec{q}_2\end{bmatrix}$ is invertible.

C7 Prove Theorem 7.2.4.

C8 If $\{\vec{v}_1,\dots,\vec{v}_k\}$ is an orthonormal basis for a subspace $\mathbb{S}$, verify that the standard matrix of $\text{proj}_{\mathbb{S}}$ can be written in the form

$$[\text{proj}_{\mathbb{S}}]=\vec{v}_1\vec{v}_1^T+\vec{v}_2\vec{v}_2^T+\cdots+\vec{v}_k\vec{v}_k^T$$

7.3 Method of Least Squares

In the sciences one often tries to find a correlation between quantities, say y and t, by collecting data from repeated experimentation. For example, y might be the position of a particle or the temperature of some body fluid at time t. Optimally, the experimenter would like to find a function f such that $y = f(t)$ for all t. However, it is unlikely that there is a simple function which perfectly matches the data. So, the experimenter tries to find a simple function which best fits the data. We now look at one common method for doing this.

Suppose that an experimenter would like to find an equation which best predicts the value of a quantity y at any time t. Further suppose that the scientific theory indicates that this can be done by an equation of the form $y = a + bt + ct^2$. The experimenter will perform an experiment in which they measure the value of y at times $t_1, t_2, \dots, t_m$ and obtains the values $y_1, y_2, \dots, y_m$.

For each data point (t_i, y_i) we get a linear equation

$$y_i = a + bt_i + ct_i^2$$

Note that t_i is a fixed number, so this equation is actually linear in variables a, b, and c. Thus, we have a system of m linear equations in 3 variables

$$\begin{aligned} y_1 &= a + bt_1 + ct_1^2 \\ \vdots &= \quad \vdots \\ y_m &= a + bt_m + ct_m^2 \end{aligned}$$

Due to experimentation error, the system is very likely to be inconsistent. So, the experimenter needs to find the values of a, b, and c which best approximates the data they collected.

To solve this problem, we observe that we can write the system in the form

$$\begin{bmatrix} y_1 \\ \vdots \\ y_m \end{bmatrix} = \begin{bmatrix} a + bt_1 + ct_1^2 \\ \vdots \\ a + bt_m + ct_m^2 \end{bmatrix} = \begin{bmatrix} 1 & t_1 & t_1^2 \\ \vdots & \vdots & \vdots \\ 1 & t_m & t_m^2 \end{bmatrix} \begin{bmatrix} a \\ b \\ c \end{bmatrix}$$

To write this more compactly we define $\vec{y} = \begin{bmatrix} y_1 \\ \vdots \\ y_m \end{bmatrix}$, $X = \begin{bmatrix} 1 & t_1 & t_1^2 \\ \vdots & \vdots & \vdots \\ 1 & t_m & t_m^2 \end{bmatrix}$, and $\vec{a} = \begin{bmatrix} a \\ b \\ c \end{bmatrix}$ to get

$$\vec{y} = X\vec{a}$$

Since we are assuming this system is inconsistent, we cannot find $\vec{a}$ such that $X\vec{a} = \vec{y}$. So, instead, we try to find $\vec{a}$ such that $X\vec{a}$ is as close as possible to $\vec{y}$. In particular, we want to minimize

$$\|\vec{y} - X\vec{a}\|$$

The Approximation Theorem tells us how to do this. In particular, since $X\vec{a}$ is a vector in the columnspace of X, the Approximation Theorem says that the vector in Col(X) that is closet to $\vec{y}$ is $\text{proj}_{\text{Col}(X)}(\vec{y})$. Hence, to determine the best $\vec{a}$ we just need to solve the system

$$X\vec{a} = \text{proj}_{\text{Col}(X)}(\vec{y}) \tag{7.1}$$

Notice that this is guaranteed to be a consistent system since $\text{proj}_{\text{Col}(X)}(\vec{y}) \in \text{Col}(X)$, so we now can find our desired vector $\vec{a}$.

Although this works, it is not particularly nice. To find the projection, we first need to find a basis for the columnspace of X, and then apply the Gram-Schmidt Procedure to make it orthogonal. After all of that work, we still need to find the projection, and then solve the system. Consequently, we try to further manipulate equation (7.1) to create an easier method.

Since we are really trying to minimize $\|\vec{y} - X\vec{a}\|$, we subtract both sides of equation (7.1) from $\vec{y}$ to get

$$\vec{y} - X\vec{a} = \vec{y} - \text{proj}_{\text{Col}(X)}(\vec{y})$$

But,

$$\vec{y} - \text{proj}_{\text{Col}(X)}(\vec{y}) = \text{perp}_{\text{Col}(X)}(\vec{y}) \in (\text{Col}(X))^{\perp}$$

By the Fundamental Theorem of Linear Algebra, the orthogonal complement of the columnspace is the left nullspace. Hence, we have that $\vec{y} - X\vec{a} \in \text{Null}(X^T)$. Therefore, by definition of the left nullspace, we have that

$$X^T(\vec{y} - X\vec{a}) = \vec{0}$$

Rearranging this gives

$$X^T X\vec{a} = X^T\vec{y}$$

This is called the **normal system**. This system will be consistent by construction. However, it need not have a unique solution. If it does have infinitely many solutions, then each of the solutions will minimize $\|\vec{y} - X\vec{a}\|$.

Wait! This is amazing! We have started with the inconsistent system $X\vec{a} = \vec{y}$, and by simply multiplying both sides by X^T, we not only get a consistent system, but the solution of the new system best approximates a solution to the original inconsistent system!

This method of finding an approximate solution is called the **method of least squares**. It is called this because we are minimizing

$$\|\vec{y} - X\vec{a}\|^2 = \|\vec{y} - (a\vec{1} + b\vec{t} + c\vec{t}^{\,2})\|^2 = \sum_{i=1}^{m} (y_i - (a + bt_i + ct_i^2))^2$$

where

$$\text{err}_i = y_i - (a + bt_i + ct_i^2)$$

measures the error between each data point y_i and the curve as shown in Figure 7.3.1.

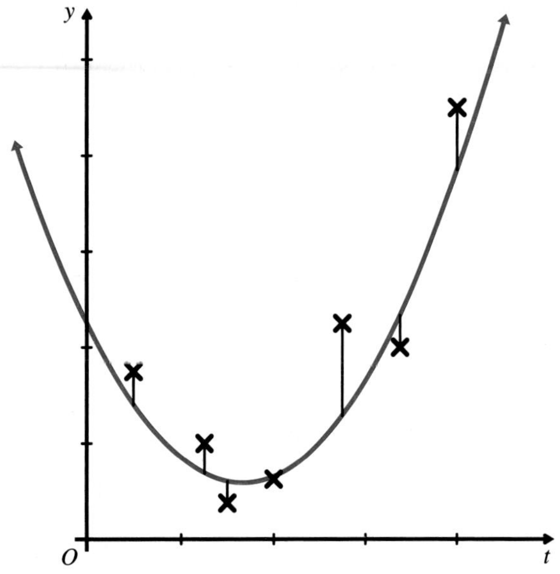

Figure 7.3.1 Some data points and a curve $y = a + bt + ct^2$. Vertical line segments measure the error err_i in the fit at each t_i.

We now demonstrate the method of least squares with a couple of examples.

EXAMPLE 7.3.1

Find a, b, c to obtain the best fitting equation of the form $y = a + bt + ct^2$ for the data points $(-1, 2), (0, 1), (1, 3)$.

Solution: We let $\vec{y} = \begin{bmatrix} 2 \\ 1 \\ 3 \end{bmatrix}$, $\vec{a} = \begin{bmatrix} a \\ b \\ c \end{bmatrix}$ and $X = \begin{bmatrix} 1 & -1 & (-1)^2 \\ 1 & 0 & 0^2 \\ 1 & 1 & 1^2 \end{bmatrix} = \begin{bmatrix} 1 & -1 & 1 \\ 1 & 0 & 0 \\ 1 & 1 & 1 \end{bmatrix}$.

The normal system gives

$$X^T X\vec{a} = X^T\vec{y}$$

$$\begin{bmatrix} 3 & 0 & 2 \\ 0 & 2 & 0 \\ 2 & 0 & 2 \end{bmatrix}\vec{a} = \begin{bmatrix} 6 \\ 1 \\ 5 \end{bmatrix}$$

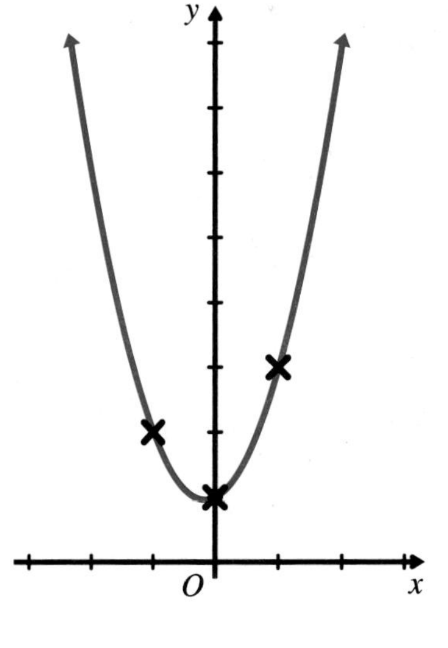

Row reducing the corresponding augmented matrix gives

$$\left[\begin{array}{ccc|c} 3 & 0 & 2 & 6 \\ 0 & 2 & 0 & 1 \\ 2 & 0 & 2 & 5 \end{array}\right] \sim \left[\begin{array}{ccc|c} 1 & 0 & 0 & 1 \\ 0 & 1 & 0 & 1/2 \\ 0 & 0 & 1 & 3/2 \end{array}\right]$$

So, the best fitting parabola is

$$p(x) = 1 + \frac{1}{2}t + \frac{3}{2}t^2$$

EXAMPLE 7.3.2

Suppose that the experimenter's data points are $(1.0, 6.1)$, $(2.1, 12.6)$, $(3.1, 21.1)$, $(4.0, 30.2)$, $(4.9, 40.9)$, $(6.0, 55.5)$. Find the values of a, b, c so that the equation $y = a + bt + ct^2$ best fits the data.

Solution: We let $\vec{y} = \begin{bmatrix} 6.1 \\ 12.6 \\ 21.1 \\ 30.2 \\ 40.9 \\ 55.5 \end{bmatrix}$, $\vec{a} = \begin{bmatrix} a \\ b \\ c \end{bmatrix}$, and $X = \begin{bmatrix} 1 & 1.0 & (1.0)^2 \\ 1 & 2.1 & (2.1)^2 \\ 1 & 3.1 & (3.1)^2 \\ 1 & 4.0 & (4.0)^2 \\ 1 & 4.9 & (4.9)^2 \\ 1 & 6.0 & (6.0)^2 \end{bmatrix} = \begin{bmatrix} 1 & 1.0 & 1.00 \\ 1 & 2.1 & 4.41 \\ 1 & 3.1 & 9.61 \\ 1 & 4.0 & 16.00 \\ 1 & 4.9 & 24.01 \\ 1 & 6.0 & 36.00 \end{bmatrix}$.

Using a computer, we can find that the solution for the system $(X^T X)\vec{a} = X^T\vec{y}$ is $\vec{a} = \begin{bmatrix} 1.63175 \\ 3.38382 \\ 0.93608 \end{bmatrix}$. The data does not justify retaining so many decimal places, so we take the best-fitting quadratic curve to be $y = 1.6 + 3.4t + 0.9t^2$.

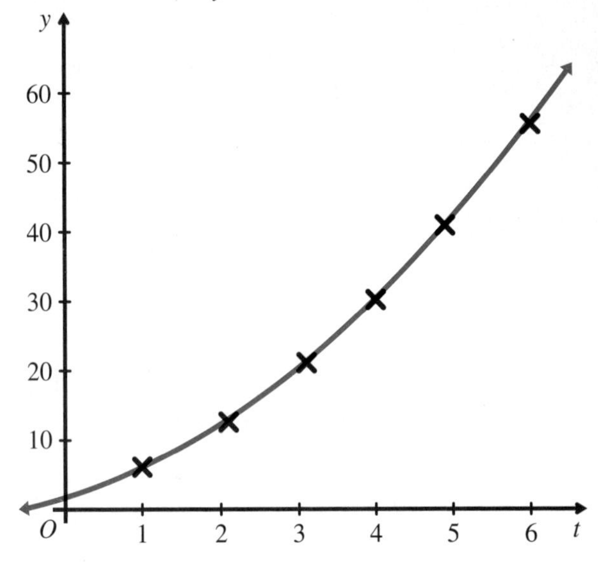

In both of these examples we find that $X^T X$ is invertible. It can be proven (see Problem C5) that $X^T X$ is invertible whenever the number of distinct t_i values is greater than or equal to the number of unknown coefficients. In this case, the method of least squares always has the unique solution

$$\vec{a} = (X^T X)^{-1} X^T \vec{y}$$

If $X^T X$ is not invertible, then the normal system will have infinitely many solutions. Each solution will be an equally good approximation. In this case, we typically take the solution with the smallest length.

So far, we have just been finding a parabola of best fit. For a more general situation, we use a similar construction. The matrix X, called the **design matrix**, depends on the desired model curve and the way the data is collected. This is demonstrated in the next example.

EXAMPLE 7.3.3

Find a and b to obtain the best-fitting equation of the form $y = at^2 + bt$ for the data points $(-1, 4), (0, 1), (1, 1)$.

Solution: We substitute each data point into the equation $y = at^2 + bt$ to get the system of linear equations

$$\begin{aligned} 4 &= a(-1)^2 + b(-1) \\ 1 &= a(0)^2 + b(0) \\ 1 &= a(1)^2 + b(1) \end{aligned}$$

We then rewrite this system into matrix form.

$$\begin{bmatrix} 4 \\ 1 \\ 1 \end{bmatrix} = \begin{bmatrix} a(-1)^2 + b(-1) \\ a(0)^2 + b(0) \\ a(1)^2 + b(1) \end{bmatrix} = \begin{bmatrix} (-1)^2 & -1 \\ 0^2 & 0 \\ 1^2 & 1 \end{bmatrix} \begin{bmatrix} a \\ b \end{bmatrix}$$

Thus, we let $\vec{y} = \begin{bmatrix} 4 \\ 1 \\ 1 \end{bmatrix}$, $\vec{a} = \begin{bmatrix} a \\ b \end{bmatrix}$, and $X = \begin{bmatrix} 1 & -1 \\ 0 & 0 \\ 1 & 1 \end{bmatrix}$.

We find that $X^T X = \begin{bmatrix} 2 & 0 \\ 0 & 2 \end{bmatrix}$ is easy to invert, and hence the method of least squares gives

$$\begin{aligned} \begin{bmatrix} a \\ b \end{bmatrix} &= (X^T X)^{-1} X^T \vec{y} \\ &= \begin{bmatrix} 2 & 0 \\ 0 & 2 \end{bmatrix}^{-1} \begin{bmatrix} 5 \\ -3 \end{bmatrix} \\ &= \begin{bmatrix} 5/2 \\ -3/2 \end{bmatrix} \end{aligned}$$

So, the equation of best fit for the given data is

$$y = \frac{5}{2}t^2 - \frac{3}{2}t$$

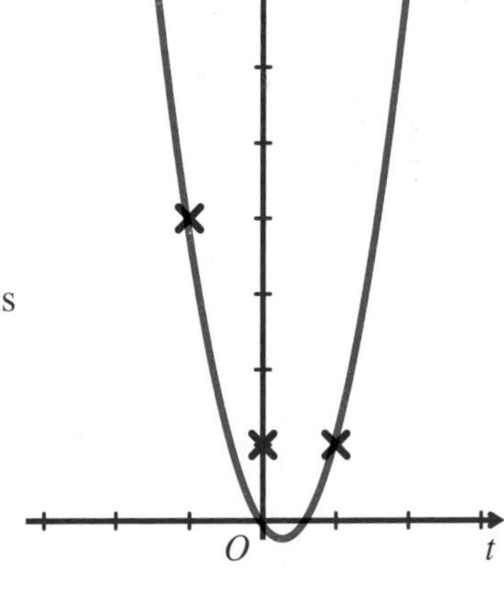

Overdetermined Systems

To get the best possible mathematical model, an experimenter will always perform their experiment many more times than the number of unknowns. This will result in a system of linear equations which has more equations than unknowns. Such a system of equations is called an **overdetermined system**.

The problem of finding the best-fitting curve can be viewed as a special case of the problem of "solving" an overdetermined system. Suppose that $A\vec{x} = \vec{b}$ is a system of m equations in n variables, where m is greater than n. With more equations than variables, we expect the system to be inconsistent unless $\vec{b}$ has some special properties.

Note that the problem in Example 7.3.2 of finding the best-fitting quadratic curve was of this form: we needed to solve $X\vec{a} = \vec{y}$ for the three variables a, b, and c, where there were 6 equations.

If there is no $\vec{x}$ such that $A\vec{x} = \vec{b}$, the next best "solution" is to find a vector $\vec{x}$ that minimizes the "error" $\|A\vec{x} - \vec{b}\|$. Using an argument analogous to that in the special case above, it can be shown that this vector $\vec{x}$ must also satisfy the normal system

$$A^T A\vec{x} = A^T\vec{b}$$

EXAMPLE 7.3.4

Determine the vector $\vec{x}$ that minimizes $\|A\vec{x} - \vec{b}\|$ for the system

$$\begin{aligned} 3x_1 - x_2 &= 4 \\ x_1 + 2x_2 &= 0 \\ 2x_1 + x_2 &= 1 \end{aligned}$$

Solution: We have $\vec{b} = \begin{bmatrix} 4 \\ 0 \\ 1 \end{bmatrix}$, $\vec{x} = \begin{bmatrix} x_1 \\ x_2 \end{bmatrix}$, and $A = \begin{bmatrix} 3 & -1 \\ 1 & 2 \\ 2 & 1 \end{bmatrix}$.

Solving the normal system gives

$$\begin{aligned} A^T A\vec{x} &= A^T\vec{b} \\ \begin{bmatrix} 14 & 1 \\ 1 & 6 \end{bmatrix}\vec{x} &= \begin{bmatrix} 3 & 1 & 2 \\ -1 & 2 & 1 \end{bmatrix}\begin{bmatrix} 4 \\ 0 \\ 1 \end{bmatrix} \\ \vec{x} &= \begin{bmatrix} 14 & 1 \\ 1 & 6 \end{bmatrix}^{-1}\begin{bmatrix} 14 \\ -3 \end{bmatrix} \\ \vec{x} &= \begin{bmatrix} 87/83 \\ -56/83 \end{bmatrix} \end{aligned}$$

So, $\vec{x} = \begin{bmatrix} 87/83 \\ -56/83 \end{bmatrix}$ is the vector that minimizes $\|A\vec{x} - \vec{b}\|$.

Geometry of the Method of Least Squares

Using our work in Section 3.4 and the Fundamental Theorem of Linear Algebra, we can get a geometric view of the Method of Least Squares.

Let $A \in M_{m\times n}(\mathbb{R})$ and $\vec{b} \in \mathbb{R}^m$ such that $A\vec{x} = \vec{b}$ is inconsistent. This implies that there exists $\vec{c}_0 \in \text{Col}(A)$ and $\vec{l}_0 \in \text{Null}(A^T)$ with $\vec{l}_0 \neq \vec{0}$ such that

$$\vec{b} = \vec{c}_0 + \vec{l}_0$$

Figure 7.3.2 makes it clear that the vector in Col(A) that is closest to $\vec{b}$ is

$$\text{proj}_{\text{Col}(A)}(\vec{b}) = \vec{c}_0$$

as indicated by the Approximation Theorem. Thus, as derived above, to find the $\vec{x}$ that minimizes $\|A\vec{x} - \vec{b}\|$ we just need to solve

$$A\vec{x} = \vec{c}_0$$

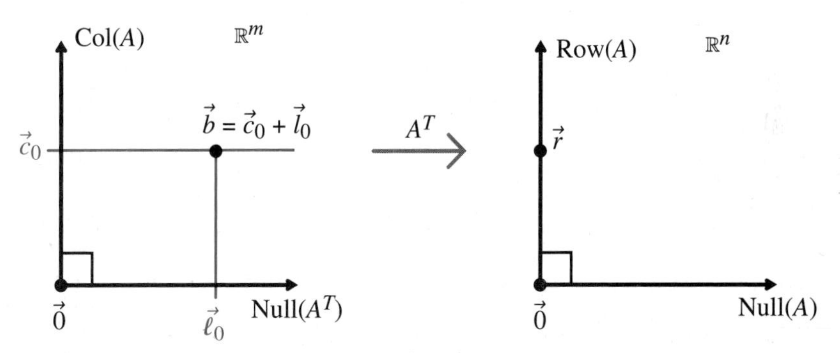

Figure 7.3.2 Geometry of the Method of Least Squares.

As we saw in Section 3.4, if we multiply any vector on the horizontal line $\vec{x} = \vec{c}_0 + \vec{l}$ for any $\vec{l} \in \text{Null}(A^T)$, by A^T, then it will be mapped to the same vector $\vec{r} \in \text{Row}(A)$. That is,

$$A^T\vec{c}_0 = A^T\vec{b}$$

But, $\vec{c}_0 = A\vec{x}$, so this gives

$$A^TA\vec{x} = A^T\vec{b}$$

which is the normal system.

If A^TA is not invertible, then the normal system will have infinitely many solutions. In this case, we want to find the solution of the normal system with the smallest length. Observe that any solution $\vec{x}$ of the normal system is a vector in $\mathbb{R}^n$. Thus, by the Fundamental Theorem of Linear Algebra, every solution has the form

$$\vec{x} = \vec{a} + \vec{n}$$

where $\vec{a} \in \text{Row}(A)$ and $\vec{n} \in \text{Null}(A)$. Since, $\vec{a}$ and $\vec{n}$ are orthogonal, we get by Theorem 7.1.3, that

$$\|\vec{x}\|^2 = \|\vec{a} + \vec{n}\|^2 = \|\vec{a}\|^2 + \|\vec{n}\|^2$$

Thus, the solution with the smallest length will be a vector in Row(A). We will look at a formula to find the solution with the smallest length in Section 8.5.

PROBLEMS 7.3

Practice Problems

For Problems A1 and A2, find a and b to obtain the best-fitting equation of the form $y = a + bt$ for the given data. Graph the data points and the best-fitting line.

A1 $(1,9), (2,6), (3,5), (4,3), (5,1)$

A2 $(-2,2), (-1,2), (0,4), (1,4), (2,5)$

A3 Find a, b, and c to obtain the best-fitting equation of the form $y = a + bt + ct^2$ for the data points $(-2,3)$, $(-1,2)$, $(0,0)$, $(1,2)$, $(2,8)$. Graph the data points and the best-fitting curve.

For Problems A4–A7, find a and b to obtain the best-fitting equation of the form $y = a + bt$ for the given data.

A4 $(-1,4), (0,1), (1,1)$

A5 $(-2,2), (1,-1), (1,-2)$

A6 $(-2,1), (-1,1), (0,2), (1,3)$

A7 $(-1,1), (0,2), (1,2), (2,3)$

For Problems A8–A11, find a and b to obtain the best-fitting equation of the form $y = a + bt + ct^2$ for the given data.

A8 $(-1,4), (0,1), (1,1)$

A9 $(-2,2), (0,-2), (2,3)$

A10 $(-1,3), (1,0), (2,-2)$

A11 $(-1,2), (-1,3), (1,1), (1,2)$

For Problems A12–A15, find the best-fitting equation of the given form for the given data points.

A12 $y = bt + ct^2$: $(-1,4), (0,1), (1,1)$

A13 $y = bt + dt^3$: $(-1,1), (0,0), (1,-1), (2,-20)$

A14 $y = a + ct^2$: $(-2,1), (-1,1), (0,2), (1,3), (2,-2)$

A15 $y = a + bt + c2^t$: $(-1,2), (-1,3), (1,1), (2,2)$

For Problems A16–A23, verify that the system $A\vec{x} = \vec{b}$ is inconsistent and then determine for the vector $\vec{x}$ that minimizes $\|A\vec{x} - \vec{b}\|$.

A16
$$\begin{aligned} x_1 + 2x_2 &= 3 \\ 2x_1 - 3x_2 &= 5 \\ x_1 - 5x_2 &= -4 \end{aligned}$$

A17
$$\begin{aligned} 2x_1 - 2x_2 &= 2 \\ 2x_1 - 3x_2 &= 1 \\ x_1 + x_2 &= 8 \end{aligned}$$

A18
$$\begin{aligned} 2x_1 + x_2 &= 5 \\ 2x_1 - x_2 &= -1 \\ 3x_1 + 2x_2 &= 3 \end{aligned}$$

A19
$$\begin{aligned} x_1 + 2x_2 &= 3 \\ x_1 \qquad &= 3 \\ -x_1 + x_2 &= 2 \end{aligned}$$

A20
$$\begin{aligned} x_1 + \qquad x_3 &= 2 \\ x_2 + 2x_3 &= 1 \\ x_1 + 2x_2 - x_3 &= 1 \\ -x_1 - x_2 - x_3 &= 2 \end{aligned}$$

A21
$$\begin{aligned} x_1 + 3x_2 &= 1 \\ -x_1 + x_2 &= 7 \\ -2x_1 + x_2 &= 6 \end{aligned}$$

A22
$$\begin{aligned} 2x_2 + x_3 &= 1 \\ x_1 + x_2 + 2x_3 &= 2 \\ x_1 - x_2 + 3x_3 &= 0 \\ x_2 \qquad &= -4 \end{aligned}$$

A23
$$\begin{aligned} x_1 + 2x_2 &= 2 \\ x_1 + 2x_2 &= 3 \\ x_1 + 3x_2 &= 2 \\ x_1 + 3x_2 &= 3 \end{aligned}$$

Homework Problems

For Problems B1 and B2, find a and b to obtain the best-fitting equation of the form $y = a + bt$ for the given data. Graph the data points and the best-fitting line.

B1 $(-2,9), (-1,8), (0,5), (1,3), (2,1)$

B2 $(1,4), (2,3), (3,4), (4,5), (5,5)$

For Problems B3–B6, find a and b to obtain the best-fitting equation of the form $y = a + bt$ for the given data

B3 $(-2,5), (-1,3), (1,-1), (2,-3)$

B4 $(-2,2), (-1,4), (1,6)$

B5 $(-2,-1), (1,-2), (2,-3)$

B6 $(-2,5), (0,2), (1,1), (2,-3)$

For Problems B7–B10, find a and b to obtain the best-fitting equation of the form $y = a + bt + ct^2$ for the given data.

B7 $(-1,2), (0,3), (1,5)$

B8 $(-2,2), (-1,2), (-1,3), (0,3)$

B9 $(-1,-2), (0,1), (1,1), (2,3)$

B10 $(-2,-3), (-1,-1), (1,-3), (2,-5)$

For Problems B11–B19, find the best-fitting equation of the given form for the given data points.

B11 $y = a + ct^2$: $(-1,2), (0,3), (1,5)$

B12 $y = bt + ct^2$: $(-1,3), (0,2), (1,2)$

B13 $y = a + ct^2$: $(-2,3), (-1,3), (0,4), (1,6)$

B14 $y = bt + dt^3$: $(-2,-1), (-1,4), (0,2), (1,6)$

B15 $y = bt + ct^2$: $(-1,6), (0,1), (1,-1), (2,3)$

B16 $y = a + bt + dt^3$: $(-1,6), (0,2), (1,-2), (2,-24)$

B17 $y = a + b2^t$: $(-1,2), (0,1), (1,-1)$

B18 $y = a\sin t + b\cos t$: $(-\pi/2,-4), (0,5), (\pi/2,3)$

B19 $y = a\sin t + b\cos t$: $(-\pi/2,-2), (0,-1), (\pi/2,1)$

For Problems B20–B25, verify that the system $A\vec{x} = \vec{b}$ is inconsistent and then determine for the vector $\vec{x}$ that minimizes $\|A\vec{x} - \vec{b}\|$.

B20
$$\begin{aligned} x_1 + x_2 &= 3 \\ x_1 + 2x_2 &= 5 \\ x_1 + x_2 &= 1 \end{aligned}$$

B21
$$\begin{aligned} x_1 + x_2 &= 7 \\ 2x_1 - x_2 &= -8 \\ x_1 + x_2 &= -4 \end{aligned}$$

B22
$$\begin{aligned} x_1 + 4x_2 &= 4 \\ 2x_1 - x_2 &= 2 \\ 2x_1 - x_2 &= 5 \end{aligned}$$

B23
$$\begin{aligned} x_2 &= -3 \\ x_1 + 2x_2 &= 1 \\ 3x_1 + 4x_2 &= 2 \end{aligned}$$

B24
$$\begin{aligned} x_1 + 2x_2 &= 1 \\ 2x_1 - x_2 &= 5 \\ 4x_1 + x_2 &= 2 \\ 2x_1 - 2x_2 &= 3 \end{aligned}$$

B25
$$\begin{aligned} x_1 + 3x_2 &= 5 \\ 2x_1 + 2x_2 &= 2 \\ x_2 &= 5 \\ -x_1 + x_2 &= 3 \end{aligned}$$

Conceptual Problems

C1 Let $A \in M_{m\times n}(\mathbb{R})$ and $\vec{b} \in \mathbb{R}^m$ such that $A\vec{x} = \vec{b}$ is a consistent system. Prove that if $\vec{y}$ is a vector in $\mathbb{R}^n$ such that $A^TA\vec{y} = A^T\vec{b}$, then $A\vec{y} = \vec{b}$.

C2 Let $A = \begin{bmatrix} \vec{a}_1 & \cdots & \vec{a}_n \end{bmatrix} \in M_{m\times n}(\mathbb{R})$. Prove if $\vec{b} \in \mathbb{R}^m$ is orthogonal to each $\vec{a}_i$, then $\vec{x} = \vec{0}$ is a solution of the normal system.

C3 Let $A \in M_{m\times n}(\mathbb{R})$ such that A^TA is invertible. Prove that for any $\vec{y} \in \mathbb{R}^n$ we have

$$\text{proj}_{\text{Col}(A)}(\vec{y}) = A(A^TA)^{-1}A^T\vec{y}$$

C4 Let $A \in M_{m\times n}(\mathbb{R})$ whose columns form a linearly independent set. We define the **pseudoinverse** of A by

$$A^+ = (A^TA)^{-1}A^T$$

Prove that A^+ has the following properties.

(a) $AA^+A = A$

(b) $A^+AA^+ = A^+$

(c) $(AA^+)^T = AA^+$

(d) $(A^+A)^T = A^+A$

In Section 8.5, we will see that the unique least squares with minimal length of $A\vec{x} = \vec{b}$ is given by $\vec{x} = A^+\vec{b}$.

C5 Let $X = \begin{bmatrix} \vec{1} & \vec{t} & \vec{t}^2 \end{bmatrix}$, where $\vec{t} = \begin{bmatrix} t_1 \\ \vdots \\ t_n \end{bmatrix}$ and $\vec{t}^2 = \begin{bmatrix} t_1^2 \\ \vdots \\ t_n^2 \end{bmatrix}$.

Show that $X^TX = \begin{bmatrix} n & \sum\limits_{i=1}^n t_i & \sum\limits_{i=1}^n t_i^2 \\ \sum\limits_{i=1}^n t_i & \sum\limits_{i=1}^n t_i^2 & \sum\limits_{i=1}^n t_i^3 \\ \sum\limits_{i=1}^n t_i^2 & \sum\limits_{i=1}^n t_i^3 & \sum\limits_{i=1}^n t_i^4 \end{bmatrix}$.

C6 Let $X = \begin{bmatrix} \vec{1} & \vec{t} & \vec{t}^2 & \cdots & \vec{t}^m \end{bmatrix}$, where $\vec{t} = \begin{bmatrix} t_1 \\ \vdots \\ t_n \end{bmatrix}$ and $\vec{t}^i = \begin{bmatrix} t_1^i \\ \vdots \\ t_n^i \end{bmatrix}$ for $1 \leq i \leq n$. Assume that at least $m+1$ of the numbers $t_1, \dots, t_n$ are distinct.

(a) Prove that the columns of X form a linearly independent set by showing that the only solution to $c_0\vec{1} + c_1\vec{t} + \cdots + c_m\vec{t}^m = \vec{0}$ is $c_0 = \cdots = c_m = 0$. (Hint: let $p(t) = c_0 + c_1t + \cdots + c_mt^m$ and show that if $c_0\vec{1} + c_1\vec{t} + \cdots + c_m\vec{t}^m = \vec{0}$, $p(t)$ must be the zero polynomial.)

(b) Use the result from part (a) to prove that X^TX is invertible. (Hint: show that the only solution to $X^TX\vec{v} = \vec{0}$ is $\vec{v} = \vec{0}$ by considering $\|X\vec{v}\|^2$.)

7.4 Inner Product Spaces

In Sections 1.3, 1.5, and 7.2, we saw that the dot product plays an essential role in the discussion of lengths, distances, and projections in $\mathbb{R}^n$. In Chapter 4, we saw that the ideas of vector spaces and linear mappings apply to more general sets, including some function spaces. If ideas such as projections are going to be used in these more general spaces, it will be necessary to have a generalization of the dot product to general vector spaces.

Inner Product Spaces

Consideration of the most essential properties of the dot product in Theorem 1.5.1 on page 60 leads to the following definition.

Definition
Inner Product
Inner Product Space

Let $\mathbb{V}$ be a vector space. An **inner product** on $\mathbb{V}$ is a function $\langle\, ,\rangle : \mathbb{V}\times\mathbb{V}\to\mathbb{R}$ such that for all $\mathbf{v},\mathbf{w},\mathbf{z}\in\mathbb{V}$ and $s,t\in\mathbb{R}$ we have

(1) $\langle \mathbf{v},\mathbf{v}\rangle \geq 0$, and $\langle \mathbf{v},\mathbf{v}\rangle = 0$ if and only if $\mathbf{v}=\mathbf{0}$ (positive definite)
(2) $\langle \mathbf{v},\mathbf{w}\rangle = \langle \mathbf{w},\mathbf{v}\rangle$ (symmetric)
(3) $\langle \mathbf{v}, s\mathbf{w}+t\mathbf{z}\rangle = s\langle \mathbf{v},\mathbf{w}\rangle + t\langle \mathbf{v},\mathbf{z}\rangle$ (right linear)

A vector space $\mathbb{V}$ with an inner product is called an **inner product space**.

Remarks

1. The notation $\langle\, ,\rangle : \mathbb{V}\times\mathbb{V}\to\mathbb{R}$ indicates that an inner product is a binary operation just like the dot product. In particular, it takes two vectors from $\mathbb{V}$ as inputs and it outputs a real number.

2. Since an inner product is right linear and symmetric, it is also **left linear**:

$$\langle s\mathbf{w}+t\mathbf{z},\mathbf{v}\rangle = s\langle \mathbf{w},\mathbf{v}\rangle + t\langle \mathbf{z},\mathbf{v}\rangle$$

   Thus, we say that an inner product is **bilinear**.

In the same way that a vector space is dependent on the definitions of addition and scalar multiplication, an inner product space is dependent on the definitions of addition, scalar multiplication, and the inner product. This will be demonstrated in the examples below.

EXAMPLE 7.4.1

The dot product is an inner product on $\mathbb{R}^n$, called the **standard inner product** on $\mathbb{R}^n$.

EXAMPLE 7.4.2

Show that the function defined by

$$\langle \vec{x}, \vec{y} \rangle = 2x_1y_1 + 3x_2y_2$$

is an inner product on $\mathbb{R}^2$.

Solution: We verify that $\langle\ ,\ \rangle$ satisfies the three properties of an inner product:

1. $\langle \vec{x}, \vec{x} \rangle = 2x_1^2 + 3x_2^2 \geq 0$. From this we also see that $\langle \vec{x}, \vec{x} \rangle = 0$ if and only if $\vec{x} = \vec{0}$. Thus, it is positive definite.

2. $\langle \vec{x}, \vec{y} \rangle = 2x_1y_1 + 3x_2y_2 = 2y_1x_1 + 3y_2x_2 = \langle \vec{y}, \vec{x} \rangle$. Thus, it is symmetric.

3. For any $\vec{x}, \vec{w}, \vec{z} \in \mathbb{R}^2$ and $s, t \in \mathbb{R}$,
$$\begin{aligned}\langle \vec{x}, s\vec{w} + t\vec{z} \rangle &= 2x_1(sw_1 + tz_1) + 3x_2(sw_2 + tz_2) \\ &= s(2x_1w_1 + 3x_2w_2) + t(2x_1z_1 + 3x_2z_2) \\ &= s\langle \vec{x}, \vec{w} \rangle + t\langle \vec{x}, \vec{z} \rangle\end{aligned}$$

So, $\langle\ ,\ \rangle$ is bilinear. Thus, $\langle\ ,\ \rangle$ is an inner product on $\mathbb{R}^2$.

Note that $\mathbb{R}^2$ with the inner product defined in Example 7.4.2 is a different inner product space than $\mathbb{R}^2$ with the dot product. However, although there are infinitely many inner products on $\mathbb{R}^n$, it can be proven that for any inner product on $\mathbb{R}^n$ there exists an orthonormal basis such that the inner product is just the dot product on $\mathbb{R}^n$ with respect to this basis. See Problem C11.

EXAMPLE 7.4.3

Verify that $\langle \mathbf{p}, \mathbf{q} \rangle = \mathbf{p}(0)\mathbf{q}(0) + \mathbf{p}(1)\mathbf{q}(1) + \mathbf{p}(2)\mathbf{q}(2)$ defines an inner product on the vector space $P_2(\mathbb{R})$ and determine $\langle 1 + x, 2 - 3x^2 \rangle$.

Solution: We first verify that $\langle\ ,\ \rangle$ satisfies the three properties of an inner product:

(1) $\langle \mathbf{p}, \mathbf{p} \rangle = (\mathbf{p}(0))^2 + (\mathbf{p}(1))^2 + (\mathbf{p}(2))^2 \geq 0$ for all $\mathbf{p} \in P_2(\mathbb{R})$. Moreover, $\langle \mathbf{p}, \mathbf{p} \rangle = 0$ if and only if $\mathbf{p}(0) = \mathbf{p}(1) = \mathbf{p}(2) = 0$. But, the only $\mathbf{p} \in P_2(\mathbb{R})$ that is zero for three values of x is the zero polynomial, $\mathbf{p}(x) = 0$. Thus $\langle\ ,\ \rangle$ is positive definite.

(2) $\langle \mathbf{p}, \mathbf{q} \rangle = \mathbf{p}(0)\mathbf{q}(0) + \mathbf{p}(1)\mathbf{q}(1) + \mathbf{p}(2)\mathbf{q}(2) = \mathbf{q}(0)\mathbf{p}(0) + \mathbf{q}(1)\mathbf{p}(1) + \mathbf{q}(2)\mathbf{p}(2) = \langle \mathbf{q}, \mathbf{p} \rangle$. So, $\langle\ ,\ \rangle$ is symmetric.

(3) For any $\mathbf{p}, \mathbf{q}, \mathbf{r} \in P_2(\mathbb{R})$ and $s, t \in \mathbb{R}$,

$$\begin{aligned}\langle \mathbf{p}, s\mathbf{q} + t\mathbf{r} \rangle &= \mathbf{p}(0)(s\mathbf{q}(0) + t\mathbf{r}(0)) + \mathbf{p}(1)(s\mathbf{q}(1) + t\mathbf{r}(1)) + \mathbf{p}(2)(s\mathbf{q}(2) + t\mathbf{r}(2)) \\ &= s(\mathbf{p}(0)\mathbf{q}(0) + \mathbf{p}(1)\mathbf{q}(1) + \mathbf{p}(2)\mathbf{q}(2)) + t(\mathbf{p}(0)\mathbf{r}(0) + \mathbf{p}(1)\mathbf{r}(1) \\ &\quad + \mathbf{p}(2)\mathbf{r}(2)) = s\langle \mathbf{p}, \mathbf{q} \rangle + t\langle \mathbf{p}, \mathbf{r} \rangle\end{aligned}$$

So, $\langle\ ,\ \rangle$ is bilinear. Thus, $\langle\ ,\ \rangle$ is an inner product on $P_2(\mathbb{R})$. That is, $P_2(\mathbb{R})$ is an inner product space under the inner product $\langle\ ,\ \rangle$.

In this inner product space, we have

$$\begin{aligned}\langle 1 + x, 2 - 3x^2 \rangle &= (1 + 0)(2 - 3(0)^2) + (1 + 1)(2 - 3(1)^2) + (1 + 2)(2 - 3(2)^2) \\ &= 2 - 2 - 30 = -30\end{aligned}$$

EXAMPLE 7.4.4

Let tr(C) represent the trace of a matrix (the definition of tr(C) is in property (4) of Theorem 6.2.1 on page 361). Then, $M_{2\times 2}(\mathbb{R})$ is an inner product space under the inner product defined by $\langle A, B\rangle = \text{tr}(A^T B)$. If $A = \begin{bmatrix} 1 & 2 \\ 3 & -1 \end{bmatrix}$ and $B = \begin{bmatrix} 4 & 5 \\ 0 & 6 \end{bmatrix}$, then under this inner product, we have

$$\left\langle \begin{bmatrix} 1 & 2 \\ 3 & -1 \end{bmatrix}, \begin{bmatrix} 4 & 5 \\ 0 & 6 \end{bmatrix} \right\rangle = \text{tr}\left(\begin{bmatrix} 1 & 3 \\ 2 & -1 \end{bmatrix} \begin{bmatrix} 4 & 5 \\ 0 & 6 \end{bmatrix} \right) = \text{tr}\left(\begin{bmatrix} 4 & 23 \\ 8 & 4 \end{bmatrix} \right) = 4 + 4 = 8$$

EXERCISE 7.4.1

Verify that $\langle A, B\rangle = \text{tr}(A^T B)$ is an inner product for $M_{2\times 2}(\mathbb{R})$. Do you notice a relationship between this inner product and the dot product on $\mathbb{R}^4$?

EXAMPLE 7.4.5

An extremely important inner product in applied mathematics, physics, and engineering is the inner product

$$\langle f, g\rangle = \int_{-\pi}^{\pi} f(x)g(x)\, dx$$

on the vector space $C[-\pi, \pi]$ of continuous functions defined on the closed interval from $-\pi$ to π. This is the foundation for Fourier Series. See Section 7.5.

Orthogonality and Length

The concepts of length and orthogonality are fundamental in geometry and have many real-world applications. Thus, we now extend these concepts to general inner product spaces. Since the definition of an inner product mimics the properties of the dot product, we can define length and orthogonality in an inner product space to match exactly with what we did in $\mathbb{R}^n$.

Definition
Norm
Unit Vector

Let $\mathbb{V}$ be an inner product space. For any $\mathbf{v} \in \mathbb{V}$, we define the **norm** (or **length**) of $\mathbf{v}$ to be

$$\|\mathbf{v}\| = \sqrt{\langle \mathbf{v}, \mathbf{v}\rangle}$$

A vector $\mathbf{v}$ in an inner product space $\mathbb{V}$ is called a **unit vector** if $\|\mathbf{v}\| = 1$.

EXAMPLE 7.4.6

Find the norm of $A = \begin{bmatrix} 1 & 0 \\ 2 & 1 \end{bmatrix}$ in $M_{2\times 2}(\mathbb{R})$ under the inner product $\langle A, B\rangle = \text{tr}(A^T B)$.

Solution: Using the result of Exercise 7.4.1 we get

$$\|A\| = \sqrt{\langle A, A\rangle} = \sqrt{\text{tr}(A^T A)} = \sqrt{1^2 + 0^2 + 2^2 + 1^2} = \sqrt{6}$$

EXAMPLE 7.4.7

Find the norm of $\mathbf{p}(x) = 1 - 2x - x^2$ in $P_2(\mathbb{R})$ under the inner product

$$\langle \mathbf{p}, \mathbf{q} \rangle = \mathbf{p}(0)\mathbf{q}(0) + \mathbf{p}(1)\mathbf{q}(1) + \mathbf{p}(2)\mathbf{q}(2)$$

Solution: We have

$$\|\mathbf{p}\| = \sqrt{(\mathbf{p}(0))^2 + (\mathbf{p}(1))^2 + (\mathbf{p}(2))^2} = \sqrt{1^2 + (1-2-1)^2 + (1-4-4)^2} = \sqrt{54}$$

EXERCISE 7.4.2

Find the norm of $\mathbf{p}(x) = 1$ and $\mathbf{q}(x) = x$ in $P_2(\mathbb{R})$ under each inner product

(a) $\langle \mathbf{p}, \mathbf{q} \rangle = \mathbf{p}(0)\mathbf{q}(0) + \mathbf{p}(1)\mathbf{q}(1) + \mathbf{p}(2)\mathbf{q}(2)$

(b) $\langle \mathbf{p}, \mathbf{q} \rangle = \mathbf{p}(-1)\mathbf{q}(-1) + \mathbf{p}(0)\mathbf{q}(0) + \mathbf{p}(1)\mathbf{q}(1)$

Of course, we have the usual properties of length in an inner product space.

Theorem 7.4.1

If $\mathbb{V}$ is an inner product space, $\mathbf{x}, \mathbf{y} \in \mathbb{V}$ and $t \in \mathbb{R}$, then

(1) $\|\mathbf{x}\| \geq 0$, and $\|\mathbf{x}\| = 0$ if and only if $\mathbf{x} = \mathbf{0}$

(2) $\|t\mathbf{x}\| = |t|\,\|\mathbf{x}\|$

(3) $|\langle \mathbf{x}, \mathbf{y} \rangle| \leq \|\mathbf{x}\|\,\|\mathbf{y}\|$, with equality if and only if $\{\mathbf{x}, \mathbf{y}\}$ is linearly dependent

(4) $\|\mathbf{x} + \mathbf{y}\| \leq \|\mathbf{x}\| + \|\mathbf{y}\|$

We now look at orthogonality in inner product spaces.

Definition

Orthogonal

Orthogonal Set

Orthonormal

Let $\mathbb{V}$ be an inner product space with inner product $\langle\ ,\ \rangle$.
Two vectors $\mathbf{v}, \mathbf{w} \in \mathbb{V}$ are said to be **orthogonal** if $\langle \mathbf{v}, \mathbf{w} \rangle = 0$.
The set of vectors $\{\mathbf{v}_1, \ldots, \mathbf{v}_k\}$ in $\mathbb{V}$ is said to be an **orthogonal set** if $\langle \mathbf{v}_i, \mathbf{v}_j \rangle = 0$ for all $i \neq j$.
An orthogonal set $\{\mathbf{v}_1, \ldots, \mathbf{v}_k\}$ is called an **orthonormal set** if we also have $\langle \mathbf{v}_i, \mathbf{v}_i \rangle = 1$ for $1 \leq i \leq k$.

With this definition, we can now repeat our arguments from Sections 7.1 and 7.2 for coordinates with respect to an orthogonal basis and projections. In particular, we get that if $\mathcal{B} = \{\mathbf{v}_1, \ldots, \mathbf{v}_k\}$ is an orthogonal basis for a subspace $\mathbb{S}$ of an inner product space $\mathbb{V}$ with inner product $\langle\ ,\ \rangle$, then for any $\mathbf{x} \in \mathbb{V}$ we have

$$\text{proj}_{\mathbb{S}}(\mathbf{x}) = \frac{\langle \mathbf{v}_1, \mathbf{x} \rangle}{\|\mathbf{v}_1\|^2}\mathbf{v}_1 + \cdots + \frac{\langle \mathbf{v}_k, \mathbf{x} \rangle}{\|\mathbf{v}_k\|^2}\mathbf{v}_k$$

Additionally, the Gram-Schmidt Procedure is also identical. If we have a basis $\{\mathbf{w}_1, \ldots, \mathbf{w}_n\}$ for an inner product space $\mathbb{V}$ with inner product $\langle , \rangle$, then the set $\{\mathbf{v}_1, \ldots, \mathbf{v}_n\}$ defined by

$$\begin{aligned}
\mathbf{v}_1 &= \mathbf{w}_1 \\
\mathbf{v}_2 &= \mathbf{w}_2 - \frac{\langle \mathbf{v}_1, \mathbf{w}_2 \rangle}{\|\mathbf{v}_1\|^2}\mathbf{v}_1 \\
&\ \vdots \\
\mathbf{v}_n &= \mathbf{w}_n - \frac{\langle \mathbf{v}_1, \mathbf{w}_n \rangle}{\|\mathbf{v}_1\|^2}\mathbf{v}_1 - \cdots - \frac{\langle \mathbf{v}_{n-1}, \mathbf{w}_n \rangle}{\|\mathbf{v}_{n-1}\|^2}\mathbf{v}_{n-1}
\end{aligned}$$

is an orthogonal basis for $\mathbb{V}$.

EXAMPLE 7.4.8

Use the Gram-Schmidt Procedure to determine an orthonormal basis for $\mathbb{S} = \text{Span}\{1, x\}$ of $P_2(\mathbb{R})$ under the inner product

$$\langle \mathbf{p}, \mathbf{q} \rangle = \mathbf{p}(0)\mathbf{q}(0) + \mathbf{p}(1)\mathbf{q}(1) + \mathbf{p}(2)\mathbf{q}(2)$$

Use this basis to determine $\text{proj}_{\mathbb{S}}(x^2)$.

Solution: Denote the basis vectors of $\mathbb{S}$ by $\mathbf{p}_1(x) = 1$ and $\mathbf{p}_2(x) = x$. We want to find an orthogonal basis $\{\mathbf{q}_1(x), \mathbf{q}_2(x)\}$ for $\mathbb{S}$. By using the Gram-Schmidt Procedure, we take $\mathbf{q}_1(x) = \mathbf{p}_1(x) = 1$ and then let

$$\mathbf{q}_2 = \mathbf{p}_2 - \frac{\langle \mathbf{q}_1, \mathbf{p}_2 \rangle}{\|\mathbf{q}_1\|^2}\mathbf{q}_1 = x - \frac{1(0) + 1(1) + 1(2)}{1^2 + 1^2 + 1^2}1 = x - 1$$

Therefore, our orthogonal basis is $\{\mathbf{q}_1, \mathbf{q}_2\} = \{1, x - 1\}$. Hence, we have

$$\begin{aligned} \text{proj}_{\mathbb{S}}(x^2) &= \frac{\langle 1, x^2 \rangle}{\|1\|^2}1 + \frac{\langle x - 1, x^2 \rangle}{\|x - 1\|^2}(x - 1) \\ &= \frac{1(0) + 1(1) + 1(4)}{1^2 + 1^2 + 1^2}1 + \frac{(-1)0 + 0(1) + 1(4)}{(-1)^2 + 0^2 + 1^2}(x - 1) \\ &= \frac{5}{3}1 + 2(x - 1) = 2x - \frac{1}{3} \end{aligned}$$

PROBLEMS 7.4

Practice Problems

For Problems A1–A5, evaluate the expression in $M_{2\times2}(\mathbb{R})$ with inner product $\langle A, B \rangle = \text{tr}(A^T B)$.

A1 $\left\langle \begin{bmatrix} 1 & 3 \\ -2 & 1 \end{bmatrix}, \begin{bmatrix} 2 & 5 \\ 1 & 1 \end{bmatrix} \right\rangle$

A2 $\left\langle \begin{bmatrix} 3 & 6 \\ 4 & -2 \end{bmatrix}, \begin{bmatrix} 3 & 0 \\ -3 & -2 \end{bmatrix} \right\rangle$

A3 $\left\| \begin{bmatrix} 1 & 0 \\ 0 & 1 \end{bmatrix} \right\|$

A4 $\left\| \begin{bmatrix} 1 & 2 \\ 2 & 0 \end{bmatrix} \right\|$

A5 $\left\| \begin{bmatrix} 3 & -1 \\ -2 & 1 \end{bmatrix} \right\|$

For Problems A6–A9, evaluate the expression in $P_2(\mathbb{R})$ with inner product $\langle \mathbf{p}, \mathbf{q} \rangle = \mathbf{p}(0)\mathbf{q}(0) + \mathbf{p}(1)\mathbf{q}(1) + \mathbf{p}(2)\mathbf{q}(2)$.

A6 $\langle x - 2x^2, 1 + 3x \rangle$

A7 $\langle 2 - x + 3x^2, 4 - 3x^2 \rangle$

A8 $\|3 - 2x + x^2\|$

A9 $\|9 + 9x + 9x^2\|$

For Problems A10–A14, determine whether $\langle \, , \rangle$ defines an inner product on $P_2(\mathbb{R})$.

A10 $\langle \mathbf{p}, \mathbf{q} \rangle = \mathbf{p}(0)\mathbf{q}(0) + \mathbf{p}(1)\mathbf{q}(1)$

A11 $\langle \mathbf{p}, \mathbf{q} \rangle = |\mathbf{p}(0)\mathbf{q}(0)| + |\mathbf{p}(1)\mathbf{q}(1)| + |\mathbf{p}(2)\mathbf{q}(2)|$

A12 $\langle \mathbf{p}, \mathbf{q} \rangle = \mathbf{p}(-1)\mathbf{q}(-1) + 2\mathbf{p}(0)\mathbf{q}(0) + \mathbf{p}(1)\mathbf{q}(1)$

A13 $\langle \mathbf{p}, \mathbf{q} \rangle = \mathbf{p}(-1)\mathbf{q}(1) + 2\mathbf{p}(0)\mathbf{q}(0) + \mathbf{p}(1)\mathbf{q}(-1)$

A14 $\langle \mathbf{p}, \mathbf{q} \rangle = \mathbf{p}(0)\mathbf{p}(0) + \mathbf{p}(1)\mathbf{p}(1) + \mathbf{q}(0)\mathbf{q}(0) + \mathbf{q}(1)\mathbf{q}(1)$

For Problems A15–A18, assume $\mathbb{S}$ is a subspace of $M_{2\times2}(\mathbb{R})$ with inner product $\langle A, B \rangle = \text{tr}(A^T B)$.

(a) Use the Gram-Schmidt Procedure to determine an orthogonal basis for the following subspaces of $M_{2\times2}(\mathbb{R})$.

(b) Use the orthogonal basis you found in part (a) to determine $\text{proj}_{\mathbb{S}}\left(\begin{bmatrix} 4 & 3 \\ -2 & 1 \end{bmatrix}\right)$.

A15 $\mathbb{S} = \text{Span}\left\{\begin{bmatrix} 1 & 0 \\ -1 & 1 \end{bmatrix}, \begin{bmatrix} 1 & 1 \\ 0 & 1 \end{bmatrix}, \begin{bmatrix} 2 & 0 \\ 1 & -1 \end{bmatrix}\right\}$

A16 $\mathbb{S} = \text{Span}\left\{\begin{bmatrix} 2 & 0 \\ 1 & -1 \end{bmatrix}, \begin{bmatrix} 1 & 0 \\ -1 & 1 \end{bmatrix}, \begin{bmatrix} 1 & 1 \\ 0 & 1 \end{bmatrix}\right\}$

A17 $\mathbb{S} = \text{Span}\left\{\begin{bmatrix} 1 & 0 \\ 2 & 2 \end{bmatrix}, \begin{bmatrix} 1 & 1 \\ 3 & 1 \end{bmatrix}, \begin{bmatrix} 3 & -1 \\ 2 & 1 \end{bmatrix}\right\}$

A18 $\mathbb{S} = \text{Span}\left\{\begin{bmatrix} 3 & 1 \\ -1 & -2 \end{bmatrix}, \begin{bmatrix} 3 & 1 \\ 0 & -1 \end{bmatrix}, \begin{bmatrix} 3 & 1 \\ 2 & 1 \end{bmatrix}, \begin{bmatrix} 1 & 1 \\ 1 & 0 \end{bmatrix}\right\}$

For Problems A19–A22, assume $\mathbb{S}$ is a subspace of $\mathbb{R}^3$ with inner product $\langle \vec{x}, \vec{y} \rangle = 2x_1y_1 + x_2y_2 + 3x_3y_3$.

(a) Use the Gram-Schmidt Procedure to determine an orthogonal basis for the following subspaces of $\mathbb{R}^3$.
(b) Use the orthogonal basis you found in part (a) to determine $\text{proj}_{\mathbb{S}}\left(\begin{bmatrix} 1 \\ 0 \\ 0 \end{bmatrix}\right)$.

A19 $\mathbb{S} = \text{Span}\left\{\begin{bmatrix} 1 \\ 1 \\ 0 \end{bmatrix}, \begin{bmatrix} -1 \\ 1 \\ 0 \end{bmatrix}\right\}$ **A20** $\mathbb{S} = \text{Span}\left\{\begin{bmatrix} 1 \\ 0 \\ 0 \end{bmatrix}, \begin{bmatrix} 1 \\ 3 \\ 2 \end{bmatrix}, \begin{bmatrix} 0 \\ 4 \\ 1 \end{bmatrix}\right\}$

A21 $\mathbb{S} = \text{Span}\left\{\begin{bmatrix} 1 \\ 1 \\ 1 \end{bmatrix}, \begin{bmatrix} 2 \\ 2 \\ 1 \end{bmatrix}\right\}$ **A22** $\mathbb{S} = \text{Span}\left\{\begin{bmatrix} 2 \\ 1 \\ 1 \end{bmatrix}, \begin{bmatrix} 1 \\ 2 \\ 1 \end{bmatrix}, \begin{bmatrix} 1 \\ 5 \\ 2 \end{bmatrix}\right\}$

For Problems A23–A26, find the projection of the given polynomials onto the given subspace $\mathbb{S}$ in $P_2(\mathbb{R})$ with inner product $\langle \mathbf{p}, \mathbf{q} \rangle = \mathbf{p}(-1)\mathbf{q}(-1) + \mathbf{p}(0)\mathbf{q}(0) + \mathbf{p}(1)\mathbf{q}(1)$.

A23 $\mathbf{p}(x) = x^2, \mathbf{q}(x) = 3 + 2x - x^2, \mathbb{S} = \text{Span}\left\{1 + 2x - x^2\right\}$

A24 $\mathbf{p}(x) = x, \mathbf{q}(x) = 1 - 3x, \mathbb{S} = \text{Span}\{1 + 3x\}$

A25 $\mathbf{p}(x) = 1 + x + x^2, \mathbf{q}(x) = 1 - x^2, \mathbb{S} = \text{Span}\{1, x\}$

A26 $\mathbf{p}(x) = 1, \mathbf{q}(x) = 2x + 3x^2, \mathbb{S} = \text{Span}\left\{1 + x, 1 - x^2\right\}$

A27 Let $\{\mathbf{v}_1, \ldots, \mathbf{v}_k\}$ be an orthogonal set in an inner product space $\mathbb{V}$. Prove that

$$\|\mathbf{v}_1 + \cdots + \mathbf{v}_k\|^2 = \|\mathbf{v}_1\|^2 + \cdots + \|\mathbf{v}_k\|^2$$

Homework Problems

For Problems B1–B5, evaluate the expression in $M_{2\times2}(\mathbb{R})$ with inner product $\langle A, B \rangle = \text{tr}(A^TB)$.

B1 $\left\langle \begin{bmatrix} 2 & 1 \\ -1 & 3 \end{bmatrix}, \begin{bmatrix} 4 & 6 \\ -7 & -3 \end{bmatrix} \right\rangle$ **B2** $\left\langle \begin{bmatrix} 2 & 5 \\ -8 & 4 \end{bmatrix}, \begin{bmatrix} 1 & -1 \\ 0 & -4 \end{bmatrix} \right\rangle$

B3 $\left\| \begin{bmatrix} 1 & -1 \\ 0 & -4 \end{bmatrix} \right\|$ **B4** $\left\| \begin{bmatrix} 1/2 & 1/2 \\ 1/2 & 1/2 \end{bmatrix} \right\|$ **B5** $\left\| \begin{bmatrix} 2 & 3 \\ -5 & 4 \end{bmatrix} \right\|$

For Problems B6–B11, evaluate the expression in $P_2(\mathbb{R})$ with inner product $\langle \mathbf{p}, \mathbf{q} \rangle = \mathbf{p}(0)\mathbf{q}(0) + \mathbf{p}(1)\mathbf{q}(1) + \mathbf{p}(2)\mathbf{q}(2)$.

B6 $\langle 1 + x - x^2, -6 - 2x^2 \rangle$ **B7** $\langle 2 + 3x, 1 + x - x^2 \rangle$

B8 $\langle 1 + x^2, 1 - x^2 \rangle$ **B9** $\|2 - x^2\|$

B10 $\|x^2\|$ **B11** $\|1 + x - x^2\|$

For Problems B12–B15, determine whether $\langle\ ,\ \rangle$ defines an inner product on $P_2(\mathbb{R})$.

B12 $\langle \mathbf{p}, \mathbf{q} \rangle = \mathbf{p}(-2)\mathbf{q}(-2) + \mathbf{p}(0)\mathbf{q}(0) + \mathbf{p}(2)\mathbf{q}(2)$

B13 $\langle \mathbf{p}, \mathbf{q} \rangle = 2\mathbf{p}(-2)\mathbf{q}(-2) + \mathbf{p}(0)\mathbf{q}(0) + 2\mathbf{p}(1)\mathbf{q}(1)$

B14 $\langle \mathbf{p}, \mathbf{q} \rangle = \mathbf{p}(-1)\mathbf{q}(-1) - \mathbf{p}(0)\mathbf{q}(0) + \mathbf{p}(1)\mathbf{q}(1)$

B15 $\langle \mathbf{p}, \mathbf{q} \rangle = \mathbf{p}(0)\mathbf{q}(2) + 2\mathbf{p}(1)\mathbf{q}(1) + \mathbf{p}(2)\mathbf{q}(0)$

For Problems B16 and B17, assume $\mathbb{S}$ is a subspace of $M_{2\times2}(\mathbb{R})$ with inner product $\langle A, B \rangle = \text{tr}(A^TB)$.

(a) Use the Gram-Schmidt Procedure to determine an orthogonal basis for the following subspaces of $M_{2\times2}(\mathbb{R})$.
(b) Use the orthogonal basis you found in part (a) to determine $\text{proj}_{\mathbb{S}}\left(\begin{bmatrix} 1 & -1 \\ 0 & 2 \end{bmatrix}\right)$.

B16 $\mathbb{S} = \text{Span}\left\{\begin{bmatrix} 1 & 0 \\ 0 & 1 \end{bmatrix}, \begin{bmatrix} 0 & 1 \\ 1 & 1 \end{bmatrix}, \begin{bmatrix} 1 & -1 \\ 1 & 1 \end{bmatrix}\right\}$

B17 $\mathbb{S} = \text{Span}\left\{\begin{bmatrix} 1 & 2 \\ 2 & 0 \end{bmatrix}, \begin{bmatrix} 1 & 0 \\ 4 & 1 \end{bmatrix}, \begin{bmatrix} 2 & 5 \\ 3 & 1 \end{bmatrix}\right\}$

For Problems B18–B21, assume $\mathbb{S}$ is a subspace of $\mathbb{R}^3$ with inner product $\langle \vec{x}, \vec{y} \rangle = 3x_1y_1 + 2x_2y_2 + x_3y_3$.

(a) Use the Gram-Schmidt Procedure to determine an orthogonal basis for the following subspaces of $\mathbb{R}^3$.
(b) Use the orthogonal basis you found in part (a) to determine $\text{proj}_{\mathbb{S}}\left(\begin{bmatrix} 1 \\ 0 \\ 1 \end{bmatrix}\right)$.

B18 $\mathbb{S} = \text{Span}\left\{\begin{bmatrix} 1 \\ 1 \\ 0 \end{bmatrix}, \begin{bmatrix} -1 \\ 1 \\ 0 \end{bmatrix}\right\}$ **B19** $\mathbb{S} = \text{Span}\left\{\begin{bmatrix} 1 \\ 1 \\ 1 \end{bmatrix}, \begin{bmatrix} 2 \\ 2 \\ 1 \end{bmatrix}\right\}$

B20 $\mathbb{S} = \text{Span}\left\{\begin{bmatrix} 3 \\ 2 \\ 1 \end{bmatrix}, \begin{bmatrix} 3 \\ 1 \\ 5 \end{bmatrix}, \begin{bmatrix} 2 \\ 1 \\ 2 \end{bmatrix}\right\}$ **B21** $\mathbb{S} = \text{Span}\left\{\begin{bmatrix} 0 \\ 2 \\ 1 \end{bmatrix}, \begin{bmatrix} 1 \\ 1 \\ 5 \end{bmatrix}\right\}$

For Problems B22–B27, find the projection of the given polynomials onto the given subspace $\mathbb{S}$ in $P_2(\mathbb{R})$ with inner product $\langle \mathbf{p}, \mathbf{q} \rangle = \mathbf{p}(-1)\mathbf{q}(-1) + \mathbf{p}(0)\mathbf{q}(0) + \mathbf{p}(1)\mathbf{q}(1)$.

B22 $\mathbf{p}(x) = 1$, $\mathbf{q}(x) = 1 + 2x^2$, $\mathbb{S} = \text{Span}\left\{1 + x + x^2\right\}$

B23 $\mathbf{p}(x) = x$, $\mathbf{q}(x) = x^2$, $\mathbb{S} = \text{Span}\{1 - x\}$

B24 $\mathbf{p}(x) = 1 - x$, $\mathbf{q}(x) = 1 + x - x^2$, $\mathbb{S} = \text{Span}\left\{x, 1 - x^2\right\}$

B25 $\mathbf{p}(x) = x$, $\mathbf{q}(x) = 2 + x + x^2$, $\mathbb{S} = \text{Span}\left\{x^2, 1 + x^2\right\}$

B26 $\mathbf{p}(x) = x$, $\mathbf{q}(x) = 2 + x + x^2$, $\mathbb{S} = \text{Span}\left\{x^2, 1 - x\right\}$

B27 $\mathbf{p}(x) = 1 - x^2$, $\mathbf{q}(x) = x^2$, $\mathbb{S} = \text{Span}\left\{1 + x, x - x^2\right\}$

Conceptual Problems

For Problems C1–C3, determine if the following statements are true or false. Justify your answer.

C1 If there exists $\mathbf{x}, \mathbf{y} \in \mathbb{V}$ such that $\langle \mathbf{x}, \mathbf{y} \rangle < 0$, then $\langle \, , \rangle$ is not an inner product on $\mathbb{V}$.

C2 For any $\mathbf{x}$ in an inner product space $\mathbb{V}$, we have $\langle \mathbf{x}, \mathbf{0} \rangle = 0$.

C3 If $\mathbf{x}$ and $\mathbf{y}$ are orthogonal in an inner product space $\mathbb{V}$, then $s\mathbf{x}$ and $t\mathbf{y}$ are also orthogonal for any $s, t \in \mathbb{R}$.

For Problems C4–C9, define the distance between two vectors $\mathbf{x}, \mathbf{y}$ in an inner product space $\mathbb{V}$ by

$$d(\mathbf{x}, \mathbf{y}) = \|\mathbf{x} - \mathbf{y}\|$$

C4 Prove $d(\mathbf{x}, \mathbf{y}) \geq 0$.

C5 Prove $d(\mathbf{x}, \mathbf{y}) = 0$ if and only if $\mathbf{x} = \mathbf{y}$.

C6 Prove $d(\mathbf{x}, \mathbf{y}) = d(\mathbf{y}, \mathbf{x})$.

C7 Prove $d(\mathbf{x}, \mathbf{z}) \leq d(\mathbf{x}, \mathbf{y}) + d(\mathbf{y}, \mathbf{z})$.

C8 Prove $d(\mathbf{x}, \mathbf{y}) = d(\mathbf{x} + \mathbf{z}, \mathbf{y} + \mathbf{z})$.

C9 Prove $d(t\mathbf{x}, t\mathbf{y}) = |t|\, d(\mathbf{x}, \mathbf{y})$ for all $t \in \mathbb{R}$.

C10 Suppose that $\mathcal{B} = \{\mathbf{v}_1, \mathbf{v}_2, \mathbf{v}_3\}$ is a basis for an inner product space $\mathbb{V}$ with inner product $\langle \, , \rangle$. Define $G \in M_{3\times 3}(\mathbb{R})$ by

$$g_{ij} = \langle \mathbf{v}_i, \mathbf{v}_j \rangle, \qquad \text{for } i, j = 1, 2, 3$$

(a) Prove that G is symmetric ($G^T = G$).

(b) Show that if $[\vec{x}]_{\mathcal{B}} = \begin{bmatrix} x_1 \\ x_2 \\ x_3 \end{bmatrix}$ and $[\vec{y}]_{\mathcal{B}} = \begin{bmatrix} y_1 \\ y_2 \\ y_3 \end{bmatrix}$, then

$$\langle \vec{x}, \vec{y} \rangle = \begin{bmatrix} x_1 & x_2 & x_3 \end{bmatrix} G \begin{bmatrix} y_1 \\ y_2 \\ y_3 \end{bmatrix}$$

(c) Determine the matrix G of the inner product $\langle \mathbf{p}, \mathbf{q} \rangle = \mathbf{p}(0)\mathbf{q}(0) + \mathbf{p}(1)\mathbf{q}(1) + \mathbf{p}(2)\mathbf{q}(2)$ for P_2 with respect to the basis $\{1, x, x^2\}$.

C11 (a) Let $\{\vec{e}_1, \vec{e}_2\}$ be the standard basis for $\mathbb{R}^2$ and suppose that $\langle \, , \rangle$ is an inner product on $\mathbb{R}^2$. Show that if $\vec{x}, \vec{y} \in \mathbb{R}^2$,

$$\begin{aligned} \langle \vec{x}, \vec{y} \rangle &= x_1y_1\langle \vec{e}_1, \vec{e}_1 \rangle + x_1y_2\langle \vec{e}_1, \vec{e}_2 \rangle \\ &\quad + x_2y_1\langle \vec{e}_2, \vec{e}_1 \rangle + x_2y_2\langle \vec{e}_2, \vec{e}_2 \rangle \end{aligned}$$

(b) For the inner product in part (a), define a matrix G, called the **standard matrix of the inner product** $\langle \, , \rangle$, by $g_{ij} = \langle \vec{e}_i, \vec{e}_j \rangle$ for $i, j = 1, 2$. Show that G is symmetric and that

$$\langle \vec{x}, \vec{y} \rangle = \sum_{i,j=1}^{2} g_{ij}x_iy_j = \vec{x}^T G \vec{y}$$

(c) Apply the Gram-Schmidt Procedure, using the inner product $\langle \, , \rangle$ and the corresponding norm, to produce an orthonormal basis $\mathcal{B} = \{\vec{v}_1, \vec{v}_2\}$ for $\mathbb{R}^2$.

(d) Define $\tilde{G}$, the $\mathcal{B}$-matrix of the inner product $\langle \, , \rangle$, by $\tilde{g}_{ij} = \langle \vec{v}_i, \vec{v}_j \rangle$ for $i, j = 1, 2$. Show that $\tilde{G} = I$ and that for $\vec{x} = \tilde{x}_1\vec{v}_1 + \tilde{x}_2\vec{v}_2$ and $\vec{y} = \tilde{y}_1\vec{v}_1 + \tilde{y}_2\vec{v}_2$,

$$\langle \vec{x}, \vec{y} \rangle = \tilde{x}_1\tilde{y}_1 + \tilde{x}_2\tilde{y}_2$$

Conclusion. For an arbitrary inner product $\langle \, , \rangle$ on $\mathbb{R}^2$, there exists a basis for $\mathbb{R}^2$ that is orthonormal with respect to this inner product. Moreover, when $\vec{x}$ and $\vec{y}$ are expressed in terms of this basis, $\langle \vec{x}, \vec{y} \rangle$ looks just like the standard inner product in $\mathbb{R}^2$. This argument generalizes in a straightforward way to $\mathbb{R}^n$; see Section 8.2 Problem C6.

7.5 Fourier Series

The Inner Product $\int_a^b f(x)g(x)\,dx$

Let $C[a,b]$ be the space of functions $f : \mathbb{R} \to \mathbb{R}$ that are continuous on the interval $[a,b]$. Then, for any $f, g \in C[a,b]$ we have that the product fg is also continuous on $[a,b]$ and hence integrable on $[a,b]$. Therefore, it makes sense to define an inner product as follows.

The inner product $\langle\ ,\ \rangle$ is defined on $C[a,b]$ by

$$\langle f, g\rangle = \int_a^b f(x)g(x)\,dx$$

The three properties of an inner product are satisfied because

(1) $\langle f, f\rangle = \int_a^b f(x)f(x)\,dx \geq 0$ for all $f \in C[a,b]$ and $\langle f, f\rangle = \int_a^b f(x)f(x)\,dx = 0$ if and only if $f(x) = 0$ for all $x \in [a,b]$.

(2) $\langle f, g\rangle = \int_a^b f(x)g(x)\,dx = \int_a^b g(x)f(x)\,dx = \langle g, f\rangle$

(3) $\langle f, sg + th\rangle = \int_a^b f(x)(sg(x) + th(x))\,dx = s\int_a^b f(x)g(x)\,dx + t\int_a^b f(x)h(x)\,dx = s\langle f, g\rangle + t\langle f, h\rangle$ for any $s, t \in \mathbb{R}$.

Since an integral is the limit of sums, this inner product defined as the *integral of the product of the values of* f *and* g *at each* x is a fairly natural generalization of the dot product in $\mathbb{R}^n$ defined as a *sum of the product of the* i*-th components of* $\vec{x}$ *and* $\vec{y}$ *for each* i.

One interesting consequence is that the norm of a function f with respect to this inner product is

$$\|f\| = \left(\int_a^b f^2(x)\,dx\right)^{1/2}$$

Intuitively, this is quite satisfactory as a measure of how far the function is from the zero function.

One of the most interesting and important applications of this inner product involves Fourier series.

Fourier Series

Let $CP_{2\pi}$ denote the space of continuous real-valued functions of a real variable that are periodic with period 2π. Such functions satisfy $f(x+2\pi) = f(x)$ for all x. Examples of such functions are $f(x) = c$ for any constant c, $\cos x$, $\sin x$, $\cos 2x$, $\sin 3x$, etc. (Note that the function $\cos 2x$ is periodic with period 2π because $\cos(2(x + 2\pi)) = \cos 2x$. However, its "fundamental (smallest) period" is π.)

In some electrical engineering applications, it is of interest to consider a signal described by functions such as the function

$$f(x) = \begin{cases} -\pi - x & \text{if } -\pi \leq x \leq -\pi/2 \\ x & \text{if } -\pi/2 < x \leq \pi/2 \\ \pi - x & \text{if } \pi/2 < x \leq \pi \end{cases}$$

This function is shown in Figure 7.5.1.

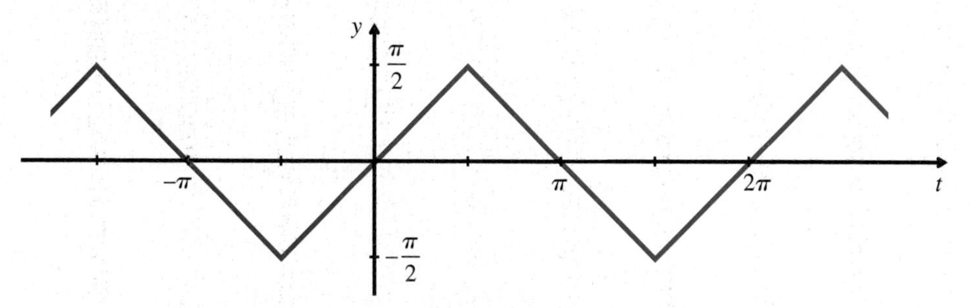

Figure 7.5.1 A continuous periodic function.

In the early nineteenth century, while studying the problem of the conduction of heat, Fourier had the brilliant idea of trying to represent an arbitrary function in $CP_{2\pi}$ as a linear combination of the set of functions

$$\{1, \cos x, \sin x, \cos 2x, \sin 2x, \ldots, \cos nx, \sin nx, \ldots\}$$

This idea developed into Fourier analysis, which is now one of the essential tools in quantum physics, communication engineering, and many other areas.

We formulate the questions and ideas as follows. (The proofs of the statements are discussed below.)

(i) For any n, the set of functions $\{1, \cos x, \sin x, \cos 2x, \sin 2x, \ldots, \cos nx, \sin nx\}$ is an orthogonal set with respect to the inner product

$$\langle f, g \rangle = \int_{-\pi}^{\pi} f(x)g(x)\, dx$$

The set is therefore an orthogonal basis for the subspace of $CP_{2\pi}$ that it spans. This subspace will be denoted $CP_{2\pi,n}$.

(ii) Given an arbitrary function f in $CP_{2\pi}$, how well can it be approximated by a function in $CP_{2\pi,n}$? We expect from our experience with distance and subspaces that the closest approximation to f in $CP_{2\pi,n}$ is $\text{proj}_{CP_{2\pi,n}}(f)$. The coefficients for Fourier's representation of f by a linear combination of $\{1, \cos x, \sin x, \ldots, \cos nx, \sin nx, \ldots\}$, called **Fourier coefficients**, are found by considering this projection.

(iii) We hope that the approximation improves as n gets larger. Since the distance from f to the n-th approximation $\text{proj}_{CP_{2\pi,n}}(f)$ is $\|\text{perp}_{CP_{2\pi,n}} f\|$, to test if the approximation improves, we must examine whether $\|\text{perp}_{CP_{2\pi,n}}(f)\| \to 0$ as $n \to \infty$.

Let us consider these statements in more detail.

(i) **The orthogonality of constants, sines, and cosines with respect to the inner product** $\langle f, g\rangle = \int_{-\pi}^{\pi} f(x)g(x)\,dx$

These results follow by standard trigonometric integrals and trigonometric identities:

$$\int_{-\pi}^{\pi} \sin nx\,dx = -\frac{1}{n}\cos nx\Big|_{-\pi}^{\pi} = 0$$

$$\int_{-\pi}^{\pi} \cos nx\,dx = \frac{1}{n}\sin nx\Big|_{-\pi}^{\pi} = 0$$

$$\int_{-\pi}^{\pi} \cos mx \sin nx\,dx = \int_{-\pi}^{\pi} \frac{1}{2}(\sin(m+n)x - \sin(m-n)x)\,dx = 0$$

and for $m \neq n$,

$$\int_{-\pi}^{\pi} \cos mx \cos nx\,dx = \int_{-\pi}^{\pi} \frac{1}{2}(\cos(m+n)x + \cos(m-n)x)\,dx = 0$$

$$\int_{-\pi}^{\pi} \sin mx \sin nx\,dx = \int_{-\pi}^{\pi} \frac{1}{2}(\cos(m-n)x - \cos(m+n)x)\,dx = 0$$

Hence, the set $\{1, \cos x, \sin x, \dots, \cos nx, \sin nx\}$ is orthogonal. To use this as a basis for projection arguments, it is necessary to calculate $\|1\|^2$, $\|\cos mx\|^2$, and $\|\sin mx\|^2$:

$$\|1\|^2 = \int_{-\pi}^{\pi} 1\,dx = 2\pi$$

$$\|\cos mx\|^2 = \int_{-\pi}^{\pi} \cos^2 mx\,dx = \int_{-\pi}^{\pi} \frac{1}{2}(1 + \cos 2mx)\,dx = \pi$$

$$\|\sin mx\|^2 = \int_{-\pi}^{\pi} \sin^2 mx\,dx = \int_{-\pi}^{\pi} \frac{1}{2}(1 - \cos 2mx)\,dx = \pi$$

(ii) **The Fourier coefficients of f as coordinates of a projection with respect to the orthogonal basis for $CP_{2\pi,n}$**

The procedure for finding the closest approximation $\text{proj}_{CP_{2\pi,n}}(f)$ in $CP_{2\pi,n}$ to an arbitrary function f in $CP_{2\pi}$ is parallel to the procedure in Sections 7.2 and 7.4. That is, we use the projection formula, given an orthogonal basis $\{\vec{v}_1, \dots, \vec{v}_n\}$ for a subspace $\mathbb{S}$:

$$\text{proj}_{\mathbb{S}}(\vec{x}) = \frac{\langle \vec{v}_1, \vec{x}\rangle}{\|\vec{v}_1\|^2}\vec{v}_1 + \cdots + \frac{\langle \vec{v}_n, \vec{x}\rangle}{\|\vec{v}_n\|^2}\vec{v}_n$$

There is a standard way to label the coefficients of this linear combination:

$$\begin{aligned}\text{proj}_{CP_{2\pi,n}}(f) = &\frac{a_0}{2}1 + a_1\cos x + a_2\cos 2x + \cdots + a_n\cos nx\\ &+ b_1\sin x + b_2\sin 2x + \cdots + b_n \sin nx\end{aligned}$$

The factor $\frac{1}{2}$ in the coefficient of 1 appears here because $\|1\|^2$ is equal to 2π, while the other basis vectors have length squared equal to π. Thus, we have

$$a_0 = \frac{1}{\pi}\int_{-\pi}^{\pi} f(x)\,dx$$

$$a_m = \frac{\langle \cos mx, f\rangle}{\|\cos mx\|^2} = \frac{1}{\pi}\int_{-\pi}^{\pi} f(x)\cos mx\,dx, \quad 1 \le m \le n$$

$$b_m = \frac{\langle \sin mx, f\rangle}{\|\sin mx\|^2} = \frac{1}{\pi}\int_{-\pi}^{\pi} f(x)\sin mx\,dx, \quad 1 \le m \le n$$

(iii) **Is** $\text{proj}_{CP_{2\pi,n}}(f)$ **equal to** f **in the limit as** $n \to \infty$**?**

As $n \to \infty$, the sum becomes an infinite series called the **Fourier series** for f. The question being asked is a question about the convergence of series—and in fact, about series of functions. Such questions are raised in calculus (or analysis) and are beyond the scope of this book. (The short answer is "yes, the series converges to f provided that f is continuous." The problem becomes more complicated if f is allowed to be piecewise continuous.) Questions about convergence are important in physical and engineering applications.

EXAMPLE 7.5.1

Determine $\text{proj}_{CP_{2\pi,3}}(f)$ for the function $f(x)$ defined by $f(x) = |x|$ if $-\pi \le x \le \pi$ and $f(x + 2\pi) = f(x)$ for all x.

Solution: We have

$$
\begin{aligned}
a_0 &= \frac{1}{\pi}\int_{-\pi}^{\pi} |x|\,dx = \pi \\
a_1 &= \frac{1}{\pi}\int_{-\pi}^{\pi} |x|\cos x\,dx = -\frac{4}{\pi} \\
a_2 &= \frac{1}{\pi}\int_{-\pi}^{\pi} |x|\cos 2x\,dx = 0 \\
a_3 &= \frac{1}{\pi}\int_{-\pi}^{\pi} |x|\cos 3x\,dx = -\frac{4}{9\pi} \\
b_1 &= \frac{1}{\pi}\int_{-\pi}^{\pi} |x|\sin x\,dx = 0 \\
b_2 &= \frac{1}{\pi}\int_{-\pi}^{\pi} |x|\sin 2x\,dx = 0 \\
b_3 &= \frac{1}{\pi}\int_{-\pi}^{\pi} |x|\sin 3x\,dx = 0
\end{aligned}
$$

Hence,

$$\text{proj}_{CP_{2\pi,3}}(f) = \frac{\pi}{2} - \frac{4}{\pi}\cos x - \frac{4}{9\pi}\cos 3x$$

The results are shown below.

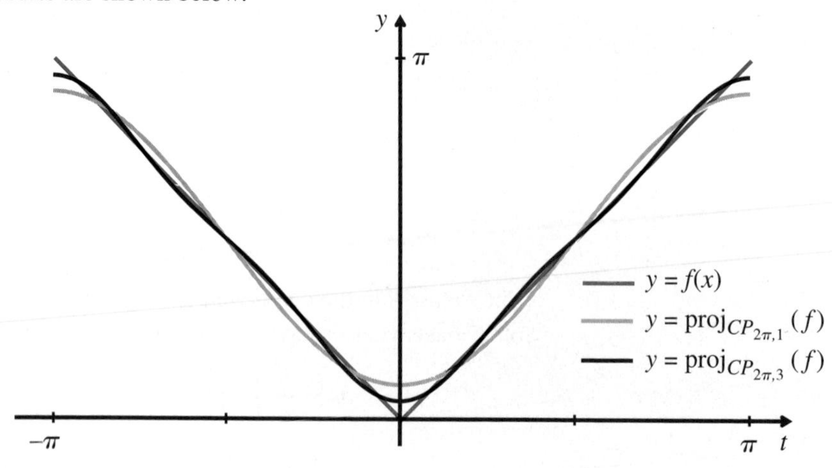

EXAMPLE 7.5.2 Determine $\text{proj}_{CP_{2\pi,3}}(f)$ for the function $f(x)$ defined by

$$f(x) = \begin{cases} -\pi - x & \text{if } -\pi \leq x \leq -\pi/2 \\ x & \text{if } -\pi/2 < x \leq \pi/2 \\ \pi - x & \text{if } \pi/2 < x \leq \pi \end{cases}$$

Solution: We have

$$\begin{aligned} a_0 &= \frac{1}{\pi}\int_{-\pi}^{\pi} f\,dx = 0 \\ a_1 &= \frac{1}{\pi}\int_{-\pi}^{\pi} f\cos x\,dx = 0 \\ a_2 &= \frac{1}{\pi}\int_{-\pi}^{\pi} f\cos 2x\,dx = 0 \\ a_3 &= \frac{1}{\pi}\int_{-\pi}^{\pi} f\cos 3x\,dx = 0 \\ b_1 &= \frac{1}{\pi}\int_{-\pi}^{\pi} f\sin x\,dx = \frac{4}{\pi} \\ b_2 &= \frac{1}{\pi}\int_{-\pi}^{\pi} f\sin 2x\,dx = 0 \\ b_3 &= \frac{1}{\pi}\int_{-\pi}^{\pi} f\sin 3x\,dx = -\frac{4}{9\pi} \end{aligned}$$

Hence,

$$\text{proj}_{CP_{2\pi,3}}(f) = \frac{4}{\pi}\sin x - \frac{4}{9\pi}\sin 3x$$

The results are shown below.

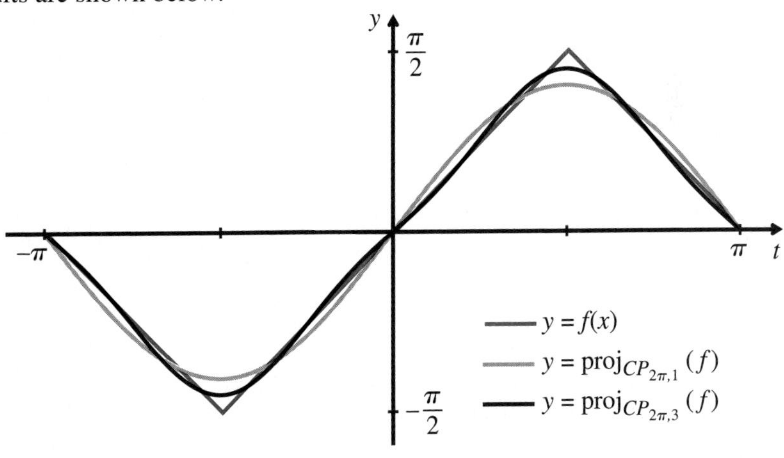

PROBLEMS 7.5

Practice Problems

For Problems A1–A3, use a computer to calculate $\text{proj}_{CP_{2\pi,n}}(f)$ for $n = 3, 7$, and, 11. Graph the function f and each of the projections on the same plot.

A1 $f(x) = x^2, -\pi \leq x \leq \pi$

A2 $f(x) = e^x, -\pi \leq x \leq \pi$

A3 $f(x) = \begin{cases} 0 & \text{if } -\pi \leq x \leq 0 \\ 1 & \text{if } 0 < x \leq \pi \end{cases}$

For Problems A4–A10, use a computer to calculate $\text{proj}_{CP_{2\pi,3}}(f)$.

A4 $f(x) = \sin^2 x, -\pi \leq x \leq \pi$

A5 $f(x) = 1 - x + x^3, -\pi \leq x \leq \pi$

A6 $f(x) = 1 + x^2, -\pi \leq x \leq \pi$

A7 $f(x) = \pi - x, -\pi \leq x \leq \pi$

A8 $f(x) = \begin{cases} 0 & \text{if } -\pi \leq x \leq 0 \\ x & \text{if } 0 < x \leq \pi \end{cases}$

A9 $f(x) = \begin{cases} -1 & \text{if } -\pi \leq x \leq 0 \\ 1 & \text{if } 0 < x \leq \pi \end{cases}$

A10 $f(x) = \begin{cases} 0 & \text{if } -\pi \leq x \leq -1 \\ 1 & \text{if } -1 \leq x \leq 1 \\ 0 & \text{if } 1 < x \leq \pi \end{cases}$

Homework Problems

For Problems B1–B4, calculate $\text{proj}_{CP_{2\pi,2}}(f)$.

B1 $f(x) = \cos^2 x, -\pi \leq x \leq \pi$

B2 $f(x) = -x, -\pi \leq x \leq \pi$

B3 $f(x) = \sin^3 x, -\pi \leq x \leq \pi$

B4 $f(x) = \begin{cases} 0 & \text{if } -\pi \leq x \leq 0 \\ x & \text{if } 0 < x \leq \pi \end{cases}$

CHAPTER REVIEW

Suggestions for Student Review

1 What is meant by *an orthogonal set of vectors* in $\mathbb{R}^n$? What is the difference between an orthogonal basis and an orthonormal basis? (Section 7.1)

2 Illustrate with a sketch why it is easy to write a vector $\vec{x} \in \mathbb{R}^2$ as a linear combination of an orthogonal basis $\{\vec{v}_1, \vec{v}_2\}$ for $\mathbb{R}^2$. (Section 7.1)

3 Define an orthogonal matrix and list the properties of an orthogonal matrix. (Section 7.1)

4 What are the essential properties of a projection onto a subspace of $\mathbb{R}^n$? How do you calculate a projection onto a subspace? What is the relationship between the formula for a projection and writing a vector as a linear combination of an orthogonal basis? (Section 7.2)

5 State the Approximation Theorem. Illustrate with a sketch in $\mathbb{R}^2$. (Section 7.2)

6 Does every subspace of $\mathbb{R}^n$ have an orthonormal basis? What about the zero subspace? How do you find an orthonormal basis? Describe the Gram-Schmidt Procedure. (Section 7.2)

7 Explain how to use the method of least squares and its relationship to the Fundamental Theorem of Linear Algebra. (Section 7.3)

8 What is an overdetermined system? How do you best approximate a solution of an inconsistent overdetermined system? What answer would our algorithm give if the overdetermined system was consistent? (Section 7.3)

9 What are the essential properties of an inner product? Give an example of an inner product on $P_2(\mathbb{R})$. Create a few different inner products for $M_{2\times2}(\mathbb{R})$. (Section 7.4)

10 Let $\mathbb{V}$ be a vector space. If $\vec{v}, \vec{w} \in \mathbb{V}$ are orthogonal with respect to one inner product on $\mathbb{V}$, will $\vec{v}$ and $\vec{w}$ be orthogonal with respect to all inner products on $\mathbb{V}$? Give some examples. (Section 7.4)

Chapter Quiz

For Problems E1–E3, determine whether the set is orthogonal or orthonormal. Show how you decide.

E1 $\left\{\frac{1}{3}\begin{bmatrix}1\\0\\1\\1\end{bmatrix}, \frac{1}{3}\begin{bmatrix}0\\1\\1\\-1\end{bmatrix}, \frac{1}{2}\begin{bmatrix}0\\1\\-1\\0\end{bmatrix}\right\}$

E2 $\left\{\frac{1}{\sqrt{3}}\begin{bmatrix}1\\0\\1\\1\end{bmatrix}, \frac{1}{\sqrt{5}}\begin{bmatrix}0\\0\\1\\-2\end{bmatrix}\right\}$

E3 $\left\{\frac{1}{\sqrt{3}}\begin{bmatrix}1\\0\\1\\1\end{bmatrix}, \frac{1}{\sqrt{3}}\begin{bmatrix}1\\1\\-1\\0\end{bmatrix}, \frac{1}{\sqrt{3}}\begin{bmatrix}0\\-1\\-1\\1\end{bmatrix}\right\}$

E4 Find an orthogonal basis for the orthogonal complement of the subspace $\mathbb{S} = \operatorname{Span}\left\{\begin{bmatrix}2\\1\\3\end{bmatrix}\right\}$ of $\mathbb{R}^3$.

E5 Let $\vec{x} = \begin{bmatrix}1\\3\\-1\\2\end{bmatrix}$. Given that $\mathcal{B} = \left\{\begin{bmatrix}1\\0\\1\\1\end{bmatrix}, \begin{bmatrix}1\\1\\-1\\0\end{bmatrix}, \begin{bmatrix}-1\\1\\0\\1\end{bmatrix}\right\}$ is an orthogonal basis for the subspace $\mathbb{S} = \operatorname{Span}\mathcal{B}$ of $\mathbb{R}^4$, use Theorem 7.1.2 to write $\vec{x}$ as a linear combination of the vectors in $\mathcal{B}$.

E6 Let $\vec{x} = \begin{bmatrix}1\\3\\-1\end{bmatrix}$ and let $\mathbb{S} = \operatorname{Span}\left\{\begin{bmatrix}1\\1\\3\end{bmatrix}, \begin{bmatrix}2\\2\\-3\end{bmatrix}, \begin{bmatrix}3\\3\\-4\end{bmatrix}, \begin{bmatrix}3\\3\\5\end{bmatrix}\right\}$. Calculate $\operatorname{proj}_{\mathbb{S}}(\vec{x})$ and $\operatorname{perp}_{\mathbb{S}}(\vec{x})$.

For Problems E7 and E8, determine whether the function $\langle\, ,\rangle$ defines an inner product on $M_{2\times 2}(\mathbb{R})$. Explain how you decide in each case.

E7 $\langle A, B\rangle = \det(AB)$

E8 $\langle A, B\rangle = a_{11}b_{11} + 2a_{12}b_{12} + 2a_{21}b_{21} + a_{22}b_{22}$

E9 Let $\mathbb{S}$ be the subspace of $\mathbb{R}^4$ defined by

$$\mathbb{S} = \operatorname{Span}\left\{\begin{bmatrix}1\\0\\1\\0\end{bmatrix}, \begin{bmatrix}1\\1\\1\\1\end{bmatrix}, \begin{bmatrix}1\\3\\3\\1\end{bmatrix}\right\}$$

(a) Apply the Gram-Schmidt Procedure to the given spanning set to produce an orthogonal basis for $\mathbb{S}$.

(b) Determine the point in $\mathbb{S}$ closest to $\vec{x} = \begin{bmatrix}1\\-2\\-1\\1\end{bmatrix}$.

E10 Create an orthogonal basis for $\mathbb{R}^3$ that contains the vector $\vec{w}_1 = \begin{bmatrix}1\\2\\2\end{bmatrix}$.

For Problems E11 and E12, find the best-fitting equation of the given form for the given set of data points.

E11 $y = a + bt$: $(-1,2), (0,3), (1,5), (1,6)$

E12 $y = a + ct^2$: $(-1,4), (0,1), (1,5)$

For Problems E13 and E14, determine the vector $\vec{x}$ that minimizes $\|A\vec{x} - \vec{b}\|$.

E13
$$\begin{aligned} x_1 + 2x_2 &= 3 \\ x_1 + x_2 &= 1 \\ x_2 &= -1 \end{aligned}$$

E14
$$\begin{aligned} x_1 - 2x_2 &= 1 \\ 2x_1 + x_2 &= 2 \\ 3x_1 + 2x_2 &= 2 \end{aligned}$$

For Problems E15–E20, determine whether the statement is true or false. Justify your answer.

E15 If $P, R \in M_{n\times n}(\mathbb{R})$ are orthogonal, then so is PR.

E16 If $P \in M_{m\times n}(\mathbb{R})$ has orthonormal columns, then P is orthogonal.

E17 If $P \in M_{m\times n}(\mathbb{R})$ has orthonormal columns, then $P^TP = I$.

E18 If $\{\vec{v}_1, \dots, \vec{v}_k\}$ is a basis for a subspace $\mathbb{S}$ of $\mathbb{R}^n$, then for any $\vec{x} \in \mathbb{R}^n$ we have

$$\operatorname{proj}_{\mathbb{S}}(\vec{x}) = \frac{\langle \vec{x}, \vec{v}_1\rangle}{\|\vec{v}_1\|^2}\vec{v}_1 + \cdots + \frac{\langle \vec{x}, \vec{v}_k\rangle}{\|\vec{v}_k\|^2}\vec{v}_k$$

E19 If $\mathbb{S}$ is a subspace for $\mathbb{R}^n$ and $\vec{x} \in \mathbb{S}$, then $\operatorname{perp}_{\mathbb{S}}(\vec{x}) = \vec{0}$.

E20 Let $\mathbb{S}$ be a subspace of $\mathbb{R}^n$ and let $\vec{x} \in \mathbb{R}^n$. Then, there is a unique vector $\vec{y} \in \mathbb{S}$ that has the property

$$\|\vec{x} - \vec{y}\| < \|\vec{x} - \vec{v}\|$$

for all $\vec{v} \in \mathbb{S}, \vec{v} \neq \vec{y}$.

E21 On $P_2(\mathbb{R})$ define the inner product $\langle \mathbf{p}, \mathbf{q}\rangle = \mathbf{p}(-1)\mathbf{q}(-1) + \mathbf{p}(0)\mathbf{q}(0) + \mathbf{p}(1)\mathbf{q}(1)$ and let $\mathbb{S} = \operatorname{Span}\{1, x - x^2\}$.

(a) Find an orthogonal basis for $\mathbb{S}$.

(b) Determine $\operatorname{proj}_{\mathbb{S}}(1 + x + x^2)$ and $\operatorname{perp}_{\mathbb{S}}(1 + x + x^2)$.

Further Problems

These exercises are intended to be challenging.

F1 Let $\mathbb{S}$ be a finite-dimensional subspace of an inner product space $\mathbb{V}$. Prove that $\text{proj}_{\mathbb{S}}$ is independent of the orthogonal basis chosen for $\mathbb{S}$. That is, for any $\mathbf{v} \in \mathbb{V}$, $\text{proj}_{\mathbb{S}}(\mathbf{v})$ will always be the same vector even if we change which orthogonal basis is being used for $\mathbb{S}$.

F2 **(Isometries of $\mathbb{R}^3$)**

(a) A linear mapping is an **isometry** of $\mathbb{R}^3$ if

$$\|L(\vec{x})\| = \|\vec{x}\|$$

for every $\vec{x} \in \mathbb{R}^3$. Prove that an isometry preserves dot products.

(b) Show that a linear mapping L is an isometry if and only if the standard matrix of L is orthogonal. (Hint: see Chapter 3 Problem F7 and Section 7.1 Problem C3.)

(c) Explain why an isometry of $\mathbb{R}^3$ must have 1 or 3 real eigenvalues, counting multiplicity. Based on Section 7.1 Problem C3 (b), the eigenvalues must be ± 1.

(d) Let A be the standard matrix of an isometry L. Suppose that 1 is an eigenvalue of A with eigenvector $\vec{u}$. Let $\vec{v}$ and $\vec{w}$ be vectors such that $\{\vec{u}, \vec{v}, \vec{w}\}$ is an orthonormal basis for $\mathbb{R}^3$ and let $P = \begin{bmatrix} \vec{u} & \vec{v} & \vec{w} \end{bmatrix}$. Show that

$$P^T AP = \begin{bmatrix} 1 & 0_{12} \\ 0_{21} & A^* \end{bmatrix}$$

where the right-hand side is a partitioned matrix, with 0_{ij} being the $i \times j$ zero matrix, and with A^* being a 2×2 orthogonal matrix. Moreover, show that the eigenvalues of A are 1 and the eigenvalues of A^*.

Note that an analogous form can be obtained for $\mathrm{P^TAP}$ *in the case where one eigenvalue is −1.*

(e) Use Chapter 3 Problem F7 to analyze the A^* of part (d) and explain why every isometry of $\mathbb{R}^3$ is the identity mapping, a reflection, a composition of reflections, a rotation, or a composition of a reflection and a rotation.

F3 A linear mapping $L : \mathbb{R}^n \to \mathbb{R}^n$ is called an **involution** if $L \circ L = \text{Id}$. In terms of its standard matrix, this means that $A^2 = I$. Prove that any two of the following imply the third.

(a) A is the matrix of an involution.

(b) A is symmetric.

(c) A is an isometry.

F4 The sum $\mathbb{S} + \mathbb{T}$ of subspaces of a finite dimensional vector space $\mathbb{V}$ is defined in the Chapter 4 Further Problems. Prove that $(\mathbb{S} + \mathbb{T})^\perp = \mathbb{S}^\perp \cap \mathbb{T}^\perp$.

F5 Finding a sequence of approximations to some vector (or function) $\mathbf{v}$ in a possibly infinite-dimensional inner product space $\mathbb{V}$ can often be described by requiring the i-th approximation to be the closest vector $\mathbf{v}$ in some finite-dimensional subspace $\mathbb{S}_i$ of $\mathbb{V}$, where the subspaces are required to satisfy

$$\mathbb{S}_1 \subset \mathbb{S}_2 \subset \cdots \subset \mathbb{S}_i \subset \cdots \subset \mathbb{V}$$

The i-th approximation is then $\text{proj}_{\mathbb{S}_i}(\mathbf{v})$. Prove that the approximations improve as i increases in the sense that

$$\|\mathbf{v} - \text{proj}_{\mathbb{S}_{i+1}}(\mathbf{v})\| \leq \|\mathbf{v} - \text{proj}_{\mathbb{S}_i}(\mathbf{v})\|$$

F6 ***QR*-factorization.** Suppose that A is an invertible $n \times n$ matrix. Prove that A can be written as the product of an orthogonal matrix Q and an upper triangular matrix R: $A = QR$.

(Hint: apply the Gram-Schmidt Procedure to the columns of A, starting with the first column.)

Note that this QR*-factorization is important in a numerical procedure for determining eigenvalues of symmetric matrices.*

CHAPTER 8

Symmetric Matrices and Quadratic Forms

CHAPTER OUTLINE

Symmetric matrices and quadratic forms arise naturally in many physical applications. For example, the strain matrix describing the deformation of a solid and the inertia tensor of a rotating body are symmetric (Section 8.4). We have also seen that the matrix of a projection is symmetric since a real inner product is symmetric. We now use our work with diagonalization and inner products to explore the theory of symmetric matrices and quadratic forms.

In this chapter we will make frequent use of the formula $\vec{x}^T\vec{y} = \vec{x} \cdot \vec{y}$ (see the bottom of page 160).

8.1 Diagonalization of Symmetric Matrices

Definition
Symmetric Matrix

A matrix $A \in M_{n\times n}(\mathbb{R})$ is said to be **symmetric** if $A^T = A$ or, equivalently, if $a_{ij} = a_{ji}$ for all $1 \leq i, j \leq n$.

EXERCISE 8.1.1

Determine whether $A = \begin{bmatrix} 3 & 1 \\ 1 & -5 \end{bmatrix}$ and/or $B = \begin{bmatrix} 2 & -3 \\ 3 & 0 \end{bmatrix}$ is symmetric.

Symmetric matrices have the following useful property.

Theorem 8.1.1

A matrix $A \in M_{n\times n}(\mathbb{R})$ is symmetric if and only if $\vec{x}\cdot(A\vec{y}) = (A\vec{x})\cdot\vec{y}$ for all $\vec{x}, \vec{y} \in \mathbb{R}^n$.

You are asked to prove Theorem 8.1.1 in Problem C2.

Our goal in this section is to look at the diagonalization of symmetric matrices. We begin with an example.

EXAMPLE 8.1.1

Diagonalize the symmetric matrix $A = \begin{bmatrix} 0 & 1 \\ 1 & -2 \end{bmatrix}$.

Solution: We have

$$C(\lambda) = \det(A - \lambda I) = \begin{vmatrix} 0 - \lambda & 1 \\ 1 & -2 - \lambda \end{vmatrix} = \lambda^2 + 2\lambda - 1$$

Using the quadratic formula, we find that the roots of the characteristic polynomial are $\lambda_1 = -1 + \sqrt{2}$ and $\lambda_2 = -1 - \sqrt{2}$. Thus, the resulting diagonal matrix is

$$D = \begin{bmatrix} -1 + \sqrt{2} & 0 \\ 0 & -1 - \sqrt{2} \end{bmatrix}$$

For $\lambda_1 = -1 + \sqrt{2}$, we have

$$A - \lambda_1 I = \begin{bmatrix} 1 - \sqrt{2} & 1 \\ 1 & -1 - \sqrt{2} \end{bmatrix} \sim \begin{bmatrix} 1 & -1 - \sqrt{2} \\ 0 & 0 \end{bmatrix}$$

Thus, a basis for the eigenspace is $\left\{ \begin{bmatrix} 1 + \sqrt{2} \\ 1 \end{bmatrix} \right\}$.

Similarly, for $\lambda_2 = -1 - \sqrt{2}$, we have

$$A - \lambda_2 I = \begin{bmatrix} 1 + \sqrt{2} & 1 \\ 1 & -1 + \sqrt{2} \end{bmatrix} \sim \begin{bmatrix} 1 & -1 + \sqrt{2} \\ 0 & 0 \end{bmatrix}$$

Thus, a basis for the eigenspace is $\left\{ \begin{bmatrix} 1 - \sqrt{2} \\ 1 \end{bmatrix} \right\}$.

Hence, A is diagonalized by $P = \begin{bmatrix} 1 + \sqrt{2} & 1 - \sqrt{2} \\ 1 & 1 \end{bmatrix}$ to $D = \begin{bmatrix} -1 + \sqrt{2} & 0 \\ 0 & -1 - \sqrt{2} \end{bmatrix}$.

Observe that the eigenvectors of the matrix A in Example 8.1.1 are orthogonal:

$$\begin{bmatrix} 1 + \sqrt{2} \\ 1 \end{bmatrix} \cdot \begin{bmatrix} 1 - \sqrt{2} \\ 1 \end{bmatrix} = (1 + \sqrt{2})(1 - \sqrt{2}) + 1(1) = 1 - 2 + 1 = 0$$

Using Theorem 8.1.1, we can prove that eigenvectors of a symmetric matrix corresponding to different eigenvalues are always orthogonal.

Theorem 8.1.2

If $\vec{v}_1, \vec{v}_2$ are eigenvectors of a symmetric matrix $A \in M_{n\times n}(\mathbb{R})$ corresponding to distinct eigenvalues λ_1, λ_2, then $\vec{v}_1$ is orthogonal to $\vec{v}_2$.

Proof: Assume that $A\vec{v}_1 = \lambda_1\vec{v}_1$ and $A\vec{v}_2 = \lambda_2\vec{v}_2$, $\lambda_1 \neq \lambda_2$. Theorem 8.1.1 gives

$$\lambda_1(\vec{v}_1 \cdot \vec{v}_2) = (\lambda_1\vec{v}_1) \cdot \vec{v}_2 = (A\vec{v}_1) \cdot \vec{v}_2 = \vec{v}_1 \cdot (A\vec{v}_2) = \vec{v}_1 \cdot (\lambda_2\vec{v}_2) = \lambda_2(\vec{v}_1 \cdot \vec{v}_2)$$

Hence, $(\lambda_1 - \lambda_2)(\vec{v}_1 \cdot \vec{v}_2) = 0$. But, $\lambda_1 \neq \lambda_2$, so $\vec{v}_1 \cdot \vec{v}_2 = 0$ as required. ■

We know that any non-zero scalar multiple of an eigenvector $\vec{v}$ of a matrix A corresponding to λ is also an eigenvector of A corresponding to λ. Consequently, by Theorem 8.1.2, if a symmetric matrix with all distinct eigenvalues is diagonalizable, then it can be diagonalized by an orthogonal matrix P (by normalizing the columns). To extend what we did in Chapter 6, we make the following definitions.

Definition
Orthogonally Similar

Two matrices $A, B \in M_{n\times n}(\mathbb{R})$ are said to be **orthogonally similar** if there exists an orthogonal matrix P such that

$$P^T AP = B$$

Remark

Since P is orthogonal, we have that $P^T = P^{-1}$ and hence if A and B are orthogonally similar, then they are similar. Therefore, all the properties of similar matrices still hold. In particular, if A and B are orthogonally similar, then $\operatorname{rank} A = \operatorname{rank} B$, $\operatorname{tr} A = \operatorname{tr} B$, $\det A = \det B$, and A and B have the same eigenvalues.

Definition
Orthogonally Diagonalizable

A matrix $A \in M_{n\times n}(\mathbb{R})$ is said to be **orthogonally diagonalizable** if there exists an orthogonal matrix P and diagonal matrix D such that

$$P^T AP = D$$

that is, if A is orthogonally similar to a diagonal matrix.

EXAMPLE 8.1.2

Find an orthogonal matrix that diagonalizes $A = \begin{bmatrix} 1 & -2 \\ -2 & 1 \end{bmatrix}$.

Solution: We have

$$C(\lambda) = \begin{vmatrix} 1-\lambda & -2 \\ -2 & 1-\lambda \end{vmatrix} = (\lambda - 3)(\lambda + 1)$$

So, the eigenvalues are $\lambda_1 = 3$, and $\lambda_2 = -1$.
For $\lambda_1 = 3$, we get

$$A - \lambda_1 I = \begin{bmatrix} -2 & -2 \\ -2 & -2 \end{bmatrix} \sim \begin{bmatrix} 1 & 1 \\ 0 & 0 \end{bmatrix}$$

Thus, a basis for the eigenspace is $\left\{ \begin{bmatrix} -1 \\ 1 \end{bmatrix} \right\}$.

For $\lambda_2 = -1$, we get

$$A - \lambda_2 I = \begin{bmatrix} 2 & -2 \\ -2 & 2 \end{bmatrix} \sim \begin{bmatrix} 1 & -1 \\ 0 & 0 \end{bmatrix}$$

Thus, a basis for the eigenspace is $\left\{ \begin{bmatrix} 1 \\ 1 \end{bmatrix} \right\}$.

As foretold by Theorem 8.1.2, the vectors $\vec{v}_1 = \begin{bmatrix} -1 \\ 1 \end{bmatrix}$ and $\vec{v}_2 = \begin{bmatrix} 1 \\ 1 \end{bmatrix}$ form an orthogonal set. Hence, if we normalize them, we find that A is diagonalized by the orthogonal matrix

$$P = \begin{bmatrix} -1/\sqrt{2} & 1/\sqrt{2} \\ 1/\sqrt{2} & 1/\sqrt{2} \end{bmatrix}$$

EXAMPLE 8.1.3

Orthogonally diagonalize the symmetric matrix $A = \begin{bmatrix} 5 & -4 & -2 \\ -4 & 5 & -2 \\ -2 & -2 & 8 \end{bmatrix}$.

Solution: We have

$$C(\lambda) = \begin{vmatrix} 5-\lambda & -4 & -2 \\ -4 & 5-\lambda & -2 \\ -2 & -2 & 8-\lambda \end{vmatrix} = -\lambda(\lambda-9)^2$$

So, the eigenvalues are $\lambda_1 = 9$ and $\lambda_2 = 0$.
For $\lambda_1 = 9$, we get

$$A - \lambda_1 I = \begin{bmatrix} -4 & -4 & -2 \\ -4 & -4 & -2 \\ -2 & -2 & -1 \end{bmatrix} \sim \begin{bmatrix} 1 & 1 & 1/2 \\ 0 & 0 & 0 \\ 0 & 0 & 0 \end{bmatrix}$$

Thus, a basis for the eigenspace of λ_1 is $\{\vec{w}_1, \vec{w}_2\} = \left\{ \begin{bmatrix} -1 \\ 1 \\ 0 \end{bmatrix}, \begin{bmatrix} -1 \\ 0 \\ 2 \end{bmatrix} \right\}$. However, observe that $\vec{w}_1$ and $\vec{w}_2$ are not orthogonal. Since we want an orthonormal basis of eigenvectors of A, we need to find an orthonormal basis for the eigenspace of λ_1. We can do this by applying the Gram-Schmidt Procedure to $\{\vec{w}_1, \vec{w}_2\}$.
Pick $\vec{v}_1 = \vec{w}_1$ and let $\mathbb{S}_1 = \operatorname{Span}\{\vec{v}_1\}$. We find that

$$\vec{v}_2 = \operatorname{perp}_{\mathbb{S}_1}(\vec{w}_2) = \vec{w}_2 - \frac{\vec{v}_1 \cdot \vec{w}_2}{\|\vec{v}_1\|^2}\vec{v}_1 = \frac{1}{2}\begin{bmatrix} -1 \\ -1 \\ 4 \end{bmatrix}$$

Then, $\{\vec{v}_1, \vec{v}_2\}$ is an orthogonal basis for the eigenspace of λ_1.
For $\lambda_2 = 0$, we get

$$A - \lambda_2 I = \begin{bmatrix} 5 & -4 & -2 \\ -4 & 5 & -2 \\ -2 & -2 & 8 \end{bmatrix} \sim \begin{bmatrix} 1 & 0 & -2 \\ 0 & 1 & -2 \\ 0 & 0 & 0 \end{bmatrix}$$

Thus, a basis for the eigenspace of λ_2 is $\{\vec{v}_3\} = \left\{ \begin{bmatrix} 2 \\ 2 \\ 1 \end{bmatrix} \right\}$.

Normalizing $\vec{v}_1, \vec{v}_2$, and $\vec{v}_3$, we find that A is diagonalized by the orthogonal matrix

$$P = \begin{bmatrix} -1/\sqrt{2} & -1/\sqrt{18} & 2/3 \\ 1/\sqrt{2} & -1/\sqrt{18} & 2/3 \\ 0 & 4/\sqrt{18} & 1/3 \end{bmatrix} \quad \text{to} \quad D = \begin{bmatrix} 9 & 0 & 0 \\ 0 & 9 & 0 \\ 0 & 0 & 0 \end{bmatrix}$$

EXERCISE 8.1.2

Orthogonally diagonalize the symmetric matrix $A = \begin{bmatrix} 2 & -1 & -1 \\ -1 & 2 & -1 \\ -1 & -1 & 2 \end{bmatrix}$.

The Principal Axis Theorem

To prove that every symmetric matrix is orthogonally diagonalizable, we use the following two results.

Theorem 8.1.3 If $A \in M_{n\times n}(\mathbb{R})$ is a symmetric matrix, then all eigenvalues of A are real.

The proof of this theorem requires properties of complex numbers and hence is postponed until Chapter 9. See Theorem 9.5.1.

Theorem 8.1.4 Triangularization Theorem
If $A \in M_{n\times n}(\mathbb{R})$ has all real eigenvalues, then A is orthogonally similar to an upper triangular matrix T.

The proof is not helpful for our purposes and so is omitted.

Theorem 8.1.5 Principal Axis Theorem
A matrix $A \in M_{n\times n}(\mathbb{R})$ is symmetric if and only if it is orthogonally diagonalizable.

Proof: We will prove if A is symmetric, then A is orthogonally diagonalizable and leave the proof of the converse as Problem C3.

Assume A is symmetric. By Theorem 8.1.3 all eigenvalues of A are real. Therefore, we can apply the Triangularization Theorem to get that there exists an orthogonal matrix P such that $P^TAP = T$ is upper triangular. Since A is symmetric, we have that $A^T = A$ and hence

$$T^T = (P^TAP)^T = P^TA^T(P^T)^T = P^TAP = T$$

Therefore, T is also a symmetric matrix. But, if T is upper triangular, then T^T is lower triangular, and so T is both upper and lower triangular. Consequently, T is diagonal. Hence, we have that $P^TAP = T$ is diagonal, so A is orthogonally similar to a diagonal matrix. ■

Remarks

1. Note that Theorem 8.1.2 applies only to eigenvectors that correspond to different eigenvalues. As we saw in Example 8.1.3, eigenvectors that correspond to the same eigenvalue do not need to be orthogonal. Thus, as in Example 8.1.3, if an eigenvalue of a symmetric matrix has algebraic multiplicity greater than 1, it may be necessary to apply the Gram-Schmidt Procedure to find an orthogonal basis for its eigenspace.

2. The eigenvectors in an orthogonal matrix that diagonalizes a symmetric matrix A are called the **principal axes** for A. We will see why this definition makes sense in Section 8.3.

PROBLEMS 8.1

Practice Problems

For Problems A1–A4, decide whether the matrix is symmetric.

A1 $A = \begin{bmatrix} 0 & 2 \\ 2 & -1 \end{bmatrix}$

A2 $B = \begin{bmatrix} 0 & 0 \\ 0 & 0 \end{bmatrix}$

A3 $C = \begin{bmatrix} 1 & 2 & 1 \\ -2 & 1 & 2 \\ -1 & -2 & 1 \end{bmatrix}$

A4 $D = \begin{bmatrix} 0 & -1 & 1 \\ -1 & 0 & -1 \\ 1 & -1 & 0 \end{bmatrix}$

For Problems A5–A16, orthogonally diagonalize the matrix.

A5 $A = \begin{bmatrix} 1 & -3 \\ -3 & 1 \end{bmatrix}$

A6 $A = \begin{bmatrix} 5 & 3 \\ 3 & -3 \end{bmatrix}$

A7 $A = \begin{bmatrix} 5 & 2 \\ 2 & 2 \end{bmatrix}$

A8 $A = \begin{bmatrix} 4 & 2 \\ 2 & 1 \end{bmatrix}$

A9 $A = \begin{bmatrix} 0 & 1 & 1 \\ 1 & 0 & 1 \\ 1 & 1 & 0 \end{bmatrix}$

A10 $A = \begin{bmatrix} 1 & 0 & -2 \\ 0 & -1 & -2 \\ -2 & -2 & 0 \end{bmatrix}$

A11 $A = \begin{bmatrix} 1 & 8 & 4 \\ 8 & 1 & -4 \\ 4 & -4 & 7 \end{bmatrix}$

A12 $A = \begin{bmatrix} 1 & 2 & 1 \\ 2 & 1 & 1 \\ 1 & 1 & 2 \end{bmatrix}$

A13 $A = \begin{bmatrix} 0 & 1 & -1 \\ 1 & 0 & 1 \\ -1 & 1 & 0 \end{bmatrix}$

A14 $A = \begin{bmatrix} 1 & 0 & -1 \\ 0 & 1 & 1 \\ -1 & 1 & 2 \end{bmatrix}$

A15 $A = \begin{bmatrix} 1 & 2 & -4 \\ 2 & -2 & -2 \\ -4 & -2 & 1 \end{bmatrix}$

A16 $A = \begin{bmatrix} -2 & 2 & -1 \\ 2 & 1 & -2 \\ -1 & -2 & -2 \end{bmatrix}$

Homework Problems

For Problems B1–B4, decide whether the matrix is symmetric.

B1 $A = \begin{bmatrix} 0 & 1 \\ 0 & 1 \end{bmatrix}$

B2 $B = \begin{bmatrix} 1 & -2 \\ 2 & 1 \end{bmatrix}$

B3 $C = \begin{bmatrix} 3 & 2 & 3 \\ 2 & 0 & 2 \\ 1 & 0 & 1 \end{bmatrix}$

B4 $D = \begin{bmatrix} -1 & 4 & 1 \\ 4 & -2 & 3 \\ 1 & 3 & 5 \end{bmatrix}$

For Problems B5–B16, orthogonally diagonalize the matrix.

B5 $A = \begin{bmatrix} 2 & 2 \\ 2 & 2 \end{bmatrix}$

B6 $A = \begin{bmatrix} 4 & -2 \\ -2 & 7 \end{bmatrix}$

B7 $A = \begin{bmatrix} 4 & 3 \\ 3 & -4 \end{bmatrix}$

B8 $A = \begin{bmatrix} 1 & -2 \\ -2 & -2 \end{bmatrix}$

B9 $A = \begin{bmatrix} 2 & -2 & -5 \\ -2 & -5 & -2 \\ -5 & -2 & 2 \end{bmatrix}$

B10 $A = \begin{bmatrix} 1 & 0 & 2 \\ 0 & 1 & 4 \\ 2 & 4 & 2 \end{bmatrix}$

B11 $A = \begin{bmatrix} 1 & 1 & 1 \\ 1 & 1 & 1 \\ 1 & 1 & 1 \end{bmatrix}$

B12 $A = \begin{bmatrix} 3 & -2 & 4 \\ -2 & 6 & 2 \\ 4 & 2 & 3 \end{bmatrix}$

B13 $A = \begin{bmatrix} 2 & -4 & -4 \\ -4 & 2 & -4 \\ -4 & -4 & 2 \end{bmatrix}$

B14 $A = \begin{bmatrix} 5 & -1 & 1 \\ -1 & 5 & 1 \\ 1 & 1 & 5 \end{bmatrix}$

B15 $A = \begin{bmatrix} 0 & 2 & -1 \\ 2 & 3 & -2 \\ -1 & -2 & 0 \end{bmatrix}$

B16 $A = \begin{bmatrix} 5 & -4 & -2 \\ -4 & 5 & 2 \\ -2 & 2 & 7 \end{bmatrix}$

Conceptual Problems

C1 Let $A, B \in M_{n\times n}(\mathbb{R})$ be symmetric. Determine which of the following is also symmetric.

(a) $A + B$ (b) $A^T A$ (c) AB (d) A^2

C2 Prove Theorem 8.1.1.

C3 Show that if A is orthogonally diagonalizable, then A is symmetric.

C4 Prove that if A is an invertible symmetric matrix, then A^{-1} is orthogonally diagonalizable.

For Problems C5 and C6, find a 2×2 symmetric matrix with the given eigenvalues and corresponding eigenvectors.

C5 $\lambda_1 = 4, \vec{v}_1 = \begin{bmatrix} 1 \\ 1 \end{bmatrix}, \lambda_2 = 6, \vec{v}_2 = \begin{bmatrix} -1 \\ 1 \end{bmatrix}$.

C6 $\lambda_1 = -1, \vec{v}_1 = \begin{bmatrix} 2 \\ 1 \end{bmatrix}, \lambda_2 = 3, \vec{v}_2 = \begin{bmatrix} -1 \\ 2 \end{bmatrix}$.

For Problems C7–C11, determine whether the statement is true or false. Justify your answer.

C7 If the matrix P diagonalizes a symmetric matrix A, then P is orthogonal.

C8 Every orthogonal matrix is orthogonally diagonalizable.

C9 If A and B are orthogonally diagonalizable, then AB is orthogonally diagonalizable.

C10 If A is orthogonally similar to a symmetric matrix B, then A is orthogonally diagonalizable.

C11 Every eigenvalue of a symmetric matrix has its geometric multiplicity equal to its algebraic multiplicity.

8.2 Quadratic Forms

In Chapter 3, we saw the relationship between matrix mappings and linear mappings. We now explore the relationship between symmetric matrices and an important class of functions called **quadratic forms**, *which are not linear. Quadratic forms appear in geometry, statistics, calculus, topology, and many other disciplines. We shall see in the next section how quadratic forms and our special theory of diagonalization of symmetric matrices can be used to graph conic sections and quadric surfaces.*

Quadratic Forms

Consider the symmetric matrix $A = \begin{bmatrix} a & b/2 \\ b/2 & c \end{bmatrix}$. If $\vec{x} = \begin{bmatrix} x_1 \\ x_2 \end{bmatrix}$, then

$$\begin{aligned} \vec{x}^T A\vec{x} &= \begin{bmatrix} x_1 & x_2 \end{bmatrix} \begin{bmatrix} a & b/2 \\ b/2 & c \end{bmatrix} \begin{bmatrix} x_1 \\ x_2 \end{bmatrix} \\ &= \begin{bmatrix} x_1 & x_2 \end{bmatrix} \begin{bmatrix} ax_1 + bx_2/2 \\ bx_1/2 + cx_2 \end{bmatrix} \\ &= ax_1^2 + bx_1x_2 + cx_2^2 \end{aligned}$$

We call the expression $ax_1^2 + bx_1x_2 + cx_2^2$ a quadratic form on $\mathbb{R}^2$ (or in the variables x_1 and x_2). Thus, corresponding to every symmetric matrix A, there is a quadratic form

$$Q(\vec{x}) = \vec{x}^T A\vec{x} = ax_1^2 + bx_1x_2 + cx_2^2$$

On the other hand, given a quadratic form $Q(\vec{x}) = ax_1^2 + bx_1x_2 + cx_2^2$, we can reconstruct the symmetric matrix $A = \begin{bmatrix} a & b/2 \\ b/2 & c \end{bmatrix}$ by choosing $(A)_{11}$ to be the coefficient of x_1^2, $(A)_{12} = (A)_{21}$ to be half of the coefficient of x_1x_2, and $(A)_{22}$ to be the coefficient of x_2^2. We deal with the coefficient of x_1x_2 in this way to ensure that A is symmetric.

EXAMPLE 8.2.1

Determine the symmetric matrix corresponding to the quadratic form

$$Q(\vec{x}) = 2x_1^2 - 4x_1x_2 - x_2^2$$

Solution: The corresponding symmetric matrix A is

$$A = \begin{bmatrix} 2 & -2 \\ -2 & -1 \end{bmatrix}$$

Notice that we could have written $ax_1^2 + bx_1x_2 + cx_2^2$ in terms of other *asymmetric* matrices. For example,

$$ax_1^2 + bx_1x_2 + cx_2^2 = \vec{x}^T \begin{bmatrix} a & b \\ 0 & c \end{bmatrix} \vec{x} = \vec{x}^T \begin{bmatrix} a & 2b \\ -b & c \end{bmatrix} \vec{x}$$

Many choices are possible. However, we agree always to choose the symmetric matrix for two reasons. First, it gives us a unique (symmetric) matrix corresponding to a given quadratic form. Second, the choice of the symmetric matrix A allows us to apply the special theory available for symmetric matrices. We now use this to extend the definition of quadratic form to n variables.

Definition
Quadratic Form

A **quadratic form** on $\mathbb{R}^n$, with corresponding $n \times n$ symmetric matrix A, is a function $Q : \mathbb{R}^n \to \mathbb{R}$ defined by

$$Q(\vec{x}) = \vec{x}^T A\vec{x}, \qquad \text{for all } \vec{x} \in \mathbb{R}^n$$

EXAMPLE 8.2.2

Let $A = \begin{bmatrix} 1 & 2 & -3 \\ 2 & -4 & 0 \\ -3 & 0 & -1 \end{bmatrix}$. Find the quadratic form $Q(\vec{x})$ corresponding to A.

Solution: We have

$$\begin{aligned}
Q(x_1, x_2, x_3) &= \begin{bmatrix} x_1 & x_2 & x_3 \end{bmatrix} \begin{bmatrix} 1 & 2 & -3 \\ 2 & -4 & 0 \\ -3 & 0 & -1 \end{bmatrix} \begin{bmatrix} x_1 \\ x_2 \\ x_3 \end{bmatrix} \\
&= \begin{bmatrix} x_1 & x_2 & x_3 \end{bmatrix} \begin{bmatrix} 1x_1 + 2x_2 - 3x_3 \\ 2x_1 - 4x_2 + 0x_3 \\ -3x_1 + 0x_2 - 1x_3 \end{bmatrix} \\
&= x_1(1x_1 + 2x_2 - 3x_3) + x_2(2x_1 - 4x_2 + 0x_3) + x_3(-3x_1 + 0x_2 - 1x_3) \\
&= x_1^2 + 4x_1x_2 - 6x_1x_3 - 4x_2^2 - x_3^2
\end{aligned}$$

Observe from the multiplication of $\vec{x}^T A\vec{x}$ in Example 8.2.2 that multiplying by $\vec{x}$ on the right of A makes the first column of A correspond to x_1, the second column of A to x_2, etc. Similarly, multiplying on the left of A by $\vec{x}^T$ makes the first row of A correspond to x_1, the second row of A to x_2, etc.

Thus, for any symmetric matrix A, the coefficient b_{ij} of x_ix_j in the quadratic form $Q(\vec{x}) = \vec{x}^T A\vec{x}$ is given by

$$b_{ij} = \begin{cases} a_{ii} & \text{if } i = j \\ 2a_{ij} & \text{if } i < j \end{cases}$$

On the other hand, given a quadratic form $Q(\vec{x}) = b_{11}x_1^2 + b_{12}x_1x_2 + \cdots + b_{nn}x_n^2$ on $\mathbb{R}^n$, we can construct the corresponding symmetric matrix A by taking

$$(A)_{ij} = \begin{cases} b_{ii} & \text{if } i = j \\ \frac{1}{2}b_{ij} & \text{if } i \neq j \end{cases}$$

EXAMPLE 8.2.3

Find the corresponding symmetric matrix for each of the following quadratic forms.

(a) $Q(\vec{x}) = 3x_1^2 + 5x_1x_2 + 2x_2^2$

Solution: The corresponding symmetric matrix $A = \begin{bmatrix} 3 & 5/2 \\ 5/2 & 2 \end{bmatrix}$.

(b) $Q(\vec{x}) = x_1^2 + 4x_1x_2 + x_1x_3 + 4x_2^2 + 2x_2x_3 + 2x_3^2$

Solution: The corresponding symmetric matrix $A = \begin{bmatrix} 1 & 2 & 1/2 \\ 2 & 4 & 1 \\ 1/2 & 1 & 2 \end{bmatrix}$.

EXERCISE 8.2.1

Find the quadratic form corresponding to each of the following symmetric matrices.

(a) $\begin{bmatrix} 4 & 1/2 \\ 1/2 & \sqrt{2} \end{bmatrix}$ (b) $\begin{bmatrix} 1 & -1 & 0 \\ -1 & 2 & 3 \\ 0 & 3 & -1 \end{bmatrix}$

EXERCISE 8.2.2

Find the corresponding symmetric matrix for each of the following quadratic forms.

(a) $Q(\vec{x}) = x_1^2 - 2x_1x_2 - 3x_2^2$

(b) $Q(\vec{x}) = 2x_1^2 + 3x_1x_2 - x_1x_3 + 4x_2^2 + x_3^2$

(c) $Q(\vec{x}) = x_1^2 + 2x_2^2 + 3x_3^2 + 4x_4^2$

Observe that the symmetric matrix corresponding to $Q(\vec{x}) = x_1^2 + 2x_2^2 + 3x_3^2 + 4x_4^2$ is in fact diagonal. This motivates the following definition.

Definition
Diagonal Form

A quadratic form $Q(\vec{x}) = b_{11}x_1^2 + b_{12}x_1x_2 + \cdots + b_{nn}x_n^2$ is in **diagonal form** if all the coefficients b_{jk} with $j \neq k$ are equal to 0. Equivalently, $Q(\vec{x})$ is in diagonal form if its corresponding symmetric matrix is diagonal.

EXAMPLE 8.2.4

The quadratic form $Q(\vec{x}) = 3x_1^2 - 2x_2^2 + 4x_3^2$ is in diagonal form.

The quadratic form $Q(\vec{x}) = 2x_1^2 - 4x_1x_2 + 3x_2^2$ is not in diagonal form.

Since each quadratic form is defined by a symmetric matrix, we should expect that diagonalizing the symmetric matrix should also diagonalize the quadratic form. We first demonstrate this with an example and then prove the result in general.

EXAMPLE 8.2.5

Consider the quadratic form

$$Q(\vec{x}) = \vec{x}^T A\vec{x} = 17x_1^2 + 12x_1x_2 + 8x_2^2$$

The corresponding symmetric matrix $A = \begin{bmatrix} 17 & 6 \\ 6 & 8 \end{bmatrix}$ is orthogonally diagonalized by $P = \begin{bmatrix} 2/\sqrt{5} & -1/\sqrt{5} \\ 1/\sqrt{5} & 2/\sqrt{5} \end{bmatrix}$ to $D = \begin{bmatrix} 20 & 0 \\ 0 & 5 \end{bmatrix}$. Use the change of variables $\vec{x} = P\vec{y}$ to express $Q(\vec{x})$ in terms of $\vec{y} = \begin{bmatrix} y_1 \\ y_2 \end{bmatrix}$.

Solution: We have

$$Q(\vec{x}) = \vec{x}^T A\vec{x} = (P\vec{y})^T A(P\vec{y}) = \vec{y}^T(P^T AP)\vec{y} = \vec{y}^T \begin{bmatrix} 20 & 0 \\ 0 & 5 \end{bmatrix} \vec{y} = 20y_1^2 + 5y_2^2$$

Recall that $P = \begin{bmatrix} \vec{v}_1 & \vec{v}_2 \end{bmatrix}$ is a change of coordinates matrix from coordinates with respect to the basis $\mathcal{B} = \{\vec{v}_1, \vec{v}_2\}$ to standard coordinates. So, in Example 8.2.5, we put $Q(\vec{x})$ into diagonal form by writing it with respect to the orthonormal basis $\mathcal{B} = \left\{ \begin{bmatrix} 2/\sqrt{5} \\ 1/\sqrt{5} \end{bmatrix}, \begin{bmatrix} -1/\sqrt{5} \\ 2/\sqrt{5} \end{bmatrix} \right\}$. The vector $\vec{y}$ is just the $\mathcal{B}$-coordinates with respect to $\vec{x}$. See Example 1.2.12 on page 27 to view this geometrically. We now prove this in general.

Theorem 8.2.1

Let $A \in M_{n\times n}(\mathbb{R})$ be a symmetric matrix and let $Q(\vec{x}) = \vec{x}^T A \vec{x}$. If P is an orthogonal matrix that diagonalizes A, then the change of variables $\vec{x} = P\vec{y}$ brings $Q(\vec{x})$ into diagonal form. In particular, we get

$$Q(\vec{x}) = \lambda_1 y_1^2 + \cdots \lambda_n y_n^2$$

where $\lambda_1, \dots, \lambda_n$ are the eigenvalues of A corresponding to the columns of P.

Proof: Since P orthogonal diagonalizes A, we have that $P^T AP = D$ is diagonal where the diagonal entries of D are the eigenvalues $\lambda_1, \dots, \lambda_n$ of A. Hence,

$$Q(\vec{x}) = \vec{x}^T A\vec{x} = (P\vec{y})^T A(P\vec{y}) = \vec{y}^T(P^T AP)\vec{y} = \vec{y}^T D\vec{y} = \lambda_1 y_1^2 + \cdots \lambda_n y_n^2$$

■

EXAMPLE 8.2.6

Let $Q(\vec{x}) = x_1^2 + 4x_1x_2 + x_2^2$. Find a diagonal form of $Q(\vec{x})$ and an orthogonal matrix P that brings it into this form.

Solution: The corresponding symmetric matrix is $A = \begin{bmatrix} 1 & 2 \\ 2 & 1 \end{bmatrix}$. We have

$$C(\lambda) = \begin{vmatrix} 1-\lambda & 2 \\ 2 & 1-\lambda \end{vmatrix} = (\lambda - 3)(\lambda + 1)$$

The eigenvalues are $\lambda_1 = 3$ and $\lambda_2 = -1$.
For $\lambda_1 = 3$, we get

$$A - \lambda_1 I = \begin{bmatrix} -2 & 2 \\ 2 & -2 \end{bmatrix} \sim \begin{bmatrix} 1 & -1 \\ 0 & 0 \end{bmatrix}$$

An eigenvector for λ_1 is $\vec{v}_1 = \begin{bmatrix} 1 \\ 1 \end{bmatrix}$, and a basis for the eigenspace is $\{\vec{v}_1\}$.

For $\lambda_2 = -1$, we get

$$A - \lambda_2 I = \begin{bmatrix} 2 & 2 \\ 2 & 2 \end{bmatrix} \sim \begin{bmatrix} 1 & 1 \\ 0 & 0 \end{bmatrix}$$

An eigenvector for λ_2 is $\vec{v}_2 = \begin{bmatrix} -1 \\ 1 \end{bmatrix}$, and a basis for the eigenspace is $\{\vec{v}_2\}$.

Therefore, we see that A is orthogonally diagonalized by $P = \frac{1}{\sqrt{2}}\begin{bmatrix} 1 & -1 \\ 1 & 1 \end{bmatrix}$ to $D = \begin{bmatrix} 3 & 0 \\ 0 & -1 \end{bmatrix}$. Thus, by Theorem 8.2.1 we get that the change of variables $\vec{x} = P\vec{y}$ brings $Q(\vec{x})$ into the form

$$Q(\vec{x}) = 3y_1^2 - y_2^2$$

EXERCISE 8.2.3

Let $Q(\vec{x}) = 4x_1x_2 - 3x_2^2$. Find a diagonal form of $Q(\vec{x})$ and an orthogonal matrix P that brings it into this form.

Classifications of Quadratic Forms

Definition
Positive Definite
Negative Definite
Indefinite
Positive Semidefinite
Negative Semidefinite

A quadratic form $Q(\vec{x})$ on $\mathbb{R}^n$ is

(1) **positive definite** if $Q(\vec{x}) > 0$ for all $\vec{x} \neq \vec{0}$.
(2) **negative definite** if $Q(\vec{x}) < 0$ for all $\vec{x} \neq \vec{0}$.
(3) **indefinite** if $Q(\vec{x}) > 0$ for some $\vec{x}$ and $Q(\vec{x}) < 0$ for some $\vec{x}$.
(4) **positive semidefinite** if $Q(\vec{x}) \geq 0$ for all $\vec{x}$.
(5) **negative semidefinite** if $Q(\vec{x}) \leq 0$ for all $\vec{x}$.

These concepts are useful in applications. For example, we shall see in Section 8.3 that the graph of $Q(\vec{x}) = 1$ in $\mathbb{R}^2$ is an ellipse if and only if $Q(\vec{x})$ is positive definite.

EXAMPLE 8.2.7

Classify the quadratic forms $Q_1(\vec{x}) = 3x_1^2 + 4x_2^2$, $Q_2(\vec{x}) = x_1^2 - x_2^2$, and $Q_3(\vec{x}) = -2x_1^2 - x_2^2$.

Solution: $Q_1(\vec{x})$ is positive definite since $Q_1(\vec{x}) = 3x_1^2 + 4x_2^2 > 0$ for all $\vec{x} \neq \vec{0}$.
$Q_2(\vec{x})$ is indefinite since $Q_2(1,0) = 1 > 0$ and $Q_2(0,1) = -1 < 0$.
$Q_3(\vec{x})$ is negative definite since $Q_3(\vec{x}) = -2x_1^2 - x_2^2 < 0$ for all $\vec{x} \neq \vec{0}$.

The quadratic forms in Example 8.2.7 were easy to classify since they were in diagonal form. The following theorem gives us an easy way to classify general quadratic forms.

Theorem 8.2.2

If $Q(\vec{x}) = \vec{x}^T A\vec{x}$ where $A \in M_{n\times n}(\mathbb{R})$ is a symmetric matrix, then

(1) $Q(\vec{x})$ is positive definite if and only if all eigenvalues of A are positive.
(2) $Q(\vec{x})$ is negative definite if and only if all eigenvalues of A are negative.
(3) $Q(\vec{x})$ is indefinite if and only if some of the eigenvalues of A are positive and some are negative.

Proof: We prove (1) and leave (2) and (3) as Problems C1 and C2.

By Theorem 8.2.1, there exists an orthogonal matrix P such that

$$Q(\vec{x}) = \lambda_1 y_1^2 + \lambda_2 y_2^2 + \cdots + \lambda_n y_n^2$$

where $\vec{x} = P\vec{y}$ and $\lambda_1, \ldots, \lambda_n$ are the eigenvalues of A. Clearly, $Q(\vec{x}) > 0$ for all $\vec{y} \neq \vec{0}$ if and only if the eigenvalues are all positive. Moreover, since P is orthogonal, it is invertible. Hence, $\vec{x} = \vec{0}$ if and only if $\vec{y} = \vec{0}$ since $\vec{x} = P\vec{y}$. Thus we have shown that $Q(\vec{x})$ is positive definite if and only if all eigenvalues of A are positive. ■

EXAMPLE 8.2.8

Classify the following quadratic forms.

$$Q_1(\vec{x}) = 4x_1^2 + 8x_1x_2 + 3x_2^2$$
$$Q_2(\vec{x}) = -2x_1^2 - 2x_1x_2 + 2x_1x_3 - 2x_2^2 + 2x_2x_3 - 2x_3^2$$

Solution: The symmetric matrix corresponding to $Q_1(\vec{x})$ is $A = \begin{bmatrix} 4 & 4 \\ 4 & 3 \end{bmatrix}$. The characteristic polynomial of A is $C(\lambda) = \lambda^2 - 7\lambda - 4$. Using the quadratic formula, we find that the eigenvalues of A are $\lambda_1 = \frac{7+\sqrt{65}}{2}$ and $\lambda_2 = \frac{7-\sqrt{65}}{2}$. Clearly $\lambda_1 > 0$. Observe that $\sqrt{65} > 7$, so $\lambda_2 < 0$. Hence, $Q_1(\vec{x})$ is indefinite.

The symmetric matrix corresponding to $Q_2(\vec{x})$ is $A = \begin{bmatrix} -2 & -1 & 1 \\ -1 & -2 & 1 \\ 1 & 1 & -2 \end{bmatrix}$. The characteristic polynomial of A is $C(\lambda) = -(\lambda+1)^2(\lambda+4)$. Thus, the eigenvalues of A are -1, -1, and -4. Therefore, $Q_2(\vec{x})$ is negative definite.

EXERCISE 8.2.4

Classify the following quadratic forms.

(a) $Q_1(\vec{x}) = 5x_1^2 + 4x_1x_2 + 2x_2^2$

(b) $Q_2(\vec{x}) = 2x_1^2 - 6x_1x_2 - 6x_1x_3 + 3x_2^2 + 4x_2x_3 + 3x_3^2$

Since every symmetric matrix corresponds uniquely to a quadratic form, it makes sense to classify a symmetric matrix by classifying its corresponding quadratic form. That is, for example, we will say a symmetric matrix A is positive definite if and only if the quadratic form $Q(\vec{x}) = \vec{x}^T A\vec{x}$ is positive definite. Observe that this implies that we can use Theorem 8.2.2 to classify symmetric matrices as well.

EXAMPLE 8.2.9

Classify the following symmetric matrices.

(a) $A = \begin{bmatrix} 3 & 2 \\ 2 & 3 \end{bmatrix}$

Solution: We have $C(\lambda) = \lambda^2 - 6\lambda + 5$. Thus, the eigenvalues of A are 5 and 1, so A is positive definite.

(b) $A = \begin{bmatrix} -2 & 2 & -4 \\ 2 & -4 & -2 \\ -4 & -2 & 2 \end{bmatrix}$

Solution: We have $C(\lambda) = -(\lambda+4)(\lambda^2 - 28)$. Thus, the eigenvalues of A are $-4, 2\sqrt{7}$, and $-2\sqrt{7}$, so A is indefinite.

PROBLEMS 8.2

Practice Problems

For Problems A1–A4, determine the quadratic form corresponding to the given symmetric matrix.

A1 $\begin{bmatrix} 1 & 3 \\ 3 & -1 \end{bmatrix}$ **A2** $\begin{bmatrix} 3 & -2 \\ -2 & 0 \end{bmatrix}$

A3 $\begin{bmatrix} 1 & 0 & 0 \\ 0 & -2 & 3 \\ 0 & 3 & -1 \end{bmatrix}$ **A4** $\begin{bmatrix} -2 & 1 & 1 \\ 1 & 1 & -1 \\ 1 & -1 & 0 \end{bmatrix}$

For Problems A5–A10, classify the symmetric matrix.

A5 $\begin{bmatrix} 4 & -2 \\ -2 & 4 \end{bmatrix}$ **A6** $\begin{bmatrix} 1 & 0 \\ 0 & 2 \end{bmatrix}$

A7 $\begin{bmatrix} 1 & 0 & 0 \\ 0 & -2 & 6 \\ 0 & 6 & 7 \end{bmatrix}$ **A8** $\begin{bmatrix} -3 & 1 & -1 \\ 1 & -3 & 1 \\ -1 & 1 & -3 \end{bmatrix}$

A9 $\begin{bmatrix} 7 & 2 & -1 \\ 2 & 10 & -2 \\ -1 & -2 & 7 \end{bmatrix}$ **A10** $\begin{bmatrix} -4 & -5 & 5 \\ -5 & 2 & 1 \\ 5 & 1 & 2 \end{bmatrix}$

For Problems A11–A19:

(a) Determine the symmetric matrix corresponding to $Q(\vec{x})$.

(b) Express $Q(\vec{x})$ in diagonal form and give the orthogonal matrix that brings it into this form.

(c) Classify $Q(\vec{x})$.

A11 $Q(\vec{x}) = x_1^2 - 3x_1x_2 + x_2^2$

A12 $Q(\vec{x}) = 5x_1^2 - 4x_1x_2 + 2x_2^2$

A13 $Q(\vec{x}) = -7x_1^2 + 4x_1x_2 - 4x_2^2$

A14 $Q(\vec{x}) = -2x_1^2 - 6x_1x_2 - 2x_2^2$

A15 $Q(\vec{x}) = -2x_1^2 + 12x_1x_2 + 7x_2^2$

A16 $Q(\vec{x}) = x_1^2 - 2x_1x_2 + 6x_1x_3 + x_2^2 + 6x_2x_3 - 3x_3^2$

A17 $Q(\vec{x}) = -4x_1^2 + 2x_1x_2 - 5x_2^2 - 2x_2x_3 - 4x_3^2$

A18 $Q(\vec{x}) = 3x_1^2 - 2x_1x_2 - 2x_1x_3 + 5x_2^2 + 2x_2x_3 + 3x_3^2$

A19 $Q(\vec{x}) = 3x_1^2 - 4x_1x_2 + 8x_1x_3 + 6x_2^2 + 4x_2x_3 + 3x_3^2$

Homework Problems

For Problems B1–B6, determine the quadratic form corresponding to the given symmetric matrix.

B1 $\begin{bmatrix} 4 & -1 \\ -1 & 3 \end{bmatrix}$ **B2** $\begin{bmatrix} 1 & -5 \\ -5 & 1 \end{bmatrix}$

B3 $\begin{bmatrix} -1 & 0 & 0 \\ 0 & -2 & 0 \\ 0 & 0 & -3 \end{bmatrix}$ **B4** $\begin{bmatrix} -1 & 1 & 1 \\ 1 & -1 & 1 \\ 1 & 1 & -1 \end{bmatrix}$

B5 $\begin{bmatrix} 3 & 6 & 5 \\ 6 & 0 & 2 \\ 5 & 2 & 1 \end{bmatrix}$ **B6** $\begin{bmatrix} 2 & 1 & 2 \\ 1 & -3 & -1 \\ 2 & -1 & 2 \end{bmatrix}$

For Problems B7–B12, classify the symmetric matrix.

B7 $\begin{bmatrix} 3 & -2 \\ -2 & 6 \end{bmatrix}$ **B8** $\begin{bmatrix} -6 & 6 \\ 6 & -11 \end{bmatrix}$

B9 $\begin{bmatrix} 3 & 0 & 6 \\ 0 & 2 & 0 \\ 6 & 0 & -2 \end{bmatrix}$ **B10** $\begin{bmatrix} 2 & 2 & 0 \\ 2 & 5 & 0 \\ 0 & 0 & 3 \end{bmatrix}$

B11 $\begin{bmatrix} 1 & 4 & 4 \\ 4 & 3 & 0 \\ 4 & 0 & -1 \end{bmatrix}$ **B12** $\begin{bmatrix} -2 & -1 & 1 \\ -1 & -2 & 1 \\ 1 & 1 & -2 \end{bmatrix}$

For Problems B13–B24:

(a) Determine the symmetric matrix corresponding to $Q(\vec{x})$.

(b) Express $Q(\vec{x})$ in diagonal form and give the orthogonal matrix that brings it into this form.

(c) Classify $Q(\vec{x})$.

B13 $Q(\vec{x}) = x_1^2 + 8x_1x_2 + x_2^2$

B14 $Q(\vec{x}) = -2x_1^2 + 12x_1x_2 - 7x_2^2$

B15 $Q(\vec{x}) = 2x_1^2 + 4x_1x_2 + 5x_2^2$

B16 $Q(\vec{x}) = -5x_1^2 + 12x_1x_2 - 10x_2^2$

B17 $Q(\vec{x}) = -3x_1^2 + 12x_1x_2 - 8x_2^2$

B18 $Q(\vec{x}) = -2x_1^2 - 6x_1x_2 + 6x_2^2$

B19 $Q(\vec{x}) = -x_1^2 + 2x_1x_2 + 4x_1x_3 - x_2^2 + 4x_2x_3 + 2x_3^2$

B20 $Q(\vec{x}) = 4x_1^2 - 2x_1x_2 + 2x_1x_3 + 4x_2^2 - 2x_2x_3 + 4x_3^2$

B21 $Q(\vec{x}) = -6x_1^2 + 4x_1x_3 - 6x_2^2 + 8x_2x_3 - 5x_3^2$

B22 $Q(\vec{x}) = -2x_1^2 - 3x_2^2 + 4x_2x_3 - 3x_3^2$

B23 $Q(\vec{x}) = -x_1^2 + 4x_1x_2 - 2x_1x_3 + 2x_2^2 - 4x_2x_3 - x_3^2$

B24 $Q(\vec{x}) = 4x_1^2 + 4x_1x_2 - 4x_1x_3 + 3x_2^2 + 5x_3^2$

Conceptual Problems

C1 Let $Q(\vec{x}) = \vec{x}^T A\vec{x}$, where A is a symmetric matrix. Prove that $Q(\vec{x})$ is negative definite if and only if all eigenvalues of A are negative.

C2 Let $Q(\vec{x}) = \vec{x}^T A\vec{x}$, where A is a symmetric matrix. Prove that $Q(\vec{x})$ is indefinite if and only if some of the eigenvalues of A are positive and some are negative.

C3 Let $Q(\vec{x}) = \vec{x}^T A\vec{x}$, where A is a symmetric matrix. Prove that $Q(\vec{x})$ is positive semidefinite if and only if all of the eigenvalues of A are non-negative.

C4 Let $A \in M_{m\times n}(\mathbb{R})$. Prove that A^TA is positive semidefinite.

For Problems C5–C8, assume that A is a positive definite symmetric matrix.

C5 Prove that the diagonal entries of A are all positive.

C6 Prove that A is invertible.

C7 Prove that A^{-1} is positive definite.

C8 Prove that P^TAP is positive definite for any orthogonal matrix P.

C9 A matrix B is called **skew-symmetric** if $B^T = -B$. Given a square matrix A, define the **symmetric part** of A to be

$$A^+ = \frac{1}{2}(A + A^T)$$

and the **skew-symmetric** part of A to be

$$A^- = \frac{1}{2}(A - A^T)$$

(a) Verify that A^+ is symmetric, A^- is skew-symmetric, and $A = A^+ + A^-$.
(b) Prove that the diagonal entries of A^- are 0.
(c) Determine expressions for typical entries $(A^+)_{ij}$ and $(A^-)_{ij}$ in terms of the entries of A.
(d) Prove that for every $\vec{x} \in \mathbb{R}^n$,

$$\vec{x}^TA\vec{x} = \vec{x}^TA^+\vec{x}$$

(Hint: use the fact that $A = A^+ + A^-$ and prove that $\vec{x}^TA^-\vec{x} = \vec{0}$.)

C10 In this problem, we show that general inner products on $\mathbb{R}^n$ are not different in interesting ways from the standard inner product.
Let $\langle\, ,\rangle$ be an inner product on $\mathbb{R}^n$ and let $\mathcal{S} = \{\vec{e}_1, \dots, \vec{e}_n\}$ be the standard basis.
(a) Verify that for any $\vec{x}, \vec{y} \in \mathbb{R}^n$,

$$\langle \vec{x}, \vec{y}\rangle = \sum_{i=1}^{n}\sum_{j=1}^{n} x_iy_j\langle \vec{e}_i, \vec{e}_j\rangle$$

(b) Let G be the $n \times n$ matrix defined by $g_{ij} = \langle \vec{e}_i, \vec{e}_j\rangle$. Verify that

$$\langle \vec{x}, \vec{y}\rangle = \vec{x}^TG\vec{y}$$

(c) Use the properties of an inner product to verify that G is symmetric and positive definite.
(d) By adapting the proof of Theorem 8.2.1, show that there is a basis $\mathcal{B} = \{\vec{v}_1, \dots, \vec{v}_n\}$ such that in $\mathcal{B}$-coordinates,

$$\langle \vec{x}, \vec{y}\rangle = \lambda_1\tilde{x}_1\tilde{y}_1 + \cdots + \lambda_n\tilde{x}_n\tilde{y}_n$$

where $\lambda_1, \dots, \lambda_n$ are the eigenvalues of G. In particular,

$$\langle \vec{x}, \vec{y}\rangle = \|\vec{x}\|^2 = \sum_{i=1}^{n} \lambda_i\tilde{x}_1{}^2$$

(e) Introduce a new basis $C = \{\vec{w}_1, \dots, \vec{w}_n\}$ by defining $\vec{w}_i = \vec{v}_i/\sqrt{\lambda_i}$. Use an asterisk to denote C-coordinates, so that $\vec{x} = x_1^*\vec{w}_1 + \cdots + x_n^*\vec{w}_n$. Verify that

$$\langle \vec{w}_i, \vec{w}_j\rangle = \begin{cases} 1 & \text{if } i = k \\ 0 & \text{if } i \neq k \end{cases}$$

and that

$$\langle \vec{x}, \vec{y}\rangle = x_1^*y_1^* + \cdots + x_n^*y_n^*$$

Thus, with respect to the inner product $\langle\, ,\rangle$, C is an orthonormal basis, and in C-coordinates, the inner product of two vectors looks just like the standard dot product.

8.3 Graphs of Quadratic Forms

In $\mathbb{R}^2$, it is often of interest to know the graph of an equation of the form $Q(\vec{x}) = k$, where $Q(\vec{x})$ is a quadratic form on $\mathbb{R}^2$ and k is a constant. If we were interested in only one or two particular graphs, it might be sensible to simply use a computer. However, by applying diagonalization to the problem of determining these graphs, we see a very clear interpretation of eigenvectors. We also consider a concrete useful application of a change of coordinates. Moreover, this approach to these graphs leads to a classification of the various possibilities; all of the graphs of the form $Q(\vec{x}) = k$ in $\mathbb{R}^2$ can be divided into a few standard cases. Classification is a useful process because it allows us to say "I really need to understand only these few standard cases." A classification of these graphs is given later in this section.

In general it is difficult to identify the shape of the graph of

$$ax_1^2 + bx_1x_2 + cx_2^2 = k$$

It is even more difficult to try to sketch the graph. However, it is relatively easy to sketch the graph of

$$ax_1^2 + cx_2^2 = k$$

Thus, our strategy to sketch the graph of a quadratic form $Q(\vec{x}) = k$ is to first bring it into diagonal form. Of course, we first need to determine how diagonalizing the quadratic form will affect the graph.

Theorem 8.3.1

Let $Q(\vec{x}) = ax_1^2 + bx_1x_2 + cx_2^2$ where $a, b,$ and c are not all zero and let P be an orthogonal matrix which diagonalizes $Q(\vec{x})$. If $\det P = 1$, then P corresponds to a rotation in $\mathbb{R}^2$.

Proof: Since A is symmetric, by the Principal Axis Theorem, there exists an orthonormal basis $\{\vec{v}, \vec{w}\}$ of $\mathbb{R}^2$ of eigenvectors of A. Let $\vec{v} = \begin{bmatrix} v_1 \\ v_2 \end{bmatrix}$ and $\vec{w} = \begin{bmatrix} w_1 \\ w_2 \end{bmatrix}$. Since $\vec{v}$ is a unit vector, we must have

$$1 = \|\vec{v}\|^2 = v_1^2 + v_2^2$$

Hence, the entries v_1 and v_2 lie on the unit circle. Therefore, there exists an angle θ such that $v_1 = \cos\theta$ and $v_2 = \sin\theta$. Moreover, since $\vec{w}$ is a unit vector orthogonal to $\vec{v}$, we must have $\vec{w} = \pm\begin{bmatrix} -\sin\theta \\ \cos\theta \end{bmatrix}$. We choose $\vec{w} = +\begin{bmatrix} -\sin\theta \\ \cos\theta \end{bmatrix}$ so that $\det P = 1$. Hence we have

$$P = \begin{bmatrix} \cos\theta & -\sin\theta \\ \sin\theta & \cos\theta \end{bmatrix}$$

This corresponds to a rotation by θ. Finally, from our work in Section 8.2, we know that this change of coordinates matrix brings Q into diagonal form. ■

Remark

If we picked $\vec{w} = -\begin{bmatrix} -\sin\theta \\ \cos\theta \end{bmatrix}$, we would find that $\det P = -1$ and that P corresponds to a rotation and a reflection.

In practice, we do not need to calculate the angle of rotation. When we orthogonally diagonalize $Q(\vec{x})$ with $P = \begin{bmatrix} \vec{v}_1 & \vec{v}_2 \end{bmatrix}$, the change of coordinates $\vec{x} = P\vec{y}$ causes a rotation of the y_1- and y_2-axes. In particular, since the y_1-axis is spanned by the first standard basis vector $\vec{e}_1$, we get that the image of the y_1-axis in the x_1x_2-plane is spanned by

$$\vec{x} = P\vec{e}_1 = \begin{bmatrix} \vec{v}_1 & \vec{v}_2 \end{bmatrix} \begin{bmatrix} 1 \\ 0 \end{bmatrix} = \vec{v}_1$$

Similarly, since the y_2-axis is spanned by the second standard basis vector $\vec{e}_2$, we get that the image of the y_2-axis in the x_1x_2-plane is spanned by

$$\vec{x} = P\vec{e}_2 = \begin{bmatrix} \vec{v}_1 & \vec{v}_2 \end{bmatrix} \begin{bmatrix} 0 \\ 1 \end{bmatrix} = \vec{v}_2$$

We demonstrate this with two examples.

EXAMPLE 8.3.1

Sketch the graph of the equation $3x_1^2 + 4x_1x_2 = 16$.

Solution: The quadratic form $Q(\vec{x}) = 3x_1^2 + 4x_1x_2$ corresponds to the symmetric matrix $A = \begin{bmatrix} 3 & 2 \\ 2 & 0 \end{bmatrix}$. Its characteristic polynomial is

$$C(\lambda) = \lambda^2 - 3\lambda - 4 = (\lambda - 4)(\lambda + 1)$$

Thus, the eigenvalues of A are $\lambda_1 = 4$ and $\lambda_2 = -1$. Thus, by an orthogonal change of coordinates, the equation can be brought into the diagonal form:

$$4y_1^2 - y_2^2 = 16$$

This is an equation of a hyperbola in the y_1y_2-plane. We observe that the y_1-intercepts are $(2, 0)$ and $(-2, 0)$, and there are no intercepts on the y_2-axis. The asymptotes of the hyperbola are determined by the equation $4y_1^2 - y_2^2 = 0$. Solving for y_2, we determine that the asymptotes are lines with equations $y_2 = 2y_1$ and $y_2 = -2y_1$. With this information, we obtain the graph in Figure 8.3.1.

Next, we need to find a basis for each eigenspace of A.
For $\lambda_1 = 4$,

$$A - \lambda_1 I = \begin{bmatrix} -1 & 2 \\ 2 & -4 \end{bmatrix} \sim \begin{bmatrix} 1 & -2 \\ 0 & 0 \end{bmatrix}$$

Thus, a basis for the eigenspace is $\{\vec{v}_1\}$, where $\vec{v}_1 = \begin{bmatrix} 2 \\ 1 \end{bmatrix}$.
For $\lambda_2 = -1$,

$$A - \lambda_2 I = \begin{bmatrix} 4 & 2 \\ 2 & 1 \end{bmatrix} \sim \begin{bmatrix} 2 & 1 \\ 0 & 0 \end{bmatrix}$$

Thus, a basis for the eigenspace is $\{\vec{v}_2\}$, where $\vec{v}_2 = \begin{bmatrix} -1 \\ 2 \end{bmatrix}$. Hence, A is orthogonally diagonalized by $P = \frac{1}{\sqrt{5}} \begin{bmatrix} 2 & -1 \\ 1 & 2 \end{bmatrix}$.

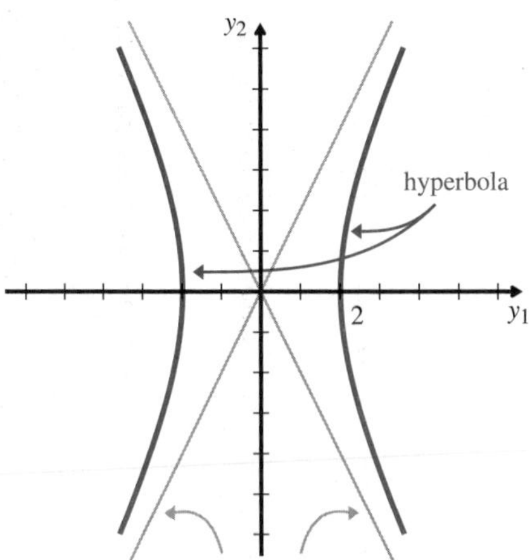

Figure 8.3.1 The graph of $4y_1^2 - y_2^2 = 16$.

EXAMPLE 8.3.1
(continued)

Now we sketch the graph of $3x_1^2 + 4x_1x_2 = 16$. In the x_1x_2-plane, we draw the y_1-axis in the direction of $\vec{v}_1$. (For clarity, in Figure 8.3.2 we have shown the vector $\begin{bmatrix} 4 \\ 2 \end{bmatrix}$ instead of $\frac{1}{\sqrt{5}}\begin{bmatrix} 2 \\ 1 \end{bmatrix}$.) We also draw the y_2-axis in the direction of $\vec{v}_2$. Then, relative to these new axes, we sketch the graph of the hyperbola $4y_1^2 - y_2^2 = 16$. The graph in Figure 8.3.2 is also the graph of the original equation $3x_1^2 + 4x_1x_2 = 16$.

In order to include the asymptotes in the sketch, we solve the change of variables $\vec{x} = P\vec{y}$ for $\vec{y}$ to get

$$\vec{y} = P^T\vec{x}$$

$$\begin{bmatrix} y_1 \\ y_2 \end{bmatrix} = \frac{1}{\sqrt{5}}\begin{bmatrix} 2 & 1 \\ -1 & 2 \end{bmatrix}\begin{bmatrix} x_1 \\ x_2 \end{bmatrix}$$

This gives

$$y_1 = \frac{1}{\sqrt{5}}(2x_1 + x_2)$$

$$y_2 = \frac{1}{\sqrt{5}}(-x_1 + 2x_2)$$

Then one asymptote is

$$y_2 = 2y_1$$

$$\frac{1}{\sqrt{5}}(-x_1 + 2x_2) = \frac{2}{\sqrt{5}}(2x_1 + x_2)$$

$$0 = x_1$$

The other asymptote is

$$y_2 = -2y_1$$

$$\frac{1}{\sqrt{5}}(-x_1 + 2x_2) = -\frac{2}{\sqrt{5}}(2x_1 + x_2)$$

$$x_2 = -\frac{3}{4}x_1$$

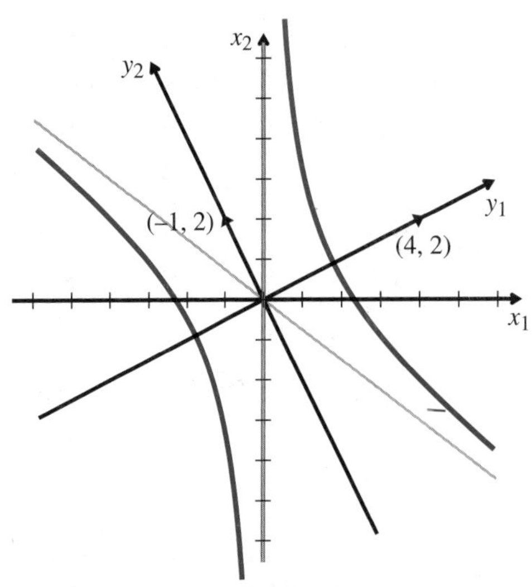

Figure 8.3.2 The graph of $3x_1^2 + 4x_1x_2 = 16$.

Remark

In Example 8.3.1, we chose $\vec{v}_2$ so that $\det P = 1$. It is a valuable exercise to see what would happen if we had chosen $\vec{v}_2 = \begin{bmatrix} 1 \\ -2 \end{bmatrix}$ instead.

EXAMPLE 8.3.2 Sketch the graph of the equation $6x_1^2 + 4x_1x_2 + 3x_2^2 = 14$.

Solution: The corresponding symmetric matrix is $A = \begin{bmatrix} 6 & 2 \\ 2 & 3 \end{bmatrix}$. We get

$$C(\lambda) = \lambda^2 - 9\lambda + 14 = (\lambda - 2)(\lambda - 7)$$

Thus, the eigenvalues are $\lambda_1 = 2$ and $\lambda_2 = 7$. Hence, the equation can be brought into the diagonal form

$$2y_1^2 + 7y_2^2 = 14$$

This is the equation of an ellipse with y_1-intercepts $(\sqrt{7}, 0)$ and $(-\sqrt{7}, 0)$ and y_2-intercepts $(0, \sqrt{2})$ and $(0, -\sqrt{2})$. We get the ellipse in Figure 8.3.3.

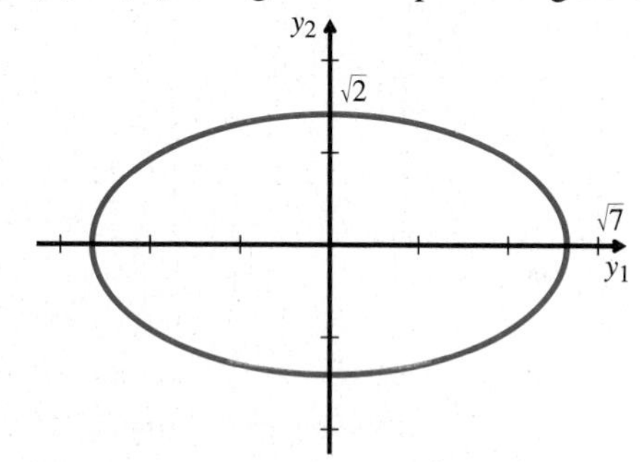

Figure 8.3.3 The graph of $2y_1^2 + 7y_2^2 = 14$.

For $\lambda_1 = 2$,

$$A - \lambda_1 I = \begin{bmatrix} 4 & 2 \\ 2 & 1 \end{bmatrix} \sim \begin{bmatrix} 1 & 1/2 \\ 0 & 0 \end{bmatrix}$$

Thus, a basis for the eigenspace of λ_1 is $\{\vec{v}_1\}$, where $\vec{v}_1 = \begin{bmatrix} 1 \\ -2 \end{bmatrix}$.

For $\lambda_2 = 7$,

$$A - \lambda_2 I = \begin{bmatrix} -1 & 2 \\ 2 & -4 \end{bmatrix} \sim \begin{bmatrix} 1 & -2 \\ 0 & 0 \end{bmatrix}$$

Thus, a basis for the eigenspace of λ_2 is $\{\vec{v}_2\}$, where $\vec{v}_2 = \begin{bmatrix} 2 \\ 1 \end{bmatrix}$.

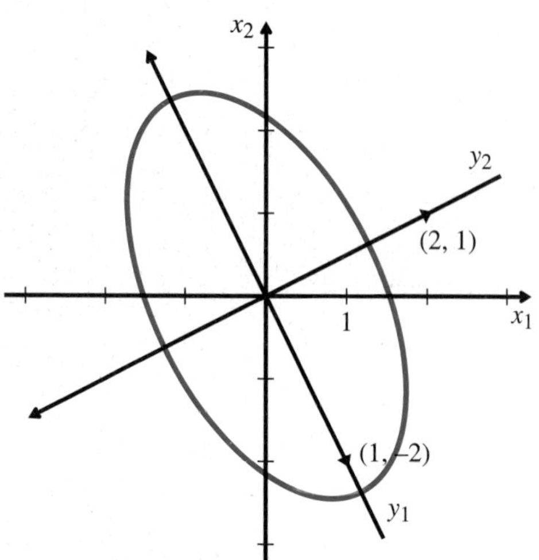

Figure 8.3.4 The graph of $6x_1^2 + 4x_1x_2 + 3x_2^2 = 14$.

In the x_1x_2-plane, we draw the y_1-axis in the direction of $\vec{v}_1$ and the y_2-axis in the direction of $\vec{v}_2$. Then, relative to these new axes, we sketch the graph of the ellipse $2y_1^2 + 7y_2^2 = 14$. This gives us the graph of $6x_1^2 + 4x_1x_2 + 3x_2^2 = 14$ in Figure 8.3.4.

Since diagonalizing a quadratic form corresponds to a rotation, to classify all the graphs of equations of the form $Q(\vec{x}) = k$, we diagonalize and rewrite the equation in the form $\lambda_1 y_1^2 + \lambda_2 y_2^2 = k$. Here, λ_1 and λ_2 are the eigenvalues of the corresponding symmetric matrix. The distinct possibilities are displayed in Table 8.3.1.

Table 8.3.1 Graphs of $\lambda_1 x_1^2 + \lambda_2 x_2^2 = k$

	$k > 0$	$k = 0$	$k < 0$
$\lambda_1 > 0, \lambda_2 > 0$	ellipse	point (0,0)	empty set
$\lambda_1 > 0, \lambda_2 = 0$	parallel lines	line $x_1 = 0$	empty set
$\lambda_1 > 0, \lambda_2 < 0$	hyperbola	intersecting lines	hyperbola
$\lambda_1 = 0, \lambda_2 < 0$	empty set	line $x_2 = 0$	parallel lines
$\lambda_1 < 0, \lambda_2 < 0$	empty set	point (0,0)	ellipse

The cases where $k = 0$ or one eigenvalue is zero may be regarded as **degenerate cases** (not general cases). The **nondegenerate cases** are the ellipses and hyperbolas, which are **conic sections**. (A conic section is a curve obtained in $\mathbb{R}^3$ as the intersection of a cone and a plane.) Notice that the cases of a single point, a single line, and intersecting lines can also be obtained as the intersection of a cone and a plane passing through the vertex of the cone. However, the cases of parallel lines (in Table 8.3.1) are not obtained as the intersection of a cone and a plane.

It is also important to realize that one class of conic sections, parabolas, does not appear in Table 8.3.1. In $\mathbb{R}^2$, the equation of a parabola is a quadratic equation, but it contains first-degree terms. Since a quadratic form contains only second-degree terms, an equation of the form $Q(\vec{x}) = k$ cannot be a parabola.

The classification provided by Table 8.3.1 suggests that it might be interesting to consider how degenerate cases arise as limiting cases of nondegenerate cases. For example, Figure 8.3.5 shows that the case of parallel lines ($y = \pm$ constant) arises from the family of ellipses $\lambda x_1^2 + x_2^2 = 1$ as λ tends to 0.

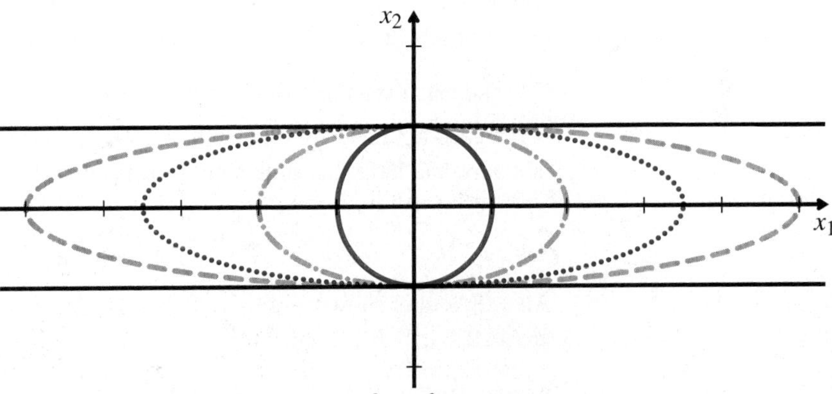

Figure 8.3.5 A family of ellipses $\lambda x_1^2 + x_2^2 = 1$. The circle occurs for $\lambda = 1$; as λ decreases, the ellipses get "fatter"; for $\lambda = 0$, the graph is a pair of lines.

Figure 8.3.6 shows that the case of intersecting lines ($k = 0$) separates the case of hyperbolas with intercepts on the x_1-axis ($x_1^2 - 2x_2^2 = k$, $k > 0$) from the case of hyperbolas with intercepts on the x_2-axis ($x_1^2 - 2x_2^2 = k$, $k < 0$).

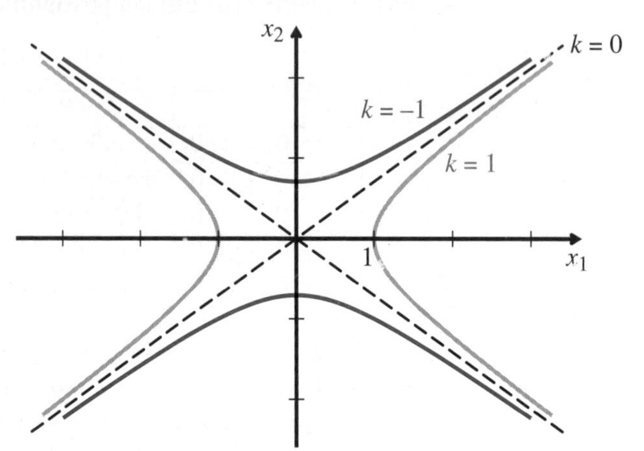

Figure 8.3.6 Graphs of $x_1^2 - 2x_2^2 = k$ for $k \in \{-1, 0, 1\}$.

EXERCISE 8.3.1

Diagonalize the quadratic form and sketch the graph of the equation $x_1^2+2x_1x_2+x_2^2 = 2$. Show both the original axes and the new axes.

Graphs of $Q(\vec{x}) = k$ in $\mathbb{R}^3$

For a quadratic equation of the form $Q(\vec{x}) = k$ in $\mathbb{R}^3$, there are similar results to what we did above. However, because there are three variables instead of two, there are more possibilities. The nondegenerate cases give ellipsoids, hyperboloids of one sheet, and hyperboloids of two sheets. These graphs are called **quadric surfaces**.

The usual standard form for the equation of an ellipsoid is $\frac{x_1^2}{a^2} + \frac{x_2^2}{b^2} + \frac{x_3^2}{c^2} = 1$. This is the case obtained by diagonalizing $Q(\vec{x}) = k$ where the eigenvalues and k are all non-zero and have the same sign. In particular, it is obtained by taking

$$a^2 = k/\lambda_1, \qquad b^2 = k/\lambda_2, \qquad c^2 = k/\lambda_3$$

An ellipsoid is shown in Figure 8.3.7. The positive intercepts on the coordinate axes are $(a, 0, 0)$, $(0, b, 0)$, and $(0, 0, c)$.

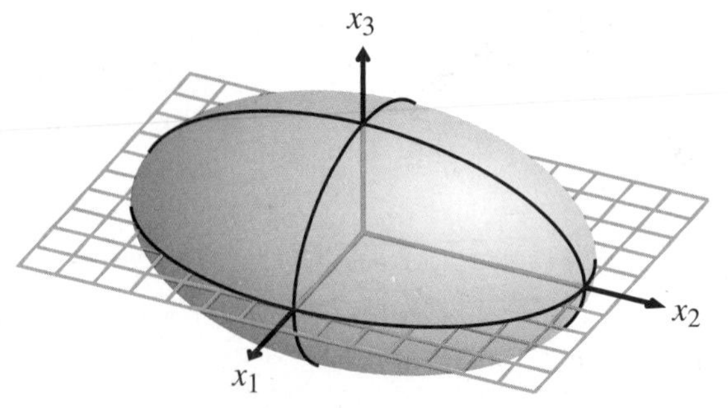

Figure 8.3.7 An ellipsoid in standard position.

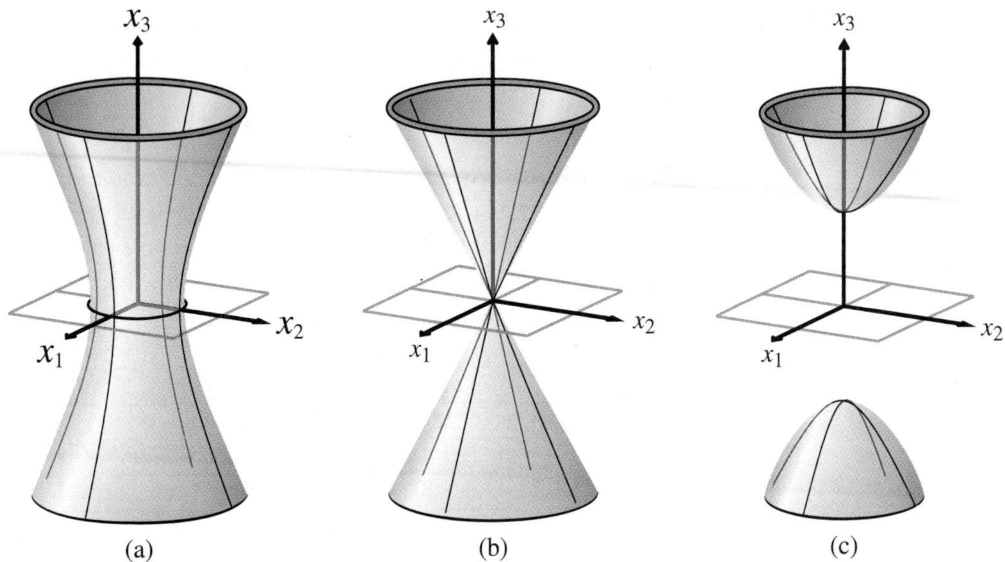

Figure 8.3.8 Graphs of $4x_1^2 + 4x_2^2 - x_3^2 = k$. (a) $k = 1$; a hyperboloid of one sheet. (b) $k = 0$; a cone. (c) $k = -1$; a hyperboloid of two sheets.

The standard form of the equation for a hyperboloid of one sheet is

$$\frac{x_1^2}{a^2} + \frac{x_2^2}{b^2} - \frac{x_3^2}{c^2} = 1$$

This form is obtained when k and two eigenvalues of the matrix of Q are positive and the third eigenvalue is negative. It is also obtained when k and two eigenvalues are negative and the other eigenvalue is positive. If the equation is rewritten as

$$\frac{x_1^2}{a^2} + \frac{x_2^2}{b^2} = 1 + \frac{x_3^2}{c^2}$$

then it is clear that for every x_3 there are values of x_1 and x_2 that satisfy the equation, so that the surface is all one piece (or one sheet). A hyperboloid of one sheet is shown in Figure 8.3.8 (a).

The standard form of the equation for a hyperboloid of two sheets is

$$\frac{x_1^2}{a^2} + \frac{x_2^2}{b^2} - \frac{x_3^2}{c^2} = -1$$

This form is obtained when k and one eigenvalue is negative and the other eigenvalues are positive, or when k and one eigenvalue are positive and the other eigenvalues are negative. Notice that if this is rewritten as

$$\frac{x_1^2}{a^2} + \frac{x_2^2}{b^2} = -1 + \frac{x_3^2}{c^2}$$

it is clear that for every $|x_3| < c$, there are no values of x_1 and x_2 that satisfy the equation. Therefore, the graph consists of two pieces (or two sheets), one with $x_3 \geq c$ and the other with $x_3 \leq -c$. A hyperboloid of two sheets is shown in Figure 8.3.8 (c).

It is interesting to consider the family of surfaces obtained by varying k in the equation

$$\frac{x_1^2}{a^2} + \frac{x_2^2}{b^2} - \frac{x_3^2}{c^2} = k$$

as in Figure 8.3.8. When $k = 1$, the surface is a hyperboloid of one sheet; as k decreases towards 0, the "waist" of the hyperboloid shrinks until at $k = 0$ it has "pinched in" to a single point and the hyperboloid of one sheet becomes a cone. As k decreases towards -1, the waist has disappeared, and the graph is now a hyperboloid of two sheets.

Table 8.3.2 Graphs of $\lambda_1 x_1^2 + \lambda_2 x_2^2 + \lambda_3 x_3^2 = k$

	$k > 0$	$k = 0$
$\lambda_1, \lambda_2, \lambda_3 > 0$	ellipsoid	point $(0, 0, 0)$
$\lambda_1, \lambda_2 > 0, \lambda_3 = 0$	elliptic cylinder	x_3-axis
$\lambda_1, \lambda_2 > 0, \lambda_3 < 0$	hyperboloid of one sheet	cone
$\lambda_1 > 0, \lambda_2 = 0, \lambda_3 = 0$	parallel planes	x_2x_3-plane
$\lambda_1 > 0, \lambda_2 = 0, \lambda_3 < 0$	hyperbolic cylinder	intersecting planes
$\lambda_1 > 0, \lambda_2, \lambda_3 < 0$	hyperboloid of two sheets	cone
$\lambda_1 = 0, \lambda_2, \lambda_3 < 0$	empty set	x_1-axis
$\lambda_1 < 0, \lambda_2 = 0, \lambda_3 = 0$	empty set	x_2x_3-plane
$\lambda_1, \lambda_2, \lambda_3 < 0$	empty set	point $(0, 0, 0)$

Table 8.3.2 displays the possible cases for $Q(\vec{x}) = k$ in $\mathbb{R}^3$. The nondegenerate cases are the ellipsoids and hyperboloids. Note that the hyperboloid of two sheets appears in the form $\frac{x_1^2}{a^2} - \frac{x_2^2}{b^2} - \frac{x_3^2}{c^2} = k, k > 0$.

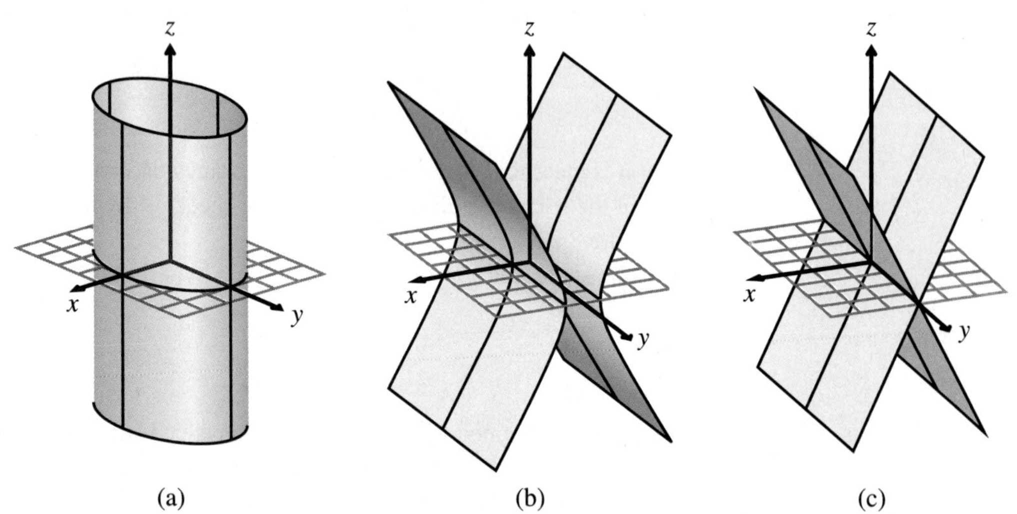

Figure 8.3.9 Some degenerate quadric surfaces. (a) an elliptic cylinder $x_1^2 + \lambda x_2^2 = 1$, parallel to the x_3-axis. (b) a hyperbolic cylinder $\lambda x_1^2 - x_3^2 = 1$, parallel to the x_2-axis. (c) intersecting planes $\lambda x_1^2 - x_3^2 = 0$.

Figure 8.3.9 shows some degenerate quadric surfaces. Note that paraboloidal surfaces do not appear as graphs of the form $Q(\vec{x}) = k$ in $\mathbb{R}^3$ for the same reason that parabolas do not appear in Table 8.3.1 for $\mathbb{R}^2$: their equations contain first-degree terms.

PROBLEMS 8.3

Practice Problems

A1 Sketch the graph of $2x_1^2 + 4x_1x_2 - x_2^2 = 6$. Show both the original axes and the new axes.

A2 Sketch the graph of $2x_1^2 + 6x_1x_2 + 10x_2^2 = 11$. Show both the original axes and the new axes.

A3 Sketch the graph of $4x_1^2 - 6x_1x_2 + 4x_2^2 = 12$. Show both the original axes and the new axes.

A4 Sketch the graph of $5x_1^2 + 6x_1x_2 - 3x_2^2 = 15$. Show both the original axes and the new axes.

A5 Sketch the graph of $x_1^2 - 4x_1x_2 + x_2^2 = 8$. Show both the original axes and the new axes.

A6 Sketch the graph of $x_1^2 + 4x_1x_2 + x_2^2 = 8$. Show both the original axes and the new axes.

A7 Sketch the graph of $3x_1^2 - 4x_1x_2 + 3x_2^2 = 32$. Show both the original axes and the new axes.

For Problems A8–A13, identify the shape of the graph of $\vec{x}^T A\vec{x} = 1$ and the shape of the graph of $\vec{x}^T A\vec{x} = -1$.

A8 $A = \begin{bmatrix} 4 & 2 \\ 2 & 1 \end{bmatrix}$

A9 $A = \begin{bmatrix} 5 & 3 \\ 3 & -3 \end{bmatrix}$

A10 $A = \begin{bmatrix} 0 & 1 & 1 \\ 1 & 0 & 1 \\ 1 & 1 & 0 \end{bmatrix}$

A11 $A = \begin{bmatrix} 1 & 0 & -2 \\ 0 & -1 & -2 \\ -2 & -2 & 0 \end{bmatrix}$

A12 $A = \begin{bmatrix} 1 & 8 & 4 \\ 8 & 1 & -4 \\ 4 & -4 & 7 \end{bmatrix}$

A13 $A = \begin{bmatrix} 8 & -2 & -2 \\ -2 & 5 & -4 \\ -2 & -4 & 5 \end{bmatrix}$

For Problems A14–A17, express $Q(\vec{x})$ in diagonal form and use the diagonal form to determine the shape of the surface $Q(\vec{x}) = k$ for $k = 1, 0, -1$.

A14 $Q(\vec{x}) = -x_1^2 + 12x_1x_2 + 4x_2^2$

A15 $Q(\vec{x}) = x_1^2 + 2x_1x_2 - 4x_1x_3 + x_2^2 - 4x_2x_3 + 4x_3^2$

A16 $Q(\vec{x}) = 2x_1^2 + 4x_1x_2 + 3x_2^2 + 4x_2x_3 + 4x_3^2$

A17 $Q(\vec{x}) = 4x_1^2 + 2x_1x_2 + 5x_2^2 - 2x_2x_3 + 4x_3^2$

Homework Problems

B1 Sketch the graph of $4x_1^2 - 24x_1x_2 + 11x_2^2 = 20$. Show both the original axes and the new axes.

B2 Sketch the graph of $2x_1^2 + 12x_1x_2 + 7x_2^2 = 11$. Show both the original axes and the new axes.

B3 Sketch the graph of $3x_1^2 - 4x_1x_2 + 6x_2^2 = 14$. Show both the original axes and the new axes.

B4 Sketch the graph of $12x_1^2 + 10x_1x_2 - 12x_2^2 = 13$. Show both the original axes and the new axes.

B5 Sketch the graph of $9x_1^2 + 4x_1x_2 + 6x_2^2 = 90$. Show both the original axes and the new axes.

B6 Sketch the graph of $x_1^2 + 6x_1x_2 - 7x_2^2 = 32$. Show both the original axes and the new axes.

For Problems B7–B14, identify the shape of the graph of $\vec{x}^T A\vec{x} = 1$ and the shape of the graph of $\vec{x}^T A\vec{x} = -1$.

B7 $A = \begin{bmatrix} 1 & 2 \\ 2 & -2 \end{bmatrix}$

B8 $A = \begin{bmatrix} -2 & -3 \\ -3 & -10 \end{bmatrix}$

B9 $A = \begin{bmatrix} 26 & 10 \\ 10 & 5 \end{bmatrix}$

B10 $A = \begin{bmatrix} 1 & 2 \\ 2 & 4 \end{bmatrix}$

B11 $A = \begin{bmatrix} -5 & -2 & 1 \\ -2 & -2 & -2 \\ 1 & -2 & -5 \end{bmatrix}$

B12 $A = \begin{bmatrix} 3 & 2 & -1 \\ 2 & 0 & 2 \\ -1 & 2 & 3 \end{bmatrix}$

B13 $A = \begin{bmatrix} -3 & 0 & 1 \\ 0 & -2 & 0 \\ 1 & 0 & -3 \end{bmatrix}$

B14 $A = \begin{bmatrix} -4 & -2 & 4 \\ -2 & -1 & 2 \\ 4 & 2 & -4 \end{bmatrix}$

For Problems B15–B21, express $Q(\vec{x})$ in diagonal form and use the diagonal form to determine the shape of the surface $Q(\vec{x}) = k$ for $k = 1, 0, -1$.

B15 $Q(\vec{x}) = 5x_1^2 + 4x_1x_2 + 2x_2^2$

B16 $Q(\vec{x}) = x_1^2 - 4x_1x_2 + x_2^2$

B17 $Q(\vec{x}) = -3x_1^2 + 8x_1x_2 - 9x_2^2$

B18 $Q(\vec{x}) = x_1^2 + 2x_1x_2 + x_2^2$

B19 $Q(\vec{x}) = x_1^2 + 6x_1x_2 + 2x_1x_3 + x_2^2 + 2x_2x_3 + 5x_3^2$

B20 $Q(\vec{x}) = x_1^2 + 4x_1x_2 + 4x_1x_3 + 5x_2^2 + 6x_2x_3 + 5x_3^2$

B21 $Q(\vec{x}) = -x_1^2 + 2x_1x_2 - 6x_1x_3 + x_2^2 - 2x_2x_3 - x_3^2$

8.4 Applications of Quadratic Forms

Some may think of mathematics as only a set of rules for doing calculations. However, a theorem such as the Principal Axis Theorem is often important because it provides a simple way of thinking about complicated situations. The Principal Axis Theorem plays an important role in the two applications described here.

Small Deformations

A small deformation of a solid body may be understood as the composition of three stretches along the principal axes of a symmetric matrix together with a rigid rotation of the body.

Consider a body of material that can be deformed when it is subjected to some external forces. This might be, for example, a piece of steel under some load. Fix an origin of coordinates $\vec{0}$ in the body; to simplify the story, suppose that this origin is left unchanged by the deformation. Suppose that a material point in the body, which is at $\vec{x}$ before the forces are applied, is moved by the forces to the point

$$f(\vec{x}) = (f_1(\vec{x}), f_2(\vec{x}), f_3(\vec{x}))$$

where we have assumed that $f(\vec{0}) = \vec{0}$. The problem is to understand this deformation f so that it can be related to the properties of the body.

For many materials under reasonable forces, the deformation is small; this means that the point $f(\vec{x})$ is not far from $\vec{x}$. It is convenient to introduce a parameter β to describe how small the deformation is. To do this, we define a function $h(\vec{x})$ by

$$f(\vec{x}) = \vec{x} + \beta h(\vec{x})$$

For many materials, an arbitrary small deformation is well approximated by its "best linear approximation," the derivative. In this case, the map $f : \mathbb{R}^3 \to \mathbb{R}^3$ is approximated near the origin by the linear transformation with matrix $\left[\frac{\partial f_j}{\partial x_k}(\vec{0})\right]$, so that in this approximation, a point originally at $\vec{v}$ is moved (approximately) to $\left[\frac{\partial f_j}{\partial x_k}(\vec{0})\right]\vec{v}$. (This is a standard calculus approximation.)

In terms of the parameter β and the function h, this matrix can be written as

$$\left[\frac{\partial f_j}{\partial x_k}(\vec{0})\right] = I + \beta G$$

where $G = \left[\frac{\partial h_j}{\partial x_k}(\vec{0})\right]$. In this situation, it is useful to write G as $G = E + W$, where

$$E = \frac{1}{2}(G + G^T)$$

is its symmetric part, and

$$W = \frac{1}{2}(G - G^T)$$

is its skew-symmetric part, as in Section 8.2 Problem C9.

The next step is to observe that we can write

$$I + \beta G = I + \beta(E + W) = (I + \beta E)(I + \beta W) - \beta^2 EW$$

Since β is assumed to be small, β^2 is very small and may be ignored. (Such treatment of terms like β^2 can be justified by careful discussion of the limit at $\beta \to 0$.)

The small deformation we started with is now described as the composition of two linear transformations, one with matrix $I + \beta E$ and the other with matrix $I + \beta W$. It can be shown that $I + \beta W$ describes a small rigid rotation of the body; a rigid rotation does not alter the distance between any two points in the body. (The matrix βW is called an *infinitesimal rotation*.)

Finally, we have the linear transformation with matrix $I + \beta E$. This matrix is symmetric, so there exist principal axes such that the symmetric matrix is diagonalized to $\begin{bmatrix} 1+\epsilon_1 & 0 & 0 \\ 0 & 1+\epsilon_2 & 0 \\ 0 & 0 & 1+\epsilon_3 \end{bmatrix}$. (It is equivalent to diagonalize βE and add the result to I, because I is transformed to itself under any orthonormal change of coordinates.) Since β is small, it follows that the numbers ϵ_j are small in magnitude, and therefore $1 + \epsilon_j > 0$. This diagonalized matrix can be written as the product of the three matrices:

$$\begin{bmatrix} 1+\epsilon_1 & 0 & 0 \\ 0 & 1+\epsilon_2 & 0 \\ 0 & 0 & 1+\epsilon_3 \end{bmatrix} = \begin{bmatrix} 1+\epsilon_1 & 0 & 0 \\ 0 & 1 & 0 \\ 0 & 0 & 1 \end{bmatrix} \begin{bmatrix} 1 & 0 & 0 \\ 0 & 1+\epsilon_2 & 0 \\ 0 & 0 & 1 \end{bmatrix} \begin{bmatrix} 1 & 0 & 0 \\ 0 & 1 & 0 \\ 0 & 0 & 1+\epsilon_3 \end{bmatrix}$$

It is now apparent that, excluding rotation, the small deformation can be represented as the composition of three stretches along the principal axes of the matrix βE. The quantities ϵ_1, ϵ_2, and ϵ_3 are related to the external and internal forces in the material by elastic properties of the material. (βE is called the infinitesimal strain; this notation is not quite the standard notation. This will be important if you read further about this topic in a book on continuum mechanics.)

The Inertia Tensor

For the purpose of discussing the rotation motion of a rigid body, information about the mass distribution within the body is summarized in a symmetric matrix N called the **inertia tensor**. The tensor is easiest to understand if principal axes are used so that the matrix is diagonal; in this case, the diagonal entries are simply the moments of inertia about the principal axes, and the moment of inertia about any other axis can be calculated in terms of these **principal moments of inertia**. In general, the **angular momentum** vector $\vec{J}$ of the rotating body is equal to $N\vec{\omega}$, where $\vec{\omega}$ is the **instantaneous angular velocity** vector. The vector $\vec{J}$ is a scalar multiple of $\vec{\omega}$ if and only if $\vec{\omega}$ is an eigenvector of N—that is, if and only if the axis of rotation is one of the principal axes of the body. This is a beginning to an explanation of how the body wobbles during rotation ($\vec{\omega}$ need not be constant) even though $\vec{J}$ is a conserved quantity. That is, $\vec{J}$ is constant if no external force is applied.

Suppose that a rigid body is rotating about some point in the body that remains fixed in space throughout the rotation. Make this fixed point the origin $(0, 0, 0)$. Suppose that there are coordinate axes fixed in space and also three reference axes that are fixed in the body (so that they rotate with the body). At any time t, these body axes make certain angles with respect to the space axes; at a later time $t + \Delta t$, the body axes have moved to a new position. Since $(0, 0, 0)$ is fixed and the body is rigid, the body axes have moved only by a rotation, and it is a fact that any rotation in $\mathbb{R}^3$ is determined by its axis and an angle. Call the unit vector along this axis $\vec{u}(t + \Delta t)$ and denote the angle by $\Delta\theta$. Now let $\Delta t \to 0$; the unit vector $\vec{u}(t + \Delta t)$ must tend to a limit $\vec{u}(t)$, and this determines the **instantaneous axis of rotation at time** t. Also, as $\Delta t \to 0$, $\frac{\Delta\theta}{\Delta t} \to \frac{d\theta}{dt}$, the **instantaneous rate of rotation about the axis**. The **instantaneous angular velocity** is defined to be the vector $\vec{\omega} = \left(\frac{d\theta}{dt}\right)\vec{u}(t)$.

(It is a standard exercise to show that the instantaneous linear velocity $\vec{v}(t)$ at some point in the body whose space coordinates are given by $\vec{x}(t)$ is determined by $\vec{v} = \vec{\omega} \times \vec{x}$.)

To use concepts such as energy and momentum in the discussion of rotating motion, it is necessary to introduce moments of inertia.

For a single mass m at the point (x_1, x_2, x_3) the **moment of inertia about the x_3-axis** is defined to be $m(x_1^2 + x_2^2)$; this will be denoted by n_{33}. The factor $(x_1^2 + x_2^2)$ is simply the square of the distance of the mass from the x_3-axis. There are similar definitions of the moments of inertia about the x_1-axis (denoted by n_{11}) and about the x_2-axis (denoted by n_{22}).

For a general axis ℓ through the origin with unit direction vector $\vec{u}$, the moment of inertia of the mass about ℓ is defined to be m multiplied by the square of the distance of m from ℓ. Thus, if we let $\vec{x} = \begin{bmatrix} x_1 \\ x_2 \\ x_3 \end{bmatrix}$, the moment of inertia in this case is

$$m\|\operatorname{perp}_{\vec{u}} \vec{x}\|^2 = m[\vec{x} - (\vec{x} \cdot \vec{u})\vec{u}]^T[\vec{x} - (\vec{x} \cdot \vec{u})\vec{u}] = m(\|\vec{x}\|^2 - (\vec{x} \cdot \vec{u})^2)$$

With some manipulation, using $\vec{u}^T\vec{u} = 1$ and $\vec{x} \cdot \vec{u} = \vec{x}^T\vec{u}$, we can verify that this is equal to the expression

$$\vec{u}^T m(\|\vec{x}\|^2 I - \vec{x}\vec{x}^T)\vec{u}$$

Because of this, for the single point mass m at $\vec{x}$, we define the **inertia tensor** N to be the 3×3 matrix

$$N = m(\|\vec{x}\|^2 I - \vec{x}\vec{x}^T)$$

(Vectors and matrices are special kinds of "tensors"; for our present purposes, we simply treat N as a matrix.) With this definition, the moment of inertia about an axis with unit direction $\vec{u}$ is

$$\vec{u}^T N \vec{u}$$

It is easy to check that N is the matrix with components n_{11}, n_{22}, and n_{33} as given above, and for $i \neq j$, $n_{ij} = -mx_ix_j$. It is clear that this matrix N is symmetric because $\vec{x}\vec{x}^T$ is a symmetric 3×3 matrix. (The term mx_ix_j is called a *product of inertia*. This name has no special meaning; the term is simply a product that appears as an entry in the inertia tensor.)

It is easy to extend the definition of moments of inertia and the inertia tensor to bodies that are more complicated than a single point mass. Consider a rigid body that can be thought of as k masses joined to each other by weightless rigid rods. The moment of inertia of the body about the x_3-axis is determined by taking the moment of inertia about the x_3-axis of each mass and simply adding these moments; the moments about the x_1- and x_2-axes, and the products of the inertia are defined similarly. The inertia tensor of this body is just the sum of the inertia tensors of the k masses; since it is the sum of symmetric matrices, it is also symmetric. If the mass is distributed continuously, the various moments and products of inertia are determined by definite integrals. In any case, the inertia tensor N is still defined, and is still a symmetric matrix.

Since N is a symmetric matrix, it can be brought into diagonal form by the Principal Axis Theorem. The diagonal entries are then the moments of inertia with respect to the principal axes, and these are called the **principal moments of inertia**. Denote these by N_1, N_2, and N_3. Let $\mathcal{P}$ denote the orthonormal basis consisting of eigenvectors of N (which means these vectors are unit vectors along the principal axes).

Suppose an arbitrary axis ℓ is determined by the unit vector $\vec{u}$ such that $[\vec{u}]_{\mathcal{P}} = \begin{bmatrix} p_1 \\ p_2 \\ p_3 \end{bmatrix}$.

Then, from the discussion of quadratic forms in Section 8.2, the moment of inertia about this axis ℓ is simply

$$\begin{aligned} \vec{u}^T N \vec{u} &= \begin{bmatrix} p_1 & p_2 & p_3 \end{bmatrix} \begin{bmatrix} N_1 & 0 & 0 \\ 0 & N_2 & 0 \\ 0 & 0 & N_3 \end{bmatrix} \begin{bmatrix} p_1 \\ p_2 \\ p_3 \end{bmatrix} \\ &= p_1^2 N_1 + p_2^2 N_2 + p_3^2 N_3 \end{aligned}$$

This formula is greatly simplified because of the use of the principal axes.

It is important to get equations for rotating motion that corresponds to Newton's equation:

The rate of change of momentum equals the applied force.

The appropriate equation is

The rate of change of angular momentum equals the applied torque.

It turns out that the right way to define the angular momentum vector $\vec{J}$ for a general body is

$$\vec{J} = N(t)\vec{\omega}(t)$$

Note that in general N is a function of time since it depends on the positions at time t of each of the masses making up the solid body. Understanding the possible motions of a rotating body depends on determining $\vec{\omega}(t)$, or at least saying something about it. In general, this is a very difficult problem, but there will often be important simplifications if N is diagonalized by the Principal Axis Theorem. Note that $\vec{J}(t)$ is parallel to $\vec{\omega}(t)$ if and only if $\vec{\omega}(t)$ is an eigenvector of $N(t)$.

PROBLEM 8.4

Conceptual Problem

C1 Show that if P is an orthogonal matrix that diagonalizes the symmetric matrix βE to a matrix with diagonal entries ϵ_1, ϵ_2, and ϵ_3, then P also diagonalizes $(I + \beta E)$ to a matrix with diagonal entries $1 + \epsilon_1$, $1 + \epsilon_2$, and $1 + \epsilon_3$.

8.5 Singular Value Decomposition

The theory of orthogonal diagonalization of symmetric matrices is extremely useful and powerful, but it has one drawback; it applies only to symmetric matrices. In many real world situations, matrices are not only not symmetric nor diagonalizable, they are not even square. So, the natural question to ask is if we can mimic orthogonal diagonalization for non-square matrices.

Let $A \in M_{m\times n}(\mathbb{R})$. To truly mimic orthogonal diagonalization we would need to find an orthogonal matrix P such that

$$P^T AP = D$$

is an $m \times n$ **rectangular diagonal matrix**; that is, a matrix where all entries are 0 except possibly entries along the main diagonal.

However, we know that this is not possible since the sizes do not work out if $n \neq m$. In particular, we require that P^T is an $m \times m$ matrix for the matrix-matrix multiplication $P^T A$ to be defined, and we need P to be an $n \times n$ matrix for AP to be defined. Thus, rather than looking for a single orthogonal matrix P, we will try to find an $m \times m$ orthogonal matrix U and an $n \times n$ orthogonal matrix V such that

$$U^T AV = \Sigma$$

is an $m \times n$ rectangular diagonal matrix. Rather than emphasizing the matrix Σ, we typically solve this equation for A to get a matrix decomposition

$$A = U\Sigma V^T$$

The difficult question is how to find such a matrix decomposition. If we are going to mimic orthogonal diagonalization, it would be helpful to relate A to a symmetric matrix. The key is to recognize that both $A^T A$ and AA^T are symmetric matrices. Indeed

$$(A^T A)^T = A^T (A^T)^T = A^T A$$

and

$$(AA^T)^T = (A^T)^T A^T = AA^T$$

Thus, if such a matrix decomposition exists, we find that

$$\begin{aligned} A^T A &= (U\Sigma V^T)^T (U\Sigma V^T) \\ &= V\Sigma^T U^T U\Sigma V^T && \text{by properties of transposes} \\ &= V\Sigma^T I\Sigma V^T && \text{since } U \text{ is orthogonal} \\ A^T A &= V\Sigma^T \Sigma V^T && (8.1) \end{aligned}$$

We observe that $\Sigma^T \Sigma$ will be a diagonal matrix, and so we can rewrite equation (8.1) as

$$V^T (A^T A)V = \text{diag}(\sigma_1^2, \ldots, \sigma_n^2)$$

where $\sigma_i = (\Sigma)_{ii}$.

Observe that this is an orthogonal diagonalization of $A^T A$. In particular, it shows us that we want the diagonal entries of Σ to be the square roots of the eigenvalues of $A^T A$.

Definition
Singular Values

Let $A \in M_{m\times n}(\mathbb{R})$. The **singular values** $\sigma_1, \dots, \sigma_n$ of A are the square roots of the eigenvalues of $A^T A$ arranged so that $\sigma_1 \geq \sigma_2 \geq \cdots \geq \sigma_n \geq 0$.

Remark

We use the convention that the singular values are always ordered from greatest to least, as it is helpful in a variety of applications to know the position of the largest singular values.

EXAMPLE 8.5.1

Find the singular values of $A = \begin{bmatrix} 1 & 1 & 1 \\ 2 & 2 & 2 \end{bmatrix}$.

Solution: We have $A^T A = \begin{bmatrix} 5 & 5 & 5 \\ 5 & 5 & 5 \\ 5 & 5 & 5 \end{bmatrix}$. The characteristic polynomial of $A^T A$ is $C(\lambda) = -\lambda^2(\lambda - 15)$. Thus, the eigenvalues (ordered from greatest to least) are $\lambda_1 = 15$, $\lambda_2 = 0$, and $\lambda_3 = 0$. So, the singular values of A are $\sigma_1 = \sqrt{15}$, $\sigma_2 = 0$, and $\sigma_3 = 0$.

EXAMPLE 8.5.2

Find the singular values of $B = \begin{bmatrix} 1 & 1 \\ -1 & 2 \\ -1 & 1 \end{bmatrix}$.

Solution: We have $B^T B = \begin{bmatrix} 3 & -2 \\ -2 & 6 \end{bmatrix}$. The characteristic polynomial of $B^T B$ is $C(\lambda) = (\lambda - 2)(\lambda - 7)$. Hence, the eigenvalues are $\lambda_1 = 7$ and $\lambda_2 = 2$. Therefore, the singular values of B are $\sigma_1 = \sqrt{7}$ and $\sigma_2 = \sqrt{2}$.

EXAMPLE 8.5.3

Find the singular values of $A = \begin{bmatrix} 2 & -1 \\ 1 & 0 \\ 1 & 2 \end{bmatrix}$ and $B = \begin{bmatrix} 2 & 1 & 1 \\ -1 & 0 & 2 \end{bmatrix}$.

Solution: We have $A^T A = \begin{bmatrix} 6 & 0 \\ 0 & 5 \end{bmatrix}$. Hence, the eigenvalues of $A^T A$ are $\lambda_1 = 6$ and $\lambda_2 = 5$. Thus, the singular values of A are $\sigma_1 = \sqrt{6}$ and $\sigma_2 = \sqrt{5}$.

We have $B^T B = \begin{bmatrix} 5 & 2 & 0 \\ 2 & 1 & 1 \\ 0 & 1 & 5 \end{bmatrix}$. The characteristic polynomial of $B^T B$ is

$$C(\lambda) = \begin{vmatrix} 5-\lambda & 2 & 0 \\ 2 & 1-\lambda & 1 \\ 0 & 1 & 5-\lambda \end{vmatrix} = -\lambda(\lambda - 5)(\lambda - 6)$$

Hence, the eigenvalues of $B^T B$ are $\lambda_1 = 6$, $\lambda_2 = 5$, and $\lambda_3 = 0$. Thus, the singular values of B are $\sigma_1 = \sqrt{6}$, $\sigma_2 = \sqrt{5}$, and $\sigma_3 = 0$.

You are asked to prove that a matrix and its transpose have the same non-zero singular values in Problem C5.

EXERCISE 8.5.1

Find the singular values of $A = \begin{bmatrix} 1 & -1 \\ 3 & 1 \\ 1 & 1 \end{bmatrix}$ and $B = \begin{bmatrix} 1 & -1 & 1 \\ 0 & 2 & 2 \end{bmatrix}$.

Theorem 8.5.1

If $A \in M_{m\times n}(\mathbb{R})$ and $\operatorname{rank}(A) = r$, then A has r non-zero singular values.

You are guided through the proof of Theorem 8.5.1 in Problem C1.

We can now define our desired decomposition.

Definition

Singular Value Decomposition (SVD)
Left Singular Vectors
Right Singular Vectors

Let $A \in M_{m\times n}(\mathbb{R})$ with $\operatorname{rank}(A) = r$ and non-zero singular values $\sigma_1, \dots, \sigma_r$. If U is an $m \times m$ orthogonal matrix, V is an $n \times n$ orthogonal matrix, and Σ is an $m \times n$ rectangular diagonal matrix with $(\Sigma)_{ii} = \sigma_i$ for $1 \leq i \leq r$ and all other entries 0, then

$$A = U\Sigma V^T$$

is called a **singular value decomposition (SVD)** of A. The orthonormal columns of U are called **left singular vectors** of A. The orthonormal columns of V are called **right singular vectors** of A.

Equation (8.1) not only indicates how to define the singular values of a matrix A, but also shows us that the desired orthogonal matrix V for a singular value decomposition of A is a matrix which orthogonally diagonalizes A^TA (ensuring that we order the columns of V so that the corresponding eigenvalues are ordered from greatest to least).

We now need to define the columns of U so that we will indeed have $A = U\Sigma V^T$. Let $V = \begin{bmatrix} \vec{v}_1 & \cdots & \vec{v}_n \end{bmatrix}$ and $U = \begin{bmatrix} \vec{u}_1 & \cdots & \vec{u}_m \end{bmatrix}$. Rewriting $A = U\Sigma V^T$ as $AV = U\Sigma$ we find that

$$\begin{aligned} A\begin{bmatrix} \vec{v}_1 & \cdots & \vec{v}_n \end{bmatrix} &= \begin{bmatrix} \vec{u}_1 & \cdots & \vec{u}_m \end{bmatrix}\begin{bmatrix} \sigma_1\vec{e}_1 & \cdots & \sigma_r\vec{e}_r & \vec{0} & \cdots & \vec{0} \end{bmatrix} \\ \begin{bmatrix} A\vec{v}_1 & \cdots & A\vec{v}_n \end{bmatrix} &= \begin{bmatrix} \sigma_1\vec{u}_1 & \cdots & \sigma_r\vec{u}_r & \vec{0} & \cdots & \vec{0} \end{bmatrix} \end{aligned}$$

Hence, we require that

$$A\vec{v}_i = \sigma_i\vec{u}_i \Rightarrow \vec{u}_i = \frac{1}{\sigma_i}A\vec{v}_i, \quad 1 \leq i \leq r$$

We now prove that the vectors $\{\vec{u}_1, \dots, \vec{u}_r\}$ indeed form an orthonormal basis.

Theorem 8.5.2

Let $A \in M_{m\times n}(\mathbb{R})$ with $\operatorname{rank}(A) = r$ and non-zero singular values $\sigma_1, \dots, \sigma_r$. Let $\{\vec{v}_1, \dots, \vec{v}_n\}$ be an orthonormal basis for $\mathbb{R}^n$ consisting of the eigenvectors of A^TA arranged so that the corresponding eigenvalues are ordered from greatest to least. If we define

$$\vec{u}_i = \frac{1}{\sigma_i}A\vec{v}_i, \quad 1 \leq i \leq r$$

then $\{\vec{u}_1, \dots, \vec{u}_r\}$ is an orthonormal basis for Col A.

Proof: Observe that for $1 \leq i, j \leq r$, we have

$$\begin{aligned}\vec{u}_i \cdot \vec{u}_j &= \frac{1}{\sigma_i}(A\vec{v}_i) \cdot \frac{1}{\sigma_j}\left(A\vec{v}_j\right) \\ &= \frac{1}{\sigma_i\sigma_j}(A\vec{v}_i)^T(A\vec{v}_j) \\ &= \frac{1}{\sigma_i\sigma_j}\vec{v}_i^T A^T A\vec{v}_j\end{aligned}$$

We have assumed that $\vec{v}_j$ is an eigenvector of A^TA corresponding to the eigenvalue $\lambda_j = \sigma_j^2$. Thus, we have

$$\begin{aligned}\vec{u}_i \cdot \vec{u}_j &= \frac{1}{\sigma_i\sigma_j}\vec{v}_i^T(\sigma_j^2\vec{v}_j) \\ &= \frac{\sigma_j}{\sigma_i}\vec{v}_i^T(\vec{v}_j) \\ &= \frac{\sigma_j}{\sigma_i}\vec{v}_i \cdot \vec{v}_j = \begin{cases} 0 & \text{if } i \neq j \\ 1 & \text{if } i = j\end{cases}\end{aligned}$$

because $\{\vec{v}_1, \ldots, \vec{v}_n\}$ is an orthonormal set. By definition, $\vec{u}_i \in \text{Col}(A)$ and, since $\dim \text{Col}(A) = r$, Theorem 4.3.7 implies that this is an orthonormal basis for Col A. ■

We now have an algorithm for finding the singular value decomposition of a matrix.

Algorithm 8.5.1

To find a singular value decomposition of $A \in M_{m\times n}(\mathbb{R})$ with $\text{rank}(A) = r$:

(1) Find the eigenvalues $\lambda_1, \ldots, \lambda_n$ of A^TA arranged from greatest to least and a corresponding orthonormal set of eigenvectors $\{\vec{v}_1, \ldots, \vec{v}_n\}$. Let $V = \begin{bmatrix}\vec{v}_1 & \cdots & \vec{v}_n\end{bmatrix}$.

(2) Define $\sigma_i = \sqrt{\lambda_i}$ for $1 \leq i \leq r$, and let Σ be the $m \times n$ matrix such that $(\Sigma)_{ii} = \sigma_i$ for $1 \leq i \leq r$ and all other entries of Σ are 0.

(3) Compute $\vec{u}_i = \dfrac{1}{\sigma_i}A\vec{v}_i$ for $1 \leq i \leq r$, and then extend the set $\{\vec{u}_1, \ldots, \vec{u}_r\}$ to an orthonormal basis $\{\vec{u}_1, \ldots, \vec{u}_m\}$ for $\mathbb{R}^m$. Take $U = \begin{bmatrix}\vec{u}_1 & \cdots & \vec{u}_m\end{bmatrix}$.

Then, $A = U\Sigma V^T$ is a singular value decomposition of A.

Remark

Since $\{\vec{u}_1, \ldots, \vec{u}_r\}$ is an orthonormal basis for $\text{Col}(A)$, by the Fundamental Theorem of Linear Algebra, one way to extend $\{\vec{u}_1, \ldots, \vec{u}_r\}$ to an orthonormal basis for $\mathbb{R}^m$ is by adding an orthonormal basis for $\text{Null}(A^T)$.

EXAMPLE 8.5.4

Find a singular value decomposition of $A = \begin{bmatrix} 1 & -1 & 3 \\ 3 & 1 & 1 \end{bmatrix}$.

Solution: Our first step is to orthogonally diagonalize $A^T A$. We find that

$$A^T A = \begin{bmatrix} 10 & 2 & 6 \\ 2 & 2 & -2 \\ 6 & -2 & 10 \end{bmatrix}$$

has eigenvalues (ordered from greatest to least) of $\lambda_1 = 16$, $\lambda_2 = 6$, and $\lambda_3 = 0$ and corresponding orthonormal eigenvectors are

$$\vec{v}_1 = \begin{bmatrix} 1/\sqrt{2} \\ 0 \\ 1/\sqrt{2} \end{bmatrix}, \quad \vec{v}_2 = \begin{bmatrix} 1/\sqrt{3} \\ 1/\sqrt{3} \\ -1/\sqrt{3} \end{bmatrix}, \quad \vec{v}_3 = \begin{bmatrix} 1/\sqrt{6} \\ -2/\sqrt{6} \\ -1/\sqrt{6} \end{bmatrix}$$

Hence, we define $V = \begin{bmatrix} 1/\sqrt{2} & 1/\sqrt{3} & 1/\sqrt{6} \\ 0 & 1/\sqrt{3} & -2/\sqrt{6} \\ 1/\sqrt{2} & -1/\sqrt{3} & -1/\sqrt{6} \end{bmatrix}$.

The non-zero singular values of A are $\sigma_1 = \sqrt{16} = 4$, $\sigma_2 = \sqrt{6}$. Hence, we take

$$\Sigma = \begin{bmatrix} 4 & 0 & 0 \\ 0 & \sqrt{6} & 0 \end{bmatrix}$$

Now, we compute

$$\vec{u}_1 = \frac{1}{\sigma_1} A\vec{v}_1 = \frac{1}{4}\begin{bmatrix} 4/\sqrt{2} \\ 4/\sqrt{2} \end{bmatrix} = \begin{bmatrix} 1/\sqrt{2} \\ 1/\sqrt{2} \end{bmatrix}$$

$$\vec{u}_2 = \frac{1}{\sigma_2} A\vec{v}_2 = \frac{1}{\sqrt{6}}\begin{bmatrix} -3/\sqrt{3} \\ 3/\sqrt{3} \end{bmatrix} = \begin{bmatrix} -1/\sqrt{2} \\ 1/\sqrt{2} \end{bmatrix}$$

Since this forms an orthonormal basis for $\mathbb{R}^2$, we take

$$U = \begin{bmatrix} 1/\sqrt{2} & -1/\sqrt{2} \\ 1/\sqrt{2} & 1/\sqrt{2} \end{bmatrix}$$

Then, $A = U\Sigma V^T$ is a singular value decomposition of A.

EXERCISE 8.5.2

Find a singular value decomposition of $A = \begin{bmatrix} 1 & -1 \\ 3 & 1 \\ 1 & 1 \end{bmatrix}$.

EXAMPLE 8.5.5

Find a singular value decomposition of $B = \begin{bmatrix} 2 & -4 \\ 2 & 2 \\ -4 & 0 \\ 1 & 4 \end{bmatrix}$.

Solution: Our first step is to orthogonally diagonalize $B^T B$. We have

$$B^T B = \begin{bmatrix} 25 & 0 \\ 0 & 36 \end{bmatrix}$$

The eigenvalues of $B^T B$ are $\lambda_1 = 36$ and $\lambda_2 = 25$ with corresponding orthonormal eigenvectors $\vec{v}_1 = \begin{bmatrix} 0 \\ 1 \end{bmatrix}$ and $\vec{v}_2 = \begin{bmatrix} 1 \\ 0 \end{bmatrix}$. Consequently, we take

$$V = \begin{bmatrix} 0 & 1 \\ 1 & 0 \end{bmatrix}$$

The singular values of B are $\sigma_1 = \sqrt{36} = 6$ and $\sigma_2 = \sqrt{25} = 5$, so we have

$$\Sigma = \begin{bmatrix} 6 & 0 \\ 0 & 5 \\ 0 & 0 \\ 0 & 0 \end{bmatrix}$$

Next, we compute

$$\vec{u}_1 = \frac{1}{\sigma_1} B\vec{v}_1 = \frac{1}{6}\begin{bmatrix} -4 \\ 2 \\ 0 \\ 4 \end{bmatrix}$$

$$\vec{u}_2 = \frac{1}{\sigma_2} B\vec{v}_2 = \frac{1}{5}\begin{bmatrix} 2 \\ 2 \\ -4 \\ 1 \end{bmatrix}$$

But, we have only two vectors in $\mathbb{R}^4$, so we need to extend $\{\vec{u}_1, \vec{u}_2\}$ to an orthonormal basis $\{\vec{u}_1, \vec{u}_2, \vec{u}_3, \vec{u}_4\}$ for $\mathbb{R}^4$. To do this we will find an orthonormal basis for Null(B^T). Row reducing B^T gives

$$B^T \sim \begin{bmatrix} 1 & 0 & -2/3 & -1/2 \\ 0 & 1 & -4/3 & 1 \end{bmatrix}$$

Hence, a basis for Null(B^T) is $\left\{ \begin{bmatrix} 1 \\ -2 \\ 0 \\ 2 \end{bmatrix}, \begin{bmatrix} 2 \\ 4 \\ 3 \\ 0 \end{bmatrix} \right\}$. Applying the Gram-Schmidt Procedure to this set and then normalizing gives

$$\vec{u}_3 = \frac{1}{3}\begin{bmatrix} 1 \\ -2 \\ 0 \\ 2 \end{bmatrix}, \quad \vec{u}_4 = \frac{1}{15}\begin{bmatrix} 8 \\ 8 \\ 9 \\ 4 \end{bmatrix}$$

Thus, we take $U = \begin{bmatrix} \vec{u}_1 & \vec{u}_2 & \vec{u}_3 & \vec{u}_4 \end{bmatrix}$ and we get $B = U\Sigma V^T$.

Applications of the Singular Value Decomposition

Pseudoinverse In Problem C4 of Section 7.3 we defined the pseudoinverse (**Moore-Penrose inverse**) of $A \in M_{m\times n}(\mathbb{R})$ as a matrix A^+ such that

$$A^+ = (A^TA)^{-1}A^T$$

However, this formula works only if A^TA is invertible. To overcome this drawback we can use the singular value decomposition.

Definition
Pseudoinverse

Let $A \in M_{m\times n}(\mathbb{R})$ with singular value decomposition $A = U\Sigma V^T$ and non-zero singular values $\sigma_1, \dots, \sigma_r$. Define Σ^+ to be the $n \times m$ matrix such that $(\Sigma^+)_{ii} = \frac{1}{\sigma i}$ for $1 \leq i \leq r$ and all other entries of Σ^+ are 0. We define the pseudoinverse A^+ of A by

$$A^+ = V\Sigma^+U^T$$

EXAMPLE 8.5.6

Find a pseudoinverse A^+ for the matrix $A = \begin{bmatrix} 1 & -1 & 3 \\ 3 & 1 & 1 \end{bmatrix}$ of Example 8.5.4. Verify that $AA^+A = A$ and $A^+AA^+ = A^+$.

Solution: We found in Example 8.5.4 that a singular value decomposition for A is $A = U\Sigma V^T$ where

$$U = \begin{bmatrix} 1/\sqrt{2} & -1/\sqrt{2} \\ 1/\sqrt{2} & 1/\sqrt{2} \end{bmatrix}, \quad \Sigma = \begin{bmatrix} 4 & 0 & 0 \\ 0 & \sqrt{6} & 0 \end{bmatrix}, \quad V = \begin{bmatrix} 1/\sqrt{2} & 1/\sqrt{3} & 1/\sqrt{6} \\ 0 & 1/\sqrt{3} & -2/\sqrt{6} \\ 1/\sqrt{2} & -1/\sqrt{3} & -1/\sqrt{6} \end{bmatrix}$$

Thus, we take

$$\Sigma^+ = \begin{bmatrix} 1/4 & 0 \\ 0 & 1/\sqrt{6} \\ 0 & 0 \end{bmatrix}$$

Then

$$A^+ = V\Sigma^+U^T = \begin{bmatrix} -1/24 & 7/24 \\ -1/6 & 1/6 \\ 7/24 & -1/24 \end{bmatrix}$$

Multiplying, we find that $AA^+ = I$, so we have $AA^+A = A$.

Computing A^+AA^+ by first multiplying A^+A gives

$$\begin{aligned} A^+AA^+ &= \begin{bmatrix} 5/6 & 1/3 & 1/6 \\ 1/3 & 1/3 & -1/3 \\ 1/6 & -1/3 & 5/6 \end{bmatrix} \begin{bmatrix} -1/24 & 7/24 \\ -1/6 & 1/6 \\ 7/24 & -1/24 \end{bmatrix} \\ &= \begin{bmatrix} -1/24 & 7/24 \\ -1/6 & 1/6 \\ 7/24 & -1/24 \end{bmatrix} \\ &= A^+ \end{aligned}$$

EXAMPLE 8.5.7

Find a pseudoinverse B^+ for the matrix $B = \begin{bmatrix} 1 & 2 \\ 1 & 2 \end{bmatrix}$.

Solution: We have that

$$B^T B = \begin{bmatrix} 2 & 4 \\ 4 & 8 \end{bmatrix}$$

The eigenvalues of $B^T B$ are $\lambda_1 = 10$ and $\lambda_2 = 0$ with corresponding orthonormal eigenvectors $\vec{v}_1 = \begin{bmatrix} 1/\sqrt{5} \\ 2/\sqrt{5} \end{bmatrix}$ and $\vec{v}_2 = \begin{bmatrix} -2/\sqrt{5} \\ 1/\sqrt{5} \end{bmatrix}$. Consequently, we take

$$V = \begin{bmatrix} 1/\sqrt{5} & -2/\sqrt{5} \\ 2/\sqrt{5} & 1/\sqrt{5} \end{bmatrix}$$

The non-zero singular value of B is $\sigma_1 = \sqrt{10}$. So, we take

$$\Sigma^+ = \begin{bmatrix} 1/\sqrt{10} & 0 \\ 0 & 0 \end{bmatrix}$$

Next, we compute

$$\vec{u}_1 = \frac{1}{\sigma_1} B\vec{v}_1 = \frac{1}{\sqrt{2}} \begin{bmatrix} 1 \\ 1 \end{bmatrix}$$

Since we only have one vector in $\mathbb{R}^2$, we need to extend $\{\vec{u}_1\}$ to an orthonormal basis $\{\vec{u}_1, \vec{u}_2\}$ for $\mathbb{R}^2$. We see that we can take $\vec{u}_2 = \begin{bmatrix} -1/\sqrt{2} \\ 1/\sqrt{2} \end{bmatrix}$. Hence,

$$U = \begin{bmatrix} 1/\sqrt{2} & -1/\sqrt{2} \\ 1/\sqrt{2} & 1/\sqrt{2} \end{bmatrix}$$

Then

$$B^+ = V\Sigma^+ U^T = \begin{bmatrix} 1/10 & 1/10 \\ 1/5 & 1/5 \end{bmatrix}$$

One useful property of the pseudoinverse is that it gives us a formula for the solution of the normal system $A^T A\vec{x} = A^T \vec{b}$ in the method of least squares with the smallest length.

Theorem 8.5.3

Let $A \in M_{m\times n}(\mathbb{R})$. The solution of $A^T A\vec{x} = A^T \vec{b}$ with the smallest length is

$$\vec{x} = A^+ \vec{b}$$

EXERCISE 8.5.3

Use Theorem 8.5.3 to determine the vector $\vec{x}$ that minimizes $\|B\vec{x} - \vec{b}\|$ for the system

$$\begin{aligned} x_1 + 2x_2 &= 1 \\ x_1 + 2x_2 &= 0 \end{aligned}$$

Effective Rank In real world applications we often have extremely large matrices and hence the computations are all performed by a computer. One important area of concern is the numerical stability of the calculations. That is, we must be careful of changes due to round-off error. This is best demonstrated with an example.

EXAMPLE 8.5.8

Consider the matrix $A = \begin{bmatrix} 1/3 & 1/2 & 1 \\ 1/7 & 1/4 & 1 \\ 10/21 & 3/4 & 2 \end{bmatrix}$. It is easy to check that the third row is the sum of the first two rows and hence $\operatorname{rank}(A) = 2$.

However, if we replace the fractions with decimals accurate to two decimals places, we would get $A' = \begin{bmatrix} 0.33 & 0.50 & 1 \\ 0.14 & 0.25 & 1 \\ 0.48 & 0.75 & 2 \end{bmatrix}$. The sum of the first two rows no longer equals the third rows, so $\operatorname{rank}(A') = 3$.

To counteract this problem, we can use Theorem 8.5.1, which says that $\operatorname{rank}(A)$ is equal to the number of non-zero singular values. In particular, we can define the **effective rank** of a matrix to be the number of singular values which are not less than some specified tolerance.

For example, the singular values for A' to two decimal places (our number of significant digits) are $\sigma_1 = 2.68$, $\sigma_2 = 0.20$, and $\sigma_3 = 0.00$ (the actual value is approximately $\sigma_3 = 0.0046$). Hence, the effective rank of A' is 2.

Image Compression Consider a digital image that is $1080x1920$ pixels in size (standard size for a high definition television). Each pixel is a value from 0 to 255 for its grayscale. Hence, a digital image is essentially an $m = 1080$ by $n = 1920$ matrix where each entry is an integer value from 0 to 255. Since it takes 8 bits to store a number from 0 to 255, such an image takes 16 588 800 bits. Transmitting large amounts of data like this is expensive, so we want to find a way to compress the information with minimal loss of quality.

It turns out that a singular value decomposition of the image matrix still perfectly represents the image. This immediately indicates that we can throw away any information corresponding to 0 singular values. Since many images contain redundant information, this amounts to substantial savings. However, we can do even better. The information corresponding to the larger singular values is the most important. That is, if k is the effective rank of the matrix, then retaining only the information corresponding to the first k singular values (since the singular values are ordered from greatest to least) will still give a very accurate representation of the original picture.

PROBLEM 8.5

Practice Problems

For Problems A1–A3, find the non-zero singular values of the matrix.

A1 $\begin{bmatrix} -2 & 0 \\ 0 & 2 \end{bmatrix}$ **A2** $\begin{bmatrix} \sqrt{3} & 2 \\ 0 & \sqrt{3} \end{bmatrix}$ **A3** $\begin{bmatrix} 1 & 1 & 1 \\ 2 & 2 & 2 \end{bmatrix}$

For Problems A4–A13, find a singular value decomposition of the matrix.

A4 $\begin{bmatrix} 2 & 2 \\ -1 & 2 \end{bmatrix}$ **A5** $\begin{bmatrix} 2 & 3 \\ 0 & 2 \end{bmatrix}$ **A6** $\begin{bmatrix} 1 & 3 \\ 2 & 1 \\ 1 & 1 \end{bmatrix}$

A7 $\begin{bmatrix} 2 & -1 \\ 2 & -1 \\ 2 & -1 \end{bmatrix}$ **A8** $\begin{bmatrix} 1 & 1 \\ 0 & 1 \\ 1 & -1 \end{bmatrix}$ **A9** $\begin{bmatrix} 1 & -1 \\ 2 & 2 \\ 1 & 1 \end{bmatrix}$

A10 $\begin{bmatrix} 1 & 0 & 1 \\ 1 & 1 & -1 \\ -1 & 1 & 0 \\ 0 & 1 & 1 \end{bmatrix}$ **A11** $\begin{bmatrix} 1 & 2 \\ 1 & 2 \\ 1 & 2 \\ 1 & 2 \end{bmatrix}$

A12 $\begin{bmatrix} 1 & 1 & 1 \\ 1 & -1 & -1 \end{bmatrix}$ **A13** $\begin{bmatrix} 1 & 1 \\ 0 & 2 \\ -2 & -1 \\ 1 & 3 \end{bmatrix}$

For Problems A14–A16, find the pseudoinverse of the matrix.

A14 $\begin{bmatrix} 2 & 2 \\ -1 & 2 \end{bmatrix}$ **A15** $\begin{bmatrix} 2 & -1 \\ 2 & -1 \\ 2 & -1 \end{bmatrix}$ **A16** $\begin{bmatrix} 1 & 1 & 1 \\ 1 & -1 & -1 \end{bmatrix}$

Homework Problems

For Problems B1–B3, find the non-zero singular values of the matrix.

B1 $\begin{bmatrix} -\sqrt{2} & 0 \\ 0 & 3 \end{bmatrix}$ **B2** $\begin{bmatrix} 2 & 2 \\ 3 & 3 \end{bmatrix}$ **B3** $\begin{bmatrix} -1 & -1 & 1 \\ 1 & 0 & 1 \end{bmatrix}$

For Problems B4–B18, find a singular value decomposition of the matrix.

B4 $\begin{bmatrix} 3 & 1 \\ 1 & 3 \end{bmatrix}$ **B5** $\begin{bmatrix} 2 & -1 \\ 2 & 1 \end{bmatrix}$ **B6** $\begin{bmatrix} 1 & 2 \\ 2 & 4 \end{bmatrix}$

B7 $\begin{bmatrix} -1 & 2 \\ 0 & 2 \\ 2 & 1 \end{bmatrix}$ **B8** $\begin{bmatrix} 1 & -1 \\ 2 & 2 \\ -1 & 1 \end{bmatrix}$ **B9** $\begin{bmatrix} 2 & 1 \\ -2 & -2 \\ 1 & 2 \end{bmatrix}$

B10 $\begin{bmatrix} 1 & 1 \\ 2 & 1 \\ -1 & 1 \end{bmatrix}$ **B11** $\begin{bmatrix} 1 & 0 \\ 0 & 1 \\ 2 & 0 \end{bmatrix}$ **B12** $\begin{bmatrix} 1 & 1 \\ 2 & 2 \\ -1 & -1 \end{bmatrix}$

B13 $\begin{bmatrix} 2 & 1 & 1 \\ 0 & 2 & 0 \\ 0 & 0 & 2 \end{bmatrix}$ **B14** $\begin{bmatrix} 1 & 1 & 1 \\ 1 & 1 & 1 \\ 2 & 2 & 2 \end{bmatrix}$

B15 $\begin{bmatrix} 1 & -1 & 1 \\ 0 & 1 & 1 \end{bmatrix}$ **B16** $\begin{bmatrix} 1 & 3 & -1 \\ 1 & 3 & -1 \end{bmatrix}$

B17 $\begin{bmatrix} 1 & 0 \\ 0 & 1 \\ 1 & 0 \\ 0 & 1 \end{bmatrix}$ **B18** $\begin{bmatrix} 1 & 2 \\ 2 & 1 \\ -1 & 1 \\ 1 & 1 \end{bmatrix}$

For Problems B19–B23, find the pseudoinverse of the matrix.

B19 $\begin{bmatrix} 2 & 0 \\ 0 & 0 \end{bmatrix}$ **B20** $\begin{bmatrix} 2 & -2 \\ 1 & 1 \end{bmatrix}$ **B21** $\begin{bmatrix} 2 & 1 \end{bmatrix}$

B22 $\begin{bmatrix} 1 & 2 \\ 3 & 1 \\ 1 & 1 \end{bmatrix}$ **B23** $\begin{bmatrix} 3 & 2 \\ 0 & 1 \\ 1 & 0 \end{bmatrix}$ **B24** $\begin{bmatrix} 4 & 0 & 1 \\ 0 & 0 & 3 \end{bmatrix}$

Conceptual Problems

C1 Let $A \in M_{m\times n}(\mathbb{R})$ with $\text{rank}(A) = r$. Prove Theorem 8.5.1 using the following steps.

(a) Prove that the number of non-zero eigenvalues of a symmetric matrix B equals $\text{rank}(B)$.

(b) Prove that $\text{Null}(A^TA) = \text{Null}(A)$.

(c) Use (b) to prove that $\text{rank}(A^TA) = \text{rank}(A)$.

(d) Conclude that A has r non-zero singular values.

C2 Let $A \in M_{m\times n}(\mathbb{R})$ and let $P \in M_{m\times m}(\mathbb{R})$ be an orthogonal matrix. Prove that PA has the same singular values as A.

C3 Let $A \in M_{n\times n}(\mathbb{R})$ be a symmetric matrix. Prove that the singular values of A are the absolute values of the eigenvalues of A.

C4 Let $U\Sigma V^T$ be a singular value decomposition for an $m \times n$ matrix A with rank r. Find, with proof, an orthonormal basis for Row(A), Col(A), Null(A), and Null(A^T) from the columns of U and V.

C5 Let $A \in M_{m\times n}(\mathbb{R})$. Prove that A and A^T have the same non-zero singular values.

C6 Let $A \in M_{m\times n}(\mathbb{R})$. Prove that the left singular vectors of A are all eigenvectors of AA^T.

C7 Let $A \in M_{m\times n}(\mathbb{R})$ with rank(A) $= r$. Show that a singular value decomposition $A = U\Sigma V^T$ allows us to write A as

$$A = \sigma_1\vec{u}_1\vec{v}_1^T + \cdots + \sigma_r\vec{u}_r\vec{v}_r^T$$

For Problems C8 and C9, let $A \in M_{n\times n}(\mathbb{R})$ with singular value decomposition $U\Sigma V^T$.

C8 Prove that if A^TA is invertible, then

$$(A^TA)^{-1}A^T = V\Sigma^+U^T$$

C9 Prove that A^+ has the following properties.
(a) $AA^+A = A$
(b) $A^+AA^+ = A^+$
(c) $(AA^+)^T = AA^+$
(d) $(A^+A)^T = A^+A$

CHAPTER REVIEW
Suggestions for Student Review

1 What does it mean for two matrices to be orthogonally similar? How does this relate to the concept of two matrices being similar? (Section 8.1)

2 List the properties of a symmetric matrix. How does the theory of diagonalization for symmetric matrices differ from the theory of diagonalization for general square matrices? (Section 8.1)

3 Explain the algorithm for orthogonally diagonalizing a symmetric matrix. When do you need to use the Gram-Schmidt procedure? (Section 8.1)

4 Explain the connection between quadratic forms and symmetric matrices. How do you find the symmetric matrix corresponding to a quadratic form and vice versa? Why did we choose to relate a quadratic form with a symmetric matrix as opposed to an asymmetric matrix? (Section 8.2)

5 How does diagonalization of the symmetric matrix enable us to diagonalize the quadratic form? What is the geometric interpretation of this? (Sections 8.2, 8.3)

6 List the classifications of a quadratic form. How does diagonalizing the corresponding symmetric matrix help us classify a quadratic form? (Section 8.2)

7 What role do eigenvalues play in helping us understand the graphs of equations $Q(\vec{x}) = k$, where $Q(\vec{x})$ is a quadratic form? How are the classifications of a quadratic form related to its graph? (Section 8.3)

8 Define the principal axes of a symmetric matrix A. How do the principal axes of A relate to the graph of $Q(\vec{x}) = \vec{x}^TA\vec{x} = k$? (Section 8.3)

9 When diagonalizing a symmetric matrix A, we know that we can choose the eigenvalues in any order. How would changing the order in which we pick the eigenvalues change the graph of $Q(\vec{x}) = \vec{x}^TA\vec{x} = k$? Explain. (Section 8.3)

10 Try to find applications of symmetric matrices and/or quadratic forms related to your chosen field of study. (Sections 8.1, 8.2, 8.3, 8.4)

11 Explain the similarities and differences between orthogonal diagonalization and the singular value decomposition. (Section 8.5)

12 Write the algorithm for finding a singular value decomposition of a matrix. Is the singular value decomposition of a matrix unique? (Section 8.5)

13 Research applications of the singular value decomposition. Why is the singular value decomposition particularly useful? (Section 8.5)

Chapter Quiz

E1 Let $A = \begin{bmatrix} 2 & -3 & 2 \\ -3 & 3 & 3 \\ 2 & 3 & 2 \end{bmatrix}$. Find an orthogonal matrix P such that $P^T AP = D$ is diagonal.

For Problems E2 and E3, write the quadratic form corresponding to the given symmetric matrix.

E2 $\begin{bmatrix} 2 & 4 \\ 4 & 5 \end{bmatrix}$

E3 $\begin{bmatrix} 1 & 0 & -2 \\ 0 & -3 & 4 \\ -2 & 4 & 1 \end{bmatrix}$

For Problems E4 and E5:

(a) Determine the symmetric matrix corresponding to $Q(\vec{x})$.

(b) Express $Q(\vec{x})$ in diagonal form and give the orthogonal matrix that brings it into this form.

(c) Classify $Q(\vec{x})$.

(d) Describe the shape of $Q(\vec{x}) = 1$ and $Q(\vec{x}) = 0$.

E4 $Q(\vec{x}) = 5x_1^2 + 4x_1x_2 + 5x_2^2$

E5 $Q(\vec{x}) = 2x_1^2 - 6x_1x_2 - 6x_1x_3 - 3x_2^2 + 4x_2x_3 - 3x_3^2$

E6 By diagonalizing the quadratic form, make a sketch of the graph of

$$5x_1^2 - 2x_1x_2 + 5x_2^2 = 12$$

in the x_1x_2-plane. Show the new and old coordinate axes.

E7 Sketch the graph of the hyperbola

$$x_1^2 + 4x_1x_2 + x_2^2 = 8$$

in the x_1x_2-plane, including the asymptotes in the sketch.

For Problems E8 and E9, find a singular value decomposition of the matrix.

E8 $A = \begin{bmatrix} 1 & 1 \\ 2 & 2 \end{bmatrix}$

E9 $B = \begin{bmatrix} 1 & 2 \\ -1 & 1 \\ -1 & -1 \end{bmatrix}$

E10 Find the pseudoinverse of the matrix $A = \begin{bmatrix} 1 & 1 \\ 0 & 0 \end{bmatrix}$.

E11 Prove that if A is a positive definite symmetric matrix, then $\langle \vec{x}, \vec{y} \rangle = \vec{x}^T A\vec{y}$ is an inner product on $\mathbb{R}^n$.

E12 Prove that if $A \in M_{4\times4}(\mathbb{R})$ is symmetric with characteristic polynomial $C(\lambda) = (\lambda - 3)^4$, then $A = 3I$.

E13 Find a 2×2 symmetric matrix such that a basis for the eigenspace of $\lambda_1 = 2$ is $\left\{\begin{bmatrix} 2 \\ -3 \end{bmatrix}\right\}$, and a basis for the eigenspace of $\lambda_2 = -1$ is $\left\{\begin{bmatrix} 3 \\ 2 \end{bmatrix}\right\}$.

E14 Prove that a square matrix A is invertible if and only if 0 is not a singular value of A.

For Problems E15–E22, determine whether the statement is true or false. Justify your answer.

E15 If A and B are orthogonally similar, then A^2 and B^2 are orthogonally similar.

E16 If A and B are orthogonally similar and A is symmetric, then B is symmetric.

E17 If $\vec{v}_1$ and $\vec{v}_2$ are both eigenvectors of a symmetric matrix A, then $\vec{v}_1$ and $\vec{v}_2$ are orthogonal.

E18 If A is orthogonally diagonalizable, then A is symmetric.

E19 If $Q(\vec{x})$ is positive definite, then $Q(\vec{x}) > 0$ for all $\vec{x} \in \mathbb{R}^n$.

E20 If A is an indefinite symmetric matrix, then A is invertible.

E21 A symmetric matrix with all negative entries is negative definite.

E22 If A is a square matrix, then $|\det A|$ is equal to the product of the singular values of A.

Further Problems

These exercises are intended to be challenging.

F1 Let $A = \begin{bmatrix} 2 & 1 & 1 \\ 0 & 2 & 0 \\ 0 & 0 & 2 \end{bmatrix}$ be the standard matrix of a linear mapping L. (Observe that A is not diagonalizable.)

(a) Find a basis $\mathcal{B}$ for $\mathbb{R}^3$ of right singular vectors of A.

(b) Find a basis C for $\mathbb{R}^3$ of left singular vectors of A.

(c) Determine ${}_C[L]_{\mathcal{B}}$.

F2 In Chapter 7 Problem F6, we saw the QR-factorization: an invertible $n \times n$ matrix A can be expressed in the form $A = QR$, where Q is orthogonal and R is upper triangular. Let $A_1 = RQ$, and prove that A_1 is orthogonally similar to A and hence has the same eigenvalues as A. (By repeating this process, $A = Q_1R_1$, $A_1 = R_1Q_1$, $A_1 = Q_2R_2$, $A_2 = R_2Q_2, \ldots$, one obtains an effective numerical procedure for determining eigenvalues of a symmetric matrix.)

F3 Suppose that $A \in M_{n\times n}(\mathbb{R})$ is a positive semidefinite symmetric matrix. Prove that A has a square root. That is, show that there is a positive semidefinite symmetric matrix B such that $B^2 = A$. (Hint: suppose that Q diagonalizes A to D so that $Q^TAQ = D$. Define C to be a positive square root for D and let $B = QCQ^T$.)

F4 (a) If $A \in M_{n\times n}(\mathbb{R})$, prove that A^TA is symmetric and positive semidefinite. (Hint: consider $A\vec{x} \cdot A\vec{x}$.)

(b) If A is invertible, prove that A^TA is positive definite.

F5 (a) Suppose that $A \in M_{n\times n}(\mathbb{R})$ is invertible. Prove that A can be expressed as a product of an orthogonal matrix Q and a positive definite symmetric matrix U, $A = QU$. This is known as a **polar decomposition** of A. (Hint: use Problems F3 and F4; let U be the square root of A^TA and let $Q = AU^{-1}$.)

(b) Let $V = QUQ^T$. Show that V is symmetric and that $A = VQ$. Moreover, show that $V^2 = AA^T$, so that V is a positive definite symmetric square root of AA^T.

For part (c), we will use the following definition. A linear mapping $L(\vec{x}) = A\vec{x}$ is said to be **orientation-preserving** if $\det A > 0$.

(c) Suppose that $A \in M_{3\times 3}(\mathbb{R})$ is the matrix of an orientation-preserving linear mapping L. Show that L is the composition of a rotation following three stretches along mutually orthogonal axes. (This result follows from part (a), facts about isometries of $\mathbb{R}^3$, and ideas in Section 8.4. In fact, this is a finite version of the result for infinitesimal strain in Section 8.4.)

CHAPTER 9

Complex Vector Spaces

CHAPTER OUTLINE

When first encountering imaginary numbers, many students wonder at the point of looking at numbers which are not real. In 1572, Rafael Bombelli was the first to show that numbers involving square roots of negative numbers could be used to solve real world problems. Currently, complex numbers are used to solve problems in a wide variety of areas. Some examples are electronics, control theory, quantum mechanics, and fluid dynamics. Our goal in this chapter is to extend everything we did in Chapters 1–8 to allow the use of complex numbers instead of just real numbers.

9.1 Complex Numbers

Introduction

The first numbers we encounter as children are the natural numbers $1, 2, 3, \ldots$ and so on. In school, we soon found out that, in order to perform certain subtractions, we had to extend our concept of number to the integers, which include the natural numbers. Then, so that division by a non-zero number could always be carried out, the concept of number was extended to the rational numbers, which include the integers. Next, the concept of number was extended to the real numbers, which include all the rational numbers.

Now, to solve the equation $x^2 + 1 = 0$ we have to extend our concept of number one more time. We define the number i to be a number such that $i^2 = -1$. The system of numbers of the form $x + yi$ where $x, y \in \mathbb{R}$ is called the **complex numbers**. Note that the real numbers are included as those complex numbers with $b = 0$. As in the case with all the previous extensions of our understanding of number, some people are initially uncertain about the "meaning" of the "new" numbers. However, the complex numbers have a consistent set of rules of arithmetic, and the extension to complex numbers is justified by the fact that they allow us to solve important mathematical and physical problems that we could not solve using only real numbers.

The Geometry of Complex Numbers

Definition
Complex Number

A **complex number** is a number of the form $z = x + yi$, where $x, y \in \mathbb{R}$ and i is an element such that $i^2 = -1$. The set of all complex numbers is denoted by $\mathbb{C}$.

EXAMPLE 9.1.1

Some examples of complex numbers are $2 - 5i$, $0 + 3i$ (usually written as $3i$), $4 + 0i$ (usually written as 4).

Definition
Real Part
Imaginary Part

If $z = x + yi \in \mathbb{C}$, then we say that the **real part** of z is x and write $\text{Re}(z) = x$. We say that the **imaginary part** of z is y (*not* yi), and we write $\text{Im}(z) = y$.

EXAMPLE 9.1.2

For $z = 2 - 5i$ we have $\text{Re}(z) = 2$ and $\text{Im}(z) = -5$.

Remarks

1. The form $z = x + yi$ is called the **standard form** of a complex number.
2. If $y \neq 0$, then $z = yi$ is said to be "purely imaginary."
3. In physics and engineering, j is sometimes used in place of i since the letter i is often used to denote electric current.
4. It is sometimes convenient to write $x + iy$ instead of $x + yi$. This is particularly common with the polar form for complex numbers, which is discussed below.

At this point it is useful to observe that we can think of any complex number $z = x + yi$ as a point (x, y) in $\mathbb{R}^2$. When doing this, we refer to the x-axis as the **real axis**, denoted Re, and the y-axis as the **imaginary axis**, denoted Im, and call the xy-plane the **complex plane**. See Figure 9.1.1. A picture of this kind is sometimes called an **Argand diagram**.

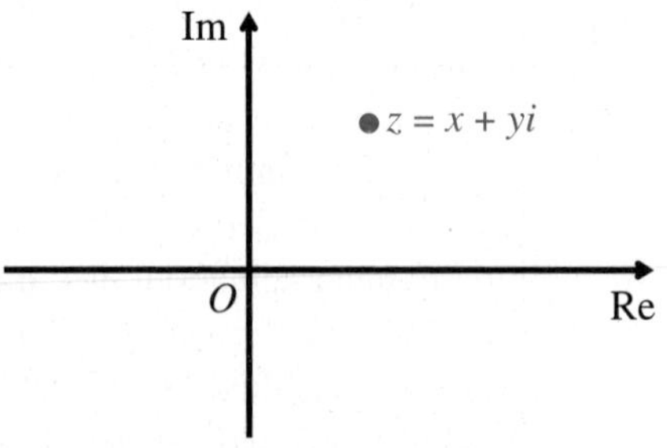

Figure 9.1.1 The complex plane.

One advantage of looking at complex numbers as points in the complex plane is that it gives us geometric motivations for properties and operations on complex numbers.

For example, we know that two points (x, y) and (u, v) in $\mathbb{R}^2$ are equal if $x = u$ and $y = v$, so we make the corresponding definition for complex numbers.

Definition
Equality

Two complex numbers $z = x + yi$ and $w = u + vi$ are **equal** if $x = u$ and $y = v$.

Similarly, since the absolute value function $|x|$ measures the distance that x is from the origin, we define the absolute value (called the modulus) of a complex number z to be the distance z is from the origin.

Definition
Modulus

For a complex number $z = x + yi$, the real number

$$|z| = \sqrt{x^2 + y^2}$$

is called the **modulus** of z.

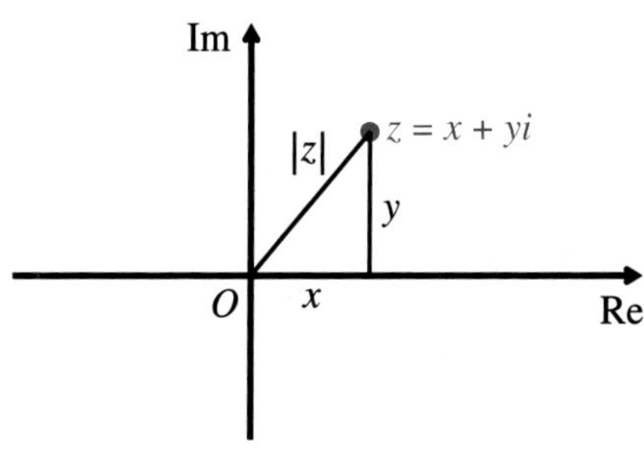

Figure 9.1.2 The Pythagorean Theorem gives the distance from z to O: $|z| = \sqrt{x^2 + y^2}$.

EXAMPLE 9.1.3

Determine the modulus (absolute value) of $z_1 = 2 - 2i$ and $z_2 = -1 + \sqrt{3}i$.

Solution: We have

$$|z_1| = |2 - 2i| = \sqrt{(2)^2 + (-2)^2} = \sqrt{8}$$
$$|z_2| = |-1 + \sqrt{3}i| = \sqrt{(-1)^2 + (\sqrt{3})^2} = \sqrt{4} = 2$$

EXERCISE 9.1.1

Determine the modulus of $z_1 = -3$ and $z_2 = -1 - i$ and illustrate with an Argand diagram.

The Arithmetic of Complex Numbers

We also use the geometry of the complex plane to define **addition** and **real scalar multiplication** of complex numbers.

Definition
Addition
Real Scalar Multiplication

For complex numbers $z_1 = x_1 + y_1 i$ and $z_2 = x_2 + y_2 i$ we define

$$z_1 + z_2 = (x_1 + x_2) + (y_1 + y_2)i$$

For any real number c we define

$$cz_1 = cx_1 + cy_1 i$$

Note that when we write $z_1 - z_2$ we mean $z_1 + (-1)z_2$.

EXAMPLE 9.1.4

Perform the following operations.

(a) $(2 + 3i) + (5 - 4i)$

Solution: $(2 + 3i) + (5 - 4i) = (2 + 5) + (3 - 4)i = 7 - i$

(b) $(2 + 3i) - (5 - 4i)$

Solution: $(2 + 3i) - (5 - 4i) = (2 - 5) + (3 - (-4))i = -3 + 7i$

(c) $4(2 - 3i)$

Solution: $4(2 - 3i) = 8 - 12i$

CONNECTION

(Section 4.7 required.) It is not difficult to show that under the operations of addition and real scalar multiplication defined above, $\mathbb{C}$ is a real vector space with basis $\{1, i\}$. Therefore, by Theorem 4.7.3, $\mathbb{C}$, as a real vector space, is in fact isomorphic to $\mathbb{R}^2$.

So far, we have only defined multiplication of a complex number by a real scalar. We, of course, also want to be able to multiply two complex numbers together. To define this, we write $z_1 = x_1 + y_1 i$ and $z_2 = x_2 + y_2 i$, and require that the product satisfy the distributive property. We get

$$\begin{aligned} z_1 z_2 &= (x_1 + y_1 i)(x_2 + y_2 i) \\ &= x_1 x_2 + x_1 y_2 i + x_2 y_1 i + y_1 y_2 i^2 \\ &= (x_1 x_2 - y_1 y_2) + (x_1 y_2 + x_2 y_1)i \end{aligned}$$

Definition
Multiplication

If $z_1 = x_1 + y_1 i, z_2 = x_2 + y_2 i \in \mathbb{C}$, then we define

$$z_1 z_2 = (x_1 x_2 - y_1 y_2) + (x_1 y_2 + x_2 y_1)i$$

EXAMPLE 9.1.5

Calculate $(3 - 2i)(-2 + 5i)$.

Solution: $(3 - 2i)(-2 + 5i) = [3(-2) - (-2)(5)] + [3(5) + (-2)(-2)]i = 4 + 19i$

EXERCISE 9.1.2

Calculate the following:

(a) $(1 - 4i) + (2 + 5i)$
(b) $(2 + 3i) - (1 - 3i)$
(c) $(2 + 2i)i$
(d) $(1 - 3i)(2 + i)$
(e) $(3 - 2i)(3 + 2i)$

In order to give a systematic method for expressing the quotient of two complex numbers as a complex number in standard form, it is useful to introduce the **complex conjugate**.

Definition
Complex Conjugate

The **complex conjugate** $\bar{z}$ of the complex number $z = x + yi$ is defined by

$$\bar{z} = x - yi$$

EXAMPLE 9.1.6

Find the complex conjugate of $2 + 5i$, $-3 - 2i$, and 4.

Solution: We have

$$\overline{2 + 5i} = 2 - 5i$$
$$\overline{-3 - 2i} = -3 + 2i$$
$$\overline{4} = 4$$

Geometrically, the complex conjugate is a reflection of $z = x + yi$ over the real axis.

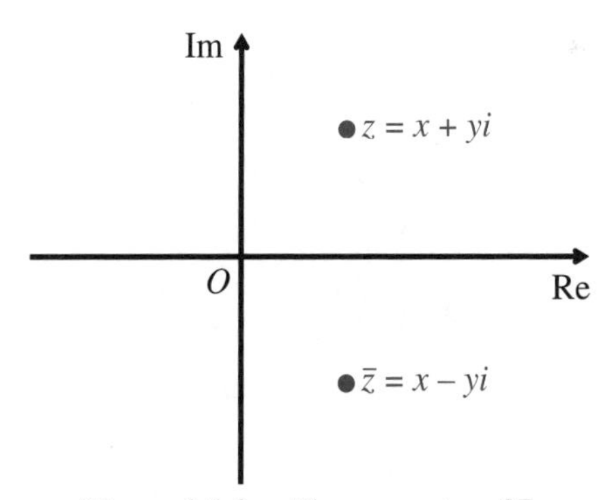

Figure 9.1.3 The geometry of $\bar{z}$.

Theorem 9.1.1 **Properties of the Complex Conjugate**

For complex numbers $z_1 = x + yi$ and z_2 with $x, y \in \mathbb{R}$, we have

(1) $\overline{\overline{z_1}} = z_1$
(2) z_1 is real if and only if $\overline{z_1} = z_1$
(3) z_1 is purely imaginary if and only if $\overline{z_1} = -z_1, z_1 \neq 0$
(4) $\overline{z_1 + z_2} = \overline{z_1} + \overline{z_2}$
(5) $\overline{z_1 z_2} = \overline{z_1}\,\overline{z_2}$
(6) $\overline{z_1^n} = \overline{z_1}^n$
(7) $z_1 + \overline{z_1} = 2\operatorname{Re}(z_1) = 2x$
(8) $z_1 - \overline{z_1} = i2\operatorname{Im}(z_1) = i2y$
(9) $z_1\overline{z_1} = x^2 + y^2 = |z_1|^2$

EXERCISE 9.1.3 Prove properties (1), (2), and (4) in Theorem 9.1.1.

The proofs of the remaining properties are left as Problem C1.

The **quotient** of two complex numbers can now be displayed as a complex number in standard form by multiplying both the numerator and the denominator by the complex conjugate of the denominator and simplifying. In particular, if $z_1 = x_1 + y_1 i$ and $z_2 = x_2 + y_2 i \neq 0$, then

$$\begin{aligned}\frac{z_1}{z_2} = \frac{z_1\overline{z_2}}{z_2\overline{z_2}} &= \frac{(x_1 + y_1 i)(x_2 - y_2 i)}{(x_2 + y_2 i)(x_2 - y_2 i)}\\ &= \frac{(x_1x_2 + y_1y_2) + (y_1x_2 - x_1y_2)i}{x_2^2 + y_2^2}\\ &= \frac{x_1x_2 + y_1y_2}{x_2^2 + y_2^2} + \frac{y_1x_2 - x_1y_2}{x_2^2 + y_2^2}i\end{aligned}$$

Notice that the quotient is defined for every pair of complex numbers z_1, z_2, provided that the denominator is not zero.

EXAMPLE 9.1.7 Write $\dfrac{2 + 5i}{3 - 4i}$ in standard form.

Solution: We have

$$\frac{2 + 5i}{3 - 4i} = \frac{(2 + 5i)(3 + 4i)}{(3 - 4i)(3 + 4i)} = \frac{(6 - 20) + (8 + 15)i}{9 + 16} = -\frac{14}{25} + \frac{23}{25}i$$

EXERCISE 9.1.4 Calculate the following quotients.

(a) $\dfrac{1 + i}{1 - i}$ (b) $\dfrac{2i}{1 + i}$ (c) $\dfrac{4 - i}{1 + 5i}$

The operations on complex numbers have the following important properties.

Theorem 9.1.2 If $z_1, z_2, z_3 \in \mathbb{C}$, then

(1) $z_1 + z_2 \in \mathbb{C}$
(2) $z_1 + z_2 = z_2 + z_1$
(3) $z_1 + (z_2 + z_3) = (z_1 + z_2) + z_3$
(4) $z_1 + 0 = z_1$
(5) For each $z \in \mathbb{C}$ there exists $(-z) \in \mathbb{C}$ such that $z + (-z) = 0$.
(6) $z_1 z_2 \in \mathbb{C}$
(7) $z_1 z_2 = z_2 z_1$
(8) $z_1(z_2 z_3) = (z_1 z_2) z_3$
(9) $1 z_1 = z_1$
(10) For each $z \in \mathbb{C}$ with $z \neq 0$, there exists $(1/z) \in \mathbb{C}$ such that $z(1/z) = 1$.
(11) $z_1(z_2 + z_3) = z_1 z_2 + z_1 z_3$

CONNECTION

Any set with two operations of addition and multiplication that satisfies properties (1) to (11) is called a **field**. Both the set of rational numbers $\mathbb{Q}$, and the set of real numbers $\mathbb{R}$ are also fields. There are other fields which play important roles in many real world applications of linear algebra.

Another important property is the familiar triangle inequality.

Theorem 9.1.3 If $z_1, z_2 \in \mathbb{C}$, then

$$|z_1 + z_2| \leq |z_1| + |z_2|$$

Polar Form

One of the reasons complex numbers are so helpful for solving problems is that they can be represented in many different ways. So far, we have seen the algebraic form $z = x + yi$ and the vector form (x, y). We now look at another way of representing complex numbers.

Rather than locating a complex number $z = x + yi$ in the complex plane using its Cartesian coordinates, we can locate it by using its modulus $r = |z|$ (distance from the origin) and any angle θ between the positive real axis and a line segment between the origin and z. See Figure 9.1.4.

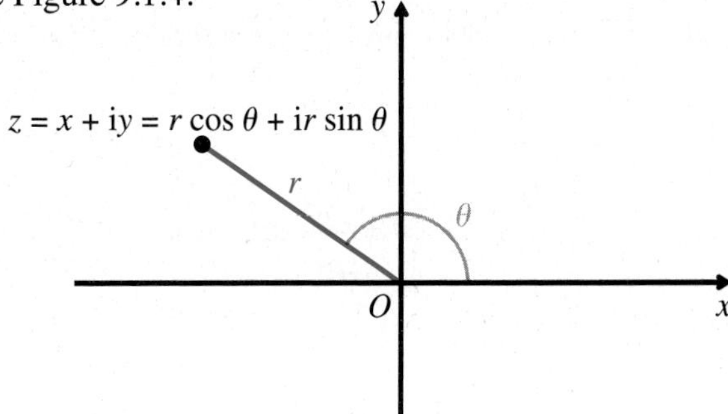

Figure 9.1.4 Polar representation of z.

We begin by making the following definition.

Definition
Argument

Let $z = x + yi \in \mathbb{C}$ with $z \neq 0$. If θ is any angle such that

$$x = |z| \cos \theta \quad \text{and} \quad y = |z| \sin \theta$$

then the angle θ is called an **argument** of z and we write

$$\theta = \arg z$$

It is important to observe that every complex number has infinitely many arguments. In particular, if θ is an argument of a complex number z, then all the values

$$\theta + 2\pi k, \quad k \in \mathbb{Z}$$

are also arguments of z.

For many applications, we prefer to choose the unique argument θ that satisfies $-\pi < \theta \leq \pi$ called the **principal argument** of z. It is denoted $\operatorname{Arg} z$.

EXAMPLE 9.1.8

Find all arguments of $z_1 = 1 - i$.

Solution: To calculate any argument of z_1, we first need to calculate the modulus.

$$|z_1| = \sqrt{1^2 + (-1)^2} = \sqrt{2}$$

Hence, any argument θ of z_1 satisfies

$$1 = \sqrt{2} \cos \theta, \quad \text{and} \quad -1 = \sqrt{2} \sin \theta$$

So, $\cos \theta = \frac{1}{\sqrt{2}}$ and $\sin \theta = -\frac{1}{\sqrt{2}}$. This gives

$$\theta = -\frac{\pi}{4} + 2\pi k, \quad k \in \mathbb{Z}$$

EXAMPLE 9.1.9

Determine the principal argument of $z_2 = \sqrt{3} + i$.

Solution: We have

$$|z_2| = \sqrt{(\sqrt{3})^2 + 1^2} = 2$$

and so, any argument θ of z_2 satisfies

$$\sqrt{3} = 2 \cos \theta, \quad \text{and} \quad 1 = 2 \sin \theta$$

So, $\cos \theta = \frac{\sqrt{3}}{2}$ and $\sin \theta = \frac{1}{2}$. This gives

$$\theta = \frac{\pi}{6} + 2\pi k, \quad k \in \mathbb{Z}$$

Therefore, the principal value is $\operatorname{Arg} z_2 = \frac{\pi}{6}$.

Remarks

1. The angles may be measured in radians or degrees. We will always use radians.
2. It is tempting but incorrect to write $\theta = \arctan(y/x)$. Remember that you need two trigonometric functions to locate the correct quadrant for z. Also note that y/x is not defined if $x = 0$.

Definition
Polar Form

The **polar form** of a complex number z is

$$z = r(\cos\theta + i\sin\theta)$$

where $r = |z|$ and θ is an argument of z.

EXAMPLE 9.1.10

Determine the modulus, an argument, and a polar form of $z_1 = 2-2i$ and $z_2 = -1+\sqrt{3}i$.

Solution: We have

$$|z_1| = |2-2i| = \sqrt{2^2+(-2)^2} = 2\sqrt{2}$$

Any argument θ of z_1 satisfies

$$2 = 2\sqrt{2}\cos\theta \quad \text{and} \quad -2 = 2\sqrt{2}\sin\theta$$

So, $\cos\theta = \frac{1}{\sqrt{2}}$ and $\sin\theta = -\frac{1}{\sqrt{2}}$, which gives $\theta = -\frac{\pi}{4} + 2\pi k$, $k \in \mathbb{Z}$. Hence, a polar form of z_1 is

$$z_1 = 2\sqrt{2}\left(\cos\left(-\frac{\pi}{4}\right) + i\sin\left(-\frac{\pi}{4}\right)\right)$$

For z_2, we have

$$|z_2| = \left|-1+\sqrt{3}i\right| = \sqrt{(-1)^2+(\sqrt{3})^2} = 2$$

Since $-1 = 2\cos\theta$ and $\sqrt{3} = 2\sin\theta$, we get $\theta = \frac{2\pi}{3} + 2\pi k$, $k \in \mathbb{Z}$. Thus, a polar form of z_2 is

$$z_2 = 2\left(\cos\left(\frac{2\pi}{3}\right) + i\sin\left(\frac{2\pi}{3}\right)\right)$$

EXERCISE 9.1.5

Determine the modulus, an argument, and a polar form of $z_1 = \sqrt{3}+i$ and $z_2 = -1-i$.

EXERCISE 9.1.6

Let $z = r(\cos\theta + i\sin\theta)$. Prove that the modulus of $\bar{z}$ equals the modulus of z and an argument of $\bar{z}$ is $-\theta$.

The polar form is particularly convenient for multiplication and division because of the trigonometric identities

$$\cos(\theta_1 + \theta_2) = \cos\theta_1 \cos\theta_2 - \sin\theta_1 \sin\theta_2$$
$$\sin(\theta_1 + \theta_2) = \sin\theta_1 \cos\theta_2 + \cos\theta_1 \sin\theta_2$$

Theorem 9.1.4 For any complex numbers $z_1 = r_1(\cos\theta_1 + i\sin\theta_1)$ and $z_2 = r_2(\cos\theta_2 + i\sin\theta_2)$, we have

$$z_1 z_2 = r_1 r_2 \left(\cos(\theta_1 + \theta_2) + i\sin(\theta_1 + \theta_2)\right)$$

In words, the modulus of a product is the product of the moduli of the factors, while an argument of a product is the sum of the arguments.

Theorem 9.1.5 For any complex numbers $z_1 = r_1(\cos\theta_1 + i\sin\theta_1)$ and $z_2 = r_2(\cos\theta_2 + i\sin\theta_2)$, with $z_2 \neq 0$, we have

$$\frac{z_1}{z_2} = \frac{r_1}{r_2}\left(\cos(\theta_1 - \theta_2) + i\sin(\theta_1 - \theta_2)\right)$$

The proofs of Theorems 9.1.4 and Theorem 9.1.5 are left for you to complete in Problem C4.

Theorem 9.1.6 If $z = r(\cos\theta + i\sin\theta)$ with $r \neq 0$, then $\frac{1}{z} = \frac{1}{r}(\cos(-\theta) + i\sin(-\theta))$.

EXERCISE 9.1.7 Describe Theorem 9.1.5 in words and use it to prove Theorem 9.1.6.

EXAMPLE 9.1.11 Calculate $(1-i)(-\sqrt{3}+i)$ and $\frac{2+2i}{1+\sqrt{3}i}$ using polar form.

Solution: We have

$$\begin{aligned}
(1-i)(-\sqrt{3}+i) &= \sqrt{2}\left(\cos\left(-\frac{\pi}{4}\right) + i\sin\left(-\frac{\pi}{4}\right)\right) 2\left(\cos\left(\frac{5\pi}{6}\right) + i\sin\left(\frac{5\pi}{6}\right)\right)\\
&= 2\sqrt{2}\left(\cos\left(-\frac{\pi}{4} + \frac{5\pi}{6}\right) + i\sin\left(-\frac{\pi}{4} + \frac{5\pi}{6}\right)\right)\\
&= 2\sqrt{2}\left(\cos\left(\frac{7\pi}{12}\right) + i\sin\left(\frac{7\pi}{12}\right)\right)
\end{aligned}$$

$$\begin{aligned}
\frac{2+2i}{1+\sqrt{3}i} &= \frac{2\sqrt{2}\left(\cos\left(\frac{\pi}{4}\right) + i\sin\left(\frac{\pi}{4}\right)\right)}{2\left(\cos\left(\frac{\pi}{3}\right) + i\sin\left(\frac{\pi}{3}\right)\right)}\\
&= \sqrt{2}\left(\cos\left(\frac{\pi}{4} - \frac{\pi}{3}\right) + i\sin\left(\frac{\pi}{4} - \frac{\pi}{3}\right)\right)\\
&= \sqrt{2}\left(\cos\left(-\frac{\pi}{12}\right) + i\sin\left(-\frac{\pi}{12}\right)\right)
\end{aligned}$$

EXERCISE 9.1.8 Calculate $(2 - 2i)(-1 + \sqrt{3}i)$ and $\dfrac{2 - 2i}{-1 + \sqrt{3}i}$ using polar form.

Powers and the Complex Exponential

From the rule for products, we find that

$$z^2 = r^2(\cos 2\theta + i \sin 2\theta)$$

Then

$$\begin{aligned} z^3 &= z^2 z = r^2 r(\cos(2\theta + \theta) + i \sin \theta(2\theta + \theta)) \\ &= r^3(\cos 3\theta + i \sin 3\theta) \end{aligned}$$

Theorem 9.1.7 de Moivre's Formula

If $z = r(\cos\theta + i\sin\theta)$ with $r \neq 0$, then for any integer n we have

$$z^n = r^n(\cos n\theta + i \sin n\theta)$$

Proof: For $n = 0$, we have $z^0 = 1 = r^0(\cos 0 + i \sin 0)$. To prove that the theorem holds for positive integers, we proceed by induction. Assume that the result is true for some integer $k \geq 0$. Then

$$\begin{aligned} z^{k+1} &= z^k z = r^k r[\cos(k\theta + \theta) + i \sin(k\theta + \theta)] \\ &= r^{k+1}[\cos((k + 1)\theta) + i \sin((k + 1)\theta)] \end{aligned}$$

Therefore, the result is true for all non-negative integers n. Then, by Theorem 9.1.4, for any positive integer m, we have

$$\begin{aligned} z^{-m} &= (z^m)^{-1} = (r^m(\cos m\theta + i \sin m\theta))^{-1} \\ &= r^{-m}(\cos(-m\theta) + i \sin(-m\theta)) \end{aligned}$$

Hence, the result also holds for all negative integers $n = -m$. ■

EXAMPLE 9.1.12

Use de Moivre's Formula to calculate $(2+2i)^3$.

Solution:

$$\begin{aligned}(2+2i)^3 &= \left[2\sqrt{2}\left(\cos\left(\frac{\pi}{4}\right)+\sin\left(\frac{\pi}{4}\right)\right)\right]^3\\ &= (2\sqrt{2})^3\left(\cos\left(\frac{3\pi}{4}\right)+i\sin\left(\frac{3\pi}{4}\right)\right)\\ &= 16\sqrt{2}\left(-\frac{1}{\sqrt{2}}+i\frac{1}{\sqrt{2}}\right)\\ &= -16+16i\end{aligned}$$

In the case where $r = 1$, de Moivre's Formula reduces to

$$(\cos\theta + i\sin\theta)^n = \cos n\theta + i\sin n\theta$$

This is formally just like one of the exponential laws, $(e^\theta)^n = e^{n\theta}$. We use this idea to define e^z for any $z \in \mathbb{C}$, where e is the usual natural base for exponentials ($e \approx 2.71828$). We begin with a useful formula of Euler.

Definition
Euler's Formula

For any $\theta \in \mathbb{R}$ we have

$$e^{i\theta} = \cos\theta + i\sin\theta$$

Observe that Euler's Formula allows us to write the polar form of a complex number z more compactly. In particular, we can now write

$$z = re^{i\theta}$$

where $r = |z|$ and θ is any argument of z. One advantage of this form is that de Moivre's Formula can be written as

$$z^n = r^n e^{in\theta}$$

Remarks

1. One interesting consequence of Euler's Formula is that $e^{i\pi} + 1 = 0$. In one formula, we have five of the most important numbers in mathematics: $0, 1, e, i$, and π.

2. One area where Euler's Formula has important applications is ordinary differential equations. There, one often uses the fact that

$$e^{(a+bi)t} = e^{at}e^{ibt} = e^{at}(\cos bt + i\sin bt)$$

EXAMPLE 9.1.13 Calculate the following using the polar form.

(a) $(2+2i)^3$ (b) $(2i)^3$ (c) $(\sqrt{3}+i)^5$

Solution:

$$(2+2i)^3 = \left(2\sqrt{2}e^{i\pi/4}\right)^3 = (2\sqrt{2})^3 e^{i(3\pi/4)} = 16\sqrt{2}(\cos(3\pi/4) + i\sin(3\pi/4)) = -16+16i$$

$$(2i)^3 = \left(2e^{i\pi/2}\right)^3 = 2^3 e^{i(3\pi/2)} = 8(\cos(3\pi/2) + i\sin(3\pi/2)) = -8i$$

$$(\sqrt{3}+i)^5 = \left(2e^{i\pi/6}\right)^5 = 2^5 e^{i5\pi/6} = 32(\cos(5\pi/6) + i\sin(5\pi/6)) = -16\sqrt{3}+16i$$

EXERCISE 9.1.9 Use polar form to calculate $(1-i)^5$ and $(-1-\sqrt{3}i)^5$.

---

n-th Roots

Using de Moivre's Formula for n-th powers is the key to finding n-th roots. Suppose that we need to find the n-th root of the non-zero complex number $z = re^{i\theta}$. That is, we need a number w such that $w^n = z$. Suppose that $w = Re^{i\phi}$. Then $w^n = z$ implies that

$$R^n e^{in\phi} = re^{i\theta}$$

Then R is the real n-th root of the positive real number r. However, because arguments of complex numbers are determined only up to the addition of $2\pi k$, all we can say about ϕ is that

$$n\phi = \theta + 2\pi k, \quad k \in \mathbb{Z}$$

or

$$\phi = \frac{\theta + 2\pi k}{n}, \quad k \in \mathbb{Z}$$

EXAMPLE 9.1.14 Find all the cube roots of 8.

Solution: We have $8 = 8e^{i(0+2\pi k)}$, $k \in \mathbb{Z}$. Thus, for any $k \in \mathbb{Z}$.

$$8^{1/3} = \left(8e^{i(0+2\pi k)}\right)^{1/3} = 8^{1/3} e^{i2k\pi/3}$$

If $k = 0$, we have the root $w_0 = 2e^0 = 2$.
If $k = 1$, we have the root $w_1 = 2e^{i2\pi/3} = -1 + \sqrt{3}i$.
If $k = 2$, we have the root $w_2 = 2e^{i4\pi/3} = -1 - \sqrt{3}i$.
If $k = 3$, we have the root $2e^{i2\pi} = 2 = w_0$.

By increasing k further, we simply repeat the roots we have already found. Similarly, consideration of negative k gives us no further roots. The number 8 has three third roots, ω_0, ω_1, and ω_2. In particular, these are the roots of equation $w^3 - 8 = 0$.

Theorem 9.1.8 If $z = re^{i\theta}$ is non-zero, then the n distinct n-th roots of z are

$$w_k = r^{1/n} e^{i(\theta+2\pi k)/n}, \quad k = 0, 1, \dots, n-1$$

EXAMPLE 9.1.15 Find the fourth roots of -81 and illustrate in an Argand diagram.

Solution: We have $-81 = 81e^{i(\pi+2\pi k)}$. Thus, the fourth roots are

$$\begin{aligned} w_0 &= (81)^{1/4} e^{i(\pi+0)/4} = 3e^{i\pi/4} \\ w_1 &= (81)^{1/4} e^{i(\pi+2\pi)/4} = 3e^{i3\pi/4} \\ w_2 &= (81)^{1/4} e^{i(\pi+4\pi)/4} = 3e^{i5\pi/4} \\ w_3 &= (81)^{1/4} e^{i(\pi+6\pi)/4} = 3e^{i7\pi/4} \end{aligned}$$

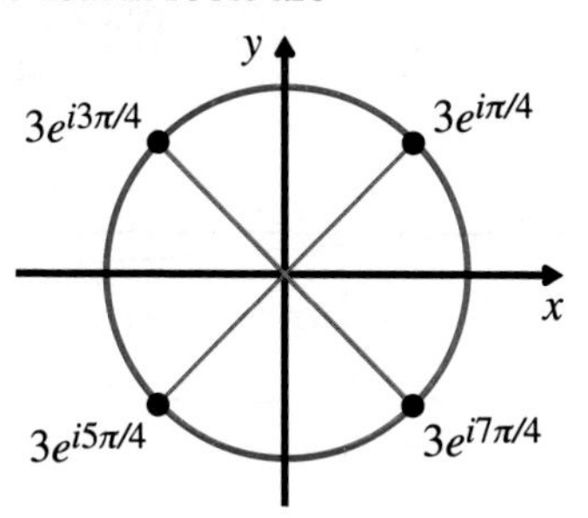

Plotting these roots shows that all four are points on the circle of radius 3 centred at the origin and that they are separated by equal angles of $\frac{\pi}{2}$.

In Examples 9.1.14 and 9.1.15, we took roots of numbers that were purely real: we were really solving $x^n - a = 0$, where $a \in \mathbb{R}$. As a contrast, let us consider roots of a number that is not real.

EXAMPLE 9.1.16 Find the third roots of $5i$ and illustrate in an Argand diagram.

Solution: $5i = 5e^{i(\frac{\pi}{2}+2k\pi)}$, so the cube roots are

$$\begin{aligned} w_0 &= 5^{1/3} e^{i\pi/6} = 5^{1/3}\left(\frac{\sqrt{3}}{2} + i\frac{1}{2}\right) \\ w_1 &= 5^{1/3} e^{i5\pi/6} = 5^{1/3}\left(\frac{-\sqrt{3}}{2} + i\frac{1}{2}\right) \\ w_2 &= 5^{1/3} e^{i3\pi/2} = 5^{1/3}(-i) \end{aligned}$$

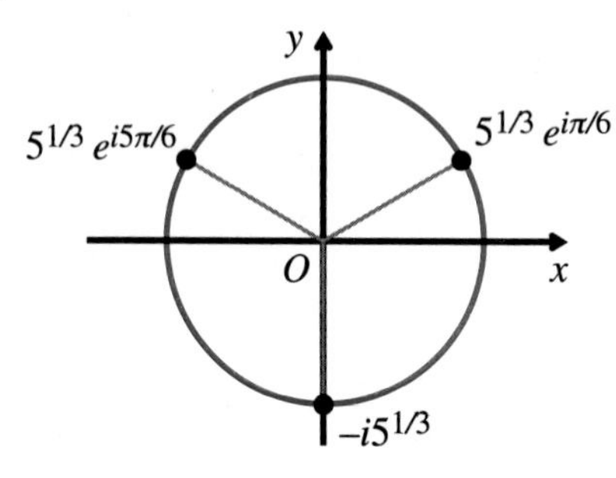

Plotting these roots shows that all three are points on the circle of radius $5^{1/3}$ centred at the origin and that they are separated by equal angles of $\frac{2\pi}{3}$.

Examples 9.1.14, 9.1.15, and 9.1.16 all illustrate a general rule: the n-th roots of a complex number $z = re^{i\theta}$ all lie on the circle of radius $r^{1/n}$, and they are separated by equal angles of $2\pi/n$.

PROBLEMS 9.1

Practice Problems

For Problems A1–A6, draw an Argand diagram for z, and determine $\bar{z}$ and $|z|$.

A1 $z = 3 - 5i$ **A2** $z = 2 + 7i$ **A3** $z = -4i$
A4 $z = -1 - 2i$ **A5** $z = 2$ **A6** $z = -3 + 2i$

For Problems A7–A12, find a polar form and the principal argument of z.

A7 $z = -3 - 3i$ **A8** $z = \sqrt{3} - i$
A9 $z = -\sqrt{3} + i$ **A10** $z = -2 - 2\sqrt{3}i$
A11 $z = -3$ **A12** $z = -2 + 2i$

For Problems A13–A18, convert z into to standard form.

A13 $z = e^{-i\pi/3}$ **A14** $z = 2e^{-i\pi/3}$
A15 $z = 2e^{i\pi/3}$ **A16** $z = 3e^{i3\pi/4}$
A17 $z = 2e^{-i\pi/6}$ **A18** $z = e^{i5\pi/6}$

For Problems A19–A26, write the number in standard form.

A19 $(2 + 5i) + (3 + 2i)$ **A20** $(2 - 7i) + (-5 + 3i)$
A21 $(-3 + 5i) - (4 + 3i)$ **A22** $(-5 - 6i) - (9 - 11i)$
A23 $(1 + 3i)(3 - 2i)$ **A24** $(-2 - 4i)(3 - i)$
A25 $(1 - 6i)(-4 + i)$ **A26** $(-1 - i)(1 - i)$

For Problems A27–A30, determine $\text{Re}(z)$ and $\text{Im}(z)$.

A27 $z = 3 - 6i$ **A28** $z = (2 + 5i)(1 - 3i)$
A29 $z = \dfrac{4}{6 - i}$ **A30** $z = \dfrac{-1}{i}$

For Problems A31–A33, express the quotient in standard form.

A31 $\dfrac{1}{2 + 3i}$ **A32** $\dfrac{2 - 5i}{3 + 2i}$ **A33** $\dfrac{1 + 6i}{4 - i}$

For Problems A34–A37, use polar form to determine z_1z_2 and $\dfrac{z_1}{z_2}$.

A34 $z_1 = 1 + i, z_2 = 1 + \sqrt{3}i$
A35 $z_1 = -\sqrt{3} - i, z_2 = 1 - i$
A36 $z_1 = -1 + \sqrt{3}i, z_2 = \sqrt{3} - i$
A37 $z_1 = -3 + 3i, z_2 = -\sqrt{3} + i$

For Problems A38–A41, use polar form to evaluate the expression.

A38 $(1 + i)^4$ **A39** $(3 - 3i)^3$
A40 $(-1 - \sqrt{3}i)^4$ **A41** $(-2\sqrt{3} + 2i)^5$

For Problems A42–A45, use polar form to determine all the indicated roots.

A42 $(-1)^{1/5}$ **A43** $(-16i)^{1/4}$
A44 $(-\sqrt{3} - i)^{1/3}$ **A45** $(1 + 4i)^{1/3}$

Homework Problems

For Problems B1–B6, draw an Argand diagram for z, and determine $\bar{z}$ and $|z|$.

B1 $z = 3 + 4i$ **B2** $z = -2 - 3i$ **B3** $z = 2 - \sqrt{3}i$
B4 $z = \frac{1}{2}i$ **B5** $z = 1 - 3i$ **B6** $z = -3 + i$

For Problems B7–B12, find a polar form and the principal argument of z.

B7 $z = 3 - 3i$ **B8** $z = -2i$
B9 $z = \sqrt{3}i$ **B10** $z = \frac{3\sqrt{3}}{2} + \frac{3}{2}i$
B11 $z = -3\sqrt{2} - 3\sqrt{2}i$ **B12** $z = -\frac{1}{4} - \frac{\sqrt{3}}{4}i$

For Problems B13–B20, convert z from polar form to standard form.

B13 $z = 3e^{i\pi/2}$ **B14** $z = 2e^{-i\pi}$
B15 $z = 4e^{i2\pi/3}$ **B16** $z = \sqrt{2}e^{i\pi}$
B17 $z = e^{-i\pi/4}$ **B18** $z = 6e^{i\pi/6}$
B19 $z = 3e^{-i2\pi/3}$ **B20** $z = e^{i\pi/3}$

For Problems B21–B34, evaluate the expression and write it in standard form.

B21 $(-1 + 2i) + (2 - 3i)$ **B22** $(3 - 4i) + (-3 - i)$
B23 $2i + (-1 - 2i)$ **B24** $(2 - i) - (3 - 3i)$
B25 $(-1 - 3i) - (-2 + i)$ **B26** $(4 + 2i) + (-2 - 3i)$
B27 $(-2 + i)(-1 - i)$ **B28** $(1 + i)(1 - i)$
B29 $(-1 - i)(1 + i)$ **B30** $(3 - i)(1 + 3i)$

B31 $(2+4i)(-1-3i)$

B32 $(-3+2i)(4-i)$

B33 $(1-2i)(-2+i)$

B34 $(1+\sqrt{3}i)(-\frac{3}{2}+\frac{3\sqrt{3}}{2}i)$

For Problems B35–B42, determine Re(z) and Im(z).

B35 $z=-2+3i$

B36 $z=\overline{3-4i}$

B37 $z=|-1+\sqrt{3}i|$

B38 $z=(1-2i)(2+2i)$

B39 $z=\dfrac{1}{2-3i}$

B40 $z=\dfrac{i}{1-i}$

B41 $z=5e^{-i\pi/2}$

B42 $z=2e^{i\pi/4}$

For Problems B43–B48, express the quotient in standard form.

B43 $\dfrac{1}{3-4i}$

B44 $\dfrac{2}{1+5i}$

B45 $\dfrac{5}{-1-2i}$

B46 $\dfrac{1-3i}{1+2i}$

B47 $\dfrac{3+5i}{-2+i}$

B48 $\dfrac{-2-3i}{4-3i}$

For Problems B49–B52, use polar form to determine z_1z_2 and $\dfrac{z_1}{z_2}$.

B49 $z_1=\sqrt{3}+i, z_2=-1-i$

B50 $z_1=2+2i, z_2=-2-2\sqrt{3}i$

B51 $z_1=2-2\sqrt{3}i, z_2=1+i$

B52 $z_1=-3\sqrt{2}+3\sqrt{2}i, z_2=1+\sqrt{3}i$

For Problems B53–B58, use polar form to convert z to standard form.

B53 $(1-i)^5$

B54 $(-2\sqrt{3}+2i)^3$

B55 $(\sqrt{(3)}+i)^4$

B56 $(2+2i)^4$

B57 $(-1+\sqrt{3}i)^4$

B58 $\left(\frac{1}{2}+\frac{\sqrt{3}}{2}i\right)^5$

For Problems B59–B64, use polar form to determine all the indicated roots.

B59 $(-1)^{1/5}$

B60 $(-16i)^{1/4}$

B61 $1^{1/5}$

B62 $(1+i)^{1/4}$

B63 $(-\sqrt{3}-i)^{1/3}$

B64 $(1+4i)^{1/3}$

Conceptual Problems

C1 Prove properties (3), (5), (6), (7), (8), and (9) of Theorem 9.1.1.

C2 If $z=r(\cos\theta+i\sin\theta)$, what is $|\overline{z}|$? What is an argument of $\overline{z}$?

C3 Use Euler's Formula to show that
(a) $\overline{e^{i\theta}}=e^{-i\theta}$
(b) $\cos\theta=\dfrac{1}{2}\left(e^{i\theta}+e^{-i\theta}\right)$
(c) $\sin\theta=\dfrac{1}{2i}\left(e^{i\theta}-e^{-i\theta}\right)$

C4 Prove Theorem 9.1.4 and Theorem 9.1.5.

C5 Let $z_1,z_2\in\mathbb{C}$ such that z_1+z_2 and z_1z_2 are each negative real numbers. Prove that z_1 and z_2 must be real numbers.

C6 Prove that if $|z|=1, z\neq 1$, then $\mathrm{Re}\left(\frac{1}{1-z}\right)=\frac{1}{2}$.

C7 Use de Moivre's Formula to prove

$$\cos 3\theta=\cos^3\theta-3\cos\theta\sin^2\theta$$
$$\sin 3\theta=3\cos^2\theta\sin\theta-\sin^3\theta$$

C8 Prove that for any $z\in\mathbb{C}$, $\mathrm{Re}(z)\leq|z|$.

C9 Derive the triangle inequality by following the steps below.
(a) Show that

$$|z_1+z_2|^2=(z_1+z_2)(\overline{z_1}+\overline{z_2})=z_1\overline{z_1}+(z_1\overline{z_2}+\overline{z_1\overline{z_2}})+z_2\overline{z_2}$$

(b) Use Theorem 9.1.1 (7) and C8 to show that

$$z_1\overline{z_2}+\overline{z_1\overline{z_2}}=2\,\mathrm{Re}(z_1\overline{z_2})\leq 2|z_1||z_2|$$

(c) Use the results above to obtain the inequality

$$|z_1+z_2|^2\leq(|z_1|+|z_2|)^2$$

9.2 Systems with Complex Numbers

In some applications, it is necessary to consider systems of linear equations with complex coefficients and complex right-hand sides. One physical application, discussed later in this section, is the problem of determining currents in electrical circuits with capacitors and inductive coils as well as resistance. We can solve systems with complex coefficients by using exactly the same elimination/row reduction procedures as for systems with real coefficients. Of course, our solutions will be complex, and any free variables will be allowed to take any complex value.

EXAMPLE 9.2.1

Solve the system of linear equations

$$\begin{aligned} z_1 + z_2 + z_3 &= 0 \\ (1-i)z_1 + z_2 &= i \\ (3-i)z_1 + 2z_2 + z_3 &= 1 + 2i \end{aligned}$$

Solution: The solution procedure is to write the augmented matrix for the system and row reduce the augmented matrix to reduced row echelon form.

$$\left[\begin{array}{ccc|c} 1 & 1 & 1 & 0 \\ 1-i & 1 & 0 & i \\ 3-i & 2 & 1 & 1+2i \end{array}\right] \begin{array}{l} \\ R_2 - (1-i)R_1 \\ R_3 - (3-i)R_1 \end{array} \sim \left[\begin{array}{ccc|c} 1 & 1 & 1 & 0 \\ 0 & i & -1+i & i \\ 0 & -1+i & -2+i & 1+2i \end{array}\right] -iR_2 \sim$$

$$\left[\begin{array}{ccc|c} 1 & 1 & 1 & 0 \\ 0 & 1 & 1+i & 1 \\ 0 & -1+i & -2+i & 1+2i \end{array}\right] \begin{array}{l} R_1 - R_2 \\ \\ R_3 + (1-i)R_2 \end{array} \sim \left[\begin{array}{ccc|c} 1 & 0 & -i & -1 \\ 0 & 1 & 1+i & 1 \\ 0 & 0 & i & 2+i \end{array}\right] \begin{array}{l} \\ \\ -iR_3 \end{array} \sim$$

$$\left[\begin{array}{ccc|c} 1 & 0 & -i & -1 \\ 0 & 1 & 1+i & 1 \\ 0 & 0 & 1 & 1-2i \end{array}\right] \begin{array}{l} R_1 + iR_3 \\ R_2 - (1+i)R_3 \\ \\ \end{array} \sim \left[\begin{array}{ccc|c} 1 & 0 & 0 & 1+i \\ 0 & 1 & 0 & -2+i \\ 0 & 0 & 1 & 1-2i \end{array}\right]$$

Hence, the solution is $\vec{z} = \begin{bmatrix} 1+i \\ -2+i \\ 1-2i \end{bmatrix}$.

We can verify that

$$\begin{aligned} (1+i) + (-2+i) + (1-2i) &= 0 \\ (1-i)(1+i) + (-2+i) &= i \\ (3-i)(1+i) + 2(-2+i) + (1-2i) &= 1 + 2i \end{aligned}$$

Remark

Due to the sheer number of calculations involved in solving a system of linear equations with complex numbers, it is highly recommended that you check your answer whenever possible.

EXAMPLE 9.2.2 Solve the system

$$\begin{aligned}(1+i)z_1 + 2iz_2 &= 1\\ (1+i)z_2 + z_3 &= \frac{1}{2} - \frac{1}{2}i\\ z_1 - z_3 &= 0\end{aligned}$$

Solution: Row reducing the augmented matrix gives

$$\left[\begin{array}{ccc|c} 1+i & 2i & 0 & 1\\ 0 & 1+i & 1 & \frac{1-i}{2}\\ 1 & 0 & -1 & 0\end{array}\right] \begin{array}{l}\frac{1}{1+i}R_1\\ \frac{1}{1+i}R_2\\ \\ \end{array} \sim \left[\begin{array}{ccc|c} 1 & 1+i & 0 & \frac{1-i}{2}\\ 0 & 1 & \frac{1-i}{2} & -\frac{1}{2}i\\ 1 & 0 & -1 & 0\end{array}\right] \begin{array}{l}R_1-(1+i)R_2\\ \\ \\ \end{array} \sim$$

$$\left[\begin{array}{ccc|c} 1 & 0 & -1 & 0\\ 0 & 1 & \frac{1-i}{2} & -\frac{1}{2}i\\ 1 & 0 & -1 & 0\end{array}\right] \begin{array}{l} \\ \\ R_3-R_1\end{array} \sim \left[\begin{array}{ccc|c} 1 & 0 & -1 & 0\\ 0 & 1 & \frac{1-i}{2} & -\frac{1}{2}i\\ 0 & 0 & 0 & 0\end{array}\right]$$

Hence, z_3 is a free variable. We let $z_3 = t \in \mathbb{C}$. Then $z_1 = z_3 = t$, $z_2 = -\frac{1}{2}i - \frac{1-i}{2}t$, and the general solution is

$$\begin{bmatrix} z_1\\ z_2\\ z_3\end{bmatrix} = \begin{bmatrix} 0\\ -i/2\\ 0\end{bmatrix} + t\begin{bmatrix} 1\\ -(1-i)/2\\ 1\end{bmatrix}, \quad t \in \mathbb{C}$$

We can verify that

$$\begin{aligned}(1+i)t + 2i\left(-\frac{1}{2}i - \frac{1-i}{2}t\right) &= 1\\ (1+i)\left(-\frac{1}{2}i - \frac{1-i}{2}t\right) + t &= \frac{1}{2} - \frac{1}{2}i\\ t - t &= 0\end{aligned}$$

EXERCISE 9.2.1 Solve the system

$$\begin{aligned}iz_1 + z_2 + 3z_3 &= -1 - 2i\\ iz_1 + iz_2 + (1+2i)z_3 &= 2 + i\\ 2z_1 + (1+i)z_2 + 2z_3 &= 5 - i\end{aligned}$$

Complex Numbers in Electrical Circuit Equations

For purposes of the following discussion only, we will denote by j the complex number such that $j^2 = -1$, so that we can use i to denote current.

In Section 2.4, we discussed electrical circuits with resistors. We now also consider capacitors and inductors, as well as alternating current. A simple capacitor can be thought of as two conducting plates separated by a vacuum or some dielectric. Charge can be stored on these plates, and it is found that the voltage across a capacitor at time t is proportional to the charge stored at that time:

$$V(t) = \frac{Q(t)}{C}$$

where Q is the charge and the constant C is called the *capacitance* of the capacitor.

The usual model of an inductor is a coil; because of magnetic effects, it is found that with time-varying current $i(t)$, the voltage across an inductor is proportional to the rate of change of current:

$$V(t) = L\frac{d\,i(t)}{dt}$$

where the constant of proportionality L is called the *inductance*.

As in the case of the resistor circuits, Kirchhoff's Laws applies: the sum of the voltage drops across the circuit elements must be equal to the applied electromotive force (voltage). Thus, for a simple loop with inductance L, capacitance C, resistance R, and applied electromotive force $E(t)$ (Figure 9.2.1), the circuit equation is

$$L\frac{d\,i(t)}{dt} + R\,i(t) + \frac{1}{C}Q(t) = E(t)$$

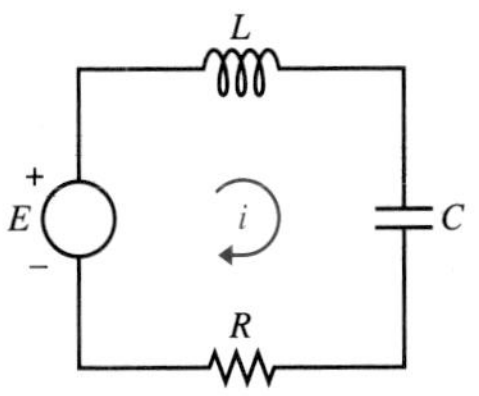

Figure 9.2.1 Kirchhoff's Voltage Law applied to an alternating current circuit.

For our purposes, it is easier to work with the derivative of this equation and use the fact that $\frac{dQ}{dt} = i$:

$$L\frac{d^2\,i(t)}{dt^2} + R\frac{d\,i(t)}{dt} + \frac{1}{C}i(t) = \frac{d\,E(t)}{dt}$$

In general, the solution to such an equation will involve the superposition (sum) of a steady-state solution and a transient solution. Here we will be looking only for the steady-state solution, in the special case where the applied electromotive force, and hence any current, is a single-frequency sinusoidal function. Thus, we can assume that

$$E(t) = Be^{j\omega t} \quad \text{and} \quad i(t) = Ae^{j\omega t}$$

where A and B are complex numbers that determine the amplitudes and phases of voltage and current, and ω is 2π multiplied by the frequency. Then

$$\frac{d\,i}{dt} = j\omega Ae^{j\omega t} = j\omega i$$

$$\frac{d^2 i}{dt^2} = (j\omega)^2 i = -\omega^2 i$$

and the circuit equation can be rewritten

$$-\omega^2 L i + j\omega R i + \frac{1}{C} i = \frac{dE}{dt}$$

Now consider a network of circuits with resistors, capacitors, inductors, electromotive force, and currents, as shown in Figure 9.2.2. As in Section 2.4, the currents are loops, so that the actual current across some circuit elements is the difference of two loop currents. (For example, across R_1, the actual current is $i_1 - i_2$.) From our assumption that we have only one single frequency source, we may conclude that the steady-state loop currents must be of the form

$$i_1(t) = A_1 e^{j\omega t}, \quad i_2(t) = A_2 e^{j\omega t}, \quad i_3(t) = A_3 e^{j\omega t}$$

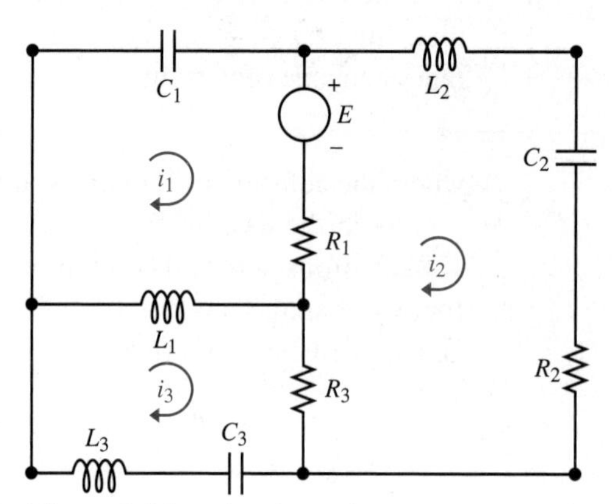

Figure 9.2.2 An alternating current network.

By applying Kirchhoff's Laws to the top-left loop, we find that

$$\left[-\omega^2 L_1(A_1 - A_3) + j\omega R_1(A_1 - A_2) + \frac{1}{C} A_1\right] e^{j\omega t} = -j\omega B e^{j\omega t}$$

If we write the corresponding equations for the other two loops, reorganize each equation, and divide out the non-zero common factor $e^{j\omega t}$, we obtain the following system of linear equations for the three variables A_1, A_2, and A_3:

$$\left[-\omega^2 L_1 + \frac{1}{C_1} + j\omega R_1\right] A_1 - j\omega R_1 A_2 + \omega^2 L_1 A_3 = -j\omega B$$

$$-j\omega R_1 A_1 + \left[-\omega^2 L_2 + \frac{1}{C_2} + j\omega(R_1 + R_2 + R_3)\right] A_2 - j\omega R_3 A_3 = j\omega B$$

$$\omega^2 L_1 A_1 - j\omega R_3 A_2 + \left[-\omega^2 (L_1 + L_3) + \frac{1}{C_3} + j\omega R_3\right] A_3 = 0$$

Thus, we have a system of three linear equations with complex coefficients for the three variables A_1, A_2, and A_3. We can solve this system by standard elimination. We emphasize that this example is for illustrative purposes only: we have constructed a completely arbitrary network and provided the solution method for only part of the problem, in a special case. A much more extensive discussion is required before a reader will be ready to start examining realistic circuits to discover what they can do. But even this limited example illustrates the general point that to analyze some electrical networks, we need to solve systems of linear equations with complex coefficients.

PROBLEMS 9.2

Practice Problems

For Problems A1–A10, determine whether the system is consistent. If it is, find the general solution.

A1
$$\begin{aligned} z_1 - iz_2 &= 3 - i \\ 2z_1 + (1 - 3i)z_2 &= 8 - 2i \end{aligned}$$

A2
$$\begin{aligned} (1 - 2i)z_1 + iz_2 &= 1 + 2i \\ (-1 - 3i)z_1 + (1 + i)z_2 &= 3 - i \end{aligned}$$

A3
$$\begin{aligned} 2z_1 - 2iz_2 &= 8 - 8i \\ (1 - 2i)z_1 - 2z_2 &= -7 - 10i \end{aligned}$$

A4
$$\begin{aligned} (1 + i)z_1 + 2z_2 &= 2 + 2i \\ (1 - i)z_1 - 2iz_2 &= 2 - 2i \end{aligned}$$

A5
$$\begin{aligned} z_1 + 3z_2 + (2 - 3i)z_3 &= 9 + i \\ 2iz_1 + 7iz_2 + (7 + 4i)z_3 &= -2 + 21i \end{aligned}$$

A6
$$\begin{aligned} z_1 + iz_2 - 2z_3 &= -2i \\ iz_1 - iz_3 + (1 + i)z_4 &= 3 \\ -2iz_1 + 2iz_3 + (-2 - 2i)z_4 &= -6 \end{aligned}$$

A7
$$\begin{aligned} iz_1 - z_2 + (-1 + i)z_3 &= -2 + i \\ z_1 + (1 + i)z_2 + z_3 &= 3 + 2i \\ -iz_1 + (1 + i)z_2 + (3 - i)z_3 &= 2 + 2i \end{aligned}$$

A8
$$\begin{aligned} z_1 - 3iz_2 - 4z_3 - 7iz_4 &= -2i \\ 3iz_2 + 3z_3 + 9iz_4 &= 3i \\ 2iz_1 + 8z_2 - 10iz_3 + 20z_4 &= 6 - i \end{aligned}$$

A9
$$\begin{aligned} z_1 + iz_2 + (1 + i)z_3 &= 1 - i \\ -2z_1 + (1 - 2i)z_2 - 2z_3 &= 2i \\ 2iz_1 - 2z_2 - (2 + 3i)z_3 &= -1 + 3i \end{aligned}$$

A10
$$\begin{aligned} z_1 + (1 + i)z_2 + 2z_3 + z_4 &= 1 - i \\ 2z_1 + (2 + i)z_2 + 5z_3 + (2 + i)z_4 &= 4 - i \\ iz_1 + (-1 + i)z_2 + (1 + 2i)z_3 + 2iz_4 &= 1 \end{aligned}$$

Homework Problems

For Problems B1–B9, determine whether the system is consistent. If it is, find the general solution.

B1
$$\begin{aligned} iz_1 + (-1 - 2i)z_2 &= i \\ 2z_1 + (-3 + 4i)z_2 &= 7 \end{aligned}$$

B2
$$\begin{aligned} z_1 + (-2 + 2i)z_2 &= 4 + 5i \\ iz_2 - 2z_2 &= -1 + 4i \end{aligned}$$

B3
$$\begin{aligned} z_1 - 2iz_2 + (1 - i)z_3 &= 2 - i \\ 3z_1 - 6iz_2 + (4 - 2i)z_3 &= 7 - 2i \end{aligned}$$

B4
$$\begin{aligned} z_1 + 3z_2 + z_3 &= 8 \\ -2z_1 - 3z_2 + z_3 &= -10 \end{aligned}$$

B5
$$\begin{aligned} z_1 - z_2 + 2z_3 &= 3 \\ 3iz_1 - 3iz_2 + (1 + 6i)z_3 &= -2 + 9i \\ iz_1 - iz_2 + 3iz_3 &= 1 + i \end{aligned}$$

B6
$$\begin{aligned} z_1 + iz_2 + 2z_3 + (-2 + 2i)z_4 &= 6 + i \\ z_1 + iz_2 + (3 + i)z_3 + (-4 + 2i)z_4 &= 9 + 4i \\ iz_1 - z_2 + 2iz_3 + (-2 - 2i)z_4 &= -1 + 6i \end{aligned}$$

B7
$$\begin{aligned} iz_1 - z_2 + 2iz_3 &= -1 - i \\ (1 + i)z_1 + 2iz_2 + (5 + 2i)z_3 &= -2 - i \\ -2iz_1 - 2iz_2 - (6 + 4i)z_3 &= 3 + 3i \end{aligned}$$

B8
$$\begin{aligned} z_1 + iz_2 + (-1 + 2i)z_3 &= -1 + 2i \\ z_2 + 2z_3 &= 2 + 2i \\ 2z_1 + (-1 + 2i)z_2 + (-6 + 4i)z_3 &= -4 \end{aligned}$$

B9
$$\begin{aligned} z_1 - iz_2 &= 1 + i \\ (1 + i)z_1 + (1 - i)z_2 + (1 - i)z_3 &= 1 + 3i \\ 2z_1 - 2iz_2 + (3 + i)z_3 &= 1 + 5i \end{aligned}$$

9.3 Complex Vector Spaces

The definition of a vector space in Section 4.2 is given in the case where the scalars are real numbers. In fact, the definition makes sense when the scalars are taken from any one system of numbers such that addition, subtraction, multiplication, and division are defined for any pairs of numbers (excluding division by 0) and satisfy the usual commutative, associative, and distributive rules for doing arithmetic. Thus, the vector space axioms make sense if we allow the scalars to be the set of complex numbers. In such cases, we say that we have a **vector space over** $\mathbb{C}$, or a **complex vector space**.

Definition
Complex Vector Space

A set $\mathbb{V}$ with operations of addition and scalar multiplication is called a **complex vector space** if for any $\mathbf{v}, \mathbf{z}, \mathbf{w} \in \mathbb{V}$ and $\alpha, \beta \in \mathbb{C}$ we have:

V1 $\mathbf{z} + \mathbf{w} \in \mathbb{V}$
V2 $\mathbf{z} + \mathbf{w} = \mathbf{w} + \mathbf{z}$
V3 $(\mathbf{z} + \mathbf{w}) + \mathbf{v} = \mathbf{z} + (\mathbf{w} + \mathbf{v})$
V4 There exists a vector $\mathbf{0} \in \mathbb{V}$ such that $\mathbf{z} + \mathbf{0} = \mathbf{z}$ for all $\mathbf{z} \in \mathbb{V}$.
V5 For each $\mathbf{z} \in \mathbb{V}$, there exists $(-\mathbf{z}) \in \mathbb{V}$ such that $\mathbf{z} + (-\mathbf{z}) = \mathbf{0}$.
V6 $\alpha\mathbf{z} \in \mathbb{V}$
V7 $\alpha(\beta\mathbf{z}) = (\alpha\beta)\mathbf{z}$
V8 $(\alpha + \beta)\mathbf{z} = \alpha\mathbf{z} + \beta\mathbf{z}$
V9 $\alpha(\mathbf{z} + \mathbf{w}) = \alpha\mathbf{z} + \alpha\mathbf{w}$
V10 $1\mathbf{z} = \mathbf{z}$

For complex vector spaces, the analog of $\mathbb{R}^n$ is $\mathbb{C}^n$ and the analog of $M_{m\times n}(\mathbb{R})$ is $M_{m\times n}(\mathbb{C})$.

Definition
$\mathbb{C}^n$

The complex vector space $\mathbb{C}^n$ is defined to be the set

$$\mathbb{C}^n = \left\{ \begin{bmatrix} z_1 \\ \vdots \\ z_n \end{bmatrix} \mid z_1, \dots, z_n \in \mathbb{C} \right\}$$

with addition and scalar multiplication defined in the expected way.

Definition
$M_{m\times n}(\mathbb{C})$

The set $M_{m\times n}(\mathbb{C})$ of all $m \times n$ matrices with complex entries is a complex vector space with standard addition and complex scalar multiplication of matrices.

All of the definitions and theorems regarding linear independence, spanning, subspaces, bases, dimension, coordinates, and determinants remain the same in complex vector spaces, with the exception that the scalars are now allowed to take any complex value.

It is instructive to look carefully at the idea of a basis for complex vector spaces.

EXAMPLE 9.3.1

Find a basis for $\mathbb{C}^1$ as a complex vector space and determine its dimension.

Solution: It is tempting for students to immediately write down $\{1, i\}$ as a basis for $\mathbb{C}^1$. However, this is not a basis since it is linearly dependent. In particular,

$$-i(1) + 1(i) = 0$$

The key is to remember that in a complex vector space we now allow the use of complex scalars. Thus, in fact, a basis for $\mathbb{C}^1$ is $\{1\}$. Indeed, every complex number z can be written in the form

$$z = \alpha 1$$

by taking $\alpha = z$.

Hence, we see that $\mathbb{C}^1$ has a basis consisting of one element, so $\mathbb{C}^1$ is a one-dimensional complex vector space.

In general, we get that the standard basis $\{\vec{e}_1, \dots, \vec{e}_n\}$ for $\mathbb{R}^n$ is also the standard basis for $\mathbb{C}^n$. Similarly, the standard basis for $M_{m\times n}(\mathbb{C})$ is the same as that for $M_{m\times n}(\mathbb{R})$.

We can also extend the definition of a linear mapping $L : \mathbb{V} \to \mathbb{W}$ to the case where $\mathbb{V}$ and $\mathbb{W}$ are both vector spaces over the complex numbers, as well as the definition of the matrix of a linear mapping with respect to bases $\mathcal{B}$ and C.

EXAMPLE 9.3.2

Let $\mathbf{z} = \begin{bmatrix} 1 \\ 2i \\ 1-i \end{bmatrix}$ and let $L : \mathbb{C}^3 \to \mathbb{C}^2$ be the linear mapping defined by

$$L(z_1, z_2, z_3) = \Big((1+i)z_1 - 2iz_2 + (1+2i)z_3, 2z_1 + (1-i)z_2 + (3+i)z_3\Big)$$

Find the standard matrix of L and use it to compute $L(\mathbf{z})$.

Solution: We have that

$$L(1,0,0) = \begin{bmatrix} 1+i \\ 2 \end{bmatrix}, \quad L(0,1,0) = \begin{bmatrix} -2i \\ 1-i \end{bmatrix}, \quad L(0,0,1) = \begin{bmatrix} 1+2i \\ 3+i \end{bmatrix}$$

Hence, the standard matrix of L is

$$[L] = \begin{bmatrix} 1+i & -2i & 1+2i \\ 2 & 1-i & 3+i \end{bmatrix}$$

Therefore,

$$L(\mathbf{z}) = [L]\mathbf{z} = \begin{bmatrix} 1+i & -2i & 1+2i \\ 2 & 1-i & 3+i \end{bmatrix} \begin{bmatrix} 1 \\ 2i \\ 1-i \end{bmatrix} = \begin{bmatrix} 8+2i \\ 8 \end{bmatrix}$$

Complex Conjugate

Since the complex conjugate is so useful in $\mathbb{C}$, we extend the definition of a complex conjugate to vectors in $\mathbb{C}^n$ and matrices in $M_{m\times n}(\mathbb{C})$.

Definition
Complex Conjugate

The **complex conjugate** of $\mathbf{z} = \begin{bmatrix} z_1 \\ \vdots \\ z_n \end{bmatrix} \in \mathbb{C}^n$ is defined to be $\overline{\mathbf{z}} = \begin{bmatrix} \overline{z_1} \\ \vdots \\ \overline{z_n} \end{bmatrix}$.

The **complex conjugate** of $Z = \begin{bmatrix} z_{11} & \cdots & z_{1n} \\ \vdots & & \vdots \\ z_{m1} & \cdots & z_{mn} \end{bmatrix} \in M_{m\times n}(\mathbb{C})$ is defined to be

$$\overline{Z} = \begin{bmatrix} \overline{z_{11}} & \cdots & \overline{z_{1n}} \\ \vdots & & \vdots \\ \overline{z_{m1}} & \cdots & \overline{z_{mn}} \end{bmatrix}$$

EXAMPLE 9.3.3

Let $\mathbf{z} = \begin{bmatrix} 1+i \\ -2i \\ 3 \\ 1-3i \end{bmatrix}$ and $W = \begin{bmatrix} 1 & 1-3i \\ -i & 2i \end{bmatrix}$. Calculate $\overline{\mathbf{z}}$ and $\overline{W}$.

Solution: We have

$$\overline{\mathbf{z}} = \begin{bmatrix} \overline{1+i} \\ \overline{-2i} \\ \overline{3} \\ \overline{1-3i} \end{bmatrix} = \begin{bmatrix} 1-i \\ 2i \\ 3 \\ 1+3i \end{bmatrix} \quad \text{and} \quad \overline{W} = \begin{bmatrix} \overline{1} & \overline{1-3i} \\ \overline{-i} & \overline{2i} \end{bmatrix} = \begin{bmatrix} 1 & 1+3i \\ i & -2i \end{bmatrix}$$

We will find the following property useful in Section 9.4.

Theorem 9.3.1

If $Z \in M_{m\times n}(\mathbb{C})$ and $\mathbf{w} \in \mathbb{C}^n$, then $\overline{Z\mathbf{w}} = \overline{Z}\,\overline{\mathbf{w}}$.

Since we will frequently need to take both the conjugate and the transpose of a vector in $\mathbb{C}^n$ or for a matrix in $M_{m\times n}(\mathbb{C})$, we invent some notation for this.

Definition
Conjugate Transpose

Let $\mathbf{z} \in \mathbb{C}^n$ and $Z \in M_{m\times n}(\mathbb{C})$. We define the **conjugate transpose** of $\mathbf{z}$ and Z by

$$\mathbf{z}^* = \overline{\mathbf{z}}^T \quad \text{and} \quad Z^* = \overline{Z}^T$$

EXERCISE 9.3.1

Let $Z = \begin{bmatrix} 1+i & 1-2i & i \\ 2 & -i & 3+i \end{bmatrix}$ and $\mathbf{z} = \begin{bmatrix} -1-i \\ 2+i \end{bmatrix}$. Find Z^* and $\mathbf{z}^*$.

The conjugate transpose has the following properties, which follow directly from the properties of the transpose and complex conjugates.

Theorem 9.3.2 If $Z, W \in M_{n\times n}(\mathbb{C})$, $\mathbf{z} \in \mathbb{C}^n$, and $\alpha \in \mathbb{C}$, then

(1) $Z^{**} = Z$
(2) $(Z + W)^* = Z^* + W^*$
(3) $(\alpha Z)^* = \overline{\alpha} Z^*$
(4) $(ZW)^* = W^* Z^*$
(5) $(Z\mathbf{z})^* = \mathbf{z}^* Z^*$

Hermitian Inner Product Spaces

We would like to have an inner product defined for complex vector spaces because the concepts of length, orthogonality, and projection are powerful tools for solving certain problems.

Our first thought would be to determine if we can extend the dot product to $\mathbb{C}^n$. Does this define an inner product on $\mathbb{C}^n$? Let $\mathbf{z} = \vec{x} + i\vec{y}$ for $\vec{x}, \vec{y} \in \mathbb{R}^n$, then we have

$$\begin{aligned}\mathbf{z} \cdot \mathbf{z} &= z_1^2 + \cdots + z_n^2 \\ &= (x_1^2 + \cdots + x_n^2 - y_1^2 - \cdots - y_n^2) + 2i(x_1y_1 + \cdots + x_ny_n)\end{aligned}$$

Observe that $\mathbf{z} \cdot \mathbf{z}$ does not even need to be a real number and so the condition $\mathbf{z} \cdot \mathbf{z} \geq 0$ does not even make sense. Thus *we cannot use the dot product as a rule for defining an inner product in* $\mathbb{C}^n$.

As in the real case, we want $\langle \mathbf{z}, \mathbf{z} \rangle$ to be a non-negative real number so that we can define the length of a vector by $\|\mathbf{z}\| = \sqrt{\langle \mathbf{z}, \mathbf{z} \rangle}$. We recall that if $z \in \mathbb{C}$, then $\overline{z}z = |z|^2 \geq 0$. Hence, it makes sense to choose

$$\langle \mathbf{z}, \mathbf{w} \rangle = \overline{\mathbf{z}} \cdot \mathbf{w}$$

as this gives us

$$\langle \mathbf{z}, \mathbf{z} \rangle = \overline{\mathbf{z}} \cdot \mathbf{z} = \overline{z_1}z_1 + \cdots + \overline{z_n}z_n = |z_1|^2 + \cdots + |z_n|^2$$

which is a non-negative real number.

Definition
Standard Inner Product on $\mathbb{C}^n$

In $\mathbb{C}^n$ the **standard inner product** $\langle\, , \rangle$ is defined by

$$\langle \mathbf{z}, \mathbf{w} \rangle = \overline{\mathbf{z}} \cdot \mathbf{w} = \overline{z_1}w_1 + \cdots + \overline{z_n}w_n, \qquad \text{for } \mathbf{z}, \mathbf{w} \in \mathbb{C}^n$$

Remark

Here we have used the definition of the standard inner product commonly used in science, engineering, and most mathematical software. It is important to note that many mathematics textbooks will use the definition

$$\langle \mathbf{z}, \mathbf{w} \rangle = \mathbf{z} \cdot \overline{\mathbf{w}}$$

for the definition of the standard inner product for $\mathbb{C}^n$.

EXAMPLE 9.3.4 Let $\mathbf{u} = \begin{bmatrix} 1+i \\ 2-i \end{bmatrix}$, $\mathbf{v} = \begin{bmatrix} -2+i \\ 3+2i \end{bmatrix}$. Determine $\langle \mathbf{v}, \mathbf{u} \rangle$, $\langle \mathbf{u}, \mathbf{v} \rangle$, and $\langle (2-i)\mathbf{u}, \mathbf{v} \rangle$.

Solution: We have

$$\begin{aligned}
\langle \mathbf{v}, \mathbf{u} \rangle &= \overline{\mathbf{v}} \cdot \mathbf{u} \\
&= \begin{bmatrix} -2-i \\ 3-2i \end{bmatrix} \cdot \begin{bmatrix} 1+i \\ 2-i \end{bmatrix} \\
&= (-2-i)(1+i) + (3-2i)(2-i) \\
&= 3 - 10i \\
\langle \mathbf{u}, \mathbf{v} \rangle &= \begin{bmatrix} 1-i \\ 2+i \end{bmatrix} \cdot \begin{bmatrix} -2+i \\ 3+2i \end{bmatrix} \\
&= 3 + 10i \\
\langle (2-i)\mathbf{u}, \mathbf{v} \rangle &= \overline{(2-i)\mathbf{u}} \cdot \mathbf{v} = \overline{(2-i)\begin{bmatrix} 1+i \\ 2-i \end{bmatrix}} \cdot \begin{bmatrix} -2+i \\ 3+2i \end{bmatrix} \\
&= (2+i)\begin{bmatrix} 1-i \\ 2+i \end{bmatrix} \cdot \begin{bmatrix} -2+i \\ 3+2i \end{bmatrix} \\
&= (2+i)(3+10i) \\
&= -4 + 23i
\end{aligned}$$

Observe that this does not satisfy the properties of the real inner product. In particular, $\langle \mathbf{u}, \mathbf{v} \rangle \neq \langle \mathbf{v}, \mathbf{u} \rangle$ and $\langle \alpha\mathbf{u}, \mathbf{v} \rangle \neq \alpha\langle \mathbf{u}, \mathbf{v} \rangle$.

EXERCISE 9.3.2 Let $\mathbf{u} = \begin{bmatrix} i \\ 1+2i \end{bmatrix}$ and $\mathbf{v} = \begin{bmatrix} 2+2i \\ 1-3i \end{bmatrix}$. Determine $\langle \mathbf{u}, \mathbf{v} \rangle$, $\langle 2i\mathbf{u}, \mathbf{v} \rangle$, and $\langle \mathbf{u}, 2i\mathbf{v} \rangle$.

We saw in Chapter 8 that the formula

$$\vec{x} \cdot \vec{y} = \vec{x}^T \vec{y}$$

was very useful. For the standard inner product on $\mathbb{C}^n$ we get

$$\langle \mathbf{z}, \mathbf{w} \rangle = \mathbf{z}^* \mathbf{w}$$

We also get the following theorem.

Theorem 9.3.3 If $A \in M_{n\times n}(\mathbb{C})$ and $\mathbf{w}, \mathbf{z} \in \mathbb{C}^n$, then

$$\langle \mathbf{w}, A\mathbf{z} \rangle = \langle A^*\mathbf{w}, \mathbf{z} \rangle$$

Proof: We have

$$\langle \mathbf{w}, A\mathbf{z} \rangle = \mathbf{w}^* A\mathbf{z} = \mathbf{w}^*(A^*)^*\mathbf{z} = (A^*\mathbf{w})^*\mathbf{z} = \langle A^*\mathbf{w}, \mathbf{z} \rangle$$

■

Properties of Complex Inner Products

Example 9.3.4 warns us that for complex vector spaces, we must modify the requirements of symmetry and bilinearity stated for real inner products.

Definition
Complex Inner Product
Complex Inner Product Space

Let $\mathbb{V}$ be a vector space over $\mathbb{C}$. A **complex inner product** on $\mathbb{V}$ is a function $\langle\,,\rangle : \mathbb{V} \times \mathbb{V} \to \mathbb{C}$ such that

(1) $\langle \mathbf{z}, \mathbf{z} \rangle \geq 0$ for all $\mathbf{z} \in \mathbb{V}$ and $\langle \mathbf{z}, \mathbf{z} \rangle = 0$ if and only if $\mathbf{z} = \mathbf{0}$
(2) $\langle \mathbf{z}, \mathbf{w} \rangle = \overline{\langle \mathbf{w}, \mathbf{z} \rangle}$ for all $\mathbf{w}, \mathbf{z} \in \mathbb{V}$
(3) For all $\mathbf{u}, \mathbf{v}, \mathbf{w}, \mathbf{z} \in \mathbb{V}$ and $\alpha \in \mathbb{C}$

$$\begin{aligned}
\text{(i)}\quad \langle \mathbf{v} + \mathbf{z}, \mathbf{w} \rangle &= \langle \mathbf{v}, \mathbf{w} \rangle + \langle \mathbf{z}, \mathbf{w} \rangle \\
\text{(ii)}\quad \langle \mathbf{z}, \mathbf{w} + \mathbf{u} \rangle &= \langle \mathbf{z}, \mathbf{w} \rangle + \langle \mathbf{z}, \mathbf{u} \rangle \\
\text{(iii)}\quad \langle \alpha\mathbf{z}, \mathbf{w} \rangle &= \overline{\alpha}\langle \mathbf{z}, \mathbf{w} \rangle \\
\text{(iv)}\quad \langle \mathbf{z}, \alpha\mathbf{w} \rangle &= \alpha\langle \mathbf{z}, \mathbf{w} \rangle
\end{aligned}$$

An complex vector space with a complex inner product is called a **complex inner product space**.

EXERCISE 9.3.3

Verify that the standard inner product on $\mathbb{C}^n$ is a complex inner product.

Remarks

1. Property (2) is the **Hermitian** property of the inner product.
2. Property (3) says that the complex inner product is not quite bilinear. However, this property reduces to bilinearity when the scalars are all real.
3. A complex inner product is sometimes called a Hermitian inner product. A complex inner product space may also be called a Hermitian inner product space or a unitary space.

EXAMPLE 9.3.5

The complex vector space $M_{m \times n}(\mathbb{C})$ can be made into a complex inner product space by adding the complex inner product defined by

$$\langle Z, W \rangle = \mathrm{tr}(Z^* W)$$

EXAMPLE 9.3.6

On the complex vector space $C[a, b]$ of complex-valued functions of a real variable x that are continuous on the closed interval $[a, b]$, we often use the complex inner product defined by

$$\langle \mathbf{f}, \mathbf{g} \rangle = \int_{\mathbf{a}}^{\mathbf{b}} \overline{\mathbf{f}(\mathbf{x})}\mathbf{g}(\mathbf{x})\, \mathbf{dx}$$

Length and Orthogonality

We can now define length and orthogonality to match the definitions in the real case.

Definition
Length
Unit Vector

Let $\mathbb{V}$ be a Hermitian inner product space. We define the **length** of $\mathbf{z} \in \mathbb{V}$ by

$$\|\mathbf{z}\| = \sqrt{\langle \mathbf{z}, \mathbf{z} \rangle}$$

If $\|\mathbf{z}\| = 1$, then $\mathbf{z}$ is called a **unit vector**.

Definition
Orthogonality

Let $\mathbb{V}$ be a Hermitian inner product space. For any $\mathbf{z}, \mathbf{w} \in \mathbb{V}$ we say that $\mathbf{z}$ and $\mathbf{w}$ are **orthogonal** if $\langle \mathbf{z}, \mathbf{w} \rangle = 0$.

Of course, these satisfy all of our familiar properties of length and orthogonality.

Theorem 9.3.4

If $\mathbb{V}$ is a Hermitian inner product space, then for any $\mathbf{z}, \mathbf{w} \in \mathbb{V}$ and $\alpha \in \mathbb{C}$ we have

(1) $\|\alpha \mathbf{z}\| = |\alpha| \, \|\mathbf{z}\|$
(2) $\frac{1}{\|\mathbf{z}\|}\mathbf{z}$ is a unit vector.
(3) $|\langle \mathbf{z}, \mathbf{w} \rangle| \leq \|\mathbf{z}\| \, \|\mathbf{w}\|$
(4) $\|\mathbf{z} + \mathbf{w}\| \leq \|\mathbf{z}\| + \|\mathbf{w}\|$

Proof: We prove (3) and leave the proof of (2) and (4) as Problem C2 and Problem C3 respectively.

If $\mathbf{w} = \mathbf{0}$, then (3) is immediate, so assume that $\mathbf{w} \neq \mathbf{0}$, and let $\alpha = \dfrac{\overline{\langle \mathbf{z}, \mathbf{w} \rangle}}{\langle \mathbf{w}, \mathbf{w} \rangle}$. Then, we get

$$\begin{aligned}
0 &\leq \langle \mathbf{z} - \alpha \mathbf{w}, \mathbf{z} - \alpha \mathbf{w} \rangle \\
&= \langle \mathbf{z}, \mathbf{z} - \alpha \mathbf{w} \rangle + \langle -\alpha \mathbf{w}, \mathbf{z} - \alpha \mathbf{w} \rangle \\
&= \langle \mathbf{z}, \mathbf{z} \rangle + \langle \mathbf{z}, -\alpha \mathbf{w} \rangle + \langle -\alpha \mathbf{w}, \mathbf{z} \rangle + \langle -\alpha \mathbf{w}, -\alpha \mathbf{w} \rangle \\
&= \langle \mathbf{z}, \mathbf{z} \rangle - \alpha \langle \mathbf{z}, \mathbf{w} \rangle - \overline{\alpha} \langle \mathbf{w}, \mathbf{z} \rangle + \overline{\alpha} \alpha \langle \mathbf{w}, \mathbf{w} \rangle \\
&= \langle \mathbf{z}, \mathbf{z} \rangle - \frac{|\langle \mathbf{z}, \mathbf{w} \rangle|^2}{\langle \mathbf{w}, \mathbf{w} \rangle} - \frac{|\langle \mathbf{z}, \mathbf{w} \rangle|^2}{\langle \mathbf{w}, \mathbf{w} \rangle} + \frac{|\langle \mathbf{z}, \mathbf{w} \rangle|^2}{\langle \mathbf{w}, \mathbf{w} \rangle} \\
&= \|\mathbf{z}\|^2 - \frac{|\langle \mathbf{z}, \mathbf{w} \rangle|^2}{\|\mathbf{w}\|^2}
\end{aligned}$$

and (3) follows. ■

EXERCISE 9.3.4

Let $\mathbb{V}$ be a Hermitian inner product space. Prove that for all $\mathbf{z} \in \mathbb{V}$ and $\alpha \in \mathbb{C}$ we have

$$\|\alpha \mathbf{z}\| = |\alpha| \, \|\mathbf{z}\|$$

Definition
Orthogonal Set
Orthonormal Set

Let $\mathcal{B} = \{\mathbf{z}_1, \ldots, \mathbf{z}_k\}$ be a set in a Hermitian inner product space. $\mathcal{B}$ is said to be an **orthogonal set** if $\langle \mathbf{z}_\ell, \mathbf{z}_j \rangle = 0$ for all $\ell \neq j$. $\mathcal{B}$ is said to be an **orthonormal set** if it is an orthogonal set and $\|\mathbf{z}_j\| = 1$ for $1 \leq j \leq k$.

9.4 Complex Diagonalization

We now extend everything we did with eigenvalues, eigenvectors, and diagonalization in Chapter 6 to allow the use of complex numbers.

Complex Eigenvalues and Eigenvectors

Definition
Eigenvalue
Eigenvector

Let $A \in M_{n\times n}(\mathbb{C})$. If there exists $\lambda \in \mathbb{C}$ and $\mathbf{z} \in \mathbb{C}^n$ with $\mathbf{z} \neq \mathbf{0}$ such that $A\mathbf{z} = \lambda\mathbf{z}$, then λ is called an **eigenvalue** of A and $\mathbf{z}$ is called an **eigenvector** of A corresponding to λ.

Since the theory of solving systems of equations, inverting matrices, and finding coordinates with respect to a basis is exactly the same for complex vector spaces as the theory for real vector spaces, the basic results on diagonalization are unchanged except that the vector space is now $\mathbb{C}^n$. A complex $n \times n$ matrix A is diagonalized by a matrix P if and only if the columns of P form a basis for $\mathbb{C}^n$ consisting of eigenvectors of A. Since the Fundamental Theorem of Algebra guarantees that every n-th degree polynomial has exactly n roots over $\mathbb{C}$, the only way a matrix cannot be diagonalizable over $\mathbb{C}$ is if it has an eigenvalue with geometric multiplicity less than its algebraic multiplicity.

EXAMPLE 9.4.1

Let $A = \begin{bmatrix} 5 & -6 \\ 3 & -1 \end{bmatrix}$. Find its eigenvectors and diagonalize over $\mathbb{C}$.

Solution: We have

$$C(\lambda) = \det(A - \lambda I) = \begin{vmatrix} 5-\lambda & -6 \\ 3 & -1-\lambda \end{vmatrix} = \lambda^2 - 4\lambda + 13$$

So, by the quadratic formula, we get that the eigenvalues of A are $\lambda_1 = 2 + 3i$ and $\lambda_2 = 2 - 3i$.
For $\lambda_1 = 2 + 3i$,

$$A - \lambda_1 I = \begin{bmatrix} 3-3i & -6 \\ 3 & -3-3i \end{bmatrix} \sim \begin{bmatrix} 1 & -(1+i) \\ 0 & 0 \end{bmatrix}$$

Hence, an eigenvector corresponding to $\lambda_1 = 2 + 3i$ is $\mathbf{z}_1 = \begin{bmatrix} 1+i \\ 1 \end{bmatrix}$.
For $\lambda_2 = 2 - 3i$,

$$A - \lambda_2 I = \begin{bmatrix} 3+3i & -6 \\ 3 & -3+3i \end{bmatrix} \sim \begin{bmatrix} 1 & -(1-i) \\ 0 & 0 \end{bmatrix}$$

Thus, an eigenvector corresponding to $\lambda_2 = 2 - 3i$ is $\mathbf{z}_2 = \begin{bmatrix} 1-i \\ 1 \end{bmatrix}$.

It follows that A is diagonalized to $\begin{bmatrix} 2+3i & 0 \\ 0 & 2-3i \end{bmatrix}$ by $P = \begin{bmatrix} 1+i & 1-i \\ 1 & 1 \end{bmatrix}$.

EXAMPLE 9.4.2

Determine whether $B = \begin{bmatrix} 2 & i \\ i & 4 \end{bmatrix}$ is diagonalizable over $\mathbb{C}$.

Solution: We have

$$C(\lambda) = \det(B - \lambda I) = \begin{vmatrix} 2-\lambda & i \\ i & 4-\lambda \end{vmatrix} = \lambda^2 - 6\lambda + 9 = (\lambda - 3)^2$$

So, the only distinct eigenvalue of B is $\lambda_1 = 3$ with algebraic multiplicity 2. For $\lambda_1 = 3$,

$$B - \lambda_1 I = \begin{bmatrix} -1 & i \\ i & 1 \end{bmatrix} \sim \begin{bmatrix} 1 & -i \\ 0 & 0 \end{bmatrix}$$

Hence, the geometric multiplicity of λ_1 is 1. Therefore, B is not diagonalizable since the geometric multiplicity of λ_1 is less than its algebraic multiplicity.

EXERCISE 9.4.1

Determine whether $A = \begin{bmatrix} 4 & 1+i \\ 1-i & 3 \end{bmatrix}$ is diagonalizable over $\mathbb{C}$. If so, diagonalize it.

EXERCISE 9.4.2

Determine whether $A = \begin{bmatrix} 0 & 1 & 0 \\ 4 & 0 & -5 \\ -2 & 1 & 2 \end{bmatrix}$ is diagonalizable over $\mathbb{C}$. If so, diagonalize it.

As usual, these examples teach us more than just how to diagonalize complex matrices. In Example 9.4.1, we see that when a matrix has only real entries, then its eigenvalues should come in complex conjugate pairs. Example 9.4.2 shows that when working with matrices with non-real entries our theory of symmetric matrices for real matrices does not apply. In particular, we had $B^T = B$, but not only did B not have real eigenvalues, it was not even diagonalizable. We will now prove our first observation that non-real eigenvalues of real matrices come in complex conjugate pairs and, moreover, that the corresponding eigenvectors will also be complex conjugates of each other. We will look at how to modify the theory of symmetric matrices to non-real matrices in Section 9.5.

Theorem 9.4.1

Let $A \in M_{n\times n}(\mathbb{R})$. If λ is a non-real eigenvalue of A with corresponding eigenvector $\mathbf{z}$, then $\overline{\lambda}$ is also an eigenvalue of A and has corresponding eigenvector $\overline{\mathbf{z}}$.

Proof: We have $A\mathbf{z} = \lambda\mathbf{z}$. Taking complex conjugates of both sides and using Theorem 9.3.1 gives

$$\overline{A\mathbf{z}} = \overline{\lambda\mathbf{z}} \Rightarrow A\overline{\mathbf{z}} = \overline{\lambda}\overline{\mathbf{z}}$$

since A is real. Hence, $\overline{\lambda}$ is an eigenvalue of A with corresponding eigenvector $\overline{\mathbf{z}}$, as required. ■

PROBLEMS 9.4

Practice Problems

For Problems A1–A10, either diagonalize the matrix over $\mathbb{C}$ or show that the matrix is not diagonalizable.

A1 $\begin{bmatrix} 2 & 1+i \\ 1-i & 1 \end{bmatrix}$

A2 $\begin{bmatrix} 3 & 5 \\ -5 & -3 \end{bmatrix}$

A3 $\begin{bmatrix} 1 & i \\ i & -1 \end{bmatrix}$

A4 $\begin{bmatrix} i & i \\ 2 & i \end{bmatrix}$

A5 $\begin{bmatrix} \cos\theta & -\sin\theta \\ \sin\theta & \cos\theta \end{bmatrix}$

A6 $\begin{bmatrix} 2 & 2 & -1 \\ -4 & 1 & 2 \\ 2 & 2 & -1 \end{bmatrix}$

A7 $\begin{bmatrix} -6-3i & -2 & -3-2i \\ 10 & 2 & 5 \\ 8+6i & 3 & 4+4i \end{bmatrix}$

A8 $\begin{bmatrix} 2 & 2 & 3 \\ 1 & 1 & -1 \\ -1 & 0 & 2 \end{bmatrix}$

A9 $\begin{bmatrix} 5 & -1+i & 2i \\ -2-2i & 2 & 1-i \\ 4i & -1-i & -1 \end{bmatrix}$

A10 $\begin{bmatrix} 2 & 1 & -1 \\ 2 & 1 & 0 \\ 3 & -1 & 2 \end{bmatrix}$

A11 $\begin{bmatrix} 0 & -i & 2+i \\ -i & 2-i & 2i \\ -i & 2-2i & 3i \end{bmatrix}$

A12 $\begin{bmatrix} 1+i & 1 & 0 \\ 1 & 1 & -i \\ 1 & 0 & 1 \end{bmatrix}$

Homework Problems

For Problems B1–B8, either diagonalize the matrix over $\mathbb{C}$ or show that the matrix is not diagonalizable.

B1 $\begin{bmatrix} 2 & -1 \\ 1 & 2 \end{bmatrix}$

B2 $\begin{bmatrix} i & 0 \\ i & i \end{bmatrix}$

B3 $\begin{bmatrix} 1+i & -1 \\ i & 0 \end{bmatrix}$

B4 $\begin{bmatrix} 2-i & i \\ -2i & 2+i \end{bmatrix}$

B5 $\begin{bmatrix} 1 & i \\ 2 & -1+2i \end{bmatrix}$

B6 $\begin{bmatrix} -1+i & 0 & -i \\ 1 & i & i \\ -2i & 0 & 2+i \end{bmatrix}$

B7 $\begin{bmatrix} 3-i & -1-i & -1+i \\ -1+i & 3+i & -1-3i \\ 0 & 0 & 2i \end{bmatrix}$

B8 $\begin{bmatrix} -i & -1 & i \\ -1+i & 1+i & 1 \\ -1+i & i & 2 \end{bmatrix}$

B9 $\begin{bmatrix} 1+2i & 3 & -2 \\ 3 & 1+2i & -2 \\ 2 & 2 & -2+2i \end{bmatrix}$

B10 $\begin{bmatrix} 5 & -2 & -i \\ -1 & 6 & i \\ i & -2i & 5 \end{bmatrix}$

B11 $\begin{bmatrix} 2i & -1+i & -1+i \\ 1-i & 2 & 1-i \\ -1-i & -1-i & 0 \end{bmatrix}$

B12 $\begin{bmatrix} 3i & -2i & -1 \\ i & 6i & 1 \\ 1 & 2 & 3i \end{bmatrix}$

Conceptual Problems

C1 Prove that if $\mathbf{z}$ is an eigenvector of $A \in M_{n\times n}(\mathbb{C})$, then $\overline{\mathbf{z}}$ is an eigenvector of $\overline{A}$.

C2 Assume that $A \in M_{n\times n}(\mathbb{C})$ is diagonalizable over $\mathbb{C}$ and has eigenvalues $\lambda_1, \dots, \lambda_n$. Prove that

$$\operatorname{tr} A = \lambda_1 + \cdots + \lambda_n$$

(Hint: see Theorem 6.2.1 on page 361.)

C3 Prove if $A \in M_{n\times n}(\mathbb{R})$ with n odd, then A has at least one real eigenvalue.

C4 Let $A \in M_{n\times n}(\mathbb{R})$ with eigenvector $\mathbf{z} = \vec{x} + i\vec{y}$, where $\vec{x}, \vec{y} \in \mathbb{R}^n$, corresponding to the non-real eigenvalue $\lambda = a + bi$.

(a) Prove that $\vec{x} \neq \vec{0}$ and $\vec{y} \neq \vec{0}$.

(b) Prove that $\vec{x} \neq k\vec{y}$ for any $k \in \mathbb{R}$.

(c) Prove that Span$\{\vec{x}, \vec{y}\}$ does not contain an eigenvector of A corresponding to an eigenvalue of A.

C5 (a) Let $A \in M_{2\times 2}(\mathbb{R})$ with eigenvector $\mathbf{z} = \vec{x} + i\vec{y}$, where $\vec{x}, \vec{y} \in \mathbb{R}^n$, corresponding to the non-real eigenvalue $\lambda = a + bi$. Prove if $P = \begin{bmatrix} \vec{x} & \vec{y} \end{bmatrix}$, then P is invertible and

$$P^{-1}AP = \begin{bmatrix} a & b \\ -b & a \end{bmatrix}$$

The matrix $\begin{bmatrix} a & b \\ -b & a \end{bmatrix}$ is called a **real canonical form** for A.

(b) Let $A \in M_{3\times 3}(\mathbb{R})$ with eigenvector $\mathbf{z} = \vec{x} + i\vec{y}$, where $\vec{x}, \vec{y} \in \mathbb{R}^n$, corresponding to the non-real eigenvalue $\lambda = a + bi$. Find a real canonical form for A.

9.5 Unitary Diagonalization

Since the complex equivalent of an orthogonal matrix is a unitary matrix, the complex equivalent of orthogonal diagonalization is unitary diagonalization.

Definition
Unitarily Similar

Let $A, B \in M_{m\times n}(\mathbb{C})$. If there exists a unitary matrix U such that $U^*AU = B$, then we say that A and B are **unitarily similar**.

If A and B are unitarily similar, then they are similar. Consequently, all of our properties of similarity still apply.

Definition
Unitarily Diagonalizable

A matrix $A \in M_{n\times n}(\mathbb{C})$ is said to be **unitarily diagonalizable** if it is unitarily similar to a diagonal matrix.

The Principal Axis Theorem says that a real matrix A is orthogonally diagonalizable if and only if it is symmetric. We observe that if A is a real symmetric matrix, then the condition $A^T = A$ is equivalent to $A^* = A$. Hence, in the complex case, the condition $A^* = A$ should take the place of the condition $A^T = A$.

Definition
Hermitian Matrix

A matrix $A \in M_{n\times n}(\mathbb{C})$ is called **Hermitian** if $A^* = A$.

EXAMPLE 9.5.1

Which of the following matrices are Hermitian?

$$A = \begin{bmatrix} 2 & 3-i \\ 3+i & 4 \end{bmatrix}, \quad B = \begin{bmatrix} 1 & 2i \\ -2i & 3-i \end{bmatrix}, \quad C = \begin{bmatrix} 0 & i & i \\ -i & 0 & i \\ -i & i & 0 \end{bmatrix}$$

Solution: We have $A^* = \begin{bmatrix} 2 & 3-i \\ 3+i & 4 \end{bmatrix} = A$, so A is Hermitian.

$$B^* = \begin{bmatrix} 1 & 2i \\ -2i & 3+i \end{bmatrix} \neq B, \text{ so } B \text{ is not Hermitian.}$$

$$C^* = \begin{bmatrix} 0 & i & i \\ -i & 0 & -i \\ -i & -i & 0 \end{bmatrix} \neq C, \text{ so } C \text{ is not Hermitian.}$$

Observe that if A is Hermitian, then we have $\overline{(A)_{ij}} = A_{ji}$, so the diagonal entries of A must be real, and for $i \neq j$ the ij-th entry must be the complex conjugate of the ji-th entry.

Remark

A linear operator $L : \mathbb{V} \to \mathbb{V}$ is called Hermitian if $\langle \mathbf{x}, L(\mathbf{y})\rangle = \langle L(\mathbf{x}), \mathbf{y}\rangle$ for all $\mathbf{x}, \mathbf{y} \in \mathbb{V}$. A linear operator is Hermitian if and only if its matrix with respect to any orthonormal basis of $\mathbb{V}$ is a Hermitian matrix. Hermitian linear operators play an important role in quantum mechanics.

Theorem 9.5.1 If $A \in M_{n\times n}(\mathbb{C})$ is Hermitian, then

(1) All eigenvalues of A are real.
(2) Eigenvectors corresponding to distinct eigenvalues are orthogonal to each other.

Proof: We will prove (1). The proof of (2) is left as Problem C1.

Suppose that λ is an eigenvalue of A with corresponding unit eigenvector $\mathbf{z}$. Using Theorem 9.3.3 and the fact that $\langle \mathbf{z}, \mathbf{z}\rangle = 1$ we get

$$\lambda = \lambda\langle \mathbf{z}, \mathbf{z}\rangle = \langle \mathbf{z}, \lambda\mathbf{z}\rangle = \langle \mathbf{z}, A\mathbf{z}\rangle = \langle A^*\mathbf{z}, \mathbf{z}\rangle = \langle A\mathbf{z}, \mathbf{z}\rangle = \langle \lambda\mathbf{z}, \mathbf{z}\rangle = \overline{\lambda}\langle \mathbf{z}, \mathbf{z}\rangle = \overline{\lambda}$$

Thus, λ is real. ■

Since a real symmetric matrix A is Hermitian, Theorem 9.5.1 (1) implies Theorem 8.1.3: that all eigenvalues of a real symmetric matrix are real. Moreover, from this result, we expect to get something very similar to the Principal Axis Theorem for Hermitian matrices. We first consider an example.

EXAMPLE 9.5.2

Unitarily diagonalize $A = \begin{bmatrix} 2 & 1+i \\ 1-i & 3 \end{bmatrix}$.

Solution: The characteristic polynomial of A is

$$C(\lambda) = \begin{vmatrix} 2-\lambda & 1+i \\ 1-i & 3-\lambda \end{vmatrix} = \lambda^2 - 5\lambda + 4 = (\lambda - 4)(\lambda - 1)$$

Hence, the eigenvalues of A are $\lambda_1 = 4$ and $\lambda_2 = 1$.
For $\lambda_1 = 4$,

$$A - \lambda_1 I = \begin{bmatrix} -2 & 1+i \\ 1-i & -1 \end{bmatrix} \sim \begin{bmatrix} 1 & -(1+i)/2 \\ 0 & 0 \end{bmatrix}$$

Thus, a corresponding eigenvector is $\mathbf{z}_1 = \begin{bmatrix} 1+i \\ 2 \end{bmatrix}$. For $\lambda_2 = 1$,

$$A - \lambda_2 I = \begin{bmatrix} 1 & 1+i \\ 1-i & 2 \end{bmatrix} \sim \begin{bmatrix} 1 & 1+i \\ 0 & 0 \end{bmatrix}$$

Thus, a corresponding eigenvector is $\mathbf{z}_2 = \begin{bmatrix} 1+i \\ -1 \end{bmatrix}$.

To unitarily diagonalize A, we need an orthonormal basis for $\mathbb{C}^2$ of eigenvectors of A. Hence, we normalize $\mathbf{z}_1$ and $\mathbf{z}_2$ and take

$$U = \begin{bmatrix} (1+i)/\sqrt{6} & (1+i)/\sqrt{3} \\ 2/\sqrt{6} & -1/\sqrt{3} \end{bmatrix}$$

to get

$$U^*AU = \operatorname{diag}(4, 1)$$

To prove that every Hermitian matrix is unitarily diagonalizable we will repeat what we did in Section 8.1. To do this we need the following very important extension of the Triangularization Theorem.

Theorem 9.5.2 Schur's Theorem

If $A \in M_{n\times n}(\mathbb{C})$, then A is unitarily similar to an upper triangular matrix T. Moreover, the diagonal entries of T are the eigenvalues of A.

Theorem 9.5.3 Spectral Theorem for Hermitian Matrices

If $A \in M_{n\times n}(\mathbb{C})$ is Hermitian, then it is unitary diagonalizable.

You are asked to prove Theorem 9.5.3 as Problem C2.

In the real case, we found that the only matrices that are orthogonally diagonalizable are symmetric matrices. Because of the power of complex numbers, there are more matrices than just Hermitian matrices that are unitarily diagonalizable.

EXERCISE 9.5.1

Show that $A = \begin{bmatrix} 0 & 1 \\ -1 & 0 \end{bmatrix}$ is unitarily diagonalizable but not Hermitian.

Normal Matrices

To look for a necessary condition for a matrix to be unitarily diagonalizable, we work backwards.

Assume $A \in M_{n\times n}(\mathbb{C})$ is unitarily diagonalizable. Let U be a unitary matrix such that $U^*AU = D$, where $D = \text{diag}(\lambda_1, \dots, \lambda_n)$. Observe that

$$DD^* = \text{diag}(|\lambda_1|^2, \dots, |\lambda_n|^2) = D^*D$$

Using this and the fact that U is unitary, we get

$$\begin{aligned} AA^* &= (UDU^*)(UDU^*)^* = UDU^*UD^*U^* = UDD^*U^* \\ &= UD^*DU^* = UD^*U^*UDU^* = (UDU^*)^*(UDU^*) = A^*A \end{aligned}$$

Consequently, if A is unitarily diagonalizable, then we must have $AA^* = A^*A$.

Definition Normal Matrix

A matrix $A \in M_{n\times n}(\mathbb{C})$ is called **normal** if $AA^* = A^*A$.

Theorem 9.5.4 Spectral Theorem for Normal Matrices

A matrix A is normal if and only if it is unitarily diagonalizable.

Of course, normal matrices are very important. We now look at some useful properties of normal matrices.

Theorem 9.5.5 If $A \in M_{n\times n}(\mathbb{C})$ is normal, then

(1) $\|A\mathbf{z}\| = \|A^*\mathbf{z}\|$, for all $\mathbf{z} \in \mathbb{C}^n$.
(2) $A - \lambda I$ is normal for every $\lambda \in \mathbb{C}$.
(3) If $A\mathbf{z} = \lambda\mathbf{z}$, then $A^*\mathbf{z} = \overline{\lambda}\mathbf{z}$.
(4) If $\mathbf{z}_1$ and $\mathbf{z}_2$ are eigenvectors of A corresponding to distinct eigenvalues λ_1 and λ_2 of A, then $\mathbf{z}_1$ and $\mathbf{z}_2$ are orthogonal.

Notice that property (4) shows us that the procedure for unitarily diagonalizing a normal matrix is exactly the same as the procedure for orthogonally diagonalizing a real symmetric matrix, except that the calculations are a little more complex.

EXAMPLE 9.5.3

Unitarily diagonalize $A = \begin{bmatrix} 4 & 1 & -i \\ 1 & 4 & -i \\ i & i & 4 \end{bmatrix}$.

Solution: We find that eigenvalues of A are $\lambda_1 = 3$ with algebraic multiplicity 2 and $\lambda_2 = 6$ with algebraic multiplicity 1. We have

$$A - 3I = \begin{bmatrix} 1 & 1 & -i \\ 1 & 1 & -i \\ i & i & 1 \end{bmatrix} \Rightarrow \mathbf{z}_1 = \begin{bmatrix} -1 \\ 1 \\ 0 \end{bmatrix}, \mathbf{z}_2 = \begin{bmatrix} i \\ 0 \\ 1 \end{bmatrix}$$

Since $\mathbf{z}_1$ and $\mathbf{z}_2$ are not orthogonal, we need to apply the Gram-Schmidt Procedure to $\{\mathbf{z}_1, \mathbf{z}_2\}$. Pick $\mathbf{w}_1 = \mathbf{z}_1$ and let $\mathbb{S}_1 = \operatorname{Span}\{\mathbf{w}_1\}$. We find that

$$\operatorname{perp}_{\mathbb{S}_1}(\mathbf{z}_2) = \mathbf{z}_2 - \frac{\langle \mathbf{w}_1, \mathbf{z}_2\rangle}{\|\mathbf{w}_1\|^2}\mathbf{w}_1 = \begin{bmatrix} i/2 \\ i/2 \\ 1 \end{bmatrix}$$

We take $\mathbf{w}_2 = \begin{bmatrix} i \\ i \\ 2 \end{bmatrix}$. We also find that

$$A - 6I = \begin{bmatrix} -2 & 1 & -i \\ 1 & -2 & -i \\ i & i & -2 \end{bmatrix} \Rightarrow \mathbf{z}_3 = \begin{bmatrix} -i \\ -i \\ 1 \end{bmatrix}$$

Hence, taking

$$U = \begin{bmatrix} -1/\sqrt{2} & i/\sqrt{6} & -i/\sqrt{3} \\ 1/\sqrt{2} & i/\sqrt{6} & -i/\sqrt{3} \\ 0 & 2/\sqrt{6} & 1/\sqrt{3} \end{bmatrix}$$

gives

$$U^*AU = \begin{bmatrix} 3 & 0 & 0 \\ 0 & 3 & 0 \\ 0 & 0 & 6 \end{bmatrix}$$

EXAMPLE 9.5.4

Unitarily diagonalize $A = \begin{bmatrix} 4i & 1+3i \\ -1+3i & i \end{bmatrix}$.

Solution: We have $C(\lambda) = \lambda^2 - 5i\lambda + 6$. Using the quadratic formula, we get eigenvalues $\lambda_1 = 6i$ and $\lambda_2 = -i$. We find that

$$A - 6iI = \begin{bmatrix} -2i & 1+3i \\ -1+3i & -5i \end{bmatrix} \Rightarrow \mathbf{z}_1 = \begin{bmatrix} 3-i \\ 2 \end{bmatrix}$$

$$A + iI = \begin{bmatrix} 5i & 1+3i \\ -1+3i & 2i \end{bmatrix} \Rightarrow \mathbf{z}_2 = \begin{bmatrix} 1+3i \\ -5i \end{bmatrix}$$

Hence, taking

$$U = \begin{bmatrix} (3-i)/\sqrt{14} & (1+3i)/\sqrt{35} \\ 2/\sqrt{14} & -5i/\sqrt{35} \end{bmatrix}$$

gives

$$U^*AU = \begin{bmatrix} 6i & 0 \\ 0 & -i \end{bmatrix}$$

PROBLEMS 9.5

Practice Problems

For Problems A1–A4, determine whether the matrix is normal.

A1 $\begin{bmatrix} 3 & 1-i \\ 1+i & 5 \end{bmatrix}$

A2 $\begin{bmatrix} 2 & -i \\ i & 1+i \end{bmatrix}$

A3 $\begin{bmatrix} 0 & 1-i \\ -1-i & 0 \end{bmatrix}$

A4 $\begin{bmatrix} 1-i & 2i \\ 2 & 3 \end{bmatrix}$

For Problems A5–A12, unitarily diagonalize the matrix.

A5 $\begin{bmatrix} 1+2i & -1 \\ -1 & 1+2i \end{bmatrix}$

A6 $\begin{bmatrix} 5i & -1-i \\ 1-i & 4i \end{bmatrix}$

A7 $\begin{bmatrix} 3 & 5 \\ -5 & 3 \end{bmatrix}$

A8 $\begin{bmatrix} 2 & 1+i \\ 1-i & 3 \end{bmatrix}$

A9 $\begin{bmatrix} 0 & i \\ i & 0 \end{bmatrix}$

A10 $\begin{bmatrix} 4 & \sqrt{2}+i \\ \sqrt{2}-i & 2 \end{bmatrix}$

A11 $\begin{bmatrix} 1 & 0 & 1+i \\ 0 & 2 & 0 \\ 1-i & 0 & 0 \end{bmatrix}$

A12 $\begin{bmatrix} i & 0 & 0 \\ 0 & -1 & 1-i \\ 0 & 1+i & 0 \end{bmatrix}$

Homework Problems

For Problems B1–B6, determine whether the matrix is normal.

B1 $\begin{bmatrix} 1+i & 1-i \\ 1-i & i \end{bmatrix}$

B2 $\begin{bmatrix} 2 & 2-i \\ 2+i & -2 \end{bmatrix}$

B3 $\begin{bmatrix} i & 1 \\ 1 & i \end{bmatrix}$

B4 $\begin{bmatrix} i & 2i \\ 2 & i \end{bmatrix}$

B5 $\begin{bmatrix} 0 & 1+i \\ 1+i & 1 \end{bmatrix}$

B6 $\begin{bmatrix} 0 & 3i \\ -3i & 1 \end{bmatrix}$

For Problems B7–B12, unitarily diagonalize the matrix.

B7 $\begin{bmatrix} 1 & 1-2i \\ 1+2i & 5 \end{bmatrix}$

B8 $\begin{bmatrix} i & 1 \\ 1 & i \end{bmatrix}$

B9 $\begin{bmatrix} -1 & -2 \\ 2 & -1 \end{bmatrix}$

B10 $\begin{bmatrix} 4i & 1+3i \\ -1+3i & i \end{bmatrix}$

B11 $\begin{bmatrix} 0 & 1 & 0 \\ -1 & 0 & 1 \\ 0 & -1 & 0 \end{bmatrix}$

B12 $\begin{bmatrix} 1 & i & -i \\ -i & -1 & i \\ i & -i & 0 \end{bmatrix}$

Conceptual Problems

C1 Let $U \in M_{n\times n}(\mathbb{C})$ be a unitary matrix.
(a) Show that $\|U\mathbf{z}\| = \|\mathbf{z}\|$ for all $\mathbf{z} \in \mathbb{C}^n$.
(b) Show that all of its eigenvalues satisfy $|\lambda| = 1$.
(c) Give a 2×2 unitary matrix such that none of its eigenvalues are real.

C2 Assume that A is Hermitian. Prove that if $\mathbf{v}$ and $\mathbf{z}$ are eigenvectors of A corresponding to distinct eigenvalues, then $\mathbf{v}$ and $\mathbf{z}$ are orthogonal.

C3 Prove that every Hermitian matrix is unitarily diagonalizable.

C4 Prove that if A is Hermitian, then $\det A$ is real.

C5 Let $A \in M_{n\times n}(\mathbb{C})$ satisfy $A^* = iA$.
(a) Prove that A is normal.
(b) Show that every eigenvalue λ of A must satisfy $\lambda = -\overline{\lambda}i$.

C6 Suppose that $A, B \in M_{n\times n}(\mathbb{C})$ are Hermitian matrices and that A is invertible. Determine which of the following are Hermitian.
(a) AB (b) A^2 (c) A^{-1}

C7 A general 2×2 Hermitian matrix can be written as $A = \begin{bmatrix} a & b+ci \\ b-ci & d \end{bmatrix}$, $a, b, c, d \in \mathbb{R}$.
(a) What can you say about $a, b, c,$ and d if A is unitary as well as Hermitian?
(b) What can you say about $a, b, c,$ and d if A is Hermitian, unitary, and diagonal?
(c) What can you say about the form of a 3×3 matrix that is Hermitian, unitary, and diagonal?

C8 Let $\mathbb{V}$ be a complex inner product space. Prove that a linear operator $L : \mathbb{V} \to \mathbb{V}$ is Hermitian ($\langle \mathbf{x}, L(\mathbf{y})\rangle = \langle L(\mathbf{x}), \mathbf{y}\rangle$) if and only if its matrix with respect to any orthonormal basis of $\mathbb{V}$ is a Hermitian matrix.

CHAPTER REVIEW

Suggestions for Student Review

1 What is the complex conjugate of a complex number? List some properties of the complex conjugate. How does the complex conjugate relate to division of complex numbers? How does it relate to the length of a complex number? (Section 9.1)

2 Define the polar form of a complex number. Explain how to convert a complex number from standard form to polar form. Is the polar form unique? How does the polar form relate to Euler's Formula? (Section 9.1)

3 List some of the similarities and some of the differences between complex vector spaces and real vector spaces. Discuss the differences between viewing $\mathbb{C}$ as a complex vector space and as a real vector space. (Section 9.3)

4 Discuss the standard inner product in $\mathbb{C}^n$. How are the essential properties of an inner product modified in generalizing from the real case to the complex case? (Section 9.3)

5 Define the conjugate transpose of a matrix. List some similarities between the conjugate transpose of a complex matrix and the transpose of a real matrix. (Section 9.3)

6 Explain how diagonalization of matrices over $\mathbb{C}$ differs from diagonalization over $\mathbb{R}$. (Section 9.4)

7 What is a Hermitian matrix? State what you can about diagonalizing a Hermitian matrix. (Section 9.5)

8 What is a normal matrix? List the properties of a normal matrix. (Section 9.5)

Chapter Quiz

For Problems E1–E6, let $z_1 = 3 + 4i$, $z_2 = 1 + 2i$, and $z_3 = 1 - 2i$. Evaluate the expression.

E1 $z_1 + z_2$ **E2** $2z_1 - iz_2$ **E3** $z_1 z_3$

E4 $z_1 z_2 z_3$ **E5** $\frac{z_1}{z_2}$ **E6** $\frac{z_2}{z_3}$

E7 Let $z_1 = 1 - \sqrt{3}i$ and $z_2 = 2 + 2i$.
(a) Find a polar form of z_1 and a polar form of z_2.
(b) Use the polar forms to determine $z_1 z_2$ and $\frac{z_1}{z_2}$.

E8 Use the polar form to determine all values of $(i)^{1/2}$.

For Problems E9 and E10, determine whether the system is consistent. If it is, find the general solution.

E9
$$\begin{aligned} z_1 - iz_2 - z_3 &= 1 \\ -2z_1 + 4iz_2 &= -2 + 6i \\ -iz_1 + 2z_2 + 4iz_3 &= 9 \end{aligned}$$

E10
$$\begin{aligned} (1-i)z_1 + (1+i)z_2 + 2z_3 &= 1 + i \\ iz_1 + 0z_2 - z_3 &= 2i \\ (1-i)z_1 + (2-i)z_2 - iz_3 &= 6 + i \end{aligned}$$

For Problems E11–E16, let $\mathbf{u} = \begin{bmatrix} 3-i \\ i \\ 2 \end{bmatrix}$ and $\mathbf{v} = \begin{bmatrix} 1 \\ 3 \\ 4-i \end{bmatrix}$.

Evaluate the expression.

E11 $2\mathbf{u} + (1+i)\mathbf{v}$ **E12** $\overline{\mathbf{u}}$

E13 $\langle \mathbf{u}, \mathbf{v} \rangle$ **E14** $\langle \mathbf{v}, \mathbf{u} \rangle$

E15 $\|\mathbf{v}\|$ **E16** $\text{proj}_{\mathbf{u}}(\mathbf{v})$

E17 Let $\mathcal{B} = \left\{ \begin{bmatrix} 1 \\ i \\ i \end{bmatrix}, \begin{bmatrix} 1 \\ 1+i \\ 1+i \end{bmatrix}, \begin{bmatrix} 1-2i \\ i \\ 1 \end{bmatrix} \right\}$. Use the Gram-Schmidt Procedure on $\mathcal{B}$ to find an orthogonal basis for $\mathbb{S} = \text{Span}\,\mathcal{B}$.

E18 Determine the projection of $\mathbf{z} = \begin{bmatrix} i \\ 1+i \\ 2-i \end{bmatrix}$ onto the subspace

$$\mathbb{S} = \text{Span}\left\{ \begin{bmatrix} i \\ -1 \\ 1+i \end{bmatrix}, \begin{bmatrix} 1-i \\ 1+i \\ 2 \end{bmatrix} \right\}$$

E19 Show that $U = \frac{1}{\sqrt{3}} \begin{bmatrix} 1-i & -i \\ 1 & -1+i \end{bmatrix}$ is unitary.

For Problems E20 and E21, either diagonalize the matrix over $\mathbb{C}$ or show that the matrix is not diagonalizable.

E20 $\begin{bmatrix} 1-3i & -1 \\ -1 & 1-i \end{bmatrix}$ **E21** $\begin{bmatrix} 3+i & 1 & i \\ 1 & 3-i & 1 \\ -1+i & 1+i & 2+i \end{bmatrix}$

E22 Let $A = \begin{bmatrix} 0 & 3+ki \\ 3+i & 3 \end{bmatrix}$.
(a) Determine k such that A is Hermitian.
(b) With the value of k as determined in part (a), unitarily diagonalize A.

APPENDIX A

Answers to Mid-Section Exercises

Section 1.1 Exercises

1 (a) $\begin{bmatrix}1\\0\end{bmatrix}$ (b) $\begin{bmatrix}-2\\-1\end{bmatrix}$ (c) $\begin{bmatrix}-1\\-1\end{bmatrix}$

2 $\vec{x} = t\begin{bmatrix}3\\-1\end{bmatrix}, t \in \mathbb{R}$

3 $\vec{x} = \begin{bmatrix}1\\1\end{bmatrix} + t\begin{bmatrix}-3\\1\end{bmatrix}, t \in \mathbb{R}$

4 $\begin{bmatrix}12\\0\\-14\end{bmatrix}$

5 $\vec{x} = \begin{bmatrix}1\\2\\2\end{bmatrix} + t\begin{bmatrix}0\\-4\\1\end{bmatrix}, t \in \mathbb{R}; \begin{cases}x_1 = 1\\x_2 = 2-4t\\x_3 = 2+t\end{cases}, t \in \mathbb{R}$

6 There are many possible answers

$\begin{bmatrix}2\\2\\1\end{bmatrix}, \begin{bmatrix}2\\1\\2\end{bmatrix}$ are in the plane. $\begin{bmatrix}1\\0\\0\end{bmatrix}$ is not in the plane.

7 $\vec{u} = \begin{bmatrix}1\\1\\1\end{bmatrix}, \vec{v} = \begin{bmatrix}2\\2\\2\end{bmatrix}$

Section 1.2 Exercises

1 (a) $\left\{\begin{bmatrix}1\\1\end{bmatrix}, \begin{bmatrix}-1\\1\end{bmatrix}\right\}$, (b) $\left\{\begin{bmatrix}-1\\2\\1\end{bmatrix}, \begin{bmatrix}1\\-2\\0\end{bmatrix}\right\}$

2 (a) A line through the origin.
(b) If $\{\vec{u}, \vec{v}\}$ is linearly dependent, then a line through the origin. Otherwise, a plane through the origin.
(c) A plane through the origin.
(d) If $\vec{u} = s\vec{v} = t\vec{w}$, then it is a line through the origin. If one of the three vectors is a linear combination of the other two (but they are not all scalar multiples of each other), then it is a plane through the origin. Otherwise, it is all of $\mathbb{R}^3$.

Section 1.3 Exercises

1 $\cos\theta = \dfrac{\vec{v} \cdot \vec{w}}{\|\vec{v}\|\|\vec{w}\|} = 0$, so $\theta = \dfrac{\pi}{2}$ rads

2 $x_1 - 3x_2 - 2x_3 = 1(1) + (-3)(2) + (-2)(3) = -11$

3 $\vec{x} = x_2\begin{bmatrix}-3/2\\1\\0\end{bmatrix} + x_3\begin{bmatrix}1/2\\0\\1\end{bmatrix}, x_2, x_3 \in \mathbb{R}$

4 $\begin{bmatrix}3\\-2\\1\end{bmatrix} \times \begin{bmatrix}2\\3\\7\end{bmatrix} = \begin{bmatrix}-17\\-19\\13\end{bmatrix}$

5 The six cross products are easily checked.

6 $4x_1 - 5x_2 + 3x_3 = -5$

7 Area$= \|\vec{u} \times \vec{v}\| = \left\|\begin{bmatrix}-2\\-1\\-2\end{bmatrix}\right\| = \sqrt{9} = 3$

8 $\vec{x} = \begin{bmatrix}0\\1\\0\end{bmatrix} + t\begin{bmatrix}3\\0\\3\end{bmatrix}, t \in \mathbb{R}.$

Section 1.4 Exercises

1 (V5) Let $\vec{x} = \begin{bmatrix} x_1 \\ \vdots \\ x_n \end{bmatrix}$. Then $-\vec{x} = \begin{bmatrix} -x_1 \\ \vdots \\ -x_n \end{bmatrix}$ since

$$\vec{x} + (-\vec{x}) = \begin{bmatrix} x_1 + (-x_1) \\ \vdots \\ x_n + (-x_n) \end{bmatrix} = \begin{bmatrix} 0 \\ \vdots \\ 0 \end{bmatrix} = \vec{0}$$

(V7) $s(t\vec{x}) = s\begin{bmatrix} tx_1 \\ \vdots \\ tx_n \end{bmatrix} = \begin{bmatrix} stx_1 \\ \vdots \\ stx_n \end{bmatrix} = st\begin{bmatrix} x_1 \\ \vdots \\ x_n \end{bmatrix} = (st)\vec{x}$

2 $S \subseteq \mathbb{R}^2$. $\vec{0} \in S$, since $2(0) = 0$. If $\vec{x}, \vec{y} \in S$, then $s\vec{x} + t\vec{y} = \begin{bmatrix} sx_1 + ty_1 \\ sx_2 + ty_2 \end{bmatrix} \in S$ since $2(sx_1 + ty_1) = s(2x_1) + t(2y_1) = sx_2 + ty_2$.
T is not a subspace since $\vec{0} \notin T$.

3 $P \subseteq \mathbb{R}^3$. Taking $a = b = 0$ gives $\vec{0} \in P$.
If $a_1\vec{v}_1 + b_1\vec{v}_2, a_2\vec{v}_1 + b_2\vec{v}_2 \in P$, then

$$\begin{aligned} s(a_1\vec{v}_1 + b_1\vec{v}_2) + t(a_2\vec{v}_1 + b_2\vec{v}_2) \\ = (sa_1 + ta_2)\vec{v}_1 + (sb_1 + tb_2)\vec{v}_2 \in P \end{aligned}$$

4 The standard basis for $\mathbb{R}^4$ is $\left\{\begin{bmatrix} 1 \\ 0 \\ 0 \\ 0 \end{bmatrix}, \begin{bmatrix} 0 \\ 1 \\ 0 \\ 0 \end{bmatrix}, \begin{bmatrix} 0 \\ 0 \\ 1 \\ 0 \end{bmatrix}, \begin{bmatrix} 0 \\ 0 \\ 0 \\ 1 \end{bmatrix}\right\}$.

Consider $\begin{bmatrix} x_1 \\ x_2 \\ x_3 \\ x_4 \end{bmatrix} = t_1\begin{bmatrix} 1 \\ 0 \\ 0 \\ 0 \end{bmatrix} + t_2\begin{bmatrix} 0 \\ 1 \\ 0 \\ 0 \end{bmatrix} + t_3\begin{bmatrix} 0 \\ 0 \\ 1 \\ 0 \end{bmatrix} + t_4\begin{bmatrix} 0 \\ 0 \\ 0 \\ 1 \end{bmatrix} = \begin{bmatrix} t_1 \\ t_2 \\ t_3 \\ t_4 \end{bmatrix}$.

Thus, for every $\vec{x} \in \mathbb{R}^4$ we have a solution $t_i = x_i$ for $1 \leq i \leq 4$. Hence, the set is a spanning set for $\mathbb{R}^4$. Moreover, if we take $\vec{x} = \vec{0}$, we get that the solution is $t_i = 0$ for $1 \leq i \leq 4$, so the set is also linearly independent.

Section 1.5 Exercises

1 $\|\vec{x}\| = \sqrt{1 + 4 + 1} = \sqrt{6}$, $\|\vec{y}\| = \sqrt{\frac{1}{6} + \frac{4}{6} + \frac{1}{6}} = 1$

2 By definition $\hat{x}$ is a scalar multiple of $\vec{x}$ so it is parallel. Using Theorem 1.5.2 (2) we get

$$\|\hat{x}\| = \left\|\frac{1}{\|\vec{x}\|}\vec{x}\right\| = \left|\frac{1}{\|\vec{x}\|}\right| \|\vec{x}\| = \frac{\|\vec{x}\|}{\|\vec{x}\|} = 1$$

3

$$\text{proj}_{\vec{v}}(\vec{u}) = \frac{\vec{v} \cdot \vec{u}}{\|\vec{v}\|^2}\vec{v} = \frac{1}{14}\begin{bmatrix} 3 \\ 1 \\ 2 \end{bmatrix}$$

$$\text{perp}_{\vec{v}}(\vec{u}) = \vec{u} - \text{proj}_{\vec{v}}(\vec{u}) = \begin{bmatrix} 1 \\ -2 \\ 0 \end{bmatrix} - \begin{bmatrix} 3/14 \\ 1/14 \\ 2/14 \end{bmatrix} = \begin{bmatrix} 11/14 \\ -29/14 \\ -1/7 \end{bmatrix}$$

4 (L1):

$$\begin{aligned} \text{proj}_{\vec{x}}(s\vec{y} + t\vec{z}) &= \frac{\vec{x} \cdot (s\vec{y} + t\vec{z})}{\|\vec{x}\|^2}\vec{x} \\ &= \frac{\vec{x} \cdot (s\vec{y}) + \vec{x} \cdot (t\vec{z})}{\|\vec{x}\|^2}\vec{x} \\ &= s\frac{\vec{x} \cdot \vec{y}}{\|\vec{x}\|^2}\vec{x} + t\frac{\vec{x} \cdot \vec{z}}{\|\vec{x}\|^2}\vec{x} \\ &= s\,\text{proj}_{\vec{x}}(\vec{y}) + t\,\text{proj}_{\vec{x}}(\vec{z}) \end{aligned}$$

(L2):

$$\begin{aligned} \text{proj}_{\vec{x}}(\text{proj}_{\vec{x}}(\vec{y})) &= \frac{\vec{x} \cdot (\text{proj}_{\vec{x}}(\vec{y}))}{\|\vec{x}\|^2}\vec{x} \\ &= \frac{1}{\|\vec{x}\|^2}\left(\frac{\vec{x} \cdot \vec{y}}{\|\vec{x}\|^2}(\vec{x} \cdot \vec{x})\right)\vec{x} \\ &= \frac{\vec{x} \cdot \vec{y}}{\|\vec{x}\|^2}\vec{x} = \text{proj}_{\vec{x}}(\vec{y}) \end{aligned}$$

Section 2.1 Exercises

1 Observe the lines intersect at the point $(-3, 2)$.

2
$$\begin{aligned}(7+2t)-2(2+t)&=3\\(7+2t)+(2+t)+3(-t)&=9\end{aligned}$$

3 There are many possible answers.

(2.1.1) (a)
$$\begin{aligned}x_1+0x_2+0x_3&=-1\\-x_1+x_2+0x_3&=\ \ 0\\x_1+x_2+0x_3&=\ \ 0\end{aligned}$$

(2.1.1) (b)
$$\begin{aligned}x_1+x_2+x_3&=1\\x_1+x_2+x_3&=2\\-x_2+x_3&=1\end{aligned}$$

(2.1.2)
$$\begin{aligned}x_1+x_2+x_3&=1\\x_2+x_3&=1\\x_3&=1\end{aligned}$$

(2.1.3)
$$\begin{aligned}x_1+x_2+x_3&=1\\x_2+x_3&=1\\x_1+2x_2+2x_3&=2\end{aligned}$$

4 $x_1 = 15, x_2 = -19/3$.

5 $\begin{bmatrix}x_1\\x_2\\x_3\end{bmatrix}=\begin{bmatrix}6-2t\\t\\2\end{bmatrix}=\begin{bmatrix}6\\0\\2\end{bmatrix}+t\begin{bmatrix}-2\\1\\0\end{bmatrix},\quad t\in\mathbb{R}$

6 $\left[\begin{array}{ccc|c}2&4&0&12\\1&2&-1&4\end{array}\right]\ R_2-\frac{1}{2}R_1\ \sim\ \left[\begin{array}{ccc|c}2&4&0&12\\0&0&-1&-2\end{array}\right]$

7 $\begin{bmatrix}1&2&-1&1\\0&2&1&-1\\0&0&0&5\end{bmatrix}$

Section 2.2 Exercises

1 (a) Not in RREF (b) In RREF
(c) Not in RREF (d) Not in RREF

2 $\begin{bmatrix}1&0&1&1\\0&1&-1&-1\\0&0&0&0\end{bmatrix}$

3 (a) $\text{rank}(A) = 2$ (b) $\text{rank}(B) = 2$

4 $\vec{x}=r\begin{bmatrix}0\\-1\\1\\0\\0\end{bmatrix}+s\begin{bmatrix}-1\\0\\0\\1\\0\end{bmatrix}+t\begin{bmatrix}-2\\-1\\0\\0\\1\end{bmatrix},\quad r,s,t\in\mathbb{R}$

Section 2.3 Exercises

1 Yes, $(-2)\begin{bmatrix}1\\-3\\-3\end{bmatrix}+\frac{19}{4}\begin{bmatrix}2\\-2\\1\end{bmatrix}+\frac{13}{4}\begin{bmatrix}-2\\2\\-3\end{bmatrix}=\begin{bmatrix}1\\3\\1\end{bmatrix}$.

2 The set is linearly independent.

Section 3.1 Exercises

1 $\mathcal{B}$ is linearly dependent. $X \in \operatorname{Span}\mathcal{B}$.

2 For any 2×2 matrix $\begin{bmatrix} x_1 & x_2 \\ x_3 & x_4 \end{bmatrix}$ we have

$$x_1\begin{bmatrix} 1 & 0 \\ 0 & 0 \end{bmatrix} + x_2\begin{bmatrix} 0 & 1 \\ 0 & 0 \end{bmatrix} + x_3\begin{bmatrix} 0 & 0 \\ 1 & 0 \end{bmatrix} + x_4\begin{bmatrix} 0 & 0 \\ 0 & 1 \end{bmatrix} = \begin{bmatrix} x_1 & x_2 \\ x_3 & x_4 \end{bmatrix}$$

Hence, $\mathcal{B}$ spans $M_{2\times2}(\mathbb{R})$. Moreover, if we take $x_1 = x_2 = x_3 = x_4 = 0$, then the only solution is the trivial solution, so the set is linearly independent.

3 $\vec{a}_1 = \begin{bmatrix} 1 \\ 3 \\ 1 \end{bmatrix}, \vec{a}_2 = \begin{bmatrix} -4 \\ 0 \\ 2 \end{bmatrix}, \vec{a}_3 = \begin{bmatrix} 5 \\ 9 \\ -3 \end{bmatrix}$

4 $A^T = \begin{bmatrix} 2 & -1 \\ 3 & 0 \\ 1 & 5 \end{bmatrix}$ and $(A^T)^T = \begin{bmatrix} 2 & 3 & 1 \\ -1 & 0 & 5 \end{bmatrix} = A$

$3A^T = \begin{bmatrix} 6 & -3 \\ 9 & 0 \\ 3 & 15 \end{bmatrix} = (3A)^T$

5 (a) $\begin{bmatrix} 11 \\ 24 \end{bmatrix}$ (b) $[10]$ (c) $\begin{bmatrix} -4 \\ 1 \\ -2 \end{bmatrix}$

6 $A = \begin{bmatrix} 1 & 1 \\ 2 & 0 \\ -2 & -3 \end{bmatrix}, \vec{x} = \begin{bmatrix} x_1 \\ x_2 \end{bmatrix}, \vec{b} = \begin{bmatrix} 3 \\ 15 \\ -5 \end{bmatrix}$

7 (a) $\begin{bmatrix} 11 \\ 24 \end{bmatrix}$ (b) $[10]$ (c) $\begin{bmatrix} -4 \\ 1 \\ -2 \end{bmatrix}$

8 (a) AB is not defined since A has 3 columns and B has 2 rows.

(b) $BA = \begin{bmatrix} 4 & 7 & -1 \\ 1 & 2 & -1 \end{bmatrix}$

(c) $A^TA = \begin{bmatrix} 5 & 8 & 1 \\ 8 & 13 & 1 \\ 1 & 1 & 2 \end{bmatrix}$

(d) $BB^T = \begin{bmatrix} 5 & 2 \\ 2 & 1 \end{bmatrix}$

9 (a) Let $\vec{x}_i$ be a solution of $A\vec{x} = \vec{e}_i$, $1 \leq i \leq m$. We can write $\vec{y} = y_1\vec{e}_1 + \cdots + y_m\vec{e}_m$, so

$$\begin{aligned} A(y_1\vec{x}_1 + \cdots + y_m\vec{x}_m) &= y_1A\vec{x}_1 + \cdots + y_mA\vec{x}_m \\ &= y_1\vec{e}_1 + \cdots + y_m\vec{e}_m = \vec{y} \end{aligned}$$

(b) By the System-Rank Theorem, $\operatorname{rank}(A) = m$.

(c) Take $B = \begin{bmatrix} \vec{x}_1 & \cdots & \vec{x}_m \end{bmatrix}$.

(d) $B = \begin{bmatrix} 1 & -2 \\ 0 & 1 \\ 0 & 0 \end{bmatrix}$

Section 3.2 Exercises

1 $f_A(-1,1,1,0) = \begin{bmatrix} 0 \\ 0 \\ 0 \end{bmatrix}, f_A(-3,1,0,1) = \begin{bmatrix} 0 \\ 0 \\ 0 \end{bmatrix}$

2 $f_A(1,0) = \begin{bmatrix} 1 \\ 3 \end{bmatrix}$, $f_A(0,1) = \begin{bmatrix} 2 \\ -1 \end{bmatrix}$, $f_A(2,3) = \begin{bmatrix} 8 \\ 3 \end{bmatrix}$

$f_A(2,3) = 2f_A(1,0) + 3f_A(0,1)$

3 (a) f is not linear. $f(1,0) + f(2,0) \neq f(3,0)$

(b) G is linear.

4 $[H] = \begin{bmatrix} 0 & 0 & 1 & 1 \\ 1 & 0 & 0 & 0 \end{bmatrix}$

Section 3.3 Exercises

1 $[R_{\pi/4}] = \begin{bmatrix} \sqrt{2}/2 & -\sqrt{2}/2 \\ \sqrt{2}/2 & \sqrt{2}/2 \end{bmatrix}, R_{\pi/4}(1,1) = \begin{bmatrix} 0 \\ \sqrt{2} \end{bmatrix}$

2 $[S] = \begin{bmatrix} 1 & 0 \\ 0 & 3 \end{bmatrix}, S(\vec{x}) = \begin{bmatrix} 2 \\ 3 \end{bmatrix}$

3 $[T] = \begin{bmatrix} 3 & 0 \\ 0 & 3 \end{bmatrix}, T(\vec{x}) = \begin{bmatrix} 6 \\ 3 \end{bmatrix}$

4 $[H] = \begin{bmatrix} 1 & 0 \\ 2 & 1 \end{bmatrix}$

5 $[F] = \begin{bmatrix} -1 & 0 \\ 0 & 1 \end{bmatrix}, F(\vec{x}) = \begin{bmatrix} -2 \\ 1 \end{bmatrix}$

6 x_1x_3-plane: $\begin{bmatrix} 1 & 0 & 0 \\ 0 & -1 & 0 \\ 0 & 0 & 1 \end{bmatrix}$. x_2x_3-plane $\begin{bmatrix} -1 & 0 & 0 \\ 0 & 1 & 0 \\ 0 & 0 & 1 \end{bmatrix}$

Section 3.4 Exercises

1 $\text{Range}(L) = \text{Span}\left\{\begin{bmatrix} 1 \\ -2 \end{bmatrix}, \begin{bmatrix} 0 \\ 1 \end{bmatrix}\right\} = \mathbb{R}^2$

2 $\text{Null}(L) = \text{Span}\left\{\begin{bmatrix} 1 \\ 1 \\ 0 \end{bmatrix}\right\}$.

3 $\left\{\begin{bmatrix} 1 \\ 1 \\ -1 \end{bmatrix}, \begin{bmatrix} 1 \\ -1 \\ 2 \end{bmatrix}, \begin{bmatrix} 3 \\ -3 \\ -2 \end{bmatrix}\right\}$

4 $\text{Col}(A)$: $\left\{\begin{bmatrix} 1 \\ 2 \\ 0 \end{bmatrix}, \begin{bmatrix} 1 \\ 3 \\ 1 \end{bmatrix}, \begin{bmatrix} 1 \\ 4 \\ 3 \end{bmatrix}\right\}$ $\quad$ $\text{Null}(A^T)$: $\{\}$

$\text{Row}(A)$: $\left\{\begin{bmatrix} 1 \\ 0 \\ -1 \\ 0 \end{bmatrix}, \begin{bmatrix} 0 \\ 1 \\ -2 \\ 0 \end{bmatrix}, \begin{bmatrix} 0 \\ 0 \\ 0 \\ 1 \end{bmatrix}\right\}$ $\quad$ $\text{Null}(A)$: $\left\{\begin{bmatrix} 1 \\ 2 \\ 1 \\ 0 \end{bmatrix}\right\}$

Section 3.5 Exercises

1 $K^{-1} = \begin{bmatrix} 7/26 & 2/13 \\ 2/13 & 3/13 \end{bmatrix}, x_1 = \frac{45}{13}, x_2 = \frac{35}{13}$

2 $A^{-1} = \begin{bmatrix} -1 & 1 & 1 \\ -1 & 0 & 1 \\ 3 & -1 & -2 \end{bmatrix}, \vec{x} = A^{-1}\vec{b} = \begin{bmatrix} 2 \\ -3 \\ 5 \end{bmatrix}$

3 (a) $[R^{-1}] = \begin{bmatrix} 0 & 1 \\ 1 & 0 \end{bmatrix}$

(b) $[S^{-1}] = \begin{bmatrix} 1 & -t \\ 0 & 1 \end{bmatrix}$

Section 3.6 Exercises

1 If $E_1 = \begin{bmatrix} 1 & 0 & 0 \\ -2 & 1 & 0 \\ 0 & 0 & 1 \end{bmatrix}, E_2 = \begin{bmatrix} 1 & 0 & 0 \\ 0 & 1 & 0 \\ -3 & 0 & 1 \end{bmatrix}$

$E_3 = \begin{bmatrix} 1 & 0 & 0 \\ 0 & 1 & 0 \\ 0 & 0 & -1 \end{bmatrix}, E_4 = \begin{bmatrix} 1 & 0 & 0 \\ 0 & 0 & 1 \\ 0 & 1 & 0 \end{bmatrix}$

$E_5 = \begin{bmatrix} 1 & -1 & 0 \\ 0 & 1 & 0 \\ 0 & 0 & 1 \end{bmatrix}$, and $R = \begin{bmatrix} 1 & 0 \\ 0 & 1 \\ 0 & 0 \end{bmatrix}$,

then $E_5E_4E_3E_2E_1A = R$.

Section 3.7 Exercises

1 $A = LU = \begin{bmatrix} 1 & 0 & 0 \\ -4 & 1 & 0 \\ 3 & -2 & 1 \end{bmatrix}\begin{bmatrix} -1 & 1 & 2 \\ 0 & 3 & 5 \\ 0 & 0 & 5 \end{bmatrix}$

2 $\vec{x}_1 = \begin{bmatrix} 10/3 \\ -11/3 \\ 5 \end{bmatrix}, \vec{x}_2 = \begin{bmatrix} 17/3 \\ -1/3 \\ 7 \end{bmatrix}$

Section 4.1 Exercises

1 $\mathcal{B}$ is linearly independent. $\mathbf{p} \in \operatorname{Span}\mathcal{B}$.

2 Clearly any polynomial $a + bx + cx^2 + dx^3 \in \operatorname{Span}\mathcal{B}$. If we consider $0 = t_1(1) + t_2x + t_3x^2 + t_4x^3$ we get that $t_1 = t_2 = t_3 = t_4 = 0$.

Section 4.2 Exercises

1 The set $\mathbb{S}$ is not closed under scalar multiplication.

2 Axioms V2, V3, V7, V8, V9, and V10 hold by Theorem 3.1.1. Show the other axioms hold.

3 Axioms V2, V3, V7, V8, V9, and V10 hold by Theorem 4.1.1. Show the other axioms hold.

4 $\mathbb{U} \subseteq P_2(\mathbb{R})$. Taking $a = b = c = 0$, we get $0 \in \mathbb{U}$. If $\mathbf{p}(x) = a + bx + cx^2, \mathbf{q}(x) = d + ex + fx^2 \in \mathbb{U}$, and $s, t \in \mathbb{R}$, then
$s\mathbf{p}(x) + t\mathbf{q}(x) = (sa + td) + (sb + te)x + (sc + tf)x^2$
and
$(sb + te) + (sc + tf) = s(b + c) + t(e + f) = sa + td$
So, $s\mathbf{p} + t\mathbf{q} \in \mathbb{U}$. Hence, $\mathbb{U}$ is a subspace of $P_2(\mathbb{R})$.

5 By definition, $\{\mathbf{0}\}$ is a non-empty subset of $\mathbb{V}$. If $\mathbf{x}, \mathbf{y} \in \{\mathbf{0}\}$ and $s, t \in \mathbb{R}$, then $s\mathbf{x} + t\mathbf{y} = s\mathbf{0} + t\mathbf{0} = \mathbf{0}$. Hence, $\{\mathbf{0}\}$ is a subspace of $\mathbb{V}$ and therefore is a vector space under the same operations as $\mathbb{V}$.

Section 4.3 Exercises

1 $\{1 - x, 2 + 2x + x^2, x + x^2\}$

2 $\dim \mathbb{S} = 3$

3 There are many possible answers. One possibility is
$\left\{\begin{bmatrix} 1 \\ 1 \\ 1 \end{bmatrix}, \begin{bmatrix} 1 \\ 0 \\ 0 \end{bmatrix}, \begin{bmatrix} 0 \\ 1 \\ 0 \end{bmatrix}\right\}$.

4 There are many possible answers.
$\left\{\begin{bmatrix} 1 \\ 1 \\ 0 \\ 0 \end{bmatrix}, \begin{bmatrix} 0 \\ 1 \\ 1 \\ 0 \end{bmatrix}, \begin{bmatrix} 2 \\ 0 \\ 0 \\ 1 \end{bmatrix}\right\}$ is a basis for the hyperplane.
$\left\{\begin{bmatrix} 1 \\ 1 \\ 0 \\ 0 \end{bmatrix}, \begin{bmatrix} 0 \\ 1 \\ 1 \\ 0 \end{bmatrix}, \begin{bmatrix} 2 \\ 0 \\ 0 \\ 1 \end{bmatrix}, \begin{bmatrix} 1 \\ 0 \\ 0 \\ 0 \end{bmatrix}\right\}$ is a basis for $\mathbb{R}^4$.

Section 4.4 Exercises

1 $\begin{bmatrix}0\\2\\2\end{bmatrix}_{\mathcal{B}} = \begin{bmatrix}6\\4\end{bmatrix}$

2 If $\vec{x} = \begin{bmatrix}x_1\\\vdots\\x_n\end{bmatrix}$, then we have $\vec{x} = x_1\vec{e}_1 + \cdots + x_n\vec{e}_n$, so $[\vec{x}]_{\mathcal{S}} = \vec{x}$.

3 $Q = \begin{bmatrix}1&1&2\\1&2&2\\2&4&3\end{bmatrix}, P = \begin{bmatrix}2&-5&2\\-1&1&0\\0&2&-1\end{bmatrix}$

Section 4.5 Exercises

1
$$\begin{aligned}
L(s\vec{x}+t\vec{y}) &= L\left(\begin{bmatrix}sx_1+ty_1\\sx_2+ty_2\\sx_3+ty_3\end{bmatrix}\right)\\
&= \begin{bmatrix}sx_1+ty_1 & sx_1+ty_1+sx_2+ty_2+sx_3+ty_3\\0 & sx_2+ty_2\end{bmatrix}\\
&= s\begin{bmatrix}x_1 & x_1+x_2+x_3\\0 & x_2\end{bmatrix} + t\begin{bmatrix}y_1 & y_1+y_2+y_3\\0 & y_2\end{bmatrix}\\
&= sL(\vec{x}) + tL(\vec{y})
\end{aligned}$$

2
$$\begin{aligned}
\mathcal{D}(s\mathbf{p}+t\mathbf{q}) &= (s\mathbf{p}+t\mathbf{q})' = s\mathbf{p}' + t\mathbf{q}'\\
&= s\mathcal{D}(\mathbf{p}) + t\mathcal{D}(\mathbf{q})
\end{aligned}$$

3 A basis for Null(L) is $\left\{\begin{bmatrix}0\\1\\-1\end{bmatrix}\right\}$.

A basis for Range(L) is $\left\{\begin{bmatrix}1&0\\0&1\end{bmatrix}, \begin{bmatrix}0&1\\1&0\end{bmatrix}\right\}$.

Section 4.6 Exercises

1 $[\text{refl}_{\vec{n}}]_{\mathcal{B}} = \begin{bmatrix}-1&0&0\\0&1&0\\0&0&1\end{bmatrix}$

2 $[L]_{\mathcal{B}} = \begin{bmatrix}2&1&0&0\\-2&-2&0&0\\1&2&1&0\\2&1&0&1\end{bmatrix}$

Section 4.7 Exercises

1 If $\mathbf{0} = t_1L(\mathbf{u}_1) + \cdots + t_kL(\mathbf{u}_k) = L(t_1\mathbf{u}_1 + \cdots t_k\mathbf{u}_k)$, then $t_1\mathbf{u}_1 + \cdots + t_k\mathbf{u}_k \in \text{Null}(L)$. Thus, $t_1\mathbf{u}_1 + \cdots + t_k\mathbf{u}_k = \mathbf{0}$ by Theorem 4.7.1. Hence, $t_1 = \cdots = t_k = 0$ as required.

2 Let $\mathbf{v} \in \mathbb{V}$. Since L is onto, there exists $\mathbf{x} \in \mathbb{U}$ such that $L(\mathbf{x}) = \mathbf{v}$. Hence, we have

$$\mathbf{v} = L(\mathbf{x}) = L(t_1\mathbf{u}_1 + \cdots + t_k\mathbf{u}_k) = t_1L(\mathbf{u}_1) + \cdots + t_kL(\mathbf{u}_k)$$

Section 5.1 Exercises

1 (a) $\begin{vmatrix} 3 & 2 \\ 2 & 1 \end{vmatrix} = 3(1) - 2(2) = -1$

(b) $\begin{vmatrix} 1 & 3 \\ 0 & -2 \end{vmatrix} = 1(-2) - 3(0) = -2$

(c) $\begin{vmatrix} 2 & 4 \\ 1 & 2 \end{vmatrix} = 2(2) - 4(1) = 0$

2 $C_{11} = 3, C_{12} = -8, C_{13} = 4$
$\det A = 1C_{11} + 2C_{12} + 3C_{13} = -1$

3 (a) $\det A = 1C_{11} + 0C_{21} + 3C_{31} + (-2)C_{41} = 112$
(b) $\det A = 0C_{21} + 0C_{22} + (-1)C_{23} + 2C_{24} = 112$
(c) $\det A = 0C_{14} + 2C_{24} + 0C_{34} + 0C_{44} = 112$

Section 5.2 Exercises

1 rA is obtained by multiplying each row of A by r. Thus, by Theorem 5.2.1,

$$\det(rA) = (r)(r)(r)\det A = r^3 \det A$$

2 $\det A = 20$

3 $\det A = 156$

Section 5.3 Exercises

1 $A^{-1} = \frac{1}{\det A}(\operatorname{cof} A)^T = \begin{bmatrix} 3 & 2 & -2 \\ -1 & -1 & 1 \\ 3 & 3 & -2 \end{bmatrix}$

Section 5.4 Exercises

1 We have $S(\vec{e}_1) = \begin{bmatrix} t \\ 0 \end{bmatrix}$ and $S(\vec{e}_2) = \begin{bmatrix} 0 \\ 1 \end{bmatrix}$. Hence,

$$\text{Area}(S(\vec{e}_1), S(\vec{e}_2)) = \left| \det \begin{bmatrix} t & 0 \\ 0 & 1 \end{bmatrix} \right| = |t| = t$$

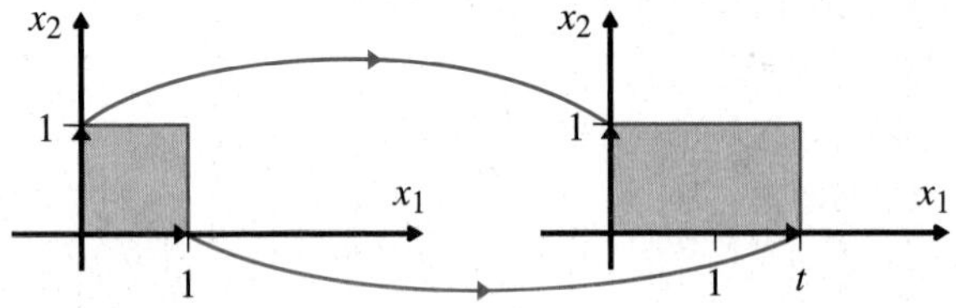

Alternately,

$$\text{Area}(S(\vec{e}_1), S(\vec{e}_2)) = |\det A|\text{Area}(\vec{e}_1, \vec{e}_2) = \left| \det \begin{bmatrix} t & 0 \\ 0 & 1 \end{bmatrix} \right| \left| \det \begin{bmatrix} 1 & 0 \\ 0 & 1 \end{bmatrix} \right| = t(1) = t$$

Section 6.1 Exercises

1 The eigenvalues are $\lambda_1 = 0$ and $\lambda_2 = 5$.

A basis for E_{λ_1} is $\left\{\begin{bmatrix} -2 \\ 1 \end{bmatrix}\right\}$.

A basis for E_{λ_2} is Span $\left\{\begin{bmatrix} 1 \\ 2 \end{bmatrix}\right\}$.

2 $\lambda_1 = 5$ has algebraic multiplicity 1 and geometric multiplicity 1. $\lambda_2 = -2$ has algebraic multiplicity 2 and geometric multiplicity 1.

3 The only real eigenvalue is $\lambda_1 = 1$ with

$$E_{\lambda_1} = \operatorname{Span}\left\{\begin{bmatrix} 0 \\ 0 \\ 1 \end{bmatrix}\right\}.$$

Section 6.2 Exercises

1 $P = \begin{bmatrix} 1 & -1 & 1 \\ 1 & -3 & -1 \\ 1 & 1 & 2 \end{bmatrix}, D = \begin{bmatrix} 0 & 0 & 0 \\ 0 & 2 & 0 \\ 0 & 0 & -1 \end{bmatrix}$

2 A is not diagonalizable since $\lambda_1 = 2$ has algebraic multiplicity 2, but the geometric multiplicity is 1.

Section 6.3 Exercises

1 $A^{100} = \begin{bmatrix} 2 - 2^{100} & -1 + 2^{100} \\ 2 - 2^{101} & -1 + 2^{101} \end{bmatrix}$

2 (a) A is not a Markov matrix.

(b) B is a Markov matrix. $\begin{bmatrix} 1/2 \\ 1/2 \end{bmatrix}$

3 We find that $T\vec{s} = \begin{bmatrix} s_2 \\ s_1 \end{bmatrix}, T^2\vec{s} = \begin{bmatrix} s_1 \\ s_2 \end{bmatrix}, T^3\vec{s} = \begin{bmatrix} s_2 \\ s_1 \end{bmatrix}$, etc.

On the other hand, the fixed-state vector is $\begin{bmatrix} 1/2 \\ 1/2 \end{bmatrix}$.

Section 7.1 Exercises

1 $\left\{\begin{bmatrix} 1/\sqrt{6} \\ 1/\sqrt{6} \\ 2/\sqrt{6} \end{bmatrix}, \begin{bmatrix} -2/\sqrt{5} \\ 0 \\ 1/\sqrt{5} \end{bmatrix}, \begin{bmatrix} 1/\sqrt{30} \\ -5\sqrt{30} \\ 2\sqrt{30} \end{bmatrix}\right\}$

2 It is easy to verify that $\mathcal{B}$ is orthonormal. We have

$$\begin{bmatrix} 3 \\ 1 \\ 4 \end{bmatrix} = \frac{8}{\sqrt{3}}\begin{bmatrix} 1/\sqrt{3} \\ 1/\sqrt{3} \\ 1/\sqrt{3} \end{bmatrix} - \frac{1}{\sqrt{2}}\begin{bmatrix} 1/\sqrt{2} \\ 0 \\ -1/\sqrt{2} \end{bmatrix} + \frac{5}{\sqrt{6}}\begin{bmatrix} 1/\sqrt{6} \\ -2/\sqrt{6} \\ 1/\sqrt{6} \end{bmatrix}.$$

3 The result is easily verified.

Section 7.2 Exercises

1 $\mathbb{S}^\perp = \operatorname{Span}\left\{\begin{bmatrix} 1 \\ -1 \\ 0 \\ 1 \end{bmatrix}\right\}$

2 $\operatorname{proj}_{\mathbb{S}}(\vec{x}) = \begin{bmatrix} 5/2 \\ 1 \\ 5/2 \end{bmatrix}, \operatorname{perp}_{\mathbb{S}}(\vec{x}) = \begin{bmatrix} -1/2 \\ 0 \\ 1/2 \end{bmatrix}$

Section 7.4 Exercises

1 The result is easily verified. We observe that $\langle A, B\rangle = a_1b_1+a_2b_2+a_3b_3+a_4b_4$, so it matches the dot product on the isomorphic vectors in $\mathbb{R}^4$.

2 (a) $\|1\| = \sqrt{3}, \|x\| = \sqrt{5}$
(b) $\|1\| = \sqrt{3}, \|x\| = \sqrt{2}$

Section 8.1 Exercises

1 $A^T = A$, so A is symmetric. $B^T = \begin{bmatrix} 2 & 3 \\ -3 & 0 \end{bmatrix}$, so B is not symmetric.

2 $P = \begin{bmatrix} 1/\sqrt{3} & -1/\sqrt{2} & 1/\sqrt{6} \\ 1/\sqrt{3} & 1/\sqrt{2} & 1/\sqrt{6} \\ 1/\sqrt{3} & 0 & -2/\sqrt{6} \end{bmatrix}, D = \begin{bmatrix} 0 & 0 & 0 \\ 0 & 3 & 0 \\ 0 & 0 & 3 \end{bmatrix}$

Section 8.2 Exercises

1 (a) $Q(\vec{x}) = 4x_1^2 + x_1x_2 + \sqrt{2}x_2^2$
(b) $Q(\vec{x}) = x_1^2 - 2x_1x_2 + 2x_2^2 + 6x_2x_3 - x_3^2$

2 (a) $\begin{bmatrix} 1 & -1 \\ -1 & -3 \end{bmatrix}$ (b) $\begin{bmatrix} 2 & 3/2 & -1/2 \\ 3/2 & 4 & 0 \\ -1/2 & 0 & 1 \end{bmatrix}$

(c) $\begin{bmatrix} 1 & 0 & 0 & 0 \\ 0 & 2 & 0 & 0 \\ 0 & 0 & 3 & 0 \\ 0 & 0 & 0 & 4 \end{bmatrix}$

3 $P = \begin{bmatrix} -1/\sqrt{5} & 2/\sqrt{5} \\ 2/\sqrt{5} & 1/\sqrt{5} \end{bmatrix}$ and $Q(\vec{x}) = -4y_1^2 + y_2^2$.

4 (a) $Q_1(\vec{x})$ is positive definite.
(b) $Q_2(\vec{x})$ is indefinite.

Section 8.3 Exercises

1

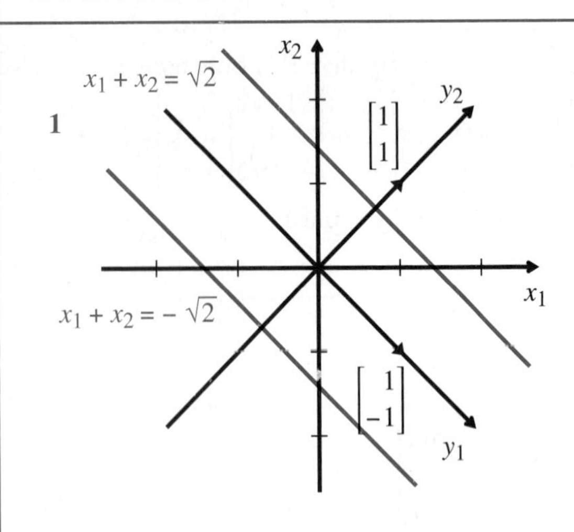

Section 8.5 Exercises

1 A has singular values $\sigma_1 = \sqrt{12}, \sigma_2 = \sqrt{2}$.
B has singular values $\sigma_1 = \sqrt{8}, \sigma_2 = \sqrt{3}, \sigma_3 = 0$.

2 $U = \begin{bmatrix} 1/\sqrt{30} & -2/\sqrt{5} & 1/\sqrt{6} \\ 5/\sqrt{30} & 0 & -1/\sqrt{6} \\ 2/\sqrt{30} & 1/\sqrt{5} & 2/\sqrt{6} \end{bmatrix}, \Sigma = \begin{bmatrix} \sqrt{12} & 0 \\ 0 & \sqrt{2} \\ 0 & 0 \end{bmatrix},$
$V = \begin{bmatrix} 3/\sqrt{10} & -1/\sqrt{10} \\ 1/\sqrt{10} & 3/\sqrt{10} \end{bmatrix}$

3 We have $B = \begin{bmatrix} 1 & 2 \\ 1 & 2 \end{bmatrix}$. From Example 8.5.7, we have that $B^+ = \begin{bmatrix} 1/10 & 1/10 \\ 1/5 & 1/5 \end{bmatrix}$. Thus, the vector $\vec{x}$ of the smallest length that minimizes $\|B\vec{x} - \vec{b}\|$ is

$$\vec{x} = B^+ \begin{bmatrix} 1 \\ 0 \end{bmatrix} = \begin{bmatrix} 1/10 \\ 1/5 \end{bmatrix}$$

Section 9.1 Exercises

1 $|-3| = 3, |-1-i| = \sqrt{2}$.

2 (a) $3+i$ (b) $1+6i$ (c) $-2+2i$
(d) $5-5i$ (e) 13

3 (1) $\overline{\overline{z_1}} = \overline{x - yi} = x + yi = z_1$
(2) $x + iy = z_1 = \overline{z_1} = x - iy$ if and only if $y = 0$.
(4) Let $z_2 = a + ib$. Then

$$\begin{aligned} \overline{z_1 + z_2} &= \overline{x + iy + a + ib} = \overline{x + a + i(y+b)} \\ &= x + a - i(y+b) = x - iy + a - ib \\ &= \overline{z_1} + \overline{z_2} \end{aligned}$$

4 (a) i (b) $1+i$ (c) $-\frac{1}{26} - \frac{21}{26}i$

5 $|z_1| = 2$. $\theta = \frac{\pi}{6} + 2\pi k, k \in \mathbb{Z}$. Hence

$$z_1 = 2\left(\cos\frac{\pi}{6} + i\sin\frac{\pi}{6}\right)$$

$|z_2| = \sqrt{2}$. $\theta = \frac{5\pi}{4} + 2\pi k, k \in \mathbb{Z}$. Hence

$$z_2 = \sqrt{2}\left(\cos\frac{5\pi}{4} + i\sin\frac{5\pi}{4}\right)$$

6

$$\begin{aligned} |\overline{z}| &= |r\cos\theta - ir\sin\theta| = \sqrt{r^2\cos^2\theta + r^2\sin^2\theta} \\ &= \sqrt{r^2} = |r| = |z| \end{aligned}$$

Using the trigonometric identities $\cos\theta = \cos(-\theta)$ and $-\sin\theta = \sin(-\theta)$ gives

$$\begin{aligned} \overline{z} &= r\cos\theta - ir\sin\theta = r\cos(-\theta) + ir\sin(-\theta) \\ &= r(\cos(-\theta) + i\sin(-\theta)) \end{aligned}$$

Hence, an argument of $\overline{z}$ is $-\theta$.

7 Theorem 9.1.5 says the modulus of a quotient is the quotient of the moduli of the factors, while the argument of the quotient is the difference of the arguments. Taking $z_1 = 1 = 1(\cos 0 + i\sin 0)$ and $z_2 = z$ in Theorem 9.1.5, gives

$$\begin{aligned} \frac{1}{z_2} &= \frac{1}{r}(\cos(0-\theta) + i\sin(0-\theta)) \\ &= \frac{1}{r}(\cos(-\theta) + i\sin(-\theta)) \end{aligned}$$

8 $(2-2i)(-1+\sqrt{3}i) = 4\sqrt{2}\left(\cos\frac{5\pi}{12} + i\sin\frac{5\pi}{12}\right)$, $\dfrac{2-2i}{-1+\sqrt{3}i} = \sqrt{2}\left(\cos\frac{-11\pi}{12} + i\sin\frac{-11\pi}{12}\right)$

9 $(1-i)^5 = -4+4i, (-1-\sqrt{3}i)^5 = -16+16\sqrt{3}i$.

Section 9.2 Exercises

1 $\vec{z} = \begin{bmatrix} 1+i \\ -3i \\ 0 \end{bmatrix} + \alpha \begin{bmatrix} i \\ -2 \\ 1 \end{bmatrix}, \alpha \in \mathbb{C}$

Section 9.3 Exercises

1 $Z^* = \begin{bmatrix} 1-i & 2 \\ 1+2i & i \\ -i & 3-i \end{bmatrix}, \mathbf{z}^* = \begin{bmatrix} -1+i & 2-i \end{bmatrix}$

2 $\langle \mathbf{u}, \mathbf{v} \rangle = -3-7i, \langle 2i\mathbf{u}, \mathbf{v} \rangle = -14+6i, \langle \mathbf{u}, 2i\mathbf{v} \rangle = 14-6i$

3 The result is easily verified.

4 $\|\alpha \mathbf{z}\|^2 = \langle \alpha \mathbf{z}, \alpha \mathbf{z} \rangle = \overline{\alpha}\alpha \langle \mathbf{z}, \mathbf{z} \rangle = |\alpha|^2 \, \|\mathbf{z}\|^2$

5 We have $\mathbf{z} = c_1\mathbf{v}_1 + \cdots + c_n\mathbf{v}_n$. Taking the inner product of both sides with $\mathbf{v}_i$ gives

$$\begin{aligned} \langle \mathbf{v}_i, \mathbf{z} \rangle &= \langle \mathbf{v}_i, c_1\mathbf{v}_1 + \cdots + c_n\mathbf{v}_n \rangle \\ &= c_1 \langle \mathbf{v}_i, \mathbf{v}_1 \rangle + \cdots + c_n \langle \mathbf{v}_i, \mathbf{v}_n \rangle \\ &= c_i \langle \mathbf{v}_i, \mathbf{v}_i \rangle \end{aligned}$$

Thus, for any $1 \leq i \leq n$, we have $c_i = \dfrac{\langle \mathbf{v}_i, \mathbf{z} \rangle}{\|\mathbf{v}_i\|^2}$.

Section 9.4 Exercises

1 A is diagonalized by $P = \begin{bmatrix} 1+i & -1-i \\ 1 & 2 \end{bmatrix}$ to $D = \begin{bmatrix} 5 & 0 \\ 0 & 2 \end{bmatrix}$.

2 A is diagonalized by $P = \begin{bmatrix} 1 & 1 & 1 \\ i & -i & 2 \\ 1 & 1 & 0 \end{bmatrix}$ to $D = \begin{bmatrix} i & 0 & 0 \\ 0 & -i & 0 \\ 0 & 0 & 2 \end{bmatrix}$.

Section 9.5 Exercises

1 Observe that $A^* = \begin{bmatrix} 0 & -1 \\ 1 & 0 \end{bmatrix} \neq A$, so A is not Hermitian.

A is unitarily diagonalized by $U = \begin{bmatrix} -i/\sqrt{2} & i/\sqrt{2} \\ 1/\sqrt{2} & 1/\sqrt{2} \end{bmatrix}$.

APPENDIX B

Answers to Practice Problems and Chapter Quizzes

CHAPTER 1

Section 1.1 Practice Problems

A1

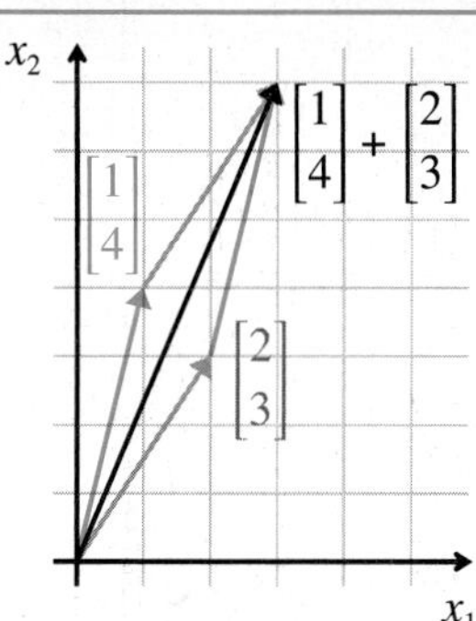

A2

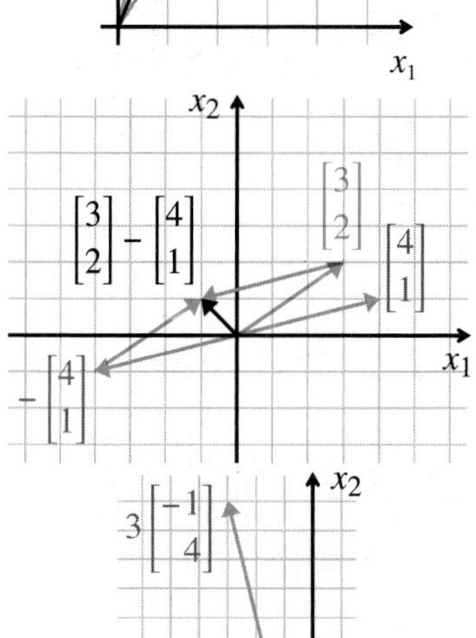

A3

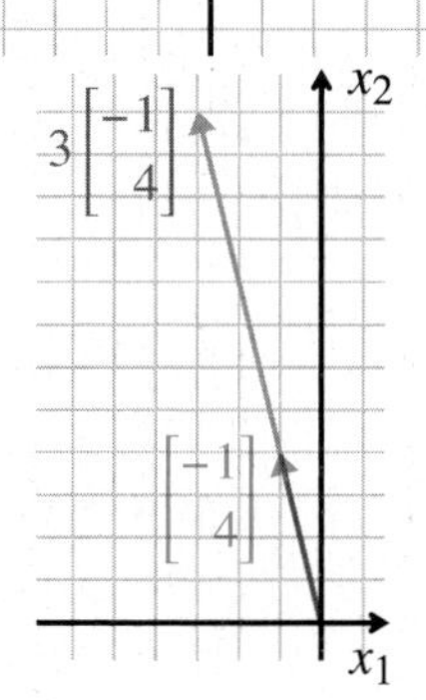

A4

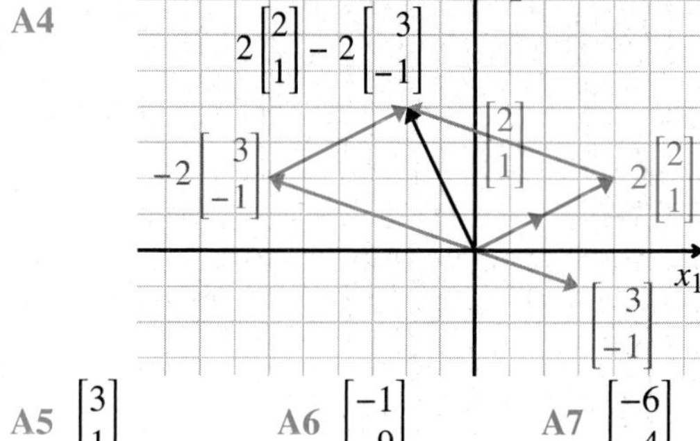

A5 $\begin{bmatrix}3\\1\end{bmatrix}$ A6 $\begin{bmatrix}-1\\-9\end{bmatrix}$ A7 $\begin{bmatrix}-6\\4\end{bmatrix}$

A8 $\begin{bmatrix}7/3\\4\end{bmatrix}$ A9 $\begin{bmatrix}3/2\\0\end{bmatrix}$ A10 $\begin{bmatrix}5\\4\sqrt{6}\end{bmatrix}$

A11 $\begin{bmatrix}-3\\2\\6\end{bmatrix}$ A12 $\begin{bmatrix}-1\\2\\-10\end{bmatrix}$ A13 $\begin{bmatrix}-24\\30\\36\end{bmatrix}$

A14 $\begin{bmatrix}7\\-2\\-5\end{bmatrix}$ A15 $\begin{bmatrix}7/3\\-4/3\\13/3\end{bmatrix}$ A16 $\begin{bmatrix}\sqrt{2}-\pi\\\sqrt{2}\\\sqrt{2}+\pi\end{bmatrix}$

A17 (a) $\begin{bmatrix}-4\\7\\-13\end{bmatrix}$ (b) $\begin{bmatrix}-10\\10\\-22\end{bmatrix}$

(c) $\vec{u}=\begin{bmatrix}-1/2\\-7/2\\9/2\end{bmatrix}$ (d) $\vec{u}=\begin{bmatrix}-3\\-6\\6\end{bmatrix}$

A18 (a) $\begin{bmatrix}4\\0\\-1/2\end{bmatrix}$ (b) $\begin{bmatrix}25\\-5\\-10\end{bmatrix}$

(c) $\begin{bmatrix}-1\\-3\\-4\end{bmatrix}$ (d) $\begin{bmatrix}8\\-8/3\\-14/3\end{bmatrix}$

A19 $\vec{PQ} = \begin{bmatrix}1\\-2\\-3\end{bmatrix}$, $\vec{PR} = \begin{bmatrix}-1\\1\\-1\end{bmatrix}$, $\vec{PS} = \begin{bmatrix}-7\\-2\\4\end{bmatrix}$, $\vec{QR} = \begin{bmatrix}-2\\3\\2\end{bmatrix}$,
$\vec{SR} = \begin{bmatrix}6\\3\\-5\end{bmatrix}$, $\vec{PQ} + \vec{QR} = \begin{bmatrix}-1\\1\\-1\end{bmatrix} = \vec{PS} + \vec{SR}$

A20 $\vec{x} = \begin{bmatrix}3\\4\end{bmatrix} + t\begin{bmatrix}-5\\1\end{bmatrix}, t \in \mathbb{R}$ **A21** $\vec{x} = \begin{bmatrix}2\\3\end{bmatrix} + t\begin{bmatrix}-4\\-6\end{bmatrix}, t \in \mathbb{R}$

A22 $\vec{x} = \begin{bmatrix}2\\0\\5\end{bmatrix} + t\begin{bmatrix}4\\-2\\-11\end{bmatrix}, t \in \mathbb{R}$ **A23** $\vec{x} = \begin{bmatrix}4\\1\\5\end{bmatrix} + t\begin{bmatrix}-2\\1\\2\end{bmatrix}, t \in \mathbb{R}$

Other correct answers are possible for problems A24–A28. **A24** $\vec{x} = \begin{bmatrix}-1\\2\end{bmatrix} + t\begin{bmatrix}3\\-5\end{bmatrix}, t \in \mathbb{R}$

A25 $\vec{x} = \begin{bmatrix}4\\1\end{bmatrix} + t\begin{bmatrix}-6\\-2\end{bmatrix}, t \in \mathbb{R}$

A26 $\vec{x} = \begin{bmatrix}1\\3\\-5\end{bmatrix} + t\begin{bmatrix}-3\\-2\\5\end{bmatrix}, t \in \mathbb{R}$ **A27** $\vec{x} = \begin{bmatrix}-2\\1\\1\end{bmatrix} + t\begin{bmatrix}6\\1\\1\end{bmatrix}, t \in \mathbb{R}$

A28 $\vec{x} = \begin{bmatrix}1/2\\1/4\\1\end{bmatrix} + t\begin{bmatrix}-3/2\\3/4\\-2/3\end{bmatrix}, t \in \mathbb{R}$

A29 $\begin{cases} x_1 &= -1 + 3t \\ x_2 &= 2 - 5t, \end{cases}$ $t \in \mathbb{R}$; $x_2 = -\frac{5}{3}x_1 + \frac{1}{3}$.

A30 $\begin{cases} x_1 &= 1 + t \\ x_2 &= 1 + t, \end{cases}$ $t \in \mathbb{R}$; $x_2 = x_1$.

A31 $\begin{cases} x_1 &= 1 + 2t \\ x_2 &= 0 + 0t, \end{cases}$ $t \in \mathbb{R}$; $x_2 = 0$.

A32 $\begin{cases} x_1 &= 1 - 2t \\ x_2 &= 3 + 2t, \end{cases}$ $t \in \mathbb{R}$; $x_2 = -x_1 + 4$.

A33 (a) Three points P, Q, and R are collinear if $\vec{PQ} = t\vec{PR}$ for some $t \in \mathbb{R}$.
(b) Since $-2\vec{PQ} = \begin{bmatrix}-6\\2\end{bmatrix} = \vec{PR}$, the points P, Q, and R must be collinear.
(c) The points S, T, and U are not collinear because $\vec{SU} \neq t\vec{ST}$ for any t.

A34 For V2: $\vec{x} + \vec{y} = \begin{bmatrix}x_1 + y_1\\x_2 + y_2\end{bmatrix} = \begin{bmatrix}y_1 + x_1\\y_2 + x_2\end{bmatrix} = \vec{y} + \vec{x}$

For V8: $(s + t)\vec{x} = \begin{bmatrix}(s + t)x_1\\(s + t)x_2\end{bmatrix} = \begin{bmatrix}sx_1 + tx_1\\sx_2 + tx_2\end{bmatrix}$
$= \begin{bmatrix}sx_1\\sx_2\end{bmatrix} + \begin{bmatrix}tx_1\\tx_2\end{bmatrix} = s\vec{x} + t\vec{x}$

A35 $\vec{F} = \begin{bmatrix}475\\25\sqrt{3}\end{bmatrix}$

Section 1.2 Practice Problems

A1 $\vec{x} \in \operatorname{Span} \mathcal{B}$
A2 $\vec{x} \in \operatorname{Span} \mathcal{B}$
A3 $\vec{x} \notin \operatorname{Span} \mathcal{B}$
A4 $\vec{x} \in \operatorname{Span} \mathcal{B}$
A5 $\vec{x} \in \operatorname{Span} \mathcal{B}$
A6 $\vec{x} \notin \operatorname{Span} \mathcal{B}$
A7 linearly dependent
A8 linearly independent
A9 linearly independent
A10 linearly dependent
A11 linearly independent
A12 linearly dependent
A13 linearly dependent
A14 linearly independent

A15 A line. $\vec{x} = s\begin{bmatrix}1\\0\end{bmatrix}, s \in \mathbb{R}$.

A16 A line. $\vec{x} = s\begin{bmatrix}-1\\1\end{bmatrix}, s \in \mathbb{R}$.

A17 A line. $\vec{x} = s\begin{bmatrix}1\\-3\\1\end{bmatrix}, s \in \mathbb{R}$.

A18 Two points in $\mathbb{R}^3$. $\vec{x} = \begin{bmatrix}1\\-3\\1\end{bmatrix}$ or $\vec{x} = \begin{bmatrix}-2\\6\\-2\end{bmatrix}$.

A19 A plane. $\vec{x} = s\begin{bmatrix}1\\0\\-2\end{bmatrix} + t\begin{bmatrix}2\\1\\-1\end{bmatrix}, s, t \in \mathbb{R}$.

A20 The origin. $\vec{x} = \vec{0}$.
A21 Not a basis
A22 A basis
A23 Not a basis
A24 Not a basis
A25 A basis
A26 Not a basis
A27 Not a basis
A28 A basis
A29 A basis
A30 A basis

A31 (a) Show that $\mathcal{B}$ spans $\mathbb{R}^2$ and is linearly independent.
(b) The coordinates of $\vec{e}_1$ with respect to $\mathcal{B}$ are $c_1 = 1$ and $c_2 = 0$.
The coordinates of $\vec{e}_2$ with respect to $\mathcal{B}$ are $c_1 = -1$ and $c_2 = 1$.
The coordinates of $\vec{x}$ with respect to $\mathcal{B}$ are $c_1 = -2$ and $c_2 = 3$.

A32 (a) Show that $\mathcal{B}$ spans $\mathbb{R}^2$ and is linearly independent.
(b) The coordinates of $\vec{e}_1$ with respect to $\mathcal{B}$ are $c_1 = 1/2$ and $c_2 = 1/2$.
The coordinates of $\vec{e}_2$ with respect to $\mathcal{B}$ are $c_1 = 1/2$ and $c_2 = -1/2$.
The coordinates of $\vec{x}$ with respect to $\mathcal{B}$ are $c_1 = ?$ and $c_2 = -1$.

A33 (a) Show that $\mathcal{B}$ spans $\mathbb{R}^2$ and is linearly independent.
(b) Tthe coordinates of $\vec{e}_1$ with respect to $\mathcal{B}$ are $c_1 = -1$ and $c_2 = -2$.
The coordinates of $\vec{e}_2$ with respect to $\mathcal{B}$ are $c_1 = 1$ and $c_2 = 1$.
The coordinates of $\vec{x}$ with respect to $\mathcal{B}$ are $c_1 = 2$ and $c_2 = 1$.

A34 Assume that $\{\vec{v}_1, \vec{v}_2\}$ is linearly independent. For a contradiction, assume without loss of generality that $\vec{v}_1 = t\vec{v}_2$. Hence, $\vec{v}_1 - t\vec{v}_2 = \vec{0}$. This contradicts the fact that $\{\vec{v}_1, \vec{v}_2\}$ is linearly independent.
On the other hand, assume that $\{\vec{v}_1, \vec{v}_2\}$ is linearly dependent. Then there exists $c_1, c_2 \in \mathbb{R}$ not both zero such that $c_1\vec{v}_1 + c_2\vec{v}_2 = \vec{0}$. Without loss of generality assume that $c_1 \neq 0$. Then $\vec{v}_1 = -\frac{c_2}{c_1}\vec{v}_2$ and hence $\vec{v}_1$ is a scalar multiple of $\vec{v}_2$.

A35 To prove this, we will prove that both sets are a subset of the other.
Let $\vec{x} \in \operatorname{Span}\{\vec{v}_1, \vec{v}_2\}$. Then there exists $c_1, c_2 \in \mathbb{R}$ such that $\vec{x} = c_1\vec{v}_1 + c_2\vec{v}_2$. Since $t \neq 0$ we get

$$\vec{x} = c_1\vec{v}_1 + \frac{c_2}{t}(t\vec{v}_2)$$

so $\vec{x} \in \operatorname{Span}\{\vec{v}_1, t\vec{v}_2\}$. Thus,

$$\operatorname{Span}\{\vec{v}_1, \vec{v}_2\} \subseteq \operatorname{Span}\{\vec{v}_1, t\vec{v}_2\}$$

If $\vec{y} \in \operatorname{Span}\{\vec{v}_1, t\vec{v}_2\}$, then there exists $d_1, d_2 \in \mathbb{R}$ such that

$$\vec{y} = d_1\vec{v}_1 + d_2(t\vec{v}_2) = d_1\vec{v}_1 + (d_2t)\vec{v}_2 \in \operatorname{Span}\{\vec{v}_1, \vec{v}_2\}$$

Hence, we also have $\operatorname{Span}\{\vec{v}_1, t\vec{v}_2\} \subseteq \operatorname{Span}\{\vec{v}_1, \vec{v}_2\}$. Therefore, $\operatorname{Span}\{\vec{v}_1, \vec{v}_2\} = \operatorname{Span}\{\vec{v}_1, t\vec{v}_2\}$.

Section 1.3 Practice Problems

A1 $\sqrt{29}$ **A2** 1 **A3** $\sqrt{2}$
A4 $\sqrt{17}$ **A5** $\sqrt{251}/5$ **A6** 1

A7 $2\sqrt{10}$ **A8** 5
A9 $\sqrt{170}$ **A10** $\sqrt{38}$
A11 Orthogonal **A12** Orthogonal
A13 Not orthogonal **A14** Orthogonal
A15 Orthogonal **A16** Not orthogonal
A17 $k = 6$ **A18** $k = 0, 3$
A19 $k = -3$ **A20** $k \in \mathbb{R}$
A21 $2x_1 + 4x_2 - x_3 = 9$ **A22** $3x_1 + 5x_3 = 26$
A23 $3x_1 - 4x_2 + x_3 = 8$

A24 $\begin{bmatrix} -27 \\ -9 \\ -9 \end{bmatrix}$ **A25** $\begin{bmatrix} -31 \\ -34 \\ 8 \end{bmatrix}$ **A26** $\begin{bmatrix} 4 \\ 5 \\ -4 \end{bmatrix}$

A27 $\begin{bmatrix} 0 \\ 0 \\ -1 \end{bmatrix}$ **A28** $\begin{bmatrix} 0 \\ 0 \\ 0 \end{bmatrix}$ **A29** $\begin{bmatrix} 0 \\ 0 \\ 0 \end{bmatrix}$

A30 (a) $\begin{bmatrix} 0 \\ 0 \\ 0 \end{bmatrix}$ (b) $\begin{bmatrix} -6 \\ 5 \\ -13 \end{bmatrix}$ (c) $\begin{bmatrix} 6 \\ 9 \\ -15 \end{bmatrix}$
(d) $\begin{bmatrix} -4 \\ 8 \\ -18 \end{bmatrix}$ (e) -14 (f) 14

A31 $x_1 - 4x_2 - 10x_3 = -85$ **A32** $2x_1 - 2x_2 + 3x_3 = -5$
A33 $-5x_1 - 2x_2 + 6x_3 = 15$ **A34** $-17x_1 - x_2 + 10x_3 = 0$

Other correct answers are possible for problems A35–A40.

A35 $\begin{bmatrix}x_1\\x_2\\x_3\end{bmatrix} = x_1\begin{bmatrix}1\\0\\-2\end{bmatrix} + x_2\begin{bmatrix}0\\1\\3\end{bmatrix}, \quad x_1, x_2 \in \mathbb{R}$

A36 $\begin{bmatrix}x_1\\x_2\\x_3\end{bmatrix} = \begin{bmatrix}0\\5\\0\end{bmatrix} + x_1\begin{bmatrix}1\\-4\\0\end{bmatrix} + x_3\begin{bmatrix}0\\2\\1\end{bmatrix}, \quad x_1, x_3 \in \mathbb{R}$

A37 $\begin{bmatrix}x_1\\x_2\\x_3\end{bmatrix} = \begin{bmatrix}1\\0\\0\end{bmatrix} + x_2\begin{bmatrix}-2\\1\\0\end{bmatrix} + x_3\begin{bmatrix}-2\\0\\1\end{bmatrix}, \quad x_2, x_3 \in \mathbb{R}$

A38 $\begin{bmatrix}x_1\\x_2\\x_3\end{bmatrix} == \begin{bmatrix}7/3\\0\\0\end{bmatrix} + x_2\begin{bmatrix}-5/3\\1\\0\end{bmatrix} + x_3\begin{bmatrix}4/3\\0\\1\end{bmatrix}, \quad x_2, x_3 \in \mathbb{R}$

A39 $\begin{bmatrix}x_1\\x_2\\x_3\end{bmatrix} = x_1\begin{bmatrix}1\\2\\0\end{bmatrix} + x_3\begin{bmatrix}0\\3\\1\end{bmatrix}, \quad x_2, x_3 \in \mathbb{R}$

A40 $\begin{bmatrix}x_1\\x_2\\x_3\end{bmatrix} = \begin{bmatrix}0\\3\\0\end{bmatrix} + x_1\begin{bmatrix}1\\-2\\0\end{bmatrix} + x_3\begin{bmatrix}0\\-3\\1\end{bmatrix}, \quad x_1, x_3 \in \mathbb{R}$

A41 $39x_1 + 12x_2 + 10x_3 = 140$

A42 $11x_1 - 21x_2 - 17x_3 = -56$

A43 $-12x_1 + 3x_2 - 19x_3 = -14$

A44 $x_2 = 0$

A45 $x_1 + x_2 + 2x_3 = 4$

A46 $14x_1 - 4x_2 - 5x_3 = 9$

A47 $2x_1 - 3x_2 + 5x_3 = 6$ A48 $x_2 = -2$

A49 $x_1 - x_2 + 3x_3 = 2$

A50 $\vec{x} = \begin{bmatrix}46/11\\3/11\\0\end{bmatrix} + t\begin{bmatrix}-2\\-3\\-11\end{bmatrix}, \quad t \in \mathbb{R}$

A51 $\vec{x} = \begin{bmatrix}7/2\\4\\0\end{bmatrix} + t\begin{bmatrix}3\\-4\\2\end{bmatrix}, \quad t \in \mathbb{R}$

A52 $\vec{x} = \begin{bmatrix}7/5\\1/5\\0\end{bmatrix} + t\begin{bmatrix}-2\\4\\10\end{bmatrix}, \quad t \in \mathbb{R}$

A53 $\vec{x} = t\begin{bmatrix}-2\\4\\10\end{bmatrix}, \quad t \in \mathbb{R}$

A54 $\sqrt{35}$ A55 $\sqrt{11}$ A56 13

A57 $\vec{u} \cdot (\vec{v} \times \vec{w}) = 0$ means that $\vec{u}$ is orthogonal to $\vec{v} \times \vec{w}$. Therefore, $\vec{u}$ lies in the plane through the origin that contains $\vec{v}$ and $\vec{w}$. We can also see this by observing that $\vec{u} \cdot (\vec{v} \times \vec{w}) = 0$ means that the parallelepiped determined by $\vec{u}, \vec{v}$, and $\vec{w}$ has volume zero; this can happen only if the three vectors lie in a common plane.

A58 We have

$$\begin{aligned}(\vec{u} - \vec{v}) \times (\vec{u} + \vec{v}) &= \vec{u} \times (\vec{u} + \vec{v}) - \vec{v} \times (\vec{u} + \vec{v})\\ &= \vec{u} \times \vec{u} + \vec{u} \times \vec{v} - \vec{v} \times \vec{u} - \vec{v} \times \vec{v}\\ &= \vec{0} + \vec{u} \times \vec{v} + \vec{u} \times \vec{v} - \vec{0}\\ &= 2(\vec{u} \times \vec{v})\end{aligned}$$

Section 1.4 Practice Problems

A1 $\begin{bmatrix}5\\9\\0\\1\end{bmatrix}$ A2 $\begin{bmatrix}10\\-7\\10\\-5\end{bmatrix}$

A3 $\begin{bmatrix}6\\0\\4\\5\\3\end{bmatrix}$ A4 $\begin{bmatrix}0\\0\\1\\1\\-3\end{bmatrix}$

A5 Subspace A6 Not a subspace

A7 Subspace A8 Subspace

A9 Not a subspace A10 Subspace

A11 Subspace A12 Not a subspace

A13 Not a subspace A14 Not a subspace

A15 Subspace A16 Not a subspace

Other correct answers are possible for problems A17–A20.

A17 $1\begin{bmatrix}0\\0\\0\\0\end{bmatrix}+0\begin{bmatrix}1\\0\\1\\2\end{bmatrix}+0\begin{bmatrix}2\\1\\-1\\-3\end{bmatrix}=\begin{bmatrix}0\\0\\0\\0\end{bmatrix}$

A18 $0\begin{bmatrix}2\\-1\\3\\2\end{bmatrix}-2\begin{bmatrix}0\\2\\1\\-1\end{bmatrix}+1\begin{bmatrix}0\\4\\2\\-2\end{bmatrix}=\begin{bmatrix}0\\0\\0\\0\end{bmatrix}$

A19 $1\begin{bmatrix}1\\1\\0\\2\end{bmatrix}+1\begin{bmatrix}1\\1\\1\\1\end{bmatrix}-1\begin{bmatrix}2\\2\\1\\3\end{bmatrix}=\begin{bmatrix}0\\0\\0\\0\end{bmatrix}$

A20 $(-3)\begin{bmatrix}1\\1\\-2\\3\end{bmatrix}+2\begin{bmatrix}1\\2\\1\\3\end{bmatrix}+\begin{bmatrix}1\\-1\\-8\\3\end{bmatrix}=\begin{bmatrix}0\\0\\0\\0\end{bmatrix}$

A21 Linearly dependent **A22** Linearly dependent

A23 Linearly independent **A24** Linearly independent

A25 Show that B is linearly independent and spans P.

A26 Show that B is linearly independent and spans P.

A27 Show that B is linearly independent and spans P.

A28 Show that B is linearly independent and spans P.

A29 A plane in $\mathbb{R}^4$ with basis $\left\{\begin{bmatrix}1\\0\\1\\1\end{bmatrix},\begin{bmatrix}1\\2\\1\\3\end{bmatrix}\right\}$.

A30 A hyperplane with basis $\left\{\begin{bmatrix}1\\0\\0\\0\end{bmatrix},\begin{bmatrix}0\\1\\0\\0\end{bmatrix},\begin{bmatrix}0\\0\\0\\1\end{bmatrix}\right\}$.

A31 A line with basis $\left\{\begin{bmatrix}3\\1\\-1\\0\end{bmatrix}\right\}$.

A32 A plane with basis $\left\{\begin{bmatrix}1\\1\\0\\2\end{bmatrix},\begin{bmatrix}1\\0\\0\\-1\end{bmatrix}\right\}$.

A33 If $\vec{x}=\vec{p}+t\vec{d}$ is a subspace of $\mathbb{R}^n$, then it contains the zero vector. Hence, there exists t_1 such that $\vec{0}=\vec{p}+t_1\vec{d}$. Thus, $\vec{p}=-t_1\vec{d}$ and so $\vec{p}$ is a scalar multiple of $\vec{d}$. On the other hand, if $\vec{p}$ is a scalar multiple of $\vec{d}$, say $\vec{p}=t_1\vec{d}$, then we have $\vec{x}=\vec{p}+t\vec{d}=t_1\vec{d}+t\vec{d}=(t_1+t)\vec{d}$. Hence, the set is Span$\{\vec{d}\}$ and thus is a subspace.

A34 Assume there is a non-empty subset $\mathcal{B}_1=\{\vec{v}_1,\dots,\vec{v}_\ell\}$ of $\mathcal{B}$ that is linearly dependent. Then there exists c_i not all zero such that

$$\vec{0}=c_1\vec{v}_1+\cdots+c_\ell\vec{v}_\ell=c_1\vec{v}_1+\cdots+c_\ell\vec{v}_\ell+0\vec{v}_{\ell+1}+\cdots+0\vec{v}_n$$

which contradicts the fact that $\mathcal{B}$ is linearly independent. Hence, $\mathcal{B}_1$ must be linearly independent.

A35 (a) Assume Span$\{\vec{v}_1,\dots,\vec{v}_k\}$ = Span$\{\vec{v}_1,\dots,\vec{v}_{k-1}\}$. Then there exists $b_1,\dots,b_{k-1}\in\mathbb{R}$ such that

$$\vec{v}_k=b_1\vec{v}_1+\cdots+b_{k-1}\vec{v}_{k-1}$$

So, $\vec{v}_k$ is a linear combination of $\vec{v}_1,\dots,\vec{v}_{k-1}$.

(b) If $\vec{v}_k$ can be written as a linear combination of $\vec{v}_1,\dots,\vec{v}_{k-1}$, then, there exist $c_1,\dots,c_{k-1}\in\mathbb{R}$ such that $c_1\vec{v}_1+\cdots+c_{k-1}\vec{v}_{k-1}=\vec{v}_k$. For any $\vec{x}\in$ Span$\{\vec{v}_1,\dots,\vec{v}_k\}$ there exist $d_1,\dots,d_k\in\mathbb{R}$ such that

$$\begin{aligned}\vec{x}&=d_1\vec{v}_1+\cdots+d_{k-1}\vec{v}_{k-1}+d_k\vec{v}_k\\&=d_1\vec{v}_1+\cdots+d_{k-1}\vec{v}_{k-1}+d_k(c_1\vec{v}_1+\cdots+c_{k-1}\vec{v}_{k-1})\\&=(d_1+d_kc_1)\vec{v}_1+\cdots+(d_{k-1}+d_kc_{k-1})\vec{v}_{k-1}\\&\in \text{Span}\{\vec{v}_1,\dots,\vec{v}_{k-1}\}\end{aligned}$$

On the other hand, if $\vec{y}\in$ Span$\{\vec{v}_1,\dots,\vec{v}_{k-1}\}$, then there exists $a_1,\dots,a_{k-1}\in\mathbb{R}$ such that

$$\begin{aligned}\vec{y}&=a_1\vec{v}_1+\cdots+a_{k-1}\vec{v}_{k-1}\\&=a_1\vec{v}_1+\cdots+a_{k-1}\vec{v}_{k-1}+0\vec{v}_k\end{aligned}$$

Thus, $\vec{y}\in$ Span$\{\vec{v}_1,\dots,\vec{v}_k\}$.

A36 The linear combination represent how much material is required to produce 100 thingamajiggers and 250 whatchamacallits.

Section 1.5 Practice Problems

A1 −3 **A2** −4 **A3** 0

A4 $\sqrt{6}$ **A5** 1 **A6** $\sqrt{15}$

A7 $\frac{1}{\sqrt{30}}\begin{bmatrix}1\\2\\5\end{bmatrix}$ **A8** $\frac{1}{\sqrt{15}}\begin{bmatrix}3\\-2\\-1\\1\end{bmatrix}$ **A9** $\frac{1}{\sqrt{6}}\begin{bmatrix}-2\\1\\0\\1\end{bmatrix}$

A10 $\frac{1}{\sqrt{39}}\begin{bmatrix}1\\2\\5\\-3\end{bmatrix}$ **A11** $\begin{bmatrix}1/2\\1/2\\1/2\\1/2\end{bmatrix}$ **A12** $\frac{1}{\sqrt{3}}\begin{bmatrix}1\\0\\1\\0\\1\end{bmatrix}$

A13 $2\sqrt{22} \approx 9.38 \leq \sqrt{26} + \sqrt{30} \approx 10.58$, $16 \leq \sqrt{26(30)} \approx 27.93$

A14 $\sqrt{41} \approx 6.40 \leq \sqrt{6} + \sqrt{29} \approx 7.83$, $3 \leq \sqrt{6(29)} \approx 13.19$

A15 $3x_1 + x_2 + 4x_3 = 0$

A16 $x_2 + 3x_3 + 3x_4 = 1$

A17 $3x_1 - 2x_2 - 5x_3 + x_4 = 4$

A18 $2x_1 - 4x_2 + x_3 - 3x_4 = -19$

A19 $x_1 - 4x_2 + 5x_3 - 2x_4 = 0$

A20 $x_2 + 2x_3 + x_4 + x_5 = 5$

A21 $\begin{bmatrix}2\\1\end{bmatrix}$ **A22** $\begin{bmatrix}3\\-2\\3\end{bmatrix}$ **A23** $\begin{bmatrix}-4\\3\\-5\end{bmatrix}$

A24 $\begin{bmatrix}1\\-1\\2\\-3\end{bmatrix}$ **A25** $\begin{bmatrix}1\\1\\-1\\2\\-1\end{bmatrix}$

A26 $\text{proj}_{\vec{v}}(\vec{u}) = \begin{bmatrix}0\\-5\end{bmatrix}, \text{perp}_{\vec{v}}(\vec{u}) = \begin{bmatrix}3\\0\end{bmatrix}$

A27 $\text{proj}_{\vec{v}}(\vec{u}) = \begin{bmatrix}36/25\\48/25\end{bmatrix}, \text{perp}_{\vec{v}}(\vec{u}) = \begin{bmatrix}-136/25\\102/25\end{bmatrix}$

A28 $\text{proj}_{\vec{v}}(\vec{u}) = \begin{bmatrix}0\\5\\0\end{bmatrix}, \text{perp}_{\vec{v}}(\vec{u}) = \begin{bmatrix}-3\\0\\2\end{bmatrix}$

A29 $\text{proj}_{\vec{v}}(\vec{u}) = \begin{bmatrix}-4/9\\8/9\\-8/9\end{bmatrix}, \text{perp}_{\vec{v}}(\vec{u}) = \begin{bmatrix}40/9\\1/9\\-19/9\end{bmatrix}$

A30 $\text{proj}_{\vec{v}}(\vec{u}) = \begin{bmatrix}0\\0\\0\\0\end{bmatrix}, \text{perp}_{\vec{v}}(\vec{u}) = \begin{bmatrix}-1\\-1\\2\\-1\end{bmatrix}$

A31 $\text{proj}_{\vec{v}}(\vec{u}) = \begin{bmatrix}-1/2\\0\\0\\-1/2\end{bmatrix}, \text{perp}_{\vec{v}}(\vec{u}) = \begin{bmatrix}5/2\\3\\2\\-5/2\end{bmatrix}$

A32 $\text{proj}_{\vec{v}}(\vec{u}) = \begin{bmatrix}0\\0\end{bmatrix}, \text{perp}_{\vec{v}}(\vec{u}) = \begin{bmatrix}3\\-3\end{bmatrix}$

A33 $\text{proj}_{\vec{v}}(\vec{u}) = \begin{bmatrix}-2/17\\-3/17\\2/17\end{bmatrix}, \text{perp}_{\vec{v}}(\vec{u}) = \begin{bmatrix}70/17\\-14/17\\49/17\end{bmatrix}$

A34 $\text{proj}_{\vec{v}}(\vec{u}) = \begin{bmatrix}14/3\\-7/3\\7/3\end{bmatrix}, \text{perp}_{\vec{v}}(\vec{u}) = \begin{bmatrix}1/3\\4/3\\2/3\end{bmatrix}$

A35 $\text{proj}_{\vec{v}}(\vec{u}) = \begin{bmatrix}3/2\\3/2\\-3\end{bmatrix}, \text{perp}_{\vec{v}}(\vec{u}) = \begin{bmatrix}5/2\\-1/2\\1\end{bmatrix}$

A36 $\text{proj}_{\vec{v}}(\vec{u}) = \begin{bmatrix}1/3\\-2/3\\-1/3\\1\end{bmatrix}, \text{perp}_{\vec{v}}(\vec{u}) = \begin{bmatrix}5/3\\-1/3\\7/3\\0\end{bmatrix}$

A37 $\text{proj}_{\vec{v}}(\vec{u}) = \begin{bmatrix}-1/3\\0\\-1/6\\-1/6\end{bmatrix}, \text{perp}_{\vec{v}}(\vec{u}) = \begin{bmatrix}-2/3\\2\\-5/6\\13/6\end{bmatrix}$

A38 (a) $\begin{bmatrix}2/7\\6/7\\3/7\end{bmatrix}$ (b) $\begin{bmatrix}220/49\\660/49\\330/49\end{bmatrix}$ (c) $\begin{bmatrix}270/49\\222/49\\-624/49\end{bmatrix}$

A39 (a) $\begin{bmatrix}3/\sqrt{14}\\1/\sqrt{14}\\-2/\sqrt{14}\end{bmatrix}$ (b) $\begin{bmatrix}24/7\\8/7\\-16/7\end{bmatrix}$ (c) $\begin{bmatrix}-3/7\\69/7\\30/7\end{bmatrix}$

A40 $R(5/2, 5/2), \|\text{perp}_{\vec{d}}(\vec{PQ})\| = 5/\sqrt{2}$

A41 $R(58/17, 91/17), \|\text{perp}_{\vec{d}}(\vec{PQ})\| = 6/\sqrt{17}$

A42 $R(17/6, 1/3, -1/6), \|\text{perp}_{\vec{d}}(\vec{PQ})\| = \sqrt{29/6}$

A43 $R(5/3, 11/3, -1/3), \|\text{perp}_{\vec{d}}(\vec{PQ})\| = \sqrt{6}$

A44 $2/\sqrt{26}$ **A45** $13/\sqrt{38}$ **A46** $4/\sqrt{5}$

A47 $\sqrt{6}$ **A48** $3/\sqrt{11}$ **A49** $13/\sqrt{21}$

A50 $5/\sqrt{3}$

A51 $R(1/7, 3/7, -3/7, 4/7)$ **A52** $R(15/14, 13/7, 17/14, 3)$

A53 $R(0, 14/3, 1/3, 10/3)$ **A54** $R(-12/7, 11/7, 9/7, -9/7)$

A55 1 **A56** 126 **A57** 5

A58 35 **A59** $k \approx 3.07$

Chapter 1 Quiz

E1 $\begin{bmatrix}1\\2\end{bmatrix}$

E2 $\frac{1}{\sqrt{54}}\begin{bmatrix}2\\-1\\7\end{bmatrix}$

E3 $\begin{bmatrix}4/5\\4/15\\8/15\\-4/15\end{bmatrix}$, $\text{perp}_{\vec{u}}(\vec{v}) = \begin{bmatrix}1/5\\-4/15\\22/15\\49/15\end{bmatrix}$

E4 $\vec{x} = \begin{bmatrix}-2\\1\\-4\end{bmatrix} + t\begin{bmatrix}7\\-3\\5\end{bmatrix}, \quad t \in \mathbb{R}$

E5 $\vec{x} = \begin{bmatrix}3\\0\\0\end{bmatrix} + x_2\begin{bmatrix}0\\1\\0\end{bmatrix} + x_3\begin{bmatrix}2\\0\\1\end{bmatrix}, x_2, x_3 \in \mathbb{R}.$

E6 $8x_1 - x_2 + 7x_3 = 9$

E7 $\mathcal{B} = \left\{\begin{bmatrix}2\\6\\4\end{bmatrix}, \begin{bmatrix}1\\3\\3\end{bmatrix}\right\}$

E8 The set is linearly independent.

E9 (a) Show B spans $\mathbb{R}^2$ and that it is linearly independent.

(b) $t_1 = \frac{11}{4}, t_2 = -\frac{1}{4}$

(c) $t_1 = \frac{11}{2}, t_2 = -\frac{1}{2}$

E10 S is not a subspace.

E11 If $d \neq 0$, then $a_1(0) + a_2(0) + a_3(0) = 0 \neq d$, so $\vec{0} \notin S$ and thus, S is not a subspace of $\mathbb{R}^3$.
On the other hand, assume $d = 0$. Show that $s\vec{x} + t\vec{y}$ satisfies the condition of S.

E12 Show that B is linearly independent and spans P.

E13 $Q(18/11, -12/11, 18/11)$

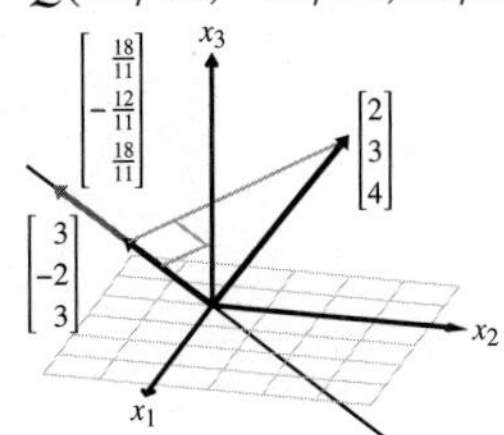

E14 Let $R(5/2, -5/2, -1/2, 3/2)$, $\|\vec{PR}\| = 1$

E15 The volume determined by $\vec{u} + k\vec{v}$, $\vec{v}$, and $\vec{w}$ is

$$\begin{aligned}|(\vec{u} + k\vec{v}) \cdot (\vec{v} \times \vec{w})| &= |\vec{u} \cdot (\vec{v} \times \vec{w}) + k(\vec{v} \cdot (\vec{v} \times \vec{w}))| \\ &= |\vec{u} \cdot (\vec{v} \times \vec{w}) + k(0)|\end{aligned}$$

which equals the volume of the parallelepiped determined by $\vec{u}$, $\vec{v}$, and $\vec{w}$.

E16 FALSE. The points $P(0,0,0)$, $Q(0,0,1)$, and $R(0,0,2)$ lie in every plane of the form $t_1x_1 + t_2x_2 = 0$ with t_1 and t_2 not both zero.

E17 TRUE. This is the definition of a line reworded in terms of a spanning set.

E18 TRUE. By definition of the plane $\{\vec{v}_1, \vec{v}_2\}$ spans the plane. If $\{\vec{v}_1, \vec{v}_2\}$ is linearly dependent, then the set would not satisfy the definition of a plane, so $\{\vec{v}_1, \vec{v}_2\}$ must be linearly independent. Hence, $\{\vec{v}_1, \vec{v}_2\}$ is a basis for the plane.

E19 FALSE. The dot product of the zero vector with itself is 0.

E20 FALSE. Let $\vec{x} = \begin{bmatrix}1\\0\end{bmatrix}$ and $\vec{y} = \begin{bmatrix}1\\1\end{bmatrix}$.
Then, $\text{proj}_{\vec{x}}\,\vec{y} = \begin{bmatrix}1\\0\end{bmatrix}$, while $\text{proj}_{\vec{y}}\,\vec{x} = \begin{bmatrix}1/2\\1/2\end{bmatrix}$.

E21 FALSE. If $\vec{y} = \vec{0}$, then $\text{proj}_{\vec{x}}\,\vec{y} = \vec{0}$. Thus, $\{\text{proj}_{\vec{x}}(\vec{y}), \text{perp}_{\vec{x}}(\vec{y})\}$ contains the zero vector so it is linearly dependent.

E22 TRUE. We have

$$\|\vec{u}\times(\vec{v}+3\vec{u})\| = \|\vec{u}\times\vec{v}+3(\vec{u}\times\vec{u})\| = \|\vec{u}\times\vec{v}+\vec{0}\| = \|\vec{u}\times\vec{v}\|$$

so the parallelograms have the same area.

CHAPTER 2

Section 2.1 Practice Problems

A1 $\vec{x} = \begin{bmatrix} 17 \\ 4 \end{bmatrix}$

A2 $\vec{x} = \begin{bmatrix} 13 \\ 0 \\ 6 \end{bmatrix} + t\begin{bmatrix} -2 \\ 1 \\ 0 \end{bmatrix}, t \in \mathbb{R}$

A3 $\vec{x} = \begin{bmatrix} 32 \\ -8 \\ 2 \end{bmatrix}$

A4 $\vec{x} = \begin{bmatrix} -1 \\ -3 \\ 2 \\ 0 \end{bmatrix} + t\begin{bmatrix} -1 \\ 1 \\ -1 \\ 1 \end{bmatrix}, t \in \mathbb{R}$

A5 A is in row echelon form.

A6 B is in row echelon form.

A7 C is not in row echelon.

A8 D is not in row echelon.

Other correct answers are possible for problems A9–A14.

A9 $\begin{bmatrix} 1 & -3 & 2 \\ 0 & 13 & -7 \end{bmatrix}$

A10 $\begin{bmatrix} 1 & -1 & 2 & 3 \\ 0 & 0 & 1 & 2 \\ 0 & 0 & 0 & 1 \end{bmatrix}$

A11 $\begin{bmatrix} 1 & -1 & -1 \\ 0 & 1 & 0 \\ 0 & 0 & 5 \\ 0 & 0 & 0 \end{bmatrix}$

A12 $\begin{bmatrix} 2 & 0 & 2 & 0 \\ 0 & 2 & 2 & 4 \\ 0 & 0 & 4 & 8 \\ 0 & 0 & 0 & 0 \end{bmatrix}$

A13 $\begin{bmatrix} 1 & 2 & 1 & 1 \\ 0 & 1 & 2 & 1 \\ 0 & 0 & 7 & 5 \\ 0 & 0 & 0 & 2 \end{bmatrix}$

A14 $\begin{bmatrix} 1 & 0 & 3 & 0 & 1 \\ 0 & 1 & -1 & 2 & 1 \\ 0 & 0 & 24 & 1 & 11 \\ 0 & 0 & 0 & 0 & 1 \end{bmatrix}$

A15 Inconsistent.

A16 Consistent. $\vec{x} = \begin{bmatrix} 2 \\ 0 \\ 3 \end{bmatrix} + t\begin{bmatrix} 0 \\ 1 \\ 0 \end{bmatrix}, t \in \mathbb{R}.$

A17 Consistent. $\vec{x} = \begin{bmatrix} 1 \\ -1 \\ 0 \\ 3 \end{bmatrix} + t\begin{bmatrix} -1 \\ -1 \\ 1 \\ 0 \end{bmatrix}, t \in \mathbb{R},$

A18 Consistent. $\vec{x} = \begin{bmatrix} 19/2 \\ 0 \\ 5/2 \\ -2 \end{bmatrix} + t\begin{bmatrix} -1 \\ 1 \\ 0 \\ 0 \end{bmatrix}, t \in \mathbb{R}.$

A19 Consistent. $\vec{x} = s\begin{bmatrix} -1 \\ 0 \\ 1 \\ 0 \end{bmatrix} + t\begin{bmatrix} 1 \\ 0 \\ 0 \\ 1 \end{bmatrix}, s, t \in \mathbb{R}.$

A20 Inconsistent.

A21 (a) $\left[\begin{array}{cc|c} 3 & -5 & 2 \\ 1 & 2 & 4 \end{array}\right]$

(b) $\left[\begin{array}{cc|c} 1 & 2 & 4 \\ 0 & -11 & -10 \end{array}\right]$

(c) Consistent.

(d) $\vec{x} = \begin{bmatrix} 24/11 \\ 10/11 \end{bmatrix}.$

A22 (a) $\left[\begin{array}{ccc|c} 1 & 2 & 1 & 5 \\ 2 & -3 & 2 & 6 \end{array}\right]$

(b) $\left[\begin{array}{ccc|c} 1 & 2 & 1 & 5 \\ 0 & -7 & 0 & -4 \end{array}\right]$

(c) Consistent.

(d) $\vec{x} = \begin{bmatrix} 27/7 \\ 4/7 \\ 0 \end{bmatrix} + t\begin{bmatrix} -1 \\ 0 \\ 1 \end{bmatrix}, t \in \mathbb{R}.$

A23 (a) $\left[\begin{array}{ccc|c} 1 & 2 & -3 & 8 \\ 1 & 3 & -5 & 11 \\ 2 & 5 & -8 & 19 \end{array}\right]$

(b) $\left[\begin{array}{ccc|c} 1 & 2 & -3 & 8 \\ 0 & 1 & -2 & 3 \\ 0 & 0 & 0 & 0 \end{array}\right]$

(c) Consistent.

(d) $\vec{x} = \begin{bmatrix} 2 \\ 3 \\ 0 \end{bmatrix} + t\begin{bmatrix} -1 \\ 2 \\ 1 \end{bmatrix}, t \in \mathbb{R}.$

A24 (a) $\left[\begin{array}{ccc|c} -3 & 6 & 16 & 36 \\ 1 & -2 & -5 & -11 \\ 2 & -3 & -8 & -17 \end{array}\right]$

(b) $\left[\begin{array}{ccc|c} 1 & -2 & -5 & -11 \\ 0 & 1 & 2 & 5 \\ 0 & 0 & 1 & 3 \end{array}\right]$

(c) Consistent.

(d) $\vec{x} = \begin{bmatrix} 2 \\ -1 \\ 3 \end{bmatrix}.$

A25 (a) $\left[\begin{array}{ccc|c} 1 & 2 & -1 & 4 \\ 2 & 5 & 1 & 10 \\ 4 & 9 & -1 & 19 \end{array}\right]$

(b) $\left[\begin{array}{ccc|c} 1 & 2 & -1 & 4 \\ 0 & 1 & 3 & 2 \\ 0 & 0 & 0 & 1 \end{array}\right]$

(c) Inconsistent.

A26 (a) $\left[\begin{array}{cccc|c} 1 & 2 & -3 & 0 & -5 \\ 2 & 4 & -6 & 1 & -8 \\ 6 & 13 & -17 & 4 & -21 \end{array}\right]$

(b) $\left[\begin{array}{cccc|c} 1 & 2 & -3 & 0 & -5 \\ 0 & 1 & 1 & 4 & 9 \\ 0 & 0 & 0 & 1 & 2 \end{array}\right]$

(c) Consistent.

(d) $\vec{x} = \begin{bmatrix} -7 \\ 1 \\ 0 \\ 2 \end{bmatrix} + t\begin{bmatrix} 5 \\ -1 \\ 1 \\ 0 \end{bmatrix}, t \in \mathbb{R}.$

A27 (a) $\left[\begin{array}{ccccc|c} 0 & 2 & -2 & 0 & 1 & 2 \\ 1 & 2 & -3 & 1 & 4 & 1 \\ 2 & 4 & -5 & 3 & 8 & 3 \\ 2 & 5 & -7 & 3 & 10 & 5 \end{array}\right]$

(b) $\left[\begin{array}{ccccc|c} 1 & 2 & -3 & 1 & 4 & 1 \\ 0 & 2 & -2 & 0 & 1 & 2 \\ 0 & 0 & 1 & 1 & 0 & 1 \\ 0 & 0 & 0 & 1 & 3/2 & 2 \end{array}\right]$

(c) Consistent.

(d) $\vec{x} = \begin{bmatrix} -4 \\ 0 \\ -1 \\ 2 \\ 0 \end{bmatrix} + t\begin{bmatrix} 0 \\ 1 \\ 3/2 \\ -3/2 \\ 1 \end{bmatrix}, t \in \mathbb{R}.$

A28 $y = 3 - x + x^2$

A29 $y = 2 - 3x^2$

A30 $y = 1 + 2x + 3x^2$

A31 $y = -3 - x + 3x^2$

A32 If $a \neq 0$, $b \neq 0$, this system is consistent, and the solution is unique. If $a = 0$, $b \neq 0$, the system is consistent, but the solution is not unique. If $a \neq 0$, $b = 0$, the system is inconsistent. If $a = 0, b = 0$, this system is consistent, but the solution is not unique.

A33 If $c \neq 0$, $d \neq 0$, the system is consistent and has no free variables, so the solution is unique. If $c = d = 0$, then the system is consistent and x_3 is a free variable, so there are infinitely many solutions. If $c \neq 0$ and $d = 0$, then the last row is $\left[\begin{array}{ccc|c} 0 & 0 & 0 & c \end{array}\right]$, so the system is inconsistent. If $c = 0$ and $d \neq 0$, then the system is consistent and x_4 is a free variable so there are infinitely many solutions.

A34 600 apples, 400 bananas, and 500 oranges.

A35 75% in algebra, 90% in calculus, and 84% in physics.

Section 2.2 Practice Problems

A1 The matrix is not in RREF.

A2 The matrix is not in RREF.

A3 The matrix is in RREF.

A4 The matrix is in RREF.

A5 The matrix is in RREF.

A6 The matrix is not in RREF.

A7 The matrix is not in RREF.

A8 $\begin{bmatrix} 1 & 0 \\ 0 & 1 \\ 0 & 0 \end{bmatrix}$; rank = 2

A9 $\begin{bmatrix} 1 & 0 & 0 \\ 0 & 1 & 0 \\ 0 & 0 & 1 \end{bmatrix}$; rank = 3

A10 $\begin{bmatrix} 1 & 0 & 0 \\ 0 & 1 & 0 \\ 0 & 0 & 1 \end{bmatrix}$; rank = 3

A11 $\begin{bmatrix} 1 & 0 & 0 \\ 0 & 1 & 0 \\ 0 & 0 & 1 \end{bmatrix}$; rank = 3

A12 $\begin{bmatrix} 1 & 2 & 0 \\ 0 & 0 & 1 \\ 0 & 0 & 0 \\ 0 & 0 & 0 \end{bmatrix}$; rank = 2

A13 $\begin{bmatrix} 1 & 1 & 0 & 0 \\ 0 & 0 & 1 & 0 \\ 0 & 0 & 0 & 1 \end{bmatrix}$; rank = 3

A14 $\begin{bmatrix} 1 & 0 & 0 & 0 \\ 0 & 1 & 0 & -2 \\ 0 & 0 & 1 & 3 \end{bmatrix}$; rank = 3

A15 $\begin{bmatrix} 1 & 0 & 0 & -1/2 \\ 0 & 1 & 0 & 3/2 \\ 0 & 0 & 1 & 1/2 \\ 0 & 0 & 0 & 0 \end{bmatrix}$; rank = 3

A16 $\begin{bmatrix} 1 & 0 & 0 & 0 & -56 \\ 0 & 1 & 0 & 0 & 17 \\ 0 & 0 & 1 & 0 & 23 \\ 0 & 0 & 0 & 1 & -6 \end{bmatrix}$; rank = 4

A17 There is one parameter. The general solution is

$$\vec{x} = t\begin{bmatrix} -2 \\ 1 \\ 1 \\ 0 \end{bmatrix}, \quad t \in \mathbb{R}.$$

A18 There are two parameters. The general solution is

$$\vec{x} = s\begin{bmatrix} 1 \\ 0 \\ 0 \\ 0 \end{bmatrix} + t\begin{bmatrix} 0 \\ -2 \\ 1 \\ 0 \end{bmatrix}, \quad s, t \in \mathbb{R}.$$

A19 There are two parameters. The general solution is

$$\vec{x} = s\begin{bmatrix}3\\1\\0\\0\end{bmatrix} + t\begin{bmatrix}-2\\0\\1\\0\end{bmatrix}, \quad s,t \in \mathbb{R}.$$

A20 There are two parameters. The general solution is

$$\vec{x} = s\begin{bmatrix}-2\\1\\1\\0\\0\end{bmatrix} + t\begin{bmatrix}0\\2\\0\\-1\\1\end{bmatrix}, \quad s,t \in \mathbb{R}.$$

A21 There are two parameters. The general solution is

$$\vec{x} = s\begin{bmatrix}0\\1\\0\\0\\0\end{bmatrix} + t\begin{bmatrix}-4\\0\\5\\1\\0\end{bmatrix}, \quad s,t \in \mathbb{R}.$$

A22 There is one parameter. The general solution is

$$\vec{x} = t\begin{bmatrix}0\\-1\\1\\0\\0\end{bmatrix}, t \in \mathbb{R}.$$

A23 $\left[\begin{array}{cc|c}1 & 0 & 24/11\\0 & 1 & 10/11\end{array}\right]. \vec{x} = \begin{bmatrix}24/11\\10/11\end{bmatrix}.$

A24 $\left[\begin{array}{ccc|c}1 & 0 & 1 & 27/7\\0 & 1 & 0 & 4/7\end{array}\right]. \vec{x} = \begin{bmatrix}27/7\\4/7\\0\end{bmatrix} + t\begin{bmatrix}-1\\0\\1\end{bmatrix}, \quad t \in \mathbb{R}.$

A25 $\left[\begin{array}{ccc|c}1 & 0 & 1 & 2\\0 & 1 & -2 & 3\\0 & 0 & 0 & 0\end{array}\right]. \vec{x} = \begin{bmatrix}2\\3\\0\end{bmatrix} + t\begin{bmatrix}-1\\2\\1\end{bmatrix}, \quad t \in \mathbb{R}.$

A26 $\left[\begin{array}{ccc|c}1 & 0 & 0 & 2\\0 & 1 & 0 & -1\\0 & 0 & 1 & 3\end{array}\right]. \vec{x} = \begin{bmatrix}2\\-1\\3\end{bmatrix}.$

A27 $\left[\begin{array}{ccc|c}1 & 0 & -7 & 0\\0 & 1 & 3 & 0\\0 & 0 & 0 & 1\end{array}\right]$. Inconsistent.

A28 $\left[\begin{array}{cccc|c}1 & 0 & -5 & 0 & -7\\0 & 1 & 1 & 0 & 1\\0 & 0 & 0 & 1 & 2\end{array}\right]. \vec{x} = \begin{bmatrix}-7\\1\\0\\2\end{bmatrix} + t\begin{bmatrix}5\\-1\\1\\0\end{bmatrix}, t \in \mathbb{R}.$

A29 $\left[\begin{array}{ccccc|c}1 & 0 & 0 & 0 & 0 & -4\\0 & 1 & 0 & 0 & -1 & 0\\0 & 0 & 1 & 0 & -3/2 & -1\\0 & 0 & 0 & 1 & 3/2 & 2\end{array}\right].$

$$\vec{x} = \begin{bmatrix}-4\\0\\-1\\2\\0\end{bmatrix} + t\begin{bmatrix}0\\1\\3/2\\-3/2\\1\end{bmatrix}, \quad t \in \mathbb{R}.$$

A30 $\begin{bmatrix}0 & 2 & -5\\1 & 2 & 3\\1 & 4 & -3\end{bmatrix}$; rank=3; 0 parameters. $\vec{x} = \vec{0}$.

A31 $\begin{bmatrix}3 & 1 & -9\\1 & 1 & -5\\2 & 1 & -7\end{bmatrix}$; rank =2; 1 parameter.

$$\vec{x} = t\begin{bmatrix}2\\3\\1\end{bmatrix}, \quad t \in \mathbb{R}.$$

A32 $\begin{bmatrix}1 & -1 & 2 & -3\\3 & -3 & 8 & -5\\2 & -2 & 5 & -4\\3 & -3 & 7 & -7\end{bmatrix}$; rank=2; 2 parameters.

$$\vec{x} = s\begin{bmatrix}1\\1\\0\\0\end{bmatrix} + t\begin{bmatrix}7\\0\\-2\\1\end{bmatrix}, \quad s,t \in \mathbb{R}.$$

A33 $\begin{bmatrix}0 & 1 & 2 & 2 & 0\\1 & 2 & 5 & 3 & -1\\2 & 1 & 5 & 1 & -3\\1 & 1 & 4 & 2 & -2\end{bmatrix}$; rank=3; 2 parameters.

$$\vec{x} = s\begin{bmatrix}2\\0\\-1\\1\\0\end{bmatrix} + t\begin{bmatrix}0\\-2\\1\\0\\1\end{bmatrix}, \quad s,t \in \mathbb{R}.$$

A34 $\vec{x} = \begin{bmatrix}-3\\1\\2\end{bmatrix}; \vec{x} = \begin{bmatrix}0\\0\\0\end{bmatrix}$

A35 $\vec{x} = \begin{bmatrix}-2\\1\\0\end{bmatrix} + t\begin{bmatrix}-5\\0\\1\end{bmatrix}, t \in \mathbb{R}; \vec{x} = t\begin{bmatrix}-5\\0\\1\end{bmatrix}, t \in \mathbb{R}$

A36 $\vec{x} = \begin{bmatrix}-5\\1\\0\\0\end{bmatrix} + s\begin{bmatrix}-9\\5\\1\\0\end{bmatrix} + t\begin{bmatrix}1\\-2\\0\\1\end{bmatrix}, \quad s,t \in \mathbb{R};$

$$\vec{x} = s\begin{bmatrix}-9\\5\\1\\0\end{bmatrix} + t\begin{bmatrix}1\\-2\\0\\1\end{bmatrix}, \quad s,t \in \mathbb{R}$$

A37 $\vec{x} = \begin{bmatrix} 2 \\ -1 \\ -1 \\ 0 \end{bmatrix} + t \begin{bmatrix} 0 \\ 2 \\ -1 \\ 1 \end{bmatrix}, \quad t \in \mathbb{R};$

$\vec{x} = t \begin{bmatrix} 0 \\ 2 \\ -1 \\ 1 \end{bmatrix}, \quad t \in \mathbb{R}$

A38 $\vec{x} = \begin{bmatrix} -3 \\ 14 \\ 4 \\ -5 \end{bmatrix}; \vec{x} = \begin{bmatrix} 0 \\ 0 \\ 0 \\ 0 \end{bmatrix}$

A39 $\vec{x} = \begin{bmatrix} 6 \\ 0 \\ -2 \\ 2 \\ 0 \end{bmatrix} + s \begin{bmatrix} -1 \\ 1 \\ 0 \\ 0 \\ 0 \end{bmatrix} + t \begin{bmatrix} -3 \\ 0 \\ 0 \\ -1 \\ 1 \end{bmatrix}, \quad s, t \in \mathbb{R};$

$\vec{x} = s \begin{bmatrix} -1 \\ 1 \\ 0 \\ 0 \\ 0 \end{bmatrix} + t \begin{bmatrix} -3 \\ 0 \\ 0 \\ -1 \\ 1 \end{bmatrix}, \quad s, t \in \mathbb{R}$

A40 (a) Row reducing the coefficient matrix gives $\begin{bmatrix} 1 & 1 & 1 & 0 \\ 1 & 2 & 1 & -2 \\ 1 & 4 & 2 & -7 \end{bmatrix} \sim \begin{bmatrix} 1 & 0 & 0 & 3 \\ 0 & 1 & 0 & -2 \\ 0 & 0 & 1 & -1 \end{bmatrix}$ Since there is a leading one in each row, there cannot be a row in the RREF of $\left[A \mid \vec{b}\right]$ of the form $\left[\begin{array}{ccc|c} 0 & \cdots & 0 & 1 \end{array}\right]$. Hence, the system will be consistent for any $b_1, b_2, b_3 \in \mathbb{R}$.

(b) Since $\text{rank}(A) = m$, there cannot be a row in the RREF of $\left[A \mid \vec{b}\right]$ of the form $\left[\begin{array}{ccc|c} 0 & \cdots & 0 & 1 \end{array}\right]$. So, the system will be consistent for any $\vec{b} \in \mathbb{R}^m$.

(c) $\vec{b} = \begin{bmatrix} 0 \\ 1 \\ 0 \end{bmatrix}$ makes the system inconsistent.

(d) If rank $A < m$, then the RREF R of A has a row of all zeros. Thus, $[R \mid \vec{e}_m]$ is inconsistent as it has a row of the form $\left[\begin{array}{ccc|c} 0 & \cdots & 0 & 1 \end{array}\right]$. Since elementary row operations are reversible, we can apply the reverse of the row operations needed to row reduce A to R on $[R \mid \vec{e}_m]$ to get $\left[A \mid \vec{b}\right]$ for some $\vec{b} \in \mathbb{R}^n$. Then this system is inconsistent since elementary row operations do not change the solution set. Thus, there exists some $\vec{b} \in \mathbb{R}^m$ such that $\left[A \mid \vec{b}\right]$ is inconsistent.

Section 2.3 Practice Problems

A1 (a) $2 \begin{bmatrix} 1 \\ 0 \\ 1 \\ 1 \end{bmatrix} + (-1) \begin{bmatrix} 2 \\ 1 \\ 0 \\ 1 \end{bmatrix} + 3 \begin{bmatrix} -1 \\ 1 \\ 2 \\ 1 \end{bmatrix} = \begin{bmatrix} -3 \\ 2 \\ 8 \\ 4 \end{bmatrix}$

(b) $\begin{bmatrix} 5 \\ 4 \\ 6 \\ 7 \end{bmatrix}$ is not in the span.

(c) $3 \begin{bmatrix} 1 \\ 0 \\ 1 \\ 1 \end{bmatrix} + (-1) \begin{bmatrix} 2 \\ 1 \\ 0 \\ 1 \end{bmatrix} + (-1) \begin{bmatrix} -1 \\ 1 \\ 2 \\ 1 \end{bmatrix} = \begin{bmatrix} 2 \\ -2 \\ 1 \\ 1 \end{bmatrix}$

A2 (a) $\begin{bmatrix} 3 \\ 2 \\ -1 \\ -1 \end{bmatrix}$ is not in the span.

(b) $(-2) \begin{bmatrix} 1 \\ -1 \\ 1 \\ 0 \end{bmatrix} + 3 \begin{bmatrix} -1 \\ 1 \\ 0 \\ 2 \end{bmatrix} + (-2) \begin{bmatrix} 1 \\ 1 \\ -1 \\ -1 \end{bmatrix} = \begin{bmatrix} -7 \\ 3 \\ 0 \\ 8 \end{bmatrix}$

(c) $\begin{bmatrix} 1 \\ 1 \\ 1 \\ 1 \end{bmatrix}$ is not in the span.

A3 $x_3 = 0$

A4 $x_1 - 2x_2 = 0, x_3 = 0$

A5 $x_1 + 3x_2 - 2x_3 = 0$

A6 $x_1 + 3x_2 + 5x_3 = 0$

A7 $-x_1 - x_2 + x_3 = 0, x_2 + x_4 = 0$

A8 $-4x_1 + 5x_2 + x_3 + 4x_4 = 0$

A9 A basis is the empty set. The dimension is 0.

A10 A basis is $\left\{ \begin{bmatrix} -2 \\ -3 \\ 1 \\ 0 \end{bmatrix}, \begin{bmatrix} 1 \\ -1 \\ 0 \\ 1 \end{bmatrix} \right\}$. The dimension is 2.

A11 It is a basis for the plane.

A12 It is not a basis for the plane.

A13 It is a basis for the hyperplane.

A14 Linearly independent.

A15 Linearly dependent.

$-3t \begin{bmatrix} 1 \\ 0 \\ 1 \\ 0 \end{bmatrix} - 2t \begin{bmatrix} 0 \\ 1 \\ 1 \\ 1 \end{bmatrix} - t \begin{bmatrix} 0 \\ 0 \\ 1 \\ 1 \end{bmatrix} + t \begin{bmatrix} 3 \\ 2 \\ 6 \\ 3 \end{bmatrix} = \begin{bmatrix} 0 \\ 0 \\ 0 \\ 0 \end{bmatrix}, \quad t \in \mathbb{R}$

A16 Linearly dependent.

$$2t\begin{bmatrix}1\\1\\0\\1\\1\end{bmatrix} - t\begin{bmatrix}2\\3\\1\\3\\3\end{bmatrix} + t\begin{bmatrix}0\\1\\1\\1\\1\end{bmatrix} = \begin{bmatrix}0\\0\\0\\0\\0\end{bmatrix}, \quad t \in \mathbb{R}$$

A17 Linearly independent.

A18 Linearly independent for all $k \neq -3$.

A19 Linearly independent for all $k \neq -5/2$.

A20 It is a basis.

A21 Only two vectors, so it cannot span $\mathbb{R}^3$. Therefore, it is not a basis.

A22 It has four vectors in $\mathbb{R}^3$, so it is linearly dependent. Therefore, it is not a basis.

A23 It is linearly dependent, so it is not a basis.

Section 2.4 Practice Problems

A1 $x_1 = 12, x_2 = 26$, and $x_3 = 28$

A2
$$\begin{aligned}(R_1 + R_2)i_1 - R_2 i_2 &= E_1\\ -R_2 i_1 + (R_2 + R_3)i_2 - R_3 i_3 &= 0\\ -R_3 i_2 + (R_3 + R_4 + R_8)i_3 - R_8 i_5 &= 0\\ (R_5 + R_6)i_4 - R_6 i_5 &= 0\\ -R_8 i_3 - R_6 i_4 + (R_6 + R_7 + R_8)i_5 &= E_2\end{aligned}$$

A3 $x_1 = 30 - t, x_2 = 10 - t, x_3 = 60 - t, x_4 = t, t \in \mathbb{R}$

A4 $\dfrac{4x^4 + x^3 + x^2 + x + 1}{(x-1)(x^2+1)^2} = \dfrac{2}{x-1} + \dfrac{2x+3}{x^2+1} + \dfrac{-2x-2}{(x^2+1)^2}$

A5 $2Al(OH)_3 + 3H_2CO_3 \rightarrow Al_2(CO_3)_3 + 6H_2O$

A6 The maximum value is 140 (occurring at (100, 40)).

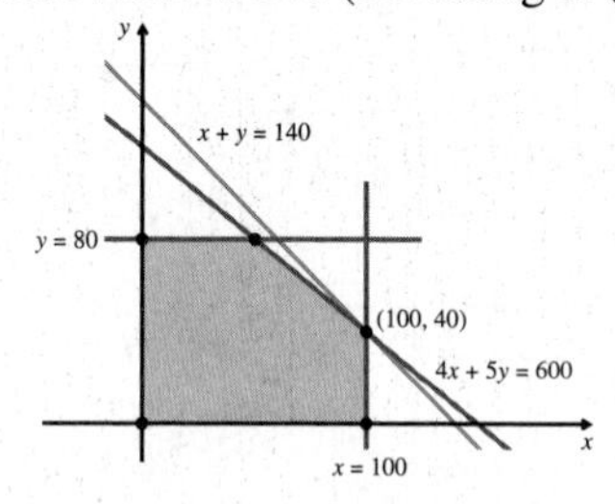

A7 To simplify writing, let $\alpha = \frac{1}{\sqrt{2}}$.
Total horizontal force: $R_1 + R_2 = 0$.
Total vertical force: $R_V - F_V = 0$.
Total moment about A: $R_1 s + F_V(2\sqrt{2}s) = 0$.
The horizontal and vertical equations at the joints are
$\alpha N_2 + R_2 = 0$ and $N_1 + \alpha N_2 + R_V = 0$;
$N_3 + \alpha N_4 + R_1 = 0$ and $-N_1 + \alpha N_4 = 0$;
$-\alpha N_2 - N_3 + \alpha N_6 = 0$ and $-\alpha N_2 + N_5 + \alpha N_6 = 0$;
$-\alpha N_4 + N_7 = 0$ and $-\alpha N_4 - N_5 = 0$;
$-N_7 - \alpha N_6 = 0$ and $-\alpha N_6 - F_V = 0$.

Chapter 2 Quiz

E1 (a) $\left[\begin{array}{cccc|c} 0 & 1 & -2 & 1 & 2\\ 2 & -2 & 4 & -1 & 10\\ 1 & -1 & 1 & 0 & 2\\ 1 & 0 & 1 & 0 & 9\end{array}\right]$.

(b) $\left[\begin{array}{cccc|c} 1 & 0 & 0 & 1/2 & 13/2\\ 0 & 1 & 0 & 0 & 7\\ 0 & 0 & -2 & 1 & -5\\ 0 & 0 & 0 & 0 & 1\end{array}\right]$. The rank is 4.

(c) The system is inconsistent.

E2 (a) $\left[\begin{array}{cccc|c} 2 & 4 & 1 & -6 & 7\\ 4 & 8 & -3 & 8 & -1\\ -3 & -6 & 2 & -5 & 0\\ 1 & 2 & 1 & -5 & 5\end{array}\right]$.

(b) $\left[\begin{array}{cccc|c} 1 & 2 & 0 & -1 & 2\\ 0 & 0 & 1 & -4 & 3\\ 0 & 0 & 0 & 0 & 0\\ 0 & 0 & 0 & 0 & 0\end{array}\right]$. The rank is 2.

(c) $\begin{bmatrix}x_1\\x_2\\x_3\\x_4\end{bmatrix} = \begin{bmatrix}2\\0\\3\\0\end{bmatrix} + x_2\begin{bmatrix}-2\\1\\0\\0\end{bmatrix} + x_4\begin{bmatrix}1\\0\\4\\1\end{bmatrix}, \quad x_2, x_4 \in \mathbb{R}$

E3 (a) $\left[\begin{array}{ccccc} 1 & 0 & 0 & 0 & 1\\ 0 & 1 & 0 & 0 & 0\\ 0 & 0 & 1 & 0 & -1/3\\ 0 & 0 & 0 & 1 & 1/3\end{array}\right]$. The rank is 4.

(b) $\vec{x} = \begin{bmatrix} x_1 \\ x_2 \\ x_3 \\ x_4 \\ x_5 \end{bmatrix} = t \begin{bmatrix} -1 \\ 0 \\ 1/3 \\ -1/3 \\ 1 \end{bmatrix}, t \in \mathbb{R}$

(c) $\mathcal{B} = \left\{ \begin{bmatrix} -1 \\ 0 \\ 1/3 \\ -1/3 \\ 1 \end{bmatrix} \right\}$. The dimension is 1.

E4 $-3x_1 - x_2 + x_3 = 0$

E5 (a) The system is inconsistent for all (a, b, c) of the form $(a, b, 1)$ or $(a, -2, c)$, and is consistent for all (a, b, c) where $b \neq -2$ and $c \neq 1$.

(b) The system has a unique solution if and only if $b \neq -2, c \neq 1$, and $c \neq -1$.

E6 (a) $\vec{x} = s \begin{bmatrix} -2 \\ -1 \\ 1 \\ 0 \\ 0 \end{bmatrix} + t \begin{bmatrix} 11/4 \\ -11/2 \\ 0 \\ -5/4 \\ 1 \end{bmatrix}, \quad s, t \in \mathbb{R}$

(b) If there exists a vector $\vec{x} \in \mathbb{R}^5$ which is orthogonal to $\vec{u}, \vec{v}$, and $\vec{w}$, then $\vec{x} \cdot \vec{u} = 0, \vec{x} \cdot \vec{v} = 0$, and $\vec{x} \cdot \vec{w} = 0$, yields a homogeneous system of three linear equations with five variables. Hence, the rank of the matrix is at most three and thus there are at least 2 parameters (# of variables - rank = $5 - 3 = 2$). So, there are in fact infinitely many vectors orthogonal to the three vectors.

E7 The set is linearly independent by Lemma 2.3.3.

E8 The set does not span $\mathbb{R}^3$.

E9 Consider $\vec{x} = t_1 \begin{bmatrix} 3 \\ 1 \\ 2 \end{bmatrix} + t_2 \begin{bmatrix} 1 \\ 1 \\ 6 \end{bmatrix} + t_3 \begin{bmatrix} 4 \\ 1 \\ 5 \end{bmatrix}$. The RREF of the corresponding coefficient matrix is I. Thus, it is a basis for $\mathbb{R}^3$ by Theorem 2.3.5.

E10 False **E11** False **E12** False

E13 False **E14** True **E15** True

CHAPTER 3

Section 3.1 Practice Problems

A1 $\begin{bmatrix} -1 & -6 & 4 \\ 6 & -4 & 2 \end{bmatrix}$

A2 $\begin{bmatrix} 13 & 4 & 11 \\ 6 & 14 & -11 \end{bmatrix}$

A3 Not defined.

A4 Not defined.

A5 $\begin{bmatrix} 10 & -12 \\ 2 & -5 \end{bmatrix}$

A6 $\begin{bmatrix} -7 & 6 & -5 \\ -4 & -13 & 5 \\ -16 & -6 & -2 \end{bmatrix}$

A7 $\begin{bmatrix} 6 & 7/3 \\ 19/2 & 19/3 \\ 21 & 34/3 \end{bmatrix}$

A8 $\begin{bmatrix} 27/2 \\ 49/3 \end{bmatrix}$

A9 10

A10 $A = \begin{bmatrix} 3 & 2 & -1 \\ 2 & -1 & 5 \end{bmatrix}, \vec{x} = \begin{bmatrix} x_1 \\ x_2 \\ x_3 \end{bmatrix}, \vec{b} = \begin{bmatrix} 4 \\ 5 \end{bmatrix}$

A11 $A = \begin{bmatrix} 1 & -4 & 1 & -2 \\ 1 & -1 & 3 & 0 \end{bmatrix}, \vec{x} = \begin{bmatrix} x_1 \\ x_2 \\ x_3 \\ x_4 \end{bmatrix}, \vec{b} = \begin{bmatrix} 1 \\ 0 \end{bmatrix}$

A12 $A = \begin{bmatrix} 1/3 & 3 & -1/4 \\ 1 & 0 & 1 \\ 1 & -1 & 0 \end{bmatrix}, \vec{x} = \begin{bmatrix} x_1 \\ x_2 \\ x_3 \end{bmatrix}, \vec{b} = \begin{bmatrix} 1 \\ 2/3 \\ 3 \end{bmatrix}$

A13 $A = \begin{bmatrix} 1 & -1 \\ 3 & 1 \\ 5 & -8 \end{bmatrix}, \vec{x} = \begin{bmatrix} x_1 \\ x_2 \end{bmatrix}, \vec{b} = \begin{bmatrix} 3 \\ 4 \\ 17 \end{bmatrix}$

A14 TRUE. Since A has 2 columns, $\vec{x}$ must have 2 entries.

A15 FALSE. Since A has 2 rows, we get that $A\vec{x} \in \mathbb{R}^2$.

A16 FALSE. IF B is the identity matrix, then $AB = A = BA$ by Theorem 3.1.7.

A17 TRUE. By definition of matrix-matrix multiplication A^TA will have n rows (since A^T has n rows) and will have n columns (since A has n columns). Thus, A^TA is $n \times n$.

A18 FALSE. Take $A = \begin{bmatrix} 0 & 1 \\ 0 & 0 \end{bmatrix}$. Then $A^2 = \begin{bmatrix} 0 & 0 \\ 0 & 0 \end{bmatrix}$.

A19 FALSE. Take $A = B = \begin{bmatrix} 0 & 1 \\ 0 & 0 \end{bmatrix}$.

A20 $(A + B)^T = \begin{bmatrix} -3 & 2 & 1 \\ -1 & 2 & 3 \end{bmatrix}, A^T + B^T = \begin{bmatrix} -3 & 2 & 1 \\ -1 & 2 & 3 \end{bmatrix}$

A21 $(AB)^T = \begin{bmatrix} -21 & -10 \\ 15 & -27 \end{bmatrix}, B^TA^T = \begin{bmatrix} -21 & -10 \\ 15 & -27 \end{bmatrix}$

A22 $\begin{bmatrix} 13 & 31 & 2 \\ 10 & 12 & 10 \end{bmatrix}$

A23 Not defined.

A24 Not defined.

A25 $\begin{bmatrix} 11 & 7 & 3 & 15 \\ 7 & 9 & 11 & 1 \end{bmatrix}$

A26 $x_1 + 2x_2 + x_3$

A27 Not defined.

A28 $\begin{bmatrix} 52 & 139 \\ 62 & 46 \end{bmatrix}$

A29 $\begin{bmatrix} 13 & 10 \\ 31 & 12 \\ 2 & 10 \end{bmatrix}$

A30 $\begin{bmatrix} 5 & -19 \\ 7 & 9 \\ 3 & 11 \\ 15 & 1 \end{bmatrix}$

A31 (a) $A\vec{x} = \begin{bmatrix} 12 \\ 17 \\ 3 \end{bmatrix}, A\vec{y} = \begin{bmatrix} 8 \\ 4 \\ -4 \end{bmatrix}, A\vec{z} = \begin{bmatrix} -2 \\ 5 \\ 1 \end{bmatrix}$

(b) $\begin{bmatrix} 12 & 8 & -2 \\ 17 & 4 & 5 \\ 3 & -4 & 1 \end{bmatrix}$

A32 $\begin{bmatrix} -13 & 16 \\ -27 & 0 \end{bmatrix}$

A33 (a) $A \in \operatorname{Span} \mathcal{B}$.

(b) $\mathcal{B}$ is linearly independent.

A34 Using the second view of matrix-vector multiplication and the fact that the i-th component of $\vec{e}_i$ is 1 and all other components are 0, we get

$$A\vec{e}_i = 0\vec{a}_1 + \cdots + 0\vec{a}_{i-1} + 1\vec{a}_i + 0\vec{a}_{i+1} + \cdots + 0\vec{a}_n = \vec{a}_i$$

Section 3.2 Practice Problems

A1 (a) $f_A : \mathbb{R}^2 \to \mathbb{R}^4$

(b) $f_A(2,-5) = \begin{bmatrix} -19 \\ 6 \\ -23 \\ 38 \end{bmatrix}, f_A(-3,4) = \begin{bmatrix} 18 \\ -9 \\ 17 \\ -36 \end{bmatrix}$

(c) $f_A(1,0) = \begin{bmatrix} -2 \\ 3 \\ 1 \\ 4 \end{bmatrix}, f_A(0,1) = \begin{bmatrix} 3 \\ 0 \\ 5 \\ -6 \end{bmatrix}$

(d) $f_A(\vec{x}) = \begin{bmatrix} -2x_1 + 3x_2 \\ 3x_1 + 0x_2 \\ x_1 + 5x_2 \\ 4x_1 - 6x_2 \end{bmatrix}$

(e) $[f_A] = \begin{bmatrix} -2 & 3 \\ 3 & 0 \\ 1 & 5 \\ 4 & -6 \end{bmatrix}$

A2 (a) The domain is $\mathbb{R}^4$. The codomain is $\mathbb{R}^3$.

(b) $f_A(2,-2,3,1) = \begin{bmatrix} -11 \\ 9 \\ 7 \end{bmatrix}, f_A(-3,1,4,2) = \begin{bmatrix} -13 \\ -1 \\ 3 \end{bmatrix}$

(c) $f_A(\vec{e}_1) = \begin{bmatrix} 1 \\ 2 \\ 1 \end{bmatrix}, f_A(\vec{e}_2) = \begin{bmatrix} 2 \\ -1 \\ 0 \end{bmatrix}, f_A(\vec{e}_3) = \begin{bmatrix} -3 \\ 0 \\ 2 \end{bmatrix},$

$f_A(\vec{e}_4) = \begin{bmatrix} 0 \\ 3 \\ -1 \end{bmatrix}$

(d) $f_A(\vec{x}) = \begin{bmatrix} x_1 + 2x_2 - 3x_3 \\ 2x_1 - x_2 + 3x_4 \\ x_1 + 2x_3 - x_4 \end{bmatrix}$

(e) $[f_A] = \begin{bmatrix} 1 & 2 & -3 & 0 \\ 2 & -1 & 0 & 3 \\ 1 & 0 & 2 & -1 \end{bmatrix}$

A3 $f : \mathbb{R}^2 \to \mathbb{R}^2$. f is not linear.

A4 $f : \mathbb{R}^3 \to \mathbb{R}^2$. f is not linear.

A5 $g : \mathbb{R}^3 \to \mathbb{R}^3$. g is not linear.

A6 $g : \mathbb{R}^2 \to \mathbb{R}^2$. g is linear.

A7 $h : \mathbb{R}^2 \to \mathbb{R}^3$. h is not linear.

A8 $k : \mathbb{R}^3 \to \mathbb{R}^3$. k is linear.

A9 $\ell : \mathbb{R}^3 \to \mathbb{R}^2$. ℓ is not linear.

A10 $m : \mathbb{R}^1 \to \mathbb{R}^3$. m is not linear.

A11 $L : \mathbb{R}^3 \to \mathbb{R}^2$. L is linear.

A12 $L : \mathbb{R}^2 \to \mathbb{R}^2$. L is not linear.

A13 $M : \mathbb{R}^3 \to \mathbb{R}^2$. M is not linear.

A14 $M : \mathbb{R}^3 \to \mathbb{R}^3$. M is linear.

A15 $N : \mathbb{R}^2 \to \mathbb{R}^3$. N is linear.

A16 $N : \mathbb{R}^3 \to \mathbb{R}^3$. N is not linear.

A17 $\begin{bmatrix} 4/5 & -2/5 \\ -2/5 & 1/5 \end{bmatrix}$

A18 $\begin{bmatrix} 16/41 & 20/41 \\ 20/41 & 25/41 \end{bmatrix}$

A19 $\begin{bmatrix} 4/9 & 4/9 & -2/9 \\ 4/9 & 4/9 & -2/9 \\ -2/9 & -2/9 & 1/9 \end{bmatrix}$ **A20** $\begin{bmatrix} 1/5 & 2/5 \\ 2/5 & 4/5 \end{bmatrix}$

A21 $\begin{bmatrix} 16/17 & -4/17 \\ -4/17 & 1/17 \end{bmatrix}$ **A22** $\begin{bmatrix} 13/14 & -1/7 & -3/14 \\ -1/7 & 5/7 & -3/7 \\ -3/14 & -3/7 & 5/14 \end{bmatrix}$

A23 $L : \mathbb{R}^2 \to \mathbb{R}^2$. $[L] = \begin{bmatrix} -3 & 5 \\ -1 & -2 \end{bmatrix}$

A24 $L : \mathbb{R}^2 \to \mathbb{R}^3$. $[L] = \begin{bmatrix} 1 & 0 \\ 0 & 1 \\ 1 & 1 \end{bmatrix}$

A25 $L : \mathbb{R}^1 \to \mathbb{R}^3$. $[L] = \begin{bmatrix} 1 \\ 0 \\ 3 \end{bmatrix}$

A26 $M : \mathbb{R}^3 \to \mathbb{R}^1$. $[M] = \begin{bmatrix} 1 & -1 & \sqrt{2} \end{bmatrix}$

A27 $M : \mathbb{R}^3 \to \mathbb{R}^2$. $[M] = \begin{bmatrix} 2 & 0 & -1 \\ 2 & 0 & -1 \end{bmatrix}$

A28 $N : \mathbb{R}^3 \to \mathbb{R}^4$. $[N] = \begin{bmatrix} 0 & 0 & 0 \\ 0 & 0 & 0 \\ 0 & 0 & 0 \\ 0 & 0 & 0 \end{bmatrix}$

A29 $L : \mathbb{R}^3 \to \mathbb{R}^2$. $[L] = \begin{bmatrix} 2 & -3 & 1 \\ 0 & 1 & -5 \end{bmatrix}$

A30 $K : \mathbb{R}^4 \to \mathbb{R}^2$. $[K] = \begin{bmatrix} 5 & 0 & 3 & -1 \\ 0 & 1 & -7 & 3 \end{bmatrix}$

A31 $M : \mathbb{R}^4 \to \mathbb{R}^4$. $[M] = \begin{bmatrix} 1 & 0 & -1 & 1 \\ 1 & 2 & 0 & -3 \\ 0 & 1 & 1 & 0 \end{bmatrix}$

A32 $\begin{bmatrix} 3 & 1 \\ 5 & -2 \end{bmatrix}$ **A33** $\begin{bmatrix} -1 & 13 \\ 11 & -21 \end{bmatrix}$

A34 $\begin{bmatrix} 1 & 1 \\ 0 & 0 \\ 1 & 1 \end{bmatrix}$ **A35** $\begin{bmatrix} 1 & 1 & 1 \\ 0 & 0 & 0 \\ 1 & 1 & 1 \end{bmatrix}$

A36 $\begin{bmatrix} 5 & 0 & 0 \\ 0 & 3 & 0 \\ 0 & 0 & 2 \end{bmatrix}$ **A37** $\begin{bmatrix} -1 & 1/2 & -1 \\ \sqrt{2} & 0 & -1 \end{bmatrix}$

A38 $\begin{bmatrix} 2 & 1 & 0 \\ 1 & -2 & 1 \\ 1 & 1 & -2 \end{bmatrix}$ **A39** $\begin{bmatrix} 3 & 2 \\ 5 & -7 \end{bmatrix}$

A40 $\begin{bmatrix} 2 & 1 \\ 1 & 1 \\ -1 & 1 \end{bmatrix}$ **A41** $\begin{bmatrix} 3 & 2 & -1 \\ 1 & 5 & -1 \end{bmatrix}$

A42 (a) $S : \mathbb{R}^3 \to \mathbb{R}^2, T : \mathbb{R}^3 \to \mathbb{R}^2$.
(b) $[S+T] = \begin{bmatrix} 3 & 3 & 2 \\ 1 & 2 & 5 \end{bmatrix}, [2S-3T] = \begin{bmatrix} 1 & -4 & 9 \\ -8 & -6 & -5 \end{bmatrix}$

A43 (a) $S : \mathbb{R}^4 \to \mathbb{R}^2, T : \mathbb{R}^2 \to \mathbb{R}^4$.
(b) $[S \circ T] = \begin{bmatrix} 6 & -19 \\ 10 & -10 \end{bmatrix}$,

$$[T \circ S] = \begin{bmatrix} -3 & 5 & 16 & 9 \\ 6 & 8 & 4 & 0 \\ -6 & -8 & -4 & 0 \\ -9 & -17 & -16 & -5 \end{bmatrix}$$

A44 $[L \circ M] = \begin{bmatrix} 11 & -4 & 1 \\ 11 & -9 & -6 \\ 3 & -2 & -1 \end{bmatrix}$

A45 $[M \circ L] = [M][L] = \begin{bmatrix} 1 & 9 \\ 8 & 0 \end{bmatrix}$

A46 The composition is not defined.

A47 The composition is not defined.

A48 The composition is not defined.

A49 $[N \circ M] = \begin{bmatrix} 5 & 0 & 3 \\ 6 & 1 & 5 \\ -3 & -3 & -6 \\ 13 & -7 & -2 \end{bmatrix}$

A50 (a) $L(\vec{x}) = (3x_1 + x_2, -x_1 - 5x_2, 4x_1 + 9x_2)$
(b) $L(x_1, x_2) = (3x_1^2 + x_2, -x_1 - 5x_2, 4x_1 + 9x_2)$

Section 3.3 Practice Problems

A1 $\begin{bmatrix} 0 & -1 \\ 1 & 0 \end{bmatrix}$

A2 $\begin{bmatrix} -1 & 0 \\ 0 & -1 \end{bmatrix}$

A3 $\begin{bmatrix} 1/\sqrt{2} & 1/\sqrt{2} \\ -1/\sqrt{2} & 1/\sqrt{2} \end{bmatrix}$

A4 $\begin{bmatrix} 0.309 & -0.951 \\ 0.951 & 0.309 \end{bmatrix}$

A5 $[V] = \begin{bmatrix} 1 & 0 \\ 3 & 1 \end{bmatrix}, [S] = \begin{bmatrix} 1 & 0 \\ 0 & 5 \end{bmatrix}$

A6 $[V \circ S] = \begin{bmatrix} 1 & 0 \\ 3 & 5 \end{bmatrix}$

A7 $[S \circ V] = \begin{bmatrix} 1 & 0 \\ 15 & 5 \end{bmatrix}$

A8 $[R_\theta \circ S] = \begin{bmatrix} \cos\theta & -5\sin\theta \\ \sin\theta & 5\cos\theta \end{bmatrix}$

A9 $[S \circ R_\theta] = \begin{bmatrix} \cos\theta & -\sin\theta \\ 5\sin\theta & 5\cos\theta \end{bmatrix}$

A10 $[H] = \begin{bmatrix} 1 & 1 \\ 0 & 1 \end{bmatrix}, [V] = \begin{bmatrix} 1 & 0 \\ -2 & 1 \end{bmatrix}$

A11 $[V \circ H] = \begin{bmatrix} 1 & 1 \\ -2 & -1 \end{bmatrix}$ **A12** $[H \circ V] = \begin{bmatrix} -1 & 1 \\ -2 & 1 \end{bmatrix}$

A13 $[F \circ H] = \begin{bmatrix} 1 & 1 \\ 0 & -1 \end{bmatrix}$ **A14** $[H \circ F] = \begin{bmatrix} 1 & -1 \\ 0 & -1 \end{bmatrix}$

A15 $\begin{bmatrix} 4/5 & -3/5 \\ -3/5 & -4/5 \end{bmatrix}$ **A16** $\begin{bmatrix} -3/5 & 4/5 \\ 4/5 & 3/5 \end{bmatrix}$

A17 $\begin{bmatrix} -15/17 & 8/17 \\ 8/17 & 15/17 \end{bmatrix}$ **A18** $\begin{bmatrix} 8/17 & 15/17 \\ 15/17 & -8/17 \end{bmatrix}$

A19 $\frac{1}{3}\begin{bmatrix} 1 & -2 & -2 \\ -2 & 1 & -2 \\ -2 & -2 & 1 \end{bmatrix}$ **A20** $\frac{1}{9}\begin{bmatrix} 1 & 8 & 4 \\ 8 & 1 & -4 \\ 4 & -4 & 7 \end{bmatrix}$

A21 $\begin{bmatrix} 0 & 0 & 1 \\ 0 & 1 & 0 \\ 1 & 0 & 0 \end{bmatrix}$

A22 $[\text{refl}_{\vec{n}}] = \begin{bmatrix} 6/7 & -2/7 & 3/7 \\ -2/7 & 3/7 & 6/7 \\ 3/7 & 6/7 & -2/7 \end{bmatrix}$

A23 $[\text{inj} \circ D] = \begin{bmatrix} 5 & 0 & 0 \\ 0 & 5 & 0 \\ 0 & 0 & 0 \\ 0 & 0 & 5 \end{bmatrix}$

A24 (a) $[P \circ S] = \begin{bmatrix} 0 & 1 & 0 \\ 2 & 0 & 1 \end{bmatrix}$

(b) There is no T.

(c) $[Q \circ S] = \begin{bmatrix} 1 & 0 & 0 \\ 0 & 1 & 0 \end{bmatrix}$

Section 3.4 Practice Problems

A1 (a) $\vec{y}_1 \in \text{Range}(L)$; $\vec{x} = \begin{bmatrix} 9 \\ -3 \\ 0 \end{bmatrix} + t\begin{bmatrix} 7 \\ -4 \\ 1 \end{bmatrix}, t \in \mathbb{R}$.

(b) $\vec{y}_2 \in \text{Range}(L)$; $\vec{x} = \begin{bmatrix} -3 \\ 2 \\ 0 \end{bmatrix} + t\begin{bmatrix} 7 \\ -4 \\ 1 \end{bmatrix}, t \in \mathbb{R}$.

(c) $\vec{v} \in \text{Null}(L)$.

A2 (a) $\vec{y}_1 \notin \text{Range}(L)$.

(b) $\vec{y}_2 \in \text{Range}(L)$; $\vec{x} = \begin{bmatrix} 2 \\ 3 \end{bmatrix}$.

(c) $\vec{v} \notin \text{Null}(L)$.

A3 (a) $\vec{y}_1 \notin \text{Range}(L)$.

(b) $\vec{y}_2 \in \text{Range}(L)$; $\vec{x} = \begin{bmatrix} 1 \\ 1 \\ -2 \end{bmatrix}$.

(c) $\vec{v} \notin \text{Null}(L)$.

A4 $C = \left\{\begin{bmatrix} 3 \\ 1 \\ 1 \end{bmatrix}, \begin{bmatrix} 5 \\ -1 \\ -1 \end{bmatrix}\right\}$ is a basis for Range(L).
The empty set is a basis for Null(L).

A5 $C = \left\{\begin{bmatrix} 2 \\ 4 \end{bmatrix}\right\}$ is a basis for Range(L).

$\mathcal{B} = \left\{\begin{bmatrix} 1/2 \\ 1 \end{bmatrix}\right\}$ is a basis for Null(L).

A6 $C = \left\{\begin{bmatrix} 1 \\ 1 \end{bmatrix}, \begin{bmatrix} -7 \\ 1 \end{bmatrix}\right\}$ is a basis for Range(L).
The empty set is a basis for Null(L).

A7 $C = \left\{\begin{bmatrix} 1 \\ 2 \\ 3 \end{bmatrix}\right\}$ is a basis for Range(L).

$\mathcal{B} = \left\{\begin{bmatrix} 0 \\ 1 \end{bmatrix}\right\}$ is a basis for Null(L).

A8 $C = \left\{\begin{bmatrix} 1 \\ 0 \end{bmatrix}\right\}$ is a basis for Range(L).

$\mathcal{B} = \left\{\begin{bmatrix} 1 \\ -1 \\ 0 \end{bmatrix}, \begin{bmatrix} 0 \\ 0 \\ 1 \end{bmatrix}\right\}$ is a basis for Null(L).

A9 A basis for Range(L) is the empty set.
The standard basis for $\mathbb{R}^3$ is a basis for Null(L).

A10 $C = \left\{\begin{bmatrix} 1 \\ 0 \end{bmatrix}, \begin{bmatrix} 0 \\ 1 \end{bmatrix}\right\}$ is a basis for Range(L).

$\mathcal{B} = \left\{\begin{bmatrix} 0 \\ 2 \\ 1 \end{bmatrix}\right\}$ is a basis for Null(L).

A11 $C = \left\{\begin{bmatrix} 1 \\ 1 \\ 1 \end{bmatrix}, \begin{bmatrix} 0 \\ 1 \\ 1 \end{bmatrix}\right\}$ is a basis for Range(L).

$\mathcal{B} = \left\{ \begin{bmatrix} -7 \\ 6 \\ 1 \end{bmatrix} \right\}$ is a basis for Null(L).

A12 $C = \left\{ \begin{bmatrix} -1 \\ 1 \\ -2 \end{bmatrix}, \begin{bmatrix} 2 \\ -4 \\ 4 \end{bmatrix} \right\}$ is a basis for Range(L).

$\mathcal{B} = \left\{ \begin{bmatrix} 1 \\ 0 \\ 0 \end{bmatrix} \right\}$ is a basis for Null(L).

A13 $C = \left\{ \begin{bmatrix} 1 \\ 0 \\ 0 \\ 0 \\ 0 \end{bmatrix}, \begin{bmatrix} 0 \\ 1 \\ 0 \\ 0 \\ -1 \end{bmatrix}, \begin{bmatrix} 0 \\ 0 \\ 0 \\ 1 \\ 1 \end{bmatrix}, \begin{bmatrix} 0 \\ 0 \\ 0 \\ 0 \\ 1 \end{bmatrix} \right\}$ is a basis for Range(L).
The empty set is a basis for Null(L).

A14 $C = \left\{ \begin{bmatrix} 1 \\ 0 \end{bmatrix}, \begin{bmatrix} -1 \\ 1 \end{bmatrix} \right\}$ is a basis for Range(L).

$\mathcal{B} = \left\{ \begin{bmatrix} 1 \\ 3 \\ 1 \\ 0 \end{bmatrix}, \begin{bmatrix} -4 \\ -3 \\ 0 \\ 1 \end{bmatrix} \right\}$ is a basis for Null(L).

A15 A basis for Col($[L]$) is $\left\{ \begin{bmatrix} 3 \\ 1 \\ 1 \end{bmatrix}, \begin{bmatrix} 5 \\ -1 \\ -1 \end{bmatrix} \right\}$.
A basis for Null($[L]$) is the empty set.
A basis for Row($[L]$) is $\left\{ \begin{bmatrix} 1 \\ 0 \end{bmatrix}, \begin{bmatrix} 0 \\ 1 \end{bmatrix} \right\}$.
A basis for Null($[L]^T$) is $\left\{ \begin{bmatrix} 0 \\ -1 \\ 1 \end{bmatrix} \right\}$.

A16 A basis for Col($[L]$) is $\left\{ \begin{bmatrix} 1 \\ 0 \end{bmatrix} \right\}$.
A basis for Null($[L]$) is $\left\{ \begin{bmatrix} 1 \\ -1 \\ 0 \end{bmatrix}, \begin{bmatrix} 0 \\ 0 \\ 1 \end{bmatrix} \right\}$.
A basis for Row($[L]$) is $\left\{ \begin{bmatrix} 1 \\ 1 \\ 0 \end{bmatrix} \right\}$.
A basis for Null($[L]^T$) is $\left\{ \begin{bmatrix} 0 \\ 1 \end{bmatrix} \right\}$.

A17 A basis for Col($[L]$) is the empty set.
A basis for Null($[L]$) is the standard basis for $\mathbb{R}^3$.
A basis for Row($[L]$) is the empty set.
A basis for Null($[L]^T$) is the standard basis for $\mathbb{R}^3$.

A18 A basis for Col($[L]$) is the standard basis for $\mathbb{R}^2$.
A basis for Null($[L]$) is $\left\{ \begin{bmatrix} 0 \\ 2 \\ 1 \end{bmatrix} \right\}$.
A basis for Row($[L]$) is $\left\{ \begin{bmatrix} 1 \\ 0 \\ 0 \end{bmatrix}, \begin{bmatrix} 0 \\ 1 \\ -2 \end{bmatrix} \right\}$.
A basis for Null($[L]^T$) is the empty set.

A19 A basis for Col($[L]$) is $\left\{ \begin{bmatrix} 1 \\ 0 \\ 0 \\ 0 \\ 0 \end{bmatrix}, \begin{bmatrix} 0 \\ 1 \\ 0 \\ 0 \\ -1 \end{bmatrix}, \begin{bmatrix} 0 \\ 0 \\ 0 \\ 1 \\ 1 \end{bmatrix}, \begin{bmatrix} 0 \\ 0 \\ 0 \\ 0 \\ 1 \end{bmatrix} \right\}$.
A basis for Null($[L]$) is the empty set.
A basis for Row($[L]$) is the standard basis for $\mathbb{R}^4$.
A basis for Null($[L]^T$) is $\left\{ \begin{bmatrix} 0 \\ 0 \\ 1 \\ 0 \\ 0 \end{bmatrix} \right\}$.

A20 A basis for Range(L) is $\left\{ \begin{bmatrix} 1 \\ -2 \\ 2 \end{bmatrix} \right\}$.
A basis for Null(L) is $\left\{ \begin{bmatrix} 2 \\ 1 \\ 0 \end{bmatrix}, \begin{bmatrix} 2 \\ 0 \\ -1 \end{bmatrix} \right\}$.

A21 A basis for Range(L) is $\left\{ \begin{bmatrix} 1 \\ -3 \\ 0 \end{bmatrix}, \begin{bmatrix} 0 \\ -2 \\ 1 \end{bmatrix} \right\}$.
A basis for Null(L) is $\left\{ \begin{bmatrix} 3 \\ 1 \\ 2 \end{bmatrix} \right\}$.

A22 A basis for Range(L) is $\{\vec{e}_1, \vec{e}_2, \vec{e}_3\}$.
A basis for Null(L) is the empty set.

A23 $[L] = \begin{bmatrix} 1 & 2 \\ -1 & 5 \\ 1 & -3 \end{bmatrix}$

A24 $[L] = \begin{bmatrix} 1 & -1 \\ 2 & -2 \\ 3 & -3 \end{bmatrix}$

A25 $[L] = \begin{bmatrix} 1 & 1/2 \\ 1 & 1/2 \\ 1 & 1/2 \end{bmatrix}$

A26 One choice is $[L] = \begin{bmatrix} 1 & 0 & -3 \\ 0 & 1 & 2 \\ 0 & 1 & 2 \end{bmatrix}$

A27 Since $\vec{r}_0$ is a solution of $A\vec{x} = \vec{b}$, that means $A\vec{r}_0 = \vec{b}$. Then we get

$$A(\vec{r}_0 + \vec{n}) = A\vec{r}_0 + A\vec{n} = \vec{b} + \vec{0} = \vec{b}$$

A28 (a) The number of variables is 4.
(b) The rank of A is 2.
(c) The dimension of the solution space is 2.

A29 (a) The number of variables is 5.
(b) The rank of A is 3.
(c) The dimension of the solution space is 2.

A30 A basis for Row(A) is $\left\{\begin{bmatrix}1\\2\end{bmatrix}\right\}$.

A basis for Col(A) is $\left\{\begin{bmatrix}1\\2\end{bmatrix}\right\}$.

A basis for Null(A) is $\left\{\begin{bmatrix}-2\\1\end{bmatrix}\right\}$.

A basis for Null(A^T) is $\left\{\begin{bmatrix}-2\\1\end{bmatrix}\right\}$.

A31 A basis for Row(A) is $\left\{\begin{bmatrix}1\\0\\-1\\0\end{bmatrix}, \begin{bmatrix}0\\1\\-2\\0\end{bmatrix}, \begin{bmatrix}0\\0\\0\\1\end{bmatrix}\right\}$.

A basis for Col(A) is $\left\{\begin{bmatrix}1\\2\\0\end{bmatrix}, \begin{bmatrix}1\\3\\1\end{bmatrix}, \begin{bmatrix}1\\4\\3\end{bmatrix}\right\}$.

A basis for Null(A) is $\left\{\begin{bmatrix}1\\2\\1\\0\end{bmatrix}\right\}$.

A basis for Null(A^T) is the empty set.

A32 A basis for Row(A) is $\left\{\begin{bmatrix}1\\0\\0\end{bmatrix}, \begin{bmatrix}0\\1\\1/2\end{bmatrix}\right\}$.

A basis for Col(A) is $\left\{\begin{bmatrix}1\\0\end{bmatrix}, \begin{bmatrix}2\\4\end{bmatrix}\right\}$.

A basis for Null(A) is $\left\{\begin{bmatrix}0\\-1/2\\1\end{bmatrix}\right\}$.

A basis for Null(A^T) is the empty set.

A33 A basis for Row(A) is $\left\{\begin{bmatrix}0\\1\\-2\\-1\end{bmatrix}\right\}$.

A basis for Col(A) is $\left\{\begin{bmatrix}-2\\3\end{bmatrix}\right\}$.

A basis for Null(A) is $\left\{\begin{bmatrix}1\\0\\0\\0\end{bmatrix}, \begin{bmatrix}0\\2\\1\\0\end{bmatrix}, \begin{bmatrix}0\\1\\0\\1\end{bmatrix}\right\}$.

A basis for Null(A^T) is $\left\{\begin{bmatrix}3/2\\1\end{bmatrix}\right\}$.

A34 A basis for Row(A) is $\left\{\begin{bmatrix}1\\0\\-1\\0\end{bmatrix}, \begin{bmatrix}0\\1\\-2\\0\end{bmatrix}, \begin{bmatrix}0\\0\\0\\1\end{bmatrix}\right\}$.

A basis for Col(A) is $\left\{\begin{bmatrix}1\\2\\0\end{bmatrix}, \begin{bmatrix}1\\3\\1\end{bmatrix}, \begin{bmatrix}1\\4\\3\end{bmatrix}\right\}$.

A basis for Null(A) is $\left\{\begin{bmatrix}1\\2\\1\\0\end{bmatrix}\right\}$.

A basis for Null(A^T) is the empty set.

A35 A basis for Row(A) is $\left\{\begin{bmatrix}1\\0\\0\end{bmatrix}, \begin{bmatrix}0\\1\\0\end{bmatrix}, \begin{bmatrix}0\\0\\1\end{bmatrix}\right\}$.

A basis for Col(A) is $\left\{\begin{bmatrix}1\\1\\1\end{bmatrix}, \begin{bmatrix}2\\1\\0\end{bmatrix}, \begin{bmatrix}8\\5\\-2\end{bmatrix}\right\}$.

A basis for Null(A) is the empty set.
A basis for Null(A^T) is the empty set.

A36 A basis for Row(A) is $\left\{\begin{bmatrix}1\\0\\2\end{bmatrix}, \begin{bmatrix}0\\1\\-1\end{bmatrix}\right\}$.

A basis for Col(A) is $\left\{\begin{bmatrix}2\\2\\4\end{bmatrix}, \begin{bmatrix}1\\-2\\3\end{bmatrix}\right\}$.

A basis for Null(A) is $\left\{\begin{bmatrix}-2\\1\\1\end{bmatrix}\right\}$.

A basis for Null(A^T) is $\left\{\begin{bmatrix}-7/3\\1/3\\1\end{bmatrix}\right\}$.

A37 A basis for Row(A) is $\left\{\begin{bmatrix}1\\2\\4\end{bmatrix}\right\}$.

A basis for Col(A) is $\left\{\begin{bmatrix}1\\1\\1\end{bmatrix}\right\}$.

A basis for Null(A) is $\left\{\begin{bmatrix}-2\\1\\0\end{bmatrix}, \begin{bmatrix}-4\\0\\1\end{bmatrix}\right\}$.

A basis for Null(A^T) is $\left\{\begin{bmatrix}-1\\1\\0\end{bmatrix}, \begin{bmatrix}-1\\0\\1\end{bmatrix}\right\}$.

A38 A basis for Row(A) is $\left\{\begin{bmatrix}1\\0\\-5\end{bmatrix}, \begin{bmatrix}0\\1\\7\end{bmatrix}\right\}$.

A basis for Col(A) is $\left\{\begin{bmatrix}1\\0\\-2\\1\end{bmatrix}, \begin{bmatrix}2\\1\\-2\\1\end{bmatrix}\right\}$.

A basis for Null(A) is $\left\{\begin{bmatrix}5\\-7\\1\end{bmatrix}\right\}$.

A basis for Null(A^T) is $\left\{\begin{bmatrix}2\\-2\\1\\0\end{bmatrix}, \begin{bmatrix}-1\\1\\0\\1\end{bmatrix}\right\}$.

A39 A basis for Row(A) is the standard basis for $\mathbb{R}^3$.

A basis for Col(A) is $\left\{\begin{bmatrix}3\\1\\1\end{bmatrix}, \begin{bmatrix}-1\\2\\3\end{bmatrix}, \begin{bmatrix}6\\5\\3\end{bmatrix}\right\}$.

A basis for Null(A) is the empty set.

A basis for Null(A^T) is the empty set.

A40 A basis for Row(A) is the standard basis for $\mathbb{R}^3$.

A basis for Col(A) is $\left\{\begin{bmatrix}1\\1\\1\\1\end{bmatrix}, \begin{bmatrix}-1\\0\\1\\2\end{bmatrix}, \begin{bmatrix}1\\0\\1\\4\end{bmatrix}\right\}$.

A basis for Null(A) is the empty set.

A basis for Null(A^T) is $\left\{\begin{bmatrix}-1\\3\\-3\\1\end{bmatrix}\right\}$.

A41 A basis for Row(A) is $\left\{\begin{bmatrix}1\\1\\0\end{bmatrix}, \begin{bmatrix}0\\0\\1\end{bmatrix}\right\}$.

A basis for Col(A) is $\left\{\begin{bmatrix}1\\2\\1\\3\end{bmatrix}, \begin{bmatrix}0\\0\\2\\4\end{bmatrix}\right\}$.

A basis for Null(A) is $\left\{\begin{bmatrix}-1\\1\\0\end{bmatrix}\right\}$.

A basis for Null(A^T) is $\left\{\begin{bmatrix}-2\\1\\0\\0\end{bmatrix}, \begin{bmatrix}-1\\0\\-2\\1\end{bmatrix}\right\}$.

A42 A basis for Row(A) is $\left\{\begin{bmatrix}1\\2\\0\\3\\0\end{bmatrix}, \begin{bmatrix}0\\0\\1\\4\\0\end{bmatrix}, \begin{bmatrix}0\\0\\0\\0\\1\end{bmatrix}\right\}$.

A basis for Col(A) is $\left\{\begin{bmatrix}1\\1\\2\\3\end{bmatrix}, \begin{bmatrix}0\\1\\0\\1\end{bmatrix}, \begin{bmatrix}0\\1\\1\\2\end{bmatrix}\right\}$.

A basis for Null(A) is $\left\{\begin{bmatrix}-2\\1\\0\\0\\0\end{bmatrix}, \begin{bmatrix}-3\\0\\-4\\1\\0\end{bmatrix}\right\}$.

A basis for Null(A^T) is $\left\{\begin{bmatrix}0\\-1\\-1\\1\end{bmatrix}\right\}$.

A43 $\vec{u} \in \text{Null}(A^T)$

A44 $\vec{v} \notin \text{Col}(A)$

A45 $\vec{w} \notin \text{Null}(A)$

A46 $\vec{z} \in \text{Row}(A)$

A47 $\left\{\begin{bmatrix}-2\\1\\1\\0\\0\end{bmatrix}, \begin{bmatrix}-3\\-1\\0\\-1\\1\end{bmatrix}\right\}$

A48 $\vec{x} = \begin{bmatrix}1\\0\\0\\0\\0\end{bmatrix} + s\begin{bmatrix}-2\\1\\1\\0\\0\end{bmatrix} + t\begin{bmatrix}-3\\-1\\0\\-1\\1\end{bmatrix}, s,t \in \mathbb{R}$

Section 3.5 Practice Problems

A1 Not invertible.

A2 $\frac{1}{34}\begin{bmatrix} 1 & -6 \\ 5 & 4 \end{bmatrix}$

A3 $\frac{1}{24}\begin{bmatrix} 3 & -6 \\ 2 & 4 \end{bmatrix}$

A4 $\frac{1}{23}\begin{bmatrix} 5 & 4 \\ -2 & 3 \end{bmatrix}$

A5 $\begin{bmatrix} 2 & 0 & -1 \\ -1 & 1 & -1 \\ -1 & 0 & 1 \end{bmatrix}$

A6 Not invertible.

A7 Not invertible.

A8 $\begin{bmatrix} 1/2 & 1/4 & -7/4 \\ 0 & 1/2 & -1/2 \\ 0 & 0 & 1 \end{bmatrix}$

A9 $\begin{bmatrix} 7 & -5 & -1 \\ 5 & -4 & -1 \\ 4 & -3 & -1 \end{bmatrix}$

A10 $\begin{bmatrix} 0 & -1 & 1 \\ -1 & 1 & 0 \\ 1 & 0 & 0 \end{bmatrix}$

A11 $\begin{bmatrix} -1 & 1 & 0 \\ -3 & 1 & -1 \\ 7/3 & -2/3 & 2/3 \end{bmatrix}$

A12 Not invertible.

A13 $\begin{bmatrix} 0 & 0 & 1 \\ 1 & 0 & 0 \\ 0 & 1 & 0 \end{bmatrix}$

A14 Not invertible.

A15 Not invertible.

A16 $\begin{bmatrix} 0 & -3 & 1/2 & -2 \\ -1 & -2 & 3/2 & -2 \\ 0 & 1 & 0 & 1 \\ 1 & 3 & -3/2 & 2 \end{bmatrix}$

A17 $\begin{bmatrix} 6 & 10 & -5/2 & -7/2 \\ 1 & 2 & -1/2 & -1/2 \\ -2 & -3 & 1 & 1 \\ 0 & -3 & 0 & 1 \end{bmatrix}$

A18 Not invertible.

A19 $\begin{bmatrix} 1 & 0 & -1 & 1 & -2 \\ 0 & 1 & 0 & -1 & 2 \\ 0 & 0 & 1 & -1 & 1 \\ 0 & 0 & 0 & 1 & -2 \\ 0 & 0 & 0 & 0 & 1 \end{bmatrix}$

A20 (a) $\vec{x} = \begin{bmatrix} 1 \\ -1 \\ 0 \end{bmatrix}$ (b) $\vec{x} = \begin{bmatrix} -3 \\ 0 \\ 2 \end{bmatrix}$

(c) $\vec{x} = \begin{bmatrix} -2 \\ -1 \\ 2 \end{bmatrix}$ (d) $\vec{x} = \begin{bmatrix} 8 \\ 0 \\ -5 \end{bmatrix}$

A21 (a) $A^{-1} = \begin{bmatrix} 2 & -1 \\ -3 & 2 \end{bmatrix}$, $B^{-1} = \begin{bmatrix} -5 & 2 \\ 3 & -1 \end{bmatrix}$

(b) $(AB)^{-1} = \begin{bmatrix} -16 & 9 \\ 9 & -5 \end{bmatrix} = B^{-1}A^{-1}$

(c) $(3A)^{-1} = \begin{bmatrix} 2/3 & -1/3 \\ -1 & 2/3 \end{bmatrix} = \frac{1}{3}A^{-1}$

(d) $(A^T)^{-1} = \begin{bmatrix} 2 & -3 \\ -1 & 2 \end{bmatrix}$

(e) $(A+B)^{-1} = \frac{1}{3}\begin{bmatrix} 7 & -3 \\ -6 & 3 \end{bmatrix}$, $A^{-1} + B^{-1} = \begin{bmatrix} -3 & 1 \\ 0 & 1 \end{bmatrix}$

A22 (a) $[R_{\pi/6}] = \begin{bmatrix} \sqrt{3}/2 & -1/2 \\ 1/2 & \sqrt{3}/2 \end{bmatrix}$,

$[R_{\pi/6}]^{-1} = [R_{-\pi/6}] = \begin{bmatrix} \sqrt{3}/2 & 1/2 \\ -1/2 & \sqrt{3}/2 \end{bmatrix}$

(b) $[S] = \begin{bmatrix} 1 & 0 \\ 2 & 1 \end{bmatrix}$, $[S^{-1}] = \begin{bmatrix} 1 & 0 \\ -2 & 1 \end{bmatrix}$

(c) $[R] = \begin{bmatrix} 0 & 1 \\ 1 & 0 \end{bmatrix}$, $[R^{-1}] = \begin{bmatrix} 0 & 1 \\ 1 & 0 \end{bmatrix} = [R]$

(d) $[(R \circ S)^{-1}] = \begin{bmatrix} 0 & 1 \\ 1 & -2 \end{bmatrix}$, $[(S \circ R)^{-1}] = \begin{bmatrix} -2 & 1 \\ 1 & 0 \end{bmatrix}$

A23 Let $\vec{v}, \vec{y} \in \mathbb{R}^n$ and $s, t \in \mathbb{R}$. Then there exists $\vec{u}, \vec{x} \in \mathbb{R}^n$ such that $\vec{x} = M(\vec{y})$ and $\vec{u} = M(\vec{v})$. Then $L(\vec{x}) = \vec{y}$ and $L(\vec{u}) = \vec{v}$. Since L is linear $L(s\vec{x} + t\vec{u}) = sL(\vec{x}) + tL(\vec{u}) = s\vec{y} + t\vec{v}$. Hence, M is linear since

$$M(s\vec{y} + t\vec{v}) = s\vec{x} + t\vec{u} = sM(\vec{y}) + tM(\vec{v})$$

A24 $\begin{bmatrix} 1 & 3 \\ 0 & 1 \end{bmatrix}$

A25 $\begin{bmatrix} 1/5 & 0 \\ 0 & 1 \end{bmatrix}$

A26 $\begin{bmatrix} 1 & 0 & 0 \\ 0 & -1 & 0 \\ 0 & 0 & 1 \end{bmatrix}$

A27 $\begin{bmatrix} 1 & 0 \\ -4 & 1 \end{bmatrix}$

A28 $\begin{bmatrix} 1 & 0 & 0 \\ 0 & 1 & -2 \\ 0 & 0 & 1 \end{bmatrix}$

A29 $\begin{bmatrix} 1 & 0 & 0 \\ 0 & 0 & 1 \\ 0 & 1 & 0 \end{bmatrix}$

A30 $L^{-1}(x_1, x_2) = \left(\frac{5}{7}x_1 - \frac{3}{7}x_2, -\frac{1}{7}x_1 + \frac{2}{7}x_2\right)$

A31 $L^{-1}(x_1, x_2, x_3) = (-3x_1 - 3x_2 + 2x_3, -2x_1 + x_3, 2x_1 + x_2 - x_3)$

A32 $X = \begin{bmatrix} -18 & -15 & 4 \\ 13 & 11 & -3 \end{bmatrix}$

A33 $X = \begin{bmatrix} -1 & 3 \\ -1 & 4 \\ -8 & 19 \end{bmatrix}$

A34 If A is invertible, then A^T is invertible by the Invertible Matrix Theorem. Let $A^T = \begin{bmatrix} \vec{a}_1 & \cdots & \vec{a}_n \end{bmatrix}$ and consider

$$\vec{0} = c_1\vec{a}_1 + \cdots + c_n\vec{a}_n = \begin{bmatrix} \vec{a}_1 & \cdots & \vec{a}_n \end{bmatrix}\begin{bmatrix} c_1 \\ \vdots \\ c_n \end{bmatrix} = A^T\vec{c}$$

Since A^T is invertible, this system has unique solution

$$\vec{c} = (A^T)^{-1}\vec{0} = \vec{0}$$

Thus, $c_1 = \cdots = c_n = 0$, so the columns of A^T form a linearly independent set.

A35 Let $A = \begin{bmatrix} 1 & 0 \\ 0 & 1 \end{bmatrix}$ and $B = \begin{bmatrix} -1 & 0 \\ 0 & -1 \end{bmatrix}$. Then, A and B are both invertible, but $A + B = \begin{bmatrix} 0 & 0 \\ 0 & 0 \end{bmatrix}$ is not invertible.

A36 Let $A = \begin{bmatrix} 1 & 1 \\ 1 & 1 \end{bmatrix}$, $X = \begin{bmatrix} 1 & 1 \\ 0 & 1 \end{bmatrix}$, and $C = \begin{bmatrix} 0 & 0 \\ 1 & 2 \end{bmatrix}$ gives $XC = AX$, but $A \neq C$.

Section 3.6 Practice Problems

A1 $E = \begin{bmatrix} 1 & -5 & 0 \\ 0 & 1 & 0 \\ 0 & 0 & 1 \end{bmatrix}$, $EA = \begin{bmatrix} 6 & -13 & -17 \\ -1 & 3 & 4 \\ 4 & 2 & 0 \end{bmatrix}$

A2 $E = \begin{bmatrix} 1 & 0 & 0 \\ 0 & 0 & 1 \\ 0 & 1 & 0 \end{bmatrix}$, $EA = \begin{bmatrix} 1 & 2 & 3 \\ 4 & 2 & 0 \\ -1 & 3 & 4 \end{bmatrix}$

A3 $E = \begin{bmatrix} 1 & 0 & 0 \\ 0 & 1 & 0 \\ 0 & 0 & -1 \end{bmatrix}$, $EA = \begin{bmatrix} 1 & 2 & 3 \\ -1 & 3 & 4 \\ -4 & -2 & 0 \end{bmatrix}$

A4 $E = \begin{bmatrix} 1 & 0 & 0 \\ 0 & 6 & 0 \\ 0 & 0 & 1 \end{bmatrix}$, $EA = \begin{bmatrix} 1 & 2 & 3 \\ -6 & 18 & 24 \\ 4 & 2 & 0 \end{bmatrix}$

A5 $E = \begin{bmatrix} 1 & 0 & 0 \\ 0 & 1 & 0 \\ 4 & 0 & 1 \end{bmatrix}$, $EA = \begin{bmatrix} 1 & 2 & 3 \\ -1 & 3 & 4 \\ 8 & 10 & 12 \end{bmatrix}$

A6 $E = \begin{bmatrix} 1 & 0 & 0 & 0 \\ 0 & 1 & 0 & 0 \\ 0 & 0 & 1 & 0 \\ 0 & 0 & -3 & 1 \end{bmatrix}$, $E^{-1} = \begin{bmatrix} 1 & 0 & 0 & 0 \\ 0 & 1 & 0 & 0 \\ 0 & 0 & 1 & 0 \\ 0 & 0 & 3 & 1 \end{bmatrix}$

A7 $E = \begin{bmatrix} 1 & 0 & 0 & 0 \\ 0 & 0 & 0 & 1 \\ 0 & 0 & 1 & 0 \\ 0 & 1 & 0 & 0 \end{bmatrix}$, $E^{-1} = \begin{bmatrix} 1 & 0 & 0 & 0 \\ 0 & 0 & 0 & 1 \\ 0 & 0 & 1 & 0 \\ 0 & 1 & 0 & 0 \end{bmatrix}$

A8 $E = \begin{bmatrix} 1 & 0 & 0 & 0 \\ 0 & 1 & 0 & 0 \\ 0 & 0 & -3 & 0 \\ 0 & 0 & 0 & 1 \end{bmatrix}$, $E^{-1} = \begin{bmatrix} 1 & 0 & 0 & 0 \\ 0 & 1 & 0 & 0 \\ 0 & 0 & -\frac{1}{3} & 0 \\ 0 & 0 & 0 & 1 \end{bmatrix}$

A9 $E = \begin{bmatrix} 1 & 0 & 0 & 0 \\ 0 & 1 & 0 & 0 \\ 2 & 0 & 1 & 0 \\ 0 & 0 & 0 & 1 \end{bmatrix}$, $E^{-1} = \begin{bmatrix} 1 & 0 & 0 & 0 \\ 0 & 1 & 0 & 0 \\ -2 & 0 & 1 & 0 \\ 0 & 0 & 0 & 1 \end{bmatrix}$

A10 $E = \begin{bmatrix} 3 & 0 & 0 & 0 \\ 0 & 1 & 0 & 0 \\ 0 & 0 & 1 & 0 \\ 0 & 0 & 0 & 1 \end{bmatrix}$, $E^{-1} = \begin{bmatrix} \frac{1}{3} & 0 & 0 & 0 \\ 0 & 1 & 0 & 0 \\ 0 & 0 & 1 & 0 \\ 0 & 0 & 0 & 1 \end{bmatrix}$

A11 $E = \begin{bmatrix} 0 & 0 & 1 & 0 \\ 0 & 1 & 0 & 0 \\ 1 & 0 & 0 & 0 \\ 0 & 0 & 0 & 1 \end{bmatrix}$, $E^{-1} = \begin{bmatrix} 0 & 0 & 1 & 0 \\ 0 & 1 & 0 & 0 \\ 1 & 0 & 0 & 0 \\ 0 & 0 & 0 & 1 \end{bmatrix}$

A12 $E = \begin{bmatrix} 1 & 1 & 0 & 0 \\ 0 & 1 & 0 & 0 \\ 0 & 0 & 1 & 0 \\ 0 & 0 & 0 & 1 \end{bmatrix}$, $E^{-1} = \begin{bmatrix} 1 & -1 & 0 & 0 \\ 0 & 1 & 0 & 0 \\ 0 & 0 & 1 & 0 \\ 0 & 0 & 0 & 1 \end{bmatrix}$

A13 $E = \begin{bmatrix} 1 & 0 & 0 & -3 \\ 0 & 1 & 0 & 0 \\ 0 & 0 & 1 & 0 \\ 0 & 0 & 0 & 1 \end{bmatrix}$, $E^{-1} = \begin{bmatrix} 1 & 0 & 0 & 3 \\ 0 & 1 & 0 & 0 \\ 0 & 0 & 1 & 0 \\ 0 & 0 & 0 & 1 \end{bmatrix}$

A14 Elementary. $5R_1$

A15 Not elementary.

A16 Elementary. $R_1 + 2R_2$

A17 Elementary. $R_3 - 4R_2$

A18 Not elementary.

A19 Not elementary.

A20 Elementary. $R_1 \updownarrow R_3$

A21 Not elementary

A22 Elementary. $1R_1$

A23 (a) $E_1 = \begin{bmatrix} 1 & 0 & 0 \\ 0 & 0 & 1 \\ 0 & 1 & 0 \end{bmatrix}$, $E_2 = \begin{bmatrix} 1 & 0 & 0 \\ 0 & 1 & 0 \\ 0 & 0 & 1/2 \end{bmatrix}$,

$E_3 = \begin{bmatrix} 1 & 0 & -4 \\ 0 & 1 & 0 \\ 0 & 0 & 1 \end{bmatrix}$, $E_4 = \begin{bmatrix} 1 & -3 & 0 \\ 0 & 1 & 0 \\ 0 & 0 & 1 \end{bmatrix}$

(b) $A^{-1} = E_4E_3E_2E_1 = \begin{bmatrix} 1 & -2 & -3 \\ 0 & 0 & 1 \\ 0 & 1/2 & 0 \end{bmatrix}$

(c) $A = \begin{bmatrix} 1 & 0 & 0 \\ 0 & 0 & 1 \\ 0 & 1 & 0 \end{bmatrix}\begin{bmatrix} 1 & 0 & 0 \\ 0 & 1 & 0 \\ 0 & 0 & 2 \end{bmatrix}\begin{bmatrix} 1 & 0 & 4 \\ 0 & 1 & 0 \\ 0 & 0 & 1 \end{bmatrix}\begin{bmatrix} 1 & 3 & 0 \\ 0 & 1 & 0 \\ 0 & 0 & 1 \end{bmatrix}$

A24 (a) $E_1 = \begin{bmatrix} 1 & 0 & 0 \\ 0 & 1 & 0 \\ -2 & 0 & 1 \end{bmatrix}, E_2 = \begin{bmatrix} 1 & 0 & 0 \\ 0 & 1 & -3 \\ 0 & 0 & 1 \end{bmatrix},$

$E_3 = \begin{bmatrix} 1 & 0 & -2 \\ 0 & 1 & 0 \\ 0 & 0 & 1 \end{bmatrix}, E_4 = \begin{bmatrix} 1 & -2 & 0 \\ 0 & 1 & 0 \\ 0 & 0 & 1 \end{bmatrix}$

(b) $A^{-1} = \begin{bmatrix} -7 & -2 & 4 \\ 6 & 1 & -3 \\ -2 & 0 & 1 \end{bmatrix}$

(c) $A = \begin{bmatrix} 1 & 0 & 0 \\ 0 & 1 & 0 \\ 2 & 0 & 1 \end{bmatrix}\begin{bmatrix} 1 & 0 & 0 \\ 0 & 1 & 3 \\ 0 & 0 & 1 \end{bmatrix}\begin{bmatrix} 1 & 0 & 2 \\ 0 & 1 & 0 \\ 0 & 0 & 1 \end{bmatrix}\begin{bmatrix} 1 & 2 & 0 \\ 0 & 1 & 0 \\ 0 & 0 & 1 \end{bmatrix}$

A25 (a) $E_1 = \begin{bmatrix} 1/2 & 0 & 0 \\ 0 & 1 & 0 \\ 0 & 0 & 1 \end{bmatrix}, E_2 = \begin{bmatrix} 1 & 0 & 0 \\ 2 & 1 & 0 \\ 0 & 0 & 1 \end{bmatrix},$

$E_3 = \begin{bmatrix} 1 & 0 & 0 \\ 0 & 1 & 0 \\ -3 & 0 & 1 \end{bmatrix}, E_4 = \begin{bmatrix} 1 & 0 & 0 \\ 0 & 1 & 0 \\ 0 & 1 & 1 \end{bmatrix},$

$E_5 = \begin{bmatrix} 1 & 0 & 2 \\ 0 & 1 & 0 \\ 0 & 0 & 1 \end{bmatrix}$

(b) $A^{-1} = \begin{bmatrix} -1/2 & 2 & 2 \\ 1 & 1 & 0 \\ -1/2 & 1 & 1 \end{bmatrix}$

(c) $A = \begin{bmatrix} 2 & 0 & 0 \\ 0 & 1 & 0 \\ 0 & 0 & 1 \end{bmatrix}\begin{bmatrix} 1 & 0 & 0 \\ -2 & 1 & 0 \\ 0 & 0 & 1 \end{bmatrix}\begin{bmatrix} 1 & 0 & 0 \\ 0 & 1 & 0 \\ 3 & 0 & 1 \end{bmatrix}$
$\begin{bmatrix} 1 & 0 & 0 \\ 0 & 1 & 0 \\ 0 & -1 & 1 \end{bmatrix}\begin{bmatrix} 1 & 0 & -2 \\ 0 & 1 & 0 \\ 0 & 0 & 1 \end{bmatrix}$

A26 (a) $E_1 = \begin{bmatrix} 1 & 0 & 0 \\ -2 & 1 & 0 \\ 0 & 0 & 1 \end{bmatrix}, E_2 = \begin{bmatrix} 1 & 0 & 0 \\ 0 & 1 & 0 \\ -1 & 0 & 1 \end{bmatrix},$

$E_3 = \begin{bmatrix} 0 & 1 & 0 \\ 1 & 0 & 0 \\ 0 & 0 & 1 \end{bmatrix}, E_4 = \begin{bmatrix} 1 & 0 & 0 \\ 0 & 1 & 0 \\ -1 & 0 & 1 \end{bmatrix},$

$E_5 = \begin{bmatrix} 1 & 0 & 0 \\ 0 & 1 & -1 \\ 0 & 0 & 1 \end{bmatrix}, E_6 = \begin{bmatrix} 1 & 0 & 0 \\ 0 & 1/3 & 0 \\ 0 & 0 & 1 \end{bmatrix}$

(b) $A^{-1} = E_6E_5E_4E_3E_2E_1 = \begin{bmatrix} -2 & 1 & 0 \\ 0 & 1/3 & -1/3 \\ 1 & -1 & 1 \end{bmatrix}$

(c) $A = \begin{bmatrix} 1 & 0 & 0 \\ 2 & 1 & 0 \\ 0 & 0 & 1 \end{bmatrix}\begin{bmatrix} 1 & 0 & 0 \\ 0 & 1 & 0 \\ 1 & 0 & 1 \end{bmatrix}\begin{bmatrix} 0 & 1 & 0 \\ 1 & 0 & 0 \\ 0 & 0 & 1 \end{bmatrix}$
$\begin{bmatrix} 1 & 0 & 0 \\ 0 & 1 & 0 \\ 1 & 0 & 1 \end{bmatrix}\begin{bmatrix} 1 & 0 & 0 \\ 0 & 1 & 1 \\ 0 & 0 & 1 \end{bmatrix}\begin{bmatrix} 1 & 0 & 0 \\ 0 & 3 & 0 \\ 0 & 0 & 1 \end{bmatrix}$

A27 (a) $E_1 = \begin{bmatrix} 1 & 0 & 0 \\ 1 & 1 & 0 \\ 0 & 0 & 1 \end{bmatrix}, E_2 = \begin{bmatrix} 1 & 2 & 0 \\ 0 & 1 & 0 \\ 0 & 0 & 1 \end{bmatrix},$

$E_3 = \begin{bmatrix} 1 & 0 & 0 \\ 0 & 1 & 0 \\ 0 & -1 & 1 \end{bmatrix}, E_4 = \begin{bmatrix} 1 & 0 & 0 \\ 0 & 1 & 0 \\ 0 & 0 & 1/2 \end{bmatrix},$

$E_5 = \begin{bmatrix} 1 & 0 & -4 \\ 0 & 1 & 0 \\ 0 & 0 & 1 \end{bmatrix}$

(b) $A^{-1} = E_5E_4E_3E_2E_1 = \begin{bmatrix} 5 & 4 & -2 \\ 1 & 1 & 0 \\ -1/2 & -1/2 & 1/2 \end{bmatrix}$

(c) $A = \begin{bmatrix} 1 & 0 & 0 \\ -1 & 1 & 0 \\ 0 & 0 & 1 \end{bmatrix}\begin{bmatrix} 1 & -2 & 0 \\ 0 & 1 & 0 \\ 0 & 0 & 1 \end{bmatrix}\begin{bmatrix} 1 & 0 & 0 \\ 0 & 1 & 0 \\ 0 & 1 & 1 \end{bmatrix}$
$\begin{bmatrix} 1 & 0 & 0 \\ 0 & 1 & 0 \\ 0 & 0 & 2 \end{bmatrix}\begin{bmatrix} 1 & 0 & 4 \\ 0 & 1 & 0 \\ 0 & 0 & 1 \end{bmatrix}$

A28 (a) $E_1 = \begin{bmatrix} 1 & 0 & 0 \\ 2 & 1 & 0 \\ 0 & 0 & 1 \end{bmatrix}, E_2 = \begin{bmatrix} 1 & 0 & 0 \\ 0 & 1 & 0 \\ 4 & 0 & 1 \end{bmatrix},$

$E_3 = \begin{bmatrix} 1 & 0 & 0 \\ 0 & -1/4 & 0 \\ 0 & 0 & 1 \end{bmatrix}, E_4 = \begin{bmatrix} 1 & 0 & 0 \\ 0 & 0 & 1 \\ 0 & 1 & 0 \end{bmatrix},$

$E_5 = \begin{bmatrix} 1 & 0 & 1 \\ 0 & 1 & 0 \\ 0 & 0 & 1 \end{bmatrix}$

(b) $A^{-1} = \begin{bmatrix} 1/2 & -1/4 & 0 \\ 4 & 0 & 1 \\ -1/2 & -1/4 & 0 \end{bmatrix}$

(c) $A = \begin{bmatrix} 1 & 0 & 0 \\ -2 & 1 & 0 \\ 0 & 0 & 1 \end{bmatrix}\begin{bmatrix} 1 & 0 & 0 \\ 0 & 1 & 0 \\ -4 & 0 & 1 \end{bmatrix}\begin{bmatrix} 1 & 0 & 0 \\ 0 & -4 & 0 \\ 0 & 0 & 1 \end{bmatrix}$
$\begin{bmatrix} 1 & 0 & 0 \\ 0 & 0 & 1 \\ 0 & 1 & 0 \end{bmatrix}\begin{bmatrix} 1 & 0 & -1 \\ 0 & 1 & 0 \\ 0 & 0 & 1 \end{bmatrix}$

Section 3.7 Practice Problems

A1 (a) An elementary matrix is lower triangular if and only if it corresponds to an elementary row operation of the form $R_i + aR_j$ where $i > j$ or of the form aR_i. In the former case, the inverse elementary matrix corresponds to $R_i - aR_j$ and hence will be lower triangular. In the latter case, the elementary matrix corresponds to $\frac{1}{a}R_i$ and hence is also lower triangular.

(b) Assume $A, B \in M_{n\times n}(\mathbb{R})$ are both lower triangular matrices. Then, by definition, we have $a_{ij} = 0$ and $b_{ij} = 0$ whenever $i < j$. Hence, for any $i < j$ we have

$$\begin{aligned}(AB)_{ij} &= a_{i1}b_{1j} + \cdots + a_{ii}b_{ij} + a_{i(i+1)}b_{(i+1)j} + \cdots + a_{in}b_{nj}\\ &= a_{i1}(0) + \cdots + a_{ii}(0) + 0b_{(i+1)j} + \cdots + 0b_{nj}\\ &= 0\end{aligned}$$

So, (AB) is lower triangular.

A2 $\begin{bmatrix}1&0&0\\2&1&0\\-1&0&1\end{bmatrix}\begin{bmatrix}-2&-1&5\\0&2&-12\\0&0&8\end{bmatrix}$

A3 $\begin{bmatrix}1&0&0\\3&1&0\\2&3/2&1\end{bmatrix}\begin{bmatrix}1&-2&4\\0&4&-8\\0&0&-1\end{bmatrix}$

A4 $\begin{bmatrix}1&0&0\\1&1&0\\1&1/3&1\end{bmatrix}\begin{bmatrix}2&-4&5\\0&9&-3\\0&0&1\end{bmatrix}$

A5 $\begin{bmatrix}1&0&0&0\\-2&1&0&0\\0&1/2&1&0\\0&0&0&1\end{bmatrix}\begin{bmatrix}1&5&3&4\\0&4&5&11\\0&0&-7/2&-13/2\\0&0&0&0\end{bmatrix}$

A6 $\begin{bmatrix}1&0&0&0\\0&1&0&0\\3&-1&1&0\\0&-4/3&17/9&1\end{bmatrix}\begin{bmatrix}1&-2&1&1\\0&-3&-2&1\\0&0&-3&-3\\0&0&0&7\end{bmatrix}$

A7 $\begin{bmatrix}1&0&0&0\\-2&1&0&0\\-3/2&3/2&1&0\\-1&-2&2&1\end{bmatrix}\begin{bmatrix}-2&-1&2&0\\0&1&2&2\\0&0&4&0\\0&0&0&0\end{bmatrix}$

A8 $\begin{bmatrix}1&0\\-1/3&1\end{bmatrix}\begin{bmatrix}3&-2\\0&13/3\end{bmatrix}$; $\vec{x}_1 = \begin{bmatrix}1\\0\end{bmatrix}, \vec{x}_2 = \begin{bmatrix}38/13\\31/13\end{bmatrix}$

A9 $\begin{bmatrix}1&0&0\\-2&1&0\\-1&4&1\end{bmatrix}\begin{bmatrix}1&0&3\\0&1&3\\0&0&-4\end{bmatrix}$; $\vec{x}_1 = \begin{bmatrix}-3\\-4\\2\end{bmatrix}, \vec{x}_2 = \begin{bmatrix}5\\2\\-1\end{bmatrix}$

A10 $\begin{bmatrix}1&0&0\\-1&1&0\\3&1&1\end{bmatrix}\begin{bmatrix}1&0&-2\\0&-4&2\\0&0&3\end{bmatrix}$; $\vec{x}_1 = \begin{bmatrix}3\\3\\2\end{bmatrix}, \vec{x}_2 = \begin{bmatrix}-4\\-2\\-3\end{bmatrix}$

A11 $\begin{bmatrix}1&0&0\\-3&1&0\\-3&2&1\end{bmatrix}\begin{bmatrix}1&0&1\\0&2&2\\0&0&1\end{bmatrix}$; $\vec{x}_1 = \begin{bmatrix}3\\2\\0\end{bmatrix}, \vec{x}_2 = \begin{bmatrix}-3\\-3\\-1\end{bmatrix}$

A12 $\begin{bmatrix}1&0&0&0\\0&1&0&0\\-3&2&1&0\\-1&0&0&1\end{bmatrix}\begin{bmatrix}-1&2&-3&0\\0&-1&3&1\\0&0&-12&0\\0&0&0&1\end{bmatrix}$; $\vec{x}_1 = \begin{bmatrix}-1\\1\\3\\-1\end{bmatrix}, \vec{x}_2 = \begin{bmatrix}-5\\-3\\-2\\0\end{bmatrix}$

Chapter 3 Quiz

E1 $\begin{bmatrix} -14 & 1 & -17 \\ -1 & 10 & -39 \end{bmatrix}$

E2 Not defined.

E3 $\begin{bmatrix} -3 & -38 \\ 0 & -23 \\ -8 & -42 \end{bmatrix}$

E4 (a) $f_A(\vec{u}) = \begin{bmatrix} -11 \\ 0 \end{bmatrix}$, $f_A(\vec{v}) = \begin{bmatrix} -16 \\ 17 \end{bmatrix}$

(b) $\begin{bmatrix} -16 & -11 \\ 17 & 0 \end{bmatrix}$

E5 (a) $[R] = [R_{\pi/3}] = \begin{bmatrix} 1/2 & -\sqrt{3}/2 & 0 \\ \sqrt{3}/2 & 1/2 & 0 \\ 0 & 0 & 1 \end{bmatrix}$

(b) $[M] = [\text{refl}_{(-1,-1,2)}] = \frac{1}{3}\begin{bmatrix} 2 & -1 & 2 \\ -1 & 2 & 2 \\ 2 & 2 & -1 \end{bmatrix}$

(c) $[R \circ M] = \frac{1}{6}\begin{bmatrix} 2+\sqrt{3} & -1-2\sqrt{3} & 2-2\sqrt{3} \\ 2\sqrt{3}-1 & -\sqrt{3}+2 & 2\sqrt{3}+2 \\ 4 & 4 & -2 \end{bmatrix}$

E6 $\vec{x} = s\begin{bmatrix} -2 \\ 1 \\ 1 \\ 0 \\ 0 \end{bmatrix} + t\begin{bmatrix} -1 \\ 0 \\ 0 \\ 1 \\ 0 \end{bmatrix}$, $s, t \in \mathbb{R}$, and

$\vec{x} = \begin{bmatrix} 5 \\ 6 \\ 0 \\ 0 \\ 7 \end{bmatrix} + s\begin{bmatrix} -2 \\ 1 \\ 1 \\ 0 \\ 0 \end{bmatrix} + t\begin{bmatrix} -1 \\ 0 \\ 0 \\ 1 \\ 0 \end{bmatrix}$, $s, t \in \mathbb{R}$.

E7 (a) $\vec{u} \notin \text{Col}(B)$, $\vec{v} \in \text{Col}(B)$

(b) $\vec{x} = \begin{bmatrix} -1 \\ -2 \\ -3 \end{bmatrix}$ (c) $\vec{y} = \begin{bmatrix} 0 \\ 1 \\ 0 \end{bmatrix}$

E8 A basis for Row(A) is $\left\{\begin{bmatrix} 1 \\ 0 \\ 1 \\ 0 \\ -1 \end{bmatrix}, \begin{bmatrix} 0 \\ 1 \\ -1 \\ 0 \\ 3 \end{bmatrix}, \begin{bmatrix} 0 \\ 0 \\ 0 \\ 1 \\ 2 \end{bmatrix}\right\}$.

A basis for Col(A) is $\left\{\begin{bmatrix} 1 \\ 2 \\ 0 \\ 3 \end{bmatrix}, \begin{bmatrix} 0 \\ 1 \\ 2 \\ 3 \end{bmatrix}, \begin{bmatrix} 1 \\ 2 \\ 1 \\ 4 \end{bmatrix}\right\}$.

A basis for Null(A) is $\left\{\begin{bmatrix} -1 \\ 1 \\ 1 \\ 0 \\ 0 \end{bmatrix}, \begin{bmatrix} 1 \\ -3 \\ 0 \\ -2 \\ 1 \end{bmatrix}\right\}$.

A basis for Null(A^T) is $\left\{\begin{bmatrix} -1 \\ -1 \\ -1 \\ 1 \end{bmatrix}\right\}$.

E9 $A^{-1} = \begin{bmatrix} 2/3 & 0 & 0 & 1/3 \\ 1/6 & 0 & 1/2 & -1/6 \\ 0 & 1 & 0 & 0 \\ -1/3 & 0 & 0 & 1/3 \end{bmatrix}$

E10 The matrix is invertible only for $p \neq 1$. The inverse is $\frac{1}{1-p}\begin{bmatrix} 1 & p & -p \\ -1 & 1-2p & p \\ -1 & -1 & 1 \end{bmatrix}$.

E11 By definition, the range of L is a subset of $\mathbb{R}^m$. We have $L(\vec{0}) = \vec{0}$, so $\vec{0} \in \text{Range}(L)$. If $\vec{x}, \vec{y} \in \text{Range}(L)$, then there exists $\vec{u}, \vec{v} \in \mathbb{R}^n$ such that $L(\vec{u}) = \vec{x}$ and $L(\vec{v}) = \vec{y}$. Hence, $L(\vec{u}+\vec{v}) = L(\vec{u}) + L(\vec{v}) = \vec{x} + \vec{y}$, so $\vec{x} + \vec{y} \in \text{Range}(L)$. Similarly, $L(t\vec{u}) = tL(\vec{u}) = t\vec{x}$, so $t\vec{x} \in \text{Range}(L)$. Thus, L is a subspace of $\mathbb{R}^m$.

E12 Consider $c_1L(\vec{v}_1) + \cdots + c_kL(\vec{v}_k) = \vec{0}$. Since L is linear, we get $L(c_1\vec{v}_1 + \cdots + c_k\vec{v}_k) = \vec{0}$. Thus, $c_1\vec{v}_1 + \cdots + c_k\vec{v}_k \in$ Null(L) and so $c_1\vec{v}_1 + \cdots + c_k\vec{v}_k = \vec{0}$. This implies that $c_1 = \cdots = c_k = 0$ since $\{\vec{v}_1, \dots, \vec{v}_k\}$ is linearly independent. Therefore, $\{L(\vec{v}_1), \dots, L(\vec{v}_k)\}$ is linearly independent.

E13 (a) $E_1 = \begin{bmatrix} 1 & 0 & 0 \\ 0 & 1/2 & 0 \\ 0 & 0 & 1 \end{bmatrix}$, $E_2 = \begin{bmatrix} 1 & 0 & 0 \\ 0 & 1 & 0 \\ 0 & 0 & 1/4 \end{bmatrix}$,

$E_3 = \begin{bmatrix} 1 & 0 & 2 \\ 0 & 1 & 0 \\ 0 & 0 & 1 \end{bmatrix}$, $E_4 = \begin{bmatrix} 1 & 0 & 0 \\ 0 & 1 & 3/2 \\ 0 & 0 & 1 \end{bmatrix}$

(b) $A = \begin{bmatrix} 1 & 0 & 0 \\ 0 & 2 & 0 \\ 0 & 0 & 1 \end{bmatrix}\begin{bmatrix} 1 & 0 & 0 \\ 0 & 1 & 0 \\ 0 & 0 & 4 \end{bmatrix}\begin{bmatrix} 1 & 0 & -2 \\ 0 & 1 & 0 \\ 0 & 0 & 1 \end{bmatrix}\begin{bmatrix} 1 & 0 & 0 \\ 0 & 1 & -3/2 \\ 0 & 0 & 1 \end{bmatrix}$

E14 $K = I_3$

E15 KM cannot equal MK for any matrix K.

E16 The range cannot be spanned by $\begin{bmatrix} 1 \\ 1 \end{bmatrix}$ because this vector is not in $\mathbb{R}^3$.

E17 The matrix of L is any multiple of $\begin{bmatrix} 1 & -2/3 \\ 1 & -2/3 \\ 2 & -4/3 \end{bmatrix}$

E18 This contradicts the Rank Theorem, so there can be no such mapping L.

E19 This contradicts Theorem 3.5.2, so there can be no such matrix.

CHAPTER 4

Section 4.1 Practice Problems

A1 $-1 - 6x + 4x^2 + 6x^3$

A2 $-3 + 6x - 6x^2 - 3x^3 - 12x^4$

A3 $-1 + 9x - 11x^2 - 17x^3$

A4 $-3 + 2x + 6x^2$

A5 $7 - 2x - 5x^2$

A6 $\frac{7}{3} - \frac{4}{3}x + \frac{13}{3}x^2$

A7 $\sqrt{2} - \pi + \sqrt{2}x + (\sqrt{2} + \pi)x^2$

A8 $0 = 0(1+x^2+x^3)+0(2+x+x^3)+0(-1+x+2x^2+x^3)$.

A9 $2 + 4x + 3x^2 + 4x^3$ is not in the span.

A10 $$-x + 2x^2 + x^3 = 2(1 + x^2 + x^3) + (-1)(2 + x + x^3) + 0(-1 + x + 2x^2 + x^3)$$

A11 $$-4 - x + 3x^2 = 1(1 + x^2 + x^3) + (-2)(2 + x + x^3) + 1(-1 + x + 2x^2 + x^3)$$

A12 $$-1 - 7x + 5x^2 + 4x^3 = (-3)(1 + x^2 + x^3) + 3(2 + x + x^3) + 4(-1 + x + 2x^2 + x^3)$$

A13 $2 + x + 5x^3$ is not in the span.

A14 The set is linearly independent.

A15 The set is linearly dependent.
$$0 = (-3t)(1 + x + x^2) + tx + t(x^2 + x^3) + t(3 + 2x + 2x^2 - x^3), t \in \mathbb{R}$$

A16 The set is linearly independent.

A17 The set is linearly dependent.
$$0 = (-2t)(1 + x + x^3 + x^4) + t(2 + x - x^2 + x^3 + x^4) + t(x + x^2 + x^3 + x^4), t \in \mathbb{R}$$

A18 Consider
$$a_1 + a_2x + a_3x^2 = t_1 1 + t_2(x-1) + t_3(x-1)^2$$
The augmented matrix is $\left[\begin{array}{ccc|c} 1 & -1 & 1 & a_1 \\ 0 & 1 & -2 & a_2 \\ 0 & 0 & 1 & a_3 \end{array}\right]$.
Since there is a leading 1 in each row, the system is consistent for all polynomials $a_1 + a_2x + a_3x^2$. Thus, $\mathcal{B}$ spans $P_2(\mathbb{R})$. Since there is a leading 1 in each column, there is a unique solution and so $\mathcal{B}$ is also linearly independent. Therefore, it is a basis for $P_2(\mathbb{R})$.

Section 4.2 Practice Problems

A1 Subspace
A2 Subspace
A3 Subspace
A4 Not a subspace
A5 Not a subspace
A6 Subspace
A7 Not a subspace
A8 Subspace
A9 Not a subspace
A10 Subspace
A11 Subspace
A12 Subspace
A13 Subspace
A14 Subspace
A15 Not a subspace
A16 Subspace
A17 Subspace
A18 Not a subspace
A19 Subspace
A20 Not a subspace

Section 4.3 Practice Problems

A1 Basis
A2 Not a basis
A3 Not a basis
A4 Not a basis
A5 Basis
A6 Not a basis
A7 Not a basis
A8 Basis
A9 $\dim \operatorname{Span} \mathcal{B} = 2$
A10 $\dim \operatorname{Span} \mathcal{B} = 3$
A11 $\dim \operatorname{Span} \mathcal{B} = 3$
A12 $\dim \operatorname{Span} \mathcal{B} = 2$
A13 $\dim \operatorname{Span} \mathcal{B} = 4$
A14 $\dim \operatorname{Span} \mathcal{B} = 3$
A15 $\dim \operatorname{Span} \mathcal{B} = 3$
A16 $\dim \operatorname{Span} \mathcal{B} = 4$

A17 Other correct answers are possible.

(a) $\left\{\begin{bmatrix}1\\2\\0\end{bmatrix}, \begin{bmatrix}1\\0\\2\end{bmatrix}\right\}$

(b) $\left\{\begin{bmatrix}1\\2\\0\end{bmatrix}, \begin{bmatrix}1\\0\\2\end{bmatrix}, \begin{bmatrix}2\\-1\\-1\end{bmatrix}\right\}$

A18 Other correct answers are possible.

(a) $\left\{\begin{bmatrix}1\\0\\0\\1\end{bmatrix},\begin{bmatrix}1\\0\\-1\\0\end{bmatrix},\begin{bmatrix}1\\1\\0\\0\end{bmatrix}\right\}$ (b) $\left\{\begin{bmatrix}1\\0\\0\\1\end{bmatrix},\begin{bmatrix}1\\0\\-1\\0\end{bmatrix},\begin{bmatrix}1\\1\\0\\0\end{bmatrix},\begin{bmatrix}1\\0\\0\\0\end{bmatrix}\right\}$

A19 $\{1+x+x^3, 1+x^2, x^2, x\}$

A20 $\left\{\begin{bmatrix}1&2\\2&1\end{bmatrix},\begin{bmatrix}1&2\\3&4\end{bmatrix},\begin{bmatrix}1&0\\0&0\end{bmatrix},\begin{bmatrix}0&1\\0&0\end{bmatrix}\right\}$

A21 $\left\{\begin{bmatrix}1&0&0\\0&0&0\end{bmatrix},\begin{bmatrix}0&1&0\\0&0&0\end{bmatrix},\begin{bmatrix}0&0&1\\0&0&0\end{bmatrix},\begin{bmatrix}0&0&0\\1&0&0\end{bmatrix},\begin{bmatrix}0&0&0\\0&1&0\end{bmatrix},\begin{bmatrix}0&0&0\\0&0&1\end{bmatrix}\right\}$

A22 $\{x, 1-x^2\}$, $\dim \mathbb{S} = 2$

A23 $\{1+x^3, x-2x^3, x^2\}$, $\dim \mathbb{S} = 3$

A24 $\left\{\begin{bmatrix}1&0\\0&0\end{bmatrix},\begin{bmatrix}0&1\\0&0\end{bmatrix},\begin{bmatrix}0&0\\0&1\end{bmatrix}\right\}$, $\dim \mathbb{S} = 3$

A25 $\left\{\begin{bmatrix}1\\0\\-1\end{bmatrix},\begin{bmatrix}0\\1\\-1\end{bmatrix}\right\}$, $\dim \mathbb{S} = 2$

A26 $\{x^2-5x+6\}$, $\dim \mathbb{S} = 1$

A27 $\left\{\begin{bmatrix}-1&-1\\1&0\end{bmatrix},\begin{bmatrix}0&0\\0&1\end{bmatrix}\right\}$, $\dim \mathbb{S} = 2$

Section 4.4 Practice Problems

A1 (a) Show that it is linearly independent and spans the plane.

(b) $\begin{bmatrix}3\\2\\1\end{bmatrix}$ and $\begin{bmatrix}5\\2\\3\end{bmatrix}$ are not in the plane. $\begin{bmatrix}3\\2\\2\end{bmatrix}_{\mathcal{B}} = \begin{bmatrix}3\\-2\end{bmatrix}$

A2 (a) $[\mathbf{q}]_{\mathcal{B}} = \begin{bmatrix}4\\1\\1\end{bmatrix}$, $[\mathbf{r}]_{\mathcal{B}} = \begin{bmatrix}2\\-1\\3\end{bmatrix}$

(b) $[2-4x+10x^2]_{\mathcal{B}} = \begin{bmatrix}6\\0\\4\end{bmatrix}$

A3 $[\mathbf{x}]_{\mathcal{B}} = \begin{bmatrix}1\\2\end{bmatrix}$, $[\mathbf{y}]_{\mathcal{B}} = \begin{bmatrix}-19/11\\-29/11\end{bmatrix}$

A4 $[\mathbf{x}]_{\mathcal{B}} = \begin{bmatrix}2\\1\end{bmatrix}$, $[\mathbf{y}]_{\mathcal{B}} = \begin{bmatrix}32/11\\35/11\end{bmatrix}$

A5 $[\mathbf{x}]_{\mathcal{B}} = \begin{bmatrix}9\\-6\\-1\end{bmatrix}$, $[\mathbf{y}]_{\mathcal{B}} = \begin{bmatrix}5\\-3\\-2\end{bmatrix}$

A6 $[\mathbf{x}]_{\mathcal{B}} = \begin{bmatrix}3\\-2\end{bmatrix}$, $[\mathbf{y}]_{\mathcal{B}} = \begin{bmatrix}2\\3\end{bmatrix}$

A7 $[\mathbf{p}]_{\mathcal{B}} = \begin{bmatrix}1\\2\end{bmatrix}$, $[\mathbf{q}]_{\mathcal{B}} = \begin{bmatrix}5/11\\3/11\end{bmatrix}$

A8 $[\mathbf{p}]_{\mathcal{B}} = \begin{bmatrix}5\\1\\-2\end{bmatrix}$, $[\mathbf{q}]_{\mathcal{B}} = \begin{bmatrix}2\\2\\-2\end{bmatrix}$

A9 $[\mathbf{p}]_{\mathcal{B}} = \begin{bmatrix}5\\-3\\4\end{bmatrix}$, $[\mathbf{q}]_{\mathcal{B}} = \begin{bmatrix}-1\\2\\-1\end{bmatrix}$

A10 $[A]_{\mathcal{B}} = \begin{bmatrix}-2\\3\\1\end{bmatrix}$, $[B]_{\mathcal{B}} = \begin{bmatrix}-2\\3\\-1\end{bmatrix}$

A11 $[A]_{\mathcal{B}} = \begin{bmatrix}1\\1\end{bmatrix}$, $[B]_{\mathcal{B}} = \begin{bmatrix}3\\-2\end{bmatrix}$

A12 $[A]_{\mathcal{B}} = \begin{bmatrix}-1/2\\1/2\\3/4\end{bmatrix}$, $[B]_{\mathcal{B}} = \begin{bmatrix}1\\2\\-1/2\end{bmatrix}$

A13 $[A]_{\mathcal{B}} = \begin{bmatrix}1\\2\\-4\end{bmatrix}$, $[B]_{\mathcal{B}} = \begin{bmatrix}1\\-1\\-2\end{bmatrix}$

A14 $Q = \begin{bmatrix}1&0\\1&2\end{bmatrix}$, $P = \begin{bmatrix}1&0\\-1/2&1/2\end{bmatrix}$

A15 $Q = \begin{bmatrix}3&0&-2\\4&1&-3\\1&0&3\end{bmatrix}$, $P = \begin{bmatrix}3/11&0&2/11\\-15/11&1&1/11\\-1/11&0&3/11\end{bmatrix}$

A16 $Q = \begin{bmatrix}1&1&1\\0&-2&-4\\0&0&4\end{bmatrix}$, $P = \begin{bmatrix}1&1/2&1/4\\0&-1/2&-1/2\\0&0&1/4\end{bmatrix}$

A17 $Q = \begin{bmatrix}1&0&1\\2&1&3\\1&1&0\end{bmatrix}$, $P = \begin{bmatrix}3/2&-1/2&1/2\\-3/2&1/2&1/2\\-1/2&1/2&-1/2\end{bmatrix}$

A18 $Q = \begin{bmatrix}1&1&0\\-2&0&1\\5&-2&1\end{bmatrix}$, $P = \begin{bmatrix}2/9&-1/9&1/9\\7/9&1/9&-1/9\\4/9&7/9&2/9\end{bmatrix}$

A19 $Q = \begin{bmatrix} 0 & 0 & 0 & 1 \\ 0 & 0 & 1 & 0 \\ 1 & 0 & 0 & 0 \\ 0 & 1 & 0 & 0 \end{bmatrix}, P = \begin{bmatrix} 0 & 0 & 1 & 0 \\ 0 & 0 & 0 & 1 \\ 0 & 1 & 0 & 0 \\ 1 & 0 & 0 & 0 \end{bmatrix}$

A20 $Q = \begin{bmatrix} 1 & 0 & 2 \\ -1 & -4 & 1 \\ -1 & -1 & 1 \end{bmatrix}, P = \begin{bmatrix} 1/3 & 2/9 & -8/9 \\ 0 & -1/3 & 1/3 \\ 1/3 & -1/9 & 4/9 \end{bmatrix}$

A21 $Q = \begin{bmatrix} -1 & 5 \\ 1 & -1 \end{bmatrix}, P = \begin{bmatrix} 1/4 & 5/4 \\ 1/4 & 1/4 \end{bmatrix}$

A22 $Q = \begin{bmatrix} 19 & 13 \\ 4 & 3 \end{bmatrix}, P = \begin{bmatrix} 3/5 & -13/5 \\ -4/5 & 19/5 \end{bmatrix}$

A23 $Q = \begin{bmatrix} 1/2 & -1/2 & 1 \\ 0 & 1/4 & -3/4 \\ 1/2 & -3/4 & 3/4 \end{bmatrix}, P = \begin{bmatrix} 3 & 3 & -1 \\ 3 & 1 & -3 \\ 1 & -1 & -1 \end{bmatrix}$

Section 4.5 Practice Problems

A1 Show that $L(s\mathbf{x} + t\mathbf{y}) = sL(\mathbf{x}) + tL(\mathbf{y})$.

A2 Show that $L(s\mathbf{x} + t\mathbf{y}) = sL(\mathbf{x}) + tL(\mathbf{y})$.

A3 Show that $\text{tr}(s\mathbf{x} + t\mathbf{y}) = s\,\text{tr}(\mathbf{x}) + t\,\text{tr}(\mathbf{y})$.

A4 Show that $T(s\mathbf{x} + t\mathbf{y}) = sT(\mathbf{x}) + tT(\mathbf{y})$.

A5 Not linear

A6 Linear

A7 Not linear

A8 Linear

A9 $\mathbf{y} \in \text{Range}(L)$. One choice is $\mathbf{x} = \begin{bmatrix} 1 \\ 2 \\ 1 \end{bmatrix}$.

A10 $\mathbf{y} \in \text{Range}(L)$. $\vec{x} = (2 - t)1 + (3 - t)x + tx^2$

A11 $\mathbf{y} \notin \text{Range}(L)$.

A12 $\mathbf{y} \notin \text{Range}(L)$.

Other correct answers are possible for problems A13–A18.

A13 A basis for Range(L) is $\left\{\begin{bmatrix} 1 \\ 1 \end{bmatrix}, \begin{bmatrix} 0 \\ 1 \end{bmatrix}\right\}$.

A basis for Null(L) is $\left\{\begin{bmatrix} 1 \\ -1 \\ 0 \end{bmatrix}\right\}$.

A14 A basis for Range(L) is $\{1 + x, x\}$.

A basis for Null(L) is $\left\{\begin{bmatrix} 1 \\ -1 \\ 0 \end{bmatrix}\right\}$.

A15 A basis for Range(L) is $\{1 + x^2\}$.
A basis for Null(L) is $\{2 + x\}$.

A16 A basis for Range(L) is $\left\{\begin{bmatrix} 1 \\ 2 \\ 1 \end{bmatrix}, \begin{bmatrix} -1 \\ 1 \\ -1 \end{bmatrix}, \begin{bmatrix} 0 \\ 0 \\ 1 \end{bmatrix}\right\}$.

A basis for Null(L) is the empty set.

A17 A basis for Range(tr) is $\{1\}$.

A basis for Null(tr) is $\left\{\begin{bmatrix} 1 & 0 \\ 0 & -1 \end{bmatrix}, \begin{bmatrix} 0 & 1 \\ 0 & 0 \end{bmatrix}, \begin{bmatrix} 0 & 0 \\ 1 & 0 \end{bmatrix}\right\}$.

A18 A basis for Range(L) is the standard basis for $M_{2\times2}(\mathbb{R})$. A basis for Null(L) is the empty set.

A19 They are not inverses of each other.

A20 They are inverses of each other.

A21 $L\left(\begin{bmatrix} a_1 \\ a_2 \\ a_3 \end{bmatrix}\right) = a_3 + (2a_2 + a_3)x + (a_1 + a_3)x^2$

A22 $L\left(a + bx + cx^2\right) = \begin{bmatrix} a & b \\ 0 & c \end{bmatrix}$

A23 One possible answer is $L\left(\begin{bmatrix} a & b \\ c & d \end{bmatrix}\right) = \begin{bmatrix} 0 \\ a \\ d \\ 0 \end{bmatrix}$

Section 4.6 Practice Problems

A1 $[L]_{\mathcal{B}} = \begin{bmatrix} 0 & 2 \\ 1 & -1 \end{bmatrix}, [L(\mathbf{x})]_{\mathcal{B}} = \begin{bmatrix} 6 \\ 1 \end{bmatrix}$

A2 $[L]_{\mathcal{B}} = \begin{bmatrix} 2 & 2 & 0 \\ 0 & 0 & 4 \\ -1 & -1 & 5 \end{bmatrix}, [L(\mathbf{x})]_{\mathcal{B}} = \begin{bmatrix} 12 \\ -4 \\ -11 \end{bmatrix}$

A3 $[L]_{\mathcal{B}} = \begin{bmatrix} -3 & 0 \\ 0 & 4 \end{bmatrix}$

A4 $[L]_{\mathcal{B}} = \begin{bmatrix} 0 & 2 \\ 1 & 0 \end{bmatrix}$

A5 $[L]_{\mathcal{B}} = \begin{bmatrix} 3 & 0 \\ 0 & 1 \end{bmatrix}, [L(\vec{x})]_{\mathcal{B}} = \begin{bmatrix} 3 \\ 1 \end{bmatrix}$

A6 $[L]_{\mathcal{B}} = \begin{bmatrix} 0 & 0 \\ -7 & 5 \end{bmatrix}, [L(\vec{x})]_{\mathcal{B}} = \begin{bmatrix} 0 \\ 6 \end{bmatrix}$

A7 $[L]_\mathcal{B} = \begin{bmatrix} 11 & 16 \\ -4 & -3 \end{bmatrix}, [L(\vec{x})]_\mathcal{B} = \begin{bmatrix} 5 \\ 1 \end{bmatrix}$

A8 $[L]_\mathcal{B} = \begin{bmatrix} 1 & -2 & -2 \\ 0 & 2 & 1 \\ 1 & 1 & 0 \end{bmatrix}, [L(\vec{x})]_\mathcal{B} = \begin{bmatrix} -3 \\ 3 \\ 2 \end{bmatrix}$

A9 $[L]_\mathcal{B} = \begin{bmatrix} 4 & 4 & 1 \\ -3 & -4 & -4 \\ 2 & 3 & 4 \end{bmatrix}, [L(\vec{x})]_\mathcal{B} = \begin{bmatrix} -15 \\ 2 \\ 4 \end{bmatrix}$

A10 $\mathcal{B} = \left\{ \begin{bmatrix} 1 \\ -2 \end{bmatrix}, \begin{bmatrix} 2 \\ 1 \end{bmatrix} \right\}, [\text{refl}]_\mathcal{B} = \begin{bmatrix} -1 & 0 \\ 0 & 1 \end{bmatrix}$

A11 $\mathcal{B} = \left\{ \begin{bmatrix} 1 \\ -2 \end{bmatrix}, \begin{bmatrix} 2 \\ 1 \end{bmatrix} \right\}, [\text{perp}_{(1,-2)}]_\mathcal{B} = \begin{bmatrix} 0 & 0 \\ 0 & 1 \end{bmatrix}$

A12 $\mathcal{B} = \left\{ \begin{bmatrix} 2 \\ 1 \\ -1 \end{bmatrix}, \begin{bmatrix} 1 \\ 0 \\ 2 \end{bmatrix}, \begin{bmatrix} 0 \\ 1 \\ 1 \end{bmatrix} \right\}, [\text{proj}]_\mathcal{B} = \begin{bmatrix} 1 & 0 & 0 \\ 0 & 0 & 0 \\ 0 & 0 & 0 \end{bmatrix}$

A13 $\mathcal{B} = \left\{ \begin{bmatrix} -1 \\ -1 \\ 1 \end{bmatrix}, \begin{bmatrix} 1 \\ 0 \\ 1 \end{bmatrix}, \begin{bmatrix} 0 \\ 1 \\ 1 \end{bmatrix} \right\}, [\text{refl}]_\mathcal{B} = \begin{bmatrix} -1 & 0 & 0 \\ 0 & 1 & 0 \\ 0 & 0 & 1 \end{bmatrix}$

A14 (a) $\begin{bmatrix} 1 \\ 2 \\ 4 \end{bmatrix}_\mathcal{B} = \begin{bmatrix} 2 \\ -1 \\ 1 \end{bmatrix}$ (b) $[L]_\mathcal{B} = \begin{bmatrix} 2 & 0 & 0 \\ -1 & 0 & 2 \\ 1 & 1 & 0 \end{bmatrix}$

(c) $L(1,2,4) = \begin{bmatrix} 4 \\ 1 \\ 6 \end{bmatrix}$

A15 (a) $\begin{bmatrix} -1 \\ 7 \\ 6 \end{bmatrix}_\mathcal{B} = \begin{bmatrix} -3 \\ 2 \\ 3 \end{bmatrix}$ (b) $[L]_\mathcal{B} = \begin{bmatrix} 0 & -2 & 2 \\ 0 & 0 & 3 \\ 1 & 0 & -3 \end{bmatrix}$

(c) $L(-1,7,6) = \begin{bmatrix} 11 \\ 6 \\ -14 \end{bmatrix}$

A16 $[L]_\mathcal{B} = \begin{bmatrix} 11 & 16 \\ -4 & -3 \end{bmatrix}$ **A17** $[L]_\mathcal{B} = \begin{bmatrix} 5 & 0 \\ 0 & -5 \end{bmatrix}$

A18 $[L]_\mathcal{B} = \begin{bmatrix} -40 & -118 \\ 18 & 52 \end{bmatrix}$ **A19** $[L]_\mathcal{B} = \begin{bmatrix} 4 & 0 \\ 0 & 6 \end{bmatrix}$

A20 $[L]_\mathcal{B} = \begin{bmatrix} 4 & 2 & 0 \\ 0 & 4 & 2 \\ 0 & 0 & 4 \end{bmatrix}$ **A21** $[L]_\mathcal{B} = \begin{bmatrix} 2 & 0 & 0 \\ 0 & -2 & 0 \\ 0 & 0 & 3 \end{bmatrix}$

A22 $[L]_\mathcal{B} = \begin{bmatrix} 2 & 1 & 0 \\ 0 & 1 & 1 \\ -2 & -3 & -2 \end{bmatrix}$ **A23** $[L]_\mathcal{B} = \begin{bmatrix} 1 & -1 & 1 \\ 0 & 2 & -1 \\ 0 & 2 & -1 \end{bmatrix}$

A24 $[D]_\mathcal{B} = \begin{bmatrix} 0 & 1 & 0 \\ 0 & 0 & 2 \\ 0 & 0 & 0 \end{bmatrix}$ **A25** $[T]_\mathcal{B} = \begin{bmatrix} -1 & -1 & -2 \\ 0 & 0 & -1 \\ 2 & 2 & 4 \end{bmatrix}$

Section 4.7 Practice Problems

A1 One-to-one, onto

A2 One-to-one, not onto

A3 Not one-to-one, onto

A4 Not one-to-one, not onto

A5 One-to-one, onto

A6 Not one-to-one, not onto

A7 Not one-to-one, not onto

A8 Not one-to-one, not onto

A9 One-to-one, onto

A10 One-to-one, not onto

A11 Not one-to-one, not onto

A12 One-to-one, onto

A13 They are not isomorphic.

A14 Define $L(a + bx + cx^2 + dx^3) = \begin{bmatrix} a \\ b \\ c \\ d \end{bmatrix}$.

A15 Define $L\left(\begin{bmatrix} a & b \\ c & d \end{bmatrix}\right) = \begin{bmatrix} a \\ b \\ c \\ d \end{bmatrix}$.

A16 Define $L(a + bx + cx^2 + dx^3) = \begin{bmatrix} a & b \\ c & d \end{bmatrix}$.

A17 If $\mathcal{B} = \{\vec{v}_1, \vec{v}_2\}$ is a basis for the plane P, then $L(a\vec{v}_1 + b\vec{v}_2) = \begin{bmatrix} a \\ b \end{bmatrix}$.

A18 $L(\begin{bmatrix} a_1 & 0 \\ 0 & a_2 \end{bmatrix}) = a_1 + a_2x$

A19 They are not isomorphic.

A20 They are not isomorphic.

A21 $L((x-1)(a_1x + a_0)) = \begin{bmatrix} a_2 & a_1 \\ 0 & a_0 \end{bmatrix}$

Chapter 4 Quiz

E1 It is a vector space.

E2 It is a vector space.

E3 It is not a vector space.

E4 It is a vector space.

E5 $\mathbb{S}$ is not a subspace.

E6 $\left\{\begin{bmatrix}-1 & 0\\ 1 & 0\end{bmatrix}, \begin{bmatrix}-2 & 0\\ 0 & 1\end{bmatrix}\right\}$ is a basis for $\mathbb{S}$.

E7 $\left\{1+x^2\right\}$ is a basis for $\mathbb{S}$.

E8 $\left\{1+x^2, 2+x\right\}$ is a basis for $\mathbb{S}$.

E9 Not a basis since it is linearly dependent.

E10 Not a basis since it is linearly dependent.

E11 Not a basis since it does not span $M_{2\times 2}(\mathbb{R})$.

E12 (a) $\mathcal{B} = \left\{\begin{bmatrix}1\\0\\1\\1\end{bmatrix}, \begin{bmatrix}1\\1\\0\\1\end{bmatrix}, \begin{bmatrix}3\\3\\1\\0\end{bmatrix}\right\}$, $\dim \mathbb{S} = 3$.

(b) $[\vec{x}]_{\mathcal{B}} = \begin{bmatrix}-2\\-1\\1\end{bmatrix}$

E13 $Q = \begin{bmatrix}-5/3 & 5/3\\ 4/3 & -1/3\end{bmatrix}$, $P = \begin{bmatrix}1/5 & 1\\ 4/5 & 1\end{bmatrix}$

E14 (a) $\left\{\begin{bmatrix}0\\1\\0\end{bmatrix}, \begin{bmatrix}1\\0\\1\end{bmatrix}\right\}$

(b) $\left\{\begin{bmatrix}0\\1\\0\end{bmatrix}, \begin{bmatrix}1\\0\\1\end{bmatrix}, \begin{bmatrix}1\\0\\-1\end{bmatrix}\right\}$

(c) $[L]_{\mathcal{B}} = \begin{bmatrix}1 & 0 & 0\\ 0 & 1 & 0\\ 0 & 0 & -1\end{bmatrix}$

(d) $[L]_S = P[L]_{\mathcal{B}}P^{-1} = \begin{bmatrix}0 & 0 & 1\\ 0 & 1 & 0\\ 1 & 0 & 0\end{bmatrix}$

E15 $[L]_{\mathcal{B}} = \begin{bmatrix}0 & 2/3 & 11/3\\ -1 & 2/3 & -10/3\\ 0 & 1/3 & 1/3\end{bmatrix}$

E16 We have $t_1\mathbf{v}_1 + \cdots + t_k\mathbf{v}_k \in \text{Null}(L)$ since

$$\mathbf{0} = L(t_1\mathbf{v}_1 + \cdots + t_k\mathbf{v}_k) = L(t_1\mathbf{v}_1 + \cdots + t_k\mathbf{v}_k)$$

Thus, $t_1\mathbf{v}_1+\cdots+t_k\mathbf{v}_k = \mathbf{0}$. Since $\{\mathbf{v}_1, \dots, \mathbf{v}_k\}$ is linearly independent, this gives $t_1 = \cdots = t_k = 0$. Hence, $\{L(\mathbf{v}_1), \dots, L(\mathbf{v}_k)\}$ is also linearly independent.

E17 FALSE. $\mathbb{R}^n$ is an n-dimensional subspace of $\mathbb{R}^n$.

E18 TRUE. The dimension of $P_2(\mathbb{R})$ is 3, so a set of 4 polynomials in $P_2(\mathbb{R})$ must be linearly dependent.

E19 FALSE. The number of components in a coordinate vector is the number of vectors in the basis. So, if $\mathcal{B}$ is a basis for a 4-dimensional subspace, then the $\mathcal{B}$-coordinate vector would have only 4 components.

E20 TRUE. Both ranks equal the dimension of Range(L).

E21 FALSE. If $L : P_2(\mathbb{R}) \to P_2(\mathbb{R})$ is a linear mapping, then the range of L is a subspace of $P_2(\mathbb{R})$, but the column space of $[L]_{\mathcal{B}}$ is a subspace of $\mathbb{R}^3$. Hence, they cannot equal.

E22 FALSE. The mapping $L : \mathbb{R} \to \mathbb{R}^2$ given by $L(x_1) = (x_1, 0)$ is one-to-one, but $\dim \mathbb{R} \neq \dim \mathbb{R}^2$.

CHAPTER 5

Section 5.1 Practice Problems

A1 38 | A2 −5 | A3 0
A4 0 | A5 64 | A6 17
A7 0 | A8 0 | A9 −20
A10 3 | A11 3 | A12 −5
A13 5 | A14 196 | A15 −136

A16 −26 | A17 98 | A18 0
A19 20 | A20 −72 | A21 72
A22 18 | A23 −90 | A24 48
A25 18 | A26 76 | A27 420
A28 −1 | A29 1 | A30 −3
A31 1

Section 5.2 Practice Problems

A1 $\det A = 30$, so A is invertible.

A2 $\det A = 1$, so A is invertible.

A3 $\det A = 8$, so A is invertible.

A4 $\det A = 0$, so A is not invertible.

A5 $\det A = -1120$, so A is invertible.

A6 $\det A = 12$, so A is invertible.

A7 0 **A8** 0 **A9** 36

A10 6 **A11** 14 **A12** -12

A13 -5 **A14** 716 **A15** -65

A16 32 **A17** 0 **A18** 448

A19 50

A20 $(b-a)(c-a)(d-a)(c-b)(d-b)(d-c)$

A21 $\det A = 4 - 2p$, so A is invertible for all $p \neq 2$.

A22 $\det A = p^2 + 2 > 0$, so A is invertible for all p.

A23 $\det A = p$, so A is invertible for all $p \neq 0$.

A24 $\det A = 6p + 32$, so A is invertible for all $p \neq -\frac{16}{3}$.

A25 $\det A = 3p - 14$, so A is invertible for all $p \neq \frac{14}{3}$.

A26 $\det A = -5p - 20$, so A is invertible for all $p \neq -4$.

A27 $\det A = 2p - 116$, so A is invertible for all $p \neq 58$.

A28 $\det A = 0$, so there is no value of p.

A29 $\det A = 13$, $\det B = 14$, $\det AB = 182$

A30 $\det A = -2$, $\det B = 56$, $\det AB = -112$

A31 $\lambda = 1, 7$ **A32** $\lambda = -\sqrt{7}, \sqrt{7}$

A33 $\lambda = -1, 0, 2$ **A34** $\lambda = 0, 2, 6$

A35 $\lambda = -1, 1$ **A36** $\lambda = 2, 8$

A37 Since rA is the matrix where each of the n rows of A must been multiplied by r, we can use Theorem 5.2.1 n times to get $\det(rA) = r^n \det A$.

A38 We have AA^{-1} is I, so

$$1 = \det I = \det AA^{-1} = (\det A)(\det A^{-1})$$

by Theorem 5.2.7. Since $\det A \neq 0$, we get $\det A^{-1} = \dfrac{1}{\det A}$.

A39 By Theorem 5.2.7., we have

$$1 = \det I = \det A^3 = (\det A)^3$$

Taking cube roots of both sides gives $\det A = 1$.

Section 5.3 Practice Problems

A1 $\begin{bmatrix} 8 \\ -10 \\ 5 \\ 1 \end{bmatrix}$ **A2** $\begin{bmatrix} 9 \\ -10 \\ 4 \\ -6 \end{bmatrix}$ **A3** $\begin{bmatrix} 4 \\ -2 \\ 2 \\ -2 \end{bmatrix}$ **A4** $\begin{bmatrix} 4 \\ -9 \\ -3 \\ -1 \end{bmatrix}$

A5 (a) $\operatorname{adj}(A) = \begin{bmatrix} 1 & 0 & -3 \\ -1 & 0 & 1 \\ 1 & 2 & -5 \end{bmatrix}$

(b) $A\operatorname{adj}(A) = \begin{bmatrix} -2 & 0 & 0 \\ 0 & -2 & 0 \\ 0 & 0 & -2 \end{bmatrix}$, $\det A = -2$,

$A^{-1} = \frac{1}{-2}\begin{bmatrix} 1 & 0 & -3 \\ -1 & 0 & 1 \\ 1 & 2 & -5 \end{bmatrix}$

A6 (a) We get $\operatorname{adj}(A) = \begin{bmatrix} 7 & -14 & -13 \\ 0 & 7 & 5 \\ 0 & 0 & 1 \end{bmatrix}$

(b) $A\operatorname{adj}(A) = \begin{bmatrix} 7 & 0 & 0 \\ 0 & 7 & 0 \\ 0 & 0 & 7 \end{bmatrix}$, $\det A = 7$,

$A^{-1} = \frac{1}{7}\begin{bmatrix} 7 & -14 & -13 \\ 0 & 7 & 5 \\ 0 & 0 & 1 \end{bmatrix}$

A7 (a) $\operatorname{adj}(A) = \begin{bmatrix} ce & 0 & -dc \\ 0 & ae-bd & 0 \\ -cb & 0 & ac \end{bmatrix}$

(b) $A\operatorname{adj}(A) = \begin{bmatrix} ace-bcd & 0 & 0 \\ 0 & ace-bcd & 0 \\ 0 & 0 & ace-bcd \end{bmatrix}$,

$\det A = ace - bcd$,

$A^{-1} = \frac{1}{ace-bcd}\begin{bmatrix} ce & 0 & -dc \\ 0 & ae-bd & 0 \\ -cb & 0 & ac \end{bmatrix}$

A8 (a) $\operatorname{adj}(A) = \begin{bmatrix} b & -2b & -13 \\ 0 & ab & 5a \\ 0 & 0 & a \end{bmatrix}$

(b) $A\operatorname{adj}(A) = \begin{bmatrix} ab & 0 & 0 \\ 0 & ab & 0 \\ 0 & 0 & ab \end{bmatrix}$, $\det A = ab$,

$A^{-1} = \frac{1}{ab}\begin{bmatrix} b & -2b & -13 \\ 0 & ab & 5a \\ 0 & 0 & a \end{bmatrix}$

A9 (a) $\operatorname{adj}(A) = \begin{bmatrix} -1 & 3+t & -3 \\ 5 & 2-3t & -2-2t \\ -2 & -11 & -6 \end{bmatrix}$

(b) $A\operatorname{adj}(A) = \begin{bmatrix} -2t-17 & 0 & 0 \\ 0 & -2t-17 & 0 \\ 0 & 0 & -2t-17 \end{bmatrix}$,

$\det A = -2t - 17$,

$A^{-1} = \frac{1}{-2t-17}\begin{bmatrix} -1 & 3+t & -3 \\ 5 & 2-3t & -2-2t \\ -2 & -11 & -6 \end{bmatrix}$

A10 (a) $\operatorname{adj}(A) = \begin{bmatrix} -1 & t & -t^2+2 \\ 1 & -t & t-3 \\ -t+1 & -1 & 3t-2 \end{bmatrix}$

(b) $A\operatorname{adj}(A) = \begin{bmatrix} -t^2+t-1 & 0 & 0 \\ 0 & -t^2+t-1 & 0 \\ 0 & 0 & -t^2+t-1 \end{bmatrix}$,

$\det A = -t^2 + t - 1$,

$A^{-1} = \frac{1}{-t^2+t-1}\begin{bmatrix} -1 & t & -t^2+2 \\ 1 & -t & t-3 \\ -t+1 & -1 & 3t-2 \end{bmatrix}$

A11 $\frac{1}{-2}\begin{bmatrix} 10 & -3 \\ -4 & 1 \end{bmatrix}$

A12 $\frac{1}{7}\begin{bmatrix} -1 & 5 \\ -2 & 3 \end{bmatrix}$

A13 $\frac{1}{8}\begin{bmatrix} -3 & 21 & 11 \\ -1 & 7 & 5 \\ 3 & -13 & -7 \end{bmatrix}$

A14 $\frac{1}{4}\begin{bmatrix} -1 & -8 & -4 \\ -2 & -12 & -4 \\ -2 & -8 & -4 \end{bmatrix}$

A15 $\begin{bmatrix} 3 & -1 & -1 \\ 2 & 0 & -1 \\ -7 & 2 & 3 \end{bmatrix}$

A16 $\frac{1}{4}\begin{bmatrix} -1 & -2 & -2 \\ -8 & -12 & -8 \\ -4 & -4 & -4 \end{bmatrix}$

A17 $\frac{1}{-56}\begin{bmatrix} 4 & -16 & 2 \\ -12 & 20 & 8 \\ 0 & -28 & -14 \end{bmatrix}$

A18 $\vec{x} = \begin{bmatrix} 51/19 \\ -4/19 \end{bmatrix}$

A19 $\vec{x} = \begin{bmatrix} 7/5 \\ -11/15 \end{bmatrix}$

A20 $\vec{x} = \begin{bmatrix} 21/11 \\ -26/11 \\ 2 \end{bmatrix}$

A21 $\vec{x} = \begin{bmatrix} 3/5 \\ -12/5 \\ -8/5 \end{bmatrix}$

A22 $\vec{x} = \begin{bmatrix} -5/9 \\ -8/3 \\ 23/9 \end{bmatrix}$

A23 $\vec{x} = \begin{bmatrix} 3 \\ 0 \\ -5/3 \end{bmatrix}$

Section 5.4 Practice Problems

A1 (a) $\text{Area}(\vec{u}, \vec{v}) = 9$ (b) $\text{Area}(A\vec{u}, A\vec{v}) = 36$

A2 (a) $\text{Area}(\vec{u}, \vec{v}) = 18$ (b) $\text{Area}(A\vec{u}, A\vec{v}) = 0$

A3 (a) $\text{Area}(\vec{u}, \vec{v}) = 11$ (b) $\text{Area}(A\vec{u}, A\vec{v}) = 286$

A4 $A\vec{u} = \begin{bmatrix} 3 \\ 1 \end{bmatrix}$, $A\vec{v} = \begin{bmatrix} 2 \\ -2 \end{bmatrix}$, $\text{Area}(A\vec{u}, A\vec{v}) = 8$

A5 (a) 63 (b) 42 (c) 2646

A6 (a) 41 (b) 78 (c) 3198

A7 Area = 2

A8 Collinear

A9 Area = 5/2

A10 Area = 22

A11 Collinear

A12 (a) 5 (b) 245

A13 The n-volume of the parallelotope induced by $\vec{v}_1, \dots, \vec{v}_n$ is $\left|\det\begin{bmatrix} \vec{v}_1 & \cdots & \vec{v}_n \end{bmatrix}\right|$. Since adding a multiple of one column to another does not change the determinant, we get that

$$\left|\det\begin{bmatrix} \vec{v}_1 & \cdots & \vec{v}_n \end{bmatrix}\right| = \left|\det\begin{bmatrix} \vec{v}_1 & \cdots & \vec{v}_n + t\vec{v}_1 \end{bmatrix}\right|$$

which is the volume of the parallelotope induced by $\vec{v}_1, \dots, \vec{v}_{n-1}, \vec{v}_n + t\vec{v}_1$.

Chapter 5 Quiz

E1 The determinant is −13, thus the matrix is invertible.

E2 The determinant is 18, thus the matrix is invertible.

E3 The determinant is 0, thus the matrix is not invertible.

E4 The determinant is −12, thus the matrix is invertible.

E5 The determinant is 180, thus the matrix is invertible.

E6 The determinant is 120, thus the matrix is invertible.

E7 The matrix is invertible for all $k \neq -\frac{7}{8} \pm \frac{\sqrt{145}}{8}$.

E8 $\det B = 21$

E9 $\det C = 7$

E10 $\det(2A) = 224$

E11 $\det A^{-1} = \frac{1}{7}$

E12 $\det(A^T A) = 49$

E13 $\lambda = -1, 0, 5$

E14 (a) $\operatorname{adj}(A) = \begin{bmatrix} 2 & -6 & 2 \\ -4 & 6 & -1 \\ 2 & -6 & -1 \end{bmatrix}$

(b) $A \operatorname{adj}(A) = \begin{bmatrix} -6 & 0 & 0 \\ 0 & -6 & 0 \\ 0 & 0 & -6 \end{bmatrix}$, $\det A = -6$

(c) $(A^{-1})_{31} = -\frac{1}{3}$.

E15 $x_2 = -\frac{1}{2}$

E16 (a) 33 (b) 792

E17 $\vec{x} = \begin{bmatrix} -8 \\ 7 \\ -1 \\ -2 \end{bmatrix}$

E18 The points are collinear.

E19 The area of the triangle is 7.

E20 The area of the triangle is 1/2.

CHAPTER 6

Section 6.1 Practice Problems

A1 $\vec{v}_1$ is an eigenvector corresponding to the eigenvalue $\lambda = 7$, and $\vec{v}_2$ is an eigenvector corresponding to the eigenvalue $\lambda = 1$.

A2 $\vec{v}_1$ is not an eigenvector, but $\vec{v}_2$ is an eigenvector corresponding to the eigenvalue $\lambda = -4$.

A3 $\vec{v}_1$ is not an eigenvector, but $\vec{v}_2$ is an eigenvector corresponding to the eigenvalue $\lambda = 0$.

A4 $\vec{v}_1$ is an eigenvector corresponding to the eigenvalue $\lambda = -6$, but $\vec{v}_2$ is not an eigenvector.

A5 $\vec{v}_1$ is not an eigenvector, but $\vec{v}_2$ is an eigenvector corresponding to the eigenvalue $\lambda = 1$.

A6 $\lambda_1 = 2, \lambda_2 = 3$. $E_{\lambda_1} = \operatorname{Span}\left\{\begin{bmatrix} 1 \\ 2 \end{bmatrix}\right\}$, $E_{\lambda_2} = \operatorname{Span}\left\{\begin{bmatrix} 1 \\ 3 \end{bmatrix}\right\}$.

A7 $\lambda = 1$. $E_{\lambda_1} = \operatorname{Span}\left\{\begin{bmatrix} 1 \\ 0 \end{bmatrix}\right\}$.

A8 $\lambda_1 = 2, \lambda_2 = 3$. $E_{\lambda_1} = \operatorname{Span}\left\{\begin{bmatrix} 1 \\ 0 \end{bmatrix}\right\}$, $E_{\lambda_2} = \operatorname{Span}\left\{\begin{bmatrix} 0 \\ 1 \end{bmatrix}\right\}$.

A9 $\lambda_1 = -1, \lambda_2 = 4$. $E_{\lambda_1} = \operatorname{Span}\left\{\begin{bmatrix} 2 \\ 5 \end{bmatrix}\right\}$, $E_{\lambda_2} = \operatorname{Span}\left\{\begin{bmatrix} 1 \\ 3 \end{bmatrix}\right\}$.

A10 $\lambda_1 = 5, \lambda_2 = -2$. $E_{\lambda_1} = \operatorname{Span}\left\{\begin{bmatrix} 3 \\ 4 \end{bmatrix}\right\}$, $E_{\lambda_2} = \operatorname{Span}\left\{\begin{bmatrix} -1 \\ 1 \end{bmatrix}\right\}$.

A11 $\lambda_1 = 0, \lambda_2 = -3$. $E_{\lambda_1} = \operatorname{Span}\left\{\begin{bmatrix} 1 \\ 1 \end{bmatrix}\right\}$, $E_{\lambda_2} = \operatorname{Span}\left\{\begin{bmatrix} 1 \\ 2 \end{bmatrix}\right\}$.

A12 $\lambda_1 = 1, \lambda_2 = 2, \lambda_3 = 3$. $E_{\lambda_1} = \operatorname{Span}\left\{\begin{bmatrix} 1 \\ 0 \\ 0 \end{bmatrix}\right\}$, $E_{\lambda_2} = \left\{\begin{bmatrix} 3 \\ 1 \\ 0 \end{bmatrix}\right\}$, $E_{\lambda_3} = \left\{\begin{bmatrix} 13 \\ 7 \\ 1 \end{bmatrix}\right\}$.

A13 $\lambda_1 = -2, \lambda_2 = 2, \lambda_3 = -1$. $E_{\lambda_1} = \text{Span}\left\{\begin{bmatrix} 0 \\ -1 \\ 1 \end{bmatrix}\right\}$, $E_{\lambda_2} = \left\{\begin{bmatrix} 2 \\ 1 \\ 1 \end{bmatrix}\right\}$, $E_{\lambda_3} = \left\{\begin{bmatrix} 2 \\ 0 \\ 1 \end{bmatrix}\right\}$.

A14 $\lambda_1 = 2$. $E_{\lambda_1} = \left\{\begin{bmatrix} -1 \\ 1 \\ 1 \end{bmatrix}\right\}$.

A15 $\lambda_1 = 1, \lambda_2 = -1$. $E_{\lambda_1} = \left\{\begin{bmatrix} -1 \\ -3 \\ 1 \end{bmatrix}\right\}$, $E_{\lambda_2} = \left\{\begin{bmatrix} -1 \\ -1 \\ 1 \end{bmatrix}\right\}$.

A16 $\lambda_1 = 2$ has algebraic multiplicity 1 and geometric multiplicity 1. $\lambda_2 = 3$ has algebraic multiplicity 1 and geometric multiplicity 1.

A17 $\lambda_1 = 2$ has algebraic multiplicity 2 and geometric multiplicity 1.

A18 $\lambda_1 = 2$ has algebraic multiplicity 2 and geometric multiplicity 1.

A19 $\lambda_1 = 1$ has algebraic multiplicity 1 and geometric multiplicity 1. $\lambda_2 = 2$ has algebraic multiplicity 1 and geometric multiplicity 1. $\lambda_3 = -2$ has algebraic multiplicity 1 and geometric multiplicity 1.

A20 $\lambda_1 = 0$ has algebraic multiplicity 2 and geometric multiplicity 2. $\lambda_2 = 6$ has algebraic multiplicity 1 and geometric multiplicity 1.

A21 $\lambda_1 = 2$ has algebraic multiplicity 2 and geometric multiplicity 2. $\lambda_2 = 5$ has algebraic multiplicity 1 and geometric multiplicity 1.

A22 $\lambda_1 = 1$ has algebraic multiplicity 1 and geometric multiplicity 1. $\lambda_2 = 3$ has algebraic multiplicity 1 and geometric multiplicity 1. $\lambda_3 = -1$ has algebraic multiplicity 1 and geometric multiplicity 1.

A23 $\lambda = 6$

A24 $\lambda = 6$

A25 $\lambda = 8$

A26 $\lambda = 10$

Section 6.2 Practice Problems

A1 $P^{-1}AP = \begin{bmatrix} 14 & 0 \\ 0 & -7 \end{bmatrix}$

A2 P does not diagonalize A.

A3 P does not diagonalize A.

A4 $P^{-1}AP = \begin{bmatrix} 1 & 0 \\ 0 & -3 \end{bmatrix}$

A5 $P^{-1}AP = \begin{bmatrix} -2 & 0 & 0 \\ 0 & -2 & 0 \\ 0 & 0 & 10 \end{bmatrix}$

A6 $P^{-1}AP = \begin{bmatrix} 3 & 0 & 0 \\ 0 & 0 & 0 \\ 0 & 0 & -1 \end{bmatrix}$

A7 $P = \begin{bmatrix} 1 & -1 \\ 0 & 5 \end{bmatrix}, D = \begin{bmatrix} 7 & 0 \\ 0 & -8 \end{bmatrix}$

A8 A is not diagonalizable.

A9 $P = \begin{bmatrix} -1 & 9 \\ 1 & 4 \end{bmatrix}, D = \begin{bmatrix} -8 & 0 \\ 0 & 5 \end{bmatrix}$

A10 $P = \begin{bmatrix} -1 & 2 \\ 1 & 5 \end{bmatrix}, D = \begin{bmatrix} 1 & 0 \\ 0 & 8 \end{bmatrix}$

A11 $P = \begin{bmatrix} -3 & 1 \\ 4 & 1 \end{bmatrix}, D = \begin{bmatrix} -6 & 0 \\ 0 & 1 \end{bmatrix}$

A12 A is not diagonalizable.

A13 A is not diagonalizable.

A14 $P = \begin{bmatrix} -1 & 1 \\ 1 & 1 \end{bmatrix}, D = \begin{bmatrix} 0 & 0 \\ 0 & 8 \end{bmatrix}$

A15 $P = \begin{bmatrix} -1 & 1 \\ 1 & 1 \end{bmatrix}, D = \begin{bmatrix} -7 & 0 \\ 0 & 3 \end{bmatrix}$

A16 $P = \begin{bmatrix} -1 & 1 & -1 \\ 1 & 2 & 0 \\ 0 & 3 & 1 \end{bmatrix}, D = \begin{bmatrix} -1 & 0 & 0 \\ 0 & 2 & 0 \\ 0 & 0 & 0 \end{bmatrix}$

A17 A is not diagonalizable.

A18 A is not diagonalizable.

A19 $P = \begin{bmatrix} 2 & 1 & 1 \\ 1 & 0 & -1 \\ 0 & 1 & 2 \end{bmatrix}, D = \begin{bmatrix} 2 & 0 & 0 \\ 0 & 2 & 0 \\ 0 & 0 & -1 \end{bmatrix}$

A20 A is not diagonalizable.

A21 $P = \begin{bmatrix} -2 & 1 & 1 \\ -2 & -2 & 1 \\ 5 & 1 & 1 \end{bmatrix}, D = \begin{bmatrix} -1 & 0 & 0 \\ 0 & 0 & 0 \\ 0 & 0 & 6 \end{bmatrix}$

A22 $P = \begin{bmatrix} -2 & 1 & 0 \\ 1 & 0 & 1 \\ 0 & 1 & 1 \end{bmatrix}, D = \begin{bmatrix} 2 & 0 & 0 \\ 0 & 2 & 0 \\ 0 & 0 & 1 \end{bmatrix}$

A23 A is not diagonalizable.

Section 6.3 Practice Problems

A1 $A^3 = \begin{bmatrix} 90 & -63 \\ -126 & 153 \end{bmatrix}$

A2 $A^{100} = \frac{1}{3}\begin{bmatrix} -5 \cdot 2^{100} + 8 & -5 \cdot 2^{101} + 10 \\ 2^{102} - 4 & 2^{103} - 5 \end{bmatrix}$

A3 $A^{100} = -\frac{1}{5}\begin{bmatrix} -6 + 4^{100} & -2 + 2 \cdot 4^{100} \\ 3 - 3 \cdot 4^{100} & 1 - 6 \cdot 4^{100} \end{bmatrix}$

A4 $A^{200} = \frac{1}{5}\begin{bmatrix} 6 - 4^{200} & -2 + 2 \cdot 4^{200} \\ 3 - 3 \cdot 4^{200} & -1 + 6 \cdot 4^{200} \end{bmatrix}$

A5 $A^{200} = \begin{bmatrix} -3(2^{200}) + 4(3^{200}) & 3(2^{201}) - 2(3^{201}) \\ -2^{201} + 2(3^{200}) & 2^{202} - 3^{201} \end{bmatrix}$

A6 $A^{100} = I$

A7 $A^{100} = \begin{bmatrix} 5 & -1 & 2 \\ 4 & 0 & 2 \\ -8 & 2 & -3 \end{bmatrix}$

A8 $A^{100} = \begin{bmatrix} -1 + 2^{101} & 1 - 2^{100} & 0 \\ -2 + 2^{101} & 2 - 2^{100} & 0 \\ 2 - 2^{101} & -1 + 2^{100} & 1 \end{bmatrix}$

A9 Not a Markov matrix.

A10 The fixed state is $\begin{bmatrix} 6/13 \\ 7/13 \end{bmatrix}$.

A11 The fixed state is $\begin{bmatrix} 4/9 \\ 5/9 \end{bmatrix}$.

A12 The fixed state is $\begin{bmatrix} 5/7 \\ 2/7 \end{bmatrix}$.

A13 Not a Markov matrix.

A14 The fixed state is $\begin{bmatrix} 1/3 \\ 1/3 \\ 1/3 \end{bmatrix}$.

A15 (a) In the long run, 25% of the population will be rural dwellers and 75% will be urban dwellers.

(b) After five decades, approximately 33% of the population will be rural dwellers and 67% will be urban dwellers.

A16 $T = \frac{1}{10}\begin{bmatrix} 8 & 3 & 3 \\ 1 & 6 & 1 \\ 1 & 1 & 6 \end{bmatrix}$. In the long run, 60% of the cars will be at the airport, 20% of the cars will be at the train station, and 20% of the cars will be at the city centre.

A17 (a) $\begin{bmatrix} J_{m+1} \\ T_{m+1} \end{bmatrix} = \frac{1}{10}\begin{bmatrix} 3 & 8 \\ 7 & 2 \end{bmatrix}\begin{bmatrix} J_m \\ T_m \end{bmatrix}$. In the long run, $\frac{8}{15} =$ 53% of the customers will deal with Johnson and $\frac{7}{15} = 47\%$ will deal with Thomson.

(b) $T^m \begin{bmatrix} 0.25 \\ 0.75 \end{bmatrix} = \frac{1}{15}\begin{bmatrix} 8 \\ 7 \end{bmatrix} + \frac{1}{15}(4.25)(-0.5)^m \begin{bmatrix} -1 \\ 1 \end{bmatrix}$

A18 $\begin{bmatrix} y \\ z \end{bmatrix} = ae^{-5t}\begin{bmatrix} -1 \\ 4 \end{bmatrix} + be^{4t}\begin{bmatrix} 2 \\ 1 \end{bmatrix}, \quad a, b \in \mathbb{R}$

A19 $\begin{bmatrix} y \\ z \end{bmatrix} = ae^{-0.5t}\begin{bmatrix} -1 \\ 1 \end{bmatrix} + be^{0.3t}\begin{bmatrix} 7 \\ 1 \end{bmatrix}, \quad a, b \in \mathbb{R}$

A20 $\begin{bmatrix} y \\ z \end{bmatrix} = ae^{-9t}\begin{bmatrix} -1 \\ 2 \end{bmatrix} + be^{3t}\begin{bmatrix} 1 \\ 1 \end{bmatrix}, \quad a, b \in \mathbb{R}$

A21 $\begin{bmatrix} y \\ z \end{bmatrix} = ae^{0.6t}\begin{bmatrix} 3 \\ 1 \end{bmatrix} + be^{0.2t}\begin{bmatrix} -1 \\ 1 \end{bmatrix}, \quad a, b \in \mathbb{R}$

A22 $\begin{bmatrix} y \\ z \end{bmatrix} = a\begin{bmatrix} -1 \\ 2 \end{bmatrix} + be^{5t}\begin{bmatrix} 2 \\ 1 \end{bmatrix}, \quad a, b \in \mathbb{R}$

A23 $\begin{bmatrix} y \\ z \end{bmatrix} = ae^{8t}\begin{bmatrix} 6 \\ 5 \end{bmatrix} + be^{-3t}\begin{bmatrix} -1 \\ 1 \end{bmatrix}, \quad a, b \in \mathbb{R}$

A24 $\begin{bmatrix} y \\ z \end{bmatrix} = ae^{5t}\begin{bmatrix} -2 \\ 1 \end{bmatrix} + be^{11t}\begin{bmatrix} 1 \\ 1 \end{bmatrix}, \quad a, b \in \mathbb{R}$

A25 $\begin{bmatrix} y \\ z \end{bmatrix} = a\begin{bmatrix} -1 \\ 1 \end{bmatrix} + be^{t}\begin{bmatrix} 2 \\ 3 \end{bmatrix}, \quad a, b \in \mathbb{R}$

A26 $\begin{bmatrix} x \\ y \\ z \end{bmatrix} = a\begin{bmatrix} 1 \\ -1 \\ 2 \end{bmatrix} + be^{-2t}\begin{bmatrix} 1 \\ 1 \\ 1 \end{bmatrix} + ce^{-4t}\begin{bmatrix} -1 \\ -3 \\ 1 \end{bmatrix}, \quad a, b, c \in \mathbb{R}$

A27 $\begin{bmatrix} x \\ y \\ z \end{bmatrix} = ae^{3t}\begin{bmatrix} 0 \\ 1 \\ 1 \end{bmatrix} + be^{6t}\begin{bmatrix} -1 \\ 0 \\ 1 \end{bmatrix} + ce^{-9t}\begin{bmatrix} 2 \\ -3 \\ 1 \end{bmatrix}, \quad a, b, c \in \mathbb{R}$

Chapter 6 Quiz

E1 (a) (i) $\begin{bmatrix}3\\1\\0\end{bmatrix}$ is not an eigenvector of A.

(ii) $\begin{bmatrix}1\\0\\1\end{bmatrix}$ is an eigenvector with eigenvalue 1.

(iii) $\begin{bmatrix}4\\1\\0\end{bmatrix}$ is an eigenvector with eigenvalue 1.

(iv) $\begin{bmatrix}2\\1\\-1\end{bmatrix}$ is an eigenvector with eigenvalue -1.

(b) $P = \begin{bmatrix}1 & 4 & 2\\0 & 1 & 1\\1 & 0 & -1\end{bmatrix}$, $D = \text{diag}(1, 1, -1)$.

E2 $\lambda_1 = 2$ has algebraic multiplicity 2 and geometric multiplicity 2. $\lambda_2 = 4$ has algebraic multiplicity 1 and geometric multiplicity 1.

E3 $\lambda_1 = 1$ has algebraic multiplicity 3 and geometric multiplicity 1.

E4 $P = \begin{bmatrix}2 & 1\\1 & 1\end{bmatrix}$, $D = \begin{bmatrix}1 & 0\\0 & 3\end{bmatrix}$

E5 A is not diagonalizable since the geometric multiplicity of $\lambda_1 = 2$ is less than its algebraic multiplicity.

E6 $P = \begin{bmatrix}-1 & -1 & 2\\1 & 0 & 2\\0 & 2 & 1\end{bmatrix}$, $D = \begin{bmatrix}9 & 0 & 0\\0 & 9 & 0\\0 & 0 & 0\end{bmatrix}$

E7 $P = \begin{bmatrix}-1 & 1 & 1\\-3 & -1 & 1\\1 & 2 & 1\end{bmatrix}$, $D = \begin{bmatrix}0 & 0 & 0\\0 & -4 & 0\\0 & 0 & -2\end{bmatrix}$

E8 $\vec{v}_1 = \begin{bmatrix}1\\2\end{bmatrix}$, so the dominant eigenvalue is $\lambda = 20$.

E9 $A^{100} = \begin{bmatrix}-3 + 4 \cdot 2^{100} & -6 + 6 \cdot 2^{100}\\2 - 2 \cdot 2^{100} & 4 - 3 \cdot 2^{100}\end{bmatrix}$

E10 (a) The solution space $A\vec{x} = \vec{0}$ is one-dimensional, since it is the eigenspace corresponding to the eigenvalue 0.

(b) 2 cannot be an eigenvalue, as we already have three eigenvalues for the 3×3 matrix A. Hence, there are no vectors that satisfy $A\vec{x} = 2\vec{x}$, so the solution space is zero dimensional in this case.

(c) Since $A\vec{x} = \vec{0}$ is one-dimensional, we could apply the Rank-Nullity Theorem, to get that

$$\text{rank}(A) = n - \text{nullity}(A) = 3 - 1 = 2$$

E11 The invariant state is $\vec{x} = \begin{bmatrix}1/4\\1/4\\1/2\end{bmatrix}$.

E12 $\begin{bmatrix}y\\z\end{bmatrix} = ae^{-0.1t}\begin{bmatrix}-1\\1\end{bmatrix} + be^{0.4t}\begin{bmatrix}2\\3\end{bmatrix}, \quad a, b \in \mathbb{R}$

E13 Since A is invertible, 0 is not an eigenvalue of A (see Section 6.2 Problem C8). Then, if $A\vec{x} = \lambda\vec{x}$ we get $\vec{x} = \lambda A^{-1}\vec{x}$, so $A^{-1}\vec{x} = \frac{1}{\lambda}\vec{x}$.

E14 Since A is diagonalizable, we have that there exists an invertible matrix P such that

$$P^{-1}AP = \text{diag}(\lambda_1, \dots, \lambda_1) = \lambda_1 I$$

Hence,

$$A = P(\lambda_1 I)P^{-1} = \lambda_1 PP^{-1} = \lambda_1 I$$

E15 Let λ be the eigenvalue corresponding to $\vec{v}$. Hence, $A\vec{v} = \lambda\vec{v}$. For any $t \neq 0$, we have that

$$A(t\vec{v}) = tA\vec{v} = t(\lambda\vec{v}) = \lambda(t\vec{v})$$

Thus, $t\vec{v}$ is also an eigenvector of A.

E16 If A is diagonalizable, then there exists an invertible matrix P such that $P^{-1}AP = D$ where D is diagonal. Since A and B are similar, there exists an invertible matrix Q such that $Q^{-1}BQ = A$. Thus, we have that

$$D = P^{-1}(Q^{-1}BQ)P = (P^{-1}Q^{-1})B(QP) = (QP)^{-1}B(QP)$$

Thus, by definition, B is also diagonalizable.

E17 This is false. The matrix $A = \begin{bmatrix}1 & 1\\0 & 1\end{bmatrix}$ is invertible, but not diagonalizable.

E18 This is false. The zero matrix is diagonal and hence diagonalizable, but is not invertible.

E19 This is true. It is Theorem 6.2.4.

E20 This is false. If $\vec{v} = \vec{0}$, then $\vec{v}$ is not an eigenvector.

E21 This is true. If the RREF of $A - \lambda I$ is I, then $(A - \lambda I)\vec{v} = \vec{0}$ has the unique solution $\vec{v} = \vec{0}$. Hence, by definition, λ is not an eigenvalue.

E22 This is true. If $L(\vec{v}) = \lambda\vec{v}$ where $\vec{v} \neq \vec{0}$, then we have $\lambda\vec{v} = L(\vec{v}) = [L]\vec{v}$ as required.

CHAPTER 7

Section 7.1 Practice Problems

A1 $\left\{\begin{bmatrix}\frac{1}{\sqrt{5}}\\ \frac{2}{\sqrt{5}}\end{bmatrix}, \begin{bmatrix}\frac{2}{\sqrt{5}}\\ \frac{-1}{\sqrt{5}}\end{bmatrix}\right\}$

A2 The set is not orthogonal.

A3 $\left\{\begin{bmatrix}4/\sqrt{18}\\ -1/\sqrt{18}\\ 1/\sqrt{18}\end{bmatrix}, \begin{bmatrix}-1/\sqrt{17}\\ 0\\ 4/\sqrt{17}\end{bmatrix}, \begin{bmatrix}4/\sqrt{306}\\ 17/\sqrt{306}\\ 1/\sqrt{306}\end{bmatrix}\right\}$

A4 The set is not orthogonal.

A5 $\left\{\begin{bmatrix}1/\sqrt{11}\\ 3/\sqrt{11}\\ -1/\sqrt{11}\end{bmatrix}, \begin{bmatrix}1/\sqrt{2}\\ 0\\ 1/\sqrt{2}\end{bmatrix}, \begin{bmatrix}3/\sqrt{22}\\ -2/\sqrt{22}\\ -3/\sqrt{22}\end{bmatrix}\right\}$

A6 $\left\{\begin{bmatrix}1/3\\ -2/3\\ 2/3\end{bmatrix}, \begin{bmatrix}-2/3\\ 1/3\\ 2/3\end{bmatrix}, \begin{bmatrix}2/3\\ 2/3\\ 1/3\end{bmatrix}\right\}$

A7 $\left\{\begin{bmatrix}1/2\\ 1/2\\ 1/2\\ 1/2\end{bmatrix}, \begin{bmatrix}-2/\sqrt{6}\\ 0\\ 1/\sqrt{6}\\ 1/\sqrt{6}\end{bmatrix}, \begin{bmatrix}0\\ 0\\ 1/\sqrt{2}\\ -1/\sqrt{2}\end{bmatrix}\right\}$

A8 The set is not orthogonal.

A9 (a) $\vec{w} = \frac{8}{9}\vec{v}_1 + \frac{19}{9}\vec{v}_2 + \frac{5}{9}\vec{v}_3$

(b) $\vec{x} = \frac{1}{9}\vec{v}_1 + \frac{2}{9}\vec{v}_2 - \frac{2}{9}\vec{v}_3$

(c) $\vec{y} = -\frac{8}{9}\vec{v}_1 - \frac{1}{9}\vec{v}_2 + \frac{4}{9}\vec{v}_3$

A10 (a) $\vec{w} = \frac{8}{3}\vec{v}_1 + \frac{19}{3}\vec{v}_2 + \frac{5}{3}\vec{v}_3$

(b) $\vec{x} = 7\vec{v}_1 - 2\vec{v}_2 - 3\vec{v}_3$

(c) $\vec{y} = \frac{22}{3}\vec{v}_1 + \frac{2}{3}\vec{v}_2 + \frac{4}{3}\vec{v}_3$

A11 (a) $\vec{x} = 2\vec{v}_1 - \frac{5}{2}\vec{v}_2 - \frac{5}{2}\vec{v}_3 - \frac{1}{2}\vec{v}_4$

(b) $\vec{y} = -\frac{5}{4}\vec{v}_1 + \frac{3}{4}\vec{v}_2 + \frac{7}{2}\vec{v}_3 + 3\vec{v}_4$

(c) $\vec{w} = \frac{5}{4}\vec{v}_1 + \frac{1}{4}\vec{v}_2 - \frac{3}{2}\vec{v}_3 + 0\vec{v}_4$

A12 It is orthogonal.

A13 It is not orthogonal. The columns of the matrix are not orthogonal.

A14 It is not orthogonal. The columns are not unit vectors.

A15 It is orthogonal.

A16 It is not orthogonal. The third column is not orthogonal to the first or second column.

A17 It is orthogonal.

A18 $\left\{\begin{bmatrix}1/\sqrt{6}\\ 1/\sqrt{3}\\ 1/\sqrt{2}\end{bmatrix}, \begin{bmatrix}-2/\sqrt{6}\\ 1/\sqrt{3}\\ 0\end{bmatrix}, \begin{bmatrix}1/\sqrt{6}\\ 1/\sqrt{3}\\ -1/\sqrt{2}\end{bmatrix}\right\}$

A19 By Theorem 7.1.1, $\mathcal{B} = \{\vec{v}_1, \dots, \vec{v}_n\}$ is linearly independent. Thus, it is a basis for $\mathbb{R}^n$ by Theorem 2.3.6.

Section 7.2 Practice Problems

A1 $\text{proj}_{\mathbb{S}}(\vec{x}) = \begin{bmatrix}-1\\ 2\end{bmatrix}$, $\text{perp}_{\mathbb{S}}(\vec{x}) = \begin{bmatrix}0\\ 0\end{bmatrix}$

A2 $\text{proj}_{\mathbb{S}}(\vec{x}) = \begin{bmatrix}11/2\\ 9\\ 11/2\end{bmatrix}$, $\text{perp}_{\mathbb{S}}(\vec{x}) = \begin{bmatrix}1/2\\ 0\\ -1/2\end{bmatrix}$

A3 $\text{proj}_{\mathbb{S}}(\vec{x}) = \begin{bmatrix}2\\ 0\\ 3\end{bmatrix}$, $\text{perp}_{\mathbb{S}}(\vec{x}) = \begin{bmatrix}0\\ \sqrt{5}\\ 0\end{bmatrix}$

A4 $\text{proj}_{\mathbb{S}}(\vec{x}) = \begin{bmatrix}4\\ 4\\ 4\end{bmatrix}$, $\text{perp}_{\mathbb{S}}(\vec{x}) = \begin{bmatrix}3\\ 0\\ -3\end{bmatrix}$

A5 $\text{proj}_{\mathbb{S}}\,\vec{x} = \begin{bmatrix}1\\ 2\\ 3\end{bmatrix}$, $\text{perp}_{\mathbb{S}}(\vec{x}) = \begin{bmatrix}0\\ 0\\ 0\end{bmatrix}$

A6 $\text{proj}_{\mathbb{S}}(\vec{x}) = \begin{bmatrix}-5/7\\ 6/7\\ 32/7\end{bmatrix}$, $\text{perp}_{\mathbb{S}}(\vec{x}) = \begin{bmatrix}12/7\\ -6/7\\ 3/7\end{bmatrix}$

A7 $\text{proj}_{\mathbb{S}}\,\vec{x} = \begin{bmatrix}2\\ 9/2\\ 5\\ 9/2\end{bmatrix}$, $\text{perp}_{\mathbb{S}}(\vec{x}) = \begin{bmatrix}0\\ -3/2\\ 0\\ 3/2\end{bmatrix}$

A8 $\text{proj}_{\mathbb{S}}(\vec{x}) = \begin{bmatrix}5/2\\ 5\\ 5/2\\ 5\end{bmatrix}$, $\text{perp}_{\mathbb{S}}(\vec{x}) = \begin{bmatrix}-1/2\\ -2\\ 5/2\\ 1\end{bmatrix}$

A9 $\left\{\begin{bmatrix}1\\ 0\\ 1\end{bmatrix}, \begin{bmatrix}-2\\ 1\\ 0\end{bmatrix}\right\}$

A10 $\left\{\begin{bmatrix}1\\ -2\\ 1\end{bmatrix}\right\}$

A11 $\left\{\begin{bmatrix}-1\\-1\\1\end{bmatrix}\right\}$

A12 $\left\{\begin{bmatrix}-1\\-3\\-5\\12\end{bmatrix}\right\}$

A13 $\left\{\begin{bmatrix}-1\\0\\1\\0\end{bmatrix}, \begin{bmatrix}1\\-3\\0\\1\end{bmatrix}\right\}$

A14 $\left\{\begin{bmatrix}1\\0\\0\end{bmatrix}, \begin{bmatrix}0\\1\\0\end{bmatrix}\right\}$

A15 $\left\{\begin{bmatrix}1\\1\\0\end{bmatrix}, \begin{bmatrix}1/2\\-1/2\\0\end{bmatrix}\right\}$

A16 $\left\{\begin{bmatrix}1\\3\\2\end{bmatrix}, \begin{bmatrix}9/14\\13/14\\-12/7\end{bmatrix}\right\}$

A17 $\left\{\begin{bmatrix}2\\1\\2\end{bmatrix}, \begin{bmatrix}1\\0\\-1\end{bmatrix}, \begin{bmatrix}-1\\4\\-1\end{bmatrix}\right\}$

A18 $\left\{\begin{bmatrix}4\\1\\2\end{bmatrix}, \begin{bmatrix}1\\2\\-3\end{bmatrix}\right\}$

A19 $\left\{\begin{bmatrix}1\\0\\0\\1\end{bmatrix}, \begin{bmatrix}0\\-1\\1\\0\end{bmatrix}, \begin{bmatrix}-1\\1\\1\\1\end{bmatrix}\right\}$

A20 $\left\{\begin{bmatrix}1\\1\\0\\1\end{bmatrix}, \begin{bmatrix}1\\-2\\3\\1\end{bmatrix}, \begin{bmatrix}1\\-2\\-2\\1\end{bmatrix}\right\}$

A21 $\left\{\begin{bmatrix}1/\sqrt{2}\\1/\sqrt{2}\\0\end{bmatrix}, \begin{bmatrix}0\\0\\1\end{bmatrix}\right\}$

A22 $\left\{\begin{bmatrix}2/3\\1/3\\-2/3\end{bmatrix}, \begin{bmatrix}4/\sqrt{45}\\2/\sqrt{45}\\5/\sqrt{45}\end{bmatrix}, \begin{bmatrix}1/\sqrt{5}\\-2/\sqrt{5}\\0\end{bmatrix}\right\}$

A23 $\left\{\begin{bmatrix}1/\sqrt{3}\\1/\sqrt{3}\\0\\1/\sqrt{3}\end{bmatrix}, \begin{bmatrix}-2/\sqrt{15}\\1/\sqrt{15}\\3/\sqrt{15}\\1/\sqrt{15}\end{bmatrix}, \begin{bmatrix}2/\sqrt{10}\\-1/\sqrt{10}\\2/\sqrt{10}\\-1/\sqrt{10}\end{bmatrix}\right\}$

A24 $\left\{\begin{bmatrix}1/\sqrt{3}\\0\\1/\sqrt{3}\\0\\1/\sqrt{3}\end{bmatrix}, \begin{bmatrix}1/\sqrt{3}\\0\\-1/\sqrt{3}\\1/\sqrt{3}\\0\end{bmatrix}, \begin{bmatrix}-1/\sqrt{15}\\3/\sqrt{15}\\1/\sqrt{15}\\2/\sqrt{15}\\0\end{bmatrix}\right\}$

A25 $\vec{y} = \frac{4}{13}\begin{bmatrix}2\\3\end{bmatrix}$

A26 $\vec{y} = \begin{bmatrix}5/2\\-1\\-3/2\end{bmatrix}$

A27 $\vec{y} = \begin{bmatrix}-1\\2\end{bmatrix}$

A28 $\vec{y} = \begin{bmatrix}9/2\\5/2\\6\end{bmatrix}$

A29 $\vec{y} = \begin{bmatrix}-1/5\\-1/5\\2/5\\3/5\end{bmatrix}$

A30 $\vec{y} = \begin{bmatrix}10/7\\3/7\\-10/7\\1/7\end{bmatrix}$

Section 7.3 Practice Problems

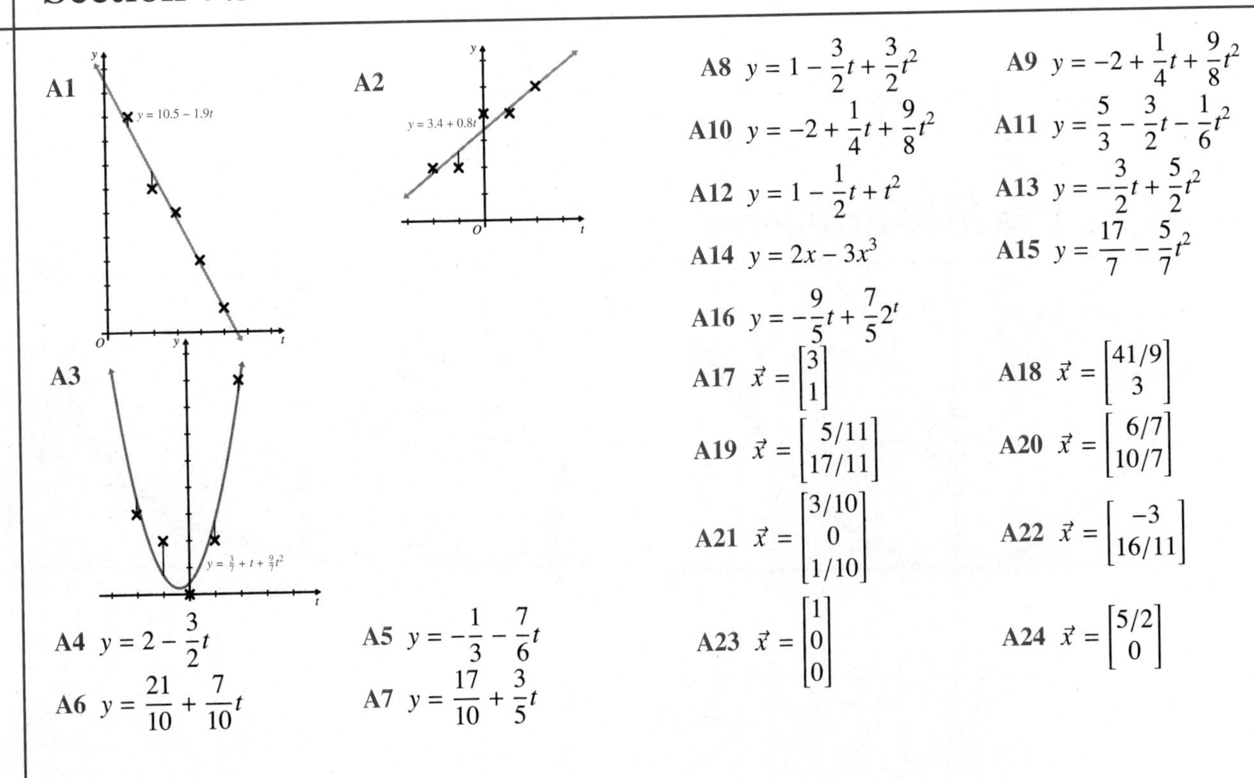

A4 $y = 2 - \frac{3}{2}t$

A5 $y = -\frac{1}{3} - \frac{7}{6}t$

A6 $y = \frac{21}{10} + \frac{7}{10}t$

A7 $y = \frac{17}{10} + \frac{3}{5}t$

A8 $y = 1 - \frac{3}{2}t + \frac{3}{2}t^2$

A9 $y = -2 + \frac{1}{4}t + \frac{9}{8}t^2$

A10 $y = -2 + \frac{1}{4}t + \frac{9}{8}t^2$

A11 $y = \frac{5}{3} - \frac{3}{2}t - \frac{1}{6}t^2$

A12 $y = 1 - \frac{1}{2}t + t^2$

A13 $y = -\frac{3}{2}t + \frac{5}{2}t^2$

A14 $y = 2x - 3x^3$

A15 $y = \frac{17}{7} - \frac{5}{7}t^2$

A16 $y = -\frac{9}{5}t + \frac{7}{5}2^t$

A17 $\vec{x} = \begin{bmatrix}3\\1\end{bmatrix}$

A18 $\vec{x} = \begin{bmatrix}41/9\\3\end{bmatrix}$

A19 $\vec{x} = \begin{bmatrix}5/11\\17/11\end{bmatrix}$

A20 $\vec{x} = \begin{bmatrix}6/7\\10/7\end{bmatrix}$

A21 $\vec{x} = \begin{bmatrix}3/10\\0\\1/10\end{bmatrix}$

A22 $\vec{x} = \begin{bmatrix}-3\\16/11\end{bmatrix}$

A23 $\vec{x} = \begin{bmatrix}1\\0\\0\end{bmatrix}$

A24 $\vec{x} = \begin{bmatrix}5/2\\0\end{bmatrix}$

Section 7.4 Practice Problems

A1 16

A2 1

A3 $\sqrt{2}$

A4 3

A5 $\sqrt{15}$

A6 -46

A7 -84

A8 $\sqrt{22}$

A9 $9\sqrt{59}$

A10 It is not positive definite.

A11 It is not bilinear.

A12 It is an inner product.

A13 It is not positive definite.

A14 It is not bilinear.

A15 (a) $\left\{\begin{bmatrix}1 & 0\\-1 & 1\end{bmatrix}, \begin{bmatrix}1 & 3\\2 & 1\end{bmatrix}, \begin{bmatrix}3 & -1\\1 & -2\end{bmatrix}\right\}$

(b) $\text{proj}_{\mathbb{S}}\left(\begin{bmatrix}4 & 3\\-2 & 1\end{bmatrix}\right) = \begin{bmatrix}4 & 5/3\\-2/3 & 7/3\end{bmatrix}$

A16 (a) $\left\{\begin{bmatrix}2 & 0\\1 & -1\end{bmatrix}, \begin{bmatrix}1 & 0\\-1 & 1\end{bmatrix}, \begin{bmatrix}0 & 2\\1 & 1\end{bmatrix}\right\}$

(b) $\text{proj}_{\mathbb{S}}\left(\begin{bmatrix}4 & 3\\-2 & 1\end{bmatrix}\right) = \begin{bmatrix}4 & 5/3\\-2/3 & 7/3\end{bmatrix}$

A17 (a) $\left\{\begin{bmatrix}1 & 0\\2 & 2\end{bmatrix}, \begin{bmatrix}0 & 1\\1 & -1\end{bmatrix}, \begin{bmatrix}2 & -1\\0 & -1\end{bmatrix}\right\}$

(b) $\text{proj}_{\mathbb{S}}\left(\begin{bmatrix}4 & 3\\-2 & 1\end{bmatrix}\right) = \begin{bmatrix}14/9 & -2/3\\4/9 & -2/9\end{bmatrix}$

A18 (a) $\left\{\begin{bmatrix}3 & 1\\-1 & -2\end{bmatrix}, \begin{bmatrix}3 & 1\\4 & 3\end{bmatrix}, \begin{bmatrix}-1 & 2\\1 & -1\end{bmatrix}\right\}$

(b) $\text{proj}_{\mathbb{S}}\left(\begin{bmatrix}4 & 3\\-2 & 1\end{bmatrix}\right) = \begin{bmatrix}4 & 1\\0 & -1\end{bmatrix}$

A19 (a) $\mathcal{B} = \left\{\begin{bmatrix}1\\1\\0\end{bmatrix}, \begin{bmatrix}-1\\2\\0\end{bmatrix}\right\}$ (b) $\text{proj}_{\mathbb{S}}(\vec{e}_1) = \begin{bmatrix}1\\0\\0\end{bmatrix}$

A20 (a) $\mathcal{B} = \left\{\begin{bmatrix}1\\0\\0\end{bmatrix}, \begin{bmatrix}0\\3\\2\end{bmatrix}, \begin{bmatrix}0\\2\\-1\end{bmatrix}\right\}$ (b) $\text{proj}_{\mathbb{S}}(\vec{e}_1) = \begin{bmatrix}1\\0\\0\end{bmatrix}$

A21 (a) $\mathcal{B} = \left\{\begin{bmatrix}1\\1\\1\end{bmatrix}, \begin{bmatrix}1\\1\\-1\end{bmatrix}\right\}$ (b) $\text{proj}_{\mathbb{S}}(\vec{e}_1) = \begin{bmatrix}2/3\\2/3\\0\end{bmatrix}$

A22 (a) $\mathcal{B} = \left\{\begin{bmatrix}2\\1\\1\end{bmatrix}, \begin{bmatrix}-2\\5\\1\end{bmatrix}\right\}$ (b) $\text{proj}_{\mathbb{S}}(\vec{e}_1) = \begin{bmatrix}8/9\\-2/9\\2/9\end{bmatrix}$

A23 $\text{proj}_{\mathbb{S}}(\mathbf{p}) = \frac{0}{9}(1+2x-x^2) = 0, \text{proj}_{\mathbb{S}}(\mathbf{q}) = \frac{11}{9}(1+2x-x^2)$

A24 $\text{proj}_{\mathbb{S}}(\mathbf{p}) = \frac{2}{7}(1+3x), \text{proj}_{\mathbb{S}}(\mathbf{q}) = \frac{-15}{21}(1+3x)$

A25 $\text{proj}_{\mathbb{S}}(\mathbf{p}) = \frac{5}{3} + x, \text{proj}_{\mathbb{S}}(\mathbf{q}) = \frac{1}{3}$

A26 $\text{proj}_{\mathbb{S}}(\mathbf{p}) = 1 + \frac{1}{2}x - \frac{1}{2}x^2, \text{proj}_{\mathbb{S}}(\mathbf{p}) = \frac{5}{2}x + \frac{5}{2}x^2$

A27 Since $\{\mathbf{v}_1, \dots, \mathbf{v}_k\}$ is orthogonal, we have $\langle \mathbf{v}_i, \mathbf{v}_j\rangle = 0$ if $i \neq j$. Hence,

$$\begin{aligned}\|\mathbf{v}_1 + \cdots + \mathbf{v}_k\|^2 &= \langle \mathbf{v}_1, \mathbf{v}_1\rangle + \cdots + \langle \mathbf{v}_k, \mathbf{v}_k\rangle \\ &= \|\mathbf{v}_1\|^2 + \cdots + \|\mathbf{v}_k\|^2\end{aligned}$$

Section 7.5 Practice Problems

A1

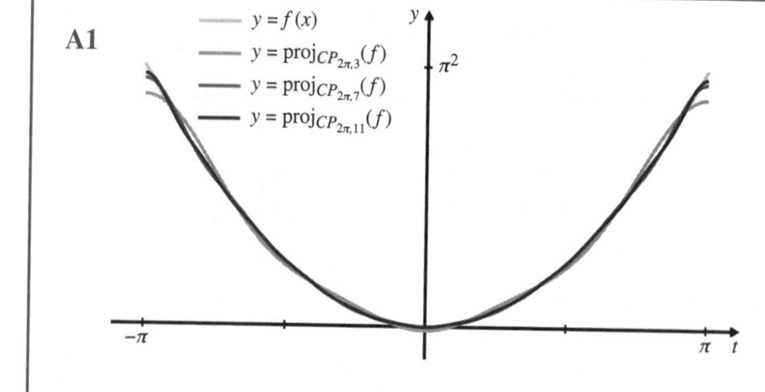

A2

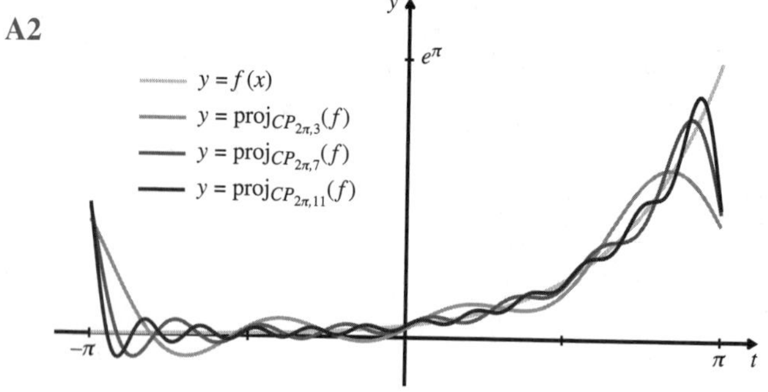

A3

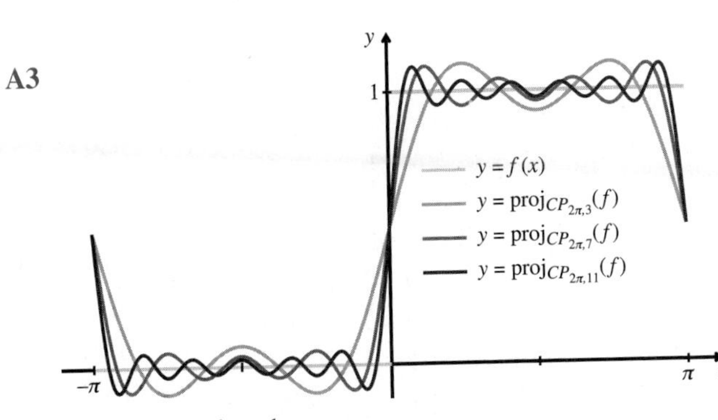

A4 $CP_{2\pi,3}(f) = \frac{1}{2} - \frac{1}{2}\cos 2x$

A5 $CP_{2\pi,3}(f) = 1 + (2\pi^2 - 14)\sin x + \left(-\pi^2 + \frac{5}{2}\right)\sin 2x + \left(\frac{2}{3}\pi^2 - \frac{10}{9}\right)\sin 3x$

A6 $CP_{2\pi,3}(f) = \frac{\pi^2}{3} + 1 - 4\cos x + \cos 2x - \frac{4}{9}\cos 3x$

A7 $CP_{2\pi,3}(f) = \pi - 2\sin x + \sin 2x - \frac{2}{3}\sin 3x$

A8 $CP_{2\pi,3}(f) = \frac{\pi}{4} - \frac{2}{\pi}\cos x + \sin x - \frac{1}{2}\sin 2x - \frac{2}{9\pi}\cos 3x + \frac{1}{3}\sin 3x$

A9 $CP_{2\pi,3}(f) = \frac{4}{\pi}\sin x + \frac{4}{3\pi}\sin 3x$

A10 $CP_{2\pi,3}(f) = \frac{1}{\pi} + \frac{2\sin 1}{\pi}\cos x + \frac{\sin 2}{\pi}\cos 2x + \frac{2\sin 3}{3\pi}\cos 3x$

Chapter 7 Quiz

E1 Neither. The vectors are not of unit length, and the first and third vectors are not orthogonal.

E2 Neither. The vectors are not orthogonal.

E3 The set is orthonormal.

E4 $\left\{\begin{bmatrix}-1\\2\\0\end{bmatrix}, \begin{bmatrix}-12\\-6\\10\end{bmatrix}\right\}$

E5 $\vec{x} = \frac{2}{3}\vec{v}_1 + \frac{5}{3}\vec{v}_2 + \frac{4}{3}\vec{v}_3$

E6 $\text{proj}_{\mathbb{S}}(\vec{x}) = \begin{bmatrix}2\\2\\-1\end{bmatrix}, \text{perp}_{\mathbb{S}}(\vec{x}) = \begin{bmatrix}-1\\1\\0\end{bmatrix}$

E7 It is not positive definite.

E8 It is an inner product.

E9 (a) $\left\{\begin{bmatrix}1\\0\\1\\0\end{bmatrix}, \begin{bmatrix}0\\1\\0\\1\end{bmatrix}, \begin{bmatrix}-1\\1\\1\\-1\end{bmatrix}\right\}$.

(b) $(5/4, -7/4, -5/4, 3/4)$

E10 $\left\{\begin{bmatrix}1\\2\\2\end{bmatrix}, \begin{bmatrix}0\\1\\-1\end{bmatrix}, \begin{bmatrix}4\\-1\\-1\end{bmatrix}\right\}$

E11 $y = \frac{39}{11} + \frac{20}{11}t$

E12 $y = 1 + \frac{7}{2}t^2$

E13 $\begin{bmatrix}2\\0\end{bmatrix}$

E14 $\begin{bmatrix}5/6\\-1/9\end{bmatrix}$

E15 PR is orthogonal since
$(PR)^T(PR) = R^TP^TPR = R^TIR = R^TR = I$

E16 One counter example is $P = \begin{bmatrix}1 & 0\\0 & 1\\0 & 0\end{bmatrix}$.

E17 Let $P = \begin{bmatrix}\vec{v}_1 & \cdots & \vec{v}_n\end{bmatrix}$. Since P has orthonormal columns, then we have

$$(P^TP)_{ij} = \vec{v}_i \cdot \vec{v}_j = 0, \quad \text{for } i \neq j (P^TP)_{ii} = \vec{v}_i \cdot \vec{v}_i = 1$$

Therefore, $P^TP = I$.

E18 The statement may be false when the basis $\{\vec{v}_1, \dots, \vec{v}_k\}$ is not an orthogonal basis for $\mathbb{S}$.

E19 We proved that $\text{perp}_{\mathbb{S}}(\vec{x}) \in \mathbb{S}^{\perp}$. By definition, $\text{proj}_{\mathbb{S}}(\vec{x}) \in \mathbb{S}$ and so, since $\vec{x} \in \mathbb{S}$ we also have that $\vec{x} - \text{proj}_{\mathbb{S}}(\vec{x}) \in \mathbb{S}$. Hence, $\text{perp}_{\mathbb{S}}(\vec{x})$ is in $\mathbb{S}$ and $\mathbb{S}^{\perp}$. So, by Theorem 7.2.2(2), $\text{perp}_{\mathbb{S}}(\vec{x}) = \vec{0}$.

E20 Assume that $\vec{z}$ is another vector in $\mathbb{S}$ such that $\|\vec{x} - \vec{z}\| < \|\vec{x} - \vec{v}\|$ for all $\vec{v} \in \mathbb{S}, \vec{v} \neq \vec{z}$. But, $\vec{y} \in \mathbb{S}$, so that would imply that $\|\vec{x} - \vec{z}\| < \|\vec{x} - \vec{y}\|$. But, since $\vec{z} \in \mathbb{S}$ we must also have that $\|\vec{x} - \vec{y}\| < \|\vec{x} - \vec{z}\|$. Hence, $\vec{z}$ cannot exist.

E21 (a) $\mathcal{B} = \{1, 2 + 3x - 3x^2\}$

(b) $\text{proj}_{\mathbb{S}}(1 + x + x^2) = -1 + \frac{1}{2}x + \frac{3}{2}x^2$

CHAPTER 8

Section 8.1 Practice Problems

A1 Symmetric

A2 Symmetric

A3 Not symmetric

A4 Symmetric

A5 $P = \begin{bmatrix} -1/\sqrt{2} & 1/\sqrt{2} \\ 1/\sqrt{2} & 1/\sqrt{2} \end{bmatrix}, D = \begin{bmatrix} 4 & 0 \\ 0 & -2 \end{bmatrix}$

A6 $P = \begin{bmatrix} -1/\sqrt{10} & 3/\sqrt{10} \\ 3/\sqrt{10} & 1/\sqrt{10} \end{bmatrix}, D = \begin{bmatrix} -4 & 0 \\ 0 & 6 \end{bmatrix}$

A7 $P = \begin{bmatrix} 2/\sqrt{5} & -1/\sqrt{5} \\ 1/\sqrt{5} & 2/\sqrt{5} \end{bmatrix}, D = \begin{bmatrix} 6 & 0 \\ 0 & 1 \end{bmatrix}$

A8 $P = \begin{bmatrix} -1/\sqrt{5} & 2/\sqrt{5} \\ 2/\sqrt{5} & 1/\sqrt{5} \end{bmatrix}, D = \begin{bmatrix} 0 & 0 \\ 0 & 5 \end{bmatrix}$

A9 $P = \begin{bmatrix} 1/\sqrt{3} & -1/\sqrt{2} & -1/\sqrt{6} \\ 1/\sqrt{3} & 1/\sqrt{2} & -1/\sqrt{6} \\ 1/\sqrt{3} & 0 & 2/\sqrt{6} \end{bmatrix}$,

$D = \begin{bmatrix} 2 & 0 & 0 \\ 0 & -1 & 0 \\ 0 & 0 & -1 \end{bmatrix}$

A10 $P = \begin{bmatrix} 2/3 & 2/3 & 1/3 \\ -2/3 & 1/3 & 2/3 \\ 1/3 & -2/3 & 2/3 \end{bmatrix}, D = \begin{bmatrix} 0 & 0 & 0 \\ 0 & 3 & 0 \\ 0 & 0 & -3 \end{bmatrix}$

A11 $P = \begin{bmatrix} 1/\sqrt{5} & 4/\sqrt{45} & -2/3 \\ 0 & 5/\sqrt{45} & 2/3 \\ 2/\sqrt{5} & -2/\sqrt{45} & 1/3 \end{bmatrix}, D = \begin{bmatrix} 9 & 0 & 0 \\ 0 & 9 & 0 \\ 0 & 0 & -9 \end{bmatrix}$

A12 $P = \begin{bmatrix} -1/\sqrt{2} & 1/\sqrt{3} & -1/\sqrt{6} \\ 1/\sqrt{2} & 1/\sqrt{3} & -1/\sqrt{6} \\ 0 & 1/\sqrt{3} & 2/\sqrt{6} \end{bmatrix}$,

$D = \begin{bmatrix} -1 & 0 & 0 \\ 0 & 4 & 0 \\ 0 & 0 & 1 \end{bmatrix}$

A13 $P = \begin{bmatrix} 1/\sqrt{2} & -1/\sqrt{6} & 1/\sqrt{3} \\ 1/\sqrt{2} & 1/\sqrt{6} & -1/\sqrt{3} \\ 0 & 2/\sqrt{6} & 1/\sqrt{3} \end{bmatrix}$,

$D = \begin{bmatrix} 1 & 0 & 0 \\ 0 & 1 & 0 \\ 0 & 0 & -2 \end{bmatrix}$

A14 $P = \begin{bmatrix} 1/\sqrt{3} & -1/\sqrt{6} & 1/\sqrt{2} \\ -1/\sqrt{3} & 1/\sqrt{6} & 1/\sqrt{2} \\ 1/\sqrt{3} & 2/\sqrt{6} & 0 \end{bmatrix}, D = \begin{bmatrix} 0 & 0 & 0 \\ 0 & 3 & 0 \\ 0 & 0 & 1 \end{bmatrix}$

A15 $P = \begin{bmatrix} 1/\sqrt{5} & 4/\sqrt{45} & -2/3 \\ -2/\sqrt{5} & 2/\sqrt{45} & -1/3 \\ 0 & 5/\sqrt{45} & 2/3 \end{bmatrix}$,

$D = \begin{bmatrix} -3 & 0 & 0 \\ 0 & -3 & 0 \\ 0 & 0 & 6 \end{bmatrix}$

A16 $P = \begin{bmatrix} -2/\sqrt{5} & 1/\sqrt{30} & -1/\sqrt{6} \\ 1/\sqrt{5} & 2/\sqrt{30} & -2/\sqrt{6} \\ 0 & 5/\sqrt{30} & 1/\sqrt{6} \end{bmatrix}$,

$D = \begin{bmatrix} -3 & 0 & 0 \\ 0 & -3 & 0 \\ 0 & 0 & 3 \end{bmatrix}$

Section 8.2 Practice Problems

A1 $x_1^2 + 6x_1x_2 - x_2^2$

A2 $3x_1^2 - 4x_1x_2$

A3 $x_1^2 - 2x_2^2 + 6x_2x_3 - x_3^2$

A4 $-2x_1^2 + 2x_1x_2 + 2x_1x_3 + x_2^2 - 2x_2x_3$

A5 Positive definite

A6 Positive definite

A7 Indefinite

A8 Negative definite

A9 Positive definite

A10 Indefinite

A11 (a) $A = \begin{bmatrix} 1 & -3/2 \\ -3/2 & 1 \end{bmatrix}$

(b) $Q(\vec{x}) = \frac{5}{2}y_1^2 - \frac{1}{2}y_2^2, P = \begin{bmatrix} -1/\sqrt{2} & 1/\sqrt{2} \\ 1/\sqrt{2} & 1/\sqrt{2} \end{bmatrix}$

(c) $Q(\vec{x})$ is indefinite.

A12 (a) $A = \begin{bmatrix} 5 & -2 \\ -2 & 2 \end{bmatrix}$

(b) $Q(\vec{x}) = y_1^2 + 6y_2^2, P = \begin{bmatrix} 1/\sqrt{5} & -2/\sqrt{5} \\ 2/\sqrt{5} & 1/\sqrt{5} \end{bmatrix}$

(c) $Q(\vec{x})$ is positive definite.

A13 (a) $A = \begin{bmatrix} -7 & 2 \\ 2 & -4 \end{bmatrix}$

(b) $Q(\vec{x}) = -8y_1^2 - 3y_2^2$, $P = \begin{bmatrix} -2/\sqrt{5} & 1/\sqrt{5} \\ 1/\sqrt{5} & 2/\sqrt{5} \end{bmatrix}$

(c) $Q(\vec{x})$ is negative definite.

A14 (a) $A = \begin{bmatrix} -2 & -3 \\ -3 & -2 \end{bmatrix}$

(b) $Q(\vec{x}) = 1y_1^2 - 5y_2^2$, $P = \begin{bmatrix} -1/\sqrt{2} & 1/\sqrt{2} \\ 1/\sqrt{2} & 1/\sqrt{2} \end{bmatrix}$

(c) $Q(\vec{x})$ is negative definite.

A15 (a) $A = \begin{bmatrix} -2 & 6 \\ 6 & 7 \end{bmatrix}$

(b) $Q(\vec{x}) = 10y_1^2 - 5y_2^2$, $P = \begin{bmatrix} 1/\sqrt{5} & -2/\sqrt{5} \\ 2/\sqrt{5} & 1/\sqrt{5} \end{bmatrix}$

(c) $Q(\vec{x})$ is indefinite.

A16 (a) $A = \begin{bmatrix} 1 & -1 & 3 \\ -1 & 1 & 3 \\ 3 & 3 & -3 \end{bmatrix}$

(b) $Q(\vec{x}) = 2y_1^2 + 3y_2^2 - 6y_3^2$,

$P = \begin{bmatrix} -1/\sqrt{2} & 1/\sqrt{3} & 1/\sqrt{6} \\ 1/\sqrt{2} & 1/\sqrt{3} & 1/\sqrt{6} \\ 0 & 1/\sqrt{3} & -2/\sqrt{6} \end{bmatrix}$

(c) $Q(\vec{x})$ is indefinite.

A17 (a) $A = \begin{bmatrix} -4 & 1 & 0 \\ 1 & -5 & -1 \\ 0 & -1 & -4 \end{bmatrix}$

(b) $Q(\vec{x}) = -3y_1^2 - 6y_2^2 - 4y_3^2$,

$P = \begin{bmatrix} -1/\sqrt{3} & -1/\sqrt{6} & 1/\sqrt{2} \\ -1/\sqrt{3} & 2/\sqrt{6} & 0 \\ 1/\sqrt{3} & 1/\sqrt{6} & 1/\sqrt{2} \end{bmatrix}$

(c) $Q(\vec{x})$ is negative definite.

(a) $A = \begin{bmatrix} 3 & -1 & -1 \\ -1 & 5 & 1 \\ -1 & 1 & 3 \end{bmatrix}$

(b) $Q(\vec{x}) = 6y_1^2 + 3y_2^2 + 2y_3^2$,

$P = \begin{bmatrix} -1/\sqrt{6} & -1/\sqrt{3} & 1/\sqrt{2} \\ 2/\sqrt{6} & -1/\sqrt{3} & 0 \\ 1/\sqrt{6} & 1/\sqrt{3} & 1/\sqrt{2} \end{bmatrix}$

(c) $Q(\vec{x})$ is positive definite.

(a) $A = \begin{bmatrix} 3 & -2 & 4 \\ -2 & 6 & 2 \\ 4 & 2 & 3 \end{bmatrix}$

(b) $Q(\vec{x}) = 7y_1^2 + 7y_2^2 - 2y_3^2$,

$P = \begin{bmatrix} 1/\sqrt{2} & -1/\sqrt{18} & -2/3 \\ 0 & 4/\sqrt{18} & -1/3 \\ 1/\sqrt{2} & 1/\sqrt{18} & 2/3 \end{bmatrix}$

(c) $Q(\vec{x})$ is indefinite.

Section 8.3 Practice Problems

A1

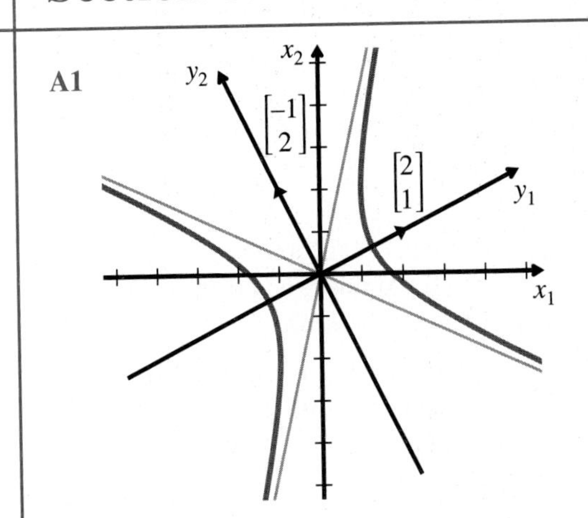

A2

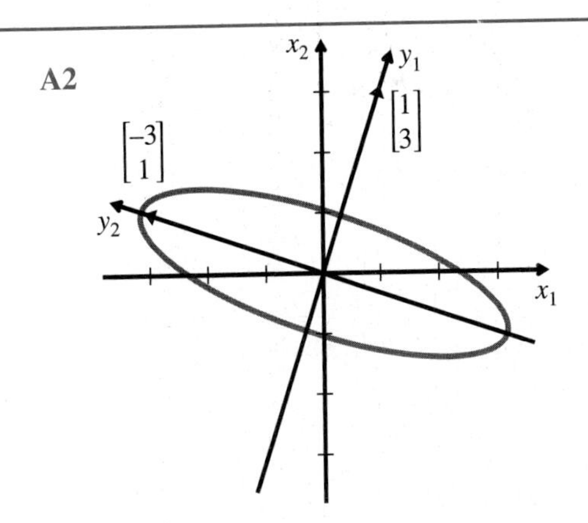

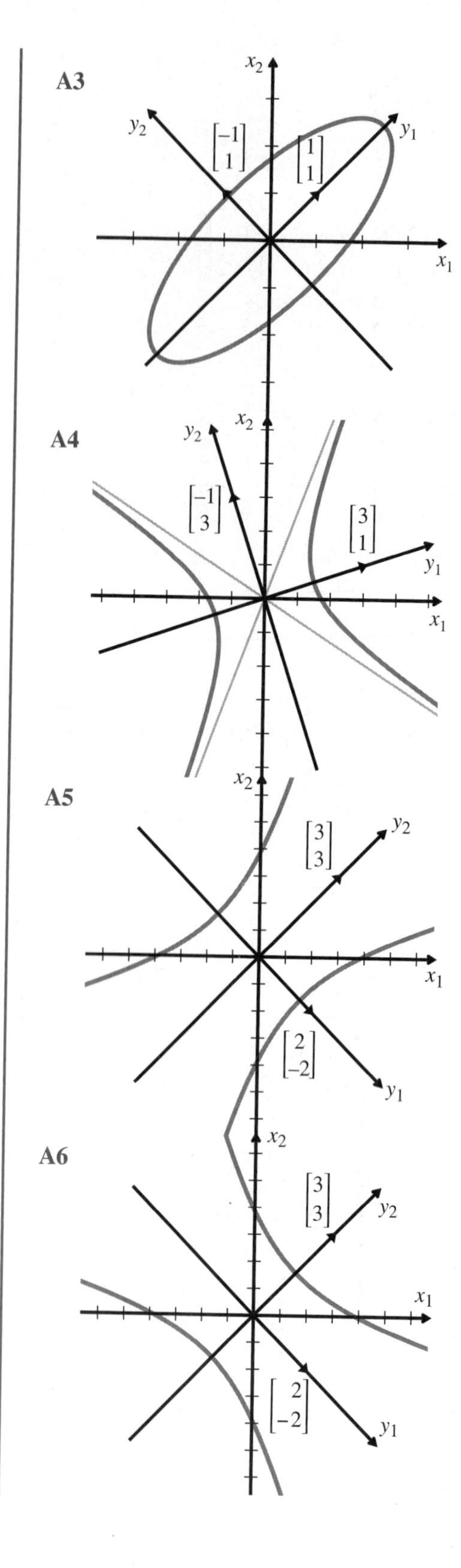

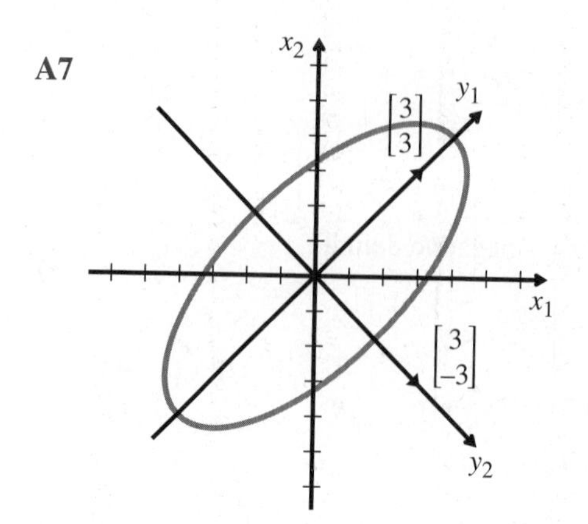

A8 The graph of $\vec{x}^T A\vec{x} = 1$ is a set of two parallel lines. The graph of $\vec{x}^T A\vec{x} = -1$ is the empty set.

A9 The graph of $\vec{x}^T A\vec{x} = 1$ is a hyperbola. The graph of $\vec{x}^T A\vec{x} = -1$ is a hyperbola.

A10 The graph of $\vec{x}^T A\vec{x} = 1$ is a hyperboloid of two sheets. The graph of $\vec{x}^T A\vec{x} = -1$ is a hyperboloid of one sheet.

A11 The graph of $\vec{x}^T A\vec{x} = 1$ is a hyperbolic cylinder. The graph of $\vec{x}^T A\vec{x} = -1$ is a hyperbolic cylinder.

A12 The graph of $\vec{x}^T A\vec{x} = 1$ is a hyperboloid of one sheet. The graph of $\vec{x}^T A\vec{x} = -1$ is a hyperboloid of two sheets.

A13 The graph of $\vec{x}^T A\vec{x} = 1$ is an elliptic cylinder. The graph of $\vec{x}^T A\vec{x} = -1$ is the empty set.

A14 The graph of $Q(\vec{x}) = 0$ is two intersecting lines. The graph of $Q(\vec{x}) = -1$ is a hyperbola opening in the y_2-direction.

A15 The graph of $Q(\vec{x}) = 1$ is two parallel lines. The graph of $Q(\vec{x}) = 0$ is a plane. The graph of $Q(\vec{x}) = -1$ is the empty set.

A16 The graph of $Q(\vec{x}) = 1$ is an elliptic cylinder. The graph of $Q(\vec{x}) = 0$ is a line. The graph of $Q(\vec{x}) = -1$ is the empty set.

A17 The graph of $Q(\vec{x}) = 1$ is an ellipsoid. The graph of $Q(\vec{x}) = 0$ is the point $(0,0,0)$. The graph of $Q(\vec{x}) = -1$ is the empty set.

Section 8.5 Practice Problems

A1 $\sigma_1 = 2, \sigma_2 = 2$

A2 $\sigma_1 = 3, \sigma_2 = 1$

A3 $\sigma_1 = \sqrt{15}$

A4 $U = \begin{bmatrix} 2/\sqrt{5} & -1/\sqrt{5} \\ 1/\sqrt{5} & 2/\sqrt{5} \end{bmatrix}, \Sigma = \begin{bmatrix} 3 & 0 \\ 0 & 2 \end{bmatrix},$
$V = \begin{bmatrix} 1/\sqrt{5} & -2/\sqrt{5} \\ 2/\sqrt{5} & 1/\sqrt{5} \end{bmatrix}$

A5 $U = \begin{bmatrix} 2/\sqrt{5} & -1/\sqrt{5} \\ 1/\sqrt{5} & 2/\sqrt{5} \end{bmatrix}, \Sigma = \begin{bmatrix} 4 & 0 \\ 0 & 1 \end{bmatrix},$
$V = \begin{bmatrix} 1/\sqrt{5} & -2/\sqrt{5} \\ 2/\sqrt{5} & 1/\sqrt{5} \end{bmatrix}$

A6 $U = \begin{bmatrix} 11/\sqrt{95} & 3/\sqrt{26} & 1/\sqrt{30} \\ 7/\sqrt{195} & -4/\sqrt{26} & 2/\sqrt{30} \\ 5/\sqrt{195} & -1/\sqrt{26} & -5/\sqrt{30} \end{bmatrix},$
$\Sigma = \begin{bmatrix} \sqrt{15} & 0 \\ 0 & \sqrt{2} \\ 0 & 0 \end{bmatrix}, V = \begin{bmatrix} 2/\sqrt{13} & -3/\sqrt{13} \\ 3/\sqrt{13} & 2/\sqrt{13} \end{bmatrix}$

A7 $U = \begin{bmatrix} -1/\sqrt{3} & -1/\sqrt{2} & -1/\sqrt{6} \\ -1/\sqrt{3} & 1/\sqrt{2} & -1/\sqrt{6} \\ -1/\sqrt{3} & 0 & 2/\sqrt{6} \end{bmatrix},$
$\Sigma = \begin{bmatrix} \sqrt{15} & 0 \\ 0 & 0 \\ 0 & 0 \end{bmatrix}, V = \begin{bmatrix} -2/\sqrt{5} & 1/\sqrt{5} \\ 1/\sqrt{5} & 2/\sqrt{5} \end{bmatrix}$

A8 $U = \begin{bmatrix} 1/\sqrt{3} & 1/\sqrt{2} & 1/\sqrt{6} \\ 1/\sqrt{3} & 0 & -2/\sqrt{6} \\ -1/\sqrt{3} & 1/\sqrt{2} & -1/\sqrt{6} \end{bmatrix},$
$\Sigma = \begin{bmatrix} \sqrt{3} & 0 \\ 0 & \sqrt{2} \\ 0 & 0 \end{bmatrix}, V = \begin{bmatrix} 0 & 1 \\ 1 & 0 \end{bmatrix}$

A9 $U = \begin{bmatrix} 0 & -1 & 0 \\ 2/\sqrt{5} & 0 & -1/\sqrt{5} \\ 1/\sqrt{5} & 0 & 2/\sqrt{5} \end{bmatrix}, \Sigma = \begin{bmatrix} \sqrt{10} & 0 \\ 0 & \sqrt{2} \\ 0 & 0 \end{bmatrix},$
$V = \begin{bmatrix} 1/\sqrt{2} & -1/\sqrt{2} \\ 1/\sqrt{2} & 1/\sqrt{2} \end{bmatrix}$

A10 $U = \begin{bmatrix} 1/\sqrt{3} & 0 & 1/\sqrt{3} & -1/\sqrt{3} \\ 1/\sqrt{3} & 1/\sqrt{3} & -1/\sqrt{3} & 0 \\ -1/\sqrt{3} & 1/\sqrt{3} & 0 & -1/\sqrt{3} \\ 0 & 1/\sqrt{3} & 1/\sqrt{3} & 1/\sqrt{3} \end{bmatrix},$
$\Sigma = \begin{bmatrix} \sqrt{3} & 0 & 0 \\ 0 & \sqrt{3} & 0 \\ 0 & 0 & \sqrt{3} \\ 0 & 0 & 0 \end{bmatrix}, V = I$

A11 $U = \frac{1}{2}\begin{bmatrix} 1 & -1 & 1 & 1 \\ 1 & -1 & -1 & -1 \\ 1 & 1 & 1 & -1 \\ 1 & 1 & -1 & 1 \end{bmatrix}, \Sigma = \begin{bmatrix} \sqrt{20} & 0 \\ 0 & 0 \\ 0 & 0 \\ 0 & 0 \end{bmatrix},$
$V = \begin{bmatrix} 1/\sqrt{5} & -2/\sqrt{5} \\ 2/\sqrt{5} & 1/\sqrt{5} \end{bmatrix}$

A12 $U = \begin{bmatrix} 1/\sqrt{2} & 1/\sqrt{2} \\ -1/\sqrt{2} & 1/\sqrt{2} \end{bmatrix}, \Sigma = \begin{bmatrix} 2 & 0 & 0 \\ 0 & \sqrt{2} & 0 \end{bmatrix},$
$V = \begin{bmatrix} 0 & 1 & 0 \\ 1/\sqrt{2} & 0 & -1/\sqrt{2} \\ 1/\sqrt{2} & 0 & 1/\sqrt{2} \end{bmatrix}$

A13 $U = \begin{bmatrix} 3/\sqrt{90} & -1/\sqrt{15} & -1/\sqrt{3} & 3/\sqrt{18} \\ 4/\sqrt{90} & 2/\sqrt{15} & -1/\sqrt{3} & -2/\sqrt{18} \\ -4/\sqrt{90} & 3/\sqrt{15} & 0 & 2/\sqrt{18} \\ 7/\sqrt{90} & 1/\sqrt{15} & 1/\sqrt{3} & 1/\sqrt{18} \end{bmatrix},$
$\Sigma = \begin{bmatrix} \sqrt{18} & 0 \\ 0 & \sqrt{3} \\ 0 & 0 \\ 0 & 0 \end{bmatrix}, V = \begin{bmatrix} 1/\sqrt{5} & -2/\sqrt{5} \\ 2/\sqrt{5} & 1/\sqrt{5} \end{bmatrix}$

A14 $A^+ = \begin{bmatrix} 1/3 & -1/3 \\ 1/6 & 1/3 \end{bmatrix}$

A15 $A^+ = \begin{bmatrix} 2/15 & 2/15 & 2/15 \\ -1/15 & -1/15 & -1/15 \end{bmatrix}$

A16 $A^+ = \begin{bmatrix} 1/2 & 1/2 \\ 1/4 & -1/4 \\ 1/4 & -1/4 \end{bmatrix}$

Chapter 8 Quiz

E1 $P = \begin{bmatrix} 1/\sqrt{6} & 1/\sqrt{3} & 1/\sqrt{2} \\ -2/\sqrt{6} & 1/\sqrt{3} & 0 \\ -1/\sqrt{6} & -1/\sqrt{3} & 1/\sqrt{2} \end{bmatrix}$ and $D = \begin{bmatrix} 6 & 0 & 0 \\ 0 & -3 & 0 \\ 0 & 0 & 4 \end{bmatrix}$.

E2 $2x_1^2 + 8x_1x_2 + 5x_2^2$

E3 $x_1^2 - 4x_1x_3 - 3x_2^2 + 8x_2x_3 + x_3^2$

E4 (a) $A = \begin{bmatrix} 5 & 2 \\ 2 & 5 \end{bmatrix}$

(b) $Q(\vec{x}) = 7y_1^2 + 3y_2^2$, $P = \begin{bmatrix} 1/\sqrt{2} & -1/\sqrt{2} \\ 1/\sqrt{2} & 1/\sqrt{2} \end{bmatrix}$

(c) $Q(\vec{x})$ is positive definite.

(d) $Q(\vec{x}) = 1$ is an ellipse, and $Q(\vec{x}) = 0$ is the origin.

E5 (a) $A = \begin{bmatrix} 2 & -3 & -3 \\ -3 & -3 & 2 \\ -3 & 2 & -3 \end{bmatrix}$

(b) $Q(\vec{x}) = -5y_1^2 + 5y_2^2 - 4y_3^2$,

$P = \begin{bmatrix} 0 & -2/\sqrt{6} & 1/\sqrt{3} \\ -1/\sqrt{2} & 1/\sqrt{6} & 1/\sqrt{3} \\ 1/\sqrt{2} & 1/\sqrt{6} & 1/\sqrt{3} \end{bmatrix}$

(c) $Q(\vec{x})$ is indefinite.

(d) $Q(\vec{x}) = 1$ is a hyperboloid of two sheets, and $Q(\vec{x}) = 0$ is a cone.

E6

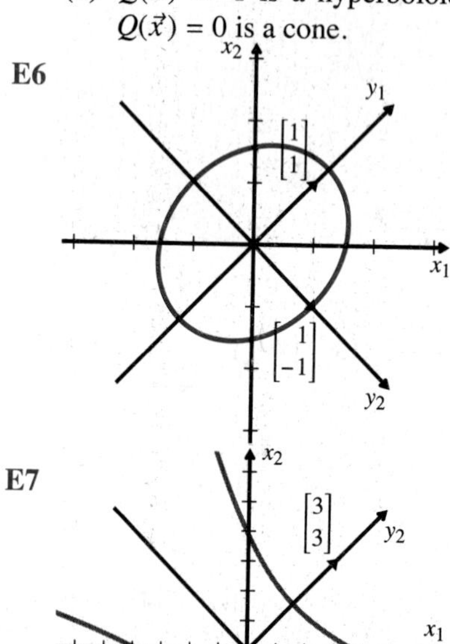

E7

E8 $U = \begin{bmatrix} /1\sqrt{5} & -2/\sqrt{5} \\ 2/\sqrt{5} & 1/\sqrt{5} \end{bmatrix}$, $\Sigma = \begin{bmatrix} \sqrt{10} & 0 \\ 0 & 0 \end{bmatrix}$,

$V = \begin{bmatrix} 1/\sqrt{2} & -1/\sqrt{2} \\ 1/\sqrt{2} & 1/\sqrt{2} \end{bmatrix}$

E9 $U = \begin{bmatrix} 5/\sqrt{35} & 0 & 2/\sqrt{14} \\ 1/\sqrt{35} & 3/\sqrt{10} & -1/\sqrt{14} \\ -3/\sqrt{35} & 1/\sqrt{10} & 3/\sqrt{14} \end{bmatrix}$,

$\Sigma = \begin{bmatrix} \sqrt{7} & 0 \\ 0 & \sqrt{2} \\ 0 & 0 \end{bmatrix}$, $V = \begin{bmatrix} 1/\sqrt{5} & -2/\sqrt{5} \\ 2/\sqrt{5} & 1/\sqrt{5} \end{bmatrix}$

E10 $A^+ = V\Sigma^+U^T = \begin{bmatrix} 1/2 & 0 \\ 1/2 & 0 \end{bmatrix}$

E11 Since A is positive definite, we have that

$$\langle \vec{x}, \vec{x} \rangle = \vec{x}^T A \vec{x} \geq 0$$

and $\langle \vec{x}, \vec{x} \rangle = 0$ if and only if $\vec{x} = \vec{0}$.
Since A is symmetric, we have that

$$\begin{aligned} \langle \vec{x}, \vec{y} \rangle &= \vec{x}^T A \vec{y} = \vec{x} \cdot A\vec{y} = A\vec{y} \cdot \vec{x} = (A\vec{y})^T \vec{x} \\ &= \vec{y}^T A^T \vec{x} = \vec{y}^T A \vec{x} = \langle \vec{y}, \vec{x} \rangle \end{aligned}$$

For any $\vec{x}, \vec{y}, \vec{z} \in \mathbb{R}^n$ and $s, t \in \mathbb{R}$ we have

$$\begin{aligned} \langle \vec{x}, s\vec{y} + t\vec{z} \rangle &= \vec{x}^T A(s\vec{y} + t\vec{z}) = \vec{x}^T A(s\vec{y}) + \vec{x}^T A(t\vec{z}) \\ &= s\vec{x}^T A\vec{y} + t\vec{x}^T A\vec{z} = s\langle \vec{x}, \vec{y} \rangle + t\langle \vec{x}, \vec{z} \rangle \end{aligned}$$

Thus, $\langle \vec{x}, \vec{y} \rangle$ is an inner product on $\mathbb{R}^n$.

E12 Since A is a 4×4 symmetric matrix, there exists an orthogonal matrix P that diagonalizes A. Since the only eigenvalue of A is 3, we must have $P^TAP = 3I$. Then multiply on the left by P and on the right by P^T and we get

$$A = P(3I)P^T = 3PP^T = 3I$$

E13 $A = \begin{bmatrix} -1/13 & -18/13 \\ -18/13 & 14/13 \end{bmatrix}$

E14 If A is invertible, then $\text{rank}(A) = n$ by the Invertible Matrix Theorem. Hence, A has n non-zero singular values by Theorem 8.5.1. But, since an $n \times n$ matrix has exactly n singular values, the matrix A cannot have 0 as a singular value.
On the other hand, if A is not invertible, then $\text{rank}(A) < n$ by the Invertible Matrix Theorem. Hence, A has less than n non-zero singular values by Theorem 8.5.1. So, A has 0 as a singular value.

E15 The statement is true. We have $P^TAP = B$, so

$$B^2 = (P^TAP)(P^TAP) = P^TAIAP = P^TA^2P$$

E16 The statement is true. We have $P^TAP = B$, so

$$B^T = (P^TAP)^T = P^TA^T(P^T)^T = P^TAP = B$$

E17 The statement is false. Take $A = I$.

E18 This is true by the Principal Axis Theorem.

E19 The statement is false since $Q(\vec{0}) = 0$.

E20 The statement is false. Take $A = \begin{bmatrix} 1 & 0 & 0 \\ 0 & -1 & 0 \\ 0 & 0 & 0 \end{bmatrix}$.

E21 The statement is false. Take $A = \begin{bmatrix} -1 & -2 \\ -2 & -1 \end{bmatrix}$.

E22 If $A = U\Sigma V^T$, then

$$\begin{aligned} |\det A| &= |\det(U\Sigma V^T)| = |\det U||\det \Sigma||\det V^T| \\ &= 1(|\det \Sigma|)(1) = \sigma_1 \dots \sigma_n \end{aligned}$$

CHAPTER 9

Section 9.1 Practice Problems

A1 $\bar{z} = 3 + 5i, |z| = \sqrt{34}$

A2 $\bar{z} = 2 - 7i, |z| = \sqrt{53}$

A3 $\bar{z} = 4i, |z| = 4$

A4 $\bar{z} = -1 + 2i, |z| = \sqrt{5}$

A5 $\bar{z} = 2, |z| = 2$

A6 $\bar{z} = -3 - 2i, |z| = \sqrt{13}$

A7 $z = \sqrt{18}e^{-i\frac{3\pi}{4}}, \operatorname{Arg} z = -\frac{3\pi}{4}$

A8 $z = 2e^{-i\frac{\pi}{6}}, \operatorname{Arg} z = -\frac{\pi}{6}$

A9 $z = 2e^{i\frac{5\pi}{6}}, \operatorname{Arg} z = \frac{5\pi}{6}$

A10 $z = 4e^{-i\frac{2\pi}{3}}, \operatorname{Arg} z = -\frac{2\pi}{3}$

A11 $z = 3e^{i\pi}, \operatorname{Arg} z = \pi$

A12 $z = \sqrt{8}e^{i\frac{3\pi}{4}}, \operatorname{Arg} z = \frac{3\pi}{4}$

A13 $\frac{1}{2} - \frac{\sqrt{3}}{2}i$

A14 $1 - \sqrt{3}i$

A15 $1 + \sqrt{3}i$

A16 $-\frac{3}{\sqrt{2}} + \frac{3}{\sqrt{2}}i$

A17 $\sqrt{3} - i$

A18 $-\frac{\sqrt{3}}{2} + \frac{1}{2}i$

A19 $5 + 7i$

A20 $-3 - 4i$

A21 $-7 + 2i$

A22 $-14 + 5i$

A23 $9 + 7i$

A24 $-10 - 10i$

A25 $2 + 25i$

A26 -2

A27 $\operatorname{Re}(z) = 3, \operatorname{Im}(z) = -6$

A28 $\operatorname{Re}(z) = 17, \operatorname{Im}(z) = -1$

A29 $\operatorname{Re}(z) = 24/37, \operatorname{Im}(z) = 4/37$

A30 $\operatorname{Re}(z) = 0, \operatorname{Im}(z) = 1$

A31 $\frac{2}{13} - \frac{3}{13}i$

A32 $-\frac{4}{13} - \frac{19}{13}i$

A33 $-\frac{2}{17} + \frac{25}{17}i$

A34 $z_1z_2 = 2\sqrt{2}\left(\cos\frac{7\pi}{12} + i\sin\frac{7\pi}{12}\right)$,

$$\frac{z_1}{z_2} = \frac{2}{\sqrt{2}}\left(\cos\frac{-\pi}{12} + i\sin\frac{-\pi}{12}\right)$$

A35 $z_1z_2 = 2\sqrt{2}\left(\cos\frac{11\pi}{12} + i\sin\frac{11\pi}{12}\right)$,

$$\frac{z_1}{z_2} = \sqrt{2}\left(\cos\frac{17\pi}{12} + i\sin\frac{17\pi}{12}\right)$$

A36 $z_1z_2 = 4i, \frac{z_1}{z_2} = -\frac{\sqrt{3}}{2} + \frac{1}{2}i$

A37 $z_1z_2 = 6\sqrt{2}\left(\cos\frac{19\pi}{12} + i\sin\frac{19\pi}{12}\right)$,

$$\frac{z_1}{z_2} = \frac{\sqrt{18}}{2}\left(\cos\frac{-\pi}{12} + i\sin\frac{-\pi}{12}\right)$$

A38 -4

A39 $-54 - 54i$

A40 $-8 - 8\sqrt{3}i$

A41 $512(\sqrt{3} + i)$

A42 The roots are $\cos\left(\frac{\pi + 2k\pi}{5}\right) + i\sin\left(\frac{\pi + 2k\pi}{5}\right), 0 \le k \le 4$.

A43 The roots are $2\left[\cos\left(\frac{-\frac{\pi}{2} + 2k\pi}{4}\right) + i\sin\left(\frac{-\frac{\pi}{2} + 2k\pi}{4}\right)\right]$, $0 \le k \le 3$.

A44 The roots are $2^{1/3}\left[\cos\left(\frac{-\frac{5\pi}{6} + 2k\pi}{3}\right) + i\sin\left(\frac{-\frac{5\pi}{6} + 2k\pi}{3}\right)\right]$, $0 \le k \le 2$.

A45 The roots are $17^{1/6}\left[\cos\left(\frac{\theta + 2k\pi}{3}\right) + i\sin\left(\frac{\theta + 2k\pi}{3}\right)\right]$, $0 \le k \le 2$, where $\theta = \arctan(4)$.

Section 9.2 Practice Problems

A1 $\mathbf{z} = \begin{bmatrix} 2 \\ 1+i \end{bmatrix}$

A2 The system is inconsistent.

A3 $\mathbf{z} = \begin{bmatrix} 1-2i \\ 2+3i \end{bmatrix}$

A4 $\mathbf{z} = \begin{bmatrix} 2 \\ 0 \end{bmatrix} + t\begin{bmatrix} -1+i \\ 1 \end{bmatrix}, t \in \mathbb{C}.$

A5 $\mathbf{z} = \begin{bmatrix} i \\ 3 \\ 0 \end{bmatrix} + t\begin{bmatrix} -2 \\ i \\ 1 \end{bmatrix}, t \in \mathbb{C}$

A6 $\mathbf{z} = \begin{bmatrix} -3i \\ 1 \\ 0 \\ 0 \end{bmatrix} + t\begin{bmatrix} 1 \\ -i \\ 1 \\ 0 \end{bmatrix} + s\begin{bmatrix} -1+i \\ -1-i \\ 0 \\ 1 \end{bmatrix}, t, s \in \mathbb{C}$

A7 $\mathbf{z} = \begin{bmatrix} 2 \\ 1 \\ i \end{bmatrix}$

A8 The system is inconsistent.

A9 $\mathbf{z} = \begin{bmatrix} 1-i \\ \frac{4}{5}+\frac{2}{5}i \\ -\frac{1}{5}-\frac{3}{5}i \end{bmatrix}$

A10 $\mathbf{z} = \begin{bmatrix} 5+i \\ -2+2i \\ -i \\ 0 \end{bmatrix} + t\begin{bmatrix} -1+2i \\ 0 \\ -i \\ 1 \end{bmatrix}, t \in \mathbb{C}$

Section 9.3 Practice Problems

A1 $\begin{bmatrix} -5-3i \\ i \end{bmatrix}$

A2 $\begin{bmatrix} 5-3i \\ 7+8i \\ -1-9i \end{bmatrix}$

A3 $\begin{bmatrix} -10+4i \\ 4+6i \end{bmatrix}$

A4 $\begin{bmatrix} -4-3i \\ -1-7i \\ -12+i \end{bmatrix}$

A5 (a) $[L] = \begin{bmatrix} 1+2i & 3+i \\ 1 & 1-i \end{bmatrix}$

(b) $L(2+3i, 1-4i) = \begin{bmatrix} 3-4i \\ -1-2i \end{bmatrix}$

(c) A basis for Range(L) is $\left\{\begin{bmatrix} 1+2i \\ 1 \end{bmatrix}\right\}$.

A basis for Null(L) is $\left\{\begin{bmatrix} -1+i \\ 1 \end{bmatrix}\right\}$.

A6 $\langle \mathbf{u}, \mathbf{v} \rangle = 2+5i, \langle \mathbf{v}, \mathbf{u} \rangle = 2-5i, \|\mathbf{u}\| = \sqrt{18}, \|\mathbf{v}\| = \sqrt{33}$

A7 $\langle \mathbf{u}, \mathbf{v} \rangle = -6i, \langle \mathbf{v}, \mathbf{u} \rangle = 6i, \|\mathbf{u}\| = \sqrt{22}, \|\mathbf{v}\| = \sqrt{20}$

A8 $\langle \mathbf{u}, \mathbf{v} \rangle = 3-i, \langle \mathbf{v}, \mathbf{u} \rangle = 3+i, \|\mathbf{u}\| = \sqrt{11}, \|\mathbf{v}\| = 2$

A9 $\langle \mathbf{u}, \mathbf{v} \rangle = 4+i, \langle \mathbf{v}, \mathbf{u} \rangle = 4-i, \|\mathbf{u}\| = \sqrt{15}, \|\mathbf{v}\| = \sqrt{5}$

A10 $(ZW)^* = \begin{bmatrix} -1-2i & 4-2i \\ -1+i & -1-i \end{bmatrix} = W^*Z^*$

A11 $(ZW)^* = \begin{bmatrix} -1-2i & 2-i \\ -1 & 2 \end{bmatrix} = W^*Z^*$

A12 Not unitary

A13 Unitary

A14 Unitary

A15 Unitary

A16 $\mathcal{B} = \left\{\begin{bmatrix} 1 \\ i \\ 1 \end{bmatrix}, \begin{bmatrix} 5 \\ -4i \\ -1 \end{bmatrix}, \begin{bmatrix} -1/2 \\ -i \\ 3/2 \end{bmatrix}\right\}$

A17 $\mathcal{B} = \left\{\begin{bmatrix} 1+i \\ 1-i \\ 1 \end{bmatrix}, \begin{bmatrix} 2 \\ -1+i \\ 2i \end{bmatrix}, \begin{bmatrix} i \\ -1+i \\ 1-i \end{bmatrix}\right\}$

A18 $\text{proj}_{\mathbb{S}}(\mathbf{z}) = \begin{bmatrix} \frac{2}{3}+i \\ 2+\frac{1}{3}i \\ 3+\frac{4}{3}i \end{bmatrix}$

A19 $\text{proj}_{\mathbb{S}}(\mathbf{z}) = \begin{bmatrix} -1+\frac{3}{4}i \\ \frac{9}{4} \\ -\frac{7}{4}-\frac{5}{4}i \end{bmatrix}$

A20 $1-2i$

A21 $3+i$

A22 5

A23 $-2-4i$

A24 (a) We have

$$1 = \det I = \det(U^*U) = \det(U^*)\det U$$
$$= \overline{\det U}\det U = |\det U|^2$$

Therefore, $|\det U| = 1$.

(b) The matrix $U = \begin{bmatrix} i & 0 \\ 0 & 1 \end{bmatrix}$ is unitary and $\det U = i$.

Section 9.4 Practice Problems

A1 $P = \begin{bmatrix} 1+i & 1+i \\ -2 & 1 \end{bmatrix}, D = \begin{bmatrix} 0 & 0 \\ 0 & 3 \end{bmatrix}$

A2 $P = \begin{bmatrix} 3+4i & 3-4i \\ -5 & -5 \end{bmatrix}, D = \begin{bmatrix} 4i & 0 \\ 0 & -4i \end{bmatrix}$

A3 The matrix is not diagonalizable.

A4 $P = \begin{bmatrix} 1+i & -1-i \\ 2 & 2 \end{bmatrix}, D = \begin{bmatrix} 1+2i & 0 \\ 0 & -1 \end{bmatrix}$

A5 If $\sin\theta = 0$, then $P = I$. Otherwise, $P = \begin{bmatrix} i & -i \\ 1 & 1 \end{bmatrix}$ and

$$D = \begin{bmatrix} \cos\theta + i\sin\theta & 0 \\ 0 & \cos\theta - i\sin\theta \end{bmatrix}.$$

A6 $P = \begin{bmatrix} 1 & 1 & 1 \\ 0 & i & -i \\ 2 & 1 & 1 \end{bmatrix}, D = \begin{bmatrix} 0 & 0 & 0 \\ 0 & 1+2i & 0 \\ 0 & 0 & 1-2i \end{bmatrix}$

A7 The matrix is not diagonalizable.

A8 $P = \begin{bmatrix} 1 & -i & i \\ -2 & -1 & -1 \\ 1 & 1 & 1 \end{bmatrix}, D = \begin{bmatrix} 1 & 0 & 0 \\ 0 & 2+i & 0 \\ 0 & 0 & 2-i \end{bmatrix}$

A9 $P = \begin{bmatrix} 1-i & -i & -2i \\ 4 & 0 & -1+i \\ 0 & 2 & 2 \end{bmatrix}, D = \begin{bmatrix} 1 & 0 & 0 \\ 0 & 1 & 0 \\ 0 & 0 & 4 \end{bmatrix}$

A10 $P = \begin{bmatrix} 0 & 1+2i & 1-2i \\ 1 & 3+i & 3-i \\ 1 & 5 & 5 \end{bmatrix}, D = \begin{bmatrix} 1 & 0 & 0 \\ 0 & 2+i & 0 \\ 0 & 0 & 2-i \end{bmatrix}$

A11 The matrix is not diagonalizable.

A12 $P = \begin{bmatrix} i & 1 & -1 \\ 0 & 1-i & 1+i \\ 1 & 1 & 1 \end{bmatrix}, D = \begin{bmatrix} 1+i & 0 & 0 \\ 0 & 2 & 0 \\ 0 & 0 & 0 \end{bmatrix}$

Section 9.5 Practice Problems

A1 Normal

A2 Not normal

A3 Normal

A4 Not normal

A5 $U = \begin{bmatrix} 1/\sqrt{2} & -1/\sqrt{2} \\ 1/\sqrt{2} & 1/\sqrt{2} \end{bmatrix}, D = \begin{bmatrix} 2i & 0 \\ 0 & 2+2i \end{bmatrix}$

A6 $U = \begin{bmatrix} (1-i)/\sqrt{6} & (-1+i)/\sqrt{3} \\ 2/\sqrt{6} & 1/\sqrt{3} \end{bmatrix}$ to $D = \begin{bmatrix} 3i & 0 \\ 0 & 6i \end{bmatrix}$

A7 $U = \begin{bmatrix} -i/\sqrt{2} & i/\sqrt{2} \\ 1/\sqrt{2} & 1/\sqrt{2} \end{bmatrix}, D = \begin{bmatrix} 3+5i & 0 \\ 0 & 3-5i \end{bmatrix}$

A8 $U = \begin{bmatrix} (1+i)/\sqrt{6} & (-1-i)/\sqrt{3} \\ 2/\sqrt{6} & 1/\sqrt{3} \end{bmatrix}, D = \begin{bmatrix} 4 & 0 \\ 0 & 1 \end{bmatrix}$

A9 $U = \begin{bmatrix} 1/\sqrt{2} & -1/\sqrt{2} \\ 1/\sqrt{2} & 1/\sqrt{2} \end{bmatrix}, D = \begin{bmatrix} i & 0 \\ 0 & -i \end{bmatrix}$

A10 $U = \begin{bmatrix} (\sqrt{2}+i)/\sqrt{12} & (\sqrt{2}+i)/2 \\ -3/\sqrt{12} & 1/2 \end{bmatrix}, D = \begin{bmatrix} 1 & 0 \\ 0 & 5 \end{bmatrix}.$

A11 $U = \begin{bmatrix} 0 & (1+i)/\sqrt{3} & (1+i)/\sqrt{6} \\ 1 & 0 & 0 \\ 0 & 1/\sqrt{3} & -2/\sqrt{6} \end{bmatrix},$

$$D = \begin{bmatrix} 2 & 0 & 0 \\ 0 & 2 & 0 \\ 0 & 0 & -1 \end{bmatrix}$$

A12 $U = \begin{bmatrix} 1 & 0 & 0 \\ 0 & (-1+i)/\sqrt{3} & (1-i)/\sqrt{6} \\ 0 & 1/\sqrt{3} & 2/\sqrt{6} \end{bmatrix},$

$$D = \begin{bmatrix} i & 0 & 0 \\ 0 & -2 & 0 \\ 0 & 0 & 1 \end{bmatrix}$$

Chapter 9 Quiz

E1 $4+6i$

E2 $8+7i$

E3 $11-2i$

E4 $15+20i$

E5 $\frac{11}{5}-\frac{2}{5}i$

E6 $-\frac{3}{5}+\frac{4}{5}i$

E7 (a) $z_1 = 2\left(\cos\frac{-\pi}{3} + i\sin\frac{-\pi}{3}\right), z_2 = 2\sqrt{2}\left(\cos\frac{\pi}{4} + i\sin\frac{\pi}{4}\right)$

(b) $z_1z_2 = 4\sqrt{2}\left(\cos\frac{-\pi}{12} + i\sin\frac{-\pi}{12}\right)$,

$\frac{z_1}{z_2} = \frac{1}{\sqrt{2}}\left(\cos\frac{-7\pi}{12} + i\sin\frac{-7\pi}{12}\right)$

E8 $w_0 = \frac{1}{\sqrt{2}} + \frac{1}{\sqrt{2}}i, w_1 = -\frac{1}{\sqrt{2}} - \frac{1}{\sqrt{2}}i$

E9 The system is inconsistent.

E10 $\mathbf{z} = \begin{bmatrix} 2 \\ 1+2i \\ 0 \end{bmatrix} + t\begin{bmatrix} -i \\ i \\ 1 \end{bmatrix}, t \in \mathbb{C}$

E11 $\begin{bmatrix} 7-i \\ 3+5i \\ 9+3i \end{bmatrix}$

E12 $\begin{bmatrix} 3+i \\ -i \\ 2 \end{bmatrix}$

E13 $11-4i$

E14 $11+4i$

E15 $\sqrt{27}$

E16 $\frac{1}{15}\begin{bmatrix} 29-23i \\ 4+11i \\ 22-8i \end{bmatrix}$

E17 $\left\{\begin{bmatrix} 1 \\ i \\ i \end{bmatrix}, \begin{bmatrix} 2i \\ 1 \\ 1 \end{bmatrix}, \begin{bmatrix} 0 \\ -\frac{1}{2}+\frac{1}{2}i \\ \frac{1}{2}-\frac{1}{2}i \end{bmatrix}\right\}$

E18 $\text{proj}_{\mathbb{S}}(\mathbf{z}) = \begin{bmatrix} 1/2 \\ i/2 \\ 2-i \end{bmatrix}$

E19 Show $UU^* = I$.

E20 A is not diagonalizable.

E21 $P = \begin{bmatrix} i & -1 & 1-i \\ 1 & 0 & -1-i \\ 0 & 1 & 2 \end{bmatrix}, D = \begin{bmatrix} 3 & 0 & 0 \\ 0 & 3 & 0 \\ 0 & 0 & 2+i \end{bmatrix}$

E22 (a) A is Hermitian if and only if $k=-1$.

(b) $U = \begin{bmatrix} (3-i)/\sqrt{14} & (3-i)/\sqrt{35} \\ -2/\sqrt{14} & 5/\sqrt{35} \end{bmatrix}, D = \begin{bmatrix} -2 & 0 \\ 0 & 5 \end{bmatrix}$

Index

Index of Notations

$\mathbb{R}^2$	2-dimensional Euclidean space, 2
$\vec{x}$	vector in $\mathbb{R}^n$, 2, 10, 48
$\vec{0}$	zero vector in $\mathbb{R}^n$, 5, 49
$\vec{PQ}$	directed line segment, 8, 12
$\mathbb{R}^3$	3-dimensional Euclidean space, 10
$\operatorname{Span} \mathcal{B}$	span of a set of vectors, 18, 52, 150, 237, 249
$\{\vec{e}_1, \dots, \vec{e}_n\}$	standard basis vectors for $\mathbb{R}^n$, 20, 23, 56
$\lVert \mathbf{v} \rVert$	length (norm) of a vector, 30, 33, 60, 412, 492
$\vec{x} \cdot \vec{y}$	dot product of vectors, 32, 33, 60
$\vec{x} \times \vec{y}$	cross product of vectors in $\mathbb{R}^3$, 38
$\mathbb{R}^n$	n-dimensional Euclidean space, 48, 241
$\operatorname{proj}_{\vec{x}}(\vec{y})$	projection of $\vec{y}$ onto $\vec{x}$, 64
$\operatorname{perp}_{\vec{x}}(\vec{y})$	projection of $\vec{y}$ perpendicular to $\vec{x}$, 66
$\left[A \mid \vec{b}\right] = \left[\vec{v}_1 \ \cdots \ \vec{v}_n \mid \vec{b}\right]$	matrix representation of a system of linear equations, 88
$A \sim B$	row equivalent matrices, 89
cR_i	elementary row operation multiply the i-th row by $c \neq 0$, 90
$R_i \updownarrow R_j$	elementary row operation swap the i-th row and the j-th row, 90
$R_i + cR_j$	elementary row operation add c times the j-th row to the i-th row, 90
$\operatorname{rank}(A)$	number of leading ones in the RREF of a matrix, 107
$\dim \mathbb{S}$	dimension of a vector space (subspace), 122, 255
$m \times n$ matrix	rectangular array with m rows and n columns, 147
$M_{m \times n}(\mathbb{R})$	vector space of $m \times n$ matrices with real entries, 147, 241
$(A)_{ij} = a_{ij}$	the ij-th entry of a matrix A, 148
$\operatorname{diag}(d_{11}, \dots, d_{nn})$	$n \times n$ diagonal matrix, 148
$O_{m,n}$	$m \times n$ zero matrix, 150
$\vec{x}^T$	transpose of a vector, 152
A^T	transpose of a matrix, 152
$A\vec{x}$	matrix-vector multiplication, 154, 157
AB	matrix multiplication, 159, 161
$I = I_n$	$n \times n$ identity matrix, 165
$\begin{bmatrix} A & B \end{bmatrix}$	block matrix, 166
$f_A(\vec{x})$	matrix mapping corresponding to A, 172
$[L]$	standard matrix of a linear mapping, 176
$M \circ L$	composition of linear mappings, 179, 279
$\operatorname{Id}(\vec{x})$	identity mapping, 180, 274
R_θ	rotation about the origin through an angle θ, 184
$\operatorname{refl}_{\vec{n}}$	reflection in the plane with normal vector $\vec{n}$, 189
$\operatorname{Range}(L)$	range of a linear mapping, 192, 274

Null(L)	nullspace of a linear mapping, 193, 274
Col(A)	column space of a matrix, 195
Null(A)	nullspace of a matrix, 198
Row(A)	row space of a matrix, 200
Null(A^T)	left nullspace of a matrix, 200
A^{-1}	inverse of the square matrix A, 207
L^{-1}	inverse of the linear mapping L, 212, 279
$P_n(\mathbb{R})$	vector space of polynomials of degree at most n, 235, 241
$\mathbf{0}(x)$	zero polynomial, 236
$\mathbf{x}$	vector in a vector space, 240
$\mathbf{0}$	zero vector in a vector space, 240
$(-\mathbf{v})$	additive inverse of a vector $\mathbf{v}$ in a vector space, 240
$\mathcal{F}(a,b)$	vector space of all functions $f:(a,b)\to\mathbb{R}$, 241
$C(a,b)$	vector space of all functions that are continuous on the interval (a,b), 241
$[\mathbf{x}]_{\mathcal{B}}$	coordinates of $\mathbf{x}$ with respect to the basis $\mathcal{B}$, 264
rank(L)	rank of a linear mapping, 276
nullity(L)	nullity of a linear mapping, 276
$[L]_{\mathcal{B}}$	matrix of a linear mapping with respect to the basis $\mathcal{B}$, 284, 288
$_{C}[L]_{\mathcal{B}}$	matrix of a linear mapping with respect to bases $\mathcal{B}$ and C, 296
$\det A$	determinant of a matrix, 307, 310, 311
C_{ij}	(i,j)-cofactor of a matrix, 308, 311
adj(A)	adjugate of the matrix A, 331
E_λ	eigenspace of the eigenvalue λ, 350
$C(\lambda)$	characteristic polynomial of a matrix, 352
$\mathbb{S}^\perp$	orthogonal complement of the subspace $\mathbb{S}$, 391
$\text{proj}_{\mathbb{S}}(\vec{y})$	projection of $\vec{y}$ onto the subspace $\mathbb{S}$, 393
$\text{perp}_{\mathbb{S}}(\vec{y})$	projection of $\vec{y}$ perpendicular to the subspace $\mathbb{S}$, 393
$\langle\,,\rangle$	inner product of a vector space, 410, 489, 491
$\vec{x}^TA\vec{x}$	quadratic form on $\mathbb{R}^n$, 432
$A = U\Sigma V^T$	singular value decomposition of the matrix A, 454
A^+	pseudoinverse (Moore-Penrose inverse) of the matrix A, 458
Re(z)	real part of a complex number, 466
Im(z)	imaginary part of a complex number, 466
$\lvert z\rvert$	modulus of a complex number, 467
$\bar{z}$	complex conjugate of a complex number, 469
$\arg z$	argument of a complex number, 472
$\mathbb{C}^n$	complex vector space of column vectors with n-entries, 486
$M_{m\times n}(\mathbb{C})$	vector space of $m\times n$ matrices with complex entries, 486
$\bar{\mathbf{z}}$	complex conjugate of $\mathbf{z}\in\mathbb{C}^n$, 488
$\bar{Z}$	complex conjugate of $Z\in M_{m\times n}(\mathbb{C})$, 488
$\mathbf{z}^*$	conjugate transpose of $\mathbf{z}\in\mathbb{C}^n$, 488
Z^*	conjugate transpose of $Z\in M_{m\times n}(\mathbb{C})$, 488

PREFACE

But God hath chosen the foolish things of the world to confound the wise; and God hath chosen the weak things of the world to confound the things which are mighty; And base things of the world, and things which are despised, hath God chosen, yea, and things which are not, to bring to nought things that are: That no flesh should glory in his presence.

I Corinthians 1: 27-29

When I thought to write this manuscript, it was to be a self-appointed task; the purpose being to record for my family reflections from my lifelong study of *the Revelation*. However, as I spent time in prayer and pondered the assignment, I realized it was not merely a personal commitment, but had become a task set by God. I felt deeply troubled because of my inadequacies. It was one thing to write for my family, but quite another to write for Divinity. Age and ill health has dulled my thinking and shortened my memory; nevertheless, it is now, when I am least capable, that He has chosen to lay upon me this assignment. While writing, I have learned that we need nothing more than a willing heart, a yielding ear and a ready pen. God equips the unequipped: He draws upon scriptures hidden away in the recesses of one's mind, stored from years of study, and makes them available to His servants.

Now that the writing is finished and I review the manuscript, I read it as a student and not as an author. I am amazed at its content. I can boast of no credentials, nor any particular presence of mind which would give me the authority or ability to comment on His great *Revelation*, graciously given through Jesus Christ to man. I willingly recognize that I am a fisherman, not a theologian; however, my lack of formal training is remediated by the guidance of the Holy Spirit. He is my source of information; my sole credential.

I have written with a degree of anxiety only in the sense of a deep awareness of the necessity to be constantly guided by God's Spirit. I have approached the work reverently, praying always that I would be given the ability to coherently unravel the threads of truth and clearly present those portions of the *Revelation* that are complex and interwoven. This task was a monumental challenge, indeed, since I had, up to the time of writing, never thoroughly understood a number of the precepts in the way they were being revealed to me. As I wrote, I recognized that some of this teaching departed notably from traditional interpretation; I realized that I may be criticized and possibly condemned. I can appreciate such an initial reaction since I, too, found it difficult to accept and surrender to this new understanding. The final writing, often startling in content, is somewhat foreign to what I previously held as truth. We grow so comfortable in our denominational doctrines that we often never challenge our current beliefs, nor look beyond them. When we are confronted by a new understanding that might differ or run deeper, we tend to hide behind the pulpit and cry heresy.

I am confident that all who continue to the end of this teaching study will applaud its truth. It is a deep study and one that should not be taken lightly. It necessitates a full reading from start to finish to grasp God's remarkable plan. Examine the work with the discernment of the Bereans who "*...searched the Scriptures every day to see if what Paul said was true*" (Acts

17: 11). The *Revelation* is a Holy document woven with fine threads, almost as a priestly garment. It must be carefully and prayerfully examined to appreciate its fabric of intricate and interwoven teachings. In it lies the key to portions of Daniel's prophecies, and it brings to light the great depth of all of God's inspired writings. The manuscript, though difficult at times, is grounded in Truth; its foundation being the scriptures.

I have reiterated a number of themes often – probably to the point of being overly repetitious. I am aware that readers, representing all stages of spiritual growth, may need the assistance of repetition to be able to receive, in one short study, a maze of teachings which were given to me over a period of possibly two years.

This has been a painful time of meditation, prayer and study; nevertheless, it has been a richly rewarding period of growth – a time of being broken and re-taught. In my unskilled way, I have attempted to share these truths with you. Be patient and persevering; in so doing, God will reward your efforts as you <u>too</u> seek to understand His *Revelation*.

I acknowledge, with pleasure, my daughters, Teresa McLeod and Deborah Robinson, who whole-heartedly encouraged me in this attempt to serve God. Especially do I recognize the tremendous sacrifice of my daughter, Deborah Melanie, who made time, despite the multitude of demands in her schedule, to word process and edit this work. She and my granddaughter, Marah Gilead Zola, have acted as readers and editors of the manuscript. Beyond these, I express my appreciation of Adrian D. Fryling of Texas for challenging me to write; to my beloved Pastor Jacques Belzile of Hinton, Alberta, who admonishes me to serve; my sister, Nancy McKinlay, who has faith in my ability; and my brothers and sisters in the faith: Debbie Christiansen, Noreen Campbell, Aileen Rudd, Walter Robinson and Robert and Georgina Gartshore -- all of whom have acted as readers and critics of this work.

TRIUMPH AND TERROR:

Revelation – the Truth for End Times

A Study in *The Revelation*

Introduction

The ***Revelation*** was written around 95 A.D. when John was banished to the Isle of Patmos by Roman rule under Emperor Domitian (81-95 A.D.). He was liberated upon Domitian's death in 96 A.D.

The Greek word for 'Revelation' is 'Apokalupsis', the uncovering of something hidden. The book is written in 'apocalyptic' form, a type of Jewish literature which uses symbolic imagery to communicate hope – the ultimate triumph of God – to those in the midst of persecution. The events are ordered according to literary patterns, rather than in strict chronological order (from Living Life Application Bible notes).[1]

The whole treatise is all the more authentic in that God chose this 'apocalyptic' form of writing for John to record the revealing, or unveiling, of things that would come to pass. It is the style of writing used by the learned, hardly that of an uneducated fisherman like John; moreover, it seems totally foreign when compared to the other books penned by him. This is confirmation, surely, that John writes under the inspiration of God.

---

[1]The Living Life Application Bible notes given for the book of the Revelation under Vital Statistics - Special Features (Wheaton, Ill.: Tyndale House Publishers, 1971), p. 1966.

CHAPTER ONE

In **Chapter One**, John is commissioned by Christ to write letters of warning to seven Christian Churches in Asia. The letters, addressed to the pastors of the early churches, speak also to the seven Church Ages throughout church history. The Study notes discuss the hidden meaning within the Church names and draw the reader's attention to the person of Christ and the promises of God.

Revelation 1

[1] The Revelation of Jesus Christ, which God gave unto him, to shew unto his servants things which must shortly come to pass; and he sent and signified it by his angel unto his servant John:
[2] Who bare record of the word of God, and of the testimony of Jesus Christ, and of all things that he saw.
[3] Blessed is he that readeth, and they that hear the words of this prophecy, and keep those things which are written therein: for the time is at hand.
[4] John to the seven churches which are in Asia: Grace be unto you, and peace, from him which is, and which was, and which is to come; and from the seven Spirits which are before his throne;
[5] And from Jesus Christ, who is the faithful witness, and the first begotten of the dead, and the prince of the kings of the earth. Unto him that loved us, and washed us from our sins in his own blood,
[6] And hath made us kings and priests unto God and his Father; to him be glory and dominion for ever and ever. Amen.
[7] Behold, he cometh with clouds; and every eye shall see him, and they also which pierced him: and all kindreds of the earth shall wail because of him. Even so, Amen.
[8] I am Alpha and Omega, the beginning and the ending, saith the Lord, which is, and which was, and which is to come, the Almighty.
[9] I John, who also am your brother, and companion in tribulation, and in the kingdom and patience of Jesus Christ, was in the isle that is called Patmos, for the word of God, and for the testimony of Jesus Christ.
[10] I was in the Spirit on the Lord's day, and heard behind me a great voice, as of a trumpet,
[11] Saying, I am Alpha and Omega, the first and the last: and, What thou seest, write in a book, and send it unto the seven churches which are in Asia; unto Ephesus, and unto Smyrna, and unto Pergamos, and unto Thyatira, and unto Sardis, and unto Philadelphia, and unto Laodicea.
[12] And I turned to see the voice that spake with me. And being turned, I saw seven golden candlesticks;
[13] And in the midst of the seven candlesticks one like unto the Son of man, clothed with a garment down to the foot, and girt about the paps with a golden girdle.
[14] His head and his hairs were white like wool, as white as snow; and his eyes were as a flame of fire;
[15] And his feet like unto fine brass, as if they burned in a furnace; and his voice as the sound of many waters.
[16] And he had in his right hand seven stars: and out of his mouth went a sharp twoedged sword: and his countenance was as the sun shineth in his strength.
[17] And when I saw him, I fell at his feet as dead. And he laid his right hand upon me, saying unto me, Fear not; I am the first and the last:
[18] I am he that liveth, and was dead; and, behold, I am alive for evermore, Amen; and have the keys of hell and of death.
[19] Write the things which thou hast seen, and the things which are, and the things which shall be hereafter;
[20] The mystery of the seven stars which thou sawest in my right hand, and the seven golden candlesticks. The seven stars are the angels of the seven churches: and the seven candlesticks which thou sawest are the seven churches.

A Revelation from God Through Jesus Christ

Notes

Revelation, Chapter 1

Revelation 1: 1-2

1 *The Revelation of Jesus Christ, which God gave unto him, to shew unto his servants things which must shortly come to pass; and he sent and signified it by his angel unto his servant John:*

2 *Who bare record of the word of God, and of the testimony of Jesus Christ, and of all things that he saw.*

The Amplified Bible gives an excellent rendition of these two verses of the Revelation. The book heading reads, "*The Revelation to John*", and the verses go on to say "*(This is) the Revelation of Jesus Christ (His unveiling of the divine mysteries). God gave it to Him to disclose and make known to His bond servants certain things which must shortly and speedily come to pass in their entirety. And He sent and communicated it through His angel (messenger) to His bond servant John. Who has testified to and vouched for all that he saw (in his visions), the word of God and the testimony of Jesus Christ.*"

The Revelation 'apokalupsis' means an uncovering of something hidden; a revealing or unveiling. According to William Barclay, the Greek word 'apokalupsis' used here simply means the disclosure of any fact.[2] He states that this same word is used in the original Greek transcript whenever God reveals His will and direction to His servants (for example, see Galations 2: 2). The Revelation is from God through Jesus Christ and not from John the Divine, as the title heading in the King James version seems to imply. William Barclay agrees it is not a revelation about Jesus Christ but, rather, a revelation given by God to Jesus Christ, who then gave it to John. Jesus, always obedient to his Father's command (John 12: 49), discloses the revelation which God gave him. John faithfully reports everything he sees and records the testimony of Jesus so that God's servants might know "*the things which must shortly come to pass.*" Jesus again assures John in Revelation 22: 6 that the purpose of the Revelation is to "*show his servants things which must shortly be done.*" God is careful to articulate this in the first and the last chapter of the Revelation.

The prophetic writing, however, not only foretells the future of man and the universe, but in its unveiling it marvelously reveals our Saviour, Jesus

---

[2]William Barclay, "The Daily Study Bible", The Revelation, Volume 1 (Burlington, Ontario: Welch Publishing Company, revised 1976), pp. 21-24.

Notes

Christ, the returning word of God (Revelation 19: 13). He is portrayed as the Magnificent, Almighty Victor, the King of kings, the Lord of lords, the Holy Son of God.

One wonders why verse one tells us these things will shortly come to pass when almost two thousand years later we are still waiting. Peter explains in II Peter 3: 8:

> 8 *But, beloved, be not ignorant of this one thing, that one day is with the Lord as a thousand years, and a thousand years as one day.*

Revelation 1: 3

> 3 *Blessed is he that readeth, and they that hear the words of this prophecy, and keep those things which are written therein: for the time is at hand.*

We are told that all will be blessed who read, hear and keep the words written in this prophecy. Revelation 22: 10 tell us this is not a sealed prophecy; that is, it is open to interpretation and understanding. Matthew 13: 10-16 teaches if we choose to be blind and deaf, we will not be healed (cleansed from sin). Read the promise of Matthew 7: 7:

> 7 *Ask, and it shall be given you; seek, and ye shall find; knock, and it shall be opened unto you;*

Revelation 1: 4

> 4 *John to the seven churches which are in Asia: Grace be unto you, and peace, from him which is, and which was, and which is to come; and from the seven Spirits which are before his throne;*

John speaks of grace first and then peace in his salutation to the seven churches in Asia. The themes of grace and peace occur repeatedly throughout the New Testament: "*By grace are you saved through faith, which is the gift of God*" (Ephesians 2:8); "*...a free gift offered to all men*" (Romans 5: 18); "*...to be carnally minded is death; but to be spiritually minded is life and peace*" (Romans 8: 6).

John tells the seven churches that grace and peace come from the Lord Jesus Christ and from the seven Spirits before his throne. The seven

Notes

Spirits are found in Isaiah 11: 2. A straightforward listing of these spirits is:

- the Spirit of the Lord – the Holy Spirit
- the Spirit of wisdom – all wisdom
- the Spirit of understanding – as God understands
- the Spirit of council – all authority
- the Spirit of might – all power
- the Spirit of knowledge – full knowledge
- the Spirit of the fear of the Lord – complete reverence and total submission to God

Revelation 1: 5

5 *And from Jesus Christ, who is the faithful witness, and the first begotten of the dead, and the prince of the kings of the earth. Unto him that loved us, and washed us from our sins in his own blood,*

So that we might be assured that the Revelation is, in fact, from Jesus Christ, we are now presented with his credentials:

- he is the faithful witness (John 14: 24; Revelation 3: 14)
- he is the first to rise from the dead (Colossians 1: 15)
- he is the Prince of the kings of the earth, King of kings and Lord of lords. (Revelation 19: 16)

This King, our Saviour, first loved us; so great is his love, he washed us from our sins in his own blood. He bled and died on our behalf.

Revelation 1: 6

6 *And hath made us kings and priests unto God and his Father; to him be glory and dominion for ever and ever. Amen.*

The declaration is that we, the believers, have already been made kings and priests in God's eyes. Revelation 5: 10 further informs us that as kings and priests, we will reign on earth. Romans 8: 17 says (paraphrased) that *we will be glorified together with Christ, having been made through him heirs of God and joint-heirs with Christ.*

Revelation 1:7

7 *Behold, he cometh with clouds; and every eye shall see him, and they also which pierced him: and all kindreds of the earth shall wail because of him. Even so, Amen.*

This verse speaks to the whole of the earth, especially to the Israelite and the Gentile Tribulation Saints. The Church will have been raptured and therefore won't be on earth at this time.

Jesus Christ Identified / John at Patmos

Notes

The first phrase, "*he cometh with clouds and every eye shall see him*" is a promise yet to be fulfilled. Recall Acts 1: 9-11, when Jesus is taken up into heaven and a cloud receives him, two men in white apparel inform his followers that he will return in the same manner as they have seen him go. The next phrase, "*And they also which pierced him; and all the kindreds of the earth shall wail because of Him*" hearkens back to the prophecy in Zechariah 12: 10-11:

> 10 *And I will pour upon the house of David* (the Israelites), *<u>and upon the inhabitants of Jerusalem</u>* (the Gentile), *the spirit of grace and of supplication; and they shall look upon me whom they have pierced, and they shall mourn for him, as one mourneth for his only son, and shall be in bitterness for him, as one is in bitterness for his firstborn.*
>
> 11 *That day shall be great mourning in Jerusalem...*

Revelation 1: 8

8 *I am Alpha and Omega, the beginning and the ending, saith the Lord, which is, and which was, and which is to come, the Almighty.*

Jesus always identifies himself. Here he says, "I am…"

- the Alpha and Omega;
- the beginning and the end;
- the Almighty; which is, which was, and which is to come.

The Alpha and Omega are the beginning and the ending letters of the Greek alphabet and so he calls himself the first and the last.

He is also qualified to call himself the beginning and the end for it is written in John 1: 1: "*In the beginning was the word, and the word was with God, and the word was God.*"

God is Almighty. God is, God was and God is to come. Revelation 21: 3 affirms this:

> 3 *...The tabernacle of God is with men, and He will dwell with them, and they shall be His people, and God himself shall be with them and be their God.*

Revelation 1: 9

9 *I John, who also am your brother, and companion in tribulation, and in the kingdom and patience of Jesus Christ, was in the isle that is called Patmos, for the word of God, and for the testimony of Jesus Christ.*

Because John is born of the Spirit, he states that he is in the Kingdom (of God - Luke 17: 20-21) and with patience endures persecution, as do his brothers in the faith. According to tradition, he has been banished by the

Emperor Domitean (in 95 A.D.) to Patmos, a desolate island in the Aegean Sea.[3] This dear old Saint is sent to labour in the quarries because, as Bishop of Ephesus, disregarding persecution, he continues to teach Jesus and Him crucified.

Notes

Revelation 1: 10

10 I was in the Spirit on the Lord's day, and heard behind me a great voice, as of a trumpet,

The early Christians referred to the day after their Sabbath (the weekly Sabbath in which they worshipped God from Friday sundown to Saturday sundown) as 'the Lord's Day'. For them, it was 'the Lord's Day' because this was the day in which they first witnessed Jesus risen from the dead. They began to commemorate the day as a celebration day in which they came together to worship and receive the blessed sacrament of the Holy Communion. They called this sacrament the 'Love Feast' and on this day they shared food with one another. To the early Christian believer, the Lord's Day soon became a replacement of the previous Holy Sabbath observed on Saturday. Read Matthew 28: 1-10:

1 In the end of the Sabbath, as it began to dawn toward the first day of the week, came Mary Magdalene and the other Mary to see the sepulchre.

9 ...behold, Jesus met them, saying, All hail. And they came and held him by the feet, and worshipped him..

Notice the reference in Acts 20: 7:

7 And upon the first day of the week, when the disciples came together to break bread, Paul preached unto them...

Thus, on the Lord's Day, while in Patmos, John is so deeply immersed in meditation and prayer that he, through the Spirit, hears from behind him the voice of Jesus. To his spiritual ears, it sounds like the call of a trumpet.

Perhaps here, at the beginning, we should understand the difference between 'the Lord's Day' and 'the Day of the Lord'. Every day of the week is 'the Lord's Day' but we, as Christians guided by the Word, honour one day as a Sabbath. We set aside this day as a special day of rest and joint worship, calling it 'the Lord's Day'. 'The Day of the Lord' is far more glorious and awesome. It is the day of the resurrection; it is also the day of

---

[3]William Smith, Bible Dictionary (Philadelphia/Toronto: John C. Winston Co., 1948), p. 486.
Charles Pfeiffer, Howard F. Vos, John Rea, eds, Wycliffe Bible Encyclopedia, Volume 2 (Chicago: Moody Bible Institute, 1975), p. 1287.
Ray Stedman, God's Final Word (Grand Rapids, MI: Discovery House Publishers, 1991), p. 12.

the Rapture, but most commonly refers to the day when Jesus returns for his Millennial reign. The final 'Day of the Lord', as pertaining to this earth, is the day of the White Throne Judgement (II Peter 3: 10-13).

Notes

Revelation 1: 11

11 Saying, I am Alpha and Omega, the first and the last; and, What thou seest, write in a book, and send it unto the seven churches which are in Asia; unto Ephesus, and unto Smyrna, and unto Pergamos, and unto Thyatira, and unto Sardis, and unto Philadelphia, and unto Laodicea.

Jesus again refers to himself as the Alpha and Omega, the first and the last. John is told to record in a book all that he sees; he is directed to send it to the seven churches which the disciples had planted in Asia.

It is no accident that the churches are recorded in the order that they appear in the scripture, nor is it by accident that these cities in Asia are named with names meaningful to church history. The wonder of it all is that God's hand can be seen in the prophetic naming of these cities which were in existence long before Christianity was brought to Asia. The awesome fact is that the Churches throughout the ages have played out, with uncanny accuracy, the truth of the predictions hidden within the church names. The spiritual condition of the first-century church of Ephesus was, indeed, as the name declares, 'desirable'. We of the last century church are of the predicted Laodicean age, continually crying 'people's rights'. God foresaw the deterioration of the church and wrote letters warning each church age. All Christians should periodically evaluate their spiritual condition and that of their church in the light of these letters. Not only do they carry a message for the believers in John's time, but speak encouragement as well as warning to all churches down through the ages. One should note the meaning within the Church name to see how it applies throughout Church history.[4]

- Ephesus: desirable
- Smyrna: myrrh (bitterness)
- Pergamos: marriage
- Thyatira: continual sacrifice
- Sardis: remnant
- Philadelphia: brotherly love
- Laodicea: people's rights

---

[4]Edward G. Dobson et. al., The Complete Bible Commentary (Nashville, Tenn.: Thomas Nelson Publishers, 1999), pp. 1782-1788.

Seven Candlesticks / Seven Stars

Notes

Revelation 1: 12

12 And I turned to see the voice that spake with me. And being turned, I saw seven golden candlesticks;

When John turns to see the voice which has spoken to him, he first sees seven golden candlesticks. Verse 20 tells us that the seven candlesticks represent the seven churches which Jesus has just listed. The first church of the seven starts off as a spiritually complete church, beloved and pleasing to the Father.

Revelation 1: 13

13 And in the midst of the seven candlesticks one like unto the Son of man, clothed with a garment down to the foot, and girt about the paps with a golden girdle.

Verses 16 to 18 affirm that the angel standing in the midst of the Churches is Jesus Christ in angelic form. We should note that Jesus wears the robes of an angel (Revelation 15: 6) and a Priest with a golden sash around his chest.

Revelation 1: 14 - 15

14 His head and his hairs were white like wool, as white as snow; and his eyes were as a flame of fire;

15 And his feet like unto fine brass, as if they burned in a furnace; and his voice as the sound of many waters.

The white hair, white as snow, denotes wisdom. He is, as is God, the '*ancient of days.*' Read Daniel 7: 9:

> *9 I beheld till the thrones were cast down, and the Ancient of days did sit, whose garment was white as snow, and the hair of his head like the pure wool: his throne was like the fiery flame, and his wheels as burning fire.*

Jesus is, as his Father, the all-knowing, all-seeing, all-consuming God. Scripture reveals Them through consistent descriptive images; for example, "*A fiery stream issued before him*" (Daniel 7: 10); "*His feet sparkled like the colour of burnished brass*" (Daniel 10: 6); "*His voice was like a noise of many waters*" (Ezekiel 43: 2).

The Two-Edged Sword / Jesus Identified / Keys of Hell and Death

Notes

Revelation 1: 16

16 And he had in his right hand seven stars: and out of his mouth went a sharp two-edged sword: and his countenance was as the sun shineth in his strength.

The seven stars in Christ's right hand represents the angels of the seven Churches (read verse 20 for a detailed explanation). Hebrews 4: 12 tells us the two-edged sword is the Word and reveals its power and purpose:

> *12 For the word of God is quick, and powerful, and sharper than any two-edged sword, piercing even to the dividing asunder of soul and spirit, and of the joints and marrow, and is a discerner of the thoughts and intents of the heart.*

John sees the countenance of Jesus as it shines with the brilliance of the sun. Matthew 17: 2 gives a similar image of Jesus as it describes him transfigured before Peter, James and John: '*his face did shine as the sun, and his raiment was white as the light.*'

Revelation 1: 17

17 And when I saw him, I fell at his feet as dead. And he laid his right hand upon me, saying unto me, Fear not; I am the first and the last:

John is slain in the presence of Christ, both from fear and through the power of the Holy Spirit. John is as a dead man when he sees Jesus in all His glory. He is overwhelmed and overpowered by the Glory of God. Jesus lays his right hand on John and tells him not to be afraid, giving assurance that the one speaking to him is the compassionate Christ.

Revelation 1: 18

18 I am he that liveth, and was dead; and, behold, I am alive for evermore, Amen; and have the keys of hell and of death.

Jesus again identifies himself. John is reminded that Jesus lives, but was dead; also, that he will live forever more. Keep in mind, when we study Revelation 9: 1, that Jesus is the one who has the keys of hell and of death. It is only through his consent that Satan is allowed, for a short moment, to use them.

The Mystery of the Stars and the Candlesticks

Notes

Revelation 1: 19

19 Write the things which thou hast seen, and the things which are, and the things which shall be hereafter;

Jesus commissions John to write not only the things which he has already seen, but also to write the things which are, as well as the future things which will come. We will see as we study that Chapter 1 speaks of the past; Chapters 2 and 3 focus on the present (all the cities and churches that John is so familiar with); and Chapters 4 to 22 look to future events which will come to pass.

Revelation 1: 20

20 The mystery of the seven stars which thou sawest in my right hand, and the seven golden candlesticks. The seven stars are the angels of the seven churches: and the seven candlesticks which thou sawest are the seven churches.

Jesus reveals the mystery of the seven stars and the seven golden candlesticks. He tells John the seven stars are the angels of the seven churches. Since the seven letters are written to the angels of the seven churches, we know that the angels represent the protective keepers of the flock: the pastors and the leadership of the Church. To establish that this refers to the pastors and not actual 'angels', remember that God does not communicate with angels through letter.

The seven candlesticks are the churches; they are the light bearers to the world, as are the two witnesses of Chapter 11.

See Appendix 1 for information on the Seven Mysteries of God.

CHAPTER TWO

The four letters of **Chapters Two** are letters of encouragement and warning. The commentary gives the approximate dates of the four Church Ages to which the letters apply. The letters promise the Crown of Life to the overcomer, and eternal punishment to the unrepentant. The Study identifies the Nicolaitans, Gnostics and Baal worshippers, and gives a brief teaching on each misleading doctrine. It shows how satanic worship slowly crept into the pure teachings of the apostles and inevitably corrupted the early Church.

Revelation 2

[1] Unto the angel of the church of Ephesus write; These things saith he that holdeth the seven stars in his right hand, who walketh in the midst of the seven golden candlesticks;
[2] I know thy works, and thy labour, and thy patience, and how thou canst not bear them which are evil: and thou hast tried them which say they are apostles, and are not, and hast found them liars:
[3] And hast borne, and hast patience, and for my name's sake hast laboured, and hast not fainted.
[4] Nevertheless I have somewhat against thee, because thou hast left thy first love.
[5] Remember therefore from whence thou art fallen, and repent, and do the first works; or else I will come unto thee quickly, and will remove thy candlestick out of his place, except thou repent.
[6] But this thou hast, that thou hatest the deeds of the Nicolaitans, which I also hate.
[7] He that hath an ear, let him hear what the Spirit saith unto the churches; To him that overcometh will I give to eat of the tree of life, which is in the midst of the paradise of God.
[8] And unto the angel of the church in Smyrna write; These things saith the first and the last, which was dead, and is alive;
[9] I know thy works, and tribulation, and poverty, (but thou art rich) and I know the blasphemy of them which say they are Jews, and are not, but are the synagogue of Satan.
[10] Fear none of those things which thou shalt suffer: behold, the devil shall cast some of you into prison, that ye may be tried; and ye shall have tribulation ten days: be thou faithful unto death, and I will give thee a crown of life.
[11] He that hath an ear, let him hear what the Spirit saith unto the churches; He that overcometh shall not be hurt of the second death.
[12] And to the angel of the church in Pergamos write; These things saith he which hath the sharp sword with two edges;
[13] I know thy works, and where thou dwellest, even where Satan's seat is: and thou holdest fast my name, and hast not denied my faith, even in those days wherein Antipas was my faithful martyr, who was slain among you, where Satan dwelleth.
[14] But I have a few things against thee, because thou hast there them that hold the doctrine of Balaam, who taught Balac to cast a stumblingblock before the children of Israel, to eat things sacrificed unto idols, and to commit fornication.
[15] So hast thou also them that hold the doctrine of the Nicolaitans, which thing I hate.
[16] Repent; or else I will come unto thee quickly, and will fight against them with the sword of my mouth.
[17] He that hath an ear, let him hear what the Spirit saith unto the churches; To him that overcometh will I give to eat of the hidden manna, and will give him a white stone, and in the stone a new name written, which no man knoweth saving he that receiveth it.
[18] And unto the angel of the church in Thyatira write; These things saith the Son of God, who hath his eyes like unto a flame of fire, and his feet are like fine brass;
[19] I know thy works, and charity, and service, and faith, and thy patience, and thy works; and the last to be more than the first.
[20] Notwithstanding I have a few things against thee, because thou sufferest that woman Jezebel, which calleth herself a prophetess, to teach and to seduce my servants to commit fornication, and to eat things sacrificed unto idols.
[21] And I gave her space to repent of her fornication; and she repented not.
[22] Behold, I will cast her into a bed, and them that commit adultery with her into great tribulation, except they repent of their deeds.
[23] And I will kill her children with death; and all the churches shall know that I am he which searcheth the reins and hearts: and I will give unto every one of you according to your works.
[24] But unto you I say, and unto the rest in Thyatira, as many as have not this doctrine, and which have not known the depths of Satan, as they speak; I will put upon you none other burden.
[25] But that which ye have already hold fast till I come.
[26] And he that overcometh, and keepeth my works unto the end, to him will I give power over the nations:
[27] And he shall rule them with a rod of iron; as the vessels of a potter shall they be broken to shivers: even as I received of my Father
[28] And I will give him the morning star.
[29] He that hath an ear, let him hear what the Spirit saith unto the churches.

Preamble for the Letter to Ephesus

Revelation, Chapter 2

Notes

Preamble: the letter to Ephesus

This is the first of seven letters to the Christian Churches in Asia. All seven were first century churches planted by the disciples after the crucifixion of Jesus. Ephesus, an actual and literal church, is also used symbolically to communicate a message of God's correction and love. The message given to Ephesus – indeed, to each one of the churches – is relevant to all churches throughout time.

The salutation 'Ephesus' is an endearing term used by a groom to his bride; when translated, it means 'desirable', which was the condition of the church in the eyes of God at that time. The message within the names of all seven churches is symbolic of a period or age in Church history, with Ephesus representing the first age. This church was, to a greater extent, pure; however, it had lost its first exuberant love. Persecution came from within the Jewish community and from the heathen – wolves had moved in, as foreseen by Paul in Acts 20: 29-31:

29 For I know this, that after my departing shall grievous wolves enter in among you, not sparing the flock.

30 Also of your own selves shall men arise, speaking perverse things, to draw away disciples after them.

31 Therefore watch, and remember, that by the space of three years I ceased not to warn every one night and day with tears.

Scriptures affirm that the Christian Church remained reasonably pure throughout the first to the third century; however, even in Paul's time, a religious dictatorship had started to show its face and love waned. There was a turning from the true faith to a religion of hypocrisy. Read Galatians 1: 6-9:

6 I marvel that ye are so soon removed from him that called you into the grace of Christ unto another gospel:

7 Which is not another; but there be some that trouble you, and would pervert the gospel of Christ.

8 But though we, or an angel from heaven, preach any other gospel unto you than that which we have preached unto you, let him be accursed.

9 As we said before, so say I now again, If any man preach any other gospel unto you than that ye have received, let him be accursed.

Revelation 2: 1

1 Unto the angel of the church of Ephesus write; These things saith he that holdeth the seven stars in his right hand, who walketh in the midst of the seven golden candlesticks;

Jesus always identifies himself plainly. He never speaks or acts in secret –

Notes

only Satan has secret ceremonies and speeches. John is commanded to write the things which Jesus is about to tell him. The words are specifically directed to the Pastor of the Church of Ephesus.

Revelation 2: 2

2 *I know thy works, and thy labour, and thy patience, and how thou canst not bear them which are evil: and thou hast tried them which say they are apostles, and are not, and hast found them liars:*

Jesus first commends the congregation for their works, for labouring patiently and continuing on in the faith. He is pleased that they test the spirit of those who call themselves apostles. Jesus sanctions their caution and applauds the exposing of liars and hypocrites.

Revelation 2: 3

3 *And hast borne, and hast patience, and for my name's sake hast laboured, and hast not fainted.*

Jesus acknowledges their patience, their labours and their faithfulness that has been maintained even under persecution for his name's sake.

Revelation 2: 4

4 *Nevertheless I have somewhat against thee, because thou hast left thy first love.*

Jesus sees beyond their works. He knows they have departed somewhat from the all-consuming love of their first conversion and, instead, are motivated by a ministry of works.

Revelation 2: 5

5 *Remember therefore from whence thou art fallen, and repent, and do the first works; or else I will come unto thee quickly, and will remove thy candlestick out of his place, except thou repent.*

The 'first works' to which Jesus refers are, primarily, that man must have a heart of love directed toward God, wanting only that all glory and honour and worship be His; and, secondly, that man must carry the message of salvation and perform works of love toward his fellow man. Recall '*the first and greatest commandment*' found in Matthew 22: 36-40:

36 *Master, which is the great commandment in the law?*
37 *Jesus said unto him, Thou shalt love the Lord thy God with all thy heart, and with all thy soul, and with all thy mind.*
38 *This is the first and great commandment.*

Notes

39 *And the second is like unto it, Thou shalt love thy neighbour as thyself.*
40 *On these two commandments hang all the law and the prophets.*

Works without love are unacceptable. We sometimes get so caught up in works, we become a stranger to the One for whom we work and forget that He is the One we love and serve.

Revelation 2: 6

6 *But this thou hast, that thou hatest the deeds of the Nicolaitanes, which I also hate.*

Jesus speaks approval to the congregation of Ephesus because they despise the doctrine of the Nicolaitanes. The word in Greek is 'nikao', to conquer, and 'laos', the people.[5] In other words,'conquerors of the people'. This probably marked the beginning of the priestly orders of the Church.

Paul chastised the believers in I Corinthians 11: 18-22 for allowing discrimination and division to creep into the Love Feast which either preceded or followed the solemn commemorative act of the Lord's Supper. Already class notability seemed to be the order of the feast: the first step toward so-called 'holy order'.

Revelation 2: 7

7 *He that hath an ear, let him hear what the Spirit saith unto the churches; To him that overcometh will I give to eat of the tree of life, which is in the midst of the paradise of God.*

Jesus tells the Ephesians to hear the Spirit's warning and return to their first love. As overcomers, they will eat of the Tree of Life which is in the midst of the paradise of God in theNew Heavens and New Earth. Read Revelation 22:1-2:

1 *And he shewed me a pure river of water of life, clear as crystal, proceeding out of the throne of God and of the Lamb.*
2 *In the midst of the street of it, and on either side of the river, was there the tree of life, which bare twelve manner of fruits, and yielded her fruit every month: and the leaves of the tree were for the healing of the nations.*

---

[5]J. Vernon McGee, Thru The Bible, Volume V (Nashville, Tenn: Thomas Nelson Publishers, 1983), p. 903.

Jack Van Impe, Revelation Bible (Troy, MI: Jack Van Impe Ministries, 1998), p. 15.
Oliver B. Greene, Revelation – Verse by Verse Study (Greenville, SC: Gospel House Inc., 1976), p. 66.

Letter to Smyrna / Gnosticism Identified

Notes

Preamble: the letter to Smyrna

Jesus addresses the second letter to the Pastor of the Church at Smyrna, the church of the catacombs. The name comes from the word 'myrrh' which means 'bitterness'. Myrrh is a fragrant ointment obtained when the tender bark of the flowering myrrh tree is pierced. The spice or perfume is associated with death. Roman persecution of the Christian Church began as early as 63 A.D. and continued in different times and places throughout Church history. This bitter affliction was linked to the crushing of myrrh. Israelite and heathen alike viciously persecuted the early Church for a period of two hundred years. The centre of the most concentrated persecution was in the city of Smyrna.

Revelation 2: 8 - 9

8 *And unto the angel of the church in Smyrna write; These things saith the first and the last, which was dead, and is alive;*

9 *I know thy works, and tribulation, and poverty, (but thou art rich) and I know the blasphemy of them which say they are Jews, and are not, but are the synagogue of Satan.*

Jesus knows their works and tribulation and poverty and praises them for enduring under persecution. He calls them rich because the faithfulness of the Saints during the crushing of the church at Smyrna brings unto God a sweet smelling savour of Christ, as spoken of in II Corinthians 2: 15-16:

> *15 For we are unto God a sweet savour of Christ, in them that are saved, and in them that perish:*
>
> *16 To the one we are the savour of death unto death; and to the other the savour of life unto life...*

Read, also, Romans 8: 18:

> *18 For I reckon that the sufferings of this present time are not worthy to be compared with the glory which shall be revealed in us.*

The synagogue of Satan to which Jesus refers may well be the foul doctrine of Gnosticism which flourished among the Greek philosophers even before the first century. It evolved in the Hebrew Christian culture (thus, the synagogue of Satan) and flourished from the second to the fourth century. Gnostics, calling themselves 'Christians', practiced five sacraments: baptism, sealing, Eucharist, anointing of oil and bridal chamber rites(free sex).[6] The Christian teachings were becoming totally defiled through the evil twist of this demonic doctrine. They taught that since the Spirit can't sin, the body – weak, sinful flesh – could participate in sexual immorality and still remain Christian. They also believed that knowledge and concentrated meditation could bring levels of spiritual enlightenment far above and beyond

---

[6]Charles F. Pfeiffer et. al., eds. Wycliffe Bible Encyclopedia, Vol. I (Chicago: Moody Press, 1975), pp. 687-688.

that provided through Christ.

Notes

The Sadducees, a Jewish political party and the opponents of the Pharisee,[7] saturated the Sanhedrin (the seat of Jewish government) with this cult doctrine. On the whole, the Sadducees were wealthy, pampered Greek intellectuals who believed themselves to be much above the foolishness of faith in Christ as the only means of entering a life of heightened spirituality pleasing to God. Their teaching confused the Christians and polluted Christianity – confusion is always companion to high-mindedness.

Revelation 2: 10

10 Fear none of those things which thou shalt suffer: behold, the devil shall cast some of you into prison, that ye may be tried; and ye shall have tribulation ten days: be thou faithful unto death, and I will give thee a crown of life.

Fear accentuates pain, so Jesus tells the Christians of Smyrna to fear not. In saying they would have tribulation for ten days, he is indicating that persecution would be for a short or limited time. Church history confirms that the tyranny was most concentrated over a ten year period. The promise Jesus makes to the Church of Smyrna is a promise he makes to all Christians of all time: to all who love his appearing, the promise is "*be thou faithful unto death and I will give thee a crown of life*" (see also James 1: 12). For a listing of the Crowns which will be awarded to the faithful in Christ, refer to Appendix 2.

Revelation 2: 11

11 He that hath an ear, let him hear what the Spirit saith unto the churches; He that over cometh shall not be hurt of the second death.

All Christians should read and hear the teachings of the scriptures, heeding the correction given by the Holy Spirit as a guide to their Christian walk. Jesus tells us that as overcomers, we won't be hurt by the second death. The second death comes to those who have once died and are called to the White Throne judgment, the final accounting at the end of the Millennium. If condemned, they are sentenced to a second death, the eternal living death.

Preamble: the letter to Pergamos

Pergamos means 'marriage': the marriage of good with evil. This Church era began around 313 A.D. when Emperor Constantine the Great (274-337 A.D.), believing that his victory in battle was the work of the Christian God,

---

[7]Charles F. Pfeiffer et. al., eds. Wycliffe Bible Encyclopedia, Vol. II (Chicago: Moody Press, 1975), pp. 1500-1502.

Notes

became the first Emperor to promote Christianity. He declared the Church the official institution of the State in His Edict of Milan, 313 A.D. He firmly established the eastern capital which he called Constantinople (Istanbul) and insisted that everyone in it worship the Christian God; however, to keep the peace, he eventually allowed them to bring into the faith their pagan ceremonies – the keeping of Christmas and Easter being two such celebrations. He elevated the Church to a place of power and bestowed honours on the bishop. Married to the world, the Church became more the Church of Babylon than that of Christ.[8] According to H. A. Ironside, Constantine presided as the acknowledged head of the Christian Church. He bore the title "Pontifex Maximus", or High Priest of the Heathen – the same title that the Pope bears at the present time.[9] The three evils which God hates most were well-established in the Christian churches of that era: Gnosticism, Nicolaitan priestly order and Baal worship. These ungodly orders are still active and accepted in some modern churches but they have hidden themselves in new names.

Revelation 2: 12

12 And to the angel of the church in Pergamos write: these things saith he which hath the sharp sword with two edges;

Just as former letters are addressed to the angel or Pastor of the Church, so also is this letter addressed to the Pastor of Pergamos. Jesus, who is '*the Word*', is speaking; it is the Word which is sharper than a two-edged sword.

Revelation 2: 13

13 I know thy works, and where thou dwellest, even where Satan's seat is: and thou holdest fast my name, and hast not denied my faith, even in those days wherein Antipas was my faithful martyr, who was slain among you, where Satan dwelleth.

Even though Jesus sees Gnosticism alive in the Church of Pergamos, he also sees that many believers are still faithful in honouring his name. Even under threat of death, they have kept his teachings and hold to the true doctrine. Jesus commends them for this.

What a treasure it is to note that Jesus has seen and kept record of the martyrdom of his faithful servant, Antipas. According to tradition, Antipas

---

[8] "Constantine" from Encyclopedia Britannica, Vol. VI (Chicago: William Brenton Publishers, 1973), pp. 385-386.

[9] H. R. Ironside, The Revelation: Ironside Commentaries (Neptune, NJ: Loizeaux Brothers Publishing, revised 1996), p. 42.

Notes

was a bishop of Pergamos who died during the first century church. Jesus knows and remembers that the Hebrew Gnostics killed Antipas and refers to this cult as 'Satan's seat' and 'where Satan dwelleth'.

Revelation 2: 14

14 But I have a few things against thee, because thou hast there them that hold the doctrine of Balaam, who taught Balac to cast a stumbling block before the children of Israel, to eat things sacrificed unto idols, and to commit fornication.

In this verse, Jesus speaks against the third cult, Baal worship, active in the Church at Pergamos. He chastises them for tolerating evil men in the church and accepting their teachings.

Baal worship[10] originated at the time of Nimrod when he and his evil queen, Semiramis I, (leaders in the tower of Babel) founded the system which the scriptures refer to as 'Mystery, Babylon'. The ziggurat of Babel, from which they ruled, devoted itself to the worship of the sun, moon and stars; satanic ceremonies; and prostitution in the name of religion. It was formed in the heart of Satan and became the 'Mother' of every heathen religion down through history. After Nimrod's death, Semiramis declared herself to be the mother of a virgin birth: through this false 'immaculate conception', she gave birth to Tammuz and presented him as the reincarnated Nimrod. She proclaimed herself Queen of heaven and Tammuz, also called Baal, was worshiped as god. The whole wicked, satanic scheme was birthed with the sole purpose of discrediting the authentic virgin birth and ministry of Jesus. This evil worship defiled Israel at the time of Moses – when Balaam, the wayward prophet, advised Balac to send the Baal-worshiping Moabite and Midianite soldiers to the Israelites camped on the plains of Moab (read Numbers, Chapters 22, 23, 25: 1-3 and, especially, Chapter 31: 16). Further information is provided in this Study in the preamble to Revelation, Chapter 17.

Revelation 2: 15

15 So hast thou also them that hold the doctrine of the Nicolaitanes, which thing I hate.

Jesus again attacks the evil doctrine of the Nicolaitanes, active in the Church of Pergamos. He has previously praised the Church of Ephesus because they hated this doctrine (see Revelation 2: 6). It is evident from Paul's correction of the Christians attending the 'love feast' and communion service in

---

[10] Wayne Carver, "The Tower of Babel" (San Antonio: The Christian Jew Foundation)

Notes

I Corinthians 11: 18 that there was discrimination between the believers. This division led to separation between the educated, affluent members and the underprivileged, unlearned believers. We know that this led to the forming of laity and clergy; Jesus refers to it as the doctrine of the Nicolaitanes. They may have held that it was lawful for them to eat things sacrificed to idols because they felt above such foolishness – their conscience giving them this freedom. They may also have been influenced to attend these feasts since it was a time when eating, or not eating, was a crucial test of faith. This questionable freedom they felt, because of high-mindedness and possibly cowardice, may have led them to mingle in the orgies of the pagan, idolatrous feasts. Having grown bold and superior, they may have brought these impurities into the 'love feast' of the Christian Church. It's difficult to know if they were, in fact, participating in a form of Baal worship. Readers will have to come to their own conclusion. One thing is obvious: their actions birthed priestly orders and from this came the hierarchy of Romanism.

Revelation 2: 16

16 Repent; or else I will come unto thee quickly, and will fight against them with the sword of my mouth.

Jesus demands the repentance of all those who have become involved with any of the evil doctrines on which he had just pronounced judgment. He says that unless they repent, he will fight against them with the sword of His mouth. Jesus is saying that through the words which God will give him, he will judge and condemn them.

Read John 12: 48-49:

48 He that rejecteth me, and receiveth not my words, hath one that judgeth him: the word that I have spoken, the same shall judge him in the last day.

49 For I have not spoken of myself; but the Father which sent me, he gave me a commandment, what I should say, and what I should speak.

Revelation 2: 17

17 He that hath an ear, let him hear what the Spirit saith unto the churches; To him that over cometh will I give to eat of the hidden manna, and will give him a white stone, and in the stone a new name written, which no man knoweth saving he that receiveth it.

Verse 16 has warned of God's wrath and now, in this verse, we see the gentle forgiveness of God. Jesus tells the Church if they will repent and cease from

The White Stone / The New Name

Notes

following evil doctrines, he will allow them the privilege of eating of the '*hidden manna*', the bread of life. Review John 6: 48-51:

48 I am that bread of life.
49 Your fathers did eat manna in the wilderness, and are dead.
50 This is the bread which cometh down from heaven, that a man may eat thereof, and not die.
51 I am the living bread which came down from heaven: if any man eat of this bread, he shall live for ever: and the bread that I will give is my flesh, which I will give for the life of the world.

Jesus also says he will give them a white stone with a new name written on it, a name that no one will know but themselves. There are several connotations associated with this message:

- At the time of Christ, in the court of law, a black stone given to the one being judged was a pronouncement of guilt and death; a white stone was an announcement of innocence, acquittal and life. Jesus is saying, of course, that in the judgment, he will give them life if they hear and receive what the Spirit is saying.
- A guest who received a special invitation would often be given a white stone[11] with a personal message meant for his eyes only. Jesus gives us a special invitation and, even now, a white stone is laid up in heaven with a new name written on it for each believer. God often gives new names; He did so in both the Old and New Testament. For example, in Genesis 17: 5, Abram (a high Father) became Abraham (a Father of many nations); in Genesis 17: 15, Sari (the princess) was changed to Sarah (a Mother of nations); in Genesis 32: 28, Jacob (trickster, heel-catcher, supplanter) was given the new name Israel (prince who prevails with God). In the New Testament, proud Saul (desired) became Paul (small, little); presumptuous Simon Barjona (son of Jona – interesting to compare his lineage with 'Jonah', the one who ran from God) was known as Peter (petros, a small stone or pebble).

When Jesus returns at the end of the Tribulation, he will bear a new name: '*a name written that no man knows*' (Revelation 19: 12). Each of his followers will also be given a new name which brings with it a personal and private message from God. It will be beautiful because He will see us through Christ and our name will reflect what He sees.

---

[11] William Barclay, "The Revelation of John, Volume 1, Chapters 1-5", The Daily Study Bible, (Burlington, Ontario: Welch Publishing Co., 1976), pp. 95-96.

Letter to Thyatira / The Spark of Reformation

Notes

Preamble: the letter to Thyatira

The church at Thyatira probably represents a period in Church history from 440 A.D. to 1517 A.D. Roman Catholicism rooted itself in the Church in 440 A.D. when Bishop Gregory proclaimed himself to be the successor of Peter and the only bishop of Rome. In the fullest sense, the Roman Catholic Church came into its own in the seventh century when the Pope was acknowledged head of all Christendom. The Church was in the darkness of the Dark Ages but, out of the remnant of true believers, God awakened men like John Wycliffe (1320-1384). This dear Saint, a British priest, began to translate the Vulgate (the Latin translation of the Bible) into English and, by so doing, saw that the doctrine of the Church had departed from its biblical mooring. Through Wycliffe's lectures, God breathed a spark of life back into the Church and, because of this, Wycliffe is called 'the morning star' of reformation.

John Huss (1369-1415), a Bohemian priest converted through translating John Wycliffe's sermons, died at the stake in 1415 rather than deny the faith. His brave words, as he stood to be burned, were: "In the truth of the gospel which I have written, taught and preached, I die willingly and joyfully today." In the flames he sang, "Jesus Christ, the Son of the Living God, have mercy on me."

William Tyndale (1494-1536), a student of Wycliffe, continued with the work of translating the scriptures and thereby advanced the fight against the bondage and suppression of the Church.[12]

Toward the close of this era, the teaching was that every time the sacrament of the Holy Communion was commemorated, the emblems became the actual body and blood of Jesus Christ. Jesus died afresh at each communion: he was forever dying and continually being sacrificed. Thus, the name Thyatira, 'continual sacrifice', is appropriate for this Church age.

Martin Luther (1483-1546), a great leader of reformation, brought with him into the church a modified form of transubstantiation and, possibly, the doctrine of 'continual sacrifice'. There are conflicting opinions on this. It may not be what Luther taught, but what some of his followers retained from their Catholic background. Even to this day, some branches of the Lutheran faith believe in the necessity of audible confession of sins; transubstantiation; and Christ's actual sacrificial death at each communion service. They believe, as the early Papists did, that Jesus repeatedly dies for them. To them, when the priest elevates the elements at the Eucharist, the bread and wine become the actual body and blood of Christ even though it still

---

[12]Historical data in this section has generally been obtained from the internet: bio10.txt at www.webzonecom.com.

Christ Identified / Recognition of Works / Baal Worship / The Woman Jezebel

Notes

resembles the bread and wine. Luther eventually taught 'consubstantiation'. In his view, because of Christ's divine presence, his substance co-exists with the bread and wine; nonetheless, the wine remains wine and the bread remains bread.

Revelation 2: 18

18 And unto the angel of the church in Thyatira write; These things saith the Son of God, who hath his eyes like unto a flame of fire, and his feet are like fine brass;

Jesus, the Son of God, once again identifies himself and addresses the letter to the Pastor of the Church of Thyatira.

Revelation 2: 19

19 I know thy works, and charity, and service, and faith, and thy patience, and thy works; and the last to be more than the first.

Jesus notes the works, charity, faith and patience, with a second reference to the works of the Church at Thyatira. Jesus focuses on their works by naming them a second time and says that their later works are greater than their first works. He is probably drawing to their attention a failing which many Christian churches, as well as individual church members, get caught up in. We sometimes become so involved in service that our works become a point of pride rather than an expression of love.

Revelation 2: 20

20 Notwithstanding I have a few things against thee, because thou sufferest that woman Jezebel, which calleth herself a prophetess, to teach and to seduce my servants to commit fornication, and to eat things sacrificed unto idols.

Jezebel was a Phoenician princess and the wife of Ahab, King of Israel. She worshipped the Phoenician god, Baal, which aroused the hostility of Elijah, God's prophet. Since Jezebel and Ahab reigned in 873 - 852 B.C., it's obvious the woman Jezebel was not present in person but her pagan teachings were very much alive in the Church of Thyatira. Jesus has previously condemned these teachings in the Church of Pergamos and again vehemently pronounces judgment against them.

Notes

Revelation 2: 21

21 And I gave her space to repent of her fornication; and she repented not.

God is long-suffering. The doctrine of Baal had polluted the worship of God well before the time of Abraham. Just as Jezebel would not repent, the churches have not rid themselves of this wickedness. Time has been given but man persists in pursuing the evil activities which God warns against, even though the 'Day of the Lord' is ever drawing nearer.

Revelation 2: 22

22 Behold, I will cast her into a bed, and them that commit adultery with her into great tribulation, except they repent of their deeds.

Very soon – for the judgment is near – the Gnostics, the Nicolaitanes and the Baal worshippers will enter the Tribulation years and face both the wrath of God and the wrath of man. We can, even now, hear the rumble of that judgment.

Revelation 2: 23

23 And I will kill her children with death; and all the churches shall know that I am he which searcheth the reins and hearts: and I will give unto every one of you according to your works.

Everyone will stand at the judgment seat and God will reveal the deeds of each; all will receive according to their works. Life everlasting is promised to any who choose to follow Jesus, but for those who do not, death and hell will be their reward. Even now, great destruction hangs over the earth. At judgement, no one will be able to hide behind the Church since every act will be exposed in the revealing light of Christ.

If Revelation 14: 9-11 brings fear to your heart, repent while you still have time. Don't delay, for without a doubt, you and probably your children (if you lead them to it) will enter into this judgment.

9 And the third angel followed them, saying with a loud voice, If any man worship the beast and his image, and receive his mark in his forehead, or in his hand,

10 The same shall drink of the wine of the wrath of God, which is poured out without mixture into the cup of his indignation; and he shall be tormented with fire and brimstone in the presence of the holy angels, and in the presence of the Lamb:

11 And the smoke of their torment ascendeth up for ever and ever: and they have no rest day nor night, who worship the beast and his image, and whosoever receiveth the mark of his name.

Command to Hold Fast / Rewards of the Overcomer

Notes

Revelation 2: 24

24 But unto you I say, and unto the rest in Thyatira, as many as have not this doctrine, and which have not known the depths of Satan, as they speak; I will put upon you none other burden.

The reference Jesus makes to '*the depths of Satan*' again speaks of the pagan doctrine of Gnosticism rampant in the early Church. Jesus tells the congregation of Thyatira to keep themselves separate from false and demonic teachings. They must continue in pure works, charity and service. He requires no more of them than this.

Revelation 2: 25

25 But that which ye have already hold fast till I come.

The Church is told to hold fast and not depart from the teachings that Jesus left with his first disciples. Jesus makes it clear that it is imperative that the Church continues to teach the pure doctrine of salvation through faith in the cleansing blood of Jesus Christ.

Revelation 2: 26 - 27

26 And he that over cometh, and keepeth my works unto the end, to him will I give power over the nations:

27 And he shall rule them with a rod of iron; as the vessels of a potter shall they be broken to shivers: even as I received of my Father.

The New Living translation offers a bold and enlightened rendition of these two verses. It reads: *"To all who are victorious, who obey me to the very end, I will give authority over all the nations. They will rule the nations with an iron rod and smash them like clay pots. They will have the same authority I received from my Father."* In John 6: 29, Jesus tells us that in order to be overcomers, we must "*believe on him whom He has sent.*" Without faith we are totally incapable of walking in the power and authority of Christ.

Daniel foresaw the rewards of the faithful and spoke of them in Daniel 7: 22:

22 ...judgment was given to the saints of the most High; and the time came that the saints possessed the kingdom.

Revelation 2: 28

28 And I will give him the morning star.

In Revelation 22: 16, Jesus is quoted as saying: "*...I am the root and*

Notes

offspring of David, the bright and morning star." Having given us his life, Jesus becomes our 'morning star': our hope and our joyous expectation for the future. Paul expresses this beautifully in II Corinthians 4: 6:

> *6 For God, who commanded the light to shine out of darkness, hath shined in our hearts, to give the light of the knowledge of the glory of God in the face of Jesus Christ.*

Peter also has a unique way of teaching this truth. Read II Peter 1: 19:

> *19 We have also a more sure word of prophecy; whereunto ye do well that ye take heed, as unto a light that shineth in a dark place, until the day dawn, and the day star arise in your hearts:*

This study would not be complete without including the words of a mighty teacher and a humble man: the beloved servant of God, Derrick Prince...

> "Jesus takes his metaphor from the planet Venus (the action of the planet Venus). At certain seasons the 'day star' or 'morning star' rises in the eastern sky immediately before the sun itself comes up over the horizon. At times, this star is so bright that it partially dispels the surrounding darkness. It thus becomes the sun's forerunner, giving assurance to all who understand its message that the sun is ready to appear.
>
> So it is with us as we give careful heed to the prophetic truth. Like the day star arising in our hearts and dispelling the surrounding darkness, there comes an unshakable inner assurance the Lord will soon appear."[13]

A special word of caution here: in Isaiah 14: 12, Lucifer is called 'the son of the morning', **not** the 'morning star'. Satan is the great deceiver. Even his name is a deception: the latin word 'Lucifer' means 'the morning star' or 'the day star.' However, he is not the morning star, nor is he the day star. He is the fallen angel, Satan, the wicked and boastful son of the morning (Isaiah 14: 12-15).

Revelation 2: 29

29 He that hath an ear, let him hear what the Spirit saith unto the churches.

Let all churches and all Christians, over all ages, hear what the Spirit says to the Churches: receive God's teachings as a personal message and heed His warning!

---

[13]Derek Prince, Appointment in Jerusalem (Lincoln, VA: Chosen Books, Inc., 1975), p. 187.

CHAPTER THREE

The final three letters of **Chapter Three** dictated to John through Jesus continue to be letters of encouragement and warning. In them we see the dying Church in Sardis; the Church of the Reformation in Philadelphia; the Church of Apostasy in Laodicea. The commentary suggests we are living in the last church age – the age of people's rights. The mind and heart of the Church is greatly defiled by worldliness.

Revelation 3

[1] And unto the angel of the church in Sardis write; These things saith he that hath the seven Spirits of God, and the seven stars; I know thy works, that thou hast a name that thou livest, and art dead.

[2] Be watchful, and strengthen the things which remain, that are ready to die: for I have not found thy works perfect before God.

[3] Remember therefore how thou hast received and heard, and hold fast, and repent. If therefore thou shalt not watch, I will come on thee as a thief, and thou shalt not know what hour I will come upon thee.

[4] Thou hast a few names even in Sardis which have not defiled their garments; and they shall walk with me in white: for they are worthy.

[5] He that overcometh, the same shall be clothed in white raiment; and I will not blot out his name out of the book of life, but I will confess his name before my Father, and before his angels.

[6] He that hath an ear, let him hear what the Spirit saith unto the churches.

[7] And to the angel of the church in Philadelphia write; These things saith he that is holy, he that is true, he that hath the key of David, he that openeth, and no man shutteth; and shutteth, and no man openeth;

[8] I know thy works: behold, I have set before thee an open door, and no man can shut it: for thou hast a little strength, and hast kept my word, and hast not denied my name.

[9] Behold, I will make them of the synagogue of Satan, which say they are Jews, and are not, but do lie; behold, I will make them to come and worship before thy feet, and to know that I have loved thee.

[10] Because thou hast kept the word of my patience, I also will keep thee from the hour of temptation, which shall come upon all the world, to try them that dwell upon the earth.

[11] Behold, I come quickly: hold that fast which thou hast, that no man take thy crown.

[12] Him that overcometh will I make a pillar in the temple of my God, and he shall go no more out: and I will write upon him the name of my God, and the name of the city of my God, which is new Jerusalem, which cometh down out of heaven from my God: and I will write upon him my new name.

[13] He that hath an ear, let him hear what the Spirit saith unto the churches.

[14] And unto the angel of the church of the Laodiceans write; These things saith the Amen, the faithful and true witness, the beginning of the creation of God;

[15] I know thy works, that thou art neither cold nor hot: I would thou wert cold or hot.

[16] So then because thou art lukewarm, and neither cold nor hot, I will spue thee out of my mouth.

[17] Because thou sayest, I am rich, and increased with goods, and have need of nothing; and knowest not that thou art wretched, and miserable, and poor, and blind, and naked:

[18] I counsel thee to buy of me gold tried in the fire, that thou mayest be rich; and white raiment, that thou mayest be clothed, and that the shame of thy nakedness do not appear; and anoint thine eyes with eyesalve, that thou mayest see.

[19] As many as I love, I rebuke and chasten: be zealous therefore, and repent.

[20] Behold, I stand at the door, and knock: if any man hear my voice, and open the door, I will come in to him, and will sup with him, and he with me.

[21] To him that overcometh will I grant to sit with me in my throne, even as I also overcame, and am set down with my Father in his throne.

[22] He that hath an ear, let him hear what the Spirit saith unto the churches.

Revelation, Chapter 3

Notes

Preamble: letter to Sardis

Sardis, when translated, means 'remnant'. The Church age which Sardis represents is probably from around 1517 A.D. to 1650 A.D. The 'remnant', protesting the traditions and teachings of the Catholic Church, were at this time referred to as 'protestants.' Jesus now addresses the Church of the Reformation.

Possibly the boldist activist of this period was Martin Luther[14] (1483-1546). He lead the 'remnant' on into the 1500's and further fueled the Reformation on October 31, 1517, when he nailed his 95 Theses to the church door in Wittenberg, Germany. Henry VIII (1509-1547) may have unwittingly helped Luther's fight against the controlling power of the hierarchy of Romanism since, at the same time and for personal reasons, he also wanted to break away from the tyranny of the Roman Catholic Church. Luther brought to light the three great truths of the New Testament which had been buried for centuries under ritual and dead formality:

- that man is justified by faith alone;
- that every believer is a priest and has direct access to God through the Lord Jesus Christ;
- that the Bible, apart from tradition, is the sole source of faith and authority for the Christian.

John Calvin[15] (1509-1564), a priest from Geneva, Switzerland, and John Knox[16] (1505-1572), a Scottish religious leader, followed Luther in the brave fight to bring light and truth back into the Church. In 1542, Satan tried to quench the rebirth of the true faith through the Inquisition. The Jesuits, priests appointed by the Roman Catholic Church, spearheaded the counter to the Reformation; however, the door was open and no man could shut it. (See Glossary of Terms for 'The Inquisition' and 'The Reformation'.)

Revelation 3: 1

1 *And unto the angel of the church in Sardis write: These things saith he that hath the seven Spirits of God, and the seven stars; I know thy works, that thou hast a name that thou livest, and art dead.*

Jesus is speaking; he is the one who has the seven Spirits of God (Isaiah 11: 2) and holds the seven churches and their pastors in his hand (Revelation 1:

[14]Historical data obtained from the internet: luther.txt at www.webzonecom.com

[15]Historical data obtained from the internet: hst_calv.txt at www.cob-net.org and calvin.txt at www.webhome.net

[16]Historical data obtained from the internet: The Reformation Online at oct311517@aol.com

The Dying Church / Call to Repent / A Promise of White Robes

Notes

20). This letter is addressed to the Pastor of the Church at Sardis. Jesus tells him the Church is alive in name only; he sees that it is almost dead spiritually.

See Appendix 3 for further information on the Seven Spirits of God.

Revelation 3: 2

2 Be watchful, and strengthen the things which remain, that are ready to die: for I have not found thy works perfect before God.

Jesus states that he has found the works of the Church at Sardis failing and in need of revival. He councils the Pastor to diligently watch over his flock and to strengthen the faith which is in danger of dying. He calls upon the Protestants (those who protest) to strengthen the great truths committed to them in the beginning (and awakened in the Reformation).

Revelation 3: 3

3 Remember therefore how thou hast received and heard, and hold fast, and repent. If therefore thou shalt not watch, I will come on thee as a thief, and thou shalt not know what hour I will come upon thee.

The Pastor is told to remember the undefiled teachings of Christ; to return to them and hold fast until he comes. He demands repentance and watchfulness, or he will come suddenly upon them in judgment and remove the Church.

Revelation 3: 4

4 Thou hast a few names even in Sardis which have not defiled their garments; and they shall walk with me in white: for they are worthy.

The few true believers, the worthy ones who have not offended the Spirit, are given assurance they will remain clothed in Christ's righteousness and walk with him throughout eternity.

Revelation 3: 5

5 He that over cometh, the same shall be clothed in white raiment; and I will not blot out his name out of the book of life, but I will confess his name before my Father, and before his angels.

Even before God created man, He had written a book listing each of His created by name, as confirmed in Psalm 139: 15-16.

15 My substance was not hid from thee, when I was made in secret, and

Notes

curiously wrought in the lowest parts of the earth.
16 Thine eyes did see my substance, yet being unperfect; and in thy book all my members were written, which in continuance were fashioned, when as yet there was none of them.

Our omniscient God has chosen not to list those who admire the beast and worship him (Revelation 13: 8 and Revelation 17: 8); He can – and will – blot names out of the Book of Life if one willingly and consciously continues to practice sin. Read Exodus 32: 33:

33 ...Whosoever hath sinned against me (continually and without sorrow), *him will I blot out of my book.*

Psalm 69: 28 also teaches the same truth:

28 Let them be blotted out of the book of the living and not written in with the righteous.

A similar teaching may be found in Revelation 20: 14-15 of this study. The scriptures make us aware that even though once born of the Spirit, it is possible to have one's name removed from the Book of Life and miss out on the Rapture – to be gathered and cast, without further judgment, into the Lake of Fire. There are conditions that must be met in order that one's name is retained in the 'Book of Life'. Jesus has promised here in Revelation 3: 5 that he will not blot a name out of the 'Book of Life' if the believer remains faithful and continues on as an overcomer. It is required that he keep his garments undefiled. One must remember that Jesus is addressing each of us, as well as the Christians in Sardis, when he makes these statements. Being committed Christians, born of the Spirit, assures us our names will remain written in the Book, referred to in Revelation 13:8 as '*the book of life of the Lamb...*'.

Revelation 20: 12 gives adequate proof that the 'Book of Life' and the 'Lamb's Book of Life' are one and the same book. We are told that the dead, small and great, will stand before God and the books (the Book of Deeds and the Book of Remembrance) will be opened. The dead are judged out of those things written in the books.

It then says another book is opened which is the 'Book of Life'. We are at no time told a second 'Book of Life' is opened with a separate set of names containing Christians only. The Father and Son are of one mind – they don't keep separate records.

Further to this, we find positive assurance in Philippians 4: 3 that the 'Book of Life' and the 'Lamb's Book of Life' is the same book. Paul requests that the Philippian Christians help those women who laboured with him in the gospel. He mentions Clement in particular, along with all of the other fellow labourers whose names are in the 'Book of Life.' If, in fact, there is

Notes

a separate book called the 'Lamb's Book of Life' (reserved for the recording of Christians only), Paul would surely have referred to their names being recorded in the 'Lamb's Book of Life' rather than simply saying the 'Book of Life.'

Revelation 20: 15 tells us that those who stand for judgment must have their names written in the 'Book of Life.' As Christians, we are refined by fire (I Corinthians 3:13-15). Our evil deeds and less-than-perfect works are destroyed and we, through the blood of Christ, are made perfect. Those whose names have been removed from the 'Book of Life' are condemned without further judgment and, without hope, are cast into the Lake of Fire (Revelation 20: 15).

Revelation 3: 6
6 He that hath an ear, let him hear what the Spirit saith unto the churches.

Everyone who is open to hearing the teachings of the Spirit are told to heed the words of Jesus as spoken to them in the letters to the early Churches. II Timothy 3: 16 impresses on us that all scriptures are written for our benefit:

> *16 All scripture is given by inspiration of God, and is profitable for doctrine, for reproof, for correction, for instruction in righteousness.*

Preamble: letter to Philadelphia

Philadelphia, being translated, means 'brotherly love'. Up until the later part of the Sardis era, the Church had disintegrated into almost non-existence, consumed by the dark cloud of Romanism. However, the Lord revitalized the faith with an outpouring of His love through lion-hearted men like John Wycliffe, John Huss and Martin Luther.

There came an even greater season of light from approximately 1650 to the early 1900s, which must surely be placed alongside the Philadelphian Church era. It was a time of brotherly love and open doors. Britain and the United States witnessed a mighty outpouring of the Holy Spirit. God had a series of great evangelists waiting in the wings. The world came alive with the love of Christ, as one brother reached out to another and the Church returned to the Spirit of Ephesus.

Many obedient and fearless men followed Luther, John Calvin and John Knox. To mention only a few[17], we might start with John Bunyan (1628-1688), an English Baptist preacher and writer, imprisoned for his faith in 1660. He wrote his immortal *Pilgrim's Progress* in 1672 while in prison.

---

[17]Biographical information obtained from the internet: www.webzonecom.com

Notes

John Wesley (1703-1791) rode horseback all over England, preaching and winning souls. The Methodist Church was founded through his teachings. Charles Wesley (1707-1788), brother to John Wesley, contributed his great ministry of song writing and praise. George Whitfield (1714-1770), a Methodist, preached against the Church of England because it followed the Roman Catholic ways. It is said that Whitfield preached like a lion. He joined Charles and John Wesley in England in 1735 and America in 1737.

George Muller (1805-1898), a Prussian who worked in Bristol and won thousands of orphans to Christ, established homes for the homeless. Charles Spurgeon (1834-1892), a Baptist, was England's best-known preacher in the last half of the 19th Century. Dwight Moody (1837-1899) evangelized in Massachusetts and England.

Andrew Murray (1828-1917) was a great promoter of missions and a highly-acclaimed writer. Billy Graham, born in 1918, possibly the most effective evangelist of all time, is known and loved the world over for his simple message of salvation.

When we have these facts before us and evidence God's faithfulness, can we ever doubt His promises? In Hebrews 13: 5, He says, "*I will never leave thee, nor forsake thee.*"

Revelation 3: 7

7 *And to the angel of the church in Philadelphia write; These things saith he that is holy, he that is true, he that hath the key of David, he that openeth, and no man shutteth; and shutteth, and no man openeth;*

When Jesus greets the Pastor of the Church of Philadelphia, He identifies himself as '*He that is holy*'. Recall the words of Isaiah 6: 3:

3 *...Holy, holy, holy, is the Lord of hosts: the whole earth is full of his glory.*

Jesus also refers to himself as true. He is '*the true bread*' (John 6: 32); '*the true judge*' (John 8: 16); '*the true vine*' (John 15: 1). He is called '*faithful and true*' (Revelation 19: 11). Jesus declares, as well, that he has '*the key of David*'. God verifies the truth of this claim in Isaiah 22: 22:

22 *And the key of the house of David will I lay upon his shoulder; so he shall open, and none shall shut; and he shall shut, and none shall open.*

God Humbles Satan's Servants / Kept from the Hour of Temptation

Notes

Revelation 3: 8

8 I know thy works: behold, I have set before thee an open door, and no man can shut it; for thou hast a little strength, and hast kept my word, and hast not denied my name.

Jesus knows the heart and the works of the Church of Philadelphia. He sees their love for him and their brethren; thus, the door was opened for the Saints of God to dispel the darkness of the Dark Ages and bring the light of Christ's love back into the Church. Reformation came from God through the power of the Holy Spirit working within His anointed. Not even the Inquisition could stop the moving of the Spirit and the revival of undefiled, virtuous faith.

Revelation 3: 9

9 Behold, I will make them of the synagogue of Satan, which say they are Jews, and are not, but do lie; behold, I will make them to come and worship before thy feet, and to know that I have loved thee.

The ones who have defiled the teachings of Christ – those claiming they are Jews who love God and do not – these will be made to worship at the feet of the Saints so that they openly confess that God loves and rewards the faithful. In Isaiah 49: 23, God says that people and kings (the whole world) will show honour to Him and will bless Israel. The Church is grafted into this promise.

23 ...they shall bow down to thee with their face toward the earth, and lick up the dust of thy feet; and thou shalt know that I am the Lord: for they shall not be ashamed that wait for me.

Revelation 3: 10

10 Because thou hast kept the word of my patience, I also will keep thee from the hour of temptation, which shall come upon all the world, to try them that dwell upon the earth

Because we have obeyed God's command to persevere, He will protect us by calling us before the time of temptation when men will be required to take the mark of the beast. This testing will come upon the whole world to establish openly those who actually belongs to Him, or to the world and its evil. Praise God! Our home is not in this world; we belong to Jesus. We will be risen in the Rapture and will escape this time of testing and sorrow. II Peter 2: 9 confirms that we will be delivered out of this temptation:

9 The Lord knoweth how to deliver the godly out of temptations, and to reserve the unjust unto the day of judgment to be punished:

Notes

Also, I Thessalonians 5: 9 says:

9 For God hath not appointed us to wrath, but to obtain salvation by our Lord Jesus Christ.

Revelation 3: 11

11 Behold, I come quickly: hold that fast which thou hast, that no man take thy crown.

We must treasure our faith and remain true to our confession. In so doing, the crown of righteousness, which is given to all those who love his appearing (II Timothy 4: 8), is secured for us, that we might lay it at the feet of Jesus (Revelation 4: 10).

Revelation 3: 12

12 Him that over cometh will I make a pillar in the temple of my God, and he shall go no more out: and I will write upon him the name of my God, and the name of the city of my God, which is new Jerusalem, which cometh down out of heaven from my God: and I will write upon him my new name.

We are overcomers when our hearts are established by grace and we are sealed by our confession: we have been transformed by the Spirit of our God; his name is written in our hearts and our faces shine forth his righteousness. We are overcomers when we walk steadfastly in the image of our Christ and cease from wavering, no longer going in and out of our role, or up and down in our spiritual emotions.

Knowing we are overcomers, Christ will make us pillars of strength in his temple (Church body), capable of imparting to others '*some spiritual gift, to the end that they may be established in the mutual faith*' (Romans 1: 11).

Having blessed us in our earthly service, Christ has made us kings and priests to rule and reign with him during the Millennium (see Revelation 1: 6). We will walk in the light of the Glory of God and the Lamb (Revelation 21: 23). We will drink of the waters of life which flow from the Holy City: the New Jerusalem which comes down from God out of Heaven (Revelation 21: 2). We will enter the Holy City bearing the name of God, the name of the Son of God and the name of the New Jerusalem (Revelation 3: 12). God identifies His own precisely and with great care!

Revelation 3: 13

13 He that hath an ear, let him hear what the Spirit saith unto the churches.

Notes

Only those born again, having received the Spirit of God, can fathom (in part) the blessing that God has prepared for us. We hear and see only by the enlightenment of the Spirit. Read I Corinthians 2: 9-10:

9 But as it is written, Eye hath not seen, nor ear heard, neither have entered into the heart of man, the things which God hath prepared for them that love him.

10 But God hath revealed them unto us by his Spirit: for the Spirit searcheth all things, yea, the deep things of God.

Preamble: letter to the Laodiceans

Laodicean means ‘people’s rights’. This final letter is written, as always, to the Pastor or Bishop of the Laodicean Church. More surely, it is written to the Laodicean age, the age of people’s rights. This is a picture of a Church age in the throws of final apostasy. It started as a whisper in the early 1900s and became a voice by 1960. It will increase in volume until the time of the Tribulation.

Revelation 3: 14

14 And unto the angel of the church of the Laodiceans write; These things saith the Amen, the faithful and true witness, the beginning of the creation of God;

The letter is from Jesus for he is ‘the Amen’. All the promises of God, through Christ, are “*yea and amen*” (II Corinthians 1: 20). He is the “*faithful and true witness*” (Isaiah 55: 3-4) and he is ‘the beginning’: “*In the beginning was the Word, and the Word was with God, and the Word was God*” (John 1: 1).

Revelation 3: 15 - 16

15 I know thy works, that thou art neither cold nor hot: I would thou wert cold or hot.

16 So then because thou art lukewarm, and neither cold nor hot, I will spue thee out of my mouth.

The Church must be on fire for God. Jesus won’t tolerate a lukewarm Church, nor the half-committed Christian. Read what Jesus says about an unresolved Christian in Matthew 6: 24:

24 No man can serve two masters: for either he will hate the one, and love the other; or else he will hold to the one, and despise the other. Ye cannot serve God and mammon.

Lovers of Self / Gold Tried by Fire

Notes

Revelation 3: 17

17 Because thou sayest, I am rich, and increased with goods, and have need of nothing; and knowest not that thou art wretched, and miserable, and poor, and blind, and naked:

Before the return of Christ, man will measure his success by the abundance of his possessions. As he becomes prosperous, he falsely places his trust in self. He looks to natural things instead of spiritual; his mind and heart are defiled by worldliness.

Paul speaks to Timothy about this very thing in II Timothy 3: 1-5:

1 This know also, that in the last days perilous times shall come.
2 For men shall be lovers of their own selves, covetous, boasters, proud, blasphemers, disobedient to parents, unthankful, unholy,
3 Without natural affection, truce breakers, false accusers, incontinent, fierce, despisers of those that are good,
4 Traitors, heady, high-minded, lovers of pleasures more than lovers of God;
5 Having a form of godliness, but denying the power thereof: from such turn away.

These are those who have chosen to enjoy the pleasures of sin for a season.

Revelation 3: 18

18 I counsel thee to buy of me gold tried in the fire, that thou mayest be rich; and white raiment, that thou mayest be clothed, and that the shame of thy nakedness do not appear; and anoint thine eyes with eye salve, that thou mayest see.

'*Gold tried in fire*' points to the divine righteousness of God and '*eye salve*' refers to spiritual enlightenment and discernment. Raiment is made white when washed in the blood of the Lamb (Revelation 7: 14). He clothes our nakedness and frees us from our shame when he covers us with his righteousness and takes away our confessed sin.

Isaiah 55: 1-3 tells us how to buy gold tried in fire so that we may be rich.

1 Ho, every one that thirsteth, come ye to the waters, and he that hath no money; come ye, buy, and eat; yea, come, buy wine and milk without money and without price.
2 Wherefore do ye spend money for that which is not bread? And your labour for that which satisfieth not? Hearken diligently unto me, and eat ye that which is good, and let your soul delight itself in fatness.
3 Incline your ear, and come unto me: hear, and your soul shall live; and I will make an everlasting covenant with you, even the sure mercies of David (life and salvation through the blood of Jesus).

Loved but Rebuked / Jesus Stands at the Door

Notes

Revelation 3: 19

19 As many as I love, I rebuke and chasten: be zealous therefore, and repent.

Hear these words of wisdom from Proverbs 3: 11-12:

11 My son, despise not the chastening of the Lord; neither be weary of his correction:

12 For whom the Lord loveth he correcteth; even as a father the son in whom he delighteth.

Our loving Father rebukes, chastens and calls for repentance even up to the end of the Tribulation; more probably, His invitation is open to the end of the Millennium, just prior to the White Throne judgment (Revelation 22: 17). All who have fallen during the release of Satan must hear the message of salvation – the gospel of forgiveness. The end can come:

- only when the gospel of the kingdom is preached in all the world (Matthew 24: 14);
- only when the angel of God flies in the midst of heaven preaching the everlasting gospel to them that dwell upon the earth – to every nation, and kindred, and tongue and people (Revelation 14: 6-7)

Only after this will the heavens and earth be consumed by fire on the day of judgement (II Peter 3: 7-13) and the 'old' is replaced by the 'new' (Revelation 21: 1).

Revelation 3: 20

20 Behold, I stand at the door, and knock: if any man hear my voice, and open the door, I will come in to him, and will sup with him, and he with me.

Jesus knocks on the door of our heart through the wooing of the Holy Spirit. If we open our heart, Christ in the form of the Holy Spirit comes to dwell within. His flow of forgiveness begins. The Holy Spirit, who helps us to understand our need of forgiveness, now grants us the assurance of forgiveness day by day. The blood of Jesus washes away the sins committed in the past and continues to cleanse us as we confess our sins and remain sincerely repentant. Through the Spirit, we continually drink from the waters of salvation which flow from the throne of God. Read Isaiah 1: 18 and rejoice!

18 Come now, and let us reason together, saith the Lord: though your sins be as scarlet, they shall be as white as snow; though they be red like crimson, they shall be as wool.

Notes

Revelation 3: 21

21 To him that over cometh will I grant to sit with me in my throne, even as I also overcame, and am set down with my Father in his throne.

There is a promise here that is so rich, so wondrous, that few have seen its truth. What does Jesus mean when he says that if we are overcomers, he will grant that we sit with him <u>in</u> his throne, even as he also overcame and is set down with his Father <u>in</u> His throne?

The wonder of it all: the silent plan of God can't be heard unless we look to Jesus and see who he is. Then we must look at ourselves in the light of his promises. Surely we must see that through him, as virtuous overcomers, we become the sanctified, glorified Sons of God – heirs and joint-heirs with Christ (Romans 8: 17).

Revelation 3: 22

22 He that hath an ear, let him hear what the Spirit saith unto the churches.

Until we study the scriptures and listen to its truth, we are rendered incapable of hearing what the Spirit is saying to the Church. A very few pertinent scriptures have been gathered and listed in the preamble to Chapter 4. Read and ponder them so that your mind and heart may grasp the truth and the magnitude of the promises of God.

Consider again I Corinthians 2: 9:

> *9 ...Eye hath not seen, nor ear heard, neither have entered into the heart of man, the things which God hath prepared for them that love him.*

Praise Him all ye people!

CHAPTER FOUR

Chapter Four gives assurance that the righteous are raptured before the Tribulation, and displays their final rewards and rejoicing in heaven. The Study teaches that the jasper and sardine stone are representative of the Father and Son, the two in One who sit upon the throne. The discussion draws attention to the submissive nature of the Raptured Saints and their ultimate inauguration into the Godhead.

Revelation 4

[**1**] After this I looked, and, behold, a door was opened in heaven: and the first voice which I heard was as it were of a trumpet talking with me; which said, Come up hither, and I will shew thee things which must be hereafter.

[**2**] And immediately I was in the spirit: and, behold, a throne was set in heaven, and one sat on the throne.

[**3**] And he that sat was to look upon like a jasper and a sardine stone: and there was a rainbow round about the throne, in sight like unto an emerald.

[**4**] And round about the throne were four and twenty seats: and upon the seats I saw four and twenty elders sitting, clothed in white raiment; and they had on their heads crowns of gold.

[**5**] And out of the throne proceeded lightnings and thunderings and voices: and there were seven lamps of fire burning before the throne, which are the seven Spirits of God.

[**6**] And before the throne there was a sea of glass like unto crystal: and in the midst of the throne, and round about the throne, were four beasts full of eyes before and behind.

[**7**] And the first beast was like a lion, and the second beast like a calf, and the third beast had a face as a man, and the fourth beast was like a flying eagle.

[**8**] And the four beasts had each of them six wings about him; and they were full of eyes within: and they rest not day and night, saying, Holy, holy, holy, Lord God Almighty, which was, and is, and is to come.

[**9**] And when those beasts give glory and honour and thanks to him that sat on the throne, who liveth for ever and ever,

[**10**] The four and twenty elders fall down before him that sat on the throne, and worship him that liveth for ever and ever, and cast their crowns before the throne, saying,

[**11**] Thou art worthy, O Lord, to receive glory and honour and power: for thou hast created all things, and for thy pleasure they are and were created.

Preamble to Chapter 4

It is imperative that you read the following study scriptures before going on with this verse-by-verse commentary of the Revelation of Jesus Christ.

Notes

Preamble: a teaching in preparation for Chapter 4

This teaching provides a close look at Jesus to help us develop faith so that with confidence we may stand on the promises of God. It is meant to help us hear and see these promises afresh, that we might grasp God's plan for us. The reading will help us accept who we are in Christ and what we become because of him.

First recall Revelation 3: 21:

> *21 To him that over cometh will I grant to sit with me* in (not on) *my throne, even as I also overcame, and am set down with my Father* in *his throne.*

Impossible, we might say, that man should ever be fit to receive this promise. How could weak and sinful man believe a promise which would have him sit with Jesus and his Father in His throne. Jesus has made other seemingly impossible promises – promises too lofty for sinful man to comprehend. Surely, man will never really reign with him as is stated in Revelation 5: 10:

> *10 And hast made us unto our God kings and priests: and we shall reign on the earth.*

Further, we ponder if we are to interpret the promise of Revelation 1: 5-6 literally. It says that he so loves us that he cleanses us and washes us from our sins in his own blood. Does this really mean that, through faith in the righteous blood of Jesus Christ, we are made worthy to reign on earth? Do we truly partake of his divine nature? (II Peter 1: 4)

To appreciate the possibility of such an Holy appointment, we must look to "*the author and finisher of our faith*" (Hebrews 12: 2): the beautiful, submissive Jesus. The scriptures tell us, in all ways, the Spirit will make us "*like unto him.*"

To begin, let's contemplate a passage from I John 3: 2-3:

> *2 Beloved, now are we the sons of God, and it doth not yet appear what we shall be: but we know that, when he shall appear, we shall be like him; for we shall see him as he is.*
>
> *3 And every man that has this hope in him purifieth himself, even as he is pure.*

With each truth about Jesus – and only a few are cited here – we may be assured that God has ordained that through the finished work of Jesus, we will be pure, as he is pure.

Jesus Victorious in Temptation / Jesus Repudiates Praise

Notes

Jesus is victorious in temptation.

Read Luke 4: 1-13. Jesus, on our behalf, overcomes the tempter. Through his faithfulness, upon our sanctification and glorification, we are never again subject to Satan's power. We are forever victorious because he is victorious. Jesus also conquered temptation which comes through the persuasion of man. Please refer to Matthew 16: 21-23.

> 21 *From that time forth began Jesus to shew unto his disciples, how that he must go unto Jerusalem, and suffer many things of the elders and chief priests and scribes, and be killed, and be raised again the third day.*
>
> 22 *Then Peter took him, and began to rebuke him, saying, Be it far from thee, Lord: this shall not be unto thee.*
>
> 23 *But he turned, and said unto Peter, Get thee behind me, Satan: thou art an offence unto me: for thou savourest not the things that be of God, but those that be of men.*

In this passage we see Jesus, short days before his crucifixion, rebuking his friend Peter and finding him an offense – even declaring Peter to be an instrument of Satan. In so doing, Jesus withstands the temptation of Satan exercised through man. After the Rapture, at our glorification, we are rendered free eternally and forever from this seductive temptation.

Jesus repudiates praise.

Recall the passage in Matthew 19: 16-17:

> 16 *And, behold, one came and said unto him, Good Master, what good thing shall I do, that I may have eternal life?*
>
> 17 *And he said unto him, Why callest thou me good? There is none good but one, that is, God...*

Jesus refuses the salutation, "Good Master". In rejecting this title, he is asserting that there is none good but God. By spurning the honour that is offered him and, instead, directing all praise to God, Jesus has given us victory for all eternity over one of the most common and, possibly, one of the most destructive weaknesses of man: that of self-gratification for talents and gifts that are given by God.

God has foreordained that He provides Jesus with a portion of His greatness and states that Jesus will share his portion with us. Read Isaiah 53: 12:

> 12 *Therefore will I* (God) *divide him* (Jesus) *a portion with the great, and he shall divide the spoil with the strong* (the faithful in Christ)*: because he hath poured out his soul unto death...*

In his death and resurrection, Jesus secured this promise for us that we may become as one in Them: the Father, the Son and the Holy Spirit. If we do not accept this truth then, in essence, we are self-righteously rejecting the

Notes

finished work of Jesus. This is not a pompous claim of a promotion to glory; rather, it is a humble acceptance of the perfect, completed plan of God as He gives us our appointment and position in holy service.

Walk patiently through the balance of this study. Open your eyes and heart to God's wondrous word and it will cause you to prostrate yourself before Him in adoration and praise. It is awe-inspiring that He should care so much for man that He and His Son will, in the end, dwell so mightily in us that He will truly make us one in Them.

One must not interpret that this study, *Triumph and Terror,* is declaring that those whom the Son has redeemed become, themselves, God. Rather, it is pointing out our absolute holy union with the Father and Son which is obtained only through the sanctity of Jesus Christ and is the result of his finished work. It is evidenced in Revelation 4: 3-4 that we receive the promises of Revelation 3: 21: we are, indeed, seated with Jesus in his throne, just as he is seated with his Father in His throne.

Stay focused in the study of Revelation 4: 3 that it might be revealed to you that the 'One' who is seated on the throne is both the Father and the Son; let the Word direct you to a full understanding of our final destiny and position in Christ. We are viewed by John as the submissive vessel prepared by God to house the third person of the Godhead (see Revelation 4: 9-11 of this study). We, the redeemed, encircling the throne of God, are embraced by the light of the Seven Spirits of God (Isaiah 11: 2) which are the seven lamps burning before the throne – these being the seven attributes of God. John sees the Godhead at its culmination: the Father, the Son and the Holy Spirit. Each of the Trinity is endowed with the Seven Spirits of God. (Please refer ahead to the commentary on Revelation 4: 5, second paragraph on.)

The submissive Jesus

We must go on and view the submissive Jesus, for as he is, so we will be. Jesus demonstrates authority in judgment and clearly recognizes and submits to the source of his holy knowledge and wisdom. He is well aware that God is truly the Judge of all (Hebrews 12: 23) and that he, Jesus, is the mouthpiece by which God speaks His judgment. Many scripture references point to this truth. See John 12: 48-49:

48 He that rejecteth me, and receiveth not my words, hath one that judgeth him: the word that I have spoken, the same shall judge him in the last day.

49 For I have not spoken of myself; but the Father which sent me, he gave me a commandment, what I should say, and what I should speak.

Jesus – the Judge, the Good Shepherd, the Bestower of Glory

Notes

Ponder the words of Jesus recorded in John 5: 30:

30 I can of mine own self do nothing: <u>as I hear, I judge</u>: and my judgment is just; because I seek not mine own will, but the will of the Father which hath sent me.

Look at John 8: 28:

28 Then said Jesus unto them, When ye have lifted up the Son of man, then shall ye know that I am he, and that I do nothing of myself; but as my Father hath taught me, I speak these things.

We are reminded in these scriptures that Jesus judges by the authority of God; the judgment is not his but is the Father's. It is through this same authority that we will judge, for it is written: '...the Saints shall judge the world' (I Corinthians 6: 2). John affirms this by saying in Revelation 20: 4:

4 And I saw thrones, and they that sat upon them, and judgment was given unto them.

Now, let's look at a few verses where Jesus shows he knows his position in the Godhead as Saviour and Son. He tells the world he is <u>as</u> God but is <u>less</u> than God (in his humanity only), since he has no power aside from God. Refer to John 6: 44:

44 No man can come to me, except the Father which hath sent me draw him: and I will raise him up at the last day.

Look at John 10: 27-30:

27 My sheep hear my voice, and I know them, and they follow me:
28 And I give unto them eternal life; and they shall never perish, neither shall any man pluck them out of my hand.
29 My Father, which gave them me, is greater than all; and no man is able to pluck them out of my Father's hand.
30 I and my Father are one (equal in power in the beginning – Genesis 1: 26 – and in the end – Revelation 22: 1).

Listen to Jesus as he prays for us in John 17: 19-22:

19 And for their sakes I sanctify myself, that they also might be sanctified through the truth.
20 Neither pray I for these alone, but for them also which shall believe on me through their word.
21 <u>That they all may be one; as thou, Father, art in me, and I in thee, that they also may be one in us</u>: that the world may believe that thou hast sent me.
22 And <u>the glory which thou gavest me I have given them</u>; that they may be one, even as we are one:

Jesus – the First Fruit; the Saviour; the Submissive One

Notes

Jesus is submissive unto the end.

Jesus models for us complete submission in his perfect relationship with God. Jesus is God, but he submits to God, that God the Father may be all in all. Read I Corinthians 15: 23-28:

> 23 *...Christ the first fruits; afterward they that are Christ's at his coming*
> (What a statement and what a glorious promise: that we will be second only to our Lord and Saviour.)
> 24 *Then cometh the end, when he* (Jesus) *shall have delivered up the kingdom to God, unto the Father; when he shall have put down all rule and all authority and power* (at the end of the thousand year reign)
> 25 *For he must reign, till he hath put all enemies under his feet.*
> 26 *The last enemy that shall be destroyed is death* (See Revelation 20:14)
> 27 *For he hath put all things under his feet. But when he saith all things are put under him, it is manifest that He* (God) *is excepted* (for it is the power of God), *which did put all things under him.*
> 28 *And when all things shall be subdued unto him, then shall the Son also himself be subject unto Him* (God), that God may be all in all.

Having now completed his commission and submitted to the Father, Jesus takes his place with the Father in the Holy City of Jerusalem, in the New Heaven and New Earth. A thousand years earlier, Jesus will have been seated in the temple at Jerusalem to reign throughout the Millennium, the Saints reigning with him. And before this, upon his resurrection from the dead, Jesus has taken his place at the right hand of the Father (Psalm 110: 1; Mark 16: 19), to intercede for the Saints. It is to this for which we are called and about which he has made promises to us. We are indeed lifted up as he is lifted up.

God helps us look beyond our doubts and enables us to see mortal man made perfect. We will become, through sanctification and glorification, sacred; holy; a reflection of God; heavenly; everlasting; and, even, divine. He plans that we will be vessels fit to house the fullness of the seven-fold Spirit of God. It would be an impossibility for us to reign with him unless we are given the attributes of Jesus. Read Ephesians 3: 19.

Hear I Corinthians 15: 42-44 and believe what the scriptures say about the resurrected:

> 42 *...it is sown in corruption; it is raised in incorruption:*
> 43 *It is sown in dishonour; it is raised in glory: it is sown in weakness; it is raised in power:*
> 44 *It is sown a natural body; it is raised a spiritual body...*

Notes

We are completely transformed: sown in dishonour but raised in glory. We become a radiant image of Jesus Christ. Never again will we be evil, dishonest or wicked; we'll be incapable of anything impure. We are sown in weakness but raised in power. We have borne the image of the earthly – prone to evil and participants in death – but will bear the image of the heavenly.

When we are called to inherit incorruption, we shall be changed forever: mortal, sinful man shall put on the sacred, holy likeness of God, becoming a reflection of the divinity of Jesus, with his mind and will. Even though man had originally chosen death, through Christ, death is swallowed up in victory.

Read Romans 8: 17-18:

> *17 And if children, then heirs; heirs of God, and joint-heirs with Christ; if so be that we suffer with him, that we may be also glorified together.*
> *18 For I reckon that the sufferings of this present time are not worthy to be compared with the glory which shall be revealed in us.*

Hear the beauty and promise of Peter's declaration: "*...he hath called us to glory and virtue: whereby are given unto us exceeding great and precious promises; that by these ye might be partakers of the divine nature...*" (II Peter 1: 3-4).

Understandably, we tremble at the magnitude of such promises; ultimately, we lack faith in God. Never, we think, could man be so purified and so changed that he could enter into such a Holy union with the Godhead. Being mortal man, we find it difficult to comprehend that mere earthen vessels could become, as it were, a chalice shaped by God and made fit through Christ to contain, in full measure, the Holy Spirit of God.

A last jewel to leave with you as we go on in the study of Revelation, Chapter 4, is found in Ephesians 2: 4-7:

> *4 But God, who is rich in mercy, for his great love wherewith he loved us,*
> *5 Even when we were dead in sins, hath quickened us together with Christ, (by grace ye are saved;)*
> *6 And hath raised us up together, and made us sit together in heavenly places in Christ Jesus:*
> *7 That in the ages to come he might shew the exceeding riches of his grace in his kindness toward us through Christ Jesus.*

As we continue on to the end of our study in Revelation, the truth of this

Notes

teaching will be soundly established. We will move closer to the person of God and embrace, with awe, the wonder of His plan.

Revelation 4: 1

1 *After this I looked, and, behold, a door was opened in heaven: and the first voice which I heard was as it were of a trumpet talking with me; which said, Come up hither, and I will shew thee things which must be hereafter.*

John sees a door open in heaven and the first voice he hears sounds like a trumpet. He is called up into heaven and is shown things which will come to pass throughout the Tribulation. He is also given a glimpse of the new creation and the New Jerusalem. As John is called into heaven, so also is the Church called to the fifth feast on the Jewish calendar – the Feast of the Trumpets.

'Pre-Tribulation' believing Saints generally acknowledge that the fifth Levitical feast (Leviticus 23: 24), the Feast of the Trumpets, is representative of the Rapture. There is adequate proof that such is the case, given that God has told us in I Thessalonians 5: 9 that He has "*not appointed us to wrath but to obtain salvation*" (deliverance from the punishment of the Tribulation years). This feast comes late in the year (September / October) and is the only feast announced by a ritual blowing of trumpets. It is a holy convocation celebrated after the harvest is over: a time of great rejoicing. The harvest metaphor, of course, extends into the spiritual realm: the Church Age harvest is complete and, as the trumpets sound, the believers, with rejoicing, meet Jesus in the air (I Thessalonians 4: 16-17).

Ten days later, indicative of only a short time after the Feast of the Trumpets, the Feast of the Atonement follows (Leviticus 23: 27). This is the time when the Messiah (Jesus) atones for the sins of the people. The Raptured will be judged, rewarded, sanctified and glorified in the wedding chambers of the Lamb (I Corinthians 3: 13-15; Matthew 8: 11 NLT). Five days later, at the Feast of the Tabernacle (Leviticus 23: 34), Jesus enthrones himself among his people as he takes his place at the right hand of his Father (Hebrews 1: 3; Revelation 3: 21).

The same steps are experienced by the Tribulation Saints:

- The trumpet will sound and the Lamb's wife will be called. Read Matthew 24: 30-31:

  30 *And then shall appear the sign of the Son of man in heaven: and then shall all the tribes of the earth mourn, and they shall see the Son of man coming in the clouds of heaven with power and great glory.*

  31 *And he shall send his angels with a great sound of a trumpet, and*

Notes

they shall gather together his elect (the House of Israel) *from the four winds, from one end of heaven to the other.*

- Those who have repented during the Tribulation will be grafted into God's promises to Israel. The living and the dead in Christ will be embraced by the atonement. See Joel 2: 32:
  32 And it shall come to pass, that whosoever shall call on the name of the Lord, shall be delivered: for in Mount Zion and in Jerusalem shall be deliverance, as the Lord hath said, and in the remnant whom the Lord shall call.

- This group will have missed the bridal feast but will attend the marriage supper of the Lamb's wife. See Revelation 19: 7-9:
  7 Let us be glad and rejoice, and give honour to him: for the marriage of the Lamb is come, and his wife hath made herself ready.
  8 And to her was granted that she should be arrayed in fine linen, clean and white: for the fine linen is the righteousness of saints.
  9 And he saith unto me, Write, Blessed are they which are called unto the marriage supper of the Lamb...

- They will be judged and rewarded; sanctified but not glorified. Jesus will then tabernacle himself among them for the thousand year reign. Read Revelation 20: 4:
  4 ...and I saw the souls of them that were beheaded for the witness of Jesus and for the word of God, and which had not worshiped the beast, neither his image, neither had received his mark upon their foreheads, or in their hands; and they lived and reigned with Christ a thousand years.

The Levitical feasts were instituted by God at the time of Moses, some 1445 years before the birth of Christ. God has precisely ordered them and timed them so that, in their entirety, they announce the purposes of Christ. It is not by some strange coincidence that Jesus shed His blood on the day of the first feast, the Feast of the Passover (Leviticus 23: 5); nor is it accidental that the Feast of the Unleavened Bread follows (Leviticus 23: 6), during which time man is purged from sin.

Then, as God has foretold, after the third full day, the Feast of the First Fruit is celebrated (Leviticus 23: 10). Jesus, of course, is the 'first fruit' of those who live.

It is not by chance that some fifty days later, at the Feast of the Pentecost (Leviticus 23: 16 and Acts 2: 1-11), the Kingdom of God is established in the heart of man. The jubilee of Pentecost is evidenced by the outpouring of

Notes

the Holy Spirit as God's Spirit joins man's spirit in power and praise. See Appendix 9, "Kingdom of Heaven / Kingdom of God".

Nor is it by chance that the next feast on the Jewish Calendar is the Feast of the Trumpets. Some September / October, at this holy convocation, the memorial blowing of the trumpets will gather the Church to her Rapture. As in the Feast of Atonement (Leviticus 23: 27), the Raptured Church will be tried by fire, for the quality of their works will be tested and revealed in the presence of the Father and Son (I Corinthians 3: 13-15).

By reason of God's promise to the Philadelphian Church (Revelation 3: 10), all Christians are assured that they will be kept from the hour of temptation. Since God's word is truth, we know that as faithful followers of Christ, we will escape the temptation of the Tribulation years. At the sounding of the trumpets, we will meet Jesus in the air (I Thessalonians 4: 16-17). This is affirmed through John, who sees us seated about the throne of God, well-before God releases His wrath upon the earth (Revelation 4: 4).

Revelation 4: 2

2 And immediately I was in the spirit: and, behold, a throne was set in heaven, and One sat on the throne.

This verse tells us John is raised as a spiritual body. I Corinthians 15: 44 informs us that we are "*sown a natural body* but *raised a spiritual body*"; I Corinthians 15: 50 tells us that "*flesh and blood cannot inherit the Kingdom of God.*" God's kingdom, the kingdom of heaven, is a kingdom of spirits (see I Corinthians 15: 35-54). In a spiritual state, having been called into heaven, John sees the throne of God with One sitting on the throne.

Before we go on to verse 3, I would like to mention that there are no grounds for believing we must have physical bodies in heaven in order that we might be attired in white robes. White robes are obviously spiritual and, therefore, are invisible to the human eye. As born-again Christians, we are, even now, robed in righteousness; yet, none of us is seen dressed as such.

Revelation 4: 3

3 And he that sat was to look upon like a jasper and a sardine (sardius) *stone: and there was a rainbow round about the throne, in sight like unto an emerald.*

John views, sitting upon the throne, the One who looks like a jasper and a sardine stone; this is representative of the Father and the Son. The jasper stone, according to Revelation 21: 11, is a brilliant stone, most precious, clear as crystal, symbolizing the pure holiness and glory of God. Without a

Notes

doubt, John is seeing the Father. He also sees the deep red stone of Sardis, symbolic of the blood atonement of Jesus; thus, he sees the Son – the two in One.

The Father and the Son are together in the throne because, though Jesus could appear as both the jasper stone and the sardine stone, God alone would not readily bear such an image since He is spirit. Only in Jesus did He become flesh and shed His blood. We know that God is sitting on the throne since Revelation 5: 7 tells us the Lamb, who was in the midst of the throne, '*came and took the book out of the right hand of Him that sat upon the throne.* It is therefore established that the One who sits upon the throne is God, the Father, and Jesus Christ, the Son. See the teachings given for Revelation 5: 6 in this Study.

Exodus 28: 17-20 affirms the authenticity of this teaching since the first and the last stone in the breastplate of the high priest was proclaimed by God to be the sardius and the jasper stone.

Jesus is acclaimed our great High Priest and declared Himself '*the first and the last*'. The voice who first speaks to John in Revelation 1: 11 heralds Himself as '*Alpha and Omega, the first and the last*' (God). He further identifies himself as the '*one which was dead and is alive*' (Christ) (See Revelation 2: 8).

John's gospel introduces this same One as '*the Word, and the Word was with God and the Word was God*' (John 1: 1).

These scriptures substantiate that the One who sits upon the throne and is likened to a jasper and a sardine stone is the Father and the Son. Absolute confirmation is that both stones are chalcedony quartz; although the sardine stone is seen in varied shades of red, the jasper stone may take on even deeper shades of red. Generally, however, it is transparent and is defined in Revelation 21: 11 as clear as crystal. See the Glossary of Terms, 'Sardine and Jasper Stone'.

We are also told there is a rainbow about the throne which looks like an emerald. Some interpret this as symbolic of God's covenant with Noah (Genesis 9: 13-16). I don't agree. The rainbow which glows as an emerald is representative of a vow made and a promise kept to Christ and his Church. In actual fact, the emerald is the stone of Judah (green denoting eternal freshness and eternal endurance). The rainbow which has the brilliance of an emerald, in this case, is a pronouncement that God has kept His promise to the Stone of Judah (Jesus), the one who reigns and will reign with God for all eternity.

The Four and Twenty Elders

Notes

The emerald is also the wedding stone of the ancient oriental times. As we read further, we see within the throne the four and twenty elders who are the risen Saints, the bride of Christ. In this we have evidence that the prayer of Jesus recorded in John 17: 21 is answered:

> *21 That they all may be one; as thou, Father, art in me, and I in thee, that they also may be one in us: that the world may believe that thou hast sent me.*

Herewith the Saints are wed to the Godhead. The emerald rainbow encircles the throne as a wedding band sealing God's promise. The promise spoken through Isaiah 53: 12 is fulfilled:

> *12 Therefore will I divide him* (Jesus) *a portion with the great* (with God), *and he shall divide the spoil with the strong* (the faithful)*; because he* (Jesus) *hath poured out his soul unto death...*

When we grasp that we are filled to the measure of all the fullness of God (Ephesians 3: 19), faith and praise will be multiplied as we begin to understand God's plan for us. We will see that even though we are given the Seven Spirits of God, as has Christ, we remain the submissive, consecrated vessel, chosen by God to house the fullness of His Holy Spirit. Through the abundance of the Spirit, He forever binds us to the Godhead. When our time of service is over and the New Heavens and New Earth replace the old, God brings His kingdom down from heaven and reinstates His original plan. We, as sanctified vessels refined by fire and totally consumed by the Holy Spirit, will return as children to the Garden of Eden. We'll bring the glory and honour of the Holy Spirit into it (Revelation 21: 24) as sanctified 'temples' (sons and daughters) filled with the glory of God.

Revelation 4: 4

> *4 And round about the throne were four and twenty seats: and upon the seats I saw four and twenty elders sitting, clothed in white raiment; and they had on their heads crowns of gold.*

The first thing we should look at in this verse is the statement '*and round about the throne were four and twenty seats.*' The New King James version has translated the Greek word 'thronos' to read 'thrones.' We may therefore visualize the elders seated on thrones around about the throne of God. We have just read in the previous verse that there is a rainbow '*round about the throne'* which looked like an emerald. The fact that both verses use the same term 'round about the throne' seems to infer that the seated elders are also encircled by the rainbow which circumscribes God's throne. This strengthens the thought that the rainbow has reference to God's promise to Jesus and his Church rather than to His Covenant with Noah.

Notes

The four and twenty elders are representative of the entire body of the risen Saints, Israelite and Gentile alike (Galatians 3: 28; Ephesians 2: 14-16). They are clothed in white garments and each is wearing awarded crowns of gold. I Peter 2: 9 gives a descriptive narration of them:

> 9 *But ye are a chosen generation, a royal priesthood, an holy nation, a peculiar people; that ye should shew forth the praises of him who hath called you out of darkness into his marvelous light:*

Some confusion could arise from the fact that under the Levitical priesthood there were twenty-four orders (I Chronicals 24: 1-19); in each order there were twenty-four priests chosen to serve an appointed length of time in the temple in a designated position which was called their 'lot.' Read Luke 1: 8-9 as an example:

> 8 *And it came to pass, that while he (*Zacharias*) executed the priest's office before God in the order of his course,*
> 9 *According to the custom of the priest's office, his lot was to burn incense when he went into the temple of the Lord.*

These Levitical priests served as representatives of the whole body of the priesthood. The priests (out of the tribe of Levi), when serving in the temple, wore linen drawers with a close-fitting cassock which came down to their feet. On their heads, they wore caps or bonnets in the form of a cup-shaped flower. All the garments, including the headpiece, were made from fine linen which would prevent them from perspiring while attending temple service. They never wore crowns (Exodus 28: 40-43; Ezekiel 44: 17-18).

For this reason, anytime we read that linen garments are worn, we may be assured that God is specifically referring to the Israelites; otherwise, it is simply said, as in the case of this verse under study, that they are clothed in white raiment. Therefore, we can be absolutely sure the twenty-four elders represent the royal priesthood, the risen Saints. By saying they are clothed in white raiment, God has chosen not to identify them as either Israelites or Gentiles. By this we know that the group about the throne includes both.

Some associate the twenty-four with the twelve patriarchs (the Wife of Christ) and the twelve apostles (the Bride of Christ).[18] If God had intended to leave this message with us, He would surely have had twelve arrayed in fine linen, in keeping with the Levitical priesthood; and twelve in white raiment, representing the Apostles and the Church.

---

[18] William Barclay, "The Revelation of John", Volume I, Daily Study Bible (Burlington, Ontario: Welch Publishing, 1976), p. 154.

The Church Baptized by the Seven Spirits of God

Notes

An example of an unidentified group which includes both Israelite and Gentile is found in Revelation 6: 9-11. These Saints are given white robes and are told to rest for a season.

When God chooses to identify Israelites specifically, He does so with linen robes; for example, in Revelation 19: 14, the armies of Old Testament Saints who follow Jesus back to earth – the armies from heaven – are clothed in fine linen, clean and white.

In John 14: 2-3, we read:

2 *In my Father's house are many mansions: if it were not so, I would have told you. I go to prepare a place for you.*

3 *And if I go and prepare a place for you, I will come again, and receive you unto myself; that where I am, there ye may be also.*

The mansions, or dwelling places, or the 'lot' at the end of our days as promised to Daniel (Daniel 12: 13), are the places in the Father's house reserved through Jesus Christ as places prepared for the Saints around the throne of God. Jesus died to secure these dwelling places for us and he will come again and receive us unto himself, that where he is – in his Father's house, we may be also. It's as if he went ahead, made a reservation for us and paid the price; in this way, when we arrive, we are expected and there will be no further price to pay. Please note that the Saints are not risen until Jesus comes again to receive them.

Read Mark 10: 37-40. James and John come to Jesus with this request:

37 *They said unto him, Grant unto us that we may sit, one on thy right hand, and the other on thy left hand, in thy glory.*

38 *But Jesus said unto them, Ye know not what ye ask: can ye drink of the cup that I drink of? and be baptized with the baptism that I am baptized with?*

39 *And they said unto him, We can. And Jesus said unto them, Ye shall indeed drink of the cup that I drink of; and with the baptism that I am baptized withal shall ye be baptized:*

40 *But to sit on my right hand and on my left hand is not mine to give; but it shall be given to them for whom it is prepared.*

Basically, in this narrative, Jesus first asks James and John if they are willing to drink of the cup of suffering that he will drink and be baptized with the baptism of death, as he will be baptized. James and John say they are willing to die that they may be with him, where he is. Jesus then states they will drink of the fullness of the cup of suffering that he drinks, and will be baptized as he is baptized – with the Seven Spirits of God (Colossians 2: 9). Jesus says he can't, however, assign to them a particular place. The places

Notes

are God's to give and they will be given without partiality to those for whom they are prepared ("*the first shall be last, and the last shall be first*" – all will be equal). Jesus prepares the places only in that he has secured them with his life; the mansions were already in the Father's house.

Without a doubt, John sees the Raptured Saints, Old Testament and New Testament alike (Matthew 8: 11), seated around the throne in the places prepared and secured for them in the Father's house. Now, at last, man is sanctified and glorified through Christ; we are made heirs and joint heirs with Christ and His glory shall be revealed in us. We are, at this point, the manifested sons of God (Romans 8: 17-19). When we receive Christ as our Saviour, we become '*sons of God.*' John says in I John 3: 2:

> 2 *Beloved, now are we the sons of God, and it doth not yet appear what we shall be: but we know that, when he shall appear, we shall be like him; for we shall see him as he is.*

We are sons of God, indeed, upon rebirth; however, the full manifestation of this appointment is not realized until we are baptized as is Jesus – baptized with the Seven Spirits of God. It is only then that we have his full likeness and are totally one in the Father and in the Son. Christ's promise is fulfilled: '*And the glory which thou gavest me I have given them; that they may be one, even as we are one*' (John 17: 22).

Praise God, for He '*hath raised us up together, and made us sit together in heavenly places in Christ Jesus*' (Ephesians 2: 6).

Revelation 4: 5

> 5 *And out of the throne proceeded lightnings and thunderings and voices: and there were seven lamps of fire burning before the throne, which are the seven Spirits of God.*

The thunderings and lightnings and voices from out of the throne speak of the awesome power, majesty and presence of God. It may well also refer to pent-up anger and judgment which God intends to release upon the earth. Please read Job 37: 2-5 NLT.

In the preamble intended as preparation for Chapter 4, teaching was given on the seven lamps burning before the throne of God. To reinforce the concept of our holy union with the Father and Son, as well as to translate the meaning of this present verse we are now looking at, review the following portion taken from earlier comments (Preamble to Chapter 4):

> We are become the submissive vessel prepared by God to house the third person of the Godhead. Encircling the throne of God and

Notes

> sheathed by the rainbow, we are embraced by, and endowed with, the Seven Spirits of God (Isaiah 11: 2) which are the seven lamps burning before the throne – these being the seven attributes of God.

To a lesser degree, upon our rebirth, we draw upon the characteristics of God. We are quickened by His Spirit and graciously empowered by His gifts (see I Corinthians 12: 4-11). Paul evidenced, during his ministry, all nine gifts of the Holy Spirit. In them, at their fullest, lies the seven Spirits of God.

Some readers may fear that man, if empowered by the Seven Spirits of God, could seek to take over the preeminence of God. God has provided, within His Seven Spirits, a protection upon man. To understand this, one must take note of the first Spirit and the last Spirit of God. See Appendix 3, "The Seven Spirits of God."

The first Spirit, the Spirit of the Lord, is the Fullness of the Holy Spirit. This is the Spirit of life, obedience and service. It is the Spirit which, at our glorification, gives us the mind and desires of Christ. Through this, we are as Christ is: totally subject and submissive to God; wanting only to yield all glory and all honour to Him (I Corinthians 15: 28; Revelation 4: 10).

The last Spirit of the Seven Spirits of God is the Spirit of the Fear of the Lord. Through this, we obtain forever, as part of our integral being, complete reverence and total submission to God our Father and Jesus Christ, His Son. With the provision and protection of God's Spirits (seven in one), man remains sealed forever as the submissive, child-like vessel sanctified by Christ; thus, we are prepared by God to house, in full measure, the Holy Spirit of God. Through the abundance of the Spirit, God weds us in holy union to the Two-in-One.

Revelation 4: 6

6 *And before the throne there was a sea of glass like unto crystal: and in the midst of the throne, and round about the throne, were four beasts full of eyes before and behind.*

John sees before the throne a sea of glass very much like crystal. This area before the throne reflects the glory of God similar to the bible description of the jasper stone (representative of God). Revelation 21: 11 says that the jasper stone is most precious and clear as crystal. The verse now under study implies that it is an area of brilliance and transparent light. It would seem that each newly converted Saint stands within this transparent light, veiled by Christ, to be seen by God and His heavenly host. They stand within the crystal clear brilliance of God for their initial moment of entrance

Notes

into a rebirthed state. The whole of heaven rejoices because they have overcome Satan through the blood of our Saviour (Luke 15: 7). Revelation 15: 2-3 supports, in part, this interpretation:

2 *And I saw as it were a sea of glass mingled with fire: and them that had gotten the victory over the beast, and over his image, and over his mark, and over the number of his name, stand on the sea of glass, having the harps of God.*

3 *And they sing the song of Moses the servant of God, and the song of the Lamb, saying, Great and marvelous are thy works, Lord God Almighty; just and true are thy ways, thou King of saints.*

These Israelite Saints appear before the throne but, being Tribulation Saints, they will have lost out on heavenly rewards. Their rewards are earthly. They appear before God and all heaven rejoices (Luke 15: 10) but, instead of entering heaven, they enter the millennial reign. At this point in time, they truly become *a kingdom of priests and a holy nation* (Exodus 19: 6).

Revelation 4: 7 - 8

7 *And the first beast was like a lion, and the second beast like a calf, and the third beast had a face as a man, and the fourth beast was like a flying eagle.*

8 *And the four beasts had each of them six wings about him; and they were full of eyes within: and they rest not day and night, saying, Holy, holy, holy, Lord God Almighty, which was, and is, and is to come.*

Isaiah 6: 1-3 talks of similar beings. He calls them 'seraphim'. Their purpose is to do God's bidding and praise Him continually night and day. They have faces and feet, and they speak as a man. Ezekiel 1: 5-10 and 10: 14-15 also refer to these creatures and calls them 'cherubim'. Both obviously are a higher order of angels. Ezekiel's four creatures have the likeness of man but each has four faces: that of a lion, an ox, an eagle and a human. John sees them as four separate beings, each having one face bearing the same characteristics of Ezekiel's creatures. They also differ in that Ezekiel's angels have only four wings full of eyes; while John sees them with six wings, as does Isaiah's vision of the beasts. It may be the number of wings which designate their order and service in the heavenly host. If this is the case, then the creatures under study here in Revelation 4: 6-8 are seraphim. These beings reflect the attributes of Christ. The lion denotes majesty; the calf or ox, humility and patience; the man, reason, intelligence, wisdom and knowledge; and the eagle is the wisest bird, flying the highest, having the keenest sight and being swift in action – the eagle stands for the sovereignty of Christ. All combined express the character and person of

Notes

God.[19] Their many eyes are a metaphor describing the all-seeing God. Their main purpose is to lead all heaven in glorious praise, acclaiming the Godhead as Holy, Almighty, Ever-being and Everlasting.

It would be well to read Ezekiel 1: 4-28 in the New Living Translation. It is easy to read and very informative; you will likely gain a deeper understanding of the above two verses.

Revelation 4: 9 - 11

9 *And when those beasts give glory and honour and thanks to him that sat on the throne, who liveth for ever and ever,*

10 *The four and twenty elders fall down before him that sat on the throne, and worship him that liveth for ever and ever, and cast their crowns before the throne, saying,*

11 *Thou art worthy, O Lord, to receive glory and honour and power: for thou hast created all things, and for thy pleasure they are and were created.*

When the living creatures, who never rest, continually give glory and honour and praise to God, the whole of heaven joins them in praise. The four and twenty elders (the bride of Christ) fall down before God the Father and the Lord Jesus Christ. They praise and worship Them and, in absolute adoration and submission, they cast their crowns of victory before the throne. The act of literally casting their crowns before God indicates that they choose not to retain any glory for themselves, but are subject unto God, giving Him total preeminence. Just as Jesus submits to the Father at the end of his Millennial reign (I Corinthians 15: 28), so also does the Church at the Rapture.

To culminate the teaching of Revelation Chapter 4, and in support of the preliminary statements and quotes found prior to this verse-by-verse commentary on this Chapter, read Revelation 21: 22. John has just viewed the new Jerusalem in the New Creation. He sees that there is "*no temple therein: for the Lord God Almighty and the Lamb are the temple of it*." Then in Revelation 21: 23, he sees the city has no need for the sun or the moon to shine in it because the "*glory of God did lighten it, and the Lamb is the light thereof*." In Revelation 22: 1, John sees "*a pure river of water of life, clear as crystal, proceeding out of the throne of God and of the Lamb*." Then in Revelation 22: 3, the scriptures state that "*there shall be no more curse: but the throne of God and of the Lamb shall be in it*..." John goes on to tell us in Revelation 22: 5 that "*there shall be no night there; and they*

---

[19]Daymond R. Duck, Revelation - God's Word for the Biblically-Inept (Lancaster, Penn.: Starburst Publishers, 1998), p. 71.

Oliver B. Greene, The Revelation (Greenville, SC: Gospel Hour Inc., 1976), p. 162.

Notes

need no candle, neither light of the sun; for the <u>Lord God giveth them light: and they shall reign for ever and ever</u>" (Father and Son).

Surprisingly, we have not seen the triune God. We see the Father and Son but the Holy Spirit seems to be non-existent. Has the Spirit returned to his source and been assimilated into the Father, leaving a two-person Godhead? I think not!

There will forever be a triune God. The explanation is the Spirit in its entirety has taken up residence in the Saints. They are endowed with the Seven Spirits of God at their sanctification when they become the <u>glorified</u> Sons of God. (Read Romans 8: 16-19.) Revelation 21: 24 records that John has seen the nations that are saved (the Tribulation Saints) and the kings of the earth (the Raptured Saints – Old and New Testament alike) walk in the light of the glory of God. One dimension of the 'light of the glory of God' is God in us. When we walk in the fullness of that light, we are not only present with Him, but He dwells within us in the totality of His Seven Spirits. According to Revelation 21: 24, the kings of the earth bring their glory and honour into the New Jerusalem. We have no glory, nor honour, aside from that which we have been given through the plentitude of the Seven Spirits of God.

As sinful, carnal beings, we find it difficult to accept God's wondrous plan for man. It requires more faith than we seem to have. How can we believe that His glory, His honour, and <u>the fullness of His Holy Spirit</u> could join the Spirit of God in us and forever mould us to the Godhead? It seems too holy, too glorious and far above our mortal understanding.

If we pause for a moment, however, and review the scriptures and the promises of God, we find that there is nothing new or frightening about this teaching. We have believed since the resurrection of Jesus and the day of Pentecost that God dwells within man through His Holy Spirit. The extent to which He <u>will</u> dwell has been overlooked or rejected by man. We somehow refuse to trust His promises. The time has come for us to open our ears and our hearts; we must humbly look ahead with acceptance, rejoicing and great expectation. Praise God, for Jesus is our hope of Glory!

Revelation 22: 17 says, "*The Spirit and the bride say Come.*" This may well be another affirmation that the bride and the Holy Spirit become one. Surely our hearts and minds must join the psalmist singing praises to His name: "*Sing unto him, sing psalms unto him: talk ye of all his wondrous works.*" (Psalm 105: 2)

CHAPTER FIVE

In **Chapter Five**, we view the Father holding a seven-sealed scroll containing the title deed to His universe and His final Will pertaining to its cleansing. All heaven rejoices as they behold 'the Christ', the Savior and Redeemer of their souls. A teaching is given on Roman law pertaining to a Last Will and Testament, and Mosaic redemption laws linking them to the righteous will of God and our Redeemer.

Revelation 5

[**1**] And I saw in the right hand of him that sat on the throne a book written within and on the backside, sealed with seven seals.
[**2**] And I saw a strong angel proclaiming with a loud voice, Who is worthy to open the book, and to loose the seals thereof?
[**3**] And no man in heaven, nor in earth, neither under the earth, was able to open the book, neither to look thereon.
[**4**] And I wept much, because no man was found worthy to open and to read the book, neither to look thereon.
[**5**] And one of the elders saith unto me, Weep not: behold, the Lion of the tribe of Juda, the Root of David, hath prevailed to open the book, and to loose the seven seals thereof.
[**6**] And I beheld, and, lo, in the midst of the throne and of the four beasts, and in the midst of the elders, stood a Lamb as it had been slain, having seven horns and seven eyes, which are the seven Spirits of God sent forth into all the earth.
[**7**] And he came and took the book out of the right hand of him that sat upon the throne.
[**8**] And when he had taken the book, the four beasts and four and twenty elders fell down before the Lamb, having every one of them harps, and golden vials full of odours, which are the prayers of saints.
[**9**] And they sung a new song, saying, Thou art worthy to take the book, and to open the seals thereof: for thou wast slain, and hast redeemed us to God by thy blood out of every kindred, and tongue, and people, and nation;
[**10**] And hast made us unto our God kings and priests: and we shall reign on the earth.
[**11**] And I beheld, and I heard the voice of many angels round about the throne and the beasts and the elders: and the number of them was ten thousand times ten thousand, and thousands of thousands;
[**12**] Saying with a loud voice, Worthy is the Lamb that was slain to receive power, and riches, and wisdom, and strength, and honour, and glory, and blessing.
[**13**] And every creature which is in heaven, and on the earth, and under the earth, and such as are in the sea, and all that are in them, heard I saying, Blessing, and honour, and glory, and power, be unto him that sitteth upon the throne, and unto the Lamb for ever and ever.
[**14**] And the four beasts said, Amen. And the four and twenty elders fell down and worshipped him that liveth for ever and ever.

God and the Seven-Sealed Scroll

Notes

Revelation, Chapter 5

Revelation 5: 1

1 *And I saw in the right hand of him that sat on the throne a book written within and on the backside, sealed with seven seals.*

John sees the person of God sitting upon the throne. He is holding, in His right hand, a book. The word '*book*' may more correctly be interpreted '*scroll*'. The fact that there is writing on both the front and the back suggests that this particular scroll contains two documents. What God is holding is a testament of His will, as well as the title deed to His creation; it is a directive to His heir in the final settlement of His universe. After all, He owns the earth and all that is within it.

Roman law required that a will be witnessed and sealed by seven seals.[20] When the document was fully written, exactly as the testator had dictated, it was fastened with threads and each witness sealed the knots with his own seal. It was to be opened only when all the witnesses were present or when a legal representative, such as an heir, was qualified to request the opening of it. Under the seals were written precise instructions which had to be followed as the will was probated.

Mosaic law required that when land was sold or transferred because of debt, the title deed was to be retained by the original owner. A scroll was written in duplicate (only one copy of which was sealed) stating that at a later date a kinsman, such as an heir, could reclaim the property if willing to pay the purchase price. He had to be first in line as a qualifying kinsman and, as such, was called the redeemer.

In the verse under study, seven predicted wraths (each being the righteous 'will' of God) are listed in the scroll and after each a seal is placed. So it is that seven seals are placed in the scroll held by the testator who, in this particular reference, is God. He has, at creation, given His universe in trust to the first Adam who sold out to Satan, thus condemning man to damnation. Praise God, this verdict is not for eternity since He has retained the title deed and has it in His hand. He has given it in trust to His Son, the second Adam, the Redeemer: the Lord Jesus Christ who has willingly paid the price. As the qualified kinsman-redeemer, he is found worthy to open the seals.

To read about the laws of redemption, you may turn to Leviticus 25: 23-28; see it applied in Ruth 4: 1-13 and Jeremiah 32: 7-14.

---

[20]William Barclay, "The Revelation of John" Volume I, The Daily Study Bible (Burlington, Ontario: Welch Publishing, Revised 1976), p. 166.

Search for the Redeemer / Jesus, the Worthy Lion of Judah

Notes

Revelation 5: 2

2 *And I saw a strong angel proclaiming with a loud voice, Who is worthy to open the book, and to loose the seals thereof?*

It is likely that the strong angel giving forth the call would be one from the higher order of angels, possibly a seraph. The call is for the one who qualifies; the one who paid the price; the one called the Redeemer.

Revelation 5: 3

3 *And no man in heaven, nor in earth, neither under the earth, was able to open the book, neither to look thereon.*

It seems that no man in heaven or earth, or under the earth – that is, among the resurrected, the living, or the dead – can be found worthy to break the seal and look into the contents of the scroll.

Revelation 5: 4

4 *And I wept much, because no man was found worthy to open and to read the book, neither to look thereon.*

Somehow, John has lost sight of the victorious Jesus. His sorrow is deep and his concern grave because he knows that unless a redeemer is found, Satan will be the victor and man is lost for eternity.

Revelation 5: 5

5 *And one of the elders saith unto me, Weep not: behold, the Lion of the tribe of Judah, the Root of David, hath prevailed to open the book, and to loose the seven seals thereof.*

One of the elders, the bride of Christ, who is seated about the throne, comforts John and tells him not to weep. This elder knows the Lion of the tribe of Judah, the Root of David. He knows him because he – Jesus – is his Redeemer. Jesus has paid the price for him. The elder knows that Jesus is worthy to open the scroll and break the seven seals found therein.

Revelation 5: 6

6 *And I beheld, and, lo, in the midst of the throne and of the four beasts, and in the midst of the elders, stood a Lamb as it had been slain, having seven horns and seven eyes, which are the seven Spirits of God sent forth into all the earth.*

This is the verse which shows us, without a doubt, that the resurrected Jesus

Notes

is in the midst of the throne with God the Father. He is there when John first beholds the One who sits upon the throne (Revelation 4: 2). Remember that Jesus is the sardine stone and God, his Father, is the jasper stone. As promised, we – the Saints – are seated about the throne. Jesus, the gentle Lamb, is risen as the Lion of Judah, the King of kings and Lord of lords. The seven Spirits of God, which is the whole Spirit of God, rest upon him. The seven horns and seven eyes speak of the mighty omnipotent, omniscient God. The 'One' who rises up out of the throne (Jesus) is the embodiment of the seven spirits of God: these being the Holy Spirit, the spirit of wisdom and understanding, the spirit of council and might, the spirit of knowledge and the spirit of fear (absolute reverence) of the Lord (Isaiah 11: 2). In Jesus *"dwelleth all the fulness of the Godhead bodily"* (Colossians 2: 9).

Revelation 5: 7

7 *And he came and took the book out of the right hand of him that sat upon the throne.*

Jesus, having entered into his full baptism, with confidence and all authority, takes the scroll from the right hand of the Father. Before this, Jesus has always sought the Father's direction; however, he now co-reigns with God. He has the total mind of God and knows full-well the will of God. The Father is in the Son. They are again One; nevertheless, although they are One, they are still the Father and the Son. Jesus no longer has to pray to learn the will of God because he is God, the Son. He is no longer part man; now, They are again the two empowered and reigning as One. On behalf of the Father, he will reign with the Saints throughout the Millennium, with the fullness of the seven Spirits of God guiding him in his reign. In the New Heavens and New Earth, in the New Jerusalem, they will forever reign together as Father and Son (Revelation 22: 1).

Revelation 5: 8

8 *And when he had taken the book, the four beasts and four and twenty elders fell down before the Lamb, having every one of them harps, and golden vials full of odours, which are the prayers of saints.*

The four creatures, having the express image and character of God (see commentary on Revelation 4: 7-8), fall down before Jesus along with the Saints who now also reflect the image and character of God. The Saints, sealed by the Seven Spirits of God, forever remain the submissive vessel filled to capacity by the third-person of the Godhead. The harps that they have are within them; that is, their hearts are rebounding the glory of God. They sing praises as a heavenly host together. The sweet music of the Spirit rises, wave upon wave, each more beautiful than the last. All-as-one and

Notes

one-as-all, love multiplies beyond human ability; music from God expounding the glory of God through man. It becomes adoration, as it were, from the throne, and praise above praise. They gesture their love as they hold up golden vessels filled with the sweet odours of prayer and praise, from time past to eternity.

Revelation 5: 9

9 *And they sung a new song, saying, Thou art worthy to take the book, and to open the seals thereof: for thou wast slain, and hast redeemed us to God by thy blood out of every kindred, and tongue, and people, and nation;*

The elders sing a new song in that the depth and dimension of their love has been enlarged. Having had a meager portion of the Spirit and now having been filled, they sing the glorious song of redemption, their hearts overflowing with adoration. They profess Jesus worthy to open the seals; they know him to be the provider and the Saviour of their soul. They confess that through his sacrifice, all nations, tongues, kindred and people are cleansed by his blood.

Revelation 5: 10

10 *And hast made us unto our God kings and priests: and we shall reign on the earth.*

The elders extol Jesus for having paid the price which has brought them back to the place of sweet fellowship with the Father. They have been raptured; raised to enter the place prepared for them; baptized with the baptism that Jesus is baptized with; endowed with the seven Spirits of God. As glorified sons of God, they worship Jesus for having saved their souls. They praise him for having bestowed upon them the fullness of the Spirit. In so doing, he has enabled them to become kings and priests unto God.

Revelation 5: 11

11 *And I beheld, and I heard the voice of many angels round about the throne and the beasts and the elders: and the number of them was ten thousand times ten thousand, and thousands of thousands;*

John sees and hears not only the Saints and the four creatures praise Jesus, but also an innumerable number of angels praising him.

Notes

Revelation 5: 12

12 Saying with a loud voice, Worthy is the Lamb that was slain to receive power, and riches, and wisdom, and strength, and honour, and glory, and blessing.

The angels are unable to sing the song of redemption – the New song – but they join in the celebration and praise of Jesus by saying, "*Worthy is the Lamb that was slain to receive power, and riches, and wisdom, and strength, and honour, and glory, and blessing.*"

Revelation 5: 13

13 And every creature which is in heaven, and on the earth, and under the earth, and such as are in the sea, and all that are in them, heard I saying, Blessing, and honour, and glory, and power, be unto him that sitteth upon the throne, and unto the Lamb for ever and ever.

John looks and hears beyond the Millennium and envisions the gathering for the White Throne judgment, as all creatures in heaven and on earth, and under the earth, and in the seas, bow their knee to the Father and Son. All will confess '*that Jesus Christ is Lord, to the glory of the Father.*' Now is Romans 14: 11 fulfilled:

> *11 ...as I live saith the Lord, every knee shall bow to me, and every tongue shall confess to God.*

So says Philippians 2: 9-11:

> *9 Wherefore God also hath highly exalted him, and given him a name which is above every name:*
> *10 That at the name of Jesus every knee should bow, of things in heaven, and things in earth, and things under the earth;*
> *11 And that every tongue should confess that Jesus Christ is Lord, to the glory of God the Father.*

Revelation 5: 14

14 And the four beasts said, Amen. And the four and twenty elders fell down and worshiped him that liveth for ever and ever.

On into eternity the Saints and heavenly host reverence and glorify the Father and Son. After the Millennium, they return to the earth with the Father (Revelation 21: 2-3) and enter the New Creation (which is as the Garden of Eden), forever obedient children of the Father and Son. Filled with the seven Spirits of God, incapable of sin, these are the beloved of the Father. The perfect union, as planned by God in His first creation, is impeccably embodied in the Father, the Son and the fullness of the Holy

Notes

Spirit culminated in man. The believers, transformed into the image of Christ, fall down and worship Him who lives forever and ever.

Now is fulfilled in its entirety II Corinthians 6: 16-18:

> 16 ...<u>*for ye are the temple of the living God*</u>*; as God hath said, I will dwell in them, and walk in them; and I will be their God, and they shall be my people.*
>
> 17 *...and I will receive you,*
>
> 18 *And will be a Father unto you, and ye shall be my sons and daughters, saith the Lord Almighty.*

Praise Him, all ye people!

CHAPTER SIX

In **Chapter Six**, we are given an overview of the Tribulation years. It devotes itself to unveiling and revealing God's judgments. In opening the seals of the scroll, Jesus is probating his Father's Will. By so doing, he opens the way for the outpouring of the wrath of heaven and releases the wrath of man. The Study teaches that the seven seals are not a set of wraths in themselves but, rather, are a revelation of the wraths to come. It establishes that the purpose of the chapter is also to warn and alert man and bring him to repentance. During the Study, the rider on the white horse is identified and the figurative nature and intent of the horse is discussed.

Revelation 6

[1] And I saw when the Lamb opened one of the seals, and I heard, as it were the noise of thunder, one of the four beasts saying, Come and see.
[2] And I saw, and behold a white horse: and he that sat on him had a bow; and a crown was given unto him: and he went forth conquering, and to conquer.
[3] And when he had opened the second seal, I heard the second beast say, Come and see.
[4] And there went out another horse that was red: and power was given to him that sat thereon to take peace from the earth, and that they should kill one another: and there was given unto him a great sword.
[5] And when he had opened the third seal, I heard the third beast say, Come and see. And I beheld, and lo a black horse; and he that sat on him had a pair of balances in his hand.
[6] And I heard a voice in the midst of the four beasts say, A measure of wheat for a penny, and three measures of barley for a penny; and see thou hurt not the oil and the wine.
[7] And when he had opened the fourth seal, I heard the voice of the fourth beast say, Come and see.
[8] And I looked, and behold a pale horse: and his name that sat on him was Death, and Hell followed with him. And power was given unto them over the fourth part of the earth, to kill with sword, and with hunger, and with death, and with the beasts of the earth.
[9] And when he had opened the fifth seal, I saw under the altar the souls of them that were slain for the word of God, and for the testimony which they held:
[10] And they cried with a loud voice, saying, How long, O Lord, holy and true, dost thou not judge and avenge our blood on them that dwell on the earth?
[11] And white robes were given unto every one of them; and it was said unto them, that they should rest yet for a little season, until their fellowservants also and their brethren, that should be killed as they were, should be fulfilled.
[12] And I beheld when he had opened the sixth seal, and, lo, there was a great earthquake; and the sun became black as sackcloth of hair, and the moon became as blood;
[13] And the stars of heaven fell unto the earth, even as a fig tree casteth her untimely figs, when she is shaken of a mighty wind.
[14] And the heaven departed as a scroll when it is rolled together; and every mountain and island were moved out of their places.
[15] And the kings of the earth, and the great men, and the rich men, and the chief captains, and the mighty men, and every bondman, and every free man, hid themselves in the dens and in the rocks of the mountains;
[16] And said to the mountains and rocks, Fall on us, and hide us from the face of him that sitteth on the throne, and from the wrath of the Lamb:
[17] For the great day of his wrath is come; and who shall be able to stand?

Preamble / The First Seal / The White Horse

Revelation, Chapter 6

Notes

Revelation, Chapter 6, is a condensed overview of the Tribulation years. In it, many windows open and God allows us to view future events in world affairs, starting with the entrance of the Antichrist. The Chapter ends with our seeing the resulting fear and devastation when the wrath of God comes down upon man. The whole of the chapter is a revealing or unveiling of God's righteous will; it assures us that God is in control, even unto the end.

It would be beneficial for those doing this study to follow along in Matthew 24: 3-36. When asked by the disciples what the signs of his coming would be, so that they might know the time of his return, Jesus actually gives a first-hand account of all we read in Revelation, Chapter 6. God always warns before he punishes.

Jesus, by opening the seals, is not at this point personally pouring out his wrath upon the earth, as it may first appear. Rather, he is opening the way for the outpouring of the wrath of God, as well as the wrath of man, upon a wayward and wicked world. Satan can do nothing without the consent of God (see Job 1: 12). The seals reveal to man God's imminent judgment on the ungodly and disobedient, so that the world may be alerted and repent.

Revelation 6: 1

1 *And I saw when the Lamb opened one of the seals, and I heard, as it were the noise of thunder, one of the four beasts saying, Come and see.*

In the process of probating the Will of God, Jesus opens the first seal and John hears the awesome might and authority of God. One of the four creatures invites him to come and see.

Revelation 6: 2

2 *And I saw, and behold a white horse: and he that sat on him had a bow; and a crown was given unto him: and he went forth conquering, and to conquer.*

John sees the counterfeit Jesus as a conqueror riding into power on a white horse; undeniably, this is the Antichrist. Jesus warns of this in Matthew 24: 4-5:

4 *...Take heed that no man deceive you.*

5 *For many shall come in my name, saying, I am Christ; and shall deceive many.*

This deceiver comes with great acclaim as a peace-maker; notice, there is no arrow in his bow. He will not shed his blood as our Saviour did, nor will he,

Notes

at the beginning of his reign, shed the blood of others. He will be looked upon as the saviour of the world and will be given world approval and complete authority. The nations will crown this most evil man on earth; he will be charming, persuasive, beguiling and evil.

Peace will follow; for a short while the world will seem at rest. Read I Thessalonians 5: 3:

> 3 *For when they shall say, Peace and safety; then sudden destruction cometh upon them, as travail upon a woman with child; and they shall not escape.*

Some believe that the earth enjoys peace for the first half of the Tribulation years. This will not be the case, as affirmed through prophecy and the opening of the seals. The second seal informs us that peace is taken from the earth. The seals declare it and the trumpet wraths and man enact it. One must remember there will be seven years Tribulation. If the first half of those years were peaceful, we could hardly call them Tribulation years. Please refer to Appendix 10, "Time Chart – The Tribulation Years."

Revelation 6: 3 - 4

3 *And when he had opened the second seal, I heard the second beast say, Come and see.*

4 *And there went out another horse that was red: and power was given to him that sat thereon to take peace from the earth, and that they should kill one another: and there was given unto him a great sword.*

When Jesus opens the second seal, the second creature invites John to come and see. The red horse which John sees is Russia. Power is given to her not once, but twice, to come down on Israel (Ezekiel, Chapters 38 and 39). Move ahead in this study to Revelation, Chapter 19: 17-18 for a full teaching on the Russian invasions of Israel. Jesus spoke of this in Matthew 24: 7:

> 7 *For nation shall rise against nation, and kingdom against kingdom: and there shall be famines, and pestilences, and earthquakes, in divers places.*

The beginning of these latter-day sorrows is even now evidenced in our world. The Tribulation is almost upon us. Take heart, dear Christians, for we will remain on earth just long enough to see the great falling away from Christian ethics and morals (which we now see) and the man of sin revealed as '*the son of perdition*' (the Antichrist) (II Thessalonians 2: 1-3). Rest assured that we will be risen before the destruction comes. For a complete teaching on this, read Revelation 11: 1 in this study.

The Third Seal / The Black Horse / The Fourth Seal / The Pale Horse

Notes

Revelation 6: 5 - 6

5 *And when he had opened the third seal, I heard the third beast say, Come and see. And I beheld, and lo a black horse; and he that sat on him had a pair of balances in his hand.*

6 *And I heard a voice in the midst of the four beasts say, A measure of wheat for a penny, and three measures of barley for a penny; and see thou hurt not the oil and the wine.*

When Jesus opens the third seal, the third beast says, "Come and see." John sees a black horse and the one who sits on the horse has a pair of balances in his hand. Black indicates starvation. Rationing is in effect; hunger and pestilence prevail. Tribes and races have risen against one another; people against people. Zechariah 14: 13 says:

> 13 *And it shall come to pass in that day, that a great tumult from the Lord shall be among them; and they shall lay hold every one on the hand of his neighbour, and his hand shall rise up against the hand of his neighbour.*

Food will be so scarce and so expensive that it will take a full day's wages to buy barely enough to keep a man alive. (A penny was the wage for a day's work in John's time. See Matthew 20: 2.) Everyone will be required to wear the number of the beast in order to buy food if, and when, it is available (Revelation 13: 17).

'The oil and the wine' generally refers to Israel. In this case, there is a warning not to hurt the 144,000 Israelites whom God is about to seal (look ahead to Revelation 7: 3-8).

Revelation 6: 7 - 8

7 *And when he had opened the fourth seal, I heard the voice of the fourth beast say, Come and see.*

8 *And I looked, and behold a pale horse: and his name that sat on him was Death, and Hell followed with him. And power was given unto them over the fourth part of the earth, to kill with sword, and with hunger, and with death, and with the beasts of the earth.*

When Jesus opens the fourth seal, John sees a pale horse. Pale, the colour of a corpse, indicates death. We are told that the beasts of the earth (the Antichrist and his followers) will kill one-fourth of the earth's population. Christ's opening of the seals is in direct obedience to God's will. God has consented to the Tribulation as just recompense for the continuous disobedience of mankind.

Jesus tells us in Matthew 24: 9-12 that the Tribulation Christians will go

Notes

through a time of great persecution and many will die for their faith. Remember, they could have spared themselves this sorrow. Had they repented prior to the Tribulation and been obedient to Him, they would have escaped God's wrath when received in the Rapture.

The horses spoken of in verses 2 to 8 should be taken figuratively since it hardly makes sense that the Antichrist factually rides into power on a white horse, nor would invasion be led by a red horse. It would also seem ridiculous that a black horse would come bearing rationing and starvation. Hell and death need no vehicle; hence, the pale horse establishes the figurative nature and intent of these verses.

The return of Jesus on a white horse (Revelation 19: 11) is probably figurative as well; however, I like to believe that Jesus and his company will gloriously return to earth on beautiful, white Arabian stallions. Such is the mind and imagination of man; thus, God often speaks in terms that man can visualize. The scriptures seem to indicate that Revelation 19: 11 should be taken figuratively since Acts 1: 11 tells us Jesus will return in the same manner as his followers had seen him go. Nevertheless, the word-picture painted in Revelation 19: 11-14 is very authentic in that we see the mighty warrior, Jesus, return victoriously to set up his kingdom.

Perhaps a more factual account of Christ's return may be viewed in Chapter 14. This chapter gives an overview of the latter part of the Tribulation years, centering in on the return of Jesus.

In the first verse of Revelation, Chapter 14, Jesus is seen as he stands on Mount Sion with 144,000 at his side, fulfilling the prophesy of Zechariah 14: 4. Later, in verse 14, John beholds him as he sits upon a cloud with a golden crown upon his head and a sickle in his hand, thus fulfilling Acts 1: 11. How glorious that God sees fit, in Revelation 19: 11-21, to depict our Saviour as the returning victorious warrior, mounted upon a white steed and crowned with many crowns. To complete the symbolic imagery and similitude to Chapter 14: 14, he bears not a sickle in his hand but a sword in his mouth.

Revelation 6: 9 - 10

9 *And when he had opened the fifth seal, I saw under the altar the souls of them that were slain for the word of God, and for the testimony which they held:*

10 *And they cried with a loud voice, saying, How long, O Lord, holy and true, dost thou not judge and avenge our blood on them that dwell on the earth?*

When Jesus opens the fifth seal, John sees those who have died because of

The Souls Under the Altar / Jesus, the Way of Escape / The Tribulation Saints Robed and Resting

Notes

their refusal to follow the beast. These Saints cry out to God asking Him how long it will be before He avenges their death. Note that the Tribulation Saints aren't sleeping, as do the Church Saints (I Thessalonians 4: 15-16); nor are they resting, as do the Old Testament Saints (Daniel 12: 13). They seem well-aware of the torment they have been through and they cry out for God to avenge them. Only when clothed in righteousness do they '*rest for a little season.*'

It is imperative that the world heed the words of Jesus as recorded in John 3: 16; even now, the shadow of death and judgment is creeping across the face of the earth:

> *16 For God so loved the world, that he gave his only begotten Son, that whosoever believeth in him should not perish, but have everlasting life.*

The way of escape has been provided by God; man has been given the freedom to accept or reject God's gracious offer.

If you have never committed your life to Christ or have fallen away from the faith, read Matthew 24: 15-22 and take warning. The '*abomination of desolation*' which will '*stand in the holy place*' is the world ruler, the Antichrist, who will set himself up as God during the Tribulation. Matthew 24: 21-22 says:

> *21 For then shall be great tribulation, such as was not since the beginning of the world to this time, no, nor ever shall be.*
> *22 And except those days should be shortened, there should no flesh be saved: but for the elect's sake those days shall be shortened.*

Revelation 6: 11

> *11 And white robes were given unto every one of them; and it was said unto them, that they should rest yet for a little season, until their fellow servants also and their brethren, that should be killed as they were, should be fulfilled.*

God comforts the deceased Tribulation Saints, Israelite and Gentile alike, and gives all of them white robes. He breathes His peace upon them and tells them to rest a little season. They are robed in white even prior to standing before Christ in judgement, for their persistent testimony has judged them. They reap the rewards of the faithful as they enter into the promises of Christ. Jesus has said, *Be thou faithful unto death, and I will give thee a Crown of Life* (Revelation 2: 10).

These Saints must wait until the full complement of their Gentile fellow servants, as well as their Israelite brethren, pass through the fire of

Tribulation and join them in death. Together they will rest under the altar; that is, under the protection of the blood of Jesus.

Notes

As difficult as it may be to believe, one must accept that all who die '*sleep*' until their resurrection in Christ. The Old Testament Israelites sleep, or rest, in the earth (Job 7: 21; Job 14: 10-14; Daniel 12: 2; Daniel 12:13; Matthew 27: 52-53; I Peter 3: 18-19; I Peter 4: 5-6). The New Testament Saints sleep until the Rapture (I Thessalonians 4: 15-16; John 14: 2-3) and the Tribulation Saints sleep (or rest) under the altar until the Second Coming of Jesus Christ (Revelation 6: 9-11; Revelation 20: 4-5). The separation from the body, to the appearing before the Lord, will be but a single, extended moment in time to those who die in Christ.

Paul states in II Corinthians 5: 10 that the resurrected '*must all appear before the judgment seat of Christ; that every one may receive the things done in his body, according to that which he has done, whether it be good or bad.*' Paul previously taught in I Corinthians 3: 13-15 that '*every man's work shall be made manifest* (shown plainly, revealed, displayed) *for the day shall declare it, because it shall be revealed by fire.*'

Through these scriptures we are aware that we are openly judged before our entrance into heaven. Hebrews 9: 27 says '*...it is appointed unto men once to die, but after this the judgment.*' Romans 14: 12 states that '*every one of us shall give account of himself to God.*' We have no scriptures which would allow us to believe that upon death we each, individually, receive a private judgment and immediate entrance into heaven. Some quote II Corinthians 5: 8 as saying "to be absent from the body is to be present with the Lord." In actuality the scripture passage reads: '*We are confident, I say, and willing rather to be absent from the body, and to be present with the Lord.*' This is a declaration of willingness and desire, not a statement of fact. Paul says in Philippians 1: 21 '*for to me, to live in Christ and to die is gain*'; he later says in Philippians 3: 11 that he willingly shares the suffering of Christ that he might attain (succeed in coming to or experience) the resurrection of the dead.

Paul teaches both the Thessalonian and the Corinthian Christians that the deceased in Christ sleep until the trumpet sounds. For affirmation, read I Corinthians 15: 51-52:

51 Behold, I shew you a mystery; we shall not all sleep, but we shall be changed,

52 In a moment, in the twinkling of an eye, at the last trump: for the trump shall sound, and the dead shall be raised incorruptible...

If those who die in Christ go immediately to heaven, who then would be raised when '*the Lord himself descends from heaven with a shout; with the*

voice of the archangel, and with the trump of God'? If the righteous dead have already risen, how then could Paul say '*the dead in Christ will rise first: then we which are alive and remain shall be caught up together with them in the clouds to meet the Lord in the air, so shall we ever be with the Lord*' (I Thessalonians 4: 16-17)?

Notes

Some believers think that the sanctified, glorified souls of the dead leave heaven to join their dead bodies as the bodies rise from the grave. They maintain that they are given new bodies in preparation for the time when they serve Christ on earth. What a useless and untimely exercise this would be since Paul teaches we are '*sown a natural body, but are raised a spiritual body*' (I Corinthians 15: 44). He goes on to say that we have '*borne the image of the earthly* (while earth-dwellers) *but in death we will bear the image of the heavenly*' (I Corinthians 15: 49). God will have to change the order of things if, at the Rapture, the Saints are clothed in a natural, glorified body to take back with them into heaven. Why would God choose to have the Saints return to heaven in their natural bodies if they have, since death, dwelt in heaven in a spiritual body? God is certainly capable of providing a natural body when there is a need for that body. The need will arise when the Saints return to earth to serve their course and stand in their lot in the Millennial years (Daniel 12: 13). See Study notes for Revelation 19: 14.

Paul had previously told the Corinthians that there is an order to the Resurrection. See I Corinthians 15: 21-23 NLT:

21 *So you see, just as death came into the world through a man, Adam, now the resurrection from the dead came through another man, Christ.*

22 *Everyone dies because all of us are related to Adam, the first man. But all who are related to Christ, the other man, will be given new life.*

23 *But there is an order to this resurrection; Christ was raised first, then when Christ comes back, all his people will be raised.*

According to John 5: 25, 28-29, Jesus – the greatest authority of all – spoke to the living and said:

25 *Verily, verily, I say unto you: the hour is coming, and now is, when the dead shall hear the voice of the son of God: they that hear shall live.*

28 *Marvel not at this; for the hour is coming, in the which all that are in the graves shall hear his voice.*

29 *And shall come forth; they that have done good unto the resurrection of life* (at the Rapture)*; they that have done evil unto the resurrection of damnation* (at the White Throne Judgment).

Notes

Then again, in John 6: 39-40, 44, 54 (NLT), when responding to the multitude, Jesus repeatedly speaks of the Saints being raised up in the Last Day. From the content of the verses, we know he is referring to the Rapture.

39 *This is the will of the Father who sent me that all He has given me I should lose nothing, but should raise it up at the last day.*

40 *This is the will of Him who sent me that everyone who sees the Son and believes in Him may have everlasting life; and I will raise him up in the last day.*

44 *No man can come to me unless the Father who sent me draws him; and I will raise him up at the last day.*

54 *Whosoever eats my flesh and drinks my blood has eternal life, and I will raise him up at the last day.*

Though quoted before, please recall John 14: 2-3, when Jesus speaks of his preparing a place for us in his Father's house. He promises that he would come again and receive us that where he is, we may be also. This surely confirms that we are not risen until the Rapture when he calls us to himself.

Revelation 6: 12 - 17

12 *And I beheld when he had opened the sixth seal, and, lo, there was a great earthquake: and the sun became black as sackcloth of hair, and the moon became as blood;*

13 *And the stars of heaven fell unto the earth, even as a fig tree casteth her untimely figs, when she is shaken of a mighty wind.*

14 *And the heaven departed as a scroll when it is rolled together; and every mountain and island were moved out of their places.*

15 *And the kings of the earth, and the great men, and the rich men, and the chief captains, and the mighty men, and every bondman, and every free man, hid themselves in the dens and in the rocks of the mountains;*

16 *And said to the mountains and rocks, Fall on us, and hide us from the face of him that sitteth on the throne, and from the wrath of the Lamb:*

17 *For the great day of his wrath is come; and who shall be able to stand?*

Jesus opens the sixth seal and John sees the very thing Jesus talked about in Matthew 24: 29:

29 *Immediately after the tribulation of those days shall the sun be darkened, and the moon shall not give her light, and the stars shall fall from heaven, and the powers of the heavens shall be shaken:*

The opening of the sixth seal is a revelation (a heavenly warning to man) rather than an outpouring of wrath at the hand of Christ, as is often thought. When the seventh seal is opened, heaven is silent (Revelation 8: 1) as they view the magnitude of the forthcoming wraths of God. The seals disclose

the ravage that will come upon man as the nations are drawn, by greed, to the plains of Meguiddo and the mountains of Israel. Isaiah spoke of these events some 734 years before Jesus was born. Read Isaiah 13: 9-13:

> 9 *Behold, <u>the day of the Lord</u> cometh, cruel both with wrath and fierce anger, to lay the land desolate: and he shall destroy the sinners thereof out of it.*
> 10 *For the stars of heaven and the constellations thereof shall not give their light; the sun shall be darkened in his going forth, and the moon shall not cause her light to shine.*
> 11 *And I will punish the world for their evil, and the wicked for their iniquity; and I will cause the arrogancy of the proud to cease, and will lay low the haughtiness of the terrible.*
> 13 *Therefore I will shake the heavens, and the earth shall remove out of her place, in the wrath of the Lord of hosts, and in the day of his fierce anger.*

Read, also, Joel 2: 30-31:

> 30 *And I will shew wonders in the heavens and in the earth, blood, and fire, and pillars of smoke.*
> 31 *The sun shall be turned into darkness, and the moon into blood, before <u>the great and the terrible day of the Lord</u> come.*

We, the believers, can forever rejoice because we have confidence that we won't be required to stand in those days. Remember the promises of God; call to mind II Peter 2: 9:

> 9 *The Lord knoweth how to deliver the godly out of temptations, and to reserve the unjust unto the day of judgment to be punished:*

Notes

CHAPTER SEVEN

In the first-half of **Chapter Seven**, we view one hundred and forty-four thousand Israelite Tribulation Saints as they are sealed with the seal of God. In the second-half, we see heaven rejoicing as a multitude of martyred Israelite and Gentile Tribulation Saints are clothed in the righteousness of Christ. The Study affirms that God, even to the very end, loves the lost. A portion of the Remnant have washed their robes and made them white in the blood of the Lamb.

Revelation 7

[**1**] And after these things I saw four angels standing on the four corners of the earth, holding the four winds of the earth, that the wind should not blow on the earth, nor on the sea, nor on any tree.
[**2**] And I saw another angel ascending from the east, having the seal of the living God: and he cried with a loud voice to the four angels, to whom it was given to hurt the earth and the sea,
[**3**] Saying, Hurt not the earth, neither the sea, nor the trees, till we have sealed the servants of our God in their foreheads.
[**4**] And I heard the number of them which were sealed: and there were sealed an hundred and forty and four thousand of all the tribes of the children of Israel.
[**5**] Of the tribe of Juda were sealed twelve thousand. Of the tribe of Reuben were sealed twelve thousand. Of the tribe of Gad were sealed twelve thousand.
[**6**] Of the tribe of Aser were sealed twelve thousand. Of the tribe of Nepthalim were sealed twelve thousand. Of the tribe of Manasses were sealed twelve thousand.
[**7**] Of the tribe of Simeon were sealed twelve thousand. Of the tribe of Levi were sealed twelve thousand. Of the tribe of Issachar were sealed twelve thousand.
[**8**] Of the tribe of Zabulon were sealed twelve thousand. Of the tribe of Joseph were sealed twelve thousand. Of the tribe of Benjamin were sealed twelve thousand.
[**9**] After this I beheld, and, lo, a great multitude, which no man could number, of all nations, and kindreds, and people, and tongues, stood before the throne, and before the Lamb, clothed with white robes, and palms in their hands;
[**10**] And cried with a loud voice, saying, Salvation to our God which sitteth upon the throne, and unto the Lamb.
[**11**] And all the angels stood round about the throne, and about the elders and the four beasts, and fell before the throne on their faces, and worshipped God,
[**12**] Saying, Amen: Blessing, and glory, and wisdom, and thanksgiving, and honour, and power, and might, be unto our God for ever and ever. Amen.
[**13**] And one of the elders answered, saying unto me, What are these which are arrayed in white robes? and whence came they?
[**14**] And I said unto him, Sir, thou knowest. And he said to me, These are they which came out of great tribulation, and have washed their robes, and made them white in the blood of the Lamb.
[**15**] Therefore are they before the throne of God, and serve him day and night in his temple: and he that sitteth on the throne shall dwell among them.
[**16**] They shall hunger no more, neither thirst any more; neither shall the sun light on them, nor any heat.
[**17**] For the Lamb which is in the midst of the throne shall feed them, and shall lead them unto living fountains of waters: and God shall wipe away all tears from their eyes.

Preamble / The Angel Holding the Winds / The Seal of God

Revelation, Chapter 7

Notes

Preamble: a teaching in preparation for Chapter 7

The first half of Chapter 7 (verses 1 through 8) records the sealing of the 144,000 Tribulation Saints from the twelve tribes of Israel; the second half (verses 9 through 17), tells us about the balance of the Israelite and Gentile Saints. This great multitude are of all nations, kindreds, peoples and tongues. The compassionate and loving God gives both groups similar rewards. They sing their song of praise before the throne of God and serve Jesus continually in the temple (Revelation 7: 15). The 144,000 follow Jesus wherever he goes; they never leave him (Revelation 14: 4). They quite probably continue to serve him on into the New Jerusalem.

We may rejoice as we study this chapter for in it we see Romans 11 fulfilled. God has kept His covenant with Israel. The remnant, according to the election of grace, is called, clothed and sealed. Through the unbelief of the Israelite, the Gentile has also obtained mercy, since God '*hath concluded them all in unbelief that He might have mercy upon all'* (Romans 11: 25-33).

Chapter seven affirms that the seals of Chapter six are not wraths ministered through Jesus. They are, instead, an unveiling of the wraths to come, a consenting to those which follow. Indeed, if they were judgments poured out by Jesus, the servants of God would have been sealed before the wraths began. It is commanded by God that the earth should not be hurt until God's elect servants are sealed (Revelation 7: 3).

Revelation 7: 1

1 And after these things I saw four angels standing on the four corners of the earth, holding the four winds of the earth, that the wind should not blow on the earth, nor on the sea, nor on any tree.

The four angels standing on the circumference of the earth are poised, ready to release the winds bearing the wrath of God.

Revelation 7: 2

2 And I saw another angel ascending from the east, having the seal of the living God: and he cried with a loud voice to the four angels, to whom it was given to hurt the earth and the sea,

John sees another angel, apparently superior, ascending from the east, bearing the seal of the living God. The angel calls a command of restraint to the four angels who have been given power to hurt the earth. I, personally, believe that the angel is Jesus, for only Jesus can place God's seal upon man; only Jesus can command the angels of God.

The 144,000 Israelites Sealed

Notes

Revelation 7: 3

3 *Saying, Hurt not the earth, neither the sea, nor the trees, till we have sealed the servants of our God in their foreheads.*

The command is that the angels must restrain the four winds bearing the wrath of God until the specially appointed servants of God are sealed in their foreheads (marked as God's protected ones). Revelation, Chapter 6: 6 has previously told us that a voice from the midst of the four creatures has commanded that none may hurt the oil and the wine. Although oil and wine can allude to healing and salvation (see Luke 10: 34), it generally speaks of God's elect.

In sealing these chosen Israelites, God protects them from harm throughout the Tribulation. We again see these elect of God in Revelation 14: 1 as they stand with Jesus on Mount Sion.

Revelation 7: 4 - 8

4 *And I heard the number of them which were sealed: and <u>there were sealed an hundred and forty and four thousand</u> of all the tribes of the children of Israel.*

5 *Of the tribe of Juda were sealed twelve thousand. Of the tribe of Reuben were sealed twelve thousand. Of the tribe of Gad were sealed twelve thousand.*

6 *Of the tribe of Aser were sealed twelve thousand. Of the tribe of Nepthalim were sealed twelve thousand. Of the tribe of Manasses were sealed twelve thousand.*

7 *Of the tribe of Simeon were sealed twelve thousand. Of the tribe of Levi were sealed twelve thousand. Of the tribe of Issachar were sealed twelve thousand.*

8 *Of the tribe of Zabulon were sealed twelve thousand. Of the tribe of Joseph were sealed twelve thousand. Of the tribe of Benjamin were sealed twelve thousand.*

John hears the number which are sealed and the number totals 144,000; all of these are from the tribes of the children of Israel. He proceeds to name the tribes: Juda, Reuben, Gad, Aser, Nepthalim, Manesseh, Simeon, Levi, Issachar, Zabulon, Joseph and Benjamin.

These Israelites are apparently virtuous men who have been sealed that they may devote their lives in service to Jesus. They are the first fruit of God and the Lamb. We assume that they, during the Tribulation, win souls to Christ. We can't be sure of this since we are told only that they are virgin and follow Jesus wherever he goes (Revelation 14: 4). We do know, however, that God in His love for man sends an angel flying in the midst of heaven, having the everlasting gospel to preach to all who dwell upon the earth. The

Notes

angel cries with a loud voice that man should fear God and give Him glory (Revelation 14: 6-7). We know also that God will appoint two witnesses who will prophesy (teach and preach) for the first three and a half years of the Tribulation. They are the '*two olive trees*', the '*two candlesticks*' standing before God (Revelation 11: 3-12).

Some readers may wonder why the tribe of Dan is not listed among the 144,000. According to Genesis 49: 17, Dan is '*a serpent by the way, an adder in the path*'; thus, Manesseh (the first-born twin son of Joseph, adopted and blessed by Jacob) is sealed instead of Dan (Genesis 48: 1-20).

Revelation 7: 9

9 *After this I beheld, and, lo, a great multitude, which no man could number, of all nations, and kindreds, and people, and tongues, stood before the throne, and before the Lamb, clothed with white robes, and palms in their hands;*

After the 144,000 are sealed, John sees a great number of others, so great that no one could number them. These are from all nations and kindreds and people and tongues – Tribulation converts who stand before the throne of God and the Lamb.

The fact that they stand before the throne does not necessarily mean that they all die. It could very well mean only that they have been converted. Upon conversion, all stand on the sea of glass before the throne of God that the whole of heaven might see them and rejoice (Luke 15:7; Revelation 14: 3). Revelation 6: 9-11 indicates that the converted of the Tribulation years are not ushered into heaven at death, but wait until all who are marked for martyrdom join them. Together they will be resurrected at Christ's return.

We know and are glad that great multitudes will be converted during the Tribulation. People from all nations will humbly stand before the Lamb, repentant and broken. One-third will live and go on into the Millennium (Zechariah 13: 8-9) but the others will temporarily die (Revelation 6: 9-11). They will be called from their graves when the whole House of Israel is called to attend the wedding feast of the Lamb's Wife (Revelation 19: 7-9). The white robes (note: not fine linen) tell us these multitudes are both Israelite and Gentile. The verse affirms this conclusion for it says they are '*of all nations, and kindreds, and people, and tongues*'. The palm boughs in their hands indicate humility and connects them with our Saviour, Jesus Christ.

Heaven Rejoices With the Repentant

Notes

Revelation 7: 10

10 And cried with a loud voice, saying, Salvation <u>to our God</u> which sitteth upon the throne, <u>and unto the Lamb</u>.

The multitude come before the throne with praise <u>to God and the Lamb</u>. This is further confirmation that in Revelation 4, verses 2 and 3, the jasper and the sardine stones are representative of the Father and the Son, the two in One who sit upon the throne.

These Saints have experienced the truth of Abraham's words: '*God will provide <u>Himself</u> a lamb for a burnt offering*' (Genesis 22: 8). Paul said, '*...Your life is hid with Christ in God.*' (Colossians 3: 3)

Revelation 7: 11

11 And all the angels stood round about the throne, and about the elders and the four beasts, and fell before the throne on their faces, and worshiped God.

All of the inhabitants of heaven fall, face down, and worship God: the angels who <u>stood</u> around the throne; the elders <u>seated</u> about the throne; and the four beasts <u>before</u> the throne.

Revelation 7: 12

12 Saying, Amen: Blessing, and glory, and wisdom, and thanksgiving, and honour, and power, and might, be unto our God for ever and ever. Amen.

The whole of heaven rejoices over the repentant multitude who stand before the throne and before the Lamb. They hear their voices praising God and they themselves fall down and worship, saying: "*Blessing, and glory, and wisdom, and thanksgiving, and honour, and power, and might, be unto our God for ever and ever Amen.*"

The reference here to the joy in heaven over the repentant affirms the words of Jesus as recorded in Luke 15: 10:

10 Likewise, I say unto you, there is joy in the presence of the angels of God over one sinner that repenteth.

Our joy is increased when we understand through the Revelation that entering the Tribulation, those who once believed but were drawn away by the world, as well as those who never believed, will not necessarily be lost for eternity. The fact is that eyes will be opened, faith will be renewed, souls

Notes

will be saved up to the end of the Millennium. The last call is given just prior to man's entrance into the new creation (Revelation 22: 17).

We can be confident of this very thing: *'He who has begun a good work will perform it until the day of Jesus Christ.'* This is the promise of Philippians 1: 6 and God always keeps His word! Because of this, we are assured that the Holy Spirit remains on earth even after the rapture of the Church. One can't be drawn to God, nor can one be born again unless the Spirit woo him and the repentant receive the indwelling Spirit (John 3: 3-18).

Revelation 7: 13 - 14

13 And one of the elders answered, saying unto me, What are these which are arrayed in white robes? And whence came they?

14 And I said unto him, Sir, thou knowest. And he said to me, These are they which came out of great tribulation and have washed their robes, and made them white in the blood of the Lamb.

Again, one of the elders, having personal knowledge of God's saving grace, tells John that all these Saints who are worshiping and praising God have come to Jesus throughout the Tribulation years. They have washed their robes and made them white through the blood of the Lamb. The act is theirs; the provision is Christ's. Praise God! He never gives up. He loves the lost.

Revelation 7: 15

15 Therefore are they before the throne of God, and serve him day and night in his temple: and he that sitteth on the throne shall dwell among them.

This verse tells us that these Saints who stand before the throne of God will serve Him, through Jesus, in the temple. Their service will begin when Jesus sits on the throne in Jerusalem and dwells among them throughout the Millennium. Revelation 20: 4 says:

> *4 And I saw thrones, and they sat upon them, and judgment was given unto them: <u>and I saw</u> the souls of them that were beheaded for the witness of Jesus, and for the word of God, and which had not worshipped the beast, neither his image, neither had received his mark upon their foreheads, or in their hands; and they lived and reigned with Christ a thousand years.*

Some serve and some reign, but only the ones who sit upon the thrones, the Raptured Saints, <u>assist with world judgment</u> during the Millennium as they reign with Christ. All have a ministry and all are rewarded.

Notes

Revelation 7: 16

16 They shall hunger no more, neither thirst any more; neither shall the sun light on them, nor any heat.

We see through these verses how merciful God is. He forgives so completely; nevertheless, these Saints suffer loss for they never actually sit with Jesus about the throne of God. They miss out entirely on this glorious time of praise and adoration. However, these 'late-comers' are loved and cared for by Jesus throughout the Millennium. They will never again hunger or thirst, nor will they be tormented by the rays of the sun or the heat of the wrath of God.

Revelation 7: 17

17 For the Lamb which is in the midst of the throne shall feed them, and shall lead them unto living fountains of waters: and God shall wipe away all tears from their eyes.

Jesus, who is in the midst of the throne (see Revelation 5: 6), shall feed them throughout the Millennium. He will lead them to the living fountains of water in the New Creation where God will wipe away their tears forever. See Revelation 22: 1- 4.

CHAPTER EIGHT

Chapter Eight verifies the seals are not a set of wraths since heaven only is affected by the awesome revelation of the seventh seal. When the seal is opened, the whole of heaven is silenced for half-an-hour. The chapter records the first four trumpet wraths, as their harsh punishment is trumpeted upon the heavens and the earth. The chapter ends with the threatening announcement of the last three woes. The Study notes provide a teaching on the angel with the censer of fire; the High Priest and the altar of incense; the star called wormwood; and the culmination of the wraths at the sounding of the third woe.

Revelation 8

[**1**] And when he had opened the seventh seal, there was silence in heaven about the space of half an hour.

[**2**] And I saw the seven angels which stood before God; and to them were given seven trumpets.

[**3**] And another angel came and stood at the altar, having a golden censer; and there was given unto him much incense, that he should offer it with the prayers of all saints upon the golden altar which was before the throne.

[**4**] And the smoke of the incense, which came with the prayers of the saints, ascended up before God out of the angel's hand.

[**5**] And the angel took the censer, and filled it with fire of the altar, and cast it into the earth: and there were voices, and thunderings, and lightnings, and an earthquake.

[**6**] And the seven angels which had the seven trumpets prepared themselves to sound.

[**7**] The first angel sounded, and there followed hail and fire mingled with blood, and they were cast upon the earth: and the third part of trees was burnt up, and all green grass was burnt up.

[**8**] And the second angel sounded, and as it were a great mountain burning with fire was cast into the sea: and the third part of the sea became blood;

[**9**] And the third part of the creatures which were in the sea, and had life, died; and the third part of the ships were destroyed.

[**10**] And the third angel sounded, and there fell a great star from heaven, burning as it were a lamp, and it fell upon the third part of the rivers, and upon the fountains of waters;

[**11**] And the name of the star is called Wormwood: and the third part of the waters became wormwood; and many men died of the waters, because they were made bitter.

[**12**] And the fourth angel sounded, and the third part of the sun was smitten, and the third part of the moon, and the third part of the stars; so as the third part of them was darkened, and the day shone not for a third part of it, and the night likewise.

[**13**] And I beheld, and heard an angel flying through the midst of heaven, saying with a loud voice, Woe, woe, woe, to the inhabiters of the earth by reason of the other voices of the trumpet of the three angels, which are yet to sound!

Notes

Revelation, Chapter 8

Revelation 8: 1

1 *And when he had opened the seventh seal, there was silence in heaven about the space of half an hour.*

When Jesus opens the seventh seal, there is an awesome silence in heaven. By opening the seventh seal, Jesus is consenting to, and ushering in, the trumpet judgments. They appear so dreadful to the company in heaven that they cease praising; it would seem as if they respond with a silent gasp of horror.

Revelation 8: 2

2 *And I saw the seven angels which stood before God; and to them were given seven trumpets.*

Seven angels stand before God, each having been given a trumpet which will be sounded to announce each of the seven woes which will someday come upon the earth through God's mighty hand.

Revelation 8: 3 - 4

3 *And another angel came and stood at the altar, having a golden censer; and there was given unto him much incense, that he should offer it with the prayers of all saints upon the golden altar which was before the throne.*

4 *And the smoke of the incense, which came with the prayers of the saints, ascended up before God out of the angel's hand.*

The altar before which the angel stands refers to the Altar of Incense (Exodus 30: 1-10). In the temple, within the Holy Place, this altar was placed in front of the veil. The veil itself divided the Holy Place from the Holy of Holies where the Arc of the Covenant rested. Yearly, on the day of atonement, the High Priest took coals from the Altar of Incense and carried them behind the veil into the Holy of Holies (known also as the Most Holy Place). The smoke from the incense covered the Mercy Seat which lay upon the Arc of the Covenant, protecting the Priest from the Glory of God (Leviticus 16: 12-13).

In the verses under study, the angel officiating in the office of the High Priest is Jesus. In his hand is a golden censer filled with incense. The smoke from the burning incense rises before the face of God and carries with it the prayers of the people. Jesus, our High Priest, our Mediator, perpetually intercedes on behalf of the believer (Hebrews 7: 25). '*For there*

Notes

is one God, and one mediator between God and men, the man Christ Jesus' (I Timothy 2: 5). Refer to the Study notes of Revelation 9: 13 for further information on this subject.

Revelation 8: 5

5 *And the angel took the censer, and filled it with fire of the altar, and cast it unto the earth: and there were voices, and thunderings, and lightnings, and an earthquake.*

Up until now, even with the opening of the seven seals, Jesus has never actively participated in the destruction of man. He is always seen as the gentle Saviour, the Prince of Peace, the good Shepherd. Now he becomes the Lion of Judah, the King of Glory, God-in-flesh, the Judge of mankind. When he takes the censer and fills it with the fire of the altar, and casts it into the earth, the '*power of heaven shall be shaken*' as he predicted in Matthew 24: 29:

29 *Immediately after the tribulation of those days shall the sun be darkened, and the moon shall not give her light, and the stars shall fall from heaven, and the powers of the heavens shall be shaken:*

We know the angel with the golden censer is Jesus for he is the executor of God's Will. He eradicates the evil of the earth through the trumpet and vial wraths. With appointed angels and, later, by his own participation, Jesus purges the earth with the fire of God.

Revelation 8: 6 - 7

6 *And the seven angels which had the seven trumpets prepared themselves to sound.*

7 *The first angel sounded, and there followed hail and fire mingled with blood, and they were cast upon the earth: and the third part of trees was burnt up, and all green grass was burnt up.*

The first angel sounds and we see an Exodus judgment trumpeted upon man. Read Exodus 9: 23 and 25:

23 *And Moses stretched forth his rod toward heaven: and the Lord sent thunder and hail, and the fire ran along upon the ground; and the Lord rained hail upon the land of Egypt.*

25 *And the hail smote throughout all the land of Egypt all that was in the field, both man and beast; and the hail smote every herb of the field, and brake every tree of the field.*

There is no question: God is the same yesterday, today and forever.

The Sounding of the Second and Third Trumpet

Notes

Revelation 8: 8 - 9

8 *And the second angel sounded, and as it were a great mountain burning with fire was cast into the sea: and the third part of the sea became blood;*

9 *And the third part of the creatures which were in the sea, and had life, died; and the third part of the ships were destroyed.*

The second angel's trumpet sounds and a mountain of fire is cast into the sea. This turns one-third of the sea into blood. It kills one-third part of all life in the sea and destroys a third part of the ships on the sea.

Again, there is a comparison to be made to Exodus; see Chapter 7: 20-21 (paraphrased):

20 *And Moses and Aaron lifted the rod as the Lord had commanded, and smote the waters, and all the waters in the river were turned to blood.*

21 *And all of the fish died, and the rivers stank and the Egyptians could not drink the water.*

There will be a great shortage of water and food in the Tribulation years. Even now, the earth's water supplies are fast becoming polluted and fish and wildlife are dying. Think of the universal sorrow and fear on earth when the second angel sounds and God's punishment comes down upon man!

Revelation 8: 10 - 11

10 *And the third angel sounded, and there fell a great star from heaven, burning as it were a lamp, and it fell upon the third part of the rivers, and upon the fountains of waters;*

11 *And the name of the star is called Wormwood: and the third part of the waters became wormwood; and many men died of the waters, because they were made bitter.*

The third angel sounds and a meteoric mass comes blazing from the sky. A third part of the rivers, as well as the fountains of water, become bitter and poisonous.

The name of the star is called wormwood which, by definition, means 'poison'. It's interesting to note that there are four kinds of wormwood found in Palestine. It's a perennial herb and the liquid from it causes mental deterioration and death. Jeremiah 9: 15 tells us:

15 *...thus saith the Lord of hosts, the God of Israel; Behold, I will feed them even this people, with wormwood, and give them water of gall to drink.*

The Fourth Trumpet Sounded – Darkness Upon the Earth

Notes

Joel 1: 15-20 (NLT) paints a vivid picture of the dreadful conditions in the world during the years of God's retribution for sin:

15 *The day of the Lord is on the way, the day when destruction comes from the Almighty. How terrible that day will be.*

16 *We watch as our food disappears before our very eyes. There are no joyful celebrations in the house of our God.* (Ezekiel's envisioned temple – Ezekiel, Chapters 40 - 42)

17 *The seeds die in the parched ground, and the grain crops fail. The barns and granaries stand empty and abandoned.*

18 *How the animals moan with hunger! The cattle wander about confused because there is no pasture for them. The sheep bleat in misery.*

19 *Lord, help us! The fire has consumed the pastures and burned up the trees.*

20 *Even the wild animals cry out to you because they have no water to drink. The streams have dried up, and the fire has consumed the pastures.*

Revelation 8: 12

12 *And the fourth angel sounded, and the third part of the sun was smitten, and the third part of the moon, and the third part of the stars; so as the third part of them was darkened, and the day shone not for a third part of it, and the night likewise.*

The fourth angel sounds and a third part of the sun, moon and stars are struck; they fail to give their light for a third part of the day and night.

Amos spoke of these days in Amos 8: 9:

9 *And it shall come to pass in that day, saith the Lord God, that I will cause the sun to go down at noon, and I will darken the earth in the clear day.*

Read also the similitude of darkness at the exodus of the Israelites from Egypt in Exodus 10: 21 and 23;

21 *And the Lord said unto Moses, Stretch out thine hand toward heaven; and there was a thick darkness in all the land of Egypt...*

23 *They saw not one another, neither rose any from his place for three days: but all the children of Israel had light in their dwellings.*

Jesus said in Matthew 24: 29 that darkness will envelope the earth for a season in the latter days of the Tribulation period; no doubt, this will contribute further to starvation and trouble among the nations. For example, one can see the possibility of Canada and the United States in bitter dispute over water rights. What with little drinkable water and shortened hours of light for growing crops, all the world – even the peaceful nations – will rise

The Three Woes / Three Trumpets Yet to Sound

up against one another.

Notes

Revelation 8: 13

13 And I beheld, and heard an angel flying through the midst of heaven, saying with a loud voice, Woe, woe, woe, to the inhibitors of the earth by reason of the other voices of the trumpet of the three angels, which are yet to sound.

Heaven cries out in woe for the inhabitants of the earth for there are yet three more trumpets to be sounded, each more bitter than the last.

- The first woe occurs when the fifth angel sounds his trumpet in Revelation 9: 1-11 and the bottomless pit is opened.
- One woe is past and, behold, there come two woes more (Revelation 9: 12).
- The second woe is administered through the sounding of the sixth trumpet, freeing evil spirits bound in the Euphrates (Revelation 9: 14-15; Revelation 16:12-13). Their leader, Satan the Beast, which is released from the bottomless pit, eventually brings about the death of the two witnesses (Revelation 11: 7).
- The second woe is past and, behold, the third woe comes quickly (Revelation 11: 14).
- The third woe is when the voice of the seventh trumpet <u>begins to sound</u>; the mystery of God, the time of love and grace which provides rebirth and salvation, is finished (Revelation 10: 7). This woe is the great day of the wrath of God (Revelation 6: 17). It is the day when the seventh trumpet <u>ends its sounding</u> (Revelation 11: 15); the day of rewarding the just and the unjust (Revelation 11: 18). It is the day that the seventh angel pours out his vial in the air and the great voice out of the temple of heaven, from the throne of God, declares: "It is done!" (Revelation 16: 17-21).
- For further clarification, consult the chart found in Appendix 5. Additional discussion on the subject of the third woe will be offered in the commentary for Revelation, Chapter 11: 15-19.

CHAPTER NINE

As **Chapter Nine** opens, the Study claims that verses one to thirteen speak of satanic forces. With God's consent, at the sounding of the fifth trumpet, Satan's angels are released from the bottomless pit. They join their Satanic king to reign affliction upon their brethren, the 'unsealed' of the earth. At their release, upon the sounding of the fifth trumpet, the first woe is past. In verses fourteen to twenty-one, the sixth trumpet presents human satanic forces (as confirmed by chapter sixteen). The Study shows that the nations – represented by the two hundred million armed horsemen, the False prophet and the Antichrist – plague the earth with cruel dictatorship, modern warfare and death.

Revelation 9

[**1**] And the fifth angel sounded, and I saw a star fall from heaven unto the earth: and to him was given the key of the bottomless pit.
[**2**] And he opened the bottomless pit; and there arose a smoke out of the pit, as the smoke of a great furnace; and the sun and the air were darkened by reason of the smoke of the pit.
[**3**] And there came out of the smoke locusts upon the earth: and unto them was given power, as the scorpions of the earth have power.
[**4**] And it was commanded them that they should not hurt the grass of the earth, neither any green thing, neither any tree; but only those men which have not the seal of God in their foreheads.
[**5**] And to them it was given that they should not kill them, but that they should be tormented five months: and their torment was as the torment of a scorpion, when he striketh a man.
[**6**] And in those days shall men seek death, and shall not find it; and shall desire to die, and death shall flee from them.
[**7**] And the shapes of the locusts were like unto horses prepared unto battle; and on their heads were as it were crowns like gold, and their faces were as the faces of men.
[**8**] And they had hair as the hair of women, and their teeth were as the teeth of lions.
[**9**] And they had breastplates, as it were breastplates of iron; and the sound of their wings was as the sound of chariots of many horses running to battle.
[**10**] And they had tails like unto scorpions, and there were stings in their tails: and their power was to hurt men five months.
[**11**] And they had a king over them, which is the angel of the bottomless pit, whose name in the Hebrew tongue is Abaddon, but in the Greek tongue hath his name Apollyon.
[**12**] One woe is past; and, behold, there come two woes more hereafter.
[**13**] And the sixth angel sounded, and I heard a voice from the four horns of the golden altar which is before God,
[**14**] Saying to the sixth angel which had the trumpet, Loose the four angels which are bound in the great river Euphrates.
[**15**] And the four angels were loosed, which were prepared for an hour, and a day, and a month, and a year, for to slay the third part of men.
[**16**] And the number of the army of the horsemen were two hundred thousand thousand: and I heard the number of them.
[**17**] And thus I saw the horses in the vision, and them that sat on them, having breastplates of fire, and of jacinth, and brimstone: and the heads of the horses were as the heads of lions; and out of their mouths issued fire and smoke and brimstone.
[**18**] By these three was the third part of men killed, by the fire, and by the smoke, and by the brimstone, which issued out of their mouths.
[**19**] For their power is in their mouth, and in their tails: for their tails were like unto serpents, and had heads, and with them they do hurt.
[**20**] And the rest of the men which were not killed by these plagues yet repented not of the works of their hands, that they should not worship devils, and idols of gold, and silver, and brass, and stone, and of wood: which neither can see, nor hear, nor walk:
[**21**] Neither repented they of their murders, nor of their sorceries, nor of their fornication, nor of their thefts.

The Fifth Trumpet Sounded / The Star Falls from Heaven / The Bottomless Pit

Notes

Revelation, Chapter 9

Revelation 9: 1

1 And the fifth angel sounded, and I saw a star fall from heaven unto the earth: and to him was given the key of the bottomless pit.

When the fifth angel sounds, John sees a star fall from heaven to earth. This star is temporarily given the key to the bottomless pit. The star is a representation of Satan.

In Luke 10: 18, when the seventy disciples tell Jesus that the devils are subject unto his name, Jesus affirms their observation:

> *18 And he said unto them, I beheld Satan as lightning fall from heaven.*

Isaiah 14: 12-15 gives a vivid picture of what Jesus saw and heard:

> *12 How art thou fallen from heaven, O Lucifer, son of the morning! how art thou cut down to the ground, which didst weaken the nations!*
>
> *13 For thou hast said in thine heart, I will ascend into heaven, I will exalt my throne above the stars of God: I will sit also upon the mount of the congregation, in the sides of the north:*
>
> *14 I will ascend above the heights of the clouds; I will be like the most High.*
>
> *15 Yet thou shalt be brought down to hell, to the sides of the pit.*

Revelation 9: 2

2 And he opened the bottomless pit; and there arose a smoke out of the pit, as the smoke of a great furnace; and the sun and the air were darkened by reason of the smoke of the pit.

The bottomless pit is a lesser hell (the abyss): the prison for the fallen angels (Jude 6). Jesus permits Satan to open the bottomless pit and, in so doing, he sets Satan against himself. Knowing that "*every kingdom divided against itself is brought to desolation; and every city or house divided against itself shall not stand*" (Matthew 12: 25), Jesus orchestrates the participation of Satan in his own destruction. Joel 2: 2 refers to this time as "*a day of darkness and of gloominess; a day of cloud and thick darkness.*"

Notes

The spirit of Satan in man is even now capable of releasing atomic blasts which, if dropped on the earth, could, as it were, open the bottomless pit and cause darkness over the face of the whole of the earth for a period of four months or more.

Revelation 9: 3

3 *And there came out of the smoke locusts upon the earth: and unto them was given power, as the scorpions of the earth have power.*

Out of the smoke come evil spirits which take on the form of locusts. They have the death sting of the scorpion; nonetheless, man is not able to escape this torment even through death, "*for death shall flee from them*" (see Revelation 6: 15-16; Revelation 9: 6).

Revelation 9: 4

4 *And it was commanded them that they should not hurt the grass of the earth, neither any green thing, neither any tree; but only those men which have not the seal of God in their foreheads.*

The locusts, or evil spirits in the form of locust, are told they are not to hurt the grass of the earth or any other green things as the locusts did in Exodus 10: 14-15:

14 *And the locusts went up over all the land of Egypt...*

15 *For they covered the face of the whole earth, so that the land was darkened; and they did eat every herb of the land, and all the fruit of the trees which the hail had left: and there remained not any green thing in the trees, or in the herbs of the field, through all the land of Egypt.*

The Tribulation locusts can torment only those without the seal of God in their foreheads. The 144,000 of God's servants, having been sealed, are spared (see Revelation 7: 3). There is some possibility that the great multitudes referred to in Revelation 7: 9 may also be sealed and, thereby, escape the torment of the locusts. We can't be sure of this since we are told only that they '*come out of the great tribulation and have washed their robes and made them white in the blood of the lamb.*' (Revelation 7: 13-14)

One may ask why Satan would be willing to torment his own. Satan has no

Notes

love or loyalty to his followers. Indeed, he delights in their destruction. If man opens his mind and heart to Satan, this wicked deceiver often demands that the captive take his own life; moreover, he requires that the lives of others be taken as well. He is not willing to die for man; however, man will die eternally, repeatedly and forever because of his allegiance to Satan.

Revelation 9: 5 - 6

5 *And to them it was given that they should not kill them, but that they should be tormented five months: and their torment was as the torment of a scorpion, when he striketh a man.*

6 *And in those days shall men seek death, and shall not find it; and shall desire to die, and death shall flee from them.*

Their pain is so extreme that men long for death but God does not allow them to escape. This torment goes on for five months (the life-span of a locust).

Revelation 9: 7 - 10

7 *And the shapes of the locusts were like unto horses prepared unto battle; and on their heads were as it were crowns like gold, and their faces were as the faces of men.*

8 *And they had hair as the hair of women, and their teeth were as the teeth of lions.*

9 *And they had breastplates, as it were breastplates of iron; and the sound of their wings was as the sound of chariots of many horses running to battle.*

10 *And they had tails like unto scorpions, and there were stings in their tails: and their power was to hurt men five months.*

These four verses give us a graphic picture of evil spirits disguised as grotesque locusts. These defiant fallen angels are the spirits who followed Lucifer and joined him in his rebellion. They are horrific to look upon and, like their master, their main purpose is to torture and utterly destroy man's relationship with God. God puts limits on what they can do, just as He put limits on Satan's torment of Job. Satan always has to submit to God.

You may find it interesting that in Italy and some other foreign countries, the locusts are called 'little horses'.

The First Woe is Past / The Sixth Trumpet Sounded

Notes

Revelation 9: 11

11 And they had a king over them, which is the angel of the bottomless pit, whose name in the Hebrew tongue is Abaddon, but in the Greek tongue hath his name Apollyon.

This verse confirms that these hideous beings are demonic and their king is Satan. His name in Hebrew is Abaddon which means 'destruction'; and in Greek, Apollyon, meaning 'destroyer'. Satan is the angel of the bottomless pit and the destroyer of mankind.

Revelation 9: 12

12 One woe is past; and, behold, there come two woes more hereafter.

In Revelation 8: 13, the angel flying in the midst of heaven says, "*Woe, woe, woe*", announcing that three woes are yet to come upon the earth when the trumpets of the last three angels are sounded. Now, in this verse, after the fifth angel sounds his trumpet (Revelation 9: 1) – which is the sounding of the first woe – there are two more woes to come.

Revelation 9: 13

13 And the sixth angel sounded, and I heard a voice from the four horns of the golden altar which is before God,

The golden altar before God refers to the Altar of Atonement (Leviticus 16: 18). It lays upon the Arc of the Covenant and is called the Mercy Seat. It is upon this that the High Priest offers the blood of the sacrificial lamb (bullock, goat); here also he lays coals taken from the Altar of Incense. Read Leviticus 16: 18 which says:

> *18 And he shall go out unto the altar that is before the Lord, and make an atonement for it; and shall take of the blood of the bullock, and of the blood of the goat, and put it upon the horns of the altar round about.*

The four horns of the golden altar are the four rings cast in gold and attached to the four corners of the Arc of the Covenant. These rings hold the carrying staffs of the Arc (Exodus 25: 12-15).

The Four Angels Released from the Euphrates / Two Hundred Million Horsemen

Notes

John hears the voice of God from the four horns of the golden altar announcing the second-last trumpet woe. This judgment comes from the very presence of God, from within the Holy of Holies. See Revelation 6: 12-17 to bring to memory how grave the judgment is.

Further enlightenment on the subject of the holy temple may be gained in reading an excellent study written by Schmitt and Laney, *Messiah's Coming Temple: Ezekiels' Prophetic Vision of the Future Temple.*[21]

Revelation 9: 14

14 Saying to the sixth angel which had the trumpet, Loose the four angels which are bound in the great river Euphrates.

The voice that comes out of the horns, the voice of God, tells the sixth angel which had the sixth trumpet to loose the four angels which are bound in the great river Euphrates. Their commission is to release human and satanic destruction upon the inhabitants of the earth. This is confirmed by looking ahead to Revelation 16: 12-13 when the sixth angel pours the sixth vial upon the Euphrates, releasing three unclean spirits out of the mouths of Satan, the Beast and the False prophet. By looking ahead, we rightly move from total satanic action, as has been the case in this chapter up until now, to human satanic involvement through the nations, the Antichrist and the False prophet.

Revelation 9: 15 - 16

15 And the four angels were loosed, which were prepared for an hour, and a day, and a month, and a year, for to slay the third part of men.

16 And the number of the army of the horsemen were two hundred thousand thousand: and I heard the number of them.

According to Oliver Greene, Greek scholars tell us that this should read, "...prepared for the hour and day and month and year."[22] That is, they have

---

[21]John W. Schmitt & J. Carl Laney, Messiah's Coming Temple (Grand Rapids, MI: Kregel Publications, 1977).

[22]Oliver B. Greene, The Revelation - Verse by Verse Study (Greenwille, SC: The Gospel Hour Publishing, Reprinted 1976), p. 261.

Notes

been preparing for this special moment in time. The four angels are loosed and will participate in the destruction of one-third of mankind. Scriptures don't clarify if these are just or unjust angels. They may be God's angels assigned to restrain Satan until this particular hour; or they could be evil angels restrained by heaven until their release so that they may work destruction upon the earth on behalf of Satan through the nations, the Beast and the False prophet (Revelation 16: 12-14). The latter would seem to be the case.

The four angels may represent four Asian nations, such as Japan, Pakistan, India and China. Millions of armed troops could very easily be amassed from within the vast population of these nations. China alone currently has over one billion citizens. In 1993, while I was on a trip to China, as we viewed the Terracotta warriors in Xian, our private tour guide from Beijing told us that China has had, for some years, a body of two hundred million armed horsemen ready at any time to go into battle.

Revelation 9: 17

17 And thus I saw the horses in the vision, and them that sat on them, having breastplates of fire, and of jacinth, and brimstone: and the heads of the horses were as the heads of lions; and out of their mouths issued fire and smoke and brimstone.

In Beijing, China, in the Forbidden City, a statue of a lion sits on each side at the top of the stairway which one climbs to enter the inner city. These lions hold a sphere representing the earth in their paws – the world in the hands of China: the future dream and hope of this great Asian power.

Here, in these verses, we read that the horses have the heads of lions and out of the heads of the horses issues fire and smoke and brimstone. Everywhere in the Forbidden City, one is greeted by the figure of the dragon. I don't remember whether any of the dragon sculptures are pictured with fire issuing forth from their mouths, but that image is certainly something one associates with dragons. Since John would have no words to describe modern warfare, he would use symbolic imagery to communicate the event. Apocalyptic form was commonly used and familiar to the writers of those times. It is more than possible John is speaking of well-equipped, modern war machines, even more advanced than those in use currently. I'm inclined to think that he's talking about modern warfare rather than demonic attack – demonic only in the sense that Satan puts the evil thought of aggression in the mind and heart of man. We learn in Revelation 16: 12-14 who the

Notes

aggressors are: God is pointing us to Eastern nations, especially China who, for many years, has had a heart of aggression toward the western world. China has been quietly preparing for the hour, the day, the month and the year, believing a massive force could bring victory and world domination. However, they have a surprise in store. They never planned on meeting Jesus. Russia, China, Iraq and other nations are, even now in 1998, aligning themselves to come against Israel. God has said He will gather all nations against Jerusalem to battle. Read Zechariah 14: 1-21, as well as Ezekiel, Chapters 38 and 39, which give a graphic, detailed account of that battle. It is called 'Armageddon'.

Revelation 9: 18

18 By these three was the third part of men killed, by the fire, and by the smoke, and by the brimstone, which issued out of their mouths.

When the four angels of verse 14 are loosed by the sixth angel with the trumpet, the second woe is administered through the mouths of the symbolic horses. At this time, one-third of the remaining population will die by these three means: by fire, by smoke and by brimstone. In Revelation 16: 12-14, when the sixth angel pours out his vial, the second woe is pictured as unclean spirits, like frogs, in the mouth of the dragon (Satan), the Beast (the Antichrist) and the False prophet (the World Church). Revelation 6, the chapter in which Jesus gives us a complete overview of the Tribulation years, describes massive devastation released upon the earth at the opening of the sixth seal. This will be accomplished through the trumpet and vial judgments and by Jesus himself when he returns to finish God's Tribulation judgments.

The same sequential structure may be used to view a complete picture for each set of judgments: the foretold judgments in the opening of the seals; the seven trumpet judgments; and the seven vial judgments. By fitting them alongside one another, we are given a complete picture of the full judgment of God. The following chart provides the scripture references for the judgments which coincide one with the other to create this complete overview:

The foretold judgments in the opening of the seals	The trumpet judgments	The vial judgment
1. Revelation 6:1	Revelation 8: 7	Revelation 16: 2
2. Revelation 6: 3	Revelation 8: 8	Revelation 16: 3
3. Revelation 6: 5	Revelation 8: 10	Revelation 16: 4
4. Revelation 6: 7	Revelation 8: 12	Revelation 16: 8
5. Revelation 6: 9	Revelation 9: 1	Revelation 16: 10

The Power of the Invading Army / Satanic Control

Notes

6. Revelation 6: 12-16	Revelation 9: 13-19	Rev. 16: 12-16
7. Revelation 6: 17; 8: 1-5	Revelation 10: 7; 11: 119	Rev. 16: 17-20

To simplify and clarify this teaching, consult Appendix 5 for a detailed Calendar of the Wraths.

Revelation 9: 19

19 For their power is in their mouth, and in their tails: for their tails were like unto serpents, and had heads, and with them they do hurt.

At first glance, it appears that this whole chapter is speaking only of satanic forces. However, Chapter 16: 12-14 sheds more light on the events and directs us to understand that the forces here in verses 15 to 19 are actually Satan working through man. The satanic spirits released previously from the pit increasingly activate intense evil in the minds of the nations, the Beast and the False prophet.

Revelation 9: 20 - 21

20 And the rest of the men which were not killed by these plagues yet repented not of the works of their hands, that they should not worship devils, and idols of gold, and silver, and brass, and stone, and of wood: which neither can see, nor hear, nor walk:

21 Neither repented they of their murders, nor of their sorceries, nor of their fornication, nor of their thefts.

Even with all the sorrows and horrors that come upon man from both Satan and God, many who remain alive refuse to repent. They, in fact, are unable to repent because Satan's angels have control of their minds and hearts. God can harden the heart of man. Continual rejection of God and on-going evil living can cause God to remove the wooing of the Holy Spirit. In Exodus 7: 3, God says:

3 And I will harden Pharaoh's heart, and multiply my signs and my wonders in the land of Egypt.

To paraphrase Paul's words in Hebrews 3: 15: "*Today, if you hear God's voice, harden not your heart as the Israelites did in the wilderness; for in so doing, you may bring upon yourself the anger of God.*"

Hardened Hearts

Notes

Paul also tells the Romans in Chapter 9: 18 (again, I paraphrase): *"God has mercy on whom He will have mercy, and He will harden the hearts of those He chooses to harden."* This is a warning to those who continually reject the calling of God. His Spirit will call for a limited time and then He allows the heart to be hardened. It's not God's doing; the fault is man's. Remember, if one continually ignores the wooing of the Holy Spirit, God's invitation may not come again, for God has said: *"I have heard thee in a time accepted, and in the day of salvation have I succoured thee: behold, now is the accepted time; behold, now is the day of salvation."* (II Corinthians 6:2; Isaiah 49: 8).

CHAPTER TEN

In **Chapter Ten**, Jesus appears, clothed in a cloud, holding an open book in his hand – the book of law and judgment, of love and grace. He takes the stance of a conqueror as he places his right foot on the sea and his left foot on the earth. Jesus, at John's request, gives the little book (the bible) to John and commands that he must digest its 'bittersweet' contents: its message of love and grace is sweet, but its message of correction and judgment is bitter.

Revelation 10

[**1**] And I saw another mighty angel come down from heaven, clothed with a cloud: and a rainbow was upon his head, and his face was as it were the sun, and his feet as pillars of fire:
[**2**] And he had in his hand a little book open: and he set his right foot upon the sea, and his left foot on the earth,
[**3**] And cried with a loud voice, as when a lion roareth: and when he had cried, seven thunders uttered their voices.
[**4**] And when the seven thunders had uttered their voices, I was about to write: and I heard a voice from heaven saying unto me, Seal up those things which the seven thunders uttered, and write them not.
[**5**] And the angel which I saw stand upon the sea and upon the earth lifted up his hand to heaven,
[**6**] And sware by him that liveth for ever and ever, who created heaven, and the things that therein are, and the earth, and the things that therein are, and the sea, and the things which are therein, that there should be time no longer:
[**7**] But in the days of the voice of the seventh angel, when he shall begin to sound, the mystery of God should be finished, as he hath declared to his servants the prophets.
[**8**] And the voice which I heard from heaven spake unto me again, and said, Go and take the little book which is open in the hand of the angel which standeth upon the sea and upon the earth.
[**9**] And I went unto the angel, and said unto him, Give me the little book. And he said unto me, Take it, and eat it up; and it shall make thy belly bitter, but it shall be in thy mouth sweet as honey.
[**10**] And I took the little book out of the angel's hand, and ate it up; and it was in my mouth sweet as honey: and as soon as I had eaten it, my belly was bitter.
[**11**] And he said unto me, Thou must prophesy again before many peoples, and nations, and tongues, and kings.

Notes

Revelation, Chapter 10

Revelation 10: 1

1 *And I saw another mighty angel come down from heaven, clothed with a cloud: and a rainbow was upon his head, and his face was as it were the sun, and his feet as pillars of fire:*

The mighty angel is Jesus. Note the following descriptive texts referring to Jesus:

• *Behold, he cometh with clouds.*	Acts 1: 11
• *He is the bow of mercy.*	Ezekiel 1: 28
• *His face shone as the sun.*	Matthew 17: 2
• *His feet were as fine brass: as if they burned in a furnace.*	Revelation 1: 15

Revelation 10: 2

2 *And he had in his hand a little book open: and he set his right foot upon the sea, and his left foot on the earth,*

Jesus takes the stance of a conqueror. The placement of his feet shows that he is about to take possession of the land and the sea: the whole of his earthly creation. His act denotes universal ownership and Godly authority, as is appropriate considering John 1: 3:

3 *All things were made by him; and without him was not any thing made that was made.*

The book he holds is a book of law and judgment, of love and grace. We, as Christians, hold that book dear to our heart.

Revelation 10: 3

3 *And cried with a loud voice, as when a lion roareth: and when he had cried, seven thunders uttered their voices.*

Jesus cries with the roar of the Lion of Judah. The Lord of lords and King of kings is about to take over his kingdom. Read Joel 3: 16:

16 *The Lord also shall roar out of Zion, and utter his voice from Jerusalem; and the heavens and the earth shall shake: but the Lord will be the hope of his people, and the strength of the children of Israel.*

The thunders echo the roaring Lion of Judah; they rumble and give warning of the coming judgments of the Creator.

The Seven Thunders / Time No Longer

Notes

Revelation 10: 4

4 *And when the seven thunders had uttered their voices, I was about to write: and I heard a voice from heaven saying unto me, Seal up those things which the seven thunders uttered, and write them not.*

The judgments spoken by the seven thunders are so terrifying and final that a voice from heaven silences John. The angel of Revelation 8: 5 (who takes the censer and fills it with the fire of the altar, casting it into the earth) is Jesus; at that time, there are voices and thunderings and lightnings and an earthquake. This was a warning of judgment upon mankind and was minor to what the seven thunders utter. Since we have a merciful God, He withholds the details of His most severe judgments until the very end; they are reserved for the supreme day and the irrevocable hour.

Revelation 10: 5 - 6

5 *And the angel which I saw stand upon the sea and upon the earth lifted up his hand to heaven,*

6 *And sware by him that liveth for ever and ever, who created heaven, and the things that therein are, and the earth, and the things that therein are, and the sea, and the things which are therein, that there should be time no longer:*

Jesus lifts his hand to heaven and swears by God that there will be no more delays; no more time before judgment. These verses agree with Revelation 22: 11 which indicates that the time will come when all conditions are permanent:

> *11 He that is unjust, let him be unjust still: and he which is filthy, let him be filthy still: and he that is righteous, let him be righteous still: and he that is holy, let him be holy still.*

There appears to be a long delay even after Jesus has sworn, in the name of God, that there will be no more delay. The reason for this is that although the trumpet and the vial judgments concur with one another, they are recorded and administered separately, each giving us a much more comprehensive view of the same judgment; that is, the judgments which Jesus previously reveals in the opening of the seals. Often, between and after each judgment, we are given other information vital to our understanding, not only of this prophetic word, but also other prophecies of the scriptures. Many windows open to us and, after viewing them all, we have a reasonable understanding of all that occurs in the Tribulation years. True to the words of Jesus, there is no delay. This detailed information by no means delays the time required to finish God's judgments. Everything fits within the confines of God's timetable.

The Mystery of God / John Asks for the Book

Notes

Revelation 10: 7

7 *But in the days of the voice of the seventh angel, when he shall begin to sound, the mystery of God should be finished, as he hath declared to his servants the prophets.*

When the voice of the seventh angel begins to sound, the mystery of God will be finished. The mystery is the critical knowledge of God's love and grace which brings about rebirth and salvation; in addition, it encompasses the mystery of Godly love of one brother for another.

Colossians 1: 26-27 leaves no doubt as to what the mystery of God is:

26 *Even the mystery which hath been hid from ages and from generations, but now is made manifest to his saints:*

27 *To whom God would make known what is the riches of the glory of this mystery among the Gentiles; which is Christ in you, the hope of glory:*

Revelation 10: 8

8 *And the voice which I heard from heaven spake unto me again, and said, Go and take the little book which is open in the hand of the angel which standeth upon the sea and upon the earth.*

The voice of God commands that John must take the little book which Jesus holds open in his hand. Jesus takes the stance of a conqueror with one foot on the sea and one on the earth.

Revelation 10: 9

9 *And I went unto the angel, and said unto him, Give me the little book. And he said unto me, Take it, and eat it up; and it shall make thy belly bitter, but it shall be in thy mouth sweet as honey.*

John, as directed, goes to Jesus and says, "Give me the little book." This seems rather disrespectful; however, it may be written in this way so that we may know that we must ask with authority, with sincerity and with commitment. When we do so, we are able to receive both the 'bitter' and 'sweet' message from Jesus.

The message of judgment is bitter and brings anguish and suffering. On the other hand, the message of grace is sweet. When we ask in sincerity, we receive the Spirit of God and can, with authority, preach salvation and judgment to mankind.

The book that John is told to eat is symbolic of the Holy Scriptures. John is to devour its contents, become thoroughly familiar with it, assimilate it and

Notes

digest its meaning. Paul, a steward of God, also tells us this in II Timothy 2: 15:

> *15 Study to shew thyself approved unto God, a workman that needeth not to be ashamed, rightly dividing the word of truth.*

Again, we receive this admonition in I Peter 3: 15:

> *15 But sanctify the Lord God in your hearts: and be ready always to give an answer to every man that asketh you a reason of the hope that is in you with meekness and fear:*

Revelation 10: 10

10 And I took the little book out of the angel's hand, and ate it up; and it was in my mouth sweet as honey: and as soon as I had eaten it, my belly was bitter.

John does as he is instructed, with the result being exactly as Jesus has told him it would be. The assurance of salvation is always sweet but scriptural chastisement is often bitter, even to the believer. In it, we are faced with change and surrender, both being difficult for sinful man.

Revelation 10: 11

11 And he said unto me, Thou must prophesy again before many peoples, and nations, and tongues, and kings.

The ancient Eastern manuscript provides a more accurate translation of this verse.[23] A literal rendering should read:

> *11 You must prophesy again about many peoples and nations, and the heads of nations and kings.*

John must prophecy of them, or about them, and not before them. He continues to do just that as he faithfully records the message of the *Revelation*. John speaks on behalf of Christ through this prophetic writing even more loudly in these last days. He speaks about and to *'many people, and nations, and tongues and kings'*. As always, the word of God is truth.

---

[23]George M. Lamsa, The Holy Bible from Ancient Eastern Manuscripts (Philadelphia: A. J. Holman Co., 1957), p. 1233.

CHAPTER ELEVEN

Chapter Eleven is probably the most difficult and most revealing of all the chapters of *the Revelation*. It introduces two witnesses who limit the activities of the Antichrist for the first three and one-half years of the Tribulation. Upon their death – which is followed by a massive earthquake – the second woe is past. As the seventh trumpet finishes sounding, it appears the time of judgment has come. The Tribulation is over and Christ has set up his kingdom. To understand the seemingly abrupt completion of the Tribulation – even before the vial wraths have been administered – the Study has drawn information from *Revelation,* chapter fifteen, verses five to eight, as well as *Daniel*, chapter twelve. Through the study of these chapters, the balance of *the Revelation* settles into a relatively uncomplicated chart of the last three and one-half years of Tribulation.

Revelation 11

[**1**] And there was given me a reed like unto a rod: and the angel stood, saying, Rise, and measure the temple of God, and the altar, and them that worship therein.

[**2**] But the court which is without the temple leave out, and measure it not; for it is given unto the Gentiles: and the holy city shall they tread under foot forty and two months.

[**3**] And I will give power unto my two witnesses, and they shall prophesy a thousand two hundred and threescore days, clothed in sackcloth.

[**4**] These are the two olive trees, and the two candlesticks standing before the God of the earth.

[**5**] And if any man will hurt them, fire proceedeth out of their mouth, and devoureth their enemies: and if any man will hurt them, he must in this manner be killed.

[**6**] These have power to shut heaven, that it rain not in the days of their prophecy: and have power over waters to turn them to blood, and to smite the earth with all plagues, as often as they will.

[**7**] And when they shall have finished their testimony, the beast that ascendeth out of the bottomless pit shall make war against them, and shall overcome them, and kill them.

[**8**] And their dead bodies shall lie in the street of the great city, which spiritually is called Sodom and Egypt, where also our Lord was crucified.

[**9**] And they of the people and kindreds and tongues and nations shall see their dead bodies three days and an half, and shall not suffer their dead bodies to be put in graves.

[**10**] And they that dwell upon the earth shall rejoice over them, and make merry, and shall send gifts one to another; because these two prophets tormented them that dwelt on the earth.

[**11**] And after three days and an half the Spirit of life from God entered into them, and they stood upon their feet; and great fear fell upon them which saw them.

[**12**] And they heard a great voice from heaven saying unto them, Come up hither. And they ascended up to heaven in a cloud; and their enemies beheld them.

[**13**] And the same hour was there a great earthquake, and the tenth part of the city fell, and in the earthquake were slain of men seven thousand: and the remnant were affrighted, and gave glory to the God of heaven.

[**14**] The second woe is past; and, behold, the third woe cometh quickly.

[**15**] And the seventh angel sounded; and there were great voices in heaven, saying, The kingdoms of this world are become the kingdoms of our Lord, and of his Christ; and he shall reign for ever and ever.

[**16**] And the four and twenty elders, which sat before God on their seats, fell upon their faces, and worshipped God,

[**17**] Saying, We give thee thanks, O Lord God Almighty, which art, and wast, and art to come; because thou hast taken to thee thy great power, and hast reigned.

[**18**] And the nations were angry, and thy wrath is come, and the time of the dead, that they should be judged, and that thou shouldest give reward unto thy servants the prophets, and to the saints, and them that fear thy name, small and great; and shouldest destroy them which destroy the earth.

[**19**] And the temple of God was opened in heaven, and there was seen in his temple the ark of his testament: and there were lightnings, and voices, and thunderings, and an earthquake, and great hail.

Command to Measure the Temple / Signs of the Rapture

Revelation, Chapter 11

Notes

Revelation 11: 1

1 *And there was given me a reed like unto a rod: and the angel stood, saying, Rise, and measure the temple of God, and the altar, and them that worship therein.*

The temple that John is told to measure is the temple that all Israel has longed for since their last temple was destroyed in 70 A.D. when under siege by Titus and the Roman army.

It would seem apparent that Israel must have her temple before the rule of the Antichrist and the second coming of Jesus. What exactly does Paul mean when he says the following in II Thessalonians 2: 1-4

1 *Now we beseech you, brethren, by the coming of our Lord Jesus Christ, and by our gathering together unto him,*

2 *That ye be not soon shaken in mind, or be troubled, neither by spirit, nor by word, nor by letter as from us, as that the day of Christ is at hand.*

3 *Let no man deceive you by any means: for that day shall not come, except there come a falling away first, and that man of sin be revealed, the son of perdition;*

4 *Who opposeth and exalteth himself above all that is called God, or that is worshipped; so that he as God sitteth in the Temple of God, shewing himself that he is God.*

Is Paul speaking of Christ's coming in the Rapture or is he referring to his coming in judgment at the end of the Tribulation? If we read the whole of II Thessalonians 2: 1-8 without careful thought, we might surely believe that the Church remains on Earth for the entire Tribulation; or, at least, to mid-Tribulation when the Antichrist has taken over the temple. If such were the case, the promises of God will have failed and our faith in Christ will have been somewhat futile. One must read this passage with care so as not to be misled. These verses are the source of the controversy among Christians who hold varied theories as to 'pre-mid- and end-Tribulation' Rapture.

It's quite clear that Paul is saying that the Thessalonian believers should not be troubled or confused as to the time of the Rapture. He tells them that the gathering together of the Church will not occur until:

a) there is a great falling away from God and the teachings of Christ;
b) the '*son of perdition*', the Antichrist, is revealed.

Paul goes on to tell his troubled followers what the '*son of perdition'* will do after he is revealed. He says that this evil dictator will exalt himself and set

Notes

himself up in the temple. He further tells them the final destiny of this Beast. When Jesus returns at his Second Coming, he consumes the Beast (the Antichrist) with '*the Spirit of his mouth*'.

Paul is not, in any way, telling the Saints that the Church will wait to be raptured until the great and glorious day of the Second Coming of our Lord. He is merely informing them as to the action of the Antichrist and revealing his final destruction. Paul states the Antichrist will be revealed only when his time comes and the power which is holding him back (either the angels, the witnesses or the Holy Spirit) steps out of the way (see II Thessalonians 2: 6-8 NLT). According to Paul, when the Antichrist is revealed, the Church will be gathered together unto Christ.

The measuring of the temple, the altar and the accounting of the worshippers therein is an act of God's claiming or taking possession of them for Himself. It is also possibly an act of evaluating the worship of His people. One must not forget that Jesus will reign in this temple after it is purified by the cleansing smoke from the glory of God (Revelation 15: 8).

Remember that the temple in which Jesus worshipped while on earth had also been defiled and partially destroyed by a previous Antichrist, Antiocus IV, and was later restored and cleansed by the Maccabees in 164 B.C. It was again rebuilt and restored by Herod in 19 B.C.

Revelation 11: 2

2 *But the court which is without the temple leave out, and measure it not; for it is given unto the Gentiles: and the holy city shall they tread under foot forty and two months.*

The angel tells John not to measure the outer court of the temple since it is set aside for the Gentiles. The Antichrist, however, will be restricted from interfering with the sacrifices or entering the temple proper (the Holy Place and the Holy of Holies) until the last forty-two months of the Tribulation because of the restraining power of the two witnesses

For affirmation, read Daniel 9: 27:

27 *And he* (the Antichrist) *shall confirm the covenant* (the Jerusalem Covenant of 1993) *with many for one week: and in the midst of the week he shall cause the sacrifice and the oblation to cease...*

Through Daniel's seventy week prophesy and II Thessalonians 2: 4, we know that the Antichrist will overrun the whole of the City after the witnesses are taken up. This will happen at the end of the first three and one-half years, half-way through the last week of Daniel's prophesy. It will be at this time that the Antichrist takes possession of the temple, thereby

Notes

breaking the nations agreement with Israel to honour the Jerusalem Covenant. For the last forty-two months, the restraining power having been lifted, the Antichrist will tread the Holy City and the Temple of Jerusalem under foot. Nothing will be holy – not even the inner sanctuary, the Holy of Holies. His unrestricted temple reign will continue for the balance of the seventieth week, a period of forty-two months.

Refer to Appendix 4 for information on the Jerusalem Covenant.

Revelation 11: 3

3 *And I will give power unto my two witnesses, and they shall prophesy a thousand two hundred and threescore days, clothed in sackcloth.*

God will give remarkable power to His chosen witnesses through the ever-present Holy Spirit. They will be the last of the earth's great evangelists. They restrain Satan (the Beast and the False prophet) for the first three and one-half years of the Tribulation, while proclaiming the imminent return of the Lord. These humble men, dressed in sackcloth, will have spiritual authority over Satan's evil forces and, thus, be free to call continually for man's repentance.

Revelation 11: 4

4 *These are the two olive trees, and the two candlesticks standing before the God of the earth.*

These two witnesses are Israelites, symbolized by the reference to the olive tree. The olive tree furnishes the basis for one of Paul's allegories, found in Romans 11: 16-25. The Gentiles are the '*wild olive tree*' grafted in upon the '*good olive tree*' to which, previously, only the Israelites belonged. The witnesses are the two candlesticks bearing the light of the Spirit to the world.

These witnesses are Elijah and Enoch, the only two Old Testament Saints who have never died. It is '*appointed unto men once to die*' and we read in verses 3 to 12 of them fulfilling a commission before keeping their appointment with death. Some feel that Moses is one of the candlesticks but we read in Deuteronomy 34: 5-6 of Moses dying and being buried:

5 *So Moses the servant of the Lord died there in the land of Moab, according to the word of the Lord.*

6 *And he buried him in a valley in the land of Moab, over against Bethpeor: but no man knoweth of his sepulchre unto this day.*

The only time a man dies twice is if he is called from the dead to stand before the White Throne judgment, finds himself condemned and is committed to the fires of hell. This becomes the second death.

Elijah and Enoch, the Chosen Evangelists

Notes

The only exceptions to one death and one resurrection are found in miracles, one of which is recorded in John 11: 38-45 when Jesus raises Lazarus from death to life. In prayer and thanksgiving to his Father, Jesus explains the purpose for this particular miracle; through it, he is showing his followers, as well as the doubting Israelites, the omnipotent God and the power of prayer. Essentially, Jesus uses this situation to enact the very essence of his ministry on earth; he demonstrates that he came to bring life to counteract the pronouncement of death.

We read in Jude, verse 9, that Michael, the archangel, contended with Satan over the body of Moses:

> 9 *Yet Michael the archangel, when contending with the devil he disputed about the body of Moses, durst not bring against him a railing accusation, but said, The Lord rebuke thee.*

Satan wanted his body so that he could tempt man to worship at his grave and give Moses glory that should instead go to God. It is generally accepted that Jude is quoting from *The Assumption of Moses*, one of the collection of apocalyptic testaments. Scholars say the original manuscript, in all probability, stated that Moses was taken immediately to God.[24] How fitting that Moses, so faithful a servant, should have risen to be with God, particularly since he was denied entrance to the Promised Land. If, in fact, he was risen at the time of his death, then Enoch, Elijah and Moses would have been received into heaven, even before the Old Testament Israelite Saints were raised with Jesus at his resurrection from the grave (Matthew 27: 52-53; Matthew 8: 11). This would seem feasible since, at the transfiguration, Elijah and Moses were seen talking with Jesus (Matthew 17: 1-8). It would be highly improbable that Moses would be called from the grave to accompany Elijah as they spoke with Jesus.

The reason some Bible scholars believe that Moses is one of the witnesses is that they misread and misunderstand Deuteronomy 18: 15-19. One must understand that Moses is speaking to the Israelites throughout this Chapter. He tells them that…

> 15 *The Lord thy God will raise up unto thee a Prophet* (Jesus) *from the midst of thee, of thy brethren* (an Israelite), *like unto me* (Moses)*; unto him ye shall hearken;*

---

[24]James H. Charlesworth, Editor, The Old Testament Pseudepigrapha, Volume I (New York: Doubleday Publishers, 1983), p.919.
The sole extant copy of The Testament of Moses is a Latin palimpsest discovered by A. M. Ceriani in the Ambrosian Library in Milan in 1861. Much of the manuscript is missing or undecipherable; however, many biblical scholars believe it to be an extract from, or a copy of, The Assumption of Moses. See also pp. 925-926 for information and listing of citations and discussions by James H. Charlesworth; R. H. Charles' The Assumption of Moses, 1897; A. M. Denis' Fragmenta; E. M. Laperrousaz' The Testament of Moses / The Assumption of Moses; C. J. Lattery's "The Messianic Expectation in the Assumption of Moses"; G. Nickelsburg's Studies on the Testament of Moses; etc.

Notes

He further relates what God has said to him:

18 *I will raise them up a Prophet from among their brethren, like unto thee (Moses), and will put my words in his mouth (see John 12: 49); and he (Jesus) shall speak unto them all that I shall command him.*

19 *And it shall come to pass, that whosoever will not hearken unto my words which he (Jesus) shall speak in my name, I will require it of him.*

We must not assume that Moses will become one of God's chosen witnesses simply because he was a faithful and obedient mouthpiece for God or because he was seen at the transfiguration.

As a finale to this discussion, please consider the following. Those who disqualify Enoch as a witness on the grounds that he is a Gentile and not an Israelite must also disqualify Abraham and Isaac as Israelites. The Hebrew tribe wasn't distinctly identified by name until Jacob prevailed with God. God was pleased and blessed Jacob for a number of reasons: because he declared himself unworthy and cried out for deliverance; because he believed God when He promised He would abundantly bless him and make his numbers as the sand of the sea (Genesis 32: 10-12); because he persevered with God and would not let the angel of God go until he received the promised blessing. It was at this point that God changed Jacob's name to Israel and lifted up Abraham's tribe above the heathen, openly identifying them as His own. God's promised blessing to Abraham and Isaac came to Israel through Jacob, the prince who prevailed with God (Genesis 32: 24-28).

Tracing the line of Israel back to the Garden of Eden, one would come upon Isaac (Genesis 21: 3); Abraham (Genesis 11: 26); Shem (Genesis 5: 32); Noah (Genesis 5: 29); Enoch, '*who walked with God, and he was not for God took him*' (Genesis 5: 19-24); Seth (Genesis 4: 25); Adam and Eve (Genesis 1: 27). As Jesus was in the loins of Abraham, so also was Jacob in the loins of Enoch; thus, Enoch is a building block, a foundation stone, in the House of Israel. For this reason, we are assured the two olive trees, the two candlesticks, are Enoch and Elijah.

Revelation 11: 5

5 *And if any man will hurt them, fire proceedeth out of their mouth, and devoureth their enemies: and if any man will hurt them, he must in this manner be killed.*

As in the days of Jeremiah, God will again empower His prophets with a message for the ungodly *"that will burn them up as if they were kindling*

Notes

wood." (NLT). In the King James translation of Jeremiah 5: 14, it reads:

14 *...I will make my words in their mouth fire; and this people wood, and it shall devour them.*

Revelation 11: 6

6 *These have power to shut heaven, that it rain not in the days of their prophecy: and have power over waters to turn them to blood, and to smite the earth with all plagues, as often as they will.*

It is understandable that the two witnesses could be thought to be Moses and Elijah. Moses called plagues down upon Egypt and turned water into blood in Exodus, Chapters 8 to 12. Elijah called fire down from heaven (read II Kings 1: 10-14) and prayed that God would withhold rain so that the land suffered drought. Though there are similarities between the gifts of Moses and the supernatural powers of the witnesses, these similarities do not in any way allow us to conclude that Moses will be a witness during the Tribulation. Keep in mind that Moses died the death appointed him and, thus, is disqualified as a possible Tribulation witness.

Revelation 11: 7

7 *And when they shall have finished their testimony, the beast that ascendeth out of the bottomless pit shall make war against them, and shall overcome them, and kill them.*

Satan, who comes up out of the bottomless pit, takes possession of a host-body which, in this case, is the Antichrist. When the witnesses have faithfully finished their appointed tasks – winning souls for Christ; restraining the Antichrist from entering the temple; and, to a great degree, limiting the works of Satan – God lifts His protective hand. Forty-two weeks into the Tribulation, with God's consent and in God's proper time, the Antichrist kills the two witnesses. They remain dead for three and one-half days. It is at this point that all three of these major prophets will have died: Moses, Elijah and Enoch.

Revelation 11: 8

8 *And their dead bodies shall lie in the street of the great city, which spiritually is called Sodom and Egypt, where also our Lord was crucified.*

In order to disgrace their memory and discredit their work, the Antichrist orders that the bodies of the two witnesses, Elijah and Enoch, be left on the streets of Jerusalem.

The Nations View the Witnesses / The Nations Rejoice

God compares the spiritual state of His beloved city to the ungodly state of Sodom and Egypt – Sodom because of gross idolatry and moral corruption; and Egypt for its spiritual darkness.

Notes

Revelation 11: 9

9 *And they of the people and kindreds and tongues and nations shall see their dead bodies three days and an half, and shall not suffer their dead bodies to be put in graves.*

Most of the world will be jubilant that the hindering influence of the witnesses is now ended. By means of global satellite communication, the entire world will view their dead bodies decaying on the streets of Jerusalem.

When I first read Revelation, some many years ago, I wondered at verses nine and ten. Through lack of understanding and, possibly, less secure faith, I thought the scriptures must be in error since, at that time, it would be impossible for all tongues and nations to look upon this sickening sight. Indeed, it is now possible through common, everyday technology. God is all truth. Praise His name!

Revelation 11: 10

10 *And they that dwell upon the earth shall rejoice over them, and make merry, and shall send gifts one to another; because these two prophets tormented them that dwelt on the earth.*

All of the earth's ungodly will rejoice and celebrate, just as we celebrate the birth of Christ. The reason they are so overjoyed is that the World Ruler (the Beast) will direct all blame for world troubles (the wrath of the trumpets) on the witnesses and the people of God. For this reason, thousands-upon-thousands of Tribulation Saints who hold the same doctrine as the witnesses will be condemned and put to death. If they somehow avoid the death sentence, they may die from starvation since they won't be able to buy or sell without the implanted number of the Beast. They won't even be able to access their bank accounts – if they still have them – nor will they be allowed to work or receive medical aid.

We have previously read about some of these believers in Revelation 6:9-11:

9 *And when he had opened the fifth seal, I saw under the altar the souls of them that were slain for the word of God, and for the testimony which they held:*

One-fourth of the world's population will, at this time, have died through the wrath of God and the wrath of man. No wonder the ungodly will be glad to

Notes

rid themselves of those who have falsely been accused of being responsible for world's turmoil.

If there are readers who think that this could never happen, then think again. Remember Hitler and his annihilation of not only the Israelites, but also many thousands from other nations. Think about the Kosovo crisis and the ethnic 'cleansing' which has happened even now in 1999. Take warning: this will happen again.

The only way to escape this fast-approaching holocaust is to walk with God and keep Jesus as your Saviour or 'life-vest'. Rest assured. We have many promises in the Scriptures which tell us that the Church will be taken up in the Rapture before this sorrow comes upon the earth. One such promise is found in Revelation 3: 10:

> *10 Because thou has kept the word of my patience, I also will keep thee from the hour of temptation* (the temptation to take the mark of the beast), *which shall come upon all the world, to try them that dwell upon the earth.*

Revelation 11: 11 - 12

> *11 And after three days and an half the spirit of life from God entered into them, and they stood upon their feet; and great fear fell upon them which saw them.*
>
> *12 And they heard a great voice from heaven saying unto them, Come up hither. And they ascended up to heaven in a cloud; and their enemies beheld them.*

God affirms the authenticity of the witnesses' ministry as the world looks on. Never have they viewed a miracle equal to this. Interesting to note that the Israelites believe the soul hovers near the body for three days;[25] thus, in their thinking, the body is not totally dead until then. Possibly, this is the reason God allows the witnesses to remain dead until this period of time has well passed.

Revelation 11: 13

> *13 And the same hour was there a great earthquake, and the tenth part of the city fell, and in the earthquake were slain of men seven thousand: and the remnant were affrighted, and gave glory to the God of heaven.*

At the same time as the witnesses are called up to heaven, a great earthquake

---

[25]Oliver B. Greene, The Revelation - Verse by Verse Study (Greenville, SC: The Gospel Hour, Inc., Revised 1976), p. 300.

destroys a tenth part of the city of Jerusalem. Seven thousand are killed. Even the Christian remnant are filled with awe and fear. They know God is shaking the earth and give Him the glory.

Notes

Revelation 11: 14

14 The second woe is past; and, behold, the third woe cometh quickly.

The second woe is past with the death of the witnesses and the devastation of the great earthquake. The third woe quickly follows.

Preamble to Revelation 11: 15-19

The following section will bring together the teaching of Revelation 11: 15-19 and Revelation 15: 5-8. Unless one sees how these two fit together (one revealing the truth of the other), a greater part of the Revelation will be misunderstood. Once the connection between the two is recognized, then suddenly the significance of Daniel's 'measures of time' also becomes apparent. Revelation 15: 5-8 almost forces their meaning upon us. So often we are confused by the literary patterns chosen by God as He presents His Revelation to man.

True to form, God first gives us a full overview of the culmination of the wraths in Revelation 11: 15-19. He later provides relevant information in Revelation 15: 5-8 and has previously given us critical details in Daniel 12: 7-12. Through these scriptures, we are able to define the time-frame of the trumpet and vial wraths and, in fact, the approximate time of the return of Jesus.

When we have completed a detailed study of these three segments of the scriptures, everything will fall into place. Armed with pertinent details gleaned from the scriptures, we will be able to recognize that the vial wraths are administered after the unhampered 1260 day temple-reign of the Antichrist is finished. By bringing these aforementioned chapters and verses together, we may obtain a crystal-clear understanding of portions that hitherto have led to misinterpretation of the scriptures. It's actually not difficult if we proceed step-by-step and deal honestly and scripturally with each segment. Satan would love to convince us that this victorious Revelation is beyond our comprehension; this is not so.

While reading this section – Revelation 11: 15-19 – **please consult Appendix 10, "Line Charts of the Tribulation and Millennial Years".** This should help clarify this section of the Study.

The Seventh Trumpet / The Termination of the Wraths

Revelation 11: 15 - 19

Notes

15 And the seventh angel sounded; and there were great voices in heaven, saying, The kingdoms of this world are become the kingdoms of our Lord, and of his Christ; and he shall reign for ever and ever.

16 And the four and twenty elders, which sat before God on their seats, fell upon their faces, and worshipped God.

17 Saying, We give thee thanks, O Lord God Almighty, which art, and wast, and art to come; because thou hast taken to thee thy great power, and hast reigned.

18 And the nations were angry, and thy wrath is come, and the time of the dead, that they should be judged, and that thou shouldest give reward unto thy servants the prophets, and to the saints, and them that fear thy name, small and great; and shouldest destroy them which destroy the earth.

19 And the temple of God was opened in heaven, and there was seen in his temple the ark of his testament: and there were lightnings, and voices, and thunderings, and an earthquake, and great hail.

Revelation 10: 7 states that when the voice of the seventh angel begins to sound, the mystery of God is finished – the mystery of love and grace which provides rebirth and salvation. Here, in these verses, the seventh angel finishes sounding and the pre-Millennial salvation draws to a close. Heaven rejoices because God's wraths are almost complete, and Christ's Millennial reign will soon begin. Those from the kingdoms of the world, who are found righteous after the Tribulation, will enter the kingdom of our Lord and His Christ. His kingdom will no longer be invisible, hidden or tucked away within the heart of man (Luke 17: 21). It will become a visible and literal kingdom over which our Lord Jesus will reign.

After an in-depth study of Revelation 11: 15-19, one must conclude that the trumpet wraths and the vial wraths, although recorded separately and administered separately as individual events, culminate together. First come the trumpet judgments and then, later, the vial judgments. Certainly, they concur with one another and collaborate with the unveiling of the seals; however, they are not, as one might think, delivered one upon the other throughout their unveiling. It is only in the final outpouring that they come together, one with the other. (This statement will be verified as we continue the Study.)

There is a period of peace after the first seal is opened (I Thessalonians 5: 3) which ends when the first trumpet wrath is sounded. The trumpet wraths accompany the seals throughout the seven years, giving substance to the revelation of that particular seal.

The vial wraths also stay true to the unsealing; however, they start in rapid succession after the end of the 1260 days temple-reign of the Antichrist. It

Notes

is at the presentation of the vial wraths that the Antichrist is evicted from the temple by the smoke from the Glory of God. Revelation, Chapter 15: 5-8, provides evidence to substantiate this truth.

In light of these last five verses of Revelation, Chapter 11, it becomes clear that the trumpet and vial judgments do, indeed, ultimately finish at the same time. Certainly, all of God's judgment must culminate *before the kingdoms of the world can become the kingdoms of our Lord*, as verse 15 proclaims. All the judgments must be meted out before *our Lord God Almighty takes His great power and reign*, as shown in verse 17. The living cannot be judged or rewarded until all the wraths are executed. None can look upon the *Holy of Holies* until this portion of judgment is completed. John could never have viewed the *ark of the testament in the temple in heaven* (the place in which God's Spirit dwells – see Revelation 11: 19) nor would he have been given ears to hear, and eyes to see the aftermath of God's retaliation for sin unless the two sets of wraths terminate together as one, and the temple is cleared of the smoke from the Glory of God. These five verses have lead many to believe that both sets of wraths are poured out together. Nothing could be further from the truth.

It is evident that as Jesus opens each seal, He consents to or opens the way for the outpouring of God's wrath (see teaching in Revelation 6: 12-17). This happens first through the trumpet judgments and is then later finalized in the vial judgments. It is as if He cracks seven lashes across the face of the earth with the vial wraths being the tail-end of the whip.

Read Chapter 15: 5-8 carefully to understand that the conferral of the vial wraths activates *the smoke from the Glory of God* which evicts the Antichrist from the temple:

5 *And after that I looked, and, behold, the temple of the tabernacle of the testimony in heaven was opened:*
6 *And the seven angels came out of the temple, having the seven plagues, clothed in pure and white linen, and having their breasts girded with golden girdles.*
7 *And one of the four beasts gave unto the seven angels seven golden vials full of the wrath of God, who liveth for ever and ever.*
8 *And the temple was filled with smoke from the glory of God, and from his power; and no man was able to enter into the temple, till the seven plagues of the seven angels were fulfilled.*

I have read these scriptures repeatedly over the years but have really never seen the significance of them. After careful reading, one realizes that it would be impossible for the vial wraths to join the trumpet wraths and begin their outpouring at the opening of the first seal; unless they were delayed, it would mean that at no time throughout the Tribulation could any man enter the temple. The two witnesses and the worshipping Israelites would be

Temples Filled With God's Glory

Notes

barred by *the smoke from the glory of God* for the first forty-two months, as would the Antichrist in the last forty-two months – this cannot be! Thus, it is clear that the vial wraths are withheld until after the Antichrist has been allowed his allotted time to reign in the temple.

Some might say the temple under discussion in Revelation 15: 5-8 is the temple in heaven; rightly so, however, it speaks also of the end-time temple in Jerusalem. The scriptures teach us that God's glory dwells within His tabernacles. His Holy Spirit or *the smoke from the Glory of God*, the Shekinah Glory, is the testimony of His presence.

In the time of the Exodus, the tabernacle of the testimony in the wilderness of the Sinai was also filled with the glory of the Lord. Moses was unable to enter the tent of the congregation because a cloud overshadowed it, as we read in Exodus 40: 35:

> *35 And Moses was not able to enter into the tent of the congregation, because the cloud abode thereon, and the glory of the Lord filled the tabernacle.*

In the time of the kings when the temple of Solomon was dedicated, the Shekinah cloud filled the Holy place so the priests could not stand and minister in the temple. Read I Kings 8: 10-11:

> *10 And it came to pass, when the priests were come out of the holy place, that the cloud filled the house of the Lord,*
> *11 So that the priests could not stand to minister because of the cloud: for the glory of the Lord had filled the house of the Lord.*

In Ezekiel, Chapters 40-42, the prophet records a vision of a glorious temple. Many speculate as to the purpose of this vision. Some say it is a memorial to Solomon's temple or that it is a more elaborate version of the Zerubbel temple built after Israel's return from exile. Others feel it is an embellishment of Herod's temple or that it may be a temple symbolic only of God's glory. The more likely truth is that Ezekiel was chosen to receive this vision that posterity might obtain specific, minute directives as to God's pattern for His end-time temple. God has always been precise when directing the building of His temples. He gives a blueprint for each, the central structure being similar for all of them (I Chronicles 28: 11-12; Acts 7: 44; Hebrews 8: 5). The point is that the glory of the Lord filled the house (Ezekiel 43: 5) Obviously, the Shekinah glory sanctifies the most Holy places and will unquestionably anoint the Tribulation temple.

Given Revelation 15: 5-8 and Revelation Chapter 4, we see that the temple in heaven (filled with the glory of God) is reserved for God's Holiness, His high-ranking angels and the sanctified, glorified sons of God. After the presentation of the vials, none – other than these glorified ones – will enter the holiness of the heavenly temple. The scriptures give no teaching or

Notes

indication that man, aside from those who are glorified, has ever entered the heavenly temple. The only temple that man enters until his glorification is the earthly temple. Thus, it can be concluded that Revelation 15: 8 refers, or has reference also to, the Tribulation temple at the time of the Antichrist.

As has been noted before, the trumpet judgments collaborate with the opening of the seals and are the means by which God's retribution is first channelled through to man. They accompany the seals down through the Tribulation years and end with the seventh trumpet wrath which joins the seventh vial wrath as it proceeds on into the 1290 days prophesied by Daniel 12: 11. The fury of the two sets of wraths now coming together is referred to as '*the great day of the wrath of God*' (Revelation 6: 17).

The fierceness of the joint wraths and the smoke from the Glory of God causes the Antichrist to vacate the temple and join the nations as they hide themselves in the dens of the rocks of the mountains. However, all power will not be taken from him until the joint-wraths end, allowing him to complete his 1290 day reign as prophesied by Daniel 12: 11.

After the eviction of the Antichrist from the temple the Tribulation will continue for another seventy-five days (Daniel 12: 11-12). The vial wraths will be poured out in rapid succession, one upon the other, during the first thirty days (Daniel's 1290 days). They culminate within the ending notes of the blast of the seventh trumpet in order that Revelation 11: 15-19 may be fulfilled. After this, the temple is no longer filled with the smoke of the glory of God (Revelation 11: 19) and the kingdoms of this world are become the kingdoms of our Lord, and His Christ. This, then, is the finalizing of the third woe as spoken of in Revelation 11: 14. In the last 45 days, Jesus will '*thrust in His sickle and gather the clusters of the vine of the earth and cast it into the great winepress of God*' (Revelation 14: 18-19). '*Blessed*', indeed, '*is he that waiteth and cometh to the 1335 days*' of Daniel 12: 12, for they are the righteous who '*attend the wedding feast of the Lamb's wife*' (Revelation 19: 7-9).

To reinforce and help clarify this teaching, I repeat here the last portion of the commentary for Revelation 8: 13... The third woe is when the voice of the seventh trumpet begins to sound. The mystery of God, the pre-millennial salvation, is finished (Revelation 10: 7). This woe is the great day of the wrath of God (Revelation 6: 17). It is the day when the seventh trumpet ends its sounding (Revelation 11: 15); the day of rewarding the just and the unjust (Revelation 11: 18). It is the day that the seventh angel pours out his vial in the air and the great voice out of the temple of heaven from the throne of God declares, "It is done!" (Revelation 16: 17-21).

To affirm the validity of this teaching, let us review the scriptures to be assured that seventy-five significant days follow the prophesy of Daniel 9:

Daniel's Prophetic Measures of Time / The Last Seventy-five Days Before the Millennium

Notes

24-27. With careful reading, it becomes apparent that Daniel's seventieth week takes us to the end of the temple reign of the Antichrist – the end of the 1260 days and the finalizing of the seventy weeks (Daniel 12: 6-7). Daniel, Chapter 12: 11-12, takes us beyond the seventy week prophesy. In all, Daniel gives us four measures of time:

- Daniel 8: 13-14 — 2300 days
- Daniel 12: 7 — 1260 days
- Daniel 12: 11 — 1290 days
- Daniel 12: 12 — 1335 days

All of these time periods begin when the daily sacrifice is taken away and the abomination of desolation (the temple reign of the Antichrist) is set up. They are, however, from two different eras. Daniel 8: 13-14, with a time measure of 2300 days, refers totally to the Syrian invasions of the Zerubbabel temple under King Antiochus IV. This is proven by comparing Daniel's prophesy (concentrating mainly on Chapter 8: 8-14 and Chapter 11: 28-45) with I Maccabees, Chapter 1: 10 to Chapter 4: 60. The Study will later verify this statement; however, first, we must view the other three measures of time, for it is in them that the period of the outpouring of the wraths and the eviction of the Antichrist is substantiated.

The last three measures of time from Daniel 12 begin in the middle of the seventieth week (see Daniel 9: 27). All but the 1260 days go beyond Daniel's seventieth week. The 1290 days exceed by 30 days, and the 1335 days exceeds by another 45 days: 75 days in all. The scriptures affirm, in several places, the time allowed the Antichrist to desecrate and defile the temple is 1260 days. Please read Revelation 13: 5:

> 5 *And there was given unto him a mouth speaking great things and blasphemies; and power was given unto him to continue forty and two months*. (1260 days)

Along with this, we have the declaration of Jesus as affirmation when he appeared in a vision to Daniel in Daniel 12: 7:

> 7 *And I heard the man clothed in linen, which was upon the waters of the river, when he held up his right hand and his left hand unto heaven, and sware by him that liveth for ever that it shall be for a time, times, and an half* (three and a half years – 1260 days)*; and when he shall have accomplished to scatter the power of the holy people, all these things shall be finished.*

We see through these scriptures that the Antichrist's temple reign is limited to the 1260 days first allotted. Jesus has sworn to this in the name of his Father. Obviously, at the termination of his appointed time, the Antichrist is driven out of the temple by the smoke from the Glory of God. This Shekinah Glory will cleanse the temple (Exodus 29: 43) and no man will

enter it until the seven vial plagues of the seven angels are fulfilled (Revelation 15: 8). The first to enter will be our Lord Jesus Christ and the anointed priests who restore and again sanctify the temple. Jesus will then rule and reign throughout the Millennium, the appointed Saints serving and reigning with Him.

Notes

The 1290 days of Daniel 12: 11 (thirty extra days above the 1260 first allotted the Antichrist) will be the measure of time taken by the seven angels with the vial wraths to administer God's judgments. They terminate the rule of the Antichrist and activate the fall of Babylon.

Now it becomes clear why the scriptures record the effects of the seventh trumpet as its sounding is announced and then, later, as it ends. In this way, we realize that there is a pause or lapse of time between the proclamation of its sounding, recorded in Revelation 10: 7, and its ending in Revelation 11: 15-19. This time lapse causes the seventh trumpet wrath to end its sounding within the 1290 days, thereby allowing the wraths to join and end simultaneously.

As soon as both sets of wraths finish, the 45 days of atonement begin – achieving the final total of the 1335 days of Daniel 12: 12:

> 12 *Blessed is he that waiteth, and cometh to the thousand three hundred and five and thirty days.*

Most blessed, indeed, are those who come through the last 45 days of judgment visited upon man by Jesus himself. These are the righteous who have been faithful to the end. A very complete record of these days of judgment may be found in Revelation 19: 7-21. We find here a perfect example of how the scriptures are not necessarily written in chronological order.

Read Revelation 19: 11-21 first to view heaven opening and Jesus riding forth in judgment:

> 11 *And I saw heaven opened, and behold a white horse; and he that sat upon him was called Faithful and True, and in righteousness he doth judge and make war.*
>
> 12 *His eyes were as a flame of fire, and on his head were many crowns; and he had a name written, that no man knew, but he himself.*
>
> 13 *And he was clothed with a vesture dipped in blood: and his name is called The Word of God.*
>
> 14 *And the armies which were in heaven followed him upon white horses, clothed in fine linen* (Israelites), *white and clean.*
>
> 15 *And out of his mouth goeth a sharp sword, that with it he should smite the nations: and he shall rule them with a rod of iron: and he treadeth the winepress of the fierceness and wrath of Almighty God.*
>
> 16 *And he hath on his vesture and on his thigh a name written, KING*

Salvation Comes to the House of Israel

OF KINGS, AND LORD OF LORDS.

17 *And I saw an angel standing in the sun; and he cried with a loud voice, saying to all the fowls that fly in the midst of heaven, Come and gather yourselves together unto the supper of the great God;*

18 *That ye may eat the flesh of kings, and the flesh of captains, and the flesh of mighty men, and the flesh of horses, and of them that sit on them, and the flesh of all men, both free and bond, both small and great.*

19 *And I saw the beast, and the kings of the earth, and their armies, gathered together to make war against him that sat on the horse, and against his army*

20 *And the beast was taken, and with him the false prophet that wrought miracles before him, with which he deceived them that had received the mark of the beast, and them that worshipped his image. These both were cast alive into a lake of fire burning with brimstone.*

21 *And the remnant were slain with the sword of him that sat upon the horse, which sword proceeded out of his mouth: and all the fowls were filled with their flesh.*

Notes

To finish here would be incomplete. One of the major events during the forty-five days would be overlooked. We need now to read of God's wondrous mercy extended to His beloved Israel, which includes, also, His kindness toward the Gentile. Read Revelation 19: 7-9:

7 *Let us be glad and rejoice, and give honour to him: for the marriage of the Lamb is come, and his wife hath made herself ready.*

8 *And to her was granted that she should be arrayed in fine linen, clean and white: for the fine linen is the righteousness of saints.*

9 *And he saith unto me, Write, Blessed are they which are called unto the marriage supper of the Lamb. And he saith unto me, These are the true sayings of God.*

Acknowledging that God's rightful anger has been appeased, we can take delight in His faithfulness as He keeps His vow to His unfaithful wife: salvation now comes to the house of Israel. To set your feet dancing as you look upon the goodness of God, you need only move ahead to the study of Revelation 19: 7-9. (**It is recommended, in fact, that you do read this section to fully comprehend God's goodness and applaud His far-reaching plan.**)

Having dealt with the measures of time spoken of in Daniel 12: 7-12, we are now ready to solve the mystery of the 2300 days of Daniel 8: 13-14.

One recognizes the omniscient God in reading Daniel – the entire book is flawless proof that the scriptures are written by an all-knowing Spirit, the Spirit of God. Through Daniel, as well as other prophets, history was written well before it happened. The consistent way that history and

Notes

prophecy correlate has made some critics conclude that the prophecies were written during or after the events described. The fact that the prophetic words pertaining to the latter days are only now being fulfilled must surely convince even the most hardened skeptics that the prophets wrote under inspiration of God.

God was totally aware that the events being revealed would remain unclear or closed until history opened the way to understanding, and the time was right for the Holy Spirit to reveal the truth to the Saints in the last days. For this reason, He told Daniel to '*shut up the word and seal the book until the time of the end.*' Nevertheless, He revealed the approximate time that understanding would come by saying the seal would be removed when '*many shall run to and fro, and knowledge shall be increased*' (Daniel 12: 4).

Daniel was completely confused by the four sets of numbers which God spoke through him. We who were born in the early twentieth century have also been puzzled by them. I, for one, have often questioned God for answers. The response has been as that given to Daniel in Daniel 12: 9:

> 9 *...Go thy way, Daniel: for the words are closed up and sealed till the time of the end.*

The time is now come when the Saints can look both forward and back, and understanding has come through the enlightenment of the Holy Spirit. Not that the Saints today are more spiritual than Daniel; rather, the time is right and the unveiling is assisted through historical records and current events. Such is the case – and the key – to understanding the measure of time given in response to the question in Daniel 8: 13-14:

> 13 *...How long shall be the vision concerning the daily sacrifice, and the transgression of desolation, to give both the sanctuary and the host to be trodden under foot?*
> 14 *And he said unto me, <u>Unto two thousand and three hundred days;</u> then shall the sanctuary be cleansed.*

The question asked is with reference to the Zerubbabel temple defiled and desecrated by Antiochus IV during his reign as king of Syria. The answer given is that there would be 2300 days from the defiling of the temple to its cleansing.

Some commentators state that there can't be an accurate fixing of exactly 2300 days since the word 'days' does not appear in Hebrew. Instead, it uses the words, 'evening / morning.' They say consideration needs to be given as to whether the prophecy refers to 2300 literal days, or to 2300 daily sacrifices – one in the evening and one in the morning, as was the practice of Moses. They say that the latter gives us two sacrifices per day; thus, 2300

sacrifices would be accomplished in 1150 days.[26]

Notes

This, I believe, should be a limited discussion. We will run into a lot of problems, especially in the Old Testament, if it is necessary to consistently replace the word 'day' with the two singular words 'morning / evening'. Should we then rewrite the whole of the scriptures, starting with Genesis, Chapter 1: 4-5?

> 4 *And God saw the light, that it was good: and God divided the light from the darkness.*
>
> 5 *And God called the light Day, and the darkness he called Night. And the evening and the morning were the first day.*

No matter how it is named, the reference in Daniel is 2300 periods of time, with twenty-four hours in each period, which includes 2300 intervals of light and darkness. The number of sacrifices during that period of time has nothing to do with the number of days or, if preferred, the number of 'morning / evenings'.

Reading the sacrificial law from Exodus 29: 38-39 will probably dispel the discussion.

> 38 *Now this is that which thou shalt offer upon the altar; two lambs of the first year day by day continually.*
>
> 39 *The one lamb thou shalt offer in the morning; and the other lamb thou shalt offer at even:*

Let's set aside unproductive discussion and use these precious hours winning souls to Christ by offering them the simple truths of God. The number of days from the time the temple is defiled and the sacrifices cease, to the time that the temple is cleansed and sacrifices are again offered, is 2300 days. I Maccabees 1: 10 to I Maccabees 4: 53 confirms that this is the case. This historical record provides dates; through the use of these dates, we determine the number of days to be 2300.

One may read Daniel 11: 21-45 alongside I Maccabees, Chapter 1 to Chapter 4, verse 60, to place dates on each movement of Antiochus IV and to list each offense against the Israelites and the Holy Temple. Though it seems, sometimes, that the scriptures are speaking of the Tribulation Antichrist, comparing scripture with the immaculate record of I Maccabees proves this is not the case. It is true, however, that Antiochus IV Epiphanies is an archetype of the Tribulation Antichrist.

The following are facts and dates gleaned through scriptures and historical

---

[26]Information obtained from the internet: Moellerhaus Homepage at http://www.ao.net/-fmoeller/2300.htm

Daniel's Prophesy Dated by the Maccabees

records. They confirm, once again, that true prophecy is always supported by fact.

Notes

- Antiochus IV Epiphanies ('Epiphanies' meaning 'the manifest god'), out of the House of Seleucus, gained succession to the throne, as King of Syria, in the year 175 B.C.
  (I Maccabees 1: 10 / Daniel 11: 21)
- He launched concentrated aggression against the Israelites in 171 B.C.; in 170 B.C. (dates gained from facts found in I Maccabees 4: 36 and confirmed through 1999 Internet search), he announced a law which required all citizens present themselves four times a year to pay formal homage to him, declaring himself a senior god of the Seleucid. He chose the Shabbat for these periodic submissions, knowing that the Israelites would be doubly breaking God's law by traveling on a Sabbath, as well as further offending God by giving honour to a false god. By deceitful promises and giving gifts, he won over some of the 'free thinking' section of the priesthood and gathered a renegade following of Jews (Daniel 11: 23-24). His first defiling of the people and the temple is not dated in I Maccabees 1: 11-15, but is certainly evident. Verse 15 states that those who followed him removed their marks of circumcision and repudiated the Holy covenant. Other historical records substantiate the dates; these dates are later gleaned through information given in I Maccabees 4: 36-60.
- On his return from the conquest of Egypt, in the year 169 B.C., Antiochus IV again marched with a strong force against Israel and Jerusalem. He entered the temple and carried off the golden altar, the lamp stand, the table for the Bread of the Presence, the sacred cups and bowls, the golden censers, the curtain and the crowns. He seized the silver, gold and precious vessels and whatever sacred treasures he found, then took them all with him when he left for his own country. He had caused much bloodshed and he gloated over what he had done.
  (quote from I Maccabees 1: 20-24 / Daniel 11: 28)
- Both sources record a full take-over of Jerusalem and total pollution of the sanctuary by the army of Antiochus IV. This happened in 167 B.C.
  (I Maccabees 1: 29-53 / Daniel 11: 30-35)
- On the fifteenth day of the month of Kisler (December, the feast of lights), also in 167 B.C., Antiochus IV joined his soldiers in Jerusalem, in the glorious land of Israel, to remain for two years, during which time he terrorized and persecuted the Israelites. Upon his arrival, he dedicated the temple to the worship of Zeus and a pig was sacrificed on the altar.
  (I Maccabees 1: 29-67 / Daniel 11: 36-43)

Daniel's Prophesy Dated by the Maccabees / Suggested Reading

Notes

- News from the east and the north alarmed Antiochus IV (news of the Maccabees' revolt) and he gathered his soldiers and left Jerusalem in 165 B.C. He left instructions with his viceroy of the territories between the Euphrates and the Egyptian fronts, to take half his armed forces to destroy the strength of Israel. (I Maccabees 3: 25-37 / Daniel 11: 44)
- Judas Maccabees and his brothers, having crushed the Syrians, entered Jerusalem to restore, cleanse and rededicate the temple. Early in the morning, on the twenty-fifth of the ninth month, the month of Kislev (again, December, the festival of lights) in the year 164 B.C., sacrifices were offered, as the law commanded, on the newly-made altar of burnt offering (I Maccabees 4: 36-60). The dedication of the temple is commemorated by the Jews at an annual festival known as Hanukkah (dedication).

The remarkable fact is that Daniel was given his first vision in 553 B.C. and his ministry ended in 536 B.C. The events just outlined occurred between the years 171 B.C. to 164 B.C. Daniel 8: 13-14 gives us, you'll remember, the information through prophesy that there would be 2300 days from the time of the defiling of the temple to the time of its cleansing.

2300 days divided by 360 days (lunar calendar) equals 6.39 years. Therefore, we add 6.39 years to the date when the priest offered the first sacrifice after the cleansing: 164 B.C. plus 6.39 years equals 170.39 B.C. This is the time-frame in which Antiochus IV first defiled the people and the temple in Jerusalem (171 - 170 B.C.)! Again, we find that history and prophesy agree. We have a living God; praise His wondrous name!

I would suggest, for your reading and research pleasure, that you either purchase a copy of *THE NEW ENGLISH BIBLE APOCRYPHA*[27] or obtain a copy through your library. Read for yourself the record of these events. They are quite detailed in I Maccabees and extremely interesting. You will be delighted, I'm sure, as I have been.

---

[27] The New English Bible Apocrypha (Oxford: Cambridge Press, 1970).

CHAPTER TWELVE

Chapter Twelve, according to the Study, effectively demonstrates the apocalyptic form of writing which was popular at the time of Christ. The Holy Spirit takes John across biblical history, both forward and back in time – a phenomenon not uncommon in bible prophecy. This chapter of *the Revelation* touches on events anywhere from the Garden of Eden to the end of the Millennium, with references not necessarily in chronological order. It speaks of the virgin birth and Lucifer's efforts to destroy our Christ. At the same time, it refers to the nation Israel awaiting her Messiah, and her birth as a nation. It talks of Satan, the Antichrist and the ten nations of the coming new order. It demonstrates God's protective hand upon His people.

Revelation 12

[1] And there appeared a great wonder in heaven; a woman clothed with the sun, and the moon under her feet, and upon her head a crown of twelve stars:
[2] And she being with child cried, travailing in birth, and pained to be delivered.
[3] And there appeared another wonder in heaven; and behold a great red dragon, having seven heads and ten horns, and seven crowns upon his heads.
[4] And his tail drew the third part of the stars of heaven, and did cast them to the earth: and the dragon stood before the woman which was ready to be delivered, for to devour her child as soon as it was born.
[5] And she brought forth a man child, who was to rule all nations with a rod of iron: and her child was caught up unto God, and to his throne.
[6] And the woman fled into the wilderness, where she hath a place prepared of God, that they should feed her there a thousand two hundred and threescore days.
[7] And there was war in heaven: Michael and his angels fought against the dragon; and the dragon fought and his angels,
[8] And prevailed not; neither was their place found any more in heaven.
[9] And the great dragon was cast out, that old serpent, called the Devil, and Satan, which deceiveth the whole world: he was cast out into the earth, and his angels were cast out with him.
[10] And I heard a loud voice saying in heaven, Now is come salvation, and strength, and the kingdom of our God, and the power of his Christ: for the accuser of our brethren is cast down, which accused them before our God day and night.
[11] And they overcame him by the blood of the Lamb, and by the word of their testimony; and they loved not their lives unto the death.
[12] Therefore rejoice, ye heavens, and ye that dwell in them. Woe to the inhabiters of the earth and of the sea! for the devil is come down unto you, having great wrath, because he knoweth that he hath but a short time.
[13] And when the dragon saw that he was cast unto the earth, he persecuted the woman which brought forth the man child.
[14] And to the woman were given two wings of a great eagle, that she might fly into the wilderness, into her place, where she is nourished for a time, and times, and half a time, from the face of the serpent.
[15] And the serpent cast out of his mouth water as a flood after the woman, that he might cause her to be carried away of the flood.
[16] And the earth helped the woman, and the earth opened her mouth, and swallowed up the flood which the dragon cast out of his mouth.
[17] And the dragon was wroth with the woman, and went to make war with the remnant of her seed, which keep the commandments of God, and have the testimony of Jesus Christ.

The Travail of the Woman

Revelation, Chapter 12

Notes

This chapter of the Revelation effectively demonstrates the apocalyptic form of writing which was popular at the time of Jesus. The Holy Spirit takes John across Biblical history, both forward and back in time – a phenomenon not uncommon in Biblical prophecy. The chapter touches on events anywhere from the Garden of Eden to the end of the millennium, with references not necessarily in chronological order.

Revelation 12: 1

1 *And there appeared a great wonder in heaven; a woman clothed with the sun, and the moon under her feet, and upon her head a crown of twelve stars:*

The woman clothed with the sun and the moon under her feet is a prophetic picture of Israel awaiting her Messiah, as well as her birth as a nation. At the same time, the image alludes to the devout Jewess, Mary, awaiting the birth of our Lord, Jesus Christ.

Revelation 12: 2

2 *And she being with child cried, travailing in birth, and pained to be delivered.*

Israel, as it were, being 'with child', struggled in birth and was delivered May 14th, 1948; similarly, the blessed virgin gave birth some two thousand years ago.

Israel has struggled from the time of Jacob and will struggle until she sees her Messiah '*coming in a cloud with power and great glory*' (Matthew 24: 30). When their Messiah, Jesus Christ, returns,"*the whole house of David and the inhabitants of Jerusalem will receive the spirit of grace and supplication; they will look upon him whom they have pierced and they shall mourn.*" (Zechariah 12: 10). Ezekiel 36: 26-27 reveals their time for mourning is over as God keeps His promise to Israel:

> 26 *A new heart also will I give you, and a new spirit will I put within you: and I will take away the stony heart out of your flesh, and I will give you an heart of flesh.*
>
> 27 *And I will put my spirit within you, and cause you to walk in my statutes, and ye shall keep my judgments, and do them.*

The 'child' is born; the travail is over – the nation, Israel, is delivered!

Satan, the Great Red Dragon / The Antichrist, Satan Incarnate

Notes

Revelation 12: 3

3 *And there appeared another wonder in heaven; and behold a great red dragon, having seven heads and ten horns, and seven crowns upon his heads.*

The red dragon, according to verse 9, is Satan who takes on bodily form in the persons of the Antichrist and False prophet. The seven heads are the seven mountains of Rome, the city in which they establish their first seat (Revelation 17: 9). The ten horns with the seven crowns are the ten nations submissive to the Antichrist, three of which have fallen through the intrigue of the False prophet, the second little horn of Daniel 7: 8; thus, leaving seven crowned nations.

Theologian William Barclay presents an interesting and informative narration on verses three and four under the heading, "Hatred of the Dragon":

> "The eastern people regard creation in the light of the struggle between the dragon of chaos and the creating God of order... The dragon, the arch-enemy of God, is a common and terrible figure in the thoughts of the east. It is the connection of the dragon and the sea which explains the river of water which the dragon emits to overcome the woman (verse 15).The dragon has seven heads and ten horns. This signifies its mighty power. It has seven royal diadems. This signifies its power over the kingdoms of the world..."

I offer these quotes from Barclay's *The Daily Study Bible*[28] for insight on eastern culture only.

Revelation 12: 4

4 *And his tail drew the third part of the stars of heaven, and did cast them to the earth: and the dragon stood before the woman which was ready to be delivered, for to devour her child as soon as it was born.*

The '*little horn*' of Daniel 8 is Satan, the dragon – the Antichrist of the Old Testament (Antiochus IV Epiphanies) and the Antichrist of the New. They come up out of the nations and *are exceeding great* (Daniel 8: 9; Revelation 13: 2). Both are granted the '*seat of Satan'* as they rule supreme upon the earth. They defile and destroy, even to the very gates of heaven. Read Daniel 8: 9-12:

9 *And out of one of them came forth a little horn, which waxed*

---

[28]William Barclay, "The Revelation of John", Vol. 2, The Daily Study Bible (Burlington, Ontario: Welch Publishing Company, Revised 1976), p. 77

exceeding great, toward the south, and toward the east, and toward the pleasant land.

10 *And it waxed great, even to the host of heaven; and it cast down some of the host and of the stars to the ground, and stamped upon them.*

11 *Yea, he magnified himself even to the prince of the host, and by him the daily sacrifice was taken away, and the place of his sanctuary was cast down.*

12 *And an host was given him against the daily sacrifice by reason of transgression, and it cast down the truth to the ground; and it practised, and prospered.*

Notes

We aren't told at what point Lucifer was cast out of heaven (Isaiah 14: 12-15; Ezekiel 28: 12-17), but we know he was on earth tempting man in the disguise of a serpent in Genesis 3: 1-5. We are well-aware that he is still here today. He was capable of presenting himself before the Lord at the testing of Job (Job 1: 6-12); however, since the victorious resurrection of Jesus Christ, he is restricted to '*going to and fro*' and '*walking up and down on the earth*'.

Ever since the nation Israel was granted autonomy, Satan has sought to devour her. She is even now being further threatened by the proposed 'grab back' of her rightful lands through the Wye River Accord.

Not only has Satan tried to devour the young nation Israel, but also he has made every effort to destroy the baby Jesus (Matthew 2: 1-16). Satan is ever-ready to destroy all 'newborns' as soon as they are birthed.

Revelation 12: 5

5 *And she brought forth a man child, who was to rule all nations with a rod of iron: and her child was caught up unto God, and to his throne.*

This verse covers a period from the birth of Jesus to the end of the Millennium. God has brought forth the *man child*, Jesus, through the virgin birth (Luke 2: 7) and *her child*, the Raptured Saints, through the new birth (Acts 2: 1-4). They (Jesus and the Saints) will '*rule all nations with a rod of iron*' (Revelation 19: 15 and 2: 26-27) and are caught up to the throne of God at the Resurrection (Jesus - Acts 1: 9) and the Rapture (the Saints - I Thessalonians 4: 16-17). God has – and will – fulfilled all of these prophesies through the person of Jesus Christ.

The Wilderness Flight / The Battle in Heaven

Revelation 12: 6

6 *And the woman fled into the wilderness, where she hath a place prepared of God, that they should feed her there a thousand two hundred and threescore days.*

Notes

When the Antichrist is no longer under the restraint of the two witnesses (the first half of the Tribulation), he takes possession of the Jerusalem temple (Daniel 9: 27). Now unhampered by the witnesses, he enjoys 1,260 days of unobstructed reign. His persecution of God's people ('*the woman*') causes them to flee into the wilderness. A place of safety has been prepared where they are nourished for the last half of the Tribulation. Where that place of safety is, only God knows. Because God is ever faithful, we may be assured there is a place prepared to receive them. God has declared it; He will provide it.

Revelation 12: 7 - 8

7 *And there was war in heaven: Michael and his angels fought against the dragon; and the dragon fought and his angels,*

8 *And prevailed not; neither was their place found any more in heaven*

Whenever and wherever there is a battle involving God's elect, Michael, the militant angel of God, will fight on their behalf. The Israeli Six Day War of 1967 seems to be valid proof of this. The very fact that Israel is still a strong, rich nation, standing proud and firm, shows that God is her keeper. As prophesied some 500 years before the birth of Christ, Israel is indeed, at this very point and time in history, '*a cup of trembling to the people round about.*' I think we can conclude that God intercedes on Israel's behalf. The nations of the earth should read Zechariah 12: 2-3 and take warning:

> 2 *Behold, I will make Jerusalem a cup of trembling unto all the people round about, when they shall be in the siege both against Judah and against Jerusalem.*
>
> 3 *And in that day will I make Jerusalem a burdensome stone for all people: all that burden themselves with it shall be cut in pieces, though all the people of the earth be gathered together against it.*

Could it be possible that when the thousand years Millennium is expired and Satan is loosed out of his prison (Revelation 20: 7-8) that he, in desperation, challenges heaven itself? Verse 7, now under study, says the war is in heaven. Since the victorious resurrection of Jesus, Satan has never been allowed audience before God in heaven to accuse His Saints. Thus, Satan '*prevailed not; neither was their place found any more in heaven*'.

Notes

Revelation 12: 9

9 *And the great dragon was cast out, that old serpent, called the Devil, and Satan, which deceiveth the whole world: he was cast out into the earth, and his angels were cast out with him.*

This may not be the original casting out of Satan from heaven (Isaiah 14: 12-13; Ezekiel 28: 13-17) but an affirmation of his final exclusion – an explanation for his furious hostility against God's people in the last days (see Revelation 12: 13-17).

Although the specifics of God's plan aren't given in detail in verses seven through ten, it is very clear as to Satan's final destination. We read of his fate in Revelation 20: 10:

10 *And the devil that deceived them was cast into the lake of fire and brimstone, where the beast and the false prophet are, and shall be tormented day and night for ever and ever.*

Revelation 12: 10

10 *And I heard a loud voice saying in heaven, Now is come salvation, and strength, and the kingdom of our God, and the power of his Christ: for the accuser of our brethren is cast down, which accused them before our God day and night.*

Since the resurrection of Jesus, the Kingdom of God has been a spiritual kingdom cherished within the heart of man. Satan has buffeted and tempted since creation and has done his utmost to hinder man's entrance into the peace and joy of God's kingdom.

When Satan, *'the accuser of our brethren* (the Israelites) *is for all eternity cast down'* and the literal Kingdom of God has come in all holiness and in the power and strength of Jesus Christ, the house of Israel (all those found written in the Book of Life) will join the Saints and rejoice in salvation. They never again will be accused before God since, once converted, their Messiah, Jesus Christ, will be their mediator.

Though Satan hasn't been allowed to accuse the believers after Jesus became their mediator; he could seduce and tempt them, but in no way could he appear before God and accuse them. When Satan is temporarily released from the pit, he may arrogantly assume he is powerful enough to viciously accuse, in the presence of God, those Israelites who had originally rejected their Messiah. He will quickly learn that Jesus is the victorious one. Just as he was cast out of heaven into the earth, Satan will be cast out of the earth into the kingdom of hell, prepared to receive him. Read and ponder the fate

of Israel as recorded in Daniel 12: 1-3:

1 *And at that time shall Michael stand up, the great prince which standeth for the children of thy people: and there shall be a time of trouble, such as never was since there was a nation even to that same time: and at that time thy people shall be delivered, <u>every one that shall be found written in the book</u>.*

2 *And many of them that sleep in the dust of the earth shall awake, some to everlasting life, and some to shame and everlasting contempt.*

3 *And they that be wise shall shine as the brightness of the firmament; and they that turn many to righteousness as the stars for ever and ever.*

Notes

Revelation 12: 11

11 And they overcame him by the blood of the Lamb, and by the word of their testimony; and they loved not their lives unto the death.

The Tribulation Saints, both Israelite and Gentile, overcome Satan and eternal death '*by the blood of the Lamb and by the word of their testimony*', the testimony being their verbal acceptance of their Saviour, Jesus Christ, and their rejection of the Beast and his mark. '*They loved not their lives unto the death*'; in other words, they willingly die for their faith. Refer to Revelation 6: 9-11; Revelation 7: 13-15; and Revelation 15: 2-3.

Revelation 12: 12

12 Therefore rejoice, ye heavens, and ye that dwell in them. Woe to the inhabitants of the earth and of the sea! for the devil is come down unto you, having great wrath, because he knoweth that he hath but a short time.

The heavens and those who dwell in them are told to rejoice that the Tribulation Saints have passed through their hour of temptation, having washed their robes and made them white in the blood of the Lamb (Revelation 7: 14). The heavens cry woe to the earth because Satan, unable to conquer the heavens, vents even greater anger upon the earth.

Revelation 12: 13

13 And when the dragon saw that he was cast unto the earth, he persecuted the woman which brought forth the man child.

Satan, having been banished from heaven, demonstrates ferocious hostility toward God's people. Ever since his first eviction at creation, Satan has tried to beguile and destroy God's children. He will intensify his persecution

Notes

during the Tribulation, working through the Antichrist and the False prophet. The focus of his greatest anger has always been toward Israel. He grows very bold and expedites his efforts when freed for a season (Revelation 20: 7-10).

Revelation 12: 14

14 And to the woman were given two wings of a great eagle, that she might fly into the wilderness, into her place, where she is nourished for a time, and times, and half a time, from the face of the serpent.

The reference to the woman given two wings to fly into the wilderness may allude to something like the exodus of the Israelites from Ethiopia in 1991; or it may be symbolic of the protective arms of God. Refer to Appendix 6, "The Ethiopian Exodus". Although we don't know the means by which God will protect one-third of the Tribulation Saints, we do know they will be nourished for the last three and one-half years of the Tribulation.

In Deuteronomy 32: 10-11, Moses sings to the Israelites, telling them they are *'the apple of God's eye'* and refers to God as an eagle:

10 He found him in a desert land, and in the waste howling wilderness; he led him about, he instructed him, he kept him as the apple of his eye.

11 As an eagle stirreth up her nest, fluttereth over her young, spreadeth abroad her wings, taketh them, beareth them on her wings:

In Exodus 19: 3-4, God tells Moses to remind the house of Jacob of His great care for them:

3 ... Thus shalt thou say to the house of Jacob, and tell the children of Israel;

4 Ye have seen what I did unto the Egyptians, and how I bare you on eagles' wings, and brought you unto myself.

We have no sure answers as to how God will protect the Tribulation Saints but we know that He will keep His promises. Zechariah 13: 9 will come to pass.

9 And I will bring the third part through the fire, and will refine them as silver is refined, and will try them as gold is tried: they shall call on my name, and I will hear them: I will say, It is my people: and they shall say, The Lord is my God.

What great love God has for His elect – these are the people who rejected Him and demanded the crucifixion of His Son (Matthew 27: 20-25). Praise God for His wondrous mercy!

The Wrath of the Dragon / The Flood of Evil / The Earth Swallows the Flood / Satan's Ferocious Hostility

Notes

Revelation 12: 15

15 And the serpent cast out of his mouth water as a flood after the woman, that he might cause her to be carried away of the flood.

Theologian William Barclay may be correct in saying that the dragon (which abides in the sea) connects the dragon and the sea; this, Barclay says, explains the river of water which the dragon casts out of his mouth.

I find this an interesting thought. Since the eastern world views the dragon as a terrible figure, an arch-enemy of God, the Holy Spirit may well have influenced John's pen to record this mythology; in this way, through this image, we can picture the Saints being swallowed up and almost drowned in the flood of the wrath of the Beast.

Revelation 12: 16

16 And the earth helped the woman, and the earth opened her mouth, and swallowed up the flood which the dragon cast out of his mouth.

In the time of Moses, the earth swallowed up the house of Korah, an opposer of righteousness, as described in Numbers 16: 30-33. In Exodus 14: 26-28, the Egyptians were swallowed up by the sea. If, in fact, the dragon casts a flood out of his mouth, God's earth is perfectly capable of swallowing up the flood since she is ever ready to respond to His commands.

It is my personal opinion that the verse under study simply means that the Saints flee to a place of safety and the earth hides them for a period of three and one-half years. The earth will, indeed, swallow up the flood that the dragon spews out of his mouth – the flood of persecution and evil – when our Saviour casts Satan, the Beast and the False prophet alive into the lake of fire burning with brimstone (Revelation 19: 20; Revelation 20: 10).

Revelation 12: 17

17 And the dragon was wroth with the woman, and went to make war with the remnant of her seed, which keep the commandments of God, and have the testimony of Jesus Christ.

The Beast (the Antichrist), which is the embodiment of Satan, persecutes and attempts to kill all those who keep the commandments of God and confess Jesus as their Lord and Saviour. Being filled with hatred inspired by Satan, he seeks out the Israelite as his main focus to vent his vengeance against God. The Beast falsely believes that in demolishing the Christians, he will destroy the power of God. Satan can never win for, even in death, the believer is victorious.

CHAPTER THIRTEEN

Chapter Thirteen, verses one to ten, relates the rise of the Antichrist, the last great political ruler of the world. The Study suggests that probably he will be a Gentile since he rises up out of the sea and has the attributes of the former Gentile nations of *Daniel* 7: 3. The Roman political and religious dictatorship, though wounded and appearing 'dead', will be revived under the leadership of the Antichrist and the False prophet. The second beast, which comes up out of the earth – probably a Jew – is the False prophet. As in the original Holy Roman Empire, the head of the World Church – the False prophet – will exercise even greater power than the Antichrist. He is the great deceiver, Satan incarnate, who gives life to the image of the Beast and orders the death of the Saints. He is the one who enforces the mark of the Beast and determines it to be the number of man – number 666.

Revelation 13

[**1**] And I stood upon the sand of the sea, and saw a beast rise up out of the sea, having seven heads and ten horns, and upon his horns ten crowns, and upon his heads the name of blasphemy.

[**2**] And the beast which I saw was like unto a leopard, and his feet were as the feet of a bear, and his mouth as the mouth of a lion: and the dragon gave him his power, and his seat, and great authority.

[**3**] And I saw one of his heads as it were wounded to death; and his deadly wound was healed: and all the world wondered after the beast.

[**4**] And they worshipped the dragon which gave power unto the beast: and they worshipped the beast, saying, Who is like unto the beast? who is able to make war with him?

[**5**] And there was given unto him a mouth speaking great things and blasphemies; and power was given unto him to continue forty and two months.

[**6**] And he opened his mouth in blasphemy against God, to blaspheme his name, and his tabernacle, and them that dwell in heaven.

[**7**] And it was given unto him to make war with the saints, and to overcome them: and power was given him over all kindreds, and tongues, and nations.

[**8**] And all that dwell upon the earth shall worship him, whose names are not written in the book of life of the Lamb slain from the foundation of the world.

[**9**] If any man have an ear, let him hear.

[**10**] He that leadeth into captivity shall go into captivity: he that killeth with the sword must be killed with the sword. Here is the patience and the faith of the saints.

[**11**] And I beheld another beast coming up out of the earth; and he had two horns like a lamb, and he spake as a dragon.

[**12**] And he exerciseth all the power of the first beast before him, and causeth the earth and them which dwell therein to worship the first beast, whose deadly wound was healed.

[**13**] And he doeth great wonders, so that he maketh fire come down from heaven on the earth in the sight of men,

[**14**] And deceiveth them that dwell on the earth by the means of those miracles which he had power to do in the sight of the beast; saying to them that dwell on the earth, that they should make an image to the beast, which had the wound by a sword, and did live.

[**15**] And he had power to give life unto the image of the beast, that the image of the beast should both speak, and cause that as many as would not worship the image of the beast should be killed.

[**16**] And he causeth all, both small and great, rich and poor, free and bond, to receive a mark in their right hand, or in their foreheads:

[**17**] And that no man might buy or sell, save he that had the mark, or the name of the beast, or the number of his name.

[**18**] Here is wisdom. Let him that hath understanding count the number of the beast: for it is the number of a man; and his number is Six hundred threescore and six.

Revelation, Chapter 13

Preamble: a teaching in preparation for Chapter 13

Notes

Before starting the study of Chapter 13, I would caution against accepting a prevalent teaching about the coming of the Beast. This theory points to verse one which says that the Beast rises up out of the sea. From this reference, the teaching goes on to conclude that the sea is the Mediterranean and the Beast, therefore, must come from one of the countries surrounding the Mediterranean. Such teachings are based purely on supposition which invariably distorts truth.

Probably, the Beast rising up out of the sea tells us he will come out of the water or from across water, possibly from a continent separate from those countries once comprising the Roman Empire.

John stands on the sands of the sea which, in past scriptures, represents multitudes of people. In Genesis 32: 12, God's promise to Jacob is that He will make his seed as the sand of the sea. His seed, as prophesied, has truly become a nation and a multitude immersed in many nations in the midst of the earth. (Genesis 28: 12-15; Genesis 48: 16-19). Pure remnants of the twelve tribes have been located and identified but, most often, the dispersed have so interbred with Gentile nations that they have almost totally lost their national identity. If the Antichrist were appointed from among them, it is unlikely that anyone but God would truly know his origin.

Daniel 7: 3 tells us that four beasts come up from the sea. Since the four beasts are representative of the four Gentile powers (Babylon, Persia-Media, Greece and Rome), can we safely conclude that God is using the same metaphor to lead us to understand the Antichrist is a Gentile? If, indeed, God had wanted us to understand that the end-time ruler would be an Israelite or someone from Europe, Asia or any of the countries bordering the Mediterranean, would He not have had him rise up out of the earth, as He does with the False prophet in Revelation 13: 11?

Some predict that the Antichrist is a Jew; one prominent theologian defines him as a Roman-Grecian Jew;[29] others merely suggest he is a Gentile.[30] All is purely speculation for we don't have scriptures which would allow us to come to a definitive conclusion. We might do well to leave this unanswerable question in the hands of God.

---

[29]Tim Lahaye, Revelation Unveiled (Grand Rapids, MI: Zondervan Publishing, 1999), p. 209.

[30]Ray C. Stedman, God's Final Word (Grand Rapids, MI: Discovery House Publishers, 1991), p. 244.

The Rise of the Antichrist / Nebuchadnezzar's Dream

Revelation 13: 1

Notes

1 *And I stood upon the sand of the sea, and saw a beast rise up out of the sea, having seven heads and ten horns, and upon his horns ten crowns, and upon his heads the name of blasphemy.*

John sees a beast rise up out of the sea. This beast is Satan in the form of the Antichrist. The seven heads are representative of the seven hills of Rome or, possibly, Constantinople (Istanbul), both of which have seven hills.

Refer to Revelation 17: 9:

9 *And here is the mind which hath wisdom. The seven heads are seven mountains, on which the woman* (the world church) *sitteth.*

The ten horns which John sees are the ten kings which submit to the rule of the Antichrist, as stated in Revelation 17: 12:

12 *And the ten horns which thou sawest are ten kings, which have received no kingdom as yet; but receive power as kings one hour with the beast.*

This will be realized with the confederation of ten kingdoms, countries of the revised Roman Empire.

One gains insight into the book of Revelation by having some knowledge of the book of Daniel. I encourage those who aren't familiar with the scriptures to read Daniel 2: 31-45, as well as Daniel 7: 1-27 before going on in this study.

In short, Daniel 2: 31-45 tells us that Daniel interprets a dream for King Nebuchadnezzar of Babylon. The dream foretells future world affairs from the time of Babylon to the Millennium. In his dream, the king sees an image of a great man. The head of fine gold depicts the Babylonian Empire, the richest and most regal of all powers. The silver breast and arms of the image stand for the Persia-Media dual rule under King Cyrus of Persia and King Darius the Mede, a more powerful but less glorious kingdom. The belly and thighs of brass are, according to history, the Grecian Empire under Alexander the Great. At his death, this kingdom was divided into four and became Thrace, Macedonia, Egypt and Syria. Next comes the Roman rule represented by two legs of iron: the western division at Rome, and the eastern division with its capital in Constantinople. This is a dreadful and long-lasting rule which, through religious and political influences, continues to rule from the seven hills of Rome, to a greater extent than is realized, even yet today. The toes and feet, partially of clay and partially of iron, are the ten major nations which submit to the coming one-world ruler to form the revised Roman Empire, the most dreadful and terrible rule of all times. The dream ends with Jesus destroying the false trinity (Satan, the Antichrist and the False prophet).

He then establishes the literal kingdom of God on earth which, when replaced, will be the final dwelling place of the Holy Trinity – the Trinity that never ends.

Notes

Revelation 13: 2

2 *And the beast which I saw was like unto a leopard, and his feet were as the feet of a bear, and his mouth as the mouth of a lion: and the dragon gave him his power, and his seat, and great authority.*

To understand this verse, we should look at Daniel 7: 1-8. Daniel is given a prophetic dream and sees the Babylonian Empire as a lion with eagle wings (royal and regal); the Persia-Media Empire as a bear (strong and aggressive); and the Grecian Empire as a leopard (swift and sleek). Lastly, Daniel sees the fourth-most dreadful and terrible beast. This represents Rome which will probably become the seat of the revised Roman Empire (the European Economic Community / the United States of Europe / the European Union). John sees that Satan gives this last ruling power (the Antichrist) the combined evil power and authority of the former four/five powers which ruled on the earth during, and following, the life of the prophet Daniel.

Revelation 13: 3

3 *And I saw one of his heads as it were wounded to death; and his deadly wound was healed: and all the world wondered after the beast.*

John sees one of the heads, the Roman Empire, as if it has been wounded to the death. It appeared that the monster empire of Rome was killed in 476 A.D.; however, although it was a '*deadly wound*', it was healed to the extent that the influence of Rome has lived and existed ever since. God is telling us through John that a power in the likeness of the Roman Empire is to be revived and raised to life to rule again under the Beast and the ten kings of the international alliance. This then will be the time of its restoration and complete healing. The whole world will stand in awe of the Beast who is '*wounded*' and then '*healed.*'

Revelation 13: 4

4 *And they worshipped the dragon which gave power unto the beast: and they worshipped the beast, saying, Who is like unto the beast? Who is able to make war with him?*

The world worships Satan who has given power to his representative, the Antichrist (Satan incarnate). They worship him because he is an accomplished orator and has shown himself as a beguiling peace-maker and

Notes

excellent administrator. He is a man of worldly wisdom and knowledge who, above all others, will bring temporary peace and prosperity to the world. One of his major assets will be that he has gained the respect and trust of Israel. It matters not that he is immoral and lacks integrity, a leader who follows in the ways of Satan. The nations idolize him and believe that only he can bring the world together as one great power. They willingly applaud him and give him full authority, worshipping him as if he were God. Though we believe this person will be a man, we must not rule out the possibility of the end-time ruler being a woman. Move ahead to Revelation 18: 7 for further discussion on this subject.

Revelation 13: 5 - 7

5 *And there was given unto him a mouth speaking great things and blasphemies; and power was given unto him to continue forty and two months.*

6 *And he opened his mouth in blasphemy against God, to blaspheme his name, and his tabernacle, and them that dwell in heaven.*

7 *And it was given unto him to make war with the saints, and to overcome them: and power was given him over all kindreds, and tongues, and nations.*

The Antichrist described here is remarkably similar to a former archetype of the Antichrist, Antiochus IV, the king of Syria from 175 B.C. - 164 B.C. Compare the extraordinary similarity as you read Daniel 11: 36:

36 *And the king shall do according to his will; and he shall exalt himself, and magnify himself above every god, and shall speak marvellous things against the God of gods, and shall prosper till the indignation be accomplished: for that that is determined shall be done.*

The Antichrist, though reigning throughout the seven years Tribulation (Daniel 9: 27), is restrained by God for the first three and one-half years through power given to God's two witnesses (Revelation 11: 5-6). Angered by the restraint, he blasphemes God and persecutes the Saints. When the two witnesses have fulfilled their commission (Revelation 11: 7) and served their appointed one thousand two hundred and three score days (Revelation 11: 3), God lifts His protective hand. The Antichrist, now having been given unrestrained power for the next forty-two months, attacks and kills the two witnesses. With the consent of God, he is allowed full authority over '*all kindred and tongues and nations*' until Daniel's seventieth week is complete.

It might seem that God is a vicious and uncaring Father in that He allows the Antichrist to overpower and persecute the Saints of the Tribulation era. Not so! He is a forgiving and loving God with patience far above that of

Notes

man. He willingly became man and allowed Himself to be rejected and crucified because His will is that none of His children should perish. Hear the words of Jesus from Matthew 18: 14:

> 14 *Even so it is not the will of your Father which is in heaven, that one of these little ones should perish.*

In His wisdom, He knows the torment of Satan will drive many to repentance, thereby saving them from eternal death. Jesus has redeemed them (all of them "*out of every kindred, and tongue, and people, and nation*") to God by his blood (Revelation 5: 9). If they are willing, he will "*lead them unto living fountains of waters: and God shall wipe away all tears from their eyes*" (Revelation 7: 17).

Revelation 13: 8

8 *And all that dwell upon the earth shall worship him, whose names are not written in the book of life of the Lamb slain from the foundation of the world.*

Only the very evil are not listed in the '*book of life of the Lamb slain from the foundation of the world*' (Revelation 17: 8). These unfortunate people will worship the Beast and seal for themselves eternal punishment. (For more information on the Book of Life, see the study notes in this commentary for Revelation 3: 5 and Revelation 20: 14-15.)

Revelation 13: 9

9 *If any man have an ear, let him hear.*

Open your ears and hear what God is saying to the end-time age, the age of 'people's rights.' Recall the letter to the Laodiceans, recorded in Revelation 3: 15-22 (NIV):

> 15 *I know your deed; that you are neither cold nor hot. I wish you were either one or the other!*
> 16 *So, because you are lukewarm – neither hot nor cold – I am about to spit you out of my mouth.*
> 17 *You say, 'I am rich; I have acquired wealth and do not need a thing.' But you do not realize that you are wretched, pitiful, poor, blind and naked.*
> 18 *I council you to buy from me gold refined in the fire, so you can become rich; and white clothes to wear, so you can cover your shameful nakedness; and salve to put on your eyes, so you can see.*
> 19 *Those whom I love I rebuke and discipline. So be earnest, and repent.*
> 20 *Here I am! I stand at the door and knock. If anyone hears my voice and opens the door, I will come in and eat with him, and he with me.*

Promise to the Overcomer / Righteous Judgment

Notes

21 To him who overcomes, I will give the right to sit with me on my throne, just as I overcame and sat down with my Father on his throne.
22 He who has an ear, let him hear what the Spirit says to the churches.

Remember God's promise: '*... if we would judge ourselves, we should not be judged'* (condemned with the world). (I Corinthians 11: 31-32).

Revelation 13: 10

10 He that leadeth into captivity shall go into captivity: he that killeth with the sword must be killed with the sword. Here is the patience and the faith of the saints.

'*He that leadeth into captivity shall go into captivity*' tells us that the false triune god who leads man into captivity shall himself go into the eternal captivity of death and hell.

'*He that killeth with the sword must be killed with the sword'* gives assurance that he who persecutes the Saints will, in the end, be killed with the sword of the Word through the judgment of Christ (Revelation 19: 15).

Revelation 14: 12-13 tells us '*the patience and faith of the saints'* is that, even in the face of death, they keep the commandments of God and retain their faith in Jesus. To paraphrase verse 13: "*blessed are those who die loving the Lord for they will rest from their labour; their good works will follow them into paradise.*"

Jesus says "*...with what measure ye mete, it shall be measured to you again"* (Matthew 7: 2). Galatians 6: 7 warns, "*Be not deceived; God is not mocked: for whatsoever a man soweth, that shall he also reap.*

Romans 2: 5-8 (NIV) reveals, in part, the message contained in verse 10:

5 But because of your stubbornness and your unrepentant heart, you are storing up wrath against yourself for the day of God's wrath when his righteous judgment will be revealed.
6 God will give each person according to what he has done.
7 To those who by persistence in doing good seek glory, honour and immortality, he will give eternal life.
8 But for those who are self-seeking and who reject the truth and follow evil, there will be wrath and anger.

Equal Authority / The Deadly Wound Healed

Notes

Revelation 13: 11

11 And I beheld another beast coming up out of the earth; and he had two horns like a lamb, and he spake as a dragon.

The beast John now sees is the False prophet who appears as a lamb. Both the Antichrist and the False prophet imitate Jesus: one rides in on a white horse (Revelation 6: 2); the other comes as a lamb. The Antichrist rises up out of the sea (Revelation 13: 1), while the False prophet comes up out of the earth. This tells us that the second Beast arises from within the boundaries of the once powerful Roman Empire. The earth is symbolic of the land of Caanan, indicating possibly that the False prophet is an Israelite. The two horns, aside from implying he comes as a lamb, show him to be equal with the Antichrist. This Beast has the appearance of Christ (the Lamb) but speaks with the mouth of Satan. He will gain power and be recognized as the dynamic leader of the World Church.

Revelation 13: 12

12 And he exerciseth all the power of the first beast before him, and causeth the earth and them which dwell therein to worship the first beast, whose deadly wound was healed.

The False prophet is the second evil beast to appear and, by consent of Satan and the Antichrist, exercises equal power. These two evil ones together deceive the world. The first beast is a civil authority, the second a religious authority. The False prophet commands that the people of the earth worship the false Christ. We see here that the wounded Beast (as it were, '*wounded unto death*' from Revelation 13: 3) is definitely healed. Verse 14 further informs us that the wound was not fatal; for the Beast wounded by the sword did live. It should be understood this means that a power similar to the Roman Empire which once appeared dead, now lives and is revived under one world rule. Those who teach that the Antichrist dies and is resurrected may be wise to search the scriptures for confirmation of this belief. I find none. If, in fact, the Antichrist of the Tribulation dies and is reincarnated, Daniel's seventieth week (Daniel 9: 27) would have recorded it. If not Daniel, then surely God would have spoken through one of the other prophets.

I recall a saying from my youth, although I don't know the source: "I was sick unto death and was wont to die; but death would not receive me." It was common, when I was a child, some seventy-plus years ago, for people to speak of being 'sick unto death', yet none of them died. I know, because they visited over tea and talked incessantly about it. If one can be sick unto death and not die, then surely one can be '*wounded unto death*' and live.

The Acts of the Deceiver / The Image of the Beast

Notes

Revelation 13: 13

13 And he doeth great wonders, so that he maketh fire come down from heaven on the earth in the sight of men,

Satan endows the False prophet with miraculous powers and, as always, he copies the acts of God. He imitates Elijah in I Kings 18: 37-38 who calls fire down from heaven, proving God's existence in the presence of His people and the prophets of Baal:

> *37 Hear me, O Lord, hear me, that this people may know that thou art the Lord God, and that thou hast turned their heart back again.*
>
> *38 Then the fire of the Lord fell, and consumed the burnt sacrifice, and the wood, and the stones, and the dust, and licked up the water that was in the trench.*

Revelation 13: 14 - 15

14 And deceiveth them that dwell on the earth by the means of those miracles which he had power to do in the sight of the beast; saying to them that dwell on the earth, that they should make an image to the beast, <u>which had the wound by a sword, and did live.</u>

15 And he had power to give life unto the image of the beast, that the image of the beast should both speak, and cause that as many as would not worship the image of the beast should be killed.

Jesus, speaking of the last days, forewarns of false prophets and false Christ's in Matthew 24: 24:

> *24 For there shall arise false Christ's, and false prophets, and shall shew great signs and wonders; inasmuch that, if it were possible, they shall deceive the very elect.*

In commanding that an image of the Antichrist be sculpted and worshipped, the False prophet boldly disregards God's second commandment, as found in Exodus 20: 4-5:

> *4 Thou shalt not make unto thee any graven image, or any likeness of any thing that is in heaven above, or that is in the earth beneath, or that is in the water under the earth:*
>
> *5 Thou shalt not bow down thyself to them, nor serve them: for I the Lord thy God am a jealous God...*

Although we are not told that such is the case, we may assume, with a fair degree of accuracy, that both the image and the man are set up in the temple of God in Jerusalem (II Thessalonians 2: 4); thus, defiling the temple as did Antiochus Epiphanies (171-164 B.C.) who placed an image of Jupiter in the holy of holies.

From all appearances, the False prophet demonstrates more power and more control than that of the Antichrist. It is the False prophet who causes man to

receive the mark of the Beast (Revelation 13: 16-18); it is he who is the satanic master of miracles. There is little wonder that, eventually, he is deposed by the nations and the Antichrist (Daniel 7: 26; Revelation 17: 16).

Notes

Revelation 13: 16 - 18

16 And he causeth all, both small and great, rich and poor, free and bond, to receive a mark in their right hand, or in their foreheads:

17 And that no man might buy or sell, save he that had the mark, or the name of the beast, or the number of his name.

18 Here is wisdom. Let him that hath understanding count the number of the beast: for it is the number of a man; and his number is Six hundred threescore and six.

The time is fast approaching when all the world will be governed by a massive computer system, in a society without use of cash or cheques. Under a Satanic power, it will be necessary that all receive the number of the Beast, number 666, in order to buy and sell.

God created man on the sixth day of creation (Genesis 1: 26-31) and, therefore, we accept six as the number of man. The False prophet, on behalf of the Antichrist, will require that man wear the number of the Beast. The Beast is a man, and 666 is the triple number of man. The number may represent the false triune god. We should take this number literally since verse 18 clearly removes the possibility of spiritualizing it. Probably it will be a prefix for bank card numbers (a PIN number) and will be necessary in order to enter an account; thus, disallowing buying or selling to those who refuse to wear the number. Cards will no longer be necessary for the citizen's number will be imprinted in the right hand or forehead, visible only to machines. The number will be man's total identification, replacing all other numbers such as Social Insurance and Health Care numbers. With a world-wide computerized identification system, the authorities will be able to trace and control all citizens in every country.

Even now, inmates in the penal system, nursing homes, and animals wear alert bracelets or computer chips for quick location. The system has long been in place waiting to be enacted internationally. When the time comes, the world as a whole will not object since society has been programmed to accept this system. The required registry for Social Insurance numbers was the first major move toward world government control.

The dual dictators will remove all religious and political freedom. Wholesale religious persecution will be the order of the day and untold numbers will die for the faith, as recorded in Revelation 6: 9-11; Revelation 7: 14; Revelation 15: 2; and, finally, Revelation 20: 4.

CHAPTER FOURTEEN

Chapter Fourteen, according to the Study, is an overview of the latter part of the Tribulation years. It is similar in character to chapter six when Christ opens the seals to alert and reveal to the world the seven Tribulation years. These chapters demonstrate the compassion of our God who consistently and repeatedly warns before He punishes. In this particular passage, He sends an angel to fly in the midst of the heavens preaching the gospel of salvation to every nation, kindred, tongue and people. Our God is an articulate God. Lest man fail to preach the gospel to the ends of the earth, He speaks its redemptive message with a mighty voice from heaven, that all might hear and repent. In this chapter, seven windows open to us. Through John's seven visions, we see:

- the 144,000 sealed Israelites;
- the angel preaching the gospel of salvation;
- the second angel announcing the fall of Babylon;
- the third angel declaring doom to those who worship the Beast;
- the assurance of blessing to those who die in the Lord;
- the earth as it is harvested;
- God's mighty wrath unleashed.

Revelation 14

[**1**] And I looked, and, lo, a Lamb stood on the mount Sion, and with him an hundred forty and four thousand, having his Father's name written in their foreheads.
[**2**] And I heard a voice from heaven, as the voice of many waters, and as the voice of a great thunder: and I heard the voice of harpers harping with their harps:
[**3**] And they sung as it were a new song before the throne, and before the four beasts, and the elders: and no man could learn that song but the hundred and forty and four thousand, which were redeemed from the earth.
[**4**] These are they which were not defiled with women; for they are virgins. These are they which follow the Lamb whithersoever he goeth. These were redeemed from among men, being the firstfruits unto God and to the Lamb.
[**5**] And in their mouth was found no guile: for they are without fault before the throne of God.
[**6**] And I saw another angel fly in the midst of heaven, having the everlasting gospel to preach unto them that dwell on the earth, and to every nation, and kindred, and tongue, and people,
[**7**] Saying with a loud voice, Fear God, and give glory to him; for the hour of his judgment is come: and worship him that made heaven, and earth, and the sea, and the fountains of waters.
[**8**] And there followed another angel, saying, Babylon is fallen, is fallen, that great city, because she made all nations drink of the wine of the wrath of her fornication.
[**9**] And the third angel followed them, saying with a loud voice, If any man worship the beast and his image, and receive his mark in his forehead, or in his hand,
[**10**] The same shall drink of the wine of the wrath of God, which is poured out without mixture into the cup of his indignation; and he shall be tormented with fire and brimstone in the presence of the holy angels, and in the presence of the Lamb:
[**11**] And the smoke of their torment ascendeth up for ever and ever: and they have no rest day nor night, who worship the beast and his image, and whosoever receiveth the mark of his name.
[**12**] Here is the patience of the saints: here are they that keep the commandments of God, and the faith of Jesus.
[**13**] And I heard a voice from heaven saying unto me, Write, Blessed are the dead which die in the Lord from henceforth: Yea, saith the Spirit, that they may rest from their labours; and their works do follow them.
[**14**] And I looked, and behold a white cloud, and upon the cloud one sat like unto the Son of man, having on his head a golden crown, and in his hand a sharp sickle.
[**15**] And another angel came out of the temple, crying with a loud voice to him that sat on the cloud, Thrust in thy sickle, and reap: for the time is come for thee to reap; for the harvest of the earth is ripe.
[**16**] And he that sat on the cloud thrust in his sickle on the earth; and the earth was reaped.
[**17**] And another angel came out of the temple which is in heaven, he also having a sharp sickle.
[**18**] And another angel came out from the altar, which had power over fire; and cried with a loud cry to him that had the sharp sickle, saying, Thrust in thy sharp sickle, and gather the clusters of the vine of the earth; for her grapes are fully ripe.
[**19**] And the angel thrust in his sickle into the earth, and gathered the vine of the earth, and cast it into the great winepress of the wrath of God.
[**20**] And the winepress was trodden without the city, and blood came out of the winepress, even unto the horse bridles, by the space of a thousand and six hundred furlongs.

Preamble / Jesus and the 144,000 / Voice from Heaven

Revelation, Chapter 14

Notes

Just as Chapter 6 is an overview of the seven Tribulation years given through Christ's opening of the seven seals, so also is Chapter 14 an overview of the last half of the Tribulation years.

Seven windows open to us through John's seven visions. We see:

1. the 144,000 sealed Israelites in verses 1-5;
2. the angel preach the everlasting gospel to the earth in verses 6-7;
3. the second angel announce the fall of Babylon in verse 8;
4. the third angel declare doom to those who worship the Beast in verses 9-11;
5. the assurance of blessing to those who die in the Lord in verse 13;
6. the earth harvested in verses 14-16;
7. God's mighty wrath unleashed in verses 17-20.

Revelation 14: 1

1 And I looked, and, lo, a Lamb stood on the mount Sion, and with him an hundred forty and four thousand, having his Father's name written in their foreheads.

In the Old Testament, Mount Moriah alone is referred to as Mount Zion; however, later, the whole of Jerusalem is known as Mount Sion. It's quite acceptable, therefore, that John should call the Mount of Olives 'Mount Sion'. The prophesy is that Jesus will return to the Mount of Olives -- Zechariah 14: 4:

4 ...his feet shall stand in that day upon the mount of Olives, which is before Jerusalem on the east...

John sees, in this vision, prophecy fulfilled. Jesus stands with the 144,000 who have the Father's name written in their foreheads. These servants of God are the 144,000 Israelites sealed by Christ in Revelation, Chapter 7, verses 1-8; these are the same for whom a voice from heaven pronounces protection, saying: "*See thou hurt not the oil and the wine.*" (Revelation 6: 6). From this we gather that, in being sealed, they have escaped death to appear with Jesus on Mount Sion.

Revelation 14: 2

2 And I heard a voice from heaven, as the voice of many waters, and as the voice of a great thunder: and I heard the voice of harpers harping with their harps:

John hears heaven rejoice and harpists play as they accompany the new song, the song that only the first fruit from the house of David will sing.

The New Song / The Faultless First Fruit

Revelation 14: 3

3 *And they sung as it were a new song before the throne, and before the four beasts, and the elders: and no man could learn that song but the hundred and forty and four thousand, which were redeemed from the earth.*

This group of Israelite Saints differ from every other group. They sing a song that none shall learn for only they have been appointed to this elite position in Christ. These Messianic Israelites stand before God and the whole of heaven while heaven rejoices; however, they never wear crowns, nor do they enter the existing heaven. Their delightful destiny is earth, in service to Jesus Christ.

Revelation 14: 4 - 5

4 *These are they which were not defiled with women; for they are virgins. These are they which follow the Lamb whithersoever he goeth. These were redeemed from among men, being the firstfruits unto God and to the Lamb.*

5 *And in their mouth was found no guile: for they are without fault before the throne of God.*

It may be difficult for some Protestants to accept this focus on celibacy; nevertheless, the Father had chosen and marked this group as undefiled, without guile and without fault, to serve Him in the Millennial temple and in the New Jerusalem. They follow Jesus wherever he goes for all eternity. They are redeemed from among men and are the '*first fruit*', the very choicest, identified exclusively for God's pleasure. It pleases Him that they are celibate, having never known a woman and having taken a vow, which they have kept, to set themselves aside from the world in service to God. God sees our sacrifices (our devotions and commitments) and is pleased by them.

I believe that any Pastor, even those who dearly love their wife and family, will agree with Paul that marriage brings with it family responsibilities. This can interrupt time with God. Read I Corinthians 7: 1-40 to hear Paul's views on marriage. In this he believes he has the mind of Christ and the Spirit of God; nevertheless, he is not dogmatic about it since, in verse 37, he simply says:

37 *...He that standeth steadfast in his heart, having no necessity, but hath power over his own will, and hath so decreed in his heart that he will keep his virgin, doeth well.*

Remember that this is also the man who states in Hebrews 13: 4:

4 *Marriage is honourable in all, and the bed undefiled...*

Notes

Notes

Before we go on, listen to what Jesus says about marriage and priesthood. In Matthew 19: 4-6, when being questioned by the hypocritical Pharisees, Jesus says:

> 4 *...Have ye not read, that he which made them at the beginning made them male and female,*
> 5 *And said, For this cause shall a man leave father and mother, and shall cleave to his wife: and they twain shall be one flesh?*
> 6 *Wherefore they are no more twain, but one flesh. What therefore God hath joined together, let no man put asunder.*

Then again, when asked by His disciples if it is not good to marry, Jesus answers in Matthew 19: 12:

> 12 *For there are some eunuchs, which were so born from their mother's womb: and there are some eunuchs, which were made eunuchs of men: and there be eunuchs, which have made themselves eunuchs for the kingdom of heaven's sake. He that is able to receive it, let him receive it.*

Jesus ends his response with the words, '*He that is able to receive it, let him receive it.*' By so saying, he is warning the world that only those who have an inner knowledge of their appointment to celibacy should pursue it.

Apparently, both marriage and celibacy are an honourable estate in the eyes of God. There is a serious warning, however, to those who take the vow of celibacy publicly and then break it privately, defiling themselves before God. To take an oath and then lie to God is to tempt the Spirit of the Lord, which can and may bring death (see Acts 5: 1-10).

Revelation 14: 6 - 7

> 6 *And I saw another angel fly in the midst of heaven, having the everlasting gospel to preach unto them that dwell on the earth, and to every nation, and kindred, and tongue, and people,*
> 7 *Saying with a loud voice, Fear God, and give glory to him; for the hour of his judgment is come: and worship him that made heaven, and earth, and the sea, and the fountains of waters.*

John sees that in the Tribulation years an angel flies in the midst of heaven with the everlasting gospel to preach to the whole of the world. This gospel has always been a message of both salvation and judgment. How gracious God is to His created that even to the end of grace, even to the time of judgment, He sends His angel to remind them and to warn them. Our God is an articulate God. Knowing that man could fail in preaching the gospel to the ends of the earth, God speaks its redemptive message with a mighty voice from heaven.

The Fall of Babylon Predicted / The Wrath of God

The world would do well to remember that the everlasting gospel is not a message of grace alone; without acceptance and repentance, there comes judgment. With urgency, the angel pronounces, in a loud voice for all to hear, the message of salvation and the message of judgment. The cry is:

> 7 *...Fear God, and give Him glory,*
> *Worship Him who made Heaven and Earth* (repent and be saved)
> *For the Hour of Judgment is come!*

Notes

Matthew 25: 31-34 Jesus warns of a day when it is too late to repent. Only his '*sheep*' will be saved:

> 31 *When the Son of man shall come in his glory, and all the holy angels with him, then shall he sit upon the throne of his glory:*
> 32 *And before him shall be gathered all nations: and he shall separate them one from another, as a shepherd divideth his sheep from the goats:*
> 33 *And he shall set the sheep on his right hand, but the goats on the left.*
> 34 *Then shall the King say unto them on his right hand, Come, ye blessed of my Father, inherit the kingdom prepared for you from the foundation of the world:*

Revelation 14: 8

> 8 *And there followed another angel, saying, Babylon is fallen is fallen, that great city, because she made all nations drink of the wine of the wrath of her fornication.*

The second angel appears announcing the fall of Babylon: the ungodly world systems ruled by the Antichrist and the False prophet. The False prophet has caused many to take the mark of the Beast and, thus, '*drink of the wine of the wrath of God*'; by so doing, he has sealed their fate and they must face the consequences of God's anger. A full study discussing the Fall of Babylon may be found in the commentary for *Revelation*, Chapters 17 and 18.

Revelation 14: 9 - 11

> 9 *And the third angel followed them, saying with a loud voice, If any man worship the beast and his image, and receive his mark in his forehead, or in his hand,*
> 10 *The same shall drink of the wine of the wrath of God, which is poured out without mixture into the cup of his indignation; and he shall be tormented with fire and brimstone in the presence of the holy angels, and in the presence of the Lamb:*
> 11 *And the smoke of their torment ascendeth up for ever and ever: and they have no rest day nor night, who worship the beast and his image, and whosoever receiveth the mark of his name.*

The Torment of the Wicked / The Ecstasy of Heaven

The third angel follows the former two angels, issuing a warning against worshiping the Beast and his image, as well as against receiving his mark in their forehead or hand. The angel proceeds to give a graphic account of the punishment God will deliver to those who do. To read of their punishment is almost more than we can bear. The tragedy is that there is no such thing as soul annihilation, for souls are indestructible – only God is able to destroy the soul (Matthew 10: 28) and He chooses never to do so. The smoke of their torment ascends up for ever and ever; they have no rest night or day.

Notes

Although there may not be a body, the soul feels itself encased within a body capable of suffering.

Paul attests to this condition of the soul when faced with the perplexity of being caught up into the third heaven into paradise. Read II Corinthians 12: 2-4:

> *2 ...(whether in the body, I cannot tell; or whether out of the body, I cannot tell: God knoweth;) such an one caught up to the third heaven.*

Paul is saying that the soul, though unclothed, seems clothed; though separated from the body, feels present in the body. In spiritual form, the emotions and sensations are multiplied beyond all and far above that which man can express. The soul feels sorrow and shame, as well as intense joy or acute loneliness. When facing terror, even the atheist has the option of praying; however, in the torment of hell, there is no God and, therefore, no hope to reach out to. (Remember, the story of the rich man and Lazarus in Luke 16: 19-31 is a parable, not a reality.) The soul thinks and feels and hears. It feels the absolute torment of hell or the blissful ecstasy of heaven: the joy of being veiled in the righteousness of Christ; love purified and multiplied a million times over; the splendour of the Alleluia chorus from Handel's 'Messiah' sung jubilantly dims in comparison to the exuberant exultation of the heavenly choir. The soul dances and sings, rejoices and praises; it lives in Spirit form and never dies.

God offers '*joy unspeakable*' to those who love and fear Him. Satan offers a sea of fire. Don't wait until after death to believe in hell.

Revelation 14: 12

12 Here is the patience of the saints: here are they that keep the commandments of God, and the faith of Jesus.

To avoid the punishment of hell, we must be patient and persevering in our walk with God. His Spirit empowers us with faith and inspires and enables us to keep His commandments.

Rest for the Righteous / Jesus in the Cloud / God Confirms Vision and Prophesy

Notes

Revelation 14: 13

13 And I heard a voice from heaven saying unto me, Write, Blessed are the dead which die in the Lord from henceforth: Yea, saith the Spirit, that they may rest from their labours; and their works do follow them.

A voice from heaven commands that John must write and assure the Saints that all who die for their faith throughout the Tribulation will be blessed. The Spirit affirms that they will rest from their labour of persecution under the Antichrist; they are blessed in that they are removed from further suffering under the false trinity. John is commanded to tell them that their good works will follow them to be recognized and rewarded at their judgment. See Revelation 6: 9-11 to view the Saints as they rest from their labour.

Revelation 14: 14

14 And I looked, and behold a white cloud, and upon the cloud one sat like unto the Son of man, having on his head a golden crown, and in his hand a sharp sickle.

When John sees Jesus in the cloud, he likely recalls when he was a young man and witnessed a similar occurrence (read Acts 1: 9-11). How could John ever forget when the two men, clothed in white, told him and the other disciples that Jesus would return in the same manner as they had seen Him go.

John sees Jesus wearing a golden, kingly crown which, surely, also reminds him of the title Pontius Pilot had placed on the cross: "Jesus of Nazareth, the King of the Jews". John probably recalls the time when they all sat together and Jesus spoke of the signs which would indicate his return; read Matthew 24: 30:

30 And then shall appear the sign of the Son of man in heaven: and then shall all the tribes of the earth mourn, and they shall see the Son of man coming in the clouds of heaven with power and great glory.

Daniel had previously seen a vision of the Son of man coming with clouds; see Daniel 7: 13:

13 I saw in the night visions, and, behold, one like the Son of man came with the clouds of heaven...

The wonderful thing about the scriptures is that we are not dependent on one man's vision. We find always that God confirms vision with vision, revelation with revelation. Learn to expect this as God guides your life. Most often, God first speaks to those who seek His will and then affirms His instructions through His servants, or in some other very definite and valid

Notes

way. Take note that God sends three angels to confirm that the time has come for Jesus to gather the vines of the earth and cast them into the winepress of God (Revelation 14: 14-20). Don't, under any circumstances, rush off into some ministry which a well-meaning brother or sister may lay upon you. God must speak to your heart and you must quietly ponder and seek His Will until He affirms His plan for your life. His instruction may start with a whisper but will end in positive assurance. Until then, stay as you are; study and pray, continuing to serve in the place that you find yourself.

Revelation 14: 15

15 And another angel came out of the temple, crying with a loud voice to him that sat on the cloud, Thrust in thy sickle, and reap: for the time is come for thee to reap; for the harvest of the earth is ripe.

An angel comes out of the temple proclaiming that the time has come for Jesus to reap the harvest of the earth. The 'fruit is ripe' for the world has reached its ultimate evil and the time of judgment is at hand.

Joel foresaw this judgment in Joel 3: 13:

13 Put ye in the sickle, for the harvest is ripe: come, get you down; for the press is full, the vats overflow; for their wickedness is great.

Revelation 14: 16

16 And he that sat on the cloud thrust in his sickle on the earth; and the earth was reaped.

Jesus, who sits on the cloud, having received confirmation from God through the angel, thrusts in his sickle and the earth is reaped.

Revelation 14: 17

17 And another angel came out of the temple which is in heaven, he also having a sharp sickle.

A second angel comes out of the temple. He has a sharp sickle but in no way helps in the reaping. At the end of the Tribulation, Jesus alone harvests the earth (Isaiah 63: 2-5).

The Earth is Harvested / The Winepress of God's Wrath

Revelation 14: 18

18 And another angel came out from the altar, which had power over fire; and cried with a loud cry to him that had the sharp sickle, saying, Thrust in thy sharp sickle, and gather the clusters of the vine of the earth; for her grapes are fully ripe.

The third angel comes out from the altar, indicating he is an angel of higher standing and greater authority, possibly closer to the mind of God. This angel has power over fire, as does the angel of Revelation, Chapter 16: 8, who pours out his vial upon the sun. The third angel comes bearing the consent of God to carry out the final crushing of the counterfeit vine of the earth. It seems that the angel at no time helps Jesus, for it is only in the harvest at the end of the world that the angels are the reapers (Matthew 13: 36-42). The angel cries out that Jesus (the one with the sharp sickle of Revelation 14: 14) should thrust in his sickle and gather the clusters of the vine of the earth, the grapes being fully ripe. Three angels have affirmed that the time has come for the earth to be harvested. God has spoken from His altar, through His angels, to His Son.

Though Jesus has the full mind of God, he awaits God's proper timing. A lesson may be learned here for those who feel they are called to serve in special ministries: be cautious not to run ahead of God's leading and God's timing. He must lead and we must follow.

Revelation 14: 19

19 And the angel thrust in his sickle into the earth, and gathered the vine of the earth, and cast it into the great winepress of the wrath of God.

We know that this angel is Jesus, for only Jesus has the power to judge and condemn on behalf of God. Only Jesus has the keys to hell (Revelation 1: 18). No one but Jesus has the authority to cast the living into the great winepress of God. Revelation 19: 13-15 supports this understanding:

13 And he was clothed with a vesture dipped in blood: and his name is called The Word of God.

15 And out of his mouth goeth a sharp sword, that with it he should smite the nations: and he shall rule them with a rod of iron: and he treadeth the winepress of the fierceness and wrath of Almighty God.

Notes

The Earth Drenched in Blood / The Six Angels

Notes

Revelation 14: 20

20 And the winepress was trodden without the city, and blood came out of the winepress, even unto the horse bridles, by the space of a thousand and six hundred furlongs.

It is said: '*the winepress was trodden without the city.*' Rightly so. Even the slaughter and burning of the red heifer was done some distance from the walls of Jerusalem, across the Kidron Valley, up on the side of the Mount of Olives. To this day, the dead are buried outside the walls, for anything that might defile is kept well away from the temple area. Only the blood of the sacrificial lamb (or other animals ascribed for atonement) is shed within the temple which is within the walls. This is representative of the holy blood shed for you and me. However, in his death, our Saviour died and shed his blood outside the walls of the Holy City, as did the red heifer whose ashes provided ritual cleansing of sins (Numbers 19: 1-19).

The sacrifice of the red heifer helps us understand God's immaculate plan of salvation; through it, He has provided a metaphor, '*a foreshadow of things to come*'. The red heifer is a pure sacrificial offering led outside the city walls and its ashes cleanse the ceremonial waters used for the purification of sin. The symbolism as it relates to Christ is obvious: as did the sacrificial red heifer, Christ died in the world, for the world. For more information, see Appendix 7, "The Location of Christ's Crucifixion and Burial".

Revelation 14: 20 further tells us that blood comes out of the winepress, even up to the horse bridles, by the space of a thousand and six hundred furlongs. This distance, equivalent to 200 miles, is precisely the length of Israel from north to south. Blood to the horse bridles, about three feet deep, is probably a symbolic description of harsh and complete judgment over all of the nations gathered in the land of Israel. One visualizes the entire land drenched in blood.

Before moving on to Chapter 15, we should review the seven angels of this chapter:

1) verse 6: the angel with the everlasting gospel
2) verse 8: the angel announcing the fall of Babylon
3) verse 9: the angel of warning
4) verse 15: the first angel out of the temple proclaiming the harvest
5) verse 17: the second angel out of the temple bearing a sickle
6) verse 18: the third angel, the angel of fire, coming out from the altar

The Seventh Angel, the Christ

This brings the total to six angels and then comes the seventh in verses 19 to 20. This angel is:

1) the true vine (John 15: 5)
2) the heir of the vineyard (Luke 20: 9-16)
3) the one who treads the winepress <u>alone</u> (Isaiah 63: 2-5)
4) the Christ, the Son of the Living God (Matthew 16: 16)

Notes

CHAPTER FIFTEEN

Chapter Fifteen gives us absolute assurance that we serve a true and righteous God. We are allowed to view His beloved Israelites as they stand – after passing through the fire of Tribulation – on the sea of glass, victorious over the Beast. We know they are Israelites and that they are born-again, for they sing the song of Moses and song of the Lamb. In this, God has kept His promise to the whole house of Israel. The Study teaches that chapter fifteen, verses five to eight, and chapter eleven, verses fifteen to nineteen, hold the key which opens *Daniel*, chapter twelve. It explains that these three portions of the scriptures are intricate and interwoven; none can be fully understood without the other. A detailed exposition is given in the Study notes to chapter eleven, unravelling the mystery of their interdependence.

Revelation 15

[**1**] And I saw another sign in heaven, great and marvellous, seven angels having the seven last plagues; for in them is filled up the wrath of God.

[**2**] And I saw as it were a sea of glass mingled with fire: and them that had gotten the victory over the beast, and over his image, and over his mark, and over the number of his name, stand on the sea of glass, having the harps of God.

[**3**] And they sing the song of Moses the servant of God, and the song of the Lamb, saying, Great and marvellous are thy works, Lord God Almighty; just and true are thy ways, thou King of saints.

[**4**] Who shall not fear thee, O Lord, and glorify thy name? for thou only art holy: for all nations shall come and worship before thee; for thy judgments are made manifest.

[**5**] And after that I looked, and, behold, the temple of the tabernacle of the testimony in heaven was opened:

[**6**] And the seven angels came out of the temple, having the seven plagues, clothed in pure and white linen, and having their breasts girded with golden girdles.

[**7**] And one of the four beasts gave unto the seven angels seven golden vials full of the wrath of God, who liveth for ever and ever.

[**8**] And the temple was filled with smoke from the glory of God, and from his power; and no man was able to enter into the temple, till the seven plagues of the seven angels were fulfilled.

The Seven Last Plagues / The Sea of Glass / The song of Victory

Revelation, Chapter 15

Notes

Revelation 15: 1

1 And I saw another sign in heaven, great and marvellous, seven angels having the seven last plagues; for in them is filled up the wrath of God.

John sees another sign in heaven which, to him, appears great and marvellous: he sees seven angels with seven vials filled with the seven last plagues of God. He knows that very soon the persecutors of the Tribulation Saints will face the vial judgments. They will drink with *'astonishment, from the hand of the Lord, the cup of His fury.'* As the song for Asaph says in Psalm 75: 8 (NLT):

8 For the Lord holds a cup in his hand;
It is filled of foaming wine mixed with spices.
He pours the wine out in judgment,
And all the wicked must drink it,
draining it to the dregs.

Revelation 15: 2

2 And I saw as it were a sea of glass mingled with fire: and them that had gotten the victory over the beast, and over his image, and over his mark, and over the number of his name, stand on the sea of glass, having the harps of God.

The sea of glass which John now sees is the same as that of Revelation 4: 6. The former sea, which was before the throne, was unoccupied and radiated peace and serenity as it reflected the glory of God. All who come to the Lord appear before God on the sea of glass; they are presented to Him that the whole of heaven may rejoice (Luke 15: 7). The same sea now shown is mingled with fire. This indicates that these dear Saints have come to the Lord through the horror and torment of the Tribulation years. Through death and faith in Christ, they are victorious over the Beast: over his image, over his mark, and over the number of his name.

One doubts that each Saint is given a harp but, rather, their throats are implanted, as it were, with the instruments of praise.

Revelation 15: 3

3 And they sing the song of Moses the servant of God, and the song of the Lamb, saying, Great and marvellous are thy works, Lord God Almighty; just and true are thy ways, thou King of saints.

The content of this verse allows us to understand that these Saints are converted Israelites. They sing the song of Moses which is a song of praise

Notes

celebrating God's mighty deliverance of His people Israel. Read Exodus 15: 1-19:

> *1 Then sang Moses and the children of Israel this song unto the Lord, and spake, saying, I will sing unto the Lord, for he hath triumphed gloriously: the horse and his rider hath he thrown into the sea.*

They also sing the song of the Lamb, celebrating salvation: redemption purchased at Calvary by the victorious Jesus. The words of this song are: "*...Great and marvelous are thy works, Lord God Almighty; just and true are thy ways, thou King of saints.*"

This group, being Israelites, will be clothed in fine linen, clean and white (Revelation 19: 6-9). Through suffering and repentance, they have made themselves ready, thereby worthy, through the blood of Christ, to attend the marriage supper of the Lamb. These Saints, being singled out from the martyrs of Revelation 6: 11 (white robes), are representative of the whole house of Israel. They have been called, converted and tested in the fire of tribulation. This group of Israelite Saints are probably those of Ezekiel 37 who are risen from their graves and brought back to Jerusalem to see Jesus plant his feet on the Mount of Olives (Zechariah 14: 4).

Revelation 15: 4

> *4 Who shall not fear thee, O Lord, and glorify thy name? For thou only art holy: for all nations shall come and worship before thee; for thy judgments are made manifest.*

The song of praise continues. The Saints declare, before the whole of heaven, that God is absolutely holy, glorious, almighty and worthy of worship. They proclaim His judgment to be just and true, and state that all nations will worship Him. (Read Zechariah 14: 16-17 and Isaiah 2: 2-3.)

Revelation 15: 5

> *5 And after that I looked, and, behold, the temple of the tabernacle of the testimony in heaven was opened:*

The intention of verses 5 to 8 has been discussed at length in the commentary for Revelation 11: 15-19. If necessary, turn back to that section of the study to refresh your memory as to the depth and significance of their teaching.

The Vials Presented / The Shekinah Glory

Notes

Revelation 15: 6-7

6 *And the seven angels came out of the temple, having the seven plagues, clothed in pure and white linen, and having their breasts girded with golden girdles.*

7 *And one of the four beasts gave unto the seven angels seven golden vials full of the wrath of God, who liveth for ever and ever.*

Dressed in priestly garments (see Ezekiel 44: 17), seven angels come out of the temple. The time of grace is drawing to a close and judgment is about to be administered through the harsh precepts of the law. The seven plagues are the vials of judgment given to the angels by one of the four creatures from the midst of the throne in the presence of God (Revelation 4: 6). The golden vials are filled with the great and terrible wrath of God as seen in Revelation 14: 10-11.

Revelation 15: 8

8 *And the temple was filled with smoke from the glory of God, and from his power; and no man was able to enter into the temple, till the seven plagues of the seven angels were fulfilled.*

One should take special note that even before the vial wraths are poured out, the temple is filled with the smoke from the Glory of God. No man is able to enter the temple until the seven plagues of the seven angels are poured out.

This verse makes us realize that the vial wraths are withheld until the Antichrist has been given his 1260 days of unimpeded reign in the temple. We see, with certainty, his eviction from the temple by the smoke from the Shekinah glory, the Glory of God.

How great must be the Glory of the Lord. How fearful and awesome when even the smoke from His glory would prevent man from entering the temple. The following readings from the scriptures demonstrate the consuming power, the protective power and the restraining power of the Shekinah glory. Exodus 24: 17 reveals God's consuming power:

> 17 *And the sight of the glory of the Lord was like devouring fire on the top of the mount* (Sinai) *in the eyes of the children of Israel.*

Exodus 33: 22 displays God's protective power. God in Christ is our '*clift of the rock*':

> 22 *And it shall come to pass, while my glory passeth by, that I will put thee in a clift of the rock, and will cover thee with my hand while I pass by.*

Praise His Holy Name!

Exodus 40: 35 confirms God's restraining power:

> *35 And Moses was not able to enter into the tent of the congregation, because the cloud abode thereon, and the glory of the Lord filled the tabernacle.*

Numbers 14: 21 gives us the awesome promise that someday we will dwell within the Glory of the Lord:

> *21 But as truly as I live, all the earth shall be filled with the glory of the Lord.*

Praise His Holy name!

Notes

CHAPTER SIXTEEN

The Study points out that **Chapter Sixteen** is devoted solely to the outpouring of the seven vial wraths. They start after the Antichrist is evicted from the temple by the smoke from the Glory of God. Their rapid distribution allows the seventh trumpet wrath and the seventh vial wrath to culminate simultaneously. An obvious lapse of time is indicated between the beginning of the sounding of the seventh trumpet and its ending. The Study teaches that the seventh trumpet must end its sounding beyond the twelve hundred and sixty days allotted to the temple reign of the Antichrist, bringing the two sets of wraths together as they finish. This authenticates chapter eleven of *the Revelation*, verses fifteen to nineteen, which states '*the kingdoms of the world are become the kingdoms of our Lord and His Christ.*' It also permits the voice from the temple to announce, '*It is done.*'

Revelation 16

[**1**] And I heard a great voice out of the temple saying to the seven angels, Go your ways, and pour out the vials of the wrath of God upon the earth.
[**2**] And the first went, and poured out his vial upon the earth; and there fell a noisome and grievous sore upon the men which had the mark of the beast, and upon them which worshipped his image.
[**3**] And the second angel poured out his vial upon the sea; and it became as the blood of a dead man: and every living soul died in the sea.
[**4**] And the third angel poured out his vial upon the rivers and fountains of waters; and they became blood.
[**5**] And I heard the angel of the waters say, Thou art righteous, O Lord, which art, and wast, and shalt be, because thou hast judged thus.
[**6**] For they have shed the blood of saints and prophets, and thou hast given them blood to drink; for they are worthy.
[**7**] And I heard another out of the altar say, Even so, Lord God Almighty, true and righteous are thy judgments.
[**8**] And the fourth angel poured out his vial upon the sun; and power was given unto him to scorch men with fire.
[**9**] And men were scorched with great heat, and blasphemed the name of God, which hath power over these plagues: and they repented not to give him glory.
[**10**] And the fifth angel poured out his vial upon the seat of the beast; and his kingdom was full of darkness; and they gnawed their tongues for pain,
[**11**] And blasphemed the God of heaven because of their pains and their sores, and repented not of their deeds.
[**12**] And the sixth angel poured out his vial upon the great river Euphrates; and the water thereof was dried up, that the way of the kings of the east might be prepared.
[**13**] And I saw three unclean spirits like frogs come out of the mouth of the dragon, and out of the mouth of the beast, and out of the mouth of the false prophet.
[**14**] For they are the spirits of devils, working miracles, which go forth unto the kings of the earth and of the whole world, to gather them to the battle of that great day of God Almighty.
[**15**] Behold, I come as a thief. Blessed is he that watcheth, and keepeth his garments, lest he walk naked, and they see his shame.
[**16**] And he gathered them together into a place called in the Hebrew tongue Armageddon.
[**17**] And the seventh angel poured out his vial into the air; and there came a great voice out of the temple of heaven, from the throne, saying, It is done.
[**18**] And there were voices, and thunders, and lightnings; and there was a great earthquake, such as was not since men were upon the earth, so mighty an earthquake, and so great.
[**19**] And the great city was divided into three parts, and the cities of the nations fell: and great Babylon came in remembrance before God, to give unto her the cup of the wine of the fierceness of his wrath.
[**20**] And every island fled away, and the mountains were not found.
[**21**] And there fell upon men a great hail out of heaven, every stone about the weight of a talent: and men blasphemed God because of the plague of the hail; for the plague thereof was exceeding great.

God Commands the Outpouring of the Vials /
The First and Second Vile

Notes

Revelation, Chapter 16

This chapter affirms the rapidity of the outpouring of God's final set of judgments. They come, one upon the other, with the seventh vial wrath joining the seventh trumpet wrath as it finishes sounding. This brings to fruition God's two sets of judgments, authenticating Revelation 11: 15-19. If needed, see previous teaching for Revelation 11: 15-19.

Revelation 16: 1

1 And I heard a great voice out of the temple saying to the seven angels, Go your ways, and pour out the vials of the wrath of God upon the earth.

As this chapter opens, there is an ominous sense of foreboding: it's as if we stand in the eye of the storm. There is no holding back God's pent-up anger; His final fury is about to rage across the earth, ending with '*lightnings, and voices, and thunderings, and an earthquake, and great hail*' (Revelation 11: 19; Revelation 16: 18).

John hears a mighty voice from within the temple, the voice of God, commanding that the seven priestly angels (Revelation 15: 6) go forth, without delay, and serve God's unrelenting wrath upon man. The unrepentant now face the second and final set of judgments.

Revelation 16: 2

2 And the first went, and poured out his vial upon the earth; and there fell a noisome and grievous sore upon the men which had the mark of the beast, and upon them which worshiped his image.

The first angel pours God's wrath upon the earth. All who worship the beast and bear his mark are afflicted with ulcerated boils, as were the Egyptians in the time of Moses (Exodus 9: 8-11)

Revelation 16: 3

3 And the second angel poured out his vial upon the sea; and it became as the blood of a dead man: and every living soul died in the sea.

The second vial, when poured out, causes the sea to become as the blood of dead men and every living thing dies in the sea. This could be literal blood but there is some possibility it should be taken figuratively since the verse says '*it became as the blood of a dead man*'.

Notes

Exodus 7: 19-20 reports that at the command of God, Aaron smote the water and the waters turn to blood. There is a definite parallel between the judgments upon Egypt and the judgments of the Tribulation. A theory is that algae along the banks of the Nile became polluted, turning the water red; the algae poisoned the fish and the pollution drove the frogs from the Nile. Thus began a chain-reaction of events, each activating the next plague, bringing God's judgments upon Egypt through His own creation.

John informs us that the waters of the sea become poisonous and all life dies in the sea. At the time of Moses, the curse upon the water was restricted to the waters of Egypt (Exodus 7: 19); however, the vial judgment will be world-wide and the curse will be upon the whole earth. Christ doesn't come to judge only those gathered at Armageddon; He will come to judge the nations of the world. None will escape except the righteous.

Before we leave this verse, recall that in the opening of the second seal (Revelation 6: 3-4), Jesus consents to the release of the red horse who brings bloodshed as the aftermath of peace. Later, we see that when the second trumpet wrath sounds, a third-part of the sea becomes blood (Revelation 8: 8). Now, in the outpouring of the second vial wrath, the whole of the sea becomes as blood. Note that the second vial wrath is more violent and far-reaching than the second trumpet wrath. The punishment grows more severe as the volume of God's wrath is released.

Revelation 16: 4

4 *And the third angel poured out his vial upon the rivers and fountains of waters; and they became blood.*

The vial of the third angel turns rivers and fountains of water into blood. Now all of the waters of the earth are unfit to drink. An obvious conclusion is that the vials must spin rapidly through their judgments since, in a matter of days, the whole of the earth will die through dehydration. The time must be extremely short or Christ would be robbed of God's full retaliation for sin.

Revelation 16: 5

5 *And I heard the angel of the waters say, Thou art righteous, O Lord, which art, and wast, and shalt be, because thou hast judged thus.*

We, as believers, are aware that God has appointed Guardian angels to watch over us. Most of us have never appreciated, however, that the whole of the universe has angels in attendance. In this verse, we learn that the angel of the waters agrees with God's divine judgment. The angel declares

God's judgment to be righteous and announces, since God has judged thus, all can be assured of this: He is the God who is, who was and ever shall be. He is the God of gods; there is no other.

Notes

Revelation 16: 6

6 For they have shed the blood of saints and prophets, and thou hast given them blood to drink; for they are worthy.

The angel of the waters sees God's wrath as righteous and justifies this belief by saying: '*For they have shed the blood of saints and prophets, and thou hast given them blood to drink...*' Jesus says in Matthew 7: 2, '*for with what measure ye mete, it shall be measured to you again.*'

Revelation 16: 7

7 And I heard another out of the altar say, Even so, Lord God Almighty, true and righteous are thy judgments.

An angel from the altar (that is, an angel from the very presence of God) agrees with the angel of the waters, saying: '*Even so, Lord God Almighty, true and righteous are thy judgments.*'

Revelation 16: 8

8 And the fourth angel poured out his vial upon the sun; and power was given unto him to scorch men with fire.

When the fourth angel pours out his vial upon the sun, power is given to the sun to burn the earth with tremendous heat. Jesus predicts '*...there shall be signs in the sun, and in the moon, and in the stars...*'(Luke 21: 25).

Malachi 4: 1 paints a vivid picture of the burning power of the sun:

1 For, behold, the day cometh, that shall burn as an oven; and all the proud, yea, and all that do wickedly, shall be stubble: and the day that cometh shall burn them up, saith the Lord of hosts, that it shall leave them neither root nor branch.

Deuteronomy 32: 24 also speaks of this devastating time:

24 They shall be burnt with hunger, and devoured with burning heat, and with bitter destruction...

Isaiah 13: 13 helps us to understand how God administers His wrath through the sun:

13 ...I will shake the heavens, and the earth shall remove out of her place, in the wrath of the Lord of hosts, and in the day of his fierce anger.

The Unrepentant Lost

Notes

We are aware the sun is a variable output star and that minor fluctuations in solar conditions cause weather changes. Imagine the catastrophic events which will occur when God shakes His universe: the sun and the solar system will be in total disarray! We know through scientific calculations that the sun is slowly shrinking and giving off less light. Scientists foresee that the sun eventually darkens and, just prior to its extinction, it will become as a 'nova' – in a great burst of light and heat, it will burn itself out. Thus, the sun enacts the fourth trumpet wrath (it grows dim - Revelation 8: 12) and the fourth vial wrath (it burns with a great heat - Revelation 16: 8). God has complete control over the heavens and its galaxy. All things are done in His time. Consult the Calendar of Seals and Wraths, Appendix 5.

God didn't need man to administer the plagues on Egypt, nor will He need him as He directs the power of the sun. He will allow men to destroy one another as all nations gather at Armageddon (Ezekiel 38: 21); however, it is God alone who sends His final wrath to purge and purify the earth.

Revelation 16: 9

9 *And men were scorched with great heat, and blasphemed the name of God, which hath power over these plagues: and they repented not to give him glory.*

These unrepentant are those who have heard the truth and may have, at one time, walked in it, but they have sold-out to sin. They accepted the Antichrist (the ways of the world) instead of following the ways of Christ. Denial can't remove the truth; however, the truth can be removed from the heart of man through shunning God's grace. This, in turn, hardens the heart. There comes a point when the Spirit no longer woos and the soul is lost to the mark of the beast. Romans 1: 24 affirms this when it says:

24 *Wherefore God also gave them up to uncleanness through the lusts of their own hearts...*

Paul again declares this truth in II Thessalonians 2: 10-12:

10 *...they received not the love of the truth, that they might be saved.*

11 *And for this cause God shall send them strong delusion, that they should believe a lie:*

12 *That they all might be damned who believed not the truth, but had pleasure in unrighteousness.*

The Fifth Vial Wrath / Man Blasphemes God

Notes

Revelation 16: 10

10 And the fifth angel poured out his vial upon the seat of the beast; and his kingdom was full of darkness; and they gnawed their tongues for pain,

The fifth angel pours out his vial upon '*the seat of the beast*'. The '*seat of the beast*' is the world (Job 1: 7; Isaiah 14: 12; I Peter 5: 8; Revelation 13: 2).

The darkness which comes upon the face of the earth through the outpouring of the fifth vial was foretold by Jesus in Matthew 24: 29 and several of the prophets, two of whom are Ezekiel and Joel. Read Ezekiel 32: 7-8 (a curse upon Egypt but a foreshadowing of things to come):

7 ...I will cover the heaven, and make the stars thereof dark; I will cover the sun with a cloud, and the moon shall not give her light.

8 All the bright lights of heaven will I make dark over thee, and set darkness upon the land, saith the Lord God.

Similarly, Joel 2: 31 speaks of this:

31 The sun shall be turned into darkness, and the moon into blood, before the great and the terrible day of the Lord come.

Probably God causes this darkness to fall upon the earth to bring relief. Without His compassion, the whole of mankind would be consumed by the scorching heat of the sun at the outpouring of the fourth vial.

Revelation 16: 11

11 And blasphemed the God of heaven because of their pains and their sores, and repented not of their deeds.

The Beast's kingdom (the earth) '*was full of darkness; and they gnawed their tongues for pain, and blasphemed God... because of their pains and their sores...*'. They are unable to repent of their deeds because their time for repentance is past and their souls are filled with anger. Obviously, the seven vials are poured out in rapid succession. The people still have their ulcerated boils inflicted upon them at the outpouring of the first vial. As always, scriptures confirm scriptures. Revelation 18: 10 says, '*for in one hour is thy judgment come*'; and Revelation 18: 19 says, '*for in one hour she is made desolate.*'

The Sixth Vial Wrath / Three Unclean Spirits / The Great Day of God Almighty

Notes

Revelation 16: 12

12 And the sixth angel poured out his vial upon the great river Euphrates; and the water thereof was dried up, that the way of the kings of the east might be prepared.

The sixth angel pours out his vial on the Euphrates (recall Revelation 9: 14 where the sixth angel with the trumpet loosed the four angels bound in the Euphrates). As a result, the waters dry up so that the '*kings of the east*' (the eastern nations) may freely attend their inauguration into hell at the battle of Armageddon. These are they who have been prepared for this special moment in God's timetable. The '*hour, and day, and month, and year*' of Revelation 9: 15 has now arrived when God gathers all nations in the valley of Megiddo and the mountains of Israel. God always opens the way that His will may be executed. The act of drying up the waters is evidence of the power and foreknowledge of God.

Revelation 16: 13

13 And I saw three unclean spirits like frogs come out of the mouth of the dragon, and out of the mouth of the beast, and out of the mouth of the false prophet.

In Leviticus 11: 10-12, the frog is included, by definition, with the unclean things in the waters; thus, it stands for those characterized as unclean, having the power of darkness and being the allies of evil. The stench of evil and death comes out of the mouth of the dragon (Satan), the Beast (the Antichrist) and the False prophet – the three members of the false trinity.

Revelation 16: 14

14 For they are the spirits of devils, working miracles, which go forth unto the kings of the earth and of the whole world, to gather them to the battle of that great day of God Almighty.

The evil spirit of Satan, acting through the Antichrist and the False prophet, incites hate in the heart of the nations against Israel and her God. Ultimately, it is the Spirit of God which directs the gathering of the kings of the nations to the great battle for the victorious judgment of the Lord (Joel 3: 12). Read Psalm 2: 1-12 for a word picture of the great gathering, when the armies of the Antichrist meet Christ and the host of heaven in the valley of Esdraelon – the great plain of Jezreel, under the mountain of Megiddo:

2 The kings of the earth set themselves, and the rulers take counsel together, against the Lord, and against his anointed, saying,

Notes

3 *Let us break their bands asunder, and cast away their cords from us.*
4 *He that sitteth in the heavens shall laugh: the Lord shall have them in derision.*
5 *Then shall he speak unto them in his wrath, and vex them in his sore displeasure.*

9 *Thou shalt break them with a rod of iron: thou shalt dash them in pieces like a potter's vessel.*

Revelation 16: 15

15 Behold, I come as a thief. Blessed is he that watcheth, and keepeth his garments, lest he walk naked, and they see his shame.

The garments referred to are the garments of righteousness received by grace through faith in Jesus Christ. Without Christ's covering, our true condition would be revealed: we would be seen laden with sin and filled with shame.

This verse may be recorded in this particular place in the *Revelation* for the benefit of the '*third-part*' who are brought through the fire of the Tribulation years. Its purpose is possibly to keep them strong in their time of great torment. Read Zechariah 13: 8-9:

8 *And it shall come to pass, that in all the land, saith the Lord, two parts therein shall be cut off and die; but the third shall be left therein.*
9 *And I will bring the third part through the fire, and will refine them as silver is refined, and will try them as gold is tried: they shall call on my name, and I will hear them: I will say, It is my people: and they shall say, The Lord is my God.*

Revelation 16: 15 is written also as a warning to all who read the *Revelation*, reminding them that they must constantly be covered by the righteousness of Christ and be ready for their call to the Rapture. In this way, only, are they able to escape these horrendous hours of judgment.

Revelation 16: 16

16 And he gathered them together into a place called in the Hebrew tongue Armageddon.

How great God is that He has absolute command of people and nations! He wills that they be '*gathered into a place called... Armageddon.*' God wills and it is done; none can oppose the will of God.

Notes

The only exception to this statement is in the case of salvation. Although God's will is that none should perish (II Peter 3: 9), man is left with the personal responsibility of deciding his own destiny. He has been given the freedom to choose whom he will serve. God has provided the way: the free gift of eternal life through Jesus Christ. Man may accept life through faith (a gift of God) in the saving power of the shed blood and resurrection of Christ; or he may choose eternal death and torment by following Satan, the god of this world.

Revelation 16: 17 - 18

17 And the seventh angel poured out his vial into the air; and there came a great voice out of the temple of heaven, from the throne, saying, It is done.

18 And there were voices, and thunders, and lightnings; and there was a great earthquake, such as was not since men were upon the earth, so mighty an earthquake, and so great.

This angel is one of the seven previously viewed in Revelation 15: 6-8 as they came out of the temple, '*clothed in pure and white linen, and having their breasts girded with golden girdles*'. You will recall that they receive seven golden vials filled with the wrath of God; whereupon, the temple, from which they have come, fills with the smoke from the glory of God. We now see, through John, this seventh angel as he pours out the last of the seven vials into the air. The mighty voice of God, from within the temple, declares: "*It is done.*" The pouring of God's judgment into the air, the very thing that gives man life, indicates the extent and severity of His judgment against the whole of the universe. Only those bearing the protection of God will escape.

Verses 17 to 21 coincide with verses 15 to 19 of Revelation 11. They confirm that the seventh trumpet wrath of Revelation 11: 15, and the seventh vial wrath, here in verse 17, finish together. Both end with the sound of '*voices and thunders and lightnings and... hail*' and each records a massive earthquake. Verse 15 of Chapter 11 declares, '*The kingdoms of this world are become the kingdoms of our lord and his Christ, and he shall reign for ever and ever.*' Revelation 16: 17 here agrees: '*It is done!*'

Revelation 16: 19

19 And the great city was divided into three parts, and the cities of the nations fell: and great Babylon came in remembrance before God, to give unto her the cup of the wine of the fierceness of his wrath.

In the wake of the earthquake, Jerusalem, the great city, is divided into three. The cities of the earth fall as the fault lines over the whole of the earth break open. The upheaval is universal; none will escape this wrath.

Notes

In this way, God remembers and avenges the sins of Babylon (the sins of the ungodly) and gives her to drink *'the cup of the wine of the fierceness of his wrath.'*

Revelation 16: 20

20 And every island fled away, and the mountains were not found.

The result of the earthquake is that the existing islands disappear and *the mountains and hills are made waste* (Isaiah 42: 15).

Revelation 16: 21

21 And there fell upon men a great hail out of heaven, every stone about the weight of a talent: and men blasphemed God because of the plague of the hail; for the plague thereof was exceeding great.

Hail falls from heaven, *'every stone about the weight of a talent'*, with a talent being anywhere from the Greek's talent of 56 pounds to Antioch's talent of 390 pounds. Even with this display of God's awesome power, unrepentant man responds by blaspheming God, being bound by their suffering and unbelief.

Many prophets speak of these days, none as poetically as Jeremiah 4: 23-26:

23 I beheld the earth, and, lo, it was without form, and void; and the heavens, and they had no light.

24 I beheld the mountains, and, lo, they trembled, and all the hills moved lightly.

25 I beheld, and, lo, there was no man, and all the birds of the heavens were fled.

26 I beheld, and, lo, the fruitful place (the Promised Land) *was a wilderness, and all the cities thereof were broken down at the presence of the Lord and by his fierce anger.*

Further detailed records of the last forty-five days of Judgment may be found in Ezekiel, Chapter 39, and Zechariah, Chapters 12 and 14. It is amazing how well-informed we are, and how explicit God is, if we take time to read His word.

CHAPTER SEVENTEEN

In **Chapter Seventeen**, an angel reveals to John the annihilation of the end-time World Church. The Study explains that the woman – the harlot – represents the leader of this religious system. She will gather a conglomeration of false religions, teaching that there are many ways to god, and that sin can be disassociated from the spirit. The woman is defined as a city and the city is Rome, the religious capital of the world. The waters on which she sits are peoples, multitudes and nations. Thus, John is seeing the fall of the satanic World Church and the removal of the False prophet as her administrator.

Revelation 17

[**1**] And there came one of the seven angels which had the seven vials, and talked with me, saying unto me, Come hither; I will shew unto thee the judgment of the great whore that sitteth upon many waters:

[**2**] With whom the kings of the earth have committed fornication, and the inhabitants of the earth have been made drunk with the wine of her fornication.

[**3**] So he carried me away in the spirit into the wilderness: and I saw a woman sit upon a scarlet coloured beast, full of names of blasphemy, having seven heads and ten horns.

[**4**] And the woman was arrayed in purple and scarlet colour, and decked with gold and precious stones and pearls, having a golden cup in her hand full of abominations and filthiness of her fornication:

[**5**] And upon her forehead was a name written, MYSTERY, BABYLON THE GREAT, THE MOTHER OF HARLOTS AND ABOMINATIONS OF THE EARTH.

[**6**] And I saw the woman drunken with the blood of the saints, and with the blood of the martyrs of Jesus: and when I saw her, I wondered with great admiration.

[**7**] And the angel said unto me, Wherefore didst thou marvel? I will tell thee the mystery of the woman, and of the beast that carrieth her, which hath the seven heads and ten horns.

[**8**] The beast that thou sawest was, and is not; and shall ascend out of the bottomless pit, and go into perdition: and they that dwell on the earth shall wonder, whose names were not written in the book of life from the foundation of the world, when they behold the beast that was, and is not, and yet is.

[**9**] And here is the mind which hath wisdom. The seven heads are seven mountains, on which the woman sitteth.

[**10**] And there are seven kings: five are fallen, and one is, and the other is not yet come; and when he cometh, he must continue a short space.

[**11**] And the beast that was, and is not, even he is the eighth, and is of the seven, and goeth into perdition.

[**12**] And the ten horns which thou sawest are ten kings, which have received no kingdom as yet; but receive power as kings one hour with the beast.

[**13**] These have one mind, and shall give their power and strength unto the beast.

[**14**] These shall make war with the Lamb, and the Lamb shall overcome them: for he is Lord of lords, and King of kings: and they that are with him are called, and chosen, and faithful.

[**15**] And he saith unto me, The waters which thou sawest, where the whore sitteth, are peoples, and multitudes, and nations, and tongues.

[**16**] And the ten horns which thou sawest upon the beast, these shall hate the whore, and shall make her desolate and naked, and shall eat her flesh, and burn her with fire.

[**17**] For God hath put in their hearts to fulfil his will, and to agree, and give their kingdom unto the beast, until the words of God shall be fulfilled.

[**18**] And the woman which thou sawest is that great city, which reigneth over the kings of the earth.

Mystery Babylon

Revelation, Chapter 17

Notes

Preamble: a teaching in preparation for Chapter 17

Before beginning the study of Chapters 17 and 18, we should take a brief look at the reason the scriptures refer to the fall of the end-time rule as 'the fall of Babylon'. Here I take the liberty of borrowing from radio sermons delivered by Wayne Carver and published by the Christian Jew Foundation.[31] I've generally paraphrased his sermons rather than taking them word-for-word and, in some cases, have either excluded portions or included additional information; nevertheless, I attribute a great deal of my understanding to his teachings and those of this remarkable Foundation.

Through Bible passages relating to Babylon, it becomes clear that this name stands for both a great system of religious error, as well as an abyss of political folly. The religious system's fall is recorded in Revelation 17. The political system's end is seen in Revelation 18. To properly understand the significance of the events in Revelation 17 and 18, we need to know the origin and background of this dual system God calls 'Mystery, Babylon'.

The passages of Scripture that deal with ecclesiastical (religious) Babylon show that the name 'Mystery, Babylon' refers to the counterfeit religion that plagued Israel in the Old Testament, as well as the Church of the New Testament. Since the apostolic days, this system has tremendously influenced the Church's move from Biblical simplicity to apostate confusion. In keeping with Satan's principle of offering a poor substitute for God's perfect plan, 'Mystery, Babylon' became the source of all counterfeit religions. Sometimes it takes the form of pseudo-Christianity; sometimes, it becomes an obvious form of paganism. Its most deceptive form, however, was found in the old religious system known as 'Romanism' and is now found in the current ecumenical 'New Age' movement.

In its political form, 'Mystery, Babylon' describes a plan for a one-world empire which will control all people of the earth from a single capital. This world system is to be governed by one man whom Satan will place upon the world throne. The first manifestation of this aspect of 'Mystery, Babylon' is found in Genesis 11: 1-9, where the descendents of Ham from the early post-flood era gathered together to build Nimrod's capital.

Nimrod (great grandson of Noah), the founder of Babel, engineered a ziggurat in which he would be worshiped as god, with his wife, Semiramis I, honoured as the queen of heaven. Later, she proclaimed herself the 'mother

---

[31] Wayne Carver, "Mystery Babylon" from Radio Sermons (San Antonio, TX: The Christian Jew Foundation Publications)

of god', as she gave birth to the bastard child, Tammuz (Baal). He was born well after the death of Nimrod, and was presented as a virgin birth of the reincarnated Nimrod. The 'Babylonian Mystery' system recognized him as the 'messiah'.

Notes

The religious system that Semiramis founded included many heathen sacred rites, one very prominent act being that of 'consecrated' prostitution. When God created many languages because of His displeasure with Nimrod's attempt to build the tower of Bab-El (Gate of God), He dispersed people all over the earth and they carried this false religion with them (see Jeremiah 7: 17-18). The so-called 'holy mother' and her illegitimate son were known by many names; for example, in Egypt, they became 'Isis and Osiris'; in Rome, 'Venus and Cupid'; in Greece, 'Aphrodite and Pan.' Pagan China knew the mother as 'Sing-Moo' long before Christianity came to China. Satan's plan was to discredit the true virgin birth which would, in reality, occur some 3000 years later.[32]

These two evil systems, joined together in end times, will form the revised Roman Empire governed by the appointed world ruler, the Antichrist. The ecumenical Church will be ruled by the False prophet, supposedly submissive to the Antichrist.

Another important point to understand before going on in this study is that God often gives us several different perspectives of the same event. This sometimes can be confusing and misleading, especially if we read the *Revelation* expecting things to follow in chronological order. We must recognize that God gives a full overview of forthcoming events and then, along the way, He opens windows and provides details of the same event. Once this is understood, everything falls automatically into place and the *Revelation* opens to us.

Such is the case for Chapters 17 and 18. The fall of Babylon is announced in Revelation, Chapter 14: 8, and the details of her judgment unfold in Chapters 17 and 18. We must not suppose that the subject matter of these chapters follow on from the outpouring of the seventh vial. This account is not chronological. The fact is, the fall of religious Babylon occurs before the outpouring of the vials. The great Mother of Harlotry is the first to fall. She falls at the hand of the Antichrist and the disgruntled nations (Daniel 7: 24-26; Revelation 17:16). Political Babylon's fall begins with the presentation of the vials (Revelation 15: 6-8) and ends with the outpouring of the seventh vial – in unison with the seventh trumpet wrath as it finishes sounding (Revelation 11: 15-19). To confirm the validity of this statement, if needed, re-read the Study notes for Revelation 11: 15-19.

---

[32]Further teaching on this subject can be found in the following sources: Alexander Hislop, The Two Babylons (Neptune, NJ: Loizeaux Brothers, Inc., 1858); A. S. Kapelrud, Baal in the Ras Shamra (1950); W. F. Albright, Yahweh and the gods of Canaan (1968); N. C. Habel, Yahweh Versus Baal (1964).

An Unholy Alliance / The Woman and the Scarlet Beast

Revelation 17: 1

1 And there came one of the seven angels which had the seven vials, and talked with me, saying unto me, Come hither; I will shew unto thee the judgment of the great whore that sitteth upon many waters:

Notes

One of the seven angels who participates in the judgment of Babylon, through the outpouring of the seven vials (Revelation 15: 6-7), invites John to come and see '*the judgment of the great whore that sitteth upon many waters*'. The angel reveals to John, throughout Chapters 17 and 18, the whole of the evil network of Satan and its annihilation through the wraths and, later, at the hand of Jesus, the Lamb (Revelation 17: 14). This woman, the '*great whore*', represents the religious system which has inherited a conglomeration of false religions of the world. The woman is defined as a city (verse 18) and the city is Rome (verse 9), the religious capital of the world. The many waters on which she sits are people -- multitudes and nations and tongues (verse 15). Thus, John is about to see the satanic False prophet, and Rome the centre of the World Church.

Revelation 17: 2

2 With whom the kings of the earth have committed fornication, and the inhabitants of the earth have been made drunk with the wine of her fornication.

An unholy alliance between the Church and State during this period has turned the hearts of kings and nations and multitudes and people to false gods with immoral and adulterous rites and rituals. These replace what God has sanctioned in His word. The False prophet woos and beguiles man and makes him forget the ten commandments, God's righteous will for man. For this reason, the city and its fiendish dictators are destroyed, as Jeremiah 51: 6-8 prophesied:

6 ...this is the time of the Lord's vengeance; he will render unto her a recompense.
7 Babylon hath been a golden cup in the Lord's hand (a cup of iniquity), *that made all the earth drunken: the nations have drunken of her wine; therefore the nations are mad.*
8 Babylon is suddenly fallen and destroyed: howl for her; take balm for her pain...

Revelation 17: 3

3 So he carried me away in the spirit into the wilderness: and I saw a woman sit upon a scarlet coloured beast, full of names of blasphemy, having seven heads and ten horns.

In his body, John remains on Patmos; however, in spirit, he is carried away by the angel. He sees a woman sitting on a scarlet-coloured beast. The

Notes

Beast is an unholy and ungodly creature for his heart is full of blasphemy. He is the Antichrist with his head in the seven hills of Rome (verse 9) and his horns decked with ten kings from the nations (verse 12). He is the embodiment of Satan; the great red dragon of Revelation 12: 3, having seven heads and ten horns and seven crowns upon his head. The Beast has been identified previously as the Antichrist in Chapter 13: 1-2 where it is stated that he has '*seven heads and ten horns, and upon his horns ten crowns, and upon his heads the name of blasphemy*.' He is there likened to all the previous great powers before him. John states that the dragon gave the Beast his power, and his seat, and his great authority (Revelation 13: 4). Without a doubt, John is seeing the False prophet riding on the Beast. This act of riding on the Beast demonstrates that he is in control and asserts authority equal to, if not greater than, the Antichrist himself as the false trinity is set up. Another connotation is that the False prophet comes into power with the consent of the Antichrist; thus, he rides in on the power and strength of the Beast.

Revelation 17: 4 - 5

4 *And the woman was arrayed in purple and scarlet colour, and decked with gold and precious stones and pearls, having a golden cup in her hand full of abominations and filthiness of her fornication:*

5 *And upon her forehead was a name written, MYSTERY, BABYLON THE GREAT, THE MOTHER OF HARLOTS AND ABOMINATIONS OF THE EARTH.*

The False prophet is arrayed with the merchandise of her trade: she is extravagantly and lustfully girded with gold and precious stones and pearls. In so doing, she appears as a prostitute and a high priestess. She follows the example of Semiramis I, the so-called queen of heaven (Jeremiah 7: 17-18; Jeremiah 44: 15-19), as she displays her trade wearing her name upon her forehead and frontlet (the custom at the time to identify a consecrated prostitute). Her names are written: Mystery, Babylon the Great, the Mother of Harlots and Abominations of the Earth. These names depict Rome as the corrupt prostitute among nations and the False prophet as her administrator.

Revelation 17: 6

6 *And I saw the woman drunken with the blood of the saints, and with the blood of the martyrs of Jesus: and when I saw her, I wondered with great admiration.*

John sees the city of Rome, the harlot in particular, intoxicated by the blood of the Saints; addicted to her acts of persecution; maddened and driven by her need to destroy all who oppose her will. The depth and depravity of her wickedness mystifies and amazes John.

Notes

Revelation 17: 7

7 *And the angel said unto me, Wherefore didst thou marvel? I will tell thee the mystery of the woman, and of the beast that carrieth her, which hath the seven heads and ten horns.*

The angel notes that John is astonished and bewildered by what he sees, and says that he will reveal to John the mystery of iniquity (II Thessalonians 2: 7-9): the Antichrist and the False prophet, both usurpers of the rights of the Father and the Son.

Revelation 17: 8

8 *The beast that thou sawest was, and is not; and shall ascend out of the bottomless pit, and go into perdition: and they that dwell on the earth shall wonder, whose names were not written in the book of life from the foundation of the world, when they behold the beast that was, and is not, and yet is.*

The Beast '*who was*' refers to the past tyranny of ancient Rome and the holy Roman Empire. The phrase '*is not*' indicates that Rome's power was broken and, by all appearances, dispensed with. However, it '*yet is*' because it simmers, almost unseen, in the facade of Christianity and democracy. It will rise again; reincarnated, as it were, with all the satanic wrath of hell under the dictatorship of the Antichrist and the False prophet. Their final destiny, of course, is hell.

Joining them will be those of the world who are designated to destruction since they have chosen evil instead of righteousness. This group wonders with great admiration and awe when they behold the revived tyranny of the Roman Empire, which '*was, and is not, and yet is.*' They have consented to live under the influence of Satan and have submitted to the controlling power of the revitalized oppression of one world rule.

Revelation 17: 9

9 *And here is the mind which hath wisdom. The seven heads are seven mountains, on which the woman sitteth.*

The seven heads are the seven hills of Rome (the first capital) from which 'Mystery, Babylon the Great', the revived Roman Empire, will rule. This verse demonstrates that God never leaves us uninformed; the scriptures always provide us with necessary information.

Seven Kings / Five Are Fallen

Revelation 17: 10

10 And there are seven kings: five are fallen, and one is, and the other is not yet come; and when he cometh, he must continue a short space.

Notes

There are seven kings in all; five are fallen. The five fallen kings are the kings of Daniel's prophecies. Daniel speaks of five powers who reign: the King of Babylon (Nebuchadnezzar); the two prominent kings of the dual Persia-Media Empire (Darius, the Mede, and Cyrus, the Persian); the King of Greece (Alexander the Great); and the Caesars of Rome. The one '*who is*' (the power of Rome which never died and will be resurrected in the Antichrist) has all the attributes of the former powers. This is envisioned by Daniel and is recorded in Daniel 7: 3-7, and again affirmed in Revelation 13: 1-2 when John receives a similar vision upon seeing the beast rise up out of the sea. The beast which he sees (the Antichrist) is as all of Daniel's beasts: he is like '*a leopard and his feet are as the feet of a bear; his mouth as the mouth of the lion; and the dragon gives him power and the seat of authority'* of Satan. This beast in bodily form is the sixth king in the person of the Antichrist.

Daniel 7: 7 sees this power as '*dreadful and terrible and exceedingly strong; and it had great iron teeth: it devoured and brake in pieces, and stamped the residue with the feet of it*'; thus, the horrible and terrible persecution of the Tribulation Saints. It differs from all the other beasts that Daniel has seen before it because it has ten horns. The horns are the ten nations who agree to follow and submit to the Antichrist.

The one who is spoken of as the one who has '*not yet come*; *and when he cometh, he must continue a short space*' is the False prophet (also known as the 'Mother of Harlots' who becomes the leader of the 'One World Church'). Daniel speaks of the rise of the False prophet in Daniel 7: 8:

> *8 ...behold, there came up among them <u>another little horn</u>, before whom there were three of the first horns plucked up by the roots: and, behold, in this horn were eyes like the eyes of man, and a mouth speaking great things.*

The False prophet, though risen last, becomes the dominant one who causes three of the nations to leave the European Confederacy. It appears that the False prophet demonstrates too much power and authority and earns the disfavour of the nations and the Antichrist; thus, he will continue as part of the ruling power for only a short period of time (Revelation 17: 16).

When the Antichrist is established in the temple and firmly supported by his followers, he will feel that he no longer needs the support of the Church and disposes of her services. This likely occurs shortly before the Antichrist is evicted from the temple toward the end of his allotted three and one-half years unhampered reign. In this way, 'Mystery, Babylon the Great', the

Notes

Mother of Harlots, falls first; soon after, the Antichrist follows. Obviously, the False prophet is the seventh of the seven kings.

Before leaving the study of verse ten, consider the following which, I believe, is worthy of your attention.

Some Christian scholars teach that the five fallen kings are the Caesars.[33] They list them as Julius Caesar, Tiberius, Caligula, Clausius and Nero. This teaching is totally unacceptable.

The Caesars are not in keeping with Daniel's vision in Daniel 7: 3-7; nor with John's vision in Revelation 13: 1-2. Clearly, verses 8 to 10 speak of a ruling power (a king or kingdom) with all the qualities of the former five fallen powers, the evil of which never died and will be reinstated under the Antichrist.

Revelation 17: 11

11 And the beast that was, and is not, even he is the eighth, and is of the seven, and goeth into perdition.

This verse is simply a play on numbers, something often done in apocalyptic writing. It deals entirely with the Beast, the one that '*was and is not, and yet is*.' He is the one that '*goeth into perdition*' (Hell) and is joined there by the False prophet (Revelation 19: 20).

Revelation 17: 12

12 And the ten horns which thou sawest are ten kings, which have received no kingdom as yet; but receive power as kings one hour with the beast.

Although the reader is likely more than familiar with the ten kings by now, it should be appreciated that God never leaves us dependent on man's interpretation of His inspired word, nor on one man's prophesy. He strengthens prophesy with prophesy, teaching with teaching; such is the case here. He tells us that the horns are kings who will receive power for a short time only.

---

[33]Tim Lahaye, Revelation Unveiled (Grand Rapids, MI: Zondervan Publishing, 1999), p. 262; Ray C. Stedman, God's Final Word (Grand Rapids, MI: Discovery House Publishers, 1999), p. 246; Oliver B. Greene, The Revelation - Verse by Verse Study (Greenville, SC: The Gospel Hour, Inc., Reprinted 1976), p. 412; J. Vernon McGee, Thru The Bible, Vol. V (Nashville: Thomas Nelson Publishers, 1983), p. 1034; Jack Van Impe, Prophecy Bible (Troy, MI: Jack Van Impe Ministries International, 1998), p. 132.

Allegiance to the Antichrist / The Victorious Jesus / The Called and the Chosen

Notes

For an additional image of the ten kingdoms, turn to Daniel 2: 31-45. Read a short passage from Nebuchadnezzar's dream as it pertains to the image with the ten toes (which are the ten kingdoms):

32 *This image's head was of fine gold* (Babylon), *his breast and his arms of silver* (Persia-Media), *his belly and his thighs of brass* (Greece),

33 *His legs of iron, his feet part of iron and part of clay* (dual capitals of the Roman Empire).

(verse 41 is more explicit; it reads...)

41 *And whereas thou sawest the feet and toes* (the Antichrist and the ten nations), *part of potters' clay, and part of iron.*

34 *Thou sawest till that a stone was cut out without hands* (Christ), *which smote the image upon his feet that were of iron and clay, and brake them to pieces.* (the fall of the Anti-christ and the ten nations)

Revelation 17: 13

13 These have one mind, and shall give their power and strength unto the beast.

The ten united nations of the European Confederacy share the carnal mind and political vision of the Beast; thus, they stand behind him with all their power and strength. Without reservation, they swear allegiance to the one world rule.

Revelation 17: 14

14 These shall make war with the Lamb, and the Lamb shall overcome them: for he is Lord of lords, and King of kings: and they that are with him are called, and chosen, and faithful.

The ensuing battle is meticulously recorded in Ezekiel, Chapters 38 and 39. It is predestined by the will of God. God, in His infinite wisdom, uses the greed of the nations to draw them to Israel so that the unrepentant may fall under the power of the victorious Jesus.

The '*called, and chosen*' spoken of in this verse must surely be the Old Testament Saints and the 144,000 sealed Israelites (Matthew 8: 11; Revelation 14: 1). The Church as a body is never directly referred to as '*the called, and the chosen.*' Thus, we may safely assume that the Church will not return with Jesus. (See also the Study notes for Revelation 19: 14.)

It should be understood that individuals are called and some are specially chosen (predestined) to appointed ministries (Romans 8: 28-30; II Timothy

1: 9); however, the only people who are called and chosen, as a nation, are the Israelites.

Notes

Revelation 17: 15

15 And he saith unto me, The waters which thou sawest, where the whore sitteth, are peoples, and multitudes, and nations, and tongues.

This verse is one which clarifies and instructs so that we are not confused by verse one when it speaks of '*the great whore that sitteth upon many waters*.' It tells us that the waters on which the whore sits are not literal waters but are, in fact, people and nations. This indicates that the false church has influence and power over the multitudes.

Revelation 17: 16

16 And the ten horns which thou sawest upon the beast, these shall hate the whore, and shall make her desolate and naked, and shall eat her flesh, and burn her with fire.

We have previously been informed in verse 12 that the ten horns upon the beast represent the ten kings. These kings so hate and resent the controlling power of the False prophet that they strip her of power, as well as excessively abuse her, causing bodily harm. We know from scriptures that they don't literally eat her flesh, nor burn her to death, for if such were the case, she would not be available for judgment when Christ returns with his Israelite armies from heaven. Revelation 19: 20 states:

> *20 And the Beast was taken, and with him the False prophet that wrought miracles before him, with which he deceived them that had received the mark of the beast, and them that worshiped his image. These both were cast alive into a lake of fire burning with brimstone.*

Revelation 17: 17

17 For God hath put in their hearts to fulfill his will, and to agree, and give their kingdom unto the beast, until the words of God shall be fulfilled.

Just as Pharaoh was putty in the hand of God, so also are the ten nations of the Tribulation years. God causes them to fulfill His will and give their kingdoms to the Beast. He allows their hearts to be filled with hatred for the False prophet, which brings about his/her loss of power. God's will is thus accomplished, as prophesied through Daniel 7: 26:

> *26 But the judgment shall sit, and they shall take away his dominion, to consume and to destroy it unto the end.*

God always prevails!

Jerusalem, the Capital of Our King

Notes

Revelation 17: 18

18 And the woman which thou sawest is that great city, which reigneth over the kings of the earth.

The woman, 'Mystery, Babylon the Mother of Harlots', is likened to the great city of Rome – the only city to continuously command the reverence and submission of a major portion of the religious world. During the Tribulation, this revitalized satanic system, namely ecclesiastical Babylon, will again actively rule over the kings of the earth from the great city of Rome. Halfway through the Tribulation, after the death of the two witnesses (Revelation 11: 3-12), a second capital will be founded in Jerusalem. At Christ's return, for his victorious Millennial reign, it will remain '*that great city, which reigneth over the kings of the earth.*'

CHAPTER EIGHTEEN

Chapter Eighteen is given over entirely to the fall of political and commercial Babylon. The Study states that Rome will be her first capital but, after the two witnesses have been put to death, a second capital will be established in Jerusalem (a reflection of the old Roman Empire which had two capitals). Babylon, the harlot (the World Church / the False prophet), is destroyed by the kings of the earth; while political Babylon (the Beast / the Antichrist), falls under the outpouring of the seven vials. When the World Church falls, the kings and the nations rejoice. When political and commercial Babylon falls, the kings and the merchants bewail and lament her passing.

Revelation 18

[1] And after these things I saw another angel come down from heaven, having great power; and the earth was lightened with his glory.
[2] And he cried mightily with a strong voice, saying, Babylon the great is fallen, is fallen, and is become the habitation of devils, and the hold of every foul spirit, and a cage of every unclean and hateful bird.
[3] For all nations have drunk of the wine of the wrath of her fornication, and the kings of the earth have committed fornication with her, and the merchants of the earth are waxed rich through the abundance of her delicacies.
[4] And I heard another voice from heaven, saying, Come out of her, my people, that ye be not partakers of her sins, and that ye receive not of her plagues.
[5] For her sins have reached unto heaven, and God hath remembered her iniquities.
[6] Reward her even as she rewarded you, and double unto her double according to her works: in the cup which she hath filled fill to her double.
[7] How much she hath glorified herself, and lived deliciously, so much torment and sorrow give her: for she saith in her heart, I sit a queen, and am no widow, and shall see no sorrow.
[8] Therefore shall her plagues come in one day, death, and mourning, and famine; and she shall be utterly burned with fire: for strong is the Lord God who judgeth her.
[9] And the kings of the earth, who have committed fornication and lived deliciously with her, shall bewail her, and lament for her, when they shall see the smoke of her burning,
[10] Standing afar off for the fear of her torment, saying, Alas, alas, that great city Babylon, that mighty city! for in one hour is thy judgment come.
[11] And the merchants of the earth shall weep and mourn over her; for no man buyeth their merchandise any more:
[12] The merchandise of gold, and silver, and precious stones, and of pearls, and fine linen, and purple, and silk, and scarlet, and all thyine wood, and all manner vessels of ivory, and all manner vessels of most precious wood, and of brass, and iron, and marble,
[13] And cinnamon, and odours, and ointments, and frankincense, and wine, and oil, and fine flour, and wheat, and beasts, and sheep, and horses, and chariots, and slaves, and souls of men.
[14] And the fruits that thy soul lusted after are departed from thee, and all things which were dainty and goodly are departed from thee, and thou shalt find them no more at all.
[15] The merchants of these things, which were made rich by her, shall stand afar off for the fear of her torment, weeping and wailing,
[16] And saying, Alas, alas, that great city, that was clothed in fine linen, and purple, and scarlet, and decked with gold, and precious stones, and pearls!
[17] For in one hour so great riches is come to nought. And every shipmaster, and all the company in ships, and sailors, and as many as trade by sea, stood afar off,
[18] And cried when they saw the smoke of her burning, saying, What city is like unto this great city!
[19] And they cast dust on their heads, and cried, weeping and wailing, saying, Alas, alas, that great city, wherein were made rich all that had ships in the sea by reason of her costliness! for in one hour is she made desolate.
[20] Rejoice over her, thou heaven, and ye holy apostles and prophets; for God hath avenged you on her.
[21] And a mighty angel took up a stone like a great millstone, and cast it into the sea, saying, Thus with violence shall that great city Babylon be thrown down, and shall be found no more at all.
[22] And the voice of harpers, and musicians, and of pipers, and trumpeters, shall be heard no more at all in thee; and no craftsman, of whatsoever craft he be, shall be found any more in thee; and the sound of a millstone shall be heard no more at all in thee;
[23] And the light of a candle shall shine no more at all in thee; and the voice of the bridegroom and of the bride shall be heard no more at all in thee: for thy merchants were the great men of the earth; for by thy sorceries were all nations deceived.
[24] And in her was found the blood of prophets, and of saints, and of all that were slain upon the earth.

Preamble / An Angel Announces the Fall of the Antichrist

Notes

Revelation, Chapter 18

Preamble: a teaching in preparation for Chapter 18

Chapter 18 is given over entirely to the destruction of political/commercial Babylon. Throughout the chapter, Babylon the Great refers to the revived Roman Empire, with Rome being the first capital for both religious and political Babylon. After the death of the two witnesses, the second capital will be established in Jerusalem. The confirmation of this is found in II Thessalonians 2: 4. In it, we are told that the abomination of desolation (the Antichrist) will oppose God and exalt himself above God. He will sit in the temple of God, claiming himself to be God. The temple will be found in the Holy City. By this we are assured the Antichrist will establish a second capital in God's beloved city of Jerusalem (Micah 4: 1-2).

Naming two capitals is a reflection of the old Roman Empire which had dual capitals: the western capital in Rome and the eastern capital in Constantinople (Istanbul). Babylon, the harlot (the World Church / the False prophet), is destroyed by the kings of the earth (Revelation 17: 16); while political Babylon (the Beast / the Antichrist) falls under the outpouring of the seven vials, as well as at the hand of the victorious Jesus. When religious Babylon falls, the kings rejoice. When political and commercial Babylon falls, the kings and merchants bewail her and lament her passing.

Revelation 18: 1 - 2

1 And after these things I saw another angel come down from heaven, having great power; and the earth was lightened with his glory.

2 And he cried mightily with a strong voice, saying, Babylon the great is fallen, is fallen, and is become the habitation of devils, and the hold of every foul spirit, and a cage of every unclean and hateful bird.

After John has witnessed the fall of religious Babylon, the Mother of Harlots, he sees an angel come down from heaven, shining forth the Glory of God and heralding loudly the fall of Babylon (as foretold in Revelation 14: 8). Obviously, the vials have been poured out and the stage is set for its final actor: the one who sits upon a white horse. We read of him in Revelation 19: 11-16 as the one who has '*eyes as a flame of fire, and on his head were many crowns.*' He is '*clothed with a vesture dipped in blood: and his name is called the Word of God* (Jesus).'

Notes

Revelation 18: 3

3 *For all nations have drunk of the wine of the wrath of her fornication, and the kings of the earth have committed fornication with her, and the merchants of the earth are waxed rich through the abundance of her delicacies.*

'Mystery, Babylon the Great', the abomination of the earth – the Antichrist – has learned his lessons well in co-existing with Babylon, the Mother of Harlots – the False prophet. The Antichrist has become the sole epitome of evil, the living Satan, for the False prophet shed her mantle of immorality upon him in her passing. Nations thirst after the wares of the False prophet because they have become drunken with '*the wrath of her fornication*' and they lust after her evil. The brokers of the earth grow fat and rich through the abundance of their addiction to her '*delicacies'*.

Revelation 18: 4 - 5

4 *And I heard another voice from heaven, saying, Come out of her, my people, that ye be not partakers of her sins, and that ye receive not of her plagues.*

5 *For her sins have reached unto heaven, and God hath remembered her iniquities.*

God in His bountiful mercy reaches out in love by His Spirit, warning the hearts of those who have not yet fallen to separate themselves from the world. His Spirit is compassionate for He knows the power of Satan and the iniquities of 'Mystery, Babylon' – her sins have touched the very gates of heaven.

Revelation 18: 6

6 *Reward her even as she rewarded you, and double unto her double according to her works: in the cup which she hath filled fill to her double.*

We who love the Lord are spared the judgment of this verse. We have One who has taken our punishment when he hung on a cross in Jerusalem on our behalf. It is painful to think that there are those who have been promised double retribution for their evil deeds. Surely, verses like this spur us on to win souls to Christ.

Torment and Sorrow / Rapid Destruction of Babylon

Revelation 18: 7

7 *How much she hath glorified herself, and lived deliciously, so much torment and sorrow give her: for she saith in her heart, I sit a queen, and am no widow, and shall see no sorrow.*

Notes

Babylon the Great, the Antichrist, will reap what he has sown. He believes himself above the punishment meted out to the False prophet for he sees himself as both king and queen of heaven – a transvestite, amphierotic – similar to the bisexual mind-set of our present generation. As with the Antichrist in Daniel's prophecy, Antiochus Epiphanies, the end-time Antichrist probably practices deviant sexual acts. Daniel speaks of Antiochus Epiphanies' sin in Daniel 11: 37 when he says:

37 *Neither shall he regard the God of his fathers, nor the desire of women, nor regard any god: for he shall magnify himself above all.*

His statement that he is '*no widow*' reveals that he has a husband and is a practicing member of the doctrine of Nimrod and Semiramis – unless, of course, the Antichrist is a woman. If such is the case, she sits as a queen and is no widow (she has a partner) and arrogantly states she will see no sorrow.

Some may question the way this study changes the female gender, used in this verse, to male gender and then back again into female. This is done because all of these verses are referring to the Antichrist who, as he arose up out of the seas in Revelation 13: 1, is spoken of in male gender terms, as is the False prophet when he comes up out of the earth in Revelation 13: 11. The scriptures, as well as our culture, seem always to refer to nations and cities in the female gender. For this reason, and in this case, the gender is totally interchangeable.

The apostle James, writing by inspiration of God, relates the rewards to those who follow in the path of 'Mystery, Babylon'. Read James 5: 1-3:

1 *...rich men, weep and howl for your miseries that shall come upon you.*
2 *Your riches are corrupted, and your garments are moth-eaten.*
3 *Your gold and silver is cankered; and the rust of them shall be a witness against you, and shall eat your flesh as it were fire...*

Revelation 18: 8

8 *Therefore shall her plagues come in one day, death, and mourning, and famine; and she shall be utterly burned with fire: for strong is the Lord God who judgeth her.*

Rapid is the destruction of Babylon the great. The vial wraths are poured out quickly, one upon the other, within the framework of Daniel's prophetic

Notes

1290 days (Daniel 12: 11). The Antichrist is allowed to reign in the temple in Jerusalem for 1260 of these days (Daniel 9: 27; Daniel 12: 6-7). At the end of this time, he is evicted from the temple by the smoke from the Glory of God (Revelation 15: 6-8).

It is during the first 30 of the 75 days which follow his eviction, that the vile judgments are poured out. Though still reigning until the end of his allotted 1290 days (Daniel 12: 11), he and the nations – those who have taken the mark of the Beast – hide from the face of God; the wrath of the Lamb; and the fury of the vile judgments (Revelation 6: 15-16). Having been given his allotted 1290 days reign, he now faces the return of Jesus. Jesus then administers the last 45 days of judgment, thereby fulfilling the 1335 days of Daniel's prophetic prophecy (Daniel 12: 12). For more detail, see the Study notes for Revelation 11: 15-19.

All of these plagues of misery come down rapidly upon the Antichrist. When Christ's 45 days of retaliation are finished, John sees Jesus lay hold of the Beast and the False prophet and cast them alive into the lake of fire burning with brimstone (Revelation 19: 20).

Blessed are they who survive the judgments of those days and come through to the 1335 days, for they are the ones who will attend the wedding feast of the Lamb's wife (Revelation 19: 7-9) and enter the Millennial reign.

It's interesting to read what Daniel 7: 11-12 has to say about this judgment and, more particularly, about the fate of the rest of the beasts:

> 12 *As concerning the rest of the beasts* (the nations), *they had their dominion taken away: yet their lives were prolonged for a season and time.*

If they have taken the mark of the Beast, we must conclude that they remain until they have viewed the victory of Jesus and then fall at his hand, as recorded in Revelation 19: 21. Those nations which have not taken the mark of the Beast will go on into the Millennium (Zechariah 14: 16-19)

Revelation 18: 9 - 15

9 *And the kings of the earth, who have committed fornication and lived deliciously with her, shall bewail her, and lament for her, when they shall see the smoke of her burning,*

10 *Standing afar off for the fear of her torment, saying, Alas, alas that great city Babylon, that mighty city! for in one hour is thy judgment come.*

11 *And the merchants of the earth shall weep and mourn over her; for no man buyeth their merchandise any more:*

Notes

12 *The merchandise of gold, and silver, and precious stones, and of pearls, and fine linen, and purple, and silk, and scarlet, and all thyine wood, and all manner vessels of ivory, and all manner vessels of most precious wood, and of brass, and iron, and marble,*
13 *And cinnamon, and odours, and ointments, and frankincense, and wine, and oil, and fine flour, and wheat, and beasts, and sheep, and horses, and chariots, and slaves, and souls of men.*
14 *And the fruits that thy soul lusted after are departed from thee, and all things which were dainty and goodly are departed from thee, and thou shalt find them no more at all.*
15 *The merchants of these things, which were made rich by her, shall stand afar off for the fear of her torment, weeping and wailing,*

Verses 9 to 15 are self-explanatory and need no detailed explanation. It's obvious that the effects of the vial wraths seriously upset the whole commercial and social life of the entire world. The sadness is that souls have been eternally lost for the majority of the world, collectively, believes that the cause of such devastation upon the earth is simply natural forces having gone astray. To a degree, they may be right; however, they fail to recognize the One who is the author of these forces, nor do they acknowledge the holy hand at the helm.

Revelation 18: 16 - 19

16 *And saying, Alas, alas that great city, that was clothed in fine linen, and purple, and scarlet, and decked with gold, and precious stones, and pearls!*
17 *For in one hour so great riches is come to nought. And every shipmaster, and all the company in ships, and sailors, and as many as trade by sea, stood afar off,*
18 *And cried when they saw the smoke of her burning, saying, What city is like unto this great city!*
19 *And they cast dust on their heads, and cried, weeping and wailing, saying, Alas, alas that great city, wherein were made rich all that had ships in the sea by reason of her costliness! for in one hour is she made desolate.*

Great is the mourning of those who have profited through industrial and commercial Babylon. '*The ships stand afar off*', waiting in the ports of the world. The ship masters and sailors bewail the loss of their livelihood. Note the thrice-mentioned rapid destruction of the revived Babylonian system (verses 10, 17, 19) during the outpouring of the vial judgments, an obvious confirmation that the vials are poured out in rapid succession.

One should visit an area recently struck by either earthquake, cyclone, or hurricane winds and flooding to get a deeper sense of the sadness, confusion and fear in the hearts of the people who have so suddenly lost everything

Notes

they have trusted in and built their lives on. Within hours, the sea and the land can rise in violent upheaval and lives are broken in the wake of such destruction.

Not once, in secular news broadcasts, has it been suggested that God might be speaking. So it is and so it will be, magnified many times over, as the world enters the darkness of the Tribulation years.

Revelation 18: 20

20 *Rejoice over her, thou heaven, and ye holy apostles and prophets; for God hath avenged you on her.*

The earth is filled with weeping and wailing; with confusion and mourning; with hunger and death; with fire and fornication; and the end is judgment. These are the rewards for those who scorn salvation and harm God's holy people.

Heaven is filled with rejoicing and praise; with angels and heavenly creatures; with apostles and prophets; with joy and peace; with the song of the Alleluia chorus in the presence of righteousness and the glory of our Lord. These are the rewards of the avenged, the holy people of God.

Revelation 18: 21 - 23

21 *And a mighty angel took up a stone like a great millstone, and cast it into the sea, saying, Thus with violence shall that great city Babylon be thrown down, and shall be found no more at all.*

22 *And the voice of harpers, and musicians, and of pipers, and trumpeters, shall be heard no more at all in thee; and no craftsman, of whatsoever craft he be, shall be found any more in thee; and the sound of a millstone shall be heard no more at all in thee;*

23 *And the light of a candle shall shine no more at all in thee; and the voice of the bridegroom and of the bride shall be heard no more at all in thee: for thy merchants were the great men of the earth; for by thy sorceries were all nations deceived.*

The mighty angel referred to is probably Jesus, for it is he who activates the violence against the great city of Babylon. Though similar to the prophesies against literal Babylon, these verses have to do with figurative Babylon, the evil of which will be totally destroyed when Jesus casts the Antichrist and False prophet into the Lake of Fire (Revelation 19: 20). The Spirit chooses to pen these verses to impress on the reader the total cleansing of the earth from the wickedness of 'Mystery, Babylon the Great' – probably also that

The Fall of Literal Babylon

Notes

man might compare it to the destruction of literal Babylon by fire. God's universe and temple will be purged by the violence of the wraths and the smoke from the Glory of God (Revelation 15: 8). Never again will Babylon's demonic music be heard on earth; her crafts and her enterprises will be silenced. The voice of evil and the sorceries of the wicked will no longer deceive the nations.

The activities around the Persian Gulf and the final end of literal Babylon is recorded in Jeremiah, Chapters 50 to 51. Even now, Jesus is consenting to and directing affliction upon literal Babylon in the present Persian Gulf situation. This area has already been visited by the destroying wind (Jeremiah 51:1) which man called 'Operation Desert Storm'; and has again been dredged by 'Operation Desert Fox'. The end of this conflict will see literal Babylon (the area ruled over or threatened by Saddam Hussein) engulfed by fire, as was Sodom and Gomorrah. The prophecy of Jeremiah 50: 39-40 says:

> *39 ...and it shall be no more inhabited for ever; neither shall it be dwelt in from generation to generation.*
>
> *40 As God overthrew Sodom and Gomorrah and the neighbour cities thereof, saith the Lord; so shall no man abide there, neither shall any son of man dwell therein.*

After Jeremiah had received the prophecies pertaining to the evil that would come upon literal Babylon, he wrote to a prince or chamberlain, named Seraiah. He told Seraiah what God had shown him as to the final destruction of literal Babylon. To hear what he wrote to his friend, Seraiah, and to hear and see Seraiah's response, read Jeremiah 51: 62-64:

> *62 Then shalt thou* (Seraiah) *say, O Lord, thou hast spoken against this place, to cut it off, that none shall remain in it, neither man nor beast, but that it shall be desolate for ever.*
>
> *63 And it shall be, when thou hast made an end of reading this book* (this letter), *that thou shalt bind a stone to it, and cast it into the midst of Euphrates:*
>
> *64 And thou shalt say, Thus shall Babylon sink, and shall not rise from the evil that I* (God) *will bring upon her...*

Jeremiah's prophesy pertaining to the fall of literal Babylon is similar to the fall of the end-time figurative Babylon spoken of in these three verses of Revelation now under study. Each has cast a millstone (a burdensome prediction) into the waters; each proclaims desolation to the areas referred to as Babylon. It seems almost obvious that the troubles in the east (the former Babylon) are a foreshadowing of the final destruction of figurative Babylon which will arise in these last days.

Martyred Prophets and Saints

Notes

It would seem that the Persian Gulf and the Palestinian unrest will eventually overflow with full force into Israel. Inevitably, all nations will gather on Israeli soil to fight the great battle of Armageddon. Please read Appendix 8, 'Warning Against False Teaching', to clarify further this portion of our study.

Revelation 18: 24

24 And in her was found the blood of prophets, and of saints, and of all that were slain upon the earth.

From the time that Mystery, Babylon the Great, the Mother of Harlots and Abominations of the Earth, was birthed by the disobedience of Nimrod, until the day of judgment, many prophets and Saints will have shed their blood in defense of their faith in God and His Son, Jesus Christ.

CHAPTER NINETEEN

Chapter Nineteen contains a detailed report of Christ's second coming. It records all that will occur within Daniel's last prophetic forty-five days – the time between the twelve hundred and ninety days of *Daniel* 12: 11 and the thirteen hundred and thirty-five days of *Daniel* 12: 12. The chapter records the joy of the alleluia chorus; the marriage of the Lamb to Israel, his wife; the battle of Armageddon; the feast of the fowl; the annihilation of the Beast and the False prophet; and the remnant of evil man slain.

Revelation 19

[**1**] And after these things I heard a great voice of much people in heaven, saying, Alleluia; Salvation, and glory, and honour, and power, unto the Lord our God:
[**2**] For true and righteous are his judgments: for he hath judged the great whore, which did corrupt the earth with her fornication, and hath avenged the blood of his servants at her hand.
[**3**] And again they said, Alleluia. And her smoke rose up for ever and ever.
[**4**] And the four and twenty elders and the four beasts fell down and worshipped God that sat on the throne, saying, Amen; Alleluia.
[**5**] And a voice came out of the throne, saying, Praise our God, all ye his servants, and ye that fear him, both small and great.
[**6**] And I heard as it were the voice of a great multitude, and as the voice of many waters, and as the voice of mighty thunderings, saying, Alleluia: for the Lord God omnipotent reigneth.
[**7**] Let us be glad and rejoice, and give honour to him: for the marriage of the Lamb is come, and his wife hath made herself ready.
[**8**] And to her was granted that she should be arrayed in fine linen, clean and white: for the fine linen is the righteousness of saints.
[**9**] And he saith unto me, Write, Blessed are they which are called unto the marriage supper of the Lamb. And he saith unto me, These are the true sayings of God.
[**10**] And I fell at his feet to worship him. And he said unto me, See thou do it not: I am thy fellowservant, and of thy brethren that have the testimony of Jesus: worship God: for the testimony of Jesus is the spirit of prophecy.
[**11**] And I saw heaven opened, and behold a white horse; and he that sat upon him was called Faithful and True, and in righteousness he doth judge and make war.
[**12**] His eyes were as a flame of fire, and on his head were many crowns; and he had a name written, that no man knew, but he himself.
[**13**] And he was clothed with a vesture dipped in blood: and his name is called The Word of God.
[**14**] And the armies which were in heaven followed him upon white horses, clothed in fine linen, white and clean.
[**15**] And out of his mouth goeth a sharp sword, that with it he should smite the nations: and he shall rule them with a rod of iron: and he treadeth the winepress of the fierceness and wrath of Almighty God.
[**16**] And he hath on his vesture and on his thigh a name written, KING OF KINGS, AND LORD OF LORDS.
[**17**] And I saw an angel standing in the sun; and he cried with a loud voice, saying to all the fowls that fly in the midst of heaven, Come and gather yourselves together unto the supper of the great God;
[**18**] That ye may eat the flesh of kings, and the flesh of captains, and the flesh of mighty men, and the flesh of horses, and of them that sit on them, and the flesh of all men, both free and bond, both small and great.
[**19**] And I saw the beast, and the kings of the earth, and their armies, gathered together to make war against him that sat on the horse, and against his army.
[**20**] And the beast was taken, and with him the false prophet that wrought miracles before him, with which he deceived them that had received the mark of the beast, and them that worshipped his image. These both were cast alive into a lake of fire burning with brimstone.
[**21**] And the remnant were slain with the sword of him that sat upon the horse, which sword proceeded out of his mouth: and all the fowls were filled with their flesh.

Preamble / The Alleluia Chorus

Revelation, Chapter 19

Notes

Chapter 19 contains a detailed report of Christ's Second Coming. It records all that will occur within Daniel's prophetic 45 days – the time between the 1290 days of Daniel 12: 11 and the 1335 days of Daniel 12: 12.

Revelation 19: 1 - 2

1 *And after these things I heard a great voice of much people in heaven, saying, Alleluia; Salvation and glory, and honour, and power, unto the Lord our God:*

2 *For true and righteous are his judgments: for he hath judged the great whore, which did corrupt the earth with her fornication, and hath avenged the blood of his servants at her hand.*

John hears many heavenly voices rejoicing. The Saints magnify the Lord because they have witnessed the fall of the great harlot of Babylon: both political and religious Babylon have been destroyed. The Tribulation Saints are avenged as implied in Revelation 6: 9-11. The whole of heaven declares the judgments of our Lord God to be true and righteous and call out praises to Jehovah. With great exultation, they worship the Lord, crying: '*Alleluia, Salvation and Glory and Honour and Power unto the Lord our God.*'

Revelation 19: 3

3 *And again they said, Alleluia. And her smoke rose up for ever and ever.*

Again they say '*Alleluia*' – "praise ye the Lord." Revelation 14: 11 has previously foretold that the smoke of their torment (those who fall with Babylon) will ascend forever and they have no rest.

Revelation 19: 4

4 *And the four and twenty elders and the four beasts fell down and worshipped God that sat on the throne, saying, Amen; Alleluia.*

The four and twenty elders (the Raptured Saints) and the four beasts (part of the heavenly host; identified in Isaiah 6: 1-3 as seraphim) fall down and worship God, crying '*Praise ye the lord, Amen*'. They, too, rejoice that God has avenged the blood of His servants.

A Command to Praise / The Marriage of the Lamb to His Wife

Notes

Revelation 19: 5

5 *And a voice came out of the throne, saying, Praise our God, all ye his servants, and ye that fear him, both small and great.*

God himself commands that we, His servants, must reverence and praise Him for His true and righteous judgments. Our Lord takes pleasure in our praises and has created us that we might '*clap our hands and shout unto God with the voice of triumph*' (Psalm 47:1). The doxology of the Psalms, among other things, must surely be a guide to perfect praise. Significantly, it is the last chapter of the psalms, Psalm 150: 1-6, that instructs us how to praise; by what means and to what extent we should glorify God:

1 *Praise ye the Lord. Praise God in his sanctuary;: praise him in the firmament of his power.*
2 *Praise him for his mighty acts: praise him according to his excellent greatness.*
3 *Praise him with the sound of the trumpet: praise him with the psaltery and harp.*
4 *Praise him with the timbrel and dance: praise him with stringed instruments and organs.*
5 *Praise him upon the loud cymbals: praise him upon the high sounding cymbals.*
6 *Let every thing that hath breath praise the Lord. Praise ye the Lord.*

Revelation 19: 6

6 *And I heard as it were the voice of a great multitude, and as the voice of many waters, and as the voice of mighty thunderings, saying, Alleluia: for the Lord God omnipotent reigneth.*

The voices of the Saints of all ages sound forth their praise and joy: their time of waiting will soon be over. The prayers of the millions throughout the centuries have been answered. There is little wonder that their glad rejoicing should be heard '*as the voice of many waters and mighty thunderings.*' Man has longed for many generations to cry '*Alleluia, our Lord God all-powerful reigneth.*'

Revelation 19: 7 - 9

7 *Let us be glad and rejoice, and give honour to him: for the marriage of the Lamb is come, and his wife hath made herself ready.*
8 *And to her was granted that she should be arrayed in fine linen, clean and white: for the fine linen is the righteousness of saints.*
9 *And he saith unto me, Write, Blessed are they which are called unto the marriage supper of the Lamb. And he saith unto me, These are the true sayings of God.*

The Bride and the Wife Become One

Notes

These verses, in my heart, are among the most wondrous of all the many marvellous revelations in the whole of the scriptures. I see here the deliverer removing ungodliness from the house of Jacob. I see the Elect (Israel) being the beloved of the Father. I hear God saying, “This is my covenant unto them, when I shall forgive their sins.” (II Chronicles 7: 14; Ezekiel, Chapters 36 and 37). I rejoice that the gifts and calling of God are without repentance (Romans 11: 25- 33). I see God having mercy upon the whole House of Israel. I see all of this in the marriage of the Lamb to his wife who, having made herself ready, is now granted that she should be arrayed, not in white robes as a bride, but in fine linen – the robes of the Levitical priesthood. She now has the righteousness of Christ and is as all the Saints before her: the old Testament Saints risen with Christ at his resurrection; the New Testament Saints risen in the Rapture; and, now, the deceased Tribulation Saints and the whole house of Israel are risen and invited to the marriage of the Lamb to become, once again, his wife. She has, at last, made herself ready!

The scriptures always refer to Israel being the wife of Christ and to the Church as the bride. Paul makes reference to the Church being espoused to one husband but concludes with the statement that his hope is that he may present the Church as chaste virgins to Christ (II Corinthians 11: 2). The only time the bride is called the Lamb’s ‘wife’ is in Revelation 21: 9 when John is taken by an angel to view the holy Jerusalem descending out of heaven. This is when the ‘bride’ and the ‘wife’ become one. At this point, God gathers up His bride and comes to earth to dwell with His wife (Revelation 21: 2-3). They are both espoused to the same husband and dwell together as one in the presence of the Father and Son. Read Ephesians 1: 10:

> *10 That in the dispensation of the fullness of times he might gather together in one all things in Christ, both which are in heaven, and which are on earth; even in him:*

The Father, in His mercy, has opened the way and prepared a place for His wife – blessed is the house of Israel and most privileged. She is, as the prodigal son, received by the compassionate Father. She is robed as a levitical bride and a great feast is made ready: the wedding feast of the husband to his wife. Refer ahead to the Study notes for Revelation 21: 2.

In Matthew 25: 1-13, Jesus relates the parable of the wise and foolish virgins. The five wise virgins (the Church) become the bride of Christ. When the bridegroom comes, they are ready and go in with him to the marriage breakfast (he comes at midnight and so this marriage feast comes very early in the day). The five foolish virgins are turned away and the door is shut.

The door remains relatively shut until the Tribulation years. In Revelation 15: 2-4, we read that a great multitude of Israelite converts rejoice and praise

Notes

before the Lord God. They give Him honour for He has relented: He has opened the door for the five foolish virgins. The marriage of the Lamb to His wife has come; to her is granted that she should make herself ready. She is arrayed in fine linen in keeping with the Levitical priesthood (Revelation 19: 8). Now the bridegroom knows her and receives her, for she has the required righteousness of the Saints. Revelation 19: 9 states that blessed are they who are called to the marriage supper (held late in the day) of the Lamb. Blessed, indeed, that God is so merciful. He has extended an invitation to the whole house of Israel. She has joyously accepted His summons to the marriage feast of the Lamb. How good God is; even though it is late, it's not too late! The prophecies are fulfilled. Redemption has come to Israel. God has allowed her, even unto the end, (through suffering and death) to make herself ready. No wonder John is so overjoyed that he falls at the feet of the messenger bearing the good news. These, after all, are his people.

John sees the prophecy of Ezekiel, Chapter 37: 16-28, fulfilled: **the house of Judah (out of which comes the Church) and the house of Israel (the Israelites) are joined into one '*stick*' as the whole house of Israel is united in salvation.** God has foretold of this appointed time when He would restore and unite His beloved Israel, again bringing the twelve tribes together as a nation (read Jeremiah 3: 11-18). He foresaw them submissive and devoted to one King, their Messiah, the Lord Jesus Christ.

Isolated remnants of the House of Judah and the House of Israel still exist in scattered tribes and small clans throughout the earth. God is rapidly drawing them back to their promised land for He has placed a longing in their souls, driving them home to Israel. According to Ezekiel and Matthew, the whole House of Israel (both living and dead) will be literally gathered together at Christ's second coming. Scripture tells us "*And then shall appear the sign of the Son of man in heaven: and then shall all the tribes of the earth mourn, and they shall see the Son of man coming in the clouds of heaven with power and great glory. And he shall send his angels with a great sound of a trumpet, and they shall gather together his elect from the four winds, from one end of heaven to the other.*" (Matthew 24: 30-31).

These elect of God (Israelites) will be sifted and sorted at the judgment seat of Christ. Daniel, in prophecy, says: "*And many of them that sleep in the dust of the earth shall awake, some to everlasting life, and some to shame and everlasting contempt.*" (Daniel 12: 2)

Misguided Worship / Jesus on the White Horse

Notes

Revelation 19: 10

10 And I fell at his feet to worship him. And he said unto me, See thou do it not: I am thy fellowservant, and of thy brethren that have the testimony of Jesus: worship God; for the testimony of Jesus is the spirit of prophecy.

John is rebuked by a brother and fellowservant because he is misdirecting his worship. One would expect that since, most often, angels are the channel through which God makes His proclamations, He would also use an angel to reveal His wondrous goodness and faithfulness to John's own people, the Israelites. Instead, He uses a brother, a fellowservant.

John is so overjoyed at seeing his beloved brethren called to the marriage supper of the Lamb and, thus, spared from eternal damnation, he falls down to worship the fellowservant. This sometimes happens to honest believers who are somehow beguiled by the bearer of the good news of salvation. Occasionally, man gives his worship to the messenger rather than the message, the created rather than the creator; the cult-inspired mass suicide of the Jonestown tragedy is an example of such misguided worship.

This image may appear in the Revelation as a warning to the Saints. I doubt that it was given to portray John as imperfect in his love for God, but more to caution man that all praise and worship belongs to God; even the very elect can fall into error. We must be ever mindful of the counsel given in I Peter 5: 8:

8 Be sober, be vigilant: because your adversary the devil, as a roaring lion, walketh about, seeking whom he may devour.

As verse 10 ends, the fellowservant tells John to worship God; he also tells John that the Spirit of prophecy (one of the nine gifts of the Holy Spirit) proclaims the coming events concerning the real conqueror – Christ, the Son of the living God.

Revelation 19: 11

11 And I saw heaven opened, and behold a white horse; and he that sat upon him was called Faithful and True, and in righteousness he doth judge and make war.

Chapter 6: 13-17, Chapter 11: 13-19 and Chapter 14: 1-20 are precursors to this great event: the coming of Jesus in authority and victory. John sees heaven open as Jesus rides forth on a white horse from heaven to earth. Jesus came, the first time, in humility, as he rode on the foal of an ass; this time, he is depicted as coming in strength and might, power and glory, on a white horse. Jesus is called '*Faithful and True*.' He presents quite a different picture from the first rider on a white horse (Revelations 6: 2). The first rider comes forth conquering and to conquer, but he fails. He

comes to set up Satan's reign and to destroy the righteous among men; while Jesus, the victorious Saviour, comes to set up a visible Kingdom of God on earth (Colossians 1: 13) and to reward the righteous who have remained faithful under Satan's attack. This time, Jesus comes in power and judgement; he comes to make war upon evil (Joel 3: 11-16). If necessary, refer to Revelation 6: 7-8 to recall the writer's personal view as to the figurative nature of this verse.

Revelation 19: 12

12 His eyes were as a flame of fire, and on his head were many crowns; and he had a name written, that no man knew, but he himself.

Jesus comes with infinite knowledge, the all-righteous, all-knowing, all-seeing Son of God. He comes wearing many diadems as King of kings, the omnipotent Lord. God has written upon him a name that no man knows, but he himself. Jesus has promised in Revelation 3: 12 that he will write that name upon us, along with the name of God and the City of God. When Jesus prayed, just hours before he died, he said, '*And the glory which thou gavest me, I have given them; that they may be one, even as we are one.*' (John 17: 22).

Revelation 19: 13

13 And he was clothed with a vesture dipped in blood: and his name is called The Word of God.

Read Isaiah 63: 2-6 (NLT) for a clear interpretation of this verse:

Why are your clothes so red, as if you have been treading out grapes? "I have trodden the winepress alone; no one was there to help me. In my anger I have trampled my enemies as if they were grapes. In my fury I have trampled my foes. It is their blood that has stained my clothes. For the time has come for me to avenge my people, to ransom them from their oppressors. I looked, but no one came to help my people. I was amazed and appalled at what I saw. So I executed vengeance alone; unaided, I passed down judgment. I crushed the nations in my anger and made them stagger and fall to the ground."

Revelation 19: 14

14 And the armies which were in heaven followed him upon white horses, clothed in fine linen, white and clean.

The armies that come from heaven and follow Jesus are pictured as riding white horses. These Saints are clothed in fine linen. (They are not the whole body of the Raptured Saints seen in Revelation 4: 4, but are an Old Testament Israelite extract from that body.) As always, '*fine linen*' refers

Notes

to Israel and, in this case, the Raptured Israelites. This army is made up of the Old Testament Saints whom Jesus preached to in prison (I Peter 3: 19; I Peter 4: 6). The graves were opened and those who loved God and believed on Jesus came up out of their graves. These Saints, according to Hebrews 11: 13 *'all died in faith, not having received the promises, but having seen them afar off, and were persuaded of them, <u>and embraced them</u>, and confessed that they were strangers and pilgrims on earth.'* When Jesus was crucified, they *'arose and went into the Holy City and appeared unto many'*, as is recorded in Matthew 27: 52-53.

Matthew 8: 11 (NLT) assures us that these Old Testament Saints, having embraced their Messiah, are risen from sleep and received into glory. Jesus states that when the Gentiles come from all over the world, they will sit down and feast with them in the Kingdom of Heaven. These Saints, a special people with a different resurrection from that of the Church, will be given a different commission. Having been risen from sleep (almost two thousand years ago), they accompany Jesus, both in the Rapture and at his Second Coming.

We can be sure of this since God has said He would bring both the living and the dead back to the land of Israel (Ezekiel 37: 12). According to Ezekiel 36:10, God speaks this promise to '*<u>all</u> the house of Israel, even all of them.*' The risen Old Testament Saints would, therefore, be included in this promise.

I Thessalonians 4: 14-17 attests to the fact that the risen Old Testament Saints follow Jesus when he comes to call us in the Rapture. Verse 14 says '*if we believe that Jesus died and rose again, even so them which sleep in Jesus will God bring with him.*' The first group who sleep in Jesus, according to Matthew 27: 52-53, are the Old Testament Saints who died in faith. Having risen with Christ at his resurrection, they no longer sleep but are alive and feasting around the throne of God (Matthew 8: 11 NLT). According to Paul, they will accompany Jesus when he comes to call us in the Rapture. Paul goes on to say that when the Lord descends from heaven with the voice of an archangel and with the trump of God, the dead who sleep in Jesus will rise first. This group who sleep in Jesus are the New Testament Saints. They were born a natural body but will be raised a spiritual body (I Corinthians 15: 35-44). Then those which are alive will be caught up together with them in the clouds to meet the Lord in the air (I Thessalonians 4: 15-17).

To consolidate and review the above teaching, remember Matthew 8: 11 tells us that the Saints, when gathered in the Rapture, sit down with Abraham, Isaac and Jacob. Remember also that Paul says the new Testament Saints sleep until the Lord descends to awaken the dead. They rise with the living to join Jesus and the Old Testament Saints in the air. Matthew records that they will feast together in the Kingdom of Heaven.

Notes

We can be assured the Church will have no part in establishing the free gift of salvation that Jesus has won for them. Some hold that Jesus brings the Church with him but, surely, the Church who has barely arrived and only just begun their supreme time of glorious, unlimited worship, won't be torn away from the presence of God to view the destruction of man. Clearly, heaven is not vacated, leaving the Father with only His heavenly host to praise Him.

Our home and our rewards are heavenly. Jesus secured this for us. All those who come to Christ after the Rapture are limited to earthly rewards. They never rejoice <u>around</u> the throne; instead, they rejoice on the sea of glass <u>before</u> the throne (Revelation 4: 6; 15: 2-3) so that all of heaven may rejoice with them (Luke 15: 7). Because God loves us all so much, He eventually brings the whole of heaven down to dwell as one on earth (Ephesians 1: 10). God, being a consistent God, will no doubt administer the royal priesthood as He did the Levitical priesthood with order of courses and lots to serve in. For the most part, our Millennial years will be caught up in heavenly praise far above, and much beyond, anything experienced on earth. Our cup, forever, will be filled and running over. (If necessary, review the Study notes for Revelation 6: 11.)

Revelation 19: 15

15 And out of his mouth goeth a sharp sword, that with it he should smite the nations: and he shall rule them with a rod of iron: and he treadeth the winepress of the fierceness and wrath of Almighty God.

He, Jesus, is the victorious one who shall do all these things: smite the nations with the words of his mouth; rule with a rod of iron; and tread the winepress of the wrath of God. In this, the prophetic words of Psalm 2: 9 are fulfilled, '*thou shalt break them with a rod of iron: thou shalt dash them to pieces like a potter's vessel.*'

Revelation 19: 16

16 And he hath on his vesture and on his thigh a name written, KING OF KINGS, AND LORD OF LORDS.

The Apostle Paul foresaw this day when he, under the inspiration of God, stated: '*Which in His* (God's) *time He shall shew, who is the blessed and only Potentate, the King of kings, and the Lord of lords*' (I Timothy 6: 15).

The Feast of the Fowls / The Sons of Japeth

Revelation 19: 17 - 18

Notes

17 *And I saw an angel standing in the sun* (Jesus)*; and he cried with a loud voice, saying to all the fowls that fly in the midst of heaven, Come and gather yourselves together unto the supper of the great God;*
18 *That ye may eat the flesh of kings, and the flesh of captains, and the flesh of mighty men, and the flesh of horses, and of them that sit on them, and the flesh of all men, both free and bond, both small and great.*

To see how these two verses fit into the final judgment on evil (before Jesus sets up his Millennial reign), we should briefly examine the prophesy of Ezekiel, Chapters 38 and 39.

First of all, we should understand Russia and her bands attack Israel twice during the seven year Tribulation period. However, before we establish this truth, we should find scriptures which will confirm that the one who attacks is, in fact, Russia.

Ezekiel gives the names of the aggressors. They are the descendants of Japheth (Noah's son) as found in Genesis 10: 2-4:

2 *The sons of Japheth; Gomer, and Magog, and Madai, and Javan, and Tubal, and Meshech, and Tiras.*
3 *And the sons of Gomer; Ashkenaz, and Riphath, and Togarmah.*
4 *And the sons of Javan; Elishah, and Tarshish, Kittim, and Dodanim.*

In Ezekiel 38: 2, God tells the prophet to face toward the north and prophesy against them:

2 *Son of man, set thy face against Gog, the land of Magog, the chief prince of Meshech and Tubal, and prophesy against him.*

Ezekiel, prophesying between 595 BC to 574 BC, is told to prophesy against Gog, a mighty warrior who is a descendent of Japheth and Chief Prince of Russia. (The Hebrew word translated 'chief' is 'Rosh', the root of the name 'Russia' – thus, 'a Russian prince.') This end-time Gog probably has distant connections to the early warrior, Gyges. The ancient historian Pliny records the Syrians thought the name Gog was the same as Gyges, since his country was called Gygea or Gog's Land. Gyges was the Chief of a Lydian princely family who took possession of the throne of Lydia around 700 BC. He went on to conquer the Scythian tribes around the Caspian and the Black Sea, a land where the wild, barbarian sons of Japheth had either settled or roamed. Since Ezekiel is proclaiming an end-time revelation, the Gog to whom he directs his prophecy can be none other than a latter-day, high-ranking Russian leader – probably a Field Marshal with the same aggressive spirit as Gyges. As prophesied through Ezekiel, Chapters 38 and 39, he will lead the descendants of Japheth's sons

(in particular, the offspring of Magog, Meshech and Tubal) in a dual attack on Israel during the Tribulation.

Notes

Ethnologists tell us that Japhethites migrated to the north beyond the Caucasus Mountains and established their civilization beyond the Black Sea and the Caspian Sea in the area of Rosh (Rosh is probably the first of the three great Scythian tribes of which Magog was the head.) This land, today, is Russia. Josephus, the historian, identifies the tribe of Magog as the Scythians, a savage wandering people living north of the Crimea, a peninsula in the South-West Soviet Union on the north coast of the Black Sea. Moscow is a modern form of the Hebrew word 'Meshech'. The Russian province called 'Tobalsk' is derived from the Hebrew word 'Tubal'. Clearly, there is assurance through scriptures and historians that the chief aggressor is Russia.[34]

To establish that Russia attacks Israel twice, we need only continue reading from Ezekiel 38: 3-4:

> *3 ...Thus saith the Lord God: Behold, I am against thee, O Gog, the chief prince of Meshech and Tubal:*
> *4 And I will turn thee back* (back into Russia from the first attack), *and put hooks into thy jaws, and I will bring thee forth* (compel them to attack again)...

The first attack will probably come soon after the world ruler, the Antichrist, is appointed by the nations (Revelation 6: 2). The world ruler comes with the appearance of a peace-maker but is later revealed as the Beast and the conqueror. For a short period, the whole of the world will enjoy a time of peace (I Thessalonians 5: 3). When all the world is resting in peace, Russia (the red horse of Revelation 6: 4) and her allies will attack Israel. They don't expect to meet resistance since Israel has taken down her defenses (see Ezekiel 38: 11) believing that all nations will honour their covenant with one another. She will feel added security in that the Antichrist will have signed an agreement which honours the Jerusalem Covenant (Daniel 9: 27). See Appendix 4, 'The Jerusalem Covenant'.

The nations will question Russia (Ezekiel 38: 13) but seem not to move to Israel's defense. The nations who question are Sheba and Dedan and the

---

[34]Wayne Carver, Radio Sermons: World War III in Prophecy. (San Antonio, TX: The Christian Jew Foundation).

Charles F. Pfeiffer, Howard F. Vos, John Rea, eds. Wycliffe Bible Encylopedia, Volume 2 (Chicago: Moody Press, 1987), pp. 1068-1069.

William Smith, Bible Dictionary (Philadelphia: John C. Winston Co., 1948), p. 375; p. 572.

J. D. Davis, Davis Dictionary (Grand Rapids, MI: Baker Books, 1980), p. 277 (Gog); p. 278 (Gomer II); p. 490 (Magog); p. 517 (Meshech); p. 835 (Tubal).

Notes

merchants of Tarshish. Sheba and Dedan first appear in Genesis 10: 7. They are the sons of Raamah, who is the son of Cush, and grandson of Ham, one of the sons of Noah. Out of this lineage came Nimrod and the nations of Iraq, Iran and Afganistan, probably also Ethiopia, Libya and Egypt. The merchants of Tarshish (the son of Javan and grandson of Japheth) could represent the merchants of the world. More than likely, those who object will be the Arabs and the nations of the European Confederacy, with heads of the nations represented in Rome or Istanbul (both cities are of the old Roman Empire and both have seven hills).

When God sees that none come to the aid of Israel (Isaiah 63: 5-6), He turns His fury upon the attackers (Ezekiel 38: 18). With the devastating loss of all but one-sixth of her forces, Russia and her allies retreat into the north (Ezekiel 38: 4; Ezekiel 39: 2).

In the second attack, God causes Russia and her bands to again rise up out of the north and come to the mountains of Israel (Ezekiel 38: 4; Ezekiel 39: 2). His purpose is that they will be present to view Jesus as the victor; also, that they might die on Israeli land at the battle of Armageddon, in the plains of Meguiddo. Ezekiel names some of the attacking nations but adds that, above the ones named, there will be many other people joining Russia in the attack. Read Ezekiel 38: 5-6:

> 5 *Persia, Ethiopia, and Libya with them; all of them with shield and helmet;*
>
> 6 *Gomer, and all his bands; the house of Togarmah of the north quarters, and all his bands: and many people with thee.*

God lists the following nations: Russia; Persia (Iran/Iraq/Afganistan); Ethiopia and Libya (adjacent states in the Arabian Peninsula at the time of Moses – or Ezekiel could mean present day Ethiopia and Libya); Gomar (Germanic people); and Togarmah (Turkey). One of the other nations will, no doubt, be China and, possibly, Japan, and other eastern nations (see Revelation 9: 14-20 and Revelation 16: 12). Read Isaiah 41: 25:

> 25 *I have raised up one from the north* (Russia), *and he shall come: from the rising of the sun* (China) *shall he call upon my name* (in cursing and slander - see Revelation 13: 6)*: and he shall come upon princes* (the now defending seven nations under the Antichrist) *as upon mortar, and as the potter treadeth clay.*

A third major world war will be raging on the plains of Meguiddo when Jesus returns to the earth. The nations will not only be fighting Israel, but will also be warring against one another for control of Israel. Ezekiel 38: 21 says:

> 21 *And I will call for a sword against him throughout all my mountains, saith the Lord God: every man's sword shall be against his brother.*

Notes

It will be at this time – just prior to Christ's return – the whole house of Israel will be risen from their graves and brought back to Jerusalem as promised: "*...Behold, O my people, I will open your graves, and cause you to come up out of your graves, and bring you into the land of Israel. And you shall know that I am the Lord, when I have opened your graves, and brought you up out of your graves.*" (Ezekiel 37: 12-13; Matthew 24: 30-31). These Israelites will look upon Jesus as he plants his feet on the Mount of Olives and recognize him as their Saviour. In that day, all of the tribes of the earth will mourn, sorrowing that they have rejected and crucified their Messiah (Zechariah 12: 9-14; Zechariah 14: 4).When Israel's aggressors fall upon the mountains and plains of Israel, the carnivorous birds of every sort and the beasts of the field will be called to devour them (Ezekiel 39: 4). Ezekiel 39: 17-20 adds more detail to the gruesome events which coincide with the two verses we are now studying.

17 ...Speak unto every feathered fowl, and to every beast of the field, Assemble yourselves, and come; gather yourselves on every side to my sacrifice that I do sacrifice for you, even a great sacrifice upon the mountains of Israel, that ye may eat flesh, and drink blood.

18 Ye shall eat the flesh of the mighty, and drink the blood of the princes of the earth, of rams, of lambs, and of goats, of bullocks, all of them fatlings of Bashan.

19 And ye shall eat fat till ye be full, and drink blood till ye be drunken, of my sacrifice which I have sacrificed for you.

20 Thus ye shall be filled at my table with horses and chariots, with mighty men, and with all men of war, saith the Lord God.

Revelation 19: 19

19 And I saw the beast, and the kings of the earth, and their armies, gathered together to make war against him that sat on the horse, and against his army.

When Jesus comes against the nations whom he has made to gather in the valley of Megiddo, all of the nations will rally together to battle against him. Though accompanied by his mighty army, Jesus fights this battle alone (Isaiah 63: 2-6). We read the power of God comes down and elements assist him. So confused are the nations that every man's sword is against his brother. Through his victory at Armageddon, Jesus will magnify and sanctify himself in the eyes of the nations. They will recognize him and know that he is the Lord (Ezekiel 38: 15-23).

The Antichrist and the False Prophet Cast Into Hell / World Peace Under the Reign of Jesus Christ

Notes

Revelation 19: 20

20 And the beast was taken, and with him the false prophet that wrought miracles before him, with which he deceived them that had received the mark of the beast, and them that worshipped his image. These both were cast alive into a lake of fire burning with brimstone.

None can come against the power of God and win. The world will learn that not even the united might of the nations can overturn His omnipotence and authority. The Beast (Antichrist) and the False prophet (the leader of the world church) are now cast alive into the lake of fire burning with brimstone. They are condemned for all eternity by their own acts. Indeed, they will have lost their right to be judged at the White Throne judgement for their names have long-been removed from the Book of Life.

Revelation 19: 21

21 And the remnant were slain with the sword of him that sat upon the horse, which sword proceeded out of his mouth: and all the fowls were filled with their flesh.

The remnant of evil men are slain through the power and judgment of Jesus; the fowl are filled with their flesh. We have seen this same power demonstrated in the garden of Gethsemane when the band of officers came from the chief priests to take Jesus to be judged and crucified. John 18: 6 records that when Jesus spoke the words, "*I am he*", the whole of the party fell to the ground under the power of God. They were temporarily slain by the Holy Spirit. This same power will cleanse the earth of evil, bind Satan for one thousand years, and establish the Millennial reign (Revelation 20: 2-3). After this great day of judgment, when Christ conquers the armies opposing him, he will usher in the Millennium of Peace. He will teach all nations – those who have never accepted the mark of the Beast – the way of peace (Micah 4: 2-4). Only then will the Arabs and Israelis be permanently reconciled. Read Isaiah 19: 23-25.

23 In that day shall be a highway out of Egypt to Assyria, and the Assyrians shall come into Egypt, and the Egyptians into Assyria, and Egypt shall serve with the Assyrians.
24 In that day shall Israel be the third with Egypt and with Assyria, even a blessing in the midst of the land;
25 Whom the Lord of hosts shall bless, saying, Blessed be Egypt my people, and Assyria the works of my hands, and Israel mine inheritance.

CHAPTER TWENTY

As **Chapter Twenty** opens, Satan is bound for the thousand year millennial reign of Jesus. When the thousand years are past, Satan is released for a 'season', during which time the millennial Saints are tested. Satan and his angels stage a second Armageddon as they viciously attempt to deceive the nations. At the height of Satan's campaign, "*...fire came down from God out of heaven, and devoured them. And the devil that deceived them was cast into the lake of fire and brimstone where the beast and the false prophet are...*'. As the chapter ends, all those who have never been called to a previous resurrection are now called to the White Throne Judgment, where Jesus judges on behalf of the Father. All those whose names remain written in the Book of Life are judged according to their deeds. Those whose names have been removed are deprived of judgment and are cast directly into the lake of fire. In the study notes for this chapter an emphasis is placed on the compassion of God. The probability of levels of hell and lengths of punishment are examined. Attention is drawn to the fact that God doesn't predestine man to heaven or hell. Man makes that choice himself. By his own free will, man determines his destiny. The only exception is in the case of the 'Chosen'. They are pre-destined even before birth to fulfill a special commission assigned by our Lord.

Revelation 20

[1] And I saw an angel come down from heaven, having the key of the bottomless pit and a great chain in his hand.
[2] And he laid hold on the dragon, that old serpent, which is the Devil, and Satan, and bound him a thousand years,
[3] And cast him into the bottomless pit, and shut him up, and set a seal upon him, that he should deceive the nations no more, till the thousand years should be fulfilled: and after that he must be loosed a little season.
[4] And I saw thrones, and they sat upon them, and judgment was given unto them: and I saw the souls of them that were beheaded for the witness of Jesus, and for the word of God, and which had not worshipped the beast, neither his image, neither had received his mark upon their foreheads, or in their hands; and they lived and reigned with Christ a thousand years.
[5] But the rest of the dead lived not again until the thousand years were finished. This is the first resurrection.
[6] Blessed and holy is he that hath part in the first resurrection: on such the second death hath no power, but they shall be priests of God and of Christ, and shall reign with him a thousand years.
[7] And when the thousand years are expired, Satan shall be loosed out of his prison,
[8] And shall go out to deceive the nations which are in the four quarters of the earth, Gog and Magog, to gather them together to battle: the number of whom is as the sand of the sea.
[9] And they went up on the breadth of the earth, and compassed the camp of the saints about, and the beloved city: and fire came down from God out of heaven, and devoured them.
[10] And the devil that deceived them was cast into the lake of fire and brimstone, where the beast and the false prophet are, and shall be tormented day and night for ever and ever.
[11] And I saw a great white throne, and him that sat on it, from whose face the earth and the heaven fled away; and there was found no place for them.
[12] And I saw the dead, small and great, stand before God; and the books were opened: and another book was opened, which is the book of life: and the dead were judged out of those things which were written in the books, according to their works.
[13] And the sea gave up the dead which were in it; and death and hell delivered up the dead which were in them: and they were judged every man according to their works.
[14] And death and hell were cast into the lake of fire. This is the second death.
[15] And whosoever was not found written in the book of life was cast into the lake of fire.

Jesus, the Keeper of the Keys

Revelation, Chapter 20

Notes

Revelation 20: 1

1 *And I saw an angel come down from heaven, having the key of the bottomless pit and a great chain in his hand.*

The angel who comes down from heaven and has the key to the abyss is Jesus. The very first words that Jesus says to John are words to describe his mandate, his person, his commission. Recall the words of Jesus recorded in Revelation 1: 17-18:

17 *...Fear not; I am the first and the last:*
18 *I am he that liveth, and was dead; and, behold, I am alive for evermore, Amen; and have the keys of hell and of death.*

In these three great claims, Jesus is saying: "I am the creator; I am the victorious Saviour; I am the final judge."

There seems to be conflicting opinions as to the identity of the angel, even though it is evident that the angel is Jesus. He would hardly allow any other than himself to do what he was born to do: to bind Satan and shut him up for a thousand years; to release him and then, after a short while, cast him into the lake of fire and brimstone – the same place into which he had already cast the Beast and the False prophet (Revelation 19: 20). It's highly unlikely that an angel would do the task so clearly appointed to Jesus.

Since the powerful archangel Michael lacked authority when contending with the devil (Jude 1: 9), we can be positive that a mere angel could never measure up to so great a task as binding and casting Satan into hell. Only Jesus has this authority. We should always compare scriptures with scriptures and, in this way, receive sound answers – answers that are scriptural!

John's free use of the term 'angel' is rather confusing. In Revelation 9: 1, we read when the fifth angel sounded, a fallen star is given the key to the bottomless pit (see Isaiah 14: 12). He opens the pit and releases his followers so that they might spew sorrow and destruction upon man. We are told that their leader is the destroyer which is, of course, Satan. Because Satan has temporary access to the keys, it may seem to imply that Jesus allows free use of his keys; however, this is not so. In Job 1: 12, Satan in all his wickedness was allowed to try Job. In the end, he became a tool in the hands of God. The testing strengthened Job's faith and furthered the purposes of God. In Revelation 9: 1-11, God again uses Satan as a tool. With God's sanction, Satan releases his satanic angels who then come against his (Satan's) own earthly followers – Satan divided against himself cannot stand (Matthew 12: 26).

Satan Bound for A Thousand Years

Notes

Revelation 20: 2 - 3

2 *And he laid hold on the dragon, that old serpent, which is the Devil, and Satan, and bound him a thousand years,*

3 *And cast him into the bottomless pit, and shut him up, and set a seal upon him, that he should deceive the nations no more, till the thousand years should be fulfilled: and after that he must be loosed a little season.*

The verses under study inform us that Jesus will lay hold of Satan, bind him, cast him into the bottomless pit, shut him up and set a seal upon him. Only Jesus can and may break the seal. The people and nations in the Millennium will be free from Satan's evil influence until the end of the thousand years when they are tested by God during Satan's temporary release. Isaiah 24: 21-23 (NLT) informs us that Satan's advocates, seen released in Revelation 9: 1-3, are also imprisoned in the bottomless pit. Isaiah says, '*In that day the Lord will punish the fallen angels in the heavens and the proud rulers of the nations on earth. They will be rounded up and put in prison until they are tried and condemned.*'

Some may wonder why Satan is loosed for a little season. The reason, of course, is that the decision to follow Jesus is left to man's free will. The sincerity of that choice must be tested. Matthew 4: 1 reminds us that even Jesus was tested:

1 *Then was Jesus led up of the Spirit into the wilderness to be tempted of the devil.*

This short period will be the testing time for the offspring of the Tribulation Saints. They will attend the White Throne judgement along with millions of others who will be called from the dead to be judged (Revelation 20: 12). These are those who have never been resurrected or judged, as are the following three groups: the Old Testament Saints (judged in prison – I Peter 3: 19; I Peter 4: 6 – and called after Christ's resurrection – Matthew 27: 52-53); the Church (the bride of Christ, clothed in righteousness and called by her bridegroom – Matthew 25: 1-12; I Thessalonians 4: 16-17; John 14: 2-3); and the Tribulation Saints (the wife, judged and called by her husband – Matthew 25: 31-34; Revelation 19: 7-9).

Souls Under the Altar

Notes

Revelation 20: 4

4 *And I saw thrones, and they sat upon them, and judgment was given unto them: and I saw the souls of them that were beheaded for the witness of Jesus, and for the word of God, and which had not worshipped the beast, neither his image, neither had received his mark upon their foreheads, or in their hands; and they lived and reigned with Christ a thousand years.*

John sees two groups who will live and reign with Christ: the Raptured Saints and the Tribulation Saints. The Saints who sit upon the thrones are the Raptured Saints who judge and reign with Jesus throughout the Millennium. They will, at this point, have the mind of Christ and the purpose of Christ which is solely to give all glory to God. They will judge (I Corinthians 6: 2) as Jesus judges. The words that they speak and the judgments that they make as they rule on earth will not be their own words or judgments, but will be the words and commandments of God. They will follow in the obedience of Jesus, as modelled in John 12: 49:

49 *For I have not spoken of myself; but the Father which sent me, he gave me a commandment, what I should say, and what I should speak.*

The second group which John sees are the Tribulation Saints. These Saints will be rewarded with eternal life as promised by Revelation 3: 5. The scriptures say that they will serve in the temple, ever present with Christ (Revelation 7: 13-15); however, their rewards are earthly rewards, while the Church and Old Testament Saints are given heavenly rewards. God has promised in Exodus 19: 5-6 that they will be '*a kingdom of priests*' and not '*kings and priests*' as is promised to His Raptured Saints (Revelation 1: 6). My sense is that they will reign as priests with Christ over their own house, to keep it in order, while the Church Saints will reign with Jesus over the nations of the earth as they stand in their 'lot' and serve their course. We are told in Zechariah 12: 11-14 that there will be great mourning in Jerusalem; every family apart will mourn because they have rejected the Messiah. Possibly, we are given this information so that we may conclude that each family unit is accountable to its own house and that house is accountable to God. In other words, each family reigns over its own clan.

Among the Tribulation Saints, there will be a third body of people saved, multitudes of them. This group is the whole house of Israel who has died since Jesus was risen from the dead – all the Israelites who were not risen at the victorious resurrection of Jesus (Matthew 27: 52-53). These are referred to as the dead in '*the valley of dry bones*' (Ezekiel 36: 17-38 and Ezekiel 37: 1-28). They are risen that they might see Jesus plant his feet upon the Mount of Olives (Zechariah 14: 4) and mourn for him whom they have pierced (Zechariah 12: 10). God saves them for His own Holy name's sake which Israel had profaned among the heathen (Ezekiel 36: 22). He

forgives them because He had promised He would never forsake them or leave them, nor forget the covenant He had made with their fathers (Deuteronomy 4: 31). These reclaimed Israelites, as well as the Israelites who will resist allegiance to the Beast during the Tribulation, stand on the Sea of Glass mingled with fire. They sing the Song of Moses and the Song of the Lamb (refer to Study notes of Revelation 15: 2-3).

Notes

Revelation 20: 5

5 *But the rest of the dead lived not again until the thousand years were finished. This is the first resurrection.*

The rest of the dead, the unsaved and the uncommitted, will await the White Throne judgement. This will be their first resurrection but their second death if judged unrighteous.

Revelation 20: 6

6 *Blessed and holy is he that hath part in the first resurrection: on such the second death hath no power, but they shall be priests of God and of Christ, and shall reign with him a thousand years.*

From the preceding studies, we must conclude that there are four separate and distinct resurrections from the dead. We have seen they include the Old Testament Saints, the Raptured Saints, the Tribulation Saints and, finally, those who are risen to the White Throne Judgment (Revelation 20: 11-15). Blessed are those who have been resurrected prior to the final judgment: having attended their first resurrection, the resurrection of life, they are protected from the second death at the resurrection of judgment.

Before going on with this study, it might be beneficial to take another look at the last two resurrected groups – the Tribulation Saints and those who are called to the White Throne Judgment. It may help believers to be consistent and determined in their walk with Christ.

The Tribulation Saints suffer tremendous agony and, most often, death. "*These are they who have come out of great tribulation and have washed their robes, and made them white in the blood of the Lamb*' (Revelation 7: 9-14). Take note that the effort is theirs; the act of washing their robes being accomplished through their own bravery and commitment to Christ. God provides the way and gives them the will, but they are the ones who wash their robes and make them white. When they appear before the Lamb to become his wife, '*they have made themselves ready*' (Revelation 19: 7) by availing themselves of the cleansing power of the blood, willingly confessing Christ and him crucified and resurrected. '*It is* (then) *granted that they be arrayed in fine linen, clean and white, for fine linen is*

Notes

the righteousness of the saints' (Revelation 19: 8). '*Blessed are they who are called unto the marriage supper of the lamb*' (Revelation 19: 9). Through great duress, these Saints, along with the 144,000 will become priests of God and of Christ in the earthly temple; they will serve night and day throughout the Millennial reign (Revelation 7: 15). Upon their entrance into the New Creation, they will become one with the bride of Christ.

The fourth and final resurrection is attended by the balance of God's created. It comprises all who have missed out in the other resurrections. They are the '*dead, small and great* (who) *appear before God at the White Throne judgment*' (Revelation 20: 12). This group will include the Millennial-birthed, for all must stand before the judgement seat of God (Matthew 25: 31-34). The books will be opened and all will be judged out of those things written in the books. No longer is mercy extended on the basis of the blood of Jesus; none will mediate on their behalf. Theirs will be a righteous judgment based solely on cold facts as recorded in the books. Even so, they are privileged above those whose names are not found written in the Book of Life, for these will face total condemnation without judgment. Their actions have judged them and their names have been removed from the Book of Life. See the teachings on the 'Book of Life' in the Study notes for Revelation 3: 5 and Revelation 20: 14-15.

Revelation 20: 7

7 *And when the thousand years are expired, Satan shall be loosed out of his prison,*

At the end of one thousand years, Satan is released that God might test the sincerity and commitment of the Tribulation Saints, especially those born throughout the Millennium. Indeed, we know there are children born during the Millennium because Isaiah 11: 8 (NLT) says: '*Babies will crawl safely among poisonous snakes. Yes, a little child will put its hand in a nest of deadly snakes and pull it out unharmed.*' Isaiah 65: 20-22 (NLT) says, '*No longer will babies die when only a few days old. No longer will adults die before they have lived a full life. No longer will people be considered old at one hundred! ... for my people will live as long as trees and will have time to enjoy their hard-won gains.*' For further information on the Millennial years, read Isaiah 11: 6-10; Isaiah 65: 19-25; Micah 4: 1-5 (NLT).

Jeremiah gives us the following information about the Millennial reign. Jeremiah 31: 33-34 tells us:

33 *But this shall be the covenant that I will make with the house of Israel; After those days, saith the Lord, I will put my law in their*

The Second Armageddon

Notes

inward parts, and write it in their hearts; and will be their God, and they shall be my people.

34 *<u>And they shall teach no more every man his neighbour, and every man his brother, saying, Know the Lord: for they shall all know me, from the least of them unto the greatest of them, saith the Lord;</u> for I will forgive their iniquity, and I will remember their sin no more.*

Even with full knowledge of the Lord, the most devout Christians have often been beguiled by Satan. Some have even turned away from serving God to follow after this evil deceiver. It is devastating, for example, when a beautiful child of God succumbs and walks in the ways of the world. Certainly it will be possible for those born in the Millennium to fall even though they have been taught by God.

Satan, while bound, can't tempt them; however, not being glorified, they will still have the sinful nature of man and may fall in this period of temptation once he is loosed. If these Millennial Christians were incapable of sin, then we, the Church, would not be appointed to reign over them with Christ. Nor would Jesus need to rule with a rod of iron, as described in Revelation 2: 27 (see also Revelation 19: 15).

Revelation 20: 8

8 *And shall go out to deceive the nations which are in the four quarters of the earth, Gog and Magog, to gather them together to battle: the number of whom is as the sand of the sea.*

Once released, Satan will work with great vigour because he knows his time is short; he may even know his end but is determined to change it if he can. He never gives up because he knows and fears the horror prepared for the wicked. He has tasted the torment of a lesser hell while separated from God and man during his thousand year captivity.

We now see a repeat of Satan's attack similar to that at the battle of Armageddon. The opposing force may not be Russia, as such; rather, it may be satanic power in the likeness of Russia and the nations during the first Armageddon at the end of the Tribulation years. The verse probably suggests a mystic personage akin in character to the Gog and Magog of Ezekiel's prophecy. This will be the final war in human history.

Satan Cast Into Hell / The Great White Throne / Earth and Heaven Flee Away

Notes

Revelation 20: 9

9 *And they went up on the breadth of the earth, and compassed the camp of the saints about, and the beloved city: and fire came down from God out of heaven, and devoured them.*

Satan and his angels pass over the whole of the earth in a last desperate effort to gather the saints of God, even the Saints at the very heart of God – those so close that these scriptures refer to them as '*about the beloved city*'. Jesus, the King and the Conqueror, the compassionate mediator, intercedes and fire comes down out of heaven to devour Satan's efforts. From the content of this verse, it appears that the duration of this temptation is short and final.

Revelation 20: 10

10 *And the devil that deceived them was cast into the lake of fire and brimstone, where the beast and the false prophet are, and shall be tormented day and night for ever and ever.*

The devil, the deceiver, is cast into the lake of fire and brimstone – the very depths of hell. As with the Beast and False prophet, judgment is immediate (Revelation 19: 20). This evil one has been judged by his rebellious and demonic deeds. There is no reprieve. Heaven has deliberated.

Revelation 20: 11

11 *And I saw a great white throne, and him that sat on it, from whose face the earth and the heaven fled away; and there was found no place for them.*

John is next given a vision of the Great White Throne with the almighty Judge sitting upon it. He who judges is Christ, on behalf of God. Jesus is no longer the loving Saviour; he is the mighty Judge, God in flesh. All creation face his anger and awesome authority; so great is he that even the earth and heaven flee away. Imagine the paralyzing fear of standing before the throne of God, the two-in-One, without a Redeemer.

In Matthew 13: 41-42 (NLT), Jesus indicates there will be a great harvest of the unjust at the end of the world. He says "*I, the Son of Man, will send my angels, and they will remove from my Kingdom everything that causes sin and all who do evil*."

Paul says in II Thessalonians 1: 7-9 that when Jesus comes in judgement, he will come with flaming fire to take vengeance on the wicked. He

Notes

teaches that the ungodly will be "*punished with everlasting destruction from the presence of the Lord and from the glory of his power.*" A great part of their agony will be the intense loneliness they feel because of total banishment and absolute separation from God and man.

Be assured, you who doubt, there definitely is a hell burning with fire and brimstone (Revelation 14: 10). Social or 'political' correctness can't change this truth; all unsaved will face shame and everlasting contempt (Daniel 12: 2). The fire is literal (Matthew 18: 8-9); it burns forever (Isaiah 33: 14). The judgment is final (Hebrews 6: 2) and the torment unending (Revelation 14: 11).

Praise God that Jesus set us free. He rescued us and saved us from this horrible moment in time! Still, we who love Christ should take care to remember his awesome role. We must live our lives accordingly, reverencing God and ever mindful of the truth of Hebrews 4: 12-13:

> *12 For the word of God is quick, and powerful, and sharper than any two-edged sword, piercing even to the dividing asunder of soul and spirit, and of the joints and marrow, and is a discerner of the thoughts and intents of the heart.*
>
> *13 Neither is there any creature that is not manifest in his sight: but all things are naked and opened unto the eyes of him with whom we have to do.*

Before leaving Revelation 20: 11, it should be mentioned that there is a commentary out in a 1999 publication which holds that God will destroy the heavens and the earth before the beginning of the Millennium; that is, before the literal kingdom of God is established on earth.

The teaching declares that God, on two occasions, causes the earth and heavens to be destroyed, once before the Millennium and once after.[35] It sites II Peter 3: 4-14 and Isaiah 65: 17-20 as supportive evidence of the first destruction; and Revelation 21: 1 as the second.

In order to analyze the credibility of this teaching, we must look at the passages given in support of this unusual concept. Certainly, II Peter 3: 4-14 speaks of two universal judgements: the first by flood played havoc with the universe and all but annihilated mankind. This judgement renewed the earth but did not destroy the universe. The great and only total destruction will come by fire. II Peter 3: 7 says '*the heavens and the earth which are now, are kept in store, reserved for the day of fire and judgment*' (paraphrased). The day of Judgement, to which Peter is referring, is when the sea and death and Hades give up their dead

---

[35]Tim Lahaye, Revelation Unveiled (Grand Rapids, MI: Zondervan Publishing, 1999), pp. 344-345; pp. 355-356.

Notes

(Revelation 20: 13). It is then, and only then, when Satan has been cast into the Lake of Fire (Revelation 20: 10) and Jesus is seated upon the great White Throne of Judgement that the heavens and earth flee from the face of Jesus (Revelation 20: 11).

John sees the New Heaven and New Earth, for the first heavens (atmospheric and astrological) and the first earth have passed away. John then sees the Holy City, the New Jerusalem, coming down from God out of heaven. There is absolutely no scriptural evidence of there being more than one White Throne Judgement and one total replacement of the heavens and earth. Only once are we told that the heavens and earth flee from the face of Jesus; only once does God declare, '*Behold, I make all things new*' (Revelation 21: 5).

We should now look at Isaiah 65: 17-19 since it is also used as a supporting reference for this unusual teaching. In Isaiah 65: 17-19 (NLT) God is speaking and says:

> *17 Look, I am creating new heavens and a new earth – so wonderful that no man will ever think about the old ones anymore.*
>
> *18 Be glad; rejoice forever more in my creation! And look! I will create Jerusalem as a place of happiness. Her people will be a source of joy.*
>
> *19 I will rejoice in Jerusalem and delight in my people and the sound of weeping and crying will be heard no more.*

Following these three verses, the reading goes on to describe the Millennium. If God does, in fact, create wonderful New Heavens and a New Earth at the end of the Tribulation, is it reasonable to believe that He will destroy them and again create another at the end of the Millennium? These scriptures suggest otherwise.

God says, '*Be glad; rejoice forever in my creation.*' He doesn't state the joy should last only until He later creates another. God goes on to say that He (God Himself) will rejoice in Jerusalem and delight in His people. Does God bring His Holy City down twice from heaven? Does He dwell on earth during the Millennium and reign with Christ? If, in fact, God causes the earth and heavens to pass away after Armageddon and before the Millennium, whatever will we do with the prophecy in Ezekiel Chapter 39?

Ezekiel 39: 9-15 tells us the Israelites will burn the weapons of war for seven years and bury the dead for seven months in order to cleanse the land. Ezekiel 39: 17-20 talks of the feathered fowls and the beasts of the field assisting as undertakers. The greater part of Ezekiel, Chapter 39, is devoted to a description of the cleansing of the land after Armageddon.

The Books are Opened / The White Throne Judgment

Notes

Why would God record this cleansing if He intends to '*wrap up the old creation*' at the end of the Tribulation and start with the New?

It is quite plausible, when reading Isaiah 65: 17-25, for one to mistakenly conclude that the whole section has reference to the Millennium. However, as is often the case, prophesy is hidden within prophecy. Isaiah, in this particular revelation, is taken beyond the Millennium and beyond the White Throne Judgement. He sees on into the New Creation and then back into the Millennium, a phenomenon not at all unusual in prophesy. An excellent example of this can be found in Revelation 12. The Holy Spirit takes John across Biblical history, both forward and back in time. One should be constantly aware that information revealed through prophecy does not necessarily follow chronological order.

Revelation 20: 12

12 And I saw the dead, small and great, stand before God; and the books were opened: and another book was opened, which is the book of life: and the dead were judged out of those things which were written in the books, according to their works.

Only those whose names are found written in the Book of Life are judged (Revelation 20: 15). Records are kept in heaven and, when the record books are opened, those who stand before God are judged out of those things written in the books, according to their works. The books that will be opened are the Book of Works (deeds) – II Corinthians 5: 10; the Book of Remembrance – Malachi 3: 16-18; and the Book of Life – Revelation 20: 12.

The Bible seems to indicate that there are degrees of punishment. It appears those whose names have not been removed from the Book of Life may suffer lesser punishment than that of the lake of fire. Jesus alludes to this in Luke 12: 47-48:

47 And that servant, which knew his lord's will, and prepared not himself, neither did according to his will, shall be beaten with many stripes.

48 But he that knew not, and did commit things worthy of stripes, shall be beaten with few stripes. For unto whomsoever much is given, of him shall be much required: and to whom men have committed much, of him they will ask the more.

Could Jesus be saying there are varied degrees and durations of punishment in hell; are there lesser levels of hell? Read Matthew 5: 25-26:

25 Agree with thine adversary quickly, whiles thou art in the way with him; lest at any time the adversary deliver thee to the judge, and the judge deliver thee to the officer, and thou be cast into prison.

Judgment According to Works / Degrees of Punishment / Levels of Hell

> 26 *Verily I say unto thee, Thou shalt by no means come out thence, till thou hast paid the uttermost farthing.*

Notes

Certainly, Revelation 14: 9-11 says that none shall escape the very depth of hell; they shall be tormented with fire and brimstone for ever and ever. Since God is a righteous God and a compassionate God, could it be possible he is saying that at the White Throne judgment, those who are cast into prison (a lesser hell) may eventually be redeemed? Read Job 33: 27-30:

> 27 *He looked upon men, and if any say, I have sinned, and perverted that which was right, and it profited me not*
> 28 *He will deliver his soul from going into the pit, and his life shall see the light.*
> 29 *Lo, all these things worketh God often times with man,*
> 30 *To bring back his soul from the pit, to be enlightened with the light of the living.*

Read I Corinthians 3: 13-15:

> 13 *Every man's work shall be made manifest: for the day shall declare it, because it shall be revealed by fire; and the fire shall try every man's work of what sort it is.*
> 14 *If any man's work abide which he hath built thereupon, he shall receive a reward.*
> 15 *If any man's work shall be burned, he shall suffer loss: but he himself shall be saved; yet so as by fire.*

This may refer to the Raptured and Tribulation Saints only, but it may also include some of those who are called to the White Throne judgment. Without a doubt, the 'tares' of Matthew 13: 24-40 are gathered and burned; however, consider Matthew 10: 42:

> 42 *And whosoever shall give to drink unto one of these little ones a cup of cold water only in the name of a disciple, verily I say unto you, he shall in no wise lose his reward.*

We have definite, pre-conceived notions of how God will judge man It seems apparent from the words of Jesus recorded in Matthew 25: 31-40 that the 'good Samaritans' – those who express love of God through generous acts but have never conformed to stereotypical Christianity – will obviously be rewarded at judgement. (This is not to say that some enter eternity without being covered by the blood of Jesus Christ. All must confess and repent, but scriptures seem to indicate some may do this even at the very gates of heaven – Revelation 22: 17; Isaiah 19: 18-25.) The more one studies the word, the more one realizes that God's judgement is complex and much more diverse from anything man may devise. What would be the point of the White Throne judgement – the reason for records being kept – unless man, in the final judgement, is given a just and righteous judgement befitting his acts?

Notes

We hear Jesus saying, "*Come ye, blessed of my Father... Inasmuch as ye have done it unto one of the least of these my brethren, ye have done it unto me.*" We also hear him saying, "*Depart from me, ye cursed, into everlasting fire, prepared for the devil and his angels.*" (Read Matthew 25: 31-46.)

We who have wrapped salvation and judgement into a neat little package may find, in the end, that we stand more in judgement than do those we have condemned.

Revelation 20: 13

13 And the sea gave up the dead which were in it; and death and hell delivered up the dead which were in them: and they were judged every man according to their works.

This verse gives us the information that none shall escape the judgment: the sea gives up the dead in it; and death (that is, the grave) and hell (Hades – those waiting for judgment) deliver up the dead in them. All of these are equitably judged according to what they have done on earth.

Revelation 20: 14 - 15

14 And death and hell were cast into the lake of fire. This is the second death.

15 And whosoever was not found written in the book of life was cast into the lake of fire.

All those who are not found written in the Book of Life are gathered and cast, without further judgement, into the Lake of Fire. This is their second death. As the final judgement ends and the New Creation dawns, there will no longer be death nor hell; death and hell are cast into the Lake of Fire.

In the very beginning, even before God created man, He recorded His created in the Book of Life (Psalm 139: 15-16). Only those whom He sees as very wicked are not recorded therein (Revelation 13: 8 and Revelation 17: 8). Hideous and evil crimes against God and man will cause Him to remove names which were once recorded in His book (Psalm 69: 28 and Exodus 32: 33). Similar teaching may be found in the Study notes for Revelation 3: 5. Those who commit the unforgivable sin, the sin of blasphemy against the Holy Ghost, may also have committed a mortal sin – a sin unto death. Jesus says in Mark 3: 29:

29 But he that shall blaspheme against the Holy Ghost hath never forgiveness, but is in danger of eternal damnation:

Notes

When Jesus says that such a one is <u>in danger</u> of eternal damnation, it could be possible that he is saying that some may not suffer a <u>permanent and eternal damnation</u>. This may mean that the punishment, in some cases, is limited in duration after the sinner is duly tested and cleansed by fire (I Corinthians 3: 15). Whatever the meaning may be, we can be confident the judgment will be fair and righteous.

Paul gives the following warning in Hebrews 6: 4-6 to those who are careless and ungrateful for salvation – those who walk with one foot in the world and one in the Church:

4 *For it is impossible for those who were once enlightened, and have tasted of the heavenly gift, and were made partakers of the Holy Ghost,*
5 *And have tasted the good word of God, and the powers of the world to come,*
6 *If they shall fall away, to renew them again unto repentance; seeing they crucify to themselves the Son of God afresh, and put him to an open shame.*

These are fearful warnings for those who, having once been saved, later turn their back on the Christ who died for them.

May I offer hope to all who wonder if they rank with those who have blasphemed against the Holy Ghost: if you wonder, then you have not yet committed the unforgivable sin. Those who have committed this sin are cold and uncaring. They are dead inside. The Spirit never woos them again. The very fact that your heart and mind are troubled is a clear sign that the Holy Spirit is still alive within you and calling you back. Don't delay; the Spirit will not always call if you continue to ignore his urgings. Satan is a willing companion. He would happily dwell with you forever. He covers you with perpetual shame and rewards you with eternal life in hell. The Holy Spirit won't dwell forever if the heart becomes hardened and the house in which he dwells becomes the house of Satan.

It is apparent that only those who still have their name in the Book of Life are judged; the others are cast into the lake of fire (Revelation 20: 15). According to scriptures, all those who fear the Lord and ponder His name will be graciously and fairly judged. Read Malachi 3: 16-18:

16 *Then they that feared the Lord spake often to one another; and the Lord hearkened and heard it and a <u>book of remembrance</u> was written before him for them that feared the Lord, and that thought upon his name.*

Notes

17 And they shall be mine, saith the Lord of hosts, in that day when I make up my jewels; and I will spare them, as a man spareth his own son that serveth him.
18 Then shall ye return, and discern between the righteous and the wicked, between him that serveth God and him that serveth him not.

Possibly God, through Malachi, is not only speaking to Israel but to all nations and people who genuinely love and fear God but have not embraced Jesus Christ as their Saviour – especially to the Arab nations, descendants of Abraham. In order to save themselves grave sorrow and deep suffering, it is imperative that these sincere souls take note that Jesus, on behalf of his Father, has said, "*I am the way, the truth and the life: no man can come to the Father, except through me.*" (John 14: 6 NLT). God has also spoken through Paul and said, "*At the name of Jesus, every knee will bow, in heaven and on earth and under the earth; and every tongue will confess that Jesus Christ is Lord, to the glory of God the Father.*" (Philippians 2: 10-11 NLT)

The whole truth is that God knows, even at the beginning, the names of those who will be His. God doesn't exclude man; man excludes himself. We are not predestined to hell; we make that decision ourselves. By our own free will, we determine our destiny; however, our future choices are not hidden from the eyes of God. He knows us so well that He is able to anticipate our decisions even before we make them.

Matthew 22: 14 tells us that '*many are called but few are chosen*'. The *chosen*, as individuals, are the elect, pre-appointed before conception, to fulfill a special commission of our Lord (see Jeremiah 1: 5). The chosen and elect, as a race, are God's chosen people, the Israelites. The *called* are free to accept or reject the love of the Father.

CHAPTER TWENTY-ONE

Chapter Twenty-one helps us to accept that the heaven in which God dwells is every bit as real as the atmospheric and astrological heavens. We see, through John, this sinless level of the heavens never passes away. It instead is brought down out of heaven to become the New Jerusalem and remains the dwelling place of the Father and Son. There are many significant truths taught in this chapter, but the most outstanding truth is revealed by the total absence of the Holy Spirit. Though we see the Father and Son, there is no visible evidence, or even mention, of the third Person of the Godhead. Since there must always be a triune God, it becomes obvious (through the finished work of Jesus Christ), the Saints become the submissive childlike vessel prepared by God to house, in all fullness, the Holy Spirit – the third Person of the Godhead.

Revelation 21

[**1**] And I saw a new heaven and a new earth: for the first heaven and the first earth were passed away; and there was no more sea.
[**2**] And I John saw the holy city, new Jerusalem, coming down from God out of heaven, prepared as a bride adorned for her husband.
[**3**] And I heard a great voice out of heaven saying, Behold, the tabernacle of God is with men, and he will dwell with them, and they shall be his people, and God himself shall be with them, and be their God.
[**4**] And God shall wipe away all tears from their eyes; and there shall be no more death, neither sorrow, nor crying, neither shall there be any more pain: for the former things are passed away.
[**5**] And he that sat upon the throne said, Behold, I make all things new. And he said unto me, Write: for these words are true and faithful.
[**6**] And he said unto me, It is done. I am Alpha and Omega, the beginning and the end. I will give unto him that is athirst of the fountain of the water of life freely.
[**7**] He that overcometh shall inherit all things; and I will be his God, and he shall be my son.
[**8**] But the fearful, and unbelieving, and the abominable, and murderers, and whoremongers, and sorcerers, and idolaters, and all liars, shall have their part in the lake which burneth with fire and brimstone: which is the second death.
[**9**] And there came unto me one of the seven angels which had the seven vials full of the seven last plagues, and talked with me, saying, Come hither, I will shew thee the bride, the Lamb's wife.
[**10**] And he carried me away in the spirit to a great and high mountain, and shewed me that great city, the holy Jerusalem, descending out of heaven from God,
[**11**] Having the glory of God: and her light was like unto a stone most precious, even like a jasper stone, clear as crystal;
[**12**] And had a wall great and high, and had twelve gates, and at the gates twelve angels, and names written thereon, which are the names of the twelve tribes of the children of Israel:
[**13**] On the east three gates; on the north three gates; on the south three gates; and on the west three gates.
[**14**] And the wall of the city had twelve foundations, and in them the names of the twelve apostles of the Lamb.
[**15**] And he that talked with me had a golden reed to measure the city, and the gates thereof, and the wall thereof.
[**16**] And the city lieth foursquare, and the length is as large as the breadth: and he measured the city with the reed, twelve thousand furlongs. The length and the breadth and the height of it are equal.
[**17**] And he measured the wall thereof, an hundred and forty and four cubits, according to the measure of a man, that is, of the angel.
[**18**] And the building of the wall of it was of jasper: and the city was pure gold, like unto clear glass.
[**19**] And the foundations of the wall of the city were garnished with all manner of precious stones. The first foundation was jasper; the second, sapphire; the third, a chalcedony; the fourth, an emerald;
[**20**] The fifth, sardonyx; the sixth, sardius; the seventh, chrysolite; the eighth, beryl; the ninth, a topaz; the tenth, a chrysoprasus; the eleventh, a jacinth; the twelfth, an amethyst.
[**21**] And the twelve gates were twelve pearls; every several gate was of one pearl: and the street of the city was pure gold, as it were transparent glass.
[**22**] And I saw no temple therein: for the Lord God Almighty and the Lamb are the temple of it.
[**23**] And the city had no need of the sun, neither of the moon, to shine in it: for the glory of God did lighten it, and the Lamb is the light thereof.
[**24**] And the nations of them which are saved shall walk in the light of it: and the kings of the earth do bring their glory and honour into it.
[**25**] And the gates of it shall not be shut at all by day: for there shall be no night there.
[**26**] And they shall bring the glory and honour of the nations into it.
[**27**] And there shall in no wise enter into it any thing that defileth, neither whatsoever worketh abomination, or maketh a lie: but they which are written in the Lamb's book of life.

Preamble / New Heaven and a New Earth

Revelation, Chapter 21

Notes

<u>Preamble: a teaching in preparation for Chapter 21</u>

As we begin this portion of our study, we should be aware that there are three levels of heaven. Two of these levels can be seen by man: the atmospheric heaven in which the birds fly; and the astrological heaven which supports the galaxies, the sun, the moon and the stars. These are the heavens which pass away (II Peter 3: 10-13). The scripture verse where we begin our study tells us that God will create a New Heaven and a New Earth; that is, He will create a new atmospheric and astrological heaven (New Heavens), as well as a new earth to replace that which has passed away. Because of this, Paul could correctly say '*world without end*' (Ephesians 3: 21) since the old is immediately replaced by the new.

The third heaven, even now and in the New Jerusalem, has no need of the sun, moon and stars which are part of the other two levels. The Glory of God and the Lamb lighten it (Revelation 21: 23). By faith, we accept the existence of a third level which is paradise, the dwelling place of God and His heavenly host. Heaven is a reality, as are the atmospheric and astrological heavens. Obviously, this sinless and undefiled level of heaven never passes away, for the second verse of our present study tells us that John sees the Holy city, the New Jerusalem, coming down from God out of heaven (Revelation 21: 2). To affirm John's vision of paradise, an angel carries him away in the Spirit and again shows him the great city, the Holy Jerusalem, descending out of heaven from God (Revelation 21: 9-10). With these facts in mind, we are now ready to continue our study.

Revelation 21: 1

1 And I saw a new heaven and a new earth: for the first heaven and the first earth were passed away; and there was no more sea.

John is shown the New Heaven (atmospheric and astrological) and New Earth for the first heaven and earth have passed away. He sees that there is no more sea: no more division or purpose for division on earth. God has made all things new. Jesus speaks of the heaven and earth passing away when he gives an account of the signs of the time: "*Heaven and earth shall pass away, but my words shall not pass away*" (Matthew 24: 35).

Peter offers confirmation and reassuring words about this event in II Peter 3: 7-13 (NLT):

7 And God has also commanded that <u>the heavens</u> and the earth will be consumed by fire on the day of judgment, when ungodly people will perish.

8 But you must not forget, dear friends, that a day is like a thousand years to the Lord, and a thousand years is like a day.

Notes

9 *The Lord isn't really being slow about his promise to return, as some people think. No, he is being patient for your sake. He does not want anyone to perish, so he is giving more time for everyone to repent.*
10 *But the day of the Lord will come as unexpectedly as a thief. Then the heavens will pass away with a terrible noise, and everything in them will disappear in fire, and the earth and everything on it will be exposed to judgment.*
11 *Since everything around us is going to melt away, what holy, godly lives you should be living!*
12 *You should look forward to that day and hurry it along – the day when God will set the heavens on fire and the elements will melt away in the flames.*
13 *But we are looking forward to the new heavens and new earth he has promised, a world where everyone is right with God.*

Peter gives this excellent advice which we would do well to heed. And so, dear friends, while we are waiting for these things to happen, we must make every effort to live a pure and blameless life. We will then be at peace with God and with man. A comforting reading is found in Isaiah 65: 17-19 (NLT). Though discussed previously, it is worthy of re-reading.

17 *Look! I am creating new heavens and a new earth – so wonderful that no one will even think about the old ones anymore.*
18 *Be glad; rejoice forever in my creation! And look! I will create Jerusalem as a place of happiness. Her people will be a source of joy.*
19 *I will rejoice in Jerusalem and delight in my people. And the sound of weeping and crying will be heard no more.*

Revelation 21: 2

2 *And I John saw the holy city, new Jerusalem, coming down from God out of heaven, prepared as a bride adorned for her husband.*

How wonderful this verse is! John sees the Holy City, the New Jerusalem, coming down from God out of heaven, prepared as a bride adorned for her husband. Praise God that the Holy City is retained as a dwelling place for our Father, God, and for our Saviour Jesus Christ. Praise Him, also, that the Saints (through Christ, His sinless children) will forever dwell within His presence in His newly created Heavens and Earth.

In reading John's description of the New Jerusalem as a bride adorned for her husband, we begin to see the harmony of the new creation. Our thoughts are directed toward the 'bride' and the 'wife' as one in Christ. This presents a perfect picture of beauty; of peace and love; of joy and oneness. In the New Jerusalem, the 'wife' and the 'bride' will be wed together in Him. The

Notes

wife is received as the bride, and the bride as the wife. If necessary, refer back to the Study notes for Revelation 19: 7-9.

God's acceptance of Israel, his wife, is reflected in Isaiah 54: 4-8 as God delivers 'a toast' to His belated bride:

4 *Fear not; for thou shalt not be ashamed; neither be thou confounded; for thou shalt not be put to shame: for thou shalt forget the shame of thy youth, and shalt not remember the reproach of thy widowhood any more.*

5 *For thy Maker is thine husband; the Lord of hosts is his name; and thy Redeemer the Holy One of Israel; The God of the whole earth shall he be called.*

6 *For the Lord hath called thee as a woman forsaken and grieved in spirit, and a wife of youth, when thou wast refused, saith thy God.*

7 *For a small moment have I forsaken thee; but with great mercies will I gather thee.*

8 *In a little wrath I hid my face from thee for a moment; but with everlasting kindness will I have mercy on thee, saith the Lord thy Redeemer.*

It's quite lovely to read Isaiah's account of the rejoicing of the Lamb's wife; Isaiah 61: 10:

10 *I will greatly rejoice in the Lord, my soul shall be joyful in my God; for he hath clothed me with the garments of salvation, he hath covered me with the robe of righteousness, as a bridegroom decketh himself with ornaments, and as a bride adorneth herself with her jewels.*

God is always faithful; He keeps His word. He will call His bride and He will call His wife. There will be no division between them – the first shall be last and the last shall be first.

Revelation 21: 3

3 *And I heard a great voice out of heaven saying, Behold, the tabernacle of God is with men, and he will dwell with them, and they shall be his people, and God himself shall be with them, and be their God.*

This verse means exactly what it says. The Holy City won't hover above us, nor will we journey to the place where the Father dwells. He will not be above us "reigning from a satellite city" as Jack Van Impe teaches.[36] According to the verse now under study, as well as Isaiah 65: 17-19, God will delight in us and dwell with us. He will be our God and we will be His

---

[36]Jack Van Impe, Prophecy Bible (Troy, MI: Jack Van Impe Ministries International, 1998), p. 164.

people.

Notes

Micah 4: 1-2 tells us that, in the last days, '*the mountain of the house of the Lord shall be established in the top of the mountains and exalted above the hills; and the people will flow unto it.*'

Even now, as Jerusalem is today, we go up to the Holy City. Scriptures indicate that after the great upheaval at the return of Jesus, when he plants his feet on the Mount of Olives, Jerusalem will be lifted even higher. Read Zechariah 14: 4:

> 4 *And his feet shall stand in that day upon the mount of Olives, which is before Jerusalem on the east, and the mount of Olives shall cleave in the midst thereof toward the east and toward the west, and there shall be a very great valley; and half of the mountain shall remove toward the north, and half of it toward the south.*

Indeed, a fault line runs through the Mount of Olives, down through the Jordan valley and on down through Elat. This will break open during the great earthquakes which will occur universally at Christ's return ('*and the cities of the nations fell'* – Revelation 16: 18-19). There shall be a vast chasm where there is now but a small valley between the Mount of Olives and the city of Jerusalem.

Isaiah, ministering as a contemporary to Amos, Hosea and Micah, prophesied with Micah that the Lord's house '*shall be exalted above the hills; and all nations shall flow unto it.*' (Isaiah 2: 2). It will be, as it were, above the hills on top of Mount Moriah. This may be the reason some believe that the Holy City, the new Jerusalem, will hover over the earth. What a disappointment if this were the case. We have been separated too long from our God. God, the Son, came to save us; God, the Spirit, came to teach and comfort us; God, the Father, will come to dwell and fellowship with us, as He did in the Garden of Eden (Genesis 3: 8).

Revelation 21: 4

> 4 *And God shall wipe away all tears from their eyes; and there shall be no more death, neither sorrow, nor crying, neither shall there be any more pain: for the former things are passed away.*

In the New Creation, there will be sweet, sweet fellowship for our Father will have wiped away our tears. There will be '*no more death, neither sorrow, nor crying, neither shall there be any more pain'* nor will there be any marriage or giving in marriage for we '*are equal to the angels; and are children of God, being the children of the resurrection.'* (Luke 20: 35-36).

Mark 12: 25 records Jesus' words in this manner:

> *25 For when they shall rise from the dead, they neither marry, nor are given in marriage; but are as the angels which are in heaven.*

Notes

All former things will have passed away and we will live in complete harmony, one with another, as we dwell under the New Heavens and on the New Earth. God has said, '*Behold, I create new heavens and a new earth; and the former things shall not be remembered nor come into mind.'* (Isaiah 65: 17). There will be no need for memories or connections with the past, for God will be our superb, Holy love, and the dwellers of heaven our family. In this glorious state of perfection, we will be complete in the Father and Son. They will be ever present and available to us in the New Jerusalem.

Revelation 21: 5

5 *And he that sat upon the throne said, Behold, I make all things new. And he said unto me, Write: for these words are true and faithful.*

He that sits upon the throne is God, the Father, and Jesus Christ, the Son. Recall the image of Revelation 4: 3: the jasper stone (God) and the sardine stone (Jesus). God, in His splendour, sits on the throne and says, "*I have made all things new.*" In so saying, He refers not only to His creation, but also to His created. We, His children, are restored to the image of God (Genesis 1: 26). Never again will we reflect the image of man and the personality of Satan.

Revelation 21: 6

6 *And he said unto me, It is done. I am Alpha and Omega, the beginning and the end. I will give unto him that is athirst of the fountain of the water of life freely.*

Jesus tells John that it is done. He who knows the beginning of creation and knows man's destiny, also knows the end of the first creation and the end of sin in fallen man. He knows that, in the new creation, man will drink freely from the river of Life (see Revelation 22: 1). He also knows that man will never again be tempted for there will be no tempter in the new creation. Man has finished his 'journey in the wilderness', brought about by disobedience, and has returned to God, having been redeemed through His Son.

Notes

Revelation 21: 7

7 *He that overcometh shall inherit all things; and I will be his God, and he shall be my son.*

God affirms that overcomers shall inherit all the blessings of the new creation. He says, "*I will be his God and he shall be my son.*" Having received our inheritance – all the promises of God – we are forever overcomers when we become joint heirs with Christ as glorified sons of God. For a complete overview of promises made to the overcomer, read Revelation 2: 7; 2: 11; 2: 17; 2: 26; 3: 5; 3: 12; and 3: 21.

Revelation 21: 8

8 *But the fearful, and unbelieving, and the abominable, and murderers, and whoremongers, and sorcerers, and idolaters, and all liars, shall have their part in the lake which burneth with fire and brimstone: which is the second death.*

God names those who will miss out on eternal paradise. Read I Timothy 1: 9-10 (NLT) as Paul also speaks of those who will be banished from the earth and the glory of the Lord:

But they (the Laws) *were not made for people who do what is right. They are for people who are disobedient and rebellious, who are ungodly and sinful, who consider nothing sacred and defile what is holy, who murder their father or mother or other people. These laws are for people who are sexually immoral, for homosexuals and slave traders, for liars and oath breakers, and for those who do anything else that contradicts the right teaching that comes from the glorious Good News entrusted to me by our blessed God.*

Among other sins, God condemns homosexuality; however, remember, it is for all sinners that Christ died. Everyone who overcomes through Christ will inherit eternal life and become a son of God. Read John 3: 15-18:

15 *That whosoever believeth in him should not perish, but have eternal life.*

16 *For God so loved the world, that he gave his only begotten Son, that whosoever believeth in him should not perish, but have everlasting life.*

17 *For God sent not his Son into the world to condemn the world; but that the world through him might be saved.*

18 *He that believeth on him is not condemned: but he that believeth not is condemned already, because he hath not believed in the name of the only begotten Son of God.*

The way to be an overcomer is through repentance and acceptance of Jesus Christ, the Saviour of all mankind. Our final victory will come when we arebaptized with the baptism that Jesus is baptized with (Matthew 20: 23);

Notes

when we are, as he is, endowed with the Seven Spirits of God (refer to Appendix 3, 'The Seven Spirits of God'). If concerned by this statement, re-read the commentary provided for Revelation 4: 5 in this Study.

Revelation 21: 9 - 10

9 *And there came unto me one of the seven angels which had the seven vials full of the seven last plagues, and talked with me, saying, Come hither, I will shew thee the bride, the Lamb's wife.*

10 *And he carried me away in the spirit to a great and high mountain, and shewed me that great city, the holy Jerusalem, descending out of heaven from God.*

When John is called to see the bride, the Lamb's wife, he is not shown a virgin bride, nor a reclaimed wife. He is shown the holy Jerusalem descending out of heaven from God. This is the city in which both the 'bride' and the 'wife' will enter the glorious liberty of the children of God (Romans 8: 21) with neither Israelite nor Gentile identified separately. Man will return to the tranquillity and sweet fellowship of the garden of Eden, with unlimited access to the Father and Son in the New Jerusalem.

Revelation 21: 11

11 *Having the glory of God: and her light was like unto a stone most precious, even like a jasper stone, clear as crystal;*

This verse reflects back to Chapter 4: 3-6, where the scriptures refer to the jasper stone as a stone of brilliance and representative of pure holiness, clear as crystal. This indicates that the Holy Jerusalem, like the jasper stone, is radiant with the glory of God. See the Glossary of Terms for 'Sardine and Jasper Stone'.

Revelation 21: 12 - 13

12 *And had a wall great and high, and had twelve gates, and at the gates twelve angels, and names written thereon, which are the names of the twelve tribes of the children of Israel:*

13 *On the east three gates; on the north three gates; on the south three gates; and on the west three gates.*

Ezekiel 48: 31-34 lists each of the three gates to the North, South, East and West, as well as naming to which tribe each gate is allotted.

> 31 *And the gates of the city shall be after the names of the tribes of Israel: three gates northward; one gate of Reuben, one gate of Judah, one gate of Levi.*

Notes

> 32 *And at the east side ... three gates; and one gate of Joseph, one gate of Benjamin, one gate of Dan.*
> 33 *And at the south side ... three gates; one gate of Simeon, one gate of Issachar, one gate of Zebulun.*
> 34 *At the west side ... three gates; one gate of Gad, one gate of Asher, one gate of Naphtali.*

It's thought-provoking to note that the gates on the east are of Joseph, Benjamin and Dan; Joseph and Benjamin are the beloved of Israel (Jacob), and Israel, the beloved of God; while Dan, according to Genesis 49: 17 is a '*serpent by the way, an adder in the path*' He is representative of Satan, the one who would hinder the return of Jesus.

It was through the golden gate in the eastern wall that Jesus made his triumphal entry into the city of Jerusalem,[37] his followers celebrating what we call 'Palm Sunday'. During the Millennium, he will enter and exit the city through this same gate which is reserved for the 'Prince' (Ezekiel 43: 1-7; Ezekiel 44: 1-3). Fulfilling God's prophesy through Ezekiel, this gate was bricked shut by the Turks in 1530 A.D. and remains shut to this day.[38] Nothing, however, will stop the return of Jesus – not Satan, nor brick walls, nor nations, nor atomic warfare. In God's time, Jesus will return to judge and reign from 'Ezekiel's envisioned temple' in Jerusalem.

The twelve gates are a perpetual reminder that the Father is forever faithful, and is approachable to all from the north, south, east and the west. They are also a reminder, even to us who read this great Revelation, that salvation is of the Jews (John 4: 22). Through the unbelief of the twelve tribes of Israel, the Gentiles obtain mercy. God '*concluded them all* (Israel) *in unbelief that He might have mercy upon all.*' We stand in awe at the wonder of God's plan, as does Paul in Romans 11: 25-33:

> 33 *O the depth of the riches both of the wisdom and knowledge of God! how unsearchable are his judgments, and his ways past finding out!*

---

[37]Israeli government approved guide information: *The Golden Gate (eastern gate) is a 7th century Byzantine structure. It stands over the original site of the Gate of Mercy of the Second temple (Zerubbabel temple); part of the side wall belongs to the primitive structure. It was through this gate that Jesus entered the city at his triumphal entry on Palm Sunday. Christians also attach to this gate the memory of the Beautiful Gate where St Peter healed the lame (Acts 3). The gate was completely walled up in 1530 by the Turks.*

[38]Charles F. Pfeiffer et. al., eds. Wycliffe Bible Encyclopedia, Vol. I (Chicago: Moody Press, 1975), p. 656.

The Twelve Foundations / The Twelve Apostles

Notes

Revelation 21: 14

14 And the wall of the city had twelve foundations, and in them the names of the twelve apostles of the Lamb.

In Luke 22: 30, Jesus tells the twelve apostles, chosen as foundation stones, that they would eat and drink at the Lord's table in His kingdom and sit on thrones judging the twelve tribes of Israel. Honoured are those who are chosen as a foundation stone. This honour can be ours also.

The rock or foundation stone on which God builds His Church is the truth that Jesus is the Christ, the Son of the living God. This truth was preached by the apostles and has been carried on, down through the ages, even to us. The truth we declare to those who follow becomes the foundation on which they will build their faith. Let our teaching and life be pure, grounded in the Word and, along with the apostles, we too are become foundation stones.

Please take note that the city has twelve foundations, yet one of the first disciples chosen was a betrayer. I believe that impetuous Peter moved ahead of God when he encouraged the disciples to chose another to take the place of Judas. In Acts 1: 20, Peter quotes the Psalms as his authority:

20 For it is written in the book of Psalms, Let his habitation be desolate, and let no man dwell therein: and his bishopric let another take.

Peter, probably quoting from Psalm 69: 25, is somewhat over-zealous. According to a number of translations – the King James, the Peshitta (Aramaic), the New International and the New Living – Peter has altered the quote. He has melded two separate references from the Psalms, Psalm 69: 25 and Psalm 109: 8, the last one being hardly applicable. In it, David is speaking a curse upon his enemy. The Peshitta, an ancient translation of the scriptures from Hebrew into Syriac, reads: *"Let their days be few, and let others take what they have saved."* In that Peter here alters scriptural quotes, God allows us to again see how human Peter was – not to 'put down' Peter but to uplift us.

When Jesus chose the twelve, he knew that one would betray him. The Bible teaches that God had already chosen Paul as the replacement. In Acts 9: 15, Ananias is told by Jesus that Paul is a chosen vessel unto Him:

*15 But the Lord said unto him (*Ananias*), Go thy way: for he (*Paul*) is a chosen vessel unto me, to bear my name before the Gentiles, and kings, and the Children of Israel:*

In Acts 22: 14-15, Ananias declares God's appointment to Paul:

14 And he said, The God of our fathers hath chosen thee, that thou shouldest know his will, and <u>see that Just One, and shouldest hear the voice of his mouth</u>.

Notes

15 For thou shalt be his witness unto all men of what thou hast seen and heard.

Read Acts 9: 1-22, Acts 22: 1-21 and Acts 26: 13-18; in these verses, Paul is personally called by Jesus (as were the first twelve) and told that he is to be made a minister and a witness of the things he has seen and of things which would appear to him.

13 At midday, O king, I saw in the way a light from heaven, above the brightness of the sun, shining round about me and them which journeyed with me.

14 And when we were all fallen to the earth, I heard a voice speaking unto me, and saying in the Hebrew tongue, Saul, Saul, why persecutest thou me? It is hard for thee to kick against the pricks.

15 And I said, Who art thou, Lord? And he said, I am Jesus whom thou persecutest.

16 But rise, and stand upon thy feet: for I have appeared unto thee for this purpose, to make thee a minister and a witness both of these things which thou hast seen, and of those things in the which I will appear unto thee;

17 Delivering thee from the people, and from the Gentiles, unto whom now I send thee,

18 To open their eyes, and to turn them from darkness to light, and from the power of Satan unto God, that they may receive forgiveness of sins, and inheritance among them which are sanctified by faith that is in me.

It's quite apparent throughout these scriptures that Paul had an appointment equal to and, possibly, even above that of the other apostles. Read Paul's opening greetings to the Romans. He calls himself a servant of Jesus Christ, called to be an apostle. Again in I Corinthians and II Corinthians, he says that he is called by the will of God to be an apostle. In Galatians, he says he is an apostle appointed "*not of men, neither by man, but by Jesus Christ and God the Father*." In I Corinthians 9: 16, Paul speaks of the clear calling he has; he understands his appointment to apostleship:

16 For though I preach the gospel, I have nothing to glory of: for necessity is laid upon me; yea, woe is unto me, if I preach not the gospel!

The New Jerusalem Measured / The Glorious City

Notes

Revelation 21: 15 - 17

15 *And he that talked with me had a golden reed to measure the city, and the gates thereof, and the wall thereof.*

16 *And the city lieth foursquare, and the length is as large as the breadth: and he measured the city with the reed, twelve thousand furlongs. The length and the breadth and the height of it are equal.*

17 *And he measured the wall thereof, an hundred and forty and four cubits, according to the measure of a man, that is, of the angel.*

The New Jerusalem is measured with a golden reed and determined to be 1500 miles wide, by 1500 miles deep and 1500 miles high: 2,250,000 square miles in all. As were all the temples before, so also is the New Jerusalem a faultless cube, the Greek symbol of perfection.

The faultless cube can be seen as a symbol of the triune nature of God: the reality of God eternally existing in three persons. The length and breadth and height are equal, as are the Father, Son and Holy Spirit co-equal in power and glory.

Revelation 21: 18 - 20

18 *And the building of the wall of it was of jasper: and the city was pure gold, like unto clear glass.*

19 *And the foundations of the wall of the city were garnished with all manner of precious stones. The first foundation was jasper; the second, sapphire; the third, a chalcedony; the fourth, an emerald;*

20 *The fifth, sardonyx; the sixth, sardius; the seventh, chrysolyte; the eighth, beryl; the ninth, a topaz; the tenth, a chrysoprasus; the eleventh, a jacinth, the twelfth, an amethyst.*

These three verses are tremendously intriguing. In verse 18, we see that the walls of the city in the New Jerusalem are of jasper. The jasper stone is representative of God who radiates the Shekinah Glory. Thus, we are seeing the New Jerusalem encased by the Glory of God. Some day, we will enter the New Jerusalem and so, as promised, we will walk in the light of God's Glory (Revelation 21: 23-24).

The foundation of the walls of the city are garnished with all manner of precious stones, as is the ephod of the Jewish High priest (see the King James translation of Exodus 28: 17-21). In the King James translation, the first stone in the first row of the ephod is the sardius stone (Jesus) and the last stone in the last row is the jasper stone (God). In this case, the reference is that Jesus, our High Priest, takes preeminence over God in the ephod. We see the Son and the Father, the First and the Last, the sardius and the jasper.

Notes

To counter this vision, verse 19 places God, the jasper stone, as the first in the foundation of the wall in the New Jerusalem. The sixth stone, not surprisingly, is the sardius stone. The stone, as well as the number of its placement, is representative of the <u>man</u>, Jesus Christ, who humbled himself (although he, as High Priest, was first) and became the sin offering on behalf of man. Thus, it is re-established and reinforced that the Father and Son, separate but together, are the First and the Last, the 'One' who sits on the throne (Revelation 4: 2-3).

Revelation 21: 21

21 And the twelve gates were twelve pearls: every several gate was of one pearl: and the street of the city was pure gold, as it were transparent glass.

The twelve gates are twelve pearls: the door of salvation, the pearl of great price. Recall the parable in Matthew 13: 45-46:

45 Again, the kingdom of heaven is like unto a merchant man, seeking goodly pearls:

46 Who, when he had found one pearl of great price, went and sold all that he had, and bought it.

Praise God that we have found the pearl of great price – salvation paid for by our Lord, Jesus Christ. Through him, the gates of heaven are open to us. Praise Him also that by grace, through faith, we have assurance of eternal life.

The verse under study also directs our attention to the streets of gold which appear as transparent glass. Recall from Revelation 4: 6 that John, when taken up into heaven, views the throne and sees before the throne a sea of glass like crystal.

Revelation 21: 22 - 23

22 And I saw no temple therein: for the <u>Lord God Almighty and the Lamb</u> are the temple of it.

23 And the city had no need of the sun, neither of the moon, to shine in it: for the <u>glory of God</u> did lighten it, and <u>the Lamb</u> is the light thereof.

John, looking within the New Jerusalem, sees there is no temple in it, for the Lord God and the Lamb are the temple of it. Together they are the Holiness, and the Glory, and the Majesty of the temple. There is no need for the sun nor the moon, for they are the light and the brightness of it.

The Nations Walk in the Light / The Sanctified Kings and Priests / The Eternal Day

Notes

In reading these two verses, we are struck by the absence of the third-person of the Godhead. Though we see the Father and Son, there is no visible evidence, or even mention, of the Holy Spirit.

There must always be a triune God and so it becomes evident, through the finished work of our Saviour, Jesus Christ, we become the submissive, sanctified vessel prepared by God to house, in all fullness, the Holy Spirit – referred to by the Church as the third-person of the Godhead. If more information is needed, refer to the teachings in this Study for Revelation 4: 3-5.

Revelation 21: 24

24 And the nations of them which are saved shall walk in the light of it: and the kings of the earth do bring their glory and honour into it.

The multitude of Saints which are saved walk in the light of the glory of the Father and the Son. '*The nations*' are representative of the Tribulation Saints, while '*the kings of the earth*' are the Raptured Saints (Revelation 5: 10). Together they bring the glory and honour of Christ into the new creation. Through Jesus, they have become sanctified kings and priests, and glorified sons of God. They are, as the seven lamps burning before the throne, endowed with the Seven Spirits of God; and, thereby, baptized with the baptism that Jesus is baptized with (Matthew 20: 23). Jesus has completed his mission. He has reconciled God and man.

Revelation 21: 25

25 And the gates of it shall not be shut at all by day: for there shall be no night there.

The gates of heaven will never be shut for there is no night there, neither are they shut at all by day. The reason the gates are never shut is that Jesus paid the price to open those 'pearly gates' of heaven and they will remain open for all eternity. The Father and Son will ever be available and eager to receive us. Gracious as God is, He places an angel at each gate to welcome us and usher us into His Holy presence (Revelation 21: 12; Revelation 22: 14).

Revelation 21: 26

26 And they shall bring the glory and honour of the nations into it.

Having been lifted up and reinstated by Christ, the Saints will bring the glory and honour of the nations (which is Christ in us) into the New Creation. We, the ransomed and redeemed, will take our appointed place as submissive and joyous children, consummated by the fullness of the Holy Spirit (complete in every detail – perfect). We are grafted in and made one in the Father and Son through the rich abundance of the Holy Spirit. Then is our Saviour's prayer fulfilled, as is recorded in John 17: 21-22:

> *21 That they all may be one; as thou, Father, art in me, and I in thee, that they also may be one in us; that the world may believe thou hast sent me.*
>
> *22 And the glory which thou gavest me I have given them; that they may be one, even as we are one.*

Revelation 21: 27

27 And there shall in no wise enter into it any thing that defileth, neither whatsoever worketh abomination or maketh a lie: but they which are written in the Lamb's book of life.

Only those who's names remain (after judgment) written in the Lamb's Book of Life will enter God's sinless new creation. The ungodly and those who lie have been judged and condemned. The Beast, the False prophet, Satan and all other evil beings will never again corrupt the Kingdom of God.

'*Praise the Lord; for the Lord is good: sing praises unto his name!*' (Psalm 135: 3)

Notes

CHAPTER TWENTY-TWO

Chapter Twenty-two personalizes heaven and confirms our position in the Godhead. The curse has been lifted and we have returned as sanctified children to the Garden of Eden. The Study shows that we no longer hunger after righteousness for we are endowed with Christ's Holy righteousness through the Seven Spirits of God. The glory which God has given to Jesus, Jesus has given to us, that we may be One in Them, as they are One. The glory which we bring into the New Jerusalem is the glory of the Father and Son. The chapter closes with a strong warning against adding to or taking away from truths revealed in *the Revelation.* After ending the inspired writing, John prays a blessing of grace and peace upon all who seek to understand and heed God's Holy *Revelation.*

Revelation 22

[1] And he shewed me a pure river of water of life, clear as crystal, proceeding out of the throne of God and of the Lamb.

[2] In the midst of the street of it, and on either side of the river, was there the tree of life, which bare twelve manner of fruits, and yielded her fruit every month: and the leaves of the tree were for the healing of the nations.

[3] And there shall be no more curse: but the throne of God and of the Lamb shall be in it; and his servants shall serve him:

[4] And they shall see his face; and his name shall be in their foreheads.

[5] And there shall be no night there; and they need no candle, neither light of the sun; for the Lord God giveth them light: and they shall reign for ever and ever.

[6] And he said unto me, These sayings are faithful and true: and the Lord God of the holy prophets sent his angel to shew unto his servants the things which must shortly be done.

[7] Behold, I come quickly: blessed is he that keepeth the sayings of the prophecy of this book.

[8] And I John saw these things, and heard them. And when I had heard and seen, I fell down to worship before the feet of the angel which shewed me these things.

[9] Then saith he unto me, See thou do it not: for I am thy fellowservant, and of thy brethren the prophets, and of them which keep the sayings of this book: worship God.

[10] And he saith unto me, Seal not the sayings of the prophecy of this book: for the time is at hand.

[11] He that is unjust, let him be unjust still: and he which is filthy, let him be filthy still: and he that is righteous, let him be righteous still: and he that is holy, let him be holy still.

[12] And, behold, I come quickly; and my reward is with me, to give every man according as his work shall be.

[13] I am Alpha and Omega, the beginning and the end, the first and the last.

[14] Blessed are they that do his commandments, that they may have right to the tree of life, and may enter in through the gates into the city.

[15] For without are dogs, and sorcerers, and whoremongers, and murderers, and idolaters, and whosoever loveth and maketh a lie.

[16] I Jesus have sent mine angel to testify unto you these things in the churches. I am the root and the offspring of David, and the bright and morning star.

[17] And the Spirit and the bride say, Come. And let him that heareth say, Come. And let him that is athirst come. And whosoever will, let him take the water of life freely.

[18] For I testify unto every man that heareth the words of the prophecy of this book, If any man shall add unto these things, God shall add unto him the plagues that are written in this book:

[19] And if any man shall take away from the words of the book of this prophecy, God shall take away his part out of the book of life, and out of the holy city, and from the things which are written in this book.

[20] He which testifieth these things saith, Surely I come quickly. Amen. Even so, come, Lord Jesus.

[21] The grace of our Lord Jesus Christ be with you all. Amen.

Living Water / The Saints Married to the Godhead

Revelation, Chapter 22

Notes

Revelation 22: 1

1 *And he shewed me a pure river of water of life, clear as crystal, proceeding out of the throne of God and of the Lamb.*

Zechariah 14: 8-9 talks about living water flowing out from Jerusalem in all seasons:

8 *And it shall be in that day, that living waters shall go out from Jerusalem; half of them toward the former sea, and half of them toward the hinder sea: in summer and in winter shall it be.*

9 *And the Lord shall be king over all the earth: in that day shall there be one Lord, and his name one.*

Zechariah may well be referring to both the Millennial years (after the great earth quake) and that of the New Jerusalem. In John 4: 14, Jesus says the water that he will give to man *"shall be in him a well of water springing up unto everlasting life.*" Here on earth, we seem to drink, but are never satisfied; we always hold out our cups for more. The time will come, in the everlasting life, that there will be a well of water springing up and man will never thirst again. I love to read Isaiah 35 in the New Living translation because, although I know it speaks of the Millennial years, it gives me a wondrous taste of the even more marvelous time when all the Saints will drink from those healing and cleansing waters which proceed out of the throne of God. In so doing, we are acting out the fact that we are purified, not only by new birth, but also by abundant new life.

Take special note that the water of Life flows out of the throne of God and of the Lamb. There is no mention of there being water of life flowing from a throne of the Holy Spirit. Of course not, dear friends, for in the New Creation, the Saints are inhabited by the fullness of the Holy Spirit and, thereby, are sanctified and married to the Godhead (Revelation 19: 7); however, life never flows through them, for life flows only from the Father and Son.

God has answered the prayer that Jesus prayed on our behalf just hours before he was crucified. Please re-read the quote from John 17: 19-23:

19 *And for their sakes I sanctify myself, that they also might be sanctified through the truth.*

20 *Neither pray I for these alone, but for them also which shall believe on me through their word;*

21 *That they all may be one; as thou, Father, art in me, and I in thee, that they also may be one in us: that the world may believe that thou hast sent me.*

22 *And the glory which thou gavest me I have given them; that they may be one, even as we are one:*

Notes

23 I in them, and thou in me, that they may be made perfect in one; and that the world may know that thou hast sent me, and hast loved them, as thou hast loved me.

Revelation 22: 2

2 In the midst of the street of it, and on either side of the river, was there the tree of life, which bare twelve manner of fruits, and yielded her fruit every month: and the leaves of the tree were for the healing of the nations.

As with the first garden, man will find two trees: one on either side of the river which flows from the throne of God. We have reason to rejoice for both trees are the tree of life. Man has only to choose abundant life and not death.

What a devastating choice man made in the first garden when he ate from the tree of the knowledge of good and evil. Conscience and lust awakened then and man became carnal, which separated him from the intimacy of God. He was ashamed and hid from God; thus, mankind was opened to the condemnation of Satan. How gracious God is that He heals us and restores our sweet innocence, our confidence and our peace. As in the beginning, He sets before us bountiful, sinless eternal life (Romans 8: 21).

The trees of life yield provision for man each month of every year and a fruit for every month. This is representative of the abundance of God. We will never again live by the sweat of our brow. On our behalf, Jesus has overcome the curse declared in Genesis which, ultimately, has become a blessing. The agony of independent living has brought us back to Him.

All man (all nations) will walk in the healing (health) of God's sheltering hand. We will eat from the Tree of Life in remembrance of God's provision and love. Being dwellers of Eden and ceasing from labour, we will have no need for meat; we will again eat only of the fruits and the herbs of the earth. As in the Millennium, there will be no fear in man or beast, nor fowl, nor fish (Genesis 9: 2-3; Isaiah 11: 6-8). The whole of God's universe will rest in love.

Revelation 22: 3

3 And there shall be no more curse: but the throne of God and of the Lamb shall be in it; and his servants shall serve him:

The curse of sin and separation from God will be lifted when Satan is destroyed and Jesus has completed his mission. At that time, both God and the Lamb will tabernacle themselves among us in the New Jerusalem which crowns the New Earth. God's specially chosen servants, the

Notes

144,000 of Revelation 14: 1-5, will serve the Father and Son in the glorious City of God. They follow Jesus where ever He goes, and '*sing a song that no man can learn*'. They are redeemed as the first-fruit of God and the Lamb.

Revelation 22: 4

4 *And they shall see his face; and his name shall be in their foreheads.*

We may not all see the face of God but we know the 144,000 will, for they bear the seal of God in their foreheads (Revelation 7: 3). It seems highly probable that we will all be granted this blessing, since Revelation 3: 12 states that Jesus will write upon us the name of God, and the name of the city of God, and he will write upon us his new name. Possibly, God will (as Christ did) evidence Himself in the same bodily form as He has created us.

Surely we have seen His face when we have seen the compassionate Jesus. That face is indelibly written in our hearts. Jesus said, '*He that hath seen me has seen the Father.*' (John 14: 9).

Revelation 22: 5

5 *And there shall be no night there; and they need no candle, neither light of the sun; for the Lord God giveth them light: and they shall reign for ever and ever.* (Father and Son)

This verse allows us to conclude that aside from the 144,000 chosen ones (Revelation 14: 4), the Saints will never dwell within the walls of the New Jerusalem. If they did, there would be no need for the twelve gates in the walls (Revelation 21: 12) since the Saints would go neither in nor out. It would also be unnecessary for God to create the New Heavens or the New Earth, for none would dwell there. If we all dwelt within the confines of the New Jerusalem, we would definitely not need the re-created Heavens, for in the Holy City there is no sun since 'Elohim' (the plural God) is the light thereof.

A joyous and significant part of this verse is that it says they – Father and Son – shall reign forever and ever. The awesome responsibility of reigning with Christ will be lifted and we – as submissive, sanctified and glorified children – will become one in Them. By looking deeply into this Revelation, we see the triune Godhead intact. Through the provision of Christ, consumed by the Holy Spirit, man again walks in the presence and image of God for all eternity.

The Future Revealed / The Blessed Faithful / John Corrected

Notes

Revelation 22: 6

6 *And he said unto me, These sayings are faithful and true: and the Lord God of the holy prophets sent his angel to shew unto his servants the things which must shortly be done.*

The angel affirms that this prophetic *Revelation* is faithful and true and will come to pass. John had seen much of the Old Testament prophecies fulfilled in Jesus. He is well aware that the church which he pastors, the Church of Ephesus, has almost, in fact, lost its first love, its sweet purity. He knows that Baal worship and Gnosticism is active in the Church, and the doctrine of the Nicolaitanes has conquered many of God's people. John has already seen the priestly order evolving and he is here viewing the wrath and punishment brought about by sin. John has been walking in the beginning truth of this Revelation even before it is given to him through Jesus.

Revelation 22: 7

7 *Behold, I come quickly: blessed is he that keepeth the sayings of the prophecy of this book.*

From this verse we understand that when we see the signs being fulfilled which are spoken of by Jesus in Matthew 24, we know that Jesus is about to enter into this, his latter-day ministry. All will be blessed, indeed, who are ready to be received by him and have heeded the admonishment of the *Revelation*. Our lives will be filled with joy and abundance, with godliness and glory. We will sing forever, along with the Saints: '*Blessing, and glory, and wisdom, and thanksgiving, and honour, and power, and might be unto our God, forever and ever. Amen!*' (Revelation 7: 11-12).

Revelation 22: 8 - 9

8 *And I John saw these things, and heard them. And when I had heard and seen, I fell down to worship before the feet of the angel which shewed me these things.*

9 *Then saith he unto me, See thou do it not: for I am thy fellowservant, and of thy brethren the prophets, and of them which keep the sayings of this book: worship God.*

A fellowservant in the form of an angel continues to speak with John. In verse 7, he reveals that Jesus is coming soon. He states that all who keep the sayings of the *Revelation* will be blessed. John is again so overjoyed at what he sees and hears that he falls in absolute awe before the fellowservant sent by God (see also Revelation 19: 10). The object of his worship is not so much for the servant as for what the servant has revealed to him; nevertheless, he is given another warning not to fall before anyone but God.

Notes

Revelation 22: 10

10 And he saith unto me, Seal not the sayings of the prophecy of this book: for the time is at hand.

Probably from this verse on, the spokesperson is again the Spirit of Jesus speaking through his angel. He identifies himself in the verses which follow. Only Jesus has the foreknowledge of God; he, only, is the Alpha and Omega – none but he has the authority to reward the just and the unjust. The Godhead alone can command the sealing or unsealing of the prophecies.

According to prophecy, all nations will submit to one World Church and Government with all the power – and atrocities – of the old Roman Empire. *Revelation* is a blow-by-blow account of the rise and fall of this kingdom. The world is soon to see Daniel's seventieth week played out.

I believe that shortly after the year 2000, the Wye River Accord will honour the Jerusalem covenant and allow Israel to build her temple on Mount Moriah. The world ruler will confirm the covenant but will break it after three and one-half years. At that time, he will take over the temple and set himself up as god. The false church will assist him and agree with his actions. God has not left us in the dark. We need only to read Daniel and *Revelation* to see the final end of this evil kingdom and witness God's holy rule being set up for Christ's Millennial reign. Beyond this, we see the New Creation with the Father and Son in the New Jerusalem.

Revelation 22: 11

11 He that is unjust, let him be unjust still: and he which is filthy, let him be filthy still: and he that is righteous, let him be righteous still: and he that is holy, let him be holy still.

Those who have neglected to commit their life to Christ should take heed for there comes a time when it is too late to repent. Soon the condition man is found in will be permanent. When the eternal day has dawned, the unjust will be left to remain unjust; the filthy will remain filthy; the righteous, if it be that of Christ, will keep righteousness; those who have the holiness of Jesus will retain his holiness.

Revelation 22: 12

12 And, behold, I come quickly; and my reward is with me, to give every man according as his work shall be.

God is warning mankind that when Christ returns, he comes with judgment and rewards. Evil has 'filled the cup' and God's wrath is about to run over. The just and the unjust will receive according to their works.

The Just Rewarded / The Unjust Condemned / The Morning Star

Notes

Revelation 22: 13

13 I am Alpha and Omega, the beginning and the end, the first and the last.

We who know him say, "Amen. Come, Lord Jesus!"

Revelation 22: 14

14 Blessed are they that do his commandments, that they may have right to the tree of life, and may enter in through the gates into the city.

The reward for the just, from the beginning of creation to the end of the Millennium, will be abundant and sinless eternal life in the New Heavens and New Earth – the whole of the earth will be as the Garden of Eden. The blessed are those who have washed their robes in the blood of the Lamb and, with perseverance, have walked in the commandments of God. All these will eat from the tree of life and freely enter the Holy City to rejoice before the Father and Son. There will be no separation between God and man. His children will bask eternally in His love.

Revelation 22: 15

15 For without are dogs, and sorcerers, and whoremongers, and murderers, and idolaters, and whosoever loveth and maketh a lie.

All those who do evil and serve wicked Babylon (Satan) will be killed at the end of the Tribulation and judged at the end of the Millennium. In the White Throne judgment, their final destination will be declared. Many will join their father, Satan, in the fires of hell; others shall receive lesser judgments. Multitudes will hear Jesus say, "*Come, ye blessed of my Father, inherit the kingdom prepared for you from the foundation of the world.*" (Matthew 25: 34).

Before we leave the study of this verse, we should note that God identifies lies as a mortal sin. We must all monitor our words to avoid such temptation. Lying is included among those listed as vicious sins, totally unacceptable by God's standards (Revelation 21: 8; Revelation 21: 27).

Revelation 22: 16

16 I Jesus, have sent mine angel to testify unto you these things in the churches. I am the root and the offspring of David, and the bright and morning star.

After a long silence, Jesus now speaks to the Churches through John. He indicates that he has sent his angel to tell the churches – and the world – what horror and punishment will overcome them unless they come to the

Notes

light of the glorious gospel. Jesus was – and is – the bright star on the horizon which disperses darkness from the heart of man. He is the light that shines in dark places and has become to us the day star which arose in our hearts. Refer to the words of II Peter 1: 19:

> *19 We have also a more sure word of prophecy; whereunto ye do well that ye take heed, as unto a light that shineth in a dark place, until the day dawn, and the day star arise in your hearts:*

Also consider II Corinthians 4: 4 - 6:

> *4 In whom the god of this world hath blinded the minds of them which believe not, lest the light of the glorious gospel of Christ, who is the image of God, should shine unto them.*
> *5 For we preach not ourselves, but Christ Jesus the Lord; and ourselves your servants for Jesus' sake.*
> *6 For God, who commanded the light to shine out of darkness, has shined in our hearts, to give the light of the knowledge of the glory of God in the face of Jesus Christ.*

Revelation 22: 17

> *17 And <u>the Spirit and the bride say, Come</u>. And let him that heareth say, Come. And let him that is athirst come. And whosoever will, let him take the water of life freely.*

As this marvelous Revelation ends, God gives us a final confirmation that the Spirit and the bride become one. Together, they say '*Come*'. Then, with the Saviour's heart, Jesus encourages all who hear to say '*Come*'. He assures all mankind that if they <u>will</u>, they may come and drink of Him who is the giver of eternal life. Even to the very end, God – in His bountiful love – continues to call by His Spirit, through His Son.

Revelation 22: 18

> *18 For I testify unto every man that heareth the words of the prophecy of this book, If any man shall add unto these things, God shall add unto him the plagues that are written in this book:*

As this Revelation closes, God issues a clear warning against adding to this prophesy; for that matter, adding to any portion of His inspired word. He speaks to you and to me and, most definitely, to the latter day seers who have added to the scriptures and the prophesies of the Bible. False prophets abound in these troubled days and many hungry souls seeking truth are deceived by their lies. Beware – heed them not.

God's great prophets have already spoken all that He would have us hear. If, in these latter days, a true prophet speaks, he will do so only in

Notes

illumination and confirmation of past prophecies (Galations 1: 8-9). Take care that every teaching you hear or read is safely and carefully supported by the scriptures. Hearken to God's warning from Deuteronomy 18: 20-22 (NLT) and use it as a guide:

> 20 *Any prophet who claims to give a message from another god or who falsely claims to speak for me must die.*
> 21 *You may wonder 'How will we know whether the prophecy is the Lord or not?'*
> 22 *If the prophet predicts something in the Lord's name and it does not happen, the Lord did not give the message. That prophet has spoken on his own and need not be feared (honoured).*

Revelation 22: 19

19 *And if any man shall take away from the words of the book of this prophecy, God shall take away his part out of the book of life, and out of the holy city, and from the things which are written in this book.*

Those who may attempt to take away from the words of this prophecy (the intent, instruction, direction) are strongly warned that their names will be taken out of the Book of Life. This is a fearful thing that must be weighed thoughtfully by anyone who dares to comment on this God-inspired *Revelation*. It's a warning that we must all seriously consider. We must continuously pray whenever we expound on this, His most marvelous book. Truly, it is a fearful thing to fall into the hands of the Lord God! Clearly, '*no prophecy of the scripture is of any private interpretation,*' for true prophecy came not by man but by the Holy Spirit who moved the prophets to speak for God (II Peter 1: 20-21). Reverencing the holy origin and purpose of the prophecies, I have spent innumerable hours comparing scriptures with scriptures. Hopefully, I have heard what God is saying through the prophets and offer you only truth. The *Revelation* is written by God but the interpretation may, in part, be that of man. For this reason, I sincerely invite correction of this Study if the reader is so directed by the Holy Spirit and the correction is well documented through the Word.

Revelation 22: 20

20 *He which testifieth these things saith, Surely I come quickly. Amen. Even so, come, Lord Jesus.*

In this verse, we hear the final words of Jesus who testifies to John of these things. He says, "*Surely, I come quickly. Amen.*" Then John breathes the prayer, '*Even so, come, Lord Jesus.*'

A Blessing

Revelation 22: 21

21 The grace of our Lord Jesus Christ be with you all. Amen.

John offers a final blessing upon all who read this prophesy.

Notes

Appendix 1

Seven Mysteries of God

1. Mystery of God's Will: He will draw all men unto Him. Read Ephesians 1: 6-10:
   6 *To the praise of the glory of his grace, wherein he hath made us accepted in the beloved.*
   7 *In whom we have redemption through his blood, the forgiveness of sins, according to the riches of his grace;*
   9 *Having made known unto us the <u>mystery of his will</u>, according to his good pleasure which he hath purposed in himself.*
   10 *That in the dispensation of the fullness of time he might gather together in one all things in Christ, both which are in heaven, and which are on earth; even in him.*

2. Mystery of iniquity – Satan in the flesh (II Thessalonians 2: 7):
   7 *For <u>the mystery of iniquity</u> doth already work: only he who now letteth will let, until he be taken out of the way.*

3. Mystery of Godliness (I Timothy 3: 16):
   16 *And without controversy great is <u>the mystery of godliness</u>: God was manifest in the flesh, justified in the Spirit, seen of angels, preached unto the Gentiles, believed on in the world, received up into glory.*

4. Mystery of God – the knowledge of God's love and grace which brings about rebirth and salvation; it encompasses not only the mystery of God's love, but also the love of one brother for another (Revelation 10: 7; Colossians 1: 26-27):
   26 *Even the mystery which hath been hid from ages and from generations, but now is made manifest to his saints:*
   27 *To whom God would make known what is the riches of the glory of this mystery among the Gentiles; which is Christ in you, the hope of glory.*

5. Mystery of the Seven Stars and the Seven Golden Candlesticks – the stars are the angels of the churches: the pastors, bishops and priests. The Golden Candlesticks are the Churches, the light-bearers of the world (Revelation 1: 20).

6. Mystery of the woman on the beast – mystery of Babylon – is, Satan, the Beast, the False prophet and the great city who reigneth over the kings of the earth (Revelation 17: 3-18). Read the preamble to Revelation, Chapter 17.

Appendix 1 – Seven Mysteries of God (continued)

7. Mystery of Israel – blindness, in part, has come upon Israel until the fullness of the Gentiles is come (Romans 11: 25-33).

   25 *For I would not, brethren, that ye should be ignorant of this mystery, lest ye should be wise in your own conceits; that blindness in part is happened to Israel, until the fulness of the Gentiles be come in.*
   26 *And so all Israel shall be saved; as it is written, There shall come out of Sion the Deliverer, and shall turn away ungodliness from Jacob:*
   27 *For this is my covenant unto them, when I shall take away their sins.*
   28 *As concerning the gospel, they are enemies for your sakes: but as touching the election, they are beloved for the fathers' sakes.*
   29 *For the gifts and calling of God are without repentance.*
   30 *For as ye in times past have not believed God, yet have now obtained mercy through their unbelief:*
   31 *Even so have these also now not believed, that through your mercy they also may obtain mercy.*
   32 *For God hath concluded them all in unbelief, that he might have mercy upon all.*
   33 *O the depth of the riches both of the wisdom and knowledge of God! How unsearchable are his judgments, and his ways past finding out!*

Appendix 2

The Crowns:

the Rewards earned through Faithful Stewardship

If any man's work abide which he hath built thereupon, he shall receive a reward.
(I Corinthians 3: 14)

- Crown of Life: promised to all those who have withstood tribulation, including martyrdom for Christ (Revelation 2: 10) (James 1: 12)

- Crown of Glory: promised to those who have served faithfully in the role of elders and pastors, also laymen of devotion (I Peter 5: 1-4)

- Crown of Rejoicing: promised to those who have won others to faith in Jesus Christ and nurtured the children of God (I Thessalonians 2: 19-20)

- Crown of Righteousness: promised to all those who love His appearing (II Timothy 4: 8)

- The Incorruptible Crown: promised to those who strive for self-discipline and are temperate in all things (I Corinthians 9: 25)

Appendix 3

The Seven Spirits of God

(Isaiah 11: 2)

- The Spirit of the Lord: the fullness of the Holy Spirit
- The Spirit of Wisdom: all wisdom
- The Spirit of Understanding: as Jesus understands
- The Spirit of Council: full authority
- The Spirit of Might: all power
- The Spirit of Knowledge: complete knowledge
- The Spirit of the Fear of the Lord: absolute reverence and total submission to God the Father and Jesus Christ the Son

Appendix 4

The Jerusalem Covenant

In May of 1993, the Israeli Parliament met to prepare a document in which they would mutually agree that the City of Jerusalem could never be used in trade for peace. All of the fourteen parties agreed as one and signed the document which is called 'the Jerusalem Covenant'. They made its existence known before they met with the United Nations and the Palestine Liberation Organization in September 1993. In the peace talks, the nations were aware that the Covenant obligated Israeli negotiators to reject any peace treaty which would include the return of the City of Jerusalem to the Palestinians.

During the talks, the PLO demanded that Israel return sections of the Gaza Strip and the West Bank; Israel painfully agreed to these demands and signed the 1993 Oslo Accord.

More recently, in the Wye River Accord of October 23, 1998, Israel again was required to trade land for peace. If the Accord is enacted, Israel will lose more of the land which she gained in the 1967 War with the Palestinians. At this point, both parties are restless and unhappy and the agreement is on hold. On the day of its signing, the Wye River Accord caused much anxiety and distress in the Israeli religious community.

I speculate that the only way Israel will settle peaceably into the latest proposed 'grab back' will be if the United Nations insists the Arabs honour Israel's land rights and allow her to build her longed-for temple on top of Mount Moriah, next to the Dome of the Rock. The existing Moslem holy place of worship will not be disturbed since the central structure of the end-time temple will be similar in size to the former temples, thus allowing Ezekiel's envisioned temple to be built on this site. The courts will be enlarged when Jesus and his sanctified priests enter the cleansed and restored temple, at that time fulfilling the specifications given by God through Ezekiel. This is the only agreement that could possibly pacify the Israelites who would suffer almost anything to once again worship and sacrifice in their own temple in Jerusalem.

The Jerusalem Covenant will play a prominent part in a world agreement in the imminent One World Rule. Israel will demand the freedom to build, worship and sacrifice forever in her temple on Mount Moriah. The Antichrist will confirm the Jerusalem Covenant and honour it for the first three and one-half years of the Tribulation. He will then break the covenant and enter and defile the temple for the balance of his temple reign (Daniel 9: 27), a period of three and one-half years.

Appendix 5
Calendar of the Seals and Wraths

The foretold judgments in the opening of the seals	The Trumpet Judgments	The Vial Judgments
Rev. 6: 2 - *I beheld a white horse: and he who sat upon him had a bow; and a crown was given unto him; and he went forth conquering and to conquer.*	Rev. 8: 7 - *When the first trumpet sounds, hail and fire mingled with blood was cast upon the earth: a third part of trees was burned up, and all green grass was burned up.*	Rev. 16: 2 - *When the first vial was poured out, grievous sores came upon all men who had the mark of the beast and those who worshipped his image.*
Rev. 6: 3-4 - *There went out another horse that was red: and power was given him to take peace from the earth, that they should kill one another; he was given a great sword.*	Rev. 8: 8-9 - *When the second trumpet sounds, a mountain burning with fire was cast into the sea: and a third part of the sea became blood; a third part of sea life died, and a third of the ships were destroyed.*	Rev. 16: 3 - *Second angel poured his vial on the sea; and it became the blood of dead men: and every living soul died in the sea.*
Rev. 6: 5-6 - *A black horse and he that sat upon him had a pair of balances in his hand. A measure of wheat for a penny and three measures of barley. See thou hurt not the oil and the wine.*	Rev. 8: 10 - *When the third trumpet sounds, the star wormwood fell from heaven on a third part of rivers and fountains and they became bitter and many died from the water.*	Rev. 16: 4 - *Third angel poured out his vial upon the rivers and fountains of water; and they became blood.*
Rev. 6: 7-8 - *A pale horse ridden by death, and hell followed with him. Power was given him to kill with a sword a fourth part of the earth, and with hunger, and death and with the beasts of the earth.*	Rev. 8: 12 - *When the fourth trumpet sounds, a third part of the sun and moon and stars were smitten so that a third part of them was darkened, and the day shone not for a third part of it, and the night likewise.*	Rev. 16: 8 - *Fourth angel poured his vial on the sun; and power was given him to scorch men with fire; but they repented not.*
Rev. 6: 9 - *On opening the fifth seal, I saw under the altar the souls of them that were slain for the word of God and for the testimony which they held.*	Rev. 9: 1-11 - *When the fifth trumpet sounds, a star fell from heaven and was given the key to the bottomless pit. He opened it and smoke from the furnace darkened the sun and the air: and locusts came out as scorpions and had power. They hurt only those without God's seal.*	Rev. 16: 10 - *Fifth angel poured his vial upon the seat of the beast; and his kingdom was full of darkness, and they gnawed their tongues for pain. They blasphemed God because of their pain and sores and repented not of their deeds.*
Rev. 6: 12-16 - *Saw an earthquake. Sun turned black and moon as blood. Stars fell. Heaven departed and mountains and islands moved. All men hid in dens and called to be hidden from God and the wrath of the Lamb.*	Rev. 9: 13-19 - *When the sixth trumpet sounds, a voice from the golden altar commanded four angels be loosed from the river Euphrates. Two hundred thousand thousand horsemen killed a third part of men; but the men who were not killed repented not.*	Rev. 16: 12-16 - *Sixth angel poured his vial on the Euphrates & the water dried and the way was prepared for the kings of the east. Three unclean spirits came out of the mouths of the dragon, the beast and the false prophet to gather all men to the battle of Armageddon.*
Rev. 6: 17 & 8: 1 - *The great day of wrath is come; and who is able to stand. When Jesus opens the seventh seal, there will be silence in heaven the space of half an hour.*	Rev. 10: 7 & 11: 15-19 - *When the seventh trumpet begins to sound, the mystery of God is finished. When it finished sounding, the kingdoms of the world become the kingdoms of our Lord. There were lightnings, voices, thunderings, hail and an earthquake.*	Rev. 16: 17-20 - *Seventh angel poured his vial in the air and a great voice from the temple said It is done. There were lightnings and earthquakes and Jerusalem was divided into three parts; and the cities of the nations fell and mountains and islands fled away.*

Appendix 6

The Ethiopian Exodus

1) **1984** -- The massive airlift known as **Operation Moses** begins on November 18th and ends on January 5th, 1985. During those six weeks, some 6,500 Ethiopian Jews are flown from Sudan to Israel. Attempts are made to keep the rescue effort secret, but public disclosure forces an abrupt end. In the end, an estimated 2.000 Jews die en route to Sudan or in Sudanese refugee camps.

2) **1984-1988** -- With the abrupt halting of **Operation Joshua** in 1985, the Ethiopian Jewish community is split in half, with some 15,000 souls in Israel, and more than 15,000 still stranded in Ethiopia. For the next five years, only very small numbers of Jews reach Israel.

3) **1986** -- The United States Congressional Caucus for Ethiopian Jewry is established with over 140 representatives currently listed.

4) **1987** -- The Ethiopian leaders in Israel organize an assembly at Binyanei Ha'uma in Jerusalem, where the Israeli public comes together in solidarity for reunification of Ethiopian Jewry.

5) **1988** -- Israeli Ambassador to the United States, Pinchas Eliav, makes a formal statement at the United Nations Human Rights Commission for the reunification of Ethiopian Jews in Israel.

6) **1989** -- Ethiopia and Israel renew diplomatic relations. This creates high hopes among Jewry for the reunification of Ethiopian Jews in Israel.

7) **1990** -- Ethiopia's ruler, Colonel Mengistu Haile Mariam, makes a public statement expressing desire to allow Ethiopian Jews to be reunited with family members in Israel.

8) **1991** -- with Eritrean rebels advancing on the capital each day, Colonel Mengistu flees Ethiopia. Israel asks the United States to urge rebels to allow a rescue operation for Ethiopian Jews. Spanning the 24th and 25th of May, **Operation Solomon** airlifts 14,324 Jews to Israel aboard thirty-four El Al jets in just over thirty-six hours. And the story of the Ethiopian Jews continues...

Sources cited: "From Addis to Jerusalem." Jewish Agency for Israel, Jerusalem Israel, 1991
"Reunify Ethiopian Jewry: Top Priority." World Union of Jewish Students, Jerusalem, Israel, 1989

Appendix 7

The Location of Christ's Crucifixion and Burial

There are conflicting opinions as to the crucifixion and burial place of Jesus. Some claim the Church of the Sepulchre is the place, while others say the Garden Tomb bordering Golgotha is where Joseph of Armethia and Nicodemus laid our Lord.

Many who accept the Church of the Holy Sepulchre as the burial place of Jesus also claim he was crucified on a high rock found within the confines of the Church. According to the Government approved tourist information, the site on which the sepulchure was later built was located within the walls at the Old City of Jerusalem. The Catholic appointed Via Dolorosa (the fourteen stations of the cross) start at the Chapel of the Flagellation and end at the Church of the Holy Sepulchure. All fourteen stations are contained within the walls of the Old City. The last five stations (shown as numbers X to XIV) claim that Jesus was stripped, crucified and buried at the site where the Church of the Holy Sepulchure stands today. According to the scriptures, Jesus was led outside the walls of the Old City of Jerusalem and crucified at Golgotha. It was there that they buried him; indeed, no one has ever been crucified or buried within the walls – not even Jesus (see Study notes for Revelation 14: 20).

Hebrews 13: 11-13 NIV tells us '*the high priest carried the blood of the animals into the most Holy Place as a sin offering, but the bodies are burned outside the camp. And so Jesus also suffered outside the city gates to make the people holy through his own blood. Let us then go to him outside the camp, bearing the disgrace he bore.*'

Some claim the Old City walls of Jerusalem were extended to include the place of the Holy Sepulchre after the death of Jesus. Even if this could be proven true, Matthew 27: 33, Mark 15: 22 and John 19: 17 all state that Jesus was taken to a place called Golgotha which, being interpreted, is 'the place of a skull.' There is only one skull in the sandstone (even yet today) and that unbelievable site borders the Garden Tomb. This is well beyond the extended walls of Jerusalem, a goodly distance from the celebrated Church of the Holy Sepulchre.

The authentic crucifixion and burial site is beside the only highway which ran from the extreme north to the far south of Canaan. This allowed those who passed to wag their heads and call abuses to Jesus (Matthew 27: 39-40). There has never been a highway within the walls of Jerusalem; none run past the Church of the Sepulchre.

VIA DOLOROSA
THE FOURTEEN STATIONS OF THE CROSS

Via Dolorosa, "Sorrowful Way", is considered by Christians the holiest road in the world, for along its uneven path Jesus was led from the place of His condemnation to that of His crucifixion and death.

1. Jesus is condemned to death.
2. Jesus receives the cross.
3. Jesus falls under the cross for the first time.
4. Jesus meets His mother Mary.
5. The cross is taken over by Simon of Cyrene.
6. Veronica wipes the seat from Jesus' face.
7. Jesus falls for the second time.
8. Jesus consoles the women of Jerusalem.
9. Jesus falls for the third time.
10. Jesus is stripped of His garments.
11. Jesus is nailed to the cross.
12. Jesus dies on the cross.
13. Jesus' body is taken off the cross.
14. Jesus' body is laid into the Sepulchre.

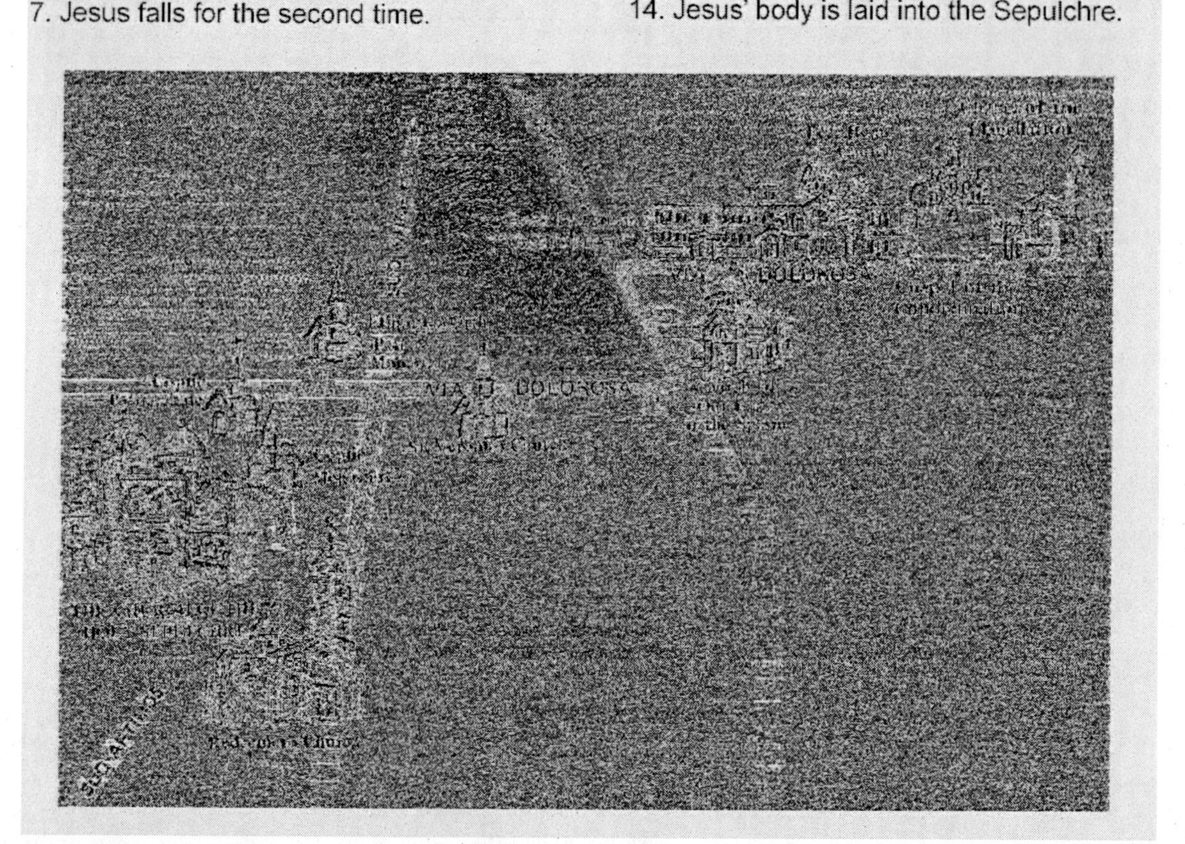

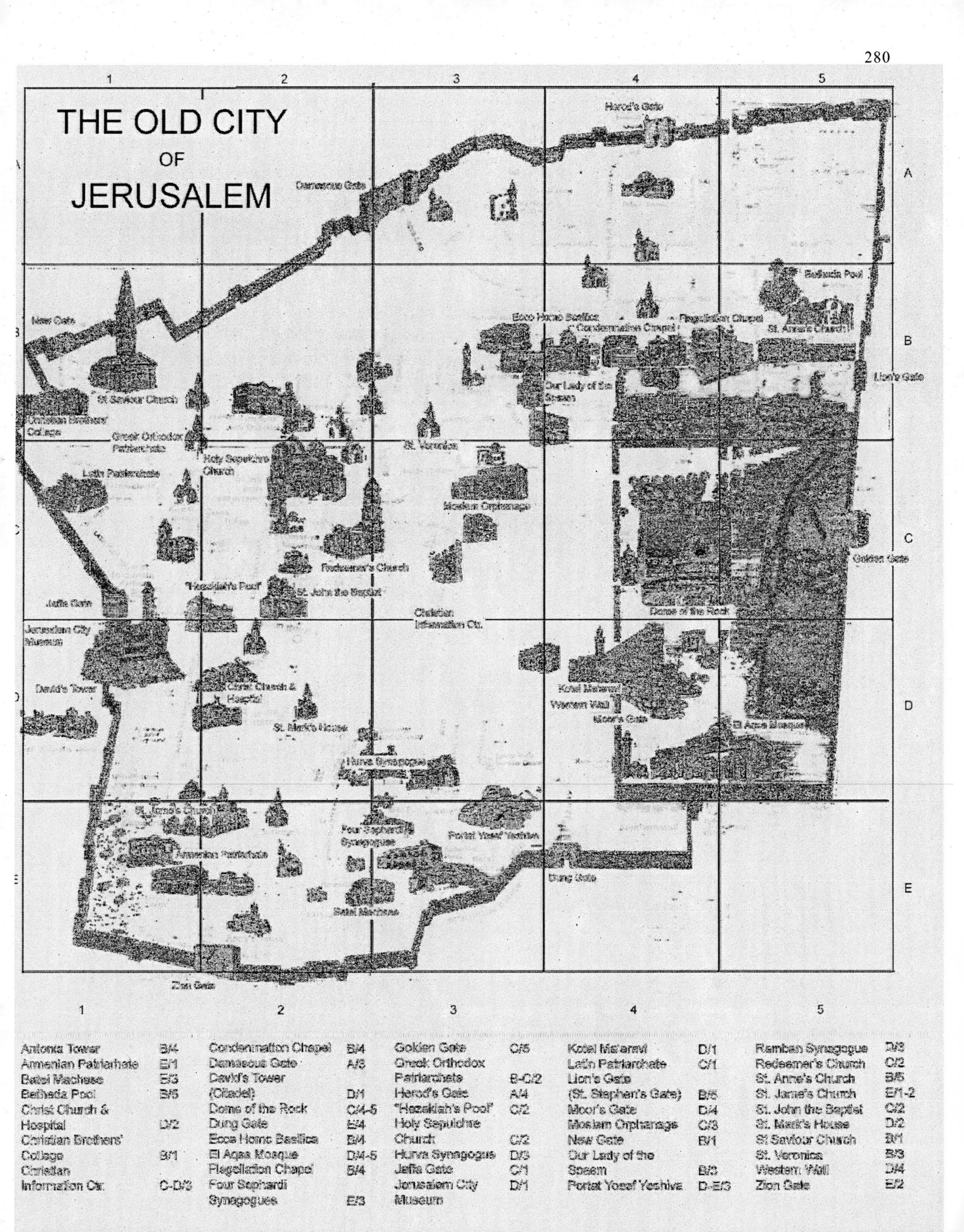
THE OLD CITY
OF
JERUSALEM
1
2
3
4
5
A
B
C
D
E
Herod's Gate
Damascus Gate
New Gate
St Saviour Church
Christian Brothers' College
Greek Orthodox Patriarchate
Latin Patriarchate
Holy Sepulchre Church
St. Veronica
Moslem Orphanage
Redeemer's Church
"Hezekiah's Pool"
St. John the Baptist
Jaffa Gate
Christian Information Ctr.
Jerusalem City Museum
David's Tower
Christ Church & Hospital
St. Mark's House
Hurva Synagogue
St. James's Church
Four Sephardi Synagogue
Portat Yosef Yeshiva
Armenian Patriarchate
Batei Machase
Dung Gate
Zion Gate
Ecce Homo Basilica
Condemnation Chapel
Flagellation Chapel
Bethesda Pool
St. Anne's Church
Lion's Gate
Our Lady of the Spasm
Golden Gate
Dome of the Rock
Kotel Ma'aravi
Western Wall
Moor's Gate
El Aqsa Mosque
Antonia Tower B/4
Armenian Patriarchate E/1
Batei Machase E/3
Bethesda Pool B/5
Christ Church & Hospital D/2
Christian Brothers' College B/1
Christian Information Ctr. C-D/3
Condemnation Chapel B/4
Damascus Gate A/3
David's Tower (Citadel) D/1
Dome of the Rock C/4-5
Dung Gate E/4
Ecce Homo Basilica B/4
El Aqsa Mosque D/4-5
Flagellation Chapel B/4
Four Sephardi Synagogues E/3
Golden Gate C/5
Greek Orthodox Patriarchate B-C/2
Herod's Gate A/4
"Hezekiah's Pool" C/2
Holy Sepulchre Church C/2
Hurva Synagogue D/3
Jaffa Gate C/1
Jerusalem City Museum D/1
Kotel Ma'aravi D/1
Latin Patriarchate C/1
Lion's Gate (St. Stephen's Gate) B/5
Moor's Gate D/4
Moslem Orphanage C/3
New Gate B/1
Our Lady of the Spasm B/3
Portat Yosef Yeshiva D-E/3
Ramban Synagogue D/3
Redeemer's Church C/2
St. Anne's Church B/5
St. James's Church E/1-2
St. John the Baptist C/2
St. Mark's House D/2
St Saviour Church B/1
St. Veronica B/3
Western Wall D/4
Zion Gate E/2

Appendix 7

<u>The Location of Christ's Crucifixion and Burial</u> (continued)

Rome would hardly recognize the Garden Tomb as the authentic place of the entombment since the land was owned by Joseph. Though powerful, the Roman Catholic Church would stop short of expropriating land from the descendants of the rich, highly respected Jew from Armethia. Being controllers of the people and wanting to claim ownership of all the holy places, the Church chose another site. They prepared a shallow tomb in the rock on which the Church of the Holy Sepulchre would later stand. They encased it in an ornate structure and proceeded to hallow it as the genuine Holy Sepulchre of Christ.

The four denominations who share and maintain the Church of the Holy Sepulchre recently invited archeologists, Martin and Berthe Biddle, to Israel. Their hope was that these professionals would authenticate this site as the burial place of Jesus. On February 5, 2002, the History Channel aired a documentary on the Biddle's research efforts. After close scrutiny of both sites, the Biddle's discredited the Garden Tomb for two reasons only.

Firstly, they felt the Christians of that era (274-313 AD) would have directed Emperor Constantine to the true location of the Holy Sepulchre. Probably the Biddles did not take into consideration that many Christians under the thumb of Rome and the threat of death worshiped as the Church of Rome dictated. Constantine became a staunch Catholic and was, therefore, influenced by the Church when he built the Church of the Sepulchre over the false tomb. (See Study note under "Preamble: the Letter to Pergamous", which follows the Study notes for Revelation Chapter 2: 11.)

The second reason the Biddles rejected the Garden Tomb was that although they recognized the site met all the requirements of the proper burial place, they concluded the Tomb itself did not meet the design of the time. They felt the Tomb in the Church of the Holy Sepulchre was more in keeping with other tombs located outside the walls of Jerusalem. They suggested the Garden Tomb was constructed much later than that of the Tomb of the Holy Sepulchre.

The Bible challenges the Biddle's conclusion, for Mattew 27: 57-60 tell us that when Joseph of Armathia claimed the body of Jesus, he and Nicodemas (John 19: 39-40) '*wrapped it in a clean linen cloth <u>and laid it in his own new tomb which he had hewn out of the rock</u>: and rolled a great stone to the door of the sepulchre, and departed.*'

John 19: 41-42 also discredits their conclusion. These scriptures say, '*Now in the place where he was crucified there was a garden: in the garden a new sepulchre, wherein man was never yet laid. There laid they Jesus therefore because of the Jews preparation day; <u>for the sepulchre was nigh at hand</u>.*' The Garden Tomb is, indeed, 'nigh at hand' for the Garden in which the Tomb is found borders Golgotha. One can stand at the edge of the Garden and see the skull in the sandstone. The Church of the Holy Sepulchre is quite some distance from Golgotha and the tomb located there is not set within a garden, nor is there a natural skull in the rock.

Appendix 7

The Location of Christ's Crucifixion and Burial (continued)

Matthew 28: 2 records that '*there was a great earthquake: for the angel of the Lord descended from heaven and came and rolled back the stone from the door, and sat upon it.*' Even yet today, the stone lies separate from the Tomb. The channel across the front of the Tomb which guided the rock remains and bears witness that the Garden Tomb is the sepulchre where Joseph and Nicodemus laid the body of Jesus.

The Garden Tomb itself authenticates the burial site, for it is finished in the manner of a rich man's tomb. Isaiah 53: 9 tells us that Jesus, in his death, made his grave with the rich. Upon entering the chamber and looking to the right, one sees a separate open chamber. This chamber contains two stone slabs (somewhat like a shallow box), one to the right and one to the left. The one to the left testifies that it was used by someone who was taller than the one for whom it had been prepared. The stone slab on which the body had been lain was obviously chiseled away to allow space for the feet of our Lord Jesus Christ. Mark 16: 5 tells us that '*when the women entered the sepulchre they saw a young man sitting on the right side, clothed in a long white garment.*' The Garden Tomb alone, designed and crafted by Joseph of Armathea, with a resting place to the left for himself and the right for his wife, verifies the feasibility of the women's testimony. In this Tomb the angel could very well sit across from the place that Jesus had lain.

Upon viewing the inner chamber of both the Church of the Sepulchre and the Garden Tomb, one must immediately conclude that the Garden Tomb meets the demands of the scriptures; while the Tomb at the Holy Sepulchre fails in all respects. The one was obviously prepared by a rich man as a burial place for himself and his wife; the other is rough and unfinished, more in keeping with a commoner's grave. It defies the ordinance of the Law by being located within the walls of the Old City of Jerusalem. The dead, according to the Law, defile the place in which the body lays. Anyone who touched the body or the grave of the dead were declared by God to be unclean for seven days (Numbers 19: 11-19). For this reason, the dead were buried outside the walls of Jerusalem. Only a Gentile bishop from Rome would dare knowingly instigate such a bold lie and commit such a brazen act against the ordinance of the Law.

One has but to visit the Garden Tomb to know that this is surely the place where they laid our Saviour. The peace and presence of Christ envelopes the Garden, assuring the heart that this is the hill where he died, was buried and resurrected.

Appendix 8

Warning Against False Teaching

This appendix is included as a special warning against a current teaching citing the book of *Zechariah* to have us believe that all references to Babylon in the *Revelation* are literal and should be attributed to Iraq and the rebuilt city of Babylon under Saddam Hussein.

The teaching is that the rebuilt city of Babylon will be the seat of the last world Empire ruled over by the Antichrist. The theory is that while the Antichrist is over in Israel summoning all nations to battle, Jesus will return.

The basis of this teaching comes out of Zechariah's seventh vision as recorded in Zechariah 5: 5-11 which has connections with Jeremiah's prophecy against the Chaldeans on the plain of Shinar and the city once known as Babel (Babylon). This area will, in truth, be annihilated, either to end the conflict with Iraq or to be used as a stepping stone to the battle of Armageddon on the plains of Meguiddo, in the land of Israel. Jeremiah, Chapters 50 and 51, records its final destruction. It, however, is not the ruling seat of the end-time Antichrist.

The Babylon mentioned in the study of *Revelation* is not the city built in the land called Shinar which was located on the plains between the Tigris and the Euphrates – commonly known in ancient times as Babylonia. It is not the city built by Nimrod and later occupied by Nebuchadnezzar (recently rebuilt by Saddam Hussein). Instead, it represents a rule likened to Babylon in which the Antichrist will flourish for a season.

II Thessalonians 2: 4 attests to the inaccuracy of such teaching since verse 4 states that the Antichrist poses as God and <u>sits in the temple of God</u>, showing himself to be God. From all scriptural indications, the temple from which the Antichrist reigns will be located in Jerusalem (Isaiah 2: 2-3; Micah 4: 1-2) and not in the land once known as Shinar (Genesis 11: 2), more commonly known as Babylon.

The battle of Armageddon, fought in the hills and plains of Israel (Ezekiel 38-39) and the battle in the area of the original Babylon are two different battles. One, no doubt, is precursory to the other. When the battle in the area of Iraq is fully finished, '*it shall be no more inhabited forever*' (Jeremiah 50: 39); while the destruction and evidence of war left at Armageddon will be eradicated by ravenous birds and the people of the land (Ezekiel 39: 4-20; Revelation 19: 17-18). According to Ezekiel 39: 25-29, Israel will dwell safely in the whole of her land and never again be afraid.

Appendix 9

Kingdom of Heaven / Kingdom of God

The Kingdom of Heaven is the third heaven, the heaven in which God dwells; it is that kingdom which never passes away (Revelation 21: 2). The Kingdom of God is the Spirit of that heavenly kingdom established through Christ on Earth (I Colossians 1: 13). When speaking to the Pharisees, Jesus stated that man could not observe the coming of the Kingdom of God because it was not a tangible kingdom. He explained it was a spiritual kingdom which, when willingly received, would abide in the heart of man (Luke 17: 20-21) and bind him to his creator, God. He stated that the Spirit of this kingdom would enter the believers and endow them with the Holy Ghost. The wisdom and knowledge of this indwelling Comforter would teach them all things pertaining to God and bring them life and peace (John 14: 26-27). Jesus explained that man could not see, nor enter into, the Kingdom of God unless he were born again – born of the Spirit of God (John 3: 1-18). Through this rebirth, at death they would be assured eternal life and entrance into the Kingdom of Heaven.

One wonders why Matthew almost consistently uses the phrase 'the Kingdom of Heaven' when referring to both the heavenly kingdom and the earthly kingdom of God. In so doing, Matthew is exercising great wisdom as he subtly introduces the earthly kingdom of God to the Israelite. Being a Hebrew, he speaks to the Hebrews in terms they understand. He aptly demonstrates this wisdom by first recording the generations of Jesus Christ. He starts with Abraham and ends with Joseph, the husband of Mary of whom Jesus was born (Matthew 1: 1-17). By starting with their Hebrew forefather, Abraham, and ending with the birth of Christ, Matthew forms a connecting link between the Old Covenant of Law and the New Covenant of Grace. He makes wide and varied uses of familiar scriptures drawn from the Old Testament to form a bridge between the Old and the New. In this way, Matthew binds the Old Testament and the New Testament together, showing the New Covenant of Grace fulfills the prophesies of the Old.

Matthew realizes the Israelite would be comforted by these familiar scriptures and, no doubt, pacified by them. Through them he very gently leads his brethren to an understanding and, possibly, to an acceptance of the Kingdom of God. Since Israelites have always known God and believed in His heavenly kingdom, they could more readily be convinced that the Kingdom of Heaven would – through their Messiah – become available to them upon their entrance into the Kingdom of God on Earth.

Matthew alone, when documenting the parables, likens the Kingdom of Heaven to the message found within the parable. At the same time, even in the beginning of his gospel, he speaks of the living Earthly kingdom and mentions their need to enter into it. He accomplishes this by recording Jesus as saying, "*Seek ye first the kingdom of God.*" (Matthew 6: 33). Again, toward the middle of his teaching, Matthew reveals the immediate availability of that Kingdom by quoting Jesus when he said, "*The Kingdom of God is come unto you.*" (Matthew 12: 28). Later, in his ardent desire to see his brethren reconciled to God, he attempts to provoke them to jealousy (Romans 11: 11) by quoting Christ's warning to the chief priests and the pharisees. Jesus spoke to them and said, "*The Kingdom of God shall be taken from you and given to a*

Appendix 9

Kingdom of Heaven / Kingdom of God (continued)

nation bringing forth fruit thereof." (Matthew 21: 43). As Matthew ends his gospel, he narrates the events of the Last Supper, the Crucifixion and the Resurrection. In so doing, he discloses the purpose and the victory of the Passover. He intentionally exposes the suffering Jesus (their Messiah) as he dies to establish the Kingdom of God on earth. With the skill of a master teacher, and the mind of an evangelist, Matthew carefully relates the words of Jesus when he said, "*I will not drink henceforth of this fruit of the vine until that day when I drink it with you in my Father's Kingdom.*" (Matthew 26: 29). By quoting Christ's words, Matthew makes them seem very personal (as if they were spoken precisely to the Israelite), and the kingdom attainable while they dwelt within their human bodies here on earth. He helps them to understand that Jesus opened the way for them to enter into the joy and freedom of that kingdom.

Matthew finishes his gospel with solid evidence that Jesus came to bring life, not only to the living, but also to the dead (Matthew 27: 52-53). Matthew's hope is that his brethren, suffering always under a load of sin, would gladly hear that through the death of Christ, their Messiah, the veil of sin separating them from God has been removed (Matthew 27: 51). He expects that after years of continual sacrifice, the Israelites will joyously accept that Jesus has set up God's Earthly Kingdom, the Kingdom of God, and presented himself to God as their mediator and redemptive Lamb.

In conclusion, both Israelite and Gentile can readily understand that the Kingdom of Heaven is God's heavenly kingdom which has existed even before creation, while the Kingdom of God is God's Earthly Kingdom. The gospels and the Acts effectively teach that this Kingdom was established in the heart of man as a living, spiritual Kingdom after the crucifixion and resurrection of Jesus Christ.

Appendix 10: Line Chart of the Tribulation Years - Chart One

TIME CHART OF THE TRIBULATION YEARS

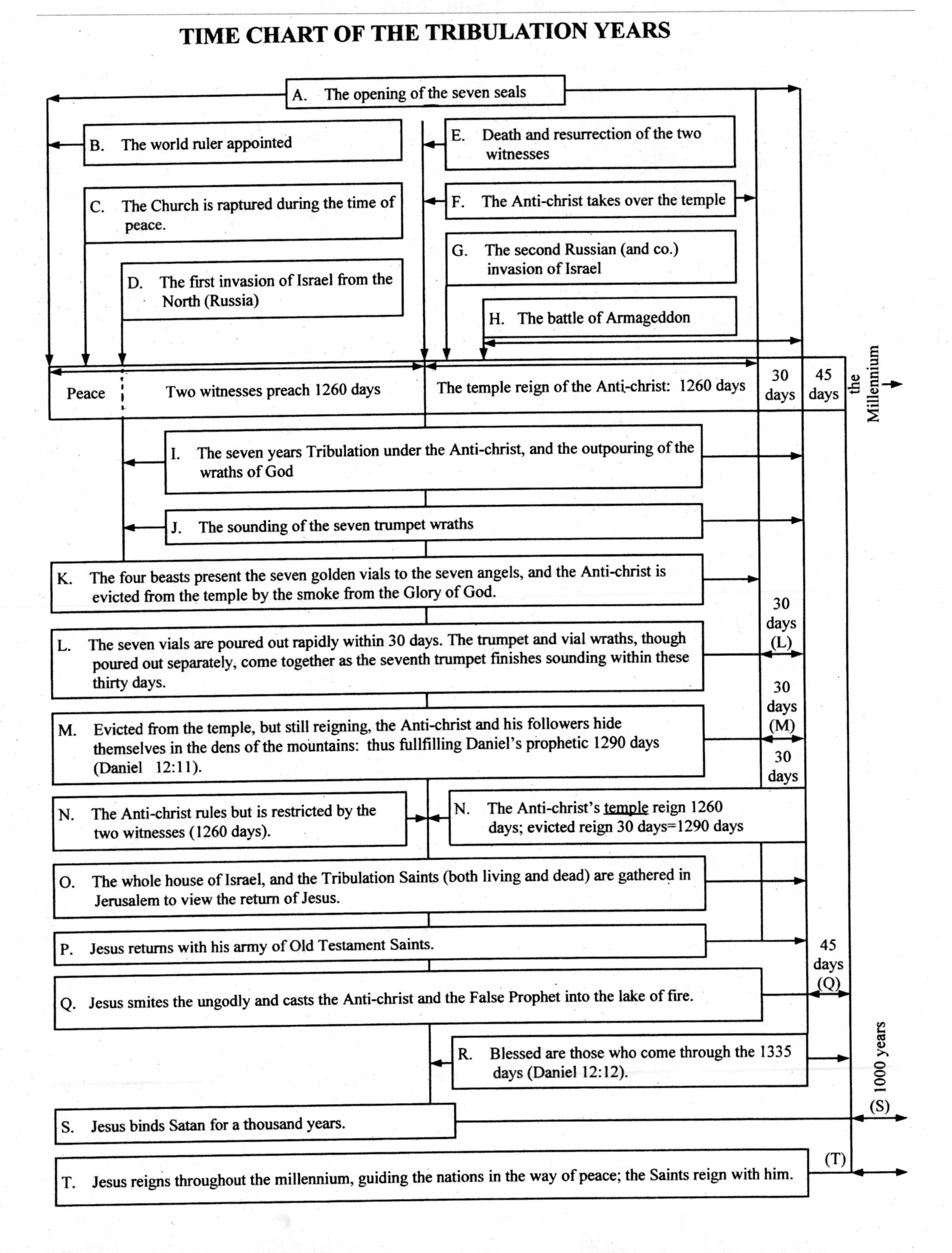

Appendix 10: Line Chart of the Millennial Years - Chart Two

TIME CHART—MILLENNIUM TO ETERNITY

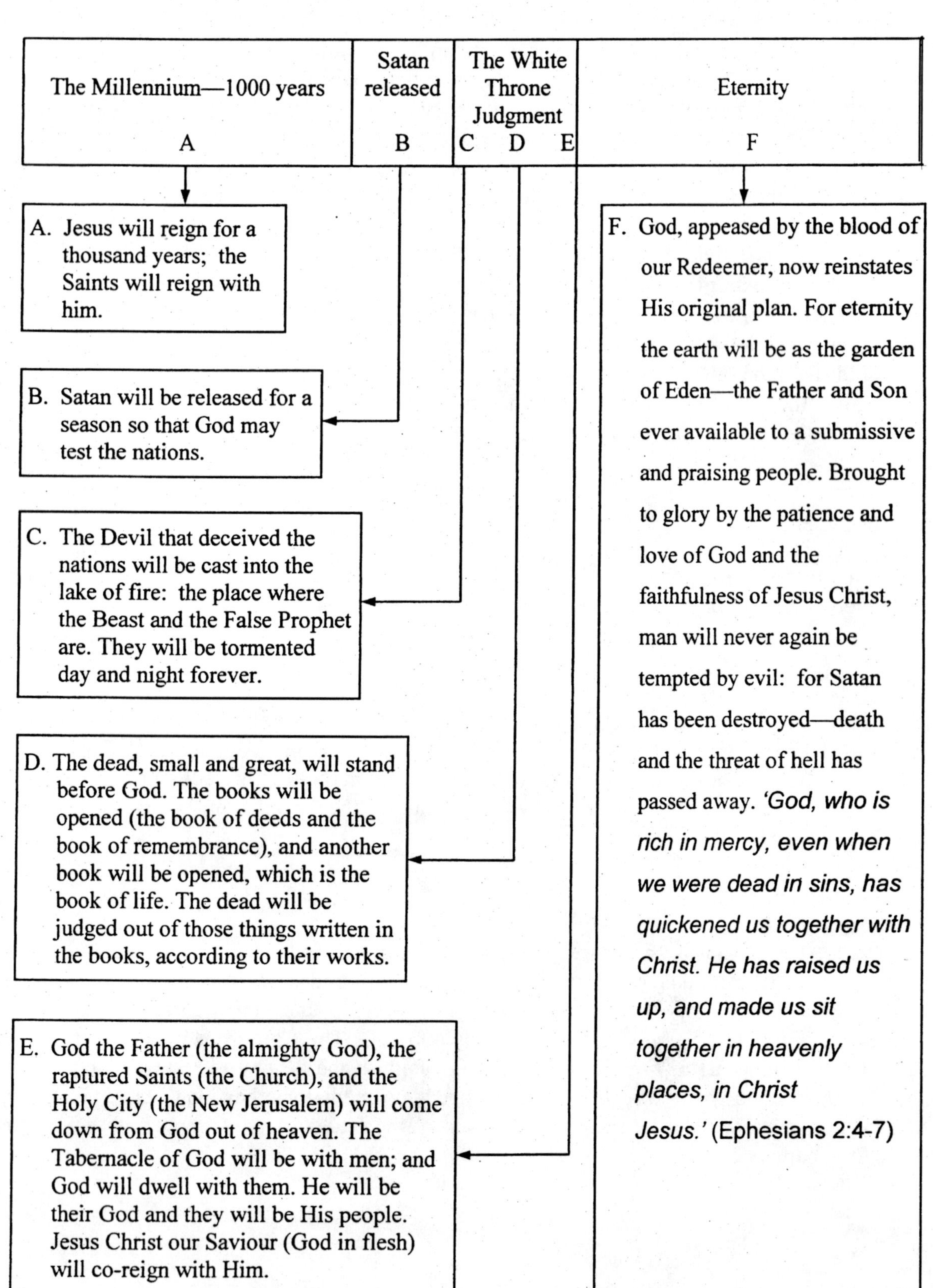

The Millennium—1000 years
A
Satan released
B
The White Throne Judgment
C D E
Eternity
F
A. Jesus will reign for a thousand years; the Saints will reign with him.
B. Satan will be released for a season so that God may test the nations.
C. The Devil that deceived the nations will be cast into the lake of fire: the place where the Beast and the False Prophet are. They will be tormented day and night forever.
D. The dead, small and great, will stand before God. The books will be opened (the book of deeds and the book of remembrance), and another book will be opened, which is the book of life. The dead will be judged out of those things written in the books, according to their works.
E. God the Father (the almighty God), the raptured Saints (the Church), and the Holy City (the New Jerusalem) will come down from God out of heaven. The Tabernacle of God will be with men; and God will dwell with them. He will be their God and they will be His people. Jesus Christ our Saviour (God in flesh) will co-reign with Him.
F. God, appeased by the blood of our Redeemer, now reinstates His original plan. For eternity the earth will be as the garden of Eden—the Father and Son ever available to a submissive and praising people. Brought to glory by the patience and love of God and the faithfulness of Jesus Christ, man will never again be tempted by evil: for Satan has been destroyed—death and the threat of hell has passed away. 'God, who is rich in mercy, even when we were dead in sins, has quickened us together with Christ. He has raised us up, and made us sit together in heavenly places, in Christ Jesus.' (Ephesians 2:4-7)

GLOSSARY OF TERMS

Antichrist / False prophet / the dragon

There have been, and will continue to be, many antichrists and false prophets until the second coming of Jesus. The Antichrist and False prophet spoken of in the *Revelation* are the predicted political and religious world rulers. They come as unified peace-makers but end up as supreme dictators. As their names indicate, they are the opposite of, and are against Christ; they are opponents of Christianity. Together they form the triune evil godhead: Satan, the Antichrist and the False prophet. The following are a few of the names, used in scriptures, identifying these evil beings:

1) **The Antichrist**: the Beast; Satan in flesh; the mystery of iniquity; the abomination of desolation; the desolator; the little horn; the conqueror on the white horse.
2) **The False prophet**: the woman on the beast; the second beast; the second little horn; the mother of harlots; the queen of heaven; the widow. Together with the Antichrist, she would share the titles: Mystery Babylon; Babylon the Great; the Abomination of the Earth.
3) **Satan**: the great red dragon; the old serpent; the devil; the deceiver; the fallen angel; Lucifer; the wicked one; Abaddon and Apollyon; the prince of the air.

Apostasy / Apostate

A person who completely forsakes his religion, faith, political party or principles. As pertaining to scriptures, those who defect from the teachings of Jesus Christ; those who completely forsake biblical teachings and turn to false doctrines.

Church

The Churches of the New Testament age were the Christian believers who followed the teaching of Christ; they fellowshipped with the apostles. Being committed to Christ, 'the Church', a body of born-again believers, will be taken up in the Rapture before the seven years Tribulation. A second Christian body will form during the Tribulation, but is not generally referred to as 'the Church'.

Dark Ages

The Dark Ages began as the priestly orders were formed and the hierarchy of Romanism gained power. The foundation of the Dark Ages was grounded as early as the first century church in the philosophy of the Nicolaitans, Baal worship and Gnosticism. Their satanic doctrines and practices were well-established by 600 A.D. The Church had departed from the truth of salvation as a free gift from God through faith in the death and resurrection of Jesus Christ; almost gone was the reality that man needs no mediator other than Jesus. For more detailed information, read the preamble for each of the letters to the Churches found in this Study for Chapters 2 and 3.

GLOSSARY OF TERMS

Day of the Lord / The Lord's Day

1) **The Day of the Lord**: Though the Resurrection and the Rapture may certainly be referred to as the victorious Day of the Lord, it is generally accepted that the Day of the Lord is when Jesus reveals his might and authority. These are the days at the end of the Tribulation when Jesus passes out God's judgment upon man and casts the Beast and the False prophet into the lake of fire. These days are recorded in Revelation 19: 11-21. See also Joel 2: 1-11.

   The Day of the Lord is also the time of the **White Throne Judgment**. It is when the books are opened and man is judged for eternity. It is the final day when the devil is cast into the Lake of Fire. A record of these days may be found in Revelation 20: 9-15. Read what Peter says about the Day of the Lord in II Peter 3: 10-13.

2) **The Lord's Day**: The Lord's Day is any day set aside solely to worship the Lord and take rest for the body. The Christians, on the whole, have observed Sunday as their day of rest and worship. The Israelites, and other denominations who keep the Law, honour Saturday as their weekly worship day and celebrate numerous other feast days as a Sabbath unto the Lord. Actually, every day of the week is the Lord's Day.

The Godhead

The **Holy Trinity** or **Triune God** (Three-in-One) is the union of the Father (God), the Son (Jesus) and the Holy Ghost (the Spirit of God which dwells within man from rebirth on).

The **unholy trinity**, or the **false trinity**, has Satan as God, the Antichrist in place of Jesus and the False prophet as representative of the Holy Spirit.

House of Israel and the House of Judah

These two houses consist of the twelve tribes which came out of the loins of Jacob, the man God called 'Israel'. Each of the twelve tribes ruled over his own clan and all submitted to one king. However, the kingdom divided in 930 B.C. The ten tribes to the north appointed Jeroboam (out of the house of Ephraim) as their king and Samaria became their capital. These ten tribes are often referred to as the 'lost tribes of the House of Israel'.

The tribes of Judah and Benjamin (the southern tribes), along with the Levites, remained loyal to king Rehoboam, the son of Solomon (I Kings 12: 17-21). These two tribes were called 'the house of Judah' or 'the house of David'. The first biblical reference to them as 'Jews' may be found in II Kings 16: 6. Only these two tribes, along with the Levites, should be referred to as Jews. Once this truth is established, the Old Testament readings become ultimately clearer. One should remember that all Jews are Israelites; however, not all Israelites are Jews. (In the same way as all Albertans are Canadians, yet all Canadians are not Albertans.)

GLOSSARY OF TERMS

The Inquisition

The Inquisition was a court, a Roman Catholic tribunal founded in 1233 by a monk named Dominic. It fell into disuse, a victim of its own success. Its purpose was to discover and suppress heresy and to punish heretics. All who challenged the authority of the Pope and the Roman Catholic Church were pronounced heretic. Customarily, they were excommunicated or burned at the stake; many died in other horrific ways.

Rome's answer to the Reformation was to bring back the office of the Inquisition. It was updated and refurbished, made fit for this modern emergency. The Papal 'Bull' of July 21, 1542, re-established the Inquisition (**Papal Bull**: Papal Bulls written by the Pope were called 'thunders' and had the power to paralyze whole nations, even bringing kings to their knees.) The so-called 'holy office' of the revised Inquisition was headed by a Spaniard named Ignatius Loyala (1491-1556) who founded the priestly order of the Jesuits. They became the spearhead of the Counter Reformation. Loyola believed the 'queen of heaven' had appeared to him and commissioned him to raise up an army to defend the papacy from reformation.

Justification / Sanctification / Glorification

- **Justification**: Man is justified upon his personal acceptance of the cleansing power of the shed blood of Jesus Christ and is freed from guilt and the penalty of sin.

- **Sanctification**: The believer, instructed by the word and empowered at rebirth by the indwelling Holy Spirit grows into sanctification in his pursuit of holiness. Wooed by the Holy Spirit, the renewed man strives always to keep the commandments and do things pleasing to God. Total sanctification comes at man's **glorification**. It is then when the Saints take on the full sinless, holy character of Jesus Christ.

- **Glorification**: Man is glorified at the Rapture; having been judged and rewarded, he is given the full likeness of Christ through the gift of the Seven Spirits of God. It is at this time that man becomes the fully sanctified, glorified sons of God.

Mystery, Babylon

For a complete teaching on the religious and political system referred to in the book of *Revelation* as 'Mystery, Babylon the Great, the Mother of Harlots and Abomination of the Earth', please read the preamble to Chapter 17.

Reformation

The great religious, social and political movement in Europe which started in the sixteenth century. It improved conditions for mankind generally. The religious aspect of the Reformation actually started when a British priest, John Wycliffe (1320-1384) began to translate the Vulgate (the Latin translation of the Bible) into English. This eventually led to the establishment of the Protestant Church which freed the people from the hold of the Roman Catholic Church. For further information, see Study notes following Revelation 2: 18 Preamble, "The Letter to Thyatira"; Revelation, Chapter 3 Preamble, "The Letter to Sardis"; and Revelation 3: 6 Preamble, "The Letter to Philadelphia".

GLOSSARY OF TERMS

Saints

The Saints refer to those who open their hearts and minds to their Saviour, Jesus Christ. They accept him, his death and his resurrection as their only means to obtain forgiveness and cleansing from sin. Jesus then becomes their salvation from judgment and eternal punishment. All born-again, committed followers of Christ are Saints. It is not a Holy order bestowed upon a select few and is not ordained by man, but by the Holy Spirit of God.

Sardine Stone and **Jasper Stone**

1) **Sardine Stone**: The Sardine stone, representative of the Son of God, has a variety of spellings. It is called Sardios, Sardius, Sardine, Sardion and Sardin. It was found near Sardis from whence it derived its name; however, the finest specimens were brought from Babylon. It is a chalcedony quartz ranging from deep orange-red to brownish-red and is the first stone listed in the breastplate of the High Priest. Rays of sunlight on the Sardis chalcedony quartz (the sardine stone) produce a deep red colour by affecting the iron salts included as impurities in the mineral. It's astonishing that the jasper stone (also a chalcedony quartz), representative of God, may sometimes, but not generally, appear blood-red, even deeper than that of the sardine.

2) **Jasper Stone**: the Jasper stone is listed last in the breastplate of the High priest; however, it is listed first in the foundation stone of the walls of the New Jerusalem. It is also chalcedony quartz, mostly rendered opaque by the inclusion of brightly-coloured iron oxides with shades of brown, yellow, red and green. It sometimes appears green as an emerald. The Jasper stone of the New Testament is defined as clear as crystal (Revelation 21: 11). How very intriguing that Revelation 4: 3 says, "*And he that sat was to look upon like a jasper and a sardine stone: and there was a rainbow round about the throne, in sight like an emerald.*" Can we doubt that the one that John sees, who sat upon the throne, is the Father and the Son?

Second Coming of Jesus

Jesus first came to earth as a baby born of the virgin Mary in the town of Bethlehem of Judea (Matthew 2: 1). His second coming to the earth will be as the two men in white apparel declared. He will come in the same manner as the disciples saw him go. A cloud received him as they watched him risen toward heaven (Acts 1: 9-11). Revelation 14: 1, Revelation 14: 14-20 and Revelation 19: 11-21 give us three aspects and a full report of his second coming. One must not confuse the **Rapture**, as recorded in I Thessalonians 4: 13-17, with the Second Coming. In the Rapture, the Church, both living and dead, is caught up together to meet our Lord in the air. Jesus won't, at this time, touch the earth.

Tabernacle

Tabernacle is both a noun and a verb: it means a dwelling place, as well as to abide with or among. As a noun, it is a place of worship, a temple; in times past, it was often a dwelling place for the Shekinah Glory of God. As a verb, it tells us that God dwells with His people.

GLOSSARY OF TERMS

Tabernacle (continued)
The tabernacle of the wilderness was a portable substitute temple made with a timber frame overlaid by animal skin coverings (curtains of black goat's hair and ram's skin dyed red; the outermost curtain was made of the skin of some marine animal, probably a dolphin). The inner curtain was made of fine-twined linen. The tabernacle was used as a place of worship by the Israelites on their journey from Egypt to the promised land of Caanan.

At the Feast of the Tabernacle (Leviticus 23: 34-43), the Israelites were required to dwell seven days within a booth made from boughs of trees, branches of palm and willows from brooks (Leviticus 23: 40). The purpose of the exercise was that they were to wait before the Lord for a time of renewal; a time of sorrowing for their sins; and a time of rejoicing within His presence.

As God 'tabernacled' Himself among the Israelites in the form of a cloud by day and a fire by night (Exodus 13: 21); as He 'tabernacled' Himself in the newly-built tent of the congregation in the form of His glory (Exodus 40: 35); if invited, so also does He 'tabernacle' Himself in the human body, the temporary dwelling place of the Spirit and soul.

At the Rapture, Christ will dwell with his Church in heaven and, later, with his Church on earth (the Tribulation Saints). He will 'tabernacle' himself in the temple, among his people, for a period of a thousand years. He and his Father will finally 'tabernacle' themselves for eternity in the New Jerusalem, ever available to the Saints (Revelation 21: 3).

Translations

1) **Septuagint (LXX)**: a pre-Christian Greek rendering of the Old Testament Hebrew scriptures.

2) **Vulgate**: the Latin version of the Bible authorized and used by the Roman Catholic Church.

3) **Peshitta**: the authorized Bible of the Church of the east, translated in ancient times from Hebrew into a branch of the Aramaic language (Syriac). These scriptures were received from the hands of the apostles themselves (in the Aramaic original) and were used by the Christians from earliest times. The Jewish commoner was more familiar with the Aramaic language and used it more readily than Hebrew. Ezra, Daniel, and one verse in Jeremiah (10: 11) were composed in Aramaic and preserved in that ancient form of the language in the midst of the Hebrew Old Testament. In the first century, Jesus and his followers generally spoke Aramaic although they knew Hebrew. From the Mediterranean east into India, the Peshitta is still the Bible of preference among Christians; however, today nearly all who use it speak Arabic, or one of the tongues of south India. (This information is drawn from the preface to the George M. Lamsa translation of the Peshitta into English; published in 1957 by A. J. Holman Company.)

GLOSSARY OF TERMS

Transubstantiation

According to the Roman Catholic Church and Eastern Orthodox dogma, the elements of the communion, at their consecration, actually become the body and blood of Christ while keeping only the appearance of bread and wine. This is called the doctrine of transubstantiation – the changing of one substance into another.

Martin Luther defended this doctrine and engaged himself in controversial disputes with Swiss theologians in the famous conference at Marburg in October 1529. He held to his Catholic background and rejected the elements of the communion as emblems only. He emphatically maintained his view that the bread and the wine, at the Eurcharist, actually become the body and blood of Jesus. He eventually concluded, after much deliberation, that the presence of Christ is in the wine and the bread; nonetheless, the wine remains wine and the bread remains bread. This revised position in Luther's theology is known as **consubstantiation**, a view recognized by many of the Lutheran bodies today.

The doctrine of transubstantiation has been seen by some as a violation of and an offense to the **secondary principles** of Christianity. Since the teaching poses no threat to the **primary principles** of the faith, we should not allow differing opinions to break the harmony of the Christian Church.

To believe the bread and the wine actually become the body and blood of Christ may well be an acceptable doctrine. Scriptures can be found which both support and oppose this position (see Matthew 26: 26-28; Mark 14: 22-24; John 6: 48-58). However, the teaching upheld within the Catholic Eucharist, which we absolutely must not embrace is that of '**continual sacrifice**'. This is the theory held by the Catholic Church and by some branches of the Lutheran churches that Christ dies afresh at each receiving of the Communion. This teaching is absolutely unacceptable since it offends the primary principles of the Church, as stated in the Apostle's Creed. The scriptures tell us that, through Christ's one righteous act, the free gift of salvation was open to all men (Roman 5: 18). Scriptures further tell us that "*Jesus died once to defeat sin and now he lives for the glory of God.*" (Romans 6: 10 NLT).

Bibliography

Barclay, William. “The Revelation of John”, The Daily Study Bible, Volume 1. Burlington, Ontario: Welch Publishing Company, revised 1976.

Barclay, William. “The Revelation of John”, The Daily Study Bible, Volume 2. Burlington, Ontario: Welch Publishing Company, revised 1976.

Carver, Wayne. “The Tower of Babel” from Radio Sermons. San Antonio, TX: The Christian Jew Foundation Publications.

Carver, Wayne. “Mystery Babylon” from Radio Sermons. San Antonio, TX: The Christian Jew Foundation Publications.

Carver, Wayne. “World War III in Prophecy” from Radio Sermons. San Antonio, TX: The Christian Jew Foundation Publications.

Charlesworth, James H., Editor. The Old Testament Pseudepigrapha, Volume 1. New York: Doubleday Publishers, 1983.

Davis, J. D. Davis Dictionary. Grand Rapids, MI: Baker Books, 1980.

Dobson, Edward G. et. al., The Complete Bible Commentary. Nashville, Tenn.: Thomas Nelson Publishers, 1999.

Duck, Daymond R. Revelation - God’s Word for the Biblically-Inept. Lancaster, Penn.: Starburst Publishers, 1998.

Encyclopedia Britannica, Vol. VI. Chicago: William Brenton Publishers, 1973.

Greene, Oliver B. Revelation – Verse by Verse Study. Greenville, SC: Gospel House Inc., 1976.

Hislop, Alexander. The Two Babylons. Neptune, NJ. Loizeaux Brothers, Inc., 1858.

Ironside, H. R. The Revelation: Ironside Commentaries. Neptune, NJ: Loizeaux Brothers Publishing, revised 1996.

Lahaye, Tim. Revelation Unveiled. Grand Rapids, MI: Zondervan Publishing, 1999.

Lamsa, George M. The Holy Bible from Ancient Eastern Manuscripts. Philadelphia: A. J. Holman Co., 1957.

Living Life Application Bible. Wheaton, Ill.: Tyndale House Publishers, 1971.

McGee, J. Vernon. Thru The Bible, Volume V. Nashville, Tenn: Thomas Nelson Publishers, 1983.

Bibliography

Pfeiffer, Charles, et. al., Editors, Wycliffe Bible Encyclopedia, Volume 1. Chicago: Moody Bible Institute, 1975.

Pfeiffer, Charles, et. al., Editors, Wycliffe Bible Encyclopedia, Volume 2. Chicago: Moody Bible Institute, 1975.

Prince, Derek. Appointment in Jerusalem. Lincoln, VA: Chosen Books, Inc., 1975.

Schmitt, John W. & Laney, J. Carl. Messiah's Coming Temple. Grand Rapids, MI: Kregel Publications, 1977.

Smith, William. Bible Dictionary. Philadelphia/Toronto: John C. Winston Company.

Stedman, Ray. God's Final Word. Grand Rapids, MI: Discovery House Publishers, 1991.

The New English Bible Apocrypha. Oxford: Cambridge Press, 1970.

Van Impe, Jack. Revelation Bible. Troy, MI: Jack Van Impe Ministries, 1998.

ISBN 155369615-8
9781553696155